PHYSICAL CHEMISTRY OF ORGANIC SOLVENT SYSTEMS

Edited by

A. K. Covington

and

T. Dickinson

Department of Physical Chemistry,
University of Newcastle upon Tyne,
Newcastle, England

PLENUM PRESS • LONDON AND NEW YORK • 1973

Library of Congress Catalog Card Number: 72–77042

ISBN 0–306–30569–0

Plenum Publishing Company Ltd
Davis House
8 Scrubs Lane
Harlesden
London NW10 6SE
Telephone 01–969 4727

U.S. Edition published by
Plenum Publishing Corporation
227 West 17th Street
New York, New York 10011

PRINTED IN GREAT BRITAIN
AT THE PITMAN PRESS, BATH

Preface

We believe this to be the first monograph devoted to the physicochemical properties of solutions in organic solvent systems. Although there have been a number of books on the subject of non-aqueous solvents[1-4], they have been devoted, almost entirely, to inorganic solvents such as liquid ammonia, liquid sulphur dioxide, etc. A variety of new solvents such as dimethylformamide, dimethylsulphoxide and propylene carbonate have become commercially available over the last twenty years. Solutions in these solvents are of technological interest in connection with novel battery systems and chemical synthesis, while studies of ion solvation and transport properties have fostered academic interest.

This monograph is primarily concerned with electrolytic solutions although discussion of non-electrolyte solutions has not been excluded. We have deliberately omitted consideration of the important area of solvent extraction, since this has been adequately covered elsewhere.

Our contributors were asked to review and discuss their respective areas with particular reference to *differences* in technique necessitated by use of non-aqueous solvents while not reiterating facts well-known from experience with aqueous solutions. We have striven to build their contributions into a coherent and consistent whole. We thank our contributors for following our suggestions so ably and for their forebearance in the face of our editorial impositions. In including a substantial quantity of numerical results in the form of appendices, we hope that this monograph will become a handbook for all working in the field of organic systems. It will have served its purpose admirably if it achieves our intention of stimulating further progress, for its initiation arose out of our own difficulties of starting research activities in the area.

We are grateful to Professor R. A. Robinson for compiling Table 1.3.1 and the other Appendices of Chapter 1, and for reading the whole of the text at the proof stage. We also wish to thank Miss Jennifer M. Thain for helping with the onerous task of preparing the index.

Newcastle upon Tyne

A. K. Covington
T. Dickinson.

1. A. K. Holliday and A. G. Massey, Inorganic Chemistry in Non-aqueous Solvents, Pergamon, Oxford 1965.
2. J. J. Lagowski (Ed.) The Chemistry of Non-aqueous Solvents. Vols I, II. Academic Press New York. 1966, 1967.
3. T. C. Waddington (Ed.). Non-aqueous Solvent Systems, Academic Press London 1965.
4. T. C. Waddington, Non-aqueous Solvents, Nelson, London 1969.
5. J. Jander and C. Lafrenz, Ionizing Solvents, Verlag Chemie-Wiley, Weinheim, (1970).

Contributors

O. R. Brown	Electrochemistry Research Laboratory, Department of Physical Chemistry, University of Newcastle upon Tyne, England.
A. K. Covington	Department of Physical Chemistry, University of Newcastle upon Tyne, England.
C. M. Criss	Department of Chemistry, University of Miami, Coral Gables, Florida 33124, U.S.A.
T. Dickinson	Electrochemisty Research Laboratory, Department of Physical Chemistry, University of Newcastle upon Tyne, England.
R. Fernández-Prini	Department of Chemistry, University of Maryland, College Park, Maryland 20742, U.S.A.
R. Garnsey	Central Electricity Generating Board, Research Laboratories, Kelvin Avenue, Leatherhead, Surrey, England.
T. E. Gough	Department of Chemistry, University of Waterloo, Waterloo, Ontario, Canada.
D. E. Irish	Department of Chemistry, University of Waterloo, Waterloo, Ontario, Canada.
E. J. King	Department of Chemistry, Barnard College, Columbia University, New York, U.S.A.
I. R. Lantzke	Department of Physical Chemistry, University of Newcastle upon Tyne, England
R. Payne	Air Force Cambridge Laboratories, L. G. Hanscom Field, Bedford, Massachusetts, U.S.A.
(the late) J. E. Prue	Department of Chemistry, The University, Reading, Berkshire, England.
M. Salomon	National Aeronautics and Space-Administration, Electronics Research Center, Cambridge, Massachusetts, U.S.A.
M. Spiro	Department of Chemistry, Imperial College of Science and Technology, London S.W.7, England.
D. W. Watts	School of Chemistry, University of Western Australia, Nedlands, W.A. 6009, Australia.

Contents

Chapter 1

Introduction and Solvent Properties

A. K. Covington and T. Dickinson
Department of Physical Chemistry,
University of Newcastle upon Tyne.

1.1 THE IMPORTANCE OF ORGANIC SOLVENTS

Physical chemists interested in electrolyte solutions have often been accused of being preoccupied with the properties of aqueous solutions. In view of the importance of water as a solvent essential to life, and its abundance on this planet, this attention may be understood if preoccupation is not. On the other hand the term non-aqueous solvents has often been interpreted as meaning various inorganic solvent systems such as those based on the oxides of nitrogen or sulphur, liquid ammonia, hydrogen fluoride and perhaps fused salts. Over the last decade or so, however, considerable interest has been generated in organic non-aqueous solvents particularly new ones like dimethylsulphoxide and propylene carbonate. Organic solvents show a wide range of dielectric constant ranging from that of *p*-dioxan (2.2) to those exceeding that of water like *N*-methylacetamide (165). However, the properties of the resulting solutions are not determined simply by the magnitude of the solvent dielectric constant, and attempts have been made recently to steer thoughts away from regarding the bulk solvent simply as a continuum characterised only by its dielectric constant.

Different sequences of solubility, differences in solvating power and possibilities of chemical or electrochemical reactions unfamiliar in aqueous chemistry have opened new vistas for physical chemists, and a large amount of quantitative data has now been accumulated. It is with the quantitative aspects we are concerned here although interest in these organic solvents transcends the traditional boundaries of inorganic, physical organic, analytical, physical chemistry and electrochemistry.

Whilst a discussion of separate solvent systems or solvent types was a possible organisation for this book, most workers in the field have tended to use more than one solvent and tend to be technique oriented. This latter arrangement has therefore been chosen.

1.2 SCOPE

We are predominantly concerned in this volume with electrolytic solutions although some mention will be found of gaseous and other non-electrolytic solutions where this helps in interpreting the properties of ionic solutions. The scope of this volume is therefore perhaps more restrictive than the title suggests; the important areas of binary non-electrolyte mixtures, solvent extraction and polymer solutions are also not covered. Adequate coverage of these may be found in excellent monographs and reviews elsewhere.

In the next section we shall discuss the physical properties and structure of the more important of the organic solvents mentioned in later chapters. In Chapter 2, the thermodynamic properties of solutions are discussed. Firstly, C. M. Criss discusses solubility and calorimetric measurements. In planning new work with unfamiliar solvents the extent of solubility of compounds is a primary consideration. Viscosity is also considered under this heading since although it is a transport property it is often used in the same way as thermodynamic measurements, to infer structural properties. An important technique for obtaining thermodynamic quantities of certain solutions is cryoscopy, the measurement of freezing point depressions, and this is discussed by R. Garnsey and J. E. Prue. Measurement of the e.m.f.s of suitable cells has provided much data on the free energies, enthalpies and entropies of solutes and of reactions. M. Salomon in sect. 2.5–2.7 reviews these data and includes consideration of polarographic half-wave potentials as well as standard electrode potentials from reversible cell measurements. Finally Criss and Salomon joined forces at our suggestion and have discussed the theoretical considerations underlying the interpretation of the thermodynamic results. This section contains discussion of medium effects expressed variously as free energies of transfer, medium effect activity coefficients or solvent activity coefficients, a theme which recurs in later chapters. We have tried to help the reader by imposing a uniform nomenclature throughout in spite of the fact that this may differ from that used in some original papers.

Thermodynamic aspects of acids and bases are deferred until Chapter 3 in which E. J. King provides an account of acid-base behaviour in organic solvents. Recent developments in spectroscopic techniques have resulted in considerable advances being made in understanding solvent-solute interactions in solution and Chapter 4 is devoted to spectroscopic measurements. Starting with sections on U.V. and visible spectroscopy by I. R. Lantzke, we progress to I.R. and Raman spectroscopy discussed together, because of their complementary nature, by D. E. Irish. The solvent systems and indeed the solutes discussed by T. E. Gough, in a section devoted to electron spin resonance spectroscopy, are rather different but the application of nuclear magnetic resonance spectroscopy reviewed by I. R. Lantzke is almost universal. Very real advances are being made in this latter area as improvements in instrumental design and technique become available. A spectroscopic technique, albeit not electromagnetic, which we have not included is that of ultrasonics. The principles of this technique and some applications have been very adequately reviewed by Blandamer and Waddington.[A1] Attention should be drawn to the work of Petrucci[A2] on tetrabutylammonium bromide in nitrobenzene-methanol mixtures and of Blandamer, Symons *et al.*[A3,A4] on tetraalkylammonium ions in various solvents. Petrucci[A5] and co-workers have also studied zinc and copper sulphates in glycol-water mixtures and Chen and Valleau[A6] have attributed a large relaxation in HCl solutions in DMF to acid association. The amount of work reported is not extensive enough to justify an additional review here.

In Chapter 5 we turn to transport properties. R. Fernandez-Prini compares and rationalises the recent theoretical advances concerning the conductance of dilute electrolyte solutions and its dependence on concentration and solvent properties. M. Spiro discusses transference number determinations in a second section, and finally reviews the information available on limiting ionic mobilities in the various solvents. The final two chapters are devoted to kinetics. Firstly in Chapter 6 D. W. Watts discusses reaction kinetics and mechanisms, predominantly for inorganic systems, since Parker has reviewed organic aspects recently.[A7] Finally O. R. Brown discusses electrochemical reactions and electrode kinetics in organic solvents; his contribution being preceded by a short review by R. Payne of the current knowledge concerning the electrical double layer in organic solvents.

1.3 CLASSIFICATION, PHYSICAL PROPERTIES, STRUCTURE AND PURIFICATION OF SOLVENTS

Classification is useful if it serves to focus attention on similarities and differences. Several classifications of organic solvent systems are possible according to dielectric constant ranges, organic group type, acid-base properties, etc. One convenient division is into three main groups:

(1) *Protic* (contain protons which they can lose)
  (a) amphiprotic, e.g. water, alcohols;
  (b) acidic or protogenic (giving protons), e.g. acetic acid, sulphuric acid;
(2) *Aprotic* (may contain protons which they are unable to lose, but may gain protons), e.g. dimethylsulphoxide, propylene carbonate.
(3) *Inert*, e.g. hexane, benzene.

Such terms have frequently been used within this book.

One disadvantage of classification schemes such as this, is that they are not clearly restrictive. Also, the terms have been used in the past in slightly different senses, for example the original usage of *aprotic* was for the *inert* hydrocarbons. (For further discussion of solvent classification, see Chapter 3.)

In the following we shall adopt an organic-group type classification outlining the essential properties of the more important and frequently used solvents. Primary considerations in choice of solvent will be its liquid range and dielectric constant followed by solubility considerations. In Table 1.3.1 compiled with Prof. R. A. Robinson's assistance, have been listed the most important physical properties of certain important solvents. Included in this Table are values calculated for the Debye–Hückel constants (A and B) and the Onsager limiting conductance equation constants (α and β). Selection of the best values for certain physical constants is no easy task in the face of discrepant values. Two monumental contributions to this problem have been 'Organic Solvents' by Riddick and Bunger,[A8a] which lists, besides purification and other details, physical properties for over 300 solvents, and the first of a new series 'Non-aqueous Electrolytes Handbook' compiled by Janz and Tomkins.[A8b] These valuable reference sources make extensive tabulation here unnecessary.

That we have not given the sources of the data in Table 1.3.1, we hope will act as a deterrent to those who might be inclined to quote them

Table 1.3.1

Physical Properties of Some Important Solvents

	Abbreviation	Molecular weight	F.p.	B.p.	Density[a]	Dielectric Constant[b]	Viscosity[c]	Dipole Moment[d]	Constants of Debye-Hückel eqn. (2.5.4)[e] A	10^{-8} B	Constants of Onsager eqn. (5.2.3)[f] α	β
water	H_2O	18.02	0.0	100.0	0.9971	78.30	0.8903	1.85	0.5115	0.3291	0.2300	60.65
methanol	MeOH	32.04	−97.8	64.6	0.7866	32.70	0.5445	1.70	1.895	0.5093	0.8522	153.4
ethanol	EtOH	46.07	−114.5	78.3	0.7851	24.55	1.089	1.69	2.914	0.5878	1.310	88.55
n-propanol	*n*-PrOH	60.10	−126.5	97.2	0.7995	20.33	2.004	1.68	3.866	0.6459	1.738	52.88
iso-propanol	*i*-PrOH	60.10	−89.5	82.4	0.7810	19.41	2.079	1.66	4.144	0.6611	1.863	52.16
n-butanol	*n*-BuOH	74.12	−89.5	117.5	0.8057	17.51	2.56	1.66	4.837	0.6960	2.175	44.60
t-butanol (30°)	*t*-BuOH	74.12	25.82	82.5	0.7762	11.62	3.32	1.66	8.727	0.8473	3.924	41.87
ethylene glycol		62.07	−12.6	197.4	1.1098	40.7	16.9	2.28	1.365	0.4565	0.6137	4.431
acetone	AC	58.08	−94.9	56.2	0.7850	20.70	0.3040	2.88	3.673	0.6401	1.692	345.4
acetophenone		120.15	20	201.5	1.0236	17.39	1.66	3.02	4.880	0.6986	2.198	69.04
tetrahydrofuran	THF	72.11	−108.5	66	0.8811	7.58	0.460	1.75	16.98	1.058	7.636	377.3
1,4 dioxan	D	88.11	11.8	101.3	1.0269	2.209	1.196	0.45	107.9	1.959	48.54	268.8
ethylene carbonate (40°)	EC	88.06	36.41	248	1.321	89.6	1.850	4.91	0.3882	0.3002	0.1746	26.62
propylene carbonate	PC	102.09	−48.8	242	1.198	64.4	2.530	4.98	0.6858	0.3629	0.3084	23.53
acetonitrile	AN	41.05	−44.9	81.6	0.7768	35.95	0.3409	3.92	1.644	0.4857	0.7393	233.8
ethylenediamine	EN	60.10	11.3	117.3	0.8922	12.9	1.54	1.99	7.650	0.8109	3.440	86.38
formamide	F	45.04	2.55	210.5	1.1292	109.5	3.302	3.73	0.3093	0.2783	0.1391	13.83
N-methylformamide	NMF	59.07	−3.8	180.5	0.9976	182.4	1.65	3.83	0.1442	0.2156	0.06484	20.95
NN-dimethylformamide	DMF	73.10	−61	153	0.9443	36.71	0.796	3.86	1.593	0.4807	0.7165	99.07
acetamide (94°)		59.07	81.1	221.2	0.9867	60.6	1.63	3.44	0.5498	0.3371	0.2472	33.93
N-methylacetamide (40°)	NMA	73.10	30.56	206	0.9420	165.5	3.020	3.73	0.1546	0.2209	0.06954	12.00
NN-dimethylacetamide	DMA	87.12	−20	165.5	0.9366	37.78	0.919	3.81	1.526	0.4738	0.6862	84.58
N-methylpropionamide	NMP	87.12	−30.9	148	0.9308	176	5.25	3.61	0.1518	0.2195	0.06825	6.860
NN-dimethylpropionamide	DMP	101.15	−43	175	0.9205	32.9	0.935	–	1.878	0.5078	0.8445	89.09
1,1,3,3,-tetramethylurea	TMU	116.16	−1.2	176.5	0.9619	23.45	1.401	3.47	3.121	0.6014	1.403	70.42
hexamethylphosphotriamide	HMPT	179.24	7.2	233	1.0202	29.60	3.25	5.39	2.201	0.5353	0.9896	27.02
dimethylsulphoxide	DMSO	78.13	18.54	189	1.0958	46.7	1.96	3.96	1.108	0.4262	0.4993	35.67
sulpholane (30°)	$TMSO_2$	120.17	28.45	285	1.2623	43.32	10.29	4.81	1.210	0.4388	0.5439	6.996

Footnotes: Values refer to 25°C unless otherwise stated.

[a] g cm^{-3}

[b] relative permittivity (ε_r)

[c] centi-poise (1 cP $\equiv 10^{-3}$ Nm^{-2} s)

[d] Debye units (1 D $= 10^{-18}$ e.s.u. cm $\equiv 3.33564 \times 10^{-3}$C m)

[e] A has the units $mol^{-1/2}$ $l^{1/2}$ and B the units cm^{-1} $mol^{-1/2}$ $l^{1/2}$

α has the units $mol^{-1/2}$ $l^{1/2}$ and β the units Ω^{-1} cm^2 $mol^{-3/2}$ $l^{1/2}$

as authoritatively selected or 'best' values. We feel much more work would have been required to produce such a table. The general sources of the data can be found amongst the references of Chapter 5. Dipole moments are gas-phase values, where available, derived from McClelland's compilation.[A9] Values for the dielectric constants of the alcohols from various sources are not in good agreement, and the values quoted are those of Dannhauser and Bahe.[A10] In selecting certain other values we have been guided by our contributors for the following: ethylene glycol[A11,A12] (M.S.), acetonitrile[A13] (M.S.), 2-propanol[A14] (J.E.P.), tetramethyl urea[A15] (J.E.P.) and HMPT[A16] (J.E.P.).

Discrepancies in physical properties and other measurements in organic solvents have often rightly been ascribed to impurities in the solvent material. Small traces of water can have a profound effect on certain properties and as pointed out in sect. 2.1.1 an apparently small amount of water expressed as a percentage is a large amount on a molar basis. Solvent purification is often no easy matter. The method to be adopted depends on the use to which the solvent is to be put. An impurity which may be anathema in one sort of physical method may be tolerated in another; for example surface active impurities must be eliminated in measurements involving interfaces, fluorescent impurities in Raman spectroscopy and ionic impurities in conductance methods. Below, only an indication of the most profitable purification method is given for solvents. The compilations[A8] should be consulted for reviews of purification methods appropriate to each solvent. Valuable work in this connection is being done by IUPAC Commissions, some reports from which are mentioned below. Another recent IUPAC publication[A17] deals with methods of assessing purity of organic substances.

In an Appendix to this chapter, some properties of selected mixed solvent systems are listed (Appendices 1.3.1–6).

1.3.1 Hydrocarbons

None of the hydrocarbons are good solvents for inorganic substances and all have low mutual solubility for water.

$CH_2CH_2CH_2CH_2CH_2CH_2$ (ring)

Cyclohexane is a commonly employed solvent in U.V. and I.R. spectroscopy. Its liquid range is 6.5 to 80.7°C. It is immiscible with water in the range 0.0055 to 99.990 wt % cyclohexane.

$CH_3(CH_2)_4CH_3$

Hexane is familiar in the laboratory as the principal component of petroleum ether. It is effectively immiscible with water.

C_6H_6

Benzene is a very extensively studied, low dielectric constant solvent. It is one of the easiest substances to purify and maintain in high state of purity. The freezing point is the simplest criterion of purity to employ. That it is a cumulative poison is often disregarded in its frequent use in the laboratory.

$C_6H_5CH_3$

Toluene is more toxic than benzene. Its freezing point is too low for recrystallisation to be a convenient method of purification although it has been used as a criterion of purity.

$CHCl_3$

Chloroform like other chlorinated hydrocarbons is unstable under storage. It reacts slowly with air and oxidising agents in the presence of light to give phosgene and other products. Commercial grade material contains stabilisers, often ethanol.

CCl_4

Carbon Tetrachloride is readily available in high purity grade and is a popular solvent. In spite of its high toxicity it is carelessly used. It may be absorbed by the body through the skin and by inhalation. The impurities present are usually other chlorinated or brominated hydrocarbons.

1.3.2 Alcohols

CH_3OH

Methanol was one of the earliest organic solvents used in physico-chemical studies. It is obtainable commercially in adequate purity for most purposes, the principal impurity being up to 0.05% water which tends to increase in amount with normal handling operations unless stringent precautions are taken. Water may be removed by distillation, or use of molecular sieves and calcium hydride.

CH_3CH_2OH

Ethanol has been less used as a solvent than methanol in quantitative studies. 'Absolute' alcohol usually contains about 0.01 vol % water, which can be removed using molecular sieves or metallic calcium.

$CH_3CH_2CH_2OHCH_3CHOHCH_3$

1-Propanol and *2-Propanol* are available anhydrous of high purity. One of the principal impurities of 1-propanol is 2-propene-2-ol.

$CH_3CH_2CH_2CH_2OH$

1-Butanol is the most toxic of the lower alcohols but fortunately its vapour pressure is low. It is not available in very high purity.

$CH_3CH_2CHOHCH_3$

2-Butanol is less toxic and may contain 1–2% impurity in commercial grade.

$(CH_3)_3COH$

t-Butanol (2 methyl-2 propanol) has been much studied particularly in recent years. Its melting point is 25.82°C but commercial grades often show melting points up to 2° lower. It is usually purified by refluxing over lime or calcium hydride followed by distillation.

$CH_2CH_2CH_2CH_2CH_2CHOH$ (ring: first CH_2 bonded to $CHOH$)

Cyclohexanol is extremely hygroscopic and has a low vapour pressure at room temperature. It melts at 25.12°C and may be purified by fractional crystallisation or by distillation. Its purity may be determined from its freezing point.

$CH_2OHCHOHCH_2OH$

Glycerol is also very hygroscopic and can absorb 50% of its weight of water. The water content can be estimated from the infrared band at 1.93 μm. Its high viscocity accounts for most of the interest in it.

$(CH_2OH)_2$

Ethylene glycol or 1,2-ethanediol is a colourless hygroscopic liquid with low vapour pressure. Neither it nor propylene glycol (1,2-propanediol) is toxic.

The structure of liquid alcohols and of alcohol-water mixtures has been reviewed by Franks and Ives.[A18] The pure liquids are associated by hydrogen bonding and dielectric relaxation time studies suggest rapid equilibrium between short-lived polymeric chains of various lengths.

1.3.3 Ethers

Unless ethers are stored cooled, in a nitrogen atmosphere or in contact with iron or copper filings, peroxides may form.

$(C_2H_5)_2O$

Ethyl ether is a familiar solvent, the most common impurities in which, besides peroxides, are ethanol, water and aldehydes. Its anaesthetic properties and the inflammability of ether-air mixtures are well known. Its higher homologues are more toxic.

$CH_3OCH_2CH_2OCH_3$

Dimethyl Cellosolve, monoglyme or 1,2 dimethoxyethane (DME) is an odourless liquid boiling at 85°C. Its dielectric constant is 7.2 and its dipole moment 1.71 D. It thus resembles tetrahydrofuran in most of its properties. It is a good solvent for the alkali metals with which it forms blue conducting solutions. It also dissolves organometallics.[A19]

$OCH_2CH_2CH_2CH_2$ (ring)

Tetrahydrofuran (*THF*) *or tetramethylene oxide* is a cyclic ether of moderate toxicity. It boils at 66°C and has a high v.p. at room temperature. It has been widely used as a solvent for organometallics, and forms blue solutions with the alkali metals. Most inorganic salts except lithium and sodium perchlorates are insoluble.

CH_2CH_2

O O

CH_2CH_2

1,4 or p-Dioxan was one of the first substances used in conjunction with water to obtain a range of mixtures of varying dielectric constant. Its faint smell suggests an inoffensive substance but in fact its vapours irritate eyes and skin, and inhalation can cause severe liver and kidney damage. Clearly it is not a substance to take liberties with. It is very hygroscopic and miscible in all proportions with water. Its dipole moment is zero at room temperature but rises with increase in temperature suggesting formation of the boat form.

$CH_2(OCH_3)_2$

Methylal or dimethoxymethane has a smell similar to chloroform and has similar toxic properties to *p*-dioxan. Its principal impurity is methanol. Little work has been reported with this solvent, perhaps because its dielectric constant is very low (3.5). It is miscible with water and a good solvent for organometallics but solubility for inorganic substances is low.

1.3.4 Ketones

CH_3COCH_3

Acetone (*AC*) is sufficiently familiar to require little description here. Its freezing point (−94.9°C) and complete miscibility with water have led to its use as an 'inert' additive to aqueous solutions to lower their freezing temperature for some n.m.r. studies.[A20] It does, however, form solvation complexes with both cobalt and copper. It has a high vapour pressure and may be readily purified by fractional distillation. It is a good solvent for organic compounds but less good for inorganic salts. $LiNO_3$, $LiClO_4$, $AgClO_4$ and $NaClO_4$ are soluble.

$CH_3CH_2COCH_3$ $\underline{CH_2CH_2CH_2CH_2CH_2}CO$ $C_6H_5COCH_3$

2-Butanone or methylethylketone, cyclohexanone and acetophenone have limited solubility in water. Their dielectric constants are similar to that of acetone (20).

1.3.5 Acids

HCOOH

Formic Acid slowly decomposes to carbon monoxide and water at room temperature, and this restricts its application as a solvent. It is extremely hygroscopic and corrosive to skin. The vapour is also dangerous. Its liquid range is 8.3 to 100°C and its dielectric constant is high (56). It is a good solvent for both inorganic and organic substances.

CH_3COOH

Acetic acid has been extensively used as a solvent medium for titration of bases. Familiar as 'glacial' acetic acid it is available in high purity, the melting point being the best criterion of purity. Like formic acid it is corrosive to skin. It is quite a good solvent for inorganic salts including chlorides and perchlorates.

$(CH_3CO)_2O$

Acetic anhydride has a large liquid range (−73.1 to 140°C) and is a good ionising solvent with high solubility for acetates with which it forms solvates, and perchlorates. It is hygroscopic reacting slowly with water to give acetic acid, which it resembles in its toxic properties except that it is even more unpleasant to use.

CF_3COOH

Trifluoroacetic acid (TFA) is a highly stable, strong acid which is very hygroscopic. It fumes in air and like acetic acid has an astringent vapour.

1.3.6 Esters

```
H2C—O
 |    \
 |     CO
 |    /
H2C—O
```

Ethylene carbonate (EC) or *1,3 dioxalan-2-one* has a high dielectric constant close to that of water and hence gives electrolytic solutions with inorganic salts. The alkali metal chlorides are, however, insoluble. Its liquid range is 36.4 to 248°C, and it is miscible with water above 40°C.

```
       H2C—O
        |    \
        |     CO
        |    /
CH3—HC—O
```

Propylene carbonate (PC) *or 4 methyldioxol-2-one* first attracted the attention of physical chemists in the early 1960s when it became commercially available. It is one of the most interesting of the newer dipolar aprotic solvents, with a high dielectric constant (64) and a convenient liquid range (−49 to 242°C although decomposition occurs above 150°C). It is an efficient solvent for both inorganic and organic compounds. It is stable, easily purified by vacuum distillation and of low toxicity. It is immiscible with water[A21] in the range 17.5 to 93 wt% PC. A study of proton chemical shifts in PC-H_2O mixtures has been interpreted in terms of the polymeric association of water.[A22] In 80% EC-PC mixture the dielectric constant is increased to 87 but the viscosity is effectively unchanged. A recent review of the applications of PC in electrochemistry is available,[A23] and the report from a IUPAC commission.[A24]

1.3.7 Nitrohydrocarbons

CH_3NO_2

Nitromethane is a liquid in the range −28.5 to 101.2°C. Like the other nitro-alkanes it is a fairly reactive solvent with considerable mutual solubility for water. The tautomeric aci-form, a pseudo acid, is more soluble in water and reacts with alkalis to form a water-soluble salt. Nitromethane has been used as a differentiating solvent for the titration of weak bases. It is not a good solvent for inorganic salts although $LiClO_4$ and $Mg(ClO_4)_2$ are soluble. It may be purified by distillation or fractional crystallisation but the purity deteriorates on storage and exposure to light.

$C_6H_5NO_2$

Nitrobenzene (NB) when very pure is a straw-coloured liquid with characteristic smell which darkens on exposure to light. It is readily absorbed by the skin and leads to systemic injuries.

1.3.8 Nitriles

CH_3CN

Acetonitrile (AN) or Methyl cyanide has been commercially available for 20 years. Its favourable dielectric constant, optical properties and its useful solubility for certain substances have led to it being a very widely used solvent. The impurities present in commercial grade material are usually water, unsaturated nitriles, toluene and perhaps aldehydes and amines. O'Donnell, Ayres and Mann[A25] have discussed purification methods and Coetzee[A26] has reported on this solvent on behalf of a IUPAC commission. Acetonitrile is a relatively inert solvent but may be hydrolysed by strong alkalis or dilute acids. It is a liquid in the range −44.9 to 81.6°C and is miscible with water over the whole composition range. It has a high vapour pressure and characteristic odour. Acetonitrile is available in special purified grades for spectroscopy and its u.v. cut-off is used as a criterion of its purity. The large amount of information available on purification procedures makes this a good example of the need to match the purification technique to the type of measurement to be made. Acetonitrile is only moderately toxic and normal handling precautions are adequate. Solubilities of inorganic salts in acetonitrile have been tabulated.[A27] Halides and nitrates are soluble but sulphates are almost insoluble.

C_6H_5CN

Benzonitrile is an excellent solvent for many anhydrous inorganic salts, organometallic compounds and a variety of organic substances. It is strongly polar with a low vapour pressure and low mutual solubility for water. It appears not to be unduly toxic.

1.3.9 Amines

Amines are basic and readily form covalent compounds with acids.

$H_2NCH_2CH_2NH_2$

Ethylenediamine is a colourless liquid of low dielectric constant (12) with a smell similar to ammonia. It is soluble in water and very hygroscopic and should always be protected from atmospheric moisture. Protection from carbon dioxide is also essential since this forms a carbonate. Ethylenediamine has found application as an ionising solvent for ionic reactions and in various aspects of electrochemistry. Exposure to its vapours can lead to dermatitis and other allergic reactions.

Readily soluble salts include KI, NaI, NaBr, NaSCN, $NaNO_3$, $NaClO_4$ and $KClO_4$. Chlorides are only slightly soluble. Many salts form solvates and there are numerous complexes formed with transition metal ions. It also dissolves the alkali metals giving blue solutions (cf. liquid ammonia).

```
HNCH2CH2CH2CH2CH2
 |_______________|
```

Piperidine is a colourless liquid boiling at 106°C. It has a characteristically unpleasant, sharp amine smell.

The principal impurities are water and carbon dioxide which can be removed by distillation from potassium hydroxide pellets. Like ethylenediamine and pyridine it has been used as a solvent for non-aqueous potentiometric titrations of very weak acids. Contact with skin leads to severe burns. The vapours cause irritation if inhaled.

```
      H
      C
    /   \\
  HC     CH
  ||     |
  HC     CH
    \   //
      N
```

Pyridine (Py) is a liquid from −41 to 115°C, with a moderate vapour pressure and distinctive, unpleasant smell.

It is only a weak base in water, with which it is completely miscible. It is very hygroscopic. As mentioned above it has often been used as a solvent or co-solvent for acid-base potentiometric titrations. It is a chronic poison.

It is a good solvent for lithium, sodium and silver salts and the alkaline earth halides. KCl, NaCl, NH_4Cl, KNO_3, KBr and KI are amongst those practically insoluble.

The principal impurities in commercial material are water and amines such as picolines and lutidines. It may be purified by treatment with solid caustic potash followed by fractional distillation. Lindauer and Mukherjee have reported on its properties and purification on behalf of IUPAC.[A28]

1.3.10 Amides

Amides make excellent solvents for a wide range of inorganic and organic materials. Some amides have higher dielectric constants than water indicating a high degree of association. All are good cation solvators and are miscible with water. A comprehensive review of electrochemical aspects[A29] and a more general review,[A30] are available.

$HCONH_2$

Formamide was used as an ionising solvent by Walden, amongst others, sixty years ago. It is a weak base easily hydrolysed in the presence of both acids and bases. It is very hygroscopic and is usually characterised in purity by a freezing point of 2.55°C and a specific conductance of less than 10^{-6} ohm^{-1} cm^{-1}. Notley and Spiro[A31] have described a method involving both molecular sieves and ion exchange resins to obtain a material of specific conductance 2×10^{-7} ohm^{-1} cm^{-1}, but low freezing point.

It boils at 210°C and has a dielectric constant higher than that of water (109). It is a good solvent for chlorides, nitrates and sulphates. It will also dissolve polar organic substances.

$HCONH(CH_3)$

N-methylformamide (NMF) has a liquid range of −4 to 180°C and a very high dielectric constant (182). It may be purified by vacuum distillation from phosphorus pentoxide. The purified solvent is not stable as shown by a change in specific conductance with time. French and Glover,[A32] who measured the conductance of alkali metal halides in NMF, concluded that it is a highly dissociating solvent; it has also great levelling action.

$HCON(CH_3)_2$

N,N-dimethylformamide (DMF) is a colourless liquid, with a liquid range of −60 to 153°C, low vapour pressure and good solubility for a wide range of substances. Its slight smell results from hydrolysis and it is only slightly basic. Water is not easy to remove. Prue and Sherrington[A33] removed most of the water from reagent grade material as the benzene azeotrope. It was then treated with phosphorus pentoxide and potassium hydroxide successively and distilled at low pressure under nitrogen. It is a popular solvent in visible and near u.v. spectrometry ($>$270 nm) and for polarographic work. It is toxic and inhalation of vapour leads to stomach and liver complaints.

It is a good solvent for the alkali metal and alkaline earth perchlorates and iodides. Lithium chloride is soluble but other chlorides are only slightly soluble. Nitrites are soluble but the solutions are not stable.

CH_3CONH_2

Acetamide freezes at 81.0°C but an extensive study of liquid acetamide as an ionising solvent has been made. The solid is colourless, odourless and deliquescent; the familiar mousy smell is indicative of impure material. It is a good solvent for many inorganic and organic substances and forms several solvates.[A34]

$CH_3CONH(CH_3)$

N-methylacetamide (NMA) has a liquid range of 30.5 to 206°C and is notable for its very high dielectric constant (191.3 at 32°C), hence the many salts which are soluble are completely ionised in it. It is relatively easy to purify by fractional distillation under nitrogen. Its purity may be gauged from its specific conductance, usually lower than 5×10^{-7} ohm^{-1} cm^{-1}. The high dielectric constant is attributed to the liquid being highly structured with polymeric chains linked by hydrogen bonding. A recent IUPAC report is available.[A35]

$CH_3CON(CH_3)_2$

N,N-dimethylacetamide (DMA) has a liquid range of −20 to 165°C and a dielectric constant of 37.8 at 25°C. It is miscible with water and is generally similar in its properties to DMF.

$CH_3CH_2CONH(CH_3)$

N-methylpropionamide (NMP) has a liquid range of −30.9 to 148°C and a dielectric constant of 176 at 25°C. It has been much less studied than NMA, DMA, etc. Hoover[A36] has presented density and dielectric constant data of NMP-H_2O at 20 to 40°C, and discussed the effects of these two highly structured solvents on each other.

$(CH_3)_2NCON(CH_3)_2$

1,1,3,3, Tetramethylurea (TMU) has a low dielectric constant for an amide. It is a good solvent for organic substances but solubility of inorganic salts is limited. Its liquid range is −1.2 to 176.5°C and water can be removed by distillation. It has a slight not unpleasant smell and is non-toxic.

$[(CH_3)_2N]_3PO$

Hexamethylphosphotriamide or tris(dimethylamino)phosphate (HMPT or HMPA) is a colourless, fairly viscous, basic liquid, miscible with water in all proportions and with many other solvents. It is a good solvent for many inorganic salts, gases, organic compounds and the alkali metals, but as yet has not been as extensively studied as some of the other dipolar aprotic solvents.[A37]

It has a very low vapour pressure and its liquid range is 7.2 to 233°C, with a dielectric constant of intermediate value (29.60). It may be purified by vacuum distillation, and dried with molecular sieves.

Pyrrolidones. 2-pyrrolidone and *N*-methyl pyrrolidone are an interesting pair of solvents having closely similar dielectric constants (27.4 and 32.0 respectively) but one is protic and the other dipolar aprotic.

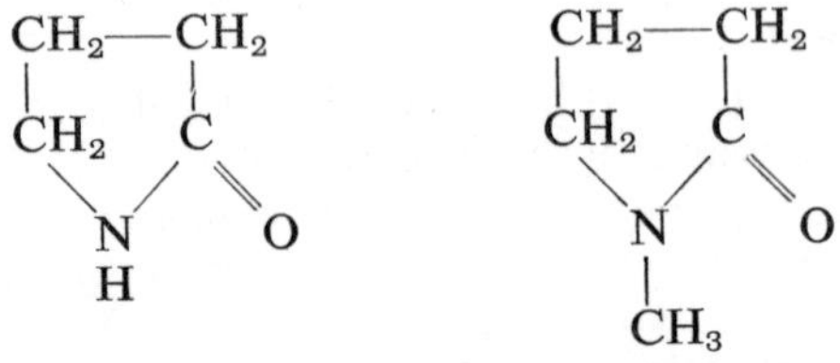

$$\begin{array}{c} CH_2-CH_2 \\ | \quad\quad | \\ CH_2 \quad C \\ \diagdown \;\; \diagup \;\; \diagdown\!\!\diagdown \\ N \quad\quad O \\ H \end{array} \qquad\qquad \begin{array}{c} CH_2-CH_2 \\ | \quad\quad | \\ CH_2 \quad C \\ \diagdown \;\; \diagup \;\; \diagdown\!\!\diagdown \\ N \quad\quad O \\ | \\ CH_3 \end{array}$$

2-pyrrolidone *N*-methylpyrrolidone

2-pyrrolidone has a liquid range of 25 to 245°C, dipole moment 3.7 D and density 1.107 g cm^{-3} at 25°C. It is a good solvent for both organic and inorganic solutes and is completely miscible with water, alcohols, ketones and aromatic hydrocarbons. A first electrochemical study has recently been reported.[A38] The chemical and electrochemical properties of *N*-methyl pyrrolidone have been surveyed.[A39]

1.3.11 Sulphur Compounds

CS_2

Carbon disulphide is extremely volatile but the pure substance has only a little smell though unpleasant. It is toxic and with its high inflammability it must be classed as hazardous.

$(CH_3)_2SO$

Dimethylsulphoxide (DMSO) has proved to be a highly versatile and useful solvent since it became commercially available. The first chapter of a recent book, otherwise devoted to biomedical aspects, gives a useful review[A40] of the chemistry of DMSO. It is an odourless, hygroscopic liquid with a bitter taste. Its clinical applications, for example to dissolve substances for direct absorption through the skin, were curtailed after reports of toxic effects. These have been disputed but it would seem wise, because of its skin-penetrating properties, always to carry out handling operations with rubber gloves. Some explosions in mixtures involving DMSO have been reported.

The purification of DMSO for electrochemical studies has been surveyed by Reddy[A41] on behalf of a IUPAC Commission. Water is the principal impurity and the purified solvent readily absorbs moisture from the atmosphere. The recommended procedure is to remove water and low boiling impurities with molecular sieve type 5A, followed by vacuum distillation. Water may be detected by its n.m.r. signal at 3.25 ppm downfield of TMS.

Its physical properties suggest that DMSO is highly associated in both liquid and solid states. This is confirmed by spectroscopic studies.[A42] The similarity of the intermolecular frequencies in the low frequency spectra of both liquid and solid DMSO suggests a high degree of dipole association. X-ray and neutron inelastic scattering studies indicate[A42] in mixed aqueous solution the formation of hydrogen-bonded DMSO-H_2O complexes, for which supporting evidence is available from other measurements.

DMSO is a good solvent for the alkali metal halides, nitrates, thiocyanates and perchlorates. It has strong solvating properties for transition metal ions such as Fe^{3+} and Ni^{2+}. The pure liquid is odourless but reagent grade material often has a sickly penetrating smell.

```
H2C      CH2
 |        |
H2C      CH2
   \    /
     S
   //  \\
  O      O
```

Sulpholane, Tetramethylenesulphone or Tetrahydrothiophene-1,1-dioxide, ($TMSO_2$) is a stable liquid in the range 28.45 to 220°C. It is a dipolar aprotic solvent of moderately high dielectric constant and viscosity. It is a good solvent for many organic compounds, and its high cryoscopic constant has led to freezing point depression studies of solutions (see sects. 2.8–2.10). It is usually purified by distillation from solid potassium hydroxide since it commonly contains an acid impurity. Its low vapour pressure is probably the reason for no toxic effects being reported.

Sulpholane resembles sulphur dioxide in its good solvent properties for inorganic compounds. Its solvating properties are weaker than those of DMSO.

$(CH_3)_2SO_2$

Dimethylsulphone ($DMSO_2$) does not melt until 109°C and hence has been little used as a solvent although some high temperature polarographic studies have reported. It is highly associated,[A42] and complexes slightly with inorganic cations.

APPENDICES 1.3.1–6

In the following tables, data are given for six important binary systems. The choice of these has been arbitrary as it is clearly impossible to be comprehensive in a book of this nature.

Table 1.3.1

Methanol-Water* (25°C)

Wt% Methanol	Mol Fraction Methanol	Density g cm^{-3}	Dielectric Constant	Viscosity cpoise
0	0	0.9971	78.30	0.8903
10	0.0588	0.9799	74.16	1.158
20	0.1232	0.9644	69.95	1.400
40	0.2725	0.9316	60.93	1.593
50	0.3599	0.9119	56.30	1.533
60	0.4574	0.8907	51.61	1.403
80	0.6922	0.8425	42.58	1.006
90	0.8350	0.8150	37.92	0.767
100	1.0000	0.7866	32.70	0.545

* Based on R. G. Bates and R. A. Robinson, *in* Chemical Physics of Ionic Solutions, p. 211. B. E. Conway and R. G. Barradas, (eds.), Wiley, N.Y., 1966,

Table 1.3.2

Ethanol-Water* (25°C)

Wt% Ethanol	Mol Fraction Ethanol	Density g cm^{-3}	Dielectric Constant	Viscosity cpoise
0	0	0.9971	78.30	0.8903
10	0.0416	0.9804	72.80	1.328
20	0.0890	0.9664	66.99	1.808
40	0.2068	0.9315	55.02	2.374
60	0.3697	0.8870	43.40	2.232
80	0.6100	0.8391	32.84	1.738
100	1.000	0.7851	24.55	1.089

* Based on T. Shedlovsky and H. O. Spivey, *J. Phys. Chem.*, **71,** 2165 (1967).

Table 1.3.3

***p*-Dioxan-Water* (25°C)**

Wt% Dioxan	Mol Fraction Dioxan	Density g cm^{-3}	Dielectric Constant	Viscosity cpoise
0	0	0.9971	78.3	0.8903
20	0.0486	1.0143	63.5	1.30
40	0.1200	1.0284	44.4	1.74
60	0.2347	1.0360	27.5	1.98
80	0.4498	1.0350	12.1	1.73
100	1.0000	1.0269	2.209	1.196

* Interpolated from T. L. Fabry and R. M. Fuoss, *J. Phys. Chem.*, **68,** 971 (1964).

Table 1.3.4

Dimethylsulphoxide-Water* (25°C)

Wt% DMSO	Mol Fraction DMSO	Density g cm^{-3}	Dielectric Constant	Viscosity cpoise
0	0	0.9971	78.3	0.8903
10	0.0250	1.0105	78.0	1.06
20	0.0545	1.0242	77.4	1.34
30	0.0899	1.0387	76.7	1.69
40	0.1333	1.0535	75.9	2.22
50	0.1874	1.0682	74.5	2.83
60	0.2570	1.0823	72.4	3.42
70	0.3499	1.0926	69.5	3.73
80	0.4798	1.0960	65.3	3.45
90	0.6749	1.0960	59.0	2.70
100	1.0000	1.0958	46.7	1.96

* Based on R. G. LeBel and D. A. I. Goring, *J. Chem. Eng. Data*, **7,** 100 (1962); J. J. Lindberg and J. Kenttämaa, *Suomen Kem. B*, **33,** 104 (1960); E. Tommila and A. Pajunen, *Suomen Kem. B*, **41,** 172 (1968); A. D. Pethybridge (University of Reading), private communication (1971).

Table 1.3.5

Sulpholane-Water*† (25°C)

Wt% Sulpholane	Mol Fraction Sulpholane	Density g cm^{-3}	Dielectric Constant	Viscosity cpoise
0	0	0.9971	78.3	0.8903
10	0.0164	1.0208	75.9	1.010
20	0.0361	1.0450	73.3	1.152
40	0.0909	1.0954	68.3	1.53
50	0.1304	1.1216	65.6	1.86
60	0.1836	1.1489	61.8	2.30
80	0.3749	1.2049	54.7	3.82
90	0.5744	1.2341	50.1	5.38
95	0.7402	1.2492	47.6	6.60

* Based on E. Tommila, E. Lindell, M-L. Virtalaine and R. Laakso, *Suomen Kem. B*, **42,** 95 (1969) (which also gives activity coefficients).

† Pure sulpholane is a solid at 25°C.

Table 1.3.6

Methanol-Sulpholane*† (25°C)

Wt% Sulpholane	Mol Fraction Sulpholane	Density g cm^{-3}	Dielectric Constant	Viscosity cpoise
0	0	0.7866	32.70	0.545
10	0.0229	0.8219	33.2	0.576
20	0.0625	0.8614	34.2	0.631
40	0.1509	0.9403	36.1	0.792
50	0.2105	0.9848	37.1	0.938
60	0.2856	1.0367	38.3	1.163
80	0.5160	1.1404	40.7	2.32
90	0.7057	1.1210	42.3	4.25

* Based on E. Tommila *et al., Suomen Kem. B*, **42,** 95 (1969).

† Pure sulpholane is a solid at 25°C.

REFERENCES

A1 M. J. Blandamer and D. Waddington, *Adv. in Mol. Relaxation Processes*, **2,** 1 (1970)

A2 S. Petrucci and M. Battistini, *J. Phys. Chem.*, **71,** 1181 (1967)

A3 M. J. Blandamer, M. J. Foster, N. J. Hidden and M. C. R. Symons, *Trans. Faraday Soc.*, **64,** 3247 (1968)

A4 M. J. Blandamer, D. E. Clarke, N. J. Hidden and M. C. R. Symons, *Trans. Faraday Soc.*, **64,** 2683 (1968)

A5 S. Petrucci *et al.*, *J. Phys. Chem.*, **71,** 1174, 3087, 3414 (1967)

A6 S. C. Chen and J. P. Valleau, *Can. J. Chem.*, **46,** 853 (1968)

A7 A. J. Parker, *Chem. Rev.*, **69,** 1 (1969)

A8a J. A. Riddick and W. B. Bunger, Organic Solvents, 3rd edition, Techniques of Chemistry, Vol. II (A. Weissburger, ed.), Wiley, New York (1970)

A8b G. J. Janz and R. P. T. Tomkins, Non-aqueous Electrolytes Handbook, Vol. 1. Academic Press, New York (1972)

A9 A. L. McClelland, Tables of Experimental Dipole Moments, Freeman, San Francisco, (1963)

A10 W. Dannhauser and L. H. Bahe, *J. Chem. Phys.*, **40,** 3058 (1964)

A11 F. Accascina and S. Petrucci, *Ric. Sci.*, **30,** 808 (1960)

A12 N. Koizumi and T. Hanai, *J. Phys. Chem.*, **60,** 1496 (1956)

A13 G. P. Cunningham, G. A. Vidulich and R. L. Kay, *J. Chem. Eng. Data*, **12,** 336 (1967)

A14 M. A. Matesich, J. A. Nadas and D. F. Evans, *J. Phys. Chem.*, **74,** 4568 (1970)

A15 B. J. Barker and J. A. Caruso, *J. Amer. Chem. Soc.*, **93,** 1341 (1971)

A16 Private communication from J. E. Prue (1971)

A17 L. A. K. Staveley (ed.), The Characterisation of Chemical Purity of Organic Compounds, Butterworths, London (1971)

A18 F. Franks and D. J. G. Ives, *Quart. Rev.*, **20,** 1 (1966)

A19 C. Agami, *Bull. Soc. Chim. France*, 1205 (1968)

A20 A. Fratiello, R. E. Lee, V. M. Nishida and R. E. Schuster, *J. Chem. Phys.*, **48,** 3705 (1968)

A21 N. F. Catherall and A. G. Williamson, *J. Chem. Eng. Data*, **16,** 335 (1971)

A22 D. R. Cogley, J. N. Butler and E. Grunwald, *J. Phys. Chem.*, **75,** 1477 (1971)

A23 R. Jasinski, *in* Adv. Electrochem. and Electrochem. Eng., Vol. 8, pp. 253–332 (P. Delahay and C. W. Tobias, eds.) (1971)

A24 T. Fujinaga and K. Izutsu, *Pure Appl. Chem.*, **25,** 273 (1971)

A25 J. F. O'Donnell, J. T. Ayres and C. K. Mann, *Anal. Chem.*, **37,** 1161 (1968)

A26 J. F. Coetzee, *Pure Appl. Chem.*, **13,** 429 (1966)

A27 J. Timmermans, I. N. LaFontaine and R. Phillippe, *Mem. Acad. Roy. Belg.*, **37,** 7–90 (1966)

A28 R. Lindauer and L. M. Mukherjee, *Pure Appl. Chem.*, **25,** 265 (1971)

A29 D. S. Reid and C. A. Vincent, *J. Electroanal. Chem.*, **18,** 427 (1968)

A30 J. W. Vaughn, *in* The Chemistry of Non-Aqueous Solvents, Vol. II, pp. 191–264 (J. J. Lagowski, ed.), Academic Press, N.Y. (1967)

A31 J. M. Notley and M. Spiro, *J. Chem. Soc. B*, 362 (1966)

A32 C. M. French and K. H. Glover, *Trans. Faraday Soc.*, **51,** 1418 (1955)

A33 J. E. Prue and P. J. Sherrington, *Trans. Faraday Soc.*, **57,** 1795 (1961)

A34 G. Jander and G. Winkler, *J. Inorg. Nucl. Chem.*, **9,** 24, 32, 39 (1959)

A35 L. A. Knecht, *Pure Appl. Chem.*, **25,** 281 (1971)

A36 T. B. Hoover, *J. Phys. Chem.*, **73,** 57 (1969)

A37 H. Normant, *Angew. Chem. Int. Ed.*, **6,** 1046 (1967); *Bull. Soc. Chim. France*, 791 (1968)

A38 C. Sinicki and M. Porteix, *J. Electroanal. Chem.*, **34,** 439 (1972)
A39 M. Breant, *Bull. Soc. Chim. France*, 725 (1971)
A40 H. H. Szmant, *in* Dimethyl Sulfoxide, pp. 1–87 (S. W. Jacob, E. E. Rosenbaum and D. C. Wood, eds.), Dekker, New York (1971)
A41 T. B. Reddy, *Pure Appl. Chem.*, **25,** 457 (1971)
A42 G. J. Safford, P. C. Schaffer, P. S. Leung, G. F. Doebbler, G. W. Brady and E. F. X. Lyden, *J. Chem. Phys.*, **50,** 2140 (1969)

Chapter 2

Thermodynamic Measurements

Part 1

Solubility, Calorimetry, Volume Measurements, and Viscosities

C. M. Criss
Dept. of Chemistry, University of Miami, Coral Gables, Florida, 33124, U.S.A.

NOTATION

a_i Activity of component i.
$a_\pm$ Mean activity of an electrolyte.
a Ion-size parameter.

A	Debye-Hückel limiting slope constant for activity coefficients (eqn. 2.3.11).
A_H, A_V	Theoretical limiting slope for heats and volumes respectively.
A_η	Experimental limiting slope for viscosities.
b	Empirical constant in eqn. 2.3.67.
B	Constant in eqn. 2.3.50.
B_η	Empirical constant in eqn. 2.3.74.
c	Molar concentration.
C	Empirical constant in eqn. 2.3.50.
C_p	Heat capacity at constant pressure.
$\bar{C}_{pi}$	Partial molal heat capacity of component *i*.
$\bar{C}^0_{pi}$	Standard partial molal heat capacity of component *i*.
d	Density of solution.
d_0	Density of solvent.
D	Empirical constant in eqn. 2.3.50.
D_η	Empirical constant in eqn. 2.3.75.
e	Electronic charge.
ε_r	Dielectric constant (relative permittivity).
$\bar{G}_i$	Partial molal free energy of component *i*.
$\bar{G}^0_i$	Standard partial molal free energy of component *i*.
H	Total enthalpy.
$\bar{H}_i$	Partial molal enthalpy of component *i*.
$\bar{H}^0_i$	Standard partial molal enthalpy of component *i*.
I	Ionic strength, defined by eqn. 2.3.12.
$\bar{J}_i$	Relative partial molal heat capacity of component *i*.
k	Boltzmann constant.
K	Constant used in eqn. 2.3.68 and defined by eqn. 2.3.70.
K_{s0}	Solubility product constant.
K_V	Empirical constant in eqn. 2.3.71.
$\bar{L}_i$	Relative partial molal enthalpy of component *i*.
m	Molal concentration.
m_i	Molal concentration of component *i*.
$m_\pm$	Mean molal concentration of an electrolyte.
M_i	Molecular weight of component *i*.
n_i	Number of moles of component *i*.
N	Avogadro number.
p	Pressure.
R	Molar gas constant.
S_V	Experimental limiting slope for volume (including valence factor).
$\bar{S}_i$	Partial molal entropy of component *i*.
$\bar{S}^0_i$	Standard partial molal entropy of component *i*.
T	Temperature.
T_s	Standard temperature.

V	Volume.
$\bar{V}_i$	Partial molal volume of component i.
$\bar{V}_i^0$	Standard partial molal volume of component i.
w	Valence factor defined by eqn. 2.3.69.
x_i	Mole fraction of component i.
z_+, z_-	Charge on indicated ion.
ω	Special function defined by eqn. 2.3.49.
$\alpha(m)$	Degree of dissociation at molality m.
α	Temperature coefficient of expansion.
β	Coefficient of compressibility.
$\gamma_\pm$	Mean activity coefficient of an electrolyte.
ΔG	Change in free energy.
ΔG^0	Standard change in free energy.
ΔG^0_{soln}, ΔH^0_{soln}, ΔS^0_{soln}	Standard change of indicated quantity upon dissolution of solute.
ΔH_f^0	Standard enthalpy of fusion.
ΔH_a	Enthalpy of association.
η	Viscosity of solution.
η_0	Viscosity of solvent.
κ	Function defined by eqn. 2.3.73.
Λ	Equivalent conductance.
Λ_0	Equivalent conductance at infinite dilution.
ν	Total number of ions produced by one molecule of electrolyte.
ν_+, ν_-	Number of cations and anions in one molecule of electrolyte.
ρ	Special function defined by eqn. 2.3.48.
$\sigma(x)$	Special function in eqn. 2.3.43.
ϕ_{C_p}, ϕ_H, ϕ_L, ϕ_V	Apparent molal quantity indicated.
Ω_V	Function defined by eqn. 2.3.72.

2.1 INTRODUCTION

Thermodynamic properties of solutions are not only useful for estimating the feasibility of reactions in solution, but they also offer one of the better methods of investigating the theoretical aspects of solution structure. This is particularly true for the standard partial molal entropy, heat capacity, and volume of the solutes, values of which are sensitive to the arrangement of solvent molecules around a solute molecule. They have been examined extensively[B1–4] in aqueous solution for the purpose of structure interpretation and more recently in non-aqueous solutions.[B5–8] Enthalpies and free energies of solvation and transfer between solvents have also been valuable in the testing of theoretical relationships such as the Born equation[B9] and modifications of the Born equation,[B10] as well as predicting solvation effects in reaction kinetics.[B11,B12]

Numerical values for these functions come from a broad range of sources, including equilibrium studies, direct calorimetric measurements, density determinations, and electrochemical measurements. The first three of these will be discussed in sects. 2.1–2.4. Electrochemical methods are treated separately in sects. 2.5–2.7.

Of the various equilibrium studies excluding cell measurements, solubilities present the most abundant source of data for non-aqueous solutions. They are most useful for the determination of free energies, and if obtained with sufficient accuracy over a temperature range they can be used to estimate heats and entropies of solution as well. Unfortunately, many of the reported data have been gathered for purposes other than calculating thermodynamic properties and being imprecise are of very little use for this purpose. The most common omission in such data is the state of solvation of the solid phase. Seidell and Linke[B13] have published a comprehensive compilation of the solubilities of inorganic elements and compounds in both water and non-aqueous solvent systems. Volume I of this two-volume set includes data abstracted through 1956, and Volume II contains data through the 'early sixties'. A planned third volume which was to list the solubilities of organic materials apparently has not been published. Solubility data reviewed in these sections (2.1–2.4) will include only that data which have appeared since the publication of this comprehensive work and which, in the opinion of the author, is potentially useful for thermodynamic calculations.

In order to use solubility data for salts of moderate solubility in the calculation of thermodynamic values, one must also have the corresponding activity coefficients. Such data, particularly at the higher concentrations, are exceedingly scarce. Most activity coefficient data which now exist are primarily for dilute solutions and have been derived from electrochemical measurements. This subject is covered elsewhere in this chapter, and some of the activity coefficients derived from this source are listed in the appendices. Some interesting data obtained from other sources, particularly from freezing point measurements, are now beginning to appear.[B14–16]

Calorimetrically measured heats of solution are generally much more reliable than those obtained from the temperature coefficient of the solubility, and should be weighted more heavily when there is a choice. The first heat measurements of thermodynamic significance in non-aqueous solvents are those of de Forcrand[B17] in 1885 and Pickering[B18] in 1888. The first of these investigators confined his measurements to a series of alcohols, while the latter investigated electrolytes in ethanol and water and some non-electrolytes in a variety of solvents. Only within the past ten years has there been any significant heat measurements for electrolytes in solvents of intermediate or high dielectric constant.[B19-28] The meagre amount of earlier work in non-aqueous solvents has been compiled by Rossini and co-workers.[B29] More recently Mishchenko and Poltoratskii[B30] have compiled additional heat data for electrolytes in non-aqueous solvents at various temperatures. These data are confined almost entirely to organic solvents of low dielectric constant.

Density measurements, like solubilities, have also frequently been obtained for purposes other than calculating thermodynamic properties, and consequently many of these are not useful for evaluating partial molal volumes. There are, however, some excellent recent measurements in this area which have been used to evaluate partial molal volumes. Since Millero[B31] has thoroughly examined all volume data for electrolytes in both aqueous and non-aqueous solutions in a recent review, only some of the more interesting results, from a structural point of view, will be included here.

While the viscosity of electrolytic solutions is not a thermodynamic property, such data frequently are reported with partial molal volumes, and viscosities are often valuable in elucidating the structure of solutions. Therefore, a few viscosity data which have a direct bearing on the discussion of the nature of these solutions are included.

In general, the discussion is limited to solutions made from pure organic solvents or mixed solvents containing no more than 50% water, except in those few cases where data are available for the entire range of solvent composition.

Thermodynamic measurements in non-aqueous solutions frequently present experimental and theoretical problems not encountered in aqueous solutions. In order to aid those not familiar with such measurements, sections are included on experimental techniques and treatment of data. The latter is of particular importance since one is generally interested in thermodynamic values at infinite dilution and these quantities are very sensitive to the method used in extrapolation.

2.2 EXPERIMENTAL TECHNIQUES AND RELATED PROBLEMS

The general techniques employed in determining thermodynamic values in non-aqueous systems are usually similar to those used in aqueous solution. These methods have been discussed elsewhere, and since space considerations preclude a detailed account, only pertinent general references and special problems related to non-aqueous solvents will be considered here.

The volumes edited by Weissberger,[B32] comprehensively cover techniques in solubility, density, calorimetric and viscosity measurements. Reilly and Rae[B33] deal with a variety of topics including a useful general discussion of solubility measurements. Other more specific treatments exist for solubilities,[B34–36] for calorimetric measurements,[B27,B37–42] (including descriptions of calorimeters suitable for non-aqueous work[B27,B41,B42] and for density measurements by various methods.[B43–47] Millero's[B47] magnetic float densitometer, having a precision of 0.3 ppm is particularly suitable for non-aqueous systems. Robinson and Stokes[B48] have reviewed methods of determining activity coefficients from measurements of the vapour pressure of the solvent, and Soldano *et al.*[B49] have described in detail a high temperature isopiestic unit which could be useful for studying organic solvent systems. Experimental methods used for obtaining accurate freezing points to be used in evaluating activity coefficients have also been described.[B50,B51] Further details are given in sects. 2.8–2.13. Precise viscosity measurements are deceptively simple, but excellent descriptions of accurate viscometers and required precautions exist.[B52–54]

In addition to the procedures and precautions required for aqueous systems, others inherent to non-aqueous systems must be observed. These are often ignored by the research worker unfamiliar with the nature of organic solvent systems. They can be divided into five categories: (1) solvent purity, (2) solubility of solutes and nature of dissolved species, (3) rate of dissolution of solutes, (4) solvent vapour pressure and (5) measurements in dilute solution. While these same items arise in work with aqueous solutions, they do not occur with such frequency and cause such practical difficulties as with non-aqueous solvents.

Obtaining and maintaining a pure solvent is probably the most difficult single problem facing the researcher in this area. To maintain purity is frequently as difficult as the initial purification process. An added problem is that it is generally not clear what degree of purity is required for reliable measurements. Nevertheless one should constantly be on guard for impurities such as water, grease or other materials being introduced by the solvent handling system. Solvents of marginal stability frequently must be vacuum-distilled at low temperature. To keep the solvent from becoming contaminated with grease and/or rubber the liquid must never come in direct contact with greased joints; vacuum valves in the system should be of the greaseless variety, constructed so that the solvent does not contact rubber O-rings. Fortunately, vacuum-tight teflon valves, which meet these requirements, are now commercially available. Most of the solvents are somewhat hygroscopic so that the solvent should be transferred to the measuring vessel (calorimeter, solubility cell, etc.) without allowing it to contact the air. Using this approach it is possible to maintain water contents of one or two thousandths of a per cent as indicated by Karl-Fischer titration.[B55] While this may appear to be a trace amount, it may be comparable, on a molal basis, with the concentration of electrolyte in dilute solutions. For example, the mole ratio of water to salt in a 0.001*m* solution of salt in a solvent containing 0.001 % water is 0.56. Furthermore, unless the vessel, used in the measurements, has been rigorously dried before admission of the solvent, the water content can easily increase to 0.01% by desorption of water from the surface of the glass. To prevent water from entering calorimetric vessels, they should be of the bomb-type construction[B42,B56] or have a vapour barrier through which the stirrer passes. While water is probably the most difficult impurity to avoid, one must also consider impurities introduced through spontaneous decomposition of the solvent. Several such cases have been reported.[B51,B57,B58]

The degree of solubility of the solute and nature of the dissolved species are of prime consideration for thermodynamic purposes. Many substances considered to be strong electrolytes in water are insoluble in organic solvents or dissolve to form an undissociated species. Ignorance of the nature of the dissolved solute can lead to considerable error in thermodynamic values calculated from such data. Ion-pair formation, particularly in solvents of lower dielectric constant can cause difficulties in the extrapolation of data to infinite dilution.

In calorimetry, not only is the degree of solubility and degree of dissociation important, but the slow rate of dissolution of many electrolytes in organic solvents must also be considered. It is advisable to make an initial study of the nature of the system by first observing the dissolution rate of the solute and then determining the degree of dissociation by conductance measurements.[B25] Much time can be saved by these preliminary examinations. The slow rate of solution of many salts may also pose problems in solubility measurements. It is advisable, as in the case of aqueous solutions, that solubility studies be conducted from both super- and under-saturation to ensure that equilibration has been attained.

Additional precautions are often required in calorimetry involving solvents with high vapour pressures. Since a significant quantity of solvent may vapourise either to fill the void above the liquid during a temperature change, or to fill the void when the ampoule containing the sample is broken. The heat required for the vaporisation process may introduce an unacceptable amount of error in the heat of solution of the solute. On the other hand, solvents having high vapour pressures are advantageous in isopiestic measurements of osmotic coefficients, since equilibrium is reached quickly; one must of course prevent evaporation of the solutions while weighing the dishes.

A reliable basis for extrapolating heat and volume data to infinite dilution is necessary to obtain standard state functions. In aqueous solutions the Debye-Hückel theory provides such a basis, since it has repeatedly been shown to be valid in dilute solutions of strong electrolytes. In most organic solvent systems, however, the Debye-Hückel theory has not been thoroughly tested, so that its use in extrapolation should be regarded with caution. In a few cases it has been reported that the experimental slopes for heat data are not in agreement with the limiting law.[B12,B25,B27] While in some cases this was caused by ion association,[B25] in other cases the causes are unknown.[B12,B27] Consequently, to avoid large uncertainties in extrapolation, one should make measurements at very low concentrations. Techniques for making accurate heat and volume measurements at low concentrations now exist,[B27,B47] but unfortunately, at these low concentrations, results are most affected by trace impurities in the solvent. Even impurities of less than 0.001% are frequently many times more concentrated than the solute being studied. It appears that in general the best approach is to work in a concentration range of 0.001 to 0.01m.

2.3 TREATMENT OF EXPERIMENTAL DATA

2.3.1 Solubility Data

The equilibrium between a crystalline electrolyte and its ions in a solvent can be expressed by the equation

$$M_{\nu_+}X_{\nu_-}(c) \rightleftharpoons \nu_+ M^{Z+}(\text{solv}) + \nu_- X^{Z-}(\text{solv}) \tag{2.3.1}$$

The free energy change accompanying the dissolution of one mole of electrolyte is given by

$$\Delta G = \Delta G^0 + RT \ln \frac{a_M^{\nu_+} a_X^{\nu_-}}{a_{MX(c)}} \tag{2.3.2}$$

By the choice of standard state, the activity of a solid is unity, consequently eqn. 2.3.2 is generally written as

$$\Delta G = \Delta G^0 + RT \ln a_M^{\nu_+} a_X^{\nu_-} \tag{2.3.3}$$

When the solute is in equilibrium with its ions, $\Delta G = 0$ and one obtains the standard free energy of solution

$$\Delta G^0_{\text{soln}} = -RT \ln K_{s0} \tag{2.3.4}$$

where the thermodynamic solubility product, K_{s0}, is given by

$$K_{s0} = a_M^{\nu_+} a_X^{\nu_-} \tag{2.3.5}$$

and the activities are for the saturated solution.

The solubility product is given by

$$K_{s0} = m_M^{\nu_+} m_X^{\nu_-} \gamma_\pm^{\nu} = \nu_+^{\nu_+} \nu_-^{\nu_-} m^{\nu} \gamma_\pm^{\nu} = (m_\pm \gamma_\pm)^{\nu} = a_\pm^{\nu} \tag{2.3.6}$$

If the solubility and activity coefficient data exist over a range of temperatures, one can obtain the standard enthalpy of solution by dividing both sides of eqn. 2.3.4 by the temperature and differentiating with respect to temperature, at constant pressure

$$\left[\frac{\partial \ln K_{s0}}{\partial T}\right]_p = -\frac{1}{R}\left[\frac{\partial\left(\frac{\Delta G^0}{T}\right)}{\partial T}\right]_p = \frac{\Delta H^0_{\text{soln}}}{RT^2} \tag{2.3.7}$$

Over moderate temperature ranges one can generally consider ΔH^0_{soln} to be constant. Upon integration and conversion to decadic logarithms eqn. 2.3.7 becomes

$$\log K_{s0} = \frac{-\Delta H^0_{\text{soln}}}{R \ln 10}\left(\frac{1}{T}\right) + C \tag{2.3.8}$$

where C is the integration constant. Consequently a plot of $\log K_{s0}$ vs. $1/T$ should be a straight line, from the slope of which, ΔH^0_{soln} can be evaluated. The entropy of solution, ΔS^0_{soln}, is readily calculated through the relation

$$\Delta S^0_{\text{soln}} = \frac{\Delta H^0_{\text{soln}} - \Delta G^0_{\text{soln}}}{T} \tag{2.3.9}$$

2.3.2 Activity Coefficients

For electrolyte solutions that are sufficiently dilute, the activity coefficient can be calculated from the limiting Debye-Hückel equation

$$\log \gamma_\pm = -A|z_+z_-|I^{\frac{1}{2}} \tag{2.3.10}$$

where

$$A = \frac{1.8246 \times 10^6}{(T\varepsilon_r)^{\frac{3}{2}} d_0^{-\frac{1}{2}}} \text{ mol}^{-\frac{1}{2}} \text{ kg}^{\frac{1}{2}} \text{ K}^{\frac{3}{2}} \tag{2.3.11}$$

is the Debye-Hückel limiting slope for activity coefficients. (For further discussion see sect. 2.5.2.) The ionic strength, I, is given by

$$I = \tfrac{1}{2}\sum_i m_i z_i^2 \tag{2.3.12}$$

In calculating mean activity coefficients by eqn. 2.3.10, one must have available accurate values of the dielectric constant (ε_r) of the solvent. These are frequently not available.

Several approaches exist for evaluating activity coefficients. For non-aqueous systems the most common method has been from electrochemical cells, (sects. 2.5–2.7). Of the remaining approaches available, the freezing point technique is most commonly employed and is considered in sects. 2.8–2.10. This gives osmotic coefficients and activity coefficients are obtained by applying the Gibbs-Duhem equation (see sect. 2.9.2).

Activity coefficients at temperatures other than that at which they were determined may be obtained using the relative partial molal enthalpy, $\bar{L}_2$, and heat capacity, $\bar{J}_2$. Differentiation with respect to temperature of the equation $\bar{G}_2 - \bar{G}_2^0 = RT \ln a_2$ for the partial molal free energy of the solute gives

$$\left(\frac{\partial \bar{G}_2}{\partial T}\right)_p + \left(\frac{\partial \bar{G}_2^0}{\partial T}\right)_p = RT\left(\frac{\partial \ln a_2}{\partial T}\right)_p + R \ln a_2 \qquad (2.3.13)$$

Remembering that

$$\left(\frac{\partial \bar{G}_2}{\partial T}\right)_p = - \bar{S}_2 \quad \text{and} \quad \left(\frac{\partial \ln a_2}{\partial T}\right)_p = \nu\left(\frac{\partial \ln \gamma_\pm}{\partial T}\right)_p \qquad (2.3.14)$$

eqn. 2.3.13 becomes, after several steps,

$$\left(\frac{\partial \ln \gamma_\pm}{\partial T}\right)_p = - \frac{\bar{H}_2 - \bar{H}_2^0}{\nu RT^2} = - \frac{\bar{L}_2}{\nu RT^2} \qquad (2.3.15)$$

For not too wide a temperature range, $\bar{J}_2$ can be considered constant. Consequently

$$\bar{L}_{2(T)} = \bar{L}_{2(T_s)} + \bar{J}_2(T - T_s) \qquad (2.3.16)$$

where T_s is the chosen standard temperature and T is the variable temperature. Substitution of eqn. 2.3.16 into eqn. 2.3.15 and subsequent integration from T to T_s gives

$$\log \gamma_{\pm(T_s)} = \log \gamma_{\pm(T)} - \frac{\bar{L}_{2(T_s)}}{\ln 10\, \nu R}\left[\frac{T_s - T}{T T_s}\right] - \frac{\bar{J}_2}{\nu R} \log \frac{T_s}{T} + \frac{\bar{J}_2 T_s}{\ln 10\, \nu R}\left[\frac{T_s - T}{T T_s}\right] \qquad (2.3.17)$$

Unfortunately $\bar{L}_2$ and $\bar{J}_2$ data are exceedingly scarce for electrolytes in non-aqueous solvents. A few values have been compiled by Mischenko and Poltoratskii.[B30]

2.3.3 Calorimetry

Heats of Solution and Dilution

Comparison between thermodynamic values is generally made with standard state functions. To obtain the standard enthalpy of solution, ΔH^0_{soln}, it is necessary to extrapolate directly measured enthalpies of solution at finite concentrations to infinite dilution; some form of the Debye-Hückel theory is generally used in this extrapolation (see sect. 2.5.2).

Most frequently, the extrapolations are made on the basis of the simple theory as expressed by eqn. 2.3.10. Differentiation of this equation and substitution into eqn. 2.3.15 gives

$$\bar{L}_2 = \ln 10\, \nu RT^2|z_+z_-|I^{\frac{1}{2}} \left(\frac{\partial A}{\partial T}\right)_p \tag{2.3.18}$$

The quantity A, defined by eqn. 2.3.11 contains two temperature dependent terms, ε_r and d_0, along with T itself. Differentiation of A, holding the pressure constant leads to

$$\left(\frac{\partial A}{\partial T}\right)_p = -\frac{A}{2} \cdot 3\left[\frac{\alpha}{3} + \frac{1}{T} + \left(\frac{\partial \ln \varepsilon_r}{\partial T}\right)_p\right] \tag{2.3.19}$$

where α is the temperature coefficient of expansion. Eqn. 2.3.18 can then be written as

$$\bar{L}_2 = A_H \frac{\nu}{2} |z_+z_-|I^{\frac{1}{2}} \tag{2.3.20}$$

where

$$A_H = -\ln 10\, RT^2 A \cdot 3\left[\frac{\alpha}{3} + \frac{1}{T} + \left(\frac{\partial \ln \varepsilon_r}{\partial T}\right)_p\right] \tag{2.3.21}$$

is the limiting slope for heats.

In order to relate $\bar{L}_2$ to the enthalpy of dilution, one must resort to the relative apparent molal heat content

$$\phi_L = \phi_H - \phi_H^0 \tag{2.3.22}$$

where the apparent molal heat content, ϕ_H, is defined by

$$\phi_H = \frac{H - n_1\bar{H}_1^0}{n_2} \tag{2.3.23}$$

At infinite dilution

$$\phi_H = \phi_H^0 = \bar{H}_2^0 \tag{2.3.24}$$

The change in heat content of a system from some finite concentration to infinite dilution is

$$H - H^0 = n_1\bar{H}_1 + n_2\bar{H}_2 - (n_1\bar{H}_1^0 + n_2\bar{H}_2^0)$$
$$= n_1(\bar{H}_1 - \bar{H}_1^0) + n_2(\bar{H}_2 - \bar{H}_2^0)$$

which upon introduction of eqn. 2.3.22 and 2.3.23

$$= n_2(\phi_H - \phi_H^0) = n_2\phi_L \tag{2.3.25}$$

By definition

$$\begin{aligned} L &= H - H^0 \\ \bar{L}_1 &= \bar{H}_1 - \bar{H}_1^0 \\ \bar{L}_2 &= \bar{H}_2 - \bar{H}_2^0 \end{aligned} \tag{2.3.26}$$

consequently eqn. 2.3.25 becomes

$$L = n_1\bar{L}_1 + n_2\bar{L}_2 = n_2\phi_L \tag{2.3.27}$$

From the definition of partial molal quantities and eqn. 2.3.27

$$\bar{L}_2 = \left(\frac{\partial L}{\partial n_2}\right)_{T,p,n_1} = \left(\frac{\partial(n_2\phi_L)}{\partial n_2}\right)_{T,p,n_1} \tag{2.3.28}$$

For constant n_1, molality is proportional to n_2, and

$$\bar{L}_2 = \left(\frac{\partial(m\phi_L)}{\partial m}\right)_{T,p,N_1} \tag{2.3.29}$$

Integration of eqn. 2.3.29 from 0 to molality m gives

$$\phi_L = \frac{1}{m}\int_0^m \bar{L}_2 \mathrm{d}m \tag{2.3.30}$$

For a pure electrolyte

$$I = K'm \tag{2.3.31}$$

where K' is a proportionality constant. Substitution of this and eqn. 2.3.20 into eqn. 2.3.30, with subsequent integration gives

$$\phi_L = \frac{2}{3} A_H \frac{\nu}{2} |z_+z_-| I^{\frac{1}{2}} \tag{2.3.32}$$

The molar heat of dilution to infinite dilution is given by

$$\Delta H_D = -\left(\frac{H - H^0}{n_2}\right) \tag{2.3.33}$$

Hence, by eqns. 2.3.25 and 2.3.32

$$\Delta H_D = -\frac{2}{3} A_H \frac{\nu}{2} |z_+z_-| I^{\frac{1}{2}} \tag{2.3.34}$$

or since

$$\Delta H_D = \Delta H^0_{\text{soln}} - \Delta H_{\text{soln}} \tag{2.3.35}$$

the heat of solution at finite concentration, ΔH_{soln}, is given by

$$\Delta H_{\text{soln}} = \frac{2}{3} A_H \frac{\nu}{2} |z_+z_-| I^{\frac{1}{2}} + \Delta H^0_{\text{soln}} \tag{2.3.36}$$

Consequently, a plot of measured molar heats of solution against $I^{\frac{1}{2}}$, or $m^{\frac{1}{2}}$ for a 1:1 electrolyte, will lead to a straight line over the range for which eqn. 2.3.10 is valid. Extrapolation to zero concentration gives ΔH^0_{soln}, the standard heat of solution. Most heat data in non-aqueous solutions have been treated in this manner.

A more sophisticated treatment, first suggested by Guggenheim and Prue[B59] for aqueous solutions of 1:1 electrolytes makes use of an expanded form of the Debye-Hückel theory. For a pure electrolyte of any type, this equation can be written as[B60]

$$\log \gamma_\pm = -A|z_+z_-| \frac{I^{\frac{1}{2}}}{1+I^{\frac{1}{2}}} + \frac{2\nu_+\nu_-}{\nu} Bm \qquad (2.3.37)$$

Differentiation of this equation with respect to temperature, holding the pressure constant, and substitution into eqn. 2.3.15 gives

$$\bar{L}_2 = \nu RT^2 \ln 10\, |z_+z_-| \frac{I^{\frac{1}{2}}}{1+I^{\frac{1}{2}}} \left(\frac{\partial A}{\partial T}\right)_p - 2(\ln 10)\nu_+\nu_- RT^2 \left(\frac{\partial B}{\partial T}\right)_p m \qquad (2.3.38)$$

Introduction of eqn. 2.3.19 leads to

$$\bar{L}_2 = A_H \frac{\nu}{2} |z_+z_-| \frac{I^{\frac{1}{2}}}{1+I^{\frac{1}{2}}} - 2\,(\ln 10)\,\nu_+\nu_- RT^2 \left(\frac{\partial B}{\partial T}\right)_p m \qquad (2.3.39)$$

One form of the Gibbs-Duhem equation is

$$\int d\bar{L}_1 = -\int \frac{n_2}{n_1}\, d\bar{L}_2 \qquad (2.3.40)$$

For a solution containing 1 kg of solvent

$$n_1 = \frac{1}{M_1}$$

where M_1 is the molecular weight of the solvent in kg, and $n_2 = m$. Upon substitution of eqn. 2.3.39 and 2.3.31 into eqn. 2.3.40, one obtains

$$\int \mathrm{d}\bar{L}_1 = -\int A_H \frac{\nu}{2} |z_+z_-| \frac{M_1 2I}{K'}$$
$$\mathrm{d}\left(\frac{I^{\frac{1}{2}}}{1+I^{\frac{1}{2}}}\right) + \int 2(\ln 10)(\nu_+\nu_-)M_1RT^2 \left(\frac{\partial B}{\partial T}\right)_p m\, \mathrm{d}m \qquad (2.3.41)$$

Recalling that

$$\mathrm{d}\left(\frac{I^{\frac{1}{2}}}{1+I^{\frac{1}{2}}}\right) = \frac{\mathrm{d}I^{\frac{1}{2}}}{(1+I^{\frac{1}{2}})^2} \qquad (2.3.42)$$

and that $dm^2 = 2\,m\,dm$, and integrating eqn. 2.3.41, $\bar{L}_1$ is evaluated

$$\frac{\bar{L}_1}{m} = -\frac{A_H\left(\frac{\nu}{2}\right)|z_+z_-|M_1 I^{\frac{1}{2}}\sigma(I^{\frac{1}{2}})}{3} + (\ln 10)(\nu_+\nu_-)M_1RT^2\left(\frac{\partial B}{\partial T}\right)_p m \tag{2.3.43}$$

where σ is a function defined by

$$\sigma(x) = \frac{3}{x^3}\left[1 + x - \frac{1}{1+x} - 2\ln(1+x)\right] \quad \text{(see 2.9.17)}$$

On a molal basis eqn. 2.3.27 becomes

$$m\phi_L = \frac{1}{M_1}\bar{L}_1 + m\bar{L}_2 \tag{2.3.44}$$

Substitution of eqns. 2.3.39 and 2.3.43 into eqn. 2.3.44 leads to

$$\phi_L = A_H\frac{\nu}{2}|z_+z_-|I^{\frac{1}{2}}\left[\frac{1}{1+I^{\frac{1}{2}}} - \frac{\sigma(I^{\frac{1}{2}})}{3}\right] - (\ln 10)\,\nu_+\nu_-RT^2\left(\frac{\partial B}{\partial T}\right)_p m \tag{2.3.45}$$

and upon substitution of eqn. 2.3.35, remembering that $\Delta H_D = -\phi_L$,

$$\Delta H_{\text{soln}} = A_H\frac{\nu}{2}|z_+z_-|I^{\frac{1}{2}}\left[\frac{1}{1+I^{\frac{1}{2}}} - \frac{\sigma(I^{\frac{1}{2}})}{3}\right] - (\ln 10)\nu_+\nu_-RT^2\left(\frac{\partial B}{\partial T}\right)_p m + \Delta H^0_{\text{soln}} \tag{2.3.46}$$

This equation can be rewritten as

$$\rho = \Delta H^0_{\text{soln}} - (\ln 10)\,\nu_+\nu_-RT^2\left(\frac{\partial B}{\partial T}\right)_p m \tag{2.3.47}$$

where

$$\rho = \Delta H_{\text{soln}} - A_H\frac{\nu}{2}|z_+z_-|I^{\frac{1}{2}}\omega \tag{2.3.48}$$

and

$$\omega = \left[\frac{1}{1+I^{\frac{1}{2}}} - \frac{\sigma(I^{\frac{1}{2}})}{3}\right] \tag{2.3.49}$$

A table of values for $\omega I^{\frac{1}{2}}$ is given in Appendix 2.3.1.

For a given temperature, a plot of ρ vs. m should result in a straight line. Guggenheim and Prue,[B59] and later Pitzer and Brewer,[B60] using a revised value of A_H, used this equation to evaluate the experimental data of Gulbransen and Robinson[B61] on aqueous NaCl. The conclusion is that it works well up to $0.05m$ but that deviations are observed at $0.1m$. Cobble and co-workers[B62,B63] successfully used eqn. 2.3.47 to extrapolate to infinite dilution aqueous heat data for higher valence type electrolytes over a wide range of temperature. In view of the fact that electrolytes in organic solvents, particularly of low dielectric constant, frequently indicate a deviation from the simple Debye-Hückel equation at low concentrations, it would appear that eqn. 2.3.47 would be a most useful means of extrapolation to infinite dilution. Held and Criss[B27,B28] employed this equation for several 1:1 electrolytes in *N*-methylformamide and *N,N*-dimethylformamide, and observed that electrolytes with low charge-density cations exhibited heats in agreement with the equation over the concentration studied ($<0.01m$). While large deviations were observed at very low concentrations for electrolytes having high charge-density cations, unpublished results in the author's laboratory indicate that these are caused by experimental problems connected with measurements in very dilute solutions.

Wood, Smith and Jongenburger[B64] have suggested a method of extrapolation using an extended form of the Debye-Hückel equation (see sect. 2.5.2 for discussion of forms of the Debye-Hückel equation)

$$\log \gamma_{\pm} = -\frac{A|z_+z_-|I^{\frac{1}{2}}}{1 + BaI^{\frac{1}{2}}} + \text{higher terms}$$

to give

$$\phi_L = A_H \frac{\nu}{2} |z_+z_-| c^{\frac{1}{2}} \left[\frac{1}{1 + Bac^{\frac{1}{2}}} - \frac{\sigma(Bac^{\frac{1}{2}})}{3}\right] + Cc + Dc^{\frac{3}{2}} \quad (2.3.50)$$

They employed this equation to extrapolate data obtained at high concentrations ($>0.01m$) for a wide variety of aqueous electrolyte solutions and observed that the extrapolated values were in good agreement with those obtained from data at very low concentrations. The author is unaware of this equation being employed for non-aqueous systems, but it appears to have potential usefulness.

Frequently electrolytes will be associated in solution. Wu and Friedman[B25] used an approximate method of obtaining standard heats of solution in such cases. They confined all measurements to low concentrations ($<0.01m$) and neglected long-range interionic effects, such as those covered by the Debye-Hückel theory. They further assumed that there is only one (unspecified) association process occurring. The heat of solution is then given by

$$\Delta H_{\text{soln}} = \Delta H^0_{\text{soln}} + [1 + \alpha(m)^{\frac{1}{2}}]\Delta H_a \quad (2.3.51)$$

where ΔH_a is the heat of association,

$$\alpha(m) = \frac{\Lambda}{\Lambda_0}$$

the degree of dissociation, and Λ and Λ_0 are the equivalent conductivities at a given concentration and infinite dilution respectively. According to eqn. 2.3.51 a plot of ΔH_{soln} vs. Λ/Λ_0 should result in a straight line with a slope of ΔH_a and an intercept of $\Delta H^0_{\text{soln}} + \Delta H_a$. The method was used successfully to evaluate ΔH^0_{soln} for lithium trifluoroacetate in propylene carbonate.

2.3.4 Heat Capacities

The standard partial molal heat capacity of electrolytes, $\bar{C}^0_{p_2}$, is probably the best thermodynamic property for providing information on ion-solvent interactions, but until very recently only two publications[B65,B66] reported this quantity for electrolytes in non-aqueous systems. The reasons for the scarcity of data become obvious when one considers the formidable experimental and theoretical problems in obtaining this function.

The traditional method of obtaining $\bar{C}^0_{p_2}$ for electrolytic solutions makes use of the apparent molal heat capacity

$$\phi_{C_p} = \frac{C_p - n_1\bar{C}^0_{p1}}{n_2} \tag{2.3.52}$$

Differentiation of eqn. 2.3.52 and conversion to molal units gives

$$\bar{C}_{p_2} = \phi_{C_p} + m\left(\frac{\partial\phi_{C_p}}{\partial m}\right) \tag{2.3.53}$$

For graphical evaluation of $\bar{C}_{p_2}$ of electrolytes it is generally better to use $m^{\frac{1}{2}}$. Recalling the relationship $dm = 2m^{\frac{1}{2}}dm^{\frac{1}{2}}$, eqn. 2.3.53 becomes

$$\bar{C}_{p_2} = \phi_{C_p} + \frac{m^{\frac{1}{2}}}{2}\left(\frac{\partial\phi_{C_p}}{\partial m^{\frac{1}{2}}}\right) \tag{2.3.54}$$

By plotting ϕ_{C_p} against $m^{\frac{1}{2}}$ and evaluating $\frac{\partial\phi_{C_p}}{\partial m^{\frac{1}{2}}}$ from the tangent to the curve, one obtains $\bar{C}_{p_2}$ for the concentration of interest. At infinite dilution $m^{\frac{1}{2}} = 0$ and eqn. 2.3.54 reduces to

$$\bar{C}^0_{p_2} = \phi^0_{C_p} \tag{2.3.55}$$

The primary problem in obtaining $\bar{C}^0_{p_2}$ is the extrapolation of ϕ_{C_p} to infinite dilution. Even with the most precise calorimeters now in existence ($\sim$0.01%) it is not practical to make measurements below about $0.1m$.[B67]

In aqueous solutions at concentrations above $0.1m$, ϕ_{C_p} is a linear function of $m^{\frac{1}{2}}$ for most electrolytes up to about $2.5m$.[B67] On the other hand, the slope of the ϕ_{C_p} vs. $m^{\frac{1}{2}}$ plots does not agree with the theoretically predicted limiting slope,[B59] obtained by differentiating the expression for A_H (eqn. 2.3.21). The rather complicated resultant expression involves, among other things, the second temperature derivative of the dielectric constant. Even for water, this quantity is known only to about 25%.[B59]

In non-aqueous solvent systems every problem is compounded. Generally dielectric constants have not been measured with sufficient accuracy to obtain useful second derivatives with respect to temperature. In addition, as in the case of aqueous solutions, the impracticability of making measurements below $0.1m$ precludes the determination of experimental data at low enough concentrations to use the limiting slope for extrapolation. Furthermore, for the few ϕ_{C_p} data which do exist in non-aqueous solutions[B65,B66] this function appears linear with $m^{\frac{1}{2}}$ to a maximum concentration of $1m$ and in some cases only to about $0.25m$. Consequently, it does not appear promising to use directly measured heat capacities for the purpose of evaluating $\bar{C}^0_{p_2}$.

Fortunately another method exists that does appear promising. Cobble and co-workers[B62,B63,B68–70] have used an *integral heat method* of evaluating $\bar{C}^0_{p_2}$ for aqueous solutions over a rather wide temperature range. For the dissolution of one mole of an electrolyte to form an infinitely dilute solution

$$MX(c) \rightarrow MX(\text{soln})$$

the standard heat of solution may be written as

$$\Delta H^0_{\text{soln}} = H^0 - n_1 H^0_1 - n_2 H^0_2 \tag{2.3.56}$$

where

$$H^0 = n_1 \bar{H}^0_1 + n_2 \bar{H}^0_2$$

$$H^0_1 = \bar{H}^0_1$$

and

$$n_2 = 1 \text{ mole}$$

Substitution of these into eqn. 2.3.56 leads to

$$\Delta H^0_{\text{soln}} = \bar{H}^0_2 - H^0_2 \tag{2.3.57}$$

Differentiation of eqn. 2.3.57 with respect to temperature at constant pressure gives

$$\left(\frac{\partial \Delta H^0_{\text{soln}}}{\partial T}\right)_p = \left(\frac{\partial \bar{H}^0_2}{\partial T}\right)_p - \left(\frac{\partial H^0_2}{\partial T}\right)_p \tag{2.3.58}$$

or

$$\Delta C^0_p = \bar{C}^0_{p_2} - C^0_{p_2}$$

which, upon rearrangement becomes

$$\bar{C}^0_{p_2} = \Delta C^0_p + C^0_{p_2} \tag{2.3.59}$$

where $C^0_{p_2}$ is the heat capacity of the pure anhydrous solute. One can expect that ΔC^0_p will be a slowly varying function of temperature so that for not too wide temperature differences

$$\Delta C^0_p = \left(\frac{\partial \Delta H^0_s}{\partial T}\right)_p \approx \left[\frac{\Delta(\Delta H^0_{\text{soln}})}{\Delta T}\right]_p \tag{2.3.60}$$

Consequently by measuring ΔH^0_{soln} at several temperatures one can obtain rather accurate values of ΔC^0_p. Since the heat capacities of most common pure electrolytes have been tabulated,[B71] one can easily calculate $\bar{C}^0_{p_2}$.

The tremendous advantages of the integral heat method over the directly measured heat capacities are that heat measurements can be made down to 10^{-3} to $10^{-4}m$ in contrast to $0.1m$, and that the theoretical relationships for extrapolating heat data are much better than those for extrapolating heat capacity data. It should be observed that this method is most accurate when the absolute value of ΔH^0_{soln} is small (<10 kcal mol^{-1}) and when data are obtained at a large number of temperatures. In aqueous solutions five or ten degree temperature intervals are convenient. Mastroianni and Criss[B8,B72] have used the method for evaluating $\bar{C}^0_{p_2}$ for $NaClO_4$ and $(CH_3)_4NBr$ in anhydrous methanol from about -5 to 50°C.

2.3.5 Molal Volumes

Evaluation of partial molal volumes of electrolytes, and in particular the standard partial molal volume, $\bar{V}^0_2$, is similar to the determination of partial molal heat capacities. As in the case of heat capacities, the apparent molal quantity is of interest. Corresponding to eqns. 2.3.52, 2.3.54 and 2.3.55 we have

$$\phi_V = \frac{V - n_1 V^0_1}{n_2} \tag{2.3.61}$$

where V^0_1 is the molar volume of the solvent

$$\bar{V}_2 = \phi_V + \frac{m^{\frac{1}{2}}}{2}\left(\frac{\partial \phi_V}{\partial m^{\frac{1}{2}}}\right) \tag{2.3.62}$$

$$\bar{V}^0_2 = \phi^0_V \tag{2.3.63}$$

For the determination of ϕ_V from experimental data it is more convenient to use the molarity concentration scale (c) and densities. For $V = 1000$ ml of solution, n_2, the number of moles of solute is numerically equal to c and eqn. 2.3.61 can be written as

$$\frac{\phi_V}{\text{ml mol}^{-1}} = \frac{1000}{cd_0}(d_0 - d) + \frac{M_2}{d_0} \tag{2.3.64}$$

where d_0 and d are the densities of the pure solvent and solute respectively. Also, on a molarity basis, eqn. 2.3.62 becomes[B73]

$$\frac{\bar{V}_2}{\text{ml mol}^{-1}} = \phi_V + \left[\frac{1000 - c\phi_V}{2000 + c^{\frac{3}{2}}(\partial\phi_V/\partial c^{\frac{1}{2}})}\right] c^{\frac{1}{2}} \frac{\partial\phi_V}{\partial c^{\frac{1}{2}}} \tag{2.3.65}$$

Consequently, one can evaluate $\bar{V}_2$ by substitution of eqn. 2.3.64 into eqn. 2.3.65.

It was observed by Masson[B74] that for aqueous electrolytic solutions ϕ_V could be expressed as

$$\phi_V = \phi_V^0 + S_V c^{\frac{1}{2}} \tag{2.3.66}$$

where S_V is the experimental slope, which varies with electrolyte, and is only approximately equal to the theoretical limiting slope. The majority of ϕ_V data in water[B75] and nearly all ϕ_V data in non-aqueous solvents[B76–81] have been extrapolated to infinite dilution through the use of eqn. 2.3.66. However, Redlich[B82] and Redlich and Meyer[B83] have shown that an equation of the form of 2.3.66 cannot be any more than a limiting law, where for a given solvent and temperature the slope, S_V, should depend only upon valence type. They suggest representing ϕ_V by

$$\phi_V = \phi_V^0 + A_V c^{\frac{1}{2}} + bc \tag{2.3.67}$$

where

$$A_V = Kw^{\frac{3}{2}} \tag{2.3.68}$$

is the theoretical limiting slope, based on molar concentration, including the valence factor

$$w = \tfrac{1}{2}\sum_i \nu_i z_i^2 \tag{2.3.69}$$

and

$$K = N^2 e^3 \left(\frac{8\pi}{1000\varepsilon_r^3 RT}\right)^{\frac{1}{2}} \left[\left(\frac{\partial \ln \varepsilon_r}{\partial p}\right)_T - \frac{\beta}{3}\right] \tag{2.3.70}$$

where β is the compressibility of the solvent. Until recently[B83,B84] the variation of dielectric constant with pressure was not known accurately enough, even in water, to calculate accurate values of the theoretical limiting slope and in organic solvents accurate data of this type are almost totally lacking.

Redlich and Meyer[B83] have shown eqn. 2.3.67 to be valid for a few salts in water for which very accurate density data exist and have shown it to be compatible with sodium acetate data in aqueous ethanol, using an estimated theoretical slope for that solvent. Unfortunately, even if data for calculating the theoretical limiting slopes in organic solvents were available, there are not at present enough accurate density data to test the relationship thoroughly. In the case of aqueous solutions, the evaluation of K, and therefore A_V from ϕ_V data requires density measurements with a precision better than 10^{-6}.[B83] Presumably systems of lower dielectric constant, in which deviations from the Debye-Hückel theory occur at lower concentrations, would require even more accurate density measurements.

Owen and Brinkley[B85] have suggested a more sophisticated equation for the extrapolation of volume data, in which an ion-size parameter, a, is included. By assuming that the parameter is temperature and pressure independent, their resultant equation, most convenient for extrapolation is

$$\phi_V = \phi_V^0 + \tfrac{2}{3}A_V\,\Omega_V c^{\frac{1}{2}} + \tfrac{1}{2}K_V c \tag{2.3.71}$$

where

$$\Omega_V = \left[\frac{1}{1+\kappa a}\,\frac{\partial \ln \varepsilon_r}{\partial p} - \frac{\sigma(\kappa a)\beta}{3}\right]\left[\frac{\partial \ln \varepsilon_r}{\partial p} - \frac{\beta}{3}\right]^{-1} \tag{2.3.72}$$

and

$$\kappa a = \left(\sum_i \nu_i z_i^2\right)^{\frac{1}{2}} \left(\frac{4\pi N e^2}{1000 k \varepsilon_r T}\right)^{\frac{1}{2}} a c^{\frac{1}{2}} \tag{2.3.73}$$

The function σ appeared in eqn. 2.3.43, and K_V is an empirical parameter. To the knowledge of the author, eqn. 2.3.71 has not been employed for the treatment of results from non-aqueous solutions.

In conclusion, it appears that both the Redlich and Meyer equation[B83] or the Owen and Brinkley treatment[B85] would be much more satisfactory than the simple Masson equation.[B74] Thorough tests of these new relationships, however, will require more accurate data on densities and also on the variation of dielectric constant of the pure solvents with pressure.

2.3.6 Viscosity

Although the viscosity of an electrolytic solution is not a thermodynamic function, such data are frequently found in conjunction with the thermodynamic quantity, $\bar{V}_2$, and the information which can be derived from accurate viscosity measurements is often useful in gaining an insight into the structure of electrolytic solutions.[B2, B86] Consequently, it seems appropriate to digress momentarily and include a brief discussion of the subject in this section.

The viscosity of aqueous solutions has been studied extensively; some of the more pertinent data, including that for a few non-aqueous solutions, are discussed by Harned and Owen.[B75] In most cases the data can be adequately represented by the Jones-Dole equation[B87]

$$\frac{\eta}{\eta_0} = 1 + A_\eta c^{\frac{1}{2}} + B_\eta c \tag{2.3.74}$$

or an extended form of this equation,

$$\frac{\eta}{\eta_0} = 1 + A_\eta c^{\frac{1}{2}} + B_\eta c + D_\eta c^2 \tag{2.3.75}$$

where η and η_0 are the viscosity of the solution and pure solvent respectively, and A_η, B_η and D_η are constants. The latter equation was suggested by Kaminsky[B88] and used by Feakins and Lawrence[B89] for *N*-methylformamide solutions. For most aqueous solutions eqn. 2.3.74 is valid up to $0.1m$.[B53,B88,B90] At higher concentrations one must employ eqn. 2.3.75. The A_η coefficient has significant theoretical interest in connection with ion–ion interaction, and can be calculated.[B91] Unfortunately, its calculation requires among other things, a knowledge of limiting ionic conductivities, which again are frequently not available for non-aqueous solutions. While there is no satisfactory way of calculating the B_η coefficient, it is a useful parameter for qualitatively interpreting ion-solvent interactions,[B2,B86] and is an approximately additive quantity for ions.[B52]

The coefficients A_η and B_η can be evaluated by rearranging eqn. 2.3.74 to

$$\frac{\frac{\eta}{\eta_0} - 1}{c^{\frac{1}{2}}} = A_\eta + B_\eta c^{\frac{1}{2}} \tag{2.3.76}$$

and plotting the left hand side against $c^{\frac{1}{2}}$. This treatment has been employed for most of the non-aqueous systems that have been investigated.[B92–96]

2.4 RESULTS

2.4.1 Non-electrolytes

The thermodynamics of solutions of non-electrolytes covers an area of knowledge at least equal in magnitude to that of electrolyte solutions. Hildebrand and Scott[B97] have thoroughly discussed the subject, mostly for non-polar solvents, in their book on Regular Solutions. For the polar solvent, water, entropies and heat capacities of non-electrolytes have proved helpful in understanding the structure of this solvent.[B98,B99] Consequently, one would expect that similar data in other polar solvents would yield useful information. In particular, thermodynamic values for solutes which do not have specific interactions, such as hydrogen bonding, with the solvent would be particularly useful. Many gases fulfil this requirement admirably, and through accurate solubility measurements over a temperature range it is possible to evaluate the partial molal entropies. Battino and Clever[B100] have reviewed gas solubilities in both polar and non-polar solvents up to 1966, and while a large amount of data exist for organic solvents, not many have been used to evaluate partial molal entropies. Only a few results, including those published since 1966, which are of particular significance in the discussion of the structure of organic solvent systems, will be considered here.

The most extensive series of studies in this area are those of Ben-Naim and co-workers.[B101] They have measured the solubilities of argon in various mixed solvents as a function of temperature, and calculated free energies, enthalpies, and entropies of solution. All data are presented in graphical form, and include the solvent systems, water-ethanol, water-*p*-dioxan, and water-methanol. The last two of these systems were studied over the entire concentration range, and while the primary interest was in aqueous and highly aqueous mixed solution, useful information is given on organic systems.

Wood and DeLaney[B102] have measured the solubility of He, N_2, Ar, and ethane in pure *N*-methylacetamide from 35 to 70°C and used the data to calculate the free energy, heat and entropy of solution, as well as the change in heat capacity for the dissolution process. (See Appendix 2.3.2.) Non-polar solutes appear to be much more soluble in *N*-methylacetamide than in water. Entropies of solution are not as negative as for aqueous solutions, but the change in heat capacity on dissolution is much more negative than in water. These results lead to some interesting conclusions as to the structural effects of non-polar solutes on this solvent.

Eley,[B99] using data from the literature, has compared the energies and entropies of solution and partial molal volumes of several gases in water with the corresponding values in some organic solvents. The data are used to explain the structure of the solutions in terms of cavity formation, and his paper includes a discussion of the effect of temperature on the thermodynamic properties. Giona and Pfeifer[B103] have measured the solubility of SO_2 in *N,N*-dimethylformamide and aqueous solutions of this solvent from 20 to 60°C and at various pressures up to slightly greater than one atmosphere. The data were used to evaluate heats of solution. Compared to the highly aqueous solvents, heats of solution in pure DMF were a slowly varying function of the concentration of SO_2. Entropies of solution were not evaluated. Brande and co-workers[B104] have measured the solubilities of acetylene, carbon dioxide, and higher acetylenic hydrocarbons in DMF and DMF-H_2O systems, from 25 to 140°C and up to 23% water. Pressures range from 50 to 760 mm. Heats of solution of the various gases in the anhydrous DMF are given.

Dymond[B105] has reported the solubility of a series of gases in cyclohexane and dimethylsulphoxide. (See Appendix 2.3.3.) Measurements on the cyclohexane systems were made from near 20°C to approximately

37°C and entropies of solution calculated, but solubilities were obtained only at 25°C for dimethylsulphoxide and no thermodynamic values were calculated.

One notable recent study of non-electrolytes in various solvents has been reported by Krishnan and Friedman[B105a] in connection with their work on solution structure. Enthalpies of solution were reported for several non-electrolytes, mostly alcohols and hydrocarbons, in the solvents water, propylene carbonate, and dimethylsulphoxide. They have also made similar measurements for some non-electrolytes in CH_3OH and CH_3OD and determined the enthalpy of transfer between the two solvents.[B105b] These data are tabulated in Appendices 2.3.4 and 2.3.5. The measurements were made at sufficiently low concentrations that solute-solute interactions are insignificant and therefore the data can be considered to be for the standard state. In the cases where the data overlap, their results are essentially in agreement with those reported earlier by Arnett and McKelvey.[B22]

2.4.2 Electrolyte Solutes

Results from five different types of measurements will be reviewed for each solvent, namely solubilities, activity coefficients, calorimetry, volumes and viscosities.

Solubility data have been thoroughly reviewed up to 1956 and less completely up to the early sixties by Seidell and Linke.[B13] Consequently, only data of more recent origin will be reviewed here. Furthermore, solubility data which result primarily from electrochemical cell investigations have been omitted since they are covered in sects. 2.5–2.7. For similar reasons only activity coefficients originating from non-electrochemical sources, primarily from freezing point studies, are considered. Often densities and viscosities have been reported but no attempt has been made to evaluate $\bar{V}_2^0$ or viscosity coefficients; in general, these sources are not included.

To review comprehensively all the data on topics covered in this section is not possible here, and therefore, only those results of most general interest which have a direct bearing on the discussion of the structure of organic solvent systems, are included. In most cases the data taken from these sources are included in the appropriate appendix, usually in the form in which they appear in the original publications.

It is generally convenient to discuss electrolytic solutions in terms of protic and aprotic solvents. With the exception of acetone and dioxan, most older investigations are confined to protic solvents having relatively low dielectric constants as compared to water. However, in recent years there has been an increasing number of papers reporting results in aprotic solvents, and in protic solvents having very high dielectric constants. Protic solvent systems will be considered first.

(i) *Protic Solvents*

(a) *Alcohols.* Methanol has probably been examined more extensively as a solvent medium for electrolytes than any solvent other than water. Its dielectric constant is sufficiently high that many simple electrolytes are soluble, although there is considerable evidence that with some electrolytes significant ion-pairing occurs in the resulting solutions.[B106] Several solubility studies have been reported in recent years;[B36,B107–115] the published data are included in Appendices 2.4.1–2.4.3. The solid phases have been indicated where these have been reported. In some cases there is a considerable unexplained difference in the reported solubility for a given salt. For example, Harner and co-workers[B108] have reported the solubility of NaI in methanol at 25°C as 625.1 g (kg of solvent)$^{-1}$. The composition of the solid phase is not given. On the other hand, Brusset and Lecoq[B109] have reported the solubility of NaI from −0.2 to 35.4°C and the solubility at 25°C, interpolated from their data at 24 and 26.5°C, is 850 g (kg of solvent)$^{-1}$. They indicate the solid phase to be NaI . 3MeOH up to 26.5°C and the non-solvated salt at higher temperatures. The former workers stated their methanol to be 99.8% pure while the latter stated the water content to be 0.01%. If the 0.2% impurity in the methanol used by Harner and co-workers[B108] was water, they would have been expected to observe a greater solubility than that observed by Brusset and Lecoq.[B109] In both investigations the solubilities were determined by analysing the saturated solution in equilibrium with the solid phase. The discrepancy exhibited in this example is by no means unique.

Accurate solubility measurements are deceptively simple. The problem is essentially knowing whether equilibrium has been established. To minimise uncertainties it would appear that whenever possible, solubilities should be obtained by approaching the equilibrium state from both under- and super-saturated solutions. For the data to be thermodynamically useful, it is essential not only to obtain accurate solubilities, but also to determine the nature of the solid phase present.

Without a knowledge of densities of solutions, it is not possible to compare the data reported in terms of molarity with that reported on a molality basis, and consequently these data are listed separately in the appendix. Some data, reported as solubility products, are also listed separately.

Activity coefficient data available for this solvent appear to be derived entirely from electrochemical studies.

Heats of solution of electrolytes in methanol and water-methanol mixtures from calorimetric measurements are numerous. The earlier work includes that of Forcrand,[B17] Lemoine[B116] and Walden,[B117] but the first extensive heat measurements in non-aqueous systems are those of Wolfenden and co-workers who measured heats of solution for 17 univalent salts[B118] at 20°C as well as the heats of dilution of a few electrolytes[B119] in several solvents, including methanol. They observed that the heats of dilution, ΔH_D, of electrolytes in most of the solvents studied were generally negative over the entire concentration range. This is in contrast to aqueous solutions in which the heats of dilution are generally negative at lower concentrations and positive at higher concentrations.[B60,B75] Furthermore, the experimental limiting slopes for the methanol solutions exceed the theoretical limiting slope by a considerable amount in the more dilute solutions. The heats of solution were generally measured at about 0.01 molar and extrapolated to infinite dilution using a theoretical correction rather than the experimental one. The authors indicate that their treatment leads to an uncertainty of about 0.1 kcal.

Moss and Wolfenden[B120] and Slansky[B121] have reported heats for several alkali metal halides and HCl in water-methanol mixtures ranging from 0 to 100% alcohol. In every case there is a maximum in the heat of solution at a solvent composition of approximately 20 mol per cent methanol. The data for the pure methanol solutions from these investigations and those of earlier workers have been compiled as heats of formation by Rossini and co-workers.[B29] Only in a few cases are the data given at infinite dilution.

Drakin and Chang[B122] have measured heats of solution of several perchlorates and some halides of the alkali metals and alkaline earths. Their results for LiCl are compatible with those of Wolfenden, *et al.*,[B118] but are not in good agreement for NaI. Their results for both LiCl and NaI are not in agreement with the results of Slansky,[B121] which they show to be internally inconsistent. Although the data of Wolfenden were obtained at 20°C, they used this data with their own to evaluate standard heats of solution at 25°C. This is probably not a serious approximation, in the light of the known heat capacities of electrolytes in methanol[B8] and the experimental errors involved. The experimental limiting slopes for sodium and silver perchlorates are in agreement with theory, but the slopes for LiCl and NaI are approximately twice the theoretical slope, which agrees qualitatively with Wolfenden *et al.*,[B119] for the halides. Mastroianni and Criss[B8] have measured the heat of solution of $NaClO_4$ in methanol at concentrations ranging from approximately 0.01 to 0.001m and observed the slope to be in agreement with theory, but the resultant standard heat of solution is somewhat more negative than that reported by Drakin and Chang.[B122]

In the last decade Mishchenko and co-workers[B123–135] have made extensive calorimetric measurements, primarily in solvents of low dielectric constant, and many of these data have been compiled by Mishchenko and Poltoratskii.[B30] Several of the papers deal with methanol or methanol-water solutions.[B125,B126,B128,B131,B134] For the most part their investigations cover wide concentration ranges, from dilute to nearly saturated solutions, but unfortunately only a few data were obtained in dilute solutions and they did not employ theoretically based extrapolations to infinite dilution. Consequently, there appears to be considerable uncertainty in standard heats of solution derived from their data.

Mishchenko and co-workers[B30,B134] and Krestov and Klopov[B136,B137] have measured heats of solution for several alkali and alkaline earth metal halides and nitrates in water-methanol mixtures. The data, which are given only in graphical form, indicate the same functional dependence upon alcohol concentrations as those reported by Slansky,[B121] and Moss and Wolfenden.[B120]

Fuchs and co-workers[B12,B23] have reported heats of solution for several electrolytes, many of which contain very large ions, in a number of solvents, including methanol.[B12] Their measurements involved concentrations as low as $5 \times 10^{-5} m$. In some cases they found a large concentration effect which was not attributed to the presence of trace amounts of water in the solvent, since some salts which might be expected to be preferentially hydrated, showed no concentration dependence. No convincing explanation for these concentration effects was suggested.

Jakuszewski and co-workers have reported heats of solution of several univalent[B138] and divalent[B139] electrolytes in anhydrous methanol. They performed their measurements at concentrations of 20,000 to 40,000 mol of CH_3OH/mol of salt and considered the values obtained to equal the standard state heats within the limits of experimental error. On a molal concentration scale their concentrations were 0.0008 to 0.0016m, and assuming the Debye-Hückel limiting law to be valid, the correction to infinite dilution from these concentrations is about 0.1 kcal for a 1:1 electrolyte. They consider their value for NaBr of -3.87 kcal mol^{-1} to be in satisfactory agreement with that of -4.00 kcal mol^{-1} reported by Slansky.[B121] Other values reported[B30,B118] for this salt in methanol are -4.13 kcal mol^{-1} and -4.30 kcal mol^{-1}.

Krishnan and Friedman[B105b] have reported enthalpies of solution of several electrolytes in methanol and deuterated methanol at 25°C, and have calculated enthalpies of transfer between the two solvents. All measurements were made at sufficiently low concentrations to avoid ion association. In the few cases where the data overlap with those of Slansky,[B121] the agreement is generally satisfactory. Their results show that for transfers to CH_3OH from CH_3OD, salts containing small ions or $CF_3CO_2^-$ have negative heats of transfer while for those with large ions the opposite is true.

While directly measured specific heats of organic solutions of electrolytes are numerous,[B30,B65,B66,B130,B131] they generally have not been measured with an accuracy high enough to evaluate apparent molal heat capacities. Only two such studies have been reported;[B65,B66] both are for methanol solutions with the most dilute measurements being about 0.2*m*, which precludes accurate extrapolation to infinite dilution. The data generally show a sharp change in slope at concentrations between 0.2 and 1.0*m*, and attempts to extrapolate the data were made in only one of these publications.[B66]

A few heats of solution at different temperatures in methanol solutions have been reported by Mishchenko and co-workers[B30] and $\Delta C^0_{p_2}$ for the dissolution process evaluated. As mentioned previously, however, their extrapolations to infinite dilution are very uncertain, and since the calculation of $\Delta C^0_{p_2}$ involves differences in the standard heats of solution, these values can only be approximate at best.

The importance of making measurements in dilute solutions and using theory to help extrapolate data to infinite dilution cannot be overemphasised. Mastroianni and Criss[B8] have used the integral heat method (see sect. 2.3.4) to obtain $\bar{C}^0_{p_2}$ for $NaClO_4$ from below 0° to around 50°C. Their data indicate that the temperature coefficient of the heat of solution changes sign at concentrations of about 0.003*m*, as a result of the marked effect of temperature on the limiting slope for methanolic solutions. This is illustrated in Fig. 2.4.1. They have also used the technique to determine $\bar{C}^0_{p_2}$ for Me_4NBr over approximately the same temperature range.[B72] In the opinion of this contributor the integral heat method, with measurements made in dilute solutions, offers the only satisfactory means currently available for determining $\bar{C}^0_{p_2}$. Leung and Grunwald[B140] have obtained $\Delta C^0_{p_2}$ for the self-ionisation of methanol and for the acid dissociation of benzoic acid in methanol, from their measurements of the heat of neutralisation at various temperatures.

As in the case of heats and heat capacities, ionic entropies have been examined more extensively in methanol and methanol-water mixtures than in other solvents. The earliest study was that of Latimer and Slansky[B141] who report entropies for NaCl, KCl, KBr, and HCl in water-methanol mixture from 0 to 100% methanol. Jakuszewski and Taniewska-Osinska[B142–144] have determined the entropies of numerous halide salts and HCl at 25°C, and more recently Franks and Reid,[B6] using data from the literature, have calculated standard ionic entropies for several species in water-methanol mixtures covering the whole concentration range. It should be observed that these authors chose the mol fraction standard state for the solute in solution and ideal ionic gas as the standard state for the pure solute instead of the conventional hypothetical one molal solution and pure solid at 0K. The former standard states are convenient when comparing entropies of solvation for the various species. Ionic entropies in methanol are considerably more negative than in water.

Millero[B7,B31] has reviewed most of the pertinent molal volume

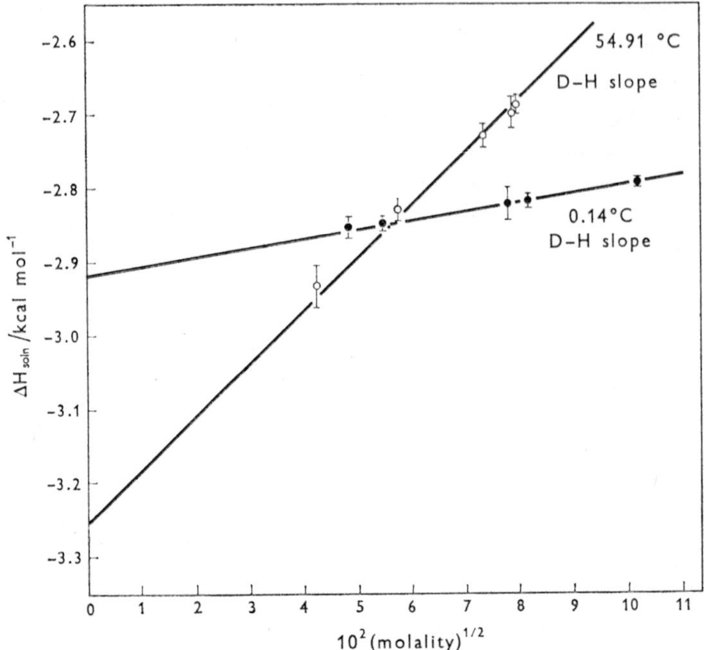

FIG. 2.4.1. Effect of concentration and temperature on the heat of solution of $NaClO_4$ in methanol, resulting in a change in sign of the temperature coefficient for the dissolution process.

and density data for methanol solutions of alkali metal and tetraalkylammonium halides and, whenever possible, obtained $\bar{V}_2^0$ values. In addition, Stark and Gilbert[B145] have graphically reported $\bar{V}_2^0$ for KBr and KI. Their values are from 0.5 to 1.1 $cm^3\ mol^{-1}$ different from the 'best' values given by Millero.[B7] Sobkowski and Minc[B80] have measured ϕ_V for methanolic HCl solutions and extrapolated the data by Masson's equation[B74] to obtain $\bar{V}_2^0$. For all electrolytes studied, $\bar{V}_2^0$ in methanol is considerably more negative than in aqueous solutions.

Viscosity coefficients, A_η and B_η, have been reported for several salts in methanol. Jones and Fornwalt[B92] reported values for KCl, KBr, KI, and NH_4Cl, and DeMaine and co-workers[B95] have reported data for several higher-charged species, primarily transition metal chlorides. Unfortunately, most of the latter data are for hydrated salts so that the resultant solutions are mixed water-methanol solvents. Kay and co-workers[B96] have evaluated B_η coefficients for several tetraalkylammonium salts in methanol, at temperatures ranging from 10 to 45°C. Tuan and Fuoss[B146] have reported B_η coefficients for tetramethyl- and tetrabutylammonium bromides in methanol. Except for the larger tetraalkylammonium salts, B_η is considerably larger in methanol than in water.

Ethanol has also received considerable attention as a solvent over a long period of time. Data on this solvent, however, are rather few compared to methanol and very few systematic studies exist. Several solubility studies have been made since the publication of Seidell and Linke.[B13] Thomas[B147] has reported solubilities for the alkali metal iodides at 20 and 25°C, and observed a decrease in solubility with an increase in ionic radius of the cation. Deno and Berkheimer[B148] have reported the solubilities of several tetraalkylammonium perchlorates. In every case the solid phase was the pure salt. Solubilities for several rare earth compounds have been reported.[B110,B112,B113] Since all of these salts form solvates in the solid phase, the results cannot be used in thermodynamic calculations without the corresponding thermodynamic values for the solid phases. Solubilities of silver chloride, caesium chloride,[B111] silver benzoate, silver salicylate[B115] and caesium nitrate[B114] have been measured in ethanol, using radioactive tracer techniques. Burgaud[B149] has measured the solubility of LiCl from 10.2 to 57.6°C and observed that there is a transition from the four-solvated solid phase to the non-solvated phase at 20.4°C.

The history of calorimetric measurements in ethanol, parallels that of measurements in methanol. The first heat measurements of any significance are those of de Forcrand[B17] and Pickering,[B18] who measured the heats of a large number of salts. The important earlier heat data, extending through the measurements of Wolfenden and co-workers[B118] have been compiled as heats of formation by Rossini and co-workers.[B29] Wolfenden and co-workers[B119] were about the first to measure accurately heats of dilutions in organic solvents. As indicated earlier, they observed that heats of dilution in ethanol, as in other organic solvents, were generally negative, in contrast to aqueous solutions which exhibit negative heats of dilution at lower concentrations but usually positive values at higher concentrations. They extrapolated their data to infinite dilution from about 0.01*m* using a theoretical correction rather than the experimentally observed slope, and estimated the uncertainty by this technique to be about 0.1 kcal.

In the last decade several papers have appeared reporting heats of solution of electrolytes in ethanol and ethanol-rich aqueous solutions. Jakuszewski and Taniewska-Osinka[B150] have measured heats for several alkali metal halides and HCl in ethanol, and Erbanova and co-workers[B151] have reported heats for the perchlorates of several alkali and alkaline earth metals and lead perchlorate. Both investigations were made in dilute solutions and in the latter the data were extrapolated to infinite dilution by means of the simple Debye-Hückel equation. The limiting slope was observed to be much greater than for aqueous solution. Mishchenko and co-workers[B129] measured the heat of solution of NaI over the entire concentration range up to saturation, at 25 and 45°C. However, as in the case of methanol, very few data were reported at low concentrations and the accuracy of the extrapolation to infinite dilution is doubtful; the reported heat of solution at infinite dilution for NaI at 25°C is nearly a kcal more positive than that of Jakuszewski and Taniewska-Osinska.[B150] Krestov and Klopov[B136,B137] have measured heats of solution of some alkali metal halides and nitrates, and calcium and strontium nitrates, in ethanol-water mixtures. They observed, as reported for methanol-water mixtures[B30,B120,B121,B134,B136,B137] that there is a maximum in the heats at about 20 mol per cent ethanol. Unfortunately these data are presented only in graphical form.

No heat capacity data have been reported for electrolytes in ethanol. However, from the temperature coefficients of the heats of dissolution of NaI in ethanol,[B129] an average ΔC_p^0 for the dissolution process may be estimated at 16 cal mol^{-1} K^{-1}. For aqueous solutions of simple electrolytes ΔC_p^0 is generally negative.[B62,B63]

Partial molal entropy data in ethanol are nearly as sparse as the heat capacity data. The only comprehensive entropy data in this solvent are those of Jakuszewski and Taniewska-Osinska,[B150] who report $\bar{S}_2^0$ for HCl and several alkali metal halides in ethanol. Ionic entropies have been calculated for the alkali metals[B152] from free energies and enthalpies of solvation, but since extra-thermodynamic assumptions were necessary, the meaning of the values is questionable. Ionic entropies in ethanol are somewhat more negative than in methanol and considerably more negative than in water.

Although density measurements of varying degrees of accuracy have been reported for ethanolic solutions, standard state partial molal volumes in ethanol have been evaluated for only a few electrolytes. Vosburgh, Connell and Butler[B153] reported $\bar{V}_2^0$ for LiCl in water and a series of alcohols, including ethanol. They observed that the salt had a much smaller value of $\bar{V}_2^0$ in the alcohols than in water, and that for all the systems studied it was smallest in ethanol. Sobkowski and Minc[B80] have reported $\bar{V}_2^0$ for HCl in water and the three lower alcohols and also observe $\bar{V}_2^0$ to be smaller in the alcohols than in water, but it is smallest in methanol rather than ethanol. Lee and Hyne[B78] have reported $\bar{V}_2^0$ at 50.25°C for the tetraalkylammonium chlorides in ethanol-water mixtures up to 0.4 mol fraction of ethanol. With the tetramethyl and tetraethyl salts, the volumes are all very positive in water but decrease rapidly with an increase in alcohol content and appear to be at a minimum around 0.3 to 0.4 mol fraction of ethanol. The higher tetraalkyl salts are not entirely consistent with this pattern.

Cox and Wolfenden[B52] have reported viscosity coefficients for NaI in ethanol. While the absence of data for this salt in methanol prevents a direct comparison, it appears that B_η is much more positive in ethanol than it would be in methanol. Padova[B154] has reported accurate relative viscosities for several electrolytes in water-ethanol mixtures containing up to 80% ethanol. However, he does not evaluate the A_η or B_η coefficients.

The quantity of thermodynamic data for the higher alcohols is severely limited. Some recent solubility studies of interest have been reported for $CsNO_3$,[B114] CsCl,[B111] and AgCl,[B111] and a few data exist for rare earth salts.[B110,B112,B113] Although there are some exceptions, solubilities of electrolytes in alcohols, as one would expect, tend to decrease as the number of carbon atoms increase.

Calorimetric measurements in dilute electrolytic solutions of the higher alcohols exist primarily for perchlorates. Erbanova and co-workers[B151] have measured heats of solution of $LiClO_4$, the alkaline earth perchlorates and $Pb(ClO_4)_2$ in ethanol, propanol and butanol, and extrapolated the data to infinite dilution to obtain ΔH^0_{soln}. The perchlorates which are among the most fully dissociated, show similar limiting slopes in all alcohols studied. The resultant ΔH^0_{soln} values of a given salt in the alcohols increase in the order EtOH $<$ PrOH $<$ BuOH.

Vosburgh and co-workers[B153] have reported $\bar{V}^0_2$ for LiCl in some aliphatic alcohols up to butanol and Sobkowski and Minc[B80] have reported the same quantity for HCl in alcohols up to *n*-propanol. For both electrolytes $\bar{V}^0_2$ increases as the number of carbons increase in the alcohol. Venkatasetty and Brown[B155] have measured the viscosities of LiI, NH_4I and Bu_4NI in butanol at 0, 25 and 50°C and attempted to fit the data to the Jones-Dole equation.[B87] Although measurements were made in relatively dilute solutions, deviations from linearity were observed over the concentration range studied and viscosity coefficients were not evaluated.

(b) *Glycols*. Ethylene glycol is the only member of this series for which a significant amount of data exist. Kraus and co-workers[B156] have recently measured the solubilities of NaCl, KCl, and $Ba(NO_3)_2$ in mixtures of water and ethylene glycol ranging from pure water to almost pure glycol, and used the data to calculate activity coefficients of the salts in the saturated solutions. The activity coefficients are based on the same reference states as for solutions of the salts in water. Similar results are reported for mono- and diacetates of ethylene glycol.

Mischenko and Tungusov[B123] have reported heats of solution of NaI at 2.5 and 25°C and KI at 25°C in ethylene glycol from low to very high concentrations. Stern and Nobilione[B157,B158] have measured the heats of transfer of HCl and $HClO_4$ from water to aqueous ethylene glycol from 10 mol per cent to nearly pure glycol. The acid concentrations ranged from 1 to 4×10^{-3} mol l^{-1}, and no attempt was made to extrapolate the data to infinite dilution. (Known heats of dilution of electrolytes in pure ethylene glycol[B159] indicate that corrections to infinite dilution are somewhat larger than the uncertainties in the data.) The results[B157,B158] are interesting since both electrolytes show maxima at approximately 10 to 20 mol per cent and minima around 70 to 80 mol per cent glycol. Similar maxima are observed in water-alcohol mixtures at about the same composition.

Wallace, Mason and Robinson[B159] have reported heats of dilution of NaCl in pure ethylene glycol. The experimental limiting slope agrees with the rather uncertain theoretical value within the limits of error. These authors list relative partial molal heat contents for both the solvent and solute.

Tungusov and Mishchenko[B130] have measured the heat capacities of pure ethylene glycol and solutions of NaI and KI in ethylene glycol and calculated the partial molal heat capacity from the data. They report a value of 21.5 cal mol^{-1} K^{-1} for NaI and conclude that $\bar{C}_{p_2}$ is independent of concentration. As indicated earlier, $\bar{C}^0_{p_2}$ values obtained through direct heat capacity measurements on organic solutions are doubtful because of the impossibility of making sufficiently accurate measurements at low concentrations.

Crickard and Skinner[B94] have reported viscosity B_η coefficients for KI and CsI in ethylene glycol. The latter salt exhibits a negative value for B_η, a phenomenon not usually observed for electrolytes in non-aqueous solvents.

(c) *Amides.* A large number of papers have appeared in recent years on various thermodynamic properties of electrolytes in amides. This interest stems primarily from the fact that amides generally have high dielectric constants, are highly structured and are usually good solvents for both electrolytes and organic species.

Formamide has probably been examined most thoroughly. Several solubility studies have been reported since the publication of Seidell and Linke.[B13] Gopal and Husain[B160] have determined the solubilities of over twenty electrolytes in formamide and at various temperatures, and Gopal and co-workers[B161] used the data to evaluate heats of solution in this solvent. Berardelli and co-workers[B162] have also reported heats of solution for several electrolytes, including some alkaline earth and transition metal halides. Alexander and co-workers[B107] have reported solubilities, in terms of the log of the solubility product, for numerous electrolytes in several solvents including some amides. Povarov and co-workers[B163] have measured the solubility of AgCl by radioactive tracer techniques in pure formamide and in solutions of NaCl and CsCl in formamide. Their results show AgCl to be considerably more soluble in formamide than indicated by the results of Alexander and coworkers.[B107] However, the latter workers point out that solubilities of silver salts in formamide and some other solvents, when determined by analysing saturated solutions, will generally be meaningless if the salts are less soluble than $10^{-2}m$. They base this on the fact that these solvents will always contain some trace impurities (e.g. ammonia or amines) which will complex with the silver, increasing its solubility.

Activity coefficients for several alkali metal halides in formamide, derived from freezing point data, have been reported in a series of papers by Vasenko.[B164–166] Other activity coefficient data on the alkali metal halides have been reported by Dawson and Griffith[B167] and by Mostkova and co-workers.[B168] Gopal and Husain[B16] have measured activity coefficients for several tetraalkylammonium and trialkylammonium halides.

Somsen and Coops[B19] and Somsen[B169] have made a comprehensive calorimetric study of the heats of solution of all the alkali metal halides in formamide. Their measurements were in dilute solution and they used the Debye-Hückel equation to correct the data to infinite dilution. Jain and co-workers[B170] have also reported calorimetrically measured heats for some alkali metal halides and a few alkaline earth halides. Unfortunately they evaluated the infinite dilution heats by extrapolating a plot of ΔH vs. m rather than the recommended ΔH vs. $m^{\frac{1}{2}}$. The original measurements are not reported so the data cannot be re-evaluated. In cases where the data of these two groups of investigators overlap, agreement is not as good as one would hope. In some cases there is a discrepancy of approximately 1 kcal mol^{-1}. Finch and co-workers[B26] have measured heats for a few alkaline earth chlorates and bromates in high dielectric constant solvents, including formamide. Their data were obtained between 0.01 and $0.05m$ and in the case of formamide were not extrapolated to infinite dilution. More recently they have reported standard heats of solution for some of the alkaline earth halides in this solvent.[B171]

Criss and co-workers,[B5] using data from the literature, calculated the entropies of several alkali metal halides in formamide. It should be observed that their published entropies are based on the mol fraction standard state instead of the more common hypothetical one molal standard state. Entropies of electrolytes in this solvent are generally more negative than in water, but more positive than in other solvents that have been examined.

Gopal and Srivastava[B81,B172] have measured the densities of a large number of uni-univalent electrolytes in formamide and extrapolated the calculated apparent molal volumes to infinite dilution by Masson's equation[B74] to obtain $\bar{V}_2^0$.

N-methylformamide (NMF) is an interesting solvent because of its extremely high dielectric constant ($\varepsilon_r \sim 180$). Strack, Swanda and Bahe[B173] measured the solubilities of several uni-univalent electrolytes in this solvent from 0 to 35°C and observed that the solubilities, expressed as molal concentrations, varied linearly with temperature. Except with LiCl, NaI, and KI, which appeared to form solvates, the data were used to evaluate heats of solution for the corresponding salts. Berardelli and co-workers[B162] have reported solubilities for several 1:1 and 2:1 electrolytes, including some of the transition metal salts in NMF. Povarov and co-workers[B163] measured the solubility of AgCl in pure NMF and in NMF-NaCl and NMF-CsCl solutions. The data have been used to evaluate the instability constant for AgCl in NMF. In view of the marked effect of trace impurities in this solvent on the solubilities of silver salts,[B107] these solubilities may be much greater than those obtained by electrochemical methods.

Heat data for electrolytes in NMF are confined primarily to uni-univalent electrolytes. Weeda and Somsen[B20] have measured heats for several alkali metal halides at low concentrations and corrected the data to infinite dilution by an extended Debye-Hückel equation. Held and Criss[B27] have also reported heats for some 1:1 electrolytes in NMF. They observed that LiCl showed a large concentration dependence at very low concentrations ($<0.001m$). Similar concentration effects were observed by Held and Criss for certain electrolytes in *N*,*N*-dimethylformamide,[B28] but recent data by Tsai and Criss[B174] indicate that this is an experimental artifact, probably caused by trace impurities in the solvent. A more reliable standard heat for LiCl can probably be obtained by correcting the data at higher concentrations ($>0.001m$) to infinite dilution by the Debye-Hückel equation. Finch and co-workers[B26,B171] have reported standard heats of solution for some alkaline earth halides, $Sr(ClO_3)_2$ and $Ba(ClO_3)_2$ in NMF. These are the only heat data that have been published for 2:1 electrolytes in NMF.

Criss, Luksha and Held[B5] have reported entropies on a mol fraction standard state for several alkali metal halides in NMF. Since the heat data employed for two of the salts, LiCl and LiBr, have been revised, new entropy values have been calculated and are listed in Appendix 2.4.41.

Feakins and Lawrence[B89] measured the relative viscosities of sodium and potassium chlorides and bromides in NMF from 25 to 45°C and expressed the data by an expanded Jones-Dole equation.[B87] The viscosity coefficients, A_η, B_η, and D_η were evaluated. While both A_η and B_η have positive values for every electrolyte studied in NMF, they are much smaller than the corresponding quantities in other organic solvents. The difference between the theoretical and experimental values of A_η may be either positive or negative.

N-methylacetamide (NMA) is another solvent which has received considerable interest in recent years because of its high degree of association and dielectric constant. Activity coefficients for a large number of uni-univalent electrolytes in NMA, obtained from freezing point depressions, have been reported at concentrations up to 1*m* by Wood and co-workers.[B14] Bonner and co-workers[B15] have also reported activity coefficients for a few iodide salts in this solvent. In the few cases where the data can be compared, the results of the latter investigators are considerably lower than those reported by Wood.

Weeda and Somsen[B21] measured heats of solution of several alkali metal halides in NMA at 35°C and used the data to evaluate solvation energies in this solvent. Except for LiI, all the salts exhibit more positive heats in this solvent than in NMF. However, comparison of heat data in the three solvents, formamide, NMF and NMA shows no consistent pattern.

Gopal and Siddiqi[B76] have measured the densities of several tetraalkylammonium iodides, KI and LiCl in NMA from 35 to 80°C and used the data to evaluate apparent molal volumes. The data have been extrapolated to infinite dilution by means of Masson's equation.[B74] The experimental slope, S_V, was observed to be positive for salts containing the smaller cations, but negative for salts having the larger tetraalkylammonium ions. The temperature dependence of $\bar{V}_2^0$ is negative for LiCl over the entire range studied while KI exhibits a maximum in the curve around 50°C. The tetraalkylammonium salts exhibit a positive temperature dependence over the whole range, with the larger ions showing the most pronounced temperature dependence.

Only a few data exist for the higher monosubstituted amides. Fuchs, Bear and Rodewald[B12] have measured heats of solution of several uni-univalent electrolytes in *N*-ethylacetamide (NEA) including some tetraalkylammonium salts and others containing complex anions. They observe that for some salts, particularly tetraphenylarsonium chloride and sodium tetraphenylborate, the heat of solution varies markedly with concentration. While they suggest that the cause may be the result of ion-pair formation, they offer no convincing evidence.

Millero[B175] has measured the densities of several uni-univalent electrolytes from 15 to 40°C in *N*-methylpropionamide (NMP) and used Masson's equation[B74] to evaluate partial molal volumes at infinite dilution. The experimental slopes, S_V, given only for $NaNO_3$ over the whole temperature range and for NaBr at 25°C, were observed to be negative in contrast to the positive values generally found for the common electrolytes in water and other solvents. Unfortunately, S_V cannot be compared to the theoretical slope because data on the pressure dependence of the dielectric constant and the compressibility of NMP are lacking. Millero[B175] also reported the viscosity coefficients A_η and B_η for $NaNO_3$ from 15 to 40°C. The reported A_η coefficients were negative in every case, but within experimental error the values are zero. Theoretically A_η can never have a negative value and most experimental results confirm this. Hoover[B176] has reported B_η coefficients for KCl at temperatures ranging from 20 to 40°C. With both $NaNO_3$ and KCl, the B_η values decrease as the temperature increases. Except for a few salts such as Li_2SO_4, $BeSO_4$ and $MgSO_4$, this is opposite from the trend observed for electrolytes in water over the same temperature range.[B86]

(*d*) *Amines*. Thermodynamic data are relatively scarce for electrolytes at low concentrations in amines. Apparent molal volumes of LiCl and $NaNO_3$ in monomethylamine from 0 to 25°C have been reported by Kelso and Felsing.[B177] The data do not obey Masson's equation[B74] and consequently were expressed by an extended form of this equation. Partial molal volumes at infinite dilution for LiCl were quite negative while those for $NaNO_3$ were positive, but in both cases $\bar{V}_2^0$ decreased as the temperature increased. Kraft and Bittrich[B178] qualitatively examined the solubility of 52 salts in diethylamine and indicated whether addition compounds are formed, but they give no numerical values for the solubilities.

Schmidt and co-workers[B179] have measured heats of solution of several electrolytes in anhydrous ethylenediamine at 25°C. Most of their data were for mol ratios of solvent to salt from 300 to 3000 with one as large as 15,000. The data were not extrapolated to infinite dilution, but the heats do not appear to be strongly dependent upon concentration.

Schmidt and co-workers[B180] have also measured apparent molal volumes and viscosities of several electrolytes in anhydrous ethylenediamine at 25°C and extrapolated the data to obtain $\bar{V}_2^0$. They observed that Masson's equation (eqn 2.3.66) was valid over the concentration range studied and their data indicate that S_V is negative in this solvent. A theoretical value for the limiting slope, A_V, is not available. The viscosity data are in agreement with the Jones-Dole equation over the concentration range studied, but the coefficients of this equation were not reported.

(*e*) *Acids*. Heats of solution of a few salts have been determined in formic acid. Kotlyarova and Ivanova[B181] measured ΔH_{soln} in formic acid from about 0.02 to 0.8*m* for KCl and to 3.6*m* for RbCl. No theoretical approach was used in the extrapolation of the data to infinite dilution, but the ΔH_{soln} vs. *m* curves used in the extrapolations are considerably different for the two salts. In another publication[B24] these authors present heat data for NaCl, CsCl and CsI in formic acid at 25°C and combine the heats with free energy data, taken from the literature, to calculate entropies of solution for NaCl, KCl, RbCl, CsCl and CsI. These data have been combined with absolute entropies of the crystalline salts to obtain partial molal entropies of the respective salts, which are listed in Appendix 2.4.41. The resulting values show $\bar{S}_2^0$ for these salts in formic acid to be more negative than for any other solvent for which data exist.[B5] Furthermore, the data appear to be inconsistent with data in other solvents.

Ivanova and Fesenko[B182] have measured the heats of solution of KBr in water-formic acid mixtures from 1.2 mol per cent acid to pure acid. Data were obtained over a wide concentration range and ΔH^0_{soln} evaluated. ΔH^0_{soln} decreases as the acid content is increased up to about 69 mol per cent acid, after which the heats increase again. Fesenko, Ivanova and Kotlyarova[B77] have measured densities for some alkali metal halides in formic acid and evaluated apparent molal volumes up to 3.2*m*. The data at low concentrations obey Masson's equation[B74] and this was employed to extrapolate the data to infinitive dilution. The value of S_V was positive in every case and $\bar{V}^0_2$ for the electrolytes investigated in this solvent are all lower than in water.

Stern and Nobilione[B158,B183] measured the heats of transfer of HCl and $HClO_4$ from water to aqueous acetic acid by determining enthalpies of dilution of aqueous HCl and $HClO_4$ with aqueous acetic acid. The resultant solvent composition range runs from pure water to 99.7 mol per cent acetic acid-water mixtures. The acid concentrations were around 0.003*m*. The correction to infinite dilution, for the highly aqueous solutions appears to be comparable to the experimental error. The limiting heat of dilution is not known for pure acetic acid.

(f) *Glycerols.* Glycerol itself and three of its derivatives, the mono-, di- and tri-acetates are the only members of this group for which data exist. Kraus and co-workers[B156] measured the solubilities of NaCl, KCl and $Ba(NO_3)_2$ in glycerol-water mixtures from 23 to 99.8 wt% glycerol and in glycerol triacetate-water mixtures containing more than 96 wt% of the organic component. They have also reported solubilities of NaCl in glycerol monoacetate-water mixtures from 25 to 99.7 wt% and glycerol diacetate-water mixtures from 78 to 99.9 wt% of organic component. Activity coefficients, based on the aqueous salt solution reference state, are also reported for the NaCl solutions. The solubilities drastically decrease in the order glycerol > mono- > di- > tri-acetate. Krickard and Skinner[B94] have measured the B_η viscosity coefficients for KI and CsI in glycerol and compared the values with those for a few other solvents. The coefficients in this solvent are more negative than those reported for any other solvent, including water.

(ii) *Aprotic Solvents*

(a) *Amides.* *N,N*-dimethylformamide has been studied far more extensively than any other amide. A careful study of the solubilities of several alkali and alkaline earth metal halides and some transition metal halides has been made by Pistoia and co-workers.[B162,B184] The salts containing either large anions or small cations tend to be most soluble. Paul and Sreenathan[B185] have made the most extensive solubility study in DMF, and give values for approximately 45 salts and acids, but Pistoia and co-workers[B184] found these data to be unreliable. Except for the simple electrolytes they observed that most species form solvates. Alexander and co-workers[B107] reported solubility products for several silver and alkali metal salts and tetraethylammonium picrate. Their data for the silver salts were obtained potentiometrically. This method avoids the errors, inherent in direct measurements of solubility, resulting from trace impurities of amines which form silver complexes. Criss and Luksha[B186] measured the solubilities of several alkali metal and silver halides analysing the saturated solutions. The solubilities of the silver salts are considerably greater than those reported by Alexander and co-workers,[B107] probably because of complexation by trace impurities. Solubilities obtained potentiometrically and a comparison of the data are discussed in detail in sect. 2.7.2.

A considerable amount of calorimetric data exists for DMF solutions. Held and Criss[B28] have measured heats of solution of several alkali metal halides and magnesium and gadolinium chloride in DMF. Data for salts having high charge-density cations show a large concentration effect in very dilute solutions ($<0.001m$). Furthermore, $MgCl_2$ shows a two-stage heat effect upon dissolution in dilute solution, indicating that an additional reaction is occurring. Recent evidence[B174] indicates that these phenomena are probably associated with uncontrollable traces of impurities in the solvent which become significant at salt concentrations of less than $0.001m$. Consequently, only the measurements above $0.001m$ can be considered reliable. Weeda and Somsen[B187] have measured heats of solution of the lithium halides and other alkali metal halides for dilute solutions and corrected the data to infinite dilution to obtain ΔH^0_{soln}. Bhatnager and Criss[B188] measured heats of solution of some tetra-alkylammonium iodides in DMF for the purpose of obtaining heats of transfer from water to DMF. Finch and co-workers[B26,B171] have measured heats of several alkaline earth halides, chlorates and bromates. They evaluated the standard heats of solution for the halide salts and two of the complex salts, $Sr(ClO_3)_2$ and $Ba(ClO_3)_2$. Krestov and Zverev[B189] measured heats of solution of several alkali metal halides and nitrates and used both the Debye-Hückel limiting equation and an extended form, eqn. 2.3.47, to extrapolate the data to infinite dilution. Their measurements ranged from about 0.01 to $0.1m$ and in some cases the observed slopes were significantly greater than that predicted by theory. Except for NaI, the two methods of extrapolation generally agreed within 0.30 kcal mol^{-1}. Fuchs and co-workers[B12] measured heats for several uni-univalent compounds, many containing complex ions, at very low concentration. In contrast to Held and Criss[B28] they found no concentration effect for LiCl in DMF containing 6 ppm water. However, for some of the electrolytes examined they did observe a strong concentration dependence, for which they offered no adequate explanation. Choux and Benoit[B190] have measured heat data at 30°C for some electrolytes having complex ions, mostly tetraethylammonium salts. The measurements were made in a concentration range of 3×10^{-3} to 5×10^{-4} and for $NaBPh_4$ they observed a strong concentration effect below $10^{-3}m$. Since conductance measurements indicate that ion pairing is insignificant for this electrolyte at the concentrations studied, they ascribed the effect to impurities in the solvent, and the data were extrapolated to infinite dilution without considering this anomalous effect. Fuchs and co-workers[B12] do not mention a strong concentration effect for $NaBPh_4$ in this solvent. There is considerable overlap in the data reported by these several groups of workers and in general the agreement is satisfactory, considering the systems involved. There are, however, three significant discrepancies: (1) The results for LiI[B12,B174,B187] disagree by about 7 kcal mol^{-1}, but the first two[B12,B174] agree within 0.6 kcal mol^{-1}, and it appears that these two are the more reliable; (2) measurements of ΔH^0_{soln} for $NaBPh_4$ disagree by about 8 kcal mol^{-1}.[B12,B190] This may be associated with the strong concentration effect observed by one group[B190] while the other group[B12] apparently made no such observation. These data should be re-examined in the light of this discrepancy; (3) the data for KBr disagree by about 1.7 kcal mol^{-1},[B12,B28,B189] but again two of the sources[B12,B28] are in substantial agreement and are to be preferred. Partial molal entropies in DMF are relatively scarce. Criss and Luksha,[B186] using free energies obtained from solubilities and calorimetrically measured heats[B28] have calculated $\bar{S}^0_2$ for several alkali metal halides. However, in view of revisions in some of the heat and solubility data, the entropies for a few of the salts have been recalculated; these $\bar{S}^0_2$ values are listed in Appendix 2.4.41.

Scattered data for several other aprotic amides exist. Pistoia and Scrosati[B191] have reported the solubilities for several alkali and alkaline earth metal halides and transition metal halides in *N,N*-dimethylacetamide. The solubilities increase in the order Cl < Br < I. Alexander and co-workers[B107] have reported solubility product values for numerous silver salts in DMA. They have also reported values for silver salts, KBr and NaCl in hexamethylphosphorotriamide (HMPT). The solubilities of the silver halides in both these solvents follow the same trend as they do in water. Coleman[B192] has measured the solubilities and activity coefficients of sodium and potassium chlorides in several dialkylamides containing water. Distribution coefficients of the salt between water and amide have also been calculated.

Heat measurements in the higher dialklyamides are confined to *N*-methylpyrrolidone and HMPT. Fuchs and co-workers[B12] measured the enthalpy of solution of several alkali metal and tetraalkylammonium halides in the first of these solvents and the lithium halides in HMPT. With the exception of lithium chloride, heats are less negative in *N*-methylpyrrolidone than in DMF but more negative than in the monosubstituted amides or formamide. The heat of solution of lithium chloride does not vary much with solvent, but it appears to be smallest in *N*-methylpyrrolidone.

(b) *Esters*. Thermodynamic data for esters appear to be confined mainly to solubility data. Some of the more recent solubility studies have included measurements by Biktimirov[B114] for caesium nitrate in several esters and by Addison and co-workers[B193] for copper nitrate in ethyl acetate. The former workers relate the log of the solubilities in the several esters to the dielectric constant of the solvent through an equation which is a linear function of $1/\varepsilon_r$.

(c) *Ketones*. Acetone is the only ketone for which any significant thermodynamic data exist. Some of the more recent solubility studies in this solvent are those of Biktimirov[B114] who reported the solubility of $CsNO_3$ and Izmailov and Chernyi[B111] who measured the solubility of AgCl and CsCl. Radioactive tracer techniques were used in all these measurements. These investigators also report solubility data for the same salts in some of the higher ketones and, as in the case of the esters, express the log of the solubility as a linear function of the reciprocal of the dielectric constant of the ketone.

Walden[B117] reported some of the earliest calorimetric data but his measurements were carried out at salt-solvent mole ratios of about 1:200 and consequently they are of little use in obtaining standard state functions. The most extensive heat measurements in this solvent have been reported by Mishchenko and co-workers.[B30,B126,B131–133] Mishchenko and Sokolov[B132,B133] measured heats of solution of $NaClO_4$, NaSCN, KSCN, NH_4SCN and $Mg(ClO_4)_2$ at 25°C and NaI at 10, 25 and 40°C. The NaI and $NaClO_4$ data were obtained over a wide concentration range, and extrapolated to infinite dilution. Following their usual treatment of heat data obtained over broad concentration ranges, the extrapolations were made as a function of m rather than the more common $m^{1/2}$. Relative partial molal enthalpies of both the solvent and solutes were calculated and the non-ideal part of the relative partial molal entropy for the solvent evaluated. The heats of solution for the several species increase in the order $Mg(ClO_4)_2 < NaI < NaClO_4 < NaSCN < KSCN < NH_4SCN$. Zhilina and Mishchenko[B126] measured the heat of solution of picric acid in acetone at 10 and 25°C and for concentrations from 0.005m to saturation. A plot of ΔH_{soln} as a function of m shows a very marked concentration dependence at low concentration and the sign of this dependence is opposite at the two temperatures investigated. The authors ascribe this concentration dependence largely to incomplete dissociation. Because of insufficient data, they give no detailed explanation for the opposite signs in the heats of dilution at the two temperatures. However, heats of dilution of weak electrolytes in water show similar changes in sign as one varies the temperature[B194] and this is due to the equilibrium constant passing through a maximum somewhere in the temperature interval. Novoselov and Mishchenko[B131] have measured the heat capacities of NaI in acetone from very dilute solutions to about 2.4m and have compared the data with similar measurements in water and methanol. No ϕ_{C_p} or $\bar{C}_{p_2}$ values were evaluated, but their treatment of the data clearly indicates that the concentration effect on $\bar{C}_p$ data is very different in aqueous solution than in electrolytic solutions of methanol or acetone.

Hood and Hohlfelder[B195] measured the viscosities of solutions of LiCl in acetone at 18 and 25°C and evaluated the A_η and B_η coefficients in the Jones-Dole equation.[B87] The value of 0.024 evaluated for A_η at 25°C agrees well with the theoretical value of 0.0237. More recently Feakins and Lawrence[B89] re-evaluated the data of Hood and Hohlfelder using an expanded Jones-Dole equation. Their A_η coefficient does not agree quite as well with the theoretical value, but the disagreement is small compared with the standard deviation of the data. The B_η coefficient evaluated by Feakins and Lawrence is slightly higher than the old value, but in view of present theory this is not significant.

(d) *Dimethylsulphoxide* (*DMSO*). DMSO has been of great interest to organic chemists as a solvent medium and partly as a result of this a considerable amount of thermodynamic data for electrolytes in this solvent has been reported. Furthermore, like DMF, the solvent is aprotic with a relatively high dielectric constant which makes it interesting from a theoretical point of view. Alexander and co-workers[B107] measured the solubility of several silver salts and tabulated additional data taken from the literature. Rumbaut and Peeters[B196] and Luehrs and co-workers[B197] reported the solubility of silver halides in DMSO. Synnott and Butler[B198] have thoroughly studied the solubility and complex formation of AgCl in DMSO-water mixtures and have critically examined the data of previous workers. The data of Luehrs and co-workers[B197] and Rumbaut and Peeters[B196] recalculated by Synnott and Butler[B198] give results in good agreement with their measurements. Addition of water to DMSO causes K_{s0} for AgCl to decrease until a water concentration of 1.8 molar is obtained at which point a plateau is reached with K_{s0} being about one order of magnitude lower than in the pure solvent. Electrochemical methods were employed in all the work which is superior[B107] to analysing saturated solutions of the salt because trace amounts of dimethylsulphide in the solvent complex with the silver ion. In addition to forming complexes with impurities, silver also forms complexes with halogens in DMSO to form AgX(soln) and AgX_2^- which lead to an even greater solubility. The same effect is seen for silver chloride in DMF.[B199] All the studies described above agree that the order of the solubilities of the silver halides in DMSO is AgCl > AgBr > AgI, as observed in aqueous solutions.

Gasser and co-workers[B200,B201] have obtained activity coefficients from the freezing point method for LiCl in DMSO up to about 0.4*m* and of CsI up to 0.25*m*. Skerlak and co-workers[B202] have calculated activity coefficients for several uni-univalent electrolytes in DMSO by the freezing point method. The measurements reach concentrations greater than 1*m*, but are presented only in graphical form. (For further discussion see sect. 2.10.3.)

Heats of solution of electrolytes in DMSO have been reported by several investigators, usually for the purpose of obtaining heats of transfer from DMSO to water or to other solvents. Arnett and McKelvey[B22] measured heats for several uni-univalent electrolytes, including alkali metal iodides and some tetraalkylammonium halides. Fuchs and co-workers[B12,B23] have made similar studies with a greater variety of electrolytes, and Choux and Benoit[B190] have measured the heat of solution of Et_4NClO_4 in DMSO at 30°C. Most recently Krishnan and Friedman[B203,B203a] made an extensive series of heat measurements on many uni-univalent electrolytes containing complex ions. The data of all of these investigators were obtained at low concentrations and were corrected to infinite dilution, except in a few cases where the experimental errors were considered to be larger than the correction to the reference state. Clark and Bear[B204] measured heats of solution of several rare earth chlorides in DMSO at concentrations of about $3 \times 10^{-4}m$. No attempt was made to correct the data to infinite dilution, even though such a correction would probably be significant since the solutes are 3:1 electrolytes. In the few cases where the published data overlap, agreement is generally satisfactory. The largest discrepancy exists for Am_4 NBr for which the two reported values[B12,B203] disagree by 1.0 kcal. As a general rule the heats of solution for the simple electrolytes are considerably more negative in DMSO than in water while the reverse is true for salts containing complex ions. However, there are some notable exceptions to this observation, particularly for electrolytes containing ions with phenyl groups.

Archer and Gasser[B201] have measured the viscosity of DMSO-CsI solutions up to about 0.7 molar and employed the extended Jones-Dole equation[B88] to fit the data. The viscosities were not sufficiently accurate to evaluate A_η, consequently, this term was calculated theoretically and the B_η and D_η coefficients evaluated by plotting $[(\eta/\eta_0) - A_\eta C^{\frac{1}{2}} - 1]/C$ against C. The resultant B_η value is 0.68, in marked contrast to the negative value observed for this quantity in water, ethylene glycol and glycerol,[B94] but consistent with B_η coefficients for electrolytes in most other organic solvents.

(e) *Propylene Carbonate* (*PC*). PC is another aprotic solvent which has been the subject of extensive thermodynamic investigations in recent years. Electrochemical studies, and the resulting solubility data are examined in sect. 2.7.3d.

Friedman and co-workers[B25,B203,B203a,B205] have measured heats of solution of numerous electrolytes in PC at 25°C, primarily to obtain ionic solvation enthalpies, and enthalpies of transfer between this and other solvents. Because many of the alkali metal halides dissolve insufficiently fast for accurate heat measurements, most of the data are for uni-univalent electrolytes containing complex ions. The only additional heat data in this solvent, reported by Choux and Benoit,[B190] are for Ph_4AsI. All measurements were made in dilute solution and most were corrected to infinite dilution using the Debye-Hückel equation. In general, enthalpies of solution of electrolytes in PC follow the same trend as in DMSO; that is, the alkali metal salts exhibit more negative heats than in water while the reverse is generally true for salts with complex cations.

(f) *Sulpholane.* Only a few thermodynamic data of interest have been published for this solvent. Choux and Benoit[B190] measured heats of solution for several salts including alkali metal and tetraalkylammonium halides and perchlorates. All measurements were made at 30°C and in dilute solutions. The data were corrected to infinite dilution using the simple Debye-Hückel theory, with the correction being about 0.03 kcal mol^{-1}. (Cryoscopic studies with solutions in this solvent are described in sect. 2.10.9.)

(g) *Acetonitrile.* A few solubility data have been reported for electrolytes in this solvent since the publication of Seidell and Linke.[B13] Luehrs and co-workers[B197] measured the solubility of the silver halides electrochemically and Alexander and co-workers[B107] measured the solubility of the silver halides and several other silver salts. In addition they have tabulated solubilities for some of the alkali metal salts taken from the literature. The values of pK_{s0}, for AgCl and AgBr, given by these two groups of workers disagree by no more than 0.5. Addison and co-workers[B193] measured the solubility of copper nitrate in acetonitrile. In general electrolytes are much less soluble in acetonitrile than in water. Exceptions appear to arise when the electrolyte contains an ion with organic groups attached. As in water and most other solvents the order of solubility of the silver halides is AgCl $>$ AgBr $>$ AgI.

Calorimetric data for electrolytes in acetonitrile are scarce. Walden[B117] measured the heats of solution of potassium iodide and tetraethyl- and tetrapropylammonium iodides in acetonitrile and, for the latter two salts, in propionitrile. The concentrations employed were high, approximately 1:200 to 1:400 mol ratio of salt to solvent, and no attempt was made to extrapolate the data to infinite dilution. Recently Kushchenko and Mishchenko[B127] have measured heats of solution of NaI and

$NaClO_4$ in acetonitrile at 25 and 50°C and over wide concentration ranges. They extrapolated the data to infinite dilution using plots of ΔH_{soln} vs. m. The temperature coefficient of their data indicates that ΔC_p^0 for the dissolution of these electrolytes is positive. However, as pointed out earlier, because of the change in limiting slope with temperature, it is possible as in the case of methanol, for ΔC_p to change sign at concentrations of less than 0.003m. In water ΔC_p^0 for $NaClO_4$ is significantly negative over the temperature range of 25 to 50°C. For methanol, ΔC_p^0 of dissolution of this electrolyte is negative, but not nearly as negative as in water and actually becomes positive at temperatures below 25°C.

Tuan and Fuoss[B146] have evaluated B_η coefficients in acetonitrile for several tetraalkylammonium salts. They observe that the B_η coefficients are additive as in other solvents, and that for the larger ions the experimental value of B_η, used in conjunction with the Einstein model, gives radii consistent with other methods of measurement. As in most other non-aqueous solvents all the B_η coefficients are positive.

(h) *Nitrocompounds.* Nitromethane has been the most commonly studied compound of this group, but even for this solvent the data are very limited. Walden[B117] measured the heat of solution of tetraethyl- and tetrapropylammonium iodides in nitromethane, but as in his other measurements these were at mol ratios of 1:200 to 1:400 and no attempt was made to extrapolate the data to infinite dilution. Wolfenden and co-workers[B118] measured heats for a few tetraalkylammonium salts and evaluated the standard heats of solution. Wolfenden and co-workers[B119] also measured heats of dilution of $NaClO_4$ and Et_4NPic in nitromethane. In both cases, the theoretical limiting slope is exceeded by approximately a factor of two.

The only published B_η coefficients for an electrolyte in nitromethane is for Bu_4NBr.[B146] The value is consistent with B_η coefficients observed for other aprotic non-aqueous solvents.

Solubilities of the silver halides have been reported in nitroethane by Luehrs and co-workers.[B197] The order of solubility follows the same trend as in water and other organic solvents.

Wolfenden and co-workers[B118] have measured heats of solution of Et_4NClO_4 and Et_4NPic in nitrobenzene and corrected the data to infinite dilution. Heats of solution of both these salts in nitrobenzene and nitromethane are considerably smaller than they are in hydrogen-bonded solvents. Wolfenden and co-workers[B119] measured heats of dilution of the same two electrolytes in nitrobenzene. Considering the system being studied, the slopes are in fair agreement with the theoretical value.

(i) *Miscellaneous Data.* Kapustinskii and co-workers[B206] have reported heats of solution of KCl in dioxan-water mixtures from pure water to 70 wt% dioxan. All data are at about 0.058*m* and values corresponding to infinite dilution were not evaluated. Mishchenko and Shadskii[B124] have measured heats of solution of NaI in water-dioxan mixtures from 46 to 68 wt% dioxan and at 2 and 25°C. The measurements covered wide concentration ranges and were extrapolated to infinite dilution. The temperature coefficient of the integral heat of solution for the 46% dioxan solvent is zero. At the higher salt concentrations the heat of solution is independent of concentration. For both KCl and NaI, ΔH_{soln} decreases as the dioxan content of the solvent is increased.

Mishchenko and Poltoratskii[B135] measured heats of solution of Bu_4NSCN over a wide concentration range in benzene at 20 and 25°C, and estimated a value of 2.5 kcal mol^{-1} for the standard heat of solution at 25°C. At the higher concentrations the heats are nearly independent of the salt concentration.

Solubility data exist for several miscellaneous systems. Deno and Berkheimer[B148] have measured solubilities of several quaternary ammonium perchlorates in benzene. Except for $(C_6H_{13})_4NClO_4$ all the salts dissolve sparingly. Kraft and Bittrich[B178] have qualitatively investigated the solubility of approximately 50 salts in triethylamine. Fleischer and Freiser[B207] measured the solubilities of copper and nickel dimethylglyoximes in water, chloroform, benzene and *n*-heptane as a function of temperature and evaluated standard heats and entropies of solution of these species in the various solvents. Both salts are only slightly soluble in all the solvents investigated, so that the equilibrium concentration can be considered approximately equal to the activity. Both the heats and entropies of solution for the nickel compound are in the order $CHCl_3 < C_6H_6 < H_2O < C_7H_{16}$ while for the copper compound the order is $H_2O < CHCl_3 < C_6H_6 < C_7H_{16}$. The authors explain the latter order in terms of structure in the solvent, but they ignore the fact that the order of the solvents is different for the two salts.

Glubokov and co-workers[B208] have published solubilities for some alkali and alkaline earth metal chlorides and aluminium chloride in tributylphosphate and di-isopentylmethylphosphonate. Both the anhydrous and hydrated salts were investigated, and while the same general trends in solubility were observed, there are also a number of differences.

Acknowledgment

The author wishes to thank Drs F. J. Millero, R. H. Wood, H. L. Friedman, and others for making their manuscripts available prior to publication. Grateful acknowledgment is also given to the National Science Foundation for partially supporting this work.

(Manuscript completed July 1971.)

REFERENCES

B1 H. S. Frank and W. Wen, *Disc. Faraday Soc.*, **24,** 133 (1957)
B2 R. W. Gurney, Ionic Processes in Solution, McGraw-Hill Book Co., Inc., New York (1953)
B3 T. Ackermann, *Disc. Faraday Soc.*, **24,** 180 (1957)
B4 G. Nemethy and H. A. Scheraga, *J. Chem. Phys.*, **36,** 3382, 3401 (1962)
B5 C. M. Criss, R. P. Held and E. Luksha, *J. Phys. Chem.*, **72,** 2970 (1968)
B6 F. Franks and D. S. Reid, *J. Phys. Chem.*, **73,** 3152 (1969)
B7 F. J. Millero, *J. Phys. Chem.*, **73,** 2417 (1969)
B8 M. Mastroianni and C. M. Criss, *J. Chem. Eng. Data*, **17,** 222 (1972)
B9 M. Born, *Z. physik*, **1,** 45 (1920)
B10 W. M. Latimer, K. S. Pitzer and C. M. Slansky, *J. Chem. Phys.*, **7,** 108 (1939)
B11 E. M. Arnett, W. G. Bentrude, J. J. Burke and P. McC. Duggleby, *J. Amer. Chem. Soc.*, **87,** 1541 (1965)
B12 R. Fuchs, J. L. Bear and R. F. Rodewald, *J. Amer. Chem. Soc.*, **91,** 5797 (1969)
B13 A. Seidell and W. F. Linke, Solubilities of Inorganic and Metal-Organic Compounds, 4th Ed., Vol. 1 (1958); Vol. 2 (1965). D. Van Nostrand Co., Inc., Princeton, N.J.
B14 R. W. Kreis and R. H. Wood, *J. Phys. Chem.*, **75,** 2319 (1971); R. H. Wood, R. K. Wicker and R. W. Kreis, *J. Phys. Chem.*, **75,** 2313 (1971)
B15 O. D. Bonner, S. J. Kim and A. L. Torres, *J. Phys. Chem.*, **73,** 1968 (1969)
B16 R. Gopal and M. M. Husain, *J. Indian Chem. Soc.*, **43,** 204 (1966)
B17 M. de Forcrand, *Compt. rend.*, **101,** 318 (1885)
B18 S. U. Pickering, *J. Chem. Soc.*, **53,** 865 (1888)
B19 G. Somsen and J. Coops, *Rec. Trav. Chim.*, **84,** 985 (1965)
B20 L. Weeda and G. Somsen, *Rec. Trav. Chim.*, **85,** 159 (1966)
B21 L. Weeda and G. Somsen, *Rec. Trav. Chim.*, **86,** 263 (1967)
B22 E. M. Arnett and D. R. McKelvey, *J. Amer. Chem. Soc.*, **88,** 2598 (1966)
B23 R. F. Rodewald, K. Mahendran, J. L. Bear and R. Fuchs, *J. Amer. Chem. Soc.*, **90,** 6698 (1968)
B24 G. P. Kotlyarova and E. F. Ivanova, *Zh. Fiz. Khim.*, **40,** 997 (1966)
B25 Y. C. Wu and H. L. Friedman, *J. Phys. Chem.*, **70,** 501 (1966)
B26 A. Finch, P. J. Gardner and C. J. Steadman, *J. Phys. Chem.*, **71,** 2996 (1967)
B27 R. P. Held and C. M. Criss, *J. Phys. Chem.*, **69,** 2611 (1965)
B28 R. P. Held and C. M. Criss, *J. Phys. Chem.*, **71,** 2487 (1967)
B29 F. D. Rossini, D. D. Wagman, W. H. Evans, S. Levine and I. Jaffe, Selected Values of Chemical Thermodynamic Properties, Circular 500, Nat. Bur. Standard (U.S.) (1952)
B30 K. P. Mishchenko and G. M. Poltoratskii, Examination of the Thermodynamics and Structure of Aqueous and Non-aqueous Electrolytic Solutions, Chemical Publisher, Leningrad Division (1968)
B31 F. J. Millero, *Chem. Rev.*, **71,** 147 (1971)
B32 A. Weissberger, (ed.), Physical Methods of Organic Chemistry, Vol. 1, Part 1. Interscience, New York (1959)
B33 J. Reilly and W. N. Rae, Physico-Chemical Methods, 5th Ed., Van Nostrand, New York (1953)
B34 I. M. Kolthoff and P. J. Elving (eds.), Treatise on Analytical Chemistry, Vol. I, Part I, Chapters 17–19. Interscience, New York (1959)
B35 O. Popovych, *Anal. Chem.*, **38,** 558 (1966)
B36 O. Popovych and R. M. Friedman, *J. Phys. Chem.*, **70,** 1671 (1966)
B37 W. P. White, The Modern Calorimeter, Chemical Catalog Co., New York (1928)

B38 H. A. Skinner (ed.), Experimental Thermochemistry, Vol. II, Interscience, New York (1962)
B39 O. Kubaschewski and E. L. Evans, Metallurgical Thermochemistry, 3rd Ed., Pergamon, New York (1958)
B40 S. Sunner and I. Wadsö, *Acta Chem. Scand.*, **13,** 97 (1959)
B41 H. L. Friedman and Y. C. Wu, *Rev. Sci. Inst.*, **36,** 1236 (1965)
B42 S. R. Gunn, *Rev. Sci. Inst.*, **29,** 377 (1958)
B43 D. A. MacInnes, M. O. Dayhoff and B. R. Ray, *Rev. Sci. Inst.*, **22,** 642 (1951)
B44 H. A. Bowman and R. M. Schoonover (with Appendix by M. W. Jones), *J. Res. Nat. Bur. Stand.*, **71C,** 179 (1967)
B45 L. G. Hepler, J. M. Stokes and R. H. Stokes, *Trans. Faraday Soc.*, **61,** 20 (1965)
B46 P. Hidnert and E. L. Peffer, Density of Solids and Liquids, Nat. Bur. Stand. (U.S.), Cir. 487 (1950)
B47 F. J. Millero, *Rev. Sci. Inst.*, **38,** 1441 (1967)
B48 R. A. Robinson and R. H. Stokes, Electrolyte Solutions, Butterworths, London (1959)
B49 B. A. Soldano, R. W. Stoughton, R. J. Fox and G. Scatchard, *in* The Structure of Electrolytic Solutions, Chapter 14 (W. J. Hamer, ed.), Wiley, New York (1959)
B50 G. Scatchard, P. T. Jones and S. S. Prentiss, *J. Amer. Chem. Soc.*, **54,** 2676 (1932)
B51 O. D. Bonner, C. F. Jordan and K. W. Bunzl, *J. Phys. Chem.*, **68,** 2450 (1964)
B52 W. M. Cox and J. H. Wolfenden, *Proc. Roy. Soc. London*, **145A,** 475 (1934)
B53 M. Kaminsky, *Z. Phys. Chem. (Frankfurt)*, **5,** 154 (1955)
B54 R. C. Hardy, NBS Viscometer Calibrating Liquids and Capillary Tube Viscometers, Monograph 55, Nat. Bur. Stand. (1962)
B55 J. Mitchell and D. M. Smith, Aquametry, Interscience, New York (1948)
B56 F. A. Askew, N. S. Jackson, O. Gatty and J. H. Wolfenden, *J. Chem. Soc.*, 1362 (1934)
B57 J. M. Notley and M. Spiro, *J. Chem. Soc. B.*, 362, (1966)
B58 C. M. French and K. H. Glover, *Trans. Faraday Soc.*, **51,** 1418 (1955)
B59 E. A. Guggenheim and J. E. Prue, *Trans. Faraday Soc.*, **50,** 710 (1954)
B60 G. N. Lewis and R. Randall, revised by K. S. Pitzer and L. Brewer, Thermodynamics, McGraw-Hill, New York (1961)
B61 E. A. Gulbransen and A. L. Robinson, *J. Amer. Chem. Soc.*, **56,** 2637 (1934)
B62 C. M. Criss and J. W. Cobble, *J. Amer. Chem. Soc.*, **83,** 3223 (1961)
B63 E. C. Jekel, C. M. Criss and J. W. Cobble, *J. Amer. Chem. Soc.*, **86,** 5404 (1964)
B64 R. H. Wood, R. W. Smith and H. S. Jongenburger, unpublished results
B65 S. I. Drakin, L. V. Lantukhova and M. K. Karapet'yants, *Zh. Fiz. Khim.*, **40,** 451 (1966)
B66 M. K. Karapet'yants, S. I. Drakin and L. V. Lantukhova, *Zh. Fiz. Khim.*, **41,** 2653 (1967)
B67 F. T. Gucker, Jr., *Ann. N.Y. Acad. Sci.*, **51,** 680 (1949)
B68 J. C. Ahluwalia and J. W. Cobble, *J. Amer. Chem. Soc.*, **86,** 5377 (1964); R. E. Mitchell and J. W. Cobble, *J. Amer. Chem. Soc.*, **86,** 5401 (1964)
B69 W. L. Gardner, E. C. Jekel and J. W. Cobble, *J. Phys. Chem.*, **73,** 2017 (1969)
B70 W. L. Gardner, R. E. Mitchell and J. W. Cobble, *J. Phys. Chem.*, **73,** 2025 (1969)
B71 K. K. Kelley, Contribution to the Data on Theoretical Metallurgy XIII. High-Temperature Heat-Content, Heat Capacity, and Entropy Data for the Elements and Inorganic Compounds, Bul. 584, U.S. Bureau of Mines (1960)
B72 M. Mastroianni, and C. M. Criss, *J. Chem. Thermodyn.*, **4,** 321 (1972)
B73 F. T. Gucker, Jr., *J. Phys. Chem.*, **38,** 307 (1934). See also F. T. Gucker, Jr., F. W. Gage and C. E. Moser, *J. Amer. Chem. Soc.*, **60,** 2582 (1938)

B74 D. O. Masson, *Phil. Mag.*, **8**, 218 (1929)
B75 H. S. Harned and B. B. Owen, The Physical Chemistry of Electrolytic Solutions, 3rd Ed. Reinhold, New York (1958)
B76 R. Gopal and M. A. Siddiqi, *J. Phys. Chem.*, **73**, 3390 (1969)
B77 V. N. Fesenko, E. F. Ivanova and G. P. Kotlyarova, *Zh. Fiz. Khim.*, **42**, 2667 (1968)
B78 I. Lee and J. B. Hyne, *Can. J. Chem.*, **46**, 2333 (1968)
B79 J. Padova and I. Abrahamer, *J. Phys. Chem.*, **71**, 2112 (1967)
B80 J. Sobkowski and S. Minc, *Rocz. Chem.*, **35**, 1127 (1961)
B81 R. Gopal and R. K. Srivastava, *J. Ind. Chem. Soc.*, **40**, 99 (1963)
B82 O. Redlich, *J. Phys. Chem.*, **44**, 619 (1940)
B83 O. Redlich and D. M. Meyer, *Chem. Rev.*, **64**, 221 (1964)
B84 O. Redlich, *J. Phys. Chem.*, **67**, 496 (1963)
B85 B. B. Owen and S. R. Brinkley, Jr., *Ann. N.Y. Acad. Sci.*, **51**, 753 (1949)
B86 M. Kaminsky, *Disc. Faraday Soc.*, **24**, 171 (1957)
B87 G. Jones and M. Dole, *J. Amer. Chem. Soc.*, **51**, 2950 (1929)
B88 M. Kaminsky, *Z. Phys. Chem. (Frankfurt)*, **12**, 206 (1957)
B89 D. Feakins and K. G. Lawrence, *J. Chem. Soc. (A)*, 212 (1966)
B90 M. Kaminsky, *Z. Phys. Chem. (Frankfurt)*, **8**, 173 (1956)
B91 H. Falkenhagen and E. L. Vernon, *Physik. Z.*, **33**, 140 (1932)
B92 G. Jones and H. J. Fornwalt, *J. Amer. Chem. Soc.*, **57**, 2041 (1935)
B93 J. C. Lafanechere and J. P. Morel, *Compt. Rend.*, **268C**, 1222 (1969)
B94 J. Crickard and J. F. Skinner, *J. Phys. Chem.*, **73**, 2060 (1969)
B95 P. A. D. deMaine, E. R. Russell, D. O. Johnston and M. M. deMaine, *J. Chem. Eng. Data*, **8**, 91 (1963)
B96 R. L. Kay, T. Vituccio, C. Zawoyski and D. F. Evans, *J. Phys. Chem.*, **70**, 2336 (1966)
B97 J. H. Hildebrand and R. L. Scott, Regular Solutions, Prentice-Hall, Englewood Cliffs (1962)
B98 E. M. Arnett, W. B. Kover and J. V. Carter, *J. Amer. Chem. Soc.*, **91**, 4028 (1969)
B99 D. D. Eley, *Trans. Faraday Soc.*, **35**, 1281, 1421 (1939)
B100 R. Battino and H. L. Clever, *Chem. Rev.*, **66**, 395 (1966)
B101 A. Ben-Naim and S. Baer, *Trans. Faraday Soc.*, **60**, 1736 (1964); A. Ben-Naim and G. Moran, *Trans. Faraday Soc.*, **61**, 821 (1965); A. Ben-Naim, *J. Phys. Chem.*, **71**, 4002 (1967)
B102 R. H. Wood and D. E. DeLaney, *J. Phys. Chem.*, **72**, 4651 (1968)
B103 A. R. Giona and G. Pfeifer, *Chim. Ind. (Milan)*, **44**, 1354 (1962)
B104 G. E. Braude, I. L. Leites and I. V. Dedova, *Khim. Prom.*, 232 (1961)
B105 J. H. Dymond, *J. Phys. Chem.*, **71**, 1829 (1967)
B105 (*a*) C. V. Krishnan and H. L. Friedman, *J. Phys. Chem.*, **73**, 1572 (1969)
B105 (*b*) C. V. Krishnan and H. L. Friedman, *J. Phys. Chem.*, **75**, 388 (1971)
B106 R. L. Kay, C. Zawoyski and D. F. Evans, *J. Phys. Chem.*, **69**, 4208 (1965)
B107 R. Alexander, E. C. F. Ko, Y. C. Mac and A. J. Parker, *J. Amer. Chem. Soc.*, **89**, 3703 (1967)
B108 R. E. Harner, J. B. Sydnor and E. S. Gilreath, *J. Chem. Eng. Data*, **8**, 411 (1963)
B109 H. Brusset and J. C. Lecoq, *Compt. Rend.*, **259**, 814 (1964)
B110 E. M. Kirmse, *Z. Chem.*, **1**, 332 (1961)
B111 N. A. Izmailov and V. S. Chernyi, *Zh. Fiz. Khim.*, **34**, 127 (1960)
B112 Z. I. Grigorovich, *Zh. Neorg. Khim.*, **8**, 986 (1963)
B113 F. R. Hartley and A. W. Wylie, *J. Chem. Soc.*, 679 (1962)
B114 R. S. Biktimirov, *Russ. J. Phys. Chem.*, **37**, 1276 (1963)
B115 V. S. Chernyi and A. P. Krasnoperova, *Zh. Fiz. Khim.*, **39**, 430 (1965)

B116 G. Lemoine, *Compt. Rend.*, **125,** 603 (1897)
B117 P. Walden, *Z. Phys. Chem.*, **58,** 479 (1907)
B118 F. A. Askew, E. Bullock, H. T. Smith, R. K. Tinkler, O. Gatty and J. H. Wolfenden, *J. Chem. Soc.*, 1368 (1934)
B119 N. S. Jackson, A. E. C. Smith, O. Gatty and J. H. Wolfenden, *J. Chem. Soc.*, 1376 (1934)
B120 R. L. Moss and J. H. Wolfenden, *J. Chem. Soc.*, 118 (1939)
B121 C. M. Slansky, *J. Amer. Chem. Soc.*, **62,** 2430 (1940)
B122 S. I. Drakin and Y. M. Chang, *Zh. Fiz. Khim.*, **38,** 2800 (1964)
B123 K. P. Mishchenko and V. P. Tungusov, *Teor. i Eksp. Kkim., Akad. Nauk Ukr. SSR*, **1,** 55 (1965)
B124 K. P. Mishchenko and S. V. Shadskii, *Teor. i Eksp. Khim., Akad. Nauk Ukr. SSR*, **1,** 60 (1965)
B125 K. P. Mishchenko and M. L. Klyueva, *Teor. i Eksp. Khim., Akad. Nauk Ukr. SSR*, **1,** 201 (1965)
B126 L. P. Zhilina and K. P. Mishchenko, *Teor. i Eksp. Khim., Akad. Nauk Ukr. SSR*, **1,** 361 (1965)
B127 V. V. Kushchenko and K. P. Mishchenko, *Teor. i Eksp. Khim., Akad. Ukr. SSR*, **4,** 403 (1968)
B128 E. P. Prosviryakova, K. P. Mishchenko and G. M. Poltoratskii, *Teor. i Eksp. Khim., Akad. Nauk. Ukr. SSR*, **5,** 129 (1969)
B129 K. P. Mishchenko, V. V. Subbotina and B. S. Krumgal'z, *Teor. i Eksp. Khim., Akad. Nauk Ukr. SSR*, **5,** 268 (1969)
B130 V. P. Tungusov and K. P. Mishchenko, *Russ. J. Phys. Chem.*, **39,** 1585 (1965)
B131 N. P. Novoselov and K. P. Mishchenko, *Zh. Obshch. Khim.*, **38,** 2129 (1968)
B132 K. P. Mishchenko and V. V. Sokolov, *Zh. Strukt. Khim.*, **4,** 184 (1963)
B133 K. P. Mishchenko and V. V. Sokolov, *Zh. Strukt. Khim.*, **5,** 819 (1964)
B134 G. V. Karpenko, K. P. Mishchenko and G. M. Poltoratskii, *Zh. Strukt. Khim.*, **8,** 413 (1967)
B135 K. P. Mishchenko and G. M. Poltoratskii, *Tr. Leningr. Tekhnol. Inst. Tsellyulozn-Bumazhn. Prom.*, No. **16,** 146 (1965)
B136 G. A. Krestov and V. I. Klopov, *Zh. Strukt. Khim.*, **5,** 829 (1964)
B137 G. A. Krestov and V. I. Klopov, *Izv. Vysshikh. Uchebn. Zavedenii, Khim. i Khim. Tekhnol.*, **9,** 34 (1966)
B138 B. Jakuszewski, S. Taniewska-Osinska and R. Logwinienko, *Bull. Acad. Polon. Sci., Ser. Sci. Chim.*, **9,** 127 (1961)
B139 B. Jakuszewski and S. Taniewska-Osinska, *Bull. Acad. Polon. Sci., Ser. Sci. Chim.*, **9,** 133 (1961)
B140 C. S. Leung and E. Grunwald, *J. Phys. Chem.*, **74,** 696 (1970)
B141 W. M. Latimer and C. M. Slansky, *J. Amer. Chem. Soc.*, **62,** 2019 (1940)
B142 B. Jakuszewski and S. Taniewska-Osinska, *Lodz. Towarz. Nauk. Soc. Sci. Lodziensis, Acta Chim.*, **4,** 17 (1959)
B143 B. Jakuszewski and S. Taniewska-Osinska, *Lodz. Towarz. Nauk. Soc. Sci. Lodziensis, Acta Chim.*, **7,** 31 (1961)
B144 B. Jakuszewski and S. Taniewska-Osinska, *Zeszyty Nauk. Uniw. Lodz. Ser. II*, **13,** 137 (1962)
B145 J. B. Stark and E. C. Gilbert, *J. Amer. Chem. Soc.*, **59,** 1818 (1937)
B146 D. F. Tuan and R. M. Fuoss, *J. Phys. Chem.*, **67,** 1343 (1963)
B147 J. D. R. Thomas, *J. Inorg. Nucl. Chem.*, **24,** 1477 (1962)
B148 N. C. Deno and Henry E. Berkheimer, *J. Org. Chem.*, **28,** 2143 (1963)
B149 J. L. Burgaud, *Compt. Rend.*, **254,** 3870 (1962)
B150 B. Jakuszewski and S. Taniewska-Osinska, *Lodz. Towarz. Nauk. Soc. Sci. Lodziensis, Acta Chim.*, **8,** 11 (1962)

B151 L. N. Erbanova, S. I. Drakin and M. K. Karapetyants, *Zh. Fiz. Khim.*, **38,** 2670 (1964)
B152 E. K. Zolotarev, *Russ. J. Phys. Chem.*, **39,** 343 (1965)
B153 W. C. Vosburgh, L. C. Connell and J. A. V. Butler, *J. Chem. Soc.*, 933 (1933)
B154 J. Padova, *J. Chem. Phys.*, **38,** 2635 (1963)
B155 H. V. Venkatasetty and G. H. Brown, *J. Ind. Chem. Soc.*, **40,** 647 (1963)
B156 K. A. Kraus, R. J. Raridon and W. H. Baldwin, *J. Amer. Chem. Soc.*, **86,** 2571 (1964)
B157 J. H. Stern and J. M. Nobilione, *J. Phys. Chem.*, **72,** 3937 (1968)
B158 J. H. Stern and J. M. Nobilione, *J. Phys. Chem.*, **73,** 928 (1969)
B159 W. E. Wallace, L. S. Mason and A. L. Robinson, *J. Amer. Chem. Soc.*, **66,** 362 (1944)
B160 R. Gopal and M. M. Husain, *J. Ind. Chem. Soc.*, **40,** 272 (1963)
B161 R. Gopal, M. M. Husain and O. N. Bhatnagar, *J. Ind. Chem. Soc.*, **44,** 1005 (1967)
B162 M. L. Berardelli, G. Pistoia and A. M. Polcaro, *Ric. Sci.*, **38,** 814 (1968)
B163 Y. M. Povarov, V. E. Kazarinov, Y. M. Kessler and A. I. Gorbanev, *Zh. Neorg. Khim.*, **9,** 1008 (1964)
B164 E. N. Vasenko, *Zh. Fiz. Khim.*, **21,** 361 (1947)
B165 E. N. Vasenko, *Zh. Fiz. Khim.*, **22,** 999 (1948)
B166 E. N. Vasenko, *Zh. Fiz. Khim.*, **23,** 959 (1949)
B167 L. R. Dawson and E. J. Griffith, *J. Phys. Chem.*, **56,** 281 (1952)
B168 R. I. Mostkova, Y. M. Kessler and I. V. Safonova, *Elektrokhimiya*, **5,** 409 (1969)
B169 G. Somsen, *Rec. Trav. Chim.*, **85,** 517 (1966)
B170 D. V. S. Jain, B. S. Lark, S. P. Kochar and V. K. Gupta, *Ind. J. Chem.*, **7,** 256 (1969)
B171 A. Finch, P. J. Gardner and C. J. Steadman, *J. Phys. Chem.*, **75,** 2325 (1971)
B172 R. Gopal and R. K. Srivastava, *J. Phys. Chem.*, **66,** 2704 (1962)
B173 G. A. Strack, S. K. Swanda and L. W. Bahe, *J. Chem. Eng. Data*, **9,** 416 (1964)
B174 Y. A. Tsai and C. M. Criss, *J. Chem. Eng. Data* **18,** 51 (1973)
B175 F. J. Millero, *J. Phys. Chem.*, **72,** 3209 (1968)
B176 T. B. Hoover, *J. Phys. Chem.*, **68,** 876 (1964)
B177 E. A. Kelso and W. A. Felsing, *J. Amer. Chem. Soc.*, **60,** 1949 (1938)
B178 M. Kraft and H. J. Bittrich, *Wiss. Z. Tech. Hochsch. Chem. Leuna-Merseburg*, **4,** 137 (1961/62)
B179 F. C. Schmidt, S. Godomsky, F. K. Ault and J. C. Huffman, *J. Chem. Eng. Data*, **14,** 71 (1969)
B180 F. C. Schmidt, W. E. Hoffman and W. B. Schaap, *Proc. Indiana Acad. Sci.*, **72,** 127 (1962)
B181 G. P. Kotlyarova and E. F. Ivanova, *Zh. Fiz. Khim.*, **38,** 423 (1964)
B182 E. F. Ivanova and V. N. Fesenko, *Zh. Fiz. Khim.*, **43,** 1006 (1969)
B183 J. H. Stern and J. Nobilione, *J. Phys. Chem.*, **72,** 1064 (1968)
B184 G. Pistoia, G. Pecci and B. Scrosati, *Ric. Sci.*, **37,** 1167 (1967)
B185 R. C. Paul and B. R. Sreenathan, *Ind. J. Chem.*, **4,** 382 (1966)
B186 C. M. Criss and E. Luksha, *J. Phys. Chem.*, **72,** 2966 (1968)
B187 L. Weeda and G. Somsen, *Rec. Trav. Chim.*, **86,** 893 (1967)
B188 O. N. Bhatnagar and C. M. Criss, *J. Phys. Chem.*, **73,** 174 (1969)
B189 G. A. Krestov and V. A. Zverev, *Izv. Vyssh. Ucheb. Zaved., Khim., Khim Tekhnol.*, **11,** 990 (1968)
B190 G. Choux and R. L. Benoit, *J. Amer. Chem. Soc.*, **91,** 6221 (1969)
B191 G. Pistoia and B. Scrosati, *Ric. Sci.*, **37,** 1173 (1967)
B192 C. F. Coleman, *J. Phys. Chem.*, **69,** 1377 (1965)

B193 C. C. Addison, B. J. Hathaway, N. Logan and A. Walker, *J. Chem. Soc.*, 4308 (1960)
B194 E. Lange, *in* The Structure of Electrolytic Solutions, Chapter 9 (W. J. Hamer, ed.), Wiley, New York (1959)
B195 G. R. Hood and L. P. Hohlfelder, *J. Phys. Chem.*, **38,** 979 (1934)
B196 N. A. Rumbaut and H. L. Peeters, *Bull. Soc. Chim., Belg.*, **76,** 33 (1967)
B197 D. C. Luehrs, R. T. Iwamoto and J. Kleinberg, *Inorg. Chem.*, **5,** 201 (1966)
B198 J. C. Synnott and J. N. Butler, *J. Phys. Chem.*, **73,** 1470 (1969)
B199 J. N. Butler, *J. Phys. Chem.*, **72,** 3288 (1968)
B200 J. S. Dunnett and R. P. H. Gasser, *Trans. Faraday Soc.*, **61,** 922 (1965)
B201 M. D. Archer and R. P. H. Gasser, *Trans. Faraday Soc.*, **62,** 3451 (1966)
B202 T. Skerlak, B. Ninkov and V. Sislov, *Glasnik Drustva Hemicara Tehnol. SR Bosne Heregovine*, **11,** 39 (1962)
B203 C. V. Krishnan and H. L. Friedman, *J. Phys. Chem.*, **73,** 3934 (1969)
B203a C. V. Krishnan and H. L. Friedman, *J. Phys. Chem.*, **74,** 3900 (1970)
B204 M. E. Clark and J. L. Bear, *J. Inorg. Nucl. Chem.*, **31,** 2619 (1969)
B205 Y. C. Wu and H. L. Friedman, *J. Phys. Chem.*, **70,** 2020 (1966)
B206 A. F. Kapustinskii, A. I. Maier and S. I. Drakin, *Tr. Mosk. Khim-Tekhnol. Inst.*, **38,** 10 (1962)
B207 D. Fleischer and H. Freiser, *J. Phys. Chem.*, **66,** 389 (1962)
B208 Y. M. Glubokov, S. S. Korovin, I. A. Apraksin, N. G. Kirilova and K. I. Petrov, *Zh. Neorg. Khim.*, **14,** 530 (1969)

APPENDIX 2.3.1*

$$I^{\frac{1}{2}}\omega = I^{\frac{1}{2}}\left[\frac{1}{1 + I^{\frac{1}{2}}} - \frac{\sigma(I^{\frac{1}{2}})}{3}\right]$$

where

$$\sigma(I^{\frac{1}{2}}) = \frac{3}{I^{\frac{3}{2}}}\left[1 + I^{\frac{1}{2}} - \frac{1}{1 + I^{\frac{1}{2}}} - 2\ln(1 + I^{\frac{1}{2}})\right]$$

I	$I^{\frac{1}{2}}\omega$
0.0001	0.007
0.0004	0.013
0.0009	0.020
0.0016	0.026
0.0025	0.032
0.0036	0.038
0.0049	0.044
0.0064	0.050
0.0081	0.056
0.0100	0.062
0.0144	0.073
0.0196	0.085
0.0256	0.095
0.0324	0.106
0.0400	0.116
0.0450	0.122
0.0484	0.126
0.0526	0.131
0.0576	0.136
0.0626	0.141
0.0676	0.145
0.0726	0.150
0.0784	0.155
0.0850	0.160
0.0900	0.164
0.0950	0.167
0.100	0.171

* Acknowledgment is gratefully given to Martin J. Mastroianni for calculating the values in this table.

APPENDIX 2.3.2

Solubility and Changes in Free Energy, Enthalpy, Entropy and Heat Capacity for Dissolution of Gases in *N*-Methylacetamide at 35°C*

Gas	$10^4 x_2$ Mol Fraction	$\Delta G_{soln}/$ kcal mol^{-1}	$\Delta H_{soln}/$ kcal mol^{-1}	$\Delta S_{soln}/$ cal mol^{-1} K^{-1}	$\Delta C_{psoln}/$ cal mol^{-1} K^{-1}
He	0.557	6.00	2.3	−12.0	0
N_2	2.461	5.09	0.5	−14.9	0
Ar	4.444	4.73	0.1	−15.0	−2
C_2H_6	41.91	3.35	−2.1	−17.8	−5

* R. H. Wood and D. E. DeLaney, *J. Phys. Chem.*, **72,** 4651 (1968).

APPENDIX 2.3.3

Solubility and Entropy of Solution of Gases in Cyclohexane at 1 Atmosphere and 25°C*

Gas	$10^4 x$ Mol Fraction	$\Delta S_{soln}/$ cal mol^{-1} K^{-1}
Ne	1.90	6.8
Ar	15.20	0.0
Kr	47.3	−3.1
Xe	210.0	−8.2
H_2	4.14	4.2
N_2	7.68	2.0
CO_2	77.1	−5.2
C_2H_6	236.0	−8.7
cyclo-C_3H_6	1395	−14.4

* J. H. Dymond, *J. Phys. Chem.*, **71,** 1829 (1967).

APPENDIX 2.3.4

Enthalpies of Solution of Various Non-electrolytes in Propylene Carbonate and Dimethylsulphoxide*

Solute	ΔH^0_{soln}/kcal mol^{-1}	
	DMSO	PC
H_2O	−1.27 (−1.28 ± 0.03)†	2.00‡
CH_3OH	−0.34 (−0.34 ± 0.03)†	1.50
C_2H_5OH	0.29 (0.28 ± 0.02)†	2.02
n-C_3H_7OH	0.61 (0.61 ± 0.04)†	2.27
n-C_4H_9OH	0.95 (0.99 ± 0.03)†	2.53
n-$C_5H_{11}OH$	1.29	2.77
i-C_3H_7OH	0.87	2.36
t-C_4H_9OH	1.19 (1.21 ± 0.04)†	2.51
t-$C_5H_{11}OH$	1.16	2.47
$C_6H_5CH_2OH$	−0.67	1.22
n-C_5H_{12}	2.75	2.19
n-C_6H_{14}	3.18	2.52
n-C_7H_{16}	3.62	2.84
cyclo-C_6H_{12}	2.72	2.12
C_6H_6	0.63 (0.65 ± 0.03)†	0.37
$C_6H_5CH_3$	0.90 (0.89 ± 0.03)†	0.56
$C_6H_5CH_2CH_3$	1.12	0.76
$C_6H_5CH(CH_3)_2$	1.36 (1.30 ± 0.04)†	0.89
m-$C_6H_4(CH_3)_2$	1.16	0.81
$C_6H_5C_6H_5$	4.98	4.53
C_6H_5F	0.31	0.13
C_6H_5Cl	0.52	0.41
C_6H_5Br	0.52	0.54
C_6H_5I	0.31	0.82
$C_6H_5NO_2$	0.55	0.27
CCl_4	0.00 ± 0.03†	
$CHCl_3$	−1.32 ± 0.09†	
CH_3COCH_3	0.37 ± 0.03†	

* Data are from C. V. Krishnan and H. L. Friedman, *J. Phys. Chem.*, **73**, 1572 (1969) unless otherwise indicated.
† E. M. Arnett and D. R. McKelvey, *J. Amer. Chem. Soc.*, **88**, 2598 (1966).
‡ Y. C. Wu and H. L. Friedman, *J. Phys. Chem.*, **70**, 501 (1966).

APPENDIX 2.3.5

Enthalpies of Solution of Some Nonelectrolytes in CH_3OH and CH_3OD*

Solute	ΔH^0_{soln}/kcal mol^{-1}	
	CH_3OH	CH_3OD
C_5H_{12}	0.91	0.95
C_6H_{14}	1.12	1.20
C_7H_{16}	1.32	1.46
C_6H_6	0.37	0.43
$C_6H_5CH_3$	0.45	0.57
$C_6H_5C_2H_5$	0.57	0.69

* C. V. Krishman and H. L. Friedman, *J. Phys. Chem.*, **75**, 388 (1971).

APPENDIX 2.4.1

Solubilities of Electrolytes in Anhydrous Methanol at 25°C

Electrolyte	$-\log K_{so}$*	$g(l\ soln)^{-1}$	$g(kg\ solv)^{-1}$§
AgCl	13.1 (13.1)		1.16×10^{-4}‖
AgBr	15.2 (15.2)		
AgI	18.3 (18.2)		
AgN_3	11.2		
AgSCN	13.9 (13.7)		
AgAc	(6.1)		
$Ag(p-SO_2OC_6H_4CH_3)$	3.2		
$AgBPh_4$	13.2		
LiF			0.176
LiCl			209.8
LiBr			342.9
$LiNO_3$			429.5
Li_2CO_3			0.555
Li_2SO_4			1.261
NaF			0.231
NaCl	1.5		14.01
NaBr			160.9
NaI			625.1
$NaNO_3$			29.36
Na_2CO_3			3.109
Na_2SO_4			0.113
$Na(p-NO_2C_6H_4O)$	−0.2		
KF			22.86
KCl	2.5		5.335
KBr	1.7		20.80
KI	0.2		170.7
$KClO_4$	4.5		
KNO_3			3.795
K_2CO_3			61.65
K_2SO_4			0.005
$K(p-NO_2C_6H_4O)$	0.2		
$KBPh_4$	5.0	1.12†	
KPic	4.2	2.29†	
CsCl	1.7		36.21‖
CsBr	2.2		
CsI	1.9		
$CsNO_3$		1.850‡	
CsPic	4.2		
Et_4NPic	2.0		

APPENDIX 2.4.1 (*contd*)

Electrolyte	$-\log K_{s0}$*	g(l soln)$^{-1}$	g(kg solv)$^{-1}$§
Bu_4NPic		410.8†	
Bu_4NBPh_4		1.45†	
$(i\text{-Amyl})_3$ BuN Pic		201.0†	
$(i\text{-Amyl})_3$ BuN BPh_4		2.17†	
CaF_2			0.145
$CaCl_2$			232.6
$CaBr_2$			558.3
CaI_2			673.7
$Ca(NO_3)_2$			1271
$CaCO_3$			0.012
$CaSO_4$			0.046
SrF_2			0.142
$SrCl_2$			180.5
$Sr(NO_3)_2$			10.61
$SrCO_3$			0.014
$SrSO_4$			0.074
BaF_2			0.044
$BaCl_2$			13.79
$Ba(NO_3)_2$			0.480
$BaCO_3$			0.064
$BaSO_4$			0.063
$CeCl_3$			650¶
$PrCl_3$			1159**
$NdCl_3$			1119**
$ScCl_3$			834.9††

* Values taken from measurements and compilation of a number of literature data by R. Alexander, E. C. F. Ko, Y. C. Mac and A. J. Parker, *J. Amer. Chem. Soc.*, **89,** 3703 (1967). The solubility product, K_{s0}, is based on the molar concentration scale. Values in parentheses are from L. G. Sillen and A. E. Martell, Stability Constants of Metal-Ion Complexes, The Chemical Soc., London (1964), as reported by Alexander *et al.*

† O. Popovych and R. M. Friedman, *J. Phys. Chem.*, **70,** 1671 (1966). The data have been converted from mol l^{-1} to g l^{-1}.

‡ R. S. Biktimirov, *Russian J. Phys. Chem.*, **37,** 1276 (1963).

§ R. E. Harner, J. B. Sydnor and E. S. Gilreath, *J. Chem. Eng. Data*, **8,** 411 (1963), except where indicated.

|| N. A. Izmailov and V. S. Chernyi, *Russian J. Phys. Chem.*, **34,** 58 (1960).

¶ F. R. Hartley and A. W. Wylie, *J. Chem. Soc.*, 679 (1962). Data have been converted from g(kg soln)$^{-1}$. Probable solid phase is $CeCl_3 \cdot 4\ CH_3OH$.

** Z. I. Grigorovich, *Zh. Neorg. Khim.*, **8,** 986 (1963). Solid phases are $PrCl_3 \cdot 3.5$ MeOH and $NdCl_3 \cdot 3$ MeOH. Concentrations converted from wt.%.

†† E. M. Kirmse, *Z. Chem.*, **1,** 332 (1961). Solid phase is $ScCl_3 \cdot 3$ MeOH. Data have been converted from mg(100 mg soln)$^{-1}$.

APPENDIX 2.4.2

Solubility of NaI in Anhydrous Methanol at Various Temperatures*

t(°C)	g(kg solv)$^{-1}$	Solid Phase
−0.2	625	$NaI \cdot 3\ CH_3OH$
+3.4	665	$NaI \cdot 3\ CH_3OH$
8.2	696	$NaI \cdot 3\ CH_3OH$
13.3	711	$NaI \cdot 3\ CH_3OH$
17.1	765	$NaI \cdot 3\ CH_3OH$
19.8	779	$NaI \cdot 3\ CH_3OH$
21.0	806	$NaI \cdot 3\ CH_3OH$
22.1	815	$NaI \cdot 3\ CH_3OH$
23.5	826	$NaI \cdot 3\ CH_3OH$
24.0	834	$NaI \cdot 3\ CH_3OH$
26.5	880	Decomposition
29.6	907	NaI
31.5	928	NaI
33.9	947	NaI
35.4	970	NaI

* H. Brusset and J. C. Lecoq, *Compt Rend.*, **259C**, 814 (1964).

APPENDIX 2.4.3

Solubility of Some Electrolytes in Methanol at Various Temperatures

t(°C)	Silver Benzoate* activity × 10^3	Silver Salicylate* activity × 10^4	$PrCl_3$† g(kg solv)$^{-1}$	$NdCl_3$† g(kg solv)$^{-1}$
0			818.8	839.2
25	1.35	7.09	1159	1119
30	1.39	7.78		
35	1.51	11.3		
40	1.68	12.5	1496	1286
45	2.04	14.9		
50	2.51	17.6		
55	2.69	22.4		
60	3.45	26.6		

* V. S. Chernyi and A. P. Krasnoperova, *Russian J. Phys. Chem.*, **39**, 220 (1965).

† Z. I. Grigorovich, *Zh. Neorg. Khim.*, **8**, 986 (1963). Solid phases are $PrCl_3 \cdot 3.5\ CH_3OH$ and $NdCl_3 \cdot 3\ CH_3OH$ at all temperatures. Data have been converted from wt. %.

APPENDIX 2.4.4

Solubilities of Electrolytes in Anhydrous Ethanol at 25°C

Electrolyte	$g(kg\ solv)^{-1}$	Ref.	$g(l\ soln)^{-1}$	Ref.
KI	19.0	(a)		
	(17.5)	(a)		
RbI	9.8	(a)		
	(9.5)	(a)		
CsCl	3.89	(b)		
CsI	3.31	(a)		
	(2.90)	(a)		
$CsNO_3$			0.439	(c)
AgCl	6.35×10^{-5}	(b)		
Me_4NClO_4			0.155	(d)
Et_4NClO_4			2.62	(d)
Pr_4NClO_4			42.6	(d)
Ph_4NClO_4			0.430	(d)
Am_4NClO_4			14.7	(d)
Hx_4NClO_4			133.0	(d)
$CeCl_3$	381.0	(e)		
$PrCl_3$	778.1	(f)		
$NdCl_3$	771.8	(f)		
$ScCl_3$	595.0	(g)		

(a) J. D. R. Thomas, *J. Inorg. Nucl. Chem.*, **24**, 1477 (1962). Values in parentheses are at 20°C.

(b) N. A. Izmailov and V. S. Chernyi, *Zh. Fiz. Khim.*, **34**, 127 (1960). The data have been converted from mol $(kg\ solv)^{-1}$.

(c) R. S. Biktimirov, *Russ. J. Phys. Chem.*, **37**, 1276 (1963).

(d) N. C. Deno and H. E. Berkheimer, *J. Org. Chem.*, **28**, 2143 (1963). Data have been converted from mol $(l\ soln)^{-1}$. Solid phase is the pure salt.

(e) F. R. Hartley and A. W. Wylie, *J. Chem. Soc.*, 679 (1962). Solid phase is reported to be $CeCl_{3.00}(EtOH)_{3.11}$. Data have been converted from $g(kg\ soln)^{-1}$.

(f) Z. I. Grigorovich, *Zh. Neorg. Khim.*, **8**, 986 (1963). Solid phases are $PrCl_3 \cdot 2.5$ EtOH and $NdCl_3 \cdot 2$ EtOH. Concentrations have been converted from wt.%.

(g) E. M. Kirmse, *Z. Chem.*, **1**, 332 (1961). Solid phase is $ScCl_3 \cdot 3EtOH$. Data converted from $mg(100\ mg\ soln)^{-1}$.

APPENDIX 2.4.5

Solubility of Silver Benzoate and Silver Salicylate in Ethanol at Various Temperatures*

t(°C)	Silver Benzoate Activity × 10^4	Silver Salicylate Activity × 10^4
25	9.04	5.37
30	9.15	5.88
35	10.3	6.60
40	12.2	8.03
45	13.0	9.61
50	16.5	13.2
55	19.1	16.4
60	22.5	20.3
65	25.5	23.8

* V. S. Chernyi and A. P. Krasnoperova, *Russ. J. Phys. Chem.*, **39**, 220 (1965).

APPENDIX 2.4.6

Solubilities of Electrolytes in Higher Alcohols at 25°C

Electrolyte	$g(kg\ solv)^{-1}$	$g(l\ soln)^{-1}$	Solid Phase	Ref.
		n-propanol		
$CsNO_3$		0.095		(a)
$CeCl_3$	340.0		$CeCl_3 \cdot 3\ PrOH$	(b)
$PrCl_3$	471.5		$PrCl_3 \cdot 2\ PrOH$	(c)
$NdCl_3$	397.8		$NdCl_3 \cdot 2\ PrOH$	(c)
$ScCl_3$	353.0		$ScCl_3 \cdot 4\ PrOH$	(d)
		n-butanol		
CsCl	0.291			(e)
$CsNO_3$		0.032		(a)
AgCl	1.45×10^{-5}			(e)
$CeCl_3$	526.0		$CeCl_3 \cdot 3\ BuOH$	(b)
$PrCl_3$	309.2		$2\ PrCl_3 \cdot 3\ BuOH$	(c)
$NdCl_3$	214.1		$2\ NdCl_3 \cdot 3\ BuOH$	(c)
$ScCl_3$	337.0		$ScCl_3 \cdot 3\ BuOH$	(d)
		n-pentanol		
$CeCl_3$	420.0		$CeCl_3 \cdot 3\ AmOH$	(b)
$PrCl_3$	321.8		$2\ PrCl_3 \cdot 3\ AmOH$	(c)
$NdCl_3$	180.6		$NdCl_3 \cdot\ AmOH$	(c)
$ScCl_3$	311.0		$ScCl_3 \cdot 4\ AmOH$	(d)
		iso-pentanol		
CsCl	0.123			(e)
AgCl	1.23×10^{-5}			(e)
		n-hexanol		
CsCl	0.139			(e)
$CeCl_3$	366.0			(b)
$ScCl_3$	274.0		$ScCl_3 \cdot 3\ HxOH$	(d)
		n-heptanol		
CsCl	0.109			(e)
$ScCl_3$	239.0		$ScCl_3 \cdot 4\ HpOH$	(d)
		n-octanol		
$ScCl_3$	203.0		$ScCl_3 \cdot 4\ OcOH$	(d)
		n-nonanol		
$ScCl_3$	156.0		$ScCl_3 \cdot 3\ NnOH$	(d)
		t-propanol		
$CeCl_3$	44.0		$CeCl_3 \cdot 2\ t\text{-}PrOH$	(b)
		t-butanol		
$CeCl_3$	5.0		$CeCl_3$	(b)

(a) R. S. Biktimirov, *Russ. J. Phys. Chem.*, **37**, 1276 (1963).
(b) F. R. Hartley and A. W. Wylie, *J. Chem. Soc.*, 679 (1962). Data converted from $g(100\ g\ soln)^{-1}$.
(c) Z. I. Grigorovich, *Zh. Neorg. Khim.*, **8**, 986 (1963). Data converted from wt. %.
(d) E. M. Kirmse, *Z. Chem.*, **1**, 332 (1961). Data converted from $mg(100\ mg\ soln)^{-1}$.
(e) N. A. Izmailov and V. S. Chernyi, *Zh. Fiz. Khim.*, **34**, 127 (1960).

APPENDIX 2.4.7

Solubilities of Some Electrolytes in Aqueous Solutions of Ethylene Glycol, Glycerol and Some of Their Derivatives at 25°C*

Wt% Glycol or Glycerol	Solubility/g(kg solv)$^{-1}$			$\gamma_{\pm(0)}$† NaCl
	NaCl	KCl	$Ba(NO_3)_2$	
		Ethylene Glycol		
0.0	360.0	360.0	102.6	
25.06	260.2	241.9	77.4	
50.14	177.2	152.2	67.1	
75.06	112.4	88.3	66.7	
85.08	90.8	70.0	69.6	
95.09	75.5	56.5	72.7	
99.97	70.0	51.3	76.0	
		Ethylene Glycol Monoacetate		
24.93	242.4			1.12
49.82	136.0			1.33
74.78	46.62			1.94
84.66	21.44			2.54
94.69	5.68			3.18
99.68	1.49			
		Ethylene Glycol Diacetate		
94.27			3.45×10^{-2}	
94.95			1.95×10^{-2}	
95.58			1.01×10^{-2}	
96.59			3.8×10^{-3}	
97.58	1.45×10^{-2}	8.6×10^{-3}	1.0×10^{-3}	
98.44	2.8×10^{-3}	2.4×10^{-3}	3.0×10^{-4}	
99.21	3.3×10^{-4}	5.2×10^{-4}	5.0×10^{-5}	
99.98	7.5×10^{-5}	8.5×10^{-5}		
		Glycerol		
22.69	284.7	265.3	88.2	
50.10	191.0	176.0	75.1	
75.07	122.6	109.2	70.0	
85.18	100.8	87.7	69.0	
94.97	83.6	70.9	69.3	
99.84	76.0	64.2	69.8	

APPENDIX 2.4.7 (*contd*)

Wt% Glycol or Glycerol	Solubility/g(kg solv)$^{-1}$ NaCl	KCl	$Ba(NO_3)_2$	$\gamma_{\pm(0)}$† NaCl
	Glycerol Monoacetate			
24.91	251.3			1.08
59.86	155.08			1.16
74.84	74.4			1.21
84.72	49.6			1.09
94.65	29.1			0.62
99.65	20.8			
	Glycerol Diacetate			
77.90	38.8			2.06
86.05	15.5			3.23
94.93	1.74			10.4
99.91	0.12			
	Glycerol Triacetate			
95.92			9.0×10^{-3}	
96.76			3.4×10^{-3}	
97.46			1.4×10^{-3}	
98.00	6.7×10^{-3}	5.2×10^{-3}	5.8×10^{-4}	
98.50	2.1×10^{-3}	2.1×10^{-3}		
99.00	5.0×10^{-4}		5.6×10^{-5}	
99.37	1.6×10^{-4}	3.3×10^{-4}		
99.95		8.6×10^{-5}		

* K. A. Kraus, R. J. Raridon and W. H. Baldwin, *J. Amer. Chem. Soc.*, **86,** 2571 (1964).

† $\gamma_{\pm(0)}$ is the mean activity coefficient based on the salt in the aqueous solution as the reference state.

APPENDIX 2.4.8

Solubilities of Electrolytes in Protic Amides at 25°C

Electrolyte	Formamide			N-methylformamide	
	g(kg solv)$^{-1}$	pK*	Ref.	g(kg solv)$^{-1}$	Ref.
LiCl	282.0		(a)	239.0	(a)
				214.0	(c)
LiBr	738.0		(a)	540.0	(a)
NaCl	93.8		(a)	32.9	(a)
	93.2		(b)	31.8	(c)
NaBr	353.0		(a)	308.0	(a)
	358.4		(b)	263.0	(c)
NaI	850.0		(a)	575.0	(a)
	566.2		(b)	637.0	(c)
$NaNO_3$	404.6		(b)		
KCl	61.8		(a)	21.4	(a)
	63.0		(b)	20.4	(c)
KBr	216.0		(a)	102.0	(a)
	213.8		(b)	96.6	(c)
KI	678.0		(a)	496.0	(a)
	692.2		(b)	427.0	(c)
KNO_3	155.5		(b)		
K_2SO_4	3.4		(b)		
CsCl		0.53	(d)		
CsBr		0.29	(d)		
CsI		0.23	(d)		
AgCl		9.4	(d)		
AgBr		11.4	(d)		
AgI		14.5	(d)		
AgN_3		7.7†	(d)		
AgSCN		9.9	(d)		
$AgBPh_4$		10.3	(d)		
NH_4Cl	110.2		(b)	48.5	(c)
NH_4Br	361.1		(b)	216.0	(c)
NH_4I	1042.0		(b)		
$MgCl_2$	84.0		(a)	88.0	(a)
$CaCl_2$	200.0		(a)	186.0	(a)
$CaBr_2$	434.0		(a)	303.0	(a)
$SrCl_2$	159.0		(b)		
$SrBr_2$	190.5		(b)		

APPENDIX 2.4.8 (*contd*)

Electrolyte	Formamide			N-methylformamide	
	g(kg solv)$^{-1}$	pK*	Ref.	g(kg solv)$^{-1}$	Ref.
$BaCl_2$	117.6		(b)		
$BaBr_2$	298.5		(b)		
$CuCl_2$	49.3		(a)	269.0	(a)
$ZnCl_2$	434.0		(a)	303.0	(a)
$ZnBr_2$	772.0		(a)	346.0	(a)
ZnI_2	1220.0		(a)	1160.0	(a)
$CdCl_2$	8.3		(a)	18.5	(a)
$CdBr_2$	25.8		(a)	329.0	(a)
CdI_2	158.0		(a)	2010.0	(a)
$MnCl_2$	54.8		(a)	100.0	(a)
$MnBr_2$	268.0		(a)	328.0	(a)
$PbCl_2$	56.23		(b)		
$PbBr_2$	120.2		(b)		

* $pK = -\log K_{s0} = -\log [M^+][X^-]$ (concentration quotients).
† At ionic strength of 0.10 – 0.05 *m*.
(a) M. L. Berardelli, G. Pistoia and A. M. Polcaro, *Ric. Sci.*, **38,** 814 (1968).
(b) R. Gopal and M. M. Husain, *J. Ind. Chem. Soc.*, **40,** 272 (1963).
(c) G. A. Strack, S. K. Swanda and L. W. Bahe, *J. Chem. Eng. Data*, **9,** 416 (1964). Data have been converted from molality.
(d) R. Alexander, E. C. F. Ko, Y. C. Mac and A. J. Parker, *J. Amer. Chem. Soc.*, **89,** 3703 (1967). Solid phases are the unsolvated salt, except for LiCl, NaI and KI.

APPENDIX 2.4.9.

Solubilities of Electrolytes in Aprotic Amides at 25°C

Electrolyte	*N,N*-Dimethylformamide				*N,N*-Dimethylacetamide				Hexamethylphosphoro-triamide		
	g(kg solv)$^{-1}$	pK*	Solid Phase†	Ref.	g(kg solv)$^{-1}$	pK*	Solid Phase†	Ref.	pK*	Solid Phase†	Ref.
LiF	1.38×10^{-3}		n.s.	(a)							
LiCl	113.0		—	(b)	86.0		—	(d)			
	275.3		LiCl·DMF	(c)							
LiBr	166.0		—	(b)	262.0		—	(d)			
NaF	1.965×10^{-3}		n.s.	(a)							
NaCl	0.355		n.s.	(a)	0.20		—	(d)			
	0.42		—	(b)							
	0.5		n.s.	(c)							
NaBr	103.0		—	(b)	65.1		—	(d)			
	32.3		n.s.	(c)							
NaI	63.5		—	(b)	346.0		—	(d)			
	37.2		n.s.	(c)							
$NaClO_3$	234.0		n.s.	(c)							
$NaClO_4$	396.0		$NaClO_4$·DMF	(c)							
$NaNO_2$	93.71		n.s.	(c)							
$NaNO_3$	130.7		n.s.	(c)							
NaSCN	299.0		—	(b)							
	213.2		NaSCN·DMF	(c)							
NaCN	187.6		n.s.	(c)							
NaAc	72.1		n.s.	(c)							

NaN_3		1.9	n.s.	(e)			
$Na(p\text{-}NO_2C_6H_4O)$		−0.58	n.s.	(e)			
$Na_2S_2O_8$	150.2		n.s.	(c)			
KCl	0.20		—	(b)	0.11	—	(d)
	0.170		n.s.	(a)			
	0.5		n.s.	(c)			
		5.4	n.s.	(e)			
KBr	8.2		—	(b)	3.9	—	(d)
	127.9		n.s.	(c)			
		2.4	n.s.	(e)			
KI	416.0		—	(b)	15.4	—	(d)
	137.1		n.s.	(c)			
		−0.5	n.s.	(e)			
$KClO_3$	181.0		n.s.	(c)			
$KClO_4$		0.1	n.s.	(e)			
	196.0		$KClO_4 \cdot DMF$	(c)			
KNO_2	13.42		n.s.	(c)			
KNO_3	22.74		n.s.	(c)			
KSCN	159.7		$KSCN \cdot DMF$	(c)			
KAc	53.2		n.s.	(c)			
$K(p\text{-}NO_2C_6H_4O)$		0.4**	n.s.	(e)			
KPic		−0.2	n.s.	(e)			
$K_2S_2O_8$	9.5		n.s.	(c)			
CsCl	0.515		n.s.	(a)			
		4.9	n.s.	(e)			
CsBr	5.57		n.s.	(a)			
		3.3	n.s.	(e)			
CsI		1.7	n.s.	(e)			
CsPic		0.5	n.s.	(e)			
NH_4Cl	0.5		n.s.	(c)			
NH_4SCN	133.7		$NH_4SCN \cdot DMF$	(c)			
Et_4NPic		0.0	n.s.	(e)			

APPENDIX 2.4.9 (*contd*)

Electrolyte	*N,N*-Dimethylformamide				*N,N*-Dimethylacetamide				Hexamethylphosphoro-triamide		
	g(kg solv)$^{-1}$	pK*	Solid Phase†	Ref.	g(kg solv)$^{-1}$	pK*	Solid Phase†	Ref.	pK*	Solid Phase†	Ref.
AgCl		14.5	n.s.	(e)		14.3	n.s.	(e)	11.9	n.s.	(e)
AgBr		15.0	n.s.	(e)		14.5	n.s.	(e)	12.3	n.s.	(e)
AgI		15.8	n.s.	(e)		14.7	n.s.	(e)			
AgN_3		11.0	n.s.	(e)		10.8	n.s.	(e)	8.5§	n.s.	(e)
AgSCN		11.5	n.s.	(e)		10.5	n.s.	(e)	7.4§	n.s.	(e)
AgAc		10.2	n.s.	(e)		9.7	n.s.	(e)			
$Ag(p\text{-}SO_2OC_6H_4CH_3)$		1.3‡	n.s.	(e)							
$AgBPh_4$		6.7§	n.s.	(e)		5.9§	n.s.	(e)	4.7‡	n.s.	(e)
$MgCl_2$	125.0		—	(b)	35.4		—	(d)			
	80.4		$MgCl_2\cdot 2DMF$	(c)							
MgI_2					570.0		—	(d)			
$Mg(NO_3)_2$	164.1		$Mg(NO_3)_2\cdot 2DMF$	(c)							
$Mg(Ac)_2$	106.0		n.s.	(c)							
$CaCl_2$	60.1		—	(b)	41.3		—	(d)			
	19.8		$CaCl_2\cdot 2DMF$	(c)							
$CaBr_2$	192.0		—	(b)	99.5		—	(d)			
$BaCl_2$	10.71		n.s.	(c)							
$Ba(NO_3)_2$	29.38		n.s.	(c)							
$CuCl_2$	171.0		—	(b)	354.0		—	(d)			

$ZnCl_2$	232.0	—	(b)	85.4	—	(b)
	63.2	$ZnCl_2 \cdot 2DMF$	(c)			
$ZnBr_2$	553.0	—	(b)	328.0	—	(b)
ZnI_2	1210.0	—	(b)	436.0	—	(b)
$Zn(NO_3)_2$	127.1	$Zn(NO_3)_2 \cdot 2DMF$	(c)			
$Zn(Ac)_2$	92.0	n.s.	(c)			
$CdCl_2$	8.8	—	(b)	25.8	—	(b)
	173.4	$2CdCl_2 \cdot 3DMF$	(c)			
$CdBr_2$	798.0	—	(b)	781.0	—	(b)
CdI_2	1470.0	—	(b)	1160.0	—	(b)
$NiCl_2$	74.9	$NiCl_2 \cdot 2DMF$	(c)			
$CoCl_2$	93.7	$CoCl_2 \cdot 2DMF$	(c)			
$FeCl_3$	114.3	$FeCl_3 \cdot 6DMF$	(c)			
$AlCl_3$	37.42	$AlCl_3 \cdot 6DMF$	(c)			
$SnCl_4$	63.46	$SnCl_4 \cdot 2DMF$	(c)			
$SnBr_4$	94.3	$SnBr_4 \cdot 2DMF$	(c)			
$SbCl_5$	41.2	$SbCl_5 \cdot DMF$	(c)			

* $pK = -\log K_{s0} = -\log [M^+][X^-]$ (concentration quotients).

† n.s.—No solvate.

‡ Analysis of saturated solution at ionic strength corresponding to solubility.

§ At ionic strength 0.10–0.05 molar.

** At 0°C.

(a) C. M. Criss and E. Luksha, *J. Phys. Chem.*, **72**, 2966 (1968). Data have been converted from mol(kg solv)$^{-1}$.

(b) M. L. Berardilli, G. Pistoia and A. M. Polcaro, *Ric. Sci.*, **38**, 814 (1968); G. Pistoia, G. Pecci and B. Scrosati, *Ric. Sci.*, **37**, 1167 (1967).

(c) R. C. Paul and B. R. Sreenathan, *Ind. J. Chem.*, **4**, 382 (1966).

(d) G. Pistoia and B. Scrosati, *Ric. Sci.*, **37**, 1173 (1967).

(e) R. Alexander, E. C. F. Ko, Y. C. Mac and A. J. Parker, *J. Amer. Chem. Soc.*, **89**, 3703 (1967).

APPENDIX 2.4.10

Solubilities of Electrolytes in Acetonitrile at 25°C

Electrolyte	pK*	Ref.
AgCl	12.9	(a)
	12.4†	(b)
AgBr	12.9	(a)
	13.2†	(b)
AgI	14.2†	(b)
AgN_3	9.6	(a)
AgSCN	10.0	(a)
AgAc	7.4‡	(a)
$AgBPh_4$	7.2‡	(a)
KPic	4.55	(a)
Na(p-$NO_2C_6H_4O$)	5.42	(a)

* $pK = -\log K_{s0} = -\log [M^+][X^-]$ (concentration quotient).
† At ionic strength 0.100 molar. 23°C.
‡ At ionic strength 0.10 − 0.05 molar.
(a) R. Alexander, E. C. F. Ko, Y. C. Mac and A. J. Parker, *J. Amer. Chem. Soc.*, **89,** 3703 (1967).
(b) D. C. Luehrs, R. T. Iwamoto and J. Kleinberg, *Inorg. Chem.*, **5,** 201 (1966).

APPENDIX 2.4.11

Solubilities of Electrolytes in Dimethylsulphoxide at 25°C

Electrolyte	pK*	Ref.
AgCl	10.4	(a)
	9.7	(b)
	10.4†	(c)
	10.28†	(d)
AgBr	10.6	(a)
	10.0	(b)
	10.6†	(c)
AgI	11.4	(a)
	11.5	(b)
	12.0†	(c)
AgN_3	6.5‡	(a)
AgSCN	7.1‡	(a)
AgAc	4.4‡	(a)
$AgB(C_6H_5)_4$	4.6§	(a)
KBr	0.60	(a)‖
$KClO_4$	−0.80	(a)‖
NaN_3	0.64	(a)

* $pK = -\log K_{so} = -\log [M^+][X^-]$ (concentration quotients).
† At ionic strength 0.100 molar.
‡ At ionic strength 0.10 − 0.05 molar.
§ From analysis of saturated solutions at ionic strength corresponding to the solubility.
‖ Tabulated from literature by ref. (a).
(a) R. Alexander, E. C. F. Ko, Y. C. Mac and A. J. Parker, *J. Amer. Chem. Soc.*, **89**, 3703 (1967).
(b) N. A. Rumbaut and H. L. Peeters, *Bull. Soc. Chim., Belges*, **76**, 33 (1967).
(c) D. C. Luehrs, R. T. Iwamoto and J. Kleinberg, *Inorg. Chem.*, **5**, 201 (1966).
(d) J. C. Synnott and J. N. Butler, *J. Phys. Chem.*, **73**, 1470 (1969).

APPENDIX 2.4.12

Solubility of Electrolytes in Tributylphosphate and Diisopentylmethylphosphonate at 25°C*

Electrolyte	$g(kg\ solv)^{-1}$	Density/$g\ cm^{-3}$
	TBP	
LiCl	117.2	1.028
$CaCl_2$	173.0	1.084
$AlCl_3$	11.1	0.988
	DPMP	
LiCl	136.1	1.023
$CaCl_2$	196.9	1.072
$AlCl_3$	18.7	0.986

* Y. M. Glubokov, S. S. Korovin, I. A. Asprakin, N. G. Kirilova and K. I. Petrov, *Zh. Neorg. Khim.*, **14,** 530 (1969). Data converted from $g(kg\ soln)^{-1}$.

APPENDIX 2.4.13

Solubility of Tetraalkylammonium Perchlorates in Benzene at 25°C*

Electrolytes	$mol\ l^{-1}$
Me_4NClO_4	0.000165
Et_4NClO_4	0.000113
Pr_4NClO_4	0.000075
Ph_4NClO_4	0.000176
Am_4NClO_4	0.00111
Hx_4NClO_4	0.871

* N. C. Deno and H. E. Berkheimer, *J. Org. Chem.*, **28,** 2143 (1963).

APPENDIX 2.4.14

Solubilities of Electrolytes in Various Solvents at 25°C

Solvent	AgCl* g(kg solv)$^{-1}$	CsCl* g(kg solv)$^{-1}$	$Cu(NO_3)$† g(kg solv)$^{-1}$	$CsNO_3$‡ g(l soln)$^{-1}$
Acetone	1.33×10^{-5}	4.27×10^{-2}		0.0564§
Methyl ethyl ketone	1.13×10^{-5}	3.18×10^{-2}		0.0162§
Acetophenone	6.83×10^{-6}	2.09×10^{-2}		
Methyl propyl ketone	2.98×10^{-6}	1.23×10^{-2}		
Acetylacetone				0.0402§
Propiophenone				0.0075§
Methyl acetate				0.00594§
Ethyl acetate			1510	0.00346§
n-Butyl acetate				0.00167§
n-Amyl acetate				0.00204§
Nitromethane			51§	
Acetonitrile			337§	

* N. A. Izmailov and V. S. Chernyi, *Zh. Fiz. Khim.*, **34**, 127 (1960). Data converted from mol (kg solv)$^{-1}$.

† C. C. Addison, B. J. Hathaway, N. Logan and A. Walker, *J. Chem. Soc.*, 4308 (1960).

‡ R. S. Biktimirov, *Russ. J. Phys. Chem.*, **37**, 1276 (1963).

§ Data at 20°C.

APPENDIX 2.4.15

Mean Activity Coefficients of Some Alkali Metal Halides and Potassium Nitrate in Formamide at the Freezing Point

m	LiCl*	NaCl*	KCl†	RbCl†	CsCl†	KNO_3‡	KBr‡	KI‡
0.02	0.942	0.936	0.936	0.929	0.922	0.912	0.927	0.929
0.05	0.930	0.919	0.917	0.903	0.888	0.868	0.902	0.904
0.1	0.929	0.909	0.907	0.883	0.858	0.821	0.883	0.887
0.2	0.940	0.903	0.899	0.866	0.826	0.761	0.869	0.887
0.3	0.959	0.906	0.896	0.859	0.804	0.721	0.870	0.888
0.4	0.982	0.913	0.898	0.854	0.797	0.688	0.876	0.898
0.5	1.009	0.924	0.898		0.792	0.662	0.881	0.909
0.6	1.039	0.936	0.900			0.640	0.895	0.921
0.7	1.072	0.951	0.903			0.621_5	0.901	0.935
0.8	1.107	0.967	0.909			0.604	0.915	0.953
0.9						0.589		
1.0						0.575		
1.1						0.563		

* E. N. Vasenko, *Zh. Fiz. Khim.*, **21,** 361 (1947).
† E. N. Vasenko, *Zh. Fiz. Khim.*, **22,** 999 (1948).
‡ E. N. Vasenko, *Zh. Fiz. Khim.*, **23,** 959 (1949).

APPENDIX 2.4.16

Mean Activity Coefficients of Some Electrolytes in Formamide at the Freezing Point

m	$\gamma_{\pm}$	m	$\gamma_{\pm}$	m	$\gamma_{\pm}$
		Me_4NI*			
0.0490	0.799	0.0747	0.733	0.0971	0.711
		Et_4NBr*			
0.1006	0.874	0.3057	0.759	0.4984	0.725
0.1999	0.809	0.4032	0.744		
		Et_4NI*			
0.1004	0.820	0.3016	0.638	0.5042	0.551
0.2010	0.710	0.4057	0.586		
		Me_3NHI*			
0.1008	0.833	0.3024	0.665	0.5027	0.585
0.2010	0.736	0.4003	0.624		
		Et_3NHCl*			
0.1005	0.849	0.3014	0.687	0.5009	0.617
0.2018	0.752	0.4043	0.640		
		Me_3PhNI*			
0.0980	0.806	0.3001	0.628	0.4037	0.581
0.1990	0.695				
		KCl†			
0.0398	0.926	0.1043	0.871	0.1584	0.840
0.0634	0.901	0.1206	0.870		
		NH_4Cl†			
0.0511	0.943	0.1120	0.877	0.1314	0.870
0.0717	0.917	0.1188	0.870	0.1475	0.855
0.0906	0.885				

* R. Gopal and M. M. Husain, *J. Ind. Chem. Soc.*, **43,** 204 (1966).
† L. R. Dawson and E. J. Griffith, *J. Phys. Chem.*, **56,** 281 (1952).

APPENDIX 2.4.17

Log $\gamma_\pm$ for Some Electrolytes in Formamide at the Freezing Point*

m	$-\log \gamma_\pm$		
	NaCl	KCl	CsCl
0.01	0.0272	0.0273	0.0282
0.02	0.0354	0.0354	0.0374
0.05	0.0472	0.0491	0.0526
0.1	0.0615	0.0658	0.0709
0.2	0.0823	0.0895	0.1004
0.3	0.0987	0.1069	0.1190

* R. I. Mostkova, Y. M. Kessler and I. V. Safonova, *Elektrokhimiya*, **5**, 409 (1969).

APPENDIX 2.4.18

Mean Activity Coefficients of Some Electrolytes in Dilute Solutions of *N*-Methylacetamide at the Freezing Point*

m	$\gamma_\pm$		
	NaI	KI	NH_4I
0.005	0.787	0.767	0.901
0.01	0.706	0.691	0.829
0.02	0.640	0.625	0.753
0.03	0.609	0.593	0.711
0.04	0.585	0.569	0.684
0.05	0.570	0.552	0.666
0.06	0.556	0.540	0.653
0.07	0.549	0.530	0.642
0.08	0.542	0.521	0.634
0.09	0.531	0.514	0.625
0.10	0.526	0.509	0.618
0.12	0.516	0.497	

* O. D. Bonner, S. J. Kim and A. L. Torres, *J. Phys. Chem.*, **73**, 1968 (1969).

APPENDIX 2.4.19

Mean Activity Coefficients of Electrolytes in *N*-Methylacetamide at the Freezing Point

Molality	LiCl	NaCl	KCl	CsCl	LiBr	NaBr	KBr	CsBr	NaI	KI	CsI	$LiNO_3$	$NaNO_3$	KNO_3
0.01	0.982	0.976	0.974	0.973	0.977	0.977	0.974	0.977	0.979	0.973	0.974	0.975	0.970	0.969
0.05	0.993	0.966	0.958	0.952	0.974	0.974	0.957	0.968	0.983	0.956	0.956	0.966	0.937	0.934
0.10	1.020	0.971	0.955	0.943	0.988	0.988	0.955	0.966	1.005	0.954	0.952	0.974	0.915	0.907
0.20	1.073	0.996			1.033	1.028	0.964	0.962	1.065	0.968	0.953	1.007	0.886	0.864
0.30	1.120	1.031			1.088	1.075	0.980		1.135	0.993	0.959	1.051	0.864	
0.40					1.151	1.126	1.000		1.213	1.025	0.966	1.102	0.847	
0.50					1.220	1.179			1.296	1.062		1.160	0.832	
0.60					1.295	1.235			1.386	1.105		1.224	0.818	
0.70						1.293			1.482	1.154			0.805	
0.80									1.584				0.794	

APPENDIX 2.4.19 (*contd*)

Molality	Me_4NCl	Et_4NCl	Pr_4NCl	Bu_4NCl	Me_4NBr	Et_4NBr	Pr_4NBr	Bu_4NBr	Et_4NI	Pr_4NI	Bu_4NI
0.01	0.971	0.973	0.974	0.975	0.967	0.969	0.968	0.969	0.962	0.960	0.959
0.05	0.941	0.953	0.956	0.962	0.917	0.932	0.928	0.932	0.899	0.894	0.893
0.10	0.920	0.945	0.951	0.963		0.904	0.898	0.906	0.842	0.837	0.835
0.20	0.888	0.940	0.952	0.974		0.863	0.856	0.870		0.750	0.749
0.30	0.859	0.939	0.959	0.990		0.829	0.823	0.843		0.683	0.683
0.40	0.831	0.939	0.967	1.006		0.799	0.796	0.822		0.628	0.630
0.50	0.803	0.939	0.976	1.022		0.771	0.773	0.804		0.582	0.586
0.60		0.938	0.984	1.038		0.744	0.752	0.788			0.549
0.70		0.937	0.993	1.052		0.719	0.734	0.775			0.518
0.80		0.934	1.001	1.066		0.694	0.717	0.763			0.492
0.90			1.008	1.078		0.671	0.702	0.752			0.469
1.00							0.688	0.743			0.450

Molality	LiForm*	NaForm*	LiAc	NaAc	KAc	LiProp*	NaProp*	KProp*
0.01	0.961	0.969	0.966	0.967	0.972	0.969	0.971	0.974
0.05	0.898	0.931	0.919	0.924	0.946	0.931	0.939	0.958
0.10	0.847	0.896	0.879	0.888	0.932	0.901	0.917	0.953
0.20	0.776	0.831	0.819	0.833	0.916	0.855	0.881	0.951
0.30	0.726		0.769	0.787	0.906	0.815	0.848	0.951
0.40	0.691		0.726	0.745	0.898	0.779		0.950
0.50					0.891			0.947
0.60					0.884			
0.70					0.878			
0.80					0.872			

* Form = Formate. Prop = Propionate.

R. H. Wood, R. K. Wicker II and R. W. Kreis, *J. Phys. Chem.*, **75**, 2313 (1971).

R. W. Kreis and R. H. Wood, *J. Phys. Chem.*, **75**, 2319 (1971).

APPENDIX 2.4.20

Standard Integral Heats ot Solution of Electrolytes in Methanol at 25°C

Electrolyte	ΔH^0_{soln}/kcal mol^{-1}	Ref.
HCl	−19.70	(a)
LiCl	−12.4	(b)
	−11.4 ± 0.2	(c)
	-12.3_8	(d)
	−11.50	(a)
LiBr	−13.4 ± 0.2	(c)
LiI	−16.9 ± 0.2	(c)
$LiClO_4$	−12.2	(b)
$LiNO_3$	−4.51	(e)
$Li(CF_3COO)$	−9.37	(f)
LiPic	−1.30	(d)
NaCl	−2.50	(d)
	−2.0	(a)
NaBr	−3.87	(e)
	−4.13	(g)
	−4.30	(d)
	−4.0	(a)
	−4.05	(f)
NaI	−8.1	(b)
	−7.25	(h)
	−7.66	(g)
	−7.52	(d)
	−7.0	(a)
	−7.14	(f)
$NaClO_4$	−2.4	(b)
	−2.86	(i)
	−2.6 ± 0.1	(c)
	−2.62	(d)
$NaNO_3$	1.12	(e)
NaSCN	−6.0	(g)
$NaBPh_4$	−3.9	(c)
	−8.02	(j)
$Na(CF_3COO)$	−4.78	(f)
NaPic	0.61	(g)
	1.9_6	(d)
KF	−5.25	(f)
KCl	1.4_1	(d)
	1.08	(a)
KBr	1.0_7	(d)
	0.87	(a)

APPENDIX 2.4.20 (*contd*)

Electrolyte	ΔH^0_{soln}/kcal mol^{-1}	Ref.
KI	-0.6_5	(d)
	0.175	(a)
	−0.15	(f)
KNO_3	4.1_9	(d)
$K(CF_3COO)$	−1.07	(f)
RbCl	1.28	(a)
$Rb(CF_3COO)$	0.33	(f)
CsCl	2.4	(b)
	1.85	(a)
CsI	4.2	(b)
$Cs(CF_3COO)$	0.41	(f)
$Mg(ClO_4)_2$	−46.6	(b)
$CaCl_2$	−25.28	(k)
$Ca(ClO_4)_2$	−28.2	(b)
$SrCl_2$	−17.66	(k)
$Sr(ClO_4)_2$	−24.6	(b)
$BaCl_2$	−8.56	(k)
$Ba(ClO_4)_2$	−14.3	(b)
$Ag(ClO_4)$	−3.4	(b)
$AgNO_3$	0.73	(e)
$CoCl_2$	−17.01	(l)
$NiCl_2$	−16.24	(l)
$PbClO_4$	−24.9	(b)
NH_4Br	0.62	(e)
NH_4NO_3	2.58	(e)
NH_4SCN	−0.59	(g)
Me_4NCl	3.18	(d)
Me_4NBr	6.98	(d)
	6.95	(m)
Et_4NCl	−0.31	(d)
Et_4NBr	4.38	(d)
Et_4NClO_4	8.24	(d)
Et_4NPic	8.2_6	(d)
Bu_4NBr	5.0 ± 0.2	(c)
Bu_4NI	10.4 ± 0.2	(c)
Bu_4NBBu_4	14.1	(c)
Bn_4NBr	7.8 ± 0.1	(c)
Am_4NCl	−2.37	(f)
Am_4NI	14.1 ± 0.3	(c)
Am_4NBAm_4	19.5	(c)
Ph_4AsCl	5.7	(c)
	−0.98	(f)
Ph_4AsI	8.65*	(j)

* Data were obtained at 30°C.

(a) C. M. Slansky, *J. Amer. Chem. Soc.*, **62,** 2430 (1940).

(b) S. I. Drakin and Y. M. Chang, *Zh. Fiz. Khim.*, **38,** 2800 (1964).

(c) R. Fuchs, J. L. Bear and R. F. Rodewald, *J. Amer. Chem. Soc.*, **91,** 5797 (1969).

(d) F. A. Askew, E. Bullock, H. T. Smith, R. K. Tinkler, O. Gatty and J. H. Wolfenden, *J. Chem. Soc.*, 1368 (1934). Values are at 20°C.

(e) B. Jakuszewski, S. Taniewska-Osinska and R. Logwinienko, *Bull. Acad. Polon. Sci., Ser. Sci. Chim.*, **9,** 127 (1961).

(f) C. V. Krishnan and H. L. Friedman, *J. Phys. Chem.*, **75,** 388 (1971).

(g) K. P. Mishchenko and G. M. Poltoratskii, Examination of the Thermodynamics and Structure of Aqueous and Nonaqueous Electrolytic Solutions, Chemical Publisher, Leningrad Division (1968).

(h) K. P. Mishchenko and M. L. Klyueva, *Teor. i Eksperim. Khim., Akad. Nauk. Ukr. SSR*, **1,** 201 (1965).

(i) M. Mastroianni and C. M. Criss, *J. Chem. Eng. Data*, **17,** 222 (1972).

(j) G. Choux and R. L. Benoit, *J. Amer. Chem. Soc.*, **91,** 6221 (1969).

(k) B. Jakuszewski and S. Taniewska-Osinaka, *Bull. Acad. Polon. Sci., Ser. Sci. Chim.*, **9,** 133 (1961).

(l) B. Jakuszewski and S. Taniewska-Osinska, *Zeszyty Nauk. Uniw. Lodz. Ser. II*, **13,** 137 (1962)

(m) M. Mastroianni and C. M. Criss, *J. Chem. Thermodyn.*, **4,** 321 (1972).

APPENDIX 2.4.21

Standard Integral Heats of Solution of Electrolytes in Deuterated Methanol at 25°C*

Electrolyte	ΔH^0_{soln}/kcal mol^{-1}
$Li(CF_3COO)$	−9.65
NaBr	−3.97
NaI	−6.98
$Na(CF_3COO)$	−4.94
KF	−5.35
KI	0.09
$K(CF_3COO)$	−1.12
$Rb(CF_3COO)$	0.33
$Cs(CF_3COO)$	0.39
Ph_4AsCl	−0.75
Am_4NCl	−2.07

* C. V. Krishnan and H. L. Friedman, *J. Phys. Chem.*, **75**, 388 (1971).

APPENDIX 2.4.22

Integral Heats of Solution of Electrolytes in Methanol at Various Concentrations and Temperatures

	ΔH_{soln}/kcal mol^{-1}					
m	NaBr*		NaI†		KI†	
	25°C	40°C	25°C	50°C	25°C	50°C
0.00	−4.13	−4.09	−7.25 (−7.66)*	−6.95		
0.01	−4.04	−3.98	(−7.58)*			
0.05	−3.70	−3.59	(−7.14)*			
0.1	−3.32	−3.17	−7.20 (−6.69)*	−6.85	0.15	0.0
0.2	−2.85	−2.71	−6.90 (−6.33)*	−6.50	0.63	0.48
0.3	−2.68	−2.54	−6.70 (−6.21)*	−6.30	0.95	0.85
0.4	−2.58	−2.45	−6.50 (−6.12)*	−6.20	0.97	0.95
0.5	−2.54	−2.38	−6.45 (−6.03)*	−6.10	0.99	1.05
0.6					1.01	1.15
0.7					1.03	1.25
0.8					1.05	1.35
0.9					1.07	1.42
1.0	−2.30	−2.02	−5.85 (−5.70)*	−5.40	1.08	1.45
1.5			−5.35	−4.80		
2.0			−5.05 (−5.15)*	−4.30		
2.5			−4.75	−4.0		
3.0			−4.50 (−4.58)*	−3.70		
3.5			−4.20	−3.30		
4.0			−3.95 (−3.97)*	−3.0		
4.5			−3.75	−2.80		
5.0			−3.55 (−3.50)*	−2.60		

APPENDIX 2.4.22 (*contd*)

m	ΔH_{soln}/kcal mol^{-1}				
	NaPic*‡		NaSCN*§		
	10°C	25°C	11°C	25°C	50°C
0.0	−10.30	0.61	−6.13	−6.0	−5.75
0.01	−5.05	1.28	−5.58	−5.40	−5.30
0.05	−2.37	1.16	−4.64	−4.54	−4.33
0.1	−2.17	0.78	−4.36	−4.26	−4.02
0.2	−2.02	0.48	−4.18	−3.98	−3.67
0.3	−1.86	0.35	−4.0	−3.80	−3.42
0.4	−1.72	0.28	−3.86	−3.68	−3.25
0.5		0.24	−3.73	−3.55	−3.11
0.6		0.24			
0.7		0.23			
1.0			−3.23	−2.96	−2.60
2.0			−2.52	−2.27	−1.87
3.0			−2.05	−1.82	−1.37
4.0			−1.70	−1.51	−0.98
5.0				−1.29	−0.69

m	ΔH_{soln}/kcal mol^{-1}			
	NH_4SCN*§		Picric acid*	
	11°C	25°C	25°C	50°C
0.0	−0.59	−0.59	2.60	3.60
0.01	−0.50	−0.50	2.32	3.35
0.05	+0.02	+0.02	2.58	3.53
0.10	0.24	0.24	2.60	3.58
0.20	0.42	0.42	2.60	3.61
0.30	0.57	0.57	2.60	3.59
0.40	0.69	0.69		3.63
0.50	0.76	0.79	2.60	3.62
0.85			2.60	
1.0	1.04	1.15		3.60
1.85				3.60
2.0	1.36	1.68		
3.0	1.60	2.0		
4.0	1.78	2.24		
5.0	1.95	2.44		
6.0	2.10	2.63		
7.0		2.80		

* K. P. Mishchenko and G. M. Poltoratskii, Examination of the Thermodynamics and Structure of Aqueous and Nonaqueous Electrolytic Solutions, Chemical Publisher, Leningrad Division (1968).

† K. P. Mishchenko and M. L. Klyueva, *Teor. i Eksp. Khim., Akad. Nauk. Ukr. SSR*, **1**, 201 (1965).

‡ L. P. Zhilina and K. P. Mishchenko, *Teor. i Eksp. Khim., Akad. Nauk. Ukr. SSR*, **1**, 361 (1965). Unsmoothed data are reported. The signs on the data at 25° are opposite that in Ref. *.

§ E. P. Prosviryakova, K. P. Mishchenko and G. M. Poltoratski, *Teor. i Eksp. Khim., Akad. Nauk. Ukr. SSR*, **5**, 129 (1969).

APPENDIX 2.4.23

Standard Integral Heats of Solution of Various Electrolytes in Water-Methanol Mixtures (25°C)*

Wt% Methanol	Mol % Methanol	ΔH^0_{soln}/kcal mol^{-1}				
		HCl	LiCl	NaCl	KCl	KI
0	0.0	−17.88	−8.90	0.92	4.12	4.87
10	5.88	−17.55	−8.0	1.40	4.41	5.20
	10.0			1.6_1†		
20	12.32	−17.32	−7.60	1.68	4.61	5.30
30	19.41	−17.20	−7.80	1.75	4.62	5.20
	20.0			1.8_5†		
	25.0			1.7_8†		
40	27.25	−17.24	−8.05	1.72	4.54	5.0
	35.0			1.5_2†		
50	36.08	−17.44	−8.40	1.59	4.28	4.62
60	45.74	−17.74	−8.80	1.375	3.87	4.10
	50.0			1.0_2†		
70	56.73	−18.03	−9.35	1.03	3.40	3.47
	65.0			0.6_0†		
80	69.41	−18.50	−9.95	0.500	2.79	2.70
	75.0			-0.1_0†		
90	83.50	−19.03	−10.65	−0.350	2.06	1.70
	85.0			-0.6_8†		
	95.0			-1.8_2†		
100	100.0	−19.70	−11.50	−2.0	1.08	0.175

Wt% Methanol	Mol % Methanol	ΔH^0_{soln}/kcal mol^{-1}				
		NaI	KBr	NaBr	RbCl	CsCl
0	0.0	−1.82	4.78	−0.030	4.0	4.24
10	5.88	−1.25	5.05	0.450	4.17	4.40
20	12.32	−1.0	5.23	0.650	4.29	4.45
30	19.41	−1.15	5.15	0.660	4.34	4.42
40	27.25	−1.40	4.95	0.550	4.35	4.34
50	36.08	−1.75	4.55	0.300	4.29	4.16
60	45.74	−2.15	4.07	−0.040	4.15	3.92
70	56.73	−2.80	3.53	−0.600	3.91	3.60
80	69.41	−3.75	2.82	−1.32	3.45	3.15
90	83.50	−5.05	2.02	−2.40	2.63	2.58
100	100.0	−7.0	0.87	−4.0	1.28	1.85

* Data from C. M. Slansky, *J. Amer. Chem. Soc.*, **62,** 2430 (1940), unless otherwise indicated.

† R. L. Moss and J. H. Wolfenden, *J. Chem. Soc.*, 118 (1939).

APPENDIX 2.4.24

Standard Integral Heats of Solution of Electrolytes in Some Higher Alcohols at 25°C

Electrolyte	Ethanol		Propanol		Butanol	
	ΔH^0_{soln}/kcal mol^{-1}	Ref.	ΔH^0_{soln}/kcal mol^{-1}	Ref.	ΔH^0_{soln}/kcal mol^{-1}	Ref.
HCl	−21.28	(a)				
LiCl	−11.90	(a)				
	-12.9_3	(b)				
LiBr	−15.75	(a)				
LiI	−20.71	(a)				
$LiClO_4$	−11.2	(c)	−10.8	(c)	−10.8	(c)
LiPic	−2.97	(b)				
NaBr	−4.61	(a)				
	−2.65	(b)				
NaI	−7.75	(a)				
	−5.86	(d)				
	-5.8_0	(b)				
$NaClO_4$	-0.8_9	(b)				
NaPic	0.4_2	(b)				
KI	−0.90	(a)				
	0.0_7	(b)				
Me_4NCl	3.52	(b)				
Me_4NBr	7.16	(b)				
Et_4NCl	0.52	(b)				
Et_4NBr	5.26	(b)				
Et_4NPic	8.95	(b)				
$Mg(ClO_4)_2$	−45.53	(c)	−40.58	(c)	−38.5	(c)
$Ca(ClO_4)_2$	−21.8	(c)	−18.2	(c)	−16.6	(c)
$Sr(ClO_4)_2$	−15.6	(c)	−13.4	(c)	−12.4	(c)
$Ba(ClO_4)_2$	−5.40	(c)	−4.21	(c)	−3.61	(c)
$Pb(ClO_4)_2$	−18.3	(c)	−16.6	(c)	−15.7	(c)

(a) B. Jakuszewski and S. Taniewska-Osinka, *Lodz. Towarz. Nauk. Soc. Sci. Lodziensis, Acta Chim.*, **8**, 11 (1962).
(b) F. A. Askew, E. Bullock, H. T. Smith, R. T. Tinkler, O. Gatty and J. H. Wolfenden, *J. Chem. Soc.*, 368 (1934).
(c) L. N. Erbanova, S. I. Drakin and M. K. Karapetyants, *Zh. Fiz. Khim.*, **38**, 2670 (1964).
(d) K. P. Mishchenko, V. V. Subbotina and B. S. Krumgal'z, *Teor. i Eksp. Khim., Akad. Nauk. Ukr. SSR*, **5**, 268 (1969).

APPENDIX 2.4.25

Integral Heats of Solution of Sodium Iodide in Ethanol at Various Concentrations at 25 and 45°C*

m	ΔH_{soln}/kcal mol^{-1}	
	25°C	45°C
0.0	−5.86	−5.54
0.01	−5.65	−5.36
0.02	−5.35	−5.20
0.03	−5.20	−5.09
0.04	−5.12	−4.97
0.05	−5.04	−4.85
0.10	−4.72	−4.40
0.20	−4.31	−3.96
0.30	−4.03	−3.66
0.40	−3.83	−3.44
0.50	−3.70	−3.25
1.0	−3.42	−2.67
1.5	−3.14	−2.33
2.0	−2.85	−2.03
2.5	−2.60	−1.70

* K. P. Mishchenko, V. V. Subbotina and B. S. Krumgal'z, *Teor. i Eksp. Khim., Akad. Nauk. Ukr. SSR*, **5**, 268 (1969).

APPENDIX 2.4.26

Integral Heats of Solution of Sodium and Potassium Iodides in Ethylene Glycol at Various Concentrations and Temperatures*

m	ΔH_{soln}/kcal mol^{-1}		
	NaI		KI
	25°C	2.5°C	25°C
0.0	−7.60	−8.10	−1.040
0.005		−7.60	
0.02	−7.14	−7.18	−0.940
0.04	−7.00	−7.18	
0.05			−0.835
0.07			−0.795
0.08	−6.84	−7.16	
0.1	−6.78	−7.09	−0.750
0.2			−0.630
0.4			−0.480
0.5	−6.40	−6.74	
0.6			−0.380
0.7	−6.31	−6.65	
0.8			−0.302
1.0	−6.18	−6.52	−0.228
1.2			−0.158
1.5	−5.96	−6.32	
1.7			0.000
2.0	−5.74	−6.10	0.084
2.2			0.138
2.5	−5.53	−5.90	0.220
2.7			0.272
3.0	−5.31	−5.69	
3.5	−5.09	−5.48	
4.0	−4.88	−5.28	
4.5	−4.66	−5.07	
5.0	−4.45		
5.3	−4.32		

* K. P. Mishchenko and V. P. Tungusov, *Teor. i Eksp. Khim., Akad. Nauk. Ukr. SSR*, **1**, 55 (1965).

APPENDIX 2.4.27

Enthalpies of Transfer of HCl and $HClO_4$ from Water to Aqueous Ethylene Glycol at 25°C*

Solvent Composition (mol fraction)	$\Delta H_{(EG \leftarrow w)}$/ kcal mol^{-1} HCl	Solvent Composition (mol fraction)	$\Delta H_{(EG \leftarrow w)}$/ kcal mol^{-1} $HClO_4$
0.1	0.40	0.102	0.33
0.2	0.25	0.278	−0.98
0.3	−0.11	0.490	−3.08
0.4	−0.60	0.690	−4.58
0.5	−1.07	0.807	−5.05
0.6	−1.50	0.896	−4.72
0.7	−1.78	0.997	−2.09
0.8	−1.70		
0.9	−1.00		
1.0	1.74		

* J. H. Stern and J. M. Nobilione, *J. Phys. Chem.*, **72**, 3937 (1968); **73**, 928 (1969).

APPENDIX 2.4.28

Standard Integral Heats of Solution of Electrolytes in Formamide and *N*-Methylformamide at 25°C

Electrolyte	Formamide		*N*-Methylformamide	
	ΔH^0_{soln}/ kcal mol^{-1}	Ref.	ΔH^0_{soln}/ kcal mol^{-1}	Ref.
LiF	4.9	(a)		
LiCl	−9.42	(a)	−10.0	(d)
	−8.4	(b)		
LiBr	−13.39	(a)		
	−12.1	(b)		
LiI	−18.25	(a)	−21.11	(e)
NaF	1.4	(a)		
NaCl	−2.10	(a)	−1.244	(d)
NaBr	−4.41	(a)	−4.15	(d)
	−3.9	(b)		
NaI	−7.43	(a)	−8.258	(d)
	−7·3	(b)	−8.261	(e)
KF	−3.17	(a)	−2.600	(e)
KCl	0.82	(a)	0.308	(d)
			0.374	(e)

APPENDIX 2.4.28 *(contd)*

KBr	0.23	(a)	−0.825	(e)
	0.2	(b)		
KI	−1.02	(a)	−3.223	(e)
	−1.9	(b)		
RbF	−5.27	(a)		
RbCl	0.71	(a)		
	0.6	(b)		
RbBr	0.75	(a)		
RbI	0.23	(a)	−1.644	(e)
	0.2	(b)		
CsF	−7.60	(a)		
CsCl	0.95	(a)	0.890	(d)
	0.6	(b)		
CsBr	1.81	(a)		
	1.3	(b)		
CsI	2.22	(a)	0.707	(e)
	1.7	(b)		
$CaCl_2$	−19.1	(b)	−24.3	(c)
	−22.7	(c)		
$CaBr_2$	−26.8	(b)	−34.05	(c)
	−29.9	(c)		
$SrCl_2$	−17.7	(b)	−16.2	(c)
	−17.6	(c)		
$SrBr_2$	−26.0	(c)	−28.0	(c)
$Sr(ClO_3)_2$			−9.96	(f)
$BaCl_2$	−11.4	(b)		
	−10.0	(c)		
$BaBr_2$	−14.4	(b)	−18.0	(c)
	−16.3	(c)		
$Ba(ClO_3)_2$			−7.68	(f)

(a) G. Somsen and J. Coops, *Rec. Trav. Chim.*, **84**, 985 (1965). G. Somsen, *Rec. Trav. Chim.*, **85**, 517 (1966).
(b) D. V. S. Jain, B. S. Lark, S. P. Kochar and V. K. Gupta, *Ind. J. Chem.*, **7**, 256 (1969).
(c) A. Finch, P. J. Gardner and C. J. Steadman, *J. Phys. Chem.*, **75**, 2325 (1971).
(d) R. P. Held and C. M. Criss, *J. Phys. Chem.*, **69**, 2611 (1965). ΔH^0_{soln} for LiCl and NaBr data have been re-evaluated by correcting to infinite dilution only those heats obtained at concentrations greater than 0.001 m.
(e) L. Weeda and G. Somsen, *Rec. Trav. Chim.*, **85**, 159 (1966).
(f) A. Finch, P. J. Gardner and C. J. Steadman, *J. Phys. Chem.*, **71**, 2996 (1967).

APPENDIX 2.4.29

Standard Integral Heats of Solution of Electrolytes in *N*-Methyl- and *N*-Ethylacetamide*

Electrolyte	*N*-Methylacetamide		*N*-Ethylacetamide (NEA)	
	ΔH^0_{soln}/ kcal mol^{-1}	Ref.	ΔH^0_{soln}/ kcal mol^{-1}	Ref.
LiCl			−10.0	(a)
LiBr			−12.0	(a)
LiI	−21.30	(b)	−15.8	(a)
NaI	−7.10	(b)	−6.5	(a)
$NaClO_4$			−1.6	(a)
$NaBPh_4$			−2.5	(a)
KF	−1.96	(b)		
KCl	1.32	(b)		
KBr	0.34	(b)		
KI	−2.20	(b)		
CsI	1.94	(b)		
Bu_4NBr			3.8	(a)
Bu_4NI			9.1	(a)
Bu_4NBBu_4			8.5	(a)
Am_4NBr			6.4	(a)
Am_4NI			12.1	(a)
Am_4NBAm_4			13.6	(a)
Ph_4AsCl			2.1	(a)

* Data for NMA are at 35°C; those for NEA are at 25°C.

(a) L. Weeda and G. Somsen, *Rec. Trav. Chim.*, **86**, 263 (1967).

(b) R. Fuchs, J. L. Bear and R. F. Rodewald, *J. Amer. Chem. Soc.*, **91**, 5797 (1969).

APPENDIX 2.4.30

Standard Integral Heats of Solution of Electrolytes in Some Aprotic Amides at 25°C

Electrolyte	*N,N*-Dimethylformamide		*N*-Methyl-pyrrolidone(a)	Hexa-methyl-phosphor-triamide(a)
	ΔH^0_{soln}/ kcal mol^{-1}	Ref.	ΔH^0_{soln}/ kcal mol^{-1}	ΔH^0_{soln}/ kcal mol^{-1}
LiCl	−11.8	(a)	−8.2	−12.3
	−11.82	(b)		
	−11.42	(c)		
	−11.25	(d)		
LiBr	−18.4	(a)	−14.5	−19.7
	−18.52	(b)		
	−18.07	(c)		
LiI	−25.5	(a)	−20.1	−24.2
	−19.1	(b)		
	−26.06	(e)		
$LiNO_3$	−8.1	(d)		
NaBr	−7.39	(c)		
	−6.9	(d)		
NaI	−12.8	(a)	−11.0	
	−13.15	(b)		
	−12.62	(c)		
	−12.4	(d)		
$NaClO_4$	−10.0	(a)	−8.8	
$NaBPh_4$	−19.8	(a)	−19.0	
	−12.05	(f)		
$NaNO_3$	−2.15	(d)		
KBr	−3.6	(a)		
	−3.89	(c)		
	−2.1	(d)		
KI	−8.1	(a)	−6.1	
	−8.04	(b)		
	−7.4	(d)		
	−8.06	(e)		
KNO_3	0.2	(d)		
RbI	−6.64	(b)		
CsI	−4.25	(b)		
	−4.25	(c)		

APPENDIX 2.4.30 (*contd*)

Electrolyte	*N,N*-Dimethylformamide		*N*-Methylpyrrolidone(a)	Hexamethylphosphortriamide(a)
	ΔH^0_{soln}/ kcal mol^{-1}	Ref.	ΔH^0_{soln}/ kcal mol^{-1}	ΔH^0_{soln}/ kcal mol^{-1}
$CaCl_2$	−23.8	(g)		
$CaBr_2$	−39.6	(g)		
$SrCl_2$	−15.8	(g)		
$SrBr_2$	−34.1	(g)		
$Sr(ClO_3)_2$	−14.96	(h)		
$Ba(ClO_3)_2$	−12.80	(h)		
Me_4NI	4.01	(i)		
Et_4NCl	1.34	(f)		
Et_4NI	3.42	(i)		
	2.95	(f)		
Et_4NClO_4	1.80	(f)		
Pr_4NI	2.00	(i)		
Bu_4NBr	0.0	(a)	3.9	
Bu_4NI	2.1	(a)	6.8	
Bu_4NBBu_4	2.1	(a)	4.8	
Am_4NBr	2.7	(a)	6.8	
Am_4NI	6.6	(a)	10.4	
Am_4NBAm_4	8.0	(a)	10.1	
Ph_4AsCl	−3.5	(a)	−1.0	
Ph_4AsI	0.69	(f)		

(a) R. Fuchs, J. L. Bear and R. F. Rodewald, *J. Amer. Chem. Soc.*, **91,** 5797 (1969).
(b) L. Weeda and G. Somsen, *Rec. Trav. Chim.*, **86,** 893 (1967).
(c) R. P. Held and C. M. Criss, *J. Phys. Chem.*, **71,** 2487 (1967). Data for LiCl, LiBr and NaI were obtained by correcting heats at concentrations >0.001*m* to infinite dilution.
(d) G. A. Krestov and V. A. Zverev, *Izv. Vyssh. Ucheb. Zaved., Khim., Khim Tekhnol.*, **11,** 990 (1968).
(e) Y. A. Tsai and C. M. Criss, *J. Chem. Eng. Data*, **18,** 51 (1973).
(f) G. Choux and R. L. Benoit, *J. Amer. Chem. Soc.*, **91,** 6221 (1969).
(g) A. Finch, P. J. Gardner and C. J. Steadman, *J. Phys. Chem.*, **75,** 2325 (1971).
(h) A. Finch, P. J. Gardner and C. J. Steadman, *J. Phys. Chem.*, **71,** 2996 (1967).
(i) O. N. Bhatnagar and C. M. Criss, *J. Phys. Chem.*, **73,** 174 (1969).

APPENDIX 2.4.31

Standard Integral Heats of Solution of Electrolytes in Ethylenediamine and Formic Acid at 25°C

Electrolyte	ΔH^0_{soln}/kcal mol^{-1}	
	Ethylenediamine*	Formic Acid†
LiI	−28.80	
NaCl		−0.20
NaBr	−10.00	
NaI	−15.70	
$NaNO_3$	−3.27	
KCl		−0.30
KI	−5.90	
KNO_3	0.97	
RbCl		−0.90
CsCl		−1.25
CsI	−2.47	
AgCl	−12.70	
AgBr	−9.97	
AgI	−8.90	
$AgNO_3$	−25.00	
HgI_2	−28.60	
$Hg(CN)_2$	−17.30	

* F. C. Schmidt, S. Godomsky, F. K. Ault and J. C. Huffman, *J. Chem. Eng. Data*, **14,** 71 (1969). The data are at concentrations in the 'dilute region', but the exact concentration range is not given.

† G. P. Kotlyarova and E. F. Ivanova, *Zh. Fiz. Khim.*, **40,** 997 (1966); **38,** 423 (1964).

APPENDIX 2.4.32

Integral Heats of Solution (ΔH^0_{soln}/kcal mol^{-1}) of KBr in Water-Formic Acid Mixtures at 25°C*

Molality	Solvent Composition (Mol % Acid)										
	1.2	2.9	6.5	14	28	48	61	69	78	88	100
0.0	4.46	4.02	3.12	1.80	0.89	0.33	0.12	0.05	0.09	0.15	0.20
0.005	4.49	4.12	3.14	1.82	0.90	0.35	0.13	0.06	0.10	0.16	0.21
0.01	4.51	4.14	3.18	1.84	0.92	0.37	0.14	0.07	0.14	0.18	0.23
0.02	4.52	4.25	3.30	1.87	0.97	0.38	0.15	0.08	0.16	0.20	0.25
0.05	4.64	4.41	3.42	1.93	1.08	0.50	0.20	0.10	0.24	0.28	0.34
0.1	4.70	4.41	3.60	2.07	1.17	0.65	0.26	0.14	0.36	0.40	0.46
0.2	4.70	4.30	3.59	2.34	1.28	0.90	0.40	0.22	0.60	0.56	0.70
0.3	4.60	4.20	3.58	2.48	1.31	0.98	0.54	0.28	0.72	0.80	0.84
0.4	4.56	4.18	3.56	2.52	1.32	1.00	0.58	0.32	0.77	0.86	0.90
0.6	4.50	4.14	3.54	2.52	1.36	1.02	0.62	0.37	0.86	0.95	1.02
0.8	4.61	4.11	3.51	2.53	1.39	1.04	0.66	0.40	0.95	1.08	1.15
1.0	4.42	4.08	3.48	2.52	1.42	1.10	0.70	0.42	1.02	1.17	1.26
1.2	4.37	4.04	3.46	2.51	1.45	1.14	0.73	0.44	1.10	1.27	1.39
1.4	4.32	4.01	3.43	2.52	1.49	1.19	0.76	0.46	1.20	1.38	1.51
1.6	4.28	3.99	3.40	2.52	1.51	1.23	0.80	0.48	1.28	1.46	1.63
1.8	4.23	3.95	3.38	2.52	1.54	1.28	0.84	0.51	1.36	1.57	1.75
2.0	4.18	3.90	3.35	2.53	1.57	1.30	0.88	0.53	1.44	1.65	1.87
2.4	4.13	3.85	3.31	2.52	1.65	1.34					
2.6	3.99	3.82	3.28	2.51	1.68						
3.0	3.96	3.76	3.22	2.52							

* E. F. Ivanova and V. N. Fesenko, *Zh. Fiz. Khim.*, **43**, 1006 (1969)

APPENDIX 2.4.33

Integral Heats of Solution for Some Electrolytes in Acetone at Various Concentrations and Temperatures*

m	ΔH_{soln}/kcal mol^{-1}					
	NaI			$NaClO_4$	Picric acid	
	10°C	25°C	40°C	25°C	10°C	25°C
0.00	−10.82	−10.54	−10.26	−8.50	−0.17	7.10
0.01	−10.48	−10.17	−9.91		0.24	3.00
0.05	−9.73	−9.41	−8.90		0.83	1.59
0.10	−9.26	−8.80	−7.88	−6.98	0.94	1.50
0.20	−8.94	−8.22	−7.52	−6.70	1.06	1.43
0.30	−8.70	−7.98	−7.30	−6.58	1.11	1.39
0.40	−8.52	−7.78	−7.12	−6.45	1.14	1.39
0.50	−8.36	−7.62	−6.98	−6.30	1.15	1.40
0.60	−8.20	−7.48	−6.82			
0.70	−8.06	−7.34	−6.70			
0.80	−7.90	−7.20	−6.56			
0.90	−7.76	−7.06	−6.40			
1.00	−7.62	−6.92	−6.24	−5.68	1.19	1.53
1.10	−7.48	−6.78	−6.08			
1.20	−7.32	−6.64	−5.94			
1.30	−7.18	−6.50	−5.80			
1.40	−7.04	−6.34	−5.70			
1.50	−6.90	−6.22	−5.58	−5.05		
1.60	−6.76	−6.10	−5.46			
1.70		−6.0	−5.36			
1.80		−5.90	−5.24			
1.90		−5.80	−5.12			
2.00		−5.72	−5.02	−4.70	1.29	1.82
2.10		−5.62	−4.92			
2.20		−5.54				
2.50				−4.25		
2.60		−5.24				
2.68		−5.24				
3.00				−3.80	1.38	1.95
3.50				−3.40		
4.00				−2.95	1.48	2.03
4.50				−2.53		
4.577				−2.40		
4.85					1.56	
5.00						2.17
5.88						2.27

* K. P. Mishchenko and V. V. Sokolov, *Zh. Strukt. Khim.*, **5**, 819 (1964); and K. P. Mishchenko and G. M. Poltoratskii, Examination of the Thermodynamics and Structure of Aqueous and Nonaqueous Electrolytic Solutions, Chemical Publisher, Leningrad Division (1968).

APPENDIX 2.4.34

Standard Integral Heats of Solution of Electrolytes in Dimethylsulphoxide at 25°C

Electrolyte	ΔH^0_{soln}/kcal mol^{-1}	Ref.
LiCl	-10.9 ± 0.3	(a), (b)
LiBr	-17.1 ± 0.1	(a), (b)
LiI	-24.2 ± 0.3	(a), (b)
NaI	-11.5 ± 0.1	(a)
	-11.53 ± 0.13	(c)
$NaClO_4$	-7.9 ± 0.1	(a)
$NaBPh_4$	-14.23 ± 0.14	(c)
	-14.2 ± 0.2	(a)
KBr	-2.7 ± 0.1	(a), (b)
KI	-6.5 ± 0.1	(a), (b)
	-6.15 ± 0.11	(c)
$KBPh_4$	-1.71	(d)
$RbBPh_4$	0.92	(d)
CsI	-2.84 ± 0.08	(c)
$CsBPh_4$	1.72	(d)
Me_4NBr	3.05	(d)
Me_4NI	3.30	(d)
Me_4NClO_4	3.45	(d)
Et_4NCl	2.42 ± 0.05	(c)
Et_4NBr	3.27 ± 0.07	(c)
Et_4NI	4.86 ± 0.10	(c)
Et_4NClO_4	3.75	(d)
	3.90*	(e)
Pr_4NCl	3.13	(d)
Pr_4NBr	3.83	(d)
Pr_4NI	3.76	(d)
Pr_4NClO_4	5.90	(d)
Bu_4NCl	3.22	(d)
	3.7 ± 0.1	(a)
Bu_4NBr	4.95	(d)
	5.1 ± 0.1	(a)
Bu_4NI	7.27	(d)
	7.2 ± 0.2	(a)
Bu_4NClO_4	3.95	(d)
Bu_4NBBu_4	9.9 ± 0.1	(a)
Am_4NCl	2.13	(d)

APPENDIX 2.4.34 (*contd*)

Electrolyte	ΔH^0_{soln}/kcal mol^{-1}	Ref.
Am_4NBr	8.60	(d)
	9.6 ± 0.1	(a)
Am_4NI	12.07	(d)
	12.1 ± 0.3	(a)
Am_4NClO_4	12.44	(d)
Am_4NBAm_4	18.4 ± 0.3	(a)
Hx_4NBr	3.99	(d)
Hx_4NClO_4	14.20	(d)
Bu_4PBr	3.92	(d)
Ph_4AsCl	−0.92	(d)
	−1.4 ± 0.2	(a)
Ph_4AsBr	−0.03	(d)
Ph_4AsI	3.44 ±0.05	(c)
Ph_4AsClO_4	6.22	(d)
Ph_4PCl	0.07	(d)
Ph_4SbBr	0.25	(d)
Me_3PhNI	1.93	(d)
$PrPh_3PBr$	3.14	(d)
$BuPh_3PBr$	4.95	(d)
$LaCl_3$	−39.0 ± 0.5 †	(f)
$NdCl_3$	−43.7 ± 1.2 †	(f)
$SmCl_3$	−49.2 ± 1.1 †	(f)
$GdCl_3$	−49.8 ± 1.5 †	(f)
$DyCl_3$	−51.8 ± 0.8 †	(f)
$HoCl_3$	−52.3 ± 0.9 †	(f)
$ErCl_3$	−52.8 ± 1.0 †	(f)
$YbCl_3$	−54.0 ± 1.2 †	(f)

* Data are at 30°C.
† Data are at concentration of 3 × 10^{-4}*m*.
(a) R. Fuchs, J. L. Bear and R. F. Rodewald, *J. Amer. Chem. Soc.*, **91**, 5797 (1969).
(b) R. F. Rodewald, K. Mahendran, J. L. Bear and R. Fuchs, *J. Amer. Chem. Soc.*, **90**, 6698 (1968).
(c) E. M. Arnett and D. R. McKelvey, *J. Amer. Chem. Soc.*, **88**, 2598 (1966).
(d) C. V. Krishnan and H. L. Friedman, *J. Phys. Chem.*, **73**, 3934 (1969).
(e) G. Choux and R. L. Benoit, *J. Amer. Chem. Soc.*, **91**, 6221 (1969).
(f) M. E. Clark and J. L. Bear, *J. Inorg. Nucl. Chem.*, **31**, 2619 (1969).

APPENDIX 2.4.35

Standard Integral Heats of Solution of Electrolytes in Propylene Carbonate at 25°C

Electrolyte	ΔH^0_{soln}/kcal mol^{-1}	Ref.
LiI	−15.05	(a)
$LiClO_4$	−9.51	(a)
$Li(CF_3COO)$	2.36 ± 0.4	(a)
NaI	−5.04	(a)
$NaClO_4$	−3.05	(a)
$Na(CF_3COO)$	3.32	(a)
$NaBPh_4$	−10.72	(a)
$K(CF_3COO)$	3.73	(a)
$KBPh_4$	0.74	(a)
$Rb(CF_3COO)$	3.78	(a)
$RbBPh_4$	2.41	(a)
$Cs(CF_3COO)$	3.04	(a)
$CsBPh_4$	2.38	(a)
Me_4NBr	5.27	(b)
Me_4NI	5.34	(b)
Me_4NClO_4	3.90	(c)
Et_4NCl	3.30	(b)
Et_4NBr	4.90	(b)
Et_4NI	6.08	(b)
Et_4NClO_4	3.58	(c)
Pr_4NCl	3.78	(c)
Pr_4NBr	5.02	(c)
Pr_4NI	4.74	(c)
Pr_4NClO_4	5.37	(c)
Bu_4NCl	3.38	(c)
Bu_4NBr	5.70	(c)
Bu_4NI	7.85	(c)
Bu_4NClO_4	3.03	(c)
Am_4NCl	1.92	(c)
Am_4NBr	8.95	(c)
Am_4NI	12.25	(c)
Am_4NClO_4	11.13	(c)
Hx_4NBr	4.10	(c)
Hx_4NClO_4	12.41	(c)
Bu_4PBr	4.80	(c)
Ph_4AsCl	0.25	(c)

APPENDIX 2.4.35 (*contd*)

Electrolyte	ΔH^{0}_{soln}/kcal mol^{-1}	Ref.
Ph_4AsBr	1.75	(c)
Ph_4AsI	5.74*	(d)
Ph_4AsClO_4	6.18	(c)
Ph_4PCl	0.90	(c)
Ph_4PClO_4	5.98	(c)
Ph_4SbBr	3.79	(c)
Me_3PhNI	4.68	(c)

* Data are at 30°C.
(a) Y. C. Wu and H. L. Friedman, *J. Phys. Chem.*, **70,** 501 (1966).
(b) Y. C. Wu and H. L. Friedman, *J. Phys. Chem.*, **70,** 2020 (1966).
(c) C. V. Krishnan and H. L. Friedman, *J. Phys. Chem.*, **73,** 3934 (1969).
(d) G. Choux and R. L. Benoit, *J. Amer. Chem. Soc.*, **91,** 6221 (1969).

APPENDIX 2.4.36

Heats of Solution at 25°C of Alkylammonium Ions in Dimethylsulphoxide and Propylene Carbonate (kcal mol^{-1})*

Electrolyte	DMSO	PC
$MeNH_2HCl$	−0.93	
$EtNH_2HCl$	0.78	
$PrNH_2HCl$	−0.49	
$BuNH_2HCl$	−1.20	
$AmNH_2HCl$	−0.73	
$HxNH_2HCl$	−0.18	
$HpNH_2HCl$	0.50	
$OctNH_2HCl$	2.57	
$MeNH_2HTFA$†	1.93	6.12
$EtNH_2HTFA$	1.69	5.80
$PrNH_2HTFA$	2.75	6.76
$BuNH_2HTFA$	2.33	6.27
$AmNH_2HTFA$	3.87	7.93
$HxNH_2HTFA$	4.24	7.95
$HpNH_2HTFA$	5.45	9.03
$OctNH_2HTFA$	6.50	10.21
Me_2NHHCl	0.46	
$Me_2NHHTFA$	−0.37	3.45
Et_2NHHBr	0.65	
$Et_2NHHTFA$	4.28	7.87
Pr_2NHHBr	−0.78	
$Pr_2NHHTFA$	5.47	8.95
Bu_2NHHCl	2.37	
$Bu_2NHHTFA$	5.49	8.72
$Am_2NHHTFA$	6.28	9.41
Me_3NHCl	2.02	
Me_3NHBr	2.14	5.80
Me_3NHTFA	3.22	5.16
Et_3NHBr	3.36	6.86
Pr_3NHBr	1.48	4.64
Bu_3NHBr	3.83	6.63
Et_3CNH_3I	1.80	3.30
$Et_3C_3H_7NH_3I$	3.25	4.48
Bu_3CNH_3I	3.05	3.93
$Bu_3C_3H_4NH_3I$	1.94	2.70
NH_4TFA	1.08	6.60

* C. V. Krishnan and H. L. Friedman, *J. Phys. Chem.*, **74**, 3900 (1970).
† HTFA—Trifluoroacetic Acid.

APPENDIX 2.4.37

Standard Integral Heats of Solution of Electrolytes in Sulpholane at 30°*

Electrolyte	ΔH^{0}_{soln}/ kcal mol^{-1}
$LiClO_4$	−5.62
NaI	−7.41
$NaClO_4$	−5.40
$NaBPh_4$	−10.90
$KClO_4$	1.10
$RbClO_4$	2.02
$CsClO_4$	2.21
Et_4NCl	3.94
Et_4NBr	4.99
Et_4NI	5.53
Et_4NClO_4	3.63
Ph_4AsI	3.87

* G. Choux and R. L. Benoit, *J. Amer. Chem. Soc.*, **91,** 6221 (1969).

APPENDIX 2.4.38

Integral Heats of Solution of Sodium Iodide and Perchlorate in Acetonitrile at Various Concentrations and Temperatures*

m	ΔH_{soln}/kcal mol^{-1}			
	NaI		$NaClO_4$	
	25°C	50°C	25°C	50°C
0.00	−6.875	−6.376	−4.152	−3.740
0.02	−6.807	−6.300	−4.075	−3.650
0.04	−6.743	−6.250	−4.000	−3.560
0.08	−6.615	−6.075	−3.850	−3.380
0.10	−6.550	−6.000	−3.776	−3.295
0.30	−6.010	−5.375	−3.275	−2.860
0.50	−5.635	−4.990	−3.030	−2.575
0.70	−5.360	−4.750	−2.795	−2.312
1.00	−5.040	−4.360	−2.615	−2.080
1.20	−4.870	−4.150	−2.505	−1.910
1.238		−4.110		
1.40	−4.720		−2.400	−1.740
1.60	−4.565		−2.310	−1.570
1.66	−4.520			
1.80			−2.215	−1.400
2.00			−2.120	−1.240
2.12				−1.150
2.50			−1.870	
3.00			−1.650	
3.10			−1.600	

* V. V. Kushchenko and K. P. Mishchenko, *Teor. i Eksp. Khim., Akad. Nauk. Ukr. SSR*, **4**, 403 (1968). See also K. P. Mishchenko and G. M. Poltoratskii, Examination of the Thermodynamics and Structure of Aqueous and Non-aqueous Electrolytic Solutions, Chemical Publisher, Leningrad Division (1968).

APPENDIX 2.4.39

Integral Heats of Solution of NaI in Water-Dioxan Mixtures at 25°C*

m	ΔH_{soln}/kcal mol^{-1}		
	45.95% Dioxan*†	52.55% Dioxan	68.15% Dioxan
0.000	−4.99	−5.44	−6.63
0.025	−4.76	−5.32	−6.49
0.050	−4.60	−5.21	−6.35
0.100	−4.50	−5.01	−6.16
0.150	−4.44	−4.93	−6.07
0.200	−4.41	−4.89	−6.02
0.300	−4.36	−4.85	−5.96
0.400	−4.32	−4.82	−5.93
0.500	−4.30	−4.80	−5.91
0.600	−4.29	−4.78	−5.90
0.700	−4.28	−4.77	−5.89
0.800	−4.275	−4.76	−5.89
0.900	−4.27	−4.75	−5.89
1.000	−4.265	−4.75	−5.89
1.100	−4.26	−4.75	−5.89
1.570	−4.26	−4.75	−5.89
2.653	−4.26	−4.75	
2.955	−4.26		

* K. P. Mishchenko and S. V. Shadskii, *Teor. i Eksp. Khim., Akad. Nauk. Ukr. SSR*, **1**, 60 (1965).

† Data at 2°C are identical to those at 25°C, within limits of experimental error.

APPENDIX 2.4.40

Standard Integral Heats of Solution of Some Tetraalkylammonium Salts in Nitromethane and Nitrobenzene at 20°C*

Electrolyte	ΔH^0_{soln}/kcal mol^{-1}	
	Nitromethane	Nitrobenzene
Me_4NBr	4.12	
Et_4NCl	−0.59	
Et_4NBr	2.59	
Et_4NClO_4	2.43	2.40
Et_4NPic	5.7_8	5.20

* F. A. Askew, E. Bullock, H. T. Smith, R. K. Tinkler, O. Gatty and J. H. Wolfenden, *J. Chem. Soc.*, 1368 (1934).

APPENDIX 2.4.41

Standard Partial Molal Entropies of Electrolytes in Various Solvents at 25°C (cal mol^{-1} K^{-1})

Electrolyte	Methanol*	Ethanol‡	Formic Acid§	Formamide‖	*N*-Methyl-formamide‖	*N,N*-Dimethyl-formamide**	Dimethyl-sulphoxide††	Propylene Carbonate‡‡
HCl	−10.7	−22.1		6.79	−0.4			
	−10.5†							
LiCl	−6.4	−18.1		11.4	2.1¶	−11.0	−6.9	−9.8
LiBr		−13.5			6.9¶	−4.5	−4.2	−4.7
LiI		−7.4					1.0	−4.3
NaF				3.5				
NaCl	0.8		−6.2	12.5	10.3			
	0.9†							
NaBr	6.1	−6.6			13.8	−6.3		
NaI	10.3	−0.3					20.5	7.1
KCl	8.6		−33.94	21.3	15.1		13.6	
	3.3†							
KBr	14.0			26.4		−1.2	12.8	
	9.7†							
KI	20.1	7.2		28.5			23.2	4.7
RbCl	14.3		−17.7	24.9				
CsCl	16.5		−20.3	27.1	21.3			
CsBr						4.2		
CsI			−4.9	41		12.7		
$MgCl_2$	−45.2							
$CaCl_2$	−31.3							
$SrCl_2$	−28.2							
$BaCl_2$	−19.3							
$CoCl_2$	−44.4							
$NiCl_2$	−44.7							

* B. Jakuszewski and S. Taniewska-Osinska, *Lodz. Towarz. Nauk., Wydzial. III, Acta Chim.*, **7,** 31 (1961).

† W. M. Latimer and C. M. Slansky, *J. Amer. Chem. Soc.*, **62,** 2019 (1940).

‡ B. Jakuszewski and S. Taniewska-Osinska, *Lodz. Towarz. Nauk., Wydzial. III, Acta Chim.*, **8,** 11 (1962).

§ G. P. Kotlyarova and E. F. Ivanova, *Zh. Fiz. Khim.*, **40,** 997 (1966). Data for ΔS^{0}_{soln} were combined with the entropies of the crystalline salts from W. M. Latimer, Oxidation Potentials, Prentice-Hall, Englewood Cliffs, N.J. (1952), to obtain $\bar{S}_2{}^{0}$.

|| C. M. Criss, R. P. Held and E. Luksha, *J. Phys. Chem.*, **72,** 2970 (1968). The data have been converted from the mol fraction to the hypothetical one molal standard state.

¶ R. P. Held and C. M. Criss, *J. Phys. Chem.*, **69,** 2611 (1965); R. P. Held, Ph.D. Thesis, University of Vermont, August 1965. The heat data from these sources were re-extrapolated from concentrations $>0.001m$ and combined with free-energy data from E. Luksha and C. M. Criss, *J. Phys. Chem.*, **70,** 1496 (1966).

** C. M. Criss and E. Luksha, *J. Phys. Chem.*, **72,** 2966 (1968). The data for LiCl, LiBr and CsBr have been re-evaluated, using revised heat and solubility data. See text.

†† Calculated from the solubility data of Alexander, *et al.*, *J. Amer. Chem. Soc.*, **89,** 3703 (1967); **90,** 3313 (1968); cell data from Appendix 2.7.16A and heat data from Appendix 2.4.34. Thermodynamic data for the solid phases used in the calculations are from P. Gray and T. C. Waddington, *Proc. Roy. Soc.*, **A235,** 106 (1956); W. M. Latimer, Oxidation Potentials, Prentice-Hall, Englewood Cliffs, N.J. (1952).

‡‡ Calculated from data in Appendicies 2.4.35 and 2.7.18A.

APPENDIX 2.4.42

Standard Partial Molal Ionic Entropies in Water-Methanol Solutions at 25°C (Mol Fraction and Ideal Ionic Gas Standard States) (cal mol^{-1} K^{-1})*

Ion	Wt % Methanol					
	0	10	20	43.1	68.3	100
H^+	39.9	38.8	37.0	35.4	35.8	47.2
Li^+	42.3	39.9	40.1	40.4	43.6	49.6
Na^+	34.9	34.3	34.3	35.7	37.4	46.1
$K^+ = Cl^-$	26.1	26.3	26.7	28.6	33.2	40.7
Rb^+	22.8					36.5
Br^-	22.3	22.3	23.0	25.0	29.9	38.0
Cs^+	22.0					35.7
I^-	17.0	16.6	16.9	19.5	22.9	29.0

* F. Franks and D. S. Reid, *J. Phys. Chem.*, **73**, 3152 (1969).

APPENDIX 2.4.43

Standard Partial Molal Volumes of Electrolytes in Various Solvents at 25°C (ml mol^{-1})

Electrolyte	Meth-anol†	Form-amide‖	*N*-Methyl-propion-amide¶	Formic Acid**	Ethylene-diamine††
HCl	−1.5‡				
LiCl	−3.8	19.60			13.5
LiBr	3.5				
NaCl	−3.3	21.10	30.7	15.5	
NaBr		28.00	35.8	20.4*	
NaI	12.3	39.85		31.5	
$NaNO_3$		33.55	39.6		35.5
NaBz			103.6		
HBz			100.2		
KCl	7.3	32.00	35.5	18.5	
KBr	15.2 15.7§	38.90	40.9	23.4	
KI	21.7 20.8§	50.75		34.5	
KNO_3		44.10			
RbCl		35.90		28.3	
RbBr		42.60		33.2*	
RbI		54.65		44.5	
$RbNO_3$		48.25			
CsCl		42.30		33.4	
CsBr		49.30		38.4*	
CsI		61.05		49.5	
$CsNO_3$		54.65			
$AgNO_3$					20.0
HgI_2					112.0
NH_4Cl		37.20			
NH_4Br	20.8	44.05			
NH_4NO_3		49.24			
Me_4NCl	83.0				
Et_4NCl	140.7				
Et_4NBr	148.0				
Pr_4NBr	220.0				
Bu_4NBr	286.2				
Bu_4NI					325.0

APPENDIX 2.4.43 (*contd*)

Electrolyte	Ethanol	*n*-Propanol	*n*-Butanol	*iso*-Butanol	Mono-methyl-amine
HCl	3.0‡	9.6‡			
LiCl	−4.4‡‡	0.1‡‡	2.5‡‡	1.1‡‡	−19.21§§
$NaNO_3$					12.32§§

† Unless otherwise indicated, data are from a compilation of 'best' values by F. J. Millero, unpublished data; see also F. J. Millero, *J. Phys. Chem.*, **73**, 2417 (1969).
‡ J. Sobkowski and S. Minc, *Rocz. Chem.*, **35**, 1127 (1961).
§ J. B. Stark and E. C. Gilbert, *J. Amer. Chem. Soc.*, **59**, 1818 (1937).
|| R. Gopal and R. K. Srivastava, *J. Phys. Chem.*, **66**, 2704 (1962); *J. Ind. Chem. Soc.*, **40**, 99 (1963).
¶ F. J. Millero, *J. Phys. Chem.*, **72**, 3209 (1968).
** V. N. Fesenko, E. F. Ivanova and G. P. Kotlyarova, *Zh. Fiz. Khim.*, **42**, 2667 (1968). Values marked by asterisk are calculated on the basis of the additivity principle.
†† F. C. Schmidt, W. E. Hoffman and W. B. Schaap, *Proc. Indiana Acad. Sci.*, **72**, 127 (1962).
‡‡ W. C. Vosburgh, L. C. Connell and J. A. V. Butler, *J. Chem. Soc.*, 933 (1933).
§§ E. A. Kelso and W. A. Felsing, *J. Amer. Chem. Soc.*, **60**, 1949 (1938).

APPENDIX 2.4.44

Standard Partial Molal Volumes and Experimental Limiting Slopes of $NaNO_3$ in *N*-Methylpropionamide at Various Temperatures*

Temperature (°C)	$\bar{V}_2^0$/ml mol^{-1}	S_V/ml mol^{-1} (mol l^{-1})$^{-\frac{1}{2}}$
15	40.6	−2.7
20	40.3	−3.0
25	39.6	−2.0
30	40.1	−3.2
35	39.9	−2.7
40	39.7	−3.4

* F. J. Millero, *J. Phys. Chem.*, **72**, 3209 (1968).

APPENDIX 2.4.45

Standard Partial Molal Volumes of Some Electrolytes in *N*-Methylacetamide at Various Temperatures*

Electrolyte	$\bar{V}_2^0$/ml mol^{-1}						
	35°	40°C	45°C	50°C	60°C	70°C	80°C
LiCl	20.5	20.2	20.0	19.7	19.2	18.4	17.8
KI	45.9	46.2	46.5	46.7	46.4	45.4	44.8
Et_4NI	175.4	177.1	178.2	179.9	181.9	182.8	183.7
Pr_4NI	247.5	249.2	251.0	252.5	254.8	256.0	256.9
Bu_4NI	322.9	324.6	325.8	327.0	329.0	330.6	331.8
Am_4NI	391.1	393.4	395.6	397.8	402.1	406.2	409.9
Hx_4NI	461.2	463.1	464.7	466.4	468.7	471.1	473.4
Hp_4NI	527.3	530.0	532.4	535.4	539.4	543.5	547.6

* R. Gopal and M. A. Siddiqi, *J. Phys. Chem.*, **73,** 3390 (1969).

APPENDIX 2.4.46

Standard Partial Molal Volume of LiCl and $NaNO_3$ in Monomethylamine at Various Temperatures (ml mol^{-1})*

Temperature/ °C	LiCl	$NaNO_3$
0	−12.36	17.45
10	−15.59	14.80
17.5	−17.51	
18		13.42
25	−19.21	12.32

* E. A. Kelso and W. A. Felsing, *J. Amer. Chem. Soc.*, **60,** 1949 (1938).

APPENDIX 2.4.47
Viscosity Coefficients of Electrolytes in Various Solvents*

Electrolyte	Methanol			N-Methylformamide (e)			Acetonitrile (d)
	$A_\eta/\text{mol}^{-\frac{1}{2}}\,\text{l}^{\frac{1}{2}}$	$B_\eta/\text{mol}^{-1}\,\text{l}$	Ref.	$A_\eta/\text{mol}^{-\frac{1}{2}}\,\text{l}^{\frac{1}{2}}$	$B_\eta/\text{mol}^{-1}\,\text{l}$	$D_\eta/\text{mol}^{-2}\,\text{l}^2$	$B_\eta/\text{mol}^{-1}\,\text{l}$
NaCl				0.0045	0.599	0.17	
NaBr				0.0050	0.567	0.25	
KCl	0.0151	0.7635	(a)	0.0065	0.615	0.14	
KBr	0.0142	0.7396	(a)	0.0039	0.584	0.21	
KI	0.0159	0.6747	(a)				
$CaCl_2$	−0.041	1.44	(b)				
NH_4Cl	0.0183	0.6610	(a)				
Me_4NBr		0.35	(c)				
		0.42	(d)				
Me_4NPic							0.78
Et_4NBr		0.48	(c)				0.69
Et_4NPic							0.85
Pr_4NBr		0.66	(c)				
Pr_4NI							0.71
Pr_4NPic							0.90
Pr_4NBPh_4							1.24
Bu_4NBr		0.85	(c)				0.93
		0.84	(d)				0.74 (c)
Bu_4NI		0.80	(c)				0.87
							0.75 (c)
Bu_4NPic							1.13
Bu_4NBPh_4							1.35

APPENDIX 2.4.47 (*contd*)

Electrolyte	Methanol			*N*-Methylformamide (e)			Acetonitrile (d)
	$A_\eta/\text{mol}^{-\frac{1}{2}}\,\text{l}^{\frac{1}{2}}$	$B_\eta/\text{mol}^{-1}\,\text{l}$	Ref.	$A_\eta/\text{mol}^{-\frac{1}{2}}\,\text{l}^{\frac{1}{2}}$	$B_\eta/\text{mol}^{-1}\,\text{l}$	$D_\eta/\text{mol}^{-2}\,\text{l}^2$	$B_\eta/\text{mol}^{-1}\,\text{l}$
Pr_3NHPic							0.88
Bu_3NHPic							0.99
$ZnCl_2$	−0.085	0.650	(b)				
$CrCl_3 \cdot 6\ H_2O$	−0.045	2.27	(b)				
$MnCl_2 \cdot 4\ H_2O$	−0.018	1.84	(b)				
$CoCl_2 \cdot 6\ H_2O$	0.000	2.19	(b)				
$CoCl_2 \cdot 2\ H_2O$	0.102	1.26	(b)				
$NiCl_2 \cdot 6\ H_2O$	0.000	2.26	(b)				
$NiCl_2 \cdot 2\ H_2O$	0.028	1.66	(b)				
$CuCl_2 \cdot 2\ H_2O$	−0.011	1.09	(b)				
$SnCl_2 \cdot 2\ H_2O$	−0.016	0.557	(b)				
$SnCl_4 \cdot 5\ H_2O$	−0.64	2.20	(b)				

Electrolyte	Ethanol (f)		*N*-Methylpropionamide		Acetone			Dimethylsulphoxide (k)		
	$A_\eta/\text{mol}^{-\frac{1}{2}}\,\text{l}^{\frac{1}{2}}$	$B_\eta/\text{mol}^{-1}\,\text{l}$	$A_\eta/\text{mol}^{-\frac{1}{2}}\,\text{l}^{\frac{1}{2}}$	$B_\eta/\text{mol}^{-1}\,\text{l}$	$A_\eta/\text{mol}^{-\frac{1}{2}}\,\text{l}^{\frac{1}{2}}$	$B_\eta/\text{mol}^{-1}\,\text{l}$	$D_\eta/\text{mol}^{-2}\,\text{l}^2$	$A_\eta/\text{mol}^{-\frac{1}{2}}\,\text{l}^{\frac{1}{2}}$	$B_\eta/\text{mol}^{-1}\,\text{l}$	$D_\eta/\text{mol}^{-2}\,\text{l}^2$
LiCl					0.024 (i) 0.0145 (j)	0.382 (i) 0.476 (j)	 0.04 (j)			
NaI	0.027	1.15								
$NaNO_3$			−0.034 (g)	1.243 (g)						
KCl				1.37 (h)						
CsI								0.012	0.68	0.47

Electrolyte	Ethylene Glycol (l)	Glycerol (l)	Methyl Ethyl Ketone (d)	Nitromethane (d)
	B_η/mol^{-1} l	B_η/mol^{-1} l	B_η/mol^{-1} l	B_η/mol^{-1} l
KI	0.0327	−0.185 (−0.176)		
CsI	−0.080	−0.408		
Bu_4NBr			1.01	0.75

* The data are at 25°C except where indicated.

(a) G. Jones and H. J. Fornwalt, *J. Amer. Chem. Soc.*, **57**, 2041 (1935).

(b) P. A. D. deMaine, E. R. Russell, D. O. Johnston and M. M. deMaine, *J. Chem. Eng. Data*, **8**, 91 (1963). Except for $SnCl_2 \cdot 2\,H_2O$ and $SnCl_4 \cdot 5\,H_2O$, the data are at 20°C.

(c) R. L. Kay, T. Vituccio, C. Zawoyski and D. F. Evans, *J. Phys. Chem.*, **70**, 2336 (1966).

(d) D. F. Tuan and R. M. Fuoss, *J. Phys. Chem.*, **67**, 1343 (1963).

(e) D. Feakins and K. G. Lawrence, *J. Chem. Soc.* (A), 212 (1966).

(f) W. M. Cox and J. H. Wolfenden, *Proc. Roy. Soc.*, **A145**, 475 (1934).

(g) F. J. Millero, *J. Phys. Chem.*, **72**, 3209 (1968).

(h) T. B. Hoover, *J. Phys. Chem.*, **68**, 876 (1964).

(i) G. R. Hood and L. P. Hohlfelder, *J. Phys. Chem.*, **38**, 979 (1934).

(j) Ref. (e), recalculated value from ref. (i).

(k) M. D. Archer and R. P. H. Gasser, *Trans. Faraday. Soc.*, **62**, 3451 (1966). The value of A_η was calculated theoretically.

(l) K. Crickard and J. F. Skinner, *J. Phys. Chem.*, **73**, 2060 (1969). Value in parentheses is calculated from data of H. T. Briscoe and W. T. Rinehart, *J. Phys. Chem.*, **46**, 387 (1942).

APPENDIX 2.4.48

Viscosity Coefficients of $NaNO_3$ and KCl in *N*-Methylpropionamide at Various Temperatures

Temperature (°C)	$NaNO_3$*		KCl†
	A_η/ $mol^{-\frac{1}{2}} l^{\frac{1}{2}}$	B_η/ $mol^{-1} l$	B_η/ $mol^{-\frac{1}{2}} l$
15	−0.021	1.296	
20	−0.039	1.277	1.39
25	−0.034	1.243	1.37
30	−0.049	1.234	1.35
35	−0.029	1.204	1.33
40	−0.069	1.182	1.31

* F. J. Millero, *J. Phys. Chem.*, **72**, 3209 (1968).
† T. B. Hoover, *J. Phys. Chem.*, **68**, 876 (1964).

Chapter 2

Thermodynamic Measurements

Part 2

Electrochemical Measurements

M. Salomon*
National Aeronautics and Space Administration,
Electronics Research Center, Cambridge, Mass.

* Present address: U.S. Army Electronics Command, Fort Monmouth, New Jersey 07703, U.S.A.

2.5.1 Introduction

In a galvanic cell where no current is flowing, the cell reactions are occurring under conditions of thermodynamic reversibility, and at constant temperature and pressure the e.m.f. is equal to the decrease in free energy in the cell reaction.[C1] Since the temperature dependence of the e.m.f. yields the entropy and enthalpy terms, e.m.f. measurements represent an important method for obtaining thermodynamic data.

Of all the solvent systems available for study, aqueous systems have received by far the greatest attention (e.g. see the recent review of Conway[C2]). Relatively little comparable work has been reported for non-aqueous solvents. Many of the theoretical treatments of aqueous systems can be applied to non-aqueous media and within the framework of classical thermodynamics, a great deal of information can be obtained concerning the physical chemistry of these organic systems.

2.5.2 Definitions and Conventions

At constant temperature and pressure, the relations between the e.m.f. E, of a reversible cell, and the free energy, entropy and enthalpy are[C1]

$$\begin{aligned} \Delta G &= -zFE \\ \Delta S &= zF(\partial E/\partial T)_p \\ \Delta H &= -zFE + zFT(\partial E/\partial T)_p \end{aligned} \qquad (2.5.1)$$

where z is the number of equivalents of reactants converted into products and F is the Faraday. Since activities, a, can be expressed as the product of concentration and activity coefficient terms, it follows that the choice of different standard states will lead to different values of ΔG^0 (and E^0). For example, the chemical potential of an ionic species i is independent of the concentration units used

$$\begin{aligned} \mu_i &= \mu_m^0 + RT \ln (m\gamma_i) \\ &= \mu_c^0 + RT \ln (cy_i) \\ &= \mu_N^0 + RT \ln (Nf_i) \end{aligned} \qquad (2.5.2)$$

Here γ_i, y_i and f_i are the molal, molar, and rational activity coefficients corresponding, respectively, to the molality (m), molarity (c), and mole fraction (N). The reference state for any solute is unit concentration and at infinite dilution in a given solvent the activity coefficient approaches unity.

The relations between the molal ($\gamma_\pm$), molar ($y_\pm$), and rational ($f_\pm$) activity coefficients are given by[C3,C4]

$$\gamma_\pm = \frac{c}{md_0} y_\pm$$

$$f_\pm = \gamma_\pm(1 + 0.001\,\nu\,W_A m)$$

$$f_\pm = y_\pm \frac{[d + 0.001c(\nu\,W_A - W_B)]}{d_0} \tag{2.5.3}$$

where ν = number of moles of ions formed by the ionisation of one mole of solute, and

W_A, W_B = molecular weights of solvent and solute respectively,
d, d_0 = densities of solution and solvent respectively.

For mixed solvents the quantity W_A is replaced by

$$\left\{\frac{x}{W_1} + \frac{1-x}{W_2}\right\}^{-1}$$

where x is the weight fraction of solvent 1 and W_1, W_2 the molecular weights of solvents 1 and 2.

The Debye–Hückel theory[C3,C4] gives, strictly, a relation for $f_\pm$ in terms of the ionic strength (I) defined on the molar scale

$$I = \tfrac{1}{2}\sum_i c_i z_i^2$$

$$\log f_\pm = -\frac{A|z_1 z_2|I^{\frac{1}{2}}}{1 + Ba\,I^{\frac{1}{2}}} \tag{2.5.4}$$

where

$$A = \frac{1.8246 \times 10^6}{(T\varepsilon_r)^{\frac{3}{2}}}\ \mathrm{mol^{-\frac{1}{2}}\,l^{\frac{1}{2}}K^{\frac{3}{2}}}$$

$$B = \frac{50.29 \times 10^8}{(T\varepsilon_r)^{\frac{1}{2}}}\ \mathrm{cm^{-1}\,mol^{-\frac{1}{2}}\,l^{\frac{1}{2}}K^{\frac{1}{2}}}$$

and z_1, z_2 are the valences of the ions; a is the ion-size parameter.

Often the distinction between scales is ignored, which is a reasonable approximation for water as a solvent but not for some non-aqueous solvents. Clearly, the reasonableness of the approximation depends on eqns. 2.5.3.

Most e.m.f. work is carried out on a molal basis, when eqns. 2.5.4 and 2.5.3 give, strictly

$$\log \gamma_{\pm} = -\frac{A d_0^{\frac{1}{2}} |z_1 z_2| I^{\frac{1}{2}}}{1 + B d_0^{\frac{1}{2}} a I^{\frac{1}{2}}} - \log (1 + 0.001\, \nu W_A m) \qquad (2.5.5)$$

The last term can be included in empirically added linear terms which take into account specific (non-electrostatic) interactions in solution. Thus eqn. 2.5.5 is often found in the form,[C5] for a 1:1 electrolyte

$$\log \gamma_{\pm} = -\frac{A' m^{\frac{1}{2}}}{1 + B' a m^{\frac{1}{2}}} + Cm \qquad (2.5.6)$$

where $A' = A d_0^{\frac{1}{2}}$, $B' = B d_0^{\frac{1}{2}}$. In writing eqn. 2.5.6 for a 1:1 electrolyte we have emphasised the calculation of ionic strength as $I = \frac{1}{2} \sum_i m_i z_i^2$ on the molal and not the molar scale. Eqn. 2.5.6 is used with C, as well as a, as adjustable parameters in the fitting of e.m.f. data for extrapolation and interpolation purposes.

Other forms of eqn. 2.5.6 are the Guggenheim approximation[C6]

$$\log \gamma_{\pm} = -\frac{A' m^{\frac{1}{2}}}{1 + m^{\frac{1}{2}}} + C' m \qquad (2.5.7)$$

and the Davies equation[C7]

$$\log \gamma_{\pm} = -\frac{A' m^{\frac{1}{2}}}{1 + m^{\frac{1}{2}}} + 0.3\, A' m \qquad (2.5.8)$$

which is an equation, without adjustable parameters, that has been found to fit data for a variety of salts in aqueous solution.

The standard potentials on the molal (E_m), molar (E_c) and rational scales (E_N) are related by

$$E_N^0 = E_m^0 + \frac{2RT}{F} \ln W_A$$

and

$$E_c^0 = E_m^0 + \frac{2RT}{F} \ln d_0 \qquad (2.5.9)$$

2.6 TYPES OF MEASUREMENTS

2.6.1 E.m.f. Measurements in Reversible cells

Throughout these sections, the IUPAC (Stockholm) conventions for cells and electrode potentials are used.

The cell (I), is a cell without transference

$$\text{Li}|\text{LiCl, solvent}|\text{AgCl}|\text{Ag} \qquad \text{(I)}$$

and the e.m.f. is given by

$$E = E_m^0 - \frac{2RT}{F}\ln 10\,(\log m + \log \gamma_\pm)$$

The standard potential (E_m^0) may be evaluated from the e.m.f. measurements as a function of m by extrapolation methods using a form of the Debye–Hückel relation (eqn. 2.5.6).

Other cells without transference such as redox cells follow the same general principles as discussed above and further details are given elsewhere.[C8] Some description of concentration cells with and without transference is necessary, however, since measurements of this type are often made in organic solvents. A concentration cell without transference is illustrated by cell (II) which is a combination of two cells of type I

$$\text{Li}|\text{Li}^+, \text{Cl}^-, S_1|\text{AgCl}|\text{Ag}{-}{-}{-}\text{Ag}|\text{AgCl}|\text{Cl}^-, \text{Li}^+, S_2|\text{Li} \qquad \text{(II)}$$

and the e.m.f. of this (double) cell without transference is given by

$$\Delta E = E_2^0 - E_1^0 - \frac{2RT}{F}\ln\frac{(a_\pm)_2}{(a_\pm)_1} \tag{2.6.1}$$

In the case where the two solvents S_1 and S_2 are identical, the term $E_2^0 - E_1^0$ is zero and the mean ionic activity $(a_\pm)_2$ may be evaluated from the measured ΔE if $(a_\pm)_1$ is known. If S_1 and S_2 are two different solvents (e.g. H_2O and an organic solvent), then $E_2^0 - E_1^0$ is not zero, but is related to the standard free energy of transfer of LiCl from S_2 to S_1; $(a_\pm)_2$ may then be evaluated from the observed ΔE only if E_2^0, E_1^0, and $(a_\pm)_1$ are known.

Consider the cell (III)

$$\text{Li}|\text{Li}^+, \text{Cl}^-, S_1\vdots\text{Cl}^-, \text{Li}^+, S_2|\text{Li} \qquad \text{(III)}$$

where the dotted line represents a liquid junction. If solvents S_1 and S_2 are identical, then the e.m.f. of this cell may be written as[C8,C9]

$$E = \frac{-RT}{F}\ln\frac{(a_\pm)_2}{(a_\pm)_1} + E_l \tag{2.6.2}$$

where the liquid junction potential is given by

$$E_{l_j} = (1-2t_-)\frac{RT}{F}\ln\frac{(a_\pm)_2}{(a_\pm)_1} \tag{2.6.3}$$

(t_- is the transference number of the anion). By proper choice of the anion and/or supporting electrolyte so that $t_- \approx 0.5$, the liquid junction potential may be made effectively zero. However, a serious problem arises for those situations where the solvents S_1 and S_2 are different. First, the relation for E_{l_j} is not given by eqn. 2.6.3 but a new, and significantly more complex, relation is required involving t_- in each solvent. A given anion, say Cl^-, need not necessarily have equal t_- values in two very different solvents. Also since t_- will vary as a function of concentration, the experimental attainment of zero liquid junction potential is highly questionable (even when S_1 and S_2 are identical). Another point is concerned with the replacement of the electrode on the RHS of cell (III) with, say, a saturated calomel electrode. In this case, a theoretical expression for the liquid junction potential will be even more complex because three ions are now involved (Li^+, K^+, Cl^-).

For the case where $S_1 = H_2O$ and $S_2 =$ organic solvent, the e.m.f. of cell (III) contains a term $E_2^0 - E_1^0$ which is related to the standard free energy of transfer of LiCl from the organic solvent to water. The same consideration applies if the reference electrode is reversible to the anion but this must be the same in both solvents if $E_2^0 - E_1^0$ is to have this significance.

Let us now consider cells of the type (IV)

$$\mathrm{Ag}|\mathrm{AgCl}|M_1^+, \mathrm{Cl}^-, \mathrm{H_2O}|M_2{}^+, X^-, S_2|M_2 \qquad \text{(IV)}$$

where M_1^+ and M_2^+ are cations, X^- is an anion and S_2 is the organic solvent. The anion X^- is often not Cl^- since many chlorides may be insoluble in the organic solvent and M_1^+ is not necessarily the same as M_2^+. For cells of this type, even *assuming* that the liquid junction potential is zero, the exact significance of the $E_2^0 - E_1^0$ term (eqn. 2.6.1) is uncertain. For these reasons, thermodynamic data based on cells containing an aqueous-organic solvent liquid junction with more than two types of ions must be regarded with caution.

For cells involving the halogen acids

$$\mathrm{Pt, H_2}|\mathrm{H}X|\mathrm{Ag}X|\mathrm{Ag} \qquad \text{(V)}$$

standard potentials for $X =$ Cl or Br are obtained fairly directly in most (protic) organic solvents. However, solutions of HI in several solvents may be easily oxidised; to avoid this difficulty such cells are often buffered to high pH values. Buffered cells take the general form

$$\mathrm{Pt, H_2}|\mathrm{H}A(m_1), MA(m_2), MX(m_3)|\mathrm{Ag}X|\mathrm{Ag} \qquad \text{(VI)}$$

where HA is a weak acid, M is usually an alkali metal, and X is a halide. The e.m.f. of this cell is given by

$$E = E^0 - \frac{RT}{F}\ln K_a - \frac{RT}{F}\ln\frac{m_{\mathrm{H}A}m_{X^-}}{m_{A^-}} - \frac{RT}{F}\ln\frac{\gamma_{\mathrm{H}A}\gamma_{X^-}}{\gamma_{A^-}} \qquad (2.6.4)$$

where K_a is the thermodynamic dissociation constant for HA. If K_a is known from either conductivity or other e.m.f. measurements, then E^0 can be easily obtained by the standard extrapolation methods; the activity coefficient ratio $\gamma_{X^-}/\gamma_{A^-}$ varies very little as a function of concentration and a plot of

$$E + \frac{RT}{F}\left(\ln K_a + \ln \frac{m_{HA}m_{X^-}}{m_{A^-}}\right)$$

versus ionic strength gives E^0 when extrapolated to zero ionic strength. On the other hand, if E^0 is known, the pK_a's of weak acids can be obtained (see sect. 3.5).

An elegant method for evaluating solubility products, and at the same time eliminating the liquid junction potential, has been tested by Owen and Brinkley[C10] and consists of e.m.f. measurements on the cell

$$\mathrm{Ag}|\mathrm{AgCl}\left|\begin{matrix}\mathrm{KCl}\ xm \\ \mathrm{KNO_3}\ (1-x)m \\ 1\end{matrix}\right|\begin{matrix}\mathrm{KNO_3}m \\ \text{salt} \\ \text{bridge}\end{matrix}\left|\begin{matrix}\mathrm{AgNO_3}\ xm \\ \mathrm{KNO_3}\ (1-x)m \\ 2\end{matrix}\right|\mathrm{AgCl}|\mathrm{Ag} \qquad \text{(VII)}$$

Here m is the total ionic strength and x is the fraction of KCl or $AgNO_3$. The e.m.f. of this cell is given by

$$E = \frac{RT}{F}\ln\frac{a_2(\mathrm{Ag}^+)}{a_1(\mathrm{Ag}^+)} + E_l \qquad (2.6.5)$$

The subscripts denote the compartment being referred to. Substituting for $a_1(\mathrm{Ag}^+)$ from $K_{s0} = a_1(\mathrm{Ag}^+)a_1(\mathrm{Cl}^-)$ we have

$$E = -\frac{RT}{F}\ln K_{s0} + \frac{2RT}{F}\ln xm + \frac{RT}{F}\ln \gamma_2(\mathrm{Ag}^+)\gamma_1(\mathrm{Cl}^-) + E_{l}, \qquad (2.6.6)$$

from which it follows that

$$\lim_{x\to 0}\left[E - \frac{2RT}{F}\ln xm\right] = -\frac{RT}{F}\ln K_{s0} + \left[\frac{RT}{F}\ln \gamma_2(\mathrm{Ag}^+)\gamma_1(\mathrm{Cl}^-)\right]_{x=0} \qquad (2.6.7)$$

The liquid junction potential is eliminated at $x = 0$ and we can now put $\gamma_2(\mathrm{Ag}^+)\cdot\gamma_1(\mathrm{Cl}^-) = \gamma_\pm^2$ where $\gamma_\pm$ can be evaluated from any of the eqns. 2.5.6–2.5.8. Using eqn. 2.5.7 as an example, K_{s0} can be evaluated by repeating the experiment for several values of m and extrapolating the left hand side of eqn. 2.6.8. to $m = 0$ since

$$[E - 2k \log xm]_{x=0} + 2k \frac{A'm^{\frac{1}{2}}}{1 + m^{\frac{1}{2}}} = -k \log K_{s0} + 2kC'm \qquad (2.6.8)$$

where $k = RT \ln 10/F$.

Well-behaved reversible reference electrodes are a prerequisite for good thermodynamic and kinetic measurements. There exist two excellent publications[C11,C12] concerned with reference electrodes in non-aqueous solvents, which should be consulted for details. Some aspects of electrode performance are discussed in Chapter 3. Many of the reference electrodes used in aqueous systems are useful in non-aqueous solvents with some notable exceptions. In many aprotic solvents the mercurous halide electrodes have been found to be unreliable due to the disproportionation reaction[C12] (which occurs to a small extent in water)

$$Hg_2X_2 + X^- \rightleftharpoons HgX_3^- + Hg$$

where X is a halide (Cl, Br, or I).

The glass electrode has found extensive use in non-aqueous solvents and seems to function quite well.[C11,C12] On the other hand silver halide electrodes appear to be soluble in most aprotic solvents but behave reasonably well in protic organic solvents such as the alcohols. Butler[C12] has tabulated a fairly complete listing of solubility and stability constants for the various polynuclear silver halide species in many solvents.

The hydrogen electrode behaves reversibly in many of the protic solvents. Some problems are encountered in the amides where the presence of platinum may lead to catalytic decomposition of the solvent, especially in the presence of HCl.[C13] Feakins and French[C14] report that the platinum-hydrogen gas electrode in acetone-water mixtures functions well on a mechanically roughened platinum surface. The use of platinum-black was found to be deleterious.

2.6.2 Polarography

(a) *Significance of the Half-wave Potential,* $E_{\frac{1}{2}}$

Polarography is a useful method for studying the chemical and electrochemical properties of various solvent systems.[C15] In this section, our interest lies only in the application of this method for the evaluation of thermodynamic properties. As an example, we shall consider first the case of the dropping mercury electrode (d.m.e.) and the reduction of an alkali metal, M, according to

$$M^+ + e^- + Hg = M(Hg) \qquad (2.6.9)$$

The total cell can be written as

$$\text{reference electrode}|M^+(\text{solution})|M(\text{Hg}) \qquad (\text{VIII})$$

If reaction 2.6.9 is diffusion controlled, then by use of the Ilkovic equation[C16] it can easily be shown[C15,C17] that the potential half-way up the polarographic wave, $E_{\frac{1}{2}}$, is given by

$$E_{\frac{1}{2}} = E_a^0 + \frac{RT}{F} \ln a_{Hg} - \frac{RT}{F} \ln \frac{y_a}{y_s} \cdot \frac{k_s}{k_a} \qquad (2.6.10)$$

where y_a and y_s are the activity coefficients of the amalgam and reducible species M^+, respectively, at the surface of the mercury drop and k_s, k_a are constants involving the square root of the diffusion coefficient for ions in solution (D_s) and metal atoms in the amalgam (D_a).

E_a^0 is the standard reversible potential of the amalgam electrode and we can refer this value to the solid electrode by reference to the cell

$$M|M^+|M(\text{Hg}) \qquad (\text{IX})$$

the e.m.f. of which, under standard conditions, is given by

$$E^0(\text{IX}) = E_a^0 - E_M^0 \qquad (2.6.11)$$

Hence (assuming $a_{Hg} = 1$)

$$E_{\frac{1}{2}} = E_m^0 + E^0(\text{IX}) - \frac{RT}{F} \ln \left\{ \frac{y_a}{y_s} \cdot \left(\frac{D_s}{D_a} \right)^{\frac{1}{2}} \right\} \qquad (2.6.12)$$

It should be noted that E^0(IX) is independent of the nature of the solvent. Thus, the half-wave potential offers a method for evaluation of standard electrode potentials if the logarithmic term in eqn. 2.6.12 can be eliminated. It is common practice to carry out polarographic measurements in solutions containing a large excess of supporting electrolyte. Extrapolating $E_{\frac{1}{2}}$ to zero ionic strength, however, does not eliminate the logarithmic term in eqn. 2.6.12. Consequently the reported $E_{\frac{1}{2}}$ values are dependent upon the nature of the supporting electrolyte. They are also dependent on the reference electrode used.

Thermodynamic measurements in non-aqueous solvents are often compared to those in the aqueous system. An important quantity is the free energy of transfer of a pair of ions from water to the organic solvent, ΔG_t^0. Here ΔG_t^0 is defined as (see also sect. 2.11.3 and eqn. 2.6.35).

$$\Delta G_t^0 = \Delta G_{\text{solv}}^0(\text{solvent}) - \Delta G_{\text{solv}}^0(\text{H}_2\text{O}) = -zF[{}^sE^0 - {}^wE^0] \qquad (2.6.13)$$

where ΔG_{solv}^0 is the solvation energy of the pair of ions. From eqn. 2.6.12 we have, for any two solvents in general (assuming the liquid junction potential has been eliminated)

$$(E_{\frac{1}{2}})_2 - (E_{\frac{1}{2}})_1 = E_2^0 - E_1^0 + \frac{RT}{zF} \ln \frac{y_{s1}}{y_{s2}} - \frac{RT}{zF} \ln \left(\frac{D_2}{D_1} \right)^{\frac{1}{2}} \qquad (2.6.14)$$

It is common practice to assume that both logarithmic terms in eqn. 2.6.14 are, essentially, zero. Since the half-wave potential, theoretically, is related to a standard free energy term, it should be possible to determine the corresponding entropy and enthalpy terms from the temperature coefficient of $E_{\frac{1}{2}}$. There are, however, very little data of this type for either organic or aqueous solvent systems.[C15]

It has been suggested that the ratio of the magnetic susceptibility to the ionic (crystal) radius, χ/r_i, is a major factor influencing the shift of $E_{\frac{1}{2}}$ upon changing solvents.[C18]

(b) *Complex Ions*

The polarographic method allows us to calculate equilibrium constants (K) for complexation reactions such as

$$M^{n+} + pX^{-b} = MX_p^{(n-pb)+} \qquad (2.6.15)$$

When a complex ion of a metal is reduced at the dropping mercury electrode, the reaction occurring may conveniently be considered as the sum of the two partial reactions

$$MX_p^{(n-pb)} = M^{n+} + pX^{-b} \qquad (2.6.9)$$

and

$$M^{n+} + ne^- + \text{Hg} = M(\text{Hg}) \qquad (2.6.9)$$

The half-wave potential is given by[C15]

$$E_{\frac{1}{2}} = E_a^0 + \frac{RT}{nF}\ln a_{Hg} + \frac{RT}{nF}\ln\frac{y_{MX}}{y_a}\cdot\frac{k_a}{k_c}\cdot K - \frac{pRT}{nF}\ln c_X y_X \qquad (2.6.16)$$

where k_c is analogous to k_a and corresponds to the complex metal ion whose molar activity coefficient is y_{MX}. c_X is the concentration of the ligand (X^{-b}) and y_X its molar activity coefficient. It is often easier to evaluate K from the difference between the $E_{\frac{1}{2}}$ value of the complex ion and that of corresponding simple metal ion using eqn. 2.6.17.

$$\Delta E_{\frac{1}{2}} = (E_{\frac{1}{2}})_c - (E_{\frac{1}{2}})_s = \frac{RT}{nF}\ln\frac{y_{MX}y_c}{y_s}\cdot\frac{k_s}{k_c}K - \frac{pRT}{nF}\ln c_X y_X \qquad (2.6.17)$$

Here the subscripts c and s refer, respectively, to the complex and simple metal ions. If the total ionic strength of the solutions is kept constant, eqn. 2.6.17 becomes, approximately,

$$\Delta E_{\frac{1}{2}} = \text{const} + \frac{RT}{nF}\ln K - p\frac{RT}{nF}\ln c_X y_X \qquad (2.6.18)$$

and K in eqn. 2.6.18 is obtained from a plot of $\Delta E_{\frac{1}{2}}$ vs. $\ln c_X$ at constant ionic strength. However, the value of K obtained is dependent upon the ionic strength (I) and the true thermodynamic value could be obtained by repeating this experiment for several ionic strengths and extrapolating K to $I = 0$. Allowance must still be made for the $(D_s/D_c)^{\frac{1}{2}}$ factor, and in dilute solutions activity coefficients may be calculated from eqns. 2.5.4–2.5.7. It should be noted that

$$\frac{RT}{nF} \ln K = E_c^0 - E_M^0 \tag{2.6.19}$$

where E_c^0 is the standard e.m.f. for the reaction

$$MX_p^{(n-pb)} + ne^- = M(\text{solid}) + pX^{-b} \tag{2.6.20}$$

Details of the evaluation of stepwise equilibrium constants by this method can be found elsewhere.[C19,C20] The techniques involved are similar to those discussed below for potentiometric titrations.

2.6.3 Stability Equilibria from Potentiometric Titrations

In organic solvents, the possibility of complex formation appears to be enhanced over that in aqueous solutions. Such equilibria are conveniently studied by means of potentiometric titrations. Although most of the published data for organic systems involve silver halides, the method is quite general and is applicable to any system providing that the concentration of at least one ion can be independently monitored, say by e.m.f. measurements. (Since a detailed analysis of this method has been given by Butler[C21] only an outline will be given here.)

Consider the common case of cation complexation where the concentration of M^+ is given from eqn. 2.6.21

$$E = \text{const.} + \frac{RT}{nF} \ln [M^+] \tag{2.6.21}$$

In saturated solutions we have the equilibrium

$$MX(s) \rightleftharpoons M^+ + X^- \qquad K_{s0} = [M^+][X^-]$$

while in both saturated and unsaturated solutions we have

$$M^+ + X^- \rightleftharpoons MX(\text{solution}) \qquad [MX] = K_1[M^+][X^-]$$

$$MX + X^- \rightleftharpoons MX_2^- \qquad [MX_2^-] = K_2[MX][X^-]$$

$$MX_2^- + X^- \rightleftharpoons NX_3^{2-} \qquad [MX_3^{2-}] = K_3[MX_2^-][X^-]$$

Additional complexes could be considered. Utilising the mass balance requirement, it can be shown that the total concentrations of M and X

in unsaturated solutions, are given by

$$c_M = [M^+]\sum_{n=0}^{3}\beta_n[X^-]^n$$

$$c_X = [X^-] + [M^+]\sum_{n=1}^{3}n\beta_n[X^-]^n \qquad (2.6.22)$$

where $\beta_0 = 1, \beta_1 = K_1, \beta_2 = K_1K_2$, and $\beta_3 = K_1K_2K_3$. $[M^+]$ is obtained experimentally from the e.m.f. measurements and the constants (β_n) can, most easily, be evaluated by a Gaussian non-linear least squares method (cf. C21–23, C24). Once these constants (β_n) have been evaluated, the solubility product K_{s0}, can be obtained from the potentiometric titration curve in the region where solid MX is present; i.e. in saturated solutions we have

$$\{c_X - c_M + [M^+]\}[M^+]^2 = K_{s0}(1 + \beta_2K_{s0})[M^+] + 2\beta_3K_{s0}^3 \qquad (2.6.23)$$

It should be pointed out that these equilibrium constants neglect activity coefficients. To correct for this effect, one can extrapolate the various constants to zero ionic strength by using some form of the Debye–Hückel relations to obtain $y_\pm$ (e.g. see eqn. 2.5.4). Ion-pairing effects, if present, would have to be taken into account.[C22,C23]

2.6.4 Volta Potentials and Real Free Energies

An important quantity required for analysis of ion-solvent interactions and structural properties is the absolute free energy (or enthalpy and entropy) of solvation. Most methods of obtaining these quantities involve some extra-thermodynamic assumption such as the extrapolation of solvation energies versus some function of crystal radii (see sect. 2.11.4). The method based on measurements of volta potential differences avoids the controversy involving the significance of these radii. This method has been used by Frumkin,[C25] Klein and Lange,[C26] Randles[C27] and Parsons *et al.*[C28,C29]

Experimentally one directs a jet of liquid to flow axially down the centre of a tube whose inner surface is covered by a stream of another liquid. An example of such a cell is

$$\underset{\beta}{\text{Hg(jet)}}|N_2|\underset{\delta}{M^+\ (m,\ \text{in solution})}|\underset{\gamma}{M}|\underset{\beta'}{\text{Hg}}$$

where the symbols β, δ, γ and β' denote the phase. A high impedance potentiometer is used to back-off the e.m.f. of this cell until no current is flowing (as indicated by a sensitive galvanometer). Under these conditions the e.m.f. of this cell is given by (cf. Parsons[C30])

$$-FE = \mu_M^\gamma - \alpha_{M^+}^\delta - \alpha_e^\beta \tag{2.6.24}$$

where E is the observed e.m.f., μ is the chemical potential and α is the real free energy ($-\alpha_e^\beta$ is the work function of mercury). If the two phases β and β' are brought into contact, the e.m.f. of the cell is zero and it follows that (cf. ref. C30, page 116)

$$^\delta\Delta^\beta\psi F = \mu_M^\gamma - \alpha_{M^+}^\delta - \alpha_e^\beta \tag{2.6.25}$$

so that E does in fact measure the volta p.d. between the phases β and δ.

Consider now the aqueous cell (X)

$$\text{Hg(jet)}|\text{N}_2|\text{KCl}(0.01m)|\text{Hg}_2\text{Cl}_2|\text{Hg} \tag{X}$$

for which the observed e.m.f. is 0.3290 V[C27] (this value is, of course, the volta p.d. $\Delta\psi$). The absolute single electrode potential (or inner potential) of mercury relative to a chloride solution of molality m_{Cl^-} and chloride ion activity a_{Cl^-} is

$$\Delta\phi(\text{Hg}, m_{\text{Cl}^-}) = \Delta\phi^\circ(\text{Hg}, \text{Cl}^-) - \frac{RT}{F}\ln a_{\text{Cl}^-} \tag{2.6.26}$$

Since $\phi = \psi + \chi$ we have

$$\Delta\psi(\text{Hg}, m_{\text{Cl}^-}) = \Delta\psi^0(\text{Hg}, \text{Cl}^-) + [\chi(m_{\text{Cl}^-}) - \chi(\text{aq})] - \frac{RT}{F}\ln a_{\text{Cl}^-} \tag{2.6.27}$$

Assuming the terms in surface potential for the KCl solution ($\chi(m_{\text{Cl}^-})$) and water ($\chi(\text{aq})$) are equal, then

$$\Delta\psi(\text{Hg}, m_{\text{Cl}^-}) = \Delta\psi^0(\text{Hg}, \text{Cl}^-) - \frac{RT}{F}\ln a_{\text{Cl}^-} \tag{2.6.28}$$

Using the limiting Debye–Hückel law to calculate a_{Cl^-} (i.e. assuming $a_{\text{Cl}^-} \equiv a_\pm$) one finds the standard volta potential difference between mercury and a chloride solution $\Delta\psi^0(\text{Hg}|\text{Cl}^-) = 0.208$ V. Once the value of $\Delta\psi^0$ (Hg|Cl$^-$) is known, we can calculate the corresponding value for any other electrode from the standard e.m.f.'s discussed above. For example, for $\Delta\psi^0(\text{Hg}|\text{Hg}_2^{2+})$ we have

$$\begin{aligned}\Delta\psi^0(\text{Hg}|\text{Hg}_2^{2+}) &= E^0(\text{Hg}|\text{Hg}^{2+}) - E^0(\text{Hg}_2\text{Cl}_2|\text{Hg}) + \Delta\psi^0(\text{Hg}|\text{Cl}^-)\\ &= 0.789 - 0.268 + 0.208 = 0.729\text{ V}\end{aligned}$$

From the cell

$$\text{K}|\text{Hg}|\text{Hg}_2\text{Cl}_2|\text{K}^+, \text{Cl}^-(a = 1)|\text{K}$$

corresponding to the cell reaction

$$K + \tfrac{1}{2}Hg_2Cl_2 = K^+_{aq} + Cl^-_{aq} + Hg \quad \Delta E^0(Hg, Cl^- - K)$$

we have

$$-\alpha^{aq}_{K^+} = \Delta G^0_f(K^+) + F\Delta E^0(Hg, Cl^- - K) - F\Delta\psi^0(Hg, Cl^-) + \alpha^{Hg}_e \tag{2.6.29}$$

where ΔG^0_f (K^+) is the free energy of formation of gaseous K^+ ions from the metal. A value of -80.6 kcal mol^{-1} for the absolute (real) free energy of solvation of the K^+ ion is obtained from this relation.[C27] Unfortunately this method of evaluating thermodynamic properties is not free from criticisms. For example, it is well known that the adsorption of water vapour on Hg can alter both the work function and surface potential. This effect has been discounted by Randles but has been raised again recently.[C31,C32]

An important aspect of this work is that it clearly defines the quantity being measured. The standard (real) free energy of solvation of an ion may be divided into a 'chemical' term, μ^0_i and an electrical term, χ, due to the electrical double layer at the surface; i.e.

$$\alpha^0_i = \mu^0_i + z_iF\chi \tag{2.6.30}$$

The surface potential is immeasurable and, at the present time, not calculable.

Case and Parsons[C28] used the volta potential measurements directly to measure real free energies of transfer of an ion from water to a given solvent, i.e.

$$\Delta\alpha_i = \alpha_i(\text{solvent}) - \alpha_i(H_2O) \tag{2.6.31}$$

Once $\Delta\alpha_i$ is measured, α_i(solvent) can be evaluated from Randles' values for $\alpha_i(H_2O)$. Another method involves the use of eqn. 2.6.30 and is based upon the assumption that the values of μ^0_i for large ions of the same size and shape are nearly equal,[C29] i.e. for monovalent hydrocarbon cations (R^+) and anions (R^-) it is assumed that

$$\lim_{r_i \to \infty} [\alpha_{R^+} - \alpha_{R^-}) \to 2F\chi \tag{2.6.32}$$

From the observed α_i and the extrapolated value of χ, the standard real free energy of solvation is evaluated from

$$\alpha_i = \alpha^0_i + RT \ln a_i \tag{2.6.33}$$

It should also be pointed out that direct measurements of α_i can be compared with those based on the splitting of experimental $\Delta\mu^0_{MX}$

values (or ΔG_t^0 values—see next section). This is easily shown since we have from eqn. 2.6.30

$$\Delta\mu_{MX}^0 = \mu_{M^+}^0 + \mu_{X^-}^0 = \alpha_{M^+}^0 + \alpha_{X^-}^0 \qquad (2.6.34)$$

because $z_{M^+} + z_{X^-} = 0$.

2.6.5 Treatment of Data

Standard free energies, enthalpies and entropies are important because they relate only to solvent-solute interactions. Activity coefficients are of interest since they relate to the departure of the system from ideal conditions.

An important concept in dealing with non-aqueous solutions is the energy of transfer of a mole of salt from water to the solvent under study. (See sect. 2.11.3.) From the standard e.m.f. values, the standard free energy of transfer from water, w, to solvent, s, is defined.

$$\Delta G_t^0 = zF({}^wE^0 - {}^sE^0) \qquad (2.6.35)$$

From the temperature dependence of ΔG_t^0, the corresponding ΔH_t^0 and ΔS_t^0 terms are obtained. Since the ΔG_t^0 terms are additive for ions they can be used to obtain other data from

$$\Delta G^0(s) = \Delta G_t^0 + \Delta G^0(w) \qquad (2.6.36)$$

Here $\Delta G^0(w)$ and $\Delta G^0(s)$ may be either the standard free energies of solution (ΔG_{soln}^0) or solvation (ΔG_{solv}^0) corresponding to the processes

$$MX(\text{solid}) = M_{(\text{soln})}^+ + X_{(\text{soln})}^- \quad \Delta G_{\text{soln}}^0 \qquad (2.6.37)$$

$$M^+(\text{gas}) + X^-(\text{gas}) = M_{(\text{soln})}^+ + X_{(\text{soln})}^- \quad \Delta G_{\text{solv}}^0$$

(cf. 2.11.1 and 2.11.2)

These data may also be evaluated directly from the e.m.f. results; using cell (I) as an example ($\Delta G_{\text{I}}^0 = -FE^0(\text{I})$) we have

$$\Delta G_{\text{soln}}^0 = \Delta G_{\text{I}}^0 + \Delta G_f^0(\text{AgCl}) - \Delta G_f^0(\text{LiCl})$$

$$\Delta G_{\text{solv}}^0 = \Delta G_{\text{soln}}^0 + \Delta G_{\text{lat}}^0(\text{LiCl})$$

$$\Delta G_{\text{lat}}^0\,(\text{LiCl}) = \Delta G_f^0(\text{Li, gas}) + \Delta G_f^0(\text{Cl, gas}) - \Delta G_f^0(\text{LiCl}) + I - A \qquad (2.6.38)$$

In these equations, ΔG_f^0 is the energy of formation of the indicated species, ΔG_{lat}^0 is the crystal lattice energy, I is the ionisation potential of the metal and A is the electron affinity of the halide. In sect. 2.11.2 the calculation of ΔG_{lat}^0 and ΔH_{lat}^0 (at 298.15 K) is discussed and in Appendix 2.11.1 values are given for the alkali metal halides and a few other selected salts.

In a recent review, Strehlow[C33] has pointed out that e.m.f. measurements in organic solvents may involve large systematic errors even when they are very reproducible. For most of the organic solvent systems discussed here, there are usually enough data available from other sources so that an independent check can be made on the e.m.f. results. For example, we can use the heats of solution from calorimetric experiments and heats of transfer to compare with the results of e.m.f. studies. The free energy of solution may be evaluated independently from solubility data (sect. 2.3.1) using the relation (for a 1:1 electrolyte)

$$\Delta G^0_{\text{soln}} = -2RT \ln (m_{\text{sat}}\gamma_\pm) \tag{2.6.39}$$

(cf. **2.3.4** and 2.3.6)

providing $\gamma_\pm$ for the saturated solution is known or calculable from eqns. 2.5.4–2.5.7.

It is often of interest to obtain E^0 values of cells such as (XI)

$$\text{Li}|\text{Li}^+, M^+|M \tag{XI}$$

where M is a metal (such as Ag or Tl) which forms highly insoluble halides. Thus consider cells of Type (I)

$$\text{Li}|\text{Li}^+, \text{Cl}^-(\text{solvent})|\text{AgCl}|\text{Ag} \tag{I}$$

$$\text{Li}|\text{Li}^+, \text{Cl}^-(\text{solvent})|\text{TlCl}|\text{Tl} \tag{I}$$

In aprotic solvents such as PC, DMSO, sulpholane, and DMF, silver chloride appears to be fairly soluble and forms complex ions such as $AgCl_2^-$ whereas thallous chloride does not (see below). From the E^0(I) values in water and in the aprotic solvent, the free energy of transfer (eqn. 2.6.35) gives the value for the transfer of 1 mole of LiCl. Once ΔG^0_t(LiCl) is known, the E^0 value for any cell involving the transfer of LiCl, such as cell (I), can be evaluated from eqn. 2.6.35 since E^0(I) is known when the solvent is water. The standard potential for cell (XI) (i.e. for M = Ag) may then be found from

$$E^0(\text{XI}) = E^0(\text{I}) - \frac{RT}{F} \ln K_{s0} \tag{2.6.40}$$

K_{s0} must be known from independent experiments such as direct solubility determination, potentiometric titrations (sect. 2.6.3) or from e.m.f. measurements on cells similar to the Owen cell (VII).

The chemical potential of an ion i can be written as

$$\mu_i = \mu_{\text{ideal}} + \mu_{\text{non-ideal}} \tag{2.6.41}$$

Differentiating with respect to temperature we have, for the non-ideal (n.i.) part,

$$\frac{d(\mu_{\text{n.i.}}/T)}{d(1/T)} = d(R \ln \gamma_i)/d(1/T) = \Delta H_D \tag{2.6.42}$$

where ΔH_D is the heat of dilution. The reasons for the departure from ideal behaviour can be attributed to ion-ion interactions and solvation effects.[C2–4,C7,C32] According to Guggenheim[C6] the constant C' in eqn. 2.5.7 if negative, may be taken as evidence in support of strong ion-association. In the absence of ion-association, $C' = 0.3A'$ (eqn. 2.5.8) according to Davies.[C7] Also the departure of log $\gamma_\pm$ from the Debye–Hückel relation, eqn. 2.5.5, can be associated with solvation effects.[C4] Hence the $\gamma_\pm$ terms are complex and interpretation is difficult without additional information from studies of partial molal enthalpies, compressibilities, osmotic coefficients, partial molal volumes, and salting-out constants.[C32,C34,C35] In a few cases it is, however, possible to arrive at some approximate values for solvation numbers and ion-association constants if we assume that at low concentrations, solvation effects are negligible and at high concentrations, solvation effects predominate.

In a solvent of molecular weight W_s, a solution containing m moles of solute per kg of solvent contains $(1000/W_s) - hm$ moles of free solvent (h is the solvation number). The true molality, m', is therefore given by

$$m' = m/(1 - 0.001\,W_s hm) \tag{2.6.43}$$

The observed activity coefficient, $\gamma_\pm$, is related to the true activity coefficient $\gamma'_\pm$ by[C4]

$$\log \gamma_\pm = \log \gamma'_\pm - \frac{h}{\nu} \log a_1 - \log\{1 + 0.001\,W_s(\nu - h)m\} \tag{2.6.44}$$

where a_1 is the solvent activity. It is also assumed[C4] that log $\gamma'_\pm$ is given by the Debye–Hückel eqn. 2.5.5.

In the absence of vapour pressure or osmotic coefficient data, the solvent activity can be obtained from the Gibbs-Duhem relation

$$\ln a_1 = -\int \frac{N_2}{1 - N_2}\, \mathrm{d} \ln a_2 \tag{2.6.45}$$

where N_2 is the mole fraction of the solute whose activity is a_2. Taking the ionic radii term a in eqn. 2.5.5 as[C4,C36]

$$a = [3V(h - \Delta)/4\pi]^{\frac{1}{3}} + r_+ + r_- \tag{2.6.46}$$

the average solvation number can be estimated by iteration procedures.[C36] In eqn. 2.6.46, V is the volume of a solvent molecule in Å^3, r_+ and r_- are the crystal radii (in Å) of the cation and anion, respectively, and Δ is an adjustable parameter called the 'penetration distance' by Robinson and Stokes.[C4] When strong ion-association effects are present, the significance of solvation numbers calculated in this manner is unclear.

For the ion-association process

$$M^+_{(\text{soln})} + X^-_{(\text{soln})} \rightleftharpoons (M^+X^-)_{\text{soln}}$$

the equilibrium constant can be written in terms of the fraction of free ions, α, by[C7,C37]

$$K_{\text{assoc}} = (1 - \alpha)/\alpha^2 m \gamma_\pm^2 \qquad (2.6.47)$$

where

$$\alpha = \gamma_\pm/\gamma'_\pm$$

In this equation $\gamma_\pm$ is the (mean) experimental activity coefficient and $\gamma'_\pm$ is that for the free ions, i.e. in the absence of the ion-pairing effect. $\gamma'_\pm$ is usually calculated from one of the Debye–Hückel or extended eqns. 2.5.5–2.5.8.

In reporting the thermodynamic data for various solvent systems, we shall avoid the duplication of data tabulated elsewhere. A great deal of information on alcohols, sugar solutions, glycols, and acetone can be found in tabular form in several places[C3,C4,C33,C38,C39] and emphasis will be on more recent data. Most of the data reported below are based on molal units with a few exceptions which are carefully noted.

The appendices have been split into two sections. Appendix 2.7A contains all thermodynamic data except the activity coefficients which are to be found in Appendix 2.7B.

2.7 EXPERIMENTAL RESULTS

2.7.1 Protic Solvents

Methanol (MeOH) and ethanol (EtOH) have for many years been the most popular of all protic solvents with the exception of water. The cell (Va)

$$\text{Pt}|\text{H}_2|\text{HCl} - S|\text{AgCl}|\text{Ag} \qquad \text{(Va)}$$

has received most attention in these solvents. This cell has been studied by a number of investigators[C40–C43] and the accepted value of E^0_m in methanol is -0.0101 V at 25°C. The temperature dependence of E^0_m has been determined by Austin *et al.*[C42] and is

$$E^0_m(t) = -0.0103 - 1.208 \cdot 10^{-3}(t - 25) - 4.00 \cdot 10^{-6}(t - 25)^2$$

where t is the temperature in °C. It should be noted that the value of E^0_m of 0.0711 V quoted by Buckley and Hartley[C41] is in error and their values for the solubility products of the silver halides (see below) may also involve some error. The activity coefficient of HCl in MeOH can be found in several reference works.[C3,C4,C38,C39]

In ethanol solutions, the E^0_m value of cell (Va) has proved to be less reproducible as discussed by Strehlow.[C33] In Table 1 of Strehlow's chapter, significant variation in E_m^0 is evident depending upon the source

Table 2.7.1

E_m^0/V	Investigators	Year	Ref.
−0.0883	Woolock and Hartley	1928	C44
−0.0759	MacFarlane and Hartley	1932	C45
−0.0760	Butler and Robertson	1929	C46
+0.02190	Mukherjee	1954	C47
−0.0763	Siekman and Grunwald	1954	C48
−0.08138	Taniguchi and Janz	1957	C49
−0.079	LeBas and Day	1960	C50
−0.0723	Teze and Schaal	1961	C51

of data. From Table 2.7.1 it is seen that the value of −0.076 V appears to be the most consistent. The result of Taniguchi and Janz[C49] should be considered the most accurate since it takes into account incomplete dissociation which is evidently important in ethanol. In this solvent the dissociation constant for HCl is[C49] 0.0113 and for HBr is[C52] 0.0187 mol kg^{-1} at 25°C. In setting up a potential scale in these solvents, each solvent will be considered individually.

(a) *Methanol*

Although the cells

$$\text{Pt, } H_2|HBr - S|AgBr|Ag \qquad \text{(Vb)}$$

$$\text{Pt, } H_2|HI - S|AgI|Ag \qquad \text{(Vc)}$$

have not received as much attention as the chloride cell, (Va) some recent data exist which seem to be fairly accurate. Buckley and Hartley[C41] made some of the earliest measurements on these cells but their E_m^0 values appear to be in error. Alfenaar, DeLigny and Remijnse[C53] made measurements on cells of type (II) and arrive at values of −0.1387 and −0.3176 V for E_m^0 for cells (Vb) and (Vc) respectively. Unfortunately, activity coefficients were not given. McIntyre[C54,C55] has made a detailed study of cell (Vc) in unbuffered solutions and found $E_m^0 = -0.31786$ V, in good agreement with the value found by Alfenaar *et al.* Appendix 2.7.1A lists the results for cell (Vc) and Appendix 2.7.1B gives the activity coefficients of HI.

Cells of the type (VIII)

$$M_X(Hg)|MX - S|AgX, Ag \qquad \text{(VIII)}$$

where M = alkali metal and X = Cl, Br, or I, have been studied for many years.[C41,C56–60] Here the subscript x in M_x denotes the mole fraction of M in the amalgam.

Using the e.m.f's for cell (IX), results for the above cell can be converted to those for the general type (I)

$$M|MX - S|\mathrm{Ag}X|\mathrm{Ag} \tag{I}$$

Since we know the E_m^0 values for the silver-silver halide electrodes versus the standard hydrogen electrode (see above), we can easily set up a series of standard potentials vs. H_2 assuming $E_m^0(H^+|H_2) = 0.0$ V. Recently Scrosati *et al.*[C61] made measurements on double amalgam cells in pure methanol. Their results for the alkali metal chlorides and bromides are in good agreement with others but the iodide systems appear to involve some error since a value of −0.2993 V is obtained for E_m^0 of cell (Vc). Table 2.7.2 lists the standard potentials of the alkali-metal electrodes.

Table 2.7.2

E_m^0 values of the cell H_2 |MX in MeOH |M at 25°C

Electrode	E_m^0/V	Year	Ref.
$Li^+|Li$	−3.045	1935	C58
	−3.0845	1966	C60
	−3.0653	1968	C61
$Na^+|Na$	−2.697*	1927	C56
	−2.728	1929	C41
	−2.7132	1961	C62
	−2.7152	1968	C61
$K^+|K$	−2.912*	1958	C59
	−2.8773	1961	C62
	−2.9362	1966	C60
	−2.9116	1968	C61

* Denotes E_c^0 value.

It should be pointed out that most of the early data of Buckley and Hartley[C41] and Koch[C57] were based on cells with liquid junctions and it is difficult to assess both the accuracy of the estimation of the liquid junction potential and the quality of the experimental work in general. In evaluating the E_m^0 values[C61] the data for methanol-iodide systems were ignored. Appendix 2.7.2B lists the activity coefficients of LiCl and KI in pure MeOH.[C60]

One other study in pure methanol by Kanning and Bowman[C63] involved e.m.f. measurements on the cell

$$\text{Pt, } H_2|H_2SO_4 - S|Hg_2\,SO_4|Hg \qquad \text{(XII)}$$

The results for E_m^0 at 20, 25, 30 and 35°C are, respectively, 0.5443, 0.5392, 0.5351, and 0.5318 V. The mean molal stoichiometric activity coefficients of H_2SO_4 in methanol are given in Appendix 2.7.3B.

There has been considerable work done with mixtures of methanol and water over the years. Studies on cell (Va) in these mixed solvents[C43,C64–69] have been summarised and tabulated in several places.[C3,C4,C38,C39,C70] Data for cell (Vb) in these mixed solvents are scarce.[C53,C71,C72] The data quoted[C72] by Bates[C70] are not available to the present author. Studies on cells (Vc) are reported.[C53–55,C71,C72,C73] In no case involving cells of type (V) was there any indication of significant incomplete dissociation of the halogen acids. The standard e.m.f.'s of cell (Vc) reported by McIntyre[C54] are consistently 2–6 mV more negative than those found by others.[C53,C71,C72,C73] This could possibly be attributed to the oxidation of the solvent by HI at the platinum electrode as suggested by Feakins.[C71,C73] E_m^0 values for the silver halide cells (V) are given in Appendix 2.7.2A, and in addition data for the calomel cells analogous to cell (V)

$$\text{Pt,}H_2|HX - H_2O\,(1 - x), \text{MeOH}(x)|Hg_2X_2|Hg \qquad (V')$$

are given.[C74,C75] Here x is the wt% of methanol and X = Cl, Br or I. The thermodynamic properties of the HBr cell (V′)[C75] are summarised in Appendix 2.7.3A and the activity coefficients are given in Appendix 2.7.4B. The activity coefficients for HI in these mixtures are given in Appendix 2.7.5B and the thermodynamic quantities derived from cell (Vc) are given in Appendix 2.7.4A (data from ref. C54). The activity coefficients of Li, Na, and K-chlorides[C76] have been tabulated by Parsons.[C39]

Feakins *et al.*[C77] used the Owen cell (VII) to obtain the solubility product of AgCl in 10 and 43.12% MeOH solutions. The standard (molal) potentials of the $Ag^+|Ag$ couple were calculated from eqn. 2.6.40 and the results are shown in Table 2.7.3.

Similar data for pure methanol do not exist. The data of Buckley and Hartley[C41] which were suggested above to involve some error, give a value of 0.764 V for the $Ag^+|Ag$ couple in pure methanol, which seems to be too low.

(b) *Ethanol*

The standard e.m.f. of cell (Va) in pure ethanol has been discussed above. The thermodynamic quantities for this cell in various EtOH-H_2O mixtures[C46] are given in Appendix 2.7.5A. The activity coefficients of

Table 2.7.3

wt% MeOH	pK_{s0}(AgCl)	E_m^0/V
0.00	9.747	0.7991
10.00	9.963	0.8047
43.12	10.780	0.8334

HCl in these mixtures[C46–48,C78–80] are given elsewhere[C3,C4,C38,C39]. Cell (Vb) has been studied by Mukherjee[C81] and by Nunez and Day.[C52] Mukherjee[C81] also reported results for cell (Vc) but there appears to be a serious systematic error in these studies. Popovych *et al.*[C82] have reported measurements on the cell of type (VIII)

$$\text{K(Hg)}|\text{KCl} - S|\text{AgCl}|\text{Ag} \qquad \text{(VIII)}$$

where S has been varied from 15 to 100% EtOH at 25°C. Appendix 2.7.6A lists the standard potentials of the K^+|K couple as a function of the wt% of EtOH. Also included in this appendix are the constants for calculating the mean molar activity coefficients from the smoothing equation

$$-\log y_\pm = Ac^{\frac{1}{2}} + Bc$$

where c is the molarity of KCl. Aleksandrov and Shikhova[C83] studied the cell

$$\text{Na(Hg)}|\text{NaI} - \text{EtOH}|\text{AgI}|\text{Ag} \qquad \text{(VIII)}$$

over the temperature range of 20 to 60°C. Their results are given in Appendices 2.7.7A and 2.7.6B. Also listed in Appendix 2.7.6B are the mean molal activity coefficients of NaBr and NaI obtained from measurements on cell (VIII) where M = K or Na and X = Br or I. Table 2.7.4 summarises the data for various cells in EtOH at 25°C.

The values in parentheses in Table 2.7.4 were obtained by combining cells 6 and 7 with cell 2 and the results indicate the uncertainties involved in these standard e.m.f's. Although the data obtained by Izmailov and Ivanova[C84] and Aleksandrov and Shikhova[C83] were not treated in a manner to account for incomplete dissociation of HBr and HI in ethanol, the difference in E^0 values due to this omission should not exceed several mV. The fact that differences in E^0 of the order of 40 mV are observed in Table 2.7.4 suggests that an error is present in one of the papers.[C82–84]

In addition to the discrepancies in E^0 values, there are also large differences in the activity coefficients for NaI as indicated in Appendix

2.7.6B. These differences are large enough to warrant additional measurements on cell (Vc) and of K_{s0} for the silver halides.

Aleksandrov *et al.*[C85] studied cell (Va) in a 25% ethanol—75% benzene mixture at 25°C. The results give $E_m^0 = -0.335$ V and for the transfer of HCl from the EtOH/benzene mixture to pure EtOH, they obtained $\Delta G_t^0 = 6.10$ kcal mol^{-1}. The value of pK_a for HCl in this mixture, was determined as 5.51.

Table 2.7.4

Standard E.m.f.'s of Various Cells in EtOH at 25°C

Cell	E_m^0	Ref.
1. H_2|HCl|AgCl|Ag	−0.08138	C49
2. H_2|HBr|AgBr|Ag	−0.1816	C52
3. H_2|H^+, Li^+|Li	−3.042	C45
4. H_2|H^+, Na^+|Na	−2.657	C45
	(−2.646)	(C84)
5. H_2|H^+, K^+|K	−2.865	C82
	(−2.825)	(C84)
6. K|KBr|AgBr|Ag	2.6317	C84
7. Na|NaBr|AgBr|Ag	2.4635	C84
8. Na|NaI|AgI|Ag	2.2599	C84
9. K|KI|AgI|Ag	2.5925	C83

(c) *Propanols*

The thermodynamic properties of HCl in both *n*- and *i*-propanol have been reviewed by others,[C3,C4,C38,C39] but recently new measurements have been reported[C86] which are summarised in Appendices 2.7.7B and 2.7.8B.

(d) *Ethylene Glycol*

Studies in ethylene glycol-water mixtures[C62,C87–93] and pure ethylene glycol[C91–93] appear to be fairly consistent although some slight differences are apparent in water-rich mixtures in the work of Knight *et al.*[C87,C88] The detailed results of Kundu *et al.*[C91–93] are summarised in Appendices 2.7.8A and 2.7.9B. For any given wt% of glycol in H_2O, the standard e.m.f. for any temperature between 5 and 45°C is given by

$$E_m^0(t) = E_m^0(25^0) - A(t - 25) - B(t - 25)^2$$

Here t is the temperature in °C and the constants A and B are given in Appendix 2.7.8A. This relation is accurate to within ±0.3 mV of the observed value.[C92,C93] Table 2.7.5 lists the standard e.m.f.'s of several cells at 25°C in 100% ethylene glycol.

Table 2.7.5

Standard E.m.f.'s in Ethylene Glycol at 25°C

Cell	E_m^0	Ref.
$H_2\|HCl\|AgCl\|Ag$	0.0235	C92
$H_2\|HBr\|AgBr\|Ag$	−0.098	C93
$H_2\|HI\|AgI\|Ag$	−0.2928	C94
$Na\|NaCl\|AgCl\|Ag$	2.8131	C62
$K\|KCl\|AgCl\|Ag$	3.0298	C62

The e.m.f.'s of the cell (V′); $H_2|HCl|Hg_2Cl_2|Hg$ (c.f. Ref. C90) can be calculated from the tabulated data for cell (Va) in Appendix 2.7.8A from the relation

$$E_m^0(\text{XIII}) = E_m^0(\text{Va}) - E_m^0(\text{VI}) = -0.04594 \text{ V at } 25°\text{C}$$

Here E_m^0(XIII) is the e.m.f. of the cell

$$Hg|Hg_2Cl_2|HCl - S|AgCl|Ag \qquad \text{(XIII)}$$

which is independent of the nature of the solvent S. The data for cell (Va) with propylene or 2,3-butylene glycol-water mixtures[C89] has been reviewed elsewhere[C4], but recently new data have been reported for the standard potentials of silver bromide and iodide electrodes in propylene glycol[C94] and the autoprotolysis constants of this and ethylene glycol.[C95] Cell (XII) has been studied by French and Hussain[C96] and the results can be found in Robinson and Stokes' book.[C4] The earlier available data for cell (Va) in glycerol solutions[C97–99] have been tabulated elsewhere.[C3,C4,C38,C39] Some very recent measurements are tabulated in Appendix 2.7.10B.

(e) *Acetic Acid*

Much of the available data for cell (Va)[C100,C101,C103] and for cell (XII)[C102] have been discussed elsewhere.[C11] The most recent papers on cells of type (V) was published in 1966[C104] in which the actual cells studied were

$$\text{Pt, } H_2|KX(m), HA(x), H_2O(100 - x)|AgX|Ag \qquad \text{(cf. V)}$$

where HA refers to acetic acid and X = Cl, Br, or I. Appendix 2.7.9A lists the results for several cells in the mixed and pure solvents. The value listed for cell (XII) was taken from LaMer and Eichelburger[C102] and differs from that of Hutchinson and Chandlee[C105] who found E_m^0 = 0.181 V. The major reason for this large difference is the fact that LaMer and Eichelburger considered sulphuric acid to be singly ionised in glacial acetic acid, i.e.

$$H_2SO_4 \rightleftharpoons H^+ + HSO_4^-$$

Hills[C11] has reviewed the data for formic acid solutions.

(f) *Other Measurements and Systems*

There are several other electrochemical measurements in protic solvents which are of interest and have not been discussed above. In pure methanol, the data of Buckley and Hartley[C41] were noted to contain some errors and the values of K_{s0} for the silver halides must therefore be questioned. The solubility products of the silver halides have been redetermined more recently by potentiometric titration[C106,C107] and these results are compared in Table 2.7.6.

Table 2.7.6

$-\log K_{s0}$ Values for Silver Halides in MeOH at 25°C

Salt	Ref. C41	Ref. C106	Ref. C107	Ref. C57
AgCl	13.05	13.0	13.1	12.8
AgBr	15.24	15.2	15.2	15.0
AgI	18.22	18.2	18.3	16.4

All these data refer to the molar scale with the exception of Buckley and Hartley's work[C41] which is based on the molal scale. Luehrs *et al.*[C106] made their measurements at 23°C in a constant ionic strength medium of 1.0M $LiClO_4$ and Parker and co-workers[C107] employed 0.01M tetraethylammonium nitrate (Et_4N-NO_3) at 25°C. With the exception of Koch's work,[C57] there seems to be fairly good agreement between the various K_{s0} values and it is of interest to calculate the E_m^0 value for the $Ag^+|Ag$ couple using eqn. 2.6.40.

Combining Buckley and Hartley's values for K_{s0} with the E_m^0 values for cells V (see Appendix 2.7.2A), the average value of $E_m^0(Ag^+|Ag)$ is found to be 0.761 ± 0.002 V. This value is very close to Buckley and Hartley's original of 0.764 V despite the fact that their data for E_m^0 for cells (V) are in error. This could be the result of a cancellation of errors which, apparently, yields K_{s0} values in agreement with those determined more recently. These values may still contain some error since the work of Feakins[C77] indicates that $E^0_m(Ag^+|Ag)$ *increases* from pure water to methanol-containing solutions (cf. Table 2.7.3). Evidently the values of K_{s0} should be redetermined using cell (VII). The values of K_{s0} for the silver halides in EtOH do not lead to a constant $E^0(Ag^+|Ag)$ value and these quantities also require further study.

Recently Manahan and Iwamoto[C108] determined the stability constants of the chloride complexes of Cu(I) and Cu(II) in methanol, ethanol, 2-propanol, 2-butanol, and acetone by observing shifts in $E_{\frac{1}{2}}$ values. Although a constant ionic strength medium of 0.1*M* ($LiClO_4$ + LiCl) was used and ion-association was neglected, the results are of interest since they indicate a small decrease in solvation energy for Cu(I) in going from H_2O to the solvents in the order listed above. The half-wave potential data show that the solvation of Cu(II) is similar to Cu(I). The stability of the chloride complexes of both Cu(I) and Cu(II) increases in the order

$$H_2O < \text{MeOH} < \text{EtOH} < \text{2-propanol} < \text{2-butanol} < \text{acetone}$$

Fultz[C109] has compiled stability constant data for organic-metal ion complexes in aqueous solutions and included some data for 50 and 75 (vol) % EtOH and dioxan in water.

Mention should be made of the work of Kisza,[C110] which has been discussed by Butler.[C12] Kisza determined the standard potentials of several electrodes in fused dimethylamine hydrochloride at 180°C.

2.7.2 The Amides

Within the amides there are examples of both protic and aprotic solvents, but they will be discussed as a distinct group because of their chemical similarities. A recent review on the electrochemistry of the amides[C13] points out disagreements regarding the stability of various reference electrodes. In most of the e.m.f. work on reversible cells in these solvents, a time dependence is usually observed and the cell potentials must be extrapolated back to zero time. The problem may be related to (a) oxidation of the solvent by silver or mercuric ions;[C112] or (b) the solubility of the electrodes $AgX|Ag$ and $HgX|Hg$;[C28,C111] or (c) the removal of HCl by bubbling hydrogen gas through the solution.[C113] At the present time, it is difficult to distinguish between these three factors. However, it seems unlikely that the solubility of AgCl is a contributing factor in formamide or *N*-methylformamide. The solubility of AgCl according to

$$Cl^- + AgCl_{solid} \rightleftharpoons AgCl_2^- \qquad K_{s2} \qquad (2.7.1)$$

is shown in Table 2.7.7 along with the solubility product in various solvents.

Table 2.7.7

Solubility Product Constants in the Amides

Salt		H_2O	F	NMF	DMF	DMA
AgCl	log K_{s0}	−9.8*	−9.4‡, −8.28§	−10.39§	−15.18‖	−14.3‡
	log K_{s2}	−4.4*	−2.12§	−1.47§	1.80‖	2.9‡
AgBr	log K_{s0}	−12.3†	−11.4‡	−11.93**	−15.0‡	−14.5‡
	log K_{s2}	−4.7†			1.6‡	2.4‡
AgI	log K_{s0}	−16.0‡	−14.5‡		−15.8‡, −16.4¶	−14.7‡
	log K_{s2}	−4.8‡			2.0‡, 1.7¶	2.6‡

* C114. † C115. ‡ C107. § C116. ‖ C117. ¶ C118. ** See text below.

It should be pointed out that most of the equilibrium constants listed in this table refer to solutions of constant ionic strength with the exception of Butler's data[C117] which have been corrected to zero ionic strength. The effect of ion association has also been neglected. The data of Pavarov *et al.*[C116] were obtained by radiotracer techniques at 18 ± 2°C.

(a) *Formamide (F)*

Pavlopoulos and Strehlow[C112] chose the Cd|$CdCl_2$ reference electrode as the basis of their work since they believed that the Ag|AgCl electrode was unstable in formamide. They studied cells of the types

$$\left.\begin{array}{ll} \text{Pt, } H_2|\text{HCl} - S|\text{CdCl}_2|\text{Cd} & \text{(cf. V)} \\ \text{Rb(Hg)}|\text{RbCl} - S|\text{CdCl}_2|\text{Cd} & \text{(cf. VIII)} \\ \text{K(Hg)}|\text{KCl} - S|\text{CdCl}_2|\text{Cd} & \text{(cf. VIII)} \end{array}\right\} \text{(XIV)}$$

In addition to these cells, other couples were studied at 18 and 25°C and the results are given in Appendix 2.7.10A. The results for cell (XIV) are not in agreement with those of Broadbank and co-workers,[C119] who studied the cells

$$\text{Pt, } H_2|\text{HCl} - S|\text{AgCl}|\text{Ag} \qquad \text{(Va)}$$

$$\text{Cd}|\text{CdCl}_2|\text{Cl}^-, S|\text{AgCl}|\text{Ag} \qquad \text{(cf. XIII) (XV)}$$

$$\text{Cd(Hg)}|\text{CdCl}_2|\text{Cl}^-, S|\text{AgCl}|\text{Ag} \qquad \text{(XVI)}$$

Cell (Va) has also been studied by Agarwal and Nyak[C113] and by Mandel and Decroly.[C120] De Rossi and co-workers[C121] have reported studies on cell (XV).

Considering cell (Va), agreement between the results of Agarwal and Nyak[C113] and of Broadbank *et al.*[C119] is fairly good as shown in Appendix 2.7.11A. The activity coefficients for HCl in formamide[C113] are given in

Appendix 2.7.11B. Broadbank *et al.* found a value of 0.7603 V for cell XVI and combining this value with 0.0505 V, which is the e.m.f. of the cell[C122] of the general type (IX)

$$\underset{\text{2 phase}}{Cd(Hg)}|Cd^{2+}, \text{aq.}|Cd$$

found the e.m.f. of cell (XV) to be 0.8103 V. De Rossi[C121] obtained a value of 0.8155 V for this cell. Combining the results of cells (Va) and (XV), Broadbank *et al.* obtained values of −0.601, −0.606 and −0.612 V for E^0_m of cell (XIV) at 5, 15 and 25°C, respectively. This last value is considerably different from Strehlow's value of −0.623 V (molal scale) or −0.617 V (molar scale) at 25°C. De Rossi[C121] incorrectly used Strehlow's E^0_c for cell (XIV) with his E^0_m for cell (XV) to obtain E^0_m(Va) = 0.198 V which is fortuitously in good agreement with the accepted value (see Appendix 2.7.11A). The confusion between the molar and molal scales is evident in Strehlow's chapter[C33] where the E^0_m values in his Table 2 are the same as the E^0_c values in the original paper.[C112] Using the K_{s0} data of Pavarov,[C116] the standard potential of the $Ag^+|Ag$ electrode in formamide is 0.6903 V.[C119a] Combining these data with 0.0967 V for E^0_m (Vb) in formamide,[C123] it is found that log K_{s0}(AgBr) = −11.4.

Pavrov *et al.*[C124–6] have reported results on double amalgam cells of the type (VIII)

$$Ag|AgCl|MCl(m, \text{in F})|M(Hg)|MCl(0.1m \text{ in F})|AgCl|Ag$$

where M = Na, K, or Cs. The results of these studies are shown in Table 2.7.8, where

Table 2.7.8

Salt	ΔE_c^0 (25°)		E_m^0 (M vs. AgX|Ag)		E_m^0 ($H^+\|H_2$ vs. M)	
	F	NMF	F	NMF	F	NMF
NaCl	0.0802	0.1259	2.8449	2.8094	−2.6509	−2.601
KCl		0.1260		3.0237		−2.816
CsCl	0.0840	0.1463	3.055	2.999	−2.856	−2.791

$\Delta E^0_c = E^0_c(H_2O) - E^0_c$(F or NMF). The standard e.m.f.'s on the molal scale for cell (I) and the standard potentials of the M^+/M couples were calculated using densities given[C13] elsewhere. In calculating the values listed in Table 2.7.8, we have taken E^0_m(Va) = 0.1990 V which is the average of those found by Agarwal and Nyak and by Broadbank *et al.*

Appendix 2.7.10A summarises these results and includes the E_m^0 for the $Li^+|Li$ electrode which has been calculated from solubility data[C127], i.e. the free energy of solution of LiCl is -8.90 kcal mol^{-1} in formamide, which leads to a value of $\Delta G_t^0(\text{LiCl}) = 1.8$ kcal mol^{-1} (or 0.0887 V) at 25°C. From eqn. 2.6.35 it is found that E_m^0 for the $Li^+|Li$ electrode in formamide is -2.9748 V at 25°C. The activity coefficients for several of the alkali metal halides are given in Appendix 2.7.12B.

Finally, it should be mentioned that De Rossi's results[C121] for cell (XIII) in formamide give a value of -0.0465 V, which is in good agreement with the theoretical e.m.f. of -0.04592 V and indicates that the calomel electrode is reversible in this solvent. Further support is given by very recent work in which a value of -0.0469 V is reported.[C128]

(b) *N-Methylformamide* (*NMF*)

Relatively few e.m.f. studies are available for this solvent.[C124-6,C129-131] Table 2.7.9. summarises the results for cell (I). With the exception of the Cs cell, agreement between the work of Luksha and Criss[C129] and of Povarov *et al.*[C124-6] appears to be fairly good, but the activity coefficients show considerable disagreement. The values for the activity coefficients of these salts have been tabulated in Appendix 2.7.12B. A

Table 2.7.9

E_m^0 Values for Cells of Type (I) in NMF at 25 and 5°C

Cell	Ref. C129 (25°C)	Ref. C124–126 (25°C)	Ref. C130 (5°C)
Li|LiCl|AgCl|Ag	3.1237		
Na|NaCl|AgCl|Ag	2.8067	2.8094	2.8203
K|KCl|AgCl|Ag	3.0212	3.0237	
Cs|CsCl|AgCl|Ag	2.987	2.999	
Na|NaBr|AgBr|Ag	2.7135		

recent paper by Mostkova and Kessler[C130] refers to the data of Luksha and Criss 'as being of dubious reliability' but offers no evidence for this statement. The measurements of Mostkova and Kessler[C130] lead to an E_m^0 value of 2.8203 V for the Na cell at 5°C and the activity coefficients are recorded in Appendix 2.7.13B. It has been observed[C130] that the minimum in the log $\gamma_\pm$ vs. $m^{\frac{1}{2}}$ plot shifts to lower m values when the solvent is changed from H_2O to F to NMF (i.e. in order of increasing dielectric constant). This was interpreted as an increase in ion-association in the order $H_2O < F < NMF$.

The only investigation of cell (Va) in NMF is that of Berardelli and co-workers[C131] who found E_m^0 (Va) in NMF $= 0.208 \pm 0.003$ V at 25°C. The standard potentials for the various couples in NMF given in Appendix 2.7.10A are based on Berardelli's result for cell (Va). The e.m.f. of the $Ag^+|Ag$ couple was calculated from eqn. 2.6.40 using log K_{s0} (AgCl) $= -10.39$.[C116] This value was then used to calculate log K_{s0} (AgBr) which is given in Table 2.7.7.

(c) *N,N-Dimethylformamide* (*DMF*)

The number of e.m.f. measurements in DMF is fewer[C132,C133] than in F and NMF, but combined with solubility data[C134] and the results of potentiometric titration,[C107,C117,C118] a considerable amount of thermodynamic data can be obtained. Kumar and Pantony[C132] investigated cell (Va) in this solvent and Butler and Synnott[C133] investigated the following cell of type (VIII)

$$\text{Li(Hg)}|\text{LiCl} - S|\text{TlCl}|\text{Tl(Hg)} \qquad \text{(VIII)}$$

Since the free energies of solvation of Li in Hg[C135] and Tl in Hg[C136,C137] are known, measurements on the above cell can be directly converted to values for the cell

$$\text{Li}|\text{LiCl} - S|\text{TlCl}|\text{Tl} \qquad \text{(I)}$$

The TlCl|Tl electrode was chosen by Butler and Synnott because the AgCl|Ag electrode is soluble in solutions containing excess chloride (cf. Table 2.7.7 and Appendix 2.7.12A). TlCl is insoluble in DMF and does not form complex ions in the presence of excess Cl^-.[C138] The e.m.f. results for cells of types (Va) and (I) are given in Table 2.7.10. The

Table 2.7.10

Type of cell (25°C)	E_c^0/V	E_m^0/V	ΔH_m^0/ kcal mol^{-1}	ΔS_m^0/ cal mol^{-1} K^{-1}
Va	0.065	0.068		
I		2.4302	-59.5 ± 1.1	-12 ± 4

potentials of cell of type I were found to be time dependent and had to be extrapolated to zero time. It was suggested[C133] that the small amount of

$TlCl_2^-$ that does form is responsible for this time dependence by establishing a diffusion potential (see e.g. C139). Kumar and Pantony used the relation

$$\ln y_{\pm} = -Ac^{\frac{1}{2}} + \beta c \qquad (2.7.2)$$

for their extrapolation to E_c^0 and found $\beta = -6.19$ for HCl. Butler and Synnott used the Guggenheim eqn. 2.5.7 and found $C = 0.136$ for LiCl. These low values suggest significant ion-association which is supported by conductivity[C140] and heat of solution[C141] data. The value of K_{assoc}(LiCl) found by Butler and Synnott is smaller than those found earlier.[C140,C141] Mean activity coefficient values for HCl and LiCl are tabulated in Appendix 2.7.14B.

The e.m.f. studies give the free energy of transfer of LiCl from water to DMF as 1.23 kcal mol^{-1}. From the solubility of LiCl in DMF, eqn. 2.6.39 can be used to calculate $\Delta G^0_{soln} = 1.06$ kcal mol^{-1} which leads to a value of 10.82 kcal mol^{-1} for ΔG_t^0. The difference between the values from the e.m.f. and solubility data (-9.6 kcal mol^{-1}) was attributed to the formation of a strong LiCl-DMF solvate.[C133]

Combining these e.m.f. results with solubility data,[C134] a table of free energies of transfer and e.m.f.'s can be compiled. (Table 2.7.11).

Table 2.7.11

ΔG_t^0 and E_m^0 Values of the Cell $M|MX|AgX|Ag$

MX	$-\Delta G^0_{solv}$(DMF)	ΔG_t^0	$E_m^0(H_2O)$	E_m^0(DMF)
LiCl	196.95	1.23	3.2624	3.2093
LiBr	194.53	−2.57	3.1114	3.2223
LiI	190.19	−7.73	2.8879	3.2231
NaCl	164.20	8.52	2.9365	2.5672
KCl	146.66	8.58	3.1497	2.7774
CsCl	137.79	9.25	3.145	2.743

Converting Butler's value of log K_{s0} for AgCl in DMF[C117] to molal units (log K_{s0}(molal) $= -15.06$), the standard potential for the $Ag^+|Ag$ electrode is calculated from eqn. 2.6.40 using Kumar and Pantony's results for cell (Va). We find $E_m^0(Ag^+|Ag) = 0.959$ V at 25°C which does not agree with Kumar and Pantony's value of 0.852 V for E_c^0. Kumar and Pantony's results for the silver electrode are based on log K_{s0}(AgCl) $= -13.3$ which is probably too large in the light of the values given in Table 2.7.7. Using the present results for the silver electrode, we can now calculate the solubility products of AgBr and AgI in DMF. The calculations yield log $K_{s0} = -14.83$ and -14.81 for AgBr and AgI, respectively, which indicates that the solubility of the silver halides in DMF increases in the order

$$\text{AgCl} < \text{AgBr} \approx \text{AgI}$$

This is opposite to the behaviour in the protic solvents H_2O, MeOH, and EtOH. Similar behaviour is found for the TlX salts in propylene carbonate and dimethylsulphoxide (see below) and is attributed to the increase in stability of the larger anions in passing from water to the aprotic solvent. The data in Table 2.7.11 were combined with the results for cell (Va) to arrive at a set of standard potentials in DMF (Appendix 2.7.10A).

Treatment of the data in this manner naturally prevents the evaluation of activity coefficients, and the standard potentials calculated from solubility data should be verified by precise e.m.f. measurements.

The polarographic behaviour of inorganic salts in these solvents has been reviewed recently by Takahashi[C142] and Mann.[C143]

(d) *N-Methylacetamide* (*NMA*) *and N,N-Dimethylacetamide* (*DMA*)

The only studies carried out in these two solvents have been concerned with cell (Va). Dawson and co-workers[C144,C145] have made a detailed study on this cell with NMA. Using eqn. 2.7.2 for the activity coefficient, standard potentials were obtained in 5° intervals from 35 to 70°C. Although no data are available on the solubility of AgCl in NMA, it is probably safe to assume that K_{s2} is small, as it is for NMF (see Table 2.7.7). However, K_{s2} for AgCl in DMA may, by a similar analogy, be quite large. The thermodynamic data for cell (Va) in NMA are given in Appendix 2.7.13A. Scrosati, Pecci and Pistoia[C146] found E_m^0(Va in DMA) = 0.1500 V at 25°C and the mean molal activity coefficients for $m < 0.06$ were given by

$$\ln \gamma_\pm = -1.526m^{\frac{1}{2}} + 5.410m$$

2.7.3 Aprotic Solvents

(a) *Acetone and Dioxan*

Everett and Rasmussen[C147] studied cell (Va) with acetone and found $E_m^0 = -0.53$ V at 25°C assuming the dissociation constant of HCl to be 10^{-8}. They also calculated $E_m^0(\text{Ag}^+|\text{Ag})$ assuming $K_{s0}(\text{AgCl}) = 4.10^{-18}$ (no reference was given for K_{s0}). More recent work[C106] yields $K_{s0} = 10^{-16.4}$. Other e.m.f. measurements on this cell have employed acetone/H_2O mixtures[C14,C149] and the results have been tabulated elsewhere.[C4,C39,C150] An AgI cell similar to (VII) was studied by Mackor[C151] in several water/acetone mixtures. Mackor's data for the solubility product of AgI are summarised in Table 2.7.12 (25°C).

Table 2.7.12

Mol % Acetone	log K_{s0}*	log K_{s2}*
19.13	−16.37	−0.061
46.4	−17.17	0.474
84.55	−19.71	0.851
100.0	−21.96	1.624

* Molar basis.

Although no work has been carried out in pure dioxan, there are many papers dealing with the thermodynamic properties of HCl in water-dioxan mixtures. The results have been thoroughly reviewed elsewhere.[C3,C4,C38,C39,C70] Some recent work is tabulated in Appendices 2.7.14A and 2.7.15B.

Feakins and Turner[C152] used buffered cells of the type (VI) to obtain the standard potentials for cells (Vb) and (Vc) in 20 and 45 wt % acetone-water mixtures. The results for E_m^0(Vb) at 25°C are 0.05991 and 0.03183 V, respectively, for the 20 and 45 % solutions, and E_m^0(Vc) = −0.15136 and −0.16025 V for the 20 and 45 % solutions, respectively. Although the activity coefficients were not tabulated, they can be calculated from data in the paper.

(b) *Acetonitrile* (*AN*)

Acetonitrile has been a popular solvent as evidenced by its coverage in many reviews.[C11,C12,C33,C143,C153] Our discussion here will therefore be relatively brief.

The basis for accurate activity coefficients and standard potentials are cells without liquid junctions. Studies on cells (XIII)[C154] and

$$\text{Tl(Hg)}|\text{TlCl}|M\text{Cl} - S|\text{AgCl}|\text{Ag}^{\text{C155}} \qquad \text{(cf. XVI)}$$

for both of which the e.m.f. is independent of the nature of the solvent, did not give the theoretical e.m.f. but showed considerable time dependence. There is no doubt that AgCl is soluble in solutions containing excess chloride[C106,C107] and that calomel is both soluble and disproportionates.[C12,C155] Spiegel and Ulich's attempt[C156] to measure activity coefficients for the cell

$$\text{Li(Hg)}|\text{LiCl} - S|\text{AgCl}|\text{Ag} \qquad \text{(cf. XVI)}$$

has therefore been seriously questioned.[C12] Pleskov[C157,C158] made measurements on cells of the type

$$M|MX(0.01M) - S\,\vdots\,\text{AgNO}_3(0.01M) - S|\text{Ag} \qquad \text{(XVII)}$$

(cf. Cell IV)

and the results have been reviewed by others. It appears that most of the work carried out in AN involve cells with liquid junctions. The aqueous calomel[C162,C163] or the $0.01M$ $AgNO_3|Ag$ electrode[C159-161] have proved to be popular reference electrodes. Providing that the liquid junction is well defined, differences in half-wave potentials ($E_{\frac{1}{2}}$) between any two ions, such as Li^+ and Cs^+, should be independent of the type of reference electrode used. This does not appear to be the case, as this difference varies up to 10 mV or more depending upon whether the reference electrode is the saturated calomel, SCE,[C162,C163] the mercury-pool reference electrode,[C164] or the $0.01M$ $AgNO_3|Ag$ electrode[C159] (cf. discussion in ref. C153). Popov and Geske[C159] actually used the $AgCl|Ag$ reference electrode but report their data versus the $AgNO_3|Ag$ electrode. Kratochvil and co-workers[C165] observed the $AgNO_3|Ag$ electrode to be very stable in AN and found $K_{assoc}(AgNO_3) = 10^{1.87}$ mol^{-1} l at 25°C.

Kolthoff and Thomas[C160] measured e.m.f.'s of cells of the type (XVII) where $M|M^+$ represents a redox couple such as *tris*(*o*-phenanthroline) Fe(III)|*tris*(*o*-phenanthroline)Fe(II) couple (abbreviated $Fe(phen)_3^{3+}|Fe(phen)_3^{2+}$), or the ferrocene-ferricinium couple (abbreviated $Fe(C_5H_5)_2^+|Fe(C_5H_5)_2$). Measurements were made in the presence and absence of supporting electrolyte (Et_4NClO_4). E_c' values were obtained from

$$E_{obs} = E_c' + \frac{RT}{F} \ln \frac{[M^{n-1}]}{[M^n]}$$

and the standard potentials were obtained as the intercept of the plot of E_c' vs. I (at zero ionic strength). In addition, E_c^0 for the $H^+|H_2$ electrode (SHE) was determined versus the $AgNO_3|Ag$ electrode using an H_2SO_4-Et_4NHSO_4 buffer. In all cases the liquid junction potential was assumed to be negligible at low concentrations. Assuming that the single electrode potential of the $Fe(phen)_3^{3+}|Fe(phen)_3^{2+}$ couple is the same in water as in AN, the E^0 of the cell

$$\text{Pt, } H_2|HX(\text{in } H_2O)\,\vdots\,ClO_4^-,\ Fe(phen)_3^{3+}|Fe(phen)_3^{2+}(\text{in AN})|\text{Pt}$$

is 1.120 V at 25°C (as for the complete aqueous cell). Thus a basis was established for relating potentials in AN to the SHE in water. An important quantity determined by Kolthoff and Thomas is the e.m.f. of the cell

$$\text{SHE}(\text{in } H_2O)\,\vdots\,AgNO_3(0.01M \text{ in AN})|Ag \qquad \text{(XVIII)}$$

which was found to equal 0.274 V at 25°C.

Coetzee and Campion[C161] recalculated the formal potential for the cell

$$Rb|RbI(a_\pm = 1 \text{ in AN})\,\vdots\,AgNO_3(0.01M \text{ in AN})|Ag$$

using Pleskov's original data[C158] and converted the results to the molal scale. These authors then assumed that the free energy of transfer of Rb^+ is given by the difference in solvation energies calculated from the Born equation

$$\Delta G^0_{\text{solv}} = -\frac{z^2e^2N}{2}\left\{1 - \frac{1}{\varepsilon_r}\right\}\left\{\frac{1}{r_+ + \delta}\right\} \qquad (2.7.3)$$

Here ε_r is the dielectric constant, N the Avogadro number, r_+ the crystal radius of the cation, and δ an adjustable parameter. For water, $\delta = 0.72$ and for AN, $\delta = 0.81$ which was evaluated[C161] by fitting eqn. 2.7.3 to the observed differences in $E_{\frac{1}{2}}$ values between Li^+ and Cs^+; Li^+ and K^+; and Na^+ and K^+. On this basis, Coetzee and Campion[C161] found that the $Rb^+|Rb$ electrode is 0.149 V more positive in AN than in water (see column 3 in Appendix 2.7.15A) and the e.m.f. of cell (XVIII) was estimated to be 0.503 V compared to Kolthoff and Thomas' value of 0.274 V and Strehlow's value[C166] of ~0.24 V. In addition to the cells studied by Coetzee and Campion, the data of Rao and Murthy[C167] and Kratochvil[C168] are also included in Appendix 2.7.15A.

It is difficult to assess the origins of the discrepancies between these investigations since much of the data is based on $E_{\frac{1}{2}}$ values which differ considerably between the various authors. More recently, Takahashi[C169] found the difference in $E_{\frac{1}{2}}$ between Rb^+ and Tl^+ to be 1.66 ± 0.01 V in H_2O-AN mixtures up to 40 mole per cent AN with the implication that this difference should hold in pure AN. Coetzee *et al.*[C163] find a value of 1.71 V for this difference in AN. In addition to these uncertainties, there is also the question of the validity of eqn. 2.7.3 which has no real advantage over the assumption* that $\Delta G^0_t(Rb^+) = 0.0$. There also appears to be some confusion over the quantities actually required in order to relate e.m.f.'s in one solvent to another. Parsons has, on several occasions,[C28,C170] pointed out that the quantity required is the real free energy. A straight-forward application of the Born equation neglects the surface potential effect as well as other important ion-solvent and ion-ion interaction energies.

(c) *Dimethylsulphoxide* (*DMSO*)

DMSO has recently proved to be a popular solvent for polarographic measurements.[C171-3] Much of the polarographic literature has been reviewed by Butler[C174] and by Takahashi.[C142] Burrus[C173] has studied the complex equilibria

$$Cd^{2+} + jCl^- \rightleftharpoons CdCl_j^{(2-j)}$$

* The author is not recommending this procedure.

by recording shifts in $E_{\frac{1}{2}}$ as a function of [Cl$^-$]. At a constant ionic strength of 0.1M $KClO_4$, formation constants were obtained and are summarised in Table 2.7.13.

Table 2.7.13

Formation Constants of Chlorocadmium and Lead Complexes

j	β_j(Cd)	K_j(Cd)	β_j(Pb)
1	$2.5 \cdot 10^3$	$2.5 \cdot 10^3$	$7.58 \cdot 10^3$
2	$3.65 \cdot 10^6$	$1.46 \cdot 10^3$	$2.14 \cdot 10^6$
3	$7.75 \cdot 10^8$	$2.11 \cdot 10^2$	
4	$8.30 \cdot 10^{10}$	$1.07 \cdot 10^2$	
5	$6.1 \cdot 10^{12}$	$7.36 \cdot 10^1$	
6	$9.2 \cdot 10^{14}$	$1.51 \cdot 10^2$	

In Table 2.7.13, β_j is the overall formation constant and K_j is the stepwise formation constant. Although activity coefficients were neglected, the results do indicate that several of these complexes are quite stable in DMSO.

The solubility of silver halides has been reported by several workers.[C106,C107,C175–177] It is well established that these salts are readily soluble in excess halide. The effect of water on the AgCl solubility has been studied by Synnott and Butler,[C177] whose results are given in Table 2.7.14.

Table 2.7.14

Equilibrium Constants for AgCl and $AgCl_2^-$ at 25°C (molar scale)

c_{H_2O}	$\log K_{s0}$	$\log K_{s1}$	$\log K_{s2}$
0.00	−11.78	−3.5	1.45
0.07	−10.234	−5.0	1.391
0.55	−10.179		1.233
5.55	−9.711		0.033
13.18	−9.558		−1.38
41.63	−9.530		−1.94
pure H_2O	−9.42	−6.6	−4.7

It is interesting to note that the value of log K_{s1} (for the solvated AgCl species) is larger in DMSO than in water indicating that the silver ion is more strongly solvated in the non-aqueous solvent.[C106,C177] In addition, the magnitude of K_{s2} for the reaction

$$AgCl_{solid} + Cl^- = AgCl_2^-$$

indicates that $AgCl_2^-$ is solvated more strongly than the Cl^- ion. There seems to be a trend in aprotic solvents indicating that large anionic species are more strongly solvated than smaller, less polarisable ones, which is the reverse of the aqueous behaviour. Parker[C179] has related these effects to the increased rates of S_N2 reactions in aprotic solvents.

Initial e.m.f. studies were concerned with the evaluation of E^0(Va). Kolthoff and Reddy[C180] obtained the formal potential of this cell from measurements with the glass electrode. The formal potential, $E_c^{0\prime}$, was established as -0.392 V at 25°C. Due to the neglect of the liquid junction potential and the fact that $(E - E_c^{0\prime})$/pH was not 0.059 V, but rather 0.070 V, it is difficult to relate this formal potential to E_c^0(Va). According to these authors the Pt, H_2 electrode in HCl-DMSO solutions reacts with the solvent and is therefore unreliable. Morel[C181] made e.m.f. measurements on the cell

$$\text{glass electrode}|\text{HCl}(m), \text{DMSO}(x)\text{-H}_2\text{O}(1 - x)|\text{AgCl}|\text{Ag},$$

and, making the questionable assumption that the potential of the glass electrode is independent of the weight fraction of DMSO up to $x = 0.80$, a series of E_m^0(Va) values were obtained (Appendix 2.7.16A). Kumar and Pantony[C132] made direct measurements on cell (Va) in pure DMSO and the results are also given in this appendix. The mean molar activity coefficients for HCl are given in Appendix 2.7.16B.

The thermodynamic properties of the lithium halides have been reported in detail.[C182–186] In all cases, the E_m^0 values of the cell

$$\text{Li}|\text{Li}X - S|\text{Tl}X|\text{Tl} \qquad \text{(I)}$$

were evaluated using the Guggenheim relation, eqn. 2.5.7. The results for $X =$ Cl, Br, and I are summarised in Appendix 2.7.17A along with the C' values (eqn. 2.5.7) and free energy and entropy data. From this appendix, it is seen that C' for LiCl at 25°C has a value of 0.650 kg mol^{-1} which is close to the value of $0.3A'$ proposed by Davies (eqn. 2.5.8). This is taken as additional evidence for the absence of ion-association.

The β values for LiBr and LiI increase too rapidly to be accounted for simply in terms of an increase in the ionic radius (cf. the term a in eqn. 2.5.4), but is easily accounted for in terms of increasing solvation. This is shown in Fig. 2.7.1, where $\ln \gamma_\pm$ is plotted vs. $m^{\frac{1}{2}}$ for LiCl, LiBr and LiI. The fact that no minimum appears in the LiCl curve up to $m \sim 1$ molal is extremely significant because unlike a 1 molal aqueous solution which contains 55.5 moles of solvent, a 1 molal DMSO solution contains only 12.8 moles of solvent. Thus the absence of a minimum in these

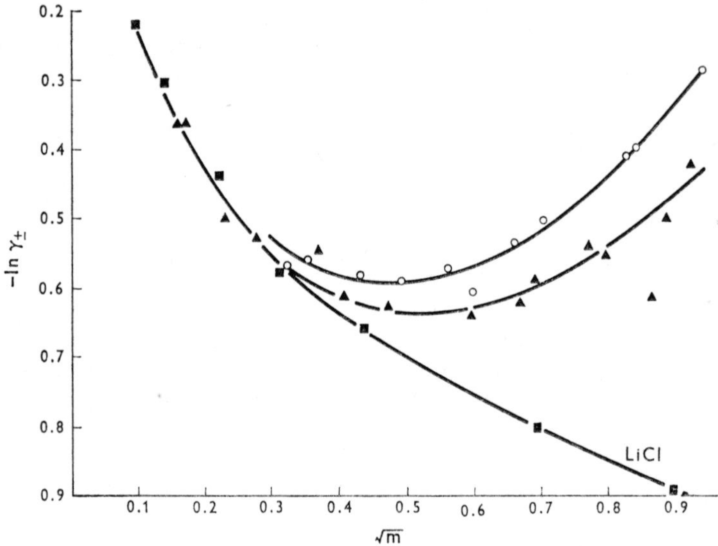

FIG. 2.7.1. Plot of $\ln \gamma_\pm$ vs. $m^{\frac{1}{2}}$ for lithium halides in DMSO at 25°C
■ LiCl
▲ LiBr
○ LiI.

very concentrated solutions supports the view[C177,C179,C185,C186] that the order of increasing solvation in DMSO is

$$Cl^- < Br^- < I^-$$

The experimental activity coefficients for LiCl, LiBr and LiI are given in Appendices 2.7.17B, 18B and 19B. An attempt has also been made to obtain solvation numbers for these salts using eqns. 2.6.43–2.6.46.[C185] The results of these calculations for several values of the 'penetration distance' (cf. eqn. 2.6.46) are shown in Table 2.7.15.

While these results are by no means of quantitative significance, they are reasonable and indicate that these salts are less solvated in DMSO than in water, which is to be expected.[C174,C179]

Table 2.7.15

Solvation Numbers in H_2O and DMSO at 25°C

Salt	H_2O		DMSO	
	$\Delta = 0.1$	$\Delta = 1.0$	$\Delta = 0.1$	$\Delta = 1.0$
LiCl	4.2	4.5	0.2	1.0
LiBr	4.5	5.0	0.2	1.0
LiI	7.5	7.7	1.5	1.8

It is also of interest to look at the equilibria

$$M_h^+ + X_h^- = MX_s$$

where the subscripts h and s refer, respectively, to the solvated and solid phase species. On the basis of solubilities and rate data, Parker[C179] concluded that the degree of solvation follows the order $Cl^- < Br^- < I^-$. The K_{s0} values for AgCl, AgBr and AgI ($10^{-10.4}$, $10^{-10.6}$ and $10^{-11.8}$, respectively, from ref. C106, C107) are not strictly consistant with this concept. However, these values of K_{s0} are not corrected for activity coefficient effects; e.g. compare Butler's result[C177] for log K_{s0} (AgCl) = −11.78 at zero ionic strength. The data for the thallium halides are more dramatic in that log K_{s0} increases markedly in the order of increasing solvation above. These data are given in Appendix 2.7.18A.

Rodewald *et al.*[C187] arrived at an opposite conclusion regarding the solvation of the halides. Their conclusions are based on calorimetric heats of solution and the data are reviewed in Table 2.7.16. Also included in this table are results derived from e.m.f. measurements.[C183,C185,C36]

Table 2.7.16

Salt	ΔH_t^0		$-\Delta H^0_{solv}$ (DMSO)		$-\Delta H^0_{solv}$ (H_2O)	
	Ref. C187	Refs. C183 C185, C36	Ref. C187	Refs. C183 C185, C36	Ref. C187	Ref. C36
LiCl	−2.05	−3.12	214.3	217.9	212.2	214.8
LiBr	−5.43	−5.67	208.4	213.2	203.0	207.5
LiI	−9.07	−8.69	201.2	205.9	192.1	197.2

The differences between the two sets of data are too large to be attributed to differences in the heats of transfer of the salts from water to DMSO. Rodewald *et al.* argue that since the solvation enthalpies in DMSO are less than in water and follow the order LiCl > LiBr > LiI, the degree of solvation is the same as it is in water (contrast[C4]); i.e.

$$Cl^- > Br^- > I^-$$

The difference between the two sets of data is due primarily to differences in ΔH^0_{solv} values and we attribute this to different lattice enthalpies used by Rodewald (cf. ref. (d) in Table I of their paper).

(d) *Propylene Carbonate* (*PC*)

Jasinski[C148] has recently reviewed much of the literature available for this solvent up to 1968. The major emphasis in this section will be on the literature published since Jasinski's review was written.

Butler *et al.*[C177,C188,C189] have made detailed studies on the solubilities of the silver halides in PC and its mixtures with water. The results follow the general pattern of aprotic solvents in that the salt AgX is soluble in excess halide, the predominant species being AgX_2^-. Polynuclear silver species were not found. A summary of these data (molar units) is given in Table 2.7.17.

Table 2.7.17

Equilibrium Constants for Silver Halides in PC at 25°C

	AgCl		AgBr	AgI
	(a)	(b)	(c)	(c)
$\log K_{s0}$	−19.87	−20.18	−20.25	−20.5
$\log K_{s1}$	−4.7	−4.7		
$\log K_{s2}$	1.0	1.0	0.7	1.3
$\log K_{s3}$	3.52	3.2	1.7	

(a) Constant ionic strength of 0.1 M Et_4NClO_4[C177].
(b) Zero ionic strength without ion-pairing[C177].
(c) Constant ionic strength of 0.1 M $LiClO_4$[C189].

These results for AgBr and AgI are in good agreement with those of Courtot-Corpez and L'Her.[C190] The solubility of AgCl has been studied over the temperature range −40 to 80°C[C189] and the results at 25°C are given in Table 2.7.18.

Table 2.7.18

Thermodynamic Quantities (kcal mol^{-1}) for AgCl in PC

Medium	ΔG_2^0	ΔG_{s0}^0	ΔG_{s2}^0	ΔH_2^0	ΔH_{s0}^0	ΔH_{s2}^0
0.1*M* Et_4NClO_4	−28.45	27.10	−1.36			
0.1*M* $LiClO_4$	−24.8	25.2	0.3	−20.1	20.8	0.8

In this table, all units are based on the molar scale and ΔH_2^0 was obtained from the temperature dependence of the overall formation constant β_2. Based on the differences in these quantities with the two supporting electrolytes, the ion-association constant for LiCl was estimated as $10^{2.4}$, which is in good agreement with the value $10^{2.75}$ obtained from conductivity data.[C191] In PC-H_2O mixtures up to $[H_2O] = 3.57$ molar, the various equilibrium constants are given by

$$\log K_{s0} = -20.0 + 3.25[H_2O]^{\frac{1}{2}}$$

$$\log K_{s2} = 1.1 - 1.62[H_2O]^{\frac{1}{2}}$$

$$\log K_{s3} = 3.75 - 4.3[H_2O]^{\frac{1}{2}} \text{ for } H_2O < 0.5M$$

E_m^0 values have been determined[C192–194] for the cells

$$M|MX - \text{PC}|\text{Tl}X|\text{Tl} \qquad \text{(I)}$$

where M = Li, Na and K and X = Cl, Br and I. The thermodynamic properties of these cells are tabulated in Appendix 2.7.19A. The mean molal activity coefficients for several salts are given in Appendices 2.7.20B to 2.7.24B. Figs. 2.7.2 and 2.7.3 are plots of $\ln \gamma_\pm$ vs. $m^{\frac{1}{2}}$ for the Li-halides and alkali metal-iodides, respectively. In Fig. 2.7.2, the Davies curve was calculated from eqn. 2.5.8 where $0.3A' = 0.511$ for PC at 25°C. The C' values (eqn. 2.5.7) for LiCl and LiBr are both negative and this is clearly seen in this figure. This behaviour is undoubtedly due to the formation of ion pairs. Employing eqn. 2.6.47, the association constants were calculated[C193,C194] and the results (molal scale) are given in Table 2.7.19.

The K_{assoc} values in parentheses are based on the conductivity data (molar scale) of Mukherjee and Boden[C191] and are probably more reliable than the e.m.f. values. Also included in Table 2.7.19 are the free energies and enthalpies of solvation calculated from eqns. 2.6.38.[C194] The RbI and CsI data were obtained from solubility data (cf. ref. C195).

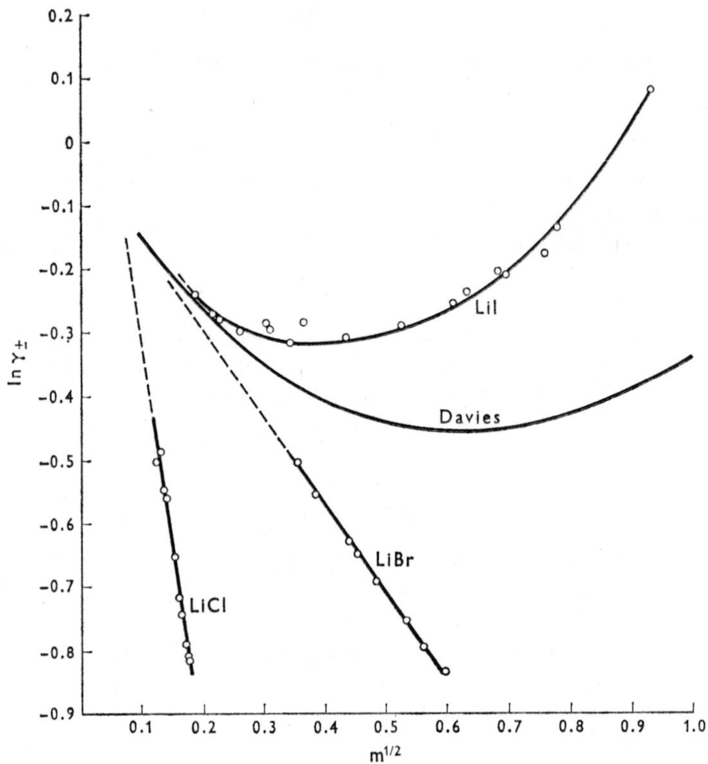

FIG. 2.7.2. Plot of ln $\gamma_{\pm}$ vs. $m^{\frac{1}{2}}$ for the lithium halides in *PC* at 25°C.

Table 2.7.19

Thermodynamic Properties of Alkali Metal Halides at 25°C

Salt	K_{assoc}	$-\Delta G^0_{solv}$ (PC)	$-\Delta H^0_{solv}$ (PC)	$-\Delta G^0_{solv}$ (H_2O)	$-\Delta H^0_{solv}$ (H_2O)
		kcal mol^{-1}		kcal mol^{-1}	
LiCl	50 ± 10 (557)	183.2	207.6	198.2	214.8
LiBr	2.5 ± 1 (19)	179.0	202.6	192.0	207.5
LiI	0.0	173.0	197.4	183.5	197.2
NaI	0.55	150.1	171.7	158.0	169.7
KI	0.0	134.9	155.6	140.5	149.7
RbI		132.5	144.4	136.4	143.2
CsI		130.7	142.8	132.4	135.8

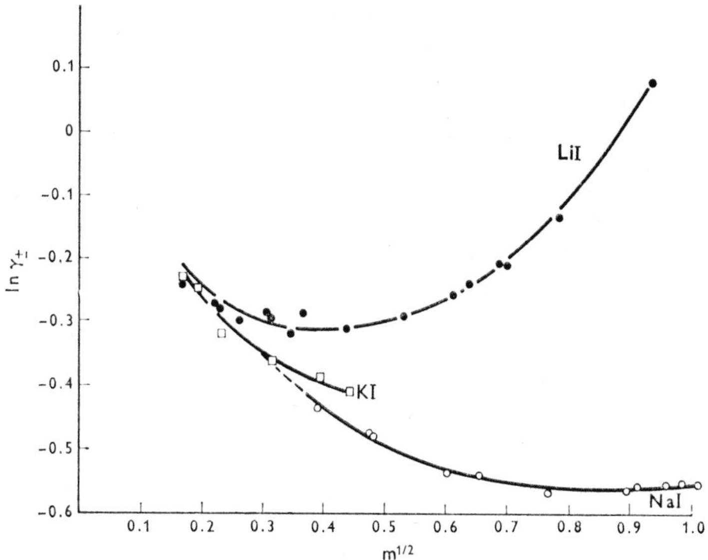

FIG. 2.7.3. Plot of ln $\gamma_\pm$ vs. $m^{\frac{1}{2}}$ for the alkali metal iodides in PC at 25°C.

The ΔH^0_{solv} values for the iodides are seen to be more negative in PC than in water whereas the ΔG^0_{solv} values for all the salts are more negative in water than they are in PC. This is due largely to the entropy of solvation which is usually much more negative in the organic solvent than in water. Table 2.7.20 lists the energies, enthalpies, and entropies of transfer of

Table 2.7.20

Energetics of Transfer of LiCl from H_2O to an Organic Solvent at 25°C (molal scale)

Solvent	ΔG^0_t/ kcal mol^{-1}	ΔH^0_t/ kcal mol^{-1}	ΔS^0_t/ cal mol^{-1} K^{-1}
MeOH	4.11	−2.63	−22.6
EtOH	6.9	−4.08	−36.8
F	0.9	−0.57	−4.9
NMF	3.13	−1.21	−14.6
DMF	1.23	−1.8	−11.0
DMSO	4.87	−3.12	−26.8
PC	14.96	7.14	−26.2

LiCl from water to several organic solvents. The large negative values of ΔS_t^0 suggests that an increase in the structure of the organic solvent accompanies the transfer.[C148] The evidence suggests that most salts which are classified as 'structure breakers' in water[C4] are structure-makers in the organic solvent.

The solubility products for TlBr and TlI given in Appendix 2.7.18A were calculated from eqn. 2.6.40 by first evaluating E_m^0 for the cell $Tl|Tl^+\vdots Li^+|Li$. As found for the DMSO system, the experimental evidence relating to the thermodynamics of the PC system support the concept that the degree of solvation of anions follows the order $Cl^- < Br^- < I^-$.

The stabilities of cation solvation can be qualitatively discussed in terms of relative free energies of transfer, $\delta\Delta_t^0$, where

$$\delta\Delta_t^0 = \Delta G_t^0(MX) - \Delta G_t^0(\mathrm{Na}X) \tag{2.7.4}$$

In eqn. 2.7.4, X is a given anion (e.g. below we take $X = \mathrm{I}$) and the free energy of transfer of the sodium ion is taken as zero. Table 2.7.21 compares these $\delta\Delta G_t^0$ values based on the e.m.f data of cell (I)[C192–195] and the half-wave potentials of the alkali metals versus the aqueous SCE.[C178,C196,C197] The entries for E_m^0 for Rb and Cs were estimated from solubility data using eqn. 2.6.35 and the ΔG_t^0 values based on the $E_{\frac{1}{2}}$ data were calculated from eqns. 2.6.13 and 2.6.14. This latter calculation neglects the activity coefficient terms and the liquid junction potential. The e.m.f. data support the view that the larger cations are more strongly solvated in the organic solvent whereas the polarographic data conflict with each other and, in part, with the e.m.f. data. It is interesting to note that the polarographic experiments give an irreversible wave for Li^+ while the waves for the other alkali metals are reversible.

Table 2.7.21

E_m^0, $E_{\frac{1}{2}}$, and $\delta\Delta G_t^0$ for the Alkali Metals in PC at 25°C*

	E_m^0/V	$E_{\frac{1}{2}}$/V		$\delta\Delta G_t^0$/kcal mol^{-1}		
	Refs.	Refs.		Refs.		
Ion	C192–C195	C178	C196, C197	C192–C195	C178	C196, C197
Li^+	1.8452	2.071	1.99	2.30	2.1	5.1
Na^+	1.6188	1.916	1.96	0.0	0.0	0.0
K^+	1.934	2.020	1.84	−2.36	−1.8	3.5
Rb^+	2.004	2.039	1.97	−4.02	−2.3	0.5
Cs^+	2.097	2.031	1.97	−6.22	−2.7	−0.2

* E_m^0 values are for the TlI/Tl cell I. All data based on $E_{\frac{1}{2}}$ refer to molar units.

(e) *Sulpholane (tetramethylene sulphone, $TMSO_2$)*

Sulpholane has only recently attracted attention and the amount of published data is therefore meagre. Benoit *et al.*[C23] and Della Monica *et al.*[C198] have studied the solubility of AgCl and AgBr at 30°C and their results are given in Table 2.7.22. Indications of strong ion-association effects (K_{assoc}(LiCl) = $10^{3.5}$) were apparent. Ion-association of LiBr was not discussed but by analogy to the PC system, it is probably important.

Table 2.7.22

Thermodynamic Properties of AgCl and AgBr in $TMSO_2$ at 30°C*

	$\log K_{s0}$	$\log K_{s2}$	$\Delta G_t^0(AgX)$/ kcal mol^{-1}
AgCl	−18.1	1.7	11.7
AgBr	−18.4 (−18.2)	1.3 (1.1)	11.9

* Data from Table III of ref. C23; and those in parentheses are derived from ref. C198. ΔG_t^0 values combines 30° $TMSO_2$ data with aqueous, 25°C, data (ref. C23). All units are on the molar scale.

Although e.m.f. data on cells without liquid junctions are not available, several papers have recently been published on the polarographic behaviour of several cations.[C199–202,C186] Half-wave potentials for the alkali metals are given in Table 2.7.23. The results are difficult to compare

Table 2.7.23

Half Wave Potentials (V) in Sulpholane

Ion	C199*	Refs. C200, C201†	C202‡	$\Delta G_t^0(M^+)$§/ kcal mol^{-1}
Li^+		−2.75	−2.67	7.1
Na^+	−1.44	−2.63	−2.56	4.8
K^+	−1.55	−2.71	−2.66	3.2
Rb^+	−1.53		−2.67	2.5
Cs^+			−2.66	2.1

* vs. AgCl|Ag electrode in sat. Et_4NCl solution at 40°C.
† vs. $AgNO_3$ (0.01 *M*)|Ag electrode at 25°C.
‡ vs. $AgNO_3$ (0.01 *M*)|Ag electrode at 30°C.
§ For single ion transfer from water (25°) to $TMSO_2$ (30°), ref. C202.

because of (a) the temperature differences; (b) the unknown differences in the liquid junction potentials; (c) the waves reported in refs. C199 and C200 are more irreversible than those reported in ref. C202. Coetzee *et al.*[C202] made the assumption that the free energy of transfer of Rb^+ from water to $TMSO_2$ can be calculated from eqn. 2.7.3. For $TMSO_2$ the δ term was evaluated from differences in $E_{\frac{1}{2}}$ for the alkali metals (cf. the analogous method applied to AN solutions discussed above). The ΔE_m^0 and corresponding ΔG_t^0 terms (eqn. 2.6.35) were then evaluated by referring the Rb electrode in $TMSO_2$ and in water to the same reference electrode. The aqueous SCE was used for this purpose. Since the $E_{\frac{1}{2}}$ values given in Table 2.7.23 are referred to the aqueous SCE, Coetzee *et al.* added a value of -0.04 V to correct the measured value for the liquid junction potential and the final results for ΔG_t^0 (molal scale) for the individual ions are given in the last column in Table 2.7.23. For the transfer of Ag^+, Coetzee *et al.* found ΔG_t^0(molal) $= -4.8$ kcal mol^{-1}. Based on the assumption that the potential of the ferrocene-ferricinium couple is independent of the nature of the solvent, Benoit's data[C23,C200] yield ΔG_t^0(molar) $= -0.69$ kcal mol^{-1} for the transfer of Ag^+. Parker and Alexander[C203] estimated that ΔG_t^0(molar) $= -1.6$ kcal mol^{-1} for this ion.

(f) *Other Measurements and Systems*

Broadhead and Elving[C204] have recently reported polarographic studies on the alkali metal ions in pyridine. Previous data for these ions date back to 1897 and 1931.

Badoz-Lambling and Bardin[C205] have reported the stability constants of several silver-salts in nitromethane at 20°C. The results, obtained by the potentiometric titration method, are given in Table 2.7.24. From this table it is seen that most AgX_2^- species are stable. No soluble complexes were found for the BrO_3^- and $(C_6H_5)_4B^-$ anions.

Hall *et al.*[C206] evaluated the equilibrium constant for the isotope exchange reaction

$$^7Li_s + {}^6LiBr(soln) \rightleftharpoons {}^6Li_s + {}^7LiBr(soln)$$

Experimentally, they measured the e.m.f. of the cell

$$^7Li|^7LiBr - S|TlBr|Tl(Hg)|TlBr|^6LiBr - S|^6Li$$

In this double cell, $S =$ diglyme ($CH_3OCH_2CH_2OCH_2CH_2OCH_3$) or PC. Measurements in PC were generally less stable than those in diglyme and this was attributed to the instability of the TlBr|Tl(Hg) electrode (but see ref. C192 where this electrode was found to be very stable in PC). For equal concentrations in both sides of the cell, the activity coefficients of ^{7}LiBr and ^{6}LiBr were assumed to be equal and $E_c^0 = 0.84 \pm 0.15$ mV at 23°C (^{7}Li side is negative) in diglyme. This leads to $K = 1.003$. In PC, they gave $E_c^0 = 0.86 \pm 0.06$ mV.

Table 2.7.24

Equilibrium Constants for Silver Salts in Nitromethane at 20°C*

Anion	$\log K_{s0}$	$\log K_{s2}$	$\log \beta_2$
Cl^-	−19.2	0.3	19.5
Br^-	−19.7	0.0	19.7
I^-	−20.5	1.5	22.0
SCN^-	−16.9	−0.4	16.4
$CH_3CO_2^-$	−15.0	0.0	15.0
CN^-	−24.0	10.0	34.0
BrO_3^-	−11.1		
$(C_6H_5)_4B^-$	−13.5		

* Molar units (I = 0.1).

2.7.4 Real Free Energies

An important concept for understanding the physical chemistry of a series of solvents involves relating the E^0 value for a single electrode in one solvent to that in another. To accomplish this, we require knowledge of the real free energies of transfer of ions from one solvent to another. The Born equation (2.7.3) is unacceptable for this purpose. Even if this equation were valid, we would still be left with the problem of the χ potential and the difference in the $zF\chi$ energies between any two solvents (cf. eqn. 2.6.30). The Born equation has been the subject of critical examination[C2,C150,C195,C207] and sect. 2.11. Table 2.7.20 demonstrates the failure of this relation as there is no correlation between ΔG_t^0 and the dielectric constant. For the transfer of ions from water to $F(\varepsilon_r = 109)$ and NMF($\varepsilon_r = 182$), the Born equation requires that ΔG_t^0 is negative which is not observed (e.g. see Table 2.7.20).

Case and Parsons measured the compensation potential between an aqueous solution and an organic solvent[C28] and directly obtained ${}^w\Delta^s\alpha_i$ where

$$ {}^w\Delta^s\alpha_i^0 = {}^s\alpha_i^0 - {}^w\alpha_i^0 $$

Using Randles' results[C27] for the real free energies of ions in water (${}^w\alpha_i^0$), Case and Parsons evaluated the corresponding quantities in several non-aqueous solvents. Appendix 2.7.20A gives the real free energies of solvation of Cl^- and Ag^+ in several solvents.

The results for Ag^+ in acetonitrile indicate that $E_m^0(Ag^+ \text{ in AN}) - E_m^0(Ag^+ \text{ in } H_2O) = 0.438$ V. Referring the $Ag^+|Ag$ electrode in AN to the aqueous SHE, we have $E_m^0(Ag^+|Ag)$ in AN vs. SHE in H_2O $= 1.237$ V at 25°C. The value of this quantity listed in Appendix 2.7.15A is 0.627 V. This latter value suggests that Ag^+ is more strongly solvated in water than in AN which is not believed to be so.[C166,C208,C209]

Case *et al.*[C29] have reported the first measurements of real free energies of solvation of hydrocarbon ions. They define a real differential free energy of solvation for the hydrocarbons as

$$\delta^0_{R^-} = \alpha^0_{R^-} - \alpha^0_{R^0}$$
$$\delta^0_{R^+} = \alpha^0_{R^+} - \alpha^0_{R^0}$$

The differential real potentials, δ^0/F, for several hydrocarbon ions in AN are given in Appendix 2.7.21A. From the last column in this appendix, the surface potential of AN can be obtained (cf. eqn. 2.6.32), i.e.

$$\chi(\text{AN}) = \frac{1}{2F}\{\delta^0_{R^+} - \delta^0_{R^-}\} = -0.10 \pm 0.06 \text{ V}$$

Acknowledgments

The author wishes to thank Drs J. N. Butler and R. Jasinski for making available data prior to publication. Dr Butler has kindly made available to me his extensive reference and reprint collection.

(Manuscript received June 1970, and final revisions incorporated December 1971.)

REFERENCES

C1 J. A. V. Butler, Electrocapillarity, Chemical Publishing Co., N.Y. (1940)
C2 B. E. Conway, *Ann. Rev. Phys. Chem.*, **16,** 481 (1966)
C3 H. S. Harned and B. B. Owen, The Physical Chemistry of Electrolytic Solutions, Reinhold, N.Y. (1958)
C4 R. A. Robinson and R. H. Stokes, Electrolyte Solutions, Butterworths, London (1959)
C5 W. Hückel, *Z. Phys.*, **26,** 93 (1925)
C6 E. A. Guggenheim, *Phil. Mag.*, **19,** 588 (1935)
C7 C. W. Davies, Ion Association, Butterworths, London (1962)
C8 G. Kortüm, Treatise on Electrochemistry, 2nd ed., Elsevier, Amsterdam (1965)
C9 G. W. Castellan, Physical Chemistry, Addison Wesley, Reading, Mass. (1964)
C10 B. B. Owen and S. R. Brinkley, *J. Amer. Chem. Soc.*, **64,** 2171 (1942)
C11 D. J. G. Ives and G. J. Janz, Reference Electrodes, Theory and Practice, Academic Press, N.Y. (1961)
C12 J. N. Butler, Reference Electrodes in Aprotic Organic Solvents, *in* Advances in Electrochemistry and Electrochemical Engineering, Vol. 7, pp. 77–175 (P. Delahay, ed.), Interscience, New York (1970)
C13 D. S. Reid and C. A. Vincent, *J. Electroanal. Chem.*, **18,** 427 (1968)

C14 D. Feakins and C. M. French, *J. Chem. Soc.*, 3168 (1956)
C15 I. M. Kolthoff and J. J. Lingane, Polarography, Vol. I, Interscience, N.Y. (1952); Vol. II (1952)
C16 D. Ilkovic, *Coll. Czech. Chem. Comm.*, **6**, 498 (1934)
C17 J. Heyrovsky and D. Ilkovic, *Coll. Czech. Chem. Comm.*, **7**, 198 (1935)
C18 D. B. Bruss and T. DeVries, *J. Amer. Chem. Soc.*, **78**, 733 (1956)
C19 F. J. C. Rossotti and H. Rossotti, The Determination of Stability Constants, McGraw-Hill, N.Y. (1961)
C20 D. D. DeFord and D. N. Hume, *J. Amer. Chem. Soc.*, **73**, 5321 (1951)
C21 J. N. Butler, Ionic Equilibria, Addison Wesley, Reading, Mass. (1964)
C22 J. N. Butler, *Anal. Chem.*, **39**, 1799 (1967)
C23 R. L. Benoit, A. L. Beauchamp and M. Deneux, *J. Phys. Chem.*, **73**, 3268 (1969)
C24 R.-P. Martin, *Rev. Pure and Appl. Chem.*, **19**, 171 (1969)
C25 A. N. Frumkin, *Z. Phys. Chem.*, **109**, 34 (1924)
C26 O. Klein and E. Lange, *Z. Elektrochem.*, **43**, 570 (1937)
C27 J. E. B. Randles, *Trans. Faraday Soc.*, **52**, 1573 (1956)
C28 B. Case and R. Parsons, *Trans. Faraday Soc.*, **63**, 1224 (1967)
C29 B. Case, N. S. Hush, R. Parsons and M. E. Peover, *J. Electroanal. Chem.*, **10**, 360 (1965)
C30 R. Parsons, Modern Aspects of Electrochemistry, Vol. 1 (J. O'M. Bockris and B. E. Conway, eds.), Butterworths, London (1954)
C31 H. F. Halliwell and S. C. Nyburg, *Trans. Faraday Soc.*, **59**, 1126 (1963)
C32 J. E. Desnoyers and C. Jolicoeur, Modern Aspects of Electrochemistry, Vol. 5 (J. O'M. Bockris and B. E. Conway, eds.), Plenum Press, N.Y. (1969)
C33 H. Strehlow, The Chemistry of Non-aqueous Solvents, Vol. 1 (J. J. Lagowski, ed.), Academic Press, N.Y. (1966)
C34 B. E. Conway and J. O'M. Bockris, Modern Aspects of Electrochemistry, Vol. 1, Ch. 2, Butterworths, London (1954)
C35 B. E. Conway, J. E. Desnoyers and A. C. Smith, *Phil. Trans. Roy. Soc. Ser. A*, **256**, 389 (1964)
C36 M. Salomon, *J. Electrochem. Soc.*, **117**, 325 (1970)
C37 G. H. Nancollas, *Quart. Rev.*, **14**, 402 (1960)
C38 B. E. Conway, Electrochemical Data, Elsevier, Amsterdam (1952)
C39 R. Parsons, Handbook of Electrochemical Constants, Butterworths, London (1959)
C40 G. Nonhebel and H. Hartley, *Phil. Mag.*, **1**, 729 (1925)
C41 P. S. Buckley and H. Hartley, *Phil. Mag.*, **8**, 320 (1929)
C42 J. M. Austin, A. H. Hunt, F. A. Johnston and H. N. Parton, quoted by B. E. Conway, reference C38
C43 I. T. Oiwa, *J. Phys. Chem.*, **60**, 754 (1956)
C44 J. L. Woolock and H. Hartley, *Phil. Mag.*, **5**, 1133 (1928)
C45 A. MacFarlane and H. Hartley, *Phil. Mag.*, **13**, 425 (1932)
C46 J. A. V. Butler and C. M. Robertson, *Proc. Roy. Soc.*, **125A**, 694 (1929)
C47 L. M. Mukherjee, *J. Phys. Chem.*, **58**, 1042 (1954)
C48 E. F. Sieckman and E. Grunwald, *J. Amer. Chem. Soc.*, **76**, 3855 (1954)
C49 H. Taniguchi and G. J. Janz, *J. Phys. Chem.*, **61**, 688 (1957)
C50 C. L. LeBas and M. C. Day, *J. Phys. Chem.*, **64**, 465 (1960)
C51 A. Tèzè and R. Schaal, *Compt. Rend.*, **252**, 3995 (1961)
C52 L. J. Nunez and M. C. Day, *J. Phys. Chem.*, **65**, 164 (1961)
C53 M. Alfenaar, C. L. DeLigny and A. G. Remijnse, *Rec. Trav. Chim.*, **86**, 555 (1967)
C54 J. M. McIntyre, Ph.D. thesis, University of Arkansas, 1968, University Microfilms Number 68–9645

C55 J. M. McIntyre and E. S. Amis, *J. Chem. Eng. Data*, **13**, 371 (1968)
C56 J. H. Wolfenden, C. P. Wright, N. L. Ross Kane and P. S. Buckley, *Trans. Faraday Soc.*, **23**, 491 (1927)
C57 F. K. V. Koch, *J. Chem. Soc.*, 1551 (1930)
C58 H. Hartley and A. MacFarlane, *Phil. Mag.*, **10**, 611 (1935)
C59 K. Brauer and H. Strehlow, *Z. Phys. Chem. (Frankfurt)*, **17**, 346 (1958)
C60 A. M. Shkodin and L. Ya. Shapovalova, *Izv. Vyssh. Ucheb. Zaved Khim. Tekhnol.*, **9**, 563 (1966)
C61 B. Scrosati, S. Schiavo and G. Pecci, *Ric. Sci.*, **38**, 367 (1968)
C62 J. K. Gladden and J. C. Fanning, *J. Phys. Chem.*, **65**, 76 (1961)
C63 E. W. Kanning and M. G. Bowman, *J. Amer. Chem. Soc.*, **68**, 2042 (1946)
C64 H. S. Harned and H. C. Thomas, *J. Amer. Chem. Soc.*, **57**, 1666 (1935); **58**, 761 (1936)
C65 G. Åkerlöf, J. W. Teare and H. Turck, *J. Amer. Chem. Soc.*, **59**, 1916 (1937).
C66 D. Feakins, *J. Chem. Soc.*, 5308 (1961)
C67 R. G. Bates and D. Rosenthal, *J. Phys. Chem.*, **67**, 1088 (1963)
C68 M. Paabo, R. G. Bates and R. A. Robinson, *Anal. Chem.*, **37**, 462 (1965)
C69 L. M. Mukherjee, *J. Amer. Chem. Soc.*, **79**, 4040 (1957)
C70 R. G. Bates *in* Hydrogen-Bonded Solvent Systems, (A. K. Covington and P. Jones, eds.), Taylor and Francis, London (1968)
C71 D. Feakins and P. Watson, *J. Chem. Soc.*, 4686 (1963); 4734 (1963)
C72 R. P. T. Tomkins, Ph.D. thesis, University of London, 1966; quoted by Bates, reference C70
C73 D. Feakins and K. H. Khoo, *J. Chem. Soc. A.*, 361 (1970)
C74 K. Schwabe and S. Ziegenbalg, *Z. Elektrochem.*, **62**, 172 (1958)
C75 K. Schwabe and E. Ferse, *Ber. Bunsenges. Phys. Chem.*, **70**, 849 (1966)
C76 G. Åkerlöf, *J. Amer. Chem. Soc.*, **52**, 2353 (1930)
C77 D. Feakins, K. G. Lawrence and R. P. T. Tomkins, *J. Chem. Soc.*, *A*, 753 (1967)
C78 H. S. Harned and C. Calmon, *J. Amer. Chem. Soc.*, **61**, 1491 (1939)
C79 H. S. Harned and D. S. Allen, *J. Phys. Chem.*, **58**, 191 (1954)
C80 K. Schwabe and M. Kunz, *Z. Elektrochem.*, **64**, 1188 (1960)
C81 L. M. Mukherjee, *J. Phys. Chem.*, **60**, 974 (1956)
C82 A. J. Dill, L. M. Itzkowitz and O. Popovych, *J. Phys. Chem.*, **72**, 4580 (1968)
C83 V. V. Aleksandrov and T. M. Shikhova, *Elektrokhimiya*, **3**, 981 (1968)
C84 N. A. Izmailov and E. F. Ivanova, *Zh. Fiz. Khim.*, **29**, 1614 (1955)
C85 V. V. Aleksandrov, L. K. Osipenko and T. A. Berezhnaya, *Elektrokhimiya*, **4**, 1008 (1968)
C86 K. Schwabe and R. Mueller, *Ber. Bunsenges. Phys. Chem.*, **73**, 178 (1969); **74**, 1248 (1970). R. N. Roy and A. Bothwell, *J. Chem. Eng. Data*, **15**, 548 (1970)
C87 S. B. Knight, J. F. Masi and D. Roesel, *J. Amer. Chem. Soc.*, **68**, 661 (1946)
C88 H. D. Crockford, S. B. Knight and H. A. Staton, *J. Amer. Chem. Soc.*, **72**, 2164 (1950)
C89 B. H. Claussen and C. M. French, *Trans. Faraday Soc.*, **51**, 1124 (1955)
C90 K. Schwabe and R. Hertzsch, *Z. Elektrochem.*, **63**, 445 (1959)
C91 K. K. Kundu and M. N. Das, *J. Chem. Eng. Data*, **9**, 87 (1964)
C92 U. Sen, K. K. Kundu and M. N. Das, *J. Phys. Chem.*, **71**, 3664 (1967)
C93 S. K. Banerjee, K. K. Kundu and M. N. Das, *J. Chem. Soc.*, *A*, 161 (1967)
C94 K. K. Kundu, D. Jana and M. N. Das, *J. Phys. Chem.*, **74**, 2625 (1970)
C95 K. K. Kundu, P. K. Chattopadhyay, D. Jana and M. N. Das, *J. Phys. Chem.*, **74**, 2633 (1970)
C96 C. M. French and Ch. F. Hussain, *J. Chem. Soc.*, 2211 (1955)
C97 W. W. Lucasse, *Z. Phys. Chem.*, **121**, 254 (1926)
C98 H. S. Harned and F. H. M. Nestler, *J. Amer. Chem. Soc.*, **68**, 665 (1946)

C99 S. B. Knight, H. D. Crockford and F. W. James, *J. Phys. Chem.*, **57**, 463 (1953)
C100 B. O. Heston and N. F. Hall, *J. Amer. Chem. Soc.*, **56**, 1462 (1934)
C101 B. B. Owen, *J. Amer. Chem. Soc.*, **54**, 1758 (1932); **57**, 1526 (1935)
C102 V. K. LaMer and W. E. Eichelberger, *J. Amer. Chem. Soc.*, **54**, 2763 (1932)
C103 L. M. Mukherjee, *J. Amer. Chem. Soc.*, **79**, 4040 (1957)
C104 H. P. Bennetto, D. Feakins and D. J. Turner, *J. Chem. Soc.*, *A*, 1211 (1966)
C105 A. W. Hutchison and G. C. Chandlee, *J. Amer. Chem. Soc.*, **53**, 2763 (1932)
C106 D. C. Luehrs, R. T. Iwamoto and J. Kleinberg, *Inorg. Chem.*, **5**, 201 (1966)
C107 R. Alexander, E. C. F. Ko, Y. C. Mac and A. J. Parker, *J. Amer. Chem. Soc.*, **89**, 3703 (1967)
C108 S. E. Manahan and R. T. Iwamoto, *J. Electroanal. Chem.*, **13**, 411 (1967)
C109 E. D. Fultz, Thermodynamic Stability Constants of Organic – Metal Ion Complexes, University of California at Livermore UCRL–50200, Feb. 1967
C110 A. Kisza, *Bull. Acad. Pol. Sci.*, *Ser. Sci. Chim.*, **12**, 707, 713 (1964); **13**, 409, 415 (1965); **14**, 687 (1966); *Z. Phys. Chem.* (*Leipzig*), **237**, 97 (1968)
C111 R. Parsons, quoted in reference C13 (cf. reference 241 of that paper)
C112 T. Pavlopoulos and H. Strehlow, *Z. Phys. Chem.* (*Frankfurt*), **2**, 89 (1954)
C113 R. K. Agarwal and B. Nyak, *J. Phys. Chem.*, **70**, 2568 (1966); **71**, 2062 (1967)
C114 E. Bern and I. Leden, *Svensk. Kem. Tidskr.*, **65**, 88 (1953)
C115 L. G. Sillén and A. E. Martell, Stability Constants of Metal-Ion Complexes, The Chemical Society, London (1954)
C116 Yu. M. Pavarov, V. E. Kazirnov, Yu. M. Kessler and I. A. Gorbanev, *Zh. Neorg. Khim.*, **9**, 1008 (1964)
C117 J. N. Butler, *J. Phys. Chem.*, **72**, 3288 (1968)
C118 H. Chateau and M. C. Moncet, *J. Chim. Phys.*, **60**, 1059 (1963)
C119a R. W. C. Broadbank, S. Dhabanandanu, K. W. Morcom and B. L. Muju, *Trans. Faraday Soc.*, **64**, 3311 (1968)
C119b R. W. C. Broadbank, B. L. Muju and K. W. Morcom, *Trans. Faraday Soc.*, **64** 3318 (1968)
C120 M. Mandel and P. Decroly, *Trans. Faraday Soc.*, **56**, 29 (1960); *Nature*, **182**, 794 (1958)
C121 M. DeRossi, G. Pecci and B. Scrosati, *Ric. Sci.*, **37**, 342 (1967)
C122 N. G. Parks and V. K. LaMer, *J. Amer. Chem. Soc.*, **56**, 90 (1934)
C123 K. W. Morcom and B. L. Muju, *Nature*, **217**, 1046 (1968)
C124 Yu. M. Povarov, Yu. M. Kessler, A. I. Gorbanev and I. V. Safonova, *Dokl. Akad. Nauk. SSSR*, **155**, 172, 1411 (1964)
C125 Yu. M. Povarov, Yu. M. Kessler and A. I. Gorbanev, *Izv. Akad. Nauk. SSSR*, **10**, 1895 (1964)
C126 Yu. M. Povarov, Yu. M. Kessler and A. I. Gorbanev, *Elektrokhimiya*, **1**, 1174 (1965)
C127 C. M. Criss, Thermodynamic Properties of Ions, U.S. Atomic Energy Comm. Rept., TID–22366
C128 B. Nayak and D. K. Sahu, *Electrochim. Acta*, **16**, 1757 (1971)
C129 E. Luksha and C. M. Criss, *J. Phys. Chem.*, **70**, 1496 (1966)
C130 R. I. Mostkova and Yu. M. Kessler, *Elektrokhimiya*, **5**, 623 (1969)
C131 M. L. Berardelli, G. Pecci and B. Scrosati, *J. Electrochem. Soc.*, **117**, 781 (1970)
C132 G. P. Kumar and D. A. Pantony, Polarography 1964, Vol. 2, p. 1061 (G. J. Hills, ed.), Interscience, N.Y. (1966)
C133 J. N. Butler and J. C. Synnott, *J. Amer. Chem. Soc.*, **92**, 2602 (1970)
C134 C. M. Criss and E. Luksha, *J. Phys. Chem.*, **72**, 2966 (1968)
C135 D. R. Cogley and J. N. Butler, *J. Phys. Chem.*, **72**, 1017 (1968)
C136 T. W. Richards and F. Daniels, *J. Amer. Chem. Soc.*, **41**, 1732 (1919)
C137 G. N. Lewis and M. Randall, *J. Amer. Chem. Soc.* **43**, 233 (1921)
C138 J. C. Synnott and J. N. Butler, *Anal. Chem.*, **41**, 1890 (1969)

C139 W. H. Smyrl and C. W. Tobias, *Electrochim. Acta*, **13**, 1581 (1968); A. K. Covington *in* Ion Selective Electrodes, p. 134 (R. A. Durst, ed.), N.B.S. Special Publication No. 314 (1969)
C140 J. E. Prue and P. J. Sherrington, *Trans. Faraday Soc.*, **57**, 1795 (1961)
C141 R. P. Held and C. M. Criss, *J. Phys. Chem.*, **71**, 2487 (1967)
C142 R. Takahashi, *Talanta*, **12**, 1211 (1965)
C143 C. K. Mann, Electroanalytical Chemistry, Vol. 3, Ch. 2 (A. J. Bard, ed.), Marcel Dekker, N.Y. (1969)
C144 L. R. Dawson, R. C. Sheridan and H. C. Eckstrom, *J. Phys. Chem.*, **61**, 1829 (1961)
C145 L. R. Dawson, W. H. Zuber and H. C. Eckstrom, *J. Phys. Chem.*, **69**, 1335 (1965)
C146 B. Scrosati, G. Pecci and G. Pistoia, *J. Electrochem. Soc.*, **115**, 506 (1968)
C147 D. H. Everett and S. E. Rasmussen, *J. Chem. Soc.*, 2812 (1954)
C148 R. Jasinski, *in* Advances in Electrochemistry and Electrochemical Engineering, Vol. 8 (P. Delahay, ed.), Interscience, New York (1971)
C149 N. A. Izmailov and V. Zabara, *Zh. Fiz. Khim.*, **20**, 165 (1946)
C150 D. Feakins and C. M. French, *J. Chem. Soc.*, 2581 (1957)
C151 E. L. Mackor, *Rec. Trav. Chim.*, **70**, 457 (1951)
C152 D. Feakins and D. J. Turner, *J. Chem. Soc.*, 4986 (1965)
C153 L. M. Gedansky and K. S. Pribadi, Electrochemical Studies in Non-aqueous Solvents. I., Acetonitrile, NASA CR–84657; N67–28011
C154 K. Kruse, E. P. Goertz and H. Petermoller, *Z. Elektrochem.*, **55**, 405 (1951)
C155 H. Ulich and G. Spiegel, *Z. Phys. Chem.*, **A177**, 103 (1936)
C156 G. Spiegel and H. Ulich, *Z. Phys. Chem.*, **A178**, 187 (1937)
C157 V. A. Pleskov, *Usp. Khim.*, **16**, 254 (1947)
C158 V. A. Pleskov, *Zh. Fiz. Khim.*, **22**, 351 (1948)
C159 A. I. Popov and D. H. Geske, *J. Amer. Chem. Soc.*, **79**, 2074 (1957)
C160 I. M. Kolthoff and F. G. Thomas, *J. Phys. Chem.*, **69**, 3049 (1965)
C161 J. F. Coetzee and J. J. Campion, *J. Amer. Chem. Soc.*, **89**, 2513 (1967)
C162 I. M. Kolthoff and J. F. Coetzee, *J. Amer. Chem. Soc.*, **79**, 870 (1957); **79**, 1852 (1957)
C163 J. F. Coetzee and D. K. McGuire with J. L. Hedrick, *J. Phys. Chem.*, **69**, 1814 (1963)
C164 S. Wawzonek and M. E. Runner, *J. Electrochem. Soc.*, **99**, 457 (1952)
C165 B. Kratochvil, E. Lorah and C. Garber, *Anal. Chem.*, **41**, 1793 (1969)
C166 H.-M. Koepp, H. Wendt and H. Strehlow, *Z. Elektrochem.*, **64**, 483 (1960)
C167 G. P. Rao and A. R. V. Murthy, *J. Phys. Chem.*, **68**, 1573 (1964)
C168 B. Kratochvil and R. Long, *Anal. Chem.*, **42**, 43 (1970)
C169 R. Takahashi, Modern Aspects of Polarography (T. Kambara, ed.), Plenum, N.Y. (1966)
C170 R. Parsons, *Disc. Faraday Soc.* **39**, 102 (1965)
C171 V. Gutmann and G. Schober, *Z. Anal. Chem.*, **171**, 339 (1959); *Monats. Chem.*, **93**, 212 (1962)
C172 I. M. Kolthoff and T. B. Reddy, *J. Electrochem. Soc.*, **108**, 980 (1961)
C173 R. T. Burrus, Ph.D. thesis, University of Tennessee, 1962, University Microfilms Number 63–4101
C174 J. N. Butler, *J. Electroanal. Chem.*, **14**, 89 (1967)
C175 N. A. Rumbant and H. L. Peeters, *Bull. Soc. Chim. Belges*, **76**, 33 (1967)
C176 J.-P. Morel, *Bull. Soc. Chim. France*, 896 (1968)
C177 J. C. Synnott and J. N. Butler, *J. Phys. Chem.*, **73**, 1470 (1969)
C178 V. A. Kuznetsov, N. G. Vasil'kevich and B. B. Damaskin, *Elektrokhimiya*, **5**, 997 (1969)
C179 A. J. Parker, *Quart. Rev.*, **16**, 163 (1962); *Chem. Rev.*, **69**, 1 (1969)

C180 I. M. Kolthoff and T. B. Reddy, *Inorg. Chem.*, **1,** 189 (1962)
C181 J.-P. Morel, *Bull. Chim. Soc. France,* 1405 (1967)
C182 D. R. Cogley and J. N. Butler, *J. Electrochem. Soc.*, **113,** 1074 (1966)
C183 W. H. Smyrl and C. W. Tobias, *J. Electrochem. Soc.*, **115,** 33 (1968)
C184 G. Holleck, D. R. Cogley and J. N. Butler, *J. Electrochem. Soc.*, **116,** 952 (1969)
C185 M. Salomon, *J. Electrochem. Soc.*, **116,** 1392 (1969)
C186 J. L. Hanley and R. T. Iwamoto, *J. Electroanal. Chem.*, **24,** 271 (1970)
C187 R. F. Rodewald, K. Mahendran, J. L. Bear and R. Fuchs, *J. Amer. Chem. Soc.*, **90,** 6698 (1968)
C188 J. N. Butler, D. R. Cogley and W. Zurosky, *J. Electrochem. Soc.*, **115,** 445 (1968)
C189 J. N. Butler, D. R. Cogley, J. C. Synnott and G. Holleck, Study of the Composition of Non-aqueous Solutions, Final Report, September, 1969, U.S. Air Force Contract Number AF 19(628)6131
C190 J. Courtot-Corpez and M. L'Her, *Bull. Soc. Chim. France,* 675 (1969)
C191 L. M. Mukherjee and D. P. Boden, *J. Phys. Chem.*, **73,** 3965 (1969)
C192 M. Salomon, *J. Phys. Chem.*, **73,** 3299 (1969)
C193 M. Salomon, *J. Electroanal. Chem.*, **24,** 1 (1970)
C194 M. Salomon, *J. Electroanal. Chem.*, **26,** 319 (1970)
C195 M. Salomon, *J. Phys. Chem.*, **74,** 2519 (1970)
C196 V. Gutmann, G. Peychal-Heiling and M. Michlmayr, *Inorg. Nucl. Chem. Let.*, **3,** 501 (1967)
C197 V. Gutmann, M. Kogelnig and M. Michlmayr, *Monats. Chem.*, **99,** 693 (1968)
C198 M. Della Monica, U. Lamanna and L. Senatore, *Inorg. Chim. Acta,* **2,** 363 (1968)
C199 J. B. Headridge, D. Pletcher and M. Callingham, *J. Chem. Soc.*, 684 (1967)
C200 J. Desbarres, P. Pichet and R. L. Benoit, *Electrochim. Acta,* **13,** 1899 (1968)
C201 R. L. Benoit, *Inorg. Nucl. Chem. Lett.*, **4,** 723 (1968)
C202 J. F. Coetzee, J. M. Simon and R. L. Bertozzi, *Anal. Chem.*, **41,** 766 (1969)
C203 A. J. Parker and R. Alexander, *J. Amer. Chem. Soc.*, **90,** 3313 (1968)
C204 J. Broadhead and P. J. Elving, *Anal. Chem.*, **41,** 1814 (1969)
C205 J. Badoz-Lambling and J.-C. Bardin, *Compt. Rend. C,* **266,** 95 (1968)
C206 J. C. Hall, R. C. Murray and P. A. Rock, *J. Chem. Phys.*, **51,** 1145 (1969)
C207 D. Feakins, *in* Physico-Chemical Processes in Mixed Aqueous Solvents (F. Franks, ed.), Elsevier, N.Y. (1967)
C208 H. Strehlow and H.-M. Koepp, *Z. Elektrochem.*, **62,** 373 (1958)
C209 H. Schneider and H. Strehlow, *Z. Phys. Chem.*, **49,** 44 (1966)

APPENDIX 2.7.1A

Thermodynamic Properties of the Cell Pt, H_2|HI in MeOH|AgI|Ag

°C	E_m^0/ V	ΔG_m^0/ kcal $\mathrm{mol^{-1}}$	ΔH_m^0/ kcal $\mathrm{mol^{-}}$	ΔS_m^0/ cal $\mathrm{mol^{-1}\ K^{-1}}$
25	−0.31786	7.330	−1.701	−30.31
35	−0.33101	7.633	−3.458	−36.01
45	−0.34926	8.054	−5.333	−42.10

APPENDIX 2.7.2A

Standard E.m.f.'s of the Cells H_2|HX|AgX|Ag and H_2|HX|Hg_2X_2|Hg in MeOH-Water Mixtures at 25°C

Wt% MeOH	E_m^0/V				
x	AgCl	AgBr	AgI	Hg_2Cl_2	Hg_2Br_2
0	0.2224	0.07106	−0.15225	0.26828	0.13940
10	0.2155	0.06655			
15			−0.15306		
20	0.2088	0.0634†		0.2545¶	0.13131
30		0.05845‡	−0.15387		
43.12	0.1958	0.05600	−0.15398‖	0.2415	0.12662††
50	0.1906	0.0538			
60	0.1818		−0.15640		0.11170‡‡
70	0.1683	0.03850§	−0.16655	0.2173§	
90	0.1135	−0.01710	−0.21153		0.05825
99	0.0840*		−0.29193	0.1027**	
100	−0.0101	−0.1387	−0.31786		−0.02072§§

* $x = 94.2$. † $x = 20.22$. ‡ $x = 33.33$. § $x = 68.33$.
‖ $x = 45.0$. ¶ $x = 20.2$. ** $x = 97.29$. †† $x = 40.00$.
‡‡ $x = 65.00$. §§ $x = 99.5$.

APPENDIX 2.7.3A

Thermodynamic Properties of the Cell $H_2|HBr|Hg_2Br_2|Hg$ in MeOH-H_2O Mixtures

Wt % MeOH	ΔG^0_m/kcal mol^{-1}			ΔH^0_m/kcal mol^{-1}		
x	15°C	25°C	35°C	15°C	25°C	35°C
0	−6.49	−6.44	−6.35	−8.02	−8.62	−9.21
40	−5.98	−5.84	−5.69	−9.64	−10.35	−11.10
65	−5.39	−5.15	−4.89	−11.89	−12.65	−13.46
90	−3.04	−2.68	−2.26	−14.32	−15.02	−15.73
99.5	0.49	0.95	1.43	−12.41	−13.18	−14.01

APPENDIX 2.7.4A

Thermodynamic Properties of the Cell $H_2|HI|AgI|Ag$ in MeOH-H_2O Mixtures

Wt % MeOH	ΔG^0_m/kcal mol^{-1}			ΔH^0_m/kcal mol^{-1}		
x	25°C	35°C	45°C	25°C	35°C	45°C
0	3.487	3.621	3.735	−0.507	−0.209	0.110
15	3.530	3.626	3.777	0.670	0.103	−0.152
30	3.547	3.658	3.819	0.240	−0.506	−1.300
45	3.551	3.736	3.966	−1.962	−2.634	−3.351
60	3.607	3.876	4.181	−4.411	−4.970	−5.518
75	3.841	4.165	4.488	−5.807	−5.800	−5.797
90	4.878	5.208	5.641	−4.959	−6.490	−8.129
99	6.732	7.063	7.520	−3.133	−5.010	−7.010
100	7.330	7.633	8.054	−1.701	−3.458	−5.333

APPENDIX 2.7.5A

Thermodynamic Properties of the Cell $H_2|HCl|AgCl|Ag$ in EtOH-H_2O Mixtures

Mole Fraction EtOH	E_m^0/V 15°C	25°C	35°C	ΔH_m^0/kcal mol^{-1} 25°	ΔS_m^0/cal mol^{-1} K^{-1} 25°
25	0.2012	0.1928	0.1845	−10.180	−19.27
50	0.1676	0.1554	0.4430	−12.030	−28.38
75	0.1196	0.1053	0.0883	−13.200	−36.23
95	0.0383	0.0215	0.0028	−12.720	−41.07
100	−0.0620	−0.0760	−0.0912	−8.274	−33.69

APPENDIX 2.7.6A

Thermodynamic Data for the Cell $H_2|H^+, K^+|K$ in EtOH-H_2O Mixtures at 25°C

Wt% EtOH	A/ mol$^{-\frac{1}{2}}$ l$^{\frac{1}{2}}$	B/ mol^{-1} l	E_m^0/V
15.0	0.6286	−0.4495	−2.9136
20.3	0.6847	−0.6692	−2.9021
40.0	0.8757	−0.8008	−2.8687
60.2	1.475	−2.3922	−2.830
80.3	0.8076	−0.6900	−2.799
92.3			−2.757
100.0			−2.865

The mean *molar* activity coefficients of KCl, $y_\pm$, can be obtained from

$$-\log y_\pm = Ac^{\frac{1}{2}} + Bc$$

APPENDIX 2.7.7A

Thermodynamic Properties of the Cell $Na|NaI|AgI|Ag$ in EtOH*

°C	ΔG_t^0/ kcal mol^{-1}	ΔE_m^0/V	°C	ΔG_t^0/ kcal mol^{-1}	ΔE_m^0/V
20	6.29	0.2728	40	7.15	0.3100
25	6.53	0.2832	45	7.35	0.3187
30	6.75	0.2927	50	7.54	0.3270
35	6.98	0.3027	60	8.10	0.3512

* $\Delta E_m^0 = E_m^0(H_2O) - E_m^0(EtOH)$. ΔG_t^0 is the free energy of transfer of NaI from water to ethanol.

APPENDIX 2.7.8A

Standard Potentials of the AgCl|Ag and AgBr|Ag Electrodes in Ethylene Glycol-Water Mixtures at 25°C*

Wt% glycol	AgCl|Ag E^0_m/V	$A \cdot 10^4$	$B \cdot 10^6$	AgBr|Ag E^0_m/V	$A \cdot 10^4$	$B \cdot 10^6$
10	0.2151	6.56	3.71	0.0675	4.85	2.92
30	0.2030	6.95	3.99	0.0583	5.27	2.50
50	0.1896	7.65	4.64	0.0478	6.59	1.64
70	0.1689	9.61	3.57	0.0328	8.65	1.21
90	0.1183	12.0	1.71	−0.0115	11.3	0.928
100	0.0235	11.6	1.79	−0.098		

* $E^0_m(t) = E^0_m(25°C) - A(t - 25) - B(t - 25)^2$ where t = °C.

APPENDIX 2.7.9A

Standard Potentials of Ag*X*|Ag and Hg_2SO_4|Hg Electrodes in Acetic Acid-Water Mixtures at 25°C

Wt% HAc	E^0_m/V AgCl|Ag	AgBr|Ag‡	AgI|Ag‡	Hg_2SO_4|Hg
10	0.2105*	0.0606	−0.1609	
20	0.1968*	0.0474	−0.1719	
40	0.1621*	0.0140	−0.2019	
60	0.1115*	−0.0385	−0.2464	
100	−0.6208†			0.338§

* B. B. Owen, *J. Amer. Chem. Soc.*, **57**, 1526 (1935).
† L. M. Mukherjee, *J. Amer. Chem. Soc.*, **79**, 4040 (1957).
‡ H. P. Bennetto, D. Feakins and D. J. Turner, *J. Chem. Soc., A*, 1211 (1966).
§ V. K. LaMer and W. E. Eichelberger, *J. Amer. Chem. Soc.*, **54**, 2763 (1932).

APPENDIX 2.7.10A

Standard Potentials in the Amides at 25°C

Electrode	F		NMF	DMF
	E_c^0/V	E_m^0/V	E_m^0/V	E_m^0/V
$Li^+\|Li$		−2.975	−2.916	−3.141
$Na^+\|Na$		−2.6509	−2.599	−2.499
$K^+\|K$	−2.872		−2.813	−2.706
$Rb^+\|Rb$	−2.858			
$Cs^+\|Cs$		−2.856	−2.779	−2.675
$Zn^{2+}\|Zn$	−0.757			
$TlCl\|Tl$				−0.711
$CdCl_2\|Cd$	−0.617	−0.612		
$Cd^{2+}\|Cd$	−0.408			
$Tl^+\|Tl$	−0.344			−0.181
$Pb^{2+}\|Pb$	−0.193			
$H^+\|H_2$	0.0	0.0	0.0	0.0
$AgI\|Ag$				0.082
$AgBr\|Ag$		0.0967	0.115	0.081
$AgCl\|Ag$		0.1190	0.208	0.068
$Hg_2Cl_2\|Hg$		0.2332	0.1296	0.2152
$Cu^+\|Cu$	0.279			
$Ag^+\|Ag$		0.6903	0.823	0.959

E_c^0 values from T. Pavlopoulos and H. Strehlow, *Z. Phys. Chem.* (*Frankfurt*), **2,** 89 (1954). For E_m^0 references, see text sect. 2.7.2.

APPENDIX 2.7.11A

Standard E.m.f.'s (V) of the Cell $H_2|HCl|AgCl|Ag$ in Formamide

°C	5	15	20	25	30	35	40	45	50	55
(a)		0.208	0.207	0.204	0.198	0.191	0.181	0.172		
(b)				0.1986	0.1937	0.1888	0.1853	0.1901	0.1753	0.1715
(c)	0.2214	0.2100		0.2002						

(a) M. Mandel and P. Decroly, *Trans. Faraday Soc.*, **56,** 3311 (1960).
(b) R. K. Agarwal and B. Nyak, *J. Phys. Chem.*, **70,** 2568 (1966); **71,** 2062 (1967).
(c) R. W. C. Broadbank, S. Dhabanandanu, K. N. Morcom and B. L. Muju, *Trans. Faraday Soc.*, **64,** 3311 (1968).

APPENDIX 2.7.12A

Equilibrium Constants for Thallous Chloride in Several Aprotic Solvents

Solvent	log K_{s0}	log K_{s1}	log K_{s2}
DMF	−9.0	−4.6	−2.1
PC	−12.4	−6.4	−4.1
DMSO	−6.4	−2.95	−1.8
DMSO*	−6.26	−3.5	−2.3

* D. R. Cogley and J. N. Butler, *J. Electrochem. Soc.*, **113,** 1074 (1966). All other data from J. C. Synnott and J. N. Butler, *Anal. Chem.*, **41,** 1890 (1969). All data in molar units at 25°C.

APPENDIX 2.7.13A

Thermodynamic Properties for the Cell $H_2|HCl|AgCl|Ag$ in *N*-Methylacetamide

°C	E_m^0/V	ΔG_m^0/ kJ mol^{-1}	ΔH_m^0/ kJ mol^{-1}	ΔS_m^0/ J mol^{-1} K^{-1}
35	0.21187	20.44	43.59	75.13
40	0.20573	19.85	43.53	75.62
45	0.20091	19.39	43.61	76.13
50	0.19456	18.77	43.55	76.68
55	0.18972	18.31	43.64	77.19
60	0.18357	17.71	43.61	77.74
65	0.17786	17.16	43.62	78.25
70	0.17194	16.59	43.63	78.80

APPENDIX 2.7.14A

Standard Molal E.m.f.'s (V) of the Cell* Pt, H_2|HBr (*m*), dioxan (*x*), $H_2O(1 - x)$|AgBr|Ag in terms of $E_m^0 = A + BT + CT^2$ where *T* is in K

Wt% Dioxan (*x*)	*A*	10^2B	10^5C
5	−0.173775	0.215176	−0.449951
10	−0.111482	0.172612	−0.379930
15	−0.097614	0.167066	−0.379954
20	0.090585	0.046588	−0.190034
45	−0.028818	0.146466	−0.420066
70	0.031531	0.112653	−0.480049
82	0.213979	−0.024327	−0.359609

* T. Mussini, C. Massarani-Formaro and P. Audrigo, *J. Electroanal. Chem.*, **33**, 177 (1971).

APPENDIX 2.7.15A

Standard Potentials in Acetonitrile at 25°C from Cells with Liquid Junctions

Couple (in AN)	E_m^0/V		
	vs. 0.01*M* $AgNO_3$ Ag in AN	vs. SHE in H_2O*	Complete Aqueous Cell (vs. SHE)
Rb^+|Rb*	−3.282	−2.779	−2.928
$Co(C_5H_5)_2^+$|$Co(C_5H_5)_2$†	−0.876	−0.373	−0.92
Tl^+|Tl*	−0.648	−0.145	−0.336
Ag^+(0.01*M*)|Ag*	0.0	0.503	

APPENDIX 2.7.15A

Standard Potentials in Acetonitrile at 25°C from Cells with Liquid Junctions (*contd*)

Couple (in AN)	E^0_m/V vs. 0.01*M* $AgNO_3$|Ag in AN	vs. SHE in H_2O*	Complete Aqueous Cell (vs. SHE)
$H^+\|H_2$‡	0.024	0.527	0.0
$Fe(C_5H_5)_2^+\|Fe(C_5H_5)_2$‡	0.068	0.571	0.394
$Ag^+\|Ag$§	0.124	0.627	0.799
$Ce^{4+}\|Ce^{3+}$‖	0.749	1.252	
$Fe(phen)_3^{3+}\|Fe(phen)_3^{2+}$‡	0.840	1.343	1.120
$Fe^{3+}\|Fe^{2+}$¶	1.57	2.07	0.771

* Based on data of J. F. Coetzee and J. J. Campion, *J. Amer. Chem. Soc.*, **89,** 2513 (1967).

† H.-M. Koepp, H. Wendt and H. Strehlow, *Z. Elektrochem.*, **64,** 483 (1960).

‡ I. M. Kolthoff and F. G. Thomas, *J. Phys. Chem.*, **69,** 3049 (1965).

§ From ref. ‡ but uncorrected for ion-association (cf. B. Kratochvil, E. Lorah and C. Garber, *Anal. Chem.*, **41,** 1793 (1969).

‖ G. P. Rao and A. R. V. Murthy, *J. Phys. Chem.*, **68,** 1573 (1964); see also C12.

¶ B. Kratochvil and R. Long, *Anal. Chem.*, **42,** 43 (1970).

APPENDIX 2.7.16A

Standard E.m.f.'s of the Cell H_2|HCl|AgCl|Ag in DMSO-Water Mixtures at 25°C

Wt% DMSO	E^0_m/V	E^0_c/V
20	0.2205	0.2215
40	0.2185	0.2210
60	0.2130	0.2170
70	0.2030	0.2075
80	0.1820	0.1870
100	0.1288	0.1335

APPENDIX 2.7.17A

Thermodynamic Properties of the Cell Li|Li*X*-DMSO|Tl*X*|Tl

Salt	°C	E^0_m/V	C'‡	$-\Delta H^0_m$/ kcal mol^{-1}	$-\Delta S^0_m$/cal mol^{-1} K^{-1}	ΔG_t^0/kcal mol^{-1}	$-\Delta S_t^0$/cal mol^{-1} K^{-1}
LiCl*	25	2.2723	0.650	60.2	26.2	4.865	26.78
	30	2.2667	0.596	60.2	26.2	—	—
	35	2.2608	0.572	60.2	26.2	5.121	—
LiBr†	25	2.2730	0.970	59.91	25.15	2.515	27.46
	35	2.2621	0.920	59.87	25.00	2.792	—
	45	2.2512	0.878	59.82	24.84	3.064	—
LiI†	25	2.3152	1.162	58.55	17.30	−0.626	27.03
	35	2.3065	1.156	59.86	21.63	−0.340	—
	45	2.2965	1.116	61.21	25.96	−0.022	—

* LiCl data from W. H. Smyrl and C. W. Tobias, *J. Electrochem. Soc.*, **115**, 33 (1968).
† LiBr and LiI data from M. Salomon, *J. Electrochem. Soc.*, **116**, 1392 (1969); **117**, 325 (1970).
‡ Constant in eqn. 2.5.7.

APPENDIX 2.7.18A

Molal Solubility Products for Tl-Halides and Standard E.m.f.'s of the Cell Li|Li$^+$, Tl$^+$|Tl at 25°C

Salt	log K_{s0}		
	DMSO	PC	H_2O
TlCl	−6.34	−12.40	−3.732
TlBr	−6.33	−12.80	−5.438
TlI	−5.62	−12.38	−7.030
E^0_m	2.6474	2.5779	2.7041

APPENDIX 2.7.19A

Thermodynamic Properties of the Cell $M|MX|TlX|Tl$ in Propylene Carbonate for M = Li, Na, K, and X = Cl, Br, I

	LiCl				LiBr			
	15°C	25°C	35°C	45°C	15°C	25°C	35°C	45°C
E_m^0/V	1.8464	1.8348	1.8232	1.8114	1.8301	1.8194	1.8089	1.7983
C'/mol^{-1} kg	−17.96	−17.78	−17.92	−17.94	−0.502	−0.568	−0.618	−0.676
$-\Delta G_m^0$/kcal mol^{-1}	42.58	42.31	42.04	41.77	42.20	41.96	41.71	41.47
$-\Delta H_m^0$/kcal mol^{-1}	50.41	50.35	50.30	50.24	49.36	49.28	49.20	49.12
$-\Delta S_m^0$/cal mol^{-1} K^{-1}	27.16	26.98	26.80	26.62	24.83	24.57	24.31	24.05

	LiI				NaI				KI
	15°C	25°C	35°C	45°C	15°C	25°C	35°C	45°C	25°C
E_m^0/V	1.8562	1.8452	1.8346	1.8240	1.6266	1.6188	1.6100	1.6014	1.934
C'/mol^{-1} kg	0.537	0.511	0.503	0.489	0.173	0.150	0.122	0.094	0.535
$-\Delta G_m^0$/kcal mol^{-1}	42.81	42.55	42.31	42.06	37.51	37.33	37.13	36.93	44.61
$-\Delta H_m^0$/kcal mol^{-1}	50.22	50.01	49.81	49.59	42.80	43.02	43.24	43.47	49.31
$-\Delta S_m^0$/cal mol^{-1} K^{-1}	25.72	25.03	24.34	23.65	18.34	19.09	19.83	20.57	15.42

APPENDIX 2.7.20A

Measured Standard Molal Real Free Energies of Solvation of Ions in some Non-aqueous Solvents at 25°C*

Solvent (s)	Ion (i)	${}^{H_2O}\Delta^s\alpha_i^0 = {}^s\alpha_i^0 - {}^{H_2O}\alpha_i^0$ kcal mol^{-1}	$-{}^s\alpha_i^0$ kcal mol^{-1}
H_2O	Cl^-		70.70 ± 1.0†
D_2O	Cl^-	0.08 ± 0.05	
CH_3OH	Cl^-	11.2 ± 0.2	59.50 ± 1.25
C_2H_5OH	Cl^-	12.4 ± 0.2	58.30 ± 1.25
i-C_3H_3OH	Cl^-	12.6 ± 0.2	58.10 ± 1.25
n-C_4H_9OH	Cl^-	12.2 ± 0.2	58.50 ± 1.25
$HCONH_2$	Cl^-	3.9 ± 0.2	66.80 ± 1.25
$HCONHCH_3$	Cl^-	10.6 ± 0.5	60.10 ± 1.20
$HCOOH$	Cl^-	2.5 ± 0.1	68.20 ± 1.15
CH_3CN	Ag^+	−10.1 ± 0.2	124.50 ± 0.75

* B. Case and R. Parsons, *Trans. Faraday Soc.*, **63**, 1224 (1967).
† J. E. B. Randles, *Trans. Faraday Soc.*, **52**, 1573 (1956).

APPENDIX 2.7.21A

Differential Real Potentials, δ/F for Monovalent Hydrocarbon Ions in Acetonitrile at 25°C*

Hydrocarbon	$-\delta_{R^+}^0/F$	$-\delta_{R^-}^0/F$	$\frac{-(\delta_{R^+}^0 - \delta_{R^-}^0)}{2F}$
Anthracene	1.87	1.90	0.02
Pyrene	1.91	1.81	0.05
Chrysene	2.03	1.63	0.20
Phenanthrene	1.89	1.61	0.14
1,2-Benzanthracene	1.80	1.83	0.03
Triphenylene	2.00	1.68	0.16
			Average = 0.10

* All data are in volts from B. Case, N. S. Hush, R. Parsons and M. E. Peover, *J. Electroanal. Chem.*, **10**, 360 (1965).

APPENDIX 2.7.1B

Mean Molal Activity Coefficients of HI in Methanol

25°C		35°C		45°C	
m/ mol kg^{-1}	$-\log \gamma_\pm$	m/ mol kg^{-1}	$-\log \gamma_\pm$	m/ mol kg^{-1}	$-\log \gamma_\pm$
0.08969	0.2192	0.08448	0.1875	0.08380	0.1912
0.08141	0.2171	0.06820	0.1900	0.04692	0.1664
0.05049	0.1921	0.04269	0.1669	0.03757	0.1591
0.03405	0.1728	0.03512	0.1554	0.03430	0.1554
0.01260	0.1279	0.01395	0.1776	0.01218	0.1124
0.00671	0.1052	0.00743	0.1040	0.00635	0.0935
0.00407	0.0871	0.00426	0.0864	0.00365	0.0760

APPENDIX 2.7.2B

Mean Molal and Molar Activity Coefficients of Several Alkali Metal Halides in Methanol at 25°C

LiCl*		KI*		NaCl†		KCl‡		RbCl‡
m/mol kg^{-1}	$\gamma_\pm$	m/mol kg^{-1}	$\gamma_\pm$	m/mol kg^{-1}	$\gamma_\pm$	c/mol l^{-1}	$y_\pm$	$y_\pm$
0.0062	0.732	0.0008	0.886	0.0001	0.943	0.001	0.873	0.873
0.0125	0.653	0.0016	0.876	0.0005	0.890	0.002	0.825	0.825
0.0250	0.539	0.0031	0.817	0.001	0.856	0.005	0.738	0.738
0.0500	0.419	0.0062	0.745	0.005	0.705	0.010	0.650	0.649
0.10	0.324	0.0125	0.664	0.01	0.615	0.020	0.567	0.558
0.25	0.259	0.0250	0.566	0.05	0.408	0.040	0.477	0.468
0.50	0.268	0 050	0.460	0.10	0.367	0.056	0.455	
1.00	0.283	0.10	0.414			0.060		0.423
		0.25	0.488			0.087		0.397
		0.50	0.654					

* A. M. Shkodin and L. Ya. Shapovalova, *Izv. Ucheb. Zaved. Khim. Khim. Tekhnol.*, **9**, 563 (1966).

† J. H. Wolfenden, C. P. Wright, N. L. Ross Kane and P. S. Buckley, *Trans. Faraday Soc.*, **23**, 491 (1927).

‡ K. Bräuer and H. Strehlow, *Z. Phys. Chem. (Frankfurt)*, **17**, 346 (1958).

APPENDIX 2.7.3B

Mean Molal Stoichiometric Activity Coefficients of H_2SO_4 in Methanol

m/mol kg^{-1}	20°C	25°C	30°C	35°C
0.0005	0.924	0.912	0.905	0.894
0.001	0.894	0.879	0.867	0.854
0.002	0.855	0.835	0.821	0.803
0.005	0.783	0.757	0.738	0.712
0.01	0.712	0.682	0.659	0.627
0.02	0.626	0.594	0.568	0.531
0.05	0.494	0.465	0.438	0.397
0.10	0.393	0.369	0.344	0.304
0.20	0.301	0.282	0.263	0.280
0.40	0.236	0.209	0.199	0.181

APPENDIX 2.7.4B

Mean Molal Activity Coefficients of HBr in Methanol-Water Mixtures

40% MeOH			
m/mol kg^{-1}	15°C	25°C	35°C
0.0011		0.9543	
0.0025	0.9376		0.9333
0.0044		0.9194	
0.0074	0.9072		0.8983
0.0078		0.9012	
0.0099	0.8968		0.8870
0.0108	0.8933		0.8835

65% MeOH			
m/mol kg^{-1}	15°C	25°C	35°C
0.0023	0.9192		0.9133
0.0031		0.9078	
0.0069	0.8800	0.8754	0.8696
0.0093	0.8676	0.8610	0.8557
0.0112		0.8513	
0.0120	0.8584		0.8439

90% MeOH			
m/mol kg^{-1}	15°C	25°C	35°C
0.0021		0.9007	
0.0022	0.9043		0.8941
0.0041	0.8782		0.8642
0.0047		0.8652	
0.0062	0.8616		0.8447
0.0075		0.8429	
0.0116	0.8304		0.8063
0.0123		0.8156	

99.5% MeOH			
m/mol kg^{-1}	15°C	25°C	35°C
0.0032		0.8584	
0.0038	0.8543		0.8414
0.0060	0.8314		0.8158
0.0068		0.8141	
0.0104	0.8022		0.7798
0.0125		0.7782	
0.0151		0.7630	
0.0158	0.7716		0.7444
0.0187		0.7458	
0.0237		0.7278	

APPENDIX 2.7.5B

Mean Molal Activity Coefficients of HI in Methanol-Water Mixtures

30% MeOH

25°C		35°C		45°C	
m/mol kg^{-1}	$-\log \gamma_\pm$	m/mol kg^{-1}	$-\log \gamma_\pm$	m/mol kg^{-1}	$-\log \gamma_\pm$
0.0680	0.9795	0.0688	0.0889	0.0812	0.1059
0.0548	0.0876	0.0588	0.0907	0.0525	0.0933
0.0251	0.0676	0.0298	0.0743	0.0296	0.0795
0.0154	0.0599	0.0165	0.0621	0.0166	0.0650
0.0133	0.0551	0.0085	0.0480	0.0105	0.0547
0.0075	0.0446			0.0064	0.0548

45% MeOH

25°C		35°C		45°C	
m/mol kg^{-1}	$-\log \gamma_\pm$	m/mol kg^{-1}	$-\log \gamma_\pm$	m/mol kg^{-1}	$-\log \gamma_\pm$
0.0871	0.1056	0.0978	0.1120	0.0916	0.1135
0.0571	0.0943	0.0633	0.1003	0.0648	0.1037
0.0212	0.0704	0.0323	0.0829	0.0310	0.0845
0.0105	0.0550	0.0155	0.0653	0.0164	0.0686
0.0048	0.0408	0.0082	0.0518	0.0060	0.0474

APPENDIX 2.7.5B (*contd*)

60% MeOH

25°C		35°C		45°C	
m/mol kg^{-1}	$-\log \gamma_{\pm}$	m/mol kg^{-1}	$-\log \gamma_{\pm}$	m/mol kg^{-1}	$-\log \gamma_{\pm}$
0.0934	0.1190	0.0928	0.1356	0.1227	0.1366
0.0680	0.1181	0.0690	0.1157	0.0895	0.1284
0.0265	0.0868	0.0253	0.0909	0.0446	0.1064
0.0145	0.0747	0.0149	0.0817	0.0225	0.0863
0.0103	0.0671	0.0120	0.0703	0.0093	0.0682
0.0067	0.0331	0.0066	0.0563	0.0064	0.0527

75% MeOH

25°C		35°C		45°C	
m/mol kg^{-1}	$-\log \gamma_{\pm}$	m/mol kg^{-1}	$-\log \gamma_{\pm}$	m/mol kg^{-1}	$-\log \gamma_{\pm}$
0.0923	0.1622	0.0936	0.1676	0.0947	0.1743
0.0446	0.1306	0.0356	0.1297	0.0359	0.1347
0.0207	0.1051	0.0217	0.1156	0.0207	0.1140
0.0124	0.0887	0.0135	0.0979	0.0144	0.1009
0.0082	0.0760	0.0072	0.0765	0.0079	0.0810

90% MeOH					
25°C		35°C		45°C	
m/mol kg^{-1}	$-\log \gamma_\pm$	m/mol kg^{-1}	$-\log \gamma_\pm$	m/mol kg^{-1}	$-\log \gamma_\pm$
0.0778	0.2236	0.0776	0.2036	0.0789	0.2161
0.0579	0.2062	0.0344	0.1607	0.0406	0.1763
0.0375	0.1712	0.0211	0.1398	0.0241	0.1494
0.0184	0.1327	0.0117	0.1149	0.0125	0.1186
0.0089	0.1037	0.0076	0.0956	0.0081	0.1009
0.0041	0.0904	0.0044	0.0788	0.0054	0.0857

99% MeOH					
25°C		35°C		45°C	
m/mol kg^{-1}	$-\log \gamma_\pm$	m/mol kg^{-1}	$-\log \gamma_\pm$	m/mol kg^{-1}	$-\log \gamma_\pm$
0.1091	0.2346	0.0814	0.1872	0.0914	0.1924
0.0569	0.2007	0.0690	0.1809	0.0727	0.1856
0.0399	0.1843	0.0403	0.1590	0.0281	0.1445
0.0193	0.1520	0.0193	0.1339	0.0187	0.1403
0.0078	0.1115	0.0082	0.0994	0.0079	0.1062
0.0032	0.0811	0.0032	0.0710	0.0029	0.0660

APPENDIX 2.7.6B

Mean Molal Activity Coefficients of Alkali Metal Halides in Ethanol

m/mol kg^{-1}	HBr* (25°C)	NaBr† (25°C)	NaI† (25°C)	NaI‡					
				20°C	25°C	30°C	40°C	50°C	60°C
0.005	0.649			0.700	0.674	0.644	0.672	0.665	0.615
0.01	0.590	0.609	0.552	0.597	0.571	0.591	0.563	0.508	0.504
0.02	0.537	0.519	0.482	0.485	0.493	0.469	0.455	0.454	0.427
0.04	0.488	0.411	0.433	0.402	0.383	0.373	0.368	0.364	0.337
0.06	0.474§	0.359	0.419	0.357	0.361	0.351	0.329	0.323	0.309
0.08	0.452‖	0.333	0.403	0.364	0.363	0.338	0.317	0.327	0.301
0.10	0.431	0.314	0.407	0.394	0.379	0.359	0.332	0.339	0.311
0.14		0.290	0.416						
0.20		0.268	0.430	0.820	0.728				
0.40			0.525						
1.00			0.838						

* L. J. Nunez and M. C. Day, *J. Phys. Chem.*, **65**, 164 (1961).
† N. A. Izmailov and E. F. Ivanova, *Zh. Fiz. Khim.*, **29**, 1614 (1955).
‡ V. V. Aleksandrov and T. M. Shikhova, *Elektrokhimiya*, **3**, 981 (1968).
§ $m = 0.05$.
‖ $m = 0.07$.

APPENDIX 2.7.7B

Mean Molal Activity Coefficients of HCl in 2-PrOH-Water Mixtures*

m/mol kg^{-1}	0°C	15°C	25°C	35°C
	x = 0 Wt% 2-Propanol			
0.1	0.802	0.800	0.796	0.791
0.05	0.834	0.832	0.830	0.826
0.02	0.877	0.877	0.875	0.873
0.01	0.906	0.905	0.904	0.902
0.005	0.930	0.929	0.928	0.926
0.002	0.954	0.953	0.952	0.954
0.001	0.966	0.966	0.965	0.964
	x = 8.03 Wt% 2-Propanol			
0.1	0.756	0.635	0.633	0.625
0.05	0.785	0.676	0.670	0.664
0.02	0.839	0.694	0.691	0.683
0.01	0.877	0.727	0.725	0.716
0.005	0.908	0.769	0.769	0.760
0.002	0.939	0.828	0.828	0.821
0.001	0.956	0.867	0.867	0.862
	x = 20.76 Wt% 2-Propanol			
0.1	0.647	0.625	0.610	0.599
0.05	0.672	0.659	0.641	0.633
0.02	0.698	0.682	0.661	0.653
0.01	0.735	0.718	0.698	0.690
0.005	0.780	0.764	0.745	0.739
0.002	0.838	0.825	0.810	0.805
0.001	0.876	0.865	0.853	0.849
	x = 44.04 Wt% 2-Propanol			
0.1	0.620	0.605	0.611	0.597
0.05	0.595	0.581	0.592	0.583
0.02	0.655	0.629	0.647	0.637
0.01	0.718	0.688	0.707	0.697
0.005	0.772	0.767	0.759	0.758
0.002	0.846	0.836	0.829	0.822
0.001	0.885	0.879	0.877	0.872

APPENDIX 2.7.7B (*contd*)

m/mol kg^{-1}	0°C	15°C	25°C	35°C
	x = 70.28 Wt% 2-Propanol			
0.1	0.424	0.391	0.389	0.379
0.05	0.475	0.440	0.439	0.432
0.02	0.555	0.526	0.524	0.524
0.01	0.626	0.599	0.599	0.598
0.005	0.697	0.676	0.675	0.674
0.002	0.782	0.768	0.766	0.762
0.001	0.835	0.822	0.822	0.822
	x = 87.71 Wt% 2-Propanol			
0.1	0.364	0.335	0.319	0.300
0.05	0.428	0.400	0.382	0.366
0.02	0.515	0.492	0.478	0.462
0.01	0.590	0.571	0.561	0.544
0.005	0.666	0.651	0.644	0.629
0.002	0.758	0.747	0.743	0.730
0.001	0.816	0.807	0.805	0.794

* R. N. Roy and A. Bothwell, *J. Chem. Eng. Data*, **15**, 548 (1970).

APPENDIX 2.7.8B

Mean Molal Activity Coefficients of HBr in 2-Propanol-Water Mixtures at 15, 25 and 35°C*

Wt% 2-PrOH	m/mol kg^{-1}	15°C	25°C	35°C
20	0.00332	0.9414	0.9270	0.9257
	0.00637	0.9087	0.9058	0.9040
	0.00894	0.8965	0.8940	0.8924
	0.01260	0.8839	0.8804	0.8780
40	0.00210	0.9281	0.9253	0.9230
	0.00442	0.9077	0.9031	0.8998
	0.00644	0.8967	0.8911	0.8871
	0.00830	0.8884	0.8825	0.8794
	0.01055	0.8822	0.8762	0.8721
65	0.00323	0.8429	0.8352	0.8287
	0.00504	0.8180	0.8094	0.8029
	0.00820	0.7899	0.7798	0.7682
	0.01171	0.7684	0.7575	0.7458
90	0.00402	0.6793	0.6674	0.6486
	0.00832	0.5952	0.5887	0.5650
	0.01108	0.5613	0.5480	0.5287
	0.01431	0.5322	0.5226	0.5011
	0.01790	0.5051	0.4978	0.4715
99	0.00503	0.6598	0.6481	0.6420
	0.00703	0.6280	0.6191	0.5989
	0.00902	0.5936	0.5933	0.5765
	0.01364	0.5724	0.5652	0.5402
	0.01707	0.5529	0.5389	0.5197
	0.02013	0.5377	0.5267	0.5096

* K. Schwabe and R. Müller, *Ber. Bunsenges. Phys. Chem.*, **74**, 1248 (1970).

APPENDIX 2.7.9B

Mean Molal Activity Coefficients of HCl and HBr in Ethylene Glycol-Water Mixtures

m/mol kg^{-1}	10% glycol				30% glycol					
	5°C	25°C		45°C	5°C		25°C		45°C	
	HCl	HCl	HBr	HCl	HCl	HBr	HCl	HBr	HCl	HBr
0.005	0.928	0.926	0.920	0.922	0.918	0.922	0.915	0.915	0.910	0.910
0.01	0.902	0.899	0.900	0.894	0.893	0.899	0.887	0.890	0.881	0.884
0.02	0.872	0.868	0.875	0.861	0.860	0.865	0.853	0.855	0.854	0.847
0.03	0.852	0.847	0.855	0.839	0.839	0.849	0.831	0.835	0.822	0.827
0.05	0.827	0.820	0.830	0.810	0.813	0.825	0.805	0.810	0.793	0.801
0.07	0.810	0.802	0.815	0.791	0.798	0.812	0.787	0.795	0.775	0.785
0.10	0.795	0.785	0.804	0.774	0.778	0.803	0.768	0.785	0.755	0.773

APPENDIX 2.7.9B (*contd*)

m/mol kg^{-1}	50% glycol						70% glycol					
	5°C		25°C		45°C		5°C		25°C		45°C	
	HCl	HBr	HCl	HBr	HCl	HBr	HCl	HBr	HCl	HBr	HCl	HBr
0.005	0.907	0.905	0.902	0.895	0.895	0.890	0.880	0.880	0.872	0.872	0.864	0.865
0.01	0.877	0.882	0.870	0.870	0.862	0.863	0.855	0.855	0.845	0.845	0.835	0.836
0.02	0.843	0.849	0.835	0.835	0.825	0.827	0.818	0.820	0.806	0.805	0.794	0.795
0.03	0.821	0.825	0.812	0.810	0.801	0.802	0.795	0.798	0.782	0.778	0.765	0.765
0.05	0.795	0.797	0.786	0.780	0.773	0.770	0.765	0.770	0.747	0.745	0.730	0.730
0.07	0.780	0.785	0.768	0.765	0.755	0.755	0.743	0.753	0.725	0.728	0.705	0.712
0.10	0.758	0.782	0.745	0.760	0.731	0.748	0.722	0.738	0.702	0.712	0.680	0.695

m/mol kg^{-1}	90% glycol						100% glycol		
	5°C		25°C		45°C		5°C	25°C	45°C
	HCl	HBr	HCl	HBr	HCl	HBr	HCl	HCl	HCl
0.005	0.842	0.845	0.832	0.830	0.823	0.820	0.815	0.807	0.795
0.01	0.803	0.820	0.793	0.800	0.872	0.790	0.756	0.748	0.735
0.02	0.750	0.780	0.741	0.760	0.725	0.745	0.696	0.682	0.665
0.03	0.725	0.755	0.710	0.730	0.692	0.715	0.660	0.645	0.627
0.05	0.692	0.715	0.675	0.690	0.655	0.675	0.625	0.605	0.583
0.07	0.672	0.695	0.654	0.670	0.629	0.655	0.596	0.578	0.555
0.10	0.750	0.685	0.627	0.760	0.600	0.640	0.568	0.550	0.520

APPENDIX 2.7.10B

Mean Molal Activity Coefficients of HCl in 95 wt% Glycerol in Water*

t/°C	m/mol kg^{-1}							
	0.001	0.002	0.005	0.01	0.02	0.05	0.1	0.2
5	0.879	0.837	0.770	0.710	0.648	0.577	0.540	0.500
10	0.877	0.836	0.768	0.708	0.647	0.575	0.538	0.496
15	0.876	0.835	0.765	0.707	0.646	0.574	0.536	0.493
20	0.876	0.834	0.765	0.705	0.645	0.572	0.535	0.489
25	0.875	0.832	0.764	0.704	0.644	0.571	0.533	0.485
30	0.874	0.831	0.762	0.703	0.643	0.569	0.531	0.481
35	0.873	0.830	0.761	0.700	0.642	0.567	0.530	0.477
40	0.871	0.828	0.758	0.698	0.639	0.565	0.527	0.475
45	0.869	0.825	0.755	0.695	0.635	0.562	0.524	0.469

* R. N. Roy and W. Vernon, *J. Electroanal. Chem.*, **30**, 335 (1971).

APPENDIX 2.7.11B

Mean Molal Activity Coefficients of HCl in Formamide

m/mol kg^{-1}	25°C	30°C	35°C	40°C	45°C	50°C	55°C
0.002	0.9725						
0.005	0.9568	0.950	0.957	0.957	0.955	0.954	0.952
0.01	0.9443	0.934	0.943	0.943	0.940	0.939	0.934
0.02	0.9258	0.908	0.928	0.928	0.904	0.919	0.911
0.03	0.9156	0.904	0.920	0.919	0.909	0.906	0.894
0.04	0.9088	0.896	0.914	0.913	0.900	0.896	0.881
0.05	0.9044	0.889	0.911	0.909	0.899	0.889	0.870
0.06		0.884	0.909	0.908	0.888	0.883	0.860
0.07		0.880	0.908	0.907	0.884	0.878	0.852
0.08		0.897	0.908	0.907	0.881	0.874	0.845
0.09		0.875	0.908	0.906	0.877	0.870	0.837
0.10		0.874	0.911	0.909	0.877	0.868	0.835

APPENDIX 2.7.12B

Mean Molal Activity Coefficients of Alkali-Metal Halides in Formamide and *N*-Methylformamide at 25°C

m/mol kg^{-1}	Formamide*		*N*-Methylformamide†				
	NaCl	CsCl	LiCl	NaCl	KCl	CsCl	NaBr
0.01			0.951	0.958	0.962	0.967	0.969
0.02			0.923	0.938	0.945	0.954	0.958
0.03		0.901	0.898	0.922	0.929	0.945	0.951
0.04			0.876	0.906	0.921	0.936	0.943
0.05	0.883‡	0.865	0.854	0.890	0.906	0.929	0.938
0.06	0.869	0.856	0.834	0.878	0.895	0.922	0.934
0.07	0.863	0.853	0.815	0.883	0.885	0.916	0.929
0.08	0.857	0.843	0.796	0.881			0.925
0.10	0.845	0.832		0.885			
0.20	0.813	0.791					
0.3	0.804	0.763					
0.5	0.798	0.727					

* Yu. M. Povarov, Yu. M. Kessler and A. I. Gorbanev, *Elektrokhimiya*, **1**, 1174 (1965).
† E. Luksha and C. M. Criss, *J. Phys. Chem.*, **70**, 1498 (1966).
‡ $m = 0.047$.

APPENDIX 2.7.13B

Mean Molal Activity Coefficients of NaCl in *N*-Methylformamide at 5°C

m/mol kg^{-1}	$\log \gamma_\pm$	m/mol kg^{-1}	$\log \gamma_\pm$
0.0224	−0.013	0.09026	−0.0109
0.04191	−0.0067	0.10060	0.00167
0.05610	−0.155	0.13807	0.0004
0.07594	−0.0085	0.15697	0.0072

Appendix 2.7.14B

Mean Activity Coefficients of HCl and LiCl in Dimethylformamide at 25°C

HCl*		LiCl†	
c/mol l^{-1}	$y_{\pm}$	m/mol kg^{-1}	$-\log \gamma_{\pm}$
0.000556	0.909	0.00100	0.047
0.001112	0.879	0.00203	0.066
0.001853	0.859	0.00502	0.102
0.002779	0.815	0.00980	0.139
0.004632	0.743	0.01005	0.140
0.00600	0.703	0.0512	0.283
0.00901	0.616	0.1003	0.367
0.01460	0.518	0.5047	0.614
0.01853	0.466	1.0001	0.716
0.02400	0.386	1.9997	0.790
0.03400	0.304		

* G. P. Kumar and D. A. Pantony, Polarography 1964, **2,** 1061 (1966).
† J. N. Butler and J. C. Synnott, *J. Amer. Chem. Soc.*, **92,** 2602 (1970).

APPENDIX 2.7.15B

Mean Molal Activity Coefficients of HBr at Various Molalities in Dioxan-Water Mixtures, at 25°C

m/mol kg^{-1}	Wt% Dioxan						
	5	10	15	20	45	70	82
0.001	0.963	0.959	0.956	0.951	0.910	0.695	0.286
0.002	0.949	0.944	0.939	0.933	0.878	0.606	0.216
0.003	0.939	0.933	0.927	0.920	0.856	0.552	0.182
0.005	0.924	0.917	0.910	0.901	0.824	0.483	0.146
0.007	0.912	0.904	0.896	0.886	0.801	0.440	0.126
0.01	0.899	0.889	0.881	0.869	0.774	0.397	0.108
0.02	0.868	0.856	0.845	0.831	0.717	0.324	0.080
0.03	0.847	0.834	0.823	0.806	0.682	0.289	0.069
0.05	0.820	0.803	0.792	0.773	0.638	0.254	0.058
0.07	0.800	0.782	0.772	0.751	0.610	0.238	0.053
0.1	0.780	0.759	0.750	0.727	0.582	0.227	0.051

* T. Mussini, C. Massarani-Formaro and P. Audrigo, *J. Electroanal. Chem.*, **33,** 177 (1971).

APPENDIX 2.7.16B

Mean Molar Activity Coefficients of HCl in Dimethyl Sulphoxide at 25°C

$c/\text{mol l}^{-1}$	$y_\pm$	$c/\text{mol l}^{-1}$	$y_\pm$
0.00072	0.964	0.00436	0.839
0.00144	0.929	0.00504	0.828
0.00180	0.912	0.00576	0.815
0.00288	0.881	0.00720	0.794
0.00360	0.857	0.00774	0.781

APPENDIX 2.7.17B

Mean Molal Activity Coefficients of LiCl in Dimethyl Sulphoxide

$m/\text{mol kg}^{-1}$	$-\ln \gamma_\pm$		
	25°C	30°C	35°C
0.005	0.152	0.160	0.165
0.010	0.220	0.230	0.230
0.020	0.306	0.310	0.310
0.030	0.360	0.364	0.365
0.050	0.440	0.440	0.447
0.080	0.511	0.520	0.525
0.100	0.548	0.560	0.565
0.1200	0.582	0.592	0.600
0.1932*	0.659		
0.4870*	0.808		
0.9870*	0.956		

* From G. Holleck, D. R. Cogley and J. N. Butler, *J. Electrochem. Soc.*, **116,** 952 (1969); all other data from W. H. Smyrl and C. W. Tobias, *J. Electrochem. Soc.*, **115,** 33 (1968).

APPENDIX 2.7.18B

Mean Molal Activity Coefficients of LiBr in DMSO

m/mol kg^{-1}	15°C	25°C	35°C	45°C
0.02673	0.6815	0.6950	0.7024	0.7120
0.05606	0.5922	0.6025	0.6121	0.6195
0.07024	0.5797	0.5872	0.5959	0.6039
0.13458	0.5698	0.5779	0.5867	0.5937
0.16782	0.5373	0.5426	0.5479	0.5542
0.22445	0.5318	0.5333	0.5344	0.5369
0.35390	0.5219	0.5269	0.5323	0.5363
0.44634	0.5315	0.5360	0.5411	0.5454
0.47745	0.5521	0.5549	0.5583	0.5608
0.54635	0.5125	0.5145	0.5173	0.5188
0.59054	0.5805	0.5814	0.5833	0.5844
0.63292	0.5694	0.5736	0.5674	0.5808
0.74850	0.5422	0.5401	0.5416	0.5421
0.79123	0.6035	0.6044	0.6071	0.6076
0.86155	0.6608	0.6560	0.6536	0.6501

APPENDIX 2.7.19B

Mean Molal Activity Coefficients of LiI in DMSO

m/mol kg^{-1}	25°C	35°C	45°C
0.10270	0.5644	0.5853	0.5900
0.12677	0.5696	0.5451	0.5628
0.18630	0.5559	0.5883	0.5932
0.24380	0.5535	0.5710	0.5767
0.32250	0.5646	0.5851	0.5885
0.36076	0.5432	0.5635	0.5685
0.44181	0.5832	0.6026	0.6053
0.50181	0.6045	0.6235	0.6251
0.69333	0.6619	0.6814	0.6826
0.71890	0.6698	0.6915	0.6942
0.89290	0.7485	0.7708	0.7718

APPENDIX 2.7.20B

Mean Molal Activity Coefficients of LiCl in PC

m/mol kg^{-1}	$\gamma_{\pm}$			
	15°C	25°C	35°C	45°C
0.01631	0.6093	0.6046	0.6057	0.6038
0.01772	0.6135	0.6137	0.6099	0.6095
0.01932	0.5732	0.5776	0.5729	0.5699
0.02003	0.5554	0.5647	0.5658	0.5675
0.02439	0.5173	0.5202	0.5175	0.5157
0.02701	0.4863	0.4879	0.4842	0.4813
0.02718	0.4786	0.4801	0.4783	0.4743
0.02771	0.4726	0.4754	0.4746	0.4744
0.02956	0.4528	0.4540	0.4510	0.4493
0.03156	0.4430	0.4446	0.4418	0.4393
0.03197	0.4391	0.4420	0.4416	0.4422

APPENDIX 2.7.21B

Mean Molal Activity Coefficients for LiBr in PC

m/mol kg^{-1}	15°C	25°C	35°C	45°C
0.64829	0.3593	0.3476	0.3380	0.3275
0.35385	0.4453	0.4340	0.4241	0.4140
0.31715	0.4596	0.4515	0.4452	0.4380
0.28684	0.4790	0.4710	0.4647	0.4553
0.23449	0.5103	0.5010	0.4957	0.4881
0.20425	0.5306	0.5219	0.5159	0.5090
0.19384	0.5409	0.5325	0.5272	0.5198
0.14837	0.5781	0.5747	0.5687	0.5636
0.12840	0.6105	0.6029	0.5962	0.5884

APPENDIX 2.7.22B
Mean Molal Activity Coefficients of LiI in PC

m/mol kg^{-1}	15°C	25°C	35°C	45°C
0.8688	1.1608	1.0862	1.0683	1.0422
0.6131	0.8953	0.8737	0.8564	0.8401
0.5818	0.8532	0.8369	0.8410	0.8295
0.4890	0.8292	0.8109	0.7982	0.7872
0.4724	0.8334	0.8145	0.8050	0.7934
0.4056	0.8031	0.7893	0.7798	0.7696
0.3748	0.7887	0.7767	0.7706	0.7635
0.2803	0.7596	0.7494	0.7445	0.7360
0.1889	0.7475	0.7377	0.7359	0.7318
0.1342	0.7609	0.7539	0.7531	0.7517
0.1178	0.7341	0.7268	0.7258	0.7224
0.09644	0.7502	0.7452	0.7430	0.7406
0.09279	0.7590	0.7517	0.7423	0.7360
0.06801	0.7471	0.7421	0.7438	0.7455
0.05201	0.7621	0.7573	0.7542	0.7495
0.04780	0.7694	0.7638	0.7644	0.7638
0.02748	0.7821	0.7850	0.7861	0.7888

APPENDIX 2.7.23B
Mean Molal Activity Coefficients for NaI in PC

m/mol kg^{-1}	15°C	25°C	35°C	45°C
1.0282	0.6116	0.5763	0.5405	0.5068
0.9286	0.5986	0.5708	0.5401	0.5117
0.9205	0.5945	0.5728	0.5447	0.5175
0.8323	0.5933	0.5733	0.5474	0.5221
0.8010	0.5869	0.5704	0.5456	0.5222
0.5882	0.5806	0.5681	0.5509	0.5336
0.4312	0.5993	0.5830	0.5650	0.5485
0.3633	0.5964	0.5848	0.5728	0.5609
0.2332		0.6125		
0.2290		0.6313		
0.1523		0.6462		

APPENDIX 2.7.24B

Mean Molal Activity Coefficients of KI in PC at 25°C

m/mol kg^{-1}	$\gamma_{\pm}$
0.1953	0.6622
0.1556	0.6776
0.1002	0.6964
0.05336	0.7266
0.03768	0.7821
0.02825	0.7944

Chapter 2

Thermodynamic Measurements

Part 3

Precision Cryoscopy

R. Garnsey
Central Electricity Generating Board, Research Laboratories, Kelvin Avenue, Leatherhead, Surrey, England

and

(the late) **J. E. Prue**
Department of Chemistry, The University, Whiteknights Park, Reading, Berkshire, England

2.8 CRYOSCOPIC MEASUREMENTS

2.8.1 Introduction

The meaning of the word 'cryoscopy' is literally 'observations on cooling'. In physical chemistry it means the study of the solid-liquid transition and in particular of the depression in the temperature of this transition which results when a second component is added to the system. The determination of freezing and melting points, the study of the melting process for pure substances by, for example, volumetric and colorimetric methods,[D1] and the determination of phase diagrams[D2] can all be included in this definition. The important analytical applications of cryoscopy for the determination of solute molecular weight and solvent purity are well known. The application of cryoscopy to the determination of equivalence points, known as cryoscopic titrimetry,[D3] is a comparatively recent development. These analytical applications have recently been comprehensively reviewed by Glasgow and Ross.[D4] In these sections (2.8–2.10) we will be principally concerned with the precise measurement of freezing point depression in organic solvent systems and with the interpretation of these data. Some of the most precise cryoscopic measurements in organic solvents have been concerned with the determination of purity. These procedures will be discussed from the standpoint of obtaining precise cryoscopic data rather than their original purpose.

In a majority of analytical applications the solvent-solute system is assumed to be ideal. Cryoscopy is, however, a useful method of studying deviations from ideality. The osmotic coefficient of the solvent is obtained directly and activity coefficients of the solute may be derived by application of the Gibbs-Duhem relation. The method is particularly useful for studying electrolyte solutions as the solutes are involatile. The number of precise conductivity studies of salts in organic solvents has increased considerably in recent years, (Chapter 5) and there is a need for complementary thermodynamic data of comparable precision to facilitate a more complete elucidation of ionic behaviour in these systems.

The major advantages of the cryoscopic method are its wide applicability and the high precision possible. Its disadvantage is that measurements are restricted to a small temperature range close to the solvent melting point which is not necessarily a convenient temperature.

2.8.2 Conditions of Measurement

Cryoscopic measurements can be performed under a variety of conditions. These conditions are chosen so that the freezing transition of the solvent system is independent of the fraction of material frozen. Provided that the composition of the liquid phase does not change as solid phase is formed, the presence of a small amount of inert substance is immaterial, e.g. measurements are commonly made in solvents saturated with air at a pressure of one atmosphere. For a system which is sensitive to water, carbon dioxide or oxygen, then nitrogen or some other inert gas may be used to exclude them. Measurements may be made at the vapour pressure of the solvent or at some higher pressure. The differences which arise as a result of dissolved gases and variation in pressure are generally slight but not invariably so.

The Phase Rule

The phase rule states that

$$F = C - P + r \tag{2.8.1}$$

where F is the number of degrees of freedom, C the number of components, P the number of phases and r the number of intensive variables. These variables are pressure, temperature, and gravitational, electrical and magnetic fields. We are normally concerned only with variations of temperature and pressure so the phase rule takes the familiar form derived by Gibbs

$$F = C - P + 2 \tag{2.8.2}$$

For a system of one component with solid in equilibrium with liquid phase there will be one degree of freedom. If the pressure is fixed the temperature will be fixed and therefore independent of the fraction of material frozen. If air (assuming it to be a single component) is introduced and is dissolved there are now two components and two phases, and at constant pressure the system will still have one degree of freedom; thus the temperature of the system will vary with the amount of air which is dissolved. As pure solid phase crystallises the concentration of air in the liquid phase will increase, and there will be a corresponding decrease in temperature until the concentration reaches saturation. At this point the composition of the liquid phase is fixed and further freezing will occur at a constant temperature with the expulsion of air. This situation will occur for any solute which is partially miscible with the solvent. Water in benzene is an example of this. The presence of water in a sample of benzene can be detected[D4] by observing the freezing behaviour before and after deliberate saturation with water.

For a system of one component in which solid, liquid and vapour are in equilibrium, there is no degree of freedom. This is the triple-point condition and the pressure will be fixed at the vapour pressure of the component at the triple-point temperature.

The difference between a freezing point determined in the presence of air at a pressure of one atmosphere and the triple-point temperature will depend on the effect of dissolved air and on the pressure difference. The presence of dissolved gas in the liquid phase always lowers the transition temperature. The amount by which it does so can be calculated from the freezing point depression equations if the solubility is known and ideal behaviour is assumed. The effect of pressure can be calculated from the Clapeyron equation

$$\delta T/\delta P = T(V_L - V_S)/\Delta H_f \tag{2.8.3}$$

where V_S and V_L are the molar volumes of solid and liquid respectively and ΔH_f is the molar heat of fusion. The transition temperature may be increased or decreased depending on the relative magnitudes of V_S and V_L. For benzene the presence of air decreases the temperature below the triple-point value by 0.031 K.[D5] The pressure difference elevates the temperature by 0.028 K[D5] resulting in a net depression of 0.003 K. If very precise work is undertaken it may be necessary to add corrections for the difference in pressure between the top and bottom of the sample container caused by the sample weight.

The differences between the triple point and the normal freezing point are generally less than 0.05 K but for some solvents the difference can be almost a degree.[D6]

2.8.3 Experimental Procedure

A system of liquid solvent and non-reacting solute in equilibrium with pure solid solvent has two degrees of freedom if the solute concentration is below the saturation value. If measurements are performed at constant pressure then two intensive variables such as temperature and liquid phase composition must be measured simultaneously. In practice this is most conveniently achieved by fixing one at a known value and measuring the other. The system may be allowed to come to thermal equilibrium at a known or measurable temperature and then the composition of the liquid phase determined. The freezing point depression may be obtained directly if the temperature difference between the system, and one with no added solute, is measured with a device such as a thermocouple. This is the basis of the differential equilibrium method. The determination of the freezing point of a solution of known concentration is not so straightforward. If pure solid solvent is added to the system then solvent will in general freeze or melt and thereby change the liquid phase concentration. The freezing point of a particular solution may be obtained, as the temperature at which an infinitesimal amount of solid is present, by extrapolation of temperature measurements made as the sample is progressively frozen or melted. This is the basis of the heating- or cooling-curve method. Alternatively the true liquid phase composition at a given temperature may be calculated if the ratio of solid to liquid phase is measured. This latter procedure has not, to our knowledge, been used to study solutions of known composition, but accurate methods of determining the solid-liquid ratio have been developed for solvent purity determinations. This procedure may prove useful for some organic solvent systems.

The appropriate choice of experimental method will depend very much on the physical properties of the solvent. Thermal equilibrium methods have the advantage, from the thermodynamic point of view, that they are conducted under equilibrium conditions. A steady state temperature is not necessarily the same as an equilibrium temperature. For experimental reasons, equilibrium methods are impracticable for the majority of organic solvent systems and a method such as the heating and cooling method, which attempts to determine the freezing point of a sample of known concentration, must be used.

(a) *Thermal Equilibrium Method*

The principles of this method were first published by White.[D7] The mixture of solution and solid solvent is contained under near adiabatic conditions. Thermal equilibrium between the phases is maintained by a method of stirring which is efficient but causes the minimum of heating. When equilibrium is achieved the temperature is recorded and a solid-free sample of solution is withdrawn for analysis. In the differential method, a temperature difference is measured between the freezing solution and pure freezing solvent, measured under identical conditions of heat loss and gain.

The most convenient device for this is a thermocouple but matched sets of thermistors or platinum resistance thermometers may be used. The method is only practicable provided:

(i) Sufficient pure solvent is available, and the heat of fusion is sufficiently high to enable a stable equilibrium condition to be achieved.

(ii) The viscosity of the solvent at the freezing point is sufficiently low to enable a solid-free sample of the solution to be withdrawn without disturbing the equilibrium.

(iii) A convenient and precise method of analysis is available for the small, dilute solution samples. The differential method is especially suited to studies of aqueous electrolyte solutions and it has reached a considerable degree of sophistication.[D8] A precision of about 10^{-4} K has been obtained with the relatively simple apparatus constructed by Brown and Prue.[D9] Peterson and Rodebush[D10] used the differential method in 1928 to make the first precise cryoscopic measurements on benzene solutions. The method has been used successfully by Dunnett and Gasser[D11] to study electrolyte solutions in dimethylsulphoxide.

(b) *Heating and Cooling Curve Method*

In these methods a solution of known concentration is either frozen or melted by uniform transfer of heat, the temperature is recorded as a function of time, and the temperature at which an infinitesimal amount of solid solvent will just coexist with the solution is ascertained. If the rate of heat transfer is constant then the rate of change of temperature with time will be lower when solid phase is present because of the latent heat required to melt or freeze the material. If the process could be carried out reversibly then a graph of temperature against time would look like A′B′EF in Fig. 2.8.1 if the solution is cooled, and line XUYVZ if the solution is heated. The lines B′EF and XUY are curved because, as more solvent solidifies, the concentration of the solution increases and the freezing point decreases; if there were no solute this region of the curve would be a horizontal line. In practice it is impossible to achieve completely reversible behaviour, the curves ABCKEF and XUVZ show the type of behaviour which is usually observed. An important factor contributing to this deviation from reversible behaviour, is that the thermal conductivity of the solid is considerably less than that of the stirred liquid; another factor in the case of cooling is the phenomenon of supercooling. An extrapolation procedure must, therefore, be used to obtain the true freezing point.

(c) *Rossini Method of Extrapolation*

Rossini and co-workers[D12,D13] have developed a method of extrapolation suitable for heating and cooling curves. They proposed analytical and graphical methods for estimating the deviation of the experimental, from the reversible curve caused by supercooling.

In addition to providing a method of determining the freezing point, they describe a procedure for estimating the purity of a liquid. A critical discussion of the practical and theoretical aspects of the heating and cooling curve method may be found in papers by Mair, Glasgow and Rossini[D12] and by Taylor and Rossini.[D13] Only an outline of the procedure relevant to the determination of the freezing point of a solution will be given here.

Consider the situation where the solution is losing heat to the surroundings which are at constant temperature T_S (Fig. 2.8.1). The problem is to extrapolate that portion of the curve which corresponds to the presence of solid phase back to the point at which it cuts the curve

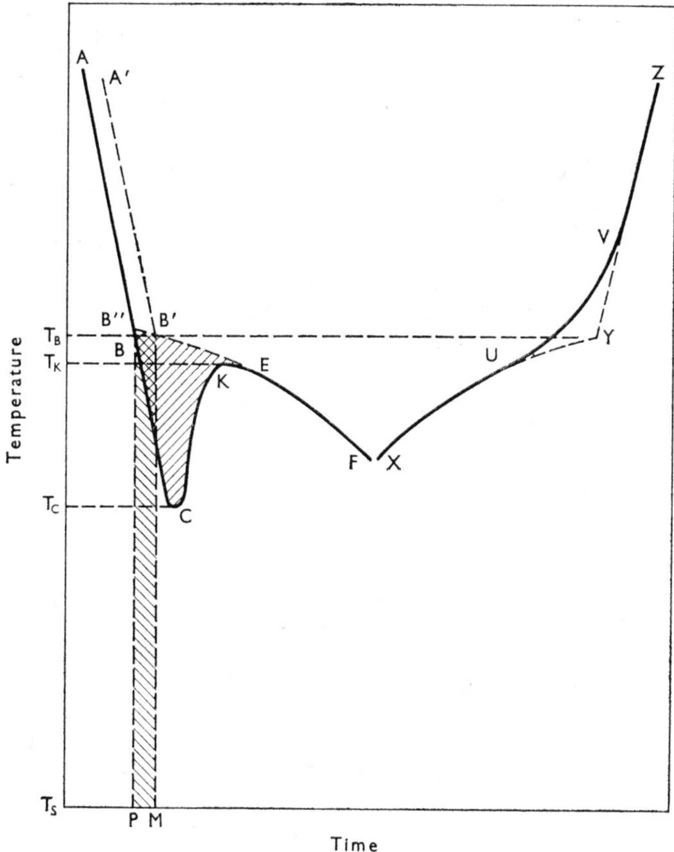

Fig. 2.8.1. Freezing and melting curves.

corresponding to liquid only. This can be done correctly only if the curvature is known. Taylor and Rossini[D13] have shown that if 'reversible' cooling is occurring then the curve is a regular hyperbola with three unknown constants. If then the temperature and time can be measured at three points on a portion of curve where 'reversible' cooling can be assumed to be occurring (portions EF in Fig. 2.8.1), then the curvature can be defined. If, however, one uses this curvature to extrapolate back to the point at which the curve cuts AB, i.e. at B″, the temperature obtained will be higher than the freezing point (point B in Fig. 2.8.1) because of the effect of supercooling. If supercooling occurs the temperature difference between the solution and the surroundings will be less than it would have been in the absence of supercooling. Therefore, the time taken to reach an equivalent point on the reversible portion of the curve, say E, will be greater when supercooling occurs. If, therefore, we extrapolate portion FE back towards line AB using the curvature for reversible behaviour, one must stop short of the line AB at some point B′ on the curve A′B′ where A′B′ is drawn parallel to AB in Fig. 2.8.1. Mair, Glasgow and Rossini[D12] showed, by equating the heat loss required to reach an equivalent point on the 'reversible' curve (point E) with and without supercooling, i.e.

$$H_{B} - H_{E} = H_{B'} - H_{E} \tag{2.8.4}$$

where H is the enthalpy in the given state, that

$$\text{area BB'MP} = \text{area BCKEB'} \tag{2.8.5}$$

Thus the position of B′ could be located. If the temperature difference between solution and surroundings is high, the position of B′ will not be significantly different from B″. Melting points can be determined from heating curves by an analogous extrapolation. The procedure is simpler because the correction for changes in rates of heating resulting from departures from 'reversible' heating is negligible, because such departures are much smaller than during supercooling.

(d) *Hoare*[D14] *Method of Extrapolation*

This is a simpler and more empirical method of determining the freezing point from cooling curves. The lowest temperature reached, T_{C}, (Fig. 2.8.1) and the highest temperature reached after supercooling, T_{K}, are measured for a series of different degrees of supercooling on the same solution. T_{K} will become higher the lower the degree of supercooling and in the limit of zero supercooling $T_{K} = T_{B}$. A plot of T_{K} against $T_{K} - T_{C}$ is found to give a straight line which is extrapolated to obtain T_{K} at $T_{K} - T_{C} = 0$. This value is taken as T_{B} the freezing point of the solution. For this method to be valid $T_{B} - T_{S}$ must be large compared with $T_{K} - T_{C}$. The method does not allow for the increase in concentration of solute as separation of solid solvent proceeds.

(e) *Methods which Reduce the Reliance on Extrapolation Procedures*

In general for the above procedures to be applicable the temperature difference between the solution and the surroundings must be large, which implies a large rate of heat extraction, and the experimental curve should follow the 'reversible' curve for an extent which is large compared with the extent of the extrapolation. These requirements are conflicting since in general the higher the rate of heat extraction the greater the deviation from reversible behaviour. Since the shape of the curve is subject to so many unknown factors it is better to reduce the reliance on extrapolation by arranging the experimental conditions so that the reversible curve is followed as closely as possible. The major requirements are then as follows:

(i) The solid must be finely divided to ensure good thermal equilibrium between the phases.

(ii) The stirring must be efficient. It is particularly important that all the solid phase is uniformly mixed with the solution.

(iii) The rate of cooling or heating must be small compared with the rate of freezing or melting, which usually means very low rates, about 10^{-3} W g^{-1}, = 1 Js^{-1}g^{-1}.

(iv) In the case of cooling, supercooling must be reduced. Cooling curves have been used extensively to study organic solvent systems and some method of initiating crystallisation and thereby reducing supercooling was normally necessary. Pettit and Bruckenstein[D15] used a platinum cold finger cooled with solid CO_2. If, however, cooling rates are low then the disruption caused by the cold finger can be as bad as that caused by the supercooling. Crystallisation may be induced by the addition of a pre-cooled inert object such as a glass bead, or by the addition of a single crystal of solvent.

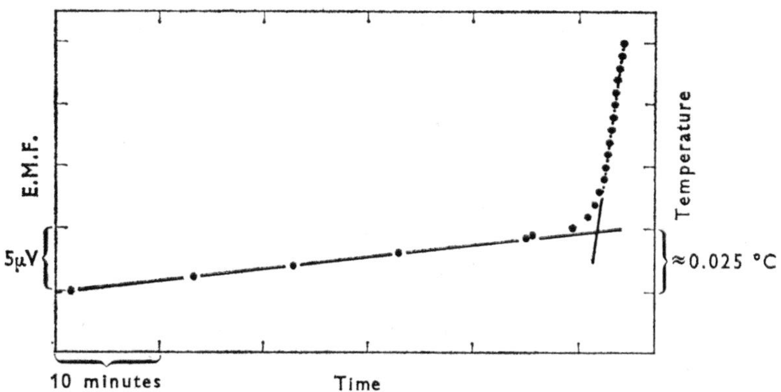

FIG. 2.8.2. Typical heating curve for dimethylsulphoxide solution.

We have used both cooling and heating curves for the study of electrolyte solutions of dimethylsulphoxide and sulpholane[D16] and found the heating-curve method more convenient and reproducible. To produce a large number of small solvent crystals, the solution was allowed to supercool and then crystallisation was induced by rapid stirring. Care was taken in the design of the stirrer to ensure that all solid was continuously agitated with the liquid. No solid should be allowed to remain on the surface of the container. Samples were heated at several different rates to check that the result was independent of heating rate. Figure 2.8.2 shows a typical heating curve for a DMSO solution. The melting point was determined with a precision of ± 0.001 K by linear extrapolation of the 'reversible' portions of the solidus and liquidus curves. The heating-curve method has been used by Munn and Kohler[D17] to study binary non-electrolyte mixtures.

Visual detection of the point at which solid finally disappears is the method adopted in most melting point apparatus[D4] used for the determination of molecular weights. A microscope and a polaroid filter make detection of solid phase easier. Visual detection has been used by Bonner, Jordan and Bunzl[D18] for precise work. The reliability of this method is suspect because without a complete melting curve it is impossible to assess whether the melting process is taking place reversibly. Visual or other direct detection of the disappearance of solid phase may be more reliable than extrapolation procedures for solvents with low heats of fusion particularly if solute concentration is high. The difference in slope between the solidus and liquidus portions of the curve at the melting point, decreases with ΔH_f because less heat is required to melt solid phase. The difference is further reduced at lower temperatures because of the curvature of the solidus line. This curvature is due to the increased depression in the freezing point caused by the increase in solute concentration as an increasing proportion of pure solvent is frozen. The curvature is greater the higher the initial solute concentration and the lower the value of ΔH_f.

(f) *Determination of Solid to Liquid Ratio*

The need for extrapolation and *in situ* analysis would be eliminated if the ratio of solid to liquid solvent could be determined. Several methods of doing this have been developed in connection with solvent purity determination.[D4] These include calorimetric, dilatometric and more recently dielectric and nuclear magnetic resonance procedures. In the calorimetric method the liquid-solid ratio is obtained by precisely monitoring the energy required to melt the sample under adiabatic conditions. The apparatus and procedures used in adiabatic calorimetry have been reviewed extensively.[D19a,b,c,D20] The apparatus is expensive but the method has the advantage that heats of fusion and heat capacities can also be obtained. Precision dilatometric procedures, first used for the study of phase change by Bunsen (calorimetry), have been developed by Plebanski[D21,D22] and Swietoslawski.[D23–D25] Measurements are usually performed at the saturation vapour pressure and the fraction F of solid melted at a given temperature, is calculated from the change in volume which accompanies the change in phase

$$F = (V_T - V_S)/(V_L - V_S) \qquad (2.8.6)$$

where V_T is the volume at the given temperature and V_L and V_S are the volumes when the sample is completely liquid and solid respectively.

Calorimetric and dilatometric studies of fusion are important in the study of the mechanism of melting where the entropy and volume of fusion are important parameters.[D1]

In the dielectric procedure developed by Ross and Frolen,[D26] the sample is made the dielectric medium between the plates of a capacitor and the change in capacitance which occurs as melting proceeds is used to determine the fraction melted

$$F = (C_T - C_S)/(C_L - C_S) \qquad (2.8.7)$$

A nuclear magnetic resonance cryoscopic technique has been developed recently by Lawrenson[D27] at the National Physical Laboratory for the determination of phase diagrams and solvent purity. The fraction melted may be determined because the N.M.R. signal from a solid is typically ten thousand times broader than for the liquid. It is possible to adjust the spectrometer to measure only the signal from the liquid. The intensity of the signal is proportional to the amount of liquid present.

2.8.4 Temperature Determination

The measurement of temperature is obviously of prime importance in all cryoscopic methods. The precise measurement of temperature and the use of thermometers is well documented.[D28–31] The choice of thermometer will depend on the precision required and the cryoscopic method adopted, which will be largely dictated by the solvent system under investigation.

(a) *Thermometers*

Temperature sensing devices which have been used for cryoscopy include mercury-in-glass, platinum resistance, thermistor, thermocouple and quartz crystal thermometers.

The mercury-in-glass thermometer of the Beckmann type is capable of measuring temperature differences with a precision of about 10^{-3} K, provided it is carefully calibrated and the necessary corrections for thermal expansion of the glass are made. These thermometers have been discussed in detail.[D32,D33] Their advantage is their low cost, but they have a high heat capacity and large immersion depth. They are generally less accurate than the other thermometers described.

Platinum resistance thermometers[D29,D34] can be used over an extensive temperature range, typically 14–750 K. They are capable of a reproducibility better than 0.001 K, which is better than the accuracy with which the thermodynamic scale has been established. They have been used to measure temperature differences at temperatures close to 273 K with a precision of about 10^{-5} K. To achieve such reproducibility and precision the resistor must be of pure platinum and be mounted in a strain-free condition. Resistivity changes can occur when the temperature is recycled because of diffusional rearrangement of impurities and lattice defects, and the relief of mechanical stresses. Very high precision thermometers are therefore rather delicate instruments. Strain-free construction is sometimes sacrificed for the sake of robustness in commercial thermometers. These high precision thermometers have the disadvantage of high heat capacity, rigidity and large immersion depth. Some small and flexible resistance thermometers are available commercially and are useful when the highest precision is not required. Like all resistance thermometers they are also electrical heaters

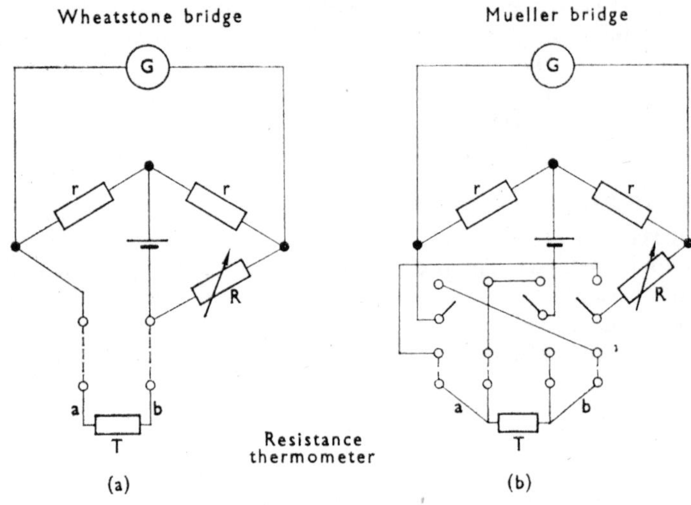

(a) For a two lead thermometer the resistance $R = T + a + b$

(b) For a four lead thermometer and reversing commutator $R_1 = T + b - a$; $R_2 = T + a - b$; $T = (R_1 + R_2)/2$

FIG. 2.8.3. Wheatstone and Mueller Bridge circuits.

and for precise work correction must be made for this electrical self-heating. The temperature coefficient of the resistance of platinum is low (*ca.* 0.0039 K^{-1}), and the apparatus required to measure such small resistance changes accurately is necessarily expensive, because all stray resistance and contact potentials must be eliminated. Correction must be made for the resistance of the thermometer leads. The resistance is usually measured with a modified D.C. Wheatstone bridge circuit known as a Mueller[D35] bridge. This bridge is compared with a simple Wheatstone bridge arrangement in Fig. 2.8.3. A four-lead thermometer is used and by reversing a commutator, the lead resistances and to some extent parasitic thermoelectric potentials are eliminated. The resistance can also be measured with a potentiometer and standard resistor. Current transformer bridges, such as the Type V.L.F.5 (Research and Engineering Controls Ltd.), eliminate the problems of contact potentials and lead resistances. These bridges use current transformers operating at 1 kHz to compare the ratio of the currents through two resistances with a modulation and demodulation device so that the actual resistance comparison is made at a frequency of 5 Hz.

The platinum resistance thermometer is used in maintaining the International Practical Temperature Scale and in consequence its temperature resistance relation is well characterised. A quadratic expression is used above 0°C

$$R_t/R_{0°C} = 1 + at + bt^2 \tag{2.8.8}$$

where t is the Celsius temperature.

From −183°C to 0°C a quartic expression is used

$$R_t/R_{0°C} = 1 + at + bt^2 + c(t - 100)t^3 \tag{2.8.9}$$

The constants a, b and c are evaluated by calibration at fixed points.

Thermistors are particularly useful resistance thermometers because of their very high negative coefficient of resistance; this is typically a 3% change in resistance per degree. Because of their high resistance and temperature coefficient, lead resistances can be ignored and relatively simple inexpensive bridge circuits can be used. Suitable circuits have been published by Faulkner, McGlashan and Stubley[D36] and others.[D37] When an A.C. bridge is used it may be necessary to operate auxiliary equipment such as heaters and stirrers on direct current to avoid interference. Thermistors are robust, small, and have a low heat capacity and rapid temperature response. They are therefore particularly useful when investigating small samples. McMullan and Corbett[D38] have written a useful article on their use in cryoscopy.

The temperature-resistance characteristics of thermistors depend upon their method of fabrication as well as their chemical composition. One consequence of this is that they are less stable than metal resistors. This instability is a serious disadvantage if a series of measurements is to be made over an extended period of time. Stable temperature characteristics are particularly desirable when making cryoscopic studies with organic solvent systems which are readily contaminated, because a determination of the freezing temperature is often the most convenient check on the solvent purity. This problem of long term stability can be alleviated by frequently checking the calibration against a thermometer which has stable characteristics, e.g. a platinum resistance thermometer. The temperature-resistance relationship for a thermistor follows the equation

$$\log R = A + B/T \tag{2.8.10}$$

where T is temperature in K, and A and B are constants. This equation is not exact and is particularly poor for a temperature range greater than 5 K. The empirical equation, $\log R = A + B/(T + \theta)$, where θ is a constant is more accurate, but is not completely satisfactory.

Thermocouples have been used extensively in cryoscopy and the use of these devices for determining temperature difference is well documented. Their advantages over platinum resistance thermometers are flexibility and small size. They can be even smaller than thermistors. Unlike resistance thermometers, they do not dissipate heat when in use. The temperature-e.m.f. relationship is more stable than with thermistors. The copper-constantan couple has been used extensively because it has a relatively high thermoelectric power and the temperature-e.m.f. characteristics are particularly linear and stable.

Thermocouples may be used to determine freezing-point depression directly by using a freezing mixture of the pure solvent as the reference temperature. It should, however, be remembered that the accuracy of a temperature depression measured in this way will be determined by the stability of the reference temperature. Should a freezing mixture of pure solvent be difficult to maintain, either because the heat of fusion is low or because the solvent is easily contaminated, it is preferable to use an alternative reference such as an ice-water mixture.

Thermoelectric potentials are quite small and should be determined at zero current so that a high precision potentiometer is required. Considerable care must be exercised to avoid parasitic e.m.f.'s which may be present at terminals and other connections. The e.m.f. produced per unit temperature may be increased by using multijunctions but the curve of current (on which sensitivity depends) versus number of junctions eventually flattens to the point that further increase in thermal bulk is no longer worthwhile.[D39] Scatchard *et al.*[D8] have used a 128 junction thermocouple to measure depressions of the triple point of water with a reported precision of 10^{-5} K. Brown and Prue[D9] obtained a precision of about 10^{-4} K with a 24 junction couple. They used the simple circuit shown in Fig. 2.8.4 to amplify the e.m.f. by a factor of 10.

The variation of e.m.f. with temperature can be represented by a power series in T

$$E = \sum_{n=1}^{n} A_n(T - T_R)^n \tag{2.8.11}$$

where A is a constant, n an integer and $T - T_R$ is the temperature difference between the arms of the thermocouple. In most cases only the first three terms need be considered.

A quartz crystal thermometer is a particularly convenient device for measuring temperatures with a precision of about 10^{-3} K. The temperature sensitive oscillations of a quartz crystal, cut in a particular plane, are monitored and compared with a standard frequency source. The quartz crystals are specially cut so that their oscillation frequency is an almost linear function of temperature. Temperature sensitivity is increased by increasing the time over which oscillations are counted.

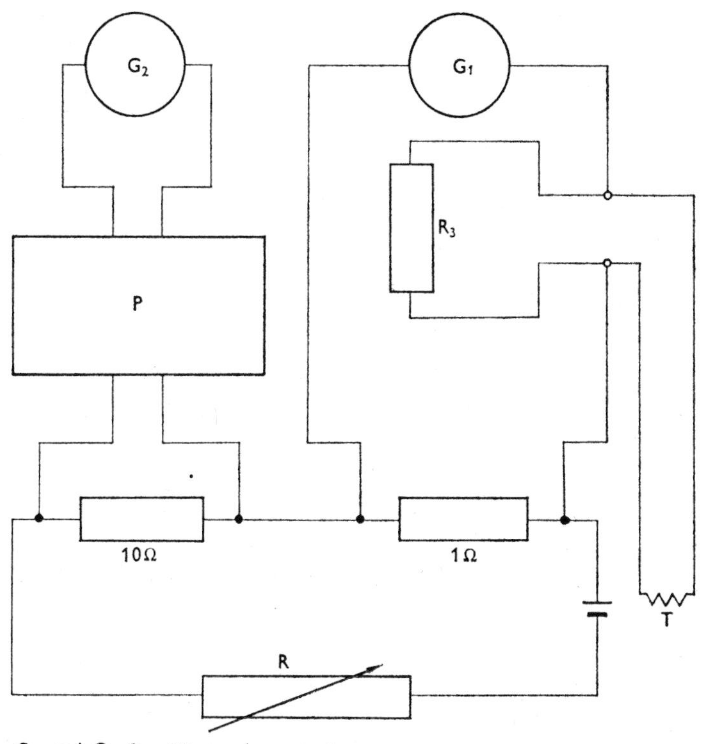

G_1 and G_2 Sensitive galvanometers
P Potentiometer
R Variable resistance ; varied to obtain null deflection on G_1
T Thermocouple
R_3 Resistance. Used as an alternative to T to check parasitic EMF's

FIG. 2.8.4. Circuit for determination of thermocouple e.m.f.

(b) *Calibration*

Primary calibration of thermometers, such as the platinum resistance thermometer with well-defined and stable characteristics, is achieved by making measurements at a series of defined temperatures or fixed points and then fitting these data to an appropriate equation. These fixed points should be selected to be as close to the proposed operating temperature range as possible. A list of Primary and Secondary Fixed points is given in Table 2.8.1. Precise calibration must be performed under closely regulated conditions and often requires specialised ancilliary equipment such as triple-point cells. An excellent calibration service is provided by institutions such as the National Physical Laboratory and the National Bureau of Standards.

Table 2.8.1

Primary and Secondary Fixed Points for the International Practical Temperature Scale*

Primary	Temperature °C	Secondary	Temperature °C
Boiling point of oxygen	−182.97	Melting point of mercury	−38.87
Triple point of water	+0.01	Ice point	0.00
Boiling point of water	100.0	Triple point of diphenyl ether	26.88
Boiling point of sulphur	444.6	Transition temperature of $Na_2SO_4.10H_2O$	32.38
Melting point of silver	960.8	Triple point of benzoic acid	122.36
Melting point of gold	1063.0	Melting point of indium	156.61
		Boiling point of naphthalene	218.0

* All at 1 atm pressure, except the triple points.

Problems often arise in the calibration of temperature-measuring devices which have been built into specialised apparatus. However, in the case of aqueous solutions sufficiently reliable values of cryoscopic depressions are available in the literature from which to calibrate the thermometer within the apparatus.

2.8.5 Design of Apparatus

The cryoscopic apparatus has three primary functions to perform which are: (i) to contain the sample and prevent its contamination during the experiment, (ii) to control the flow of heat between the sample and the surroundings, (iii) to ensure the thermal equilibrium between the phases. Typical cryoscopic assemblies for the differential equilibrium method and the heating or cooling curve method are shown in Figs. 2.8.5 and 2.8.6 respectively. The basic design for the differential equilibrium method was developed initially by White.[D7a,D7b] The apparatus is designed to provide an adiabatic environment. Heat transfer to the surroundings is minimised and the stirrers are designed so that mechanical heating is low. Errors due to heat transfer between sample and surroundings are compensated by making the sample and reference vessels identical. The design for the melting and freezing curve method was developed by Rossini and co-workers for the determination of solvent purity. The thermostat bath, set at a temperature either well above or well below the freezing point of the sample, provides the temperature gradient, and the heat transfer from the sample to the surroundings is controlled by varying the pressure between the inner and outer walls of the

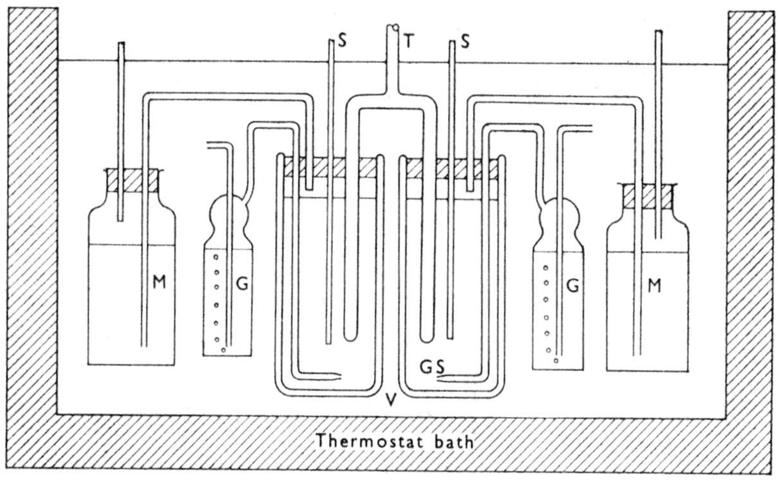

T Thermocouple
M Pre-equilibrated make-up solution
GS Gas stirrer
S Sample extraction tube connected to analysis system
G Saturator for stirrer gas
V Silvered vacuum vessels

FIG. 2.8.5. Freezing point depression apparatus; equilibrium method.

container. (For methods which call for very low rates of heat transfer it is usually more convenient completely to evacuate the double-walled container and vary the rate of heat transfer by varying the thermostat temperature.)

When studying organic solvent systems it is normally necessary to prevent contamination by rigorously excluding air. Measurements may be required at the vapour pressure of the solution, in which case the sample compartment must be evacuated. These conditions create problems of sample manipulation. A useful book by Shriver[D37] deals with these problems. One problem is stirring. It can be overcome by rocking the apparatus or operating a stirrer remotely by a solenoid or rotating magnet. Another problem is actually filling the apparatus. It may be necessary to fill the apparatus inside an inert atmosphere box or it may be necessary to deliver samples by syringe into the vessel. Where both solvent and solute are sufficiently volatile, distillation may be more convenient. Bonner *et al.*[D18] avoided the problem of atmospheric contamination by completely eliminating the vapour space by enclosing the sample in a cylinder closed by a piston. Both solvent and solutes, which were liquid, were injected into the container with syringes.

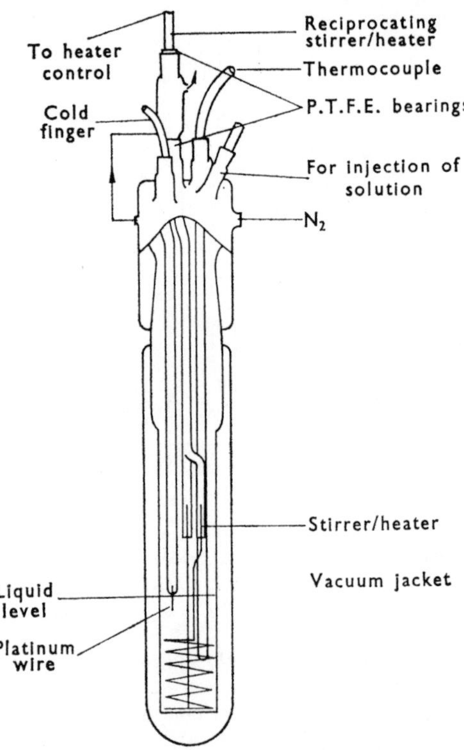

FIG. 2.8.6. Apparatus for the determination of freezing and melting curves.

To maintain equilibrium between the solid and liquid phases it is necessary to maintain efficient stirring. The appropriate type of stirrer will depend on the physical properties of the mixture to be stirred. The majority of organic solid/liquid mixtures are viscous. For a viscous mixture of components of comparable density a reciprocating stirrer is more effective than a rotary propeller type in a cylindrical container. With the latter there is a tendency for concentric laminar flow currents to be set up around the stirrer and mixing between these currents is relatively slow. The stroke of the stirrer should traverse the whole sample but not break the liquid surface. If the stirrer travels above the sample surface, solid solvent can be lifted out of solution, deposited on the walls of the container and thus change the concentration of the sample. For mixtures of a non-viscous liquid and a much denser solid, e.g. sodium sulphate in water, a stirrer with a piston-like action will simply pack solid firmly against the bottom of the container and a propeller-type stirrer is therefore more effective. For a mixture such as ice and water where the solid is less dense than the liquid, a system which lifts the liquid up and pours it over the solid, such as the air-jet circulators used by Brown and Prue,[D9] is more effective.

2.9 RESULTS

2.9.1 Thermodynamic Expression for the Depression of Freezing Point

For the equilibrium between a liquid mixture of two non-reacting substances and the pure solid phase of one of them, (1), the following thermodynamic relationship between the activity a_1 of species 1 and the depression of freezing point $\theta = T_0 - T$ may be derived.[D40]

$$-\ln a_1(T) = \int_{T_0}^{T} -\frac{\Delta_f H}{RT^2}\,\mathrm{d}T \tag{2.9.1}$$

where $\Delta_f H$ is the molal heat of fusion of the pure substance.

For non-ideal solutions

$$-\ln a_1(T) = \phi M_1 \sum_i m_i \tag{2.9.2}$$

where ϕ is the practical osmotic coefficient of the solvent, M_1 its molecular weight in kg and $\sum_i m_i$ the sum of the molalities of the solutes.

If $\Delta_f H$ can be taken as constant over the temperature range from T_0 to T then from eqns. 2.9.1 and 2.9.2

$$\phi \sum_i m_i = \frac{\Delta_f H}{RM_1T_0^2}\,\frac{T_0\theta}{(T_0 - \theta)} \tag{2.9.3}$$

The quantity $M_1RT_0^2/\Delta_f H$ is the molal cryoscopic constant λ_C of the solvent, given in the limit as $\sum_i m_i \to 0$ by

$$\lambda_C = \theta / \sum_i m_i \tag{2.9.4}$$

hence

$$\phi \sum_i m_i = T_0\theta/\lambda_C(T_0 - \theta) \tag{2.9.5}$$

In general $\Delta_f H$ is not independent of temperature, and

$$\frac{\partial \Delta_f H}{\partial T} = \Delta Cp \tag{2.9.6}$$

where ΔCp = the difference in the partial molal heat capacity of the solvent in the solution and the molal heat capacity of the solid solvent.

If this quantity is a linear function of T

$$\Delta C_p = \Delta a + \Delta bt \tag{2.9.7}$$

then the following relationships may be derived[D41] by integration of eqn. 2.9.1

$$\lambda_c \phi \sum_i m_i = \left(\frac{\Delta_f H^0 + \bar{L}_1}{TT_0}\right) \theta - (\Delta_f C^0 + \bar{J}_1)\left[\frac{\theta}{T} + \ln\left(1 - \frac{\theta}{T_0}\right)\right]$$
$$-\Delta b\left[\frac{\theta^2}{2T} - \frac{T_0\theta}{T} + T_0 \ln\left(1 - \frac{\theta}{T_0}\right)\right] \tag{2.9.8}$$

or alternatively

$$\lambda_c \phi \sum_i m_i = \left(1 + \frac{\bar{L}_1}{\Delta_f H^0}\right) + \left(\frac{1}{T_0} - \frac{\Delta_f C^0}{2\Delta_f H^0} + \frac{\bar{L}_1}{T_0 \Delta_f H^0} - \frac{\bar{J}_1}{2\Delta_f H^0}\right)\theta^2$$
$$+ \left(\frac{1}{T_0^2} - \frac{2\Delta_f C^0}{3T^0 \Delta_f H^0} + \frac{\Delta b}{6\Delta_f H^0} + \frac{\bar{L}_1}{T_0^2 \Delta_f H^0} - \frac{2\bar{J}_1}{3T_0 \Delta_f H^0}\right)\theta^3 + \ldots \tag{2.9.9}$$

where $\Delta_f H^0$ is the heat of fusion of the pure solvent $\bar{L}_1$ is the relative partial molal enthalpy of the solvent in the solution at T_0, $\Delta_f C^0$ is the change in ΔC_p on fusion of the pure solvent and $\bar{J}_1$ is the relative partial molal heat capacity of the solvent.

$\bar{L}_1$ and $\bar{J}_1$ are rarely known for solvents other than water. However, for most solvents the heat of fusion is such that the error in ϕ introduced by ignoring these terms will be less than the experimental uncertainty in θ. Likewise the term in Δb is usually negligible compared with the uncertainty in θ. Ignoring these terms, eqn. 2.9.9 reduces to

$$\lambda_c \phi \sum_i m_i = \theta + \left[\frac{1}{T_0} - \frac{\Delta_f C^0}{2\Delta_f H}\right]\theta^2 + \frac{\theta^3}{T_0^2} \tag{2.9.10}$$

2.9.2 Presentation of Results

The practical osmotic coefficient ϕ obtained directly from cryoscopic measurements gives a measure of the deviation from ideality for the solvent species. The corresponding measure of deviation from ideality for the solute species is the activity coefficient, γ. γ and ϕ are related as a consequence of the Gibbs-Duhem equation. The following expression was first derived by Bjerrum[D42]

$$\mathrm{d}(1 - \phi)\sum_i m_i = -\sum_i m_i \mathrm{d} \ln \gamma_i \tag{2.9.11}$$

The integrated form of this equation for a single solute is

$$-\ln \gamma = 1 - \phi + \int_0^m (1 - \phi)\mathrm{d} \ln m \tag{2.9.12}$$

Thus if θ is measured as a function of m so that $1 - \phi$ can be expressed as a function of m then values of the activity coefficient of the solute can be calculated. In practice this is often done in the following way. Equation 2.9.12 can be written in the form

$$-\ln \gamma = 1 - \phi + \int_0^{m_i} (1 - \phi)\mathrm{d} \ln m + \int_{m_i}^{m} (1 - \phi)\mathrm{d} \ln m \tag{2.9.13}$$

where m_i is the lowest molality at which ϕ can be obtained experimentally. The second integral may be evaluated graphically or numerically from a smoothed graph of $1 - \phi$ against $\ln m$. The first integral can only be evaluated by assuming a relationship between $1 - \phi$ and $\ln m$ for solutions more dilute than m_i; any error in this will of course become smaller, the smaller the value of m_i. Scatchard, Jones and Prentiss[D43] have used this method of representing results.

One purpose of calculating γ is to enable comparisons to be drawn between experimental data and theory. However, using Bjerrum's relation a theoretical expression for ϕ may be derived and compared directly with the experimental value. Taking this process a step further, a theoretical value of θ may be calculated and compared directly with the experimentally determined value.

Taking, as an example the Guggenheim[D44] extension of the Debye-Hückel equation for the mean ionic activity coefficient of a single 1:1 electrolyte

$$\ln \gamma_{\pm} = \frac{A(\ln 10)m^{\frac{1}{2}}}{1 + Ba\, m^{\frac{1}{2}}} + 2\beta m \tag{2.9.14}$$

where

$$A \ln 10 = (2\pi N d_0)^{\frac{1}{2}}(e^2/\varepsilon_r kT)^{\frac{3}{2}} \quad \text{(cf. 2.5.7)} \tag{2.9.15}$$

If eqn. 2.9.14 is inserted into the Bjerrum relation 2.9.11, we obtain

$$1 - \phi = \tfrac{1}{3} A\,(\ln 10)m^{\frac{1}{2}}\sigma(Ba\, m^{\frac{1}{2}}) - \beta m \tag{2.9.16}$$

where

$$\sigma(x) = \frac{3}{x^3}\int_0^x \left(\frac{x}{1+x}\right)^2 \mathrm{d}x$$

$$= \frac{3}{x^3}\left[1 + x - \frac{1}{1+x} - 2\ln(1+x)\right] \tag{2.9.17}$$

$$B = \left(\frac{8\pi N e^2}{k}\right)^{\frac{1}{2}}(d_0/\varepsilon_r T)^{\frac{1}{2}} \tag{2.9.18}$$

a is the ion size parameter

ε_r is the relative permittivity (dielectric constant) of the medium and d_0 the solvent density.

These equations have two variables a and β. Since the first term in the binominal expansion of $(1 + x)^{-1}$ is proportional to x^2, Ba can be given an arbitrary constant value and variations in this term can be accommodated by β. Guggenheim suggested setting $Ba = 1\ \text{mol}^{-\frac{1}{2}}\ \text{kg}^{\frac{1}{2}}$ for aqueous solutions.

By taking $Ba = 1\ \text{mol}^{-\frac{1}{2}}\ \text{kg}^{\frac{1}{2}}$ we can define a standard contribution to the osmotic coefficient by the equation

$$1 - \phi^{st} = \tfrac{1}{3} A\,(\ln 10)\, m^{\frac{1}{2}}\, \sigma(m^{\frac{1}{2}}) \qquad (2.9.19)$$

and a standard depression of freezing point by

$$\theta^{st} = 2\lambda_c m \phi^{st} \qquad (2.9.20)$$

Then

$$\phi - \phi^{st} = \beta m \qquad (2.9.21)$$

and for small values of θ

$$\theta - \theta^{st} = 2\lambda_c \beta m^2 \qquad (2.9.22)$$

Thus a graph of $\theta - \theta^{st}$ against m^2 should give a straight line through the origin. This method of representing results has been used by Prue and co-workers.[D9,D16]

In accordance with the Brönsted principle of specific ion interactions[D45] the parameter β refers to a specific interaction coefficient.

2.9.3 Incompletely Dissociated Electrolyte Solutions

For a partially associated 1:1 electrolyte, assuming that the free ions obey the Debye-Hückel equation and the associated part behaves ideally, the following equations relate the depression of freezing point θ to the degree of association $(1 - \alpha)$ and the association constant K.

$$\sum_i m_i = (1 - \alpha)m + 2\alpha m \phi_i \qquad (2.9.23)$$

substituting in 2.9.5

$$T_0\theta/\lambda_c(T_0 - \theta) = (1 - \alpha)m + 2\alpha m \phi_i \qquad (2.9.24)$$

$$1 - \phi_i = \tfrac{1}{3} A(\ln 10)(\alpha m)^{\frac{1}{2}} \sigma\{Bd(\alpha m)^{\frac{1}{2}}\} \qquad (2.9.25)$$

$$-\ln \gamma_i = A(\ln 10)(\alpha m)^{\frac{1}{2}}/\{1 + Bd(\alpha m)^{\frac{1}{2}}\} \qquad (2.9.26)$$

$$K = (1 - \alpha)/m\alpha^2\gamma_i^2 \qquad (2.9.27)$$

ϕ_i is the osmotic coefficient of the free ions and d is the distance of closest
ϕ_i is the osmotic coefficient of the free ions. d is the distance of closest approach of the free ions. The choice of a value of d is arbitrary, so that the product Bd may be regarded as an adjustable parameter. If θ is obtained experimentally, then for a chosen value of Bd, α can be obtained from eqns. 2.9.24 and 25 and then K is obtained from 2.9.26 and 27. A value of a, the closest distance of approach of the associated ions, may

be calculated by inserting this experimental value of K together with the chosen value of d into the Bjerrum[D46] equation (2.9.28) and then solving for a

$$K = 4\pi N \int_a^d \exp(e^2/\varepsilon_r kTr) r^2 \mathrm{d}r \tag{2.9.28}$$

This procedure has been used by Garnsey and Prue.[D16]

2.10 REVIEW OF CRYOSCOPIC STUDIES

2.10.1 Acetic Acid

Anhydrous acetic acid melts at 16.635°C and has a cryoscopic constant, $\lambda_c = 3.59$ K kg mol^{-1}.[D47] Raoult[D48] used this solvent for molecular weight determinations and it has been used by several workers[D49–52] for the investigation of hydrocarbon solutes. The dielectric constant is 6.194 at 18°C. Few investigations of electrolyte solutions have been reported and association is pronounced.[D47,D53] Lithium salts in particular, appear to polymerise in acetic acid solution. Turner and Bissett[D54] found that lithium iodide, and to a lesser extent lithium nitrate, appear to polymerise, though sodium iodide did not exhibit similar behaviour. Kenttämaa[D47] calculated equilibrium constants for the reaction

$$n\text{LiX} \rightarrow (\text{LiX})_n \tag{2.10.1}$$

The halides appear to dimerise, with the chloride being the most strongly associated and the iodide the least. For lithium nitrate the predominant species was found to be a trimer. Cryoscopic studies of solutions of cobalt and nickel acetates indicate that these salts exist mainly as dimers in acetic acid.[D55]

References to cryoscopic measurements on a variety of solutes such as amines, acetates and acids will be found in a paper by Jander and Klaus.[D56]

2.10.2 Benzene

Peterson and Rodebush[D10] made precise measurements by the differential equilibrium method on a number of non-electrolytes (toluene, methanol, ethanol, water, acetic acid and benzoic acid). Precise measurements by a similar technique on carboxylic acids[D57] and tetraalkylammonium salts[D58,D59] were made by Kraus and co-workers. The cryoscopic constant is $\lambda_c = 5.089$ K kg mol^{-1}. Measurements on the tetraalkylammonium salts were made over a wide concentration range and degrees of association as a function of concentration were evaluated. The association of phenolic compounds in benzene has been studied cryoscopically by Vanderborgh *et al.*[D60] while Brown *et al.*[D61] have investigated complex formation between ethyl-lithium and triethylamine in benzene solution.

2.10.3 Dimethylsulphoxide

Dimethylsulphoxide melts at 18.54°C and has a cryoscopic constant of about 4 K kg mol^{-1}, and is therefore a convenient solvent for cryoscopic experiments. Unfortunately, discrepant values have been reported for the cryoscopic constant. Weaver and Keim[D62] obtained $\lambda_c = 4.09$ K kg mol^{-1} from cooling curve measurements with naphthalene and urea as solutes. Skerlak and Ninkov[D63] reported $\lambda_c = 3.96 \pm 0.06$ K kg mol^{-1} from measurements on water, methanol and urea, again using a conventional Beckmann technique, and confirmed this by a calorimetric determination of the heat of fusion. An independent determination of the heat of fusion gives[D64] $\lambda_c = 3.86$ K kg mol^{-1}. Garnsey and Prue[D16] made measurements by a heating-curve technique with benzoic acid as solute up to 0.2 mol kg^{-1} and obtained $\lambda_c = 4.07 \pm 0.02$ K kg mol^{-1}. A value of $\lambda_c = 4.36$ K kg mol^{-1} was obtained[D65] from a few measurements on rather concentrated solutions of benzene as solute, and Bonner and co-workers[D66] have recently reported $\lambda_c = 4.40 \pm 0.01$ K kg mol^{-1} from measurements on several non-electrolytes. The measurements were made by a heating-curve method,[D18] which had however, earlier given a cryoscopic constant 15% too high for *N*-methylacetamide.[D67]

The first systematic studies of electrolyte solutions in dimethylsulphoxide were made by Skerlak, Ninkov and Sislov[D68] who used a conventional Beckmann technique to study lithium and potassium perchlorates; potassium, ammonium and silver nitrates; lithium chloride and potassium iodide. Measurements were made up to a molality of 1 mol kg^{-1} and the specific differences of activity coefficients among the salts showed the same qualitative patterns as in water. The work was criticised by Dunnett and Gasser[D11] who made more precise measurements, with rigorous exclusion of water, by the equilibrium technique on lithium chloride solutions; samples of solution were analysed conductimetrically. The same technique was subsequently used for measurements on solutions of caesium[D69] and rubidium iodides.[D70]

Garnsey and Prue[D16] made measurements by a heating-curve method on solutions of the perchlorates of lithium and potassium, and of lithium chloride. Measurements were made up to a molality of about 0.25 mol kg^{-1}. The first two salts behave as strong electrolytes with activity coefficients given by the equation (cf. eqn. 2.9.14)

$$\ln \gamma = \frac{-A(\ln 10)m^{\frac{1}{2}}}{1 + Bam^{\frac{1}{2}}} + \beta' m \qquad (2.10.2)$$

with $a = 5$ Å and $\beta' = 0.02$ kg mol^{-1} for potassium perchlorate and $\beta' = 0.27$ kg mol^{-1} for lithium perchlorate. The results for lithium chloride could not, however, be fitted by such an equation, and with an ion-size parameter of 5 Å for the free ions it was necessary to invoke an association constant $K_A = 3$ kg mol^{-1} in order to fit the results within ± 0.01 K which was still outside the experimental accuracy. The agreement with the results of Dunnett and Gasser[D11] for lithium chloride is only moderate, their values of the freezing point depression being 0.005–0.02 K less than those of Garnsey and Prue.[D16]

Bonner and co-workers[D66] have recently reported cryoscopic measurements up to 0.1 mol kg^{-1} on ten electrolytes, viz. LiCl, NaCl, KCl, NH_4Cl, CsCl, NaI, KI, NH_4I, NEt_4I and $(Et_3N(CH_2)_2NEt_3)I_2$. The measurements were made by a heating curve method which as mentioned earlier gave a rather high value for the cryoscopic constant of this solvent. For lithium chloride the actual values of freezing point depression appear to be in good agreement (within ± 0.005 K) with those of Dunnett and Gasser.[D11]

For lithium chloride there are also two sets[D71,D72] of e.m.f. measurements of activity coefficient values. Precise measurements[D71] at 25, 30 and 35°C on the cell

$$\text{Pt}|\text{Li}|\text{LiCl in DMSO }|\text{TlCl(s)}|\text{Tl-Hg}|\text{Pt}$$

over about the same concentration range as the cryoscopic measurements of Garnsey and Prue, lead to the conclusion that lithium chloride is a strong electrolyte with activity coefficient values given by

$$\ln \gamma = \frac{-A(\ln 10)m^{\frac{1}{2}}}{1 + m^{\frac{1}{2}}} + \beta' m \qquad (2.10.3)$$

with $\beta' = 0.325$ kg mol^{-1} at 25°C. A value of $A \ln 10 = 2.57$ kg$^{-\frac{1}{2}}$ mol$^{\frac{1}{2}}$ was used; a better value of the dielectric constant[D4] gives $A \ln 10 = 2.69$ kg$^{-\frac{1}{2}}$ mol$^{\frac{1}{2}}$, and this will lead to a small change in β'. According to eqn. 2.10.3, the osmotic coefficient of a 0.1 mol kg^{-1} solution would be 0.84. The corresponding value from the freezing point is 0.79 according to Bonner *et al.*, whilst Garnsey and Prue[D16] obtained 0.83. The e.m.f. results at molalities below 0.02 mol kg^{-1} were anomalously low and were neglected. These deviations were ascribed[D71] to the effect of the solubility of thallium (I) chloride and introduce an uncertainty into the conclusions reached from e.m.f. studies. The second set of e.m.f. measurements[D72] was made with a cell in which a lithium amalgam electrode replaced the lithium electrode. Measurements were made over a temperature range and a large temperature correction was applied. According to Holleck, Cogley and Butler,[D72] cryoscopic measurements are probably in error because of the effect of heat transfer between the reference and the solution vessel. This could make the apparent values of cryoscopic depression too low. More experimental work by both cryoscopic and e.m.f. techniques is required. (The e.m.f. work mentioned above is also discussed in sect. 2.7.3c.)

2.10.4 Ethylene Carbonate

The melting point is 36.41°C and the cryoscopic constant $\lambda_c = 5.40$ K kg mol^{-1}. Earlier values for λ_c of around 7 K kg mol^{-1} are ascribed to supercooling errors.[D73] In the work of Bonner *et al.*,[D73] the value calculated from the heat of fusion, determined calorimetrically, was confirmed by extrapolation to zero concentration of the freezing point depression results (determined from heating curves) for several non-electrolytes. For the non-electrolytes, the osmotic coefficients diminish sharply with increasing concentration, e.g. at 0.1 mol kg^{-1}, $\phi = 0.859$ for carbon tetrachloride and $\phi = 0.695$ for urea. For the alkali metal iodides the specific differences show the same sequence as in water. Surprisingly, ammonium iodide shows a much faster decrease of ϕ with increasing concentration than the alkali metal iodides. Zinc chloride and zinc iodide are weak electrolytes in this solvent.

2.10.5 Ethylenediamine

The melting point is 11.3°C and the cryoscopic constant $\lambda_c = 2.43$ K kg mol^{-1}. The cryoscopic constant was obtained by cooling-curve measurements on benzene and naphthalene solutions.[D15] In contrast to some alkali halides, silver nitrate was detectably associated over the concentration range 0.01 – 0.1 mol kg^{-1} and a dissociation constant of about 4×10^{-4} mol kg^{-1} was reported.

2.10.6 Formamide

The melting point is 2.5_5°C and the cryoscopic constant $\lambda_c = 2.56_5$ K kg mol^{-1}.[D74–76] Dawson and Griffiths obtained this cryoscopic constant by a cooling curve method with urea and sucrose as solutes. However, Mostkova, Kessler and Safonova[D76] obtained $\lambda_c = 3.548$ K kg mol^{-1} from data for urea solutions and $\lambda_c = 3.565$ K kg mol^{-1} with water as the solute and considered the latter value the more reliable. The dielectric constant is approximately 109 at room temperature and about 120 at the freezing point. It is difficult to obtain the pure solvent; Mostkova *et al.* suggest the presence of water in the solvent, possibly as a result of solvent decomposition, as a reason for the discrepancies between their own, and the earlier data.[D74,D77]

The measurements by Dawson and Griffiths on potassium and ammonium chlorides and on acetic, propionic and benzoic acids, and those by Mostkova *et al.* on sodium, potassium and caesium chlorides give results not unlike those in water. Values of the activity coefficients of alkali halides reported by Vasenko[D77] are based on a value $\lambda_c = 3.166$ K kg mol^{-1}. Some cryoscopic measurements on solutions of tetraalkylammonium salts have been reported by Gopal and Hasain.[D78]

2.10.7 Formic Acid

Formic acid melts at 8.40°C and has a cryoscopic constant $\lambda_c = 1.932$ K kg mol^{-1}.[D79] These results of J. Lange[D79] were obtained by a Beckmann cooling-curve technique in which a differential apparatus gave values of freezing point depression for which a precision of ± 0.0001 K was reported.[D80] Formic acid has a dielectric constant of 58.5 at 16°C and results for potassium chloride, potassium picrate and tetramethylammonium chloride were fitted to an extended Debye-Hückel equation for completely dissociated electrolytes. Tetramethylammonium chloride showed the smallest deviation from the limiting law and this was ascribed to the affinity of the organic cation for the solvent.

2.10.8 *N*-Methylacetamide

N-Methylacetamide melts at 30.56°C and has a cryoscopic constant of 5.77 K kg mol^{-1}. Discrepant values have been reported for the cryoscopic constant. Kreis and Wood[D67] calculated a value from the enthalpy of fusion. This was obtained by measuring the enthalpies of solution in water of crystalline *N*-methylacetamide at several temperatures below its melting point and of liquid *N*-methylacetamide at several temperatures above the melting point; both sets of results were fitted to smoothing functions and the difference in the enthalpies of solution of crystalline and liquid *N*-methylacetamide at the melting point is of course equal to the enthalpy of fusion. The value obtained in this way was about 15% higher than that reported by Bonner, Jordan and Bunzl[D18] from direct calorimetric measurements. The same authors reported confirmation of their value by cryoscopic measurements using a heating-curve technique. However, at molalities below about 0.05 mol kg^{-1} the cryoscopic depressions showed an unusual concentration dependence for both non-electrolytes and electrolytes, and the results were therefore suspect. Bonner and Woolsey[D81] have recently suggested that this solvent is unsuitable for precise work because even the dry liquid is unstable and decomposes to give appreciable concentrations of impurities.

Wood and co-workers[D82,D83] have reported an extensive series of careful measurements from 0.1 to 0.8 mol kg^{-1} on alkali halides, nitrates and carboxylates, and on tetramethylammonium halides. An equilibrium technique was used. The results were fitted to extended Debye-Hückel equations. The solvent has a high dielectric constant (178 at 30.5°C) and osmotic and activity coefficients are higher at corresponding concentrations than in water. The sequences of specific differences are, however, the same as in water. This implies that the sequences in both solvents arise from ion-solvent competition effects of a general nature, and are not closely related to structural effects. The matter is carefully discussed by Wood *et al.*

2.10.9 Sulpholane ($TMSO_2$)

Two groups of workers[D16,D84,D85] have made precise cryoscopic measurements by heating-curve methods with sulpholane which freezes at 28.45°C. The solid phase is a mesomorphic one for which dielectric constant and other evidence suggest that the molecules can rotate relatively freely in the solid. The first transition is characterised by an exceptionally low heat of fusion and a correspondingly high cryoscopic constant ($\lambda_c = 64.1$ K kg mol^{-1}). Measurements on benzoic acid by the two groups[D16,D84] lead to concordant values of the cryoscopic constant, but some solutes, e.g. carbon tetrachloride, cyclohexane and pyridine, with molar volumes close to that of sulpholane, dissolve in the solid and give abnormal results.[D84,D85] The advantage to be expected from the exceptionally high cryoscopic constant is not realised in practice. The freezing-point depressions are large, but the thermal buffering is poor and the solidus and liquidus portions of the heating curves have similar slopes, which makes the freezing-point more difficult to locate. Furthermore, measurements at different concentrations spread over a wide temperature range and should be corrected to a constant temperature; this requires enthalpy of dilution data which are not available.

Measurements on alkali metal perchlorates suggest that these salts are all detectably associated, the association constants decreasing in the sequence Li $<$ Na $<$ K $\approx$ Rb (the agreement with conductivity results[D86] is not satisfactory). The association of the perchlorates and the sequence of these constants emphasises the poor solvating power of sulpholane. Lithium chloride is a weak electrolyte which appears to be polymerised at concentrations of about 2×10^{-3} mol kg^{-1}.

2.10.10 Other Solvents

There are a number of solvents for which some cryoscopic data are available but which have not been studied extensively. The melting point, cryoscopic constant and literature references to these solvents and those discussed above are included in Table 2.10.1.

Mean activity coefficients derived from cryoscopic measurements have been tabulated by C. M. Criss in Appendices 2.4.15–19. These values should be used with caution as considerable uncertainties can arise from (a) the experimental method, (b) the value used for the cryoscopic constant, (c) integration using the Gibbs-Duhem equation (eqn. 2.8.13).

Table 2.10.1

Melting Points and Cryoscopic Constants

Solvent	Melting Point °C	Cryoscopic Constant K kg mol^{-1}	Ref.
Acetic Acid	16.635	3.59	D47
Acetamide	81.1	3.8	D87, D88
Benzene	5.50	5.089	D10, D57, D58, D59
Cyclohexane	6.5	20.0	D17, D89
Cyclohexanol	24.0	39.0	D90
Dimethylsulphoxide	18.54	{4.07, 4.4}	D16, D11, D66
Ethylene carbonate	36.41	5.40	D73
Ethylenediamine	11.3	2.43	D15
Formamide	2.55	2.57	D74, D75, D76
Formic acid	8.40	1.932	D79, D80
Hexamethylene-phosphoramide	7.2	6.93	D91, D92
Nitrobenzene	5.7	69.0	D93
N-methylacetamide	30.56	5.77	D67, D82, D83
Pyridine	−42.0	4.97	D37, D94
Sulpholane	28.45	64.1	D16, D84, D85
t-Butyl alcohol	24.0*	8.38	D95

* But see Table 1.3.1; there have been a number of reports of m.p. < 25°C.

REFERENCES

D1 A. R. Ubbelohde, Melting and Crystal Structure, Clarendon Press, Oxford (1965)

D2 R. Haase and H. Schonert, Solid-liquid Equilibrium, p. 13, Pergamon Press (1969)

D3 S. Bruckenstein and N. E. Vanderborgh, *Anal. Chem.*, **38,** 687 (1966)

D4 A. R. Glasgow, Jr. and G. S. Ross, Treatise on Analytical Chemistry, Chapter 88, Pt. 1, Vol. 8 (I. M. Kolthoff and P. J. Elving, eds.)

D5 T. W. Richards, E. K. Carver and W. C. Schumb, *J. Amer. Chem. Soc.*, **41,** 2019 (1919)

D6 F. D. Rossini, B. J. Mair and A. J. Streiff, Hydrocarbons in Petroleum, p. 253, Reinhold, N.Y. (1953)

D7 (*a*) W. P. White, *J. Phys. Chem.*, **44,** 393 (1920)

D7 (*b*) W. P. White, *J. Amer. Chem. Soc.*, **48,** 1149 (1926)

D8 G. Scatchard, B. Vonnegat and D. W. Beaumont, *J. Chem. Phys.*, **33,** 1292 (1960)

D9 P. G. M. Brown and J. E. Prue *Proc. Roy. Soc. A.*, **232,** 320 (1955)
D10 J. M. Peterson and W. H. Rodebush, *J. Phys. Chem.*, **32,** 709 (1928)
D11 J. S. Dunnett and R. P. H. Gasser, *Trans. Faraday Soc.*, **61,** 922 (1965)
D12 B. J. Mair, A. R. Glasgow, Jr. and F. D. Rossini, *J. Res. Bur. Std.*, **26,** 591 (1941)
D13 W. J. Taylor and F. D. Rossini, *J. Res. Nat. Bur. Std.*, **32,** 197 (1944)
D14 J. P. Hoare, *J. Chem. Educ.*, **37,** 146 (1960)
D15 L. D. Pettit and S. Bruckenstein, *J. Inorg. Nucl. Chem.*, **24,** 1478 (1962)
D16 R. Garnsey and J. E. Prue, *Trans. Faraday Soc.*, **64,** 1206 (1968)
D17 R. J. Munn and F. Kohler, *Monats. Chem.*, **91,** 381 (1960)
D18 O. D. Bonner, C. F. Jordan and K. W. Bunzl, *J. Phys. Chem.*, **68,** 2450 (1964)
D19 (*a*) A. R. Glasgow, Jr., G. S. Ross, A. T. Horten, D. Enagonio, H. D. Dixon, C. P. Saylor, G. T. Furukawa, M. L. Reilly and J. M. Henning, *Anal. Chim. Acta*, **17,** 54 (1957)
D19 (*b*) G. Pilcher, *Anal. Chim. Acta*, **17,** 144 (1957)
D19 (*c*) D. R. Stull, *Anal. Chim. Acta*, **17,** 133 (1957)
D20 J. M. Sturtevant, 'Calorimetry' *in* Physical Methods of Organic Chemistry, Vol. 1, 2nd Ed., p. 731 (A. Weissberger, ed.), Interscience, N.Y. (1949)
D21 T. Plebanski, *Bull. Acad. Polon. Sci. Ser. Sci. Chim.*, **8,** 117 (1960)
D22 T. Plebanski, *Bull. Acad. Polon. Sci. Ser. Sci. Chim.*, **8,** 23, 239 (1960)
D23 W. Swietoslawski, *Bull. Acad. Polon. Sci., Class A.*, 113 (1947)
D24 W. Swietoslawski, *Zh. Fiz. Khim.*, **36,** 2087 (1962)
D25 W. Swietoslawski and W. Tornassi, *Rocz. Chem.*, **22,** 105 (1948)
D26 G. S. Ross and L. J. Frolen, *J. Res. Nat. Bur. Std.*, **67A,** 607 (1963)
D27 E. F. G. Herington and I. J. Lawrenson, *J. Appl. Chem.*, **19,** 337, 341 (1969)
D28 Temperature, Its Measurement and Control in Science and Industry, Reinhold, New York (1941)
D29 Temperature. Its Measurement and Control in Science and Industry, Vols. 2 and 3, Reinhold, New York (1955)
D30 C. Halpern and R. J. Moffat, Bibliography of Temperature Measurement, January 1953 to June 1960, Nat. Bur. Std. Monograph 27 (1961)
D31 J. M. Sturtevant, Temperature Measurement *in* Physical Methods of Organic Chemistry, 2nd ed., Part 1, pp. 1–27 (Weissberger, ed.), Interscience, New York (1949)
D32 J. Busse, Liquid-in-Glass Thermometers in ref. D25, p. 228–255
D33 J. F. Swindells, Calibration in Liquid-in-Glass Thermometers, Nat. Bur. Std. Circ. 600 (1959)
D34 J. P. Pratt and D. C. Ailion, *Rev. Sci. Inst.*, **40,** 1614 (1969)
D35 H. F. Stimson, Precision Resistance Thermometry and Fixed Points, in ref. D29, p. 151–152, and Mueller, E. F., Precision Resistance Thermometry in ref. D28, p. 166–170
D36 E. A. Faulkner, M. L. McGlashan and D. Stubley, *J. Chem. Soc.*, 2837 (1965).
D37 D. F. Shriver, The Manipulation of Air-Sensitive Compounds, p. 163, McGraw-Hill, N.Y. (1969)
D38 R. K. McMullan and J. D. Corbett, *J. Chem. Educ.*, **33,** 313 (1956)
D39 W. P. White, *J. Amer. Chem. Soc.*, **36,** 2011 (1914)
D40 E. A. Guggenheim, Thermodynamics, 4th ed., p. 289, North-Holland, Amsterdam (1959)
D41 G. N. Lewis and M. Randall, Thermodynamics, 2nd ed., p. 406 (K. S. Pitzer and L. Brewer, eds.), McGraw-Hill, New York (1961)
D42 N. Bjerrum. *Z. Phys. Chem.*, **104,** 406 (1923)
D43 G. Scatchard, R. T. Jones and S. S. Prentiss, *J. Amer. Chem. Soc.*, **54,** 2676 (1932)
D44 E. A. Guggenheim, *Phil. Mag.*, **19,** 588 (1935)

D45 J. N. Brønsted, *Kgl. Danske Videnskab Selskab, Mat-fys Medd.*, **4,** No. 4 (1921)
D46 N. Bjerrum, *Kgl. Danske Videnskab Selskab, Mat-fys Medd.*, **7,** No. 9 (1926)
D47 J. Kenttämaa, *Suomen Kem.*, **32B,** 9 (1959)
D48 F. M. Raoult and A. Recoin, *J. Phys. Chem.*, **5,** 423 (1890)
D49 K. Freudenberg, E. Bruch and H. Rau, *Ber. Deut. Chem. Ges.*, **62B,** 3078 (1929)
D50 K. Freudenberg and E. Bruch, *Ber. Deut. Chem. Ges.*, **63B,** 535 (1930)
D51 K. Hess, *Ber. Deut. Chem. Ges.*, **63B,** 518 (1930)
D52 A. Garthe and K. Hess, *Ber. Deut. Chem. Ges.*, **64B,** 882 (1931)
D53 W. C. Eichelbergen, *J. Amer. Chem. Soc.*, **56,** 799 (1934).
D54 W. E. S. Turner and C. C. Bissett, *J. Chem. Soc.*, 1777 (1914)
D55 W. P. Tappmeyer and A. W. Davidson, *Inorg. Chem.*, **2,** 823 (1963)
D56 S. Jander and H. Klaus, *J. Inorg. Nucl. Chem.*, **1,** 1228 (1955)
D57 B. C. Barton and C. A. Kraus, *J. Amer. Chem. Soc.*, **73,** 4561 (1951)
D58 D. T. Copenhafer and C. A. Kraus, *J. Amer. Chem. Soc.*, **73,** 4557 (1951)
D59 H. S. Young and C. A. Kraus, *J. Amer. Chem. Soc.*, **73,** 4732 (1951)
D60 N. E. Vanderborgh, N. R. Armstrong and W. D. Spall, *J. Phys. Chem.*, **74,** 1734 (1970)
D61 T. L. Brown, R. L. Gerteis, D. A. Baffus and J. A. Ladd, *J. Amer. Chem. Soc.*, **86,** 2135 (1964)
D62 E. E. Weaver and W. Keim, *Proc. Indiana Acad. Sci.*, **70,** 123 (1960)
D63 T. Skerlak and B. Ninkov, *Glasmik Drustva Hemicara Technol., S. R. Bosne, Hercegovine*, **11,** 43 (1962)
D64 H. L. Clever and E. F. Westrum, *J. Phys. Chem.*, **74,** 1309 (1970)
D65 J. Kenttämaa, J. J. Lindberg and A. Nissema, *Suomen Kem.*, **34B,** 98 (1961)
D66 Si Joong Kim, O. D. Bonner and Doo-Soon Shin, *J. Chem. Thermodyn.*, **3,** 411 (1971)
D67 R. W. Kreis and R. H. Wood, *J. Chem. Thermodyn.*, **1,** 523 (1969)
D68 T. Skerlak, B. Ninkov and V. Sislov, *Glasmik Drustva Hemicara Technol. S.R. Bosne, Hercegovine*, **11,** 39 (1962)
D69 M. D. Archer and R. P. H. Gasser, *Trans. Faraday Soc.*, **62,** 3451 (1966)
D70 J. M. Crawford and R. P. H. Gasser, *Trans. Faraday Soc.*, **63,** 2758 (1967)
D71 W. H. Smyrl and C. W. Tobias, *J. Electrochem. Soc.*, **115,** 33 (1968)
D72 G. Holleck, D. R. Cogley and J. N. Butler, *J. Electrochem. Soc.*, **116,** 952 (1969)
D73 O. D. Bonner, Si Joong Kim and A. L. Torres, *J. Phys. Chem.*, **73,** 1968 (1969)
D74 C. R. Dawson and E. J. Griffiths, *J. Phys. Chem.*, **56,** 281 (1952)
D75 G. Somsen and J. Coops, *Rev. Trav. Chim.*, **84,** 985 (1965)
D76 R. I. Mostkova, Yu. M. Kessler and I. V. Safonova, *Elektrokhimiya*, **5,** 409 (1969)
D77 E. N. Vasenko, *Zh. Fiz. Khim.*, **21,** 361 (1947); **22,** 999 (1948); **23,** 459 (1949)
D78 Ram Gopal and M. M. Hasain, *J. Indian Chem. Soc.*, **43,** 204 (1966)
D79 J. Lange, *J. Phys. Chem. (A)*, **187,** 27 (1940)
D80 J. Lange, *J. Phys. Chem. (A)*, **186,** 291 (1940)
D81 O. D. Bonner and G. B. Woolsey, *J. Phys. Chem.*, **75,** 2879 (1971)
D82 R. H. Wood, R. K. Wicker and R. W. Kreis, *J. Phys. Chem.*, **75,** 2313 (1971)
D83 R. W. Kreis and R. H. Wood, *J. Phys. Chem.*, **75,** 2319 (1971)
D84 M. Della Monica, L. Jannelli and U. Lamanna, *J. Phys. Chem.*, **72,** 1068 (1968)
D85 L. Jannelli, M. Della Monica and A. Della Monica, *Gazz. Chim. Ital.*, **94,** 552 (1964)
D86 R. Fernández-Prini and J. E. Prue, *Trans. Faraday Soc.*, **62,** 1257 (1966)
D87 M. L. Willard and C. Marcsh, *J. Amer. Chem. Soc.*, **62,** 1253 (1940)
D88 J. C. Davis, *J. Chem. Educ.*, **43,** 611 (1966)
D89 R. Meyer and J. Metzger, *Bull. Soc. Chim. France*, 4583 (1967)
D90 R. Mikulak and O. Runquist, *J. Chem. Educ.*, **38,** 557 (1961)

D91 J. Ducom, *Compt. Rend. C*, 264, 722 (1967)
D92 H. Normant, *Angew. Chem. Internat. Ed.*, **6,** 1046 (1967)
D93 F. S. Brown and C. R. Bury, *J. Phys. Chem.*, **30,** 694 (1926)
D94 J. A. Dilts and D. F. Shriver, *J. Amer. Chem. Soc.*, **90,** 5769 (1968); *J. Amer. Chem. Soc.*, **91,** 4088 (1969)
D95 M. J. Bigelow, *J. Chem. Educ.*, **46,** 108 (1969)

Chapter 2

Thermodynamic Measurements

Part 4

Interpretation of Thermodynamic Data

C. M. Criss
Department of Chemistry, University of Miami,
Coral Gables, Florida, 33124, U.S.A.

and

M. Salomon*
National Aeronautics and Space Administration
Electronics Research Center, Cambridge, Massachusetts

* Present address: U.S. Army Electronics Command, Fort Monmouth, New Jersey, 07703, U.S.A.

2.11 THEORETICAL CONSIDERATIONS

2.11.1 Introduction

A significant portion of chemical research during the past century has been devoted to the measurement and interpretation of thermodynamic properties of electrolytes in water. While some of these studies have been made over wide temperature ranges, the majority of measurements have been limited to one temperature, usually 25°C. The results of these studies have led to considerably improved understanding of the structure of such solutions and have been useful for the estimation of thermodynamic data in aqueous solution. On the other hand, they have not been particularly useful for elucidating the structure of electrolytic solutions in general nor for estimating thermodynamic quantities in non-aqueous solutions. In view of the vast effort expended in the study of aqueous systems, it is not very sensible merely to repeat these same measurements on all the possible solvent systems of potential interest, particularly since most of these are experimentally more troublesome than aqueous solutions. Consequently, it is useful to try to systematise the data which presently exist and to examine these data in the light of present theory as well as to examine methods of estimating thermodynamic values for systems for which no data exist.

2.11.2 Ion Solvation and Lattice Energies

In considering the thermodynamics of ion solvation, two equilibria of great importance are

$$MX_{\text{solid}} = M^{+}_{\text{soln}} + X^{-}_{\text{soln}} \qquad (2.11.1)$$

$$M^{+}_{\text{gas}} + X^{-}_{\text{gas}} = M^{+}_{\text{soln}} + X^{-}_{\text{soln}} \qquad \text{(cf. 2.6.37)(2.11.2)}$$

The subscripts solid and gas refer, respectively, to the solid and gas phase species. The two processes in 2.11.1 and 2.11.2 are related to each other through the cycle

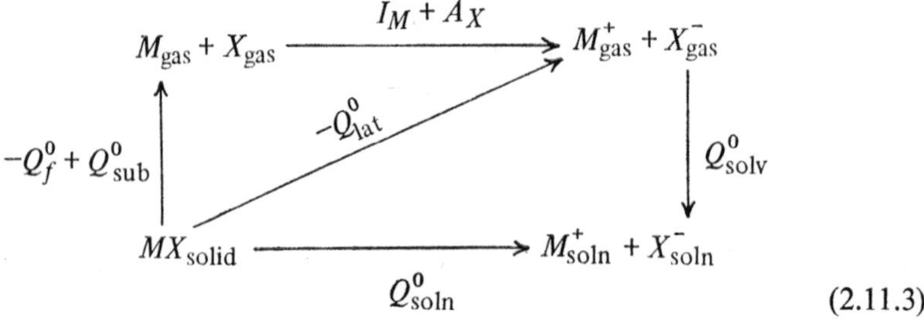

(2.11.3)

In this cycle, I and A are the ionisation potential of metal atom M and the electron affinity of the non-metal entity X, respectively. Q^0 is the energy or enthalpy of the indicated process; e.g. ΔH_f^0 is an enthalpy of formation and ΔH^0_{soln} and ΔH^0_{solv} correspond to equilibria 2.11.1 and 2.11.2, respectively. The thermodynamic quantities for reaction 2.11.1 are directly measureable as discussed in sect. 2.6.5 whereas those quantities for reaction 2.11.2 must be calculated indirectly from

$$\Delta G^0_{solv} = \Delta G^0_{soln} + \Delta G^0_{lat}$$
$$\Delta H^0_{solv} = \Delta H^0_{soln} + \Delta H^0_{lat} \qquad (2.11.4)$$

It is often preferable to work with the ΔG^0_{solv} and ΔH^0_{solv} terms because they may be divided into contributions from each ion. It should be noted that ΔG^0_{soln} and ΔH^0_{soln} are directly measureable and are therefore considerably more accurate than the corresponding solvation energies and enthalpies. It is apparent that lattice energies and enthalpies are important quantities because of their contribution to the accuracies with which ΔG^0_{solv} and ΔH^0_{solv} are known. For many years lattice energies were obtained from theoretical calculations[E1-7] based on the Born-Mayer-Madelung treatment which is applicable only to ionic crystals. An important result of these calculations is that the crystal radii of the various ions are obtained from the repulsive potential. These calculations also yield the electron affinities of the non-metallic species, which were not measureable with great accuracy in the 1930s. Crystal radii obtained from this method are often referred to as 'thermochemical' radii.[E8] They are largely dependent upon the compressibility of the crystal and an empirical repulsion constant, B, in the repulsive energy term $B \exp(-r/\rho)$ (cf. refs. E7, E8 and 2.11.4 for further details).

There are several empirical relations for the lattice energy (cf. E7, E8) and a recent one proposed by Halliwell and Nyburg[E9] is given by

$$\frac{-\Delta H^0_{lat}}{\text{kcal mol}^{-1}} = \frac{600}{(r_+ + r_-)/\text{Å}}\left\{1 - \frac{0.4}{(r_+ + r_-)/\text{Å}}\right\} \qquad (2.11.5)$$

In 2.11.5, r_+ is the cation radius of Ahrens[E10] and r_- is the Pauling[E11] anion radius. Lattice free energies and enthalpies can be determined experimentally by two methods. One method reported by Helmholz and Mayer[E12] involved high temperature equilibrium vapour pressure measurements of the reaction

$$NaCl_s \rightleftharpoons NaCl_g$$

At higher temperatures, vapour conductivity measurements were made for

$$NaCl_g \rightleftharpoons Na_g^+ + Cl_g^-$$

A second experimental method involves use of the Born-Haber cycle (cf. eqn. 2.11.3). For example the lattice free energy and enthalpy of NaCl are given by

$$\Delta H^0_{\text{lat}}(\text{NaCl}) = \Delta H^0_f(\text{Na}_g) + \Delta H^0_f(\text{Cl}_g) - \Delta H^0_f(\text{NaCl}_s) + I - A$$
$$\Delta G^0_{\text{lat}}(\text{NaCl}) = \Delta G^0_f(\text{Na}_g) + \Delta G^0_f(\text{Cl}_g) - \Delta G^0_f(\text{NaCl}_s) + I - A \quad (2.11.6)$$

A major problem associated with the use of 2.11.6 has been the lack of accurate electron affinities. However, recent measurements of A for the halides were made with great precision by Berry and Reimann[E13] by u.v. absorption spectra of alkali halide vapours heated by shock waves. With accurate A values, eqns. 2.11.6 probably represent the best method for the evaluation of ΔH^0_{lat} and ΔG^0_{lat}[E14] since the heats and free energies of formation are known for most of the common salts.[E15–17] There are several cases for which the thermodynamic data required for the calculation of ΔG^0_{lat} are incomplete, and providing ΔH^0_{lat} is known, the former quantity can be obtained from the relation

$$\Delta G^0_{\text{lat}} = \Delta H^0_{\text{lat}} - T\{\Sigma S^0_{i,g} - S^0_s\} \quad (2.11.7)$$

In this equation S^0_s is the crystal entropy for which data are fairly complete,[E15–17] or may be estimated empirically,[E15] and $\Sigma S^0_{i,g}$ represents the gaseous entropy terms, i.e.

$$\Sigma S^0_{i,g} = S^0_{\text{trans}} + S^0_{\text{rot}} + S^0_{\text{vib}} \quad (2.11.8)$$

It is assumed that the electronic term does not contribute to the entropy of the gaseous species since they are, by definition, in their ground states. These terms may be evaluated by statistical mechanics provided the necessary information on the molecular structure of the species is available.[E18–22]

Appendix 2.11.1 lists both the lattice free energies and enthalpies for a series of salts. All data are based on the standard state of 1 atm and 25°C and were calculated from eqns. 2.11.6 with the exceptions of the tetraalkylammonium halides[E23] and the alkali metal perchlorates[E24] which were calculated by statistical mechanics (eqns. 2.11.7, 2.11.8).

2.11.3 Medium Effects

The medium effect describes the energy and enthalpy changes which accompany the transfer of a chemical entity from one solvent to another. The factors which dictate the magnitude of this change are discussed below. Owen[E25] in a pioneering study, differentiated primary and secondary medium effects but more recent work has concentrated attention on the first of these. King has recently given a lucid discussion of these effects in regard to the behaviour of acids and bases.[E26] The discussion of medium effects presented below will therefore be limited to general definitions; further details may be found elsewhere[E25-28,E93].

Consider the cell (V) of sect. 2.3

$$\text{Pt, } H_2|HCl(m)\text{-solvent}|AgCl|Ag \qquad (V)$$

where the solvent may be water (w), or some pure organic solvent or mixed solvent (s). The e.m.f. of this cell is given by

$$^wE_m = {}^wE_m^0 - 2(RT/F)\{\ln m + \ln {}^w\gamma_\pm\} \qquad (2.11.9)$$

The superscript w indicates that we are considering the aqueous cell. In the pure or mixed organic solvent s we have

$$^sE_m = {}^sE_m^0 - 2(RT/F)\{\ln m + \ln {}^s\gamma_\pm\} \qquad (2.11.10)$$

It should be noted that the activity coefficients in 2.11.10 are unity at infinite dilution in the specific solvent under consideration. For the case where m is equal in both solvents,

$$^wE_m - {}^sE_m = {}^wE_m^0 - {}^sE_m^0 + 2(RT/F)\{\ln [{}^s\gamma_\pm/{}^w\gamma_\pm]\}$$

and using

$$\Delta G_t^0 = -zF({}^sE^0 - {}^wE^0) \qquad (2.6.13)$$

then

$$\Delta G_t = \Delta G_t^0 + 2RT\{\ln [{}^s\gamma_\pm/{}^w\gamma_\pm]\} \qquad (2.11.11)$$

According to Owen's definitions,[E25] the ΔG_t term in 2.11.11 is related to the total medium effect, ΔG_t^0 to the primary medium effect and the logarithmic term is the secondary medium effect. It is evident therefore that the primary medium effect (or simply the medium effect) reflects differences in ion-solvent interactions, and the secondary medium effect (or salt effect or concentration effect[E26]) reflects differences in ion-ion interactions and solvation effects; the former quantity is, of course, independent of concentration whereas the latter quantity, which is usually several orders of magnitude smaller, is defined for a constant (finite) concentration. The standard free energy of transfer is defined for the transfer of 1 mole of substance from water to the organic solvent, i.e.

$$\Delta G_t^0 = \Delta G_s^0 - \Delta G_w^0 = ({}^wE^0 - {}^sE^0)zF \qquad (2.11.12)$$

In 2.11.12, ΔG_s^0 and ΔG_w^0 may represent standard free energies of solvation, solution, or the standard free energy change of an electrochemical reaction. Thus in the example cited (cell V), ΔG_t^0 refers to the transfer of 1 mole of HCl between the solvent S and water. The chemical potential of HCl in solvent S may be written alternatively as

$$\mu_{HCl} = {}^s\mu_{HCl}^0 + 2RT \ln m\, {}^s\gamma_\pm$$

referred to a chosen standard state in the solvent S or

$$\mu_{\text{HCl}} = {}^{w}\mu^{0}_{\text{HCl}} + 2RT \ln m\, {}^{w}\gamma_{\pm}$$

referred to a chosen standard state in water, which lead to eqns. 2.11.9 and 2.11.10.
Hence

$$\ln {}^{w}\gamma_{\pm} - \ln {}^{s}\gamma_{\pm} = ({}^{s}\mu^{0}_{\text{HCl}} - {}^{w}\mu^{0}_{\text{HCl}})/2RT = \Delta G^{0}_{t}/2RT \tag{2.11.13a}$$

$$= \ln {}^{w}_{s}\gamma_{\text{HCl}}$$

which is called the medium effect.[E26] It is sometimes written as $\ln {}_{m}\gamma_{\text{HCl}}$. The nomenclature ${}^{w}_{s}\gamma$ distinguishes the solvent pair, whereas the more normally used ${}_{m}\gamma$ does not. Water is not always the reference solvent (see for example Chapter 6 where methanol is chosen and the molarity scale is used with the nomenclature ${}^{\text{MeOH}}_{s}\gamma_{i}$).

The solubility method of determining medium effects is more generally applicable. The chemical potentials of a solute MX in saturated solution in solvent S and in water are equal since each is in equilibrium with the same solid phase, provided no crystal solvates are formed (sect. 2.3.1), thus

$$\mu_{MX} = {}^{s}\mu^{0}_{MX} + 2RT \ln {}^{s}m^{\text{sat}}\, {}^{s}\gamma_{MX}$$

$$\mu_{MX} = {}^{w}\mu^{0}_{MX} + 2RT \ln {}^{w}m^{\text{sat}}\, {}^{w}\gamma_{MX} \tag{2.11.13b}$$

and

$$({}^{s}\mu^{0}{}_{MX} - {}^{w}\mu^{0}_{MX})/2RT = \Delta G^{0}_{t}/2RT = \ln \frac{{}^{w}m^{\text{sat}}\, {}^{w}\gamma_{MX}}{{}^{s}m^{\text{sat}}\, {}^{s}\gamma_{MX}}$$

$$= \ln {}^{w}_{s}\gamma_{MX} = \ln {}_{m}\gamma_{MX}$$

Hence

$$\ln {}^{w}\gamma_{MX} = \ln {}^{w}K_{s0}/{}^{s}K_{s0} \tag{2.11.13c}$$

where ${}^{w}K_{s0}$ and ${}^{s}K_{s0}$ are the solubility products of the salt MX.

For many applications in kinetics, the ratio ${}^{w}\gamma_{MX}/{}^{s}\gamma_{MX}$ has been taken as unity and the medium effect determined simply from the ratio of solubilities in the two solvents (see sect. 6.1.3).

Other names used for the medium effect are solvent activity coefficient (Chapter 6), degenerate activity coefficient and distribution coefficient.

The extra thermodynamic assumptions which have been proposed to split the medium effect for electrolytes, that is for electrically neutral combinations of ions, into values for individual ions are discussed in sect. 2.11.4b.

2.11.4 Single Ion Solvation

(a) *Individual Solvents*

In discussing methods available for the evaluation of energies and enthalpies of solvation of individual ions, it is impossible to avoid the aqueous system since it serves as the basis for models which we require for understanding the non-aqueous systems. For most ion-solvent and solvent-solvent interactions, the energies involved may be calculated from electrostatic principles. Covalent and charge-transfer forces as well as London forces require a quantum mechanical approach. The forces to be considered[E29] in ion solvation are given in Table 2.11.1.

Table 2.11.1

Forces Involved in Ion Solvation

Type of Force	Internuclear Distance Function
1. Born charging (BC)	r^{-1}
2. Ion-dipole	r^{-2}
3. Ion-quadrupole	r^{-3}
4. Dipole-dipole	r^{-3}
5. Dipole-quadrupole	r^{-4}
6. Ion-induced dipole	r^{-4}
7. Quadrupole-quadrupole	r^{-5}
8. Ion-induced quadrupole	r^{-5}
9. Induced dipole-induced dipole	r^{-6}
10. Induced dipole-induced quadrupole	r^{-8}
11. Induced quadrupole-induced quadrupole	r^{-10}
.	.
.	.
.	.
12. Repulsive forces	r^{-n}
13. Dispersion or London forces	r^{-6}

In this table we have omitted interactions of multipoles higher than the quadrupole. Since very little is known about forces 7–11, they are often neglected. The repulsive energy takes the general form B/r^n or $B\exp(-r/\rho)$. Here B is the 'repulsive constant', ρ is a constant related to the compressibility and n is a constant whose value is usually 9[E11,E30,E31] or in some cases[E32] 12, as in the Lennard-Jones potential energy function. The constant B may be evaluated theoretically by reference to a hard or soft sphere model (see below) or by differentiating the total energy, $U(r)$, with respect to r since $dU(r)/dr = 0$ at the equilibrium separation.[E5,E11,E32] Assuming the factors listed in Table 2.11.1 are additive (i.e. no perturbations), we still have to account for the work done in creating a hole in the solvent and placing the ion inside it. For aqueous solutions, this term is estimated[E29,E31,E32] to be 10–20 kcal mol^{-1} and is dependent upon geometry and the number of H-bonds which re-form in the presence of the ion. It is important to note that the terms 2–12 in Table 2.11.1 are all dependent upon the structural nature of ion solvation and this leads to the necessity of adopting a model for the ion in its solvent cage. There are considerable voids in our knowledge of the structure of aqueous solutions[E1] and the situation for non-aqueous solutions is even poorer.

The crystal radii of the ions are important quantities in these calculations and Table 2.11.2 lists those radii from several sources. The radii of Gourary and Adrian appear to be more consistent than those of Pauling or Goldschmidt[E14,E39,E40] although they are not completely free from criticism.[E14,E39]

The Born equation

$$\Delta G_{\mathrm{Born}} = -\frac{(ze)^2 N}{2R}\left\{1 - \frac{1}{\varepsilon_r}\right\}_p \tag{2.11.14}$$

and the Born-Bjerrum equation

$$\Delta H_{\mathrm{Born}} = -\frac{(ze)^2 N}{2R}\left\{1 - \frac{1}{\varepsilon_r} - \frac{T}{\varepsilon_r}\frac{\partial \ln \varepsilon_r}{\partial T}\right\}_p \tag{2.11.15}$$

have dominated the many attempts to calculate the free energy and enthalpy of solvation and the corresponding ΔG_t^0 and ΔH_t^0 terms. In eqns. 2.11.14 and 2.11.15, ε_r is the dielectric constant, R is a radius and all other terms have their usual significance. Latimer, Pitzer and Slansky[E41] assumed that

$$R = r_i + \delta \tag{2.11.16}$$

where r_i is the crystal radius of ion i and δ is a variable parameter. The Born equation actually yields the free energy of transfer of 1 mole of substance from vacuum into an infinite volume of dielectric. To obtain $\Delta G^0_{\mathrm{solv}}$ in molal units (see also ref. E65), we consider the ions to be contained in a volume, V, which will be filled with solvent later. The initial state pressure is RT/NV and the molality in the final state is $1/NV\rho$ where N is the Avogadro number and ρ is the density of the solvent.

Table 2.11.2

Crystal Radii of Some Important Ions (r_i in Å)

Ion	P	G	A	G & A	T & F	M
Li^+	0.61	0.78	0.68	0.93	0.90	
Na^+	0.96	0.98	0.97	1.17	1.21	
K^+	1.33	1.33	1.33	1.49	1.51	
Rb^+	1.48	1.49	1.47	1.64	1.65	
Cs^+	1.66	1.65	1.67	1.83	1.80	
Ag^+	1.26	1.13	1.18			
Tl^+	1.44	1.49	1.47			
NH_4^+	1.48	1.45				
$AsPh_4^+$						4.2
NMe_4^+						3.47
NEt_4^+						4.00
NPr_4^+						4.52
NBu_4^+						4.94
NAm_4^+						5.29
F^-	1.34	1.33	1.33	1.16	1.19	
Cl^-	1.81	1.81	1.81	1.64	1.65	
Br^-	1.95	1.96	1.96	1.80	1.80	
I^-	2.17	2.20	2.20	2.04	2.01	
ClO_4^-	2.00			(1.85)		2.36 (2.45)
OH^-				(1.12)		1.40 (1.47)
BPh_4^-						4.2
Pic^-						3.61

Notes to table: P = Pauling;[E11] G = Goldschmidt;[E33] A = Ahrens;[E10] G & A = Gourary and Adrian,[E34] and the values in parentheses are from ref. E24 and E36; T & F = Tosi and Fumi;[E6] M = miscellaneous sources. ClO_4^- and OH^- are from ref. E8, while the values in parentheses are from ref. E9; $AsPh_4^+$ (tetraphenylarsonium) and BPh_4^- (tetraphenyl boride) are from ref. E38. Tetraalkylammonium ions are from R. A. Robinson and R. H. Stokes, Electrolyte Solutions, Academic Press, Inc., New York (1959). Picrate ion is from D. F. Tuan and R. M. Fuoss, *J. Phys. Chem.*, **67**, 1343 (1963).

For unit fugacity ($f = p$), the free energy of the isolated ion is

$$G_g = G_g^0 + RT \ln (RT/NV)$$

and in solution

$$G_s = G_s^0 + RT \ln (1/NV\rho)$$

The Born equation yields $G_s - G_g = \Delta G_{\mathrm{Born}}$ so that

$$\Delta G_{\mathrm{solv}}^0 = \Delta G_{\mathrm{Born}} + RT \ln (RT\rho) \qquad (2.11.17)$$

According to Latimer, Pitzer and Slansky, the individual $\Delta G_{\mathrm{solv}}^0$ (ion) values are obtained by extrapolating the observed solvation energies for a series of salts, say MCl, to infinite cation radius (i.e. $1/R_+ = 0$). The restraint on this method is that, for a plot of individual ΔG^0_{solv}(ion) values vs. $1/R$, all points must lie on one line and the intercept is zero at $R = \infty$. A similar method was independently proposed by Verwey.[E42,E43] Latimer *et al.* found the adjustable parameter δ in 2.11.16 to equal 0.85 Å for cations and 0.10 Å for anions. Individual ΔG^0_{solv}(ion) values are given in Appendix 2.11.2. Latimer and Verwey found that for gegen-ions of equal crystal radius, the anion had a more negative ΔG^0_{solv}(ion) value. Verwey[E43] recognised that the δ quantities are related to structural effects and proposed one of the first models of ion-solvent structure. Verwey's structural models are shown in Figs. 2.11.1a and b.

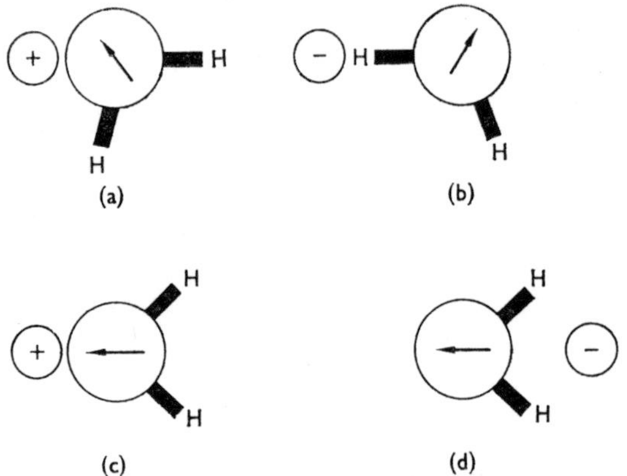

FIG. 2.11.1. The orientation of water about cations and anions.

a and *b;* after Verwey.[E43]
c and *d;* after Bernal and Fowler.[E56]

The corresponding data for the enthalpies of solvation (ΔH^0_{solv}(ion)) for aqueous solutions are given in Appendix 2.11.3. More recent applications of the Born equation to ion solvation in a series of solvents have been reported by Izmailov,[E44] Khomutov,[E45] and Criss and Luksha.[E46] The results for the non-aqueous systems are summarised in Appendices 2.11.4 (free energies) and 2.11.5 (enthalpies). It should be noted that Izmailov assumed $\delta = 0$ and used Pauling's radii; Khomutov used the crystal radii of Gourary and Adrian and assumed $\delta = 0.74$ Å for cations and 0.42 Å for anions.

Use of the Born equation for the calculation of free energies and enthalpies of solvation is based on a model of a continuous dielectric medium. Modifications of this equation are still subject to some criticism.[E1,E47–49] *A priori* adjustment of the crystal radii is not theoretically acceptable[E50] because a leading term in eqn. 2.11.14 is the gas phase radius; i.e. eqn. 2.11.14 should more accurately be written as

$$\Delta G_{\mathrm{Born}} = \frac{(ze)^2 N}{2}\left\{\frac{1}{r_s \varepsilon_r} - \frac{1}{r_g}\right\} \tag{2.11.18}$$

where r_s and r_g are, respectively, the ionic radius in solution and in the gas phase. For water the term $1/r_g$ is, at most, 80 times larger[E50] than $1/r_s\varepsilon_r$. Since ions are compressed in solution then $1/r_g < 80/r_s\varepsilon_r$.[E51] However, even though there is this decrease in radius due to compression by the large electrostatic forces in solution, it does not decrease sufficiently to warrant large δ corrections. The consideration of dielectric saturation either by reference to continuous or discontinuous models,[E50,E52] or by empirical adjustment of the dielectric constant,[E53,E54] all improve the results but offer no great advantage since the exact function for ε_r versus distance is unknown and hence ε_r becomes a catch-all, as are the δ corrections, for all the uncertainties associated with the Born equation.[E1]

Izmailov[E55] proposed that a plot of differences in conventional energies or enthalpies (defined below) for salts with both ions of equal principal quantum number, n, vs. $1/n^2$ gives $\Delta G^0_{\mathrm{solv}}(\mathrm{H}^+)$ or $\Delta H^0_{\mathrm{solv}}(\mathrm{H}^+)$ as the intercept. A list of differences in conventional energies, $\delta\Delta G^0_{\mathrm{conv}}$', for several salts in water is shown in Table 2.11.3 The data used by Izmailov

Table 2.11.3

Salt	n	$1/n^2$	$\frac{1}{2}\delta\Delta G^0_{\mathrm{conv}}$/ kcal mol^{-1}	
			Ref. E55	Ref. E36
NaF	3	0.11	260.0	263.2
KCl	4	0.063	255.2	257.7
RbBr	5	0.040	254.8	256.8
CsI	6	0.0278	255.3	254.5

are significantly different from those of Salomon;[E36] plots of $\frac{1}{2}\delta\Delta G^0_{\mathrm{conv}}$ *vs.* n^{-2} give $\Delta G^0_{\mathrm{solv}}(\mathrm{H}^+) = -256.5$ kcal mol^{-1} based on the data in ref. E55 and -251.5 kcal mol^{-1} based on the data in ref. E36. The difference is due mainly to the use of different values for lattice free energies. The results in Izmailov's two important papers[E44,E55] are very similar and we have quoted the data reported in his earlier work.[E44]

Structural theories of ion solvation begin with the classic paper of Bernal and Folwer,[E56] who assumed that the solvation enthalpies were equal for anions and cations of equal crystal radius. Using the Pauling radii, which are equal for K^+ and F^-, the experimental solvation energy for KF was split equally and then empirically corrected for the difference in polarisabilities of the two ions; the single ion enthalpies of solvation for K^+ and F^- were found to be -94 and -97 kcal mol^{-1}, respectively. Eley and Evans' statistical mechanical calculations[E57] give energies and enthalpies, which are considered to be too negative.[E58,E59] Mischenko[E60] and Vasil'ev *et al.*[E61] assumed an equal split of $\Delta H^0_{\text{solv}}(\text{CsI})$ to be a better approximation. The entropies of solvation were calculated from

$$\Delta S^0_{\text{solv}}(\text{ion}) = S^0_{\text{aq}} - S^0_g + R \ln 24.45 \qquad (2.11.19)$$

where $R \ln 24.45$ accounts for the difference in standard states of the ion in the gas phase (1 mole per 24.45 litres) and in solution (1 mole per 1000 grams of water). The entropies of solution, S^0_{aq}', were obtained from the relative values (based on $S^0_{\text{H}^+,\text{aq}}(\text{relative}) = 0$) by taking $S^0_{\text{H}^+,\text{aq}}$ (absolute) $= -3.4$ cal mol^{-1} K^{-1}. The results of Vasil'ev *et al.*, which are similar to those of Mishchenko, are given in Appendices 2.11.2 and 2.11.3.

Krestov and Klopov[E62] studied the heats of solution of sodium and potassium halides in methanol and in aqueous mixtures with methanol. To evaluate single ion contributions, they used the expression[E63]

$$\left\{\frac{\Delta H^0_{\text{solv}}(M^+)}{\Delta H^0_{\text{solv}}(X^-)}\right\}_{\text{water}} = \left\{\frac{\Delta H^0_{\text{solv}}(M^+)}{\Delta H^0_{\text{solv}}(MX) - \Delta H^0_{\text{solv}}(M^+)}\right\}_{\text{solvent}} \qquad (2.11.20)$$

Essentially this expression states that the ratio of heats of solvation in water for any pair of ions is equal to their ratio in any other solvent. They use the CsI assumption to arrive at the single ion solvation enthalpies given in Appendix 2.11.6. According to these results, a maximum in $\Delta H^0_{\text{solv}}(\text{ion})$ occurs in all cases at an alcohol mol fraction of around 0.25.

Using the crystal radii of Gourary and Adrian for which $r_{\text{Rb}^+} = r_{\text{Cl}^-} = 1.64$ Å, Blandamer and Symons[E39] split the experimental value of $\Delta G^0_{\text{solv}}(\text{RbCl})$ to obtain the single ion values (see Appendix 2.11.2). As proof of the validity of this procedure, they offer a plot of $\Delta G^0_{\text{solv}}(\text{ion})$ *vs.* $1/r^2$ in which all points for both cations and anions lie on a single line. These results indicate that $\Delta G^0_{\text{solv}}(\text{ion})$ is dependent only upon $1/r^2$ which conflicts with that discussed above where structural differences lead to greater solvation energies for anions than for cations of equal size.

Buckingham's attempt[E29] to sum all the significant terms in Table 2.11.1 has led to several new approaches to the problem. Buckingham attributed the differences in solvation enthalpies of anions and cations of equal size to the ion-quadrupole interaction energy. Referring to Fig. 2.11.1, Buckingham used models (c) and (d) as the basis of his calculations.

On this basis, the ion-dipole energy is equal for positive and negative ions of equal charge and radii. This is so because when a solvent molecule is inverted upon changing the sign on the ion, the sign of the ion-dipole energy remains unchanged since the direction of the dipole is also inverted. However, since the direction of the quadrupole is unaffected by the charge on the ion, the ion-quadrupole interaction energy will change sign when an anion is replaced by a cation. Thus the difference in solvation enthalpies for ions of equal charge and radius is given by (for monovalent ions)

$$\Delta H^0_{\text{solv}}(M^+) - \Delta H^0_{\text{solv}}(X^-) = \frac{\alpha e\Theta}{R^3} - \frac{\beta\mu\Theta}{R^4} \qquad (2.11.21)$$

Here α and β are constants which are dependent on structural effects, e is the electronic charge and μ and Θ are, respectively, the dipole and quadrupole moments of the solvent molecule. Using Verwey's data for ΔH^0_{solv}(ion) for K^+ and F^-, Buckingham obtained a value of 1.95×10^{-26} e.s.u. cm^2 for the quadrupole moment of water. Duncan and Pople's theoretical[E64] value is $0.215 \cdot 10^{-26}$ e.s.u. cm^2. In eqn. 2.11.21 the radius term R is given by eqn. 2.11.16, where δ is taken as the radius of the solvent molecule. If the solvent molecule is not axially symmetric, then the charge distribution must be represented by three principal quadrupole moments, Θ_{xx}, Θ_{yy} and Θ_{zz} instead of a single value. Having evaluated Θ as described above, Buckingham summed the electrostatic energies, ion-induced dipole forces and the dispersion energies. Assuming a 'hard sphere' model (i.e. no repulsion) in which a tetrahedral or octahedral solvent configuration is assumed, individual ionic enthalpies of solvation were obtained which are, admittedly too negative. Somsen,[E31] and Muirhead-Gould and Laidler[E65] carried out similar calculations, but included repulsion effects and quadrupole-quadrupole interactions. Somsen did not consider the energies required for hole formation and hydrogen-bonding whereas Muirhead-Gould and Laidler did.

Halliwell and Nyburg[E30] argued that the division into individual ionic contributions of ΔH^0_{solv} could be achieved semi-empirically using Buckingham's concepts. They employed enthalpies of solvation relative to the hydrogen ion (conventional enthalpies) defined by

$$\Delta H^0_{\text{conv}}(M^{z+}) = \Delta H^0_{\text{solv}}(M^{z+}) - |z|\Delta H^0_{\text{solv}}(\text{H}^+)$$

$$\Delta H^0_{\text{conv}}(X^{z-}) = \Delta H^0_{\text{solv}}(X^{z-}) + |z|\Delta H^0_{\text{solv}}(\text{H}^+)$$

For symmetrical electrolytes, the difference in conventional enthalpies, $\delta\Delta H^0_{\text{conv}}$, is

$$\delta\Delta H^0_{\text{conv}} = \{\Delta H^0_{\text{solv}}(M^{z+}) - \Delta H^0_{\text{solv}}(X^{z-})\} - 2|z|\Delta H^0_{\text{solv}}(\text{H}^+) \qquad (2.11.22)$$

Since $\delta\Delta H^0_{conv}$ is an experimental quantity, $\Delta H^0_{solv}(H^+)$ can be obtained as the intercept of a plot of $\delta\Delta H^0_{conv}$ versus some function of radius for the terms inside the brackets of eqn. 2.11.22. For the model of ion solvation used by Buckingham (Fig. 2.11.1c and 2.11.1d), the leading term for ions of equal size is the ion-quadrupole energy since the ion-dipole terms cancel. The function of R used by Halliwell and Nyburg was therefore $1/R^3$ (Pauling-Arhens radii were used) and the absolute solvation enthalpy of the proton was found to equal -260.7 kcal mol^{-1}. Morris[E14] repeated these calculations using the radii of Gourary and Adrian and employing more recent thermodynamic data and obtained -263.7 kcal mol^{-1} for $\Delta H^0_{solv}(H^+)$. The absolute solvation enthalpy of the proton obtained in this way is the *average* value of the intercepts for the plots of ΔH^0_{conv}(ion) vs. $1/R^3$. It should therefore be *independent* of the choice of radii providing the sum $r_+ + r_-$ for a given salt equals the observed sum. Morris also obtained practically identical results by plotting

$$(\Delta H^0_{solv}(MX) - \Delta H_{neut}) \text{ vs. } 1/r$$

for a series of salts MX having one common ion (e.g. taking $X = Cl^-$, the intercept at $r_+ = \infty$ is $\Delta H^0_{solv}(Cl^-)$). The non-electrostatic contribution to the enthalpy was obtained from the empirical relation

$$\frac{\Delta H_{neut}}{\text{kcal mol}^{-1}} = -0.2\left(\frac{r}{\text{Å}}\right)^4 \quad \text{(cf. Noyes, ref. E53)}$$

Somsen[E31,E66] obtained individual ΔH^0_{solv}(ion) data for the alkali metals and halides in the amides using the method of Halliwell and Nyburg. The quadrupole moments of the amides were obtained by modifying eqn. 2.11.21 to allow for repulsion effects. Considering a system of 'soft spheres', eqn. 2.11.21 becomes

$$\frac{\Delta H^0_{solv}(M^+) - \Delta H^0_{solv}(X^-)}{\text{kcal mol}^{-1}} = \frac{3.689 \times 10^4}{\text{e.s.u. cm}^{-1}}\frac{\Theta}{R^3} - \frac{5.511 \times 10^{13}}{(\text{e.s.u.})^2\text{ cm}^{-1}}\frac{\mu\Theta}{R^4} \tag{2.11.23}$$

for the tetrahedral configuration and

$$\frac{\Delta H^0_{solv}(M^+) - \Delta H^0_{solv}(X^-)}{\text{kcal mol}^{-1}} = \frac{5.533 \times 10^4}{\text{e.s.u. cm}^{-1}}\frac{\Theta}{R^3} - \frac{1.707 \times 10^{14}}{(\text{e.s.u.})^2\text{ cm}^{-1}}\frac{\mu\Theta}{R^4} \tag{2.11.24}$$

for the octahedral configuration. The values of the quadrupole moments for water and the amides based on Somsen's equation 2.11.23[E31,E66] are give in Table 2.11.4.

The use of eqn. 2.11.24 leads to smaller values for Θ but in no case[E14,E30,E31,E65,E66] is there satisfactory agreement with the theoretical value of Duncan and Pople. On the other hand, it has been suggested[E29] that the theoretical value of Θ calculated from approximate ground state electronic wave functions is not to be considered reliable. It is interesting

Table 2.11.4

Solvent	$\Theta \times 10^{26}$ e.s.u. cm^2
H_2O	2.4
F	5.6
NMF	5.0
NMA	5.3

to note that lower values of Θ can be obtained by modifying eqns. 2.11.21 or 2.11.23 to allow for an ion-dipole term. For the case where the ion-dipole terms do not cancel (see below), then eqn. 2.11.21 should be replaced by

$$\Delta H^0_{\text{solv}}(M^+) - \Delta H^0_{\text{solv}}(X^-) = \frac{\alpha\mu(1 - \cos\theta)}{R^2} + \frac{\beta e\Theta}{R^3} - \frac{\gamma\mu\Theta}{R^4} \tag{2.11.25}$$

In eqn. 2.11.25, θ is the angle between the dipole vector and the cation; it is implied that the anion lies along this vector. The constants α, β, and γ are characteristic of the geometry of the solvent molecules in the primary solvation shell and differ from those in eqns. 2.11.21 and 2.11.24.

A direct summation of all the factors involved in ion solvation has proved to be a difficult task. Even with simplifying assumptions, Muirhead-Gould and Laidler[E65] had to evaluate well over 100 interaction energies for each system. While the results appear to be in satisfactory agreement with those obtained by other methods, they are quite sensitive towards such factors as geometry and the energy of formation of hydrogen bonds. Muirhead-Gould and Laidler estimate that an uncertainty of 1 kcal mol^{-1} in the energy of formation of a hydrogen bond leads to a change of about 10 kcal mol^{-1} in the overall solvation energy. Somsen's and Laidler's results for the alkali metals in water are given in Table 2.11.5. The uncertainties which arise in attempting to include all possible interaction energies justify the continuing work on empirical methods.

Salomon[E36] has extended Halliwell and Nyburg's method for obtaining enthalpies of solvation by allowing for the possibility that the ion-dipole term is not equal for ions of equal charge and radius. Conway and

Table 2.11.5

Enthalpies of Solvation in H_2O for Tetrahedral and Octahedral Configurations Calculated from Discontinuous Models*

Ion	$-\Delta H^0_{solv}$		$-\Delta G^0_{solv}$	
	Tetra†	Octa†	Tetra ‡	Octa ‡
Li^+	103.5	154.1	130.9	
Na^+	87.3	122.5	107.9	
K^+	74.2	99.7	81.1	
Rb^+	69.7	92.3	70.8	81.0
Cs^+	64.9	84.5	63.6	70.0

* Molal units used; all data are in kcal mol^{-1} at 25°C.
† Ref. E31.
‡ Ref. E65.

Salomon[E58,E59] have shown that if Verwey's model of cation solvation is considered (see Fig. 2.11.1), then the orientation of a water molecule at the cation is 52° to the radial direction. If anions lie along this direction, then the difference in ion-dipole energy is $\mu e(1 - \cos 52°)/R^2$ which is the term included in eqn. 2.11.25. By fitting the conventional enthalpies to a least square power series

$$\Delta H^0_{conv}(\text{ion}) = z + \frac{A}{R^2} + \frac{B}{R^3} \qquad (2.11.26)$$

Salomon evaluated[E36] absolute enthalpies of solvation in water and in PC. The results of these calculations are found in Appendices 2.11.3 and 2.11.5. They differ significantly from those discussed above and this is attributed to the large difference in the ion-dipole interaction energy between the solvent and anions and cations of equal radii. Salomon also suggested that individual free energies of solvation may be obtained from a plot of differences in conventional free energies versus $1/r_i$. For ions of equal charge and radius, $\Delta G^0_{solv}(H^+)$ is obtained from

$$\Delta G^0_{conv}(M^+) - \Delta G^0_{conv}(X^-) = \frac{\text{constant}}{r_i} - 2\Delta G^0_{solv}(H^+) \qquad (2.11.27)$$

The basis for this relationship in $1/r_i$ rather than a complex power series in $1/r^n$ is attributed to a compensation between the enthalpy and entropy terms. It is generally accepted that $\Delta H_{solv}(\text{ion})$ is a complex function of $1/r^n$. The corresponding entropy effects should also involve the same functions since those factors which lead to an increase in the interaction between the ion and solvent molecule also tend to restrict the degrees of freedom of the water molecule in the primary hydration shell thereby decreasing the entropy.[E57] The $1/r$ function given by eqn. 2.11.27 is therefore an empirical one and a plot of ΔG^0_{conv}(ion) vs. $1/r$ (cf. Fig. 2.11.2 for the alkali halides in water) gives an intercept which is not to be associated with the absolute solvation energy of the reference ion. This is because it includes contributions from the other ions which have not

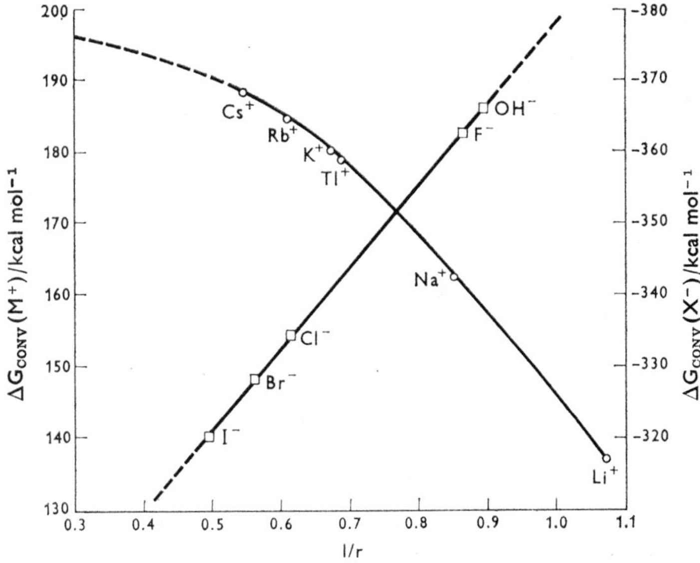

FIG. 2.11.2. Conventional free energies vs. $1/r$ in water.[E36]

cancelled due to incomplete compensation between the enthalpy and entropy. The anion and cation plots will therefore have *different* intercepts. These ion-multipole interactions which are not fully compensated are assumed to cancel when differences in ΔG^0_{conv}(ion) are taken. Figures 2.11.3 and 2.11.4 are plots of conventional free energy differences for ions of equal charge and radii versus $1/r$. From Fig. 2.11.3 for the aqueous system, the intercept at $r = \infty$ gives $\Delta G^0_{solv}(H^+) = -235.0$ kcal mol^{-1}. Taking[E35] the absolute entropy of solvation of the proton as -31.3 cal mol^{-1} K^{-1}, it is found that $\Delta H^0_{solv}(H^+) = -244.3$ kcal mol^{-1} which is in good agreement with the value of -247.3 obtained from the least squares fit using eqn. 2.11.26. It was also found that the value of $\Delta G^0_{solv}(H^+)$ is independent of the choice of radii used. While the intercepts of plots of ΔG^0_{conv}(ion) vs. $1/r$ are dependent upon which set of radii

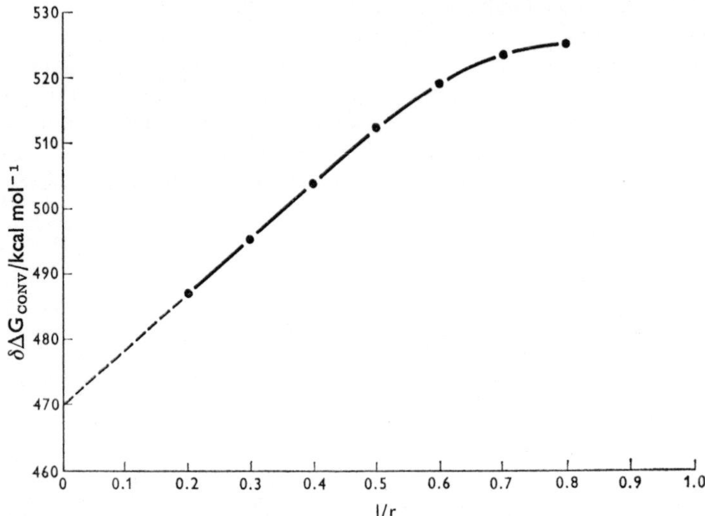

FIG. 2.11.3. Differences in conventional free energies between ions of equal radii and charge vs. $1/r$ in water.[E36]

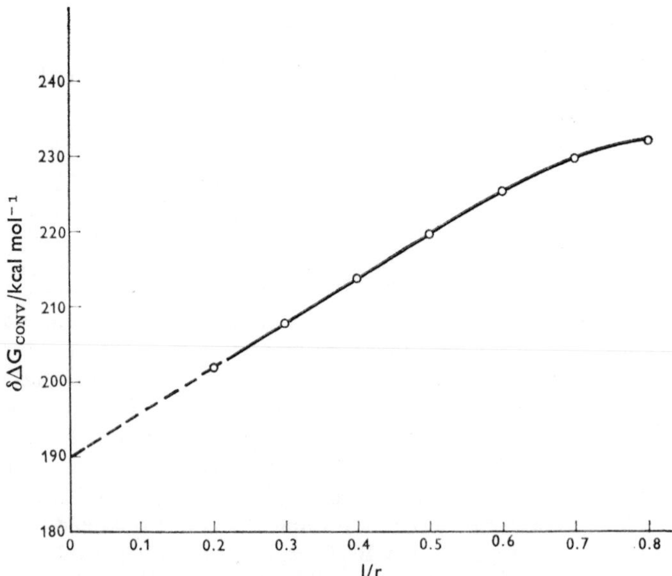

FIG. 2.11.4. Differences in conventional free energies between ions of equal radii and charge vs. $1/r$ in propylene carbonate.[E36]

are used, the difference between the anion and cation intercepts is independent of these radii as long as the sum $r_+ + r_-$ equals the observed value. The single ion solvation energies obtained from Fig. 2.11.3 for the aqueous system are given in Appendix 2.11.2 and for the PC system (Fig. 2.11.4), the data are given in Appendix 2.11.4.

An interesting application of the CNDO method of Pople *et al.*[E67,E68] has been reported by Burton and Daly.[E69] The CNDO method is based on a semi-empirical self-consistent molecular orbital theory in which it is assumed that two centre repulsion integrals are independent of the orientation of atomic orbitals; hence there is a complete neglect of differential overlap (CNDO) and the calculation of molecular properties is greatly simplified. Burton and Daly considered both tetrahedral and octahedral models of ion solvation and their results for the solvation enthalpies of several ions are shown in Table 2.11.6. It is assumed that these ΔH^0_{solv}(ion) values are based on molar units. Burton and Daly conclude that the octahedral hydration is to be preferred. However, they do point out that the results may not be sufficiently accurate to support either model. In general, the CNDO method usually leads to good values for bond angles and lengths for AB_2 and AB_3 type molecules, but it does not give satisfactory results for the total energy of a system.[E68,E69] (See Table 2.11.6.)

Table 2.11.6

ΔH^0_{solv} (Ion) Values Calculated from the CNDO Method (Data at 25°C from Ref. E69)

Ion	ΔH^0_{solv}(ion)/kcal mol^{-1}	
	Tetrahedral Model	Octahedral Model
Li^+	−150.8	−198.8
Be^{2+}	−654.4	−789.2
Na^+	−48.9	−108.9
Mg^{2+}	−171.8	−268.7

(b) *Medium Effects*

There has been considerable effort expended in attempting to evaluate the energies and enthalpies of single ion transfer between two solvents. Once again the Born equation has played a major role in dividing the free energy for the transfer of 1 mole of uni-valent salt from water *w* to solvent *s* according to

$$\Delta G_t^0 = \frac{Ne^2}{2}\left\{\frac{1}{{}^s\varepsilon_r} - \frac{1}{{}^w\varepsilon_r}\right\}\left\{\frac{1}{r_+} + \frac{1}{r_-}\right\} \tag{2.11.28}$$

Thus for a single ion, we have

$$\Delta G_t^0(\text{ion}) = \frac{Ne^2}{2r_i}\left\{\frac{1}{{}^s\varepsilon_r} - \frac{1}{{}^w\varepsilon_r}\right\} \tag{2.11.29}$$

It should be clear that if the Born equation is not corrected in any way, there is no doubt as to its failure. According to eqn. 2.11.29 it follows that ΔG_t^0(ion) will always be positive for ${}^w\varepsilon_r > {}^s\varepsilon_r$. This is not believed to be true for many solvents as discussed below. Conversely for ${}^s\varepsilon_r > {}^w\varepsilon_r$, eqn. 2.11.29 predicts a negative value for ΔG_t^0(ion), which is also questionable. Noyes[E53] chose to evaluate 'effective' dielectric constants as his variable parameter to correct the Born equation. Using Pauling radii and correcting for non-electrostatic effects, he found that the 'effective' dielectric constant, ε_{eff}, was different for each ion. Latimer *et al.*[E41] chose to vary the crystal radii by adding *constant* terms to r_+ and r_- such that $\delta_+ \neq \delta_-$. The δ corrections can be adjusted so that the sign of ΔG_t^0 in eqn. 2.11.29 may be either negative or positive. Since this original paper appeared, several authors have re-evaluated these δ corrections for both aqueous and non-aqueous solutions. Table 2.11.7 lists these δ corrections for the various solvents.

Table 2.11.7

Empirical Corrections to Pauling Crystal Radii for Use in the Modified Born Equation

Solvent	δ_+/Å	δ_-/Å	Ref.
H_2O	0.85	0.10	E41
H_2O	0.70		E70
H_2O	0.74	0.42	E45
H_2O	0.85	0.25	E71, E72
H_2O	0.72		E73
H_2O	0.64		E74
DMF	0.85	1.00	E46
DMF	0.78	0.42	E75
MeOH	0.79	0.39	E71, E72
CH_3CN	0.72	0.61	E71
HCOOH	0.78	0.38	E71
F	0.85	0.25	E71
PC	0.80	0.67	E70, E36

Strehlow and co-workers[E71,E72] evaluated δ_+ and δ_- for several solvents by considering free energies of transfer obtained from e.m.f. and solubility measurements. They first defined relative free energies of transfer, $\delta\Delta G_t^0$, by

$$\delta\Delta G_t^0 = \Delta G_t^0(\text{ion}) - \Delta G_t^0(\text{reference ion})$$

where Cs^+ was taken as the reference cation and I^- as the reference anion. For the transfer from water to methanol, Strehlow obtained the $\delta\Delta G_t^0$ values (molar scale) shown in Table 2.11.8.

Table 2.11.8

Cation	$\delta\Delta G_t^0$/ kcal mol^{-1}	Anion	$\delta\Delta G_t^0$/ kcal mol^{-1}
Na^+	−0.55	Cl^-	1.42
K^+	−0.15	Br^-	0.68
Rb^+	−0.12	I^-	0.0
Cs^+	0.0		

Correcting eqn. 2.11.29 by substituting R in eqn. 2.11.16 for r_i, we obtain the following relationship for cations

$$\delta\Delta G_t^0 = \frac{Ne^2}{2}\left\{\frac{1 - 1/{}^w\varepsilon_r}{{}^wR_+} + \frac{1 - 1/{}^s\varepsilon_r}{{}^sR_{Cs^+}} - \frac{1 - 1/{}^w\varepsilon_r}{{}^wR_{Cs^+}} - \frac{1 - 1/{}^s\varepsilon_r}{{}^sR_+}\right\} \tag{2.11.30}$$

The aqueous values for R_+ are known and for the non-aqueous solvent, R_+ is evaluated by adjusting δ_+ until the observed ΔG_t^0 is obtained. A similar relation to eqn. 2.11.30 exists for anions. Coetzee and Campion[E73] used a relation similar to eqn. 2.11.30 to evaluate δ_+ for acetonitrile (see sect. 2.7.3b). These authors obtained $\delta\Delta G_t^0$ values for the alkali metals by taking differences in half-wave potentials after correcting for the liquid junction potential. Once δ_+ and δ_- are found, the free energies of transfer of individual ions from water to any other solvent are calculated from

$$\Delta G_t^0(\text{ion}) = \frac{Ne^2}{2}\left\{\frac{1 - 1/{}^w\varepsilon_r}{{}^wR_+} - \frac{1 - 1/{}^s\varepsilon_r}{{}^sR_+}\right\} \tag{2.11.31}$$

Strehlow's results for the transfer from water to methanol are given in Appendix 2.11.7 and are compared to other results (see below).

Pleskov[E76] originally made the assumption that large ions such as Rb^+ and Cs^+ may have the same free energy of solvation in all solvents.

Strehlow argued[E71,E72] that this assumption is a first approximation and added, via eqn. 2.11.30, the small electrostatic contribution to $\Delta G_t^0(\mathrm{Cs}^+)$. Along these lines Strehlow sought a redox system such as

$$M \rightleftharpoons M^+ + e$$

where the cation M^+ is large and its charge is diffuse. The ferrocinium| ferrocene and cobalticinium|cobaltocene couples were found to be acceptable reference systems.[E71,E72] Hence the e.m.f. of a cell composed of one of these couples versus a hydrogen electrode is related to ΔG_t^0 (H^+). The value of $\Delta G_t^0(\mathrm{H}^+)$ (from water to MeOH) obtained in this manner is 0.23 kcal mol^{-1}. Lauer[E75] used Strehlow's data to arrive at the general relations

$$\frac{\delta_+}{\text{Å}} = 0.865 - 3.14/\varepsilon_r$$

$$\frac{\delta_-}{\text{Å}} = 1.81\ \varepsilon_r/(\varepsilon_r - 3.63) - 1.59$$

These two relations are based on data for H_2O, MeOH, AN, F, and HCOOH. Using the above relations, Lauer then evaluated $\Delta G^0_{\mathrm{solv}}(\mathrm{ion})$ for several other solvents and the results are given in Table 2.11.9.

Table 2.11.9

Free Energies of Solvation of Li^+ and Cl^- in Several Solvents at 25°C (Data are from Ref. E75)

Solvent	ε_r	$-\Delta G^0_{\mathrm{solv}}(\mathrm{Li}^+)$/ kcal mol^{-1}	$-\Delta G^0_{\mathrm{solv}}(\mathrm{Cl}^-)$/ kcal mol^{-1}
NMF	182.4	113.5	79.8
F	109.5	114.5	78.5
H_2O	78.5	114.9	77.6
HCOOH	57.0	115.6	75.7
$MeNO_2$	40.0	116.6	72.8
DMF	36.7	117.0	72.4
AN	36.7	117.0	72.4
MeOH	31.2	117.7	69.8
Me_2CO	20.7	120.2	64.7
THF	7.4	137.9	35.7
PhOMe	4.4	170.4	12.2
Et_2O	4.3	173.2	10.8

There are many uncertainties in these results, especially for those solvents where no experimental data exist. Lauer's results for LiCl in water give $\Delta G^0_{solv}(\text{LiCl}) = -192.5$ kcal mol^{-1} whereas the data of Latimer *et al.*[E41] and Salomon[E36] give -198 kcal mol^{-1} for this quantity.

Nelson and Iwamoto[E77] have suggested that the redox system Fe(II)$(Me_2phen)_3$/Fe(III)$(Me_2phen)_3$ be adopted as a reference reaction whose standard free energy change is independent of solvent. This suggestion has not gained much support. Benoit and co-workers have used equilibrium and e.m.f. data to obtain free energies of transfer of halides and complex silver halide species.[E78,E79] The split in free energies was made by assuming that the standard e.m.f. of the ferricinium-ferrocene electrode is independent of the solvent. Their results are given in Appendix 2.11.8. The assumptions regarding the constancy of a single electrode potential independent of solvent cannot be regarded as valid because, in addition to the uncersainty associated with liquid junction potentials (see sect. 2.6.1), one is dealing with 'real' and not chemical free energies. Case and Parsons[E80] have estimated that the difference in surface potentials between water and methanol or ethanol is

$$\chi(\text{MeOH}) - \chi(\text{H}_2\text{O}) \approx -0.3 \text{ V}$$

$$\chi(\text{EtOH}) - \chi(\text{H}_2\text{O}) \approx -0.4 \text{ V}$$

Hence the contribution to the measured real free energy of transfer by the surface potentials is about -7 kcal mol^{-1}, which exceeds all the $\Delta G^0_t(M^+)$ values obtained by various procedures as shown in Appendix 2.11.7.

Grunwald, Baughman and Kohnstam[E38] measured the derivative of the partial molar free energies of several salts with respect to water mol fraction (Z_1) in 50% dioxan-water mixture. This derivative, dG^0_t/dZ, was divided equally between the large tetraphenyl boride ion and the tetraphenyl phosphonium ion. They found positive values of this derivative for cations and negative values for anions. A positive value of dG^0/dZ corresponds to a negative value for ΔG^0_t(ion) and similarly, a negative value for dG^0/dZ, corresponds to a positive value for ΔG^0_t(ion). The significance of this work lies in the conclusion that the cations prefer the mixed solvent to water whereas the reverse is true for most anions. There is a considerable amount of evidence, based mainly on the work of Strehlow, Feakins and co-workers, Parsons and others (see below), that $\Delta G^0_t(M^+)$ is indeed likely to be negative for the transfer from water to several solvents where ${}^s\varepsilon_r < {}^w\varepsilon_r$.

Feakins and co-workers[E81–E86] have made extensive e.m.f. studies on the hydrogen halide cells without liquid junctions in various solvents and evaluated single ion contributions to the free energies of transfer by use of the relations

$$\Delta G_t^0(MX) = \Delta G_t^0(M^+) + k/r_- \tag{2.11.32}$$

$$\Delta G_t^0(MX) = \Delta G_t^0(X^-) - m/r_+ \tag{2.11.33}$$

Equations 2.11.32 and 2.11.33 are strictly empirical relations. The $1/r$ dependence is attributed to the compensation effect. Figure 2.11.5 shows plots of ΔG_t^0, ΔH_t^0, and $T\Delta S_t^0$ vs. $1/r_+$ for the alkali metal chlorides where the transfer is from water to a 20% dioxan-water

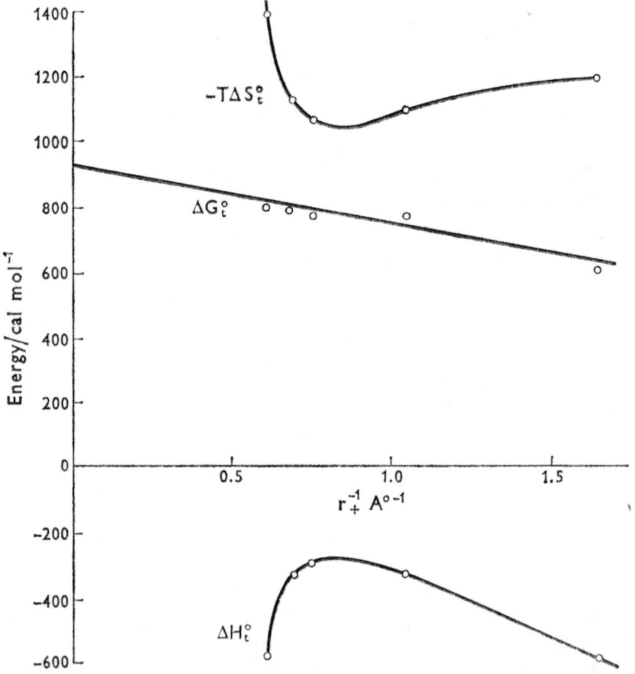

FIG. 2.11.5. The energetics of transfer of alkali metal chlorides from water to 20% dioxan-water at 25°C.[E86]

mixture. On the basis of these results, Feakins *et al.* have concluded that although ΔH_t^0 and ΔS_t^0 contain complex structural contributions, they tend to cancel in ΔG_t^0. While it does appear that much cancellation does occur, it has been pointed out above that it may not be complete. If complete cancellation of the structural effects occurs, then the value of $\Delta G_t^0(H^+)$ should be independent of its method of calculation, i.e. with respect to eqn. 2.11.32 and 2.11.33. This was not found to be true.

Table 2.11.10

$\Delta G_t^0(H^+)$/kcal mol^{-1} for the Transfer from Water to MeOH/H_2O and Dioxan/H_2O Mixtures at 25°C (Molar Scale)

	Wt% Methanol			Wt% Doxan	
	10	43.12	90	20	45
Eqn. 2.11.32	−0.435	−1.109	−5.702	−2.10	−5.70
Eqn. 2.11.33	−0.570	−3.430	−4.962	−0.50	−4.30
Ref.		E85		E82, E86	

Table 2.11.10 shows the large differences in $\Delta G_t^0(H^+)$ obtained from the two equations for the transfer from water to several mixed solvents. For ion transfer from water to acetic acid (HAc)-water mixtures,[E83] eqn. 2.11.32 does not yield a linear relation and it was suggested that if the data were available, eqn. 2.11.33 would indeed be linear. Approximate results for these systems, which were obtained from eqn. 2.11.32, are given in Appendix 2.11.9. The use of eqn. 2.11.32 for the transfer of ions from water to DMSO has not led to any reasonable results and it was concluded that this relation must be used with care.[E87] In another study[E88] the use of eqn. 2.11.33 for evaluating the free energy of transfer of halides and alkali metal ions from water to NMF was found to give a linear plot and reasonable results; for Li^+, Na^+, and K^+, $\Delta G_t^0(M^+)$ was found to be positive (see Appendix 2.11.12) which is contrary to the values predicted by the Born equation since ${}^{NMF}\varepsilon_r \gg {}^{H_2O}\varepsilon_r$.

DeLigny and Alfenaar[E89,E90] suggested that the extrapolation based on eqns. 2.11.32 and 2.11.33 could be improved by considering the 'neutral' part of the free energy of transfer. They also attempted to account for ion-dipole and ion-quadrupole interactions by use of the relations

$$\Delta G_t^0(HX) - \Delta G_{t,\text{neut}}^0(X^-) = \Delta G_t^0(H^+) + \frac{a}{r_x} + \frac{b}{r_x^2} + \frac{c}{r_x^3} + \ldots \quad (2.11.34)$$

$$\Delta G_t^0(H^+) - \Delta G_t^0(M^+) + \Delta G_{t,\text{neut}}^0(M^+) = $$

$$= \Delta G_t^0(H^+) - \frac{a}{r_M} + \frac{d}{r_M^2} + \frac{e}{r_M^3} + \ldots \quad (2.11.35)$$

They assume that $\Delta G^0_{t,\text{neut}}$ is simply proportional to r^2 so that one can obtain this quantity for any ion by plotting ΔG^0_t(inert gas) vs. r^2. Alfenaar and DeLigny's values for the free energies of transfer of neutral substances from water to methanol are given in Appendix 2.11.10. The derived values for $\Delta G^0_{t,\text{neut}}$ for the alkali metal ions and the halides are given in Appendix 2.11.11. Plots of the conventional free energies of transfer corrected for the neutral part (cf. the LHS of eqns. 2.11.34 and 2.11.35) versus $1/r$ were non-linear[E89,E90] and were therefore extrapolated to a common intercept. The constants *a*, *b*, *c*, *d*, *e* were evaluated from a least squares fit.[E90] The results of Alfenaar and DeLigny are given in Appendix 2.11.7. The data of Salomon[E91] in Appendix 2.11.7 were obtained by use of eqn. 2.11.27.

Alfenaar and DeLigny,[E89–90] Morris,[E14] and Salomon[E36,E91] have all assumed that $\Delta G^0_{t,\text{neut}}$ (or $\Delta H^0_{t,\text{neut}}$) is simply related to the ionic radius r_i. If this is true (e.g. see Noyes[E53] who equated $\Delta G^0_{t,\text{neut}}$ for ions to an iso-electronic rare gas), then the choice of crystal radii becomes important. There is no universal acceptance of a given set of radii and since they do vary considerably, it is probably more valid to take differences in $\Delta G^0_{t,\text{conv}}$(ion) since $\Delta G^0_{t,\text{neut}}$ will cancel for anions and cations of equal charge and radii. This was seen to be true for the estimation of ΔG^0_{solv}(ion) in H_2O[E36] and MeOH[E91] where the use of Pauling radii or Gourary and Adrian's radii gave identical results. However, if the method for evaluating $\Delta G^0_{t,\text{neut}}$ discussed by Noyes is correct, then the above data must be questioned.

A popular approach to the evaluation of individual ionic contributions to the medium effect involves the use of a reference ion or pair of ions. In order to assign equal ΔG^0_{solv}(ion) or ΔG^0_t(ion) values between two ions, the following specifications must apply:[E37] the ions must be large and relatively non-polarisable, have a small charge density and should not participate in specific interactions with the solvent (e.g. destroy a hydrogen-bonded structure which may or may not be re-established with the ion participating in hydrogen bonding). A variety of molecular and/or ionic species have been proposed for this purpose. Above we have discussed several cases regarding electrode reactions and the use of large reference ions such as Ph_4As^+ and Ph_4P^+ with Ph_4B^-.[E38] Popovych[E92,E93] has criticised the use of Ph_4As^+ and Ph_4P^+ as reference ions because of the uncertainty in their crystal radii. Instead he recommends the use of triisoamylbutylammonium (TAB^+) with Ph_4B^- based on the equality of the Stokes radii of these two ions in water and methanol. A comparison of ΔG^0_t(ion) values obtained by the various methods for the transfer from water to an organic solvent is given in Appendices 2.11.7 and 2.11.12. ΔG^0_t(ion) values for the transfer of Cl^-, picrate ion, K^+, TAB^+, and Ph_4B^- from water to ethanol-water mixtures are summarised in Appendix 2.11.13.[E92,E93] The data of Coetzee *et al.*,[E73,E94] which are based on half-wave potentials, were converted to e.m.f. differences for single electrode reactions and are assumed to be measures of the 'chemical' free energy of transfer of the given ion.

Grunwald and co-workers[E95–97] have proposed a method for the evaluation of $\Delta G_t^0(H^+)$ from empirical relations based on Grunwald and Winstein's earlier proposal[E98] of a linear free energy relation. Considering substituted anilinium and aliphatic acids, they proposed the following empirical relation

$$\Delta pK(HA) = \Delta G_t^0(H^+)/RT\,(\ln 10) + m_- Y_-$$
$$\Delta pK(BH^+) = \Delta G_t^0(H^+)/RT\,(\ln 10) + m_0 Y_0$$

For the transfer from water to ethanol-water mixtures, ΔpK is defined as $pK(s) - pK(w)$, and HA represents the acids formic, acetic, and benzoic, and BH^+ represents the anilinium, *p*-toluidinium, and *n*-methylanilinium ions. m_- and m_0 are substituent constants which are independent of the nature of the solvent and Y_- and Y_0 are parameters characteristic only of the solvent. The solvent parameters were found to fit the following relations

$$Y_0 = -(1 - W_{H_2O})^2$$
$$Y_- = 1 - W_{H_2O}^2$$

where W_{H_2O} is the weight fraction of water. Taking differences in ΔpK values

$$\Delta pK_0 - \Delta pK_- = m_0 Y_0 - m_- Y_-$$

and the substituent constants m_0 and m_- are obtained from a least squares analysis.[E95] Wynne-Jones[E99] has objected to this approach and has shown that the solvent parameter Y is not constant but is in fact dependent upon the nature of the substituent. Popovych[E93] has also objected to the use of these empirical constants by questioning the validity of the constancy of m_0 and m_-.

Parker[E37,E100] has made a survey of the results of the various methods for obtaining free energies of transfer of the individual ions. Assumptions involving $\Delta G_t^0(I_2) = \Delta G_t^0(I_3^-)$ were tested using data for the instability constants for the tri-iodide-iodine equilibrium;[E101,E102] double electrochemical cells were used to estimate $\Delta G_t^0(I^-)$[E101] and $\Delta G_t^0(Ag^+)$[E103] neglecting the liquid junction potential, the magnitude of which may be about one third of the observed e.m.f. A comparison of $\Delta G_t^0(Ag^+)$ from water to a series of organic solvents based on the various assumptions is shown in Appendix 2.11.14. Parker,[E103] has commented on the remarkable correspondence between the estimates of this quantity using the various assumptions. We will never know if an assumption is valid but the greater the number of independent assumptions leading to similar values, the greater the confidence in these values. In general, however, there are dramatic differences in the results obtained on the basis of different assumptions. An interesting case is that for the transfer of H^+ and the alkali metal ions from water to methanol and methanol-water mixtures (see Appendix 2.11.7). Those methods based on splitting the experimental values equally between two large ions invariably give a positive value for $\Delta G_t^0(M^+)$ whereas those methods based on an extrapolation of $1/r$ give a negative value. The positive values of $\Delta G_t^0(M^+)$ obtained by Izmailov based on the latter method cannot be regarded as being sufficiently accurate since his plots show considerable curvature thereby introducing a large uncertainty.[E44] Above we have discussed some of the likely sources of error involved in the extrapolation methods. There are also many possible sources of error in the method of assigning equal ΔG_t^0 values to the large reference ions. While it is probably safe to assume that the Born charging energies for large anions and cations are equal, it is difficult to imagine that all the specific ion-solvent interactions are identical. The choice of a given pair of reference ions based on equal (large) radii appears to be questionable due to the large uncertainties in their 'crystal' radii (e.g. see E93). It is also important to note that for most of these data, the experimental $\Delta G_t^0(MX)$ determinations for these reference salts are based on solubility data. The problems associated with using solubility data are:

(1) The possibility of micelle formation.
(2) The possible formation of complex ions.
(3) The possible formation of crystal solvates.
(4) Ion pair formation.

While arguments have been presented to discount these complicating factors,[E93] they are not based upon experimental evidence and the complications may therefore be present. There is no doubt that the data obtained from the e.m.f. of cells without liquid junctions are more reliable than the solubility data. In Table 2.11.11 below, the free energies of transfer of HCl and KCl obtained by summing the individual contributions of each ion are given. The accepted values of ΔG_t^0(HCl) and ΔG_t^0(KCl) are those based on the e.m.f. data (see E75, E81 and sect. 2.6.5).

Methods for obtaining individual contributions to the enthalpy of solvation were discussed above. For the most part, identical procedures have been used to obtain the individual ionic contributions to the standard enthalpy of transfer. Appendix 2.11.15 lists these individual ΔH_t^0(ion) values. These data are based on the following methods:

(1) $\Delta H_t^0(Ph_4B^-) = \Delta H_t^0(Ph_4As^+)$; ref. E104–107.
(2) Plots of ΔH_{solv}^0(conv) *vs.* $1/r^3$; ref. E31, E66.
(3) Plots of ΔH_{solv}^0(conv) *vs.* $1/r^2 + 1/r^3$; ref. E36.

It is apparent that there have been far fewer attempts to evaluate ΔH_t^0(ion) and ΔH_{solv}^0(ion) than the corresponding free energy terms. In several recent reviews on medium effects,[E37,E72,E93] the subject of enthalpies of transfer is completely neglected. The individual ionic contributions to ΔH_t^0 and ΔS_t^0 are of great importance because they reflect important structural differences in the behaviour of ions in various solvents which are lost in the simpler free energy terms. They are therefore more difficult to evaluate and interpret, and any questionable assumption regarding the physical model for the calculations could possibly lead to large uncertainties in the resulting ΔH_t^0(ions) values.

Table 2.11.11

Free Energies of Transfer (in kcal mol^{-1}) of HCl and KCl from Water to Pure Methanol at 25°C (Data Obtained by Summing the Individual Ionic Contributions)

Electrolyte	$\Delta G_t^0(M^+) + \Delta G_t^0(X^-)$	Experimental $\Delta G_t^0(MX)$ Method	ΔG_t^0(ion) Calc. From	Ref.
HCl	5.36	e.m.f.	$1/r$ plot	E81
	5.36	solubility + e.m.f.	reference ion	E92
	5.59	e.m.f.	$1/r$ plot	E44
KCl	6.29	e.m.f.	$1/r$ plot	E81
	5.29	solubility	reference ion	E92
	4.09	e.m.f.	$1/r$ plot	E44
	5.4	solubility	average for many methods	E37*

* Based on molar units; all other data are based on molal units.

2.11.5 SINGLE ION ENTROPIES, VOLUMES AND VISCOSITIES

The same questions which arise in the division of enthalpies and free energies of solvation and transfer into their ionic components also arise in the division of entropies, volumes, and viscosities of electrolytes into their ionic components. However, because of the inherently simpler concepts involved, a greater variety of approaches in making the divisions have been employed. Some of these methods can also be used, as discussed in sect. 2.12.3, to estimate thermodynamic properties of electrolytes in organic solutions for which no data exist.

(a) *Ionic Entropies*

Nearly all current attempts to divide partial molal entropies of electrolytes in organic solvents into their ionic components have required a knowledge of the 'absolute' entropy of the corresponding ions in water. Consequently, it appears worthwhile to examine some of the methods which have been used to evaluate absolute $\bar{S}^0_{ion}$ values in this solvent.

Several diverse techniques have been employed for making this division; all of them involve extra-thermodynamic assumptions, most of which have been the subject of criticism. The most important single factor leading to confidence in any of the values is the fact that they all give results in a small range ($\bar{S}^0_{H^+} = -1.5$ to -6.3 cal mol^{-1} K^{-1}), which fortuitously is not far from the conventional value of $\bar{S}^0_{H^+} = 0$. The methods can be divided into three groups, (1) Electrochemical, (2) Born treatment and (3) Correspondence plots.

Electrochemical Method. The electrochemical method is based on the Eastman thermocell.[E108] Essentially the method consists of measuring the e.m.f. of a cell employing two similar reversible electrodes immersed in a common electrolyte, but at different temperatures. The difference in potential divided by the temperature difference between the electrodes, is related to the entropy of transfer of the ions across the temperature gradient. By making certain assumptions it is possible to estimate the absolute entropy of an individual ion which can then be used in evaluating other ionic entropies. The details of the method have been discussed by Conway and Bockris;[E47] see also the general review of thermocells by Agar.[E109] A variation of the method has been used by Lang and Hesse.[E110]

Still another approach[E111] is to make use of the temperature coefficient of an appropriate series of cell reactions. It is necessary to assume that the potential of the electrocapillary maximum corresponds to zero metal-solution potential difference. A recent modification suggested by Ikeda[E112] makes use of the Eastman thermocell as well as the temperature coefficient of the mobility of an ion.

Born Treatment. Several attempts have been made to evaluate absolute ionic entropies by means of the Born equation (eqn. 2.11.18) or the modified Born equation.[E41] The method consists mainly of employing the temperature coefficient of the ionic free energies of solvation as discussed in sect. 2.11.2 to obtain the entropy of solvation. The entropy of the individual gaseous ions can be calculated by the methods also discussed in sect. 2.11.2 and consequently ionic entropies evaluated.

Correspondence Plot Methods. Gurney[E35] suggested a method of division which, like some of the other methods, is ultimately based in part on the temperature coefficient of the mobility of an ion, a quantity which can be indisputedly obtained on an absolute basis. In his procedure the partial molal entropy of a reference ion, for example the hydrogen ion, is initially assigned an arbitrary value. This automatically fixes the values of all other ions. These entropy values are then plotted against the ionic viscosity coefficient, $B_{\eta\ \mathrm{ion}}$, which has been previously evaluated from ionic mobility data by techniques described later in this section. Gurney[E35] observed that such a plot resulted in two straight lines, parallel to each other, one containing cations and the other anions. By making the proper assignment of entropy to the reference ion it was possible to make these two lines superimpose. Using this approach, he evaluated the entropy of aqueous hydrogen ion at 25°C to be -5.5 cal mol^{-1} K^{-1}.

Criss and Cobble[E113] have also used a correspondence method of dividing entropies into ionic components. Their division was made by plotting entropies of ions at one temperature against the entropies of the corresponding ions at another temperature. Linear results could be obtained only when the entropies at 25°C were based on an assignment of -5.0 cal mol^{-1} K^{-1} for the hydrogen ion.

Table 2.11.12

Assignments of the Entropy of the Hydrogen Ion in Water at 25°C

$S^0_{H^+}(aq)$/cal mol^{-1} K^{-1}	Method Employed	Investigator	Date
−5.0*	Thermocell	Eastman[E114]	1928
−4.7	Isothermal Thermocell	Lang and Hesse[E110]	1933
−1.5†	Born Treatment	Latimer, Pitzer and Slansky[E41]	1939
−5.4	Thermocell	Lee and Tai[E111]	1941
−2.1	Thermocell	Goodrich *et al.*[E115]	1950
−6.3‡	Thermocell	Crockford and Hall[E116]	1950
−5.5	Correspondence Plot	Gurney[E35]	1953
−2.1	Born Treatment	Latimer[E117]	1955
−4.48	Thermocell	deBethune, Licht and Swendeman[E118]	1959
−3.3	Born Treatment	Noyes[E53]	1962
−5.0	Correspondence Plot	Criss and Cobble[E113]	1964
−5.7	Thermocell	Breck and Lin[E119]	1965
−5.3	Thermocell and ion mobility	Ikeda[E112]	1965

* 15°C.
† The reported value for Cl^-(aq) was combined with $\bar{S}^0$ for HCl(aq) from Selected Values of Chemical Thermodynamic Properties, NBS Technical Note 270–3, January, 1968.
‡ 30°C. Calculated from $S^0_{Cl^-}$(aq) at 30°C.

A summary of the methods employed and the resultant values is given in Table 2.11.12. In view of the wide variety of techniques employed in these divisions it is remarkable that the results are in such close agreement. Since the majority of the values fall close to a value of -5 cal mol^{-1} K^{-1} it appears worthwhile, in the absence of more precise values, to assign arbitrarily a value of $\bar{S}^0_{H^+}(aq) = -5.0$ cal mol^{-1} K^{-1}.

As mentioned previously, attempts at dividing experimental partial molal entropies of electrolytes in non-aqueous solutions into their ionic components have rested almost entirely on the ionic entropies of the ions in aqueous solution, which of course are dependent upon the assignment of $\bar{S}^0_{H^+}(aq)$. The method makes use of the correspondence principle, which involves plotting $\bar{S}^0_{ion}(aq)$ for the ions in aqueous solution against the corresponding $\bar{S}^0_{ion}(X)$ values for the ions in non-aqueous solutions. Similar to the approach used by Gurney[E35] for evaluating $\bar{S}^0_{ion}(aq)$, a proper choice for $\bar{S}^0_{H^+}(X)$ in the non-aqueous solution will result in a linear relationship with both the cations and anions falling on the same straight line. The method has been used by Latimer and Jolly[E120] to evaluate absolute ionic entropies in liquid ammonia and by Jakuszewski and Taniewska to evaluate absolute $\bar{S}^0_{ion}(X)$ values in methanol[E121] and ethanol.[E122] More recently the correspondence principle was used by Criss, Held and Luksha[E123] to evaluate ionic entropies in formamide, NMF, DMF and deuterium oxide as well as re-evaluating the data for other solvents from previous studies.

The absolute entropies for ions in DMF, evaluated by this technique, were reported[E123] to be only 1.5 cal mol^{-1} K^{-1} different from values obtained in the same solvent by Criss and Luksha,[E46] using the Born treatment as modified by Latimer, Pitzer and Slansky. Revised entropy data for electrolytes in DMF (see Appendix 2.4.41) lead to poorer agreement, but nevertheless, are satisfactory in the light of uncertainties in the experimental data. The fact that there is satisfactory agreement lends support to the view that these are indeed absolute values. Recent free energy data on lithium salts in DMF, summarised in Table 2.7.11, give entropies for these salts considerably different from those listed in Appendix 2.4.41. While the experimental technique (e.m.f. measurements) employed in obtaining the free energies should give reliable results, the entropies calculated from them are not consistent with the entropies of other salts in DMF, and have therefore not been included in Appendix 2.4.41.

Ionic entropies in all the non-aqueous solvents examined up to the present time can be expressed as a linear function of the entropies of the ions in water as:

$$\bar{S}^0_{ion}(X) = a + b\bar{S}^0_{ion}(aq) \tag{2.11.36}$$

where a and b are empirical constants.

For theoretical reasons, Criss and co-workers[E123] tabulated entropies on a mol fraction standard state, and consequently the parameters in eqn. 2.11.36 were based on that standard state. We have re-evaluated the parameters of eqn. 2.11.36 on a hypothetical one molal standard state and have listed them in Table 2.11.13. It should be noticed that the slope,

Table 2.11.13

Constants for Equation 2.11.36 (Hypothetical 1 Molal Standard State)*

Solvent	a/cal mol^{-1} K^{-1}	b	Δ†
D_2O	0.3	1.04 ± 0.02	±0.4
H_2O	0.0	1.00	0.0
F	−0.3	0.63 ± 0.03	±1.6
NMF	−4.4	0.66 ± 0.01	±0.6
DMSO	−12.1	0.92 ± 0.04	±2.1
PC	−10.8	0.75 ± 0.04	±2.0
MeOH	−10.6	0.82 ± 0.01	±0.7
DMF	−9.0	0.44 ± 0.06	±3.4
EtOH	−15.7	0.76 ± 0.01	±0.6
NH_3	−20.9	0.82 ± 0.04	±2.2

* Evaluated from the data in Table 2.11.14.
† Δ is the standard deviation between the calculated and the experimental values.

b, in eqn. 2.11.36 is independent of the choice of standard state, but that the term

$$R\left[\ln\frac{1000}{M_X} - b\ln\frac{1000}{M_{H_2O}}\right]$$

must be added to or subtracted from the parameter, a, to convert it from one standard state to the other. Conversion of the entropy of an ion from one standard state to the other can be facilitated through the equation

$$\bar{S}^0_{\text{ion}}(m) = \bar{S}^0_{\text{ion}}(N) + R\ln\frac{1000}{M_1} \tag{2.11.37}$$

where m and N represent the molal and mol fraction standard states respectively and M_1 is the molecular weight of the solvent. A summary of absolute partial molal entropies of ions based on the hypothetical one molal standard state, taken mostly from Criss and co-workers,[E123] is given in Table 2.11.14.

When there are insufficient data to make a correspondence plot, one is probably safe in setting the entropy of the potassium and chloride ions equal to each other, since their entropies are remarkably close in most solvents (Table 2.11.14). The similarity in the properties of potassium

Table 2.11.14

Standard Absolute Ionic Entropies (cal mol^{-1} K^{-1}) at 25°C (Hypothetical 1 Molal Standard State)*

Ion	$\bar{S}_2^0$									
	NH_3	EtOH†	DMF	MeOH	PC	DMSO	NMF	F†	H_2O	D_2O
H^+	−28.0	−19.9		−15.1			−8.0	−3.2	−5.0	
Li^+	−20.0	−15.9	−5.5	−10.8	−13.9	−13.7	−5.5	1.4	−1.6	−0.8
Na^+	−12.9	−9.0	−7.3	−3.6	−2.5	−5.8	2.7	2.5	9.4	10.2
K^+	−6.6	−1.4	−3.0	4.2	4.2	8.5	7.5	11.3	19.5	20.4
Rb^+	1.0			9.9				14.9	24.7	
Cs^+	1.0		3.0	12.1			13.7	17.1	26.8	28.0
F^-								1.1	2.7	2.4
Cl^-	−2.3	−2.2	−4.9	4.4	4.1	6.8	7.6	10.1	18.2	19.1
Br^-	−2.3	2.4	1.0	9.7	9.2	9.5	11.1	15.2	24.3	25.5
I^-	3.0	8.6	9.7	14.9	9.5	14.7		20.8	31.1	32.8

* Data are taken from ref. E123, except those for DMF and NMF which have been revised and PC and DMSO which are newly calculated (see Appendix 2.4.41).

† Entropies in these solvents have been reassigned from those in ref. E123.

and chloride ions is observed in other areas, such as ionic mobilities and viscosities, as will be discussed later.

Franks and Reid,[E124] using data from the literature, have evaluated partial molal entropies of ions in aqueous methanol and dioxan. The data are listed in Appendix 2.4.42. In their division they made the assumption that $\bar{S}^0_{K^+} = \bar{S}^0_{Cl^-}$ — for each solvent system, which appears reasonable in the light of the data in Table 2.11.14.

(b) *Ionic Volumes*

The division of partial molal volumes into their absolute ionic components presents the same problems as the division of other thermodynamic properties in that non-thermodynamic arguments must be invoked. Millero[E125] has reviewed the the various methods of making this division for aqueous solutions and has summarised the corresponding $\bar{V}^0_{H^+}$ values. Most of the methods yield values falling in the range of 0 to -5 cm^3 mol^{-1}. From a critical evaluation of the data, Millero[E125] suggests that a $\bar{V}^0_{H^+}$ of close to -5.0 cm^3 mol^{-1} at 25°C is the 'best' value.

Millero[E126] has used the method of Mukerjee[E127] to divide $\bar{V}^0_2$ into its ionic components for solutions of electrolytes in *N*-methylpropionamide. The method involves assigning $\bar{V}^0_{H^+}$ so that a best fit is obtained when $\bar{V}^0_{ion}$ for both cations and anions are plotted as a function of the cube of the radius. The resultant value of $\bar{V}^0_{H^+}$ for this solvent is 3.4 cm^3 mol^{-1}. $\bar{V}^0_{ion}$ for the other ions can be estimated by the expression

$$\frac{\bar{V}^0_{ion}}{\text{cm}^3\ \text{mol}^{-1}} = 4.28\left(\frac{r}{\text{Å}}\right)^3 + 3.4 \qquad (2.11.38)$$

Millero has also used the correspondence principle method[E128] to evaluate ionic volumes in NMP as well as in methanol. Similar to the case for entropies, ionic volumes in the non-aqueous system are plotted against the absolute $\bar{V}^0_{ion}(aq)$ values, and the $\bar{V}^0_{H^+}(X)$ for the non-aqueous system assigned so that values for both cations and anions fall on the same line. The volumes can be expressed, similar to eqn. 2.11.36, as

$$\bar{V}^0_{ion}(X) = a + b\bar{V}^0_{ion}(aq) \qquad (2.11.39)$$

where a and b are empirical constants. Using the previously evaluated $\bar{V}^0_{H^+} = 3.4$ cm^3 mol^{-1} in NMP, Millero[E128] observed that the correspondence principle is obeyed, to within the limits of experimental error, indicating that the two methods give essentially the same value for $\bar{V}^0_{H^+}$. The correspondence principle, at least for univalent ions, appears to work also for formamide and formic acid, (Fig. 2.11.6). Values for $\bar{V}^0_{ion}$ in these solvents, along with those from previous work are summarised in Table 2.11.15. The parameters for eqn. 2.11.39 for the corresponding solvents are given in Table 2.11.16.

Table 2.11.15

Standard Ionic Volumes at 25°C (cm^3 mol^{-1})

Ion	Water*	Formamide†	NMP‡	Formic Acid†	Methanol‡
H^+	−5.4		3.4		−16.6
Li^+	−6.3	−4.7			−16.0
Na^+	−6.6	−3.2	6.0	−4.0	−14.6
K^+	3.6	7.7	11.1	−1.0	−4.6
Rb^+	8.7	11.6		8.8	
Cs^+	15.9	18.0		13.9	
NH_4^+	12.4	12.9			
Cl^-	23.2	24.3	24.7	19.5	11.3
Br^-	30.1	32.2	29.8	24.4	18.6
I^-	41.6	43.1		35.5	26.3
NO_3^-	34.4	36.8	33.6		

* Ref. E125.
† Evaluated from data in Appendix 2.4.43.
‡ Ref. E128. Ionic volumes in methanol have been reassigned on the basis of the correspondence principle.

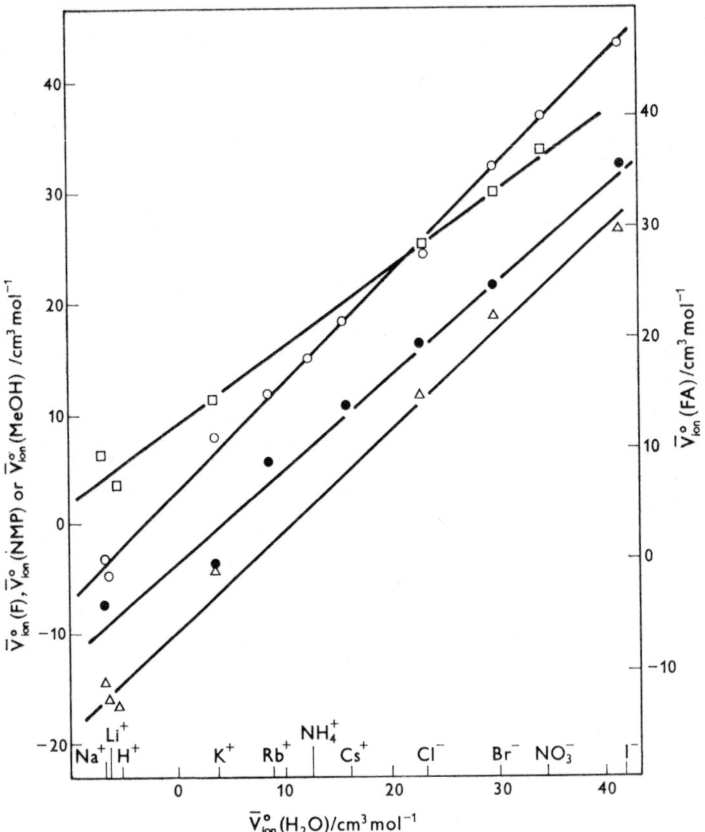

FIG. 2.11.6. Ionic partial molal volumes in formamide, O; formic acid, ●; *N*-methylpropionamide, □; and methanol, △; *vs.* the corresponding volumes in water at 25°C.

(c) *Ionic Viscosities*

Viscosity is a kinetic property rather than a thermodynamic one; however, as discussed in sect. 2.3.6, the viscosity of an electrolytic solution, and in particular the B_η coefficient of eqn. 2.3.74 is of interest because it indicates the degree of ion-solvent interaction in a solution. Viscosities, like thermodynamic properties, cannot be measured for single ions, and consequently a method must be found for dividing this quantity into its ionic components. Only a few approaches have been reported.

Table 2.11.16

Parameters and Standard Deviations for eqn. 2.11.39

Solvent	a/cm^3 mol^{-1}	b	Δ*
H_2O	0.0	1.00	0.0
Formamide	2.6	0.98 ± 0.01	± 1.0
NMP	8.8	0.70 ± 0.02	± 1.2
Formic Acid	-0.2	0.84 ± 0.03	± 1.9
Methanol	-9.7	0.90 ± 0.03	± 1.5

* Δ is the standard deviation between the calculated and the experimental values.

Cox and Wolfenden[E129] suggested a method based on absolute ionic mobilities, which can be experimentally determined. They observed that the temperature coefficient of mobility of the lithium and iodate ions conforms with Stokes law, inasmuch as the variation of the fluidity of the solvent with temperature is expected to affect the mobility of an ion. As a consequence of this they reasoned that $B_{\eta\text{ion}}$ is proportional to the ionic volume, and by making the division in the required ratio arrived at a value for $B_{\eta\text{Li}^+}$ of 0.14_6, from which the other $B_{\eta\text{ion}}$ values can be immediately obtained.

Gurney[E35] has suggested a method of division of B_η that does not require the concept of hydrated ions obeying Stokes law. He observed that a plot of the temperature coefficient of mobility versus mobility, both of which can be experimentally determined for individual ions, gives a single straight line for the simple ions and another line for the complex anions. He reasoned that since such a plot gave an indication of the behaviour of individual ions it might be possible to assign $B_{\eta\text{ion}}$ coefficients so that a similar plot would result. In particular he noticed the great similarity in behaviour of potassium and chloride. As a result of this reasoning he divided B_η for KCl equally between the potassium and chloride ions. This gives $B_{\eta\text{ion}}$ as -0.00_7 from which values for the other ions can be evaluated. A plot of the resultant $B_{\eta\text{ion}}$ values against the temperature coefficient of the mobility of the corresponding ions gives plots nearly identical with those for mobility of ions versus the temperature coefficient of their mobility. This method of division leads to a $B_{\eta\text{Li}+}$ value of 0.14_7 which is in remarkable agreement with the value of 0.14_6 suggested by Cox and Wolfenden.

In organic solvent systems mobility data for ions, and indeed even viscosity B_η coefficient data, are so scarce that up until now no systematic approach has been used to divide B_η into its ionic components. In a recent publication Criss and Mastroianni[E130] compared $B_{\eta\text{ion}}$ for simple and complex ions in water, methanol, and acetonitrile, but the division of B_η in the organic solvents was purely arbitrary.

Since results for ionic entropies in organic solvent systems are more plentiful than for ionic mobilities, it would appear that one could use the correspondence plot method, in the reverse manner from which Gurney employed it to evaluate $B_{\eta\text{ion}}$ in aqueous solutions. Plots of this type, for the very few B_η data that do exist for electrolytes of simple ions in methanol and NMF, indicate that the method is feasible. The values of $B_{\eta\text{ion}}$ obtained from such plots for both of these solvents are consistent with the values one obtains by setting $B_{\eta K^+} = B_{\eta Cl^-}$.

2.12 INTERPRETATION OF DATA

2.12.1 Introduction

It is one thing to arrive at a set of ΔG_t°(ion), ΔH_t°(ion), $\bar{S}^\circ_{\text{ion}}$, $\bar{V}^\circ_{\text{ion}}$ or $B_{\eta\text{ion}}$ values but quite another to interpret their significance. Intuition, which is often invoked in defence of a concept, can be very misleading. The structural properties of the non-aqueous solvents and conclusions based on equilibrium constants and single ion contributions to the various thermodynamic functions and transport properties require careful consideration. Incorrect data have tended to obscure several issues, since one is faced with the difficult task of choosing between several sets of data for a given system. The behaviour of ions in these solvents is indeed complex and is reflected in measurable quantities such as equilibrium constants, rates of reactions, solvation numbers, transport properties, and molar volumes and entropies to mention but a few. Our approach to these problems is to obtain correlations between the data in the various solvents, and by employing correspondence principles as developed in the previous section, attempt to explain and predict some of the properties mentioned above. A good model for ion-solvent structural interactions will do both; if our model becomes the variable parameter, however, then it becomes possible to explain everything and predict nothing.

2.12.2 Structural Aspects of Electrolytic Solutions

That there is a direct connection between the entropy of certain single systems and the amount of order or disorder in the system has led many workers to interpret entropy changes as reflecting the degree of 'structure' in liquid systems. In particular the standard entropies of ions in solution are often used for assessing the effect of the ion on the structure of the solvent.

To simplify the theoretical discussion which follows, it is convenient to convert the entropies listed in Table 2.11.14 to a standard state which will give an identical ideal entropy of mixing of the solute and solvent for every solvent system. The question arises whether one should choose the mol fraction or the molar concentration standard state. Convincing arguments can be found to support either choice. Following the convention used by Criss and co-workers[E123] and Franks and Reid,[E124] we shall adopt the mol fraction standard state for the solution, and for the solute standard state we shall use the ideal ionic gas as suggested by Franks and Reid.[E124] The latter choice removes the inherent entropy of the ion so that $\bar{S}^0_{\text{ion}}$ represents only the entropy change in the solvent because of the presence of the ion. On this basis $\bar{S}^0_{\text{ion}}$ is identical to the entropy of solvation of the ion. The $\bar{S}^0_{\text{ion}}$ values based on the new standard states are listed in Table 2.12.1.

Table 2.12.1

Standard Absolute Ionic Entropies at 25°C (Mol Fraction and Ideal Gas Standard State)*

Ion	$\bar{S}^0_{\text{ion}}$/cal mol^{-1} K^{-1}									
	NH_3	EtOH	DMF	MeOH	PC	DMSO	NMF	F	H_2O	D_2O
H^+	−62.1	−52.0		−47.9			−39.6	−35.4	−39.0	
Li^+	−59.9	−53.8	−42.5	−49.4	−50.2	−50.6	−42.9	−36.6	−41.4	−40.4
Na^+	−56.3	−50.4	−47.8	−45.7	−42.3	−46.2	−38.2	−39.0	−33.9	−32.9
K^+	−51.6	−44.4	−45.1	−39.5	−37.2	−33.5	−35.0	−31.8	−25.4	−24.3
Rb^+	−46.4			−36.2				−30.6	−22.6	
Cs^+	−47.7		−42.8	−35.3			−32.5	−29.7	−21.8	−20.4
F^-								−39.9	−40.1	−40.2
Cl^-	−47.0	−44.9	−46.7	−39.0	−37.0	−34.9	−34.6	−32.7	−26.4	−25.3
Br^-	−49.5	−42.8	−43.3	−36.2	−34.4	−34.7	−33.6	−30.1	−22.8	−21.4
I^-	−45.5	−37.9	−35.9	−32.3	−35.4	−30.8		−25.8	−17.3	−15.4

* Data are evaluated from the data of Table 2.11.14.

Three observations are apparent from a glance at the entropies listed in Table 2.12.1. Firstly, the ionic entropies are all negative, which is taken as indicating that all simple ions have a net ordering effect on every solvent listed. Consequently, the common concept that some ions in water, such as caesium or iodide, disorder the solvent is an artefact of the standard state. Secondly, in general as the sizes of the ions increase there is a decrease in the entropy, indicating that large ions have less of an ordering effect on the solvents than small ions. Thirdly, while there is a wide variation in ionic size there is, in general, an increase in ionic entropies for ions of all sizes in the various solvents in the order

$$NH_3 < \text{EtOH} \approx \text{DMF} < \text{MeOH} < \text{PC} < \text{DMSO} < \text{NMF} < \text{F} < H_2O < D_2O$$

indicating that ions cause a much greater net ordering of solvents at the liquid ammonia end than at the aqueous end of the scale.

For the practical purpose of estimating entropies in various solvents, particularly those for which no entropy data exist, it would be useful to find some physical property or combination of properties which would have values increasing proportionally in the order listed above. An examination of various properties of these solvents, summarised in Table 1.3.1, indicates that no such simple relationship exists. Because of this, any method involving only one of these properties, such as the derivative of the Born equation, is immediately doomed to failure. Use of more qualitative concepts such as basicities or solvent polarities appears equally unfruitful since agreement among investigators concerning these properties for such diverse solvents is almost totally lacking.

Criss and co-workers[E123] observed that if one considers the physical properties of the solvents collectively, along with current ideas concerning hydrogen bonding in many of these liquids, it is possible to rationalise the order listed above for the entropies. *Entropies of ions are most positive in those solvents which have the highest degree of internal order in the pure state.* Greyson[E131] has made the same observation with respect to the entropies of the alkali metal halides in water and in D_2O. By the term *internal order*, it is understood that one means an association between solvent molecules leading to an ordered structure. A system which is highly structured is not necessarily highly ordered, since it may be a disordered structure like that of an amorphous solid.

Frank and co-workers[E132,E133] and others have attributed many of the properties of water and aqueous solutions to its ability to polymerise three dimensionally to form 'flickering clusters' or 'icebergs'. One can extend these arguments to other solvents by considering the properties listed in Table 1.3.1, along with information concerning the various possibilities for hydrogen-bonding, if any. One can assume that order within a pure solvent is caused primarily by three things: (1) hydrogen bonding, (2) dipole-dipole interactions and (3) molecular shape. The first of these would generally be expected to predominate, but presumably if the dipole moment is high enough it may cause more ordering than a weakly hydrogen-bonded system. The third factor would have a minor effect. In general one would expect systems with three-dimensional polymerisation through hydrogen bonding to have a higher degree of order than systems exhibiting two- or one-dimensional polymerisation, and that solvents with large dipole moments to exhibit a higher degree of order than those with lower moments.

Detailed examination of the nature of the non-aqueous solvents under consideration is instructive. Formamide has two hydrogen atoms capable of forming hydrogen bonds. X-ray investigations on the solid show that because of the position of the hydrogen bonds it exists as large polymerised sheets, the sheets being held together by van der Waals' forces similar to those of graphite.[E134] One would expect fragments of these sheets to exist in solution giving rise to two-dimensional 'icebergs'. N.M.R.[E135] and dielectric constant[E136] studies on NMF show this solvent to form one-dimensional chain polymers in the liquid state. Presumably the degree of order would decrease accordingly. Since pure DMSO and PC are incapable of hydrogen bonding, one might expect order within these solvents to be minimal. However, both solvents exhibit extremely high dipole moments (3.96 and 4.98 D respectively) compared with the other solvents being considered. Therefore, it is not inconceivable that dipole-dipole interation leads to a greater ordering in these solvents than in MeOH which is capable of forming hydrogen bonds. On the basis of dipole moment alone one would expect the entropies to be more positive in PC than in DMSO. For some ions this appears to be the case, but in general the reverse seems to be true. DMF is also incapable of forming hydrogen bonds, but again one may expect that its relatively large dipole moment may lead to ordering to an extent about equal to that found in EtOH. Of the two alcohols one would expect a smaller degree of order in EtOH than in MeOH. While liquid ammonia has long been accepted as a hydrogen-bonded solvent, its properties indicate these bonds to be very weak. For example, its dielectric constant, viscosity, heat of vaporisation, and boiling point are amongst the lowest of all solvents indicating minor interactions. Furthermore, if for liquid ammonia the nitrogen oscillates through the plane formed by the three hydrogens, as has been proposed for gaseous ammonia,[E22] hydrogen bonding would be ruled out on theoretical grounds. The truth probably lies somewhere in between and we conclude that while there may be some hydrogen bonding in this solvent it is insufficient to lead to significant structure within the solvent.

A little reflection upon the nature of the solvents and the processes that occur during the course of measuring the various physical properties listed in Table 1.3.1 enables one to understand why no single property can be simply related to the degree of order within the solvent. Consider viscosity as an example. Among other things, an increased molecular size and non-ordered structure, such as intermingling hydrocarbon chains, will cause an increase in viscosity although neither of these cause an increased order in the liquid. Consequently, large chain hydrocarbons will be very viscous, but have little order. The dielectric constant serves as another example. This property is dependent upon several factors, including not only the dipole moment of the solvent molecules but also the normal orientation of the dipoles in the liquid state. Thus while both formamide and NMF have the same dipole moment, there is a wide divergence in their dielectric constants. This is caused by the fact that in liquid NMF the dipoles tend to be aligned in one direction while in formamide, because of the orientation of molecules from hydrogen bonding, there is partial cancellation of the dipoles leading to a lower dielectric constant. Thus, in spite of the fact that formamide is probably more ordered than NMF, its dielectric constant is very much smaller. In conclusion one can say that while a highly ordered structure will tend to exhibit high values for the physical properties listed in Table 1.3.1, the fact that a particular liquid has large values for a specific property does not necessarily mean that it is highly ordered. It is for this reason that in considering the amount of order within a solvent one must consider all of the properties collectively.

Up until now the discussion has been qualitative, but no theory is complete until it be placed on a quantitative basis. Because of the choice of standard state for the data given in Table 2.12.1, the partial molal entropy, $\bar{S}^0_{\text{ion}}$, is identical to the entropy of solvation, ΔS^0_{solv}, and this quantity is considered to be made up of three terms: (1) An entropy decrease caused by the loss of the degrees of freedom of the gaseous ions, ΔS^0_F, (2) an entropy increase caused by the disordering of the. original solvent structure, ΔS^0_D and (3) an entropy decrease resulting from the ordering of the solvent molecules around the ions, ΔS^0_0 Consequently, we can write*

$$\bar{S}^0_{\text{ion}} = \Delta S^0_D + \Delta S^0_0 + \Delta S^0_F \tag{2.12.1}$$

The last term of this equation is equal to the entropy change on the dissolution of a gas in a solvent to form an ideal solution, and because of the choice of standard state will be identical for all the solvent systems. The non-ideal part of the entropy change for formation of these solutions is included in the first and second terms on the right side of eqn. 2.12.1.

The entropy of orientation of solvent molecules around an ion will always be a negative quantity since it is associated with an increase in the order within the system. Its value may be considered to be a function of both how many and to what extent molecules are affected. As an example, a solvent consisting of molecules having a high dipole moment would exhibit strong ion-solvent interactions in the first solvation sphere but relatively weak interactions outside this sphere, where the electric field surrounding the ion has been diminished by the presence of the dipole. On the other hand, solvents having molecules with small dipole moments would exhibit weak interactions with the ion in the first solvation sphere, but the ion would exert its influence to a greater distance, since the smaller solvent dipoles would not reduce the electric field surrounding the ion as much as the larger dipoles. As a result one may imagine the net entropy of orientation of solvent molecules around an ion as being approximately constant for solvents having molecules

* Because of the change in standard state this equation has one less term than that given by Criss and co-workers.[E123]

with not too widely different dipole moments. If this is actually the case, then eqn. 2.12.1 can be written as

$$\bar{S}^0_{\text{ion}} = \Delta S^0_D + C \quad (2.12.2)$$

where C is a constant, which includes the ideal entropy of mixing and entropy of orientation of solvent molecules. It will therefore be strongly dependent upon the charge density of the ion.

The electric field surrounding an ion is sufficiently strong that one can safely assume that the original solvent structure in the vicinity of an ion is totally disrupted. In the case of water, which has about the largest degree of structure this assumption is consistent with the Frank and Wen model.[E133] With this additional assumption, eqn. 2.12.2 can be written as

$$\bar{S}^0_{\text{ion}} = kS_{\text{str}} + C \quad (2.12.3)$$

where S_{str} is the entropy of structure of the solvent, that is the decrease in the entropy of the solvent as a result of association leading to an ordered system. According to eqn. 2.12.3 plots of S_{str} *vs.* $\bar{S}^0_{\text{ion}}$ should lead to a series of parallel lines for the simple ions, one for each ion.

The problem that must be faced is how to evaluate S_{str}. On first consideration one might expect the deviation from Trouton's rule to be proportional to S_{str}, but unfortunately this describes the structure of the various liquids only at their respective boiling points, which may be markedly different from their structures at 25°C. Another approach would entail the use of the entropy of self-association of the various solvents in some 'inert solvent'. This raises the question as to what is meant by an 'inert solvent', and in any event data are not presently available to make such calculations. A third method would be to calculate the difference in the experimentally measured entropy of vaporisation at 25°C and a statistical mechanical calculation in which hydrogen bonding and dipole-dipole interactions have been purposely neglected. To the knowledge of the authors no such calculations have been performed for these systems.

In the absence of better methods, Criss and co-workers[E123] by the use of model compounds, estimated ideal boiling points for the various solvents and assumed that the deviations between these and the actual boiling points would be proportional to S_{str}. The deviations from the ideal boiling points for certain solvents are listed in Table 2.12.2. Equation 2.12.3 can then be rewritten as

$$\bar{S}^0_{\text{ion}} = k'\Delta T_{bp} + C \quad (2.12.4)$$

where k' is a proportionality constant and ΔT_{bp} is the deviation from the ideal boiling point. Plots of $\bar{S}^0_{\text{ion}}$ vs. ΔT_{bp} are shown in Fig. 2.12.1 and the values of k' and C for each ion are listed in Table 2.12.3. In addition

Table 2.12.2

Deviation Between Actual and Ideal Boiling Points of Various Solvents

Solvent	Boiling Point Deviation*/K
D_2O	173
H_2O	174
F	162
NMF	128
DMSO	135†
PC	151‡
MeOH	117
DMF	98
EtOH	97
NH_3	57

* Values are from Ref E123 unless otherwise indicated.

† The value for DMSO was obtained from Table 1.3.1 and an estimated ideal boiling point for DMSO of 54°C, obtained from the known boiling point for 2-methylpropene of −6°C (see Handbook of Chemistry and Physics, 50th ed., Chemical Rubber Co., Cleveland, Ohio, 1969–1970) plus an estimated 60°C due to the increased molecular weight. The latter estimate was based on the increase in boiling point of the compounds CH_4, $C(CH_3)_4$, $(CH_3CH_2)_3CH$ and 3,3-dimethylpentane, with molecular weight.

‡ The value for PC was obtained from Table 1.3.1 and an estimated ideal boiling point for PC of 91°C from the known boiling point of 1,3-dimethylcyclopentane (see J. P. McCullough, R. E. Pennington, J. C. Smith, I. A. Hossenlopp, and G. Waddington, *J. Amer. Chem. Soc.*, **81**, 5880 (1959).

to the solvents previously reported,[E123] data are also listed for DMSO and PC. The fact that k' is approximately the same for all ions indicates that the original solvent structure is completely disrupted in the vicinity of the ion for all ions studied. The large negative values for C indicate, not surprisingly, that solvents with no internal structure can be more highly ordered than those which have structure.

The use of ΔT_{bp}, like other physical parameters, is dependent upon more than order alone and consequently is not entirely satisfactory. However, considering the uncertain manner in which ΔT_{bp} must be estimated, the agreement with eqn. 2.12.4 is as good as can be expected.

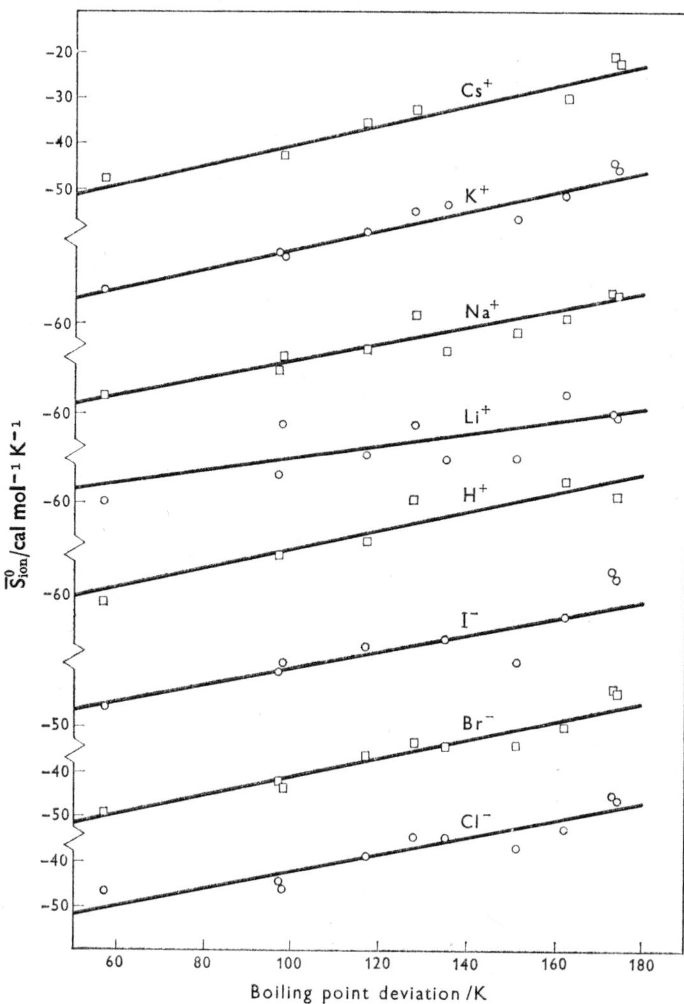

FIG. 2.12.1. Partial ionic entropies in various solvents *vs*. deviations from the ideal boiling points of the solvents.

Table 2.12.3

Constants for Eqn. 2.12.4

	k'/cal mol^{-1} K^{-2}	C/cal mol^{-1} K^{-1}
H^+	0.147 ± 0.007*	−76.0 ± 3.4†
Li^+	0.099 ± 0.007	−66.2 ± 4.4
Na^+	0.104 ± 0.007	−63.8 ± 4.5
K^+	0.131 ± 0.007	−62.6 ± 4.5
Rb^+	0.113 ± 0.012	−58.2 ± 5.3
Cs^+	0.141 ± 0.009	−62.4 ± 5.0
Cl^-	0.115 ± 0.006	−59.6 ± 4.1
Br^-	0.129 ± 0.008	−60.3 ± 4.9
I^-	0.139 ± 0.008	−57.8 ± 5.0

* Standard deviation of the slope.
† Standard deviation.

In an extension of these ideas, Franks and Reid[E124] have examined the ionic entropies in mixed water-methanol solutions (see Appendix 2.4.42) and in 20% aqueous dioxan, and have observed that the entropies of the ions for each of these systems can be expressed by an equation having the same form as eqn. 2.11.36. Similar to the pure solvent systems, the entropy of a given ion has no correlation with the solvent dielectric constant, nor is there a linear correlation of the entropy with solvent composition. Instead, the entropies reach a maximum in the vicinity of 40 mol per cent methanol. The authors explain this in terms of the solvent having the highest degree of structure near this composition. As in the pure non-aqueous solvents, the relative magnitude of the effect of ions on the solvent structure is the same for all ions, both negative and positive. This observation led the authors to conclude that there is no evidence for preferential solvation in these mixed solvent systems.

It is interesting to examine these concepts in terms of other data for electrolytic solutions. On the basis of the Frank and Wen model,[E133] Millero[E128, E137] has assumed the partial molal volume of an ion to be given by

$$\bar{V}^0_{\text{ion}} = \bar{V}^0_{\text{int}} + \bar{V}^0_{\text{elect}} + \bar{V}^0_{\text{disord}} + \bar{V}^0_{\text{caged}} \tag{2.12.5}$$

where $\bar{V}^0_{\text{int}}$ is the intrinsic partial molal volume of the ion, $\bar{V}^0_{\text{elect}}$ is the electrostriction partial molal volume, that is the volume caused by ion-solvent interactions, $\bar{V}^0_{\text{disord}}$ is the disordered partial molal volume (generally attributed to void space effects), and $\bar{V}^0_{\text{caged}}$ is the caged partial molal volume caused by a caged water structure around ions containing hydrophobic groups, such as the tetraalkylammonium ions. For the simple alkali metal and halide ions the last term can be neglected. It is difficult to determine the importance of the various contributions to $\bar{V}^0_{\text{ion}}$, and in order to gain some insight into these, Millero[E128] has examined $\bar{V}^0_{\text{ion}}$ values in water, methanol and NMP, using semi-empirical correlations which have previously been found useful for aqueous solutions.[E138,E139] One of these gives the ionic volumes by the expression

$$\bar{V}^0_{\text{ion}} = Ar^3 - Bz^2/r \tag{2.12.6}$$

where A and B are constants, r is the crystal radius of the ion and z is the ionic charge. The first term is equal to $\bar{V}^0_{\text{int}} + \bar{V}^0_{\text{disord}}$ and the second term gives $\bar{V}^0_{\text{elect}}$. Millero[E128] also examined these systems in terms of the more recent expressions[E140-142]

$$\frac{\bar{V}^0_{\text{ion}}}{\text{cm}^3\ \text{mol}^{-1}} = 2.52\left(\frac{r}{\text{Å}}\right)^3 + A'r^2 - B'z^2/r \tag{2.12.7}$$

$$\frac{\bar{V}^0_{\text{ion}}}{\text{cm}^3\ \text{mol}^{-1}} = 2.52\frac{(r+a)^3}{\text{Å}^3} - B''z^2/r \tag{2.12.8}$$

in an attempt to separate the intrinsic size of an ion from the void space effects. The first of these equations assumes the void space to be proportional to the surface of an ion (second term) while the second assumes the void space to be represented by a spherical shell around the ion (first term). The constants A, A', and a were evaluated for the three systems,[E128] and it was observed that the A constants are larger than the theoretical value in all three solvents. Furthermore, these constants, like ionic entropies, show no simple correlation to any of the common physical properties, including the compressibility of the solvents. On the other hand, there are indications that $\bar{V}^0_{\text{disord}}$ is larger in the more highly ordered solvents.

The $\bar{V}^0_{\text{elect}}$ term would be expected to be approximately related to the dielectric constant of the solvent through the equation.[E143]

$$\bar{V}^0_{\text{elect}} = -\frac{Nz^2e^2}{2\varepsilon_r r}\left[\frac{\text{d}\ln\varepsilon_r}{\text{d}p}\right] = -\frac{Bz^2}{r} \tag{2.12.9}$$

where ε_r is the dielectric constant. Using the known values of $[\text{d}\ln\varepsilon_r]/\text{d}p$ for water and methanol, Millero[E128] was able to calculate B values within a factor of two of those obtained through eqn. 2.12.6. Considering the approximate nature of this equation the divergence between the experimental and calculated values of B is considered satisfactory. If one assumes $[\text{d}\ln\varepsilon_r]/\text{d}p$ to be approximately constant for the various solvents, then B should vary linearly with $1/\varepsilon_r$, and this was observed to be the case within the limits of experimental error for water, methanol and NMP.[E128]

Using the $\bar{V}^0_{ion}$ values listed in Table 2.11.15, which were assigned by the correspondence method, we have repeated the calculations for the constants of eqn. 2.12.6 and 2.12.7 for the above-mentioned solvents and also for formamide and formic acid. These are summarised in Table 2.12.4. It can be observed that, consistent with the previous discussion,

Table 2.12.4

Constants in Eqn. 2.12.6 and 2.12.7*† for $\bar{V}^0_{ion}$/cm^3 mol^{-1}

Solvent	A/Å^{-3}	B/Å	A'/Å^2	B'/Å
Water	4.35 ± 0.07	7.5 ± 2.1	4.14 ± 0.12	11.0 ± 1.8
Formamide	4.36 ± 0.07	4.2 ± 2.0	4.18 ± 0.07	7.7 ± 1.2
NMP	3.84 ± 0.03	−2.7 ± 0.7‡	2.86 ± 0.06	−0.6 ± 0.7‡
Formic Acid	3.72 ± 0.10	7.2 ± 3.0	2.82 ± 0.17	10.2 ± 2.7
Methanol	3.21 ± 0.08	14.8 ± 2.4	1.58 ± 0.15	16.0 ± 2.2

* Ionic radii used in the calculations are identical to those used in ref. E128.
† The errors are for the standard deviation of the slope or the standard deviation.
‡ This value is significantly different from that reported by Millero[E128] even though the volume assignments are identical.

the A and A' constants tend to be most positive for the most structured liquids. Consequently, one would expect a direct correlation between A or A' and the constant, a, in eqn. 2.11.36 representing the entropies. Unfortunately, there are no entropy data for NMP, and while ionic entropies have been reported for formic acid (see Appendix 2.4.41), they are totally inconsistent with the behaviour of the entropies of electrolytes in other solvents. A comparison of A or A' with the constant of eqn. 2.11.36 for the remaining three solvents indicates that such a correlation does exist.

As mentioned above, the B constant of eqn. 2.12.6 should be approximately linearly related to the reciprocal of the dielectric constant. Values of $1/\varepsilon_r$ have been plotted against the B and B' values listed in Table 2.12.4. The results are shown in Fig. 2.12.2. With the exception of NMP these fall on a straight line within the limits of experimental error.

Because of the increase of $\bar{V}^0_{int}$ over the theoretical value (i.e. $2.52\,r^3$/Å^3) not only in water, but in methanol and NMP, Millero[E128] suggested that this term is related to the disordered region surrounding a solvated ion.

This region, consistent with the Frank and Wen model,[E133] may be viewed as the void space caused by the solvated ion rather than improper packing in the electrostricted region.

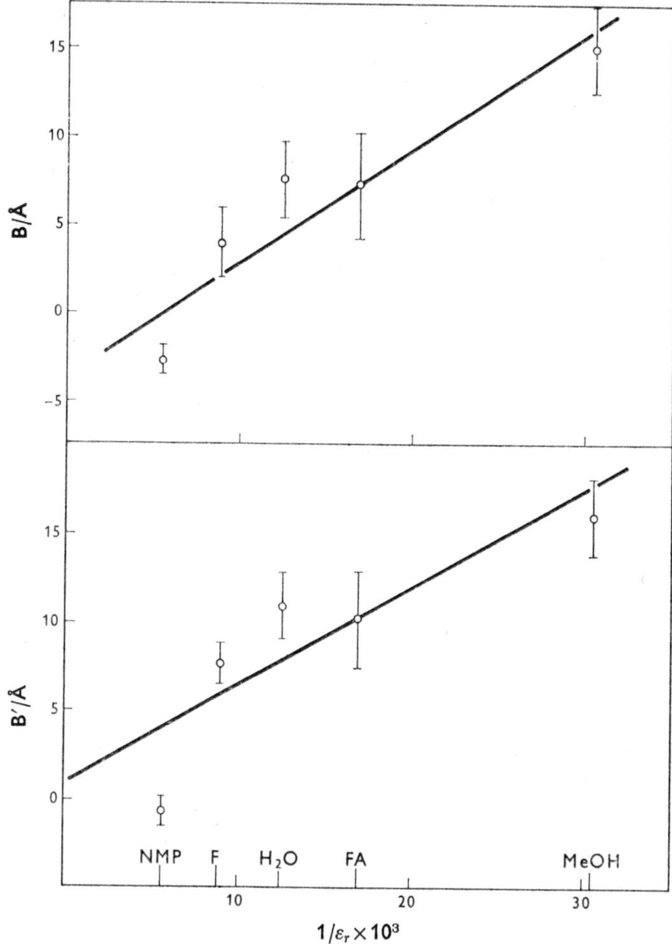

FIG. 2.12.2. Constants B and B' for eqns. 2.12.6 and 2.12.7 *vs.* $1/\varepsilon_r$ for various solvents.

This concept is entirely consistent with that deduced from ionic entropies. For solvents which are initially highly ordered, there is strong competition between the ion and other solvent molecules to increase order. On the one hand the ion is attempting to orient solvent molecules around itself (electrostrict), while on the other hand the solvent molecules are attempting to stay in the highly structured bulk liquid. This leads to a relatively large disturbed region surrounding the solvated ion, leading to large values for $\bar{V}^0_{\text{disord}}$ and $\bar{S}^0_{\text{ion}}$. For solvents with a small amount of initial structure the competition is mostly in favour of the ion. As a result more solvent molecules will be influenced by the ion, giving rise to a smaller void space region surrounding the ion, and consequently a smaller $\bar{V}^0_{\text{disord}}$ and more negative $\bar{S}^0_{\text{ion}}$.

For the same reason $\bar{V}^0_{\text{elect}}$ will tend to have large negative values because more solvent molecules are electrostricted by the ion. One would expect the actual number of solvent molecules ordered by an

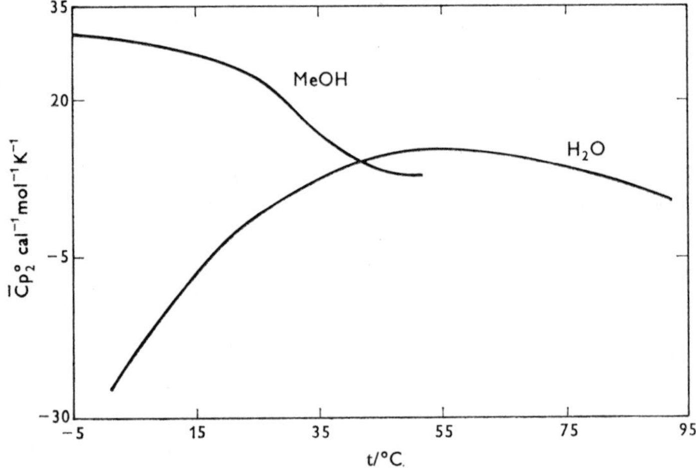

FIG. 2.12.3. Partial molal heat capacities of $NaClO_4$ in water and methanol as a function of temperature.

ion to depend upon both the degree of structure in the solvent and the dielectric constant of the solvent. Consequently, the $\bar{V}^0_{\text{elect}}$ should also depend upon both of these quantities instead of only the dielectric constant as suggested by eqn. 2.12.9. Indeed, this omission is implicit in the basic assumption made in eqn. 2.12.9 in that the dielectric is assumed to be a continuum.

These arguments should apply equally well to partial molal heat capacities in organic solvents. Data for $\bar{C}^0_{p2}$ of electrolytes in organic solvents are exceedingly rare.[E144–147] However, $\bar{C}^0_{p_2}$ for $NaClO_4$ and Me_4NBr are known as a function of temperature in both methanol and water.[E145,E146] The data for $NaClO_4$ are shown in Fig. 2.12.3. Any discussion of electrolytic solutions necessarily centres around water, so it is instructive first to examine the behaviour of $\bar{C}^0_{p_2}$ in aqueous solutions. It is observed that for $NaClO_4$, like all simple electrolytes in water, $\bar{C}^0_{p_2}$ is negative at room temperature and lower. This is generally explained in terms of the presence of the ion 'melting' the water structure, with the result that the amount of heat required to increase the temperature of the solution by a given amount is less than that required for pure water. The consequence of this is that the ion contributes a negative component to the heat capacity. The increase in $\bar{C}^0_{p_2}$ with temperature (generally up to 60 or 80°C) has been attributed by Ackermann and co-workers[E148] to the additional heat required for the removal of the outer hydration shells. At temperatures above 80°C much of this water has been removed and $\bar{C}^0_{p_2}$ decreases again. Presumably this decrease will continue until at some higher temperature additional heat will be required to remove the first hydration sheath, at which time there will again be an increase in $\bar{C}^0_{p_2}$. Mastroianni and Criss,[E145] on the other hand, attribute the increase in$\bar{C}^0_{p_2}$ with temperature to the fact that at higher temperatures there is less inherent structure in the water to be disrupted so that the ion is able to 'melt' relatively little of the structure, and thereby have a smaller negative effect. Indeed, for $NaClO_4$, $\bar{C}^0_{p_2}$ actually becomes positive because of the large inherent heat capacity of the perchlorate ion. At temperatures above 80°C so much bulk water structure has been disrupted by thermal effects that the ability of the ion to order water around itself or to electrostrict water becomes the dominating feature. This ordering effectively removes water from solution with a resultant negative contribution to $\bar{C}^0_{p_2}$. Presumably $\bar{C}^0_{p_2}$ will become more and more negative as the temperature increases and the ion is able to order more solvent molecules around itself. This will also result in more negative ionic entropies and volumes. These effects ar-actually observed to be the case for as high as the data are available (200°C for volumes and heat capacities and 300°C for entroe pies).[E125,E149,E150]

In all of these arguments, one must keep in mind that partial molal quantities of ions reflect not so much the absolute amount of structure around the ion, but rather the amount of structure around the ion relative to that of the pure solvent at that temperature. Thus, while the net amount of order in a solution decreases as the temperature increases, the effect of the ion on ordering the system, through ion-dipole interactions, relative to the pure solvent at that temperature continues to increase.

One can apply the same arguments to methanol. From Fig. 2.12.3 it appears that the $\bar{C}^0_{p_2}$ *vs.* temperature curve for $NaClO_4$ in methanol is of the same general form as that in water, except that the maximum is shifted to a lower temperature by 50 to 100°C and the value of $\bar{C}^0_{p_2}$ at the maximum is much greater than for the aqueous solution. This is what one would expect for a solvent which has significantly less inherent structure than water. At temperatures near 0°C there is relatively little inherent structure in the solvent that can be disrupted by the ion, but yet a sufficient amount that the ion cannot strongly order (electrostrict) solvent molecules around itself; consequently, there is a relatively high value for $\bar{C}^0_{p_2}$. At room temperature the electrostriction effect has become the dominating one and $\bar{C}^0_{p_2}$ is somewhat less, and at higher temperatures $\bar{C}^0_{p_2}$ should continue to decrease.

Table 2.12.5

$B_{\eta\,\mathrm{ion}}$ Coefficients in Water, Methanol and Acetonitrile at 25°C*

Ion	$B_{\eta\,\mathrm{ion}}$/mol^{-1} l		
	Water	Methanol	Acetonitrile
Li^+	0.15		
Na^+	0.09		0.44
K^+	−0.007	0.382	
Rb^+	−0.03		
Cs^+	−0.05		
NH_4^+	−0.01	0.279	
Me_4N^+	0.12	−0.03	0.25
Et_4N^+	0.38	0.12	0.32
Pr_4N^+	0.86	0.30	0.37
Bu_4N^+	1.28	0.49	0.56
Cl^-	−0.007	0.382	
Br^-	−0.03	0.358	0.37
I^-	−0.08	0.293	0.34
Pic^-	0.43†		0.53
BPh_4^-			0.87

* Data from ref. E130.
† 18°C.

The $B_{\eta\,\mathrm{ion}}$ coefficients in water, methanol and acetonitrile have been tabulated for several ionic species, including complex ions.[E130] The assigned values are given in Table 2.12.5, and a plot of $B_{\eta\,\mathrm{ion}}$ *vs.* r is shown in Fig. 2.12.4. The assignment of $B_{\eta\,\mathrm{ion}}$ in acetonitrile is purely arbitrary, but assuming that the method of division of $B_{\eta\,\mathrm{ion}}$ in water and methanol (i.e. $B_{\eta K^+} = B_{\eta Cl^-}$) to be valid, one can compare these coefficients in the two solvents in the light of the previous discussion.

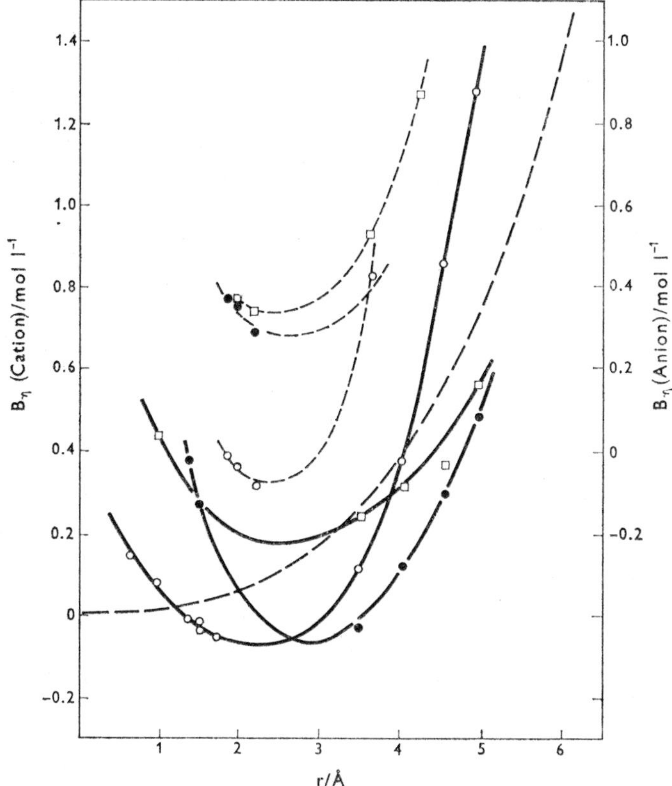

FIG. 2.12.4. Variation of $B_{\eta\text{ion}}$ coefficient with ionic radius. Cations, solid lines; Anions, dashed curves; Einstein equation, broken curve; Open circles, aqueous solutions; Solid circles, methanol solutions; Squares, acetonitrile solutions. Radii are those of Pauling (see Table 2.11.2). Data taken from ref. E130.

First it is noticed that $B_{\eta\text{ion}}$ values for all simple cations and anions are considerably greater in methanol than in water. The small values, even negative, in water have been attributed to the strong disruption effect of the ions on the bulk structure of water.[E151] Thus, while there has been a net ordering effect, as indicated by negative $\bar{S}^0_{\text{ion}}$ values, this ordering has been in the vicinity of the ion. In some cases the bulk structure of the water has been sufficiently depolymerised to result in a net decrease in viscosity.

In methanol, the $B_{\eta\text{ion}}$ coefficients of the simple ions are larger than in water, indicating a larger degree of ordering of solvent molecules around the ion. This is also consistent with the fact that $\bar{S}^0_{ion}$ and $\bar{V}^0_{elect}$ are much more negative in this solvent than in water. Indeed, the effect of electrostriction is so great that it overcomes any viscosity decrease because of disruption of the bulk solvent structure and instead causes a considerable increase in viscosity. It is observed that the smaller ions in both solvents have larger $B_{\eta\,ion}$ coefficients than the large simple ions. This is the result of greater electrostriction by the small ions in both solvents.

The complex tetraalkylammonium ions present an entirely different situation. Because of the interaction of the hydrophobic groups with water these ions tend to structure the water (probably not in the same way as a simple ion). For this reason, these ions have been considered to behave abnormally in water.[E107,E132,E133,E152,E153] Presumably in organic solvents this effect would not be present and these ions would behave like any other ion. Figure 2.12.4 shows this not to be the case. While the increase in $B_{\eta\,ion}$ values with radius for the tetraalkylammonium ions is not nearly as great in methanol and acetonitrile as it is in water, there is nevertheless a significant increase so that the difference is a matter of degree rather than of form. However, for the organic solvents this increase may be mostly caused by the expected increase in $B_{\eta\,ion}$ as calculated by the Einstein equation[E154] (see broken curve).

The relationship between ionic enthalpies of solvation, or of transfer, and structure is not nearly as clear as for the quantities just discussed. In the first place, the enthalpy of solvation is dependent not only on the degree of increase, or decrease, in the structure of a solution (i.e. the number of bonds made or broken), but also upon the energy of each bond. So while there may be some relationship between enthalpies and structural properties of a solution it certainly will not be a simple one. Furthermore, the problem is complicated by the fact that there is very little agreement on how to divide enthalpies into their ionic components. Even for aqueous solutions, for which one would expect some agreement, the division of enthaplies of solvation proposed by various investigators varies by up to 50 kcal mol^{-1} (see Appendix 2.11.3). For organic solvent systems the method of division of either solvation enthalpies or enthalpies of transfer has not been applied uniformly to all the solvents, not only because of the individual preferences of the investigators, but also because of the lack of data in some cases (e.g. those divisions made on the assumption that $\Delta H^0_t(Ph_4B^-) = \Delta H^0_t(Ph_4As^+)$). With such complications it is difficult to obtain quantitative conclusions concerning the effect on the enthalpy from structural aspects of the solutions.

Furthermore, the use of ionic solvation enthalpies raises another problem, since the enthalpies of formation of the gaseous ions are frequently not known accurately or impossible to obtain (as with some of the complex organic ions). In order to avoid this problem several investigators[E104–107,E155,E156] have resorted to comparing enthalpies of transfer from one solvent to another, but at the cost of introducing additional complications in the interpretation of the results. Friedman and co-workers[E107,E153] in their excellent investigations concerning the effects of simple and complex ions on the structure of aqueous solutions, have suggested the use of propylene carbonate as the nearly ideal solvent for ions. With this assumption any abnormalities in ΔH_t(ion) from PC to other solvents would be explained mostly in terms of structural abnormalities in the other solvent.

Krishnan and Friedman[E107] have examined enthalpies of transfer of simple and complex ions from DMSO to PC and from water to PC, using the convention that $\Delta H_t(Ph_4As^+) = \Delta H_t(Ph_4B^-)$. The results of these studies are shown in Figs. 2.12.5a and b. The $\Delta H_{PC \leftarrow DMSO}$ values for both the simple and tetraalkylammonium ions are linear with radius, but the slopes of the lines for the two types of cations are significantly different. The tetraalkylammonium ions are also linear in carbon number and exhibit negative enthalpies of transfer. On the basis of earlier work with aliphatic alcohols in PC and DMSO,[E157] the authors suggest that the solvation of the tetraalkylammonium ions in these solvents is dependent upon the additive contributions from methylene groups and that these contributions are the same whether they are in R_4N^+ or ROH. The arylsubstituted cations, on the other hand, show no correlation with ionic radius, indicating that the central atom has a large effect upon the solvation of the phenyl group. The positive enthalpies of transfer from DMSO to PC for the alkali metal ions appear to indicate that for these ions DMSO acts as a stronger Lewis base than does PC.

Enthalpies of transfer from water to PC, as seen in Fig. 2.15.5b show the same qualitative behaviour as $\Delta H^0_{PC \leftarrow DMSO}$ for the simple ions, but different interpretations are offered for the two cases,[E107] specifically a difference in basicity for the PC ← DMSO process, but specific hydration effects for the PC ← H_2O transfer. This can be justified on the basis of the difference in the single-ion values for the two systems. The basic assumption supporting this view is that the single-ion enthalpies have been correctly assigned.

For the tetraalkylammonium ions the behaviour of ΔH_t(ion) for the two systems is entirely different, in that there is an increase in $\Delta H^0_{PC \leftarrow H_2O}$ with radius, up to the NAm_4^+ ion, after which there appears to be a decrease. Because of the regularity of $\Delta H^0_{PC \leftarrow DMSO}$, the authors suggest that the curve represents a phenomenon of the aqueous solutions rather than of the PC solutions. Further evidence for this view is presented by the fact that the curve is very similar to that observed in the hydration of primary alcohols.[E157]

Choux and Benoit[E104] have compared ΔH_t(ion) for the systems

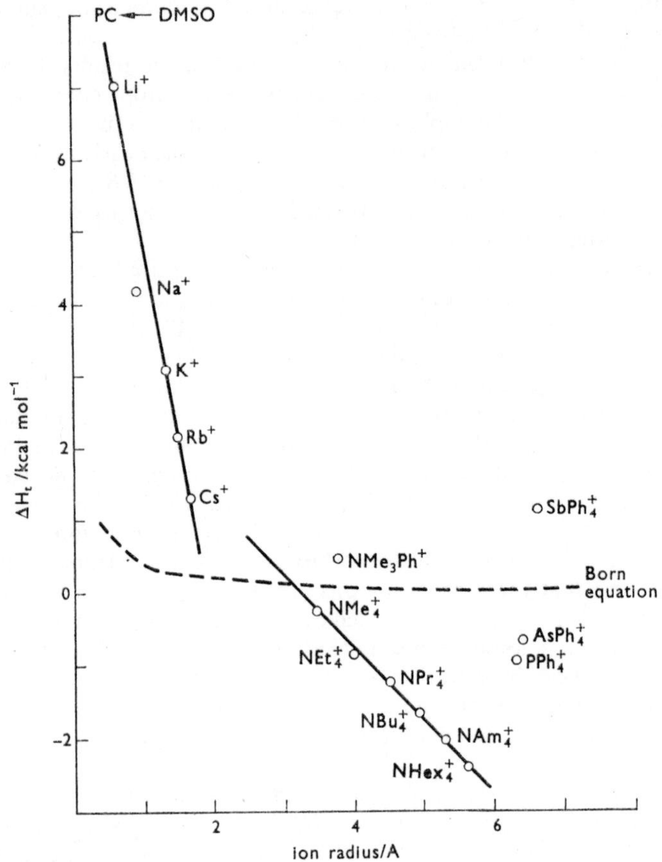

FIG. 2.12.5. (*a*) Ionic enthalpies of transfer from dimethylsulphoxide to propylene carbonate (from ref. E107 by permission of the authors).

$TMSO_2$ ← PC, DMSO ← PC, DMF ← PC, MeOH ← PC, and H_2O ← PC. In every case the division into ionic components was made on the basis that $\Delta H_t(Ph_4As^+) = \Delta H_t(BPh_4^-)$. They observed curves, in so far as data were available, similar to that in Fig. 2.12.5b for each of the systems, but the depth of the minimum of the curves decreases in the order

$$H_2O \leftarrow PC > MeOH \leftarrow PC > TMSO_2 \leftarrow PC > DMF \leftarrow PC > DMSO \leftarrow PC$$

The depth of the curves changes more as a result of the variation of the simple cations in the various solvents (up to 8 kcal mol^{-1}), rather than the tetraalkylammonium ions, which vary relatively little (~4 kcal mol^{-1} for the smaller ions). Indeed, the enthalpy of transfer between any two of the solvents, including water, appears to be less than 2 kcal mol^{-1} for tetraethylammonium ion. For the last of the systems in the above series, DMSO ← PC, data for only one tetraalkylammonium ion were available and the curve was estimated from comparison with other curves.[E104]

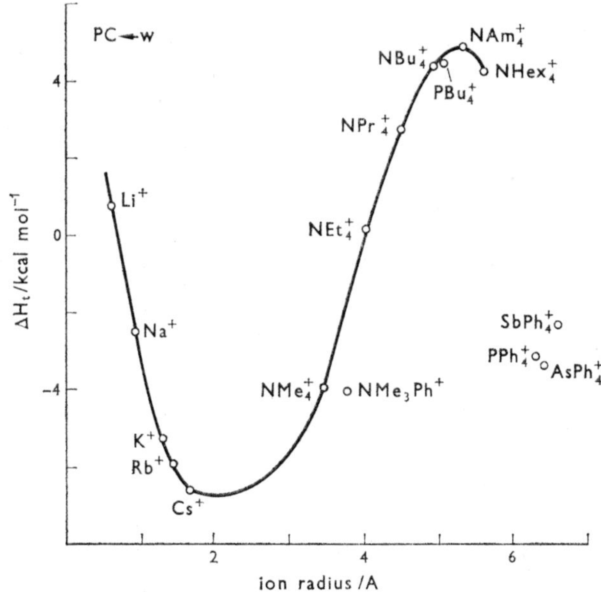

FIG. 2.12.5. (*b*) Ionic enthalpies of transfer from water to propylene carbonate (after ref. E107).

The more recent treatment by Krishnan and Friedman[E107] shows that for this system the simple ions and tetraalkylammonium ions form two distinct straight lines (see Fig. 2.12.5a). The important thing to notice is that if one accepts the view that PC is the ideal solvent for ions, then the 'water-like' behaviour in terms of enthalpies of solvation of the other solvents decreases in the order

$$H_2O > MeOH > TMSO_2 > DMF > DMSO$$

This order is essentially the *reverse* of the solvating ability, as viewed from enthalpies, of the solvents with respect to PC. For the aprotic solvents this order is in agreement with that obtained from phenol or iodine adducts,[E158] which predict the solvating abilities to be in the order

$$\text{DMSO} > \text{DMF} > \text{TMSO}_2$$

One must keep in mind that the solvating ability of a solvent is the sum effect of the energy released through ion-solvent interactions and the energy required to rupture a solvent molecule away from the bulk liquid. The magnitude of 'non-specific solvation', that is, solvation in the gas phase, may be entirely different. The charge transfer transition energy, or *Z*-value, obtained from *N*-methylpyridinium iodide has been used[E159] to evaluate non-specific solvation magnitudes, and for the solvents considered above the solvating strengths are in the order

$$\text{H}_2\text{O} > \text{MeOH} > \text{TMSO}_2 > \text{DMSO} > \text{DMF}$$

This is almost the exact opposite of that observed for the liquid phase solvation enthalpies.

Solvation of the anions in these solvents, as reflected in ΔH_t(ion), appears to be entirely different from the cations. While anionic enthalpies of transfer as a function of radius appear to follow the same general trend in the aprotic solvents $TMSO_2$, DMSO and DMF, they follow a different trend in water and methanol. This has been attributed to the stronger solvation of the smaller anions as a result of hydrogen bonding with the solvent.[E104] Choux and Benoit report that in the aprotic solvents, the order of solvation for the anions is

$$\text{DMF} > \text{DMSO} > \text{TMSO}_2 > \text{PC}$$

However, Fuchs and co-workers[E105] contend that solvating ability of these solvents is in the order

$$\text{DMF} > \text{PC} > \text{DMSO}$$

All of the above arguments have been based on the convention that the enthalpy of transfer of Ph_4As^+ and Ph_4B^- are equal for every solvent pair. In view of the extreme sensitivity of ΔH_t(ion) for the arylsubstituted ions upon the nature of the central atom,[E107] this convention seems dubious. For example, Ph_4Sb^+ and Ph_4P^+ have approximately the same radii as Ph_4As^+, but $\Delta H_{\text{PC}\leftarrow\text{DMSO}}$ for the three species is significantly different. Recent evidence found by Krishnan and Friedman[E160] indicates that Ph_4As^+ and Ph_4B^- differ markedly in solvation in water and methanol and they suggest that this convention has no particular advantage over the convention based on equal enthalpies of transfer of Cs^+ and I^-.

As one would expect, the isotope effect on enthalpies of transfer is small. Only one such study has been reported for an organic solvent system; the enthalpies of transfer of several electrolytes and non-electrolytes from CH_3OH to CH_3OD.[E160] In most cases values are positive, including those for alkyl groups, which in the corresponding aqueous systems give negative enthalpies of transfer. Krishnan and Friedman attribute this to be the dominating effect, during the transfer process, being the difference in enthalpy required to make a cavity in CH_3OD and CH_3OH, with more enthalpy required to disrupt CH_3OD due to the stronger hydrogen bonds. The alkali metal ions show the same trend in both the aqueous and methanolic systems, but $\Delta H_{D_2O \leftarrow H_2O}$ is about 0·3 kcal mol^{-1} more positive than $\Delta H_{CH_3OD \leftarrow CH_3OH}$, apparently because of the structure-breaking effect of the ion on the water structure, or conversely the lack of structure-breaking effect in the methanol system. These isotope studies have been extremely helpful in probing the relative degree of structure in the two solvents.

With the exception of water and methanol, the previous discussion has been confined to solvation of aprotic solvents of quite different types of molecules. Because hydrogen bonding leads to more obvious structural possibilities, one might expect solvation enthalpies of transfer in protic solvents to be more amenable to explanation. One might also expect solvation enthalpies of transfer to be more predictable in solvents having molecules of the same general type. Since all but one of the protic solvents listed in Appendix 2.11.15 are amides and only one of the amides is aprotic, we shall make both of these comparisons simultaneously. The data show that, in general, the simple cation enthalpies of transfer from water to various solvents, $\Delta H^0_{S \leftarrow H_2O}$, decrease in the order

$$\mathrm{F} > \mathrm{MeOH}^* > \mathrm{NEA}^* > \mathrm{DMF}^* > \mathrm{NMA} > \mathrm{NMF} > \mathrm{NMP}^*$$

Electrolytes in those solvents marked with an asterisk have been divided into ionic components by the convention that $\Delta H_t(Ph_4As^+) = \Delta H_t(Ph_4B^-)$, while those in the other solvents have been divided into their respective components by the modified Buckingham theory.[E31] The use of two methods of division confuses the issue, but assuming that both methods give approximately the same results, one can speculate on this particular order. In the first place, the enthalpies of transfer of the ions from water to all the solvents are significantly negative, except for formamide for which the enthalpy of transfer is positive. It may be recalled that in the discussion of ionic entropies and volumes it was concluded that formamide is the most highly structured of the organic liquids being almost as structured as water. The positive enthalpy of transfer observed suggests that, while ions are able to increase the overall order in formamide to a greater extent than water, the enthalpy required to create a hole in the solvent for the ion is larger than for water.

It is difficult to explain the order of the remaining solvents. Since $\Delta H^0_{S \leftarrow H_2O}$ for all of these systems is negative, it is tempting to apply the argument that less energy is required to disrupt the solvent structure in these less structured solvents, with the result that the transfer process is exothermic. However, on the basis of previous discussions, one is hard pressed to explain the presence of DMF in the centre of the above series, since DMF is considered to be relatively unstructured with respect to the *N*-monosubstituted amides. It is equally difficult to explain the order in terms of expected base strengths of the various amides.

The behaviour of the simple anions in these solvents is even more difficult to explain. The overall trend in the enthalpies of transfer from water to the various solvents appears to be opposite to that for the cations. However, one should view this with caution since the $\Delta H^0_{S \leftarrow H_2O}$ values for the three halides give three distinct orders to the above series of solvents. Until some trends are established, any theoretical explanation appears hopeless.

2.12.3 Estimation of Thermodynamic Functions in Organic Solvent Systems

The usefulness of any theory is directly proportional to its ability correctly to describe systems for which no data are available. One would hope that the correlations of the data and theoretical considerations presented in previous sections would make it possible to estimate accurately thermodynamic quantities for systems which are experimentally difficult to measure. To some extent this appears to be possible. However, the systems are extremely complex and there is a wide variety of phenomena involved, which makes extrapolation of ideas to experimentally virgin areas subject to question. As mentioned previously, ionic entropies and volumes are physically more meaningful; therefore, they should present the best opportunities for successful estimations.

For a given solvent, one should be able, with reasonable confidence, to estimate the entropy of simple univalent ions by eqn. 2.11.36, if the constants a and b for that solvent are known. In this respect the equation is useful for filling in the few holes which exist in the data listed in Table 2.11.14. It would be interesting to know whether eqn. 2.11.36 is also useful for making estimations of ionic entropies of complex and polyvalent ions. Fortunately, a few experimental entropies are available for 2:1 electrolytes for comparison. If eqn. 2.11.36 is modified to include the absolute ionic charge, z, it will account for divalent ions.[E161] With the charge included, eqn. 2.11.36 becomes

$$\bar{S}^0_{\text{ion}}(X) = az + b\bar{S}^0_{\text{ion}}(\text{aq}) \qquad (2.12.10)$$

Using the constants a and b from Table 2.11.13, which are evaluated only from univalent ions, the above equation has been used to calculate the entropies for several 2:1 electrolytes listed in Table 2.12.6. Agreement between the calculated and experimental values is amazingly good,

Table 2.12.6

Calculated and Experimental Entropies (cal mol^{-1} K^{-1}) of Divalent Cations in Methanol at 25°C

Ion	$\bar{S}^0_{ion}$(Calc)*/ cal mol^{-1} K^{-1}	$\bar{S}^0_{ion}$(Exp)†/ cal mol^{-1} K^{-1}	Difference/ cal mol^{-1} K^{-1}
Mg^{++}	−52.5	−54.0	−1.5
Ca^{++}	−42.8	−40.1	2.7
Sr^{++}	−37.1	−37.0	0.1
Ba^{++}	−26.9	−28.1	−1.2
Co^{++}	−51.5	−53.2	−1.7

* Calculated by eqn. 2.12.10, using $\bar{S}^0_{ion}$(aq) from W. M. Latimer, Oxidation Potentials, Prentice-Hall, Inc., Englewood Cliffs, N.J. (1952), after adjustment for $\bar{S}^0_{H^+} = -5.0$ cal mol^{-1} K^{-1}.

† Obtained from data in Appendix 2.4.41 and $\bar{S}^0_{Cl^-}$ obtained from Table 2.11.14.

which suggests that this equation is valid for estimating entropies of divalent cationic species and perhaps even higher charged ions in all organic solvents.

Estimations based on eqn. 2.11.36 or 2.12.10 assume that the constants a and b are known. When this is not the case, approximate estimates of a and b can probably be made, assuming that one can estimate the amount of structure in a solvent relative to those for which entropy data are already available. By interpolating between solvents one can then estimate the constant a from those listed in Table 2.11.13. With the exception of DMSO and DMF, the b constants in Table 2.11.13 for the non-aqueous systems are not too widely different, so it seems reasonable that an average of these values may give a fair estimate of b for other organic solvents. Fortunately, $\bar{S}^0_{ion}(X)$ in eqn. 2.12.10 is not extremely sensitive to the value of b when $\bar{S}^0_{ion}$(aq) is small. Substituting the average value of b into eqn. 2.12.10 gives

$$\bar{S}^0_{ion}(X) = az + 0.73\,\bar{S}^0_{ion}(\text{aq}) \qquad (2.12.11)$$

With the apparent success of predicting ionic entropies of higher charged type ions in organic solvents with eqn. 2.12.10, it is tempting to modify eqn. 2.11.39 for ionic volumes to take into account higher charged species. Similar to eqn. 2.12.10, ionic volumes should be given by

$$\bar{V}^0_{ion}(X) = a'z + b'\bar{V}^0_{ion}(\text{aq}) \qquad (2.12.12)$$

Unfortunately, we know of no accurate experimental $\bar{V}^0_2$ data for high valence type ions in organic solvents with which to compare values calculated from this equation.

Partial molal volumes of several tetraalkylammonium halides are available (see Appendix 2.4.43) which makes it possible to compare experimental values of $\bar{V}^0_{ion}$ with those calculated from eqns. 2.11.39 or 2.12.12. The comparisons are shown in Table 2.12.7. The fact that the

Table 2.12.7

Calculated and Experimental Volumes ($cm^3\ mol^{-1}$) of Tetraalkylammonium Ions in Methanol at 25°C

Ion	$\bar{V}^0_{ion}$(calc)*/ $cm^3\ mol^{-1}$	$\bar{V}^0_{ion}$(exp)†/ $cm^3\ ml^{-1}$	Difference/ $cm^3\ mol^{-1}$
Me_4N^+	66.1	71.7	5.6
Et_4N^+	119.6	129.4	9.8
Pr_4N^+	178.4	201.4	23.0
Bu_4N^+	233.6	267.6	34.0

* Calculated from eqn. 2.12.12 using constants from Table 2.11.16 and $\bar{V}_{ion}{}^0(aq)$ from F. J. Millero, The Partial Molal Volumes of Electrolytes in Aqueous Solutions, 'Chapter 13 *in* Water and Aqueous Solutions, (R. A. Horne, ed.), Wiley, New York (1972). $\bar{V}^0_{ion}$ was reassigned on the basis of $\bar{V}^0_{H^+}(aq) = -5.4\ cm^3\ mol^{-1}$.

† Calculated from the data in Appendix 2.4.43 and $\bar{V}^0_{ion}$ for Cl^- and Br^- from Table 2.11.15.

calculated and experimental values are considerably different, particularly for the larger ions, is not surprising in view of generally accepted opinion that these ions behave abnormally in water. What is surprising, however, is that for the larger ions the experimental values of $\bar{V}^0_{ion}$ in water and methanol are about the same (e.g. $\bar{V}^0_{Bu_4N^+}$ is 270.3 $cm^3\ mol^{-1}$ in water and 267.6 $cm^3\ mol^{-1}$ in methanol).
(Manuscript completed January 1971.)

REFERENCES

E1 B. E. Conway, *Ann. Rev. Phys. Chem.*, **17**, 481 (1966)
E2 M. Born and J. E. Mayer, *Z. Physik.*, **75**, 1 (1932)
E3 M. L. Huggins and J. E. Mayer, *J. Chem. Phys.*, **1**, 643 (1933)
E4 M. L. Huggins, *J. Chem. Phys.*, **5**, 143 (1937); **15**, 212 (1947)
E5 J. Sherman, *Chem. Rev.*, **11**, 93 (1932)
E6 M. P. Tosi and F. G. Fumi, *J. Phys. Chem. Solids*, **25**, 31 (1964)
E7 M. F. C. Ladd and W. H. Lee, Progress in Solid State Chemistry, Vol. 1, Ch. 2 (H. Reiss, ed.), Macmillan, N.Y. (1964); see also Vol. 2 (1965) and 3 (1967)
E8 A. F. Kapustinskii, *Quart. Rev.*, **10**, 283 (1956)

E9 H. F. Halliwell and S. C. Nyburg, *J. Chem. Soc.*, 4603 (1960)
E10 L. H. Ahrens, *Geochim. Cosmochim. Acta*, **2**, 155 (1952)
E11 L. Pauling, Nature of the Chemical Bond, Cornell University Press, Ithaca, N.Y. (1945).
E12 L. Helmholz and J. E. Mayer, *J. Chem. Phys.*, **2**, 245 (1934)
E13 R. S. Berry and C. W. Reimann, *J. Chem. Phys.*, **38**, 1540 (1963)
E14 D. F. C. Morris, Structure and Bonding, **4**, 63 (1968)
E15 W. M. Latimer, Oxidation Potentials, 2nd ed., Prentice Hall, Englewood Cliffs, N.J. (1952)
E16 D. D. Wagman, W. H. Evans, V. B. Parker, I. Halow, S. M. Bailey and R. H. Schumm, Selected Values of Thermodynamic Properties, N.B.S. Technical Note No. 270 (1968)
E17 JANAF Thermochemical Tables, Dow Chemical Co., Midland Mich., Aug. 1965 and addenda
E18 E. A. Moelwyn-Hughes, Physical Chemistry, Pergamon, London (1961)
E19 E. B. Wilson, *Chem. Rev.*, **27**, 17 (1940)
E20 E. B. Wilson, J. C. Decius and P. C. Cross, Molecular Vibrations, McGraw-Hill, N.Y. (1955)
E21 K. W. F. Kohlrausch, Der Smekal-Raman Effect, Springer, Berlin (1938)
E22 G. Herzberg, Infrared and Raman Spectra, Van Nostrand, N.Y. (1959)
E23 R. H. Boyd, *J. Chem. Phys.*, **51**, 1470 (1969)
E24 M. Salomon, J. Electrochem. Soc., **118**, 1614 (1971)
E25 B. B. Owen, *J. Amer. Chem. Soc.*, **54**, 1758 (1932)
E26 E. J. King, Acid-Base Equilibria, Pergamon, London (1965)
E27 R. A. Robinson and R. H. Stokes, Electrolyte Solutions, Butterworths, London (1959)
E28 H. S. Harned and B. B. Owen, Physical Chemistry of Electrolytic Solutions, Reinhold, N.Y. (1958)
E29 A. D. Buckingham, *Disc. Faraday Soc.*, **24**, 151 (1957)
E30 H. F. Halliwell and S. C. Nyburg, *Trans. Faraday Soc.*, **59**, 1126 (1963)
E31 G. Somsen, *Rec. Trav. Chim.*, **85**, 526 (1966)
E32 J. S. Muirhead-Gould and K. J. Laidler *in* Chemical Physics of Ionic Solutions, Ch. 6 (B. E. Conway and R. G. Barradas, eds.), Wiley, N.Y. (1966)
E33 V. M. Goldschmidt, *Chem. Ber.*, **60**, 1263 (1927)
E34 B. S. Gourary and F. J. Adrian, Solid State Physics, **10**, 127 (1960)
E35 R. W. Gurney, Ionic Processes in Solution, Dover, N.Y. (1962)
E36 M. Salomon, *J. Phys. Chem.*, **74**, 2519 (1970)
E37 A. J. Parker, *Chem. Rev.*, **69**, 1 (1969)
E38 E. Grunwald, G. Baughman and G. Kohnstam, *J. Amer. Chem. Soc.*, **82**, 5801 (1960)
E39 M. J. Blandamer and M. C. R. Symons, *J. Phys. Chem.*, **67**, 1304 (1963)
E40 J. E. Desnoyers and C. Jolicoeur *in* Modern Aspects of Electrochemistry, Vol. 5 (J. O'M. Bockris and B. E. Conway, eds.), Plenum, N.Y. (1969)
E41 W. M. Latimer, K. S. Pitzer and C. M. Slansky, *J. Chem. Phys.*, **7**, 108 (1939)
E42 E. J. W. Verwey, *Chem. Weekblad*, **37**, 530 (1940)
E43 E. J. W. Verwey, *Rec. Trav. Chim.*, **61**, 127 (1942)
E44 N. A. Izmailov, *Russ. J. Phys. Chem.*, **34**, 1142 (1960)
E45 N. E. Khomutov, *Russ. J. Phys. Chem.*, **39**, 336 (1965)
E46 C. M. Criss and E. Luksha, *J. Phys. Chem.*, **72**, 2966 (1968)
E47 B. E. Conway and J. O'M. Bockris, Modern Aspects of Electrochemistry, Vol. 1, Ch. 2, Butterworths, London (1951)
E48 D. R. Rosseinsky, *Chem. Rev.*, **65**, 467 (1965)
E49 See selected papers *in* Chemical Physics of Ionic Solutions (B. E. Conway and R. G. Barradas, eds.), Wiley, N.Y. (1966)

E50 K. J. Laidler and C. Pegis, *Proc. Roy. Soc.*, **A241,** 80 (1957)
E51 R. H. Stokes, *J. Amer. Chem. Soc.*, **86,** 979, 982, 2333 (1964)
E52 B. E. Conway, J. E. Desnoyers and A. C. Smith, *Phil. Trans. Roy. Soc.*, **A256,** 389 (1964)
E53 R. M. Noyes, *J. Amer. Chem. Soc.*, **84,** 513 (1962)
E54 E. Glueckauf, see Ch. 5 in ref. E49
E55 N. A. Izmailov, *Dokl. Akad. Nauk. SSSR*, **149,** 884 (1963)
E56 J. D. Bernal and R. H. Fowler, *J. Chem. Phys.*, **1,** 515 (1933)
E57 D. D. Eley and M. G. Evans, *Trans. Faraday Soc.*, **34,** 1093 (1938)
E58 B. E. Conway *in* Modern Aspects of Electrochemistry, Ch. 2 (B. E. Conway and J. O'M. Bockris, eds.), Butterworths, London (1964)
E59 B. E. Conway and M. Salomon, see Ch. 24 in ref. E49
E60 K. P. Mishchenko, *Zh. Fiz. Khim.*, **26,** 1736 (1952); K. P. Mishchenko and E. I. Kvyat, *Zh. Fiz. Khim.*, **28,** 1451 (1954)
E61 V. P. Vasil'ev, E. K. Zolotarev, A. F. Kapustinskii, K. P. Mishchenko, E. A. Podgornaya and K. B. Yatsimirskii, *Russ. J. Phys. Chem.*, **34,** 840 (1960)
E62 G. A. Krestov and V. I. Klopov, *Zh. Struk. Khim.*, **5,** 829 (1964)
E63 K. P. Mishchenko, *Acta Physicochim*, **3,** 693 (1935)
E64 A. B. F. Duncan and J. Pople, *Trans. Faraday Soc.*, **49,** 217 (1953)
E65 J. Muirhead-Gould and K. J. Laidler, *Trans. Faraday Soc.*, **63,** 944, 953, 958 (1967)
E66 L. Weeda and G. Somsen, *Rec. Trav. Chim.*, **86,** 263 (1967)
E67 J. A. Pople, D. P. Santry and G. A. Segal, *J. Chem. Phys.*, **43,** S129 (1965)
E68 J. A. Pople and G. A. Segal, *J. Chem. Phys.*, **43,** S136 (1965); **44,** 3289 (1966)
E69 R. E. Burton and J. Daly, *Trans. Faraday Soc.*, **66,** 1281 (1970)
E70 Y. Wu and H. L. Friedman, *J. Phys. Chem.*, **70,** 501 (1966)
E71 H. M. Koepp, H. Wendt and H. Strehlow, *Z. Elektrochem.*, **64,** 483 (1960)
E72 H. Strehlow *in* The Chemistry of Non-Aqueous Solvents (J. J. Lagowski, ed.), Academic Press, N.Y. (1966)
E73 J. F. Coetzee and J. J. Campion, *J. Amer. Chem. Soc.*, **89,** 2513 (1967)
E74 J. M. Simon, Ph.D. Thesis, Univ. of Pittsburgh quoted in ref. E93
E75 J. C. Lauer, *Electrochim. Acta*, **9,** 1617 (1964)
E76 W. A. Pleskov, *Usp. Khim.*, **16,** 254 (1947)
E77 I. V. Nelson and R. T. Iwamoto, *Anal. Chem.*, **33,** 1795 (1961)
E78 R. L. Benoit, *Inorg. Nucl. Chem. Lett.*, **4,** 723 (1968)
E79 R. L. Benoit, A. L. Beauchamp and M. Deneux, *J. Phys. Chem.*, **73,** 3268 (1969)
E80 B. Case and R. Parsons, *Trans. Faraday Soc.*, **63,** 1224 (1967)
E81 D. Feakins and P. Watson, *J. Chem. Soc.*, 4686 (1963)
E82 D. Feakins and D. J. Turner, *J. Chem. Soc.*, 4986 (1965)
E83 H. P. Bennetto, D. Feakins and D. J. Turner, *J. Chem. Soc., A*, 1211 (1966)
E84 D. Feakins, K. G. Lawrence and R. P. T. Tomkins, *J. Chem. Soc., A*, 753 (1967)
E85 D. Feakins *in* Physico-Chemical Processes in Mixed Aqueous Solvents (F. Franks, ed.), Elsevier, N.Y. (1967)
E86 H. P. Bennetto and D. Feakins *in* Hydrogen-Bonded Solvent Systems (A. K. Covington and P. Jones, eds.), Taylor and Francis, London (1968)
E87 M. Salomon, *J. Electrochem. Soc.*, **117,** 325 (1970)
E88 M. Salomon, *J. Phys. Chem.*, **73,** 3299 (1969)
E89 C. L. DeLigny and M. Alfenaar, *Rec. Trav. Chim.*, **84,** 81 (1965)
E90 M. Alfenaar and C. L. DeLigny, *Rec. Trav. Chim.*, **86,** 929 (1967)
E91 M. Salomon, *J. Electrochem. Soc.*, **118,** 1609 (1972).
E92 O. Popovych, *Anal. Chem.*, **38,** 558 (1966)

E93 O. Popovych *Crit. Rev. Anal. Chem.*, **1,** 73 (1970); O. Popovych and A. J. Dill, *Anal. Chem.*, **41,** 456 (1969)

E94 J. F. Coetzee, J. M. Simon and R. J. Bertozzi, *Anal. Chem.*, **41,** 766 (1969)

E95 J. E. Leffler and E. Grunwald, Rates and Equilibria of Organic Reactions, Wiley, New York (1963)

E96 E. Grunwald and B. J. Berkowitz, *J. Amer. Chem. Soc.*, **73,** 4939 (1951)

E97 B. Gutbezahl and E. Grunwald, *J. Amer. Chem. Soc.*, **75,** 559, 565 (1953)

E98 E. Grunwald and S. Winstein, *J. Amer. Chem. Soc.*, **70,** 846 (1948)

E99 Lord Wynne-Jones *in* Hydrogen-Bonded Solvent Systems (A. K. Covington and P. Jones, eds.), Taylor and Francis, London (1968)

E100 R. Alexander and A. J. Parker, *J. Amer. Chem. Soc.*, **89,** 5549 (1967)

E101 A. J. Parker, *J. Chem. Soc.*, *A*, 220 (1966)

E102 R. Alexander, E. C. F. Ko, Y. C. Mac and A. J. Parker, *J. Amer. Chem. Soc.*, **89,** 3703 (1967)

E103 A. J. Parker and R. Alexander, *J. Amer. Chem. Soc.*, **90,** 3313 (1968)

E104 G. Choux and R. L. Benoit, *J. Amer. Chem. Soc.*, **91,** 6221 (1969)

E105 R. Fuchs, J. L. Bear and R. F. Rodewald, *J. Amer. Chem. Soc.*, **91,** 5797 (1969)

E106 E. M. Arnett and D. R. McKelvey, *J. Amer. Chem. Soc.*, **88,** 2598 (1966)

E107 C. V. Krishnan and H. L. Friedman, *J. Phys. Chem.*, **73,** 3934 (1969)

E108 E. D. Eastman, *J. Amer. Chem. Soc.*, **48,** 1482 (1926)

E109 J. N. Agar *in* Advances in Electrochemistry and Electrochemical Engineering, Vol. 3, pp. 31–121 (P. Delahay, ed.), Interscience, New York (1963)

E110 E. Lang and T. Hesse, *Z. Elektrochem.*, **39,** 374 (1933)

E111 F. H. Lee and Y. K. Tai, *J. Chinese Chem. Soc.*, **8,** 60 (1941)

E112 T. Ikeda, *J. Chem. Phys.*, **43,** 3412 (1965)

E113 C. M. Criss and J. W. Cobble, *J. Amer. Chem. Soc.*, **86,** 5385 (1964)

E114 E. D. Eastman, *J. Amer. Chem. Soc.*, **50,** 283, 292 (1928)

E115 J. C. Goodrich, F. M. Goyan, E. E. Morse, R. G. Preston and M. B. Young, *J. Amer. Chem. Soc.*, **72,** 4411 (1950)

E116 H. D. Crockford and J. L. Hall, *J. Phys. Chem.*, **54,** 731 (1950)

E117 W. M. Latimer, *J. Chem. Phys.*, **23,** 90 (1955)

E118 A. J. deBethune, T. S. Licht and N. Swendeman, *J. Electrochem. Soc.*, **106,** 616 (1959)

E119 W. G. Breck and J. Lin, *Trans. Faraday Soc.*, **61,** 2223 (1965)

E120 W. M. Latimer and W. L. Jolly, *J. Amer. Chem. Soc.*, **75,** 4147 (1953)

E121 B. Jakuszewski and S. Taniewska-Osinska, *Lodz. Towarz. Nauk. Soc. Sci. Lodziensis, Acta Chim.*, **4,** 17 (1959)

E122 B. Jakuszewski and S. Taniewska-Osinska, *Lodz. Towarz. Nauk. Soc. Sci. Lodziensis, Acta Chim.*, **8,** 11 (1962)

E123 C. M. Criss, R. P. Held and E. Luksha, *J. Phys. Chem.*, **72,** 2970 (1968)

E124 F. Franks and D. S. Reid, *J. Phys. Chem.*, **73,** 3152 (1969)

E125 F. J. Millero, The Partial Molal Volumes of Electrolytes in Aqueous Solutions, Ch. 13 *in* Water and Aqueous Solutions (R. A. Horne, ed.), Wiley, New York (1972)

E126 F. J. Millero, *J. Phys. Chem.*, **72,** 3209 (1968).

E127 P. Mukerjee, *J. Phys. Chem.*, **65,** 740 (1961)

E128 F. J. Millero, *J. Phys. Chem.*, **73,** 2417 (1969)

E129 W. M. Cox and J. H. Wolfenden, *Proc. Roy. Soc.*, London, **145A,** 475 (1934)

E130 C. M. Criss and M. J. Mastroianni, *J. Phys. Chem.*, **75,** 2532 (1971)

E131 J. Greyson, *J. Phys. Chem.*, **66,** 2218 (1962)

E132 H. S. Frank and M. W. Evans, *J. Chem. Phys.*, **13,** 507 (1945)

E133 H. S. Frank and W. Y. Wen, *Disc. Faraday Soc.*, **24,** 133 (1957)

E134 G. C. Pimentel and A. L. McClellan, The Hydrogen Bond, Freeman, San Francisco, Calif. (1960)

E135 L. A. LaPlanche, H. B. Thompson and M. T. Rogers, *J. Phys. Chem.*, **69**, 1482 (1965)
E136 G. R. Leader and J. F. Gormley, *J. Amer. Chem. Soc.*, **73**, 5731 (1951)
E137 F. J. Millero, *J. Phys. Chem.*, **72**, 4589 (1968)
E138 P. Mukerjee, *J. Phys. Chem.*, **65**, 744 (1961)
E139 L. G. Hepler, *J. Phys. Chem.*, **61**, 1426 (1957)
E140 B. E. Conway, R. E. Verrall and J. E. Desnoyers, *Z. Phys. Chem.*, **230**, 157 (1965)
E141 E. Glueckauf, *Trans. Faraday Soc.*, **61**, 914 (1965)
E142 S. W. Benson and C. S. Copeland, *J. Phys. Chem.*, **67**, 1194 (1963)
E143 P. Drude and W. Nernst, *Z. Phys. Chem.*, **15**, 79 (1894)
E144 M. K. Karapetiyants, S. I. Drakin and L. V. Lantukhova, *Zh. Fiz. Khim.*, **40**, 451 (1966)
E145 M. J. Mastroianni and C. M. Criss, *J. Chem. Eng. Data*, **17**, 222 (1972)
E146 M. J. Mastroianni and C. M. Criss, *J. Chem. Thermodyn.*, **4**, 321 (1972)
E147 C. S. Leung and E. Grunwald, *J. Phys. Chem.*, **74**, 696 (1970)
E148 H. Rüterjans, F. Schreiner, U. Sage and T. Ackermann, *J. Phys. Chem.*, **73**, 986 (1969)
E149 W. L. Gardner, R. E. Mitchell and J. W. Cobble, *J. Phys. Chem.*, **73**, 2025 (1969)
E150 C. T. Liu and W. T. Lindsay, Jr., Scientific Paper 70-9B6-CSALT-P2, October 20, 1970, Westinghouse Research Laboratories, Pittsburgh, Pa.
E151 M. Kaminsky, *Disc. Faraday Soc.*, **24**, 171 (1957)
E152 R. L. Kay, T. Vituccio, C. Zawoyski and D. F. Evans, *J. Phys. Chem.*, **70**, 2336 (1966)
E153 H. L. Friedman, *J. Phys. Chem.*, **71**, 1723 (1967)
E154 A. Einstein, *Ann. Physik.*, **19**, 289 (1906), *Ann. Physik.*, **34**, 591 (1911)
E155 R. F. Rodewald, K. Mahendran, J. L. Bear and R. Fuchs, *J. Amer. Chem. Soc.*, **90**, 6698 (1968)
E156 O. N. Bhatnagar and C. M. Criss, *J. Phys. Chem.*, **73**, 174 (1969)
E157 C. V. Krishnan and H. L. Friedman, *J. Phys. Chem.*, **73**, 1572 (1969)
E158 R. S. Drago, B. Wayland and R. L. Carlson, *J. Amer. Chem. Soc.*, **85**, 3125 (1963)
E159 R. S. Drago and K. F. Purcell, The Coordination Model for Non-aqueous Solvent Behavior, *in* Progress in Inorganic Chemistry (F. A. Cotton, ed.), Interscience, New York (1964)
E160 C. V. Krishnan and H. L. Friedman, *J. Phys. Chem.*, **75**, 388 (1971)
E161 B. Jakuszewski and S. Taniewska-Osinska, *Lodz. Towarz. Nauk. Soc. Sei. Lodziensis, Acta Chim.*, **7**, 31 (1961)

APPENDIX 2.11.1

Standard Lattice Free Energies and Enthalpies of 1:1 Salts*

Salt	$-\Delta H^0_{lat}$ kcal mol^{-1}	$-\Delta G^0_{lat}/$ kcal mol^{-1}	Salt	$-\Delta H^0_{lat}/$ kcal mol^{-1}	$-\Delta G^0_{lat}/$ kcal mol^{-1}
LiF	248.6	230.0	CsF	174.7	161.4
LiCl	205.9	188.5	CsCl	157.6	144.9
LiBr	195.8	178.7	CsBr	152.0	140.2
LiI	182.1	165.5	CsI	143.8	132.3
$LiClO_4$	198.7	179.5	$CsClO_4$	156.6	138.4
NaF	220.7	202.1	AgCl	218.7	201.5
NaCl	188.1	170.6	AgBr	215.9	198.7
NaBr	179.8	162.6	AgI	212.3	195.4
NaI	167.9	151.1	TlCl	179.0	162.2
$NaClO_4$	180.7	161.7	TlBr	175.2	158.4
KF	196.4	178.4	TlI	169.0	152.3
KCl	171.4	154.1	NH_4Cl	161.6	144.2
KBr	164.7	147.6	Me_4NCl	132.1	
KI	154.5	138.2	Me_4NBr	130.1	
$KClO_4$	169.7	150.8	Me_4NI	123.8	
RbF	186.6	170.2	Et_4NI	109.6	
RbCl	164.4	149.0	Pr_4NI	117.9	
RbBr	158.1	142.9			
RbI	149.2	134.4			
$RbClO_4$	163.9	145.1			

* Me = CH_3
Et = $CH_3 \cdot CH_2$
Pr = $CH_3 \cdot CH_2 \cdot CH_2$.

APPENDIX 2.11.2

Single Ion Solvation Free Energies in Water, $-\Delta G^0_{solv}$(ion)/kcal mol^{-1}, at 25°C (molal scale)

Ion	Ref. 1	2	3	4	5	6	7	8	9	10
H^+	251.1	258.0	259.7	283	256	260.5			259.5	235.0
Li^+	114.6	117.0		123	121	122.1	114.8	112.1	122.9	97.8
Na^+	89.7	96.0	101.8	107	97	98.2	92.2	97.9	97.6	72.4
K^+	73.5	78.0	84.1	84	79	80.6	76.5	79.9	81.3	54.9
Rb^+	67.5	74.4	79.0	75	74	75.5	71.0	73.3	75.8	50.7
Cs^+	60.8	64.0	71.3		66	67.8	65.0	67.3	67.0	46.7
Ag^+		112.0			113			146		87.9
Tl^+				101	80			114		56.9
OH^-										131.1
F^-	113.9		83.4	81	107	99.1	98.1	100.3	103.3	128.7
Cl^-	84.2	74.0	70.0	52	79	70.7	77.6	71.8	75.8	100.3
Br^-	78.0	68.0	63.1	47	72	64.9	72.5	65.8	69.2	94.2
I^-	70.0	59.4	54.2		64	57.2	65.9	57.2	61.8	85.7

1. W. M. Latimer, K. S. Pitzer and C. M. Slansky, *J. Chem. Phys.*, **7**, 108 (1939).
2. N. A. Izmailov, *Russ. J. Phys. Chem.*, **34** 1142 (1960); *Dokl. Akad. Nauk. SSSR*, **149**, 884 (1963).
3. R. M. Noyes, *J. Amer. Chem. Soc.*, **84**, 513 (1962).
4. D. D. Eley and M. G. Evans, *Trans. Faraday Soc.*, **34**, 1093 (1938).
5. V. P. Vasil'ev, E. K. Zolotarev, A. F. Kapustinskii, K. P. Mishchenko, E. A. Podgornaya and K. B. Yatsimirskii, *Russ. J. Phys. Chem.*, **34**, 840 (1960).
6. J. E. B. Randles, *Trans. Faraday Soc.*, **52**, 1573 (1956); Randles' work involves 'real' free energies.
7. J. C. Lauer, *Electrochimica Acta*, **9**, 1617 (1964).
8. K. J. Laidler and J. Muirhead-Gould, *Trans. Faraday Soc.*, **63**, 953 (1967).
9. M. J. Blandamer and M. C. R. Symons, *J. Phys. Chem.*, **67**, 1304 (1936).
10. M. Salomon, *J. Phys. Chem.*, **74**, 2519 (1970).

APPENDIX 2.11.3

Single Ion Solvation Enthalpies in Water, $-\Delta H^0_{solv}$(ion)/kcal mol^{-1}, at 25°C (Molal Scale)

Ion	Ref. 1	2	3	4	5	6	7	8	9
H^+		276	264.0				260.7	244.3	248.2
Li^+	121.2	133	126.0	127	129.7	120.5		107.6	111.5
Na^+	94.6	115	100.0	101	103.6	94.8		80.2	84.1
K^+	75.8	90	80.0	81	83.4	74.6		60.2	64.1
Rb^+	69.2	81	74.0		77.5	68.6		53.1	57.0
Cs^+	62.0	73	66.0	67	69.6	60.2		46.1	50.0
Ag^+		104	116.0	117				96.0	
Tl^+		110		82				62.0	
OH^-							110.3	143.3	
F^-	122.6	91	118.0	116		123.2	120.8	140.3	136.4
Cl^-	88.7	59	85.0	84		89.9	86.8	106.0	102.1
Br^-	81.4	52	77.0	76		82.1	80.3	101.6	97.7
I^-	72.1	45	68.0	67		71.7	70.5	88.8	84.9

1. W. M. Latimer, K. S. Pitzer and C. M. Slansky, *J. Chem. Phys.*, **7**, 108 (1939).
2. D. D. Eley and M. G. Evans, *Trans. Faraday Soc.*, **34**, 1093 (1938).
3. N. A. Izmailov, *Dokl. Akad. Nauk. SSSR.*, **149**, 884 (1963).
4. V. P. Vasil'ev *et al.*, *Russ. J. Phys. Chem.*, **34**, 840 (1960).
5. R. M. Noyes, *J. Amer. Chem. Soc.*, **84**, 513 (1962).
6. G. Somsen, *Rec. Trav. Chim.*, **85**, 526 (1966).
7. H. F. Halliwell and S. C. Nyburg, *Trans. Faraday Soc.*, **59**, 1126 (1963).
8. M. Salomon, *J. Phys. Chem.*, **74**, 2519 (1970).
9. K. J. Laidler and J. Muirhead-Gould *in* Chemical Physics of Ionic Solutions (B. E. Conway and R. G Barradas, eds.), Wiley. N.Y. (1966); the data reported here were interpolated from Fig. 3 of this work.

APPENDIX 2.11.4

Single Ion Solvation Energies in Organic Solvents, $-\Delta G^0_{\mathrm{solv}}(\mathrm{ion})/\mathrm{kcal\ mol^{-1}}$, at 25°C (molal scale)

Ion	Solvent											
	MeOH*	MeOH‖	MeOH¶	EtOH*	EtOH‖	*n*-BuOH†	HCOOH*	HCOOH†	Me_2CO†	AN†	DMF‡	PC§
H^+	253.0		237.9	252.0		252.0	246.0	243.0	252.0	249.0		
Li^+	116.0		102.0	115.0		110.0	116.0	115.3	112.0	114.1	129.2	95.0
Na^+	93.0	101.9	75.6	90.0	100.8	90.5	99.5	95.0	91.6	93.7	102.5	71.9
K^+	76.0	81.7	57.0	73.2	82.2	70.6	73.9	74.0	72.0	73.0	84.9	56.6
Rb^+	69.0		54.8	66.5		66.0	73.2	69.7	67.5	68.8		54.5
Cs^+	60.4		49.4	58.0		60.7	65.0	63.5	61.7	62.8	71.5	52.7
Ag^+	108.0			108.0			120.8					85.4
F^-											90.6	112.0
Cl^-	71.0	84.3	91.7	71.3	84.7	75.4	78.3	79.0	77.0	77.8	64.3	88.2
Br^-	67.0	77.2	86.0	66.2	77.5	70.0		73.3	71.5	72.5	61.1	84.0
I^-	59.6	68.1	79.6	58.5	67.6	63.0		66.0	64.0	65.0	56.4	78.0

* N. A. Izmailov, *Russ. J. Phys. Chem.*, **34**, 1142 (1960); *Dokl. Akad. Nauk. SSSR*, **149**, 884 (1963).
† N. E. Khomutov, *Russ. J. Phys. Chem.*, **39**, 336 (1965).
‡ C. M. Criss and E. Luksha, *J. Phys. Chem.*, **72**, 2966 (1968).
§ M. Salomon, *J. Phys. Chem.*, **74**, 2519 (1970).
‖ G. A. Krestov and Klopov, *Zh. Struk. Khim.*, **5**, 829 (1964).
¶ M. Salomon, J. Electrochem. *Soc.*, **118**, 1609 (1971).

APPENDIX 2.11.5

Single Ion Solvation Enthalpies, $-\Delta H^0_{solv}$(ion)/kcal mol^{-1}, in Organic Solvents at 25°C (molal scale)

Ion	MeOH*	EtOH*	F†	NMF‡	DMF§	NMA‡	PC‖
H^+							
Li^+	127.0	127.0	122.9	127.9	142.6	127.6	106.2
Na^+	100.0	99.0	99.6	102.1	114.3	80.5	
K^+		79.0	79.7	83.7	95.7	82.2	64.4
Rb^+			74.0	77.4		76.0	58.1
Cs^+			65.8	68.8	80.4	67.4	51.8
F^-			116.9	112.6		113.3	
Cl^-	84.0	90.0	88.1	84.8	73.9	85.3	101.4
Br^-	79.0	78.0	81.4	78.6	69.6	78.9	96.4
I^-	70.0	69.0	72.4	70.7	64.1	71.0	91.2

* N. A. Izmailov, *Russ. J. Phys. Chem.*, **34**, 1142 (1960).
† G. Somsen, *Rec. Trav. Chim.*, **85**, 526 (1966).
‡ L. Weeda and G. Somsen, *Rec. Trav. Chim.*, **86**, 263 (1967).
§ C. M. Criss and E. Luksha, *J. Phys. Chem.*, **72**, 2966 (1968).
‖ M. Salomon, *J. Phys. Chem.*, **74**, 2519 (1970).

APPENDIX 2.11.6

Single Ion Heats of Solvation, $-\Delta H^0_{solv}$(ion)/kcal mol^{-1}, in Mixed Alcohol–Water Solvents at 25°C (molal scale)*

Ion	Mol Fraction MeOH					Mol Fraction EtOH			
	0	0.2	0.6	0.8	1.0	0.2	0.6	0.8	1.0
Na^+	100.5	99.5	99.8	100.9	101.9	99.2	99.5	100.2	100.8
K^+	80.6	80.1	80.4	81.1	81.7	79.5	80.6	81.5	82.2
Cl^-	83.5	82.9	82.9	83.5	84.3	82.3	82.8	83.7	84.7
Br^-	76.1	75.2	75.7	76.3	77.2	74.9	75.8	76.7	77.5
I^-	66.7	66.2	66.9	67.5	68.1	66.0	66.7	67.1	67.6

* From G. A. Krestov and V. I. Klopov, *Zh. Struk. Khim.*, **5**, 829 (1964).

APPENDIX 2.11.7

Single Ion Free Energies of Transfer from Water to MeOH–H_2O Mixtures at 25°C

Wt% MeOH	Ref.	$-\Delta G_t^o$(ion)/kcal mol^{-1}								
		H^+	Li^+	Na^+	K^+	Rb^+	Cs^+	Cl^-	Br^-	I^-
10	1	−0.51	−0.35	−0.21	−0.21			0.67	0.62	0.55
10	2	−0.29	−0.16	−0.03	−0.03			0.45	0.40	0.33
33.4	1	−0.95	−1.45	−1.09	−0.99			2.45	2.25	1.94
43.12	1	−2.84	−2.04	−1.44	−1.24			3.46	3.16	2.76
43.12	2	−1.74	−1.06	−0.51	−0.47	−0.52	−0.48	2.37	2.11	1.68
43.12	3	−7.90	−7.08	−6.56	−6.37	−6.56	−6.53	8.52	8.25	7.82
68.33	1	−4.69	−3.38	−2.46	−1.89			5.86	5.46	4.76
87.68	2	−3.68	−2.87	−1.66	−1.45	−1.37	−1.32	5.98	5.50	4.51
87.68	3	−8.43	−7.09	−5.81	−4.68	−4.59	−5.55	10.63	10.15	9.16
90.0	1	−5.45	−4.23	−2.86	−1.67			7.96	7.49	9.42
90.0	3	−8.29	−7.07	−5.72	−4.51			10.75		
100.0	2	−1.97		−1.83	−1.22	−1.21	−1.28	7.77	7.25	6.22
100.0	3	−5.84	−6.7	−5.9	−5.2	−5.4	−5.5	11.2	10.6	9.65
100.0	4	−3.01	−3.94	−3.04	−2.19			8.4	7.9	6.9
100.0	5	(−0.99)	(−0.94)	−0.55	−0.17	−0.08	0.01	5.46	4.81	4.03
100.0	6	5.0	1.0	3.0	2.0	5.4	3.6	3.0	1.0	−0.2
100.0	7	−2.9	−3.8	−2.8	−1.9	−4.1	−2.7	8.5	8.2	6.1
100.0	8	1.24			1.18	0.65		1.09		

1. R. P. T. Tomkins, Thesis, University of London (1966).
2. M. Alfenaar and C. L. DeLigny, *Rec. Trav. Chim.*, **86**, 929 (1967).
3. B. Case and R. Parsons, *Trans. Faraday Soc.*, **63**, 1224 (1967); these authors are reporting 'real' free energies of transfer.
4. D. Feakins and P. Watson, *J. Chem. Soc.*, 4686 (1963).
5. H. Strehlow *in* The Chemistry of Non-aqueous Solvents (J. J. Lagowski, ed.), Academic Press, N.Y. (1966); the values in parenthesis were obtained by using Feakins' $\Delta G_t^o(MX)$ data for HCl and LiCl (cf. ref. 4 above) and subtracting Strehlow's value of $\Delta G_t^o(Cl^-)$.
6. N. A. Izmailov, *Russ. J. Phys. Chem.*, **34**, 1142 (1960).
7. M. Salomon, J. Electrochem. *Soc.*, **118**, 1609 (1971).
8. O. Popovych, *Anal. Chem.*, **38**, 558 (1966).

APPENDIX 2.11.8

Free Energies of Transfer, ΔG_t^0(ion)/kcal mol^{-1}, from PC to Several Organic Solvents* (molar scale)

Ion	From PC to DMF	AN		DMSO		$TMSO_2$
Ag^+	−8.1	−9.9		−12.9		−2.3
Cl^-	0.8	0.4	−0.3	0.0	−0.9	−0.1
Br^-	0.6	−0.4	0.0	−0.6	−1.4	−0.6
$AgCl_2^-$	−0.5	1.0	0.8	−0.5	−0.8	−1.1
$AgBr_2^-$	−0.7	−0.6	0.6	−1.0	−1.3	−1.4

* From R. L. Benoit, A. L. Beauchamp and M. Deneux, *J. Phys. Chem.*, **73**, 3268 (1969). The data for $TMSO_2$ refer to 30°C, the remaining data to 25°C. Where there are two entries, Benoit *et al.* have used several sources for equilibrium data.

APPENDIX 2.11.9

Free Energies of Transfer from Water to Aqueous Acetic Acid Solvents at 25°C* (molar scale)

Wt%HAc	ΔG_t^0(ion)/kcal mol^{-1} H^+	Cl^-	Br^-	I^-
10	−0.19	0.44	0.41	0.38
20	−0.26	0.82	0.78	0.69
40	−0.12	1.48	1.41	1.23
60	0.21	2.32	2.30	1.93

* From H. P. Bennetto, D. Feakins and D. J. Turner, *J. Chem. Soc., A*, 1211 (1966).

APPENDIX 2.11.10

Free Energies of Transfer of Neutral Substances from Water to Methanol at 25°C* (molal units)

Substance	r/Å	$\Delta G^0_{t,\text{neut}}$/ kcal mol^{-1}
He	1.29	0.93
Ne	1.60	1.02
Ar	1.92	1.31
Kr	1.98	1.47
Xe	2.18	
Rn	2.3	2.08
CH_4	2.1	1.67
CCl_4	3.6	3.91
$Fe(C_2H_5)_2$	3.8	4.36
$Sn(CH_3)_4$	4.2	5.38
$Sn(C_2H_5)_4$	4.7	6.80

* From M. Alfenaar and C. L. DeLigny, *Rec. Trav. Chim.*, **86,** 929 (1967).

APPENDIX 2.11.11

Neutral Free Energies of Transfer, $\Delta G^0_{t,\text{neut}}$(ion)/kcal mol^{-1}, for Various Ions from Water to MeOH–H_2O Mixtures at 25°C* (molal units)

Ion	Wt % MeOH			
	10	43.12	87.68	100.0
H^+	−0.319	−1.868	−3.967	−2.531
Li^+	−0.183	−1.187	−3.156	
Na^+	−0.054	−0.638	−1.948	−2.387
K^+	−0.056	0.596	−1.740	−1.780
Rb^+		−0.642	−1.652	−1.767
Cs^+		−0.607	−1.062	−1.836
Cl^-	0.422	2.249	5.695	7.211
Br^-	0.369	1.987	5.212	6.692
I^-	0.300	1.558	4.222	5.665

* From M. Alfenaar and C. L. DeLigny, *Rec. Trav. Chim.*, **86,** 929 (1967).

APPENDIX 2.11.12

Single Ion Free Energies of Transfer from Water to Organic Solvents at 25°C*

Ion	ΔG_t^0(ion)/kcal mol^{-1}															
	MeOH	F	DMF	DMA	DMSO	AN	AN§	AN‖	HMPT	$MeNO_2$	$TMSO_2$	$TMSO_2$§	PC	PC¶	NMF‡	EtOH‖
H^+																2.52
Li^+			−11.0†		−6.0**		9.22					−7.15		2.8	0.52	
Na^+							6.68					−4.84		0.5	0.33	
K^+	2.0	−1.8	−3.0		−4.1	1.0	4.61	1.0	−4.5			−3.23	−0.3	−1.7	0.23	3.87
Rb^+							3.92					−2.54		−3.8		
Cs^+	1.5	−1.4	−3.0	−4.2	−4.4	−0.3	3.47					−2.08		−6.0	0.97	
Ag^+	1.1	−4.0	−5.9	−7.9	−10.4	−7.5			−12.6	3.4	1.6	−4.8	−0.3	2.5		
Cl^-	3.4	3.4	12.3	14.1	10.9	12.0	7.64	10.1	15.4	10.1			14.1	12.1	2.68	4.47
Br^-	2.8	2.7	9.6	10.9	7.8	8.6	5.05	7.5	12.6					10.2		
I^-	2.0	1.9	5.6	6.1	3.8	5.3	1.91	4.4		5.6	5.3		6.1	7.7		
Ph_4B^-	−5.6	−5.7	−9.3	−9.3	−9.1	−2.2		−7.5	−3.7		−8.3		−8.3			−5.18

* All data are from A. J. Parker, *Chem. Rev.*, **69**, 1 (1969) unless stated otherwise (molar scale).
† J. N. Butler and J. C. Synott, *J. Amer. Chem. Soc.*, **92**, 2602 (1970), molal scale.
‡ M. Salomon, *J. Phys. Chem.*, **73**, 3299 (1969), molal scale.
§ J. F. Coetzee *et al.*, *J. Amer. Chem. Soc.*, **89**, 2517 (1967); *Anal. Chem.*, **41**, 766 (1969), molal scale.
‖ O. Popovych, *Crit. Rev. Anal. Chem.*, **1**, 73 (1970), molal scale.
¶ M. Salomon, *J. Phys. Chem.*, **74**, 2519 (1970), molal scale.
** ΔG_t^0(LiCl) from W. H. Smyrl and C. W. Tobias, *J. Electrochem. Soc.*, **115**, 33 (1968); see also G. Holleck, D. R. Cogley and J. N. Butler, *J. Electrochem. Soc.*, **116**, 952 (1969).

APPENDIX 2.11.13

Free Energies of Transfer of Several Ions from Water to EtOH/H_2O Mixtures at 25°C*

Wt %EtOH	ΔG_t^0(ion)/kcal mol^{-1}				
	$TAB^+ = Ph_4B^-$	$(Picrate)^-$	K^+	Cl^-	H^+
10.0	−0.41	0.14	0.11	0.26	−0.08
20.0	−0.94	0.23	0.12	0.70	−0.35
30.0	−1.77	0.14	1.15	1.19	−0.68
40.0	−2.73	0.00	0.14	1.76	−1.12
50.0	−3.41	−0.25	0.27	2.25	−1.41
60.0	−4.04	−0.30	0.60	2.63	−1.56
70.0	−4.45	−0.37	1.13	2.89	−1.36
80.0	−4.75	−0.41	1.79	3.15	−1.01
90.0	−5.03	−0.33	2.73	3.71	−0.70
100.0	−5.18	−0.22	3.87	4.47	2.52

* O. Popovych and A. J. Dill, *Anal. Chem.*, **41**, 456 (1969) (molal scale).

APPENDIX 2.11.14

Free Energy of Transfer of Ag^+ from Water to Organic Solvents, Based on Various Assumptions*

Assumption for ΔG_t^0	$\Delta G_t^0(Ag^+)$/kcal mol^{-1}									
	MeOH	F	DMF	DMA	DMSO	AN	HMPT	$MeNO_2$	$TMSO_2$	PC
$Ph_4As^+ = Ph_4B^-$	1.77	−3.4	−5.46	−7.23	−8.87	−6.82	−8.87		−3.14	
$I_2 = I_3^-$	1.09	−3.28	−6.14	−8.05	−10.10	−7.91		3.00	−2.73	−8.73
$Fe(C_5H_5)_2^+ = Fe(C_5H_5)_2$	1.09	−2.59				−7.23				
$Rb^+ = Cs^+ = 0$	2.86	0.81	0.13			−5.73				
Strehlow	0.55	−2.04				−8.86				
no E_{l_j}	2.05	−2.86	−5.45	−7.23	−9.95	−6.54	−11.18	4.23	0.55	−8.59
S_N2 Transition State Anion	0.27	−5.19	−6.69	−8.60	−11.32	−7.64	−13.78	3.82		−10.23
Parker-Alexander†	1.09	−3.96	−5.87	−7.91	−10.10	−7.50	−12.55	3.41	−1.64	−9.00

* A. J. Parker, *Chem. Rev.*, **69**, 1 (1969).

† R. Alexander and A. J. Parker, *J. Amer. Chem. Soc.*, **90**, 3313 (1968); this is an average value which does not include the data for $\Delta G_t^0(Rb^+ \text{ or } Cs^+) = 0$.

APPENDIX 2.11.15

Standard Enthalpies of Transfer of Ions, ΔH_t^0(ion)/kcal mol^{-1}, from Water to Organic Solvents*

Ion	From Water to															
	MeOH†	DMSO†	DMSO‡	DMSO§	DMSO††	PC§	PC†	$TMSO_2$†	DMF†	DMF‡	NMP‡	NEA‡	PC‖	F¶	NMF**	NMA**
Li^+	−3.7	−6.5	−6.5	−6.31		0.73	−0.05	5.4	1.5	−5.9	−5.9	−0.2	1.4	2.4	−7.4	−7.1
Na^+	−4.8	−7.2	−4.7	−6.62	−7.15	−2.44	−3.3	−4.05	−5.9	−6.9	−7.8	−1.9	−0.3	4.8	−7.3	−5.8
K^+	−4.7	−8.5	−6.9	−8.34	−8.84	−5.24	−6.0	−6.4	−7.5	−9.2	−10.0	−5.9	−4.2	5.1	−9.1	−7.6
Rb^+	−4.7			−8.01		−5.87	−6.7	−6.8	−7.2				−5.0	5.4	−8.8	−7.4
Cs^+	−3.8	−8.3		−7.18	−7.78	−6.40	−7.4	−6.3	−6.8				−5.7	5.6	−8.6	−7.2
Ph_4As^+	0.8	−2.3	−9.6	−2.84	−2.32	−3.49	−2.6	−2.85	−2.15	−13.0	−11.3	−0.7	−0.2			
Cl^-	1.9	5.1	5.2	4.49	4.89	6.31	7.45	6.7	2.8	6.5	7.3	−0.2	4.6	1.1	5.1	4.6
Br^-	0.9	1.1	0.0	0.83	1.34	3.24	4.35	3.35	−1.35	0.9	2.0	−1.2	5.2	0.7	3.5	3.2
I^-	−0.4	−2.6	−4.9	−3.05	−2.52	−0.78	0.1	−1.6	−5.4	−4.0	−1.4	−2.8	−2.4	−0.7	1.0	0.7
ClO_4^-	−1.1	−4.8	−6.5	−4.60		−3.93	−3.05	−4.7	−7.9	−6.4	−4.3	−3.0	−3.6			
Ph_4B^-	0.8	−2.3	−9.6	−2.84	−2.31	−3.49	−2.6	−2.85	−2.15	−13.0	−11.3	−0.7	−1.6			

* All values are on the molar scale except those from refs ‖ ¶ ** which are molal scale.

† G. Choux and R. L. Benoit, *J. Amer. Chem. Soc.*, **91**, 6221 (1969).

‡ R. Fuchs, J. L. Bear and R. F. Rodewald, *J. Amer. Chem. Soc.*, **91**, 5797 (1969); the ΔH_t^0(ion) values in this reference are all referred to the reference solvent methanol. Conversion to H_2O as the reference solvent was accomplished via

$$\Delta H_t^0(H_2O \rightarrow S) = \Delta H_t^0(H_2O \rightarrow MeOH) + \Delta H_t^0(MeOH \rightarrow S)$$

where $\Delta H_t^0(H_2O \rightarrow MeOH)$ values are those from ref. †.

§ C. V. Krishnan and H. L. Friedman, *J. Phys. Chem.*, 73, 3934 (1969).

‖ M. Salomon, *J. Phys. Chem.*, **74**, 2519 (1970).

¶ G. Somsen, *Rec. Trav. Chim.*, **85**, 526 (1966).

** L. Weeda and G. Somsen, *Rec. Trav. Chim.*, **86**, 263 (1967).

†† E. M. Arnett and D. R. McKelvey, *J. Amer. Chem. Soc.*, **88**, 2598 (1966).

Chapter 3

Acid-base Behaviour

E. J. King
Department of Chemistry, Barnard College,
Columbia University, New York

3.1 INTRODUCTION

So ubiquitous is water on this planet that our view of chemistry, by custom and convenience, is strongly biased in favour of aqueous solutions. What then are we to make of other solvents where the strongest acid may be the least ionised, where a respectable acid like CH_3CO_2H behaves as a base, where an anion base achieves stabilisation only by bonding to a molecule of its conjugate acid? Examination of acids and bases in non-aqueous solvents cannot fail to be enlightening.

This chapter is primarily about acid-base equilibria in anhydrous organic solvents. Equilibria in water-organic solvent mixtures, on which much effort has been expended for both practical and theoretical reasons, have been the subject of recent authoritative reviews.[F1,F2]

We shall consider the problems of acidity and basicity of solvents, medium effects, acidity scales, homo- and hetero-conjugation, ionisation and dissociation. Experimental investigations of the most important solvent systems are covered to mid-1970. The selection of a few papers for discussion from the voluminous literature of the subject necessarily reveals the interests and prejudices of the writer. The reader is referred elsewhere for the analytical applications.[F3–7] The annual Fundamental Reviews published in Analytical Chemistry give full coverage to current developments with extensive references to both fundamental and applied work.

3.2 CLASSIFICATION OF SOLVENTS

For the purposes of this chapter it is convenient to classify solvents in such a way that we start with familiar systems and add complexities one at a time. This will be no more and no less arbitrary than other classifications which might be more appropriate in other contexts, such as a discussion of solvation effects. We first distinguish two broad classes based on dielectric constant. In solvents of high dielectric constant, often referred to loosely as *polar solvents*, ion-pairing is minimal, even negligible in dilute solutions. The strength of an acid in these solvents can be assigned a numerical value independent of the bases with which it reacts. By contrast, in solvents of low dielectric constant, loosely called *non-polar solvents*, ion-pairing is important and acid strength depends on the choice of standard base. We have to draw the line arbitrarily between these two classes and a dielectric constant (ε_r) of 30 is chosen as the dividing value. This keeps methanol with the polar solvents but relegates ethanol and the higher alcohols to the other class. By making a distinction based on dielectric constant, we postpone the complication of ion-pairing to late in the chapter.

Each broad class in turn is sub-divided into hydrogen-bonded and non-hydrogen-bonded solvents. The term *protic* is frequently used for the first sub-division, *aprotic* for the second. Molecules of an aprotic solvent, such as $(CH_3)_2SO$ or $HCON(CH_3)_2$, may contain hydrogen atoms and they may accept hydrogen bonds, but they lack the ability to donate hydrogen atoms to such bonds and cannot be hydrogen-bonded liquids. Examples of these classes and subclasses are given in Table 3.2.1.

Table 3.2.1

Classification of Solvents

High Dielectric Constant		Low Dielectric Constant	
Hydrogen-bonded	Non-hydrogen-bonded	Hydrogen-bonded	Non-hydrogen-bonded
Water	Acetonitrile	Ethanol	Acetic anhydride
Deuterium oxide	Dimethylacetamide	*iso*-Propanol	Acetone
Hydrogen peroxide	Dimethylformamide	*tert*-Butanol	Benzene
Methanol	Hexamethylphosphorotriamide	Methyl cellosolve	Carbon tetrachloride
Ethylene glycol	Dimethylsulphoxide		Chlorobenzene
	Nitrobenzene	Acetic acid	Chloroform
Sulphuric acid	Nitromethane	Trichloroacetic acid	Cyclohexane
Fluorosulphuric acid	Propylene carbonate		Dioxan
Hydrofluoric acid	Sulpholane	Ammonia	Ethyl ether
Formic acid		Ethylenediamine	*n*-Hexane
Hydrocyanic acid			Pyridine
			Sulphur dioxide
Ethanolamine			Tetrahydrofuran
Formamide			
Acetamide			
N-Methylformamide			
N-Methylacetamide			

Each subclass can be further divided according to the acidic or basic proclivities of its members. There are again rather arbitrary distinctions. Formic acid is an obvious example of an acidic or *protogenic* solvent, but it has weak basic properties too. The acidic properties of hydrocarbons are generally latent, and these solvents are not ordinarily regarded as protogenic. Ethanolamine and dimethyl sulphoxide are basic or *protophilic* solvents. When both basic and acidic properties of a solvent are frequently called into play, it is convenient to assign it to a third group, *amphiprotic* solvents, characterised by appreciable self-ionisation or *autoprotolysis*

$$SH + SH \leftrightarrows SH_2^+ + S^-$$

In the following sections we examine the acidity and basicity of solvents, and medium effects. Thereafter, we take up the principal solvents one by one according to the classification scheme that has just been presented.

3.3 ACIDITY AND BASICITY OF SOLVENTS

Although our intuitive notions of which solvents are basic or acidic often suffice, the concepts of acidity and basicity lose definition as we examine them much more closely. By basicity we have in mind the propensity of a solvent to accept protons from acids. Clearly this is not a property of an isolated solvent molecule, nor is it necessarily independent of the acid used as a probe.

3.3.1 Acidity and Basicity in the Gaseous Phase

By examining proton transfers between isolated molecules or ions in the gaseous phase we can establish an extremity of behaviour which may be of some help in understanding the nature of solvents in the liquid state. Swimming in the non-aqueous ocean, we push away any life preservers at our peril.

The direction of such gas phase reactions can be judged from the standard enthalpy change ΔH^0 because entropy changes are small.[F8] The *proton affinity* **P** of a base is defined[F8] as the energy liberated at constant pressure when a gaseous acid is formed by combining its gaseous conjugate base with gaseous protons. It is equivalent to the standard enthalpy change for $HX^z(g) \rightarrow H^+(g) + X^{z-1}(g)$. The experimental determination of proton affinities[F9,F10] is still fraught with large uncertainties, so that only gross distinctions can be made. It is plain, for example, that in the gas phase, ammonia (**P** = 870 ± 25 kJ mol^{-1} at 25°) is more basic than water (749 ± 21). Methanol (**P** = 753 ± 14 kJ mol^{-1}) is about as basic as water but less basic than ethanol (~791).

Gaseous acids may be arranged in order of their ability to transfer protons to gaseous bases by observing whether the reactions are endothermic or exothermic. The results of such a study,[F11] using the techniques of ion cyclotron resonance and pulsed double-resonance spectroscopy, are

$$t\text{-BuOH} > \text{EtOH} > \text{MeOH} > H_2O$$
$$(CH_3CO)_2CH_2 > CH_3COCN > HCN > H_2O$$
$$Et_2NH > Me_2NH > EtNH_2 > MeNH_2 > NH_3$$
$$Et_2NH > H_2O > NH_3$$

Acidity and basicity in the gaseous phase are intrinsic properties of individual molecules. For substances in the liquid state these properties belong to the entire phase and have less localised meanings. The transfer of a reaction from the gaseous to the liquid phase involves the Gibbs free energies of solvation of reactants and products

$$\Delta G^0_{\text{in solution}} = \Delta G^0_{\text{in gas}} + \Delta G^0_{\text{solv}}(\text{products}) - \Delta G^0_{\text{solv}}(\text{reactants}) \quad (3.3.1)$$

Solvation is a complex process calling into play some or all of the following interactions: ion-dipole, dipole-dipole, London dispersion, and hydrogen-bonding. Since all the quantities on the right hand side of eqn. 3.3.1 are large and hard to measure precisely, it is not yet worthwhile to pursue this approach.

3.3.2 Experimental Measures of Acidity or Basicity of Liquid Solvents

The weapons of analytical chemistry have been brought to bear on reactions in which different solvents S react with a standard acid or base.

The bases S and H_2O may be compared through the proton transfer equilibrium

$$S + H_3O^+ \leftrightarrows SH^+ + H_2O \quad (3.3.2)$$

Such equations never attempt to represent the complicated interactions and degrees of solvation of the particles in a real liquid phase. Although the protons are represented as solvated by one molecule each of S and H_2O, we are in fact ignorant of their degree of solvation. The customary convention is to take the symbol SH^+ to represent the set of all forms of the proton solvated by S and similarly for H_3O^+.

The equilibrium represented by eqn. 3.3.2 is established in solvents containing low concentrations of water and strong acid. An equilibrium constant for the reaction has the form

$${}^sK_p = \frac{[SH^+][H_2O]}{[S][H_3O^+]}\frac{({}^sy_{SH^+})({}^sy_{H_2O})}{({}^sy_S)({}^sy_{H_3O^+})} = Q_p({}^sY_p) \quad (3.3.3)$$

We shall use the presuperscript s to denote constants (K) and activity coefficients (y on the molarity scale) which refer to standard states in the solvent S. With this choice of standard state, as the concentrations of water and acid in S decrease, the equilibrium concentration Q_p approaches ${}^sK_p$ and, for any species i, ${}^sy_i$ approaches 1. Because the measurements we are discussing are made in very dilute solutions, we can make the following approximations

$$({}^sy_{SH^+})/({}^sy_{H_3O^+}) \approx 1, \quad {}^sy_S = {}^sy_{H_2O} \approx 1, \text{ and } {}^sK_p \approx Q_p$$

The results of several investigations of the distribution of protons between water and methanol are given in Table 3.3.1. Other solvents have not been investigated as frequently as methanol, and their quotients, given in Appendix 3.3.1, are less precisely known.

Table 3.3.1

The Proton Exchange Reaction at 25°C

$$\mathbf{MeOH + H_3O^+ \leftrightarrows H_2O + MeOH_2^+}$$

Method	10^3Q_p	Ref.
Indicator	9.4	L. S. Guss and I. M. Kolthoff, *J. Amer. Chem. Soc.*, **62**, 1494 (1940)
	9.0	P. Salomaa, *Acta Chem. Scand.*, **11**, 191 (1957)
Electromotive force	7.9	J. Koskikallio, *Suomen Kem.*, **30B**, 43 (1957)
Conductance	9.4	L. Thomas and E. Marum, *Z. phys. Chem.*, **143** 191 (1929)
	7.2	H. Strehlow, *Z. phys. Chem. (Frankfurt)*, **24**, 240 (1960)
Kinetics	9.2	H. A. Schmidt and C. H. Reichardt, *J. Amer. Chem. Soc.*, **63**, 605 (1941)
	8.5	J. Koskikallio, *Acta Chem. Scand.*, **13**, 671 (1959)
	8.3	C. E. Newall and A. M. Eastham, *Can. J. Chem.*, **39**, 1752 (1961)
Average	8.6 ± 0.6	

What can we learn about solvent basicity from these values? Clearly, the distribution of protons between methanol and water depends not only on Q_p but also on the concentrations of water and methanol

$$\frac{[\text{MeOH}^+]}{[\text{H}_3\text{O}^+]} = Q_p \frac{[\text{MeOH}]}{[\text{H}_2\text{O}]} \tag{3.3.4}$$

In the solutions used to determine Q_p, water was kept at low concentrations (0 to 0.6M in Guss and Kolthoff's experiments). The ratios $[\text{MeOH}_2^+]/[\text{H}_3\text{O}^+]$ ranged from greater than unity to less than unity and depended as much on the mass action effect of water and methanol as on their intrinsic basicities. If Q_p were valid for a solution containing equal concentrations of the two solvents, its small value would indicate that water was more basic than methanol. In fact, there is considerable evidence that the opposite is the case for mixtures containing substantial proportions of water.[F12] It is not proper to apply Q_p (or ${}^sK_p$) to media other than that for which it was established, namely methanol containing traces of water. Moreover, having compared two alcohols separately with water by such measurements, we cannot compare them with each other. The constant ${}^sK_p$ for ethanol is based on a standard state in ethanol and is unrelated to ${}^sK_p$ for methanol. Only when both constants refer to the same reference state are we justified in comparing them, and then we are not comparing the basicities of pure liquid methanol and ethanol.

To compare acidities of solvents we might elect to study equilibria of the type

$$S_i^- + R\text{H} \leftrightarrows R^- + S_i\text{H} \tag{3.3.5}$$

where S_iH is one of a set of solvents to be compared with a reference solvent RH. If we chose to study RH at low concentrations in S_iH, the constants we obtain have the same limitation as ${}^sK_p$. On the other hand, if we study each S_iH at low concentration in one reference solvent RH,[F13] we obtain a set of constants $_{R_\text{H}}K$ which enable us to compare the acid strengths of the various solvents S_iH with each other only at low concentration in RH but do not enable us to compare the acidities of the pure solvents.

Another possible approach to determination of solvent basicities and acidities would be to measure the change in some physical property, an infrared or ultraviolet frequency or an NMR chemical shift, of molecules of a probe substance upon transfer from some reference solvent to other solvents.[F14,F15] To cite but one example, Agami and Caillot[F16] measured the shifts in wave number $\Delta\sigma$ of the $\equiv$C—H valence vibration band of phenylacetylene upon transfer from carbon tetrachloride to nineteen other solvents. The results in Table 3.3.2 appear to establish an order of basicity from carbon tetrachloride (low) to hexamethylphosphorotri-amide (high), an order which in general agrees with that established by NMR measurements. Nevertheless, for a number of reasons it would be rash to make too much of small differences between adjacent members of the series. The shifts measure perturbations of frequencies, not proton transfers. We are not dealing with the pure solvents but with them contaminated by appreciable concentrations of phenylacetylene. The wave number shifts are sensitive to changes in concentration and temperature. It is not easy to convert these wave number shifts to respectable equilibrium constants, for that would require identification of the species involved and measurement of the absorption intensities. Even were it possible to get such constants, we would still have the problem of comparing the incomparable: constants referring to standard states in different solvents.

Table 3.3.2

Wavenumber Shifts of the ≡C—H Band of Phenylacetylene Relative to Carbon Tetrachloride

Solvent	$\Delta\sigma/\text{cm}^{-1}$
Carbon tetrachloride	0
Benzene	12
Acetonitrile	30
Methyl Methylal (dimethoxymethane)	40
Ethylene carbonate	45
Tetramethylenesulphone (sulpholane)	47
Acetone	50
Dioxan	53
Ethyl ether	55
Butyl ether	55
Tetrahydrofuran	65
1,2-Dimethoxyethane (monoglyme)	65
Diglyme	65
Tetraglyme	68
Dimethylformamide	95
Dimethylsulphoxide	110
N-Methylpyrrolidone	110
N,N,N′,N′-Tetramethylenediamine	115
Hexamethylphosphotriamide	153

3.3.3 Medium Effects on the Proton

We can now attempt to approach the problem of solvent basicity by another route: the medium effect on the proton. The chemical potential of the proton in a solvent S can be expressed in two ways, depending on whether we take the standard state to be the hypothetical one-molal solution in S

$$\mu_{\text{H}} = {}^{s}\mu_{\text{H}}^{0} + RT\ln m_{\text{H}} + RT\ln {}^{s}\gamma_{\text{H}} \tag{3.3.6}$$

or in water

$$\mu_{\text{H}} = {}^{w}\mu_{\text{H}}^{0} + RT\ln m_{\text{H}} + RT\ln {}^{w}\gamma_{\text{H}} \tag{3.3.7}$$

It is customary in this context to express concentrations m_H on the molality scale and the corresponding activity coefficients are ${}^s\gamma_H$ or ${}^w\gamma_H$ depending on the choice of standard state. By subtracting eqn. 3.3.6 from eqn. 3.3.7 we obtain

$$\ln {}^w\gamma_H - \ln {}^s\gamma_H = ({}^s\mu_H^0 - {}^w\mu_H^0)/RT \qquad \text{cf. 2.11.13a (3.3.8)}$$

The right hand side is a property of the two standard states and is independent of molality. We call it the *medium effect*[F17,F18] on the proton and represent it by

$$\text{Medium effect} = \ln ({}_m\gamma_H) = ({}^s\mu_H^0 - {}^w\mu_H^0)/RT \qquad (3.3.9)$$

or by

$$\ln ({}_m\gamma_H) = \ln ({}^w\gamma_H/{}^s\gamma_H) = (\ln {}^w_s\gamma_H) \qquad \text{cf. 2.11.13a (3.3.10)}$$

The medium effect is proportional to the reversible work of transfer of one mole of protons from infinite dilution in water to infinite dilution in solvent S. A negative value of the medium effect indicates that the proton is more stable in S, a positive value that it is more stable in water. The medium effect will thus depend on the relative basicities of water and S, basicity being understood to include not only the proton affinities of the gaseous solvent and water molecules but also the Gibbs free energies of solvation of the ions H_3O^+ in water and SH^+ in S.

Were we able to measure it directly, ${}_m\gamma_H$ would give us a way of expressing the basicity of solvents relative to some common standard.[F2] Water was chosen as the standard in our development of the concept, but if its peculiarities are judged to make it unsuitable, another solvent such as methanol or dimethylformamide could be substituted.

It was pointed out in sect. 2.11.3 that we could obtain combinations of medium effects for two ions but not the medium effect for a single ion. For example, the standard potential of cell (V) in sect. 2.6.

$$(\text{Pt})\text{H}_2(1\ \text{atm})|\text{HCl in H}_2\text{O or } S|\text{AgCl, Ag} \qquad (3.3.11)$$

can be used to obtain the sum of the medium effects on the hydrogen and chloride ions

$$(1/k)[{}^wE^0 - {}^sE^0] = \log {}_m\gamma_H + \log {}_m\gamma_{Cl} \qquad (3.3.12)$$

where k is the Nernst factor $RT(\ln 10)/F$. The various extra-thermodynamic assumptions that have been tried to separate the medium effect of one ion from that of another have been discussed in sect. 2.11.4 and will not be reviewed here.[F2,F19,F20] The results of various determinations are collected in Appendix 3.3.2 and some mean values are given in the third column of Table 3.3.3. It is evident that at the present time different assumptions give discordant results so that only gross distinctions between solvents can be recognised.

If solvent basicity is a complex of effects, we may be tempted to split ${}_m\gamma_H$ into electrostatic, hydrogen-bonding, and other contributions. We can try to estimate the electrostatic contribution by using the well-known sphere-in-continuum model of Born, the deficiencies of which are widely recognised and need not be laboured here.[F21,F22] Consider S and water to be structureless media with dielectric constants ${}^s\varepsilon_r$ and ${}^w\varepsilon_r$. Let the ions SH^+ and H_3O^+ be represented by conducting spheres of the same radius r and charge e. Their Gibbs free energies depend on the dielectric constant of the solvents. Consequently, the reversible electrical work of transferring a mole of protons from water to S, which is the electrostatic contribution to the medium effect, is given by

$$\log {}_m\gamma_{H\,el} = \frac{e^2}{2rkT(\ln 10)}\left(\frac{1}{{}^s\varepsilon_r} - \frac{1}{{}^w\varepsilon_r}\right) \tag{3.3.13}$$

where k is the Boltzmann constant. Values of this electrostatic contribution calculated for a radius of 2 Å are given in the fourth column of Table 3.3.3.

Table 3.3.3

Observed and Calculated Medium Effects on the Proton at 25°C

Solvent	ε_r	$\log {}_m\gamma_H$	
		Experimental*	Born†
MeOH	32.6	0.0 ± 2.0	1.08
EtOH	24.3	2.5 ± 1.8	1.72
n-BuOH	17.1	2.3 ± 2.0	2.76
i-AmOH	14.7	(4.2)	3.35
HCO_2H	56.1	7.9 ± 1.7	0.31
Me_2SO	46.6	−3.6 ± 2.0	0.53
CH_3CN	36.0	4.3 ± 1.5	0.91
Me_2CO	20.7	(3.3)	2.65
$HCONH_2$	109.5	−1.2 ± 0.6	−0.22
N_2H_4	51.7	−14.	0.40
NH_3	16.9	−16.	2.81

* Averages of selected values from Appendix 3.3.2 except those in parentheses which are single values.

† A radius of 2 Å is assumed for both H_3O^+ and SH^+.

The alcohols and water are evidently fairly close together in basicity. The medium effects calculated from the change in dielectric constant correspond roughly to the experimental values. For the other solvents in Table 3.3.3 it is evident that the electrostatic predictions are wide of the mark. The acidic solvent HCO_2H is much less basic than water and has a very large positive medium effect. Expressed in terms of ${}_m\gamma_H$, the failure of the Born model is spectacular: predicted value 2, observed value 8×10^7. Acetonitrile and acetone are likewise less basic than water. Formamide, hydrazine and ammonia, which are more basic than water, stabilise the proton and have negative values of the medium effect.

The present values of log ${}_m\gamma_H$ therefore merely confirm popular notions about acidity and basicity of solvents. If the future brings more accurate experimental values, we shall be able to make more subtle distinctions. Perhaps it is not amiss to add, though, that the confusion of present values is greatest for methanol, the most thoroughly investigated of the solvents.

It may be noted finally that if a precise scale of medium effects on the proton were established, we should then have a universal pH scale because of the relation

$$-\log {}_w a_H = -\log {}_s a_H - \log {}_m\gamma_H \qquad (3.3.14)$$

3.3.4 Autoprotolysis

A solvent SH capable of being both a proton donor and a proton acceptor can undergo autoprotolysis

$$SH + SH \leftrightarrows SH_2^+ + S^-$$

The ionisation of water to H_3O^+ and OH^-, or hydrated forms of these, is well known; other examples may be cited

$$2NH_3 \leftrightarrows NH_4^+ + NH_2^-$$
$$2H_2SO_4 \leftrightarrows H_3SO_4^+ + HSO_4^-$$
$$2HCONH_2 \leftrightarrows [HC(OH)NH_2]^+ + [HCONH]^-$$
$$2(CH_3)_2SO \leftrightarrows [(CH_3)_2SOH]^+ + [CH_3(CH_2)SO]^-$$

It is unwise to assume autoprotolysis without seeking confirmation. Perchloric acid, for example, does not ionise by simple proton transfer[F23]

$$3HClO_4 \leftrightarrows Cl_2O_7 + H_3O^+ + ClO_4^-; \ pK = 6.2$$

Autoprotolysis constants have the form

$$K_S = a_{SH_2^+}\, a_{S^-} \qquad (3.3.15)$$

and can be determined in several ways.[F24] For example, the ratio of the specific conductance of the pure solvent to the limiting molar conductance of the ionised solvent should give the concentration of either ion.

This is risky if the solvent is difficult to purify, a particular problem for aprotic solvents, for if the impurities ionise they may be mistaken for solvent ions. Cells with and without liquid junction are often used to determine K_S, e.g.

$$(Pt)H_2|MS \text{ and } MCl \text{ in } SH|AgCl, Ag$$

Or the acidity constant K_a of an acid HX may be determined by potentiometry or spectrophotometry and combined with the basicity constant K_b of its conjugate base X^-, measured by conductance or in other ways

$$K_a \times K_b = K_S \qquad (3.3.16)$$

Values of the autoprotolysis constants and associated thermodynamic properties of the solvent are given in Appendix 3.3.3, and some selected values are in Table 3.3.4. Most of these are concentration products

$$P_S = [SH_2^+][S^-] \qquad (3.3.17)$$

rather than thermodynamic constants. The distinction between K_S and P_S is not important for values of low precision.

Table 3.3.4

Autoprotolysis Constants at 25°C*

Solvent	pK_S	p$\boldsymbol{K_S}$
H_2SO_4	3.57	5.59
$HOC_2H_4NH_2$	5.7	8.1
H_2O	14.00	17.49
CH_3CO_2H	14.5	16.9
CH_3OH	16.6	19.6
C_2H_5OH	18.8	21.5
NH_3	29.	32.
CH_3CN	>32.	>35.

* Selected from values in Appendix 3.3.3; molal scale.

In comparing thermodynamic constants for different solvents we have to bear in mind that they refer to different standard states. To put them on a unified scale we would have to know the medium effects on the ions. It is still possible to distinguish some gross effects in examining the constants in Table 3.3.4. We should first take into account the different molalities of the pure solvents by dividing K_S by the square of the molality of SH

$$\boldsymbol{K}_S = \frac{K_S}{[S\text{H}]^2} = K_S \times M_{S\text{H}}^2 \tag{3.3.18}$$

where $M_{S\text{H}}$ is the molar mass of SH in kg. We may expect differences in K_S due to difficult dielectric constants of the solvents. The ions $H_3SO_4^+$ and HSO_4^-, for example, will be stabilised in part because of the high dielectric constant of H_2SO_4, namely 101 at 25°C. More importantly, the acidity and basicity of the solvent are both manifested in K_S. A large constant may result when one of these faculties is particularly prominent, e.g. the acidic nature of H_2SO_4 or the basic nature of ethanolamine.

3.3.5 Levelling and Differentiation[F24]

In solvents that undergo autoprotolysis the strongest acid is the solvent cation SH_2^+ and the strongest base is the solvent anion S^-. Water, as is well known, converts strong acids such as $HClO_4$, HNO_3, or HCl to the conjugate anion and the solvent cation H_3O^+, and the conversion is complete as long as the acid is not very concentrated. The strong acids are said to be *levelled* by water. Likewise, bases such as O^{2-}, NH_2^-, and CH_3O^- are completely converted to their conjugate acids and the solvent anion OH^-. They too are levelled by water. In other solvents the intrinsic differences in acid (or base) strength may not be levelled and the acids (or bases) are *differentiated.*

A strongly acid solvent such as H_2SO_4 levels virtually all bases, i.e. it converts them to their conjugate acids and the solvent anion HSO_4^-. An obvious example is dimethylsulphoxide, which is a strong base in this solvent[F25]

$$(CH_3)_2SO + H_2SO_4 \rightarrow (CH_3)_2SOH^+ + HSO_4^-$$

Substances too weak to function as bases in water may develop this faculty in such a strongly acidic solvent

$$CH_3CO_2H + H_2SO_4 \rightarrow CH_3CO_2H_2^+ + HSO_4^-$$
$$C_6H_5NO_2 + H_2SO_4 \rightarrow C_6H_5NO_2H^+ + HSO_4^-$$

In a 0.3 molal solution of nitrobenzene in sulphuric acid approximately 18% of the base is converted to its conjugate acid.

Acidic solvents also differentiate some acids levelled by water. In pure sulphuric acid the strongest acid is the solvent cation $H_3SO_4^+$ and $HClO_4$ is a weak acid with pK_a equal to about 4. The solvent molecule H_2SO_4 is a still weaker acid.

A basic solvent, such as ethanolamine, will level many acids by converting them to the solvent cation $HOC_2H_4NH_3^+$ and will differentiate many bases.

Acetonitrile, an aprotic solvent, is a very weak base and an exceptionally weak acid. Little levelling of acids and virtually no levelling of bases occurs in this solvent. Differentiation is the rule. Perchloric acid appears to be strong in acetonitrile; other acids are differentiated[F26]

$$HBr > H_2SO_4 > HNO_3 > HCl \text{ or picric acid}$$

The autoprotolysis constant sets the limits of differentiation of a solvent. In water we are usually restricted to studying acids with pK_a values between 1 and 13, or with extreme care and less precision, to values between 0 and 14. The range of acids that can be investigated is much narrower in H_2SO_4; 3.6 pK_a units, which corresponds to a range of potentials in electromotive force measurements of 200 mV. The range is likewise narrow in ethanolamine; 5.7 pK_a units. In acetonitrile, on the other hand, acids with pK_a values between 0 and 32 should be amenable to investigation. We shall see later that conjugation reactions complicate such studies in acetonitrile.

The autoprotolysis constant is also important in determining the sharpness index of a titration.[F27] For a successful titration we want the sharpness index to be large at the equivalence point; this corresponds to a large change in pH or potential per increment of titrant, indispensable for the precise location of the point. Since the sharpness index varies inversely with the autoprotolysis constant, it is advantageous to use solvents with small values of K_S. Useful titrations can therefore be made in water or acetonitrile, but they cannot be very successful in pure H_2SO_4.

Differentiating effects are not solely attributable to the acidity or basicity of the solvent. In comparing acetic acid with ammonium ion, we find that the difference in pK_a values is 4.48 in water but only 0.8 in methanol. In water the electrostatic energies of the ions are lower than in methanol because of its higher dielectric constant. This affects acetic acid, the ionisation of which creates ions, more than ammonium ion, for which ionisation merely shifts positive charge from NH_4^+ to SH. In aprotic solvents specific solvation effects may be large and result in some differentiation.

3.4 MEDIUM EFFECTS ON THE IONISATION OF WEAK ACIDS[F28]

3.4.1 Formulation

The medium effect on the proton was defined by eqn. 3.3.9. Medium effects on a molecular or ionic weak acid HX^{z+1} and conjugate base X^z, where z is the charge, can be defined by analogous relations

$$\ln {}_m\gamma_{HX} = ({}^s\mu^0_{HX} - {}^w\mu^0_{HX})/RT \tag{3.4.1}$$

and

$$\ln {}_m\gamma_X = ({}^s\mu^0_X - {}^w\mu^0_X)/RT \tag{3.4.2}$$

The reversible work done on the system in transferring a mole each of H^+ and X^z from infinite dilution in water to infinite dilution in solvent S and of a mole of HX^{z+1} in the reverse direction is given by

$$\Delta G_t^0 = RT \ln \frac{{}_m\gamma_H \, {}_m\gamma_X}{{}_m\gamma_{HX}} = RT \ln {}_m\Gamma_a \tag{3.4.3}$$

The medium effect on the ionisation of a weak acid is thus expressed by $\ln {}_m\Gamma_a$. It depends on individual interactions of H^+, X^z, and HX^{z+1} with water and solvent S but not on their interactions with each other.

The medium effect on ionisation is most directly obtained by comparing the acidity constants of the weak acid in S and water. The former constant is defined by

$${}^sK_a = Q_a \frac{{}^s\gamma_H \, {}^s\gamma_X}{{}^s\gamma_{HX}} = Q_a {}^s\Gamma_a \tag{3.4.4}$$

where the activity coefficients revert to unity at infinite dilution in S. The acidity constant ${}^sK_a$ is thus the limiting value of the equilibrium concentration quotient Q_a as the concentration of acid approaches zero. Likewise, we can define an acidity constant based on reference states in water

$${}^wK_a = Q_a \frac{{}^w\gamma_H \, {}^w\gamma_X}{{}^w\gamma_{HX}} = Q_a {}^w\Gamma_a \tag{3.4.5}$$

We may elect to study the weak acid in either water or S. Different equilibrium quotients Q_a and activity coefficient factors ${}^w\Gamma_a$ would be obtained, but the value of ${}^wK_a$, which is fixed by the choice of standard state, is invariant to changes in the medium. It necessarily differs from ${}^sK_a$ which refers to another standard state.

The activity coefficient factors are related through the medium effect

$${}^w\Gamma_a = ({}_m\Gamma_a)({}^s\Gamma_a) \tag{3.4.6}$$

By combining this with eqns. 3.4.4 and 3.4.5 we obtain a relation that enables us to find the medium effect from the difference in pK_a values

$$\log ({}_m\Gamma_a) = -\log ({}^sK_a/{}^wK_a) = \Delta pK_a \tag{3.4.7}$$

To achieve an understanding of solvent effects on the ionisation of acids we ought to have the contributions of each species to the over-all effect, $\log {}_m\Gamma_a$. The problems in estimating the contribution of the H^+ ion have been mentioned in sect. 3.3.3. Solubility and electromotive force measurements give us directly combinations of medium effects for two ions, e.g. $({}_m\gamma_{H^+})({}_m\gamma_{X^-})$ or $({}_m\gamma_{H^+})/({}_m\gamma_{HX^+})$. The medium effect on a molecular species can be obtained directly from solubility measurements in water and S or indirectly by subtracting the ionic contribution from the over-all effect.[F29]

3.4.2 Sources of the Medium Effect

An obvious but small contribution to the medium effect comes from the different mass action effects of water and solvent S. A kilogram contains a different number of moles of these two substances. We can allow for this by taking out the molal concentrations of S and water, $1000/M_s$ and $1000/M_w$, where M_s and M_w are the molar masses. This gives

$$\log {}_m\mathbf{\Gamma}_a = \log {}_m\Gamma_a - \log (M_s/M_w) \qquad (3.4.8)$$

The correction term is 0.25 for methanol and 0.74 for sulphuric acid. For a homologous series of solvents such as the alcohols, the correction term is small but increases with increasing molar mass.

Electrostatic interactions between ions of the acid and the solvents are more important. Ions are more stable in media of high dielectric constant. For an ionogenic reaction, such as the ionisation of acetic acid, we expect the total work of transfer of ions from water ($\varepsilon_r = 78.4$ at 25°C) to ethanol ($\varepsilon_r = 24.3$) to include the work required to raise the electrostatic energy of the ions. This will make a positive contribution to $\log {}_m\Gamma_a$. The electrostatic contribution will be even larger when more highly charged ions are involved, as in the acid ionisation of $H_2PO_4^-$ ions, but will be small (close to zero) for isoelectronic reactions, such as the ionisation of $CH_3NH_3^+$. This correlation between charge type of the acid and medium effect is at least qualitatively confirmed by experience.[F28]

The simplest starting point for the quantitative treatment of the electrostatic part of the medium effect is the Born model of rigid spherical ions in a continuous dielectric. For a molecular acid whose ions have a mean radius r, the Born estimate of the electrostatic contribution to the medium effect is

$$\log {}_m\Gamma_{el} = \frac{e^2}{rkT(\ln 10)}\left(\frac{1}{{}^s\varepsilon_r} - \frac{1}{{}^w\varepsilon_r}\right) \qquad (3.4.9)$$

The crudeness of the model[F30] should dissuade us from accepting this as gospel. Current disillusionment with the Born equation is understandable, but its minor triumphs should not be denied. It does predict that ΔpK for molecular acids should be a linear function of the reciprocal of the dielectric constant, and this is verified for water-alcohol mixtures of moderate to high dielectric constant.[F28]

It is sometimes forgotten that the Born equation is only applicable to the ions, and the over-all medium effect will include a contribution from the molecular acid or base. The importance of the medium effect on molecules is now widely recognised.[F31–34]

The relative basicities of S and water make an important contribution to the over-all medium effect *via* $\log {}_m\gamma_H$. Other solute-solvent interactions may be specific to the particular weak acid or its conjugate base. Weak acid anions X^- are stabilised by water and other hydrogen-bond donating solvents, such as methanol and formamide. Acetonitrile, dimethylformamide, and other solvents that are hydrogen-bond acceptors but not donors are poor solvators of anions, unless the anion is large and polarisable.[F35–37]

The medium effects for benzoic acid, displayed in Table 3.4.1, illustrate the importance of solvation. The over-all effect (column 3) is too large

Table 3.4.1

Medium Effects for the Ionisation of Benzoic Acid at 25°C

Solvent	ε_r	log $_m\Gamma_a$	log $_m\gamma_H$	log $_m(\gamma_{OBz^-}/\gamma_{HOBz})$	log $_m\gamma_{HOBz}$	Ref.
CH_3OH	32.6	5.2	0 ± 2	3 to 7	−2.1	1
CH_3CN	36.0	16.5	4.3 ± 1.5	10.5 to 13.5		2
$(CH_3)_2SO$	46.7	6.8	-3.6 ± 2	8.4 to 12.4		3

1. E. J. King, Acid-base Equilibria, p. 283, Pergamon Press, Oxford (1965).
2. J. F. Coetzee, *Progr. Phys. Org. Chem.*, **4**, 45 (1967).
3. C. D. Ritchie and R. E. Uschold, *J. Amer. Chem. Soc.*, **90**, 2821 (1968).
Values of log $_m\gamma_H$ are from Table 3.3.3.

to be accounted for by the basicity of the solvents (log $_m\gamma_H$ in column 4). We expect electrostatic effects to be largest in methanol, which has the lowest dielectric constant, yet the medium effect is smallest for this solvent. Part of the medium effects in column 5 can be attributed to the molecular benzoic acid, which is more soluble and more stable in the organic solvents than in water. This leaves for the aprotic solvents acetonitrile and dimethylsulphoxide a considerable residual effect, perhaps 6 to 11 units, due to log $_m\gamma_{OBz^-}$. The benzoate ion is stabilised in water and methanol by hydrogen bonding with the solvent; it is much less solvated in the aprotic solvents.

Solute-solvent interactions always include London dispersion forces. Since HX and X^- scarcely differ in molecular size, we expect approximate cancellation of the effect of dispersion forces in log $_m(\gamma_{X^-}/\gamma_{HX})$. Dispersion forces are equivalent to the interaction between virtual electronic oscillators.[F38] As much as a tenfold enhancement in interaction occurs if the oscillator of the solute particle is delocalised. The effect of delocalisation is equivalent to an increase in the distance between the two monopoles of the oscillator so that induced dipole-induced dipole interactions verge towards charge-dipole interactions. Because delocalised oscillators are also associated with a transition in the optical spectrum, a conjugate pair HX, X^- for which one species is delocalised and the other is not is a one-colour indicator. Picrate ion, for example, is a delocalised oscillator; picric acid, acetate ion, and acetic acid are not.

Solvent oscillators are localised but their induced dipole moments increase with molecular size. On this basis we expect dispersion forces to be greater between picrate and methanol than between picrate and water. This accounts for the change in equilibrium constant of the reaction

$$\text{HPic} + \text{OAc}^- \leftrightarrows \text{HOAc} + \text{Pic}^-$$

from 2.6×10^{-4} in water to 78×10^{-4} in methanol and 210×10^{-4} in ethanol. (Thereafter, owing to the short range of dispersion forces and spatial limitations on packing larger alcohol molecules about picrate ion, no further increase in equilibrium constant occurs.) An even more spectacular example of the occasional importance of dispersion effects has been discovered by Fong and Grunwald.[F39] The one-colour indicator 2,4,6,2′,4′,6′-hexanitrodiphenylamine in the ionised form is a strong source of dispersion interaction

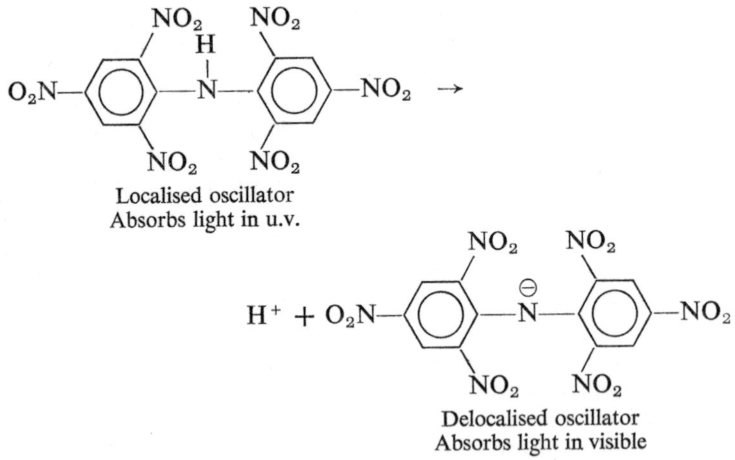

Localised oscillator
Absorbs light in u.v.

$\rightarrow H^+ +$

Delocalised oscillator
Absorbs light in visible

The equilibrium quotient for this reaction increases when small amounts of acetone are added to an aqueous solution of the indicator

Wt% Acetone	0	12	24
$10^3 Q_a$	1.8	7.4	55

This increase occurs despite a decrease in dielectric constant, which raises the electrical energy of the ions. If much more acetone is added, the dielectric constant effect becomes predominant and Q_a decreases.

Medium effects are not invariably based on water as the reference solvent. Indeed, the manifold peculiarities of water make it a poor standard of comparison for some purposes. Methanol is sometimes used in place of water, but a non-hydrogen-bonding solvent may be a better choice. Dimethylformamide,[F40] for example, is a poor solvator of anions so that its use as a reference brings out solvation effects in other solvents.

3.4.3 Uniformity of Medium Effects and Acidity Functions

If we limit ourselves to acids of the same charge type, with the same ionising group, having molecules about the same size, we often find approximate equality of medium effects in a given solvent. This is illustrated by the data in Table 3.4.2. In the two hydrogen-bonding

Table 3.4.2

Medium Effects on Weak Acids*

Acid	Methanol			Ethanol		
	ΔpK	log $_m(\gamma_H\gamma_X)$	log $_m\gamma_{HX}$	ΔpK	log $_m(\gamma_H\gamma_X)$	log $_m\gamma_{HX}$
Acetic	4.95	3.57	−1.38	5.65	4.08	−1.57
Monochloroacetic	4.95	3.58	−1.37	5.66	4.05	−1.61
Benzoic	5.20	3.58	−1.62	5.93	3.97	−1.96
Salicylic	4.89	2.85	−2.04	5.59	3.14	−2.45
o-Nitrobenzoic	5.43	3.23	−2.20	6.28	3.76	−2.52
m-Nitrobenzoic	4.81	2.41	−2.40	5.41	2.66	−2.75
p-Nitrobenzoic	4.99	2.56	−2.43	5.46	2.63	−2.83

Acid	Acetone		
	ΔpK	log $_m(\gamma_H\gamma_X)$	log $_m\gamma_{HX}$
Acetic	7.80	7.09	−0.71
Monochloroacetic	6.95	6.54	−0.41
Benzoic	7.75	6.21	−1.54
Salicylic	6.25	4.23	−2.02
o-Nitrobenzoic	7.58	5.45	−2.13
m-Nitrobenzoic	7.17	4.61	−2.56
p-Nitrobenzoic	7.18	4.58	−2.60

* O. M. Konovalov, *Zh. Fiz. Khim.*, **39**, 693 (1965).

solvents, methanol and ethanol, the medium effects of acetic and chloroacetic acids match closely, and there is fair agreement between those of benzoic, salicylic, *m*-nitrobenzoic, and *p*-nitrobenzoic acids. Even so, *o*-nitrobenzoic acid is consistently out of line. There are greater variations of the medium effect in acetone, because the acid anions can no longer be solvated by hydrogen-bonding. The component parts of the over-all medium effect given in Table 3.4.2 exhibit variations in acetone which indicate the importance of specific solvation of the anions and the molecules in that solvent.

Approximate uniformity of medium effects for similar acids can be used to set up a crude universal acidity scale,[F41] for the spacing of p^sK_a values on a linear scale will be independent of the medium. The acidity scales or order of p^sK_a values in different solvents are aligned so that they match at the p^sK_a values of some selected standard acids. This universal scale is found to be a crude approximation if it includes a variety of acids and breaks down altogether if acids of different charge types are compared.

Hammett acidity functions[F42–45] make somewhat more fastidious use of the approximate uniformity of medium effects. Equilibrium between an indicator base B^z of charge z and its conjugate acid HB^{z+1} is characterised by the constant

$$^wK_a = (^wa_{\rm H})\frac{[B^z]}{[{\rm H}B^{z+1}]}\frac{^wy_B}{^wy_{B\rm H}} \qquad (3.4.10)$$

As before, the factors ^wy_B and $^wy_{B\rm H}$ are activity coefficients on the molar concentration scale referred to unity at infinite dilution in water. The indicator is commonly used in very dilute solution. If the concentrations of other solutes are also low, we may approximate these activity coefficients by medium effects

$$\frac{^wy_B}{^wy_{B\rm H}} = \frac{^sy_B}{^sy_{B\rm H}}\frac{_my_B}{_my_{B\rm H}} \approx \frac{_my_B}{_my_{B\rm H}} \qquad (3.4.11)$$

Indicator acidity functions are operationally defined by the relation

$$\rm H_z \equiv p(^wK_a) + \log([B^z]/[BH^{z+1}]) \qquad (3.4.12)$$

where the ratio of the concentrations can be found by spectrophotometry.[F42–45] The indicator acidity constant ^wK_a can be obtained independently by conductance, potentiometric, or other measurements. In view of eqns. 3.4.10 and 3.4.11, the acidity function can also be expressed by

$$\rm H_z \simeq p(^wa_{\rm H}) - \log\frac{_my_B}{_my_{B\rm H}} \qquad (3.4.13)$$

Although such acidity functions may fly off like kites into this solvent or that, they are all tied down to the pH scale in water.

A Hammett indicator is one member of a set of molecules with the same ionising group and same charge type z and with close similarity of internal charge distribution and size. For two indicator bases C and D satisfying these requirements, we expect approximate equality of medium effects

$$\frac{{}_m y_C}{{}_m y_{CH}} \simeq \frac{{}_m y_D}{{}_m y_{DH}} \tag{3.4.14}$$

The requirements are stringent: if C is a primary amine, D cannot be a tertiary one. Substituted nitroanilines and *N,N*-dialkylnitroanilines thus define different acidity functions, and two more functions are based on alkylated indoles and amides. Within each set of closely similar substances eqn. 3.4.14 is satisfied.

Since any one indicator of a set has a limited useful pH range of 3–4 units, it is necessary to use a series of indicators with overlapping ranges to cover any considerable spread of acidities. The acidity scales constructed in this way for aqueous strong acids appear to be well established. A simple check on the consistency between H_0 scales in different media is to see whether they lead to the same value of $p({}^wK_a)$ for some test acid according to eqn. 3.4.12. For aqueous acids the range of values of $p({}^wK_a)$ obtained in different media is generally less than 0.1 unit.

Non-aqueous solvents are more of a problem than aqueous acids. We expect their medium effects to be less uniform and more unpredictable, especially in non-hydrogen-bonding solvents. It may be recalled that one-colour indicators are just those substances whose medium effects may include a sizeable contribution from London dispersion forces. For acidic solvents such as formic acid H_0 functions have been investigated and reference will be made to these later. Current interest[F46,F47] centres more on H_- acidity functions which are useful for strongly basic media, such as ethylenediamine or mixtures of methanol and sodium methoxide. More details on indicator acidity functions will be given when we consider specific solvents later in this chapter.

The interesting redox acidity function[F48,F49] $R_0(H)$ deserves further investigation. It is based on measurements of the cell

$$(Pt)H_2|HCl(a = 1), Fer(c), Fer^+(c) \text{ in } H_2O|Pt|Fer(c), \\ Fer^+(c), H^+ \text{ in } S|H_2(Pt)$$

where *Fer* stands for ferrocene and Fer^+ for the ferricinium cation. Let E be the electromotive force of the cell and k be the Nernst factor $RT(\ln 10)/F$. The redox acidity function is then defined by

$$R_0(H) = E/k \tag{3.4.15}$$

If medium effects on the ferrocene-ferricinium ion couple cancel, this is equivalent to

$$R_0(\mathrm{H}) \simeq \mathrm{p}({}^{w}a_{\mathrm{H}}) = \mathrm{p}({}^{s}a_{\mathrm{H}}) - \log {}_{m}\gamma_{\mathrm{H}} \tag{3.4.16}$$

and is the basis for a universal pH scale anchored to aqueous solutions. It is reasonable to expect medium effects to be small because ferrocene and ferricinium ion are virtually the same size and the ion has a low charge density and does not form hydrogen bonds. Unlike the H_0 function, $R_0(\mathrm{H})$ does not require a set of specially selected indicators, and a wide range of solvents can be covered without the accumulation of errors that results when several indicators are used with overlapping ranges.

3.5 HYDROGEN-BONDED SOLVENTS OF HIGH DIELECTRIC CONSTANT

3.5.1 General Characteristics

The reaction of an acid with a hydrogen-bonded solvent of high dielectric constant is represented by

$$\mathrm{H}X^{z+1} + S\mathrm{H} \leftrightarrows S\mathrm{H}_2^+ + X^z \tag{3.5.1}$$

The acidity constant corresponding to this equilibrium is

$$K_a = \frac{a_{S\mathrm{H}_2^+} a_{X^z}}{a_{HZ^{z+1}}} \tag{3.5.2}$$

It is understood that all species are solvated, principally by hydrogen bonding with the solvent. This interaction stabilises not only cations such as $C_6N_5NH_3^+$ but also anions such as CH_3COO^-. Other solvation effects peculiar to specific ions or molecules may be noticeable but are not usually prominent. The medium effect, $\log({}_{m}\Gamma_{a})$ or ΔpK, is thus often approximately constant for a given class of acids in one of these solvents.

There may be equilibria other than those expressed by eqn. 3.5.1 in concentrated solutions, but they can generally be ignored except in the most precise work. It is possible that some dimerisation of carboxylic acids occurs even in aqueous solution.[F50] Ion-pairing is negligible except in methanol, the solvent with the lowest dielectric constant (32·70 at 25°C) of this group.

Virtually all these solvents are amphiprotic and therefore they level both acids and bases. For differentiation between acids or bases it is preferable to choose a solvent from one of the other classes.

The methods of studying acid-base reactions in these solvents stem from those used for aqueous solutions.[F51,F52] Variations and adaptations will be discussed in the following sections.

3.5.2 Methanol

The most water-like of this class of solvents is methanol, for it maintains much the same nice balance of basic and acidic properties found in water. Its autoprotolysis constant is smaller than that of water (Table 3.3.4) because of its lower dielectric constant. Medium effects for transfer of ionisation equilibria from water to methanol are approximately constant for closely related acids. For six cation acids, the pyridinium ion and five methyl derivatives, the average medium effect is 0.06 ± 0.02, small because the ionisation of these cations creates no new charge field. For phenol and thirteen of its derivatives the medium effect is 4.32 ± 0.09; smaller values are obtained for nitrophenols, possibly because the anions are stabilised by dispersion interactions with methanol. For 23 carboxylic acids, aliphatic and aromatic, the average medium effect is 4.87 ± 0.15. Values of the medium effect for individual acids are collected in Appendix 3.5.5.

Conductance measurements have been used[F53] to investigate the ionisation of hydrochloric and acetic acids in pure methanol and methanol-water mixtures. Hydrochloric acid is not completely dissociated in mixtures containing more than 40 wt% methanol; in pure methanol its acidity constant is 0.059 mol l^{-1}. Values of the constant were obtained from the slope of a graph of $1/\Lambda S$ against $c\Lambda S y^2$ according to the relation[F54]

$$\frac{1}{\Lambda S} = \frac{1}{\Lambda^\circ} + \frac{1}{K_a \Lambda^{\circ 2}} c\Lambda S y^2 \qquad (3.5.3)$$

Here Λ is the molar conductance at concentration c, y is the mean ionic activity coefficient, Λ° is the molar conductance at infinite dilution, and S is a function defined by Shedlovsky.[F54]

Acetic acid is so weak in methanol ($pK_a = 9.63$ at 25°C) that it is impracticable to attempt to determine K_a from measurements on solutions of the weak acid alone. The solvent corrections for these solutions are uncertain. Moreover, a long extrapolation is required to obtain the limiting molar conductance according to eqn. 3.5.3. It is better to find Λ° independently by using Kohlrausch's law[F55]

$$\Lambda^\circ(\text{HOAc}) = \Lambda^\circ(\text{HCl}) + \Lambda^\circ(\text{NaOAc}) - \Lambda^\circ(\text{NaCl}) \qquad (3.5.4)$$

Since hydrochloric acid, sodium acetate and sodium chloride are highly dissociated in methanol, it is practicable to find their limiting conductances by extrapolation of the molar conductances of dilute solutions.

Some results of Shedlovsky and Kay's investigation[F53] are given in Appendix 3.5.1.

Electromotive force measurements can be used in several ways to investigate acid-base equilibria in methanol. The hydrogen, glass, silver-silver chloride and calomel electrodes function reversibly in this solvent but with lower precision than in water. It is doubtful if any pK_a value has been established beyond the second decimal place in pure methanol; this is equivalent to a precision of about ±0.5 mV in electromotive force. The familiar hydrogen-silver, silver chloride cell[F52]

$$\text{Pt, } H_2(g)|HX\text{, Na}X\text{, NaCl in MeOH}|\text{AgCl, Ag} \qquad (3.5.5)$$

has been used to determine acidity constants. Some of Tabagua's[F56] results are given in Appendix 3.5.5.

Grunwald's differential titration method[F57] can also give results of high precision provided the slope of the titration curve at the equivalence point is not too large. The results of Ritchie and Heffley's study[F58] of the ionisation of some picolinium ions by this method are given in Appendix 3.5.3. A direct, rather than differential, potentiometric titration had to be used for derivatives of bicyclo[2.2.2]octane-1-carboxylic acid;[F59] these results are given in Appendix 3.5.4.

The familiar technique of pH titration[F52] can be used to investigate acid-base equilibria in methanol if the cell is carefully standardised. In the usual pH meter cell

$$\begin{array}{c|c|c|l} \text{Glass} & \text{Test} & \text{Salt bridge} & \text{Reference} \\ \text{electrode} & \text{solution} & \text{solution} & \text{electrode} \end{array} \qquad (3.5.6)$$

both test and salt bridge solutions should be prepared with methanol as the solvent to minimise liquid junction potentials. Some workers[F60] have advocated use of methanol in place of water in the internal solution of the glass electrode; a faster and more reproducible response is claimed. The validity of this practice is still in doubt.[F61]

The pH meter cell must be standardised with buffers of known pH. Aqueous buffers are sometimes used when the solvent under investigation is a methanol-water mixture. One must then assume that liquid junction potentials thereby introduced are approximately constant from one test solution to another.[F62] There is also a risk that transfer of a glass electrode from aqueous to methanolic solution may cause irreproducible changes in its asymmetry potential. A better procedure is to prepare standard buffers in methanol. Three such buffers were assigned pH values[F63] by the same procedure used for standard aqueous buffers.[F63] The assigned values, designated pH*(S), should closely approximate p($^s a_H$). If an unknown solution X is now measured in the standardised cell, the operational definition of its pH is

$$\text{pH}^*(X) = \text{pH}^*(S) + \frac{E_X - E_S}{k} \qquad (3.5.7)$$

where E_X and E_S are the electromotive forces of the cell 3.5.6 when the test solutions are X and standard buffer and k is the Nernst factor $RT(\ln 10)/F$.

The derivation of pK_a values from pH measurements is too familiar an exercise to be repeated here. Since pH* values refer to a standard state for activities in methanol, the acidity constants will likewise have this reference state, i.e. they will be ${}^sK_a$ values. When measurements are made in very dilute solutions (ca. $10^{-3}M$), not only are corrections to zero ionic strength small but ion-pairing of buffer salts becomes negligible. Some results of Juillard's measurements,[F60,F65] based on the de Ligny standard buffers, are given in Appendix 3.5.5.

Spectrophotometric measurements are used to established the colour ratio R of an indicator H*In*

$$R = \frac{[In^-]}{[\mathrm{H}In]} = \frac{\varepsilon - \varepsilon_a}{\varepsilon_b - \varepsilon} \tag{3.5.8}$$

were ε_a and ε_b are the molar absorptivities characteristic of the acidic (H*In*) and basic (In^-) forms and ε is that of the equilibrium mixture. The ratio R can then be combined with stoichiometric relationships to give either the acidity constant of the indicator or, if $K_a(\mathrm{H}In)$ is known, the ratio of its acidity constant to that of another acid HX used along with its salt MX as a buffer in the same solution[F52]

$$\frac{K_a(\mathrm{H}In)}{K_a(\mathrm{H}X)} = \frac{y_{In^-}}{y_{X^-}}\frac{y_{\mathrm{H}X}}{y_{\mathrm{H}In}}\frac{[In^-]}{[X^-]}\frac{[\mathrm{H}X]}{[\mathrm{H}In]} \tag{3.5.9}$$

When very dilute solutions are used the activity coefficient ratios in this equation will be close to unity,[F66] and ion-pairing of the salts can be neglected.

It may not be possible to prepare sufficiently alkaline solutions of exceptionally weak acids to establish the value of ε_b. The method of Stearns and Wheland[F67] may then be used for an acid such as phenol in methanol (p(${}^sK_a$) = 14.4), which has appreciable optical absorptivity only in the ionised form. The absorbance A of solutions containing low concentrations of acid c_a and sodium methoxide, NaS, is represented by

$$\frac{bc_a}{A} = \frac{1}{\varepsilon_b} + \frac{K_S}{\varepsilon_b K_a}\frac{1}{[S^-]} \tag{3.5.10}$$

where b is the internal path length of the cell and K_S is the autoprotolysis constant of methanol, SH. A graph of bc_a/A against $1/[S^-]$ will then be linear, and ε_b and K_a can be calculated from the slope and intercept.

Values of the acidity constants of phenol and some of its derivatives obtained by spectrophotometry[F70,F71] are given in Appendix 3.5.2.

Acidity functions in methanol can also be established by spectrophotometry. Current interest centres on the H_- function for neutral indicator acids defined by

$$H_- \equiv pK_a + \log R \qquad (3.5.11)$$

where R is the colour ratio used in eqn. 3.5.8. For solutions of sodium methoxide in methanol very weak acids are required as indicators. One suitable set consists of selected nitro derivatives of aniline and diphenylamine, which ionise by losing a proton from their amino groups. With this set of indicators we can link the pH scale in water to the H_- scale in methanol by way of various water-methanol mixtures. When established by this stepwise procedure the acidity constant and H_- function in eqn. 3.5.11 are still based on a standard state in water.[F46] A disadvantage of such a universal scale is the accumulation of errors in the stepwise procedure. Several workers[F72,F73] have therefore advocated the use of an acidity function based on a standard state in methanol. The acidity constant in eqn. 3.5.11 is then ${}^sK_a$; the acidity function has been variously designated ${}_sH_-$, H_-^*, or H_M.

Different classes of indicators define slightly different H_- scales just as they do in water solutions: nitroanilines and nitrodiphenylamines one scale,[F72,F73] substituted phenols another.[F69,F74,F75] Polynitroaromatic hydrocarbons are used to define a J_M scale,[F76] and carboxylic acid derivatives of nitroanilines and nitrodiphenylamine a H_{2-} scale.[F77]

The acidity function H_- increases as the concentration of alkali metal methoxide in methanol is increased. Introducing the expressions for acidity and autoprotolysis constants into eqn. 3.5.11, we can write it in the form

$$H_- = pK_S + \log [MeO^-] + \log (y_{OMe^-}\, y_{HIn}/y_{In^-}) \qquad (3.5.12)$$

The activity coefficient term, though it tends towards unity at infinite dilution in water (or in methanol for the H_M function), is not negligible. The acidity function therefore increases more rapidly than the first two terms on the right hand side of eqn. 3.5.12 would lead us to expect, and it reaches quite large values for concentrated solutions of alkali methoxides. For example, H_- is 10.66 for 0.001M NaOMe, 15.77 for a 3.00M solution;[F46] H_M for the same solutions[F73] is 13.71 and 18.98. The differences between the values of H_- or H_M at these two concentrations would be 3.48 if the activity coefficient term in eqn. 3.5.12 were negligible; the actual differences are larger: 5.11 and 5.27.

In concentrated solutions the value of H_- depends on the particular alkali metal methoxide used.[F46,F75,F76] For equimolar solutions lithium methoxide gives the lowest basicity, sodium methoxide next, and potassium methoxide the highest. This may be a result of the formation of solvent-bridged ion-pairs[F46]

$$\underset{}{M^{+}}\cdots\overset{\delta-}{\underset{\displaystyle CH_3}{O}}\text{—}\overset{\delta+}{H}\cdots{}^{-}\underset{\displaystyle CH_3}{O}$$

We expect such pairings to be strongest in lithium methoxide solutions and to reduce the basicity of the methoxide ion.

Other thermodynamic functions associated with ionisation in methanol have been determined for a few acids. Heats and entropies of ionisation (ΔH^0 and ΔS^0) can be obtained from the temperature coefficient of p($^{s}K_a$). When measurements are made at only two[F58,F59] or three[F78] temperatures, rather rough values of the heats of ionisation must be expected,[F79] but these are sufficiently precise for some purposes. They reveal, for example, that heats of ionisation are generally larger in methanol than in water, whether because of greater solvation of the molecular acid by methanol or of the ions by water or both. When p($^{s}K_a$) values are measured[F70] with a precision of ± 0.02 at nine temperatures from 10 to 50°C and the data are fitted to the Clarke and Glew equation,[F71,F80] the precision of ΔH^0 may be as good as ± 0.30 kJ mol^{-1} and a rough value of ΔC_p^0 can even be obtained. These data for substituted phenols may be found in Appendix 3.5.2 and provide evidence of the importance of solute-methanol interactions.[F70] The entropy of ionisation is about 40 J K^{-1} mol^{-1} more negative in methanol than in water. Part of this loss of entropy is due to the greater electrostatic effect of ions in methanol, a solvent with less than half the dielectric constant of water. Moreover, methanol is a less ordered solvent than water,[F12] so that the entropy of methanol has farther to fall when solvent molecules become oriented about ions. A linear correlation between relative Gibbs free energy changes (or relative pK_a values) and entropies of ionisation of unhindered phenols is observed[F70] in methanol as it is in water,[F79,F81] but the quantitative predictions of the Born theory are not fulfilled. A phenol with bulky *t*-butyl groups substituted at the positions on either side of the ionising group, has a larger negative entropy change on ionisation than an unhindered phenol; compare, for example, 2,6-di-*t*-butylphenol, $\Delta S^0 = -0.212$ kJ K^{-1} mol^{-1}, with unhindered 4-*t*-butylphenol, $\Delta S^0 = -0.154$, or 3,5-di-*t*-butylphenol, $\Delta S^0 = -0.156$. This cannot be the result of a solvent exclusion effect, i.e. steric restriction on the number of methanol molecules affected by the electrostatic field of the phenoxide ion, for that would produce less decrease in entropy for the hindered phenol. Very likely the two bulky alkyl groups cause loss of librational and vibrational motion of the methanol molecule that is wedged between them to hydrogen bond with the phenoxide group.[F70]

Direct calorimetric studies of acid-base reactions in methanol are desirable if only to check heats of ionisation obtained from the temperature coefficients of equilibrium constants. An old value[F82] of the heat of ionisation of acetic acid, 18.6 kJ mol^{-1} at 17°C, is all we have to compare with the Rochester and Rossall value of 21.5 kJ mol^{-1} at 25°C. Leung and Grunwald[F83] have reported heats of neutralisation in methanol with an estimated error of only ±0.2%. The heats of ionisation of methanol and benzoic acid derived from these heats of neutralisation are given in Appendices 3.3.3 and 3.5.2. Because of the long liquid range of methanol (−97.8 to 64.65°C) and the precision of their measurements, Leung and Grunwald were able to show that the change in heat capacity associated with the ionisation of benzoic acid varies monotonically from −0.011 at −90°C to −0.225 kJ K^{-1} mol^{-1} at +32.5°C. This is in marked contrast to the change in heat capacity for ionisation of benzoic acid in aqueous solutions which goes through a maximum near 35°C. The temperature variation of ΔC_p^0 in methanol is proportional to that of the heat capacity of pure liquid methanol

$$\frac{d\Delta C_p^0}{dT} = -k\,\frac{dC_p\,(l)}{dT}$$

where k is about 15 for the self-ionisation of methanol and 16 for the ionisation of benzoic acid.

A few volume changes associated with acid-base reactions in methanol have been measured, none with high precision. The volume changes ΔV_b^0 for the ionisation of two bases[F84] are given in Table 3.5.1; for the conjugate acid $\Delta V_a^0 = \Delta V_s - \Delta V_b^0$ where ΔV_s is the volume change for

Table 3.5.1

Volume Changes for Ionisation in Methanol at 25°C

Base	Acid	In Methanol ΔV_b^0 (cm^3 mol^{-1})	In Methanol ΔV_a^0 (cm^3 mol^{-1})	In Water ΔV_b^0 (cm^3 mol^{-1})	In Water ΔV_a^0 (cm^3 mol^{-1})
Pyridine	Pyridinium ion	−53.4	10	−28.1	6
Piperidine	Piperidinium ion	−49.5	6	−24.3	2

self-ionisation of methanol,[F85] namely −43.3 cm^3 mol^{-1} at 25°C. For base ionisation reactions, which are ionogenic processes

$$B + SH \leftrightarrows BH^+ + S^-$$

the magnitude of the volume change is twice as large in methanol as it is in water. This is in accord with the less ordered structure of methanol.

The compressibility of methanol is three times that of water, so that electrostrictive compression of solvent about ions will be much larger in methanol. Millero[F86] has estimated that the partial molar volume of the hydrogen ion is 10 cm^3 mol^{-1} less in methanol than in water. The difference in volumes of acid and conjugate base must therefore be considerable. Rochester and Rossall[F85] have reported volume changes for ionisation of some of the phenols mentioned earlier, and these values are given in Appendix 3.5.2. Phenoxide ion in methanol has a much smaller apparent molar volume than it does in water whereas that of phenol is about the same in the two solvents. Electrostrictive compression is evidently larger in methanol for the phenoxide ion. Rochester and Rossall[F85] found linear relationships between pK_a and ΔV_a and between viscosity B_η coefficients of the sodium phenoxide salts and their apparent molar volumes.

3.5.3 Formic Acid

Anhydrous formic acid is an interesting acidic solvent; its H_0 value is[F87] -2.2 and the medium effect on the proton (Appendix 3.3.2) is 7.4. Its dielectric constant, 56.1 at 25°C, is large enough to ensure complete dissociation of 1–1 electrolytes in it. It is a good solvent for many organic and inorganic compounds.[F88] Yet formic acid has not achieved popularity as a solvent for the study of weak bases for two reasons. Its autoprotolysis constant is rather large (pK_S about 7), so that the range of acidities in it, though larger than that in sulphuric acid, is still limited. Potentiometric titrations can be carried out in formic acid,[F89] but the potential change at the end point is less than 200 mV and the accuracy is only $\pm 2\%$. Glass,[F90] quinhydrone,[F89] and silver-silver chloride electrodes[F91] can be used.

A more important drawback to the use of formic acid is the difficulty of preparing and maintaining absolutely pure solvent. Formic acid undergoes spontaneous decomposition to either H_2 and CO_2 or H_2O and CO. Acid solutions can be used for only a matter of hours before bubbles begin to form.[F59] The presence of traces of carbon monoxide probably accounts for poisoning of the hydrogen electrode in formic acid. The other electrodes function properly only in rigorously purified solvent.[F89]

Perchloric acid is a strong electrolyte in formic acid, but the acidity constant of hydrochloric acid[F88] is only 0.009 mol kg^{-1}. Aniline, diethylamine and a number of other bases are strong. Conductivity measurements[F88] have been used to study some substituted tetrazoles with pK_a values in the range 1.78 to 2.06.

3.5.4 Ethanolamine[F92, F93]

Ethanolamine in aqueous solution is approximately as basic as ammonia, $pK_b = 4.6$ *vs.* 4.76 for NH_3. The value of the H_- acidity function in pure ethanolamine is 15.35. Its dielectric constant of 37.7 at 25°C is high enough so that ion-pairing of salts is small in dilute solutions. Acids with $p({}^{w}K_a)$ values below 14 are completely ionised in ethanolamine. Weaker acids have been studied by the spectrophotometric method of Stearns and Wheland (sect. 3.5.2). For a typical acid, 2,4-dinitroaniline, $p({}^{s}K_a)$ is 3.15 as compared with 15.8 for $p({}^{w}K_a)$. Because of the large autoprotolysis constant of ethanolamine ($pK_S = 5.2$), the range of acids that can be investigated is limited and corresponds roughly to acids with $p({}^{w}K_a)$ values between 13.5 and 18.5.

3.5.5 Amides

The principal hydrogen-bonding amide solvents are formamide, acetamide, *N*-methylformamide (NMF), and *N*-methylacetamide (NMA). All four are basic solvents. For example, the medium effect for transfer of the proton from water to formamide (Appendix 3.3.2) is −1.2; the redox acidity function $R_0(\mathrm{H})$ is 4.7 in molten acetamide at 98°C.[F94] These solvents are most notable for their very high dielectric constants, which range from 109.5 for formamide to 182.4 for NMF at 25°C. Their autoprotolysis constants are comparable with those of water and the alcohols, e.g. pK_S of formamide is 17.0 (Appendix 3.3.3), so that a wide range of acid strengths can be investigated.

Experimental work on acid-base behaviour in these solvents has usually used conductance or electromotive force measurements. The pure solvents have satisfactorily low conductivities, $0.05\text{–}2.6 \times 10^{-6}$ $\Omega^{-1}\,\mathrm{cm}^{-1}$, but some of these solvents are difficult to purify.[F95] Reid and Vincent[F96] have reviewed electrodes that are suitable for use in amide solvents. Hydrogen electrode potentials may drift because of platinum-catalysed decomposition of the solvent in acid solutions. Agarwal and Nayak[F97] found a change of 7 mV/h in formamide, small enough to permit an accurate extrapolation to zero time. The glass electrode does not seem to have been tested thoroughly in the anhydrous solvents. Reynaud[F98] has used it in water-NMA mixtures. The calomel electrode is generally unsatisfactory, although Rossi, Pecci, and Scrosati[F99] determined its standard potential in formamide with a precision of ±0.7 mV. The silver-silver chloride electrode has been used in all four amide solvents, but there is a difference of opinion[F94,F96] about its suitability in acetamide. The standard electromotive forces of the hydrogen-silver, silver chloride cell and mean activity coefficients of hydrochloric acid can be found in Appendices 2.2.10, 11 and 13. At best the reproducibility was ±0.2 mV, at worst ±3 mV. Cadmium-cadmium chloride electrodes are also useful, but they equilibrate more slowly than silver-silver chloride electrodes.[F99]

Formamide is difficult to purify. Mandel and Decroly[F100] determined the acidity constants of formic and acetic acids in formamide at six temperatures between 15 and 45°C. The precision of their electromotive force measurements was claimed to be better than ±1 mV; their standard potentials, however, differ significantly from those of subsequent workers.[F97,F99,F101] Greater precision was achieved by Agarwal and Nayak[F102] in their determination of the acidity constant of oxalic acid. The results of these investigations are given in Appendix 3.5.6.

Acetamide melts at 81°C, so that acid-base equilibria must be studied above about 90°C. It is interesting that the dielectric constant and autoprotolysis constant of acetamide are roughly comparable with those of water in this temperature range,[F94] yet acetamide lacks the structural peculiarities of water. Molten acetamide is a good solvent for inorganic and organic compounds.[F103] In view of the high temperature it is not surprising that no measurements comparable in precision with those on formamide solutions have been reported. Trémillon[F94] has studied a large number of acids by voltammetry; the ionic strength was kept constant at $1M$ by addition of tetramethylammonium chloride.

N-Methylacetamide tenaciously holds traces of acetic acid and acetate[F95] and is difficult to purify.[F104] Dawson and co-workers[F95] have used conductance measurements to investigate the ionisation of acetic acid and its chloro-derivatives. The solvent correction to the conductivity presents a difficult problem.

3.6 NON-HYDROGEN-BONDED SOLVENTS OF HIGH DIELECTRIC CONSTANT

3.6.1 General Characteristics

Non-hydrogen-bonded liquids of high dielectric constant are often referred to as *dipolar aprotic solvents.* Nitrobenzene, *N,N*-dimethylformamide, acetonitrile, and nitromethane have dielectric constants in the range 35 to 38, close to that of methanol. The dielectric constants of sulpholane, dimethylsulphoxide and propylene carbonate are higher. Any peculiarities of acid-base behaviour in dipolar aprotic solvents as compared with methanol cannot be accounted for on the basis of the Born sphere-in-continuum model.

Dipolar aprotic solvents show little ability to donate protons to bases or hydrogen atoms to hydrogen bond acceptors. Dimethylsulphoxide, for example, has a pK_a value[F105] of about 32. Acetonitrile and nitromethane have such feeble acidic or basic properties that they are excellent differentiating solvents. All dipolar aprotic solvents are sufficiently basic to combine with and stabilise protons from acids. Going by the medium effects on the proton (Table 3.3.3), we expect dimethylsulphoxide to be more basic than water and acetonitrile to be much less basic. Dimethylformamide is between water and dimethylsulphoxide in basicity. Sulpholane and other sulphones are very weak bases.

Molecules of these aprotic solvents are more polar than those of protic solvents. Their dipole moments range from 3.4 D for acetonitrile to 4.7 D for sulpholane, as compared with values less than 2 D for water and methanol. We therefore expect stronger ion-dipole and molecule-dipole solute-solvent interactions in dipolar aprotic solvents. Molecules of these solvents are also more polarisable, so that they will have special affinity for polarisable anions like picrate. The London dispersion effects alluded to in sect. 3.4.2 will thus be more pronounced in dipolar aprotic solvents.

Medium effects for transfer of acid-base equilibria from water to a dipolar aprotic solvent may be expected to differ from those from water to methanol. The change in dielectric constant is about the same, but other solute-solvent interactions are very different. For a cation acid

$$BH^+ + S \leftrightarrows B + HS^+$$

in a basic solvent like dimethylsulphoxide the cation BH^+ will be stabilised by hydrogen-bonding with the solvent S. The molecular base B, on the other hand, cannot acquire hydrogen bonds from S, so its stability will depend on comparatively weak dipole-dipole and dispersion forces. The medium effect for cation acids will include not only a contribution from the proton but also a small positive effect from the conversion of BH^+ to B.

The medium effect for molecular acids should be larger, for the extent of reaction

$$HA + S \leftrightarrows HS^+ + A^-$$

depends on stabilisation of the anion by the solvent. None of the dipolar aprotic solvents is capable of solvating anions by hydrogen bonding, so we expect a rather large positive contribution from the conversion of HA to A^- to the medium effect for transfer of molecular acid ionisation reactions from water or methanol to dipolar aprotic solvents. Since the charge distribution of anions will change when they form hydrogen bonds with a solvent,[F106] medium effects between water and a dipolar aprotic solvent are less uniform than those from one dipolar aprotic solvent to another. The occasional importance of dispersion effects may be illustrated by the observation[F35] that the ratio of acidity constants of picric to benzoic acids is $10^{3.5}$ in water, $10^{5.3}$ in methanol, but $10^{11.9}$ in dimethylsulphoxide.

Since anions cannot achieve stability in dipolar aprotic solvents by forming hydrogen bonds with solvent, they commonly do so by bonding with any acid present. Red solutions of the 3,5-dinitrophenolate ion in either dimethylformamide[F107] or acetonitrile[F108] turn yellow when the free acid is added. Such a combination of the anion with its conjugate acid

$$A^- + HA \leftrightarrows (A \cdots HA)^-$$

is called *homoconjugation*. Combination of the anion with another acid

$$A^- + HX \leftrightarrows (A \cdots HX)^-$$

is called *heteroconjugation*. With these equilibria we associate the formation constants $K_f(AHA^-)$ and $K_f(AHX^-)$. The extent of conjugation will depend on the basicity of the solvent. We expect $K_f(AHA^-)$ to be small in the more basic solvents like dimethylsulphoxide where both A^- and solvent molecules are competing for HA. Much larger formation constants may be expected in the less basic solvents like acetonitrile.[F109] Under certain circumstances higher order complexes, such as $A(HA)_2^-$, may form.

3.6.2 Experimental Methods

Conductance,[F26,F110–112] spectrophotometric,[F111] solubility,[F112] and electromotive force measurements[F113] can be adapted to give formation constants of the conjugation complexes as well as acidity constants of the acids. The reproducibility of measurement is often lower than that found for water or methanol solutions, and it is common practice to simplify the calculation by introducing various approximations. For a given acid the results of different methods or different investigators seldom indicate a precision of better than 0.1 pK unit. Greater precision can of course be achieved in measuring relative values of acidity constants.[F114]

The *conductance method* is satisfactory only if the solvent can be rigorously purified. Through failure to appreciate this, the first values of pK_a of picric acid in acetonitrile proved to be much too small, 5.6 and 8.9 as compared with 11.0 from electromotive force measurements on buffered solutions.[F113] D'Aprano and Fuoss[F115] found that acetonitrile having a satisfactory specific conductance of about $10^{-8}\ \Omega^{-1}\ \mathrm{cm}^{-1}$ still contained a trace of ammonia. This was converted to ammonium picrate when acid was added to the solvent giving a spurious contribution to the conductance of picric acid solutions. This discovery moved them to make the flat assertion that 'dissociation constants of weak acids cannot be determined in aprotic solvents conductimetrically'. This may be an overly pessimistic view, conductance values of pK_a for acids in dimethylsulphoxide[F109] and dimethylformamide[F107] agree well with those from spectrophotometric and electromotive force measurements. Approximate values of pK_a and pK_f can be obtained from conductometric titrations of a weak acid with a weak base.[F116,F117]

The *spectrophotometric method* described in sect. 3.5.2 may be used if conjugation and ion-pairing are negligible, e.g. for 2,6-dinitrophenol in dimethylsulphoxide.[F108] More commonly, some account must be taken of these complications. The acid to be investigated is incorporated with the indicator in a buffer to reduce the effect of acidic or basic impurities from the solvent. The buffer salt is often the tetraethylammonium salt of the acid. Such salts are less subject to ion-pairing than alkali metal salts would be. The buffer contains an excess of the weak acid so that virtually all A^- ions are converted to AHA^- ions. We expect an indicator cation $In\mathrm{H}^+$ to have less tendency to undergo heteroconjugation with AHA^- than with A^-. The principal equilibrium is then

$$2\mathrm{H}A \leftrightarrows A\mathrm{H}A^- + \mathrm{H}^+;\ K = K_a(\mathrm{H}A) \times K_f(A\mathrm{H}A^-)$$

If the buffer contains an indicator with known acidity constant, spectrophotometry can be used to establish the activity of the hydrogen ion. From this and stoichiometric relationships we get the constant K but cannot separate it into K_a and K_f without further information.

Solubility measurements can be used to find the homoconjugation constant

$$K_f(A\mathrm{H}A^-) = \frac{[A\mathrm{H}A^-]}{[\mathrm{H}A][A^-]} \tag{3.6.1}$$

Let S be the molar solubility of salt MA in a solution of HA of stoichiometric concentration c_a. From the material balance and electroneutrality conditions, together with the assumption that the hydrogen ion concentration is negligible in comparison with other concentrations we can show that

$$[M^+] = [A^-](1 + K_f[\mathrm{H}A]) \tag{3.6.2}$$

and

$$[\mathrm{H}A] = c_a - S + [M^+A^-] \tag{3.6.3}$$

The concentration of the ion-pair is given by

$$[M^+A^-] = K_{s0}(MA)/K_d(MA) \tag{3.6.4}$$

where

$$K_{s0} = [M^+][A^-]y^2 \tag{3.6.5}$$

and

$$K_d = [M^+][A^-]y^2/[M^+A^-] \tag{3.6.6}$$

The dissociation constant K_d must be known from independent conductance measurements. The activity coefficient y can be estimated with sufficient precision by the Debye–Hückel limiting law. Finally, by combining these equations we obtain

$$(S - [M^+A^-])^2y^2 = K_{s0} + K_{s0}K_f[\mathrm{H}A] \tag{3.6.7}$$

From a plot of the left hand side against [HA] we obtain K_f as the ratio of slope to intercept. The values of the concentrations of M^+A^- and HA are refined by iteration.

An analogous procedure can be used to find heteroconjugation constants from solubility measurements of MA in solutions of HX.[F118]

Electromotive force measurements have given some of the best values of K_a and K_f. Butler[F119] has recently reviewed electrodes suitable for aprotic organic solvents. These include electrodes reversible with respect to either silver or chloride[F120] ions. The glass electrode behaves satisfactorily in many of these solvents. It responds more rapidly when the internal solution is made with the aprotic solvent instead of water.[F119,F121,F122] The hydrogen electrode is more temperamental, behaving well for some workers and not for others. There are disagreements about other electrodes as well, so that the preparation of well-behaved galvanic cells is evidently still something of an art.

Most electromotive force studies have relied on the establishment of a pH scale based on standard buffers. The pK_a values required for establishment of the scale have sometimes been obtained from conductance or spectrophotometric measurements that were not of first quality. For the cation of the buffer salt, tetraethylammonium ion is preferred over an alkali metal ion because it does not pair as extensively with A^- ions from the acid. It has been observed, however, that in cells with salt bridges the liquid junction potential varied less with concentration when KCl rather than Et_4NCl was used in the salt bridge.[F123]

For buffers in which no homoconjugation occurs and which contain stoichiometric concentrations c_a of acid HA and c_s of salt Et_4NA a graph of pH against log (c_s/c_a) should be a straight line of unit slope, provided the hydrogen ion concentration is negligible in comparison with c_a or c_s. When homoconjugation occurs, the plot is curved.

The acidity constant K_a can be obtained from the half-neutralisation point of titration or from the pH of a buffer with equal stoichiometric concentrations of acid and salt. The equation for the titration curve is[F108,F113]

$$y^2 c_s a_H^2 - yK_a[(c_s + c_a) + K_f(c_s - c_a)^2]a_H + K_a^2 c_a = 0 \quad (3.6.8)$$

Homoconjugation changes the slope of the titration curve at the half-neutralisation point without affecting the pH. When $c_a = c_s$ eqn. 3.6.8 reduces to

$$\text{pH}_{\frac{1}{2}} = \text{p}K_a + \log y \quad (3.6.9)$$

where y is the molarity scale activity coefficient of the A^- ion referred to unity at infinite dilution in the given solvent. If the buffer salt undergoes appreciable ion-pairing,[F124] this equation is no longer valid.[F108] The apparent pK_a value calculated from eqn. 3.6.9 is then found to vary with the concentration of the salt.

To obtain the formation constant for HA^- a second buffer mixture is required. Let $r = a_H/(a_H)_{\frac{1}{2}}$ where a_H pertains to the second buffer and $(a_H)_{\frac{1}{2}}$ to the half-neutralisation point. The titration curve (eqn. 3.6.8) then becomes[F113]

$$c_s r^2 - r[(c_a + c_s) + K_f(c_s - c_a)^2] + c_a = 0 \qquad (3.6.10)$$

which can be solved for K_f. In practice, several buffers are used and an average of the K_f values is taken. The electromotive force method works best when the product $(c_a + c_s)K_f$ is over 100; for smaller values the solubility method is preferred.[F113]

3.6.3 Dimethylsulphoxide

Dimethylsulphoxide is an excellent solvent for electrolytes. Since its dielectric constant is high (46.7 at 25°C), ion-pairing in dilute solutions can be neglected. The medium effect for transfer of a proton is approximately −3.6 (Table 3.3.3), so that dimethylsulphoxide is more basic than water. Homoconjugation is therefore slight, an acid HA is more likely to hydrogen-bond with plentiful Me_2SO than with its own anion A^-. Formation constants for AHA^- ions are only one-hundredth as large in dimethylsulphoxide as in the much less basic solvent acetonitrile.[F109] Kolthoff and Reddy[F125] reported an autoprotolysis constant of 5×10^{-18} mol^2 l^{-2}. More recent values[F105,F126] near 10^{-32} are consistent with the wide range of acidities found by Ritchie and Uschold.[F127]

Spectrophotometry[F109,F114,F125,F128,F129] and potentiometry[F109,F125,F127,F130–133] have been the most commonly used methods for investigating acid-base equilibria in dimethylsulphoxide. Few if any pK_a values have been established to better than 0.1 unit, and a complication of results would be premature. The reader will wish to hold all the pK_a values that are subsequently quoted with his fingertips so that a new breeze of facts can blow away the less substantial ones.

The glass electrode functions satisfactorily in dimethylsulphoxide[F130] up to a pH of 28. Its response is not always Nernstian.[F125] Ritchie[F127] calibrated the glass electrode with solutions of *p*-toluenesulphonic acid whereas Kolthoff[F125] used buffer mixtures of two nitrophenols and their tetraethylammonium salts. The acidity constants of the nitrophenols were obtained from conductance and spectrophotometric measurements. The hydrogen electrode is reported to give unsatisfactory results because of reduction of dimethylsulphoxide and poisoning of the platinum black,[F119] yet it has been used successfully in voltammetry.[F126] The silver-silver chloride electrode is unsuitable because silver chloride too readily forms complexes with excess chloride in dimethylsulphoxide.[F119] Silver-0.01M silver nitrate[F109,F125] and silver-0.05M silver perchlorate[F130] are satisfactory reference half-cells.

Representative values of p(^sK_a) in dimethylsulphoxide and medium effects, ΔpK, for transfer of acids from four other solvents to dimethylsulphoxide are given in Table 3.6.1. Kolthoff[F109] and Ritchie[F127,F131,F134]

Table 3.6.1

Acidity Constants in Dimethylsulphoxide and Medium Effects

$$\Delta pK = pK_a \text{ in } Me_2SO - pK_a \text{ in Solvent } S$$

Acid	$p(^SK_a)$	$S = H_2O$	$S =$ MeOH	$S =$ DMF	$S =$ AN
Acetic acid	12.6	7.8	3.1	−0.9	
Benzoic acid	11.1	6.9	1.8	−1.2	−9.6
Salicylic acid	6.8	3.8	−1.1	−1.4	−9.9
HSO_4^- ion	14.45	12.45		−2.7	−11.4
Phenol	16.4	6.4	2.0		−10.2
2-Nitrophenol	11.0	3.8			−11.2
4-Nitrophenol	11.0	3.8	−0.4	−1.6	−10.0
2,6-Dinitrophenol	4.9	1.2	−2.6	−0.9	−11.6
2,6-Dinitro-4-chlorophenol	3.5	0.5		−1.2	
Picric acid	−0.5	−1.2		−1.9	−11.5
Ammonium ion	10.5	1.3		1.0	
Ethylammonium ion	11.0	0.4			
n-Butylammonium ion	11.1	0.5		2.0	
Diethylammonium ion	10.5	−0.4			
Triethylammonium ion	9.0	−1.9		−0.2	
Anilinium ion	3.6	−1.0		−0.8	

Data from ref. F109 (Me_2SO): R. A. Robinson and R. H. Stokes, Electrolyte Solutions, Second Edition (revised), Butterworths, London (1968) (H_2O); ref. F107 (DMF = dimethylformamide); ref. F109 and F132 (AN = acetonitrile); Appendices 3.5.1–3.5.5 (MeOH).

may be consulted for longer compilations. Molecular acids are generally weaker in dimethylsulphoxide than in water, whereas cation acids have about the same strength in the two solvents. We might expect molecular acids to be stronger in the more basic of the solvents, dimethylsulphoxide. To explain why this is not so we note that anions cannot accept hydrogen bonds from dimethylsulphoxide. Water or methanol by forming hydrogen bonds with an anion disperses some of its charge and stabilises it.[F106]

Other contributions to medium effects are generally small. Approximately one unit in ΔpK arises from the decrease in dielectric constant from 78 for water to 47 for dimethylsulphoxide. Dispersion interactions are more important in dimethylsulphoxide because it is more polarisable than water. We expect to see evidence of this effect only for acids, such as picric acid or certain hydrocarbons, that form coloured anions. Picric acid, unlike many other molecular acids, is indeed stronger in dimethylsulphoxide than in water.

Anion stabilisation by hydrogen bonding is sensitive to the effect of substituents in the acid[F131] because both the substituent and the hydrogen bond alter the charge distribution of the anion.[F106] Medium effects for transfer of acids from water to dimethylsulphoxide are therefore far from uniform. The same is true for transfers from methanol, another hydrogen-bonding solvent. The data in Table 3.6.1 indicate that transfers of molecular acids, including picric acid, to dimethylsulphoxide from dimethylformamide or acetonitrile show remarkably uniform medium effects. None of these solvents hydrogen-bonds to anions and, all being polarisable, they give similar dispersion effects. The effect of solvent basicity now becomes noticeable. The more basic character of dimethylsulphoxide makes acids stronger in it than in dimethylformamide or acetonitrile. Since dimethylformamide is somewhat less basic than dimethylsulphoxide, the medium effect between the two is small and negative. Acids are much weaker in acetonitrile because the transfer of a proton alone from this solvent to dimethylsulphoxide has a medium effect $\log({}_{m}\gamma_{H})$ of -8 or -9 units.

In view of the diversity of medium effects between water and dimethylsulphoxide, we must expect any acidity function that is anchored to the pH scale in water to apply only to acids with very similar structures.[F127] Dolman and Stewart[F135] have reported values of the H_- acidity function in dimethylsulphoxide-water mixtures based on 24 substituted aniline and diphenylamine indicators. Dimethylsulphoxide containing 0.4 mol per cent water has a H_- value of 26.2.

3.6.4 *N,N*-Dimethylformamide

The general characteristics of acid-base equilibria in dimethylformamide are similar to those in dimethylsulphoxide. Neither solvent can stabilise anions by hydrogen bonding. Both are more basic than water,[F136–138] although the basicity of dimethylformamide is probably less than that of dimethylsulphoxide. Both have moderately large dielectric constants. Ion-pairing and homoconjugation[F107] will be somewhat more extensive in dimethylformamide but can be minimised by working with low concentrations of acids and salts.[F136] Potentiometry has been the most popular technique for investigating acids in dimethylformamide.[F35,F107,F119,F121,F123,F136,F137,F139–144] Occasional use is made of conductance[F107,F145,F146] and spectrophotometric[F35,F107,F123,137,F147] measurements. The glass and hydrogen electrodes function satisfactorily in this solvent, although not always with the precise Nernstian slope, 59.2 mV at 25°C.[F137,F140] Various standard buffers have been proposed for calibrating the glass electrode.[F107,F121,F142] Kolthoff, Chantooni and Smagowski[F107] chose 4-chloro-2,6-dinitrophenol and 2,6-dinitrophenol as standard acids because they do not undergo homoconjugation and used buffer mixtures of these acids and their tetraethylammonium salts. The acidity constants of the acids were determined by the conductance and spectrophotometric methods. Silver-silver chloride electrodes have been rejected[F96,F139] because of the solubility of silver chloride in dimethylformamide but were found by Petrov and Umanskii[F143,F144] to give steady potentials for a month. The calomel electrode prepared with dimethylformamide is frequently used.[F35,F96,F123,F136,F141,F142] Juillard[F123] has noted variations in liquid junction potential when the calomel electrode is used, especially with tetraethylammonium perchlorate in the salt bridge. The effect of slow deterioration of the calomel electrode[F136] can be minimised by frequent standardisation of the cell or by using the differential titration technique.

A selection of values of pK_a for acids in dimethylformamide at 25°C is given in Appendix 3.6.1, and some values of pK_f for formation of the homoconjugate ions are also included. In most cases, these values are the average of concordant results by different investigators. As noted in the preceding section, medium effects for transfer from dimethylformamide to dimethylsulphoxide are reasonably uniform. For transfer from acetonitrile to dimethylformamide the medium effect is -8.5 ± 0.5 for acids without intramolecular hydrogen bonds and -10 for acids with such bonds. Dimethylformamide, being a much stronger base than acetonitrile, is capable of breaking the intramolecular hydrogen bond.[F107]

Approximate values of ΔH_a^0 and ΔS_a^0 for ionisation of acids in dimethylformamide are given in Appendix 3.6.2. For carboxylic acids[F144] ΔH_a^0 is more positive and ΔS_a^0 is more negative in dimethylformamide than in water. These effects can be attributed in large part to the difference in dielectric constant of the two solvents, so that it is better to compare thermodynamic properties in dimethylformamide with those in methanol. For acetic acid[F144] and the picolinium ion[F136] the entropy change on ionisation in dimethylformamide is more negative by a factor of 2 than that for methanol. Since this medium effect on ΔS_a^0 appears to hold for a molecular acid as well as cation acids, it can be attributed to greater solvation of the proton in dimethylformamide, the more basic solvent. This hypothesis receives support from the observation that the net enthalpy changes for ionisation are less positive in dimethylformamide than in methanol, ΔH^0 for solvation of the proton makes a more negative contribution to ΔH_a^0 in the former solvent.[F136]

Little work has been done on acid-base equilibria in other disubstituted amides. Madic and Trémillon[F148] investigated the behaviour of various acids in hexamethylphosphorotriamide using a voltammetric technique with hydrogen and silver-silver perchlorate electrodes. Solvation effects in this solvent are very specific. Acid-base titrations in tetramethylurea have been reported by Culp and Caruso.[F149,F150]

3.6.5 Acetonitrile

An exceptionally wide range of acid and base strengths can be studied in acetonitrile because its autoprotolysis constant is so low,[F151] less than

10^{-32}. The weakest acid that has been investigated so far is phenol[F152] with $p({}^sK_a) = 26.6$. The solvent anion CH_2CN^- is unstable,[F152,F153] and there is only marginal evidence[F151,F154] for its formation by reaction of bases with the solvent, $B + CH_3CN \leftrightarrows BH^+ + CH_2CN^-$.

Acetonitrile, as noted in sect. 3.6.3, is much less basic than water, dimethylsulphoxide, or dimethylformamide. The medium effect for transfer of the proton to it from water is 4.3 ± 1.5 (Table 3.3.3). In mixtures with methanol or ethanol, acetonitrile effectively behaves as an inert component.[F155] Anions cannot be stabilised by hydrogen-bonding them to acetonitrile and therefore undergo homo- and heteroconjugation.[F156,F157] The data in Table 3.6.2 show that homoconjugation

Table 3.6.2

Homoconjugation Constants*

Acid	Solvent	K_{f1}	K_{f2}	Ref.
Benzoic acid	Acetonitrile	4,000		F118
	Dimethylformamide	250		F107
	Dimethylsulphoxide	60		F109
3,5-Dinitrobenzoic acid	Acetonitrile	10,000		F118
	Dimethylformamide	120		F107, F121
	Dimethylsulphoxide	23		F109
2,4-Dinitrobenzoic acid	Acetonitrile	19,500	7.4	F118
o-Nitrobenzoic acid	Acetonitrile	9,600	13.5	F118
m-Nitrobenzoic acid	Acetonitrile	10,700	37	F118
p-Nitrophenol	Acetonitrile	3,100		F108
	Dimethylformamide	200		F107
	Dimethylsulphoxide	40		F109
3,5-Dinitrophenol	Acetonitrile	44,000		F108
	Dimethylformamide	1,500		F107
	Dimethylsulphoxide	30		F109
o-Nitrophenol	Acetonitrile	130		F108
	Dimethylsulphoxide	60		F109
2,4-Dinitrophenol	Acetonitrile	110		F108
2,6-Dinitrophenol	All 3 solvents	~0		F107, F108, F109
Ethylammonium ion	Acetonitrile	25	2	F158, F159
Diethylammonium ion	Acetonitrile	2		F158, F159
Triethylammonium ion	Acetonitrile	~0		F158, F159

* K_{f1} is for $HA + A^- \leftrightarrows HA_2$; K_{f2} is for $HA + HA_2^- \leftrightarrows (HA)_2A^-$.

occurs to a larger extent in acetonitrile than in the more basic dimethylsulphoxide and dimethylformamide. Homoconjugation is less important for acidic cations (see the data in Table 3.6.2 for ethyl-substituted ammonium ions[F158,F159]) because both acetonitrile and the base B compete for hydrogen bonds with the acid BH^+. Clearly BHB^+ ions will form only when B is fairly strong.

Formation of higher order homoconjugate ions, e.g.

$$HA + HA_2^- \leftrightarrows (HA)_2A^-$$

is not uncommon in acetonitrile; some examples are included in Table 3.6.2.

Considering ion-solvent interactions other than hydrogen bonding, we expect ion-dipole forces to be stronger for cations, which are attracted by the exposed negative end of the CH_3CN dipole, than for anions. Acetonitrile, to judge by its refractive index, is less polarisable than dimethylformamide, more nearly comparable with methanol. Dispersion interactions in acetonitrile will therefore be appreciable but not as prominent as in dimethylformamide.

Although the dielectric constant of acetonitrile is close to those of dimethylformamide and methanol, ion-pairing can be more troublesome in it because the separate ions are less stabilised by solvation. Pairing is also large if the electrostatic attraction is reinforced by hydrogen-bonding.[F117,F132] For example, the dissociation constants for the ion pairs n-$BuNH_3^+ClO_4^-$, $Bu_2NH_2^+ClO_4^-$, and $Bu_3N^+ClO_4^-$ are, respectively, 0.015, 0.014 and 0.0095 mol l^{-1} whereas that for $Et_4N^+ClO_4^-$, in which hydrogen-bonding is absent, is 0.056 mol l^{-1}, about four times as large.[F117] Homoconjugate anions have less concentrated negative charges than simple anions. Thus $Et_3NH^+AHA^-$ ($A^- =$ 3,5-dinitrobenzoate) with a dissociation constant of 0.03 mol l^{-1} is more highly dissociated than $Et_3NH^+A^-$ with a constant of 1.2×10^{-5} mol l^{-1}.

Acetonitrile is difficult to obtain free of troublesome impurities. Mere traces of ammonia and water may give rise to quite erroneous results[F111,F115,F161] unless the solutions are well buffered. It is important to make measurements on freshly prepared solutions because changes in conductance and pH occur as they age.[F111,F153,F162–164] The change in behaviour of perchloric acid solutions, for example, is noticeable within an hour.[F164] Solutions of tetraethylammonium hydroxide in acetonitrile are also unstable.

Conductance measurements on unbuffered solutions are unreliable,[F115] but useful information can be obtained from conductometric titrations.[F108,F116,F117,F165,F166] Solubility[F108,F112,F118] and spectrophotometric[F111] measurements give more precise results. Kolthoff and his collaborators[F167,F168] have made a detailed spectrophotometric study of acid-base indicators in acetonitrile. Indicators with cationic acid forms can give trouble because of formation of ion pairs with anions.[F118] The most satisfactory indicators are ones like 2,6-dinitrophenol, 2,6-di-*t*-butyl-4-nitrophenol, and picric acid which have little tendency to form homo- or heteroconjugate ions because of steric hindrance.[F108]

Potentiometric measurements on buffered solutions in acetonitrile have given some of the best results. Experience with the hydrogen electrode is mixed;[F154,F156,F169,F170] best results were obtained by Kolthoff and Thomas.[F171] Fortunately, the glass electrode works in acetonitrile.[F113,F119,F132,F163,F169,F170] Kolthoff and Chantooni recommended that the glass electrode be calibrated with picric acid-tetrabutylammonium picrate buffers[F172] based on $pK_a = 11.0 \pm 0.1$ for picric acid. These buffer mixtures have the advantage that homoconjugation is negligible. The effect of hydrogen-bond donors and acceptors on the pH of buffers in acetonitrile has been investigated.[F173] The most popular reference electrode[F113,F153,F170,F172] is Pleskov's[F174] silver-0.01M silver nitrate in acetonitrile with a salt bridge solution of 0.1M Et_4NClO_4 in acetonitrile. The conventional aqueous calomel electrode is unsuitable for a number of reasons. Potassium chloride deposits at the junction between the aqueous and acetonitrile phases.[F153] The liquid junction potential between it and a dilute solution in acetonitrile is about 250 mV.[F171] The problem with any junction potential this large is its questionable reproducibility from one test solution to another.

The ionisation of an acid by reaction with solvent may be thought of as occurring in stages. First, the acid molecule forms a hydrogen bond with a solvent molecule

$$\mathrm{H}A + S \leftrightarrows A\mathrm{H}\cdots S$$

We expect this solvation to be weak in acetonitrile. Then the solvated acid ionises

$$A\mathrm{H}\cdots S \leftrightarrows A^-\mathrm{H}S^+$$

and the ion-pair *dissociates*

$$A^-\mathrm{H}S^+ \leftrightarrows A^- + \mathrm{H}S^+$$

Conductance and electromotive force measurements, but not spectrophotometry, will differentiate between A^- and $A^-\mathrm{H}S^+$, so we might expect to get somewhat different values of acidity constants by the different methods. The precision of measurement in acetonitrile is still too low to enable us to test this hypothesis. Kolthoff and Chantooni,[F172] for example, obtained the following pK_a values for salicylic acid, 16.9 by conductance titration, 16.7 by potentiometry, and 16.95 by spectrophotometry. Potentiometric values generally seem to be the most reliable.

Medium effects have already been discussed in sect. 3.6.3. For a majority of substituted ammonium ions the transfer from water to acetonitrile increases pK_a by between 7.2 and 7.9 units.[F159] The scatter in ΔpK_a values is still larger for molecular acids[F108,F118] because anion peculiarities are not levelled by solvation in acetonitrile. Medium effects are usually more uniform for transfers from acetonitrile to dimethylformamide (Table 3.6.1). We expect $\log {}_m\gamma_{HA}$ for this transfer to be negative since dimethylformamide is more basic than acetonitrile, and we expect $\log {}_m\gamma_{A^-}$ to be close to zero since neither solvent is a good solvator of anions. If the $\log ({}_m\gamma_{A^-}/{}_m\gamma_{HA})$ is to be positive, we can set an upper limit to $\log {}_m\gamma_H$. In view of the identity

$$\log {}_m\gamma_H = \Delta pK - \log ({}_m\gamma_{A^-}/{}_m\gamma_{HA})$$

the medium effect on the proton must be more negative than ΔpK itself, which is -10 for several nitrophenols. This agrees with the value of $\log {}_m\gamma_H = -10.95$ obtained by Strehlow's ferrocene-ferricinium electrode measurements for transfer from acetonitrile to dimethylformamide.[F137,F138]

Substituent effects on acid ionisation in acetonitrile can be expressed within a few tenths of a pK unit by linear free energy relationships. For 11 out of 12 substituted benzoic acids the Hammett relationship is[F118]

$$p({}^sK_a) = 20.65 - 2.4\sigma$$

The exception is *p*-hydroxybenzoic acid. For phenols[F157] the corresponding equation is

$$p({}^sK_a) = 27.2 - 4.76\sigma$$

The *o*-nitrophenols do not conform to this relationship because of intramolecular hydrogen bonding. There is also a linear free energy relationship between the $p({}^sK_a)$ values of aliphatic-substituted ammonium ions[F159] and Taft σ^* substituent constants. Other linear free energy relationships hold for hetero- and homoconjugation reactions in acetonitrile.[F118,F157] The ratio of slopes (Hammett ρ values) for ionisation of phenols and benzoic acids is 4.76/2.4 or 2.0. Virtually the same ratio is obtained for heteroconjugation reactions, HA with Cl^- and A^- with p-bromophenol. From this observation, Chantooni and Kolthoff[F118] concluded that the relative hydrogen-bond donating properties of the molecules HA were affected by substituents in the same way as the relative hydrogen-bond accepting properties of the anions A^-.

3.6.6 Nitromethane and Nitrobenzene

The following recital should sound familiar after the discussion of acetonitrile. Nitromethane and nitrobenzene have dielectric constants of 38 and 35 at 25°C. Their molecules have large dipole moments and are polarisable. The solvents have very low basicity and acidity. They are not good at solvating ions. Ion-pairing and homoconjugation are therefore prevalent in these solvents. Very strong acids or very weak bases can be studied in them. Far fewer quantitative investigations of acid-base behaviour have been carried out in them than in acetonitrile. Neither solvent is extensively miscible with water.

Spectrophotometric methods have been most frequently used[F175–179] to investigate acid-base behaviour. Van Looy and Hammett[F178] established an H_0 scale for solutions of sulphuric acid in nitromethane. They treated ion pairs as undissociated in this solvent. From the variation in H_0 with concentration of sulphuric acid they concluded that homoconjugation occurred not to $H_2SO_4.HSO_4^-$ but to $(H_2SO_4)_2.HSO_4^-$. Kolthoff, Stocesocá, and Lee[F176] investigated the reactions of picric acid and some Lewis acids with pyridine and aniline in nitrobenzene. In this solvent, as in acetonitrile, little if any of the homoconjugate of picric acid and picrate is formed.

Potentiometric titrations have been carried out by Hall,[F180] Feakins[F181] and Korolev and Stepanov.[F182] The glass electrode can be used in these solvents. Feakins and his co-workers[F181] prepared their calomel electrode with a saturated solution of potassium chloride in methanol and established a slowly flowing junction between it and the nitrobenzene solution.

Jasiński and Pawlak[F183,F184] showed that the analysis of conductance titration curves developed by Kolthoff and Chantooni[F116] for acetonitrile solutions holds also for nitromethane.

3.6.7 Sulphones

Sulphones, like acetonitrile and nitromethane, have low basicity and poor solvating ability, but they are easier to purify and handle. Sulpholane (tetramethylene sulphone or tetrahydrothiophene-1,1-dioxide) is the most readily available and extensively investigated sulphone, but its melting point is inconveniently high,[F187] 28.45°C. Its 3-methyl and 2,4-dimethyl derivatives melt at 0 and −3.3°C, respectively. Were it more readily available, 3-methylsulpholane, which is liquid at room temperature and easily purified, would be the preferred solvent.[F188] Very weak acids and bases can be investigated because the potential range is unusually large,[F188] namely −800 to +1200 mV. The autoprotolysis constants of the sulphones have not been determined but are evidently very small.

Titration in sulpholanes[F188,F189] have been done with the glass-calomel electrode combination. An appropriate salt bridge solution is tetramethyl- or tetraethylammonium perchlorate in 3-methylsulpholane. The calomel electrode is prepared with the same salt, rather than with potassium chloride, to avoid precipitation of potassium perchlorate.[F188] The titration curves indicate good differentiation between acids. Mixtures of perchloric acid with either nitric or hydrochloric acid are clearly resolved.[F188] Another suitable reference electrode is silver-silver perchlorate (0.01 *M*)-lithium perchlorate (0.1 *M*).[F190]

Spectrophotometry has been used to establish H_0 and H_- acidity function scales in pure sulpholane and sulpholane-water mixtures. Very strongly basic solutions ($H_- = 20$) have been prepared by dissolving phenyltrimethylammonium hydroxide in sulpholane containing 5 mol per cent water.[F191,F192] Solutions of acids in sulpholane have H_0

values[F192,F194] between −1 and −9. At the same concentration of sulphuric acid sulpholane solutions are less acidic than nitromethane solutions.[F193]

Homoconjugation has been observed both in the spectrophotometric[F193] had the potentiometric[F188] work.

3.6.8 Propylene Carbonate

Propylene carbonate (4-methyl-1,3-dioxolan-2-one) is an interesting polar solvent with the highest dielectric constant of any aprotic solvent: 66.1 at 20°C, 61.7 at 30°C.[F195] It is more stable and more easily purified than formamide,[F196] though rapidly decomposed by aqueous alkali.[F197] A steady electromotive force could not be obtained from a hydrogen-silver-silver chloride cell containing 0.02 molal hydrochloric acid in propylene carbonate.[F196] Several other electrodes are suitable for use in it.[F119,F120,F198] No systematic investigation of acids and bases in propylene carbonate has yet been reported.

3.7 HYDROGEN-BONDED SOLVENTS OF LOW DIELECTRIC CONSTANT

In this section we return to solvents capable of donating a hydrogen atom to hydrogen bonds. Solvents of this type with low dielectric constant include ethanol and the higher alcohols, ethylenediamine, and anhydrous acetic acid. Anions are more solvated and stabilised by such hydrogen bond donors than by the dipolar aprotic solvents considered in the preceding section. Homo- and heteroconjugation reactions are less frequently encountered in the hydrogen-bonded solvents. While there is some evidence for homoconjugate phenol-phenolate ions in isopropyl alcohol[F199] and ethylenediamine,[F200] this is exceptional.

Ion-pairing replaces conjugation as the important complication in acid-base behaviour in these solvents of low dielectric constant. In ethanol, which has the largest dielectric constant, 24.55 at 25°C, for this class, the force between two ions some specified distance apart is about three times that in water and twice that in dimethylsulphoxide. In acetic acid, dielectric constant 6.22, the force is four times that in ethanol. From conductance measurements we find no appreciable pairing of sodium and chloride ions in methanol,[F53] pairing with an association constant of 11 kg mol^{-1} in ethanol,[F201] and virtually complete association (association constant greater than 10^5) in acetic acid.[F202] In the alcohols association increases as the temperature rises.[F203,F204] Hydrogen-bonding between the two ions may contribute to the stability of an ion pair, but the effect is less striking than such stabilisation in dipolar aprotic solvents because of competition between solvent and ions for hydrogen bonds and because association at low dielectric constants is already extensive.

Triple ions and higher aggregates are often encountered and give rise to some peculiar effects. Kolthoff and Bruckenstein,[F205] for example, found that if water, a base, is added to a solution of the indicator *p,p'*-dimethylaminoazobenzene (*In*) in acetic acid, the colour change shows that more of the acid form (InH^+) of the indicator is produced. The water initially froms $H_3O^+OAc^-$ pairs which aggregate with InH^+ ions and stabilise them

$$InH^+OAc^- + H_3O^+OAc^- \leftrightarrows \begin{matrix} InH^+OAc^- \\ OAc^-H_3O^+ \end{matrix} \leftrightarrows InH^+OAc^-H_3O^+ + OAc^-$$

or

$$OAc^- \, InH^+OAc^- + H_3O^+$$

All the classical experimental methods have been brought to bear on acid-base reactions in these solvents. There is space to cite only a few. The conductance studies of Shedlovsky and his collaborators[F201,F204,F206,F207] on hydrochloric and acetic acids in ethanol and 1-propanol are models of careful work. Less precise but still useful results have been obtained from conductance titrations in such solvents as *t*-butyl alcohol.[F208]

For electromotive force measurements there are many suitable electrodes, and only the most novel features can be summarised here. The glass electrode is generally usable in these solvents. It has been found that in solutions of hydrochloric acid in anhydrous acetic acid, the glass electrode is subject to acid errors of as much as 70 mV relative to the chloranil electrode, owing to the incorporation of chloride ions in the surface gel layer of the glass.[F209] Since ethylenediamine reacts with calomel but not with mercury (II) chloride, a suitable reference electrode for this solvent can be constructed of a mercury pool in contact with ethylenediamine saturated with respect to both $HgCl_2$ and LiCl.[F210]

Cells without liquid junction have been used infrequently.[F209,F211,F212] Kilpi and his collaborators[F213] did differential titrations of bases in pure acetic acid and acetic acid-water mixtures. Kolthoff and Bruckenstein in a notable series of papers[F3,F202,F214–216] reported, and gave quantitative interpretations of, potentiometric titration curves in anhydrous acetic acid. They also made spectrophotometric measurements.[F3,F202,F205,F215,F216] Indicator measurements have also been reported in ethylenediamine[F200] and *t*-butyl acohol.[F217] Anhydrous acetic acid and ethylenediamine have convenient freezing points for cryoscopic work.[F200,F218,F219]

The experimental behaviour of acids in these solvents has some unusual features. The most striking contrast in behaviour is found in comparing anhydrous acetic acid with water. For example, the colour of a weak indicator base in solutions of perchloric acid does not depend on pH, as it would in aqueous solutions, but on the concentration of undissociated perchloric acid, $c_u = [HClO_4] + [H^+ClO_4^-]$, and on the formation constant of the indicator perchlorate salt $In\,H^+ClO_4^-$.[F3,F216] In aqueous solution, buffer mixtures of a base B and its perchlorate salt at equimolar concentration have a pH equal to $pK_S - pK_b(B)$. This relation is not applicable to buffers in anhydrous acetic acid.[F3,F216] Moreover, dilution of such a buffer with more acetic acid changes its pH but does not affect the colour of an indicator base in the mixture.[F3,F216]

Ion-pairing is responsible for these peculiarities. As noted in sect. 3.6.5, the reaction of an acid with solvent can be considered to occur in a series of steps. In the solvents under consideration here, the first step, solvation of the acid, is probably complete. We shall use the formula $HS \cdots\cdots HX$ or $HS\cdot HX$ to stand for the acid in its solvated condition, however many molecules of solvent HS that may involve. The shift of a proton along the axis of the hydrogen bond leads to ionisation, $HSH \cdots\cdots X \leftrightarrows SH_2^+X^-$ with an *ionisation constant*

$$K_i(HX, SH) = \frac{a_{SH_2^+X^-}}{a_{HS\cdot HX}} \simeq \frac{[SH_2^+X^-]}{[HS\cdot HX]} \tag{3.7.1}$$

Dissociation of the hydrogen-bonded ion-pair $SH_2^+X^- \leftrightarrows SH_2^+ + X^-$ is characterised by a *dissociation constant*

$$K_d(SH_2^+X^-) = \frac{a_{SH_2^+}a_{X^-}}{a_{SH_2^+X^-}} \simeq \frac{[SH_2^+][X^-]}{[SH_2^+X^-]}\,y^2 \tag{3.7.2}$$

Analogous expressions can be written for solutions of bases. Some values of ionisation and dissociation constants in anhydrous acetic acid are given in Table 3.7.1.

For solvents like acetic acid with very low dielectric constants it may be possible to approximate y, the mean activity coefficient of the SH_2^+ and X^- ions by 1. The ionic strength of solutions in such solvents will be low because of the low degree of dissociation. Yet ions, ion-pairs, and molecules may not behave ideally even though the interionic forces are negligible. The test of whether an equilibrium concentration quotient like $[SH_2^+][X^-]/[SH_2^+X^-]$ can be identified with an equilibrium constant is whether its value is independent of concentration changes.

Acid strength in solvents of low dielectric constant is measured by the value of the ionisation constant, for the ionisation step involves breaking the covalent H—X bond. Dissociation of the ion-pair depends primarily on the volumes of the ions, less characteristic and intimate details of acids than those involved in the ionisation step. The strength of an acid will depend on the base with which it reacts, whether the solvent SH or some basic solute B or In, because the degree to which ionisation occurs will vary with the stability of the ion-pair, $SH_2^+X^-$, BH^+X^-, or $In\,H^+X^-$,

Table 3.7.1

Ionisation and Dissociation Constants for Acid-Base Equilibria in Anhydrous Acetic Acid at 25°C[F202,F205]

Acid(HX)	Base(B)	$K_i(BHX)$	$K_d(BH^+X^-)$
Hydrochloric	*p*-Naphtholbenzein	1.3×10^2	3.9×10^{-6}
p-Toluenesulphonic	*p*-Naphtholbenzein	3.7×10^2	4.0×10^{-6}
Acetic	Pyridine	5.4	9.4×10^{-7}
Acetic	*p,p'*-Dimethylamino-azobenzene	0.10	5.0×10^{-6}

that is formed. *There is no unique scale of acid strengths in solvents of low dielectric constant.* Consider, for example, the data in Table 3.7.2. Urea is a weaker base than *p*-naphtholbenzein when they both react with perchloric acid but a stronger base when they react with *p*-toluene-sulphonic acid or hydrochloric acid.

Table 3.7.2

Ionisation Constants in Anhydrous Acetic Acid[F202]

Acid(HX)	Base(B)	$K_i(B.HX)$
Perchloric	*p*-Naphtholbenzein	2.0×10^5
	Urea	1.6×10^5
p-Toluenesulphonic	*p*-Naphtholbenzein	3.7×10^2
	Urea	1.2×10^3
Hydrochloric	*p*-Naphtholbenzein	1.3×10^2
	Urea	4.9×10^2

Because of ion-pairing, medium effects are not uniform for transfer of acid-base equilibria from water or methanol to hydrogen-bonded solvents of low dielectric constant. Nor are acidity functions, as Hammett clearly recognised, applicable to such solvents.[F220]

The direct result of experimental measurements is an acidity constant (K_a) or 'over-all dissociation constant'[F3] that is a combination of K_i and K_d. Conductance measurements do not distinguish between $HS \cdot HX$ and $SH_2^+X^-$ because they detect only free ions. Electromotive force measurements likewise give information about the free ions and do not distinguish between the solvated molecular acid and the ion-pair. It is therefore convenient to define the *acidity constant* as follows

$$K_a = \frac{a_{SH_2^+}a_{X^-}}{a_{HS\cdot HX} + a_{SH_2^+X^-}} = \frac{K_iK_d}{1 + K_i} \tag{3.7.3}$$

For solutions of a weak acid at stoichiometric concentration c_a this can be approximated by the familiar formula

$$K_a = \frac{[H^+]^2y^2}{c_a - [H^+]} \tag{3.7.4}$$

Strong acids have ionisation constants greater than unity. Their acidity constants are virtually equal to K_d. For very weak acids $K_i \ll 1$ and eqn. 3.7.3 reduces to $K_a = K_iK_d$. Thus, if we compare two weak acids HX and HY, the ratio of their acidity constants is not identical with their relative strengths

$$\frac{K_i(HX)}{K_i(HY)} = \frac{K_a(HX)}{K_a(HY)} \cdot \frac{K_d(HY)}{K_d(HX)} \tag{3.7.5}$$

Spectrophotometric measurements generally do not distinguish between ion-pairs ($In\,H^+X^-$) or higher aggregates and free ions ($In\,H^+$). The relation between such measurements and K_a is therefore less straightforward than that for conductance or pH measurements.[F202,F216,F221]

3.8 NON-HYDROGEN-BONDED SOLVENTS OF LOW DIELECTRIC CONSTANT

This last group of solvents includes substances as diverse as acetone and benzene. Acid-base reactions in such solvents are complicated by extensive ion-pairing and by formation of other ionic and molecular aggregates. In acetone, which has a dielectric constant of 20.7 at 25°C, sodium perchlorate at 0.01M concentration is 80% associated to ion-pairs whereas the degree of association is only 31% in nitromethane, 22% in acetonitrile, and 4% in sulpholane.[F222] In other respects acetone behaves like the dipolar aprotic solvents discussed in sect. 3.6. Jasiński and Pawlak,[F223,F224] for example, showed that conductance titrations in acetone could be treated quantitatively in the same way as those in acetonitrile.[F116] The familiar potentiometric, conductometric, and spectrophotometric methods are applicable to the ionisation of anilines in acetone and acetone-water mixtures.[F225]

Acid-base reactions in benzene and other solvents of very low dielectric constant are worth considering in more detail because of their greater complexity. Only a short summary is attempted here, for readers can turn to a lengthy review by Davis[F226] for more details. None of the solvents is sufficiently basic to remove protons from acids, so that acid-base behaviour in them is confined to association reactions between acidic and basic solutes, of which the simplest is

$$B + \mathrm{H}X \leftrightarrows B\mathrm{H}X$$

Association usually involves ionisation to produce a hydrogen-bonded ion-pair $B\mathrm{H}^+X^-$, but dissociation of the pair to free ions can be neglected in benzene. A number of combinations are known to give, not an ion-pair, but a hydrogen-bonded aggregate $B \cdots\cdots \mathrm{H}X$.[F227–230] For that reason it seems preferable to refer to the quotient

$$K(B\mathrm{H}X) = \frac{a_{\mathrm{BHX}}}{a_{\mathrm{B}}a_{\mathrm{HX}}} \simeq \frac{[B\mathrm{H}X]}{[B][\mathrm{H}X]} \quad (3.8.1)$$

as an *association constant* rather than an ionisation constant.[F226]

Useful exploratory studies of acid-base behaviour in solvents of low dielectric constant have been made by conductance[F231–234] and potentiometric[F231,F233,F235–237] titrations. Association constants are usually obtained from spectrophotometric measurements. The strengths of various bases can be compared by means of their association with an indicator acid like 2,4-dinitrophenol.[F241] If both acid and base are colourless, a competition for the base can be established between the acid and an indicator acid like bromophthalein magenta E.[F238,F242]

In solvents like benzene, other reactions than simple 1:1 association between B and $\mathrm{H}X$ may occur. Self-association of the acid or base is one such auxiliary reaction.[F226] A classic example is the dimerisation of carboxylic acids in benzene. If allowance is not made for this, constant values of the quotient $[B\mathrm{H}X]/[B][\mathrm{H}X]$ will not be obtained. (Variations in the quotient cannot be attributed to interionic forces or other non-ideal behaviour: $B\mathrm{H}X$ is scarcely dissociated into ions at all and in spectrophotometric work very low concentrations of B and $\mathrm{H}X$ can be used.) Evidence for association ratios other than 1:1 can be obtained from indicator studies. The method developed by Kolthoff and Bruckenstein[F3,F202,F205] for studying reactions in anhydrous acetic acid fails for reactions in benzene and similar solvents[F247] because more than one acid molecule reacts with the indicator to give complexes of the form $In\,\mathrm{H}^+X^-(\mathrm{H}X)_n$. In such studies it is generally a good approximation to assume that the molar absorptivity of $In\,\mathrm{H}^+$ is independent of the anion, whether X^-, $X\mathrm{H}X^-$, or $X^-(\mathrm{H}X)_n$.[F248] Phenolic indicators like bromophthalein magenta E, on the other hand, give anions capable of a wide range of colours depending on the cation with which they are paired.[F226] Davis has given an extensive review[F226] of the elucidation of the structures of homo- and heteroconjugate complexes.

In addition to ultraviolet and visible spectrophotometry, various other methods can be used to study the complexes: cryoscopy,[F249] differential vapour pressure measurements,[F228,F250–252] and infrared absorption spectrometry.[F227,F251–256]

The association constant measures acid or base strength, but because of ion association there is no unique scale in solvents of low dielectric constant (sect. 3.7). Nor are satisfactory acidity functions possible in these solvents.[F257] Different acids can be compared only with respect to their reaction with some chosen base. The stronger the proton acceptor or donor, the larger is the association constant and the shorter the $B^{+} \cdots\cdots X^{-}$ distance in the ion-pair. We therefore expect a loss in enthalpy by association which will be larger the stronger the proton donor. In forming the pair, three translational and three rotational degrees of freedom are changed to one $B^{+} \cdots\cdots X^{-}$ stretching and five deformation degrees of freedom, all low frequency modes that make large contributions to the entropy of association. The greater the strength of the $B^{+} \cdots\cdots X^{-}$ bond, the more negative the entropy of association.[F258] There is thus a correlation between ΔH and $T\Delta S$ for association. Davis and Hetzer[F259] found that at 25°C $\Delta H/T\Delta S$ for reactions in benzene was approximately constant and equal to 1.8 when B was a nitrogen base or 1.4 when it was an oxygen base

The ratio of the association constant of a given acid to that of some standard acid of the same charge and chemical type is roughly independent of the solvent.[F237,F260] If we compare phenols with carboxylic acids or primary with tertiary amines, we must expect the ratio to vary because of specific solvation effects. The acidity of carboxylic acids, for example, is enhanced when both oxygen atoms of the carboxylate group can form hydrogen bonds with the reference base or solvent.[F261] Benzene and other aromatic solvents can solvate acids by π electron donation. The association of phenols with bases is less in benzene than in heptane or cyclohexane because the π electrons of benzene attract the phenolic –OH group, stabilising the unassociated phenol.[F262] Chlorobenzene is less polarisable and a poorer π electron donor than benzene whereas toluene is a better donor. We therefore find that the association constants for reaction between 2,4-lutidine and bromphenol blue fall in the order, 438 l mol^{-1} in chlorobenzene, 179 in benzene, and 113 in toluene.[F243] Chlorobenzene solvates and stabilises the acid bromphenol blue least, so the association of it with the base occurs to the largest extent in this solvent. The base strength of amines vary from solvent to solvent in a complex manner that seems to depend on the solvating ability of the solvent.[F241,F263]

Because dissociation of the ion-pair BHX is hardly detectable in benzene and virtually complete in water, the difference in log K values for acid-base equilibria in the two solvents is not a simple medium effect but includes also the formation of BH^+X^- from the ions

$$\log {}_bK(BHX) = \log \frac{{}^wK_a(HX)}{{}^wK_a(BH^+)} - \log {}_bK_f(BH^+X^-) + \log \frac{({}_my_{BH^+})({}_my_{X^-})}{({}_my_B)({}_my_{HX})} \quad (3.8.2)$$

If ion-size were the principal effect determining stability of the ion pair or the medium effects for the ions, we might expect for a series of similar acids HX_j that a plot of $\log {}_bK(BHX_j)$ against $p({}^wK_a(HX_j))$ would give a straight line with slope equal to -1.[F264] Remarkably good linear relationships are found,[F242,F262,F265] but the slopes are less than -1. For *meta*-substituted benzoic acids, reacting in benzene with the base 1,3-diphenylguanidine, the linear relationship is

$$\log {}_bK(BHX) = 14.37 - 2.17\, p({}^wK_a)$$

Ortho and *para*-substituted acids give different straight lines owing to chelation and steric hindrance. To account for slopes less than -1, we must assume that the formation constant of the ion pair, the medium effects, or both depend on the strength of the acid. Mead[F266] found linear correlations between heats of neutralisation of amines with trichloroacetic acid in benzene and base strength in water; primary, secondary and tertiary amines fell on parallel lines.

REFERENCES

F1 R. G. Bates, *in* Hydrogen-Bonded Solvent Systems, pp. 49–86 (A.K. Covington and P. Jones, eds.), Taylor and Francis Ltd., London (1968)

F2 R. G. Bates *in* Solute-Solvent Interactions, Chaper 2 (J. F. Coetzee and C. D. Ritchie, eds.) Marcel Dekker, New York (1969); R. G. Bates, *J. Electroanal. Chem.*, **29**, 1 (1971)

F3 I. M. Kolthoff and P. J. Elving (eds.), Treatise on Analytical Chemistry, Academic Press, New York (1959)

F4 C. L. Wilson and D. W. Wilson (eds.), Comprehensive Analytical Chemistry, Vol. IB, Elsevier, Amsterdam (1960)

F5 J. S. Fritz, Acid-Base Titrations in Nonaqueous Solvents, G. Frederick Smith Chemical Company, Columbus, Ohio (1952)

F6 I. Gyenes, Titration in Non-Aqueous Media, Van Nostrand, Princeton, N.J. (1967)

F7 W. Huber, Titration in Non-Aqueous Solvents, Academic Press, New York (1967)

F8 E. J. King, Acid-Base Equilibria, Pergamon Press, Oxford (1965)

F9 V. I. Vedeneyev, L. V. Gurvich, V. N. Kondrat'yev, V. A. Medvedev and Ye. L. Frankevich, Bond Energies, Ionization Potentials and Electron Affinities, Edward Arnold Ltd., London (1966)

F10 M. DePaz, J. J. Leventhal and L. Friedman, *J. Chem. Phys.*, **51,** 3748 (1969)

F11 J. I. Brauman and L. K. Blair, *J. Amer. Chem. Soc.*, **90,** 5636, 6561 (1968); **91,** 2126 (1969)

F12 F. Franks and D. J. G. Ives, *Quart. Rev. (London)*, **20,** 1 (1966)

F13 J. Hine and M. Hine, *J. Amer. Chem. Soc.*, **74,** 5266 (1952); P. Ballinger and F. A. Long, *J. Amer. Chem. Soc.*, **82,** 795 (1960)

F14 E. M. Arnett, *Progr. Phys. Org. Chem.*, **1,** 255–8 (1963)

F15 C. Agami, *Bull. Soc. Chim. Fr.*, 2183 (1969)

F16 C. Agami and M. Caillot, *Bull. Soc. Chim. Fr.*, 1990 (1969)

F17 R. G. Bates, Determination of pH, Chapter 6, John Wiley and Sons, New York (1964)

F18 Ref. F8, Chapter 10

F19 A. J. Parker and R. Alexander, *J. Amer. Chem. Soc.*, **90,** 3313 (1968).

F20 O. Popovych and A. J. Dill, *Anal. Chem.*, **41,** 456 (1969)

F21 Ref. F8, Chapter 7, 8, 10 and 11

F22 L. Hepler, *Austr. J. Chem.*, **17,** 587 (1964)

F23 N. Bout and J. Potier, *Rev. Chim. Miner.*, **4,** 621 (1967)

F24 Ref. F8, Chapter 11

F25 S. K. Hall and E. A. Robinson, *Can. J. Chem.*, **42,** 1113 (1964)

F26 I. M. Kolthoff, S. Bruckenstein and M. K. Chantooni, Jr., *J. Amer. Chem. Soc.*, **83,** 3927 (1961)

F27 J. N. Butler, Ionic Equilibrium. A Mathematical Approach, pp. 158–161, Addison-Wesley Publishing Company, Reading, Mass. (1964)

F28 Ref. F8, Chapters 10 and 11

F29 Ref. F8, pp. 282–284

F30 R. Reynaud, *Bull. Soc. Chim. Fr.*, 4605 (1967)

F31 N. A. Izmailov and V. S. Chernyi, *Zh. Fiz. Khim.*, **34,** 319 (1960)

F32 O. M. Konovalov, *Zh. Fiz. Khim.*, **39,** 693 (1965)

F33 R. Reynaud, *Bull. Soc. Chim. Fr.*, 3945 (1968)

F34 J. Juillard, *Bull. Soc. Chim. Fr.*, 1894 (1968)

F35 B. W. Clare, D. Cook, E. C. F. Ko, Y. C. Mac and A. J. Parker, *J. Amer. Chem. Soc.*, **88,** 1911 (1966)

F36 J. F. Coetzee and J. J. Campion, *J. Amer. Chem. Soc.*, **89,** 2517 (1967)

F37 L. P. Hammett, Physical Organic Chemistry, 2nd Edn., pp. 232–235, McGraw Hill, N.Y. (1970)

F38 E. Grunwald and E. Price, *J. Amer. Chem. Soc.*, **86,** 4517 (1964)

F39 D.-W. Fong and E. Grunwald, *J. Phys. Chem.*, **73,** 3909 (1969)

F40 L. P. Hammett, Physical Organic Chemistry, 2nd ed., p. 223, McGraw-Hill Book Co., New York, N.Y. (1970)

F41 See, for example, G. Charlot and B. Trémillon, Chemical Reactions in Solvents and Melts, Pergamon Press, Oxford (1969)

F42 M. Paul and F. A. Long, *Chem. Rev.*, **57,** 1 (1957)

F43 R. H. Boyd *in* Solute-Solvent Interactions, Chapter 3 (J. F. Coetzee and C. D. Ritchie, eds.), Marcel Dekker, New York, N.Y. (1969)

F44 L. P. Hammett, Physical Organic Chemistry, 2nd ed., Chapter 9, McGraw-Hill Book Co., New York, N.Y. (1970)

F45 Ref. F8, Chapter 12

F46 K. Bowden, *Chem. Rev.*, **66,** 119 (1966)

F47 C. H. Rochester, *Quart. Rev. (London)*, **20,** 511 (1966)

F48 H.-M. Koepp, H. Wendt and H. Strehlow, *Z. Elektrochem.*, **64,** 483 (1960); H. Strehlow and H. Wendt, *Z. Phys. Chem. (Frankfurt)*, **30,** 141 (1961)

F49 J. Vedel, *Ann. Chim. (Paris)*, [14], **2,** 335 (1967)

F50 A. Katchalsky, H. Eisenberg and S. Lifson, *J. Amer. Chem. Soc.*, **73,** 5889 (1951)

F51 A. Albert and E. P. Serjeant, Ionization Constants of Acids and Bases, Methuen, London (1962)
F52 Ref. F8, Chapters 2–6
F53 T. Shedlovsky and R. L. Kay, *J. Phys. Chem.*, **60,** 151 (1956)
F54 T. Shedlovsky, *J. Franklin Inst.*, **225,** 739 (1938)
F55 G. Kortüm and H. Wenck, *Ber. Bunsenges. Phys. Chem.*, **70,** 435 (1966)
F56 I. D. Tabagua, *Tr. Sukhumsk. Gos. Ped. Inst.*, **15,** 119 (1962); *Chem. Abstr.*, **60,** 14373d (1964)
F57 E. Grunwald, *J. Amer. Chem. Soc.*, **73,** 4934 (1951)
F58 C. D. Ritchie and P. D. Heffley, *J. Amer. Chem. Soc.*, **87,** 5402 (1965)
F59 C. D. Ritchie and G. H. Megerle, *J. Amer. Chem. Soc.*, **89,** 1452 (1967)
F60 For example, J. Juillard, *Bull. Soc. Chim. Fr.*, 1727 (1966)
F61 A. E. Bottom and A. K. Covington, *J. Electroanal. Chem.*, **24,** 251 (1970)
F62 K. C. Ong, R. A. Robinson, and R. G. Bates, *Anal. Chem.*, **36,** 1971 (1964)
F63 C. L. de Ligny, P. F. M. Luykx, M. Rehbach and A. A. Wienecke, *Rec. Trav. Chim.*, **79,** 699, 713 (1960)
F64 R. G. Bates, Determination of pH, John Wiley and Sons, New York, N.Y. (1964)
F65 J. Juillard and M.-L. Dondon, *Bull. Soc. Chim. Fr.*, 2535 (1963)
F66 C. D. Rochester and B. Rossall, *J. Chem. Soc. B*, 743 (1967)
F67 R. S. Stearns and G. W. Wheland, *J. Amer. Chem. Soc.*, **69,** 2025 (1947)
F68 B. D. England and D. A. House, *J. Chem. Soc.*, 4421 (1962)
F69 C. H. Rochester, *J. Chem. Soc.*, 676 (1965)
F70 C. H. Rochester and B. Rossall, *Trans. Faraday Soc.*, **65,** 1004 (1969)
F71 P. D. Bolton, C. H. Rochester and B. Rossall, *Trans. Faraday Soc.*, **66,** 1348 (1970)
F72 R. Schaal and G. Lambert, *J. Chim. Phys.*, 1164 (1962)
F73 R. A. More O'Ferrall and J. H. Ridd, *J. Chem. Soc.*, 5030 (1963)
F74 C. H. Rochester, *J. Chem. Soc. B*, 121 (1966)
F75 F. Terrier and R. Schaal, *Compt. Rend.*, *C*, **263,** 476 (1966)
F76 F. Terrier, *Ann. Chim.* (*Paris*), **4,** 153 (1969)
F77 F. Terrier, *Bull. Soc. Chim. Fr.*, 1894 (1969)
F78 H. Schmid, A. Maschka and W. Melhardt, *Monatsh. Chem.*, **99,** 443 (1968).
F79 Ref. F8, Chapter 9
F80 E. C. W. Clarke and D. N. Glew, *Trans. Faraday Soc.*, **62,** 539 (1966)
F81 J. W. Larson and L. G. Hepler *in* Solute-Solvent Interactions, Chapter 1 (J. F. Coetzee and C. D. Ritchie, eds.), Marcel Dekker, New York, N.Y. (1969)
F82 J. H. Wolfenden, W. Jackson and H. B. Hartley, *J. Phys. Chem.*, **31,** 850 (1927)
F83 C. S. Leung and E. Grunwald, *J. Amer. Chem. Soc.*, **74,** 696 (1970)
F84 S. D. Hamann and S. C. Lim, *Australian J. Chem.*, **7,** 329 (1954)
F85 C. H. Rochester and B. Rossall, *Trans. Faraday Soc.*, **65,** 992 (1969)
F86 F. J. Millero, *J. Phys. Chem.*, **73,** 2417 (1969)
F87 R. Stewart and T. Mathews, *Can. J. Chem.*, **38,** 602 (1960)
F88 T. C. Wehman and A. I. Popov, *J. Phys. Chem.*, **72,** 4031 (1968)
F89 A. I. Popov and J. C. Marshall, *J. Inorg. Nucl. Chem.*, **19,** 340 (1961); **24,** 1667 (1962)
F90 A. M. Shkodin, N. A. Izmailov and N. P. Dzuba, *Zh. Obshch. Khim.*, **20,** 1999 (1950); **23,** 27 (1953); *Zh. Anal. Khim.*, **6,** 273 (1951)
F91 L. M. Mukherjee, *J. Amer. Chem. Soc.*, **79,** 4040 (1957)
F92 F. Masure and R. Schaal, *Bull. Soc. Chim. Fr.*, 1138 (1956)
F93 F. Masure, R. Schaal and P. Souchay, *Bull. Soc. Chim. Fr.*, 1143 (1956)
F94 S. Guiot and B. Trémillon, *J. Electroanal. Chem.*, **18,** 216 (1968)

F95 L. R. Dawson, J. W. Vaughn, M. E. Pruitt and H. C. Eckstrom, *J. Phys. Chem.*, **66**, 2684 (1962)
F96 D. S. Reid and C. A. Vincent, *J. Electroanal. Chem.*, **18**, 427 (1968)
F97 R. K. Agarwal and B. Nayak, *J. Phys. Chem.*, **70**, 2568 (1966)
F98 R. Reynaud, *Compt. Rend. C.*, **266**, 1623 (1968)
F99 M. de Rossi, G. Pecci and B. Scrosati, *Ric. Sci.*, **37**, 342 (1967)
F100 M. Mandel and P. Decroly, *Trans. Faraday Soc.*, **56**, 29 (1960); *in* Electrolytes, pp. 176–186 (B. Pesce, ed.), Pergamon Press, Oxford (1962); *Nature*, **201**, 290 (1964)
F101 R. W. Broadbank, S. Dhabanandana, K. W. Morcom and B. L. Muju, *Trans. Faraday Soc.*, **64**, 3311 (1968)
F102 R. K. Agarwal and B. Nayak, *Indian J. Chem.*, **7**, 1268 (1969)
F103 G. Jander and G. W. Winckler, *J. Inorg. Nucl. Chem.*, **9**, 24, 32 (1959)
F104 O. D. Bonner, C. F. Jordan and K. W. Bunzl, *J. Phys. Chem.*, **68**, 2450 (1964)
F105 R. Stewart and J. R. Jones, *J. Amer. Chem. Soc.*, **89**, 5069 (1967)
F106 R. C. Kerber and A. Porter, *J. Amer. Chem. Soc.*, **91**, 366 (1969)
F107 I. M. Kolthoff, M. K. Chantooni, Jr. and H. Smagowski, *Anal. Chem.*, **42**, 1622 (1970)
F108 I. M. Kolthoff, M. K. Chantooni, Jr. and S. Bhowmik, *J. Amer. Chem. Soc.*, **88**, 5430 (1966)
F109 I. M. Kolthoff, M. K. Chantooni, Jr. and S. Bhowmik, *J. Amer. Chem. Soc.*, **90**, 23 (1968)
F110 C. M. French and I. G. Roe, *Trans. Faraday Soc.*, **49**, 314 (1953)
F111 M. K. Chantooni, Jr. and I. M. Kolthoff, *J. Amer. Chem. Soc.*, **85**, 2195 (1963)
F112 I. M. Kolthoff and M. K. Chantooni, Jr., *J. Phys. Chem.*, **66**, 1675 (1962)
F113 I. M. Kolthoff and M. K. Chantooni, Jr., *J. Amer. Chem. Soc.*, **87**, 4428 (1965)
F114 J. O. Frohliger, J. E. Dziedzic and O. W. Steward, *Anal. Chem.*, **42**, 1189 (1970)
F115 A. D'Aprano and R. M. Fuoss *J. Phys. Chem.* **73**, 223 (1969)
F116 I. M. Kolthoff and M. K. Chantooni, Jr., *J. Amer. Chem. Soc.*, **87**, 1004 (1965)
F117 J. F. Coetzee and G. P. Cunningham, *J. Amer. Chem. Soc.*, **87**, 2534 (1965)
F118 M. K. Chantooni, Jr. and I. M. Kolthoff, *J. Amer. Chem. Soc.*, **92**, 7025 (1970)
F119 J. N. Butler *in* Adv. Electrochem. and Electrochem. Eng. (P. Delahay, ed.), **7**, 77 (1969)
F120 J. C. Synnott and J. N. Butler, *Anal. Chem.*, **41**, 1890 (1969)
F121 J. Juillard, *J. Chim. Phys.*, **67**, 691 (1970)
F122 J. Badoz-Lambling, J. Desbarres and J. Tacussel, *Bull. Soc. Chim. Fr.*, 53 (1962)
F123 J. Juillard, *J. Chim. Phys.*, **63**, 1190 (1966)
F124 I. M. Kolthoff and M. K. Chantooni, Jr., *J. Amer. Chem. Soc.*, **90**, 3005 (1968)
F125 I. M. Kolthoff and T. B. Reddy, *Inorg. Chem.*, **1**, 189 (1962)
F126 J. Courtot-Coupez and M. LeDemezet, *Compt. Rend. C.*, **266**, 1438 (1968)
F127 C. D. Ritchie and R. E. Uschold, *J. Amer. Chem. Soc.*, **89**, 2752 (1967)
F128 E. C. Steiner and J. M. Gilbert, *J. Amer. Chem. Soc.*, **85**, 3054 (1963)
F129 E. C. Steiner and J. M. Gilbert, *J. Amer. Chem. Soc.*, **87**, 382 (1965)
F130 C. D. Ritchie and R. E. Uschold, *J. Amer. Chem. Soc.*, **89**, 1721 (1967)
F131 C. D. Ritchie and R. E. Uschold, *J. Amer. Chem. Soc.*, **90**, 2821 (1968)
F132 I. M. Kolthoff and M. K. Chantooni, Jr., *J. Amer. Chem. Soc.*, **90**, 5961 (1968)
F133 K. K. Barnes and C. K. Mann, *Anal. Chem.*, **36**, 2502 (1964)
F134 C. D. Ritchie, *J. Amer. Chem. Soc.*, **91**, 6749 (1969)

F135 D. Dolman and R. Stewart, *Can. J. Chem.*, **45**, 911 (1967)
F136 C. D. Ritchie and G. H. Megerle, *J. Amer. Chem. Soc.*, **89**, 1447 (1967)
F137 G. Demange-Guérin, *Talanta*, **17**, 1075 (1970)
F138 G. Demange-Guérin, *Talanta*, **17**, 1099 (1970)
F139 M. Tézé and R. Schaal, *Bull. Soc. Chim. Fr.*, 1372 (1962)
F140 G. Demange-Guérin and J. Badoz-Lambling, *Bull. Soc. Chim. Fr.*, 3277 (1964)
F141 J. Juillard, *Compt. Rend.*, **260**, 1923 (1965)
F142 J. Juillard and B. Loubinoux, *Compt. Rend. C.*, **264**, 1680 (1967)
F143 S. M. Petrov and Yu. I. Umanskii, *Zh. Fiz. Khim.*, **41**, 1374 (1967)
F144 S. M. Petrov and Yu. I. Umanskii, *Zh. Fiz. Khim.*, **42**, 3052 (1968)
F145 A. B. Thomas and E. G. Rochow, *J. Amer. Chem. Soc.*, **79**, 1843 (1957)
F146 P. G. Sears, R. K. Wolford and L. R. Dawson, *J. Electrochem. Soc.*, **103**, 633 (1956)
F147 V. I. Slovetskii, A. I. Ivanov, S. A. Shevelev, A. A. Fainzil'berg and S. S. Novikov, *Zh. Fiz. Khim.*, **41**, 834 (1967)
F148 C. Madic and B. Trémillon, *Bull. Soc. Chim. Fr.*, 1634 (1968)
F149 S. Culp and J. A. Caruso, *Anal. Chem.*, **41**, 1329 (1969)
F150 S. Culp and J. A. Caruso, *Anal. Chem.*, **41**, 1876 (1969)
F151 I. M. Kolthoff and M. K. Chantooni, Jr., *J. Phys. Chem.*, **72**, 2270 (1968)
F152 J. F. Coetzee and G. R. Padmanabhan, *J. Phys. Chem.*, **69**, 3193 (1965)
F153 J. F. Coetzee and G. R. Padmanabhan, *J. Phys. Chem.*, **66**, 1708 (1962)
F154 W. S. Muney and J. F. Coetzee, *J. Phys. Chem.*, **66**, 89 (1962)
F155 A. D'Aprano and R. M. Fuoss, *J. Phys. Chem.*, **73**, 400 (1969)
F156 J. F. Coetzee, *Progr. Phys. Org. Chem.*, **4**, 45 (1967)
F157 I. M. Kolthoff and M. K. Chantooni, Jr., *J. Amer. Chem. Soc.*, **91**, 4621 (1969)
F158 J. F. Coetzee, G. R. Padmanabhan and G. P. Cunningham, *Talanta*, **11**, 93 (1964)
F159 J. F. Coetzee and G. R. Padmanabhan, *J. Amer. Chem. Soc.*, **87**, 5005 (1965)
F160 J. Pople and M. Gordon, *J. Amer. Chem. Soc.*, **89**, 4253 (1967)
F161 J. F. Coetzee and D. K. McGuire, *J. Phys. Chem.*, **67**, 1810 (1963)
F162 G. J. Janz and S. S. Danyluk, *J. Amer. Chem. Soc.*, **81**, 3846 (1959)
F163 I. M. Kolthoff and S. Ikeda, *J. Phys. Chem.*, **65**, 1020 (1961)
F164 I. M. Kolthoff M. K. Chantooni, Jr. and S. Bhowmik, *Anal. Chem.*, **39**, 1627 (1967)
F165 P. J. R. Bryant and A. W. H. Wardrop, *J. Chem. Soc.*, 895 (1957)
F166 I. M. Kolthoff and M. K. Chantooni, Jr., *J. Amer. Chem. Soc.*, **85**, 426 (1963)
F167 I. M. Kolthoff, S. Bhowmik and M. K. Chantooni, Jr., *Proc. Nat. Acad. Sci. U.S.*, **56**, 1370 (1966)
F168 I. M. Kolthoff, M. K. Chantooni, Jr. and S. Bhowmik, *Anal. Chem.*, **39**, 315 (1967)
F169 E. Römberg and K. Cruse, *Z. Elektrochem.*, **63**, 404 (1959)
F170 J. Desbarres, *Bull. Soc. Chim. Fr.*, 2103 (1962)
F171 I. M. Kolthoff and F. G. Thomas, *J. Phys. Chem.*, **69**, 3049 (1965)
F172 I. M. Kolthoff and M. K. Chantooni, Jr., *J. Phys. Chem.*, **70**, 856 (1966)
F173 I. M. Kolthoff and M. K. Chantooni, Jr., *Anal. Chem.*, **39**, 1080 (1967)
F174 V. A. Pleskov, *Zh. Fiz. Khim.*, **22**, 351 (1948)
F175 L. C. Smith and L. P. Hammett, *J. Amer. Chem. Soc.*, **67**, 23 (1945)
F176 I. M. Kolthoff, D. Stŏcesocá and T. S. Lee, *J. Amer. Chem. Soc.*, **75**, 1834 (1953)
F177 S. Aronoff, *J. Phys. Chem.*, **62**, 428 (1958)
F178 H. van Looy and L. P. Hammett, *J. Amer. Chem. Soc.*, **81**, 3872 (1959)
F179 W. Rodziewicz and H. Smagowski, *Rocz. Chem.*, **40**, 511 (1966)
F180 H. K. Hall, Jr, *J. Phys. Chem.*, **60**, 63 (1956)
F181 D. Feakins, W. A. Last and R. A. Shaw, *J. Chem. Soc.*, 2387 (1964)

F182 B. A. Korolev and B. I. Stepanov, *Izv. Vyssh. Ucheb. Zaved., Khim. Khim. Tekhnol.*, **11,** 1193 (1968); *C. A.*, **70,** 74125m
F183 T. Jasiński and Z. Pawlak, *Rocz. Chem.*, **43,** 605 (1969)
F184 T. Jasiński and Z. Pawlak, *Rocz. Chem.*, **43,** 133 (1969)
F185 R. L. Benoit and G. Choux, *Can. J. Chem.*, **46,** 3215 (1968)
F186 F. G. Bordwell, R. H. Imes and E. C. Steiner, *J. Amer. Chem. Soc.*, **89,** 3905 (1967)
F187 M. Della Monica, L. Jannelli and U. Lamanna, *J. Phys. Chem.*, **72,** 1068 (1968)
F188 D. H. Morman and G. A. Harlow, *Anal. Chem.*, **39,** 1869 (1967)
F189 A. P. Zipp, *Anal. Chem.*, **42,** 943 (1970)
F190 J. Desbarres, P. Pichet and R. L. Benoit, *Electrochim. Acta*, **13,** 1899 (1968)
F191 C. H. Langford and R. L. Burwell, Jr., *J. Amer. Chem. Soc.*, **82,** 1503 (1960)
F192 R. Stewart and J. P. O'Donnell, *J. Amer. Chem. Soc.*, **84,** 493 (1962)
F193 E. M. Arnett and C. F. Douty, *J. Amer. Chem. Soc.*, **86,** 409 (1964)
F194 R. W. Alder, G. R. Chalkley and M. C. Whiting, *Chem. Comm.*, 405 (1966)
F195 L. Simeral and R. L. Amey, *J. Phys. Chem.*, **74,** 1443 (1970)
F196 B. R. Staples, NBS Tech. Note 453 (1968)
F197 A. H. Saadi and W. H. Lee, *J. Chem. Soc. (B)*, 1 (1966)
F198 E. Kirowa-Eisner and E. Gileadi, *J. Electroanal. Chem.*, **25,** 481 (1970)
F199 L. Hummelstedt and D. N. Hume, *Anal. Chem.*, **32,** 1792 (1960)
F200 S. Bruckenstein and L. M. Mukherjee, *J. Phys. Chem.*, **66,** 2228 (1962)
F201 H. O. Spivey and T. Shedlovsky, *J. Phys. Chem.*, **71,** 2165 (1967)
F202 I. M. Kolthoff and S. Bruckenstein, *J. Amer. Chem. Soc.*, **78,** 1 (1956)
F203 M. Goffredi and T. Shedlovsky, *J. Phys. Chem.*, **71,** 2176 (1967)
F204 M. Goffredi and T. Shedlovsky, *J. Phys. Chem.*, **71,** 2182 (1967)
F205 S. Bruckenstein and I. M. Kolthoff, *J. Amer. Chem. Soc.*, **78,** 10 (1956)
F206 H. O. Spivey and T. Shedlovsky, *J. Phys. Chem.*, **71,** 2171 (1967)
F207 M. Goffredi and T. Shedlovsky, *J. Phys. Chem.*, **71,** 4436 (1967)
F208 L. W. Marple and G. J. Scheppers, *Anal. Chem.*, **38,** 553 (1966)
F209 A. T. Cheng, R. A. Howald and D. L. Miller, *J. Phys. Chem.*, **67,** 1601 (1963)
F210 S. Bruckenstein and L. M. Mukherjee, *J. Phys. Chem.*, **64,** 1601 (1960)
F211 A. M. Shkodin and L. I. Karkuzaki, *Ukrain. Khim. Zh.*, **27,** 48 (1961); *Chem. Abstr.*, **55,** 17301c
F212 A. M. Shkodin and L. I. Karkuzaki, *Zh. Fiz. Khim.*, **33,** 2795 (1959)
F213 See for example S. Kilpi and E. Lindell, *Acta Chem. Scand.*, **19,** 1420 (1965); *Ann. Acad. Sci. Fennicae*, **A136,** 3 (1967)
F214 S. Bruckenstein and I. M. Kolthoff, *J. Amer. Chem. Soc.*, **78,** 2974 (1956)
F215 I. M. Kolthoff and S. Bruckenstein, *J. Amer. Chem. Soc.*, **79,** 1 (1957)
F216 S. Bruckenstein and I. M. Kolthoff, *J. Amer. Chem. Soc.*, **79,** 5915 (1957)
F217 L. Marple and J. S. Fritz, *Anal. Chem.*, **35,** 1223 (1963)
F218 J. Kenttämaa, *Suomen Kemi.*, **B32,** 220 (1959)
F219 E. Grunwald and E. Price, *J. Amer. Chem. Soc.*, **86,** 2965 (1964)
F220 S. Bruckenstein, *J. Amer. Chem. Soc.*, **82,** 307 (1960)
F221 L. M. Mukherjee, S. Bruckenstein and F. A. K. Badawi, *J. Phys. Chem.*, **69,** 2537 (1965)
F222 R. Fernández-Prini and J. Prue, *Trans. Faraday Soc.*, **62,** 1257 (1966)
F223 T. Jasiński and Z. Pawlak, *Rocz. Chem.*, **41,** 1943 (1967)
F224 T. Jasiński and Z. Pawlak, *Rocz. Chem.*, **42,** 129, 2169 (1968)
F225 F. Aufauvre, Thesis, Faculté des Sciences de l'Université de Clermont-Ferrand, France (1969)
F226 M. M. Davis, National Bureau of Standards Monograph 105, U.S. Government Printing Office, Washington, D.C. (1968)
F227 D. F. DeTar and R. W. Novak, *J. Amer. Chem. Soc.*, **92,** 1361 (1970)

F228 L. M. Mukherjee, J. J. Kelly, W. Baranetzky and J. Sica, *J. Phys. Chem.*, **72,** 3410 (1968)
F229 J. O. Jenkins and J. W. Smith, *J. Chem. Soc.* (*B*), 1538 (1970)
F230 R. Scott, D. de Palma and S. Vinogradov, *J. Phys. Chem.*, **72,** 3192 (1968)
F231 V. K. LaMer and H. C. Downes, *J. Amer. Chem. Soc.*, **53,** 888 (1931)
F232 A. A. Maryott, *J. Res. Nat. Bur. Standards*, **38,** 527 (1947)
F233 B. P. Bruss and G. Harlow, *Anal. Chem.*, **30,** 1833, 1836 (1958)
F234 N. van Meurs and E. Dahmen, *Anal. Chim. Acta*, **19,** 64 (1958)
F235 H. B. van der Heijde, *Anal. Chim. Acta*, **16,** 392 (1957)
F236 R. R. Miron and D. M. Hercules, *Anal. Chem.*, **33,** 1770 (1961)
F237 N. A. Izmailov and L. I. Spivak, *Zh. Fiz. Khim.*, **36,** 757 (1962); *Chem. Abstr.*, **57,** 11920h
F238 M. M. Davis and P. J. Schuhmann, *J. Res. Nat. Bur. Standards*, **39,** 221 (1947)
F239 R. P. Bell and J. W. Bayles, *J. Chem. Soc.*, 1518 (1952)
F240 J. W. Bayles and A. Chetwyn, *J. Chem. Soc.*, 2328 (1958)
F241 R. G. Pearson and D. C. Vogelsong, *J. Amer. Chem. Soc.*, **80,** 1038 (1958)
F242 M. M. Davis and H. B. Hetzer, *J. Res. Nat. Bur. Standards*, **60,** 569 (1958)
F243 O. Popovych, *J. Phys. Chem.*, **66,** 915 (1962)
F244 M. Rumeau and B. Trémillon, *Bull. Soc. Chim. France*, 1049 (1964)
F245 T. Jasiński and T. Widernikowa, *Rocz. Chem.*, **43,** 1253, 1257 (1969)
F246 J. Steigman and P. M. Lorenz, *J. Amer. Chem. Soc.*, **88,** 2083, 2093 (1966)
F247 L. E. I. Hummelstedt and D. N. Hume, *J. Amer. Chem. Soc.*, **83,** 1564 (1961)
F248 J. Steigman and W. Cronkright, *J. Amer. Chem. Soc.*, **92,** 6729 (1970)
F249 S. Kaufman and C. R. Singleterry, *J. Phys. Chem.*, **56,** 604 (1952)
F250 S. Bruckenstein and A. Saito, *J. Amer. Chem. Soc.*, **87,** 698 (1965)
F251 J. F. Coetzee and R. M.-S. Lok, *J. Phys. Chem.*, **69,** 2690 (1965).
F252 S. Bruckenstein and D. F. Untereker, *J. Amer. Chem. Soc.*, **91,** 5741 (1969)
F253 G. M. Barrow and E. A. Yerger, *J. Amer. Chem. Soc.*, **76,** 5211 (1954)
F254 E. A. Yerger and G. M. Barrow, *J. Amer. Chem. Soc.*, **77,** 4474 (1955)
F255 E. A. Yerger and G. M. Barrow, *J. Amer. Chem. Soc.*, **77,** 6206 (1955)
F256 T. Shibuya and Y. I'haya, *Bull. Chem. Soc. Japan*, **37,** 896 (1964)
F257 W. N. Sanders and J. E. Berger, *Anal. Chem.*, **39,** 1473 (1967)
F258 J. W. Bayles and B. Evans, *J. Chem. Soc.*, 6984 (1965)
F259 M. M. Davis and H. B. Hetzer *J. Res. Nat. Bur. Standards*, **65A,** 209 (1961)
F260 R. P. Bell, The Proton in Chemistry, pp. 56–59, Cornell University Press, Ithaca, N.Y. (1959)
F261 M. M. Davis and M. Paabo, *J. Amer. Chem. Soc.*, **82,** 5081 (1960)
F262 M. M. Davis, *J. Amer. Chem. Soc.*, **84,** 3623 (1962)
F263 J. W. Bayles and A. F. Taylor, *J. Chem. Soc.*, 417 (1961)
F264 P. Huyskens and T. Zeegers-Huyskens, *J. Chim. Phys.*, **61,** 81 (1964)
F265 M. M. Davis and M. Paabo, *J. Org. Chem.*, **31,** 1804 (1966)
F266 T. E. Mead, *J. Phys. Chem.*, **66,** 2149 (1962)

APPENDIX 3.3.1

Proton Transfer Quotients* at 25°C

for $S + H_3O^+ \leftrightarrows SH^+ + H_2O$

$$Q_p = \frac{[SH^+][H_2O]}{[S][H_3O^+]}$$

Solvent S	$10^3 Q_p$	Refs.
H_2O	1000.0	
MeOH	8.6	Table 3.3.1
EtOH	3.9	H. Goldschmidt and P. Dahll, *Z. Phys. Chem.*, **114,** 1 (1924); L. Thomas and E. Marum, *Z. Phys. Chem.*, **A143,** 191 (1929); L. S. Guss and I. M. Kolthoff, *J. Amer. Chem. Soc.*, **62,** 1494 (1940); E. A. Braude and E. S. Stern, *J. Chem. Soc.*, 1976 (1948); C. E. Newall and A. M. Eastham, *Can. J. Chem.*, **39,** 1752 (1961)
i-PrOH	9.1	N. A. Izmailov and V. V. Alexandrov, *Zh. Fiz. Khim.*, **31,** 2619 (1957)
n-BuOH	2.8	Guss and Kolthoff, op. cit.
i-BuOH	6.0	Izmailov and Alexandrov, op. cit.
i-AmOH	11.5	Izmailov and Alexandrov, op. cit.
Benzyl alcohol	56.	Izmailov and Alexandrov, op. cit.
Acetone	85.	N. A. Izmailov and V. N. Izmailova, *Zh. Fiz. Khim.*, **29,** 1050 (1955)
Dimethylsulphoxide	160.	I. M. Kolthoff and T. B. Reddy, *Inorg. Chem.*, **1,** 189 (1962)
Acetonitrile	0.2	I. M. Kolthoff and S. Ikeda, *J. Phys. Chem.*, **65,** 1020 (1961)
Sulphuric acid	0.024	R. H. Flowers, R. J. Gillespie, and E. A. Robinson, *Can. J. Chem.*, **38,** 1363 (1960)

* No corrections have been made for association.

APPENDIX 3.3.2

Medium Effects* on the Proton at 25°C

Solvent	$\log \frac{\rho_s{}^\circ}{\rho_w{}^\circ}$	$\log {}_m\gamma_H$
MeOH	−0.103	3.1[1], 1.85[2], 0.4[3a], 0.3[3b], −0.1[3c], −1.45[4], −1.6[5], −2.3[6], −2.6[5]
EtOH	−0.104	4.6[7], 3.9[1], 1.85[2], 1.5[8], −2.0[6]
n-BuOH	−0.091	4.2[1], −1.0[6]
i-AmOH	−0.089	4.2[2]
HCO_2H	+0.088	10.0[1], 7.9[3b], 6.8[6], 4.7[3a]
CH_3CN	−0.108	7.7[9], 5.5[1], 5.2[10], 3.9[6], 2.5[3c], 2.4[3b]
$(CH_3)_2CO$	−0.103	3.3[1]
$(CH_3)_2SO$	+0.042	−1.5[1], −5.8[11]
$HCONH_2$	+0.054	−0.6[6], −0.7[3a], −1.2[3b], −2.5[3c]
N_2H_4	+0.004	−13.2[1], −15[3b]
NH_3		−15.8[1], −17[3b]

* $\log {}_m\gamma_H$ for transfer from water to solvent *S*. Values on molal scale; to convert to molar scale, use

$$\log {}_m\gamma_H - \log \rho_s{}^\circ/\rho_w{}^\circ = \log {}_m y_H$$

where $\rho_s{}^\circ$ and $\rho_w{}^\circ$ are the densities of pure *S* and water.

REFERENCES

1. N. A. Izmailov, *Dokl. Akad. Nauk. SSSR*, **149**, 1364 (1964). Extrapolation of ΔG^0 of solvation *vs.* n^{-2} where *n* is the principal quantum number of the first vacant orbital of isoelectronic pairs of ions. Earlier values for methanol of 3.3 and 3.7 were based on extrapolation *vs.* reciprocal radius
2. O. Popovych and A. J. Dill, *Anal. Chem.*, **41**, 456 (1969). Assume equal medium effects for i-Am_3BuN^+ and BPh_4^-
3. H. Strehlow *in* The Chemistry of Non-Aqueous Solvents, Vol. I (J. J. Lagowski, ed.), Academic Press, New York, N.Y. (1966). Chapter 4. Values from: (3a) H_0, (3b) Pleskow's rubidium electrode postulate, and (3c) ferrocene and cobalticene redox couples, assumed to have potentials unaffected by solvent
4. M. Alfenaar and C. L. DeLigny, *Rec. Trav. Chim.*, **86**, 929 (1967). Extrapolation *vs.* reciprocal radius
5. A. L. Andrews, H. P. Bennetto, D. Feakins, K. G. Lawrence and R. P. T. Tompkins, *J. Chem. Soc., A*, 1486 (1968). Extrapolation *vs.* reciprocal radius
6. B. Case and R. Parsons, *Trans. Faraday Soc.*, **63**, 1224 (1967). From real ΔG^0 of solvation based on $\Delta G^0(H^+)$ in water of 1101.0 kJ mol^{-1}
7. B. Gutbezahl and E. Grunwald, *J. Amer. Chem. Soc.*, **75**, 565 (1953). Grunwald-Winstein activity postulate
8. A. J. Dill, L. M. Itzkowitz and O. Popovych, *J. Phys. Chem.*, **72**, 4580 (1968). Calculated from H_0

9. I. M. Kolthoff, M. K. Chantooni, Jr. and S. Bhowmik, *J. Amer. Chem. Soc.*, **90**, 23 (1968). Comparison of acidities of protonated water and dimethylsulphoxide in acetonitrile
10. I. M. Kolthoff and F. G. Thomas, *J. Phys. Chem.*, **69**, 3049 (1965). *o*-Phenanthroline iron(II, III) redox couple
11. J. Courtet-Coupez, A. Laouenan and M. LeDemezet, *Compt. Rend. C*, **267**, 1475 (1968). Cyclic voltammetry with ferrocene-ferricinium redox couple

APPENDIX 3.3.3

Autoprotolysis Constants and Associated Thermodynamic Properties of Solvents at 25°C*

Solvent	pK_S	pK_S	ΔH_S	Other Properties	Method†	Ref.
H_2O	14.00	17.49	55.8	$\Delta S^0 = -0.081$	EMF,	1
				$\Delta C^0 = -0.22$	Cal.,	
				$\Delta V^0 = -22.11$	d.	
			55.8	$\Delta S^0 = -0.081$	Cal.	2a
				$\Delta C_p{}^0 = -0.19$		
D_2O	14.96	17.75	59.9	$\Delta S^0 = -0.085$	EMF	3
				$\Delta C_p{}^0 = -0.23$		
H_2O_2	12.7	15.6			EMF	4
CH_3OH	16.45	19.44			EMF	5
			46.9		Cal.	6
	16.72	19.71			EMF	7
	16.47	19.46			EMF	8
	16.71	19.70	45.2	$\Delta S^0 = -0.168$	EMF	9
				$\Delta C_p{}^0 = -0.25$		
	16.45	19.44	49.4	$\Delta S^0 = -0.163$	Cond.	10
				$\Delta V = -43$	d.	11
			42.8	$\Delta S^0 = -0.176$	Cal.	2b
				$\Delta C_p{}^0 = -0.237$		
C_2H_5OH			45.2	$(\Delta S^0 = -0.20)$	Cal.	12, 6
	18.9	21.6			EMF	13
	18.72	21.29			EMF	14
	18.67	21.34			EMF	15
	18.91	21.58	32.2	$\Delta S^0 = -0.25$	Cond.	10
n-C_3H_7OH	19.24	21.68			EMF	15
i-C_3H_7OH	20.58	23.02			EMF	15
n-C_4H_9OH	21.56	23.82			EMF	16
n-$C_5H_{11}OH$	20.65	22.76			EMF	16
n-$C_6H_{13}OH$	19.74	21.66			EMF	16
n-$C_8H_{17}OH$	19.44	21.24			EMF	16
Ethylene glycol	15.84	18.25	48.3	$\Delta S^0 = -0.141$	EMF	17
				$\Delta C_p{}^0 = -0.12$		
Propylene glycol	17.21	19.45	45.5	$\Delta S^0 = -0.177$	EMF	17
				$\Delta C_p{}^0 = -0.13$		
Diethylene glycol	17.4	19.4			EMF	8
Ethyl cellosolve	19.3	21.4			EMF	8
H_2SO_4	3.46	5.48	20.1	$\Delta S^0 = 0.001$	Cal.	18
	3.57	5.59	20.9	$\Delta S^0 = 0.002$	Cal.	19. 20
	3.64	5.65			H_0	21
	3.33	5.34			Cry.	22
D_2SO_4	4.34	6.34			Cond.	23
HSO_3F	7.4	9.4			Cond.	24
H_2SeO_4	~2.				Interp.	25
H_3PO_4	0.8				?	25
HF	10.5	13.9			Cond.	26
HCO_2H	6.5	10.2			Cond.	27
	6.17	9.84			H_0	28
	$\geqslant$6.83	$\geqslant$10.50			Cond.	29
CH_3CO_2H	12.55	16.11		(at 19–20°)	EMF	30
			24.0	$\Delta S^0 = -0.16$	Cal.	31
	14.45	18.05			EMF,	32
					K_aK_b	32
	13.0	16.6		(at 105.7°C)	Cond.	33

APPENDIX 3.3.3 (*contd*)

Solvent	pK_S	pK_S	ΔH_S	Other Properties	Method†	Ref.
NH_3	32.7	35.8		(at −50°C)	EMF	34
	29.8	32.6	110.0	$\Delta S^0 = -0.20$	Cal..	35
			109.0	(at −33.4°C)	ΔG_f^0	
	27.7	30.5			Cal.	36
	32.5	35.6		(at −33°C)		
$NH_2(CH_2)_2NH_2$	15.2	17.6			K_aK_b	37
$NH_2C_2H_4OH$	5.2	7.6		(at 20°C)	EMF	38
	5.7	8.1			Cond.	39
	5.15	7.55		(at 20°C)	H	40
$HCONH_2$	17.0	19.7		(at 20°C)	EMF	41
$HCON(CH_3)_2$	18.0	20.3		(at 20°C)	EMF	42
CH_3CONH_2	10.5	13.0		(at 94°C)	EMF	43
	14.6	17.1		(at 98°C)	EMF	44
CH_3CN	26.2	29.0			K_aK_b	45
	26.0	28.8			EMF	46
	>32.2	>35.0			K_aK_b	47
$(CH_3)_2SO$	17.4	19.6			K_aK_b	48

* pK_S values on molal scale; convert to molar scale with

$$pK_S(\text{molar}) = pK_S(\text{molal}) - 2 \log \rho^0$$

where ρ^0 is the density of the solvent. $\Delta H_S{}^0$ in kJ mol^{-1}; ΔS^0 and $\Delta C_p{}^0$ in kJ mol^{-1}; ΔV^0 in cm^3 mol^{-1}

† Methods: EMF, cell measurements; Cond., conductance; d., densities; Cal., calorimetry; H_0 and H., acidity functions; Cry., cryoscopy; Inter., interpolation; K_aK_b, product of acidity and basicity constants; $\Delta G_f{}^0$, standard Gibbs free energies of formation.

REFERENCES

1. J. W. Larson and L. G. Hepler *in* Solute-Solvent Interactions, Chapter 1 (J. F. Coetzee and C. D. Ritchie, eds.), Marcel Dekker, New York, N.Y. (1969)
2. C. S. Leung and E. Grunwald, *J. Phys. Chem.*, **74,** 687 (a), 696 (b) (1970)
3. A. K. Covington, R. A. Robinson and R. G. Bates. *J. Phys. Chem.*, **70,** 3820 (1966)
4. A. G. Mitchell and W. F. K. Wynne-Jones, *Trans. Faraday Soc.*, **52,** 824 (1956)
5. N. Bjerrum, A. Unmack and L. Zechmeister, *Kgl. Danske Videnskab. Selskab Mat. Fys. Medd.*, **5,** No. 11 (1925)
6. J. H. Wolfenden, W. Jackson and H. B. Hartley, *J. Phys. Chem.*, **31,** 850 (1927)
7. P. S. Buckley and H. Hartley, *Phil. Mag.*, **8,** 320 (1929)
8. M.-L. Dondon, *J. Chim. Phys.*, **48,** C34 (1951)
9. J. Koskikallio, *Suomen Kemi.*, **30B,** 111 (1957)
10. G. Briere, B. Crochon and N. Felici, *Compt. Rend.*, **254,** 4458 (1962)
11. C. H. Rochester and B. Rossall, *Trans. Faraday Soc.*, **65,** 992 (1969). Apparent molar volumes
12. C. M. van Deventer and L. T. Reicher, *Z. Phys. Chem.*, **8,** 536 (1891)
13. P. S. Danner, *J. Amer. Chem. Soc.*, **44,** 2832 (1922)
14. A. McFarlane and H. Hartley, *Phil. Mag.*, **13,** 425 (1932)
15. A. Tézé and R. Schaal, *Compt. Rend.*, **253,** 114 (1961)
16. A. R. Kreshkov, N. Sh. Aldarova and N. T. Smolova, *Zh. Fiz. Khim.*, **43,** 2846 (1969)
17. K. K. Kundu, P. K. Chattopadhyay, D. Jana and M. N. Das, *J. Phys. Chem.*, **74,** 2633 (1970); see also K. K. Kundu and M. N. Das, *J. Chem. Eng. Data*, **9,** 82 (1964)
18. B. J. Kirkbride and P. A. H. Wyatt, *Trans. Faraday Soc.*, **54,** 483 (1958)

19. R. J. Gillespie *in* Chemical Physics of Ionic Solutions (B. E. Conway and R. G. Barradas, eds.), Wiley, New York (1966)
20. B. Dacre and P. A. H. Wyatt, *J. Chem. Soc.*, 2962 (1961)
21. M. I. Vinnik and R. S. Ryabova, *Zh. Fiz. Khim.*, **38**, 606 (1964)
22. P. A. H. Wyatt, *Trans. Faraday Soc.*, **65**, 585 (1969)
23. R. J. Gillespie and E. A. Robinson *in* Non-Aqueous Solvent Systems, Chapter 4 (T. C. Waddington, ed.), Academic Press, New York, N.Y. (1965)
24. R. J. Gillespie, *Acc. Chem. Res.*, **1**, 202 (1968)
25. S. Wasif, *J. Chem. Soc.*, *A*, 118 (1970)
26. M. E. Runner, G. Balog and M. Kilpatrick, *J. Amer. Chem. Soc.*, **78**, 5183 (1956)
27. H. I. Schlesinger and R. P. Calvert, *J. Amer. Chem. Soc.*, **33**, 1924 (1911); H. I. Schlesinger and A. W. Martin, *J. Amer. Chem. Soc.*, **36**, 1589 (1914)
28. L. P. Hammett and A. J. Deyrup, *J. Amer. Chem. Soc.*, **54**, 4239 (1932)
29. T. C. Wehman and A. I. Popov, *J. Phys. Chem.*, **72**, 4031 (1968)
30. S. Kilpi and M. Puranen, *Ann. Acad. Sci. Fenn.*, **57A**, No. 10 (1941)
31. W. L. Jolly, *J. Amer. Chem. Soc.*, **74**, 6199 (1952). Value uncorrected for incomplete dissociation
32. S. Bruckenstein and I. M. Kolthoff, *J. Amer. Chem. Soc.*, **78**, 2974 (1956)
33. R. J. L. Martin, *Austr. J. Chem.*, **18**, 321 (1965)
34. V. A. Pleskow and A. M. Monossohn, *Acta Physicochim. URSS*, **1**, 713 (1935)
35. H. D. Mulder and F. C. Schmidt, *J. Amer. Chem. Soc.*, **73**, 5575 (1951); W. L. Jolly, *J. Amer. Chem. Soc.*, **74**, 6199 (1952)
36. L. V. Coulter, J. R. Sinclair, A. G. Cole and G. C. Roper, *J. Amer. Chem. Soc.*, **81**, 2986 (1959)
37. W. B. Schaap, R. E. Bayer, J. R. Siefker, J. L. Kim, P. W. Brewster and F. C. Schmidt, *Rec. Chem. Progr.*, **22**, 197 (1961)
38. F. Masure and R. Schaal, *Bull. Soc. Chim. Fr.*, 1141 (1956)
39. P. W. Brewster, F. C. Schmidt and W. B. Schaap, *J. Phys. Chem.*, **65**, 990 (1961)
40. C. Vermesse-Jacquinot, *J. Chim. Phys.*, **62**, 185 (1965)
41. F. H. Verhoek, *J. Amer. Chem. Soc.*, **58**, 2577 (1936)
42. M. Tézé and R. Schaal, *Bull. Soc. Chim. Fr.*, 1372 (1962)
43. G. Jander and G. Winkler, *J. Inorg. Nucl. Chem.*, **9**, 32 (1959)
44. S. Guiot and B. Trémillon, *J. Electroanal. Chem.*, **18**, 261 (1968)
45. J. F. Coetzee and G. R. Padmanabhan, *J. Phys. Chem.*, **66**, 1708 (1962)
46. J. Desbarres, *Bull. Soc. Chim. Fr.*, 2103 (1962)
47. I. M. Kolthoff and M. K. Chantooni, Jr., *J. Phys. Chem.*, **72**, 2270 (1968)
48. I. M. Kolthoff and T. B. Reddy, *Inorg. Chem.*, **1**, 189 (1962)

APPENDIX 3.5.1

Acid-Base Equilibria in Alcohols
Shedlovsky's Conductance Results at 25°C

K_A, association constant in kg mol^{-1}; K_a, acidity constant in mol kg^{-1}; $\Lambda°$, limiting conductance in cm^2 Ω^{-1} mol^{-1}

Solvent	Methanol	Ethanol	1-Propanol
Purity, wt %	99.99	99.67	99.99
K_A(HCl)	13.	38.	137.
$\Lambda°$(HCl)	198.5	84.65	29.81
K_A(NaCl)		11.	93.
pK_a(HOAc)	9.521	10.034	10.608
$\Lambda°$(HOAc)	185.5	48.9	20.06
Reference	1	2	3

REFERENCES

1. T. Shedlovsky and R. L. Kay, *J. Phys. Chem.*, **60**, 151 (1956). Earlier electromotive force measurements gave pK_a(HOAc) = 9.62: L. J. Minnick and M. Kilpatrick, *J. Phys. Chem.*, **43**, 259 (1939)
2. H. O. Spivey and T. Shedlovsky, *J. Phys. Chem.*, **71**, 2165, 2171 (1967). Electromotive force measurements in pure ethanol gave pK_a(HOAc) = 10.41: I. D. Tabagua, M. M. Tsurtsumiya and N. A. Izmailov, *Tr. Sukhumsk. Gos. Ped. Inst.*, **15**, 103 (1962); *Chem. Abst.*, **60**, 14370g
3. M. Goffredi and T. Shedlovsky, *J. Phys. Chem.*, **71**, 2176, 2182, 4436 (1967). Results are also given at 15°C and 35°C for HCl and at 15°C for HOAc (pK_a = 10.359)

APPENDIX 3.5.2

Thermodynamics of Ionisation of Some Acids in Methanol at 25°C

Acidity constants, K_a, in mol kg^{-1}; ΔH^0 in kJ mol^{-1}; ΔS^0 and $\Delta C_p{}^0$ in kJ K^{-1} mol^{-1}; ΔV in cm^3 mol^{-1}. Log $({}_m\Gamma_a)$ = medium effect on ionisation = $p({}^sK_a) - p({}^wK_a)$; ΔV = molar volume change associated with ionisation.

Acid	pK_a	$\log({}_m\Gamma_a)$	ΔH^0	$-\Delta S^0$	$-\Delta C_p{}^0$	$-\Delta V$	Ref.
Acetic acid	9.52	4.76	21.5	0.112	0.0		1
Benzoic acid	9.28	5.08	17.97	0.117	0.17		2
Phenol	14.36	4.44	37.1	0.150	0.29	38.5	1, 3
2-*t*-Butylphenol	16.36		49.7	0.147	0.24		1
3-Nitrophenol	12.33	4.04	34.1	0.122	0.13		1
4-Bromophenol	13.50	4.24	35.9	0.138	0.14	35.6	1, 3
4-*t*-Butylphenol	14.54	4.42	37.6	0.154	0.28	37.4	1, 3
4-Formylphenol	11.91	4.38	33.5	0.116	0.10	30.4	1, 3
4-Nitrophenol	11.40	4.35	34.5	0.102	0.09	31.7	1, 3
2,4-Di-*t*-butylphenol	16.64		49.6	0.152	0.15		1
2,6-Di-*t*-butylphenol	17.20		34.9	0.212	0.0		1
3,5-Di-*t*-butylphenol	14.76		37.7	0.156	0.26	39.3	1, 3
2,4,6-Tri-*t*-butylphenol	17.51		42.1	0.194	0.0		1
2,6-Di-*t*-butyl-4-bromophenol	15.80		42.4	0.160	0.26	54.	1, 3
2,6-Di-*t*-butyl-4-formylphenol	12.27	4.32	33.1	0.124	0.22	31.8	1, 3
2,4,6-Trinitrophenyl-amine	13.57		58.5	0.063	0.0		1

REFERENCES

1. P. D. Bolton, C. H. Rochester and B. Rossall, *Trans. Faraday Soc.*, **66,** 1348 (1970)—a recalculation of earlier results (*Trans. Faraday Soc.*, **65,** 1004 (1969)) using the Clarke and Glew equation
2. C. S. Leung and E. Grunwald, *J. Amer. Chem. Soc.*, **74,** 696 (1970). Other values of pK_a and $\Delta C_p{}^0$ from −90° to 32.5°C and values of ΔH^0 from −95° to 35°C are given
3. C. H. Rochester and B. Rossall, *Trans. Faraday Soc.*, **65,** 992 (1969). Values of the molar volumes were not extrapolated to infinite dilution

APPENDIX 3.5.3

The Ionisation of Picolinium Ions in Methanol

$$4\text{-}X\mathrm{CH_2C_5H_4NH^+} \leftrightarrows 4\text{-}X\mathrm{CH_2C_5H_4N} + \mathrm{H^+}$$

Acidity constants, K_a, in mol kg^{-1}; ΔH^0 in kJ mol^{-1}; ΔS^0 in kJ K^{-1} mol^{-1}

X	pK_a at 0.1°C	pK_a at 25°C	ΔH^0 at 25°C	ΔS^0 at 25°C
H	6.44	5.99	28	−0.022
CH_3	6.45	5.99	29	−0.020
C_2H_5	6.43	5.97	29	−0.020
C_6H_5	6.13	5.69	27	−0.020
OH	6.23	5.75	30	−0.010
CN	4.90	4.46	27	0.004
NH_3^+	4.03	3.67	23	0.005

REFERENCES

C. D. Ritchie and P. D. Heffley, *J. Amer. Chem. Soc.*, **87**, 5402 (1965)

APPENDIX 3.5.4

The Ionisation of Bicyclo[2.2.2]octane-1-carboxylic Acids in Methanol

$$XC(CH_2CH_2)_3CCOOH \rightleftarrows H^+ + XC(CH_2CH_2)_3CCOO^-$$

Acidity constants, K_a, in mol kg^{-1} with precision of ± 0.01 pK unit; ΔH^0 in kJ mol^{-1} with precision ± 4 kJ mol^{-1}; ΔS^0 the same for all acids: -0.16 ± 0.01 kJ K^{-1} mol^{-1}.

X	pK_a at 0°C	pK_a at 25°C	ΔH^0 at 25°C
H	10.29	10.12	10
COO^-	10.30*	10.16*	8
COOH	9.99†	9.77†	14
OH	10.01	9.88	8
$COOC_2H_5$	9.97	9.83	9
Br	9.84	9.65	12
CN	9.65	9.52	8
$N(CH_3)_2^+$	9.33	9.27	4

* Includes $-\log 2$ where 2 is a statistical factor.
† Includes $+\log 2$.

REFERENCES

C. D. Ritchie and G. H. Megerle, *J. Amer. Chem. Soc.*, **89,** 1452 (1967)

APPENDIX 3.5.5

Medium Effects on the Ionisation of Weak Acids in Methanol at 25°C

$$\log ({}_m\Gamma_a) = \mathrm{p}({}^sK_a) - \mathrm{p}({}^wK_a)$$

Acidity constants are on the molal scale. When possible, values of $\mathrm{p}({}^wK_a)$ were taken from R. A. Robinson and R. H. Stokes, Electrolyte Solutions, Second Edition (Revised) (Butterworths, London, 1968) Appendix 12; other values are from the references cited.

Acid	$\mathrm{p}({}^sK_a)$	$\log {}_m\Gamma_a$	Ref.
Acetic acid	9.42	4.66	1
	9.52	4.76	2
	9.68	4.92	3
	9.71	4.95	4
Monochloracetic acid	7.81	4.95	4
Propionic acid	9.61	4.75	1
n-Butyric acid	9.59	4.81	1
Phenylacetic acid	9.17	4.87	1
Trichloracetic acid	4.90	4.68	3
Benzoic acid	9.17	4.98	1
	9.25	5.04	5
	9.38	5.17	3
	9.41	5.20	4
Salicyclic acid	7.85	4.85	3
	7.89	4.89	4
o-Methylbenzoic acid	8.90	4.99	5
m-Methylbenzoic acid	9.19	4.92	5
	9.27	5.00	1
p-Methylbenzoic acid	9.35	4.98	1
	9.36	4.99	5
m-Chlorobenzoic acid	8.60	4.77	1
p-Chlorobenzoic acid	8.83	4.85	1
o-Nitrobenzoic acid	7.60	5.43	4
m-Nitrobenzoic acid	8.12	4.63	1
	8.30	4.81	4
p-Nitrobenzoic acid	8.12	4.69	1
	8.42	4.99	4
trans-Cinnamic acid	9.21	4.77	1
m-Nitrocinnamic acid	8.79	4.67	1

APPENDIX 3.5.5 (*contd*)

Acid	$p(^sK_a)$	$\log {}_m\Gamma_a$	Ref.
p-Nitrocinnamic acid	8.73	4.68	1
m-Chlorocinnamic acid	8.96	4.67	1
α-Chloropropionic acid	7.96	5.13	5
β-Chloropropionic acid	8.99	4.92	5
α-Bromopropionic acid	8.12	5.25	5
β-Bromopropionic acid	8.90	4.89	5
Thiolacetic acid	7.70	4.27	1
Thiolpropionic acid	7.68	4.40	1
Thiol-*n*-butyric acid	7.88	4.45	1
Thiolbenzoic acid	7.28	4.62	1
Benzenephosphonic acid	6.73	4.83	1
Benzenephosphinic acid	6.37	4.83	1
Chloromethylphosphonic acid	5.81	4.52	1
Diphenylphosphinic acid	5.56	3.3	1
Phenol	14.34	4.32	6
	14.36	4.34	7
o-Cresol	14.89	4.56	6
m-Cresol	14.47	4.37	6
p-Cresol	14.64	4.36	6
2,3-Xylenol	15.18	4.64	6
2,4-Xylenol	15.14	4.54	6
2,5-Xylenol	15.01	4.60	6
2,6-Xylenol	15.36	4.73	6
3,4-Xylenol	14.73	4.37	6
3,5-Xylenol	14.61	4.42	6
2,4,6-Tribromophenol	10.00	3.90	1
4-Nitrophenol	10.99	3.83	1
	11.40	4.35	7
2,6-Dibromo-4-nitrophenol	7.21	3.83	1
2,6-Dichloro-4-nitrophenol	7.30	3.75	1
2,4-Dinitrophenol	7.65	3.57	1
	7.75	3.67	5
2,5-Dinitrophenol	8.80	3.57	5
	8.88	3.65	1

APPENDIX 3.5.5 (*contd*)

Acid	$p(^{s}K_a)$	$\log {}_m\Gamma_a$	Ref.
2,6-Dinitrophenol	7.44	3.72	5
	7.50	3.78	8
	7.52	3.81	1
	7.65	3.93	8
2,4,6-Trinitrophenol	3.43	2.73	5
	3.57	2.87	1
Pyridinium ion	5.27	0.05	9
2-Picolinium ion	6.01	0.05	9
3-Picolinium ion	5.72	0.09	9
4-Picolinium ion	6.06	0.08	9
2,4-Lutidinium ion	6.66	0.03	9
2,6-Lutidinium ion	6.76	0.04	9

REFERENCES

1. J. Juillard, *Bull. Soc. Chim. Fr.*, 1727 (1966)
2. T. Shedlovsky and R. L. Kay, *J. Phys. Chem.*, **60**, 151 (1956)
3. I. D. Tabagua, *Tr. Sukhumsk. Gos. Ped. Inst.*, **15**, 119 (1962); *Chem. Abstr.*, **60**, 14373d (1964)
4. O. M. Konovalov, *Zh. Fiz. Khim.*, **39**, 693 (1965)
5. J. Juillard and M.-L. Dondon, *Bull. Soc. Chim. Fr.*, 2535 (1963)
6. C. H. Rochester, *Trans. Faraday Soc.*, **62**, 355 (1966)
7. P. D. Bolton C. H. Rochester and B. Rossall *Trans. Faraday Soc.*, **66**, 1348 (1970); see Appendix 3.5.2
8. G. Kortüm and K.-W. Koch, *Ber. Bunsenges. Phys. Chem.*, **69**, 677 (1965)
9. C. H. Rochester, *J. Chem. Soc. B*, 33 (1967)

APPENDIX 3.5.6

Acidity Constants for Acids in Formamide

Acid	$t/°C$	$p(^sK_a)$	$\Delta H^0/$ kJ mol^{-1}	Ref.
Formic	20	5.52		1
	25	5.49_5	9.0	
	30	5.46		
	35	5.48		
	40	5.52		
	45	5.55_5		
Acetic	15	6.92		1
	25	6.82	13.4	
	30	6.76		
	35	6.76_5		
	40	6.77		
	45	6.85		
Oxalic, first H^+	25	2.83		2
	30	2.02		
	35	2.32		

REFERENCES

1. M. Mandel and P. Decroly, *Trans. Faraday Soc.*, **56**, 29 (1966). Precision ±0.02 but note that E^0 values do not agree well with those of other workers (Appendix 2.2.11)
2. R. K. Agarwal and B. Nayak, *Indian J. Chem.*, **7**, 1268 (1969). Precision ±0.05 to ±0.06

APPENDIX 3.6.1

Selected Values of pK_a and pK_f for Acids in *N,N*-Dimethylformamide at 25°C

Acid	pK_a(HA)	pK_f(HA_2^-)	Refs.
Hydrochloric acid	3.4	−2.2	1, 2, 3
Formic acid	11.6		4, 5
Acetic acid	13.4		3, 5, 6
Benzoic acid	12.6_6		3–7
Salicylic acid	8.2_4	−1.7	3, 6, 7
p-Nitrobenzoic acid	10.6_2		3, 6
Picric acid	1.4		3, 8
o-Nitrophenol	12.2	−2.0	4, 6
p-Nitrophenol	12.3	−2.2	3, 4, 6, 9
2,4-Dinitrophenol	6.3_4		2–4, 6
2,6-Dinitrophenol	6.0		3, 4, 6, 7
4-Chloro-2,6-dinitrophenol	4.6_9		3

REFERENCES

1. A. B. Thomas and E. G. Rochow, *J. Amer. Chem. Soc.*, **79**, 1843 (1957)
2. B. W. Clare, D. Cook, E. C. F. Ko, Y. C. Mac and A. J. Parker, *J. Amer. Chem. Soc.*, **88**, 1911 (1966)
3. I. M. Kolthoff, M. K. Chantooni, Jr .and H. Smagowski, *Anal. Chem.*, **42**, 1622 (1970)
4. S. M. Petrov and Yu. I. Umanskii, *Zh. Fiz. Khim.*, **41**, 1374 (1967)
5. S. M. Petrov and Yu. I. Umanskii, *Zh. Fiz. Khim.*, **42**, 3052 (1968)
6. J. Juillard, *J. Chim. Phys.*, **67**, 691 (1970)
7. G. Demange-Guérin, *Talanta*, **17**, 1075 (1970)
8. P. G. Sears, R. K. Wolford and L. R. Dawson, *J. Electrochem. Soc.*, **103**, 633 (1956)
9. C. D. Ritchie and G. H. Megerle, *J. Amer. Chem. Soc.*, **89**, 1447 (1967)

APPENDIX 3.6.2

Thermodynamic Properties Associated with the Ionisation of Acids in *N,N*-Dimethylformamide*

Carboxylic Acids at 25°C†

Acid	pK_a	ΔH_a^0	ΔS_a^0
Formic acid	11.5_5	11.0	−0.18
Acetic acid	13.2_5	17.5	−0.20
Chloracetic acid	10.1_0	16.6	−0.14
Benzoic acid	12.2_0	35.9	−0.11
Salicylic acid	7.8_5	13.9	−0.10

Picolonium Ions‡

$$4\text{-}XCH_2C_5H_4NH^+ \leftrightarrows 4\text{-}XCH_2C_5H_4N + H^+$$

X	pK_a at 15°C	pK_a at 35°C	ΔH_a^0 at 25°C	ΔS_a^0 at 25°C
H	4.96	4.82	12	−0.06
CH_3	4.98	4.83	13	−0.05
C_2H_5	5.01	4.83	15	−0.05
C_6H_5	4.77	4.62	12	−0.05
OH	4.53	4.24	25	−0.03
CN	3.61	3.57	3	−0.06
NH_3^+	3.90	3.62	24	0.01

* ΔH_a^0 in kJ mol^{-1}, ΔS_a^0 in kJ K^{-1} mol^{-1}.

† S. M. Petrov and Yu. I. Umanskii, *Zh. Fiz. Khim.*, **42,** 3052 (1968). Acidity constants in moles per kilogram of solvent.

‡ C. D. Ritchie and G. H. Megerle, *J. Amer. Chem. Soc.*, **89,** 1447 (1967). Acidity constants in moles per liter of solution.

Chapter 4

Spectroscopic Measurements

Part 1

Electronic Absorption Spectroscopy

I. R. Lantzke*
Department of Physical Chemistry,
University of Newcastle upon Tyne

* Present address: Graylands Teachers' College, Mimosa Avenue, Graylands, Western Australia 6010.

NOTATION

λ	Wavelength
λ_{max}	Wavelength of absorption maximum
ε	Molar extinction coefficient. Values in $1\ mol^{-1}\ cm^{-1}$ are one tenth of the unfamiliar S.I. unit of $m^2\ mol^{-1}$
ε_{max}	Molar extinction coefficient at the absorption maximum
ε_r	Dielectric constant
$\bar{\nu}$	Wave number
$\nu_{\frac{1}{2}}$	Band width at half height expressed in wave numbers
ν	Frequency
ΔE	Energy of an optical transition
h	Planck constant
n	Refractive index; non-bonding electrons
K	Equilibrium constant
K_A	Association constant

4.1 INTRODUCTION

These sections are restricted to a discussion of electronic absorption spectroscopy, and in particular the use of such spectral measurements in the determination of ionic, and to a lesser extent molecular, interactions in organic solvents. No consideration is given to the study of electronically excited states, despite their importance and the current interest in this topic, since it is not yet particularly applicable to the physical chemistry of solution equilibria. Additionally photochemical and emission processes have been well described recently.[G1–8]

Electronic absorption spectroscopy (U.V.-visible spectroscopy) has been used by solution chemists for a number of years now, so that the theory and techniques are well established, although by no means static. There have been no recent dramatic applications to electrolyte solutions, but rather a steady increase in understanding, for example, of charge transfer and solvent effects. There are numerous texts[G9–14] and reviews[G15,G16] available on absorption spectroscopy, but they are not particularly appropriate for the chemist interested in electrolyte solutions, who is not always fully conversant with spectroscopic theory and practice. Before considering particular chemical examples, it seems desirable to summarise some pertinent fundamentals and to draw attention to some less well recognized pitfalls of electronic spectroscopy.

The chemist employing spectroscopy, does so because the observed transitions can be related (a) to the identity of atoms or molecules, (b) to atomic or molecular structure and (c) to the energetics of chemical process. Electrolyte chemists have used spectroscopy principally as a convenient method of analysis, such as in the determination of constants of association and dissociation, and for obtaining qualitative information about solvation and intermolecular forces. Spectroscopic techniques have been used also in attempts to determine the structure of species in solution.

4.2 THEORETICAL ASPECTS

Electronic absorption spectra of polyatomic molecules in solution generally show broad featureless bands. From the band maximum is obtained the average energy of the electronic transition involved, while the band area is proportional to the amount of energy absorbed. The shapes of these bands are determined by the associated vibrational changes of the absorbing species. The position of an absorption maximum is frequently referred to by its wavelenth (λ), but in considering transition energies it is better to employ frequency units, since these are directly proportional to energy. The most commonly employed (spectroscopic) measure of frequency is the wave number ($\bar{\nu} = 1/\lambda$). A specified change in a transition energy gives a constant difference in frequency regardless of the spectral region concerned, but this is not the case on a wavelength scale.

Energy absorption, i.e. an electron transition, occurs if two criteria are satisfied. Firstly the frequency of the light must match up with the energy separation between the ground and excited states, i.e.

$$\Delta E = h\nu \tag{4.2.1}$$

Secondly there must be a finite possibility of the electron being promoted. This depends upon the concentration of the absorbing species and upon its nature. The limitations on the abilities of an electron transition occurring are called selection rules. For electric dipole radiation, which provides the only interactions of sufficient intensity to be important to solution chemists, the strict selection rules are as follows. Allowed transitions involve, (i) a change of one atomic unit in the electronic orbital angular momentum; (ii) no change in multiplicity (spin). That is, electric dipole radiation will give rise to a transition only if the symmetries of the ground and excited electronic states differ by the symmetry of a dipole, and if there is no change in electronic spin during this transition. In practice there are a number of conditions that make forbidden transitions allowed to some extent, so that weak absorptions do occur.

In the case of the spin selection rule, the presence of spin-orbit coupling makes spin forbidden bands allowed. Symmetry selection rules are based upon the absence of vibrations, and in certain cases non-totally symmetric vibrations make allowed an otherwise forbidden transition. An example of this is the weak transition of an n electron of nitrate ion to a π^* orbital, occurring at about 300 nm. This is symmetry forbidden between the planar ground state and a planar excited state, but transitions are possible from a vibrationally distorted (pyramidal) ground state to its excited state. Average values of ε_{max} for various transitions are: allowed, 10^4; symmetry forbidden, spin allowed, 10^2; spin forbidden 10^{-1}, 1 mol^{-1} cm^{-1}.

Of considerable significance to any discussion of electronic spectroscopy is the Franck-Condon principle: the time taken for an electronic transition is short compared with the times taken for nuclear movements. This means that the Franck-Condon excited state of a molecule, i.e. that existing momentarily after excitation, has the same nuclear geometry as the ground state from which the transition occurred. Further, the solvent molecules around the excited molecules also remain in the same position during the transition. Only after the transition is completed will the excited molecule in the Franck-Condon excited state and its solvation sphere (if any) rearrange into their new equilibrium (excited state) positions. This has an important bearing on the interpretation of solvent effects.

4.3 SOLVENT EFFECTS

Solvents may modify the shape, position or intensity of absorption bands. They do so because of the interaction of the solute with the solvent. Thus while solvent effects have been used more often in making spectroscopic assignments, they also provide information about solute-solvent interactions.

Changes in band shape result from changes in the vibrations of the solute. The loss or blurring in solution of sharp bands occurring in vapour phase spectra results from strong solute-solvent forces (such as dipole-dipole, ion-dipole or hydrogen bonding) which tend to break down vibrational quantisation. In solutions where dispersion forces only are significant, solute spectra continue to show a banded structure if they do so in the vapour spectrum.

Solvent shifts, that is changes in the energy of maximum absorption, are difficult to predict, since they are generally small, and are usually the resultant of several individual effects which may either cancel out or reinforce. In a clear analysis of solvent effects on absorption energies of intramolecular transitions, Bayliss and McRae[G17] have shown how a qualitative interpretation of solvent shifts is possible by considering, (a) the momentary transition dipole present during the optical absorption process; (b) the difference in permanent dipole moment between ground and excited states of the solute; and (c) the Franck-Condon principle. More recently Suppan[G18] has reviewed quantitative treatments of solvent effects on electronic transitions, indicating possible improvements in the value of the cavity radius and of the solvent polarity functions used.

In any solution dispersion forces are operative, and these invariably contribute a small shift towards the red of the absorption maximum.[G19] The magnitude of this polarisation shift is a function[G19] of the solvent refractive index (n), the transition intensity, and the size of the solute molecule. When solute-solvent forces are such that the solute is solvated, another contribution is the effect of the optical transition on the solvent cage. If as a result of the transition the dipole moment of the Franck-Condon excited state is less than that of the ground state, the solvent cage will be placed under strain. This will result in a smaller contribution of solute-solvent solvation energy to the total stabilisation of the excited state, relative to the situation in the ground state. That is, it will contribute a shift towards shorter wavelengths (a blue shift). The size of this contribution will depend upon the change in solute dipole moment, and the extent and type of solvation as discussed below. The overall resultant direction of the solvent shift will depend on the relative magnitudes of the dispersion shift and the shift due to solvent cage strain.

When the dipole moment of the solute is greater in the excited state, then, although the solvent cage becomes strained, the Franck-Condon excited state is formed in an already partly oriented solvent cage, and the excited state is more solvated than the ground state.

Four limiting cases have been distinguished for intramolecular transitions,[G17]

(a) Non-polar solute in a non-polar solvent.

Here only dispersion forces contribute to the solvation of the solute, so a red shift, proportional to solvent n is expected.

(b) Non-polar solute in a polar solvent.

In the absence of a solute dipole moment there is no significant orientation of solvent about the solute, so again a red shift is expected.

(c) Polar solute in a non-polar solvent.

As the solvent is non-polar the forces contributing to solvation are dipole-solvent polarisation and dispersion forces. If the solute dipole moment increases during the transition, the Franck-Condon excited state is more solvated by dipole-solvent polarisation, and a red shift, dependent upon solvent n and change in solute dipole moment is expected. If the solute dipole moment decreases during the transition, the Franck-Condon excited state is less solvated, thus contributing a blue shift, again proportional to the above two factors. In this latter case the resultant shift may be red or blue depending on the relative magnitudes of the polarisation red shift and the blue shift.

(d) Polar solute in a polar solvent.

Here the ground state solvation energy results largely from dipole-dipole and ion-dipole forces, so there is an oriented solvent cage. If the solute dipole moment is increased during the transition, the Franck-Condon excited state is formed in a cage of already partly oriented dipoles. This will result in a red shift, its magnitude depending on the magnitude of the change in dipole moment during the transition, the value of the solvent dipole moment, and the extent to which the solute and solvent molecules interact. This last includes hydrogen-bonding interactions.

If the solute dipole moment decreases during the transition the Franck-Condon excited state is in a strained cage of oriented dipoles, and this will contribute a blue shift. The superimposed polarisation red shift will usually be less, so the resultant will be a shift to the blue.

Figure 4.3.1 depicts these possibilities.

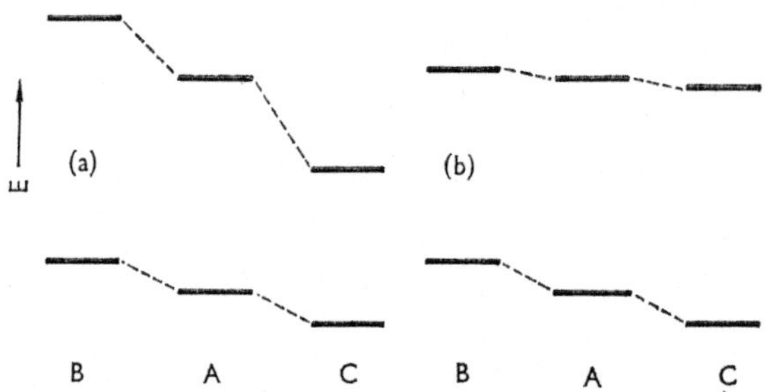

FIG. 4.3.1. Diagrammatic representations of solvent effects on the transition energies of polar solutes in polar solvents. *A* refers to a reference solvent, *B* to a less solvating solvent and *C* to a more solvating solvent. (*a*) Case where the dipole moment of the Franck-Condon excited state of the solute is larger than that of the ground state. (*b*) Case where the Franck-Condon excited state dipole moment is less than that of the ground state.

For an intermolecular transition (except charge transfer to solvent) the direction of the solvent shift is determined similarly. That is, if the Franck-Condon excited state has a larger dipole moment than the ground state, a red shift is observed with increase in solvating power of the solvent. Conversely, when the transition results in a decrease in dipole moment, there will be a blue shift contribution from solvent cage strain, which in most cases outweighs the dispersion red shift.

In the case of charge transfer to solvent spectra (of anions) the picture is more complex since these transitions do not involve a linear movement of charge and there is no simple correlation between solvent shifts and solvent polarity.[G20] In strongly solvating (and hydrogen bonding) solvents blue shifts are observed. In aprotic and poorly solvating solvents red shifts are observed. These shifts are frequently large by comparison with those observed with intramolecular transitions.

Analysis of solvent shifts requires a knowledge of the transition concerned, and how this changes the dipole moment of the absorbing species. In interpreting observed shifts it is necessary to distinguish between polarisation red shifts (proportional to solvent refractive index), and shifts dominated by solvent cage strain, when solvent dipole moment may be more important. Nevertheless, in many cases the use of a simplified, approximate treatment yields useful information about both the ground and excited states.[G18] Examples of solvent shifts are included in the discussion of Types of Transition (sect. 4.4) and in Selected Applications (sect. 4.6).

Solvent effects upon spectral intensity cannot be interpreted in a simple qualitative fashion. However, they offer propects of being a more sensitive probe of weak solute-solvent interactions when the theoretical models are more refined, since significant changes in the intensity of forbidden transitions may arise from weak intermolecular interactions which make little difference to the transition energy.

Mechanisms[G21–25] have been identified whereby the intensities of solute electronic transitions may be perturbed by the solvent. Forbidden transitions may gain intensity either from strong solute transitions, or from strong solvent transitions. Each particular case requires separate examination. Of interest among the mechanisms is the case of intensity increases in forbidden singlet-triplet transitions due to the presence of either a paramagnetic species in the solution, or a heavy atom in the solvent.[G23] Enhancements of up to 10^5 in the singlet-triplet bands of organic liquids containing oxygen (at increased pressures) have been observed.[G26]

4.4 TYPES OF TRANSITION

Deriving information about the states of molecules from their spectra requires a knowledge of the transitions being observed. To some extent the problem is systematised by regarding electronic transitions in complex molecules as localised in certain chemical groups or bonds that are present. However, unlike Raman and infra-red spectra, the assignment of an observed band to a specific transition in a polyatomic species is invariably difficult in view of the complexity of molecular orbital calculations of multi-electron species. As a result, many assignments must be regarded as tentative. Frequently, the observed spectra are analysed by using less mathematically demanding models. For example, the spectra of transition metal ions are usually separated into metal ion *d–d* transitions, intra-ligand transitions and metal-ligand charge transfer. However, if there is extensive mixing of metal and ligand wave functions in the molecular orbitals of the complex, it is necessary to describe the spectra in terms of molecular orbital theory.[G14] Such is the case for the

Table 4.4.1

Absorption Bands of some Anions in Solution

(From Reference G20 Except where Noted)

Anion	Solvent	Approx. λ_{max}(nm)	Approx. ε_{max} ($l\ mol^{-1}\ cm^{-1}$)	Assignment
Cl^-	aq.	175		CTTS
Br^-	aq.	198		CTTS
	AN	219		CTTS
I^-	aq.	226		CTTS
		193		CTTS
	AN	245		CTTS
OH^-	aq.	187	3×10^3	CTTS
SH^-	aq.	230	7.9×10^3	CTTS
S^{2-}	pH 15, aq.	360	weak	CTTS
Se^{2-}	aq.	417	unknown	unknown
CN^-	aq.	<182		
NCS^-	aq.	222	3.5×10^3	CTTS
N_3^-	aq.	189	$>1 \times 10^4$	intramolecular
		203	$\sim 4 \times 10^3$	CTTS
		233	4×10^2	$^1\Delta_u \leftarrow {}^1\Sigma_g^+$
HCO_3^-	aq.	222	<1	forbidden $n \rightarrow \pi^*$
		<192	$>4 \times 10^2$	unknown
CO_3^{2-}	aq.	270	<1	forbidden $n \rightarrow \pi^*$
		217		intramolecular
		~200	$\sim 2 \times 10^2$	forbidden $n \rightarrow \pi^*$
NO_2^-	aq.	355	22.5	$n_0 \rightarrow \pi^*$
		~287	9.4	$n_0 \rightarrow \pi^*$
		210	5.4×10^3	$\pi \rightarrow \pi^*$
		192	$\sim 5 \times 10^3$	CTTS?
	AN[G28]	370	24.2	$n_0 \rightarrow \pi^*$
		214	6.2×10^3	$\pi \rightarrow \pi^*$
	DMF[G28]	372	27.6	$n_0 \rightarrow \pi^*$
NO_3^-	aq.	303	7.0	$n \rightarrow \pi^*$
		202	9.9×10^3	$\pi \rightarrow \pi^*$
	(other solvents, see Table 4.4.2 and 4.4.3)			
HPO_4^{2-}	aq.	208	<1	intramolecular
		270	<1	intramolecular
HSO_3^-	aq.	<182	weak	unknown
SO_3^{2-}	aq.	227	$>10^3$	CTTS
HSO_4^-	aq.	<167	$>10^2$	CTTS
SO_4^{2-}	aq.	175	3×10^2	CTTS
$S_2O_3^{2-}$	aq.	<200	$>4 \times 10^3$	intramolecular
		215	4×10^3	CTTS
		244	2×10^2	intramolecular
$HSeO_3^-$	aq.[G29]	<185		CTTS
		211	1800	$n \rightarrow \pi^*$
	AN[G29]	191	5600	CTTS
		~222		$n \rightarrow \pi^*$

Anion	Solvent	Approx. λ_{max}(nm)	Approx. ε_{max} (l mol^{-1} cm^{-1})	Assignment
SeO_3^{2-}	aq.[G29]	<185		CTTS
		~200		$n \rightarrow \pi^*$
	EtOH[G29]	206	5100	$n \rightarrow \pi^*$
	aq.[G29]	~185		$n \rightarrow \pi^*$
$HSeO_4^-$	AN[G29]	190		$n \rightarrow \pi^*$
SeO_4^{2-}	aq.[G29]	<185		CTTS
		~195		$n \rightarrow \pi^*$
	EtOH[G29]	199	450	$n \rightarrow \pi^*$
ClO^-	aq.	291	3.1, 3.6 × 10^2	intramolecular
BrO^-	aq.	333	3.3 × 10^2	intramolecular
IO^-	aq.	365	31	intramolecular
		255	400	
		<238	>10^3	
ClO_3^-	aq.	<200	>10^3	intramolecular
BrO_3^-	aq.	<200	>10^3	intramolecular
IO_3^-	aq.	<200	>10^3	intramolecular
ClO_4^-	aq.	<180	>20	intramolecular
IO_4^-	aq.	~217	~10^4	intramolecular
XeO_4^{2-}	aq.	250	6 × 10^3	unknown
MnO_4^-	aq.[G30]	~500		mixed $t_1 \rightarrow 2e$
				$3_{t2} \rightarrow 2e$
		400–333		$t^1 \rightarrow 4t_2$
		303		mixed $t_1 \rightarrow 2e$
				$3t_2 \rightarrow 2e$
		229.9		mixed $3t_2 \rightarrow 4t_2$
				$1e \rightarrow 4t_2$

tetrahedral MO_4^{n-} species, e.g. MnO_4^-, CrO_4^{2-}, also ClO_4^- and SO_4^{2-}, although the appropriate molecular orbitals of many of these remain uncertain.[G27]

The polyatomic ions most commonly used in studying electrolyte solutions are among those for which the least concordant results exist. Table 4.4.1 summarises some more recent assignments for a number of these. In some cases it seems likely that other less intense bands also exist, but have not been recorded, while as mentioned in sect. 4.5 the results for short wavelength bands should be viewed with caution. There are several compilations of spectral data appertaining to organic species.[G1,G9,G15,G28–29]

While there are several conventions in use for referring to electronic transitions, for intramolecular transitions that of enumerating only the molecular orbitals involved is perhaps the most applicable for the present purpose.

4.4.1 $n \rightarrow \pi^*$ Transitions

These are intramolecular transitions associated with the presence of lone pair electrons e.g. C = O, S = O, NO_2^- and NO_3^-, the singlet-singlet (spin allowed) transition generally being observed as a weak band ($\varepsilon_{max} \sim 10$–$10^2$) at about 300 nm, and the spin forbidden triplet occurring as a very weak absorption ($\varepsilon_{max} \sim 10^{-3}$) about 3000 cm^{-1} to the red.[G15,G33] The transition intensity is low not only with molecules where it is symmetry forbidden, but vibronically allowed, but also in cases where the macrosymmetry of the species is such that the transition appears to be allowed.[G33]

In hydrocarbon solvents the spectra show a vibrational structure. Some signs of vibrational structure are discernible in the spectra of NO_2^-[G28] and NO_3^-[G34] in DMF at room temperature. The red shift of $n \rightarrow \pi^*$ transitions, observed with decrease in dielectric constant and hydrogen bonding ability of the solvent, is consistent with the Franck-Condon excited state having a smaller dipole moment than the ground state. This is as expected when an electron is promoted from a localised lone pair to a more diffuse π^* orbital.

Table 4.4.2 presents some results for the nitrate ion.[G34–36]

Table 4.4.2

Solvent Effects on the Singlet $n \rightarrow \pi^*$ Transition of Nitrate Ion

Solvent	λ_{max}(nm)	ε_{max} (l mol^{-1} cm^{-1})	Ref.
Water	301.5	7.30	G35
	301.5	7.44	G34
	301.9		G36
DMF	313.9	4.02	G35
	313.2	4.01	G34
AN	312.7	4.47	G35
	311.6	4.83	G34
MeOH	303.0		G36
	303.0	6.83	G34
EtOH	303.4	6.35	G35
	303.4	6.36	G34
	303.8		G36
Ethylene glycol	302.5		G36
	302.5	7.37	G34

On the basis of intensity measurements and low temperature crystal spectra other assignments have been suggested[G34] for the nitrate band at *ca.* 300 nm (e.g. $n \rightarrow \sigma^*$), but the general consensus of opinion[G35–36] favours $n \rightarrow \pi^*$. Solvent effects on either transition should be the same, as each will involve the transfer of negative charge from an oxygen atom to the nitrogen atom, leading to a more dispersed charge.

The intensity of forbidden $n \rightarrow \pi^*$ transitions is also solvent dependent, but interpretation is not simple. For example, in a study of some forbidden $n \rightarrow \pi^*$ transition intensities[G22,G37] neither the properties of the solvent alone, nor of the solute alone, was adequate to explain the observed perturbations. However, it may be possible to use changes in band *area* with variation in solvent composition to assess changes in ground state solvation.[G38]

4.4.2 $\pi \rightarrow \pi^*$ Transitions

These intramolecular transitions result in the characteristic intense bands of unsaturated organic compounds, observable in the usual solvent range of $\lambda > 200$ nm. They also produce the short wavelength cut-off of solvents with a π electron system. For example the intense band centred around 198 nm in DMF and DMA is considered to consist of $n \rightarrow \pi^*$ and $\pi \rightarrow \pi^*$ transitions.[G39–40]

$\pi \rightarrow \pi^*$ transitions may or may not result in the Franck-Condon excited state possessing a different dipole moment, as the transition may remain relatively localised, in which case little change occurs. Thus, the solvent shifts observed may be red, blue or negligible. Table 4.4.3 contains some selected examples of solvent effects on the wavelength of maximum absorption for several different species.

Table 4.4.3

Solvent Effects on the Singlet $\pi \rightarrow \pi^*$ Transitions of Several Species

Solvent	λ_{max}(nm)			
	NO_3^- (Ref. G35) ($E^1 \leftarrow A_1{}^1$)	*p*-NO_2PhOMe (Ref. G41)	*p*-NO_2PhOH (Ref. G41)	*p*-NO_2PhO$^-$ (Ref. G41)
H_2O	201.9	316.4	317.4	403.2
AN	205.3	307.7	307.7	
DMA		314.5	321.5	434.8
MeOH		305.8	311.5	
EtOH	202.6	304.9	314.5	
Cyclohexane		295.9	287.4	

The wavelength change of the nitrate ion maximum can be interpreted as resulting from the excited state having a smaller dipole moment,[G35] so that the electronically excited state is formed in a strained solvent cage, the extent of strain depending *inter alia* on the solvating power of the solvent.

The solvent shifts of the *p*-nitrophenyl compounds have been interpreted[G41] using *p*-nitroanisole as a reference. The spectrum of *p*-nitroanisole shows a red shift of the $\pi \rightarrow \pi^*$ band with increasing solvent polarity, indicating an increase in dipole moment with the electronic transition. The spectrum of *p*-nitrophenol shows a red shift as the solvent becomes a stronger hydrogen-bond acceptor, a result of the excited state being a stronger hydrogen-bond donor than the ground state. On the other hand the spectrum of *p*-nitrophenoxide ion (like nitrate ion) shows a blue shift with increase in solvent hydrogen-bond donor strength, interpreted as resulting from a reduction in charge around an oxygen atom during the transition and hence a decrease in solute-solvent solvation, giving rise to strain of the solvent cage.

4.4.3 $n \rightarrow \sigma^*$ Transitions

These occur at the UV end of the usual range for solution spectrophotometry, and in general have lower extinction coefficients than $\sigma \rightarrow \sigma^*$ transitions.[G15] The band due to an $n \rightarrow \sigma^*$ transition is removed by protonation or co-ordination of the lone pair. The presence of a heteroatom carrying lone pairs generally lowers the UV cut-off of a solvent to *ca.* 200 nm (e.g. H_2O and MeOH have UV cut-offs of *ca.* 180 nm and 200 nm respectively).

4.4.4 $\sigma \rightarrow \sigma^*$ Transitions

These transitions occur in the far and vacuum UV region of the spectrum and result in the intense absorptions which limit the short wavelength transparency of saturated hydrocarbon solvents (e.g. cyclohexane has a cut-off of *ca.* 180 nm).

4.4.5 *d–d* and *f–f* Transitions

These are characteristic of complexed, transition and lanthanide metal ions, and the theory relating to them is well described in inorganic texts and reviews.[G2,G5,G7,G14,G27,G42–44] Although forbidden (Laporte rule) they are allowed through asymmetric vibrations; bands with ε_{max} of 10–100 l mol^{-1} cm^{-1} being usual for octahedral complexes. Metal *d–d* transitions are comparatively insensitive to the environment of the complex, although not completely so, as shown in Table 4.4.4. They are, of course, sensitive to changes in symmetry at the metal ion, and changes in ligands, aspects frequently used in determining transition metal-ion solvation, and in kinetic (see Chapter 6) and theoretical studies.[G14,G27,G42–44]

Table 4.4.4

Data for Absorption Maxima of some *d–d* Transitions in Several Solvents (λ_{max} in nm, ϵ_{max} in l mol^{-1} cm^{-1})

Compound	Water[G45]		Methanol[G45]		Acetone[G45]		DMF[G46]		DMA[G47]	
	λ_{max}	ε_{max}	λ_{max}	ε_{max}	λ_{max}	ε_{max}	λ_{max}	ε_{max}	λ_{max}	ε_{max}
cis-$[Coen_2Cl_2]^+$	525	72.5	526	78.7			545	106	545	108
trans-$[Coen_2Cl_2]^+$	617	36.0	610	~37.0			622	41	620	39
	455	27.0	455	25.9			460†	28	460†	26
	387	36.0	387	45.0			388	48	394	47
cis-$[Coen_2(SCN)_2]^+$	487	340.0	492	408.0	492	439*				
	308	2900.0	326	2780.0	332	3310*				
$[Cr(SCN)_6]^{3+}$	564	159.0	554	114.0	556	103				
	420	129.0	555	92.0	414	83				
	309	25000.0	309	28000.0						
	234	19200.0	234	(25500)						

* Acetone with 10% CH_3OH.
† Shoulder.

Transitions between *f* orbitals give sharp lines, the number and positions of the components in each line group being very dependent on the symmetry of the metal ion. The energy levels and intensities of the solution spectra of a number of +3 lanthanides have been collated by Carnall *et al.*[G48]

4.4.6 Charge Transfer

The spectra obtained in an apparently wide variety of chemical situations are described as 'charge transfer'. In fact they all have two things in common. They involve the transfer of an electron from a donor having a high energy, filled orbital, to an acceptor having a low energy, vacant orbital, and there is orbital overlap. One group of charge-transfer spectra are those intramolecular transitions to which are attributed the intense, short wavelength absorptions of many transition metal complexes. The far UV absorption of SO_4^{2-} and ClO_4^- is also sometimes attributed to such transitions.[G20,G27]

Intermolecular charge transfer can be subdivided into three different groups:

(i) Charge-transfer complexing is well known[G49–53] and is characterised by the formation of a new absorption band. It must be considered whenever the solvent is a good electron donor, e.g. dipolar aprotic solvents. Thus contact charge-transfer complexes of dissolved oxygen and many solvents are known.[G54–58]

(ii) Charge transfer to solvent spectra (CTTS) which have been reviewed recently[G20] are somewhat different, as the promoted electron is not generally transferred to any special solvent orbital but rather resides within the same solvent cage as the donor ion.[G20] These spectra have provided some useful information about solute-solvent interaction.[G20]

(iii) Ion association, in which there is orbital overlap, also produces significant change. In the absence of orbital overlap the extent of perturbation of a spectrum by ion association is negligible (sect. 4.6.1).

4.5 EXPERIMENTAL ASPECTS

Many texts describe the applications of spectrophotometry to analytical problems and the underlying Beer-Lambert laws (e.g. ref. G59), but there are several practical points relevant to working in a wide range of solvents which are worth emphasising. It is necessary to remember that the Beer-Lambert laws fail, not only when there are changes in the absorbing species, as in ion-association or charge-transfer complexing, but also if the sample is dichroic, or contains light scattering particles. The finite width of the wave band passed also results in deviations from the laws, these increasing as the difference in absorbance of the maxima and minima increases. With broad solution bands and moderate slit widths the deviations are generally within experimental error.[G59,G60]

Generally, when determining association constants (ion-pair or charge-transfer) the spectrum of the associated species is unknown and a Benesi-Hildebrand type plot[G61] is used. Several sources of experimental error in these spectrophotometric methods were investigated by Davies and Prue[G62] who suggested modified techniques. More recently the consequences of attributing all deviations from Beer's law to the formation of one species with an unknown spectrum, have been examined.[G63–69] The occurrence of overlapping equilibria can be detected sometimes by determining K at a number of wavelengths,[G65,G70] and the dependence of K upon the concentration scale used, points to solvent interactions.[G66] A check upon the validity of a particular interpretation of the spectral changes observed, requires an examination of the system by other physical methods, although actual values of K may not be identical (see sect. 4.6).

When changing to a previously unused type of solvent it is necessary to confirm that the same species as before are actually present in solution (sect. 4.6 includes examples of solvents influencing ion association, chemical reactions such as a redox reaction, ligand exchange of a transition metal ion, and dimerisation). It is also necessary when comparing spectral intensities in different solvents to measure the width at half height ($\bar{\nu}_{\frac{1}{2}}$ in cm^{-1}) as well as ε_{max}, since the correct measure of intensity is the band area.

Computer resolution of a single, complex envelope into component bands (e.g. ref. G71) tends to mask the assumptions made. In general an unequivocal fit requires more information than the number and shape of the component curves. Electronic absorption bands are usually treated as having a Gaussian shape when plotted on a frequency scale (e.g. ε *vs.* $\bar{\nu}$), but other shapes have been suggested.[G72]

Several complications are possible when working at shorter wavelengths ($\lambda < 220$ nm) and published spectra of this region should be treated with caution. Stray light usually leads to spurious maxima as shown in Fig. 4.5.1. (Similar anomalous behaviour can be observed in the far red if stray light is present.) The detection and allowance for stray light have been described.[G59,G60,G73–75] An example of the effect of stray light would seem to be in earlier reports of a maximum at 185 nm in the spectrum of sulphate ion[G76] whereas more recent measurements show a maximum at 175 nm.[G77] Even this peak, measured as it was on the side of the water band, may be spurious for a reason discussed below.

Dissolved oxygen in the solvent results in enhanced absorption in many cases,[G54–58] due to the occurrence of contact charge transfer.[G54,G55,G57] This is in addition to the possible enhancement of triplet spectra mentioned earlier. The removal of dissolved oxygen can extend significantly the short wavelength cut-off of some solvents.[G56] Oxygen gas in the light path of a spectrophotometer also leads to spurious spectra at wavelengths shorter than *ca.* 200 nm.[G78]

It has been shown that spurious maxima or minima occur in a solute spectrum measured at the end of (or through) a solvent absorption band.[G79] They occur at points where the slopes of the absorption curves of the solute and solvent are in the ratio of their molar volumes. Their effect is more important when the solute absorption is weak.

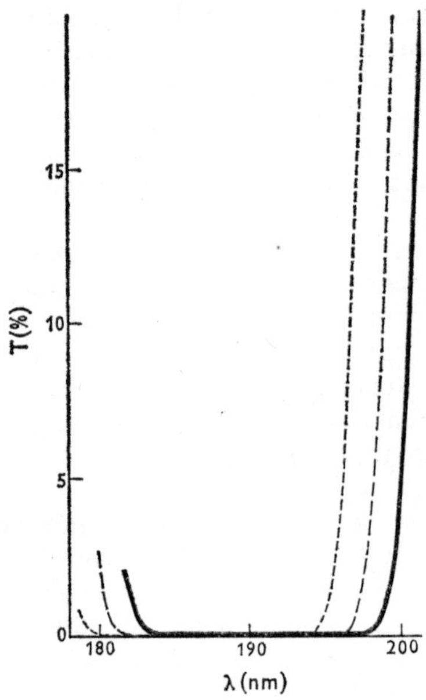

FIG. 4.5.1. Stray light effects in the low U.V. region, obtained using 0.099 *M* aqueous KCl solution, a water reference, and a nitrogen purged spectrophotometer. Cell path lengths of 10 mm ——, 5 mm — — and 2 mm - - -. (A. K. Covington and I. R. Lantzke, unpublished results.)

4.6 SELECTED APPLICATIONS

The most frequent application of electronic spectroscopy to the study of electrolyte solutions is as a method of quantitative analysis. In such analytical uses the spectroscopy of the system must be considered. In many cases it seems probable that the species 'seen' by electronic spectroscopy are not identical with those measured by other techniques. The problem as related to ion association has been recognised for several years.[G80,G81] The basic argument is that for a system of the type

$$\underset{\binom{\text{covalent}}{\text{molecule}}}{AB} \rightleftharpoons \underset{\binom{\text{ionised}}{\text{molecule}}}{A^+B^-} \rightleftharpoons \underset{\binom{\text{contact}}{\text{ion-pair}}}{A^+ \ldots B^-} \rightleftharpoons \underset{\binom{\text{solvent separated}}{\text{ion-pair}}}{A^+ \ldots S \ldots B^-}$$

$$\rightleftharpoons \underset{\binom{\text{solvated}}{\text{ions}}}{A^+(S) + B^-(S)}$$

(where S represents an unspecified number of solvent molecules), electronic spectroscopy may well 'see' all species except the solvated ions as being identical 'molecules', whereas Raman spectroscopy for example is only expected to 'see', as molecules, those species with some covalent character in the bond. Unfortunately one does not, in general, know exactly the boundary up to which each technique measures.

At the same time it is generally necessary to make some assumptions about activity coefficients, the most usual being that all types of ion pairs have the same activity coefficient as the undissociated molecules, *viz.* unity in the above case. Therefore, although it is always desirable to use several physical techniques to confirm the nature of a process under investigation, the values of the apparent thermodynamic parameters obtained by different techniques may be expected to be different.

There are cases, however, involving precise and careful work showing good agreement between techniques. For example, spectrophotometric and calorimetric determinations of ΔH for ionisation in water of a number of substituted phenols and anilines agree well[G82] (within 420 J mol^{-1}).

Spectrophotometric methods of obtaining thermodynamic parameters of ionic equilibria often require comparatively high ionic strengths. This makes desirable an extended Debye-Hückel equation, or experimental values of the relevant ionic activities. In non-aqueous solvents such information is frequently still unavailable (see Chapter 2).

Electronic spectroscopy is also used in making qualitative analyses of systems and to gain insight into the state or nature of a solute or solvent. Except on the rare occasions where the theoretical model is well developed, such studies take the traditional form of observing the spectral changes produced by variations of some suitable parameter of the system. Variation in solute spectrum may result from changes in its shape, in its nature, or just some modification of the transitions by the solvent. The following arbitrary selection of examples illustrates occasions where these structural, chemical and solvation effects modify the observed spectra.

4.6.1 Ion Association

Spectroscopy has been used widely in investigating ion association,[G83,G84] although it seems applicable only in cases where there is some orbital overlap of the associating ions. There are several systems where only electrostatic interaction is postulated and there are no detectable deviations from Beer's Law, but other evidence shows the occurrence of association. Examples of such systems are sodium and potassium tetraphenylborate,[G71] thallous hydroxide[G85] and various tetralkylammonium halides.[G86]

It might be expected that small changes in spectra, similar to those of solvent effects might result from electrostatic interactions, but if they do they are generally less than the experimental error of current practice, since Beer's Law usually holds even in concentrated solutions. However, there are occasions when a generalised electrostatic interaction produces spectral shifts. Addition of tetraethylammonium halide to solutions of *p*-nitrophenol in DMA produced red shifts in the absorption band attributed to hydrogen bonding of the halide ion by the phenolic proton, without any evidence for the formation of a specific complex.[G41] Added electrolytes and non-electrolytes also produce wavelength shifts in CTTS spectra with no apparent specific complexation.[G20]

When association does produce spectral change the form such change takes is variable. Cases are known of band shift, or of change in extinction coefficient, or perhaps only one band in the total spectrum is influenced. This latter is the case with transition metal complexes, ion association usually resulting in the appearance of a new band. With most complexes such ion-pair bands occur at shorter wavelengths than the *d–d* transitions, which thus remain unaltered. For example the ion-pair band of cobalt (III) complexes appears at *ca.* 300 nm, on the side of the metal-ligand charge transfer band. However the ion-pair band of trisethylenediamine-ruthenium(III) iodide ($[Ru\ en_3]^{3+} \ldots I^-$) occurs at 450 nm ($\varepsilon \sim 10^3$ to 10^4 l mol^{-1} cm^{-1}) which is a longer wavelength than that of the *d–d* transitions (310 nm, $\varepsilon = 360$ l mol^{-1} cm^{-1}).[G87]

Results, available to 1968, of spectrophotometric and conductance studies of ion-association of anions, with several bisethylenediamine-cobalt(III) complexes in dipolar aprotic solvents, have been summarised by Watts,[G88] who has emphasised the importance of the solvent in determining the magnitude of K_A. Of greatest significance is the ability or otherwise of the solvent to provide protons for hydrogen-bonding interactions with anions, values of K_A in aprotic solvents being up to 10^3 greater than in a protic solvent of similar dielectric constant.[G88] The agreement between conductance and spectrophotometric results is fair, but too few systems have been duplicated to determine if any trend occurs. As with the alkali metal-carbanion ion pairs described below, the concept of a preferential ion-pair geometry has been elaborated.[G88] In the case of *cis*-dichlorobisethylenediaminecobalt(III) chloride, NMR evidence has confirmed the expected structure of the chloride ion adjacent to the edge of the octahedron remote from the chloride ligands.[G89]

Mostly it is not possible to distinguish between contact and solvent separated ion pairs but, in an interesting series of experiments involving alkali metal ion-carbanion ion pairs, the difference has been demonstrated.[G90–93] This work has required solvents stable to redox reactions, and the choice has fallen on organic ethers with low dielectric constants and small dipole moments. These solvate anions poorly. Examples of the systems investigated are various alkali metal fluorenyl salts,[G91] sodium biphenyl[G94] and sodium and lithium dihydroanthracene.[G95] In these systems[G91–95] the contact ion pair absorbed at shorter wavelength than the solvent-separated ion pair ($\Delta E \sim 4$ kcal mol^{-1} (1.67×10^4 J mol^{-1})), so the equilibrium between the two types of ion pairs could be evaluated,[G91] and the effects of added complexing agents used to investigate the ion-pair structure.[G94–95] This work is particularly interesting when considered in conjunction with that on low frequency vibrations of alkali metal ions in solvent cages, described in sect. 4.10.

The transition observed is considered to be of a $\pi \rightarrow \pi^*$ type, being from the lowest unoccupied molecular orbital of the conjugated species (ion pair) to the next above.[G90] The differences between the spectra of the contact and solvent-separated ion pairs, and between different cations with the same anion, have been explained as a result of the electronic transition on the dipole moments of the two different types of ion pair. This change in dipole moment depends upon the polarisation direction of the transition and the direction of the line joining the centres of the opposing ions in the ion pair. In the cases reported, the optical transition has resulted in the excited state of the solvent-separated ion pair being formed in a more favourably oriented solvent cage by comparison with that of the contact ion pair (cf. 4.3 Solvent Effects).

An interesting variant, from determinations of ion association using spectroscopy, has been its use in exploring the properties of mixed solvents with an ion pair—free ion equilibrium as a probe.[G96–98] The system chosen was that of *bis-meso*-(stilbenediamine)nickel(II) acetate ($[Ni(stien)_2](OAc)_2$) for which there is a solvent -and anion-dependent equilibrium between a diamagnetic yellow form (probably square planar or distorted tetrahedral) and a paramagnetic blue octahedral form. The equilibrium, Yellow $\rightleftharpoons$ Blue, was followed with change in solvent composition, its variation being related via changes in solute solvation to changes in solvent structure. The solvent systems investigated were aqueous mixtures of methanol and urea,[G96] various alcohols, ketones and amides,[G97] and glycol ethers and poly(ethylene oxide).[G98]

4.6.2 Transition Metal Complexing

Changes in *d–d* transition spectra can be used to characterise and identify solvent and anion co-ordination to a transition metal ion or complex. (Unfortunately there are no equally convenient transitions available with non-transition metal ions.) While the number of transition metal-solvent complexes now known is too large to detail here, it is worth noting that sometimes these compounds have been used to further spectroscopic theory, but at other times spectroscopy has been used to assess their stability.

Precise stability constant values can be calculated from such spectroscopic data provided two conditions are met. Firstly, and in general, the spectra of all the light absorbing species present in solution must be known accurately, and be different. Secondly, activity coefficient data are needed for the solutes in the solvent concerned. As mentioned earlier, in most organic solvents this latter information is sketchy or unavailable, and more often than not, for multistage equilibria the first criterion cannot be satisfied either.

For this latter reason, for example, it was not possible to use spectroscopy to evaluate the various equilibria in the hydrolysis of uranyl ion in perchlorate media.[G99] Without assignment of all bands to their appropriate transitions it was not possible to rule out band crossing or overlapping, with pH change.

Some examples of solvents affecting spectra through changes in solvation or chemical reaction follow. However, it is often impossible to distinguish between solvation, ion association and co-ordination on observations of electronic spectra alone and recourse must be made to other evidence, e.g. conductance or NMR.

When hexaquochromium(III) perchlorate is dissolved in aqueous mixtures of DMSO the absorption maxima of the two *d–d* bands shift to lower energies (407 and 575 nm to 444 and 634 nm respectively) as the mole fraction of DMSO increases.[G100] This is interpreted as arising from changes in the average number of co-ordinated DMSO ligands, the wavelength change resulting from the different ligand field strengths of DMSO and H_2O.

Conversely no regular changes in spectra were observed with increased coordination of solvent acetonitrile when $RhCl_3.3H_2O$ or in one case $RhCl_6^{3-}$ were dissolved in acetonitrile and its aqueous mixtures.[G101] The geometry of the four chloro(acetonitrile)rhodium(III) complexes so prepared was elucidated using IR and NMR spectroscopy. From the report,[G101] the observed *d–d* transitions ($^1A_{1g} \rightarrow {}^1T_{1g}$ and $^1A_{1g} \rightarrow {}^1T_{2g}$) appeared to be sufficiently different to make some qualitative assessment of the ligand preferences for rhenium of H_2O, AN and Cl^-. Unfortunately no such assessment was made.

A qualitative assessment of the relative ligand strengths of interaction between nitrate ion, water and four organic solvents and a number of $+3$ lanthanide ions was made from changes in *f–f* transitions.[G102] The affinity series of the lanthanides for the nitrate ion and the solvents was DMF $>$ tributylphosphate $>$ NO_3^- $\sim$ H_2O $>$ EtOH $>$ dioxan. In DMF, hexadimethylformamide-lanthanides were the only complexes present, although conductivity measurements showed a major portion of the nitrate ion to be ion-paired, while in anhydrous dioxan the solvent-solute interaction is so weak that the rubidium-lanthanide nitrate double salts employed were not soluble. In water, in the absence of excess nitrate ion, hexaquo-lanthanide complexes predominate with little nitrate ion-association,[G103] in line with better nitrate solvation in protic solvents.

With more robust complexes spectral changes are more likely to result from solvent effects, rather than from ligand exchange. For example the observed shifts in the absorption maxima of spectra of neodymium acetylacetone and neodymium benzoylacetone complexes in methanol-water mixtures were due to solvent effects.[G104]

The application of spectroscopy to kinetic studies of ligand replacement is illustrated in Chapter 6.

Chemical reactions other than ligand exchange may also produce spectral changes. A number of metal ions in higher oxidation states (including Fe(III)) undergo spontaneous reduction to a lower oxidation state when dissolved in acetonitrile.[G105]

The state of aggregation of the solute may depend upon the solvent and so lead to solvent dependent spectral changes. In solutions of Fe^{3+} ions there has been uncertainty whether the species present were monomeric, dimeric or larger aggregates.[G106] In strongly co-ordinating solvents iron(III) complexes seem to be monomeric,[G107–111] except in the presence of alkali.[G111–113] Dimeiisation in such poorly co-ordinating solvents as ether, benzene, or toluene may result from the presence of traces of moisture. Scrupulously dry solutions of ferric chloride[G114] and *N*-substituted salicylaldimine complexes of iron(III)[G115] in these and similar solvents are monomeric, but traces of moisture cause the latter at least to dimerise.[G115]

Studies of the complexes of iron are important because of their relevance to the biologically important iron porphyrins. *In vivo* the haem and haemin prosthetic groups are buried in protein, so that their chemical behaviour in organic solvents is likely to be a better guide to their behaviour in the protein environment than is their behaviour in water. Spectroscopy is often used to monitor reactions of haems and haemins, their spectra being very characteristic. These consist of the Soret band, an intense allowed $\pi \rightarrow \pi^*$ transition at 380–420 nm ($\varepsilon \approx 60 - 170 \times 10^3$ l mol^{-1} cm^{-1}) and at longer wavelengths vibronically allowed $\pi \rightarrow \pi^*$ transitions ($\lambda \approx 500 - 600$ nm, $\varepsilon \approx 10 \times 10^3$ l mol^{-1} cm^{-1}) and charge transfer transitions in the far red ($\varepsilon \approx 10^3$ l mol^{-1} cm^{-1}).[G116] The number, position and intensity of the bands observed are very dependent upon the symmetry of the porphyrin ring and the spin state of the iron.[G116–118]

Since changes in the haemin moiety produce large changes in their spectra, they provide a good example with which to illustrate some influences of solvents on solute environment and solution structure.

When chloroprotohaemin (chloroprotoporhyrin iron(III)) is dissolved in DMSO, monomeric bis(dimethylsulphoxide)protohaemin is the predominant complex formed.[G110,G111] But when chloroprotohaemin is dissolved in DMA the whole spectrum is markedly different, in particular the Soret band has an extinction coefficient approximately half that observed in DMSO, although its integrated intensity remains the same. In DMA the predominant complex species is chloro(dimethylacetamide) protohaemin.[G111] The spectra of solutions of chloroprotohaemin in aqueous alkali[G113] and DMSO containing alkali[G111] are different again, the Soret band being much less intense and separated into two peaks, while the other bands are also modified. These changes result from dimerisation of the complex which distorts the porphyrin rings[G116] and modifies the spectra considerably.

Besides considering the stability of the solute in various solvents it is also necessary to consider solvent stability. Thus the amide solvents such as DMF, DMA and HMPT slowly decompose to yield amines which are readily discerned by their smell. But less obvious effects may be overlooked, for example the autoxidation of dioxan, whereby solutions of SnI_4 in dioxan show changes in spectrum on standing which are due to the formation of SnI_2 and I_2 as a result of this reaction.[G119] Oxidation reactions of DMSO have also been reviewed recently.[G120]

4.6.3 Solvent Effects

A major solvent effect, frequently influencing the choice of solvent when a solution spectrum is required, is the stability of the solute in various types of solvent. This point has been referred to earlier but a further example illustrating the stability to redox reactions of DMSO is the report of the spectrum of superoxide in this solvent.[G121] Potassium superoxide dissolved in DMSO had an absorption band at 246 nm ($\varepsilon \sim 1200$ l mol^{-1} cm^{-1}) which was stable for several hours.[G121]

Here we are more concerned with solvent shifts of band positions, that is with the effects of solvents on optical transition energies. Such shifts are usually quoted in wave numbers, although energy values are sometimes given. While solvent shifts have been used frequently as an aid in assigning transitions, their application in determinations of ion-solvent and ion-ion interaction are of more interest to the present discussion, although also of relevance are studies of factors influencing solvent shifts. Scales of solvent polarity based on solvent shifts are relevant too and are also mentioned below.

In investigations of solvent effects two experimental approaches exist. One is to observe $\bar{\nu}_{max}$ for a particular solute transition in a variety of solvents and compare the results with theoretical predictions. The other involves measuring $\bar{\nu}_{max}$ for the band in a range of solvent mixtures (usually binary) and comparing the shift with some function of the mixed solvent. Obvious experimental limitations in either approach are the need for the solute to have a transition in an accessible spectral region, not too overlapped by other bands, and for it to undergo measureable shifts with solvent change. For these reasons mostly organic ions and molecules have been studied, as there are very few inorganic ions with

$n \rightarrow \pi^*$, $\pi \rightarrow \pi^*$ or CTTS transitions not obscured in many organic solvents.

When comparing shifts in a variety of solvents the usual assumptions are that shifts arising from ion-dipole and dipole-dipole interaction are linear in some function of solvent dielectric constant[G18,G122] (ε_r) (e.g. $f(\varepsilon_r) = [2(\varepsilon_r - 1)]/(2\varepsilon_r + 1)$ or $(\varepsilon_r - 1/(\varepsilon_r + 1))$ and this is generally the case. But solvents in which additional solute-solvent interaction occurs give results which do not lie on such plots of $\bar{\nu}$ against $f(\varepsilon_r)$.

Suppan[G18], taking the case of hydrogen bonding, and using aniline and 3-amino-4-methoxyacetophenone as examples, has made the following points. In a hydrogen-bonding solvent the amount of deviation from the $\bar{\nu}$ against $f(\varepsilon_r)$ plot is different for different absorption bands of the same solute species. The most anomalous bands are those involving transitions of an electron which is specifically hydrogen-bonded in the ground state. The magnitude of this hydrogen-bonding anomaly seems to depend both on charge transfer to or from the bonded site and on the strength of the hydrogen bond.

The measurement of solvent effects in mixtures should provide information about the environment of the solute, but interpretation is not always clear cut. Change in the bulk dielectric constant can be important but with ionic solutes specific solvation is frequently more important. In ideal and regular solutions, mole fraction statistics apply to the bulk solvent properties, so that non-linear plots of $\bar{\nu}$ against solvent mol fraction show preferential solvation.[G123] But where the solution is far from ideal, either volume fractions or a more complex function may well be in order,[G124,G125] although mol fractions are still usually used.

A recent study of preferential solvation[G123] showed that although the functions $f(\varepsilon_r)$ of binary mixtures of DMF with ether, and ethanol with cyclohexane, are nearly linear in solvent mol fraction, the variation of $\bar{\nu}_{max}$ for solutions of five polar organic molecules was not. Using a thermodynamic argument and assuming an ideal solvent mixture, preferential solvation by the polar-solvent of one of the solutes studied (4-nitroaniline) was predicted down to 0.05 mol fraction DMF in the DMF-ether mixtures, agreeing well with the observed shifts. In the ethanol-cyclohexane mixtures where hydrogen bonding also occurs the experimental results were not in such good accord.

Because the nitrate ion has a readily accessible $n \rightarrow \pi^*$ transition at ~300 nm it has been the subject of a number of studies,[G34–37,G126] many in water, but some in other solvents. In order to observe solvent effects on the nitrate ion alone it is necessary that the counter ion does not interfere; tetralkylammonium ions are suitable. Provided this is the case and there is no ion association, $\bar{\nu}_{max}$ shifts to the red with decreasing solvent dielectric constant and with decreasing ability of the solvent to donate hydrogen bonds. ε_{max}, or more accurately the integrated intensity of the band, also changes, decreasing with decrease in solvent dielectric constant and with decreasing solvent hydrogen-bond donor ability. These observations are interpreted as resulting from a change in solvation of the ground state. The hydrogen-bonding solvents form weak complexes with nitrate ion, as shown by the existence of an isosbestic point in spectra in methanol-acetonitrile mixtures,[G32] stabilising the ground state. At the same time this perturbs the D_{3h} symmetry of the ion, enhancing the transition probability. Solvent effects on the vibrational spectra of nitrate ion confirm this (see sect. 4.11).

When the cation of the nitrate salt ion-pairs with the nitrate, significant shifts in $\bar{\nu}_{max}$ are observed with changes in salt concentration, or type of cation, or with solvent, since the extent and magnitude of ion-ion interaction varies. Thus the band at *ca.* 300 nm in spectra of copper(II) nitrate measured in acetonitrile-water mixtures[G127] showed negligible change on addition of acetonitrile up to $x_{AN} \sim 0.4$ ($\lambda = 302.5$ nm, $\log \varepsilon \sim 1$). Increases in x_{AN} up to 0.7 resulted in λ_{max} shifting to the blue (*ca.* 3300 cm^{-1}) and the band intensity increasing markedly. At higher concentrations of acetonitrile, λ_{max} shifted toward the red and intensity continued to increase until in anhydrous acetonitrile $\lambda \approx 323$ nm and $\log \varepsilon \approx 3$. These changes were interpreted,[G127] in conjunction with conductance and molecular weight determinations, as resulting from the formation of $CuNO_3^+$ ion at $x_{AN} > 0.4$, while the changes at $x_{AN} > 0.7$ were ascribed to changes in solvation shell composition.

Solvent effects on certain charge-transfer transitions are sufficiently large for them to be used as defining scales of solvent polarity. A scale of solvent 'Z values' has been proposed[G128] using the solvent shift (1 kcal $= 2.859 \times 10^{-3}\ \bar{\nu}/\text{cm}^{-1}$) of the charge transfer band of 1-ethyl-4-carbomethoxypyridinium iodide. This band arises from a charge-transfer transition within a contact ion pair, with a resultant decrease in the ion-pair dipole moment. An increase in solvent polarity results in a relatively larger decrease in solvation of the excited state than in the ground state, so the transition is shifted to the blue. The Z values of aprotic solvents do not correlate with those of protic solvents.[G129]

Another proposed solvent scale is based on the energy (in kcal mol^{-1}) of the maximum of the CTTS transition of iodide ion.[G20,G130] There is no simple correlation between this scale and Z values,[G131] which has been interpreted as showing that a CTTS transition does not involve a linear movement of charge.[G20]

Acknowledgments

The author wishes to thank Dr F. M. Hall for reading the manuscript, and Professor N. S. Bayliss and Dr A. K. Covington for helpful discussions.

REFERENCES

G1 J. G. Calvert and J. N. Pitts, Photochemistry, Wiley, New York (1966)
G2 E. L. Wehry, *Quart. Rev.*, **21,** 213 (1967)
G3 F. Wilkinson, *Quart. Rev.*, **20,** 403 (1966)
G4 F. McCapra, *Quart. Rev.*, **20,** 485 (1966)
G5 P. D. Fleischauer and P. F. Fleischauer, *Chem. Rev.*, **70,** 199 (1970)
G6 W. A. Noyes, G. S. Hammond and J. N. Pitts (eds.), Advances in Photochemistry, Vol. 1, Interscience, New York (1963)
G7 A. W. Adamson, W. L. Waltz, E. Zinato, D. W. Watts, P. D. Fleischauer and R. D. Lindholm, *Chem. Rev.*, **68,** 541 (1968)
G8 C. A. Parker, Photoluminescence in Solutions, Elsevier, Amsterdam (1968)
G9 G. Herzberg, Electronic Spectra of Polyatomic Molecules, Van Nostrand Co., New York (1966)
G10 R. E. Dodd, Chemical Spectroscopy, Elsevier, Amsterdam (1962)
G11 G. W. King, Spectroscopy and Molecular Structure, Holt, Rinehart and Winston, New York (1964)
G12 C. N. Banwell, Fundamentals of Molecular Spectroscopy, McGraw-Hill, London (1966)
G13 N. Mataga and T. Kubota, Molecular Interactions and Electronic Spectra, Marcel Dekker Inc., New York (1970)
G14 B. N. Figgis, Introduction to Ligand Fields, Interscience, New York (1966)
G15 S. F. Mason, *Quart. Rev.*, **15,** 287 (1961)
G16 D. A. Ramsay, *Ann. Rev. Phys. Chem.*, **12,** 255 (1961)
G17 N. S. Bayliss and E. G. McRae, *J. Phys. Chem.*, **58,** 1002 (1954)
G18 P. Suppan, *J. Chem. Soc.* (*A*), 3125 (1968)
G19 N. S. Bayliss, *J. Chem. Phys.*, **18,** 292 (1950)
G20 M. J. Blandamer and M. F. Fox, *Chem. Rev.*, **70,** 59 (1970)
G21 N. S. Bayliss, *J. Mol. Spect.*, **31,** 406 (1969)
G22 N. S. Bayliss and G. Wills-Johnson, *Spectrochim. Acta*, **24A,** 563 (1968)
G23 G. W. Robinson, *J. Chem. Phys.*, **46,** 572 (1967)
G24 W. Liptay, *Z. Naturforschung*, **21A,** 1605 (1966)
G25 O. E. Weigang, *J. Chem. Phys.*, **41,** 1435 (1964)
G26 D. F. Evans, *Proc. Roy. Soc.* (*Lond.*), **A255,** 55 (1960)
G27 T. M. Dunn *in* Modern Coordination Chemistry, pp. 229 (J. Lewis and R. G. Wilkinson, eds.), Interscience, New York (1960)
G28 S. J. Strickler and M. Kasha, *J. Amer. Chem. Soc.*, **85,** 2899 (1963)
G29 A. Treinin and J. Wilf, *J. Phys. Chem.*, **74,** 4131 (1970)
G30 J. P. Dahl and H. Johansen, *Theoret. Chim. Acta*, **11,** 8 (1968)
G31 H. M. Hershenson, Ultraviolet and Visible Absorption Spectra, Academic Press, New York (Index for 1930–1954), 1956; (Index for 1955–1959), 1961
G32 DMS, UV Atlas of Organic Compounds, Vols. 1–4, Verlag Chemie, Weinheim; Butterworths, London (1968)
G33 J. W. Sidman, *Chem. Rev.*, **58,** 689 (1958)
G34 E. Rotlevi and A. Treinin, *J. Phys. Chem.*, **69,** 2645 (1965)
G35 (*a*) S. J. Strickler, Ph.D. Thesis, Florida State University, 1961 (University Microfilms, Ann. Arbor, No. 66–1537)
(*b*) S. J. Strickler and M. Kasha *in* Molecular Orbitals in Chemistry, Physics and Biology, pp. 241 (Per-Olov Löwdin and Bernard Pullman, eds.), Academic Press, New York (1964)
G36 N. Arnal, J. Salvinien and P. Viallet, *J. Chim. Phys.*, **64,** 1174 (1967)
G37 N. S. Bayliss and G. Wills-Johnson, *Spectrochim. Acta*, **24A,** 551 (1968)

G38 A. K. Covington, I. R. Lantzke and G. A. Porthouse, unpublished data
G39 H. D. Hunt and W. T. Simpson, *J. Amer. Chem. Soc.*, **75**, 4540 (1953)
G40 E. B. Nielsen and J. A. Schellman, *J. Phys. Chem.*, **71**, 2297 (1967)
G41 A. J. Parker and D. Brody, *J. Chem. Soc.*, 4061 (1963)
G42 C. K. Jorgensen, Absorption Spectra and Chemical Bonding in Complexes, Pergamon, London (1961)
G43 F. A. Cotton and G. Wilkinson, Advanced Inorganic Chemistry, 2nd Edn., Chapter 26, Interscience, New York (1966)
G44 C. K. Jorgensen, Orbitals in Atoms and Molecules, p. 101, Academic Press, London (1962)
G45 J. Bjerrum, A. W. Adamson and O. Bostrup, *Acta Chem. Scand.*, **10**, 329 (1956)
G46 I. R. Lantzke and D. W. Watts, *Aust. J. Chem.*, **19**, 949 (1966)
G47 I. R. Lantzke and D. W. Watts. Unpublished results
G48 W. T. Carnall, P. R. Fields, K. Rajnak, U.S. Atomic Energy Commission 1968. ANL 7358
G49 R. S. Mulliken, *J. Amer. Chem. Soc.*, **72**, 600 (1950)
G50 G. Briegleb, Electronen-Donator-Acceptor Komplexen, Springer-Verlag, Berlin (1961)
G51 J. Rose, Molecular Complexes, Pergamon Press, Oxford (1967)
G52 J. N. Murrel, *Quart. Rev.*, **15**, 191 (1961)
G53 R. S. Mulliken and W. B. Person, *Ann. Rev. Phys. Chem.*, **13**, 107 (1962)
G54 J. Jortner and U. Sokolov, *J. Phys. Chem.*, **65**, 1633 (1961)
G55 M. A. Slifkin and A. C. Allison, *Nature*, **215**, 949 (1967)
G56 O. Nilsson, *Acta Chem. Scand.*, **21**, 1501 (1967)
G57 H. Tsubomura and R. S. Mulliken, *J. Amer. Chem. Soc.*, **82**, 5966 (1960)
G58 L. Paoloni and M. Cignitti, *Sci. Repts. 1st Super Sanita*, **2**, 45 (1962)
G59 G. F. Lothian, Absorption Spectrophotometry, 3rd Edn., Adam Hilger Ltd., London (1969)
G60 L. S. Goldring, R. C. Hawes, G. H. Hare, A. O. Beckman and M. E. Stickney, *Anal. Chem.*, **25**, 869 (1953)
G61 H. A. Benesi and J. H. Hildebrand, *J. Amer. Chem. Soc.*, **71**, 2703 (1949)
G62 W. G. Davies and J. E. Prue, *Trans. Faraday Soc.*, **51**, 1045 (1955)
G63 N. S. Bayliss and C. J. Brackenridge, *J. Amer. Chem. Soc.*, **77**, 3959 (1955)
G64 M. C. R. Symons, *Disc. Faraday Soc.*, **24**, 117 (1957)
G65 D. G. Johnson and R. E. Bowen, *J. Amer. Chem. Soc.*, **87**, 1655 (1965)
G66 P. J. Trotter and M. W. Hanna, *J. Amer. Chem. Soc.*, **88**, 3724 (1966)
G67 R. Foster and C. A. Fyfe, *Progress in Nuclear Magnetic Resonance Spectroscopy*, **4**, 1 (1969)
G68 E. L. Heric, *J. Phys. Chem.*, **73**, 3496 (1969)
G69 G. Norheim, *Acta Chem. Scand.*, **23**, 2808 (1969)
G70 T. V. Mal'kova, N. A. Fateeva and K. B. Yatsimirskii, *Zh. Neorg. Khim.*, **12**, 915 (1967); *Chem. Abs.*, **67**, 26380g (1967)
G71 K. Nagano and D. E. Metzler, *J. Amer. Chem. Soc.*, **89**, 2891 (1967)
G72 D. B. Siano and D. E. Metzler, *J. Chem. Phys.*, **51**, 1856 (1969)
G73 W. Slavin, *Anal. Chem.*, **35**, 561 (1963)
G74 R. E. Poulson, *Applied Optics*, **3**, 99 (1964)
G75 A. K. Covington and M. J. Tait, *Electrochimica Acta*, **12**, 123 (1967)
G76 M. Halmann and I. Platzner, *Proc. Chem. Soc.*, 261 (1964)
G77 J. T. Shapiro, Ph.D. Thesis, Bryn Mawr College, Bryn Mawr, Pa. U.S.A. 1966 quoted by M. J. Blandamer and M. F. Fox in ref. G20
G78 J. Barrett and A. L. Mansell, *Nature*, **215**, 949 (1967)
G79 N. S. Bayliss and C. J. Brackenridge, *Chem. and Ind.*, 477 (1955)
G80 T. F. Young and D. E. Irish, *Ann. Rev. Phys. Chem.*, **13**, 435 (1962)

G81 J. E. Prue *in* Proceedings of the 3rd Symposium on Coord. Chem. Debrecen, Hungary 1970, Vol. 1, p. 25 (M. T. Beck, ed.), Hungarian Academy of Sciences, Budapest (1970)
G82 J. W. Larson and L. G. Hepler *in* Solvent-Solute Interactions, p. 1 (C. Ritchie and J. Coetzee, eds.), Dekker, New York (1969)
G83 C. W. Davies, Ion Association, pp. 55, Butterworths, London (1962)
G84 J. E. Prue, Ionic Equilibria, Chapter 2, Pergamon, Oxford (1966)
G85 D. A. L. Hope, R. J. Otter and J. E. Prue, *J. Chem. Soc.*, 5226 (1960)
G86 A. I. Popov and R. E. Humphrey, *J. Amer. Chem. Soc.*, **81,** 2043 (1959)
G87 H. Elsebernd and J. K. Beattie, *Inorg. Chem.*, **7,** 2468 (1968)
G88 D. W. Watts, *Record of Chemical Progress*, **29,** 131 (1968)
G89 W. A. Millen and D. W. Watts, *J. Amer. Chem. Soc.*, **89,** 6858 (1967)
G90 H. V. Carter, B. J. McClelland and E. Warhurst, *Trans. Faraday Soc.*, **56,** 455 (1960)
G91 T. E. Hogan-Esch and J. Smid, *J. Amer. Chem. Soc.*, **88,** 307 (1966)
G92 D. Casson and B. J. Tabner, *J. Chem. Soc. (B)*, 572 (1969)
G93 P. Biloen, T. Fransen, A. Tulp and G. J. Hoytink, *J. Phys. Chem.*, **73,** 1581 (1969)
G94 R. V. Slates and M. Szwarc, *J. Amer. Chem. Soc.*, **89,** 6043 (1967)
G95 D. Nicholls and M. Szwarc, *Proc. Roy. Soc.*, **A301,** 223 (1967)
G96 R. D. Gillard and H. M. Sutton, *J. Chem. Soc. (A)*, 1309 (1970)
G97 R. D. Gillard and H. M. Sutton, *J. Chem. Soc. (A)*, 2172 (1970)
G98 R. D. Gillard and H. M. Sutton, *J. Chem. Soc. (A)*, 2175 (1970)
G99 J. T. Bell and R. E. Biggers, *J. Mol. Spect.*, **22,** 262 (1967)
G100 K. R. Ashley, R. E. Hamm and R. H. Magnuson, *Inorg. Chem.*, **6,** 413 (1967)
G101 B. D. Casikis and M. L. Good, *Inorg. Chem.*, **8,** 1095 (1969)
G102 I. Abrahamer and Y. Marcus, *J. Inorg. and Nucl. Chem.*, **30,** 1563 (1968)
G103 I. Abrahamer and Y. Marcus, *Inorg. Chem.*, **6,** 2103 (1967)
G104 N. K. Davidenko and A. A. Zholdakov, *Zh. Neorg. Khim.*, **14,** 83 (1969); (*Chem. Abstr.*, **70,** 72503j)
G105 J. Reedijk and W. L. Groeneveld, *Rec. Trav. Chim.*, **87,** 1293 (1968)
G106 F. A. Cotton and G. Wilkinson, Advanced Inorganic Chemistry, 2nd Edn., p. 858, Interscience, New York (1966)
G107 D. W. Meek and R. S. Drago, *J. Amer. Chem. Soc.*, **83,** 4322 (1961)
G108 R. S. Drago, D. M. Hart and R. L. Carlson, *J. Amer. Chem. Soc.*, **87,** 1900 (1965)
G109 R. S. Drago, R. L. Carlson and K. F. Purcell, *Inorg. Chem.*, **4,** 15 (1965)
G110 N. Ellfolk and K. Mattsson, *Suomen Kemi. B*, **42,** 319 (1969)
G111 S. B. Brown and I. R. Lantzke, *Biochem. J.*, **115,** 279 (1969)
G112 H. J. Schugar, A. T. Hubbard, F. C. Anson and H. G. Gray, *J. Amer. Chem. Soc.*, **90,** 71 (1969)
G113 S. B. Brown, T. C. Dean and P. Jones, *Biochem. J.*, **117,** 733 (1970)
G114 J. Fajer and H. Linschitz, *J. Inorg. and Nucl. Chem.*, **30,** 2259 (1968)
G115 A. Van den Bergen, K. S. Murray, M. J. O'Connor, N. Rehak and B. O. West, *Aust. J. Chem.*, **21,** 1505 (1968)
G116 R. J. P. Williams *in* Hemes and Hemoproteins, pp. 557 (B. Chance, ed.), Academic Press, New York (1967)
G117 D. W. Smith and R. J. P. Williams, *Biochem. J.*, **110,** 297 (1968)
G118 M. Kotani, *Ann. N.Y. Acad. Sci.*, **158,** 20 (1969)
G119 B. Mishra and V. Ramakrishna, *Spectrochim. Acta*, **A25,** 288 (1969)
G120 W. W. Epstein and F. W. Sweat, *Chem. Rev.*, **67,** 247 (1967)
G121 I. B. C. Matheson and J. Lee, *Spectrosc. Lett.*, **2,** 117 (1969)
G122 E. G. McRae, *J. Phys. Chem.*, **61,** 562 (1957)
G123 J. Midwinter and P. Suppan, *Spectrochim. Acta*, **25A,** 953 (1969)

G124 E. A. Guggenheim, Mixtures, Chapters 10 and 11, Oxford University Press, London (1952)
G125 J. H. Hildebrand and R. L. Scott, Solubility of Non Electrolytes, 3rd Edn., p. 328, Reinhold Publishing Corp., New York (1950)
G126 D. Meyerstein and A. Treinin, *Trans. Faraday Soc.*, **57**, 2104 (1961) and references therein
G127 C. C. Addison, B. J. Hathaway, N. Logan and A. Walker, *J. Chem. Soc.*, 4308 (1960)
G128 E. M. Kosower, *J. Amer. Chem. Soc.*, **80**, 3253 (1958); *J. Amer. Chem. Soc.*, **82** 2188 (1960)
G129 A. J. Parker, *Quart. Rev. (Lond.)*, **16**, 163 (1962)
G130 M. Smith and M. C. R. Symons, *Disc. Faraday Soc.*, **24**, 206 (1957)
G131 T. R. Griffiths and M. C. R. Symons, *Trans. Faraday Soc.*, **56**, 1125 (1960)

Chapter 4

Spectroscopic Measurements

Part 2

Infrared and Raman Spectroscopy

D. E. Irish
Department of Chemistry, University of Waterloo,
Waterloo, Ontario, Canada.

4.7 INTRODUCTION

In this section attention is focused on data obtained with Raman and infrared spectrophotometers and the interpretation of those data in so far as they may give insight into the nature and structure of solutions formed by the solution of an electrolyte in a non-aqueous solvent. The emphasis is thus on electrolyte chemistry. Even with this restriction the large literature requires selection and the absence of particular references reflects the limitations of space, the interests of the author and his failure to be knowledgeable with all important studies conducted or in progress.

Infrared spectroscopy has been a widely used tool for the study of organic solvent systems for many years and the techniques involved can be learned from many good monographs.[H1-5] Raman spectroscopy has also been employed to provide significant information about the constitution and physical properties of these systems. Until the recent advent of the laser, however, Raman facilities were only available in a few laboratories and a general lack of familiarity with the principles and the potential of this approach existed. In the immediate future we can expect many new results from this form of spectroscopy and a brief discussion of the principles and techniques is in order.

4.8 PRINCIPLES AND TECHNIQUES

Infrared and Raman are complementary forms of spectroscopy. Both provide information about the vibrational and rotational modes of motion of polyatomic molecules and crystals. The information obtainable is not redundant because of a difference in selection rules. The absorption of infrared radiation occurs when the dipole moment changes during a vibration. The frequency of the absorbed radiation is directly relatable to the frequency of the vibration. The Raman effect is observed when the polarisability of the molecule changes during a vibration. This change may occur when no dipole moment change occurs, as during a symmetric stretching vibration of a homonuclear diatomic molecule, for example. Thus signals may be obtained in the Raman spectrum which are not observed in the infrared spectrum and vice versa. Complete information about the vibrational spectrum can only be obtained from studying both types of spectrum.

When exciting radiation of frequency ν_0 is directed onto a sample which is transparent in that spectral region and the scattered radiation is analysed by means of a suitable monochromator, a spectrum is obtained which contains the frequency ν_0, resulting from Rayleigh scattering by the sample molecules, and also the frequencies v_S and v_{aS} of much lower intensity, resulting from inelastic scattering by the sample molecules. The frequency ν_S is lower than ν_0 and ν_{aS} is higher. The differences, $(\nu_0 - \nu_S)$ and $(\nu_{aS} - \nu_0)$, are independent of the magnitude of ν_0 and are characteristic of the sample; they are referred to as Stokes and anti-Stokes Raman shifts, respectively. To understand their origin imagine a molecule interacting with a photon. If some of the energy of the photon, $h\nu_0$, is transferred to the molecule causing a transition from a ground state to an excited state, the photon will be scattered with lower energy $h\nu_S$. It follows that the energy states of the molecule are then separated by an amount $h(\nu_0 - \nu_S)$. (See Fig. 4.8.1.) Some molecules, already in an excited state, will transfer energy to the photon and thus de-excite. The scattered photon will then have more energy, $h\nu_{aS}$, than the incident photon, $h\nu_0$. $(\nu_0 - \nu_S)$ must, however, equal $(\nu_{aS} - \nu_0)$. We will, in this section confine our attention to vibrational transitions and thus the quantum jump involved is from state $v = 0$ to $v = 1$ (Stokes) or vice versa (anti-Stokes). The point to note is that the interaction is only *allowed* if a change in polarisability of the molecule accompanies the vibration. We thus see the origin of a Raman spectrum.

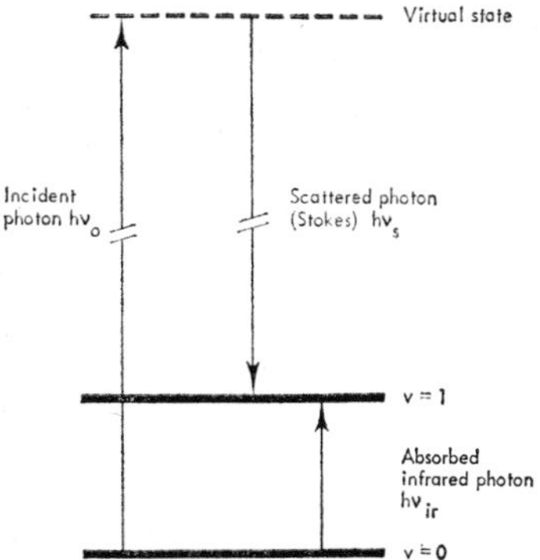

FIG. 4.8.1. Comparison of energy changes accompanying absorption of infrared radiation and excitation of a Stokes Raman line. $h\nu_{ir} \ll h\nu_0$ or $h\nu_S$. The difference $h(\nu_0 - \nu_S)$ equals $h\nu_{ir}$ and thus Raman shift and infrared absorption band are coincident.

The distinction between Raman excitation and infrared absorption is illustrated in Fig. 4.8.1. Because Raman spectra can be excited by visible radiation, samples can be contained in Pyrex glass sample cells and thus sample handling can be easier than for infrared techniques.

Because the energy states are a unique property of the molecule the Raman spectrum can be used to identify the species. A combination of knowledge of infrared and Raman lines, and their degrees of depolarisation is frequently sufficient to indicate the geometry of a molecule. The intensity of Raman lines is relatable to the number of scattering centres per unit volume and thus species concentrations can be obtained.[H6] Changes in the spectra can be related to creation of new species or to perturbation of a species by its environment. The laser provides a highly monochromatic, coherent, intense, narrow beam of exciting radiation. This convenience has resulted in new instrumentation with improved

Table 4.9.1

References to Vibrational Assignments of Non-aqueous Solvents

Solvent	Formula	Ref.
Acetaldehyde	CH_3CHO	1. (a) P. Cossee and J. H. Schachtschneider, *J. Chem. Phys.*, **44**, 97 (1966)
-acetamide, *N*,*N*,-dimethyl	$(CH_3)_2NCOCH_3$	2. (a) V. Chalapathi and K. Venkata Ramiah, *Proc. Indian Acad. Sci.*, **68A**, 109 (1968)
Acetic acid	CH_3COOH	3. (a) T. Miyazawa and K. S. Pitzer, *J. Amer. Chem. Soc.*, **81**, 74 (1959) (b) K. Nakamoto and S. Kishida, *J. Chem. Phys.*, **41**, 1554 (1964) (c) M. Haurie and A. Novak, *J. Chim. Phys.*, **62**, 137 (1965)
Acetone*	CH_3COCH_3	4. (a) G. Dellepiane and J. Overend, *Spectrochim. Acta*, **22**, 593 (1966) (b) J. R. Allkins and E. R. Lippincott, *Spectrochim. Acta*, **25A**, 761 (1969) (c) 1a
Acetonitrile*	CH_3CN	5. (a) J. C. Evans and H. J. Bernstein, *Can. J. Chem.*, **33**, 1746 (1955) (b) H. W. Thomson and R. L. Williams, *Trans. Faraday Soc.*, **48**, 502 (1952) (c) G. W. Chantry and R. A. Plane, *J. Chem. Phys.*, **35**, 1027 (1961) (d) C. C. Addison, D. W. Amos and D. Sutton, *J. Chem. Soc.*, 2285 (1968) (e) H. Johansen, *Z. Phys. Chem. (Leipzig)*, **230**, 240 (1965)
Anisole (methoxybenzene)	$C_6H_5OCH_3$	6. (a) C. V. Stephenson, W. C. Coburn, Jr. and W. S. Wilcox, *Spectrochim. Acta*, **17**, 933 (1961) (b) N. L. Owen and R. E. Hester, *Spectrochim. Acta*, **25A**, 343 (1969)
Benzene*	C_6H_6	7. (a) C. K. Ingold *et al.*, *J. Chem. Soc.*, 222–333 (1946) (b) B. L. Crawford, Jr. and F. A. Miller, *J. Chem. Phys.*, **17**, 249 (1949) (c) L. Corrsin, B. J. Fax and R. C. Lord, *J. Chem. Phys.*, **21**, 1170 (1953) (d) A. C. Albrecht, *J. Mol. Spectrosc.*, **5**, 236 (1960) (e) J. R. Scherer and J. Overend, *Spectrochim. Acta*, **17**, 719 (1961)
Benzonitrile	C_6H_5CN	8. (a) J. H. S. Green, *Spectrochim. Acta*, **17**, 607 (1961)

γ-Butyrolactone	$CH_2CH_2CH_2CO$ (ring closed through O)	9. (a) J. R. Durig, G. L. Coulter and D. W. Wertz, *J. Mol. Spectrosc.*, **27,** 285 (1968)
Carbon disulphide*	CS_2	10. (a) E. K. Plyler and C. J. Humpheys, *J. Res. Nat. Bur. Stand.*, **39,** 59 (1947) (b) J. C. Evans and H. J. Bernstein, *Can. J. Chem.*, **34,** 1127 (1956) (c) T. Wentink, Jr., *J. Chem. Phys.*, **29,** 188 (1958)
Carbon tetrachloride*	CCl_4	11. (a) H. L. Welsh, M. F. Crawford and G. D. Scott, *J. Chem. Phys.*, **16,** 97 (1948) (b) A. Müller and B. Krebs, *J. Mol. Spectrosc.*, **24,** 180 (1967)
Chloroform*	$CHCl_3$	12. (a) J. R. Madigan and F. F. Cleveland, *J. Chem. Phys.*, **19,** 119 (1951) (b) T. Shimanouchi and I. Suzuki, *J. Mol. Spectrosc.*, **6,** 277 (1961) (c) V. Galasso, G. De Alti and G. Costa, *Spectrochim. Acta*, **21,** 669 (1965) (d) A. Ruoff and H. Bürger, *Spectrochim. Acta*, **26A,** 989 (1970)
Cyclohexane*	C_6H_{12}	13. (a) C. W. Beckett, K. S. Pitzer and R. Spitzer, *J. Amer. Chem. Soc.*, **69,** 2488 (1947) (b) D. A. Dows, *J. Mol. Spectrosc.*, **16,** 302 (1965) (c) H. Takahashi, T. Shimanouchi, K. Fukushima and T. Miyazawa, *J. Mol. Spectrosc.*, **13,** 43 (1964) (d) K. B. Wiberg and A. Shrake, *Spectrochim. Acta*, **27A,** 1139 (1971)
Dimethylsulphoxide*	$CH\ SOCH_3$	14. (a) W. D. Horrocks, Jr. and F. A. Cotton, *Spectrochim. Acta*, **17,** 134 (1961) (b) J. H. Carter, J. M. Freeman and T. Henshall, *J. Mol. Spectrosc.*, **20,** 402 (1966)
1,4-Dioxan*	$OCH_2CH_2OCH_2CH_2$ (ring)	15. (a) F. E. Malherbe and H. J. Bernstein, *J. Amer. Chem. Soc.*, **74,** 4408 (1952) (b) W. R. Ward, *Spectrochim. Acta*, **21,** 1311 (1965) (c) R. G. Snyder and G. Zerbi, *Spectrochim. Acta*, **23A,** 391 (1967) (d) O. H. Ellestad, P. Klaboe and G. Hagen, *Spectrochim. Acta*, **27A,** 1025 (1971)
Ethanol	C_2H_5OH	16. (a) K. Krishnan, *Proc. Indian Acad. Sci.*, **53A,** 151 (1961) (b) C. Tanaka, *Nippon Kagaku Zasshi,* **83,** 792 (1962) (c) U. Liddel and E. D. Becker, *Spectrochim. Acta*, **10,** 70 (1957)

Table 4.9.1 (*contd*)

Solvent	Formula	Ref.
Ethers, aliphatic	ROR	17. (a) 15c. (b) J. P. Perchard, J. C. Monier and P. Dizabo, *Spectrochim. Acta*, **27A**, 447 (1971)
Formaldehyde*	HCHO	18. (a) 1a
-formamide, *N*,*N*-dimethyl*	$(CH_3)_2NCHO$	19. (a) G. Kaufmann and M. J. F. Leroy, *Bull. Soc. Chim. Fr.*, 402 (1967) (b) C. A. Indira Chary and K. Venkata Ramiah, *Proc. Indian Acad. Sci.*, **69A**, 18 (1969) (c) 2a
-formamide, *N*,*N*-dimethylthio	$(CH_3)_2NCHS$	20. (a) 19b
Formic acid	HCOOH	21. (a) G. E. Tomlinson, B. Curnutte and C. E. Hathaway, *J. Mol. Spectrosc.* **36**, 26 (1970) (b) 3a (c) 3b
Methanol*	CH_3OH	22. (a) M. Falk and E. Whalley, *J. Chem. Phys.*, **34**, 1554 (1961) (b) 16a (c) 16c
Nitrobenzene	$C_6H_5NO_2$	23. (a) 6a
Nitromethane*	CH_3NO_2	24. (a) D. C. Smith, C. Y. Pan and J. R. Nielson, *J. Chem. Phys.*, **18**, 706 (1952) (b) G. Malewski, M. Pfeiffer and P. Reich, *J. Mol. Structure*, **3**, 419 (1969)
Paraffins	C_XH_Y	25. (a) R. G. Snyder and J. H. Schachtschneider, *Spectrochim. Acta*, **19**, 85 (1963) (b) J. H. Schachtschneider and R. G. Snyder, *Spectrochim. Acta*, **19**, 117 (1963) (c) R. G. Snyder and J. H. Schachtschneider, *Spectrochim. Acta*, **21**, 169 (1965)

Phenylacetylene	$C_6H_5C{\equiv}CH$	26. (a) J. C. Evans and R. A. Nyquist, *Spectrochim. Acta*, **16**, 918 (1960)
Phenylisocyanate	C_6H_5NCO	27. (a) 6a
Phenylisothiocyanate	C_6H_5NCS	28. (a) 6a
n-Propanol	C_3H_7OH	29. (a) K. Fukushima and B. J. Zwolinski, *J. Mol. Spectrosc.*, **26**, 368 (1968) (b) 16a
Pyridine	NCHCHCHCHCH (ring)	30. (a) 7c (b) H. Takahashi, K. Mamola and E. K. Plyler, *J. Mol. Spectrosc.*, **21**, 217 (1966)
Tetrahydrofuran	$OCH_2CH_2CH_2CH_2$ (ring)	31. (a) A. Palm and E. R. Bissell, *Spectrochim. Acta*, **16**, 459 (1960) (b) M. Forel, J. Derouault, J. LeCalvé and M. Rey-Lafon, *J. Chim. Phys.*, **66**, 1232 (1969)
Thionylaniline	C_6H_5NSO	32. (a) 6a
Tributyl phosphate and other tri-substituted phosphates	$(C_4H_9O)_3PO$	33. (a) F. S. Mortimer, *Spectrochim. Acta*, **9**, 270 (1957) (b) J. R. Ferraro, D. F. Peppard and G. W. Mason, *Spectrochim. Acta*, **19**, 811 (1963) (c) L. Winand and P. Drèze, *Bull. Soc. Chim. Belg.*, **71**, 410 (1962) (d) L. J. Bellamy and L. Beecher, *J. Chem. Soc.*, 475 (1952)

* A Raman spectrum, obtained with laser excitation, is illustrated in Appendix C of a recent monograph: J. Loader, Basic Laser Raman Spectroscopy, Heyden and Son Ltd. (1970).

sensitivity which is responsible for the increased activity in the field. The techniques and principles of Raman spectroscopy can be found in a number of recent monographs.[H4–8]

4.9 PURE SOLVENTS

Many non-aqueous solvents possess a rich vibrational spectrum of their own. This presents the spectroscopist with a severe problem. Although, for electrochemical reasons, he may wish to study a particular solute in a particular solvent, the interference resulting from overlap of bands of the solvent with those generated by the solute, or species such as ion pairs or solvates formed after dissolution, may make the interpretation impossible. In some cases a spectral region can be cleared to provide a 'window' by using a solvent in which hydrogen has been replaced by deuterium. In other cases the investigator must design his experiment with components for which the interference is tolerable. The availability of computer techniques for contour analysis has made it possible to extract a certain amount of information from overlapping bands[H9] but such methods must be used with caution. Reproducibility and smooth changes of spectral parameters such as frequency, halfwidth and intensity over a wide composition range provide some confidence in such spectral analyses.

Collected in Table 4.9.1 are references to the vibrational analyses (both infrared and Raman) of some of the more common non-aqueous solvents. For substances not listed here the reader may find it helpful to refer to a book which provides aid for locating spectral data in published collections and bibliographies.[H10] Impurities, which can be a serious source of erroneous data, can be detected by comparison of the spectrum of the solvent with that listed in Table 4.9.1.

In addition to the normal modes of vibration, combinations, and overtones, many organic liquids generate low-frequency Raman lines (30 to 85 cm^{-1} region).[H11,H12] The latter were interpreted as evidence for a solid-like, but disordered (quasicrystalline) structure. The low-frequency lines shift and intensity changes occur on addition of a solute such as $AgNO_3$ or $Zn(NO_3)_2$. Such modes are also found in the far infrared spectra of polar solvents and have been attributed to dipole-dipole complexes or clusters.[H13,H14]

4.10 ALKALI METAL ION VIBRATIONS

In 1965 Edgell and co-workers[H15] reported an infrared band, from solutions of alkali metal salts of $Co(CO)_4^-$ dissolved in tetrahydrofuran (THF), which was assigned to the cation vibrating in a solvent cage. Subsequently a detailed report about the nature of the vibration of these monatomic ions in several non-aqueous solvents was given.[H16] The bands are broad and of medium intensity. Thus their properties are subject to more error in measurement than most vibrational bands. The main facts and conclusions may be summarised:

1. In dimethylsulphoxide (DMSO-d_6), a solvent of relatively high dielectric constant ($\varepsilon_r = 46.7$), a band occurs at 425 ± 3 cm^{-1} for LiCl, LiBr, LiI and $LiNO_3$. (The deuterated solvent was used to minimise band overlap with solvent bands.) Sodium salts generate a band at *ca.* 200 cm^{-1}.

2. In tetrahydrofuran (THF), a solvent of lower dielectric constant ($\varepsilon_r = 7.58$), the i.r. frequency for Li^+ and Na^+ salts is anion dependent, see Table 4.10.1.

3. From these data it is concluded that the cation is vibrating with respect to a solvent cage for DMSO-d_6 and, for solvents of lower dielectric constant in which cation–anion forces are stronger, the anion is a member of the cage thereby achieving intimate contact. A distinction is drawn between this type of ion pair and a 'diatomic species' M^+A^- in a solvent cavity.

4. The integrated absorbance of the Na^+ band of $NaCo(CO)_4$ in THF is directly proportional to concentration for $0.04 < C < 0.32M$. This suggests that the molar absorbance is independent of environment for these moieties.

5. A Raman line is expected if the interaction is significantly covalent. No such line was observed in keeping with the predominance of ionic forces.

6. Both pressure dependence and isotope dependence are consistent with a 'lattice-like' origin of the mode of vibration.

7. Force constant and potential energy (U) calculations indicate that the experimentally measured U does not vary with dielectric constant in a manner predicted by ion-pair potentials; both electrostatic *and* repulsion forces in the solution are important. Solvation measurements provide information about the former; vibration studies about the latter.

This type of study, still in its infancy, is important if it provides new insight into the short-range forces in electrolyte solutions because classical measurements can only give limited information. A new picture of the contact ion pair emerges and emphasis is placed on the role of solvent structure in interionic interactions; an emphasis which is being recognised as essential if models are to be extended beyond the limited 'sphere in continuum' picture of the very dilute solution.*

Concurrent with the first report of Edgell *et al.*, Evans and Lo[H17] reported similar bands from tetraalkylammonium salts in benzene. They were attributed to interionic modes of ion aggregates formed in a solvent of low dielectric constant ($\varepsilon_r = 2.3$) because of an anion

* A more extensive discussion of these observations has recently appeared. See W. F. Edgell *in* Ions and Ion Pairs in Organic Reactions, Vol. 1, Chapter 4 (M .Szwarc, ed.) Wiley-Interscience, New York (1972).

Table 4.10.1

Low Frequency Infrared Vibrations of Cations in Non-aqueous Solvents

Cation	Anion $Co(CO)_4^-$	Cl^-	Br^-	I^-	ClO_4^-	NO_3^-	SCN^-	BPh_4^-	BF_4^-	HSO_4^-	Solvent	Ref.
$^7Li^+$	413	387	378	373		407		412			THF	H16
		425	424	424		425					DMSO-d_6	H16
		429	429	429	429	429					DMSO	H18, H19
		416	421	420	421	421					DPSO	H19
		426	425	425	426	425					DBSO	H19
		416	418	419	419	420					Py	H22
		400	400	400	400	397	402				2PY	H20
$^7Li^+$		377	398	398	398	398	398				1M2PY	H20
$^6Li^+$		409	420	420	420	420					1M2PY	H21
$^7Li^+$		419	419	419	419	419	419				1V2PY	H20
								410			THF	H24
Na^+	192			184				198			THF	H16
	199			194		200		203			DMSO-d_6	H16
		199	199	198	200	206	200	198			DMSO	H18, H19
			216	219	217	218	221	220			DPSO	H19
				219	221	219	224	226			DBSO	H19
	180										Py	H16
				170	182		180	179			Py	H22
	183										Pip	H16
			206	207	207	207	206		206		2PY	H20
			204	204	204	206	207	205	207		1M2PY	H20
								175			Py, D, Pip, THF	H24

K^+	142									THF	H16
			153	153		154	153			DMSO	H19
				156	153		154	153		DPSO	H19
				152			155	154		DBSO	H19
				147	145		144	147		2PY	H20
				140	140		138	138		1M2PY	H20
								133		Py	H24
Rb^+			125	123	122	125				DMSO	H19
				123						DPSO	H19
				106	106			106		1M2PY	H20
Cs^+				110	109					DMSO	H19
NH_4^+		214	214	214	214	214	214			DMSO	H18, H19
		221	222	223	224	223	224			DPSO	H19
			226	225	226	226	225			DBSO	H19
			217	216	218	218	219	218	218	2PY	H20
			207	206	207	213	207		207	207	1M2PY	H20
				196	197	201	199	199	198	Py	H22
								198		Py	H24
ND_4^+				183						Py	H22
								183		Py	H24

Cation	Cl^-	Br^-	Cl HCl^-	Cl DCl^-	Br HBr^-	$HCr_2(CO)_{10}^-$	$Cr_2(CO)_{10}^{2-}$	Solvent	Ref.
$(n\text{-}C_4H_9)_4N^+$	120	80	102	102	73			B	H17
$(n\text{-}C_5H_{11})_4N^+$	119	80						B	H17
Na^+						200	200	DMSO-d_6	H16

dependence. In the interim, Popov and co-workers have contributed significantly to this field[H18-23] by studies of a wider range of anions, solvents and mixed solvent systems (Table 4.10.1). Where their data overlap with those of Edgell *et al.*, agreement is within the uncertainty claimed (*ca.* 4 cm^{-1}). To generalise, their data showed a marked dependence on the cation but no such dependence on the anion, at least until very high concentrations were reached (e.g. 7.0*M* NH_4SCN in DMSO[H19]). Also no new spectral lines of anions such as NO_3^- and SCN^-, ascribable to contact ion pairing, were discovered.[H19,H21] Support is thus provided for the assignment to cations in a solvent cage. Ancillary NMR studies suggested $M(DMSO)_2$ ($M = Li^+$ or NH_4^+)[H23] but a less well-defined solvation sphere for Na^+. For the solvent 1-methyl-2-pyrrolidone ($\varepsilon_r = 32$) (1M2PY), four molecules of 1M2PY are in the 'cage' surrounding Li^+ but the number falls as dioxan replaces 1M2PY in the mixed-solvent system.[H20]

French and Wood[H24] also studied salts of BPh_4^-. The value of 175 cm^{-1} which they report for $NaBPh_4$ in THF has not been reproduced and 194 to 198 cm^{-1} appears to be a better value.[H16,H22] Because the frequency was insensitive to solvent for $NaBPh_4$ in Py, D, Pip and THF they ascribed the band to contact ion pairs. The change of extinction coefficient with solvent, cation and concentration, and the isotopic shift were also reported while possible ion-pair potential functions were considered.

4.11 ION-PAIR DETECTION AND RELATED ION SOLVATION

Electrolyte chemists turn to non-aqueous solvents to explore the effect of a changing, and usually low, dielectric constant on the degree of ion pair formation and to observe differences attributed to a changed solvation sphere, specific solvation or changed solvent structure. The stretching vibration of an ion pair $M^+ - A^-$ may be detectable by infrared spectroscopy (as mentioned in sect. 4.10) but is undetectable by Raman spectroscopy. If A^- is a polyatomic anion, however, new lines characteristic of a 'bound' species may be detected.

Thus unperturbed nitrate ion, of D_{3h} point group, has a four line spectrum, three of which are Raman active and three infrared active. (Figure. 4.11.1.) When ion pairs are formed additional lines appear in the spectrum. When the symmetry is lowered to C_{2v}, six coincident infrared and Raman lines are predicted. Generally five can be detected; it is assumed that $\nu_3(A_1)$ and $\nu_5(B_1)$ are accidentally coincident. The frequency values given in parentheses (Fig. 4.11.1) are nominal values and depend on the cation and solvent. Usually the displacements from the values of unperturbed (dilute solution) nitrate ion are toward the values of the corresponding nitrate crystal. Thus $\nu_1(A_1')$ is greater than 1050 cm^{-1} in some cases. In aqueous solution $\nu_3(E')$ of the solvated anion appears as a doublet with about 56 cm^{-1} separation of components, but it is singlet in chloroform (see sect. 4.12). Spectra which show pairs of lines (e.g. 830 and 816 cm^{-1} in the infrared; 720 and 740 cm^{-1} in the Raman) with intensity ratios that are concentration dependent, reveal equilibria between solvated nitrate ion and contact ion pairs, as illustrated below.

Janz and co-workers have observed that the $\nu_1(A_1')$ stretching mode of nitrate ion can be resolved into two Raman active components at 1041 and 1036 cm^{-1} for $AgNO_3$-CH_3CN. Because these lines are very intense it has been possible to follow the intensity ratio down to 0.034M. Good agreement was obtained between the degree of ion-pair formation measured by spectroscopy and that obtained from conductance using the

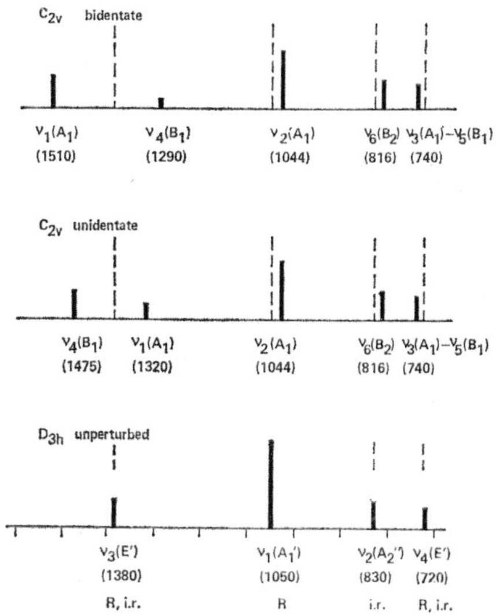

FIG. 4.11.1. Generalised scheme showing how the vibrational spectrum of nitrate ion is altered on going from 'free' nitrate ion, of D_{3h} symmetry, to ion-paired nitrate ion of C_{2v} symmetry. For C_{2v} all lines are both infrared and Raman active.

Wishaw-Stokes equation.[H25,H26] Evidence that the average number of molecules of the solvent co-ordinated by the solute decreases from four in dilute solution, to two in the range ~0.5 to ~5M, and one at higher concentrations was advanced.[H26] In solutions of $AgNO_3$-CH_3CN-H_2O, Ag^+ appears to be preferentially solvated by CH_3CN, even when the concentration of acetonitrile in the solvent has been reduced to 4.7 mol%.[H27]

Similarly, for zinc nitrate in anhydrous methanol, new lines characteristic of ion pairs have been detected.[H28] In addition to lines at 715 (R),

828 (i.r.) and 1044 (R) cm^{-1} characteristic of free nitrate, lines at 754 (R, i.r.), 817 (i.r.), 1310 (R, i.r.) and 1500 (i.r.) cm^{-1} characteristic of bound inner-sphere nitrate were observed. (Other lines are masked by the solvent.) The concentration dependence of the 817 and 828 cm^{-1} pair of infrared lines permitted calculation of association constants at 25°C, $K_1 = 2.6$ and $K_2 = 0.05$ mol l^{-1}. The lines characteristic of bound nitrate are markedly temperature dependent, the penetration of the anion into the first co-ordination sphere of zinc being favoured by high temperature. Thus below −30°C these ion pairs are not detectable by infrared spectroscopy. The decrease from 6 to 4 observed for the solvation number at −82°C by NMR spectroscopy was thus attributed to a rapid exchange between solvent, bound between cation and anion, and bulk solvent promoted by the presence of the anion in the outer sphere.[H28] In pure water, zinc nitrate does not form contact ion pairs except at concentrations so high that contact is forced by packing requirements. Vibrational spectral data for nitrate ion in water have been recently reviewed in detail.[H29] From a Raman band intensity analysis a smooth increase in K_1 and K_2 has been observed for $Ca(NO_3)_2$ as the dioxan:water ratio of mixed solvent increases.[H30] Both molar intensities and frequency positions are sensitive to the dioxan:water ratio.

In order to elucidate the mechanism of extraction of nitrates by tri-*n*-butyl phosphate (TBP) several authors have examined the infrared frequencies of metal nitrates in TBP.[H31,H32] The magnitude of the separation of component lines in the region of $\nu_3(E')$ (~1380 cm^{-1}) was considered to be a measure of the dis-symmetry induced in the nitrate group.[H31] The interaction was believed to lower the symmetry of nitrate ion from D_{3h} to C_{2v} and thus cause the $\nu_3(E')$ mode to lose its degeneracy becoming distinct $\nu_1(A_1)$ and $\nu_4(B_1)$ modes. The separation of these components was greater for TBP solvates than hydrates. Some caution was exercised in comparing this separation with 'percentage electrostatic character' as polarisation by cations can cause such splittings.[H32–34] The presence of a diagnostic metal-nitrate Raman vibration provides the strongest evidence for covalence in such an interaction.[H29] The splittings in tri-*n*-octyl phosphine oxide (TOPO) in CCl_4 and tri-*n*-octyl amine nitrate in benzene are similar to those in TBP. Intimate contact between metal and nitrate was inferred, the metal being the principal factor controlling the magnitude of the separation; the influence of medium is small.[H35] Infrared bands of inorganic nitrates in acetone have also been tabulated.[H36]

Solvation of Zn^{2+} by acetonitrile in the system $ZnCl_2$-CH_3CN-H_2O was inferred from Raman spectral studies by Evans and Lo.[H37] Bands characteristic of two distinct forms of CH_3CN, in equilibrium, were observed. The complex involved zinc-containing species and was not influenced greatly by the nature and number of other ligands attached to zinc. It was shown that complex formation caused an increase in the C-N stretching force constant. Data for zinc, cadmium and mercury(II) nitrates and halides in acetonitrile have also been obtained by Addison *et al.*[H38] Agreement with the data of Evans and Lo[H37] is good but assignments of complexed and free acetonitrile differ. A scheme which involves a 40 to 45 cm^{-1} shift to higher frequencies for complexed CH_3CN is favoured. The C-C stretching frequency was also found to increase by about 25 cm^{-1}. The average number of acetonitrile molecules bonded to zinc was obtained from the decrease of the 2251 cm^{-1} line of free CH_3CN. Thus the species present was assigned an empirical formula $Zn(CH_3CN)_2(NO_3)_2$. Absence of a line characteristic of a metal-oxygen bond suggests that zinc and nitrate are held in contact by ionic forces. Similar spectra were obtained for $Cd(NO_3)_2$. For $HgCl_2$ and $HgBr_2$ no splitting of the CH_3CN bands was found, revealing that CH_3CN was not co-ordinated to mercury. Co-ordination did occur for the nitrate, however, and a low frequency (248 cm^{-1}) line was assigned to Hg-O stretching of the associated mercuric nitrate species.

Raman spectra of $ZnCl_2$, $HgCl_2$, LiCl and HCl in methanol (L) have been investigated.[H39] A set of equilibria involving $ZnCl_2L_4$, $ZcCl_4L_2^{2-}$ and a possible $(ZnCl_2)_{2n}$ polymer were invoked for $ZnCl_2$, somewhat as in aqueous solution,[H40] although no attempt was made to measure species concentrations. Similar complex ions were proposed for mercuric chloride and evidence for ion pairs was discovered in methanol solutions of LiCl and $CaCl_2$. The vibrational spectrum of zinc chloride in tributylphosphate (TBP) diluted with benzene, acetone or pyridine has also been studied.[H41] Bands at 345 cm^{-1} and 305 cm^{-1} were attributed to a $ZnCl_2$ species of C_{2v} symmetry (bent molecule) when the TBP:$ZnCl_2$ ratio is less than 2:1. (Compare these frequencies with those from solid $ZnCl_2$ complexes.[H42]) Above this ratio, in benzene solution, evidence for polynuclear complexes was advanced.

Halocomplexes have also been isolated in non-aqueous solvents by solvent extraction and identified by spectroscopy. Woodward and Bill[H43] undertook to obtain direct structural information by Raman spectroscopy for the species $(H^+)(InBr_4^-)$ which Irving and Rossotti[H44] suggested existed in the ethyl ether phase after extraction from aqueous indium bromide-hydrobromic acid. In the absence of HBr negligible amounts of indium were extracted. In all cases the typical four line spectrum of a tetrahedral species,[H45,H46] $InBr_4^-$, was obtained. As already noted the stretching vibration of H^+-$(InBr_4^-)$ would not be detected by Raman spectroscopy if the attachment is purely electrostatic. Although, in principle, such an attachment should lower the symmetry of the tetrahedral $InBr_4^-$ ion and thus cause degenerate modes of the latter to split, the symmetrical anion is apparently not sufficiently deformable for this expectation to be observed. Thus no direct spectral evidence for the attachment of H^+ (or H_3O^+) was obtained. In contrast to the ether solutions, aqueous solutions generate spectra which suggest appreciable concentrations of lower species, $InBr^{2+}$, $InBr_2^+$ and $InBr_3$,[H47] but not $InBr_4^-$.

In subsequent ether extraction studies, Woodward and Taylor identified $InCl_4^-$, $GaCl_4^-$ and $FeCl_4^-$ (tetrahedral species)[H48] and $SnCl_3^-$ and $SnBr_3^-$ (pyramidal species).[H49] Raman spectra of $GaCl_4^-$ in TBP, present as $H^+GaCl_4^-$(2TBP) were also reported by Morris *et al.*[H50] Li^+ may also accompany $GaCl_4^-$ into di-isopropyl ether, but not NH_4^+ or Ca^{2+}.[H51] When the molar ratio of chloride to gallium is low the Raman spectrum of the extract is consistent with the tetrahedral $GaCl_3OH_2$ species.[H51] Spectral studies of extracts have also lead to the characterisation of the following species: $CuCl_2^-$ and $CuBr_2^-$ in diethyl ether,[H52] $TlCl_4^-$ in isopropyl ether,[H53] ZnX_n^{2-n} (X = Cl or Br and n = 2, 3, 4) in TBP,[H54] $TiCl_4 \cdot TBP$ and $TiCl_4 \cdot 2TBP$ in TBP,[H55] and (Li^+TBP) (HgX_3^-) for the extraction of HgX_2LiX (X = Cl, Br) with TBP.[H56] Far infrared studies of these halogeno-anions have been reported.[H57,H58] No direct evidence for association of H^+ with the anions was provided, but new data for $SbCl_6^-$, $AuCl_4^-$ and $AuBr_4^-$ were presented.[H58]

These references suffice to indicate the power of spectroscopy to identify complex ions in non-aqueous solvents. In general the species extracted is not the predominant species in the aqueous phase (but compare ref. H54). Frequently a species of high co-ordination number exists in the organic phase. Chemical analysis reveals that it is often associated with a solvated cation, for example H_3O^+, M^+TBP, etc. It is surprising that no manifestation of H_3O^+ has been reported in the vibrational spectrum, especially in the infrared spectrum. The explanation, in terms of electrolyte theory, for the nature and stability of the species in the non-aqueous phase is lacking. To date no quantitative intensity studies of the species concentrations in the extracts, similar to those performed for species in aqueous media,[H53,H6] have come to our attention. This field offers considerable scope to the spectroscopist interested in non-aqueous solutions of electrolytes.

Silver thiocyanate-pyridine solutions have been investigated with infrared spectroscopy.[H59] The intensities of the C-N stretching vibration of free SCN^- (2059 cm^{-1}) and S-bonded mononuclear complexes (2089 cm^{-1}) were measured for various compositions. Stability constants were obtained by the graphical method of Fronaeus[H60] for the species AgSCN, $Ag(SCN)_2$, $Ag(SCN)_3$. In aqueous solution evidence has been obtained[H61] for a series of polynuclear complexes $Ag_m(SCN)_{m+2}^{2-}$, $m \geqslant 2$, at high concentrations of SCN^-, but in pyridine such species are formed only when $c_M > c_L$. Probably pyridine binds strongly to silver blocking bridging SCN-groups, in contrast to weaker solvation by water. The bridging SCN-group generates infrared intensity at 2102 cm^{-1}.

The stability constants of mononuclear species are much smaller in pyridine than water, possibly because of the stronger interaction between silver ion and the solvent pyridine.

An ion pair, associated through a hydrogen bond, is formed by the reaction of protonic acid with bases in non-dissociating solvents. Thus, in chloroform solutions, acetic acid and pyridine give

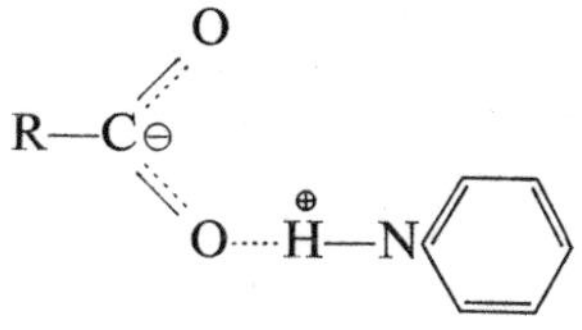

```
       O
      ⁄
R—C⊖
      ╲      ⊕
       O····H—N(pyridine ring)
```

Stronger acids give pyridine·$(HA)_2$. For sufficiently weak acids the proton is not transferred and a simple hydrogen-bonded complex exists. The extent of the acid-base reaction has been followed by infrared spectroscopy; equilibrium constants have been obtained as well as information about the number of species present in the ion pair, the manner in which they are linked and the role of the solvent.[H62–65] Base strengths of acetamide, methanol, dioxan, acetone and acetonitrile in sulphuric acid were obtained by Raman spectroscopy by Deno and Wisotsky.[H66] The ratio of bands characteristic of C—O and C—H or C—N and C—H vibrations gave the desired pK_{BH^+} values. Base strengths of methanol in aqueous HCl and 2-propanol in aqueous $HClO_4$ and HCl have also been obtained by measurement of Raman line intensities.[H67]

4.12 HYDROGEN BONDING

Infrared spectroscopy has been described as one of the most sensitive methods for the detection of hydrogen bonding.[H68] Vibrational spectra are significantly altered by hydrogen-bond formation and these changes constitute a criterion for the existence of the bond; they provide direct evidence of the role of the proton in the interaction and a measure of the degree of association and strength of the bond.

The spectral changes of the proton donor can be broadly described as follows:

(a) The stretching frequency, ν_s, shifts to lower values, broadens and increases markedly in intensity. The frequency and intensity are sensitive to changes in concentration, temperature and solvent.

(b) The bending frequency, ν_b, shifts to higher values, but only to a small extent in comparison with ν_s. No pronounced changes in half-width or intensity occur.

(c) New torsional vibrations and stretching and bending vibrations are generated by the associated species. The former are normally diffuse bands in the 400 to 800 cm^{-1} region and the latter usually are below 200 cm^{-1}. For practical reasons ν_s has been most thoroughly studied. The findings have been reviewed in detail by a number of authors.[H68–71]

Most spectral investigations have been restricted to hydrogen bonding in pure substances or in mixtures of non-electrolytes. Our concern here is with data relating to hydrogen bond formation between a donor and the ions of an electrolyte. Such interactions can be considered as one form of ion solvation and thus are of particular interest to the solution chemist. A portion of the infrared spectrum of chloroform in CCl_4 is shown in Fig. 4.12.1.[H72] The C—H stretching vibration is the intense sharp line at 3019 cm^{-1}. Small combination bands occur on either side. For solutions of tetraphenylarsonium chloride in chloroform a broad intense band at 2934 cm^{-1} increases in intensity with increasing concentration of salt.

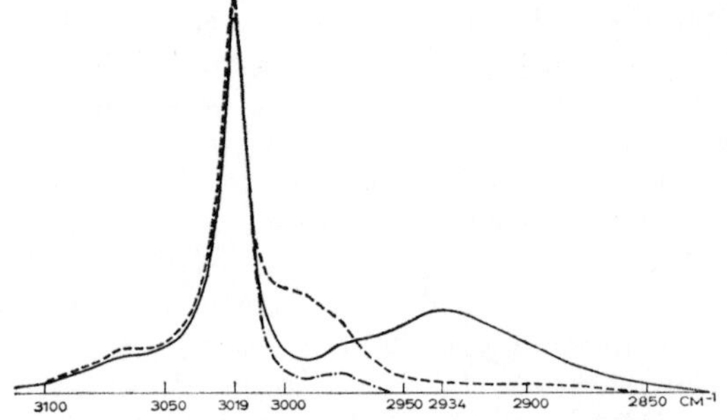

FIG. 4.12.1. A portion of the infrared spectra of chloroform.

—·—·— chloroform containing CCl_4.
———— chloroform containing Ph_4AsCl.
– – – – – chloroform containing Ph_4AsNO_3.

For tetraphenylarsonium nitrate the hydrogen bonding feature occurs as a shoulder at 2984 cm^{-1}. These bands are attributed to chloroform bonded to the anions Cl^- and NO_3^- respectively.

Ion pairing should be more favoured in chloroform than in water because of the lower dielectric constant of chloroform ($\varepsilon_r = 4.71$). Such ion pairing between the tetraphenylarsonium cation and the nitrate anion should cause changes in the vibration spectrum of nitrate ion as described in sect. 4.11. Davis *et al.*[H73] noted that the spectrum of tetraphenylarsonium nitrate in water is similar to that of sodium nitrate in water, i.e. the $\nu_3(E')$ mode is a doublet with components at 1344 and 1394 cm^{-1}. In chloroform this band is a broad singlet at 1360 cm^{-1}. It was concluded that the split of about 50 cm^{-1} of $\nu_3(E')$ was a property of the nitrate ion in water and not caused by ion pairing; otherwise a split would have been observed for enhanced ion pairing in chloroform. The absence of evidence for contact ion pairs reinforces the assignment of the hydrogen bonding C—H stretch to the chloroform-anion interaction. Judging from the greater shift of ν_s (Fig. 4.12.1) the bond to chloride ion is stronger than the bond to nitrate ion.

The O—H stretch of the methanol monomer occurs at 3682 cm^{-1} in the gas phase and 3643 cm^{-1} for dilute solutions of CH_3OH in CCl_4. Hydrogen bonding causes this mode to shift to lower frequencies (about 3380 cm^{-1}) and to broaden.[H74,H75] Electrolytes have been studied in this and other alcohols by a number of investigators. Raman spectra of saturated solutions of electrolytes in methanol were reported by Hester and Plane.[H76] Addition of salt caused the low frequency band (3380 cm^{-1}) to shift to higher values. Of all the salts studied, which included perchlorates, nitrates, sulphates, chlorides and fluorides, the perchlorates produced the largest shift to higher frequencies (about 3550 cm^{-1}). The trend to raise the frequency is in the order $ClO_4^- > NO_3^- > SO_4^{2-} > Cl^- > F^- > H_2O$; no consistent trend was detectable for the cations. Thus it was concluded that ClO_4^- is most effective in breaking the hydrogen-bonded structure of methanol. These findings are analogous to those recently reported for the effect of perchlorate ion on water structure.[H77,H78] The above anion order is also that of increasing charge density and thus ion-alcohol interactions, which increase in magnitude from left to right in the above series, are responsible for holding the O—H frequency at lower values. Presumably the hydrogen bonds between alcohol molecules have been largely broken and those between CH_3OH and anion have been created.

The CH_3 deformation mode occurs at 1460 cm^{-1}. The first overtone of this mode is thought to be in Fermi resonance with the symmetric CH_3 stretching mode giving two lines at 2943 and 2835 cm^{-1}. On addition of salts, the former line intensity increases markedly relative to the latter, resulting in an intensity reversal in most cases. However, band overlap makes any interpretation based on structure inconclusive. The intensity of the 1035 cm^{-1} C—O stretching band is also enhanced by $LiClO_4$, $Mg(ClO_4)_2$ and $ZnCl_2$, and diminished by LiCl.[H76] Previously Kecki[H39] had studied this region and concluded:

(i) The action of an anion or electron donor on the hydroxyl hydrogen of methanol causes an increase of frequency of ν(C—O) and a decrease of intensity.

(ii) The action of a cation or electron acceptor on the oxygen causes a decrease of frequency and an increase of intensity of ν(C—O).

(iii) The breaking of hydrogen bonds between methanol molecules does not influence the position and intensity of ν(C—O).

Minc and Kurowski[H79] had also noted that LiCl and $LiClO_4$ affect the molar intensity in opposite ways. Explanations in terms of ion pairs were advanced. A correlation between molar intensity of ν(C—O) and cation electric-field intensity is discussed in terms of bond polarity.[H79,H80] A new line at 1112 cm^{-1} has been tentatively assigned to a LiCl complex in methanol.[H76] Although the reported data support the conclusion that ion pairs of calcium nitrate exist in methanol, there is a lack of similar evidence for magnesium nitrate[H76] as the changes in the ν_3 nitrate region are not definitive.[H29]

Bufalini and Stern[H81,H82] have reported on the effect of a number of electrolytes on the infrared spectra of some hydrogen bonded compounds (methanol, 1-butanol, *t*-butanol, *N*-methylacetamide) when both are dissolved in dilute benzene solution. They noted that the absorption maximum of methanol associated to the electrolyte decreased with increasing anion radius, $ClO_4^- < NO_3^- < Br^- < Cl^-$. Picrate produced no shifted peak. Lund[H83] reported a similar correlation with apparent molar volume of the anions, for a series of tetrabutylammonium salts in chloroform containing 0.10 *Mp*-cresol; for the same series the apparent integrated intensity increased when the apparent molar volume decreased. Bufalini and Stern[H81] reported that above $30 \times 10^{-3}M$ methanol in benzene the 3640 cm^{-1} band departed in a negative sense from Beer's law and the ~3300 cm^{-1} band intensity became detectable. If, to a solution sufficiently dilute for the 3300 cm^{-1} band to be absent, some electrolyte was added, this band appeared and the intensity at 3640 cm^{-1} decreased. Absorptivity values of 5.07 and 10.0 were obtained for the free O—H and associated (to Bu_4NBr) O—H respectively. The intensity ratio $[MeOH]_{assoc.}/[MeOH]_{free}$ increased in the order $Bu_4NCO_2H < C_6H_5N(CH_3)_2 \cdot HCl < AgClO_4 \sim Bu_4NNO_3 < Bu_4NBr < Bu_4NCl$. For Bu_4NBr-MeOH association ΔH was found to be -28 kJ mol^{-1} and ΔS -58.6 J mol^{-1} K^{-1}. Coupling the spectral results with other data,[H84,H85] these authors[H82] conclude that on the average one alcohol molecule solvates each ion pair. The latter are preferentially solvated over quadrupoles. The interpretation was valid for the other alcohols and for *N*-methylacetamide. Allerhand and Schleyer[H86] studied the infrared spectral shifts resulting from the dissolution of quaternary ammonium halides in methanol-CCl_4 mixtures. The shifts to low frequencies were greatest for the chloride and decreased in the order $Cl^- > F^- > Br^- > I^-$. With the exception of the anomalous position of F^-, this order is the inverse of that found when covalent halides are used as proton acceptors. A sharp break in the order apparently occurs when the halogen is completely ionised; no reversion is observed as the ionic character of the halogen atom increases.[H87] A trend with cation was found for the chloride salts but not bromide and iodide.[H86] Perelygin[H88] reported a cation trend for salts dissolved in acetonitrile containing a small fixed amount of methanol. Bands at 3523, 3483 and 3388 cm^{-1} were ascribed to O—H bonds perturbed by Na^+, Li^+ and Mg^{2+} respectively. ClO_4^- had no influence but I^- was observed to be hydrogen-bonded to alcohol.

Hyne and Levy[H89] reported results of a quantitative investigation of the effect of added $(n\text{-Bu})_4NBr$ on the infrared spectra of dilute *t*-butanol in CCl_4. An absorption at 3496 cm^{-1} attributed to dimeric alcohol is swamped by very intense absorption at 3380 cm^{-1} when salt is added. However, the latter intensity does not markedly occur at the expense of the 3650 cm^{-1} band of monomeric alcohol. The absorptivity ratio of these two bands was estimated to be 20 to 1. The absorbance data are consistent with specific interaction between the alcohol and ions or ion pairs. These species are described as nucleation centres for aggregates of alcohol molecules. Comparison with the work of Bufalini and Stern[H82] suggests that changing the inert solvent, C_6H_6 to CCl_4, changes the distribution and type of associated species present.

More recently it was shown by both NMR and IR spectroscopy that the 3300 cm^{-1} band is entirely due to a 1:1, alcohol-anion complex.[H90] For the system 1-phenylethanol $(0.05M)$ – $(n\text{-}C_4H_9)_4NBr$ – CCl_4 the association constant is about 40 mol l^{-1}. Evidence that the interaction is primarily electrostatic is provided. In another study[H91] a 1:1 association constant for trihalomethanes to tetralkylammonium halides in CCl_4 or CH_3CN was measured by NMR and verified by IR. The intensity of the infrared, C—D, hydrogen-bonded stretching frequency at 2170 to 2200 cm^{-1} was found to be directly proportional to the concentration of complex, computed with the equilibrium constant deduced from NMR measurements. The equilibrium constant increases in the order $I^- < Br^- < Cl^-$ and for a given anion $CHF_3 < CHCl_3 < CHBr_3 < CHI_3$. For a given complex, the equilibrium constant in CCl_4 is about twice that in CH_3CN. No new C—D stretch characteristic of hydrogen-bonding was observed for CDI_3 in the presence of tetra-alkylammonium halide salts up to $1M$, although such a band could be seen in the presence of DMSO or Py. Infrared spectral changes in the 500 to 900 cm^{-1} region suggest that CHI_3 interacts with anions through the iodine rather than the hydrogen. For $CHCl_3$ and $CHBr_3$ the C—X stretch appears as a doublet and the concentration dependence of the intensity indicates that the two lines arise from two forms of the molecule, complexed and free. But for CHI_3, a weak, 420 cm^{-1} C—I stretch intensity was enhanced by a factor of about 100 on addition of the salt, indicating polarisation of the C—I bond by the anion which produces a larger bond dipole derivative. No second line was observed. The other trihalomethanes did not show this intensity enhancement.

Relationships between the enthalpy of reaction and the change in the O—H stretching frequency have been disputed.[H92,H93] New values for enthalpies of reaction for phenol with several Lewis bases have been determined by calorimetry. A good linear relationship between ΔH and ν_{OH} was reported.[H93] No unique relation exists between vibrational frequencies and lengths for O—H · · · O hydrogen bonds but decreasing

ν_s(OH) frequencies generally correlate with decreasing hydrogen-bond length.[H94] A relationship exists between hydrogen bond strength and the low ν(OH···O) frequency when mass effects are the same.[H95] Smooth continuous trends suggest that X—H···A interactions differ only in degree and not in type: there is no fundamental distinction between very weak dipolar interactions and very strong hydrogen-bonding interactions.[H96] Bands, arising from hydrogen bonding between acid molecules HX and acetone (in CCl_4), are shifted from the band observed in the absence of acetone by an amount which is directly proportional to the HX dipole moment. For HCl, the order with respect to the oxygen-containing substance is ethyl ether > dioxan > acetone.[H97]

4.13 SOLVATION AND MEDIUM EFFECTS

Lithium perchlorate and silver perchlorate are quite soluble in oxygen-containing solvents, more so than other alkali or alkaline earth perchlorates. $AgClO_4$ is also soluble in aromatic solvents but $LiClO_4$ is not appreciably soluble in hydrocarbon solvents. These facts suggest that association occurs between Li^+ and the oxygen atom of the solvent. The explanation is supported by spectral data. The $\nu_2(C = 0)$ band of acetone occurs at 1707 cm^{-1}, ν_4(C-C) at 1225 cm^{-1}, ν_1(C-C) at 787 cm^{-1} and $\nu_5[\delta(CCO)]$ at 530 cm^{-1}. Pullin and Pollock[H98] reported evidence for two species of acetone in the salt solutions. One form has infrared frequencies similar to pure acetone and the other form generates a lower ν_2 and a higher ν_4, ν_1 and ν_5. The growth of the intensity, of bands not found in pure acetone, with increase of solute concentration indicated solvation of the type $M^+(acetone)_2$, with partial covalency in the Ag^+ interaction.

Minc *et al.*,[H99] in a more extensive Raman study of perchlorates in acetone, show that the cation affects the line positions and integrated molar intensities in the order $Na^+ < Li^+ < Ba^{2+}$. The perchlorate ion spectrum was most perturbed for $LiClO_4$. Perchlorate ion and hexane were both found to cause destruction of the dipole structure of acetone. The acetone-cation interaction is considered to be similar to hydrogen bond formation, causing the C=O bond to weaken and the C—C—C angle to decrease. The larger size of Ba^{2+} is partly responsible for greater effects on the lines of acetone, whereas formation of Li^+-ClO_4^- ion pairs could cause the pronounced changes of the perchlorate spectrum and reduce the Li^+-acetone interaction.

Changes in the infrared spectra of acetonitrile caused by dissolved $LiClO_4$, $NaClO_4$, $Mg(ClO_4)_2$, NaI or LiI were studied by Perelygin.[H100] Perchlorate ion had no effect on the absorption bands. The blue shift of the C≡N vibration and C—C vibration and the increase of intensity of the CH_3 band were attributed to binding to cations (cf. ref. H38 in sect. 4.11). Evidence that I^- binds to the CH_3 group was also presented. Kecki[H101] has reported that the C—C and C≡N infrared stretching bands of acetonitrile are split into two components on addition of Cu^{2+}, Ni^{2+}, Co^{2+} and Na^+. One band is at the position of the pure solvent. The second, at higher frequencies, is attributed to acetonitrile solvating the cations; the shift is larger for the transition metal ions. Equilibria were recognised and treated with data obtained from the integrated band intensity.

Sodium tetrabutylaluminate ($NaAlBu_4$) is soluble in saturated hydrocarbon solvents. The solvent acts as a dispersing medium for the salt[H102] and does not compete with a co-ordinating solvent such as THF. Thus the ion-THF interactions can be followed without interference from cyclohexane, present as a major component. For THF:salt ratios greater than one, two sets of infrared lines are observed. Lines at 1071 cm^{-1} and 913 cm^{-1}, characteristic of the C—O—C asymmetric and symmetric stretching vibrations respectively, are assigned to free THF in equilibrium with a bound form which contributes intensity at 1048 cm^{-1} and ~902 cm^{-1}. The same features are observed for solutions of $NaAlEt_4$ and $NaAl(octyl)_4$; thus the interaction must be between Na^+ and THF. From intensity ratios it was concluded that complexes, $Na^+(THF)_n$, were formed in stepwise fashion with a limiting value of four for n, when the salt concentration is about $0.25M$. Stepwise stability constants were estimated.[H103]

Raman spectra of water-DMSO mixtures have been obtained over the entire concentration range.[H104] Most bands of DMSO were observed to increase in frequency slightly with increasing water content but the S=O stretching frequency (located at 1046 cm^{-1} in the spectrum of pure DMSO) shifts to lower frequencies in an approximately linear fashion with increasing dielectric constant, and would occur at 1010 cm^{-1} in an infinitely dilute solution of DMSO in water. This band did not split as was reported for DMSO-phenol mixtures[H105] and DMSO-acetic acid mixtures[H106] where the evidence for hydrogen-bond complexes is definite. The lability of the associated DMSO-water species may result in observation of a single mean spectral line. Solvent effects on S=O, P=O and N=O stretching frequencies have also been studied by Bellamy *et al.*[H107] The geometry of the solvation sphere, consisting of six DMSO molecules arranged around a cation, may be similar to that reported for solid complexes of the type $[M(DMSO)_6]^{n+}(ClO_4)_n^-$—that is, D_{3d} or S_6.[H108] The review by Reynolds[H109] can be consulted for references to other structural studies of DMSO complexes.

Vibrational spectra of many inorganic crystalline compounds, in which molecules of interest to the non-aqueous solution chemist are bonded to metal ions, have been tabulated by Adams[H110] and Nakamoto.[H111] Frequencies of similar species in solution will be displaced from those of the solids because of the change of state, but the latter provide guidance for the assignment of the former. Recent studies of solids include solvates with acetic acid,[H112], acetonitrile[H113], 1,4-dioxan,[H114] diethyl ether,[H115] DMF,[H116] ethanol,[H117] nitromethane[H118] and tetramethylene sulphoxide.[H119]

The intensities of polarised Raman bands of binary mixtures of many solvents have been studied over a wide concentration range by Fini *et al.*[H120] The scattering coefficient increases, in most cases linearly, with the refractive index of the mixture and is in good agreement with a theoretical formula, derived from Onsager's theory of dielectric polarisation.[H121] For those systems which do not conform to the theory specific intermolecular interactions are probable.

4.14 SUMMARY

The purpose of this review has been to illustrate and document the kinds of information about non-aqueous solvent systems which have been obtained by vibrational spectroscopy. We have seen that these include insight into intermolecular forces and structure of the pure solvents, the nature of the solvation shell around ions and their solvation numbers, the identification of ion pairs and complexes, measurement of mass law constants and their dependence on the polarity of the solvent, the detection and characterisation of the hydrogen bond and measurement of acid and base strengths. Little kinetic data have so far been obtained by Raman spectroscopy but recent progress in the study of ultra-fast proton transfer[H9,H122] and the detection of associated ions of type $[Br^-, n(Br_2)]$ during the bromination of acetic acid[H123] presage considerable advance in this area in the future.[H124]

Acknowledgments

This manuscript was prepared while the author was on sabbatical leave 1970–1 in the School of Chemistry, The University of Newcastle upon Tyne. He is grateful to Dr A. K. Covington and Dr I. R. Lantzke for helpful discussions.

REFERENCES

H1 G. F. Lothian, Absorption Spectrophotometry, 3rd ed., Adam Hilger Ltd., London (1969)

H2 N. L. Alpert, W. E. Keiser and H. A. Szymanski, IR-Theory and Practice of Infrared Spectroscopy, 2nd ed., Plenum Press, New York (1970)

H3 M. Davies (ed.), Infrared Spectroscopy and Molecular Structure, Elsevier Publishing Co., Amsterdam (1963)

H4 R. P. Bauman, Absorption Spectroscopy, Wiley, New York (1962)

H5 N. B. Colthup, L. H. Daly and S. E. Wiberley, Introduction to Infrared and Raman Spectroscopy, Academic Press, New York (1964)

H6 D. E. Irish and H. Chen, *Appl. Spectr.*, 25, **1** (1971)

H7 H. A. Szymanski (ed.), Raman Spectroscopy, Theory and Practice, Plenum Press, New York (1967)

H8 A. Anderson (ed.), The Raman Effect, Dekker, New York (1971)

H9 A. R. Davis, D. E. Irish, R. B. Roden and A. J. Weerheim, *Appl. Spectr.*, **26**, 384 (1972); D. E. Irish and H. Chen, *J. Phys. Chem.*, **74**, 3796 (1970)

H10 R. W. A. Oliver and M. I. Lomax, A Guide to the Published Collections and Bibliographies of Molecular Spectra, Perkin-Elmer, Beaconsfield, Bucks., England (1971)

H11 L. A. Blatz, *J. Chem. Phys.*, **47**, 841 (1967)

H12 P. Waldstein and L. A. Blatz, *J. Phys. Chem.*, **71**, 2271 (1967)

H13 R. J. Jakobsen and J. W. Brasch, *J. Amer. Chem. Soc.*, **86**, 3571 (1964)

H14 B. J. Bulkin, *Helv. Chim. Acta*, **52**, 1348 (1969)

H15 W. F. Edgell and A. T. Watts, Abstracts, Symposium on Molecular Structure and Spectroscopy, p. 85, Ohio State University, June (1965); W. F. Edgell, A. T. Watts, J. Lyford, IV and W. Risen, Jr., *J. Amer. Chem. Soc.*, **88**, 1815 (1966)

H16 W. F. Edgell, J. Lyford, IV, R. Wright, W. Risen, Jr. and A. Watts, *J. Amer. Chem. Soc.*, **92**, 2240 (1970)

H17 J. C. Evans and G. Y.-S. Lo, *J. Phys. Chem.*, **69**, 3223 (1965)

H18 B. W. Maxey and A. I. Popov, *J. Amer. Chem. Soc.*, **89**, 2230 (1967)

H19 B. W. Maxey and A. I. Popov, *J. Amer. Chem. Soc.*, **91**, 20 (1969)

H20 J. L. Wuepper and A. I. Popov, *J. Amer. Chem. Soc.*, **91**, 4352 (1969)

H21 J. L. Wuepper and A. I. Popov, *J. Amer. Chem. Soc.*, **92**, 1493 (1970)

H22 W. J. McKinney and A. I. Popov, *J. Phys. Chem.*, **74**, 535 (1970)

H23 B. W. Maxey and A. I. Popov, *J. Amer. Chem. Soc.*, **90**, 4470 (1968)

H24 M. J. French and J. L. Wood, *J. Chem. Phys.*, **49**, 2358 (1968)

H25 G. J. Janz, K. Balasubrahmanyam and B. G. Oliver, *J. Chem. Phys.*, **51**, 5723 (1969)

H26 K. Balasubrahmanyam and G. J. Janz, *J. Amer. Chem. Soc.*, **92**, 4189 (1970)

H27 B. G. Oliver and G. J. Janz, *J. Phys. Chem.*, **74**, 3819 (1970)

H28 S. A. Al-Baldawi, M. H. Brooker, T. E. Gough and D. E. Irish, *Can. J. Chem.*, **48**, 1202 (1970)

H29 D. E. Irish, *in* Ionic Interactions: From Dilute Solutions to Molten Salts, Vol. 2, Chapter 9, p. 187 (S. Petrucci, ed.), Academic Press, New York (1971).

H30 Y.-K. Sze, M.Sc. Thesis, University of Waterloo, Waterloo, Ontario, 1970; Y.-K. Sze and D. E. Irish, to be published

H31 J. R. Ferraro, *J. Inorg. Nucl. Chem.*, **10**, 319 (1959)

H32 L. I. Katzin, *J. Inorg. Nucl. Chem.*, **24**, 245 (1962)

H33 H. Brintzinger and R. E. Hester, *Inorg. Chem.*, **5**, 980 (1966)

H34 R. E. Hester and W. E. L. Grossman, *Inorg. Chem.*, **5**, 1308 (1966)

H35 J. M. P. J. Verstegen, *J. Inorg. Nucl. Chem.*, **26**, 25 (1964)

H36 G. Nortwitz and D. E. Chasan, *J. Inorg. Nucl. Chem.*, **31**, 2267 (1969)

H37 J. C. Evans and G. Y.-S. Lo, *Spectrochim. Acta*, **21**, 1033 (1965)

H38 C. C. Addison, D. W. Amos and D. Sutton, *J. Chem. Soc.* (*A*), 2285 (1968)

H39 Z. Kecki, *Spectrochim. Acta*, **18**, 1155, 1165 (1962)

H40 D. E. Irish, B. McCarroll and T. F. Young, *J. Chem. Phys.*, **39**, 3436 (1963)

H41 K. Schaarschmidt, *Z. Chem.*, **8**, 343 (1968)

H42 G. E. Coates and D. Ridley, *J. Chem. Soc.*, 166 (1964)

H43 L. A. Woodward and P. T. Bill, *J. Chem. Soc.*, 1699 (1955)

H44 H. Irving and F. J. C. Rossotti, *J. Chem. Soc.*, 1927, 1938, 1946 (1955)

H45 D. E. Irish, *in* Ref. 7. Chapter 7, p. 224

H46 F. A. Cotton, Chemical Applications of Group Theory, p. 273, Interscience Publishers, New York (1963)

H47 M. P. Hanson and R. A. Plane, *Inorg. Chem.*, **8**, 746 (1969)

H48 L. A. Woodward and M. J. Taylor, *J. Chem. Soc.*, 4473 (1960)
H49 L. A. Woodward and M. J. Taylor, *J. Chem. Soc.*, 407 (1962)
H50 D. F. C. Morris, B. D. Andrews and E. L. Short, *J. Inorg. Nucl. Chem.*, **28,** 2436 (1966)
H51 K. Schug and L. I. Katzin, *J. Phys. Chem.*, **66,** 907 (1962)
H52 J. A. Creighton and E. R. Lippincott, *J. Chem. Soc.*, 5134 (1963)
H53 T. G. Spiro, *Inorg. Chem.*, **4,** 731 (1965)
H54 D. F. C. Morris, E. L. Short and D. N. Waters, *J. Inorg. Nucl. Chem.*, **25,** 975 (1963)
H55 J. E. D. Davies and D. A. Long, *J. Chem. Soc.* (*A*), 2560 (1968)
H56 J. E. D. Davies and D. A. Long, *J. Chem. Soc.* (*A*), 2564 (1968)
H57 M. J. Taylor, *J. Chem. Soc.* (*A*), 1780 (1968)
H58 M. J. Taylor, *Inorg. Nucl. Chem. Letters*, **4,** 33 (1968)
H59 R. Larsson and A. Miezis, *Acta Chem. Scand.*, **22,** 3261 (1968)
H60 S. Fronaeus, *in* Techniques of Inorganic Chemistry, Vol. 1, Chapter 1 (A. B. Jonassen and A. Weinberger, eds.), Interscience, New York (1963)
H61 I. Leden and R. Nilsson, *Z. Naturforsch.*, **10A,** 67 (1955)
H62 G. M. Barrow and E. A. Yerger, *J. Amer. Chem. Soc.*, **76,** 5211 (1954)
H63 E. A. Yerger and G. M. Barrow, *J. Amer. Chem. Soc.*, **77,** 4474, 6206 (1955)
H64 G. M. Barrow, *J. Amer. Chem. Soc.*, **78,** 5802 (1956)
H65 J. W. Smith and M. C. Vitoria, *J. Chem. Soc.* (*A*), 2468 (1968)
H66 N. C. Deno and M. J. Wisotsky, *J. Amer. Chem. Soc.*, **85,** 1735 (1963)
H67 R. E. Weston, Jr., S. Ehrenson and K. Heinzinger, *J. Amer. Chem. Soc.*, **89,** 481 (1967)
H68 G. C. Pimentel and A. L. McClellan, The Hydrogen Bond, W. H. Freeman and Co. San Francisco (1960)
H69 H. E. Hallam in Ref. 3. Chapter 12, p. 403
H70 R. J. Jakobsen, J. W. Brasch and Y. Mikawa, *Appl. Spectr.*, **22,** 641 (1968)
H71 A. S. N. Murthy and C. N. R. Rao, *Appl. Spectr. Rev.*, **2,** 69 (1968)
H72 J. D. Riddell and D. E. Irish, unpublished work
H73 A. R. Davis, J. W. Macklin and R. A. Plane, *J. Chem. Phys.*, **50,** 1478 (1969)
H74 U. Liddel and E. D. Becker, *Spectrochim. Acta*, **10,** 70 (1957)
H75 A. V. Stuart and G. B. B. M. Sutherland, *J. Chem. Phys.*, **24,** 559 (1956)
H76 R. E. Hester and R. A. Plane, *Spectrochim. Acta*, **23A,** 2289 (1967)
H77 G. E. Walrafen, *J. Chem. Phys.*, **52,** 4176 (1970)
H78 Z. Kecki, P. Dryjanski and E. Kozlowska, *Roczniki Chem.*, **42,** 1749 (1968)
H79 S. Minc and S. Kurowski, *Spectrochim. Acta*, **19,** 339 (1963)
H80 S. Kurowski and S. Minc, *Spectrochim. Acta*, **19,** 345 (1963)
H81 J. Bufalini and K. H. Stern, *Science*, **130,** 1249 (1959)
H82 J. Bufalini and K. H. Stern, *J. Amer. Chem. Soc.*, **83,** 4362 (1961)
H83 H. Lund, *Acta Chem. Scand.*, **12,** 298 (1958)
H84 E. A. Richardson and K. H. Stern, *J. Amer. Chem. Soc.*, **82,** 1296 (1960)
H85 K. H. Stern and E. A. Richardson, *J. Phys. Chem.*, **64,** 1901 (1960)
H86 A. Allerhand and P. von R. Schleyer, *J. Amer. Chem. Soc.*, **85,** 1233 (1963)
H87 P. J. Krueger and H. D. Mettee, *Can. J. Chem.*, **42,** 288 (1964)
H88 I. S. Perelygin, *Optics and Spectrosc.*, **13,** 194 (1962)
H89 J. B. Hyne and R. M. Levy, *Can. J. Chem.*, **40,** 692 (1962)
H90 R. D. Green, J. S. Martin, W. B. McG. Cassie and J. B. Hyne, *Can. J. Chem.*, **47,** 1639 (1969)
H91 R. D. Green and J. S. Martin, *J. Amer. Chem. Soc.*, **90,** 3659 (1968)
H92 R. West, D. L. Powell, L. S. Whatley, M. K. T. Lee and P. von R. Schleyer, *J. Amer. Chem. Soc.*, **84,** 3221 (1962)
H93 T. D. Epley and R. S. Drago, *J. Amer. Chem. Soc.*, **89,** 5770 (1967)
H94 H. Ratajczak and W. J. Orville-Thomas, *J. Mol. Structure*, **1,** 449 (1967–68)

H95 J. B. Brasch, R. J. Jakobsen, N. T. McDevitt and W. G. Fateley, *Spectrochim. Acta*, **24A,** 203 (1968)
H96 H. E. Hallam, *J. Mol. Structure*, **3,** 43 (1969)
H97 I. M. Aref'ev and V. I. Malyshev, *Optics and Spectrosc.*, **13,** 112 (1962)
H98 A. D. E. Pullin and J. McC. Pollock, *Trans. Faraday Soc.*, **54,** 11 (1958)
H99 S. Minc, Z. Kecki and T. Gulik-Krzywicki, *Spectrochim. Acta*, **19,** 353 (1963)
H100 I. S. Perelygin, *Optics and Spectrosc.*, **13,** 198 (1962)
H101 Z. Kecki, *Roczniki Chem.*, **44,** 847 (1970)
H102 E. Schaschel and M. C. Day, *J. Amer. Chem. Soc.*, **90,** 503 (1968)
H103 E. G. Höhn, J. A. Olander and M. C. Day, *J. Phys. Chem.*, **73,** 3880 (1969)
H104 J. J. Lindberg and C. Majani, *Acta Chem. Scand.*, **17,** 1477 (1963)
H105 P. Biscarini and S. Ghersetti, *Gazz. Chim. Ital.*, **92,** 61 (1962)
H106 J. J. Lindberg and C. Majani, *Suomen Kem.*, **37B,** 21 (1964)
H107 L. J. Bellamy, C. P. Conduit, R. J. Pace and R. L. Williams, *Trans. Faraday Soc.*, **55,** 1677 (1959)
H108 C. V. Berney and J. H. Weber, *Inorg. Chem.*, **7,** 283 (1968)
H109 W. L. Reynolds, *Progr. Inorg. Chem.*, **12,** 1 (1970)
H110 D. M. Adams, Metal Ligand and Related Vibrations, Edward Arnold, London (1967)
H111 K. Nakamoto, Infrared Spectra of Inorganic and Co-ordination Compounds, 2nd edn., Wiley-Interscience, New York (1970)
H112 P. W. N. M. van Leeuwen and W. L. Groeneveld, *Rec. Trav. Chim.*, **87,** 86 (1968)
H113 J. Reedijk and W. L. Groeneveld, *Rec. Trav. Chim.*, **87,** 1079 (1968)
H114 G. W. A. Fowles, D. A. Rice and R. A. Walton, *Spectrochim. Acta*, **26A,** 143 (1970)
H115 H. Wieser and P. J. Krueger, *Spectrochim. Acta*, **26A,** 1349 (1970)
H116 W. Schneider, *Helv. Chim. Acta*, **46,** 1842 (1963)
H117 P. W. N. M. van Leeuwen, *Rec. Trav. Chim.*, **86,** 247 (1967)
H118 W. L. Driessen and W. L. Groeneveld, *Rec. Trav. Chim.*, **88,** 491 (1969)
H119 J. Reedijk, P. W. N. M. van Leeuwen and W. L. Groeneveld, *Rec. Trav. Chim.*, **87,** 1073 (1968)
H120 G. Fini, P. Mirone and P. Patella, *J. Mol. Spectrosc.*, **28,** 144 (1968)
H121 P. Mirone, *Spectrochim. Acta*, **22,** 1897 (1966)
H122 H. Chen and D. E. Irish, *J. Phys. Chem.*, **75,** 2672 (1971)
H123 M. Delhaye, B. Hequet, J. Landais, J.-C. Merlin and F. Wallart, *Compt. Rend.*, **271C,** 314 (1970)
H124 M. Delhaye, *Appl. Opt.*, **7,** 2195 (1968)

Chapter 4

Spectroscopic Measurements

Part 3

E.S.R. Spectroscopy

T. E. Gough

Department of Chemistry, University of Waterloo, Waterloo, Ontario, Canada.

4.15 INTRODUCTION

The first successful observations of the resonant absorption of energy by unpaired electrons experiencing a change in spin wave function were reported by Zavoisky[J1] in 1945 and by Cummerow and Halliday[J2] in 1946. Following these early papers the technique of electron spin resonance has been applied with considerable success to problems in a remarkable diversity of fields ranging from biochemistry to solid state physics. Currently, over 700 papers concerned with electron spin resonance are listed in each volume of Chemical Abstracts. However, to date, only relatively few groups of workers have consciously attempted to use electron spin resonance as a means of studying the behaviour of solvent systems. Rather, because of the wealth of information made available concerning the electronic environment of the unpaired electron, the majority of studies performed have been 'solute-oriented'. In such studies, the role of the solvent, when used, has been to provide an environment in which the individual electron spins are isolated from one another to prevent spin-exchange reactions, and in which rapid molecular tumbling can average out any anisotropies in the spectrum. Thus, studies are made in solution in order to obtain well-resolved isotropic spectra. Frequently such studies have been performed in various solvents but no attempt has been made to correlate the observed changes in the spectrum with chemical or physical properties of the solvent molecules. In this survey we concentrate upon studies which have as their prime purpose the elucidation of those properties of the solvent which can affect an electron spin resonance spectrum by way of solvation of the paramagnetic solute, and thus restrict ourselves to phenomena arising from solute-solvent interactions. Furthermore, we restrict coverage so as to exclude several studies of transition metal ions in solution in which the solvent molecules are better described as ligands; such a distinction is of course arbitrary and may have resulted in the omission of papers considered relevant by some readers.

It does not seem necessary or advisable to describe once again the basic principles of electron spin resonance, since there are now available a substantial number of reviews and books dealing with these matters. Introductory articles have been written by Carrington[J3] and Atherton[J4] and there exist several text books[J5–7] which treat the subject on a more quantitative level. Electron spin resonance papers have been regularly reviewed in the Chemical Society of London's 'Annual Reports' and these articles may be used as guides to the more significant aspects of current developments in the field. Three features of an electron spin resonance spectrum are of interest: the hyperfine splitting constants of any nuclei with non-zero spin in the molecule, the g-factor of the radical, and the widths of the various lines in the spectrum.

4.15.1 Isotropic Hyperfine Splitting Constants

Extraction of the isotropic hyperfine splitting from an experimental spectrum, of an organic free radical in solution, is normally quite straightforward with the aid of the first order Hamiltonian,

$$\mathscr{H} = g\beta H\hat{S}_z + \sum_i a_i \hat{I}_{zi}\hat{S}_z$$

where g is the time-average g value, β is the Bohr magneton, H the applied magnetic field, $\hat{S}_z$ the component of the electronic spin angular momentum operator along the direction of the applied field (z-component), and a_i and $\hat{I}_{zi}$ are respectively the hyperfine splitting constant and the z-component of the spin angular momentum operator of the ith nucleus.

Because of the large number of lines in a spectrum, overlapping of lines can lead to difficulties though these have largely been removed by the availability of computer simulation facilities. Furthermore, if one is interested in solvent effects then one obviously chooses a solute whose spectrum is capable of rapid analysis. Having obtained the isotropic hyperfine splitting constants it is necessary to relate them to the electronic structure of the radical.

It has been found possible to relate isotropic hyperfine splitting constants to the spin densities at the nucleus of interest and at each nucleus bonded to it.[J8] Thus for the fragment

$$\begin{array}{l} X_2 \\ \quad \diagdown \\ \qquad Y\text{—}X_1 \\ \quad \diagup \\ X_3 \end{array}$$

$$a^y = [S^y + \sum_i Q^y_{yx_i}]\rho^\pi_y + \sum_i Q^y_{x_i y}\rho^\pi_{x_i} \qquad (4.15.1)$$

where S and Q are constants and ρ^π_i is the spin density at the ith atom. For the case of most common interest, namely that of the aromatic C—H fragment, eqn. 4.15.1 reduces to the McConnell relationship, eqn. 4.15.2

$$a^{\mathrm{H}} = Q^{\mathrm{H}}_{\mathrm{CH}}\rho^\pi_{\mathrm{C}} \qquad (4.15.2)$$

Thus in order to calculate a value of a hyperfine splitting constant, to compare with an experimental value, it is necessary to know the appropriate spin densities and the appropriate proportionality constants. Since the spin density, rather than the unpaired electron density is required, simple Hückel theory may not be employed: McLachlan's S.C.F. theory[J9] has gained wide-spread acceptance for the calculation of spin densities because even though it is relatively simple, the accuracy of its predicted spin densities generally exceeds the precision with which the proportionality constants in eqns. 4.15.1 and 4.15.2 are known.

The problems encountered in deriving values of the various proportionality constants have been reviewed by Bolton,[J10] who has, with Colpa, proposed an extended form of the McConnell equation designed to account for the effect of excess charge upon α-proton hyperfine splitting constants (eqn. 4.15.3).

$$A^{\mathrm{H}} = [Q^{\mathrm{H}}_{\mathrm{CH}} + K^{\mathrm{H}}_{\mathrm{CH}}\xi^\pi_{\mathrm{C}}]\rho^\pi_{\mathrm{C}} \qquad (4.15.3)$$

where K_{CH}^{H} is a second proportionality constant and ξ_{C}^{π} is the excess charge on the carbon carrying the proton of interest. A recent survey of experimental data[J11] shows that there exists a similar lack of agreement on the proportionality constants governing the interaction of β protons. Although the basic features of the mechanisms of the various hyperfine interactions are now clear, it is obvious that the quantitative description of such mechanisms is the weakest link in the chain connecting experimental isotropic hyperfine splitting constants to the electronic structure of the radical. However, efforts are still being made both experimentally[J12] and theoretically[J13,J14] to improve this situation.

Fortunately, when the effects of solvent upon electron spin resonance spectra are being examined, it is seldom necessary to discuss absolute magnitudes of hyperfine splitting constants, since one is normally concerned with changes in spin density accompanying changes in solvent. Since the calculation of spin density will involve empirical constants these can be adjusted to accommodate the choice of constants in eqns. 4.15.1–3. It is worth noting that, to date, the proportionality constants in eqns. 4.15.1–3 have been assumed as independent of solvent, any variations in the hyperfine splitting constants being attributed to changes in spin densities. While this assumption is legitimate within the L.C.A.O. approximation, it should be remembered that electrostatic repulsion between electrons in different orbitals is responsible for the spin polarisations which lead to the observation of hyperfine splitting constants. Such electrostatic repulsions are sensitive functions of the spatial distributions of the constituent atomic orbitals, and should these distributions be affected by the solvent then the proportionality constants in eqns. 4.15.1–3 will be solvent dependent. Much of the theoretical background to the calculations of spin densities and of Q values has been presented, in a convenient form, by Memory,[J15] to whom the reader should refer for more quantitative discussions of these topics.

4.15.2 *g*-factors

The g-factor of an aromatic radical determines the magnetic field at which the unpaired electron will resonate with the fixed frequency microwaves of a conventional electron spin resonance spectrometer. Thus we have the familiar resonance eqn. 4.15.4

$$h\nu = g\beta H \tag{4.15.4}$$

and furthermore since g is defined by eqn. 4.15.5

$$g = \frac{3}{2} + \frac{S(S+1) - L(L+1)}{2J(J+1)} \tag{4.15.5}$$

we can see that the resonant field depends not only upon the spin angular momentum of the electron (S) but upon its orbital (L) and total (J)

angular momentum. For aromatic radicals, to a very good first approximation, the unpaired electron has no orbital angular momentum and thus from eqn. 4.15.5, $g = 2$. Relativistic corrections modify this value to $g = 2.00232$, which is known as the free spin value, since it applies to an electron unconstrained by an attractive nuclear field. The fact that the unpaired electron of a free radical has g effectively equal to the free spin, can be understood in terms of the extensive delocalisation of the molecular orbital containing the unpaired electron.

However, in the presence of spin-orbit coupling, the description of the molecular orbital containing the unpaired electron in terms of a delocalised linear combination of atomic $2p_z$ orbitals is insufficient; spin-orbit coupling, treated as a perturbation, mixes into the ground state π function contributions from localised σ and σ^* orbitals. The constraints thus placed upon the unpaired electron cause deviations from free spin behaviour and departure of g from 2.00232. Stone has presented[J16,J17] a useful approach to the calculation of g for organic radicals, in which the molecule is divided into 'groups'. The g-tensor for each group is then calculated, and these tensors summed to give the g-tensor for the molecule. In solution of course, molecular tumbling averages out the anisotropies of the g-tensor so that g is a scalar as in eqn. 4.15.5. However, when the tumbling is not sufficiently rapid, line-width effects are observed (see below), and for their interpretation knowledge of the form of the g-tensor is essential. The importance of Stone's theory to solvation studies lies in the idea of constitutive contributions to g. The magnitude of the deviation produced by a group depends upon the density of the unpaired electron at the group, and also upon the ratio of the atomic spin-orbit coupling constant to the energy gap between the π and σ or σ^* orbitals. Since the spin-orbit coupling constant increases roughly as the fourth power of the atomic number, the inclusion of heteroatoms into the framework will lead to noticeable effects which are enhanced by the fact that such heteroatoms normally carry non-bonded electrons. The energy gap involved in the perturbation is then approximately the $n \rightarrow \pi^*$, rather than the $\sigma \rightarrow \pi^*$, transition energy. Thus, should the g-value of a radical prove solvent dependent, this effect may be interpreted in terms of redistribution of the unpaired electron at the dictates of solvent-solute interactions.

4.15.3 Line-width Effects

Line-width effects in electron spin resonance spectra have been the subject of three recent reviews[J18–20] and so there seems little point in further detailed repetition of the principles at this time. Basically, line-width effects will be observed when the Hamiltonian describing the spin systems contains time-dependent elements having frequency components comparable to frequency separations in the spectrum. The mechanisms responsible for the line-width effects fall into two main categories, those which involve fluctuations in spin density and those which involve incomplete averaging, via molecular tumbling, of the anisotropies in the hyperfine and *g*-tensors. The former case has received much attention since, under suitable conditions,[J20] it leads to the phenomenon of line-width alternation. From the temperature dependence of this effect, a number of workers have determined activation energies for internal rotations and re-arrangements which are potentially of interest with respect to solvation phenomena. However, extraction of such kinetic parameters requires considerable expenditure of spectrometer and computer time and few studies have been undertaken to examine the effect of solvation upon the results. With the increased availability of computer-interfaced spectrometers it seems certain that more attention will be paid to this area in the future. The second class of mechanism, that of incomplete averaging of hyperfine and *g*-tensors, has been the subject of considerable theoretical and practical investigation.[J18] In principle, information concerning the effect of solute-solvent interactions upon the solute's tumbling rate may be obtained from an analysis of such broadening effects. However, although the general features of the phenomenon seem clear, a recent experimental study[J21] has made it obvious that a quantitative understanding has yet to be achieved. The major source of error would seem to be the assumption that the tumbling motion of the solute is isotropic. An attempt has been made[J22] to remedy this deficiency, but the resulting equations are complex, and not readily applicable to experimental situations. It is perhaps important to stress that broadening due to incomplete averaging of the hyperfine and *g*-tensors is of frequent occurrence, and should be allowed for, at least in a phenomenological fashion, when analysing line-width effects due to other mechanisms. A simple method for such an analysis has been given.[J23]

In Fig. 4.15.1 we reproduce the electron spin resonance spectrum of monoprotonated *p*-benzosemiquinone in tetrahydrofuran at −63°C, a spectrum which shows most of the features discussed above. The spectrum has been analysed in terms of four hyperfine splitting constants, as shown in Fig. 4.15.1, since the asymmetric disposition of the —OH group causes the protons *meta* to the site of protonation to be non-equivalent. The four splitting constants are readily obtained by measuring distances from the first line of the spectrum; the *g*-value may of course be obtained by simultaneous measurement of microwave frequency and magnetic field at the centre of the spectrum. Readers unfamiliar with line-width effects may care to compute the expected relative intensities of the lines and compare the results with the experimental amplitudes of the first derivative trace. In such a presentation the peak-to-peak amplitude, for a Lorentzian line, is proportional to the reciprocal of the square

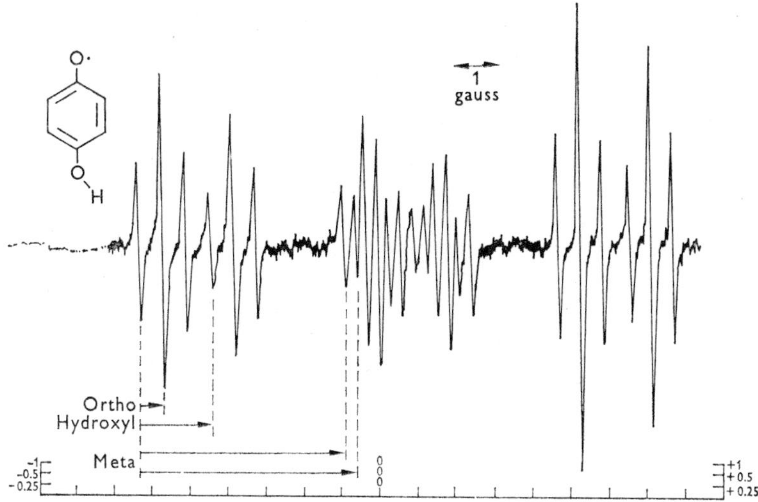

FIG. 4.15.1. Electron spin resonance spectrum of monoprotonated *p*-benzosemiquinone in tetrahydrofuran at −63°C.

of the line-width, and as such is a sensitive method for estimating relative line-widths. Figure 4.15.1 shows line-width alternation, due to out-of-phase modulation of the *meta* hyperfine splitting constant, generated by rotation of the —OH group, and in addition line-width variations due to incomplete averaging of the hyperfine and *g*-tensors.

4.16 GENERAL THEORY OF SOLVATION EFFECTS

The foundations for almost all quantitative discussions of solvent effects upon electron spin resonance spectra were laid down by the theory of Gendell, Freed and Fraenkel[J24] (G.F.F. theory), who assumed that changes in hyperfine splitting constants were solely due to re-distribution of a radical's spin density. This redistribution was assumed to accompany the formation of complexes between the radical and the solvent molecules. In formulating a model for the complexes, Gendell *et al.* restricted their attention to radicals containing polar substituents and postulated that each substituent was able to form a localised complex with one solvent molecule. Thus for a radical RX dissolved in a binary solvent mixture A:B, two solvates of the radical are postulated to be in equilibrium with each other according to eqn. 4.16.1

$$RXA + B \rightleftharpoons RXB + A \tag{4.16.1}$$

Each solvate will have its own hyperfine splitting constant, a_A and a_B for RXA and RXB, respectively. The appearance of the spectrum will now depend crucially, not only on the relative amounts of the two solvates, but also on the rates at which the solvates are interconverted. When the interconversion is slow the observed spectrum will be a superposition of the spectra of the two solvates. By slow it is meant that the lifetime of each solvate must be long compared with the time $(a_A - a_B)^{-1}$. Only one brief, and to date unconfirmed, report of such a situation has been made.[J25] Normally the interconversion between solvates is fast and so the observed spectrum will be a time-average of the spectra of the two solvates; thus, rather than containing the separate splitting constants a_A and a_B the spectrum will show one splitting, $\bar{a}$, given by eqn. 4.16.2 where p is used to denote fractional populations.

$$\bar{a} = p_A a_A + p_B a_B \tag{4.16.2}$$

From straightforward manipulation of eqns. 4.16.1–2 it may be shown that for any solvent ratio $\alpha = [B]/[A]$, $\bar{a}$ is given by eqn. 4.16.3 where K is the equilibrium constant for eqn. 4.16.1

$$\bar{a} = \tfrac{1}{2}(a_A + a_B) + \tfrac{1}{2}(a_B - a_A(K\alpha - 1)/(K\alpha + 1) \tag{4.16.3}$$

This equation reproduced very well the general form of the dependence of a_N for several nitrobenzene anion radicals dissolved in acetonitrile-water mixtures.[J26] However, agreement with the G.F.F. theory should not be taken as conclusive evidence for the existence in solution of 1:1 solvent-solute complexes, since detailed examination of alternative stoichiometries was not undertaken. Indeed, Gendell *et al.* made no attempt to fit eqn. 4.16.3 to any experimental data, other than noting the similarity discussed above.

The G.F.F. theory was put to a more quantitative test by Luckhurst and Orgel[J27] who examined the electron spin resonance spectrum of fluorenone ketyl, generated by electrochemical reduction, in mixtures of dimethylformamide and methanol. In the range of mol fractions of methanol less than 0.15, the G.F.F. theory was able to account for the variation of all four observed splitting constants on the assumption that the alcohol and the ketyl formed a 1:1 adduct. In this region of alcohol concentrations the dimethylformamide is in excess and it is not necessary to postulate a stoichiometry for the solvation of the ketyl by dimethylformamide. At concentrations of methanol in excess of 0.15 the above description of the system was found to be inadequate, presumably because of the formation of 1:2 ketyl-alcohol adducts or because of a more general form of solvent effect upon the 1:1 adduct.

An extension of the G.F.F. theory has been provided by Luckhurst in his analysis of the manner in which the sodium hyperfine coupling constant of the hexamethylacetone-sodium ion-quartet depends upon solvent and upon temperature.[G28] When hexamethylacetone is reduced by sodium, in ethereal solution, the electron spin resonance spectrum of the ketyl shows hyperfine interaction with two equivalent ^{23}Na nuclei, denoting[G29] that two ion-pairs aggregate to form an ion-quartet. In tetrahydrofuran a_{Na} is almost independent of temperature while in 2-methyltetrahydrofuran a_{Na} changes from 0.25 g at 20°C to 1.90 g at −100°C. Luckhurst postulated[G28] that the radical has n equivalent sites at which the solvent may interact. Thus in a binary solvent mixture A:B there will be n equilibria of the type expressed by eqn. 4.16.4

$$\mathrm{RXA}_{n-m}\mathrm{B}_m + \mathrm{B} \rightleftharpoons \mathrm{RXA}_{n-m-1}\mathrm{B}_{m+1} + \mathrm{A} \qquad (4.16.4)$$

and hence $(n + 1)$ distinct solvates among which it is necessary to postulate rapid equilibration so that the observed sodium coupling constant ā will be given by eqn. 4.16.5.

$$\bar{a} = \sum_{m=0}^{n} p_m a_m \qquad (4.16.5)$$

Luckhurst then assumed that a_m is a linear function of the composition of the solvent shell i.e. eqn. 4.16.6

$$a_m = \{(n - m)a_A + a_B\}/n \qquad (4.16.6)$$

and that A and B form an ideal mixture, whereupon he derived eqn. 4.16.7

$$\bar{a} = [a_A(1 - x) + a_B Kx]/[1 + x(K - 1)] \qquad (4.16.7)$$

in which x denotes the mol fraction of solvent B, and K is the equilibrium constant describing any equilibrium in which A is replaced by B in the solvent sheath. Equation 4.16.7, when recast into terms of solvent concentration ratios rather than mol fractions, is identical to eqn. 4.16.3 predicted by G.F.F. theory. It is obvious therefore, that experimental agreement with the G.F.F. theory is not diagnostic of the presence of 1:1 solute-solvent complexes.

Plots of $\bar{a}$ *vs.* x, according to eqn. 4.16.7, are given in Fig. 4.16.1 for various values of K. These plots represent the expected solvent-dependence of a hyperfine splitting constant for a system described by the G.F.F. theory.

Luckhurst was able to fit eqn. 4.16.7 to experimental results for temperatures from −60°C to +20°C, and furthermore to obtain a reasonable straight line for the plot of ln K *vs.* $1/T$, from which ΔH and ΔS for equilibrium 4.16.4 were 1.5 kcal mol^{-1} and 2.1 cal mol^{-1}°K^{-1} respectively for A = tetrahydrofuran and B = methyltetrahydrofuran.

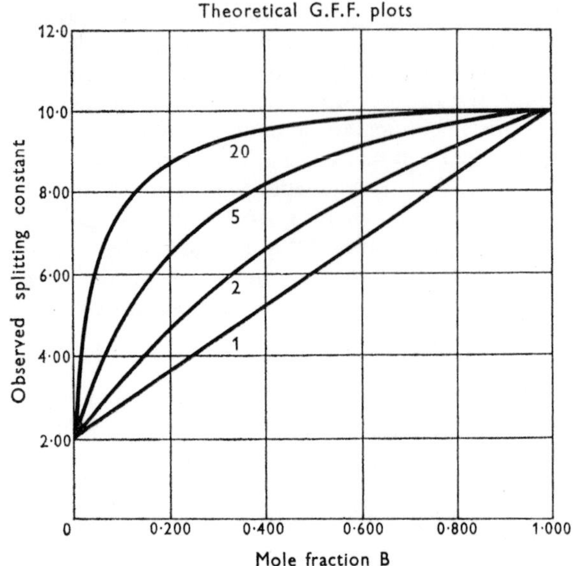

FIG. 4.16.1. Theoretical plots of eqn. 4.16.7. The numerals beside each plot indicate the value of K; for all plots $a_A = 2.0$, $a_B = 10.0$.

It should be noted that there is an important distinction to be made between the hexamethylacetone-sodium ion-quartet and the situation described by the original G.F.F. theory. In the G.F.F. theory it was assumed that the solvent dependence of hyperfine splitting constants is to be attributed to modifications in spin density distributions, whereas for the ion-quartet the spin density distribution is the same in tetrahydrofuran and in methyltetrahydrofuran. The variation in a_{Na} with solvent must be due to variation in the geometry of the ion-quartet, which will in turn vary the efficiency of the mechanism whereby spin is transferred to the alkali metal nucleus. Thus, in this case, the solvent dependence is to be attributed to variations in Q rather than ρ. The situation is common in the study of ionic association through electron spin resonance spectroscopy and has thwarted many attempts at quantitative descriptions of the effect of solvation upon such association; until the geometry of the ionic associate in solution is firmly established it is not too rewarding to discuss how the spectrum varies with change in solvent.

While some applications of the G.F.F. theory have been successful, there are several examples where it has failed even though chemically one would suspect that the system meets the requirements of the theory.

Stone and Maki examined the solvent dependence of the ^{13}C splitting of position 1 of *p*-benzosemiquinone generated electrochemically in acetonitrile-water and dimethylsulphoxide-water mixtures.[J30] Since *p*-benzosemiquinone contains two polar groups, there are effectively three solvates to be considered, RXA_2, RXAB and RXB_2. Rapid exchange of solvent molecules between RXBA and RXAB removes the asymmetry introduced by labelling only position 1 with ^{13}C. Associating the splitting constants a_1, a_2 and a_3 with the solvates RXA_2, RXAB and $RXAB_2$ respectively, a_1 and a_3 may be measured directly. From calculated spin densities, Stone and Maki deduced that a_2 is the mean of a_1 and a_3 and then applied standard G.F.F. theory. Very good agreement between experiment and theory was obtained for the dimethylsulphoxide-water solvent system but for the acetonitrile-water system the G.F.F. theory was inadequate. For the former system the ^{13}C splitting constant is almost a linear function of the mol-fraction composition of the solvent, the maximum deviation from linearity being less than 10% of (a_1-a_3); considering eqn. 4.16.7 and Fig. 4.16.1 one can see that deviations from linearity are controlled by the value of K and when $K = 1$, a linear plot of splitting constant *versus* mol fraction will be obtained. Thus for binary solvent systems, whose components have comparable solvating abilities with respect to the solute under consideration, the G.F.F. theory need only provide small corrections in order to obtain good agreement between experiment and theory. For the acetonitrile-water solvent system the deviations from linearity, for the ^{13}C splitting of position 1 of *p*-benzosemiquinone, approach approximately $0.45(a_1-a_3)$ and the G.F.F. theory does not prove satisfactory.

Gross and Symons have pointed out[J31] that G.F.F. theory is inadequate when the solvating characteristics of the components of the solvent mixture are dissimilar and caution against its application whenever ion-pairs or hydrogen-adducts are formed. They also point out that non-specific secondary solvation can also have a measurable effect upon electron spin resonance spectra.

Recently the *p*-chloronitrobenzene anion radical has been examined in a series of acetonitrile-alkyl alcohol mixtures.[J32] The results, for the solvent dependence of the nitrogen hyperfine splitting constant, were treated in terms of an extended form of G.F.F. theory, in which four solvates RXA_3, RXA_2B, $RXAB_2$ and RXB_3 are postulated to be in equilibrium with each other via a series of three solvent substitution reactions analogous to eqn. 4.16.1. Such an approach removes the necessity of postulating generalised secondary solvation but leaves five adjustable parameters (3 equilibrium constants and the nitrogen hyperfine splitting constants of RXA_2B and $RXAB_2$) with which to simulate a curve containing no maxima and only one inflection. Not surprisingly, a reasonable fit to the experimental data was obtained.

In summary, while the G.F.F. theory would seem to describe, in a qualitative fashion the variation of electron spin resonance spectra with change in solvent composition, its quantitative success has been limited. In order to improve upon the theory it seems necessary to acquire more systematic experimental information on the solvent dependence of hyperfine splitting constants.

4.17 SELECTED EXAMPLES

We now turn to a consideration of papers dealing with more specific aspects of solute-solvent interactions and the extent to which they affect electron spin resonance spectra. It is convenient to classify these papers by way of their solutes, rather than by way of their solvents.

4.17.1 Semiquinones

Semiquinones were among the first radicals to be studied in solution by electron spin resonance spectroscopy,[J33,J34] and since that time have been the subject of considerable attention. Semiquinones have been observed as anion radicals, monoprotonated neutral radicals, and diprotonated cation radicals. No systematic study of the last group as a function of solvent has been attempted, since they are formed only in strongly acidic media. *p*-Benzosemiquinone has been of particular interest because it is possible to label the carbonyl carbon with ^{13}C and the carbonyl oxygen with ^{17}O. The four remaining positions all have a hydrogen bonded to them so that it is possible to obtain three splitting constants a_H, a_C and a_O for a system with three unknown spin densities. Thus several cross-checks are available when spin-density distributions are being determined. Gendell *et al.*[J24] proposed that the effect of solvent upon the spin-density distribution could be simulated, within the framework of a Hückel or McLachlan calculation, by empirical adjustment of the Coulomb integral of the carbonyl oxygen, though they caution against a precise physical interpretation of the procedure. However, the basic features are clear; accepting the G.F.F. postulate that solvent-solute complexes are formed at the carbonyl oxygen, it can be seen that the formation of these complexes increases the effective electronegativity of oxygen, and hence modifies the most appropriate value for the Coulomb integral of this atom. The increase in electronegativity will cause an increase in the charge density on the oxygen, which is accompanied by a decrease in the spin density. In fact most proton hyperfine splittings observed for semiquinones are relatively independent of solvent because the redistribution of charge and spin is essentially localised within the carbonyl group; as the solvating power of the medium increases spin is transferred, within the carbonyl group, from oxygen to carbon. Employing the Karplus-Fraenkel relationship (eqn. 4.15.1), to translate from spin densities to hyperfine splitting constants, it can be seen that the change from an aprotic solvent to a more strongly complexing protic solvent should make both a_C and a_O more positive. The most thorough investigation of ^{17}O-labelled semiquinones is that of Broze, Luz and Silver[J35] who derived optimum values of the resonance integral of the C—O bond and of the Coulomb integral for oxygen in protic and aprotic environments. Their results were in good agreement with previous calculations[J36] performed to interpret the ^{13}C splittings of the carbonyl carbon, and also with several other experimental studies of ^{13}C and ^{17}O splittings in semiquinones.[J30,J36,J37] Since complexing by solvent, affects the spin density upon the carbonyl oxygen it should also modify the g-value; as spin is repelled from the oxygen atoms the g-value of the semiquinone should be shifted towards the free-spin value. This is born out by experimental studies[J39] but quantitatively the situation is complicated by the solvent-dependence of the $n \rightarrow \pi^*$ transition energy, which will also modify the amount of spin-orbit coupling present.

It is useful to compare these results with those from investigations of the effect of ionic association upon semiquinones, since a cation should behave somewhat like a protic solvent insofar as it will be attracted to regions of high negative charge density. The geometry of alkali metal-semiquinone ion-pairs has been established from an analysis of line-width trends, observed within the alkali-metal septets of species such as [durosemiquinone – Na_2]$^+$, which reveal that the sodium cations occupy positions along the 0–0 axis of the semiquinone.[J40] For a number of systems it has been found that the ion-pair has a g-value closer to free spin that does the free ion in an aprotic medium.[J41] Furthermore, ionic association also causes the ^{13}C splitting of the carbonyl carbon to be less negative.[J42] The addition of a second cation to form the triple ion enhances both effects: the g-value decreases further[J43] as does the carbonyl ^{13}C splitting constant.[J44] These observations confirm the hypothesis that semiquinones form solvent complexes, particularly with protic solvents and that these complexes are formed at the carbonyl oxygens. Symons[J45] has shown that both a_C and a_O for the carbonyl groups of p-benzosemiquinone can be correlated with the Z-value of the solvent used† (see Fig. 4.17.1). Since the Z-value is an empirical measure of the ion-solvating power of a solvent,[J46] being equal to the energy of the first electronic absorption band of 1-ethyl-4-methoxycarbonylpyridinium iodide in that solvent, it seems clear that the effect of solvent upon the electron spin resonance spectrum of symmetrical semiquinones is now reasonably well understood.

The results discussed above have been concerned with symmetrical semiquinones in which the carbonyl groups are equivalent. Oakes and

† In this correlation the sign of the a_{13c} axis in Fi. 5 of ref. J45 is in error.

Symons examined[J45] the electron spin resonance spectra of 2:6 dimethyl-*p*-benzosemiquinone and 2:6 di-*t*-butyl-*p*-benzosemiquinone in a wide range of solvents. The electron releasing properties of the substituents tend to make oxygen 8 (remote from the substituents) carry a higher negative charge and thus this oxygen will be preferentially solvated with respect to oxygen 7. Thus the radical is inherently asymmetric; solvation by a species seeking negative charge will accentuate the asymmetry and this effect should be revealed by observation of e.g. a_H and a_{Me} for the methyl-substituted semiquinone. Simple V.B. theory

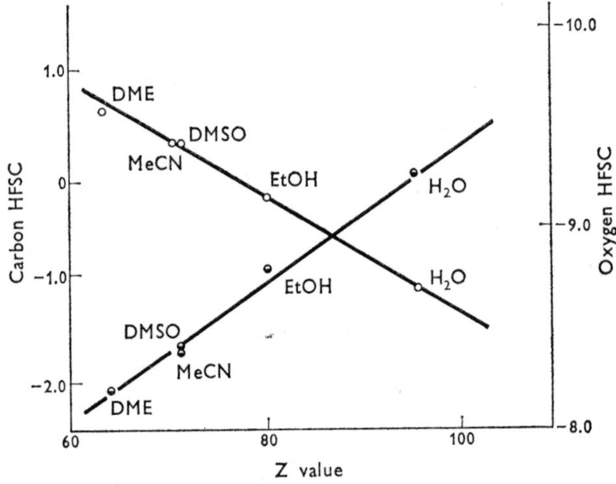

FIG. 4.17.1. Correlation of the carbonyl oxygen and carbon hyperfine splitting constants (in gauss) of *p*-benzosemiquinone with the *Z* value of the solvent; ○ = ^{13}C, ○ = ^{17}O.

shows that as the charge on oxygen 8 increases, a_{Me} should increase and a_H decrease. Oakes and Symons found that $(|a_{Me}| - |a_H|)$ increased with increasing *Z*-value of the solvent though the correlation was inferior to that discussed above for *p*-benzosemiquinone. They concluded that the molar volume of the solvent also plays a role in determining the observed splitting constant. A bulky molecule is unable to solvate oxygen 7, irrespective of its *Z*-value, since it is excluded by the adjacent methyl or *t*-butyl groups. Thus for a given *Z*-value the larger the solvent molecule the larger should be $(|a_{Me}| - |a_H|)$. As for the symmetrical semiquinones, these conclusions may be confirmed by comparison with the corresponding ion-pairs, whose spectra show that association occurs preferentially but not exclusively at the oxygen remote from the substituent groups.[J47,J48]

The polarity of the solvent has also been found to control the solvent dependence of the electron spin resonance spectra of monoprotonated semiquinones.[J49–51] Monoprotonation renders the semiquinone asymmetric but in these cases solvation opposes rather than reinforces the effect producing the asymmetry; that is, in a polar solvent protonation perturbs the radical less than it does in a non-polar solvent. As a result the ring hyperfine splitting constants vary linearly with the polarity of the solvent. The hyperfine splitting constant of the hydroxylic proton of monoprotonated durosemiquinone was not found to be a simple function of the polarity of the solvent.[J49] A correlation may be found between this splitting constant and the molar volume of the solvent molecules[J50] indicating, that when solvation occurs at a sterically crowded site, the size of the solvent molecules is an important consideration. For this particular case the hyperfine splitting constant is controlled by two mechanisms, spin polarisation from unpaired-electron density on oxygen, and hyperconjugation with spin density on the adjacent carbon atom. A quantitative estimate of the relative importance of the two mechanisms has been given[J12] and it is clear that the latter mechanism is a sensitive function of the dihedral angle between the O—H bond and the plane of the aromatic ring. Evidently solvation by bulky molecules tends to increase this angle, making the splitting constant dependent upon both the molar volume and the polarity of the solvent.

4.17.2 Phenoxyl Radicals

Since monoprotonated semiquinones may be regarded as *p*-hydroxy substituted phenoxyl radicals, it is of interest to compare the results just, presented with those obtained for more stable phenoxyl radicals. The electron spin resonance spectra of 2-6-di-*t*-butyl-4-methylphenoxyl (I) and 2-4-6-tri-*t*-butylphenoxyl (II) have been recorded in a wide range of solvents.[J52] The para-methyl and meta-hydrogen hyperfine splitting constants of the former radical were found to be solvent dependent and to be linearly related to the dipole moment of the solvent molecules, irrespective of whether the solvent was protic or aprotic. However, for the latter radical, variations in the spectrum with change in solvent were relatively minor and did not depend upon the polarity of the solvent. It was argued[J52] that the presence of three bulky *t*-butyl groups effectively insulates the π-electron system of the radical from solvent molecules, preventing the formation of solvent-solute complexes.

An interesting extension of this work is provided by the electron spin resonance spectrum of 2-6 di-*t*-butyl-4-methoxymethylphenoxyl (III), which has been recorded in several hydrocarbon solvents all having extremely low dielectric constants and dipole moments.[J53] The hyperfine splitting constant of the *para*-methylene substituent was found to

FORMULAE

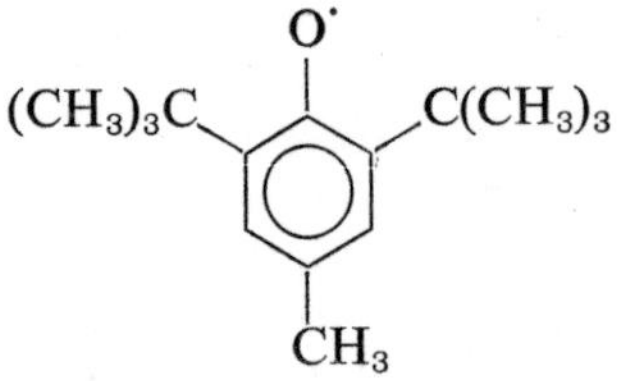

$O^{\cdot}$ $(CH_3)_3C$ $C(CH_3)_3$ CH_3

I

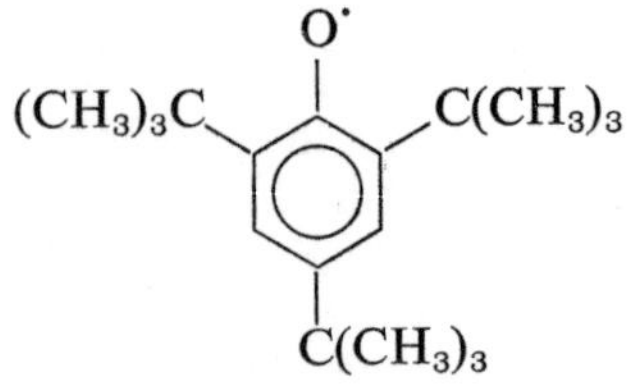

$O^{\cdot}$ $(CH_3)_3C$ $C(CH_3)_3$ $C(CH_3)_3$

II

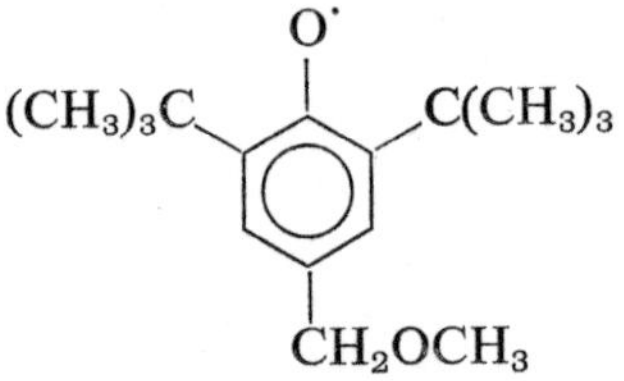

$O^{\cdot}$ $(CH_3)_3C$ $C(CH_3)_3$ CH_2OCH_3

III

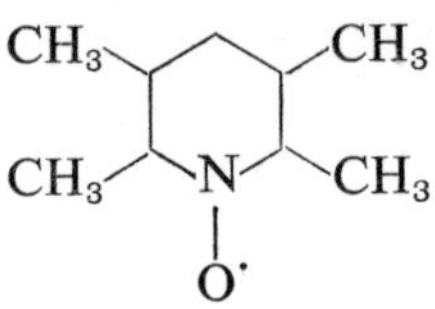

CH_3 CH_3 CH_3 N CH_3 $O^{\cdot}$

IV

O CH_3 CH_3 CH_3 N CH_3 $O^{\cdot}$

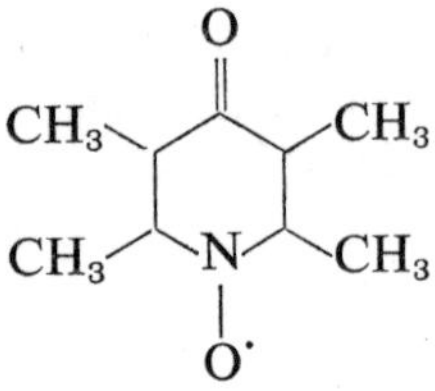

V

increase by 10% when the solvent was changed from *n*-hexane to mesitylene; in contrast the *para*-methyl hyperfine splitting constant of (I) is insensitive to such minor modifications in solvent polarity. It was found that the solvent dependence of the *para*-methylene group of (III) could be correlated with the reciprocal of the ionisation potential of the solvent which suggests[J53] that the solvent is forming a charge-transfer complex with the lone-pair electrons of the methoxymethyl group; a theoretical investigation supporting this tentative interpretation has since been published.[J54] For a correlation to exist between the methylene hyperfine splitting and the ionisation potential of the solvent it is necessary that the geometry of the methoxymethyl group be the same from solvent to solvent. In this case free rotation of the methoxymethyl ensures this prerequisite but, had the methoxymethyl group been flanked by two bulky substituents, then once again it would have proved necessary to consider the steric limitations imposed by the molecular size of the solvent.

4.17.3 Nitroxides

The nitroxide grouping >NO is isolectronic with $[>C{=}O]^-$ and hence the behaviour of nitroxide radicals is expected to be very similar to that discussed above for semiquinones. That is, polar solvents are expected to complex with the oxygen of the nitroxide group, locating charge at this centre and repelling spin towards nitrogen. The values of the appropriate constants in the Karplus-Fraenkel relationship, eqn. 4.15.1, are such that this redistribution of spin density will result in the hyperfine splitting constant of nitrogen becoming larger and more positive. Early results established this trend for the radical (IV) for which a_N was found to increase when an aprotic solvent was replaced by a protic solvent,[J55] and for diphenyl nitric oxide[J56] and di-*p*-anisyl nitric oxide[J57] both of which have nitrogen hyperfine splitting constants which correlate roughly with the dielectric constant of the solvent. A more extensive study[J52] of the latter two nitroxides showed that the polarity of the solvent could also be expressed as a function of the dipole moment, rather than the dielectric constant, of the solvent. However, solvents having —OH or —NH groups, and thus capable of forming hydrogen bonds with the solute, did not conform to the linear relationship between splitting constant and dipole moment. A remarkable illustration of the existence of hydrogen bonding between solvent and solute, was provided by the electron spin resonance spectrum of diphenyl nitric oxide dissolved in diethylamine or di-*n*-propylamine, which showed[J52] an additional doublet splitting of 2.00 gauss in the spectrum attributed to delocalisation of spin density onto the NH proton of a solvent molecule: for such a splitting to be observed, the solute-solvent complex must be stable for at least 1 μsec, which is approximately 10^4 times longer than the period between molecular collisions. Obviously such long lived complexes will not be well described in terms of non-specific solvent-solute interactions. For the radical (V) an excellent correlation has been found between the hyperfine splitting constant of the nitroxide nitrogen and the *Z*-value of the solvents,[J58] irrespective of whether the solvent is protic or aprotic. This correlation indicates that the factors controlling the empirical *Z*-value (the ability of a solvent to solvate an ionic dipole) are the same factors which vary the nitrogen hyperfine splitting constant. Therefore, a more polar solvent favours a greater degree of ionic character within the NO grouping, thereby placing greater spin density on nitrogen and increasing the hyperfine splitting constant of this nucleus.

The solvent-dependences of the electron spin resonance spectra of mono- and di-phenyl nitric oxides have been studied in a variety of solvents and the results discussed in terms of molecular orbital theory.[J59] Both simple Hückel and McLachlan S.C.F. calculations were performed; because the spin density *meta* to the nitroxide group is undoubtedly negative, the Hückel theory is unsatisfactory. Nevertheless, this theory was found to be more satisfactory than the S.C.F. theory for describing the *ortho* and *para* proton hyperfine splitting constants. When performing these empirical calculations, one must supply values for the Coulomb integrals of nitrogen and oxygen and for the resonance integrals of the carbon-nitrogen and nitrogen-oxygen bonds. Following the method adopted for semiquinones[J24] the effects of solvation were introduced by varying only the Coulomb integral of oxygen so as to make this site more electronegative in the more polar solvents. Although the results of the calculations leave something to be desired quantitatively, they nevertheless are sufficient to show that the basic postulate of the G.F.F. theory (namely that of complex formation by solvent at the electronegative oxygen) is a good description of the mechanism whereby solvents can modify the electron spin resonance spectrum of nitroxides.

Just as for the semiquinones, rearrangement of spin density within a nitroxide should lead to a modification of the *g*-values. As the polarity of the solvent is increased, the spin density on oxygen decreases and so the *g*-value should shift towards the free-spin value of 2.00232. An experimental study of the *g*-value of diphenyl nitric oxide confirms this expectation,[J60] in carbon tetrachloride g was found to be 2.00572 ± 0.00005, while in water g was 2.00502 ± 0.00005. Hückel molecular orbital calculations were used to reproduce the spin density distributions in the various solvents and hence to determine the modification in the Coulomb-integral parameter for oxygen produced by solvation. Theoretical *g*-shifts were then calculated and from the difference between the experimental and theoretical *g*-shifts, an estimate was made of the solvent dependence of the $n \rightarrow \pi^*$ transition. This estimate was in reasonable agreement with the experimentally determined variation of the energy of this band in the various solvents employed.

The high local spin density on nitrogen in a nitroxide results in a high degree of anisotropy in the hyperfine tensor of this nucleus, whose isotropic splitting constant dominates the spectrum. Furthermore, the considerable spin density on the oxygen leads to relatively large anisotropies in the *g*-tensor of the radical. Measured components of both tensors have been published[J61] for three nitroxides. Whenever molecular tumbling in solution is insufficiently rapid to average completely the anisotropies in these tensors, the components of the spectrum will exhibit differential broadening, the amount of which will be controlled by the I_Z values defining each line.[J18] In principle one can utilise this differential broadening to determine tumbling rates in solution and this has been done semi-quantitatively for macromolecules, using nitroxides as 'spin labels'.[J62,J63] It is to be anticipated that such studies will eventually provide quantitative information on solute-solvent interactions since the basic techniques for interpreting the data now seem clear.[J64,J65]

4.17.4 Nitroaromatic Anions

As discussed in the section on general theory, the solvent dependence of a_N for aromatic nitroanions was one of the phenomena which prompted the formulation of the G.F.F. theory and since that time interests in these radicals has been maintained. From an extensive compilation of experimental data, Rieger and Fraenkel have deduced optimum molecular-orbital parameters for the nitro-group, and shown that the effects of solvation can be accommodated within G.F.F. theory, by increasing the Coulomb integral of oxygen as the polarity of the solvent increases.[J66] They also deduced optimum values for the proportionality constants in eqn. 4.15.1 when it refers to the hyperfine splitting constant of the nitrogen atom of a nitro group. However, some anomalies were present in this work: steric effects can cause a nitro-group to rotate out of the plane of the aromatic ring, resulting in decreased conjugation between substituent and ring and this must be allowed for when attempting correlations between solutes. More interesting, as far as solvation is concerned, are the anomalies reported for various *meta*-dinitrobenzene anions. Such spectra showed line-width alternation which was attributed to an out-of-phase correlation in the hyperfine splitting constants of the two nitrogen atoms. The mechanism generating this out-of-phase correlation was postulated as either instantaneous solvation or rotation of one nitro-group. In the solvents employed (dimethylformamide and acetonitrile) the *para*-dinitrobenzene anion did not show this line-width alternation, though when less polar solvents are added line-width alternation may also be observed for this radical;[J67] under these conditions ionic association at one nitro-group, rather than selective solvation, is responsible for the out-of-phase correlation between the nitrogen hyperfine splitting constants.

An interesting series of experiments[J68] underline the importance of hydrogen bonding between protic solvents and nitro-anions. For the nitrobenzene anion, produced electrolytically in acetonitrile or in dimethylformamide, addition of protic solvents increases a_N, as expected from G.F.F. theory, and this increase was qualitatively correlated with the dissociation constant of the added protic solvent. Such behaviour was found to be typical of several *para* substituted aromatic nitro-anions. However, a_N for the radical anion of *o*-nitrobenzoic acid was found to be almost constant throughout a series of dimethylformamide-ethanol solvent mixtures, suggesting that the formation of a strong intramolecular hydrogen bond precludes the formation of a solvent complex with ethanol by the nitro-group. This conclusion is substantiated by the observation in the spectrum of an extra doublet splitting, of almost 12 gauss, attributed to the carboxylic proton interacting with the nitro-group.

Aromatic nitro-anions provide several examples of the importance of eliminating the effects of ionic association before interpreting variations in the spectrum in terms of solvent-solute interactions. Kitigawa *et al.* demonstrated that the electron spin resonance spectrum of the anion radical of *p*-chloronitrobenzene, generated electrochemically in dimethylformamide, is a function of the concentration and nature of the supporting electrolyte.[J69] The nitrogen hyperfine splitting constant increased with increasing concentration of supporting electrolyte, the effects being more marked the smaller the cation added. Because the spectrum was not seriously affected by the addition of tetraethylammonium perchlorate, it was concluded that ionic association did not occur between the nitro-anion and the bulky tetraalkylammonium ions. However, this conclusion should not be taken as universally applicable since it has been shown[J70] that the addition of tetra-*n*-butylammonium salts to the *p*-dinitrobenzene anion in ethanol causes a decrease in a_N. In these experiments the concentrations of added electrolyte were far above those necessary to perform an electrochemical generation of radical anions; the alkylammonium ions act rather like aprotic solvents, displacing the alcohol molecules by virtue of their abundance rather than their affinity for the anion.

4.18 SUMMARY

We have attempted to show how electron spin resonance spectra may be used to reveal the presence and characteristics of solute-solvent interactions between aromatic radicals and non-aqueous solvents. Large areas of the application of electron spin resonance spectroscopy to problems involving solvation have been omitted, partly for reasons of space, partly because of recent reviews, and partly because their interpretation is largely centred on the solute. As a partial remedy we cite recent reviews of the study of solutes in liquid crystals[J71,J72] and of the many experiments made on ionic associates.[J73–75]

The theory of localised solvent-solute complexes, proposed by Gendell, Freed and Fraenkel[J24], has been shown to provide an interpretation of solvent effects upon electron spin resonance spectra, which is acceptable to 'chemical intuition' and which correctly predicts, at least in a qualitative fashion, most of the observed trends in spectral parameters. Further progress towards a quantitative understanding would seem to await a theory, for the calculation of spin densities, which explicitly includes the spatial distribution of solvent molecules around the solute, as has indeed been recently attempted.[J76]

REFERENCES

J1 E. Zavoisky, *J. Phys. U.S.S.R.*, **9**, 211 (1945)
J2 R. L. Cummerow and D. Halliday, *Phys. Rev.*, **70**, 433 (1946)
J3 A. Carrington, *Quart. Rev.*, **17**, 67 (1963)

J4 N. Atherton, *Science Progress*, **56**, 179 (1968)
J5 M. Bersohn and J. C. Baird, An Introduction to Electron Paramagnetic Resonance, W. A. Benjamin Inc., New York (1966)
J6 P. B. Ayscough, Electron Spin Resonance in Chemistry, Methuen and Co. Ltd., London (1967)
J7 A. Carrington and A. D. McLachlan, Introduction to Magnetic Resonance, Harper and Row, New York (1967)
J8 M. Karplus and G. K. Fraenkel, *J. Chem. Phys.*, **35**, 1312 (1961)
J9 A. D. McLachlan, *Mol. Phys.*, **3**, 233 (1960)
J10 J. R. Bolton *in* Radical Ions, Chapter 1 (E. T. Kaiser and L. Kevan, eds.), Interscience Publishers, New York (1968)
J11 G. R. Underwood and V. L. Vogel, *J. Chem. Phys.*, **51**, 4323 (1969)
J12 T. E. Gough and G. A. Taylor, *Can. J. Chem.*, **47**, 3717 (1969)
J13 T. Yonezawa, T. Kawamura and H. Kato, *J. Chem. Phys.*, **50**, 3482 (1969)
J14 M. F. Chiu and B. T. Sutcliffe, *Theoretica Chimica Acta*, **16**, 331 (1970)
J15 J. D. Memory, Quantum Theory of Magnetic Resonance Parameters, McGraw-Hill Book Co., New York (1968)
J16 A. J. Stone, *Proc. Roy. Soc.*, **A271**, 424 (1963)
J17 A. J. Stone, *Mol. Phys.*, **6**, 509 (1963)
J18 G. K. Fraenkel, *J. Phys. Chem.*, **71**, 139 (1967)
J19 A. Hudson and G. R. Luckhurst, *Chem. Rev.*, **69**, 191 (1969)
J20 P. D. Sullivan and J. R. Bolton, *Adv. Mag. Res.*, **4**, 39 (1970)
J21 B. G. Segal, A. Reymond and G. K. Fraenkel, *J. Chem. Phys.*, **51**, 1336 (1969)
J22 J. H. Freed, *J. Chem. Phys.*, **41**, 2077 (1964)
J23 T. E. Gough, *Can. J. Chem.*, **47**, 331 (1969)
J24 J. Gendell, J. H. Freed and G. K. Fraenkel, *J. Chem. Phys.*, **37**, 2823 (1962)
J25 C. Corvaja and G. Giacometti, *J. Amer. Chem. Soc.*, **86**, 2736 (1964)
J26 L. H. Piette, P. Ludwig and R. N. Adams, *J. Amer. Chem. Soc.*, **84**, 4212 (1962)
J27 G. R. Luckhurst and L. E. Orgel, *Mol. Phys.*, **8**, 117 (1964)
J28 G. R. Luckhurst, *Mol. Phys.*, **9**, 179 (1965)
J29 N. Hirota and S. I. Weissman, *J. Amer. Chem. Soc.*, **86**, 2538 (1964)
J30 E. W. Stone and A. H. Maki, *J. Amer. Chem. Soc.*, **87**, 454 (1965)
J31 J. M. Gross and M. C. R. Symons, *Trans. Faraday Soc.*, **63**, 2117 (1967)
J32 M. T. Hertrich and T. Layloff, *J. Amer. Chem. Soc.*, **91**, 6910 (1969)
J33 B. Venkataraman and G. K. Fraenkel, *J. Amer. Chem. Soc.*, **77**, 2707 (1955)
J34 B. Venkataraman and G. K. Fraenkel, *J. Chem. Phys.*, **23**, 588 (1955)
J35 M. Broze, Z. Luz and B. L. Silver, *J. Chem. Phys.*, **46**, 4891 (1967)
J36 M. R. Das and G. K. Fraenkel, *J. Chem. Phys.*, **42**, 1350 (1965)
J37 W. M. Gulick and D. H. Geske, *J. Amer. Chem. Soc.*, **88**, 4119 (1966)
J38 B. L. Silver, Z. Luz and C. Eden, *J. Chem. Phys.*, **44**, 4258 (1966)
J39 P. J. Zandstra, *J. Chem. Phys.*, **41**, 3655 (1964)
J40 T. E. Gough and P. R. Hindle, *Can. J. Chem.*, **47**, 3393 (1969)
J41 P. S. Gill and T. E. Gough, *Trans. Faraday Soc.*, **64**, 1997 (1968)
J42 B. S. Prabhananda, M. P. Khakhar and M. R. Das, *J. Amer. Chem. Soc.*, **90**, 5980 (1968)
J43 T. E. Gough and P. R. Hindle, *Can. J. Chem.*, **47**, 1698 (1969)
J44 P. R. Hindle, unpublished results
J45 J. Oakes and M. C. R. Symons, *Trans. Faraday Soc.*, **64**, 2579 (1968)
J46 E. M. Kosower, *J. Amer. Chem. Soc.*, **80**, 3253 (1958)
J47 T. A. Claxton, J. Oakes and M. C. R. Symons, *Trans. Faraday Soc.*, **64**, 596 (1968)
J48 T. A. Claxton and J. Oakes, *Trans. Faraday Soc.*, **64**, 607 (1968)
J49 T. A. Claxton, T. E. Gough and M. C. R. Symons, *Trans. Faraday Soc.*, **62**, 279 (1966)

J50 T. E. Gough, *Trans. Faraday Soc.*, **62,** 2321 (1966)
J51 T. A. Claxton, J. Oakes and M. C. R. Symons, *Trans. Faraday Soc.*, **63,** 2125 (1967)
J52 K. Mukai, H. Nishiguchi, K. Ishizu, Y. Deguchi and H. Takaki, *Bull. Chem. Soc. Japan*, **40,** 2731 (1967)
J53 Y. Oishi, K. Mukai, H. Nishigushi, Y. Deguchi and H. Takaki, *Tetrahedron Letters*, 4773 (1968)
J54 S. Aono and M. Suhara, *Bull. Chem. Soc. Japan*, **41,** 2553 (1968)
J55 A. V. Il'yasov, *Zh. Strukt. Chim.*, **3,** 95 (1962)
J56 Y. Deguchi, *Bull. Chem. Soc. Japan*, **35,** 260 (1962)
J57 K. Umemoto, Y. Deguchi and H. Takaki, *Bull. Chem. Soc. Japan*, **36,** 560 (1963)
J58 R. Briere, H. Lemaire and A. Rassat, *Tetrahedron Letters*, 1775 (1964)
J59 P. B. Ayscough and F. P. Sargent, *J. Chem. Soc. B*, 907 (1966)
J60 T. Kawamura, S. Matsunami, T. Yonezawa and K. Fukui, *Bull. Chem. Soc. Japan*, **38,** 1935 (1965)
J61 O. H. Griffith, D. W. Cornell and H. M. McConnell, *J. Chem. Phys.*, **43,** 2909 (1965)
J62 J. D. Ingham, *J. Macromol. Sci., C*, **2,** 279 (1968)
J63 O. H. Griffith and A. S. Waggoner, *Accounts Chem. Res.*, **2,** 17 (1969)
J64 W. Plachy and D. Kivelson, *J. Chem. Phys.*, **47,** 3312 (1967)
J65 I. V. Alexandrov, A. N. Ivanova, N. N. Korst, A. V. Lazarev, A. I. Prikhozenko and V. B. Stryukov, *Mol. Phys.*, **18,** 681 (1970)
J66 P. H. Rieger and G. K. Fraenkel, *J. Chem. Phys.*, **39,** 609 (1963)
J67 M. J. Blandamer, J. M. Gross and M. C. R. Symons, *Nature*, **205,** 591 (1965)
J68 P. Ludwig, T. Layloff and R. N. Adams, *J. Amer. Chem. Soc.*, **86,** 4568 (1964)
J69 T. Kitigawa, T. Layloff and R. N. Adams, *Anal. Chem.*, **36,** 925 (1964)
J70 J. Oakes and M. C. R. Symons, *Chem. Comm.*, 294 (1968)
J71 A. Saupe, *Angew. Chem.*, **7,** 97 (1968)
J72 G. R. Luckhurst, *Mol. Cryst.*, **2,** 363 (1967)
J73 M. C. R. Symons, *J. Phys. Chem.*, **71,** 172 (1967)
J74 N. Hirota, *J. Phys. Chem.*, **71,** 127 (1967)
J75 M. Swarc, Carbanions, Living Polymers and Electron Transfer Processes, Interscience Publishing Co., N.Y. (1968)
J76 T. A. Claxton and D. McWilliams, *Trans. Faraday Soc.*, **65,** 3129 (1969)

Chapter 4

Spectroscopic Measurements

Part 4

N.M.R. Spectroscopy

I. R. Lantzke*
Department of Physical Chemistry, University of Newcastle upon Tyne

* Present address: Graylands Teachers' College, Mimosa Av., Graylands, Western Australia 6010.

NOTATION

I	Nuclear spin quantum number
s	Effective spin quantum number
m	Nuclear magnetic quantum number
μ	Maximum observable component of the magnetic moment
γ	Gyromagnetic (or magnetogyric) ratio
ξ	Asymmetry parameter (eqn. 4.22.4)
β	Bohr magneton
ω	Angular frequency, nuclear Larmor precession frequency (in radian s^{-1})
ω_e	Electron Larmor precession frequency (in radian s^{-1})
ν	Frequency (in Hz)
$\Delta\nu$	Chemical shift separation between two signals (in Hz)
$\nu_{\frac{1}{2}}$	Half width at half height of resonance absorption signal
H_0	Uniform static magnetic field
H_1	Oscillating magnetic field
M	Magnetisation
eq	Electric field gradient
Q	Nuclear quadrupole moment
σ	Shielding constant
a_i	Hyperfine interaction constant for nuclear spin-electron spin interaction
g	Splitting factor
δ	Chemical shift in ppm
χ_M	Molar magnetic susceptibility
χ_v	Volume magnetic susceptibility
T_1	Constant of spin-lattice or longitudinal relaxation
T_2	Constant of spin-spin or transverse relaxation
T_e	Electron spin relaxation time
T_{exch}	Characteristic exchange time for a particular process
τ_c	Correlation time for phase coherence of two spins
τ_d	Correlation time for translational jump motion
τ_r	Correlation time for rotational motion
τ_e	Correlation time for hyperfine nuclear spin-electron spin interaction
t	Time
$I_{(t)}$	Intensity of NMR signal at time 't'
J_{ij}	Coupling constant for nuclei i and j
b_{ij}	Distance between protons i and j
$\mathring{a}$	Distance of closest approach of two spins
r	Interproton distance
p_A	Fraction of specified nuclei in environment A

x_A	Mole fraction of species A
ϕ_A	Volume fraction of species A
D	Self diffusion coefficient
α	Degree of ionisation
α_{ij}	Constant of interaction of nuclei i and j
K_d	Dissociation constant
K_A	Association constant
K_a	Acid dissociation constant
k	Boltzmann constant
η	Viscosity
n	Refractive index, number of spins
ε_r	Dielectric constant
TMS	Tetramethylsilane
T	Temperature
$\hbar$	Planck constant divided by 2π
θ	Angular measure

4.19 INTRODUCTION

Nuclear magnetic resonance (NMR) spectroscopy has rapidly become a major physical technique of chemistry because of the wealth of information obtainable using it. Besides yielding information of purely spectroscopic and theoretical interest, the technique is applied to a wide range of kinetic problems, solution equilibria, solvation phenomena and intermolecular and interionic interactions, and in certain cases, the electron distribution in molecules. This list does not include the great use made of it by organic and latterly inorganic chemists in the characterisation of molecular species and determination of their structures. These are well described.[K1–4] Additionally there are a number of books and reviews of NMR spectroscopy (of which the following are only representative) ranging in coverage through the fundamentals[K5–7] and specialist series on NMR[K8,K9] to reviews of special areas and applications.

The actual recording of NMR spectra is not significantly different from one solvent to another except in the case of liquid crystal solvents, and in fact the bulk of NMR studies to date have been in organic solvents. However, most applications of NMR to physical chemistry have involved aqueous solutions. The following summary of basic NMR theory is an outline only, and the reader is referred to one of the fundamental texts for detail.

4.20 CHEMICAL SHIFTS

Nuclei with angular momentum (spin) also possess a magnetic moment and it is with these nuclei that NMR spectroscopy is possible. When placed in a static magnetic field, nuclei with spin ($I \geqslant \frac{1}{2}$) lose the degeneracy of the $(2I + 1)$ possible spin states, and if a suitable oscillating magnetic field is applied, transitions ($\Delta m = 1$) may occur between the various energy levels.

The conditions under which such a transition may occur are given by the relationship:

$$\nu_0 = \frac{\omega_0}{2\pi} = -\frac{\gamma H_0}{2\pi} \qquad (4.20.1)$$

where ν_0 is the frequency of the applied, oscillating magnetic field (H_1), ω_0 its angular velocity, γ is the gyromagnetic (or magnetogyric) ratio and H_0 is the static magnetic field.

In classical terms this can be considered as the nuclear magnets taking up preferred 'orientations' with or against the applied magnetic field, and undergoing transitions between one orientation and another when the applied oscillating field is at the resonant frequency.

For any given nuclear species once either ω or the static magnetic field at that nucleus is specified, the transition energy is also specified. Fortunately for chemists the static magnetic field experienced by a nucleus is compounded from the applied field (H_0) and all the local fields, and herein lies the value of NMR spectroscopy, as it uses the nuclear magnet as a sensitive probe for local magnetic fields. It is also important to note that the energies involved in nuclear transitions are small (e.g. for a substance with χ_M of 10^{-4} (χ_M is the molar magnetic susceptibility) at 10,000 gauss field ($\nu = 42.6$ MHz) the interaction energy is a factor of about 10^{-7} of RT[K6a]) so that the use of NMR spectroscopy does not perturb the chemistry of the system measurably.

In contrast to 'conventional' spectroscopy then, NMR (and ESR) are used not to measure the transition energy (ΔE), but rather the external applied conditions necessary to cause this transition. For experimental reasons these conditions (of ν and/or H) are not expressed in absolute values but relative to some specified standard. The choice of suitable standard substances is discussed by Laszlo.[K10]

For convenience, internal standards are most frequently used. Tetramethylsilane (TMS) is almost universally adopted for ^{1}H resonances, and is usually sufficiently soluble in non-aqueous solvents. The standard texts also describe the use of external standards and the application of bulk susceptibility corrections. Some factors influencing the susceptibilities of mixed solvents and the calculation of bulk susceptibility effects are discussed in sect. 4.24.1.

The difference between the applied field (or frequency) required for resonance of the standard (H_0) and the nucleus under observation (H) is termed the chemical shift. It can be expressed in Hertz (Hz), in which case either H_0 or ω should also be specified, or in field independent parts per million (ppm) given by:

$$\delta = \frac{H - H_c}{H_0}$$

Chemical shifts are the principle source of experimental information obtainable in an NMR experiment, and to date they have been the most used. [The other sources of information to be discussed below relate to spin-spin coupling and relaxation.] Chemical shifts are frequently characteristic of a particular environment (e.g. methyl proton resonances occur at about 1 ppm downfield from the tetramethylsilane proton resonance (TMS), while aromatic protons occur about 7 ppm downfield).

Chemical shifts are frequently modified by kinetic processes as discussed in sect. 4.22.1, the most common situation being that where the nucleus under observation (X) exchanges rapidly between two environments A and B, e.g. acid protons in aqueous solution. In such cases the measured shift δ_{obs} is given by

$$\delta_{\text{obs}} = p_{\text{A}}\delta_{\text{A}} + p_{\text{B}}\delta_{\text{B}} \qquad (4.20.2$$

where p_{A} and p_{B} are the fractions of X in environments A and B, respectively. Applications of this relationship are included in sect. 4.23.

4.21 RELAXATION

Resonance absorption involves a net gain in magnetic energy by the sample, which is only possible when a greater proportion of nuclei are in the lower spin state or states. The mechanisms whereby a Boltzmann distribution of spins tends to be, or is established or maintained, are the relaxation processes, and they are very important phenomena for three reasons.

Firstly, unless there is adequate relaxation during the resonance experiment, the $(2I + 1)$ energy levels tend to become equally populated and the signal disappears: 'saturation' occurs. Secondly, line widths are inversely proportional to the average times nuclei spend in the upper energy levels, and so depend on relaxation rates. Finally the effect of nuclear or electron spins upon the energy levels of neighbouring nuclear spins is influenced by their respective relaxation rates.

To prevent the sample from becoming 'saturated', with a consequent disappearance of the signal, it is necessary for excited nuclei to lose energy or relax. The processes of relaxation are important phenomena for two reasons. They maintain an uneven equilibrium of nuclear spins during the resonance experiment, thus preventing saturation; they also determine line widths which are inversely proportional to the average times nuclei spend in the excited state.

Kinetic applications of relaxation phenomena result from their time dependent nature. From measurements of line widths, line shapes, and direct determinations of relaxation times, many fast chemical reactions ($t_{\frac{1}{2}} <$ ca. 1 s) can be studied (sect. 4.22). Slower processes ($t_{\frac{1}{2}} >$ ca. 60 s) can also be studied by NMR using the change with time of chemical shift measurements or signal area (i.e. concentration). Relaxation measurements can also give information as to the symmetry of the immediate environment of quadrupolar nuclei ($I > \frac{1}{2}$).

The magnetic energy of excited nuclei may be lost either to the 'lattice' of surrounding atoms and molecules, or transferred from one nucleus to another. The former, variously called spin-lattice, longitudinal relaxation, or just T_1, is the process maintaining the unequal distribution of nuclear spins, but the latter, termed spin-spin, transverse relaxation or T_2, is also important since it helps determine the life time in the upper energy level of any given nucleus. Relaxation is effectively an exponential process and both longitudinal and transverse relaxation are usually characterised by their exponential half-life periods denoted by T_1 and T_2 respectively.

The interpretation of relaxation and line-width measurements requires an appreciation of the mechanisms contributing to the process. It also involves the use of relatively lengthy theoretical procedures; while frequently some approximation or approximations must be made to render the models mathematically tractable. However, NMR is unique in that an exact mathematical description of the basic phenomena is possible.

The principles of nuclear relaxation can be treated either in terms of the magnetic properties of individual nuclei or of the macroscopic measureable magnetisation resulting from an ensemble of nuclei. For a nucleus with spin $I = \frac{1}{2}$ the case is represented by Fig. 4.21.1.

In the static magnetic field (H_0) any nucleus with a magnetic moment and angular momentum precesses with frequency v_0 (see eqn. 4.20.1). Because of the relaxation processes a Boltzmann distribution of the spins is established, with slightly more nuclei in the lower energy level, i.e. that represented in Fig. 4.21.1. At resonance, ν_0 and the applied oscillating (r.f.) field H_1 are in phase and energy absorption occurs since more transitions occur from the lower to the upper energy levels

At these frequencies there is negligible probability of spontaneous loss of spin energy, so for relaxation to occur there must be a coupling between the nuclear spins and a fluctuating field with the frequency ν_0. In the case of nuclei with spin $I = \frac{1}{2}$ this must be a fluctuating magnetic field. Possible sources of such fluctuating magnetic fields arise from:

(i) Magnetic moments of other nuclei arising from either:

(a) Direct coupling of magnetic dipoles modulated by molecular translation and rotation (Brownian motion).

(b) Indirect scalar coupling modulated by chemical exchange of one or more of the nuclei involved, or by rapid relaxation of one of the spins.

[An interesting example of rapid quadrupolar relaxation of one nucleus influencing another, is the absence of spin-spin splitting between ^{35}Cl ($I = \frac{3}{2}$) and $^{17}O(I = \frac{5}{2})$ in ClO_3^-, but its occurrence in the symmetric ClO_4^- ion.[K11] The broadening of ^1H resonances of amides (and similar asymmetric ^{14}N groups) due to ^{14}N quadrupole relaxation arises from the same cause.]

(ii) The magnetic moments of unpaired electrons modulated either by direct or indirect coupling (sects. 4.22.1 and 4.25.1).

(iii) Asymmetric distributions of electronic charges, giving rise to diamagnetic anisotropy, permitting coupling of the nucleus with H_0, modulated by the rotational motion of the molecules.

In addition, nuclei with spin $I > \frac{1}{2}$ have a quadrupole moment which may interact with fluctuating electric fields. When present this coupling is often the major cause of relaxation.

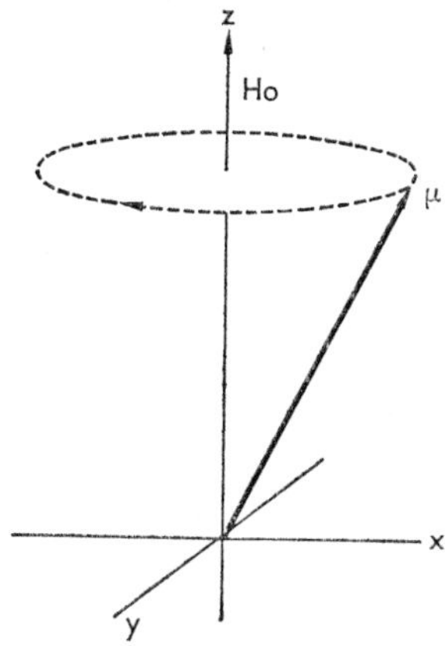

FIG. 4.21.1. Vector representation of the Larmor precession of a magnetic nucleus about the steady applied magnetic field H_0.

The most phenomenological mathematical treatment of relaxation is that of Bloch.[K12] In these equations the return of the longitudinal component of the magnetisation (the component in the direction of H_0) towards its equilibrium value, is the spin-lattice relaxation process. The decay of components of the magnetisation in the plane perpendicular to the static field H_0, is the spin-spin relaxation process since this conserves the energy of the spin system.

The equation relating the decay of nuclear magnetism in the z direction is given by

$$\frac{d\mathbf{M}_z}{dt} = \frac{\mathbf{M}_o - \mathbf{M}_z}{T_1} \qquad (4.21.1)$$

where $\mathbf{M}_o$ is the equilibrium nuclear magnetisation in the z direction and $\mathbf{M}_z$ is the instantaneous value of this magnetisation. T_1 may be estimated from measurements of the recovery of a saturated signal using

$$I_{(t)} = I_{(e)}[1 - \exp(-t/T_1)] \tag{4.21.2}$$

where $I_{(t)}$ indicates the signal intensity at time t and $I_{(e)}$ that at equilibrium. The use of a suitably phased, pulse spectrometer for spin-echo experiments and a technique using adiabatic fast passage[K13] also give T_1, and where T_1 is short, are to be preferred.

In the static field H_0 the nuclear magnets precess around the z axis, but a signal is only induced in the detector if they are in phase so that the resultant vector in the XY plane is not zero. The loss of this phase coherence is associated with T_2 such that

$$\left.\begin{aligned} \frac{d\mathbf{M}_x}{dt} &= -\frac{\mathbf{M}_x}{T_2} \\ \frac{d\mathbf{M}_y}{dt} &= -\frac{\mathbf{M}_y}{T_2} \end{aligned}\right\} \tag{4.21.3}$$

The only safe way of measuring values of T_2 is with a pulse spectrometer. For broad lines, where field inhomogeneity broadening constitutes only a small part of the line width, the relationship[K6b]

$$\nu_{\frac{1}{2}} = \frac{1}{2\pi T_2} \tag{4.21.4}$$

may also be used with a high resolution spectrum in the absorption mode. The other described method of using the decay of the 'wiggle beat', obtained on rapid passage through the resonance, is unsatisfactory. In practice the results are mostly a function of the experimental conditions used.

For diamagnetic salt solutions

$$T_1 \approx T_2 \tag{4.21.5}$$

since they are both determined by interaction with the same random field and measuring either one of them is generally sufficient.

Term by term calculations of the transition probabilities for the contributions to the relaxation time of nuclei with spin $I = \frac{1}{2}$, result in equations of the form[K5a,K6c,K14,K15]

$$\frac{1}{T_{1,2}} = \Sigma\left(\frac{1}{T_{1,2}}\right)_{\text{components}} \tag{4.21.6}$$

where each of the $(1/T_{1,2})_{\text{components}}$ is the relaxation due to one of the mechanisms listed above. Each one is an expression including terms in ω and τ_c, the correlation time (see later), and in some cases distance or magnetic field terms, together with γ and $\hbar$. To determine the approximations made and hence the conditions of validity of the derivation the reader should consult an appropriate textbook.

Equation 4.21.6 can be used to interpret time-dependent phenomena (sect. 4.22) by choosing conditions where only certain of the $(1/T_{1,2})_{\text{components}}$ terms are significant.

4.22 TIME-DEPENDENT PROCESSES

Fast time-dependent processes are of particular interest in NMR spectroscopy since it is a technique which measures the sum of nuclear behaviour over about 10^{-1} to 10^{-3} s. Kinetic processes contribute to NMR spectra in some fashion or other, frequently drastically modifying both relaxation, chemical-shift and spin-coupling measurements.

4.22.1 Use of Relaxation Measurements

Because of the approximations involved in the mathematical derivations used, it is our intention only to outline the major areas which can be investigated using relaxation measurements.

(i) *Solution Microdynamics*

From measurements of T_1 and T_2, it is possible to derive time constants for molecular rotation and translation of a pure liquid using an isotopic dilution technique, provided there is no chemical exchange of the isotopes. With solutions, the life time of solvent molecules in the solvation sphere may be determined also, as well as the life time for solute translation. The parameters derived are, τ_d the average time between two translational jumps, and τ_c the correlation time.

τ_c is a measure of the time interval during which any two spins in the system maintain a given orientation with respect to each other, and is related to the time of rotation of a molecule, or the time of diffusion into a neighbouring position. For most liquids τ_c is roughly the time of rotation of a molecule and is generally of the order of 10^{-10} s. It occurs in the relaxation expressions because the random frequency fluctuations of molecular motion contain all frequencies below that appropriate to τ_c of the Brownian motion, and except in viscous liquids this includes ν.

The form eqn. 4.21.6 takes when determining these kinetic terms from solvent relaxation measurements is

$$\frac{1}{T_1} = \left(\frac{1}{T_1}\right)_{\text{intra}} + \left(\frac{1}{T_1}\right)_{\text{inter}} \qquad (4.22.1)$$

The intra-molecular relaxation $(1/T_1)_{\text{intra}}$ is a function of the rotation of a solvent molecule in its particular environment, which in the case of extreme narrowing ($\omega^2\tau_c^2 \ll 1$; i.e. τ_c is sufficiently short for there to be a significant proportion of the fluctuating molecular motion of frequency ω), and for a system of n spins (identical nuclei) closely approximates[K15a] to

$$\left(\frac{1}{T_1}\right)_{\text{intra}} = \gamma^4\hbar^2 \frac{1}{n} \sum_{\substack{i,j \\ i<j}} 2\alpha_{ij} b_{ij}^{-6} \tau_c \qquad (4.22.2)$$

where b_{ij} is the distance between protons i and j, $\alpha_{i,j} = \frac{3}{2}$ if i and j are chemically equivalent and $\alpha_{i,j} = 1$ if they are not, and the other symbols have their usual meanings.

The intermolecular relaxation $(1/T_1)_{\text{inter}}$ is a function of the movement of neighbouring solvent molecules and, for identical spins and the case of extreme narrowing, is given by[K15b]

$$\left(\frac{1}{T_1}\right)_{\text{inter}} = \frac{4\pi}{3} \gamma^4\hbar^2 I(I+1) \frac{C_1\tau_d}{\mathring{a}^3} \left(1 + \frac{2\mathring{a}^2}{5D\tau_d}\right) \qquad (4.22.3)$$

where C_1 is the concentration of spins per cm^3, $\mathring{a}$ is the distance of closest approach between two spins, and D is the self-diffusion coefficient of the particle.

Acetone provides an example[K16,K17] of the results obtained with a pure liquid. The measured proton-relaxation times, determined in degassed mixtures (oxygen is paramagnetic) of acetone and acetone -d_6, were extrapolated to zero acetone concentration where $(1/T_1)_{\text{inter}}$ becomes negligible, to give $(1/T_1)_{\text{intra}}$. The technique gives reasonable values of $(1/T_1)_{\text{intra}}$ since, although ^{2}H has a magnetic moment ($I = 1$), its contribution to nuclear relaxation is proportional to γ and $\gamma_{1_H}/\gamma_{2_H} \simeq 7$. Values of $(1/T_1)_{\text{intra}}$ of 3.5×10^{-2} s^{-1} and 5.3×10^{-2} s^{-1} were obtained.[K16,K17]

Using an estimate of the interproton distances it was possible to calculate τ_c(0.08×10^{-11} s[K16]) using eqn. 4.22.2 and $(1/T_1)_{\text{intra}}$. Also, by making an estimate of intermolecular distances ($\mathring{a}$) and using the self-diffusion coefficient of acetone from other sources, it was possible to calculate τ_d(3.5×10^{-11} s[K16]) using eqn. 4.22.3 and $(1/T_1)_{\text{inter}}$. The significance of these and the limited other results for organic solvents are discussed by Hertz.[K15c]

To date aqueous solutions and ionic solutes have been the subjects of most NMR determinations of solution microdynamics.[K18,K19] Since they illustrate the principles and formulae involved, the reviews[K14,K15] of this topic should be consulted for further information.

The application of relaxation measurements to non-ionic solutes, also incidentally showing up solvent effects, is illustrated by some results on 1,2-dichloroethane in benzene-d_6 and carbon disulphide.[K20] In view of a considerable solvent effect on the dielectric relaxation of 1,2-dichloroethane,[K21] the longitudinal relaxation of the protons was determined using pulse techniques. By extrapolation of $(1/T_1)$ to infinite dilution of 1,2-dichloroethane to obtain $(1/T_1)_{\text{intra}}$, values of the rotational correlation τ_r were obtained (1.3×10^{-12} s and 0.6×10^{-12} s in benzene-d_6 and carbon disulphide respectively). τ_r is related to τ_c but requires further assumptions about the rotation-step constant and time constant of finite rotational-step motion. The two techniques support the argument that 1,2-dichloroethane in benzene adopts a particular gauche conformation.

For nuclei with spin $I = \frac{1}{2}$, only magnetic coupling can provide relaxation mechanisms, but for nuclei with quadrupole moments electric fields contribute the major path for relaxation. This is the case for most of the convenient magnetically active diamagnetic ions (e.g. Periodic groups I and VII) for which the relaxation expression is[K5b]

$$\frac{1}{T_1} = \frac{3}{80}\,\frac{2I + 3}{I^2(2I - 1)}\,(1 + \xi^2/3)\left(\frac{eqQ}{\hbar}\right)^2 2\tau_c \qquad (4.22.4)$$

where τ_c is a correlation time for fluctuations of the electric field gradient eq, Q is the nuclear quadrupole moment, and ξ is the asymmetry parameter usually assumed to be zero.

Although equations for specific molecular processes have been derived[K22,K23] from eqn. 4.22.4, they have not been applied to non-aqueous solutions, possibly because of the uncertainties as to the best model of the relaxation mechanism, and hence of the valid assumptions permissible in their derivation.

The binding of small molecules to larger, biologically interesting species[K14] has been studied in a similar way, by considering the components of the observed relaxation, only here changes in relaxation times are treated as arising primarily from a change in the rotational contributions. So far such work has been confined to aqueous solutions.

(ii) *Macrokinetics*

Typical of 'fast' reactions for which rate constants (and activation parameters) have been obtained using NMR relaxation measurements are;[K6,K7,K19,K24–28] hindered internal motions of molecules; exchange between bulk and solvation-sphere solvent; metal-complex ligand exchange; organometallic exchange reactions and proton and electron transfer.

These have been determined on conventional spectrometers by measuring line widths, of the broadened lines usually seen, as these are directly proportional to $1/T_2$ (eqn. 4.21.4).

Diamagnetic Systems

When a nucleus (X) can move between two or more environments there are contributions to the observed spectrum from X in each of these environments. Depending upon the average time X spends in each environment two limiting cases and an intermediate region can be distinguished.

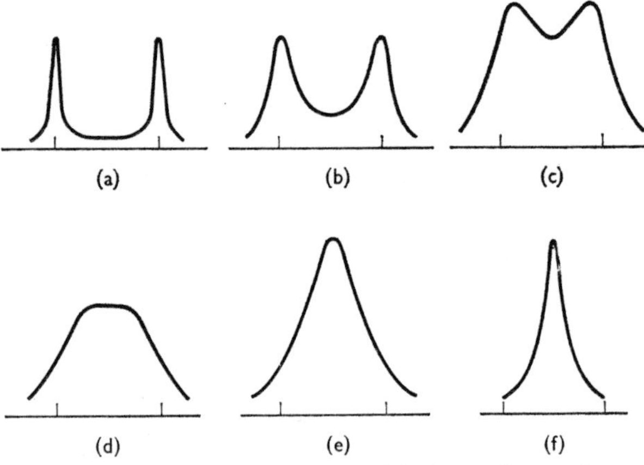

FIG. 4.22.1. Change in line shape with increasing exchange rate between two positions with equal populations for the following values of the function $2\pi\tau\Delta\nu$: (*a*) 10; (*b*) 4; (*c*) 2; (*d*) $\sqrt{2}$; (*e*) 1; (*f*) 0.5.

When species X occurs in two magnetically-different positions (A and B) and there is no exchange between these, two signals are observed, the line width of each being determined by $1/T_2$. (In practice the line width of narrow lines is determined by inhomogeneity of the steady magnetic field H_0.) If there is slow exchange of X between environment A and environment B the line widths increase and the lines eventually overlap. Around a critical exchange rate the two broadened lines coalesce. At faster rates of exchange the coalesced line becomes narrower until line width ceases to depend upon exchange rate. The critical exchange rate is related to the chemical shift separation ($\Delta\nu$) between the chemical shifts of X in A and B without exchange, such that the time for the exchange is of the order of $(2\pi\Delta\nu)^{-1}$. Figure 4.22.1 illustrates these changes.

Various analyses for the different situations have been applied[K29,K30] but the most generally used are modifications of the Bloch equations.

In the slow-exchange limit, i.e. when there are two separate resonances but each is broadened by exchange

$$T'^{-1}_{2A} = T^{-1}_{2A} + \tau^{-1}_{A} \tag{4.22.5}$$

T'_{2A} is the measured transverse relaxation of X, T_{2A} is the transverse relaxation of X in environment A with no exchange, and τ_A is the lifetime of X in environment A.

In the fast exchange limit where the line width becomes independent of exchange rate, resonance absorption occurs when

$$\omega = p_A\omega_A + p_B\omega_B \tag{4.22.6}$$

and

$$T_2^{-1} = p_A T^{-1}_{2A} + p_B T^{-1}_{2B} \tag{4.22.7}$$

Here ω $(= 2\pi\nu)$ is the radiofrequency required for resonance, and p is the fraction of X nuclei at the appropriate site. The significance of eqn. 4.22.6 is that the observed shift of the line is a concentration weighted average of the chemical shifts for the separate environments. The use of this relationship for analytical purposes is described in sect. 4.23.

When the exchange rate does contribute to the line width, eqn. 4.22.7 must be modified to

$$T_2^{-1} = p_A T^{-1}_{2A} + p_B T^{-1}_{2B} + p_A^2 p_B^2(\omega_A - \omega_B)^2(\tau_A + \tau_B) \tag{4.22.8}$$

In the intermediate range, where values of τ are such that the lines overlap and coalesce, there are several possible approaches.[K7a,K26] For the simplest case of $\tau_A = \tau_B = 2\tau$ and $p_A = p_B$ the critical lifetime for exact collapse is given by

$$\tau = \frac{1}{\sqrt{2}\pi(\nu_A - \nu_B)} \tag{4.22.9}$$

where $(\nu_A - \nu_B)$ is the frequency difference (in Hz) between the resonances at the A and B sites. In general the best method is to use curve fitting since it is impossible to estimate exact collapse.[K28]

Quadrupolar nuclei can be relaxed by fluctuating electric as well as magnetic fields, and this quadrupolar relaxation may be influenced by chemical exchange modification of the environment. If τ is the average lifetime, of the environment of the quadrupolar nucleus, between exchanges then for fast or slow exchange ($\tau \ll (1/2\pi\nu)$ or $\tau \gg (1/2\pi\nu)$ respectively) the exchange relaxation is small and line broadening is small. These conditions amount respectively to measuring (in the time of one Larmor period) either the resultant of a large number of exchanges, or else very little exchange.

When $\tau \approx (1/2\pi\nu)$, exchange relaxation is important in determining the line width of the quadrupolar nucleus.[K19,K24,K36]

Paramagnetic Systems

The following discussion deals first with the effects of a paramagnetic species on the relaxation rates of diamagnetic nuclei, then with rate determinations in which a paramagnetic species is involved.

Relaxation of a nucleus (X) in a paramagnetic compound is generally dominated by contributions from nuclear spin-electron spin magnetic dipole-dipole interaction and isotropic (scalar) nuclear-electron spin exchange. The relaxation effects of the two mechanisms are additive

$$\frac{1}{T_{1,2}} = \left(\frac{1}{T_{1,2}}\right)_{\text{dip.}} + \left(\frac{1}{T_{1,2}}\right)_{\text{exch.}} \tag{4.22.10}$$

and expressions have been derived for each mechanism in terms of nuclear and electronic parameters.[K5c,K6d,K7b,K31,K32,K33] These show that sharp nuclear resonance lines may be obtained from X when electronic relaxation is short, i.e. $\omega_e^2\tau_e^2 \ll 1$ or $(1/T_e) \gg a_i$, where ω_e is the electronic Larmor precession frequency, τ_e is the correlation time for the hyperfine interaction between the nucleus and an unpaired electron, T_e is the electron-spin relaxation time, and a_i is the hyperfine interaction constant which measures the extent to which the spins of the nucleus and an unpaired electron are coupled.

Also if $(1/T_{\text{exch.}}) \gg a_i$, sharp lines will occur in the NMR spectrum. Here $T_{\text{exch.}}$ is the characteristic exchange time for the paramagnetic species.

Under these conditions nucleus X will sense an average hyperfine magnetic field which is not zero because of the Boltzmann distribution of electron spins. The effect of this on the chemical shift of X, which is dependent upon ai and the time average of the z component of the electron spin, is discussed in sect. 4.25.1.

Conversely when $a_i \gg (1/T_e)$ and $a_i \gg (1/T_{\text{exch.}})$, nuclear resonances are greatly broadened and may be difficult to detect or resolve. The conditions which favour ESR spectroscopy of paramagnetic species are the reverse of those for NMR, so that as a first approximation 'good' NMR spectra are not expected from species giving 'good' ESR spectra.

When nuclear relaxation due to a chemical exchange process is associated with paramagnetic relaxation, it is generally possible to evaluate τ, or at least its lower limit, using modified slow or fast approximation equations.[K33,K34] In practice the spectra observed show only one line (from the exchanging bulk solvent or ligand) and it is necessary to determine experimentally whether the slow or fast exchange approximation applies. As illustrated in Fig. 4.22.1, an increase in exchange rate results in an increase in the width of each line when exchange is slow, but a decrease in the width of the averaged line when it is fast. When exchange is too fast to influence line width it is only possible to place a lower limit on the rate.

The most important case is that of fast exchange between bulk and co-ordinated solvent or ligand at a paramagnetic centre. The relationship derived when the concentration of paramagnetic ion is small is of the form[K35]

$$\frac{1}{T_2} = \frac{1}{T_{2A}} + f(M) \qquad (4.22.11)$$

where T_{2A} is the relaxation rate of solvent in the bulk environment, and $f(M)$ is the contribution to the relaxation due to the paramagnetic ion, and includes concentration, contact shift and lifetime terms.[K19,K33] The relationship for concentrated solutions of paramagnetic ions (particularly lanthanides), has also been derived.[K37]

It has been suggested that the relaxation of solvent in the bulk environment (T_{2A}) is made up of two contributions;[K38] one the normal pure-solvent relaxation (T°_{2A}); the other arising from the interaction of paramagnetic ions and the solvent molecules beyond the first solvation sphere (T^{p}_{2A})

$$\frac{1}{T_{2A}} = \frac{1}{T^{\circ}_{2A}} + \frac{1}{T^{p}_{2A}} \qquad (4.22.12)$$

Deverell[K19] has tabulated rates of solvent exchange from paramagnetic ions measured up to early 1967.

An interesting example of the determination of solvent exchange rates using solvent relaxation measurements is that of solvent DMSO from $[Ni(DMSO)_6]^{2+}$. The contribution of outer-sphere relaxation (T^{p}_{2A}) cannot be evaluated because of the high freezing point of DMSO (18.45°C), but there are now five published estimates of $k_{exchange}$ and of the activation parameters of this process, with considerable divergence between them. Blackstaffe and Dwek[K39] employed a pulse spectrometer to determine $(T_1)^{-1}$ and $(T_2)^{-1}$ by a spin-echo technique, while the four other results are based on line-width measurements.[K40a,b,K41,K42,K43] Values of $k_{exchange}$ ($= 1/\tau_{Ni}$) were calculated from the change in $(T_2)^{-1}$ of the DMSO ^{1}H resonance, using the equation derived by Swift and Connick[K35] for exchange-controlled relaxation.

Perhaps the most noteworthy aspect of these studies is the divergence between the calculated parameters, $\Delta H^{\ddagger}$ lying between 7.3[K39] and 12.1[K42] kcal mol^{-1} with $\Delta S^{\ddagger}$ and $k_{exchange}$ equally different. Since the basic experimental measurements appear to be in reasonable agreement, in so far as they can be estimated from the published data, the cause of these differences would appear to lie in the subsequent assumptions and the treatment of the data.

Electron exchange reactions may also be followed using line-width measurements. The relaxation of a diamagnetic nucleus will be modified if the species picks up or loses an electron, the magnitude of the effect depending upon a_i and the time the unpaired electron spends in the ion or radical, Such exchange reactions have been analysed using both modified Bloch equations[K7c] and a density matrix formalism.[K31]

4.22.2 Spin-spin Coupling

Spin-spin coupling or splitting is a time dependent phenomenon which has been extensively used in structural determinations.[K3,K6e,K7d] It occurs when there is isotropic coupling of nuclear spins through the bonding electrons. When the signals of two nuclei X_1 and X_2 are split by one another, the field experienced by X_1 is modified via the effects on the bonding electrons of the $(I + 1)$ orientations of X_2 and *vice versa.* If the frequency of the transitions between the $(I + 1)$ orientations of X_2 can be increased sufficiently then X_1 will sense only the average field and the spin multiplet of X_1 will collapse, in an analogous fashion to that described for single resonances (sect. 4.22.1). In practice this means increasing the relaxation rate for X_2, as often happens when X_2 has a quadrupole moment (e.g. NH_3, ClO_3^-), by using double irradiation (double resonance), and sometimes in a paramagnetic environment.

The value of spin-spin coupling results from the near constancy of the coupling constants, for a given geometric relationship between the nuclei concerned. The coupling constants (J) are a measure of the interaction energy between the two nuclei concerned.

The use of coupling constants is common in the assignment of organic molecular structures.[K1–4,K45,K46]

4.23 SOLUTION EQUILIBRIA

There have been various applications of NMR measurements to solution properties,[K18,K19,K33,K47] although non-aqueous electrolyte solutions have been neglected in contrast to aqueous solutions. While the same general principles apply in the study of solution properties by NMR as by other techniques, there is also an interesting difference because the time scale of NMR measurement is unique.

Electronic and vibrational spectroscopy involve rapid energy changes, and a species with a lifetime greater than 10^{-12}–10^{-14} s is detectable, although in liquids typical times between collisions are 10^{-10}–10^{-11} s so that these are the minimum times of duration of complexes. NMR has a slower time-scale, with the complex life time required for a separate resonance needing to be greater than the reciprocal of the frequency difference between the separate free and complex resonances (eqn. 4.22.9). This is of the order of 10^{-2} s.

The differences in complex lifetimes which make their detection using different techniques necessary have led to definitions of association which are of the form: 'complexing occurs when the aggregate exists for times long enough to be detected by the experimental technique employed'.[K10] This is the definition adopted here.

When studying solution equilibria some care is needed in choosing the concentration scale employed. The intensity of an NMR signal is proportional to solution molarity (subject to no signal saturation or molecular association etc.), but ionic and solvent interactions in more concentrated solutions are not. In ideal solutions mol fraction statistics apply, and for the most part in non-ideal solutions mol fractions are also used, although in certain circumstances volume fractions fit the data better.[K48,K49] The use of molarities and not mol fractions when working with different solvents can be misleading. The examples in Table 4.23.1 illustrate the differences between molarity and some other concentration scales, for 1 molal solutions.

Table 4.23.1

Solvent	M.W.	Approx. Density (g/cm^3)	Approx. Molarity*	Moles Solvent per Mole of Solute	Mol fraction of Solvent
H_2O	18.015	1.0	1.0	55.5	0.982
MeOH	32.04	0.793 (20°C)	0.79	31.2	0.969
DMSO	78.13	1.1	1.1	12.80	0.927
HMPT	179.20	1.02	1.02	5.58	0.848

* Molarity calculated on the assumption of no volume changes on solution.

The following discussion of solution equilibria has been divided arbitrarily into ionic and molecular sections, with that on ionic solutions sub-divided into ion-solvent and ion-ion interactions. Obviously there is overlap, and experimentally it is necessary to make measurements over a range of concentrations and extrapolate to infinite dilution, in order to determine the relative proportions of species and forces present. At the risk of appearing trite, it must be emphasised that it is highly desirable whenever possible to use relaxation measurements as well as spin-spin couplings when interpreting chemical shifts.

4.23.1 Ion-Solvent Interactions

The terms and definitions relevant to solution properties as used here are those of Bockris.[K50]

Most work on ion-solvent interactions has utilised solvent ^1H resonances, although $Me^{17}OH$ has been used.[K51] In the case of a few ions, e.g. Al^{3+}, Be^{2+} and Ga^{3+} in a number of donor solvents,[K52] the life time of the solvent protons in their solvation spheres (τ_H) is sufficiently long at room temperature for two signals to be observed. By lowering the temperature two signals have also been obtained with Mg^{2+} solutions.[K53] These correspond to free and solvation sphere solvent, and by direct integration of signal areas the solvation numbers of these ions have been obtained.[K19] In mixed solvents, the separate resonances of more than one type of solvating species have been observed.

An alternative technique, for assessing solvation numbers of cations with solvent residence times $(\tau_s) > 10^{-4}$ s, uses the large contact shift of solvent resonances when a paramagnetic ion is solvated (see sect. 4.25). The method which is applicable to paramagnetic and diamagnetic ions, is based upon the following. In a solution of dysprosium ion (Dy^{3+}) the measured solvent chemical shift is given by eqns. 4.20.2 or 4.22.6 since solvent exchange is fast

$$\delta = p_A\delta_A + p_B\delta_B \tag{4.20.2}$$

Addition of a cation (M^+) with a relatively slow solvent exchange time effectively removes the solvent around M^+ from the exchanging system, thus altering p_A and p_B. With known quantities of salts and solvent it is possible to calculate the amount of solvent 'tied up' with M^+, hence its solvation number. This method neglects the M^+ counter ion, but results agree with those from direct integrations, e.g. for ^{17}O resonances of $H_2{}^{17}O$, solvation numbers found were 5.9[K54], 5.95[K55] (Al^{3+}) and 3.8[K56], 4.14[K55] (Be^{2+}). Dy^{3+} has been used because it does not increase line widths greatly.

Covington and Lilley[K52] have tabulated the principal cation solvation number investigations to late 1969.

Equally conclusive methods for determining solvation numbers of cations with $\tau_s < 10^{-4}$ s have not been described, although several of superficial validity have been suggested. One such uses the temperature and concentration dependence of solvent chemical shifts[K56] and gives reasonable values for Na^+(4), but a high value for Al^{3+}(13). By assuming, that M^+ influences only the primary solvation sphere, and the chemical shift of this solvent (δ_{M^+}) is temperature independent, eqn. 4.23.1 was derived.[K56]

$$\delta = \frac{n_h m}{M_1}(\delta_{M^+} - \delta_0) + \delta_0 \tag{4.23.1}$$

Here δ_0 and δ_{M^+} are the chemical shifts of the solvent in the bulk solvent and in the solvation sphere, respectively, m is the molality of the solution, M_1 the molecular weight of the solvent and n_h the co-ordination or solvation number of the ion.

NMR is a useful tool for studying selective solvation of ions. Most of the work to date has been concentrated on cations, with water as one component of the solvent mixture. For the time being we will exclude the situation where the 'solvation' shell of the ion includes another ion (ion pairing).

(i) *Solvent Resonances*

When $\tau_{solvent} >$ ca. 10^{-3} s one expects to see resonances from each of the components in the solvation shell, and this has been the case with, for example, Al^{3+}, where for a number of water-donor solvent mixtures[K57-61] both species could be detected at suitable solvent compositions and temperatures, e.g. with DMSO-water mixtures separate solvating DMSO peaks were observed 0.2 ppm downfield from bulk DMSO, and solvating H_2O peaks ca. 4.2 ppm downfield from bulk H_2O.[K57]

By direct integration of areas the average number of each type of solvating molecule was determined. This showed preferential solvation of Al^{3+} by DMSO[K58] at mole fraction of DMSO (x_{DMSO}) $>$ ca. 0.25 but preferential solvation by water at $x_{H_2O} >$ ca. 0.75.

When $\tau_{solvent} < 10^{-4}$ s only one signal is observed but by using eqn. 4.20.2 deductions can be made about the fraction of solvent molecules involved in solvation. A non-linear plot of δ against solvent mol fraction is usually interpreted as showing selective solvation (although care is needed as pointed out in sect. 4.23).

For a paramagnetic ion, line-width measurements can be used, since for resonances of each solvent

$$\frac{1}{T_2} = \frac{p_D}{T_{2D}} + \frac{p_P}{T_{2P}} \tag{4.23.2}$$

where the subscripts D and P refer to diamagnetic and paramagnetic environments respectively. When the paramagnetic ion is in low concentration

$$\frac{1}{T_2} - \frac{1}{T^0_{2D}} = \frac{p_P}{T_{2P}} \tag{4.23.3}$$

where T°_{2D} is the relaxation of the solvent component in the pure solvent mixture. In these conditions it is suggested[K62] that a plot of $(1/T_{2P}) \times \phi_{solvent}$ (where ϕ is the volume fraction) against $x_{solvent}$ should be linear in the absence of preferential solvation.

(ii) *Solute Resonances (and Line Widths if $I > \frac{1}{2}$)*

Solute resonances provide less equivocal results about solute-solvent interactions, and also give information about solute-solute interactions as well. Where the observed nucleus has spin $I > \frac{1}{2}$ the symmetry of the solvation sphere is expected to influence the quadrupolar relaxation; asymmetric fields leading to broader lines,[K19] e.g. ^{27}Al line widths ($I = \frac{5}{2}$) in Al $(DMF)_6^{3+}$ (octahedral) and $KAl(OH)_4$ (tetrahedral) are 39 Hz and 200 Hz respectively.[K63] It must be pointed out that this is not always the case; the above example refers to a situation with slow exchange, but in one case of fast exchange the variation in T_2 for Cs^+ ($I = \frac{7}{2}$) in H_2O–D_2O mixtures is ascribed to changes in solvent viscosity.[K64]

To date, line-width measurements have not been used for quantitative studies of ion-solvent interactions, but the relationship between line-width and viscosity has been confirmed for ^{81}Br and ^{35}Cl resonances of lithium halides in methanol water mixtures.[K65]

The measurement of solute chemical shifts in mixed solvents cannot be used in general to determine the exact solvation shell composition, in the situation of fast exchange, since for a solvation number of n there are $(n - 1)$ compositions with unknown chemical shifts. However, the existence of preferential solvation can be shown from the analysis of the curvature of a plot of δ at infinite dilution against solvent mol fraction.

This has been done for the ion in DMSO-water mixtures using ^{133}Cs resonances.[K66] The chemical shift of Cs^+ ion is very dependent on the type and concentration of anion present.[K67] For example, NO_3^- and ClO_4^- give upfield shifts and I^- solutions give downfield shifts, but this is helpful in establishing more precisely the infinite dilution shift. A graph of Cs^+ shifts at infinite dilution against solvent mol fraction shows the negative curvature associated with preferential solvation of Cs^+ by DMSO.

Much less work has been reported on solvation of anions in organic solvent systems, using their resonances, although there seems no real reason for this. Line widths of most (magnetically active) anionic nuclei are broad, and their NMR sensitivities low, but well within the capabilities of modern apparatus.

There have been reports on ^{19}F resonance in several aqueous-organic solvent mixtures,[K68] ^{35}Cl and ^{81}Br resonances and line-widths in methanol water mixtures,[K65] and a short note on ^{35}Cl in mixtures of water with DMSO and acetonitrile.[K49]

In this last mixture, the chemical shift of ^{35}Cl is linear in water activity but not in volume fraction, observations interpreted as indicating an absence of preferential solvation. The results could, however, be interpreted as showing preferential solvation, with the preference resulting from the water-acetonitrile interaction, rather than from a preference of chloride ion for one or the other solvent.

4.23.2 Ion-Ion Interactions

(i) *Ionisation of Acids and Bases*

In protic solvents, hydrogen ion exchange between solvent and added acid or base is fast, so the chemical shift (^1H resonance) of the exchanging protons is determined by the fast exchange eqn. 4.20.2.

For the equilibrium[K69]

$$\underset{N(1-\alpha)}{HNO_3} + \underset{(M-N\alpha)}{BuOH} \rightleftharpoons \underset{N\alpha}{NO_3^-} + \underset{N\alpha}{BuOH_2^+}$$

where α is the degree of ionisation and N and M are the acid and alcohol mol fractions respectively, proton resonances give

$$\delta_{obs} = p_{HNO_3}\delta_{HNO_3} + p_{BuOH}\delta_{BuOH} + p_{BuOH_2^+}\delta_{BuOH_2^+} \qquad (4.23.4)$$

so that the relationships $p_{HNO_3} = (N(1-\alpha)/(M+N\alpha))$ and $p_{BuOH_2^+} = (2N\alpha/M + N - 2N\alpha)$ give the fractions of protons in each of these environments.

By taking chemical shifts relative to pure solvent ($\delta_{BuOH} = 0$) the values of δ_{HNO_3} and $\delta_{BuOH_2^+}$ can be obtained as $\alpha \rightarrow 0$ and $\alpha \rightarrow 1$ respectively, and so α values can be derived. There are implicit assumptions of the constancy of δ_{HNO_3} and $\delta_{BuOH_2^+}$, as well as negligible effect of the anion (NO_3^- in this case) on δ_{obs}.

Using ^{14}N, ^{13}C or possibly ^{17}O resonances (and suitably enriched BuOH) simplifies the determination of α since

$$\delta_{obs}(^{14}N) = p_{HNO_3}\delta_{HNO_3} + p_{NO_3^-}\delta_{NO_3^-}$$
$$\delta_{obs}(^{13}C \text{ or } ^{17}O) = p_{BuOH}\delta_{BuOH} + p_{BuOH_2^+}\,\delta_{BuOH_2^+} \qquad (4.23.5)$$

and measurements made relative to either NO_3^- or BuOH respectively enable α values to be determined readily. These principles were developed for aqueous solutions;[K70,K71] a summary of this work in water is given by Akitt *et al.*[K72]

In aprotic solvents at room temperature, proton exchange is fast for mineral acids, but slow for weak acids. If protonation of the solvent results in a shift of any of its resonances then, for mineral acids, relationships of the type in eqns. 4.23.4 and 4.23.5 should be applicable to the appropriate solvent resonances, as well as to the solute resonance. With weak acids (e.g. alcohols), the proton resonance of the undissociated solute is observed.[K10,K73]

The determination by NMR of *Brönsted* base strength of aprotic solvent resonances is not likely to be satisfactory because of the very low concentrations of hydrogen ions in these solvents, but the use of added (reference) acid makes possible the indirect determination of base strength.

In most non-aqueous solvents, there are likely to be side effects due to self association, complexation (solute-solvent interaction) or intramolecular hydrogen bonding.[K69,K74–76] Usually such effects can be detected by suitable variation of the concentration of solute, by extrapolating to infinite dilution, or change of solvent.

Solvent relaxation rates are only likely to provide information about ionisation in the circumstances where either,

(a) τ for residence of protons on the solvent molecule is long and so results in line broadening, e.g. at low temperatures,

(b) protonation occurs on a quadrupolar solvent nucleus, changing its symmetry and hence the field gradient, so changing the relaxation rate.

On the other hand solute relaxation measurements are sensitive to changes in the environment, and in principle afford information about the dissociation of electrolytes. They have not been used, however, to calculate dissociation constants of acids or bases, because of the complexity of the treatment required.

(ii) *Ion Association*

We discuss here only diamagnetic ion interactions. In certain circumstances the presence of a paramagnetic ion gives additional information (sect. 4.25). Ion association may modify any or all of the usually measured NMR parameters *viz.* relaxation, chemical shifts and spin-spin couplings.

While some change in solvent or solute relaxation is expected, there is a problem in interpretation. Extrapolation to infinite dilution yields the ion-solvent interaction, but in order to obtain the concentration of ion pairs in solution the relaxation rate of the ion pair is needed, and allowance for viscosity effects must also be made.

Most workers using relaxation measurements have shown the existence of ion-ion interactions but have not attempted to calculate equilibrium constants.[K46,K52] Nevertheless, observation of line widths especially for quadrupolar nuclei, provides valuable corroboration for ion association in many cases. The stability constants of several paramagnetic ions and weak ligands in aqueous solution have been calculated from solvent proton relaxation measurements.[K77]

Chemical shift measurements while less sensitive than relaxation measurements are more convenient experimentally and computationally. The usual assumption is that the measured (solute) shift is the resultant of the concentration weighted average of the shifts for the free ion and the ion pair (eqn. 4.20.2). By appropriate substitution and manipulation suitable linear equations can be derived[K33] to give either the stoichiometric K or, if activity coefficients are available, the thermodynamic K.

For example using the ^{1}H resonances of the methylene protons in the system[K78]

$$(n\text{-Bu})_4\text{N}^+ + \text{I}^- \rightleftharpoons (n\text{-Bu})_4\text{N}^+ \ldots \text{I}^- \tag{4.23.6}$$

the relationship between the observed shift (δ_{obs}) and the stoichiometric concentration (c) is

$$c = \frac{(\delta_{IP} - \delta_I)K_d(\delta_{\text{obs}} - \delta_I)}{(\delta_{IP} - \delta_{\text{obs}})^2} \tag{4.23.7}$$

where K_d is the ion pair dissociation constant and the subscripts I and IP refer to free ions and ion pairs respectively. Typically in nitrobenzene solution,[K78] $K_d = 5.3\ \text{l mol}^{-1}$ when Debye-Hückel limiting-law activity coefficients are incorporated in the calculation, whereas the conductivity value[K79] is $3.7\ \text{l mol}^{-1}$.

An alternative and preferable equation is

$$\delta'_{\text{obs}} = \frac{\delta'_{IP}}{2} - \frac{K_d}{2}\frac{\delta'_{\text{obs}}}{c\gamma_\pm^2} \tag{4.23.8}$$

(where the symbols are the same, except that the shifts are relative to the infinite-dilution shift of the measured resonance), since this avoids the reliance on δ_{IP}, which is generally estimated from a difficult extrapolation. Equation 4.23.8 is applicable only to cases of self dissociation, and neglects terms in α^2 (α = fraction associated) but similar equations can be derived for other conditions.

Coupling constants should be susceptible to ion association, but seem to be so only infrequently. Values of K have not yet been derived from such measurements.

4.23.3 Molecular Association

The study of molecular association using chemical-shift measurements has given results showing good internal consistency,[K80–2] but not always good agreement with results obtained by other methods.[K82] Although formally one may distinguish between the association of two molecular solutes in an inert solvent and one molecular solute associating with a less inert solvent, the same mechanisms (i.e. hydrogen bonding or charge transfer) are invoked, and this topic is considered in sect. 4.24 (Solvent Effects).

4.24 SOLVENT EFFECTS

The chemical shifts, line widths, and coupling constants in the high resolution spectrum of a given compound depend to some extent on the solvent. It is the purpose of this section to examine the relationships known to apply to these solvent effects. There is no consideration of the chemical effects of solvent variation, such as changes in kinetic processes and in positions of equilibria, nor of changes in the solute, such as preferred conformation or preferential solvation.

Besides their theoretical interest, solvent effects also provide information about solutions and solvents. Solvent effects are widely employed in NMR spectroscopy for such experimental reasons as the separation of otherwise overlapping resonances, or as an aid in the analysis of complex spin systems by changing their appearance. It is also necessary to choose a solvent which does not resonate in a region of other interest; this is particularly so for ^1H spectra in organic solvents.

Relaxation

There have been no widespread studies of solvent effects on relaxation rates, although various workers have reported isolated cases. In correlating solvent effects it is necessary to be sure that the same species and relaxation mechanisms are involved. For example with ionic solutes a change to a solvent of lower dielectric constant may produce significant ion association, while on changing to a solvent with small molecules some spin-rotational relaxation of the solvent may occur.

For many systems a linear relationship is observed between η/T and solute relaxation rate (η is the viscosity of the solution), which is consistent with the Debye relationship

$$\tau_c = \frac{4\pi\eta a^3}{3kT} \qquad (4.24.1)$$

(a is the radius of a rigid sphere and τ_c is its correlation time), provided the interaction energy of the relaxing species with its environment is constant. However, the linear relationship is not always observed,[K52] and its very existence for ionic nuclei has been the origin of some speculation in view of the atypical environments of ions in solutions.[K82]

An examination of the terms for intra- and inter-molecular relaxation eqns. (4.22.2 and 4.22.3) shows that $(1/T_1)_{intra}$ is proportional to the time-dependent probability of solute rotation, while $(1/T_1)_{inter}$ is proportional to solvent mobility. Thus specific solute-solvent forces will modify $(1/T_1)_{intra}$ as in the example of 1,2-dichloroethane[K20] quoted on p. 493, while $(1/T_1)_{inter}$ should correlate with the self-diffusion coefficients of the solvent.

Attempts to quantify such a relationship in aqueous solution have not been particularly successful because of the high degree of structure in water.[K83] Unstructured non-aqueous solvents offer good possibilities here.

Coupling Constants

Because spin-spin coupling is anisotropic, it has been suggested[K73] that it should provide a good probe for solvent effects. Certainly solvent effects on coupling constants have been observed frequently, but attempts to derive widespread correlations have not been very successful. When the effects of conformational changes are minimised, relationships between coupling constants and other parameters have been obtained for specific sets of compounds, but these do not hold widely.

The following are some examples of the variability of solvent effects.

1. $^1J_{CH}$ of chloroform in a series of solvents increases with proton acceptor strength of the solvent, but for phenylacetylene no such correlation is observed.[K84]

2. In a number of cases geminal coupling constants (^2J) can be explained in terms of the reaction field, but in others there is no correlation with solvent dielectric constant.[K10]

3. For a series of fluoroaromatic molecules the ^{13}CH and ^{13}CF coupling constants depend on both the substituent and the position of substitution.[K73]

For a review of results up to 1966–67 see Laszlo's article.[K10]

Chemical Shifts

The solvent shift δ_{obs}, taken as the change in chemical shift for a given solute resonance in a reference phase and at infinite dilution in a solvent, may arise from a number of mechanisms, such that

$$\delta_{obs} = \delta_b + \delta_w + \delta_E + \delta_a + \delta_s \tag{4.24.2}$$

where the contributions result from changes in bulk susceptibility of solvent, (δ_b), van der Waals' forces (δ_w), the reaction field (δ_E), magnetic anisotropy of solvent molecules (δ_a) and specific interactions (δ_s).

In general practice, it is impossible to separate out contributions from all these mechanisms, although under suitable conditions considerable simplification is possible. Assessment of contributions resulting from general effects viz. δ_b, δ_w, δ_a and δ_E is usually considered to require the use of an external reference. For specific effects an internal reference is commonly used, and contributions from general effects are disregarded by comparison with the specific effects.

4.24.1 Non-specific Solvent Effects

Theoretical treatments of chemical shifts use the shielding constant σ defined by $H_{loc} = H_0(1 - \sigma)$, where H_{loc} is the field experienced at the nucleus. Since σ is more fundamental than, but related to σ $[= (\sigma_{ref} - \sigma) = (H - H_{ref}/H_{ref})]$, shielding constants are used in the following discussion.

Bulk susceptibility corrections (δ_b) are required when using an external reference. They are required because of differences in the susceptibilities of solvents (and solutions) in cylindrical sample tubes. They are calculated using the formula.[K6f,K7f,K10,K85]

$$\delta = \delta_{obs} + \frac{2\pi}{3}(\chi_{v,ref} - \chi_v) \qquad (4.24.3)$$

Where $\chi_{v,ref}$ and χ_v are the volume susceptibilities of the reference and sample solutions respectively, and the sign of δ is negative for a downfield shift. χ_v is of course negative for diamagnetic substances. Where mixed solvents are involved it is generally assumed that

$$(\chi_v)_{mix} = \phi_1\chi_1 + \phi_2\chi_2 + \ldots \qquad (4.24.4)$$

where ϕ is the volume fraction. There is evidence that in some cases at least, when a volume change occurs on mixing, $(\chi_v)_{mix}$ is not given precisely by eqn. 4.24.3, with deviations of 5×10^{-3} in the susceptibility being typical.[K85,K86,K87]

The effect of van der Waals' forces (δ_w) is difficult to assess.[K7g,K10,K85] Experimentally σ_w has been obtained by comparing vapour and solution shifts of non-polar solutes, and making susceptibility corrections. Buckingham *et al.*,[K88] suggest the existence of two components, one due to distortion of the electronic environment of the solute nucleus by the solvent in its equilibrium configuration, the other due to a time-dependent distortion of solute electronic structure as solvent molecules move.

Theoretical approaches[K7g] show σ_w to be proportional to a function of the refractive index (n) and the ionisation potential (I) of the solvent, but comparisons of calculation with experiment (where possible) show large differences between them, while with less simple systems (where isolation of the σ_w term is not possible) no correlations are observed between functions of n and/or I and δ_{obs}.

The reaction field (δ_E)[K10,K85] or polar effect[K88] occurs with polar solutes. These polarise the surrounding solvent, only to be affected in turn by the polarised solvent. Thus δ_E is proportional to[K89] μ (dipole moment) and n of the solute and dielectric constant of the solvent. The conclusions from a number of experiments,[K10] are that the reaction field exists and contributes to NMR solvent shifts, but is more specific an effect than expected from simple theory; changing in magnitude and in direction at different positions in the same molecule.

It has been suggested that in some cases, solvent shifts which could be the result of polar effects may be better described as resulting from collision complexes.[K90] From ^1H chemical shifts of nitromethane in various solvent mixtures containing internal TMS reference, good linear plots for association constants were obtained. The values of K were small (ca. 0.1 l mol^{-1}) but their temperature dependence was considered to be consistent with association.

Many solvents are magnetically anisotropic, and in these a solute experiences a solvent effect (δ_a) due to the proximity and average orientation of the solvent molecules. Although the effect is real it is not large, and its experimental determination requires other solvent contributions to be known accurately.[K85,K90,K91] Most quantitative theoretical calculations are in poor agreement with the available experimental results.[K88,K92] Recent calculation for benzene solutions of non-polar solutes, treating the shifts as being determined primarily by molecular packing, have been encouraging.[K93]

Solvent anisotropy has been invoked on occasion to explain changes in solute resonances on changing solvents, as for instance with ^{7}Li resonances in a variety of solvents.[K94,K95]

Of relevance to studies of preferential solvation are recent determinations of σ_a in mixed solvents,[K91] which showed that anisotropy screenings of mixtures are not normally linearly additive mol or volume fraction functions of the anisotropy screenings of the pure constituents. It is suggested that only in the case of ideal mixtures is the relationship between anisotropy and mol fraction linear.

4.24.2 Specific Solvation

Under this heading come a variety of anisotropic effects which, represented by δ_s in eqn. 4.24.2, contribute significantly to the solvent shift. By and large there is other chemical or physical evidence for the existence of the mechanisms listed below.

Experimentally the chemical shift of a nucleus (usually ^{1}H) influenced by the interaction is measured as the concentration of one of the associating species is varied. The results are inserted in a suitable equation derived from the equilibrium expression for the association, and values of K obtained. By varying the temperature, ΔH and ΔS terms may be obtained.

There are two techniques for calculating K values,[K10,K82] one involving curve fitting, the other algebraic manipulation to obtain a linear equation and graphical evaluation of K. Both methods require knowledge (or an assumption) of the stoichiometry of the complex.

In the curve fitting method[K96] the observed frequency (ν) of the observed resonance is the weighted average of the monomer (ν_1) and n-mer (ν_n) populations, where only n-mer aggregates are formed, so that

$$\nu = (v_1 M + nK_n M^n)/(M + nK_n M^n) \qquad (4.24.5)$$

while

$$c = M + nK_n M^n \qquad (4.24.6)$$

K_n is the association constant of the n-mer, M is the monomer concentration at equilibrium and c is the total (free and complexed) concentration of the resonating species. By assuming arbitrary values of ν_1, ν_n and K_n values of c and ν can be calculated for any given value of M, and the theoretical curve fitted to the experimental values by trial and error.

An example of the algebraic method for evaluating K is afforded by a simple acceptor (A); donor (D) complex[K82,K97]

$$A + D \rightleftharpoons AD$$

when $K_A = [AD]/[A][D]$ and the chemical shifts (relative to an internal standard) of a suitable resonance of the acceptor species, alone, completely associated, and in the solution are δ_0^A, δ_{AD}^A and δ_{obs}^A respectively then eqn. 4.24.7 applies.

$$\frac{(\delta_{\text{obs}}^A - \delta_0^A)}{[D]} = -(\delta_{\text{obs}}^A - \delta_0^A)K_A + (\delta_{AD}^A - \delta_0^A)K_A \qquad (4.24.7)$$

A plot of $(\delta_{\text{obs}}^A - \delta_0^A)/D]$ against $(\delta_{\text{obs}}^A - \delta_0^A)$ gives a straight line of slope $-K_A$.

For either method to give meaningful values of K it is necessary that the following assumptions be true.

1. δ_{obs}^A results from molecular complexing of A and D and the fast exchange situation applies.

2. There are no significant solvent effects on the chemical shifts of the various species as the concentration of donor (in this case) is varied.

3. The solvent is inert.[K10]

Where self association of one of the solute species also occurs (e.g. in hydrogen bonding studies of water and organic molecules[K98]) this complication must be allowed for in two ways.[K10] Firstly in its effect on the concentration of free monomer, secondly from its effect on the chemical shift measured, which in the usual situation of fast exchange, is the concentration weighted average for all the environments of the nucleus under scrutiny.

Frequently, the solvent is not strictly inert and the interaction of solvent with either one or both solutes must be allowed for also.[K10] These situations where more rigorous calculational methods are required have been reviewed.[K10]

In favourable cases it is possible to decrease the rate of the dissociative process, obtain separate resonances, and use area measurements to determine the concentrations of the species present.

When $\Delta G \approx RT$ ($K \approx 0.05$ to $5\ \text{l mol}^{-1}$) it is not possible from the thermodynamic parameters, to distinguish between complexing due to hydrogen bonding, charge transfer, or electrostatic interaction. It has been suggested that this may be done using PMR spectroscopy.[K99] Hydrogen bonding of solute to dioxan produces large shifts ($\delta \geqslant 0.25$ ppm) of the donor protons, while K_A values for electron acceptor solutes and α-methylnaphthalene are significantly greater than K_A with benzene if charge transfer provides a significant associative mechanism.

(i) *Hydrogen Bonding*

Hydrogen bonding usually results in a downfield shift of the resonance of the proton involved, the solvents most effective in causing this being those with strong electron-donor properties.[K6g,K7h,K10,K82] The exceptions occur in aromatic solvents and occasionally in dilution studies.[K100] Most often, proton spectra have been measured, a wide range of donors, acceptors and 'inert' solvents being used. Donors possessing OH, NH, SH and CH functional groups have been employed. Processes ranging from self association of chloroform to inter- and intra-molecular hydrogen bonding of polyfunctional species (e.g. dicarboxylic acids and amino acids) have been examined. A variety of examples is discussed by Laszlo[K10] and Foster and Fyfe,[K82] while the relationship between the nature of hydrogen bonding and the proton shifts produced is discussed by Emsley, Feeney and Sutcliffe.[K7j]

The effect of hydrogen bonding on the resonance of the acceptor has not been greatly studied. ^{17}O resonances of methanol[K101] and acetone[K101,K102] show an upfield shift on hydrogen bonding. Downfield shifts were observed for ^{13}C resonances of a series of carbonyl compounds,[K102-4] and for ^{31}P resonances of triphenylphosphine oxide[K105] on hydrogen bonding. With both types of resonance approximately linear relationships between chemical shifts and pK values the more acidic solvents were obtained.

The question of agreement between K values for hydrogen bonding when determined by different techniques is still open, basically because too few precisely comparable studies have been made. Frequently comparison is not practicable because of the differences in the experimental conditions required and the assumption made, while unsuspected traces of water have lead to erroneous results.[K100] By utilising values of K from infra-red measurements, self-consistent chemical shifts of monomeric and hydrogen-bonded species have been obtained,[K105-108] while comparison of infra-red results with NMR results obtained in similar but different solvents, show values of K to be in good agreement.[K74,K109,K110] However, the weak association of phenols with alkyl halides, tri-*n*-butylamine and tri-*n*-butylphosphine in the same solvents have values of K two or three times larger as determined by infra-red measurements compared with those obtained from NMR.[K111]

Other correlations of NMR hydrogen-bonding shifts have been noted, e.g. with pK of proton donor,[K112] with $\Delta\nu$ for OD stretching[K113] and for C≡C stretching (of propargyl bromide),[K113] and with ΔG of hydrogen bonding.[K99] There are several reports of correlations between infra-red measurements and coupling constants.[K115-117]

(ii) *Charge-transfer Complexes*

Charge-transfer complexes have been studied by a number of techniques, particularly spectroscopic ones, but NMR observations do not completely accord with those of other methods.[K97]

In NMR work the chemical shift of a nucleus in one species, usually the acceptor (A), is determined at various low concentrations, either in the donor (D) as solvent, or in an 'inert' solvent containing donor. The experimental values of δ are averaged values for fast exchange of A between sites A and AD, and therefore depend upon the Fraction of AD, unlike electronic spectroscopy where optical density is proportional to the concentration of AD.

From δ may be derived $\delta_{A(D)}$, the chemical shift of the particular nucleus in pure A (or D), δ_{AD} its chemical shift in the complex, and Δ_{AD} the difference between these two. Values of K are calculated by one of the methods outlined earlier.

The oddities of the NMR method are most interesting. Firstly, there is negligible change in chemical shift of the resonances of the solute in many systems in which charge-transfer complexing is known to occur.[K97,K99] Only with aromatic donor molecules have significant shifts been observed, and with such a system ^{1}H and ^{19}F nuclei behave differently. Yet the same value of K within experimental error is obtained, whether donor, acceptor, ^{1}H or ^{19}F resonances, or more than one resonance in the same species is employed. Further, these values of K agree with those obtained by electronic spectroscopy provided $[D] = 2[A]$ in the latter determinations.

Values of Δ_{AD} show interesting variations. Its magnitude for ^{1}H resonances of either A or D, is independent of donor in some systems, yet decreases as K increases in other systems. The magnitude of Δ_{AD} determined for ^{19}F resonances is larger than that obtained for ^{1}H, and depends upon whether the fluorine nucleus is part of the donor or acceptor species. Values of Δ_{AD} for ^{19}F resonances of acceptors increase as the ionisation potential of the donor decreases, the shifts being upfield, while shifts for donors are downfield on complexing.

Explanations for this behaviour are tentatively[K97,K99] that the shifts observed depend upon the magnetic anisotropy and/or ring currents of aromatic systems, and the geometry of the complex. If there is charge transfer, as implied by the ^{19}F resonances, it is thought to be small in order to explain the behaviour of the ^{1}H resonances. The formation of higher aggregates, e.g. A_2D or D_2A, is not thought to be significant. At least some of the variation in values of K determined by NMR and by

UV spectroscopy may be due to anomalies in the latter method. Possible causes have been reviewed recently.[K99]

(iii) *Aromatic Solvent Shifts*

When a polar organic or metal-organic molecule is dissolved in benzene, (or other aromatic solvent) its ^1H spectrum is significantly different from that obtained in an 'inert' solvent, e.g. chloroform, as in benzene the various solute resonances are shifted upfield by different amounts. The recognition of a relationship between the substituents, the geometry of the solute and the magnitude of the upfield shift have made this a useful tool for steroid chemists.[K1,K10,K73] The mechanism of these shifts is not fully understood, but Laszlo[K10] in a careful analysis concludes that some form of chemical association provides the best model; however, the associated species must have a lifetime of less than 10^{-2} s.

It is important that the occurrence of aromatic solvent shifts be recognised as the magnitude of the shifts induced in polar molecules can be large, e.g. the ^1H resonance of chloroform moves upfield 1.5 ppm when diluted by benzene to 5%.[K7k] Even apparently non-polar solutes can be influenced by aromatic solvents, and it has been shown that TMS and the similar Sn, Ge and Pb tetramethyls suffer aromatic solvent shifts[K118] of the order of 0.10 to 0.15 ppm. For aromatic solvents it has been suggested that neopentane, cyclohexane or di-*t*-butylmethane should be used as reference for precise work.

(iv) *Liquid-crystal Solvents*

From the positions and the number of lines observed in the spectra of solutes dissolved in a liquid-crystal solvent it is possible to determine their bond angles, relative bond lengths, and the signs of spin-spin coupling constants. In some cases chemical shielding anisotropies may be determined, and the method appears to offer another technique for investigating solvent-solute interactions. Several reviews of the area have now been published.[K119–122]

'Liquid crystals' is one of the names by which certain mesomorphic phases or mesophases are known, and as the name implies refers to materials which show a high degree of structural order, yet are liquid. The topic has been reviewed recently.[K123–125] Those most used in NMR work have been the thermotropic variety which have a definite mesophase range between the melting point and a second transition producing the isotropic liquid. Both pure substances (e.g. *p*-azoxyanisole) and mixtures (e.g. butyl-*p*-(*p*-ethoxybenzoyl)phenyl carbonate and *p*-(*p*-ethoxyphenylazo)phenyl heptanoate) which provide a mesophase at room temperature, have been used.

Although other sources of ordered solutions are known, they have not yet been used widely for NMR spectroscopy. These include certain optically-active polymers in organic solvents.[K126,K127] and suitable surfactants mixed with water, or water and organic solvents.[K127]

Three types of thermotropic liquid crystals have been characterised, the most used for NMR being the nematic mesophase, in which intermolecular forces cause a long-range order by orienting the long axes of the comparatively linear molecules parallel, although the molecules remain free to move relative to each other, and 'domains' with different orientation occur.

In a magnetic field the direction of orientation is rotated so that the long axes of the molecules are all largely oriented parallel to the magnetic field. This results from the interaction of the magnetic susceptibility of the solvent molecules. The liquid crystals provide an anisotropic solvent for the compound to be studied. The effects of this on the NMR spectrum are considered shortly.

The other thermotropic mesophases (cholesteric and smectic) also result from an ordering of solvent molecules. The cholesteric occurring in solvents possessing a centre of asymmetry, so that the structure is characterised by a helical twist. The smectic mesophase consists of molecules parallel to each other with the extra order being in layers. They are generally viscous and are not so suitable for high resolution NMR spectroscopy.

An experimental difference from conventional NMR is the need to, not spin the sample, spin it slowly, or else spin it parallel to the magnetic field. This treatment is necessary because the solvent molecules are aligned parallel to the magnetic field. Frequently spectrum enhancement is needed also.

A solute dissolved in a nematic mesophase is capable of rapid motion, but because of the anisotropic environment, rotational freedom is restricted causing partial alignment. From an NMR point of view this means that sharp lines are observed and intermolecular interaction is averaged out as in an isotropic solvent. However, intramolecular dipolar coupling is no longer reduced to zero. Thus for two magnetically equivalent protons (normally resonating at a field H_0), in the condition of dipolar coupling the field experienced by each is modified by an amount ΔH depending upon the orientation of the nuclear spin of the other. The single line at H_0 is now split into two lines at $(H_0 - \Delta H)$ and $(H_0 + \Delta H)$, and the spectrum is symmetrical. The frequency separation between the two lines is given for protons by[K5d, K120]

$$\Delta\nu = \frac{3\gamma^2\hbar}{4\pi r^3}(3\cos^2\theta - 1) \qquad (4.24.8)$$

when r is the interproton distance and θ is the angle between this axis and the magnetic field.

The interpretation of any but the simplest spectrum, however, requires a detailed analysis using a computer since the spectra are complex, (e.g. the spectrum of benzene consists of about fifty lines).

The description of the spectrum by the spin Hamiltonian requires three terms;[K138] the first representing the Zeeman energy, the second the indirect spin-spin coupling, and the third dipolar coupling. The first and second terms are the usual ones required for the calculation of spectra in isotropic solvents. The dipolar coupling term includes the direct coupling constant, D_{ij} which is dependent upon the geometry of the molecule. The calculation of D_{ij} requires the averaging of $(1 - 3\cos^2\theta_{ij})$ over the motions of the molecule. In general only molecular tumbling is considered and the small intermolecular vibrations neglected. The method of calculating values of r, and using symmetry properties to simplify the problem are to be found in the papers of Saupe[K125] and the review by Buckingham and McLauchlan.[K119]

A comparison of the bond lengths and angles of a number of organic molecules as obtained by NMR in nematic solvents and by other methods showed very good agreement.[K121]

Relaxation measurements in mesophase solvents should provide insight into molecular dynamics in these systems, but as yet this aspect has been neglected. Calculations of T_1 for a pair of protons on one molecule[K130] suggest a longitudinal relaxation of about 1 s, resulting from the modulation of the dipolar coupling. They also predict a dependence of T_1 on the nuclear-resonant frequency applied.

(iv) *Optically-active Solvents*

A recent interesting application of solvent effects has been the use of optically-active solvents in the determination of the optical purity and the absolute configuration of solutes. Work so far has centred on ^1H resonances and organic solutes, covering various alcohols,[K131,K132] amines, sulphoxides, α-hydroxy- and α-amino-acids,[K131] and epoxides (the solvent here being an optically active nematic phase[K133]). There are also reports on disymmetric nickel(II) complexes,[K134] and the use of ^{19}F resonances.[K135]

The experimental observations are that the spectra of enantiomeric mixtures in certain optically-active solvents, either pure, or diluted with an optically-inactive solvent, show small splittings of some of the peaks. The separations may be up to 8 Hz (measured at 100 MHz) but are typically about one-third of this value. A typical solvent is (—) 2,2,2,-trifluorophenylethanol, but the requirements of a suitable solvent have not been fully delineated.

The observations are explained as the result of strong solvent-solute interaction (e.g. hydrogen bonding, charge transfer or dipolar attraction) to produce labile diasteriomeric solvates. In these solvates the conformation of each isomer is sufficiently different for some of the enantiomeric nuclei to be in magnetically different environments.

4.25 PARAMAGNETIC SPECIES

While the NMR study of paramagnetic species in solution is not confined to non-aqueous solvents, the bulk of the work so far has been carried out in organic solvents for reasons of stability. The results have been confined almost exclusively to transition-metal-complex solutes; much less attention has been afforded the solvents except when co-ordinated as ligands.[K34] In favourable conditions these studies provide information about NMR spectroscopic theory; metal-ligand bonding; the electronic structure of ligands, ion association, bulk susceptibilities, various kinetic processes, and molecular structures. The topic has been reviewed recently,[K31,K32,K33] and current literature is evaluated in the Specialist Reports of the Chemical Society.[K136]

The conditions required in order to obtain NMR resonances, which are not broadened beyond detection, are given in sect. 4.21. These amount to situations where the magnetic fluctuations due to the unpaired electron (which has a magnetic moment about 10^3 that of an atomic nucleus) are sufficiently fast for the nuclear-spin transitions to be affected only by an average field.

4.25.1 Electron Distribution

The spectra of paramagnetic molecules are in general quite different from those of comparable diamagnetic molecules, with the resonances shifted to quite different fields. The size and direction of the shifts depend upon the structure of the compound, the nature of the ligand, and the metal ion. Most of the spectra reported so far have been of ^1H resonances although some ^{19}F studies[K32] have been carried out.

To obtain electron-spin densities it is necessary to distinguish Fermi contact shifts from the pseudocontact shifts for each compound. This may be done sometimes by comparing two series of complexes differing only in central metal ion, if it can be shown that the modes of spin delocalisation are identical but one member is magnetically isotropic.[K137] Alternatively, the shifts for a given nucleus in the paramagnetic species are compared in solution and in the solid. Fermi contact shifts are the same in fixed and mobile phases; the ratio of pseudocontact shifts in fixed and mobile environments is related to the *g*-value anisotropies.[K32] (While theoretically generally applicable the method is restricted because of the wide lines obtained with solids.) Discrimination between the two mechanisms is also possible on the basis of the nuclear T_1 and T_2 as these have different dependences on the shift origin.[K32,K34]

The Fermi contact, or isotropic contact or isotropic hyperfine contact mechanism applies when there is finite unpaired electron density at the nucleus. This either adds to, or subtracts from the external field, depending upon the sign and magnitude of a_N (the hyperfine coupling constant). From this electron spin-nuclear spin interaction, electron-spin densities are obtained by comparing experimental results with theoretical calculations. Details of the method and derived results, which are more relevant to the inorganic chemist, are available in the reviews.[K32,K33,K34]

4.25.2 Ion Association

Pseudocontact or dipolar shifts are the results of dipolar interaction between the electronic magnetic moment and the nuclear spin which do not vanish in magnetically anisotropic systems. Like the Fermi contact shift their magnitude is proportional to a_N and the distance between the two spins.

In a solution of free ions, some being paramagnetic, any dipolar nuclear-electronic interaction between cation and anion is averaged by molecular tumbling. However, if ion pairing occurs with preferred orientation of the partners this is no longer the case when the paramagnetic ion possess significant magnetic anisotropy. From the observed dipolar shift ($\Delta\nu_D$) of appropriate nuclei in the diamagnetic ion it is possible to calculate interionic distances.

For a given complex, $\Delta\nu_D$ depends upon certain physical constants of the electron and the proton, the temperature, a function of the g-values and g-tensor anisotropy and a geometric factor which relates the relative orientation of the nucleus with respect to the g-tensor. In the case of proton resonances, and an axially symmetric octahedral Co(II) system, the relationship is[K138,K139,K140]

$$\Delta\nu_D = -[\beta^2 \nu s(s+1)/3kT][f(g)] < (3\cos^2\theta - 1)/r^3 >_{\mathrm{av}} \quad (4.25.1)$$

where S is the effective spin quantum number, β, k, T and ν have their usual significance, $f(g)$ is a function of g-values, θ is the angle made between the principal ligand field axis and a vector from the paramagnetic ion to the nucleus in question, and r is the length of this vector.

Typical associating pairs for which interionic distances were obtained using proton resonances[K141] are $(Bu_4N)^+$ and $(Ph_3P)MI_3^-$, and[K142] $[(C_4H_9)Ph_3P]^+$ and $[Ph_3PMX_3]^-$ (X = Br, I), where M = Co(II) and Ni(II) and the solvent was deuterochloroform in both studies. For R_4N^+ and MX_n^{3-} (M = Fe(III) or Cr(III)) contact ion pairs in DMSO-d_6 and solvent-separated ion pairs in D_2O were postulated.[K143]

The method gives internally consistent values of r for a given model of the ion-pair geometry. To select the most realistic model of the ion pair it has been necessary to use line-width measurements. It would seem desirable to evaluate the usefulness and accuracy of this approach to electrolyte studies, by examining some of these association equilibria using other techniques.

4.26 CONCLUSIONS

NMR spectroscopy is a powerful tool for examining solutions, but results obtained using it cannot always be taken at face value.

In the field of structural determination NMR results agree well with those of other techniques, (although there have been conflicting results as for example the site of protonation of amides,[K144] which seem now to have been resolved[K145]). This is not quite the case for determinations of primary solvation numbers of cations, where a wide range of techniques have given a wide range of values.[K146] However, values determined by different NMR techniques agree with one another, and with ultrasonic values, and are entirely reasonable.

Measurements of macroscopic kinetics by NMR agree with those made in other ways, in the few cases where another technique has been used, although a closer scrutiny may be needed. On the other hand correlation times τ_c derived[K15] by NMR can be up to an order of magnitude smaller than those calculated using the Debye relationship (eqn. 4.24.1).

In the determinations of thermodynamic functions of solution equilibria, the position is different. Values obtained by NMR are generally similar to those obtained in other ways, but are seldom the same.

While there is need for many more experimental results, especially in non-aqueous solvents, the trend is for differences to be greatest for weak interactions. The problem has been recognised and is now being examined experimentally[K112,K115,K147–150] using empirical correlations,[K151] and with attempts to assess the differences inherent in using different techniques.[K152] This latter approach has been summarised recently by Prue.[K153]

Measurement of thermodynamic parameters in organic solvent systems using spectroscopic and other techniques should clarify the problem, because of the wide range of solvent properties which can thus be utilised.

Acknowledgments

Thanks are due to Mr J. W. Akitt, Mr K. E. Newman and Dr A. K. Covington for helpful discussions and comments on the manuscript.

REFERENCES

K1 L. M. Jackman and S. Sternhell, Nuclear Magnetic Resonance Spectroscopy *in* Organic Chemistry, 2nd edn., Pergamon Press, Oxford (1969)

K2 J. D. Roberts, Nuclear Magnetic Resonance. Applications to Organic Chemistry, McGraw-Hill, New York (1959)

K3 J. D. Roberts, An Introduction to the Analysis of Spin–Spin Splitting in High-Resolution Nuclear Magnetic Resonance Spectra, W. A. Benjamin, New York (1961)

K4 D. W. Mathieson (ed.), Nuclear Magnetic Resonance for Organic Chemists, Academic Press, London (1967)

K5 A. Abragam, The Principles of Nuclear Magnetism, (a) p. 289, (b) p. 313, (c) p. 191, (d) p. 217, Oxford University Press, Oxford (1961)

K6 J. A. Pople, W. G. Schneider and H. J. Bernstein, High Resolution Nuclear Magnetic Resonance, (a) p. 13, (b) p. 37, (c) p. 202, (d) p. 207, (e) p. 103, (f) p. 80, (g) p. 400, McGraw-Hill, New York (1959)

K7 J. W. Emsley, J. Feeney and L. H. Sutcliffe, High Resoluton Nuclear Magnetic Resonance Spectroscopy, (a) p. 486, (b) p. 25, (c) p. 503, (d) p. 280 (e) p. 507, (f) p. 65 (g) p. 95, (h) p. 534, (j) p. 535, (k) p. 258, Pergamon Press, Oxford (1965)

K8 (*a*) J. W. Emsley, J. Feeney and L. H. Sutcliffe (eds.), Progress in Nuclear Magnetic Resonance Spectroscopy, Vol. 1, Pergamon Press, Oxford (1966) (*b*) J. S. Waugh (ed.), Advances in Magnetic Resonance, Vol. 1, Academic Press Inc. New York (1966)

K9 E. F. Mooney (ed.), Annual Review of NMR Spectroscopy, Vol. 1, Academic Press, N.Y. (1968)

K10 P. Laszlo, *Progress in Nuclear Magnetic Resonance Spectroscopy*, **3**, 231 (1967)

K11 B. I. Silver and Z. Luz, *Quart. Revs.*, **21**, 458 (1967)

K12 F. Bloch, *Phys. Rev.*, **70**, 460 (1946)

K13 J. G. Poweles, *Ber. Bunsen for Phys. Chem.*, **67**, 328 (1963)

K14 O. Jardetzky, *Advances in Chem. Phys.* **7**, 499 (1964)

K15 H. G. Hertz *Progress in Nuclear Magnetic Resonance Spectroscopy*, **3**, 159 (1967), (a) p. 194, (b) p. 177, (c) p. 193

K16 M. D. Zeidler, *Ber. Bunsen Phys. Chem.*, **69**, 659 (1965)

K17 G. Bonera and A. Rigamonti, *J. Chem. Phys.*, **42**, 171 (1965)

K18 H. G. Hertz, G. Stalidis and H. Versmold, *J. Chim Phys.*, **66**, 19e Réunion, 177 (1969)

K19 C. Deverell, *Progress in Nuclear Magnetic Resonance Spectroscopy*, **4**, 235 (1969)

K20 E. Bock and E. Tomchuk, *Canad. J. Chem.*, **47**, 2167 (1969)

K21 K. Chitoku and K. Higasi, *Bul.. Chem. Soc. Japan*, **40**, 773 (1967)

K22 H. G. Hertz, *Z. Elektrochem.*, **65**, 20 (1961)

K23 K. A. Valiev and B. M. Khabibullin, *Russ. J. Phys. Chem.*, **35**, 1118 (1961)

K24 A. Lowenstein and T. M. Connor, *Ber. Bunsen Phys. Chem.*, **67**, 280 (1963)

K25 C. S. Johnson, Advances in Magnetic Resonance, **1**, 33 (1965)

K26 L. W. Reeves, *Advances in Phys. Org. Chem.*, **3**, 187 (1965)

K27 R. A. Pethrick and E. Wyn-Jones, *Quart. Rev.*, **23**, 301 (1969)

K28 N. S. Ham and T. Mole, *Progress in Nuclear Magnetic Resonance Spectroscopy*, **4**, 91 (1969)

K29 F. A. L. Anet and A. J. R. Bourne, *J. Amer. Chem. Soc.*, **89**, 760 (1967)

K30 M. Rabinovitz and A. Pines, *J. Amer. Chem. Soc.*, **91**, 1585 (1969)

K31 E. de Boer and H. van Willigen, *Progress in Nuclear Magnetic Resonance Spectroscopy*, **2**, 111 (1967)

K32 D. R. Eaton and W. D. Philips, *Advances in Magnetic Resonance*, **1**, 103 (1965)

K33 J. F. Hinton and E. S. Amis, *Chem. Rev.*, **67**, 367 (1967)
K34 T. R. Stengle and C. H. Langford, *Co-ord. Chem. Rev.*, **2**, 349 (1967)
K35 T. J. Swift and R. E. Connick, *J. Chem. Phys.*, **37**, 307 (1962).
K36 A. G. Marshall, *J. Chem. Phys.*, **52**, 2527 (1970)
K37 J. Reuben and D. Fiat, *J. Chem. Phys.*, **51**, 4918 (1969)
K38 Z. Luz and S. Meiboom, *J. Chem. Phys.*, **40**, 2686 (1964)
K39 S. Blackstaffe and R. A. Dwek, *Mol. Phys.*, **15**, 279 (1968)
K40 (*a*) N. A. Matwiyoff in ref. K41
(*b*) N. A. Matwiyoff in ref. K42
K41 S. Thomas and W. L. Reynolds, *J. Chem. Phys.*, **46**, 4164 (1967)
K42 N. S. Angerman and R. B. Jordan, *Inorg. Chem.*, **8**, 2579 (1969)
K43 G. S. Vigee and P. Ng, *J. Inorg. Nucl. Chem.*, **33**, 2477 (1971)
K44 S. Sternhell, *Quart. Rev.*, **23**, 236 (1969)
K45 W. McFarlane, *Quart. Rev.*, **23**, 187 (1969)
K46 J. Burgess and M. C. R. Symons, *Quart. Rev.*, **22**, 276 (1968)
K47 J. N. Butler, D. R. Cogley, J. C. Synnott, G. Holbeck, Study of the Composition of Non-Aqueous Solutions of Potential Use in High Energy Density Batteries, Air Force Cambridge Research Laboratories, Office of Aerospace Research, U.S.A.F. Bedford, Mass., U.S.A. (1969)
K48 J. H. Hildebrand and R. L. Scott, Solubility of Non-Electrolytes, 3rd Edn., p. 328, Reinhold Publishing Corporation, N.Y., U.S.A. (1950)
K49 C. H. Langford and T. R. Stengle, *J. Amer. Chem. Soc.*, **91**, 4014 (1969)
K50 J. O.'M. Bockris, *Quart. Rev.*, **3**, 173 (1949)
K51 S. Nakemura and S. Meiboom, *J. Amer. Chem. Soc.*, **89**, 1765 (1967)
K52 A. K. Covington and T. H. Lilley, *Specialist Periodical Reports: Electrochemistry*, **1**, 48–51 (1970)
K53 A. D. Pethybridge and J. E. Prue, *Ann. Reports Chem. Soc.*, **65A**, 129 (1968)
K54 M. Alei and J. A. Jackson, *J. Chem. Phys.*, **41**, 3402 (1964)
K55 R. E. Connick and D. N. Fiat, *J. Chem. Phys.*, **39**, 1349 (1963)
K56 E. R. Malinowski, P. S. Knapp and B. Fleuer, *J. Chem. Phys.*, **45**, 4274 (1966); but see J. W. Akitt, *J. Chem. Soc.*, 2865 (1971)
K57 A. Fratiello, R. E. Lee, V. M. Nishida and R. E. Schuster, *J. Chem. Phys.*, **47**, 4951 (1967)
K58 S. Thomas and W. Reynolds, *Inorg. Chem.*, **9**, 78 (1970)
K59 D. P. Olander, R. S. Marianelli and R. C. Larson, *Anal. Chem.*, **41**, 1097 (1969)
K60 J. F. Hinton and E. S. Amis, *Spectrochim Acta A*, **25**, 709 (1969)
K61 J. F. O'Brien and M. Alei, *J. Phys. Chem.*, **74**, 743 (1970)
K62 L. S. Frankel, C. H. Langford and T. R. Stengle, *J. Phys. Chem.*, **74**, 1376 (1970)
K63 W. G. Movius and N. A. Matwiyoff, *Inorg. Chem.*, **6**, 847 (1967)
K64 D. E. O'Reilly and E. M. Peterson, *J. Chem. Phys.*, **51**, 4906 (1969)
K65 C. Hall, G. L. Haller and R. E. Richards, *Mol. Phys.*, **16**, 377 (1969)
K66 A. K. Covington, I. R. Lantzke and G. A. Porthouse, to be published
K67 J. D. Halliday, R. E. Richards and R. R. Sharp, *Proc. Roy. Soc. Lond., A*, **313**, 45 (1969)
K68 A. Carrington, F. Dravnicks and M. C. R. Symons, *Mol. Phys.*, **3**, 174 (1960)
K69 G. Mavel, *Mem. Poudres Annexe*, **43**, No. 3572 (1961); *Chem. Abstr.*, **56**, 8197h (1962)
K70 H. S. Gutowsky and A. Saika, *J. Chem. Phys.*, **21**, 1688 (1953)
K71 Y. Masuda and T. Kanda, *J. Phys. Soc. Japan*, **8**, 432 (1953)
K72 J. W. Akitt, A. K. Covington, J. G. Freeman and T. H. Lilley, *Trans. Faraday Soc.*, **65**, 2701 (1969)

K73 J. Feeney, *Ann. Reports Chem. Soc., B*, **64,** 5 (1967)
K74 M. Martin, *J. Chim. Phys.*, **59,** 736 (1962)
K75 C. Dorémieux-Morin and G. Martin, *J. Chim. Phys.*, **60,** 1341 (1963)
K76 S. Forsen, *J. Phys. Chem.*, **64,** 767 (1960)
K77 H. G. Hertz and G. Keller, *in* Nuclear Magnetic Resonance in Chemistry, p. 199 (B. Pesce, ed.), Academic Press, London (1965)
K78 R. L. Buckson and S. G. Smith, *J. Phys. Chem.*, **68,** 1875 (1964)
K79 E. Hirsch and R. M. Fuoss, *J. Amer. Chem. Soc.*, **81,** 4507 (1959)
K80 D. P. N. Satchell and R. S. Satchell, *Chem. Comm.*, 110 (1969)
K81 M. I. Foreman, R. Foster and C. A. Fyfe, *J. Chem. Soc., B*, 528 (1970)
K82 R. Foster and C. A. Fyfe, *Progress in Nuclear Magnetic Resonance Spectroscopy*, **4,** 1 (1969)
K83 C. Hall, R. E. Richards, G. N. Schulz and R. R. Sharp, *Mol. Phys.*, **16,** 529 (1969)
K84 D. F. Evans, *J. Chem. Soc.*, 5575 (1963)
K85 J. Ronayne and D. H. Williams, *Ann. Rev. N.M.R. Spectroscopy*, **2,** 83 (1969)
K86 W. R. Angus and D. V. Tilston, *Trans. Faraday Soc.*, **43,** 221 (1947)
K87 S. Broersma, *J. Chem. Phys.*, **17,** 873 (1949)
K88 A. D. Buckingham, T. Schaefer and W. G. Schneider, *J. Chem. Phys.*, **32,** 1227 (1960)
K89 R. A. Robinson and R. H. Stokes, Electrolyte Solutions, 2nd Edn., p. 9, Butterworths, London (1959)
K90 I. D. Kuntz and M. D. Johnston, *J. Amer. Chem. Soc.*, **89,** 6008 (1967)
K91 J. Homer, M. H. Everdell, E. J. Hartland and C. J. Jackson, *J. Chem. Soc. A*, 1111 (1970)
K92 B. P. Dailey, *J. Chem. Phys.*, **41,** 2304 (1964)
K93 J. K. Becconsall, T. Winkler and W. von Philipsbon, *J. Chem. Soc. D*, 430 (1969)
K94 G. E. Maciel, J. K. Hancock, L. F. Lafferty, P. A. Mueller and W. K. Musker, *Inorg. Chem.*, **5,** 554 (1966)
K95 J. W. Akitt and A. J. Downs, The Alkali Metals, Chem. Soc. Special Publication No. 22, p. 199 (1967)
K96 M. Saunders and J. B. Hyne, *J. Chem. Phys.*, **29,** 1319 (1958)
K97 R. Foster and C. A. Fyfe, *Trans. Faraday Soc.*, **61,** 1626 (1965)
K98 J. E. Bundschuh, F. Takahashi and N. C. Li, *Spectrochimica Acta A*, **24,** 1639 (1968)
K99 M. D. Johnson, F. P. Gasparro and I. D. Kuntz, *J. Amer. Chem. Soc.*, **91,** 5715 (1969)
K100 N. Muller and P. I. Rose, *J. Phys. Chem.*, **70,** 3975 (1966)
K101 J. Reuben, *J. Amer. Chem. Soc.*, **91,** 5725 (1969)
K102 H. A. Christ and P. Diehl, *Helv. Phys. Acta*, **36,** 170 (1963)
K103 G. E. Maciel and J. J. Natterstad, *J. Chem. Phys.*, **42,** 2752 (1965)
K104 G. E. Maciel and G. C. Ruben, *J. Amer. Chem. Soc.*, **85,** 3903 (1963)
K105 G. E. Maciel and R. V. James, *Inorg. Chem.*, **3,** 1650 (1964)
K106 C. J. Davis and K. S. Pitzer, *J. Phys. Chem.*, **64,** 886 (1960)
K107 M. M. Kreevoy, H. B. Charman and D. R. Vinard, *J. Amer. Chem. Soc.*, **83,** 1978 (1961)
K108 T. C. Chiang and R. M. Hammaker, *J. Mol. Spect.*, **18,** 110 (1965)
K109 J. A. Happe, *J. Phys. Chem.*, **65,** 72 (1961)
K110 C. J. Cresswell and A. L. Allred, *J. Phys. Chem.*, **66,** 1469 (1962)
K111 W. H. Brandt and J. Chojnowski, *Spectrochim. Acta A*, **25,** 1639 (1969)
K112 G. Socrates, *Trans. Faraday Soc.*, **66,** 1052 (1970)
K113 C. J. Clemmett, *J. Chem. Soc., A*, 455 (1969)

K114 E. B. Whipple, J. H. Goldstein, L. Mandell, G. S. Reddy and G. R. McClure, *J. Amer. Chem. Soc.*, **81,** 1321 (1959)

K115 H. Kriegsmann, G. Engelhardt, R. Radeglia and H. Geissler, *Z. Phys. Chem. (Leipzig)*, **240,** 294 (1969)

K116 E. V. Van den Berghe, L. Verdonck and G. P. Van der Kelen, *J. Organometallic Chem.*, **16,** 497 (1969)

K117 L. I. Vinogradov, Yu. Yu. Samitov, E. G. Yarkova and A. A. Muratova, *Opt. Spektrosk.*, **26,** 959 (1969)

K118 P. Laszlo, A. Speert, R. Ottinger and J. Reisse, *J. Chem. Phys.*, **48,** 1732 (1968)

K119 A. D. Buckingham and K. A. McLauchlan, *Progress in Nuclear Magnetic Resonance Spectroscopy*, **2,** 63 (1967)

K120 G. R. Luckhurst, *Quart. Rev.*, **22,** 179 (1968)

K121 S. Meiboom and L. C. Snyder, *Science*, **162,** 1337 (1968)

K122 L. C. Snyder and S. Meiboom, *Mol. Cryst. and Liquid Cryst.*, **7,** 181 (1969)

K123 I. G. Chistyakov, *Usp. Fiz. Nauk.*, **89,** 563 (1966); *Soviet Phys. Rev.*, **9,** 551 (1967)

K124 P. A. Winsor, *Chem. Rev.*, **68,** 1 (1968)

K125 A. Saupe, *Angew. Chem.*, **80,** 99 (1969)

K126 K. D. Lawson and T. J. Flautt, *J. Amer. Chem. Soc.*, **89,** 5489 (1967)

K127 M. Panar and W. D. Phillips, *J. Amer. Chem. Soc.*, **90,** 3880 (1968)

K128 J. M. Corkill, J. F. Goodman and J. Wyer and references therein, *Trans. Faraday Soc.*, **65,** 9 (1969)

K129 L. C. Snyder, *J. Chem. Phys.*, **43,** 4041 (1965).

K130 P. Pincus, *Solid State Communications*, **7,** 415 (1969)

K131 W. H. Pirkle and S. D. Bear, and references therein, *J. Amer. Chem. Soc.*, **91,** 5150 (1969)

K132 F. A. L. Anet, L. M. Sweeting, T. A. Whitney and D. J. Cram, *Tetrahedron Lett.*, 2617 (1968)

K133 E. Sackmann, S. Meiboom and L. C. Snyder, *J. Amer. Chem. Soc.*, **90,** 2183 (1968)

K134 R. E. Ernst, M. J. O'Connor and R. H. Holm, *J. Amer. Chem. Soc.*, **90,** 5305 (1968)

K135 T. McL. Spotswood, J. M. Evans and J. H. Richards, *J. Amer. Chem. Soc.*, **89,** 5052 (1967)

K136 N. N. Greenwood, J. W. Akitt, W. Errington, T. C. Gibb and B. P. Straughan, Specialist Periodical Reports. Spectroscopic Properties of Inorganic and Organometallic Compounds. Chemical Society, London (1969–1971)

K137 W. DeW. Horrocks, *Inorg. Chem.*, **9,** 690 (1970)

K138 H. M. McConnell and R. E. Robertson, *J. Chem. Phys.*, **29,** 1361 (1958)

K139 G. N. La Mar, *J. Chem. Phys.*, **43,** 1085 (1965)

K140 J. P. Tesson, *J. Chem. Phys.*, **47,** 579 (1967)

K141 G. N. La Mar, *J. Chem. Phys.*, **41,** 2992 (1964)

K142 R. H. Fischer and W. deW. Horrocks, *Inorg. Chem.*, **7,** 2659 (1968)

K143 D. W. Larsen, *J. Amer. Chem. Soc.*, **91,** 2920 (1969)

K144 S. J. Kuhn and J. S. McIntyre, *Canad. J. Chem.*, **43,** 995 (1965)

K145 M. Liler, *J. Chem. Soc., B*, 385 (1969); *J. Chem. Soc., B*, 334 (1971)

K146 B. E. Conway and J. O'M. Bockris *in* Modern Aspects of Electrochemistry, p. 47 (J. O'M. Bockris, ed.), Butterworths, London (1954)

K147 T. Yokoyama, G. R. Wiley and S. I. Miller, *J. Org. Chem.*, **34,** 1859 (1969)

K148 L. Odberg, E. Hogfeldt, *Acta Chem. Scand.*, **23,** 1330 (1969)

K149 S. H. Marcus and S. I. Miller, *J. Phys. Chem.*, **73,** 453 (1969)

K150 V. V. Kushchenko, *Zh. Struct. Khim.*, **11,** 140 (1970)

K151 W. Kohler, *Z. Phys. Chem.* (*Leipzig*), **242,** 220 (1969); W. Kohler, H. J. Bittrick and R. Radeglia, *J. Prakt. Chem.*, **311,** 643 (1969); W. Kohler, *Z. Chem.*, **9,** 467 (1969); W. Kohler, H. J. Bittrick and R. Radeglia, *Z. Chem.*, **9,** 72 (1969); W. Kohler, *Z. Chem.*, **9,** 392 (1969).

K152 C. Agami, *Bull. Soc. Chim. Fr.*, 2183 (1969)

K135 J. E. Prue *in* Proceedings of the 3rd Symposium on Co-ordination Chemistry, Debrecen, Hungary, 1970, Vol. I, p. 25 (M. T. Beck, ed.), Hungarian Academy of Sciences, Budapest (1970)

Chapter 5

Conductance and Transference Numbers

Part 1

Conductance

R. Fernández-Prini*
Department of Chemistry, University of Maryland,
College Park, Maryland 20742, U.S.A.

* Present address: t, Facultad de Farmacia y Bioquímica, Universidad de Buenos Aires, Argentina.

NOTATION

a	distance of closest approach of ions.
b	$(ze)^2/a\varepsilon_r k\mathrm{T}$.
c	molarity.
d	distance of closest approach of *free* ions.
e	protonic charge.
f_{ij}	$n_i n_{ij}$.
F	coefficient of the viscosity eqn. 5.2.37.
F	Faraday.
$\mathbf{F}_j$	force acting on ion j.
k	Boltzmann constant.
l	Bjerrum's distance.
K_A	association constant.
K_3	dissociation constant of triple-ions.
N_A	Avogadro number.
n_i	stoichometric concentration (ions cm^{-3}).
n_{ij}	number density of j ions near an i ion.
p_{ij}	solvent pressure on j ion.
q	$(ze)^2\kappa/\varepsilon_r k\mathrm{T}$.
R_j	hydrodynamic radius of ion j.
$\mathbf{s}_{ji}$	fluid velocity relative to ion j.
S	coefficient of Onsager limiting law.
T	absolute temperature.
$\mathbf{u}_{ij}$	local velocity of the medium in the vicinity of ion j.
$\mathbf{X}$	external electric field.
$\mathbf{v}_{ij}$	velocity of ion j near an i ion.
x	κr.
y	κa.
z_j	charge number including its sign.
z	charge number; conductance variable in Fuoss and Shedlovsky eqn. 5.4.10.
α	relaxation coefficient of the limiting law degree of dissociation.
β	electrophoretic coefficient of the limiting law.
$\gamma_\pm$	mean molar activity coefficient.
ε_r	dielectric constant.
ε_r'	dielectric coefficient.
η	solvent viscosity.
η_s	solution viscosity.
κ	reciprocal radius of the ionic atmosphere.
λ_{ij}	coefficient of eqn. 5.2.14.
Λ	molar conductance.
Λ_i	ionic molar conductance.
Λ°	molar conductance at infinite dilution.

Λ_{LL} molar conductance according to Onsager limiting law.
ζ_j coefficient of eqn. 5.2.14.
π_{ij} coefficient of eqn. 5.2.14.
Φ_{ij} $(z_j e\psi_i)/kT$.
ψ_{ij} electrostatic interionic potential acting on ion j due to ion i and its ionic atmosphere.
ρ_{ij} charge density around ion j.
σ conductivity of the solution.
ω_j mobility of ion j.

5.1 INTRODUCTION

The theory of electrolytic conductance was successfully extended during the fifties so that it could be used to predict the conductance of solutions having a concentration such that $\kappa d < 0.2$–0.5 (see Notation). In the last five years a revision of the theoretical treatments has enabled the establishment of a firmer basis for the methods employed to evaluate conductance data.

Conductance measurement, besides providing information on the mobility of ionic species in solution, is the most direct and accurate technique available at present to determine the extent to which ions associate in solution. Section 5.3 is devoted to the discussion of ionic association.

In these sects. 5.1–5.6 we shall review fundamentally the studies of conductance of electrolytes in non-aqueous solvents carried out during the last fifteen years. Fortunately many of the valuable conductance data obtained before this period have been reassessed using the more refined theoretical equations. These results are included in the five tables of Appendix 5.1. A brief account of the studies of the dependence of conductance on temperature and pressure has been included since they are a powerful means of understanding the mechanism of ionic transport.

5.2 THEORY OF CONDUCTANCE OF ELECTROLYTIC SOLUTIONS

5.2.1 The Limiting Law

The fact that conductance measurements can be carried out without sophisticated apparatus to an accuracy of 0.02% or better, has provided a large amount of very good experimental information. In order to describe in detail these accurate data, it is necessary to have an equally good theoretical treatment for electrolytic conductance. At present at least two treatments are capable of fitting the experimental results, within experimental error, in relatively dilute solutions ($\kappa d < 0.5$). A long time elapsed, however, between the first success in conductance theory and further improvements.

Shortly after Debye and Hückel[L1] had presented their momentous work on the free energy of electrolyte solutions, Onsager derived[L2] theoretically, the empirical equation proposed by Kohlrausch to represent the molar conductance of an electrolyte solution. For solutions of a single symmetrical electrolyte this equation is given by

$$\Lambda = \Lambda^\circ - S\sqrt{c} \tag{5.2.1}$$

While Debye and Hückel dealt in their work both with long range coulombic interactions and with short range repulsive interactions (introducing a distance of closest approach for two ions considered rigid), this was not the case in conductance theory. The short range repulsive interactions were successfully introduced only a score of years later.

Long range coulombic forces take no account of size or shape of the ions in solution; only their charge and concentration are important. Thus eqn. 5.2.1 is a limiting law capable only of representing experimental data in extremely dilute solutions, where specific effects may be

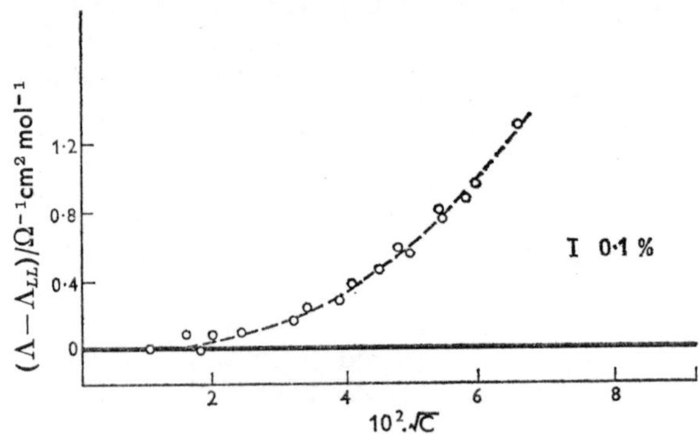

FIG. 5.2.1. Positive deviations from the limiting law for $LiClO_4$ solutions in dimethylformamide.[L3]

neglected. Onsager's eqn. 5.2.1 corresponds to Debye and Hückel's limiting law for equilibrium properties, being another example of the dependence of electrolytic properties on the square root of the ionic strength of the solution.

In Fig. 5.2.1 the molar conductance of $LiClO_4$ solutions in dimethylformamide,[L3] from which the values Λ_{LL} predicted by the limiting law have been subtracted, are plotted against $c^{\frac{1}{2}}$. It may be observed that as concentration decreases the limiting law is approached but that even at a concentration of 0.001 molar the measured conductance deviates by 0.25% (roughly ten times the experimental error) from the limiting law.

The long range coulombic interactions, which produce the decrease of the conductance described by eqn. 5.2.1, are amenable to a rather simple physical interpretation. Let us consider a central ion j immersed in a solvent. Ion j will be surrounded by an ionic atmosphere having an average charge opposite to that of the central ion. When no external fields are applied to the solution, the ionic atmosphere has spherical symmetry around the central ion. When the external electric field $\mathbf{X}$ is switched on, a force $z_j e\mathbf{X}$ acts on ion j, z_j being its charge number including the sign. The central ion and its ionic atmosphere move in opposite directions towards the electrodes. This countercurrent of ions

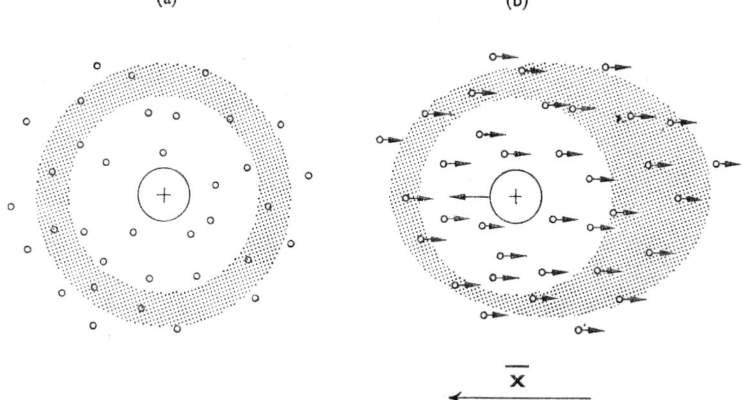

FIG. 5.2.2. Schematic illustration of the electrophoretic and relaxation effects. (*a*) No external electric field, (*b*) the central cation moves under the influence of the external electric field **X**. The shaded areas depict the ionic atmosphere having an average excess negative charge. The small circles represent solvent molecules.

produces a decrease in the effective velocity of the ion j due to two effects. The electrophoretic effect produces a local flow of solvent in the opposite direction to the displacement of the central ion. This is set up by the fact that the particles forming the ionic atmosphere collide with solvent molecules, transferring part of their momenta to the solvent. Therefore, the central ion moves in a medium which locally is moving in the opposite direction. Hence the velocity of ion j with respect to the electrodes is reduced in an amount equal to the average velocity of the solvent in its immediate neighbourhood. Simultaneously, as ion j travels along the direction of the field, its ionic atmosphere lags behind, since it is unable to reform itself sufficiently fast around the moving ion. These two effects are schematically depicted in Fig. 5.2.2.

The effective force acting upon a moving ion, which must be equal and opposite to the force of the external field, may be described in terms of three contributions,

$$-z_j e\mathbf{X} = \mathbf{F}_j^{\text{vis}} + \mathbf{F}_j' + \mathbf{F}_j'' \tag{5.2.2}$$

The first term is due to the viscous drag of the solvent on the moving ion. When interionic effects vanish, this is the only remaining force and determines the ionic conductance at infinite dilution, Λ_j°. $\mathbf{F}_j'$ is the force acting on the ion j due to the stress exerted by the local fluid on its surface. $\mathbf{F}_j''$ is the force due to electrostatic interactions with the ions surrounding j, at low concentrations, the relaxation effect stems from this force.

The value of S in eqn. 5.2.1 derived by Onsager[L2] is given for symmetrical electrolytes by

$$S = \alpha\Lambda^\circ + \beta$$

$$\alpha = \frac{(ze)^2\kappa}{3(2+\sqrt{2})\varepsilon_r\, kTc^{\frac{1}{2}}} = 82.0460 \times 10^4 \frac{z^3}{(\varepsilon_r T)^{\frac{3}{2}}} \Big/ \text{mol}^{-\frac{1}{2}}\, \text{l}^{\frac{1}{2}}$$

$$\beta = \frac{z^2 eF\kappa}{3\pi\eta c^{\frac{1}{2}}} = 82.487 \frac{z^3}{\eta(\varepsilon_r T)^{\frac{1}{2}}} \Big/ \Omega^{-1}\, \text{cm}^2\, \text{mol}^{-\frac{3}{2}}\, \text{l}^{\frac{1}{2}} \tag{5.2.3*}$$

Λ° being the molar conductance at infinite dilution, and the viscosity η is in poise.

5.2.2 Higher Order Contributions to Conductance

The neglect of short range, hard-core interactions in the derivation of eqn. 5.2.1 is the reason for the positive deviations from the limiting law illustrated in Fig. 5.2.1. The solution of the problem of electrolytic conductance, accounting for both long range and short range interactions, was worked out by Pitts[L4] in 1953 and independently by Fuoss and Onsager[L5] in 1957.

Both treatments make use of the same general equations for transport processes in fluids and of the same model to represent the electrolyte solution. However, they lead to somewhat different results due to the manner in which the problem is approached and because of the different boundary conditions employed to evaluate the constants which appear upon integration of the differential equations. We shall give here an account of the conductance theory based on the mathematical approach used by Pitts[L4] and shall point out the differences and agreements between his treatment and that of Fuoss and Onsager.[L5] The mathematical technique used by the latter authors has been given in detail by Fuoss

* Numerical factors for the conversion of units will not be used in the present chapter. The values of the universal constants are taken from F. D. Rossini, *Pure Appl. Chem.*, **8**, 95 (1964).

and Accascina.[L6] The final integrations of the equations will be omitted but may be found in the original papers.

The model used to represent the ionic solution is that of Debye and Hückel; spherical rigid ions with charges at their centres, immersed in a continuum having the properties of a macroscopic fluid with dielectric constant and viscosity identical to those of the solvent. The closest distance at which two ions may approach will be denoted by a.

The centre of coordinates is an arbitrary point in the bulk of the solution, always remaining stationary (point O in Fig. 5.2.3). Ion i is at a distance $\mathbf{r}$ from ion j. The time-average number-density of j ions at Q when there is an i ion at P, is denoted by n_{ij}. A similar meaning is

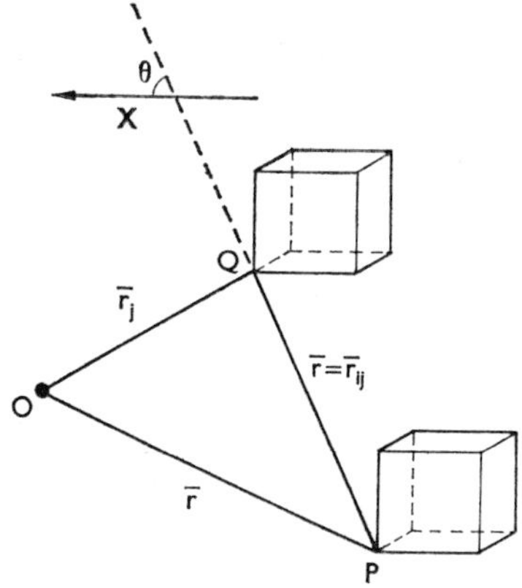

FIG. 5.2.3

attached to n_{ji}. Then if n_i and n_j represent the stoichiometric number density of ions i and j, it is obvious that $f_{ij}(\mathbf{r}) = n_i n_{ij}(\mathbf{r}) = n_j n_{ji}(\mathbf{r}) = f_{ji}(\mathbf{r})$. The functions f_{ij} and f_{ji} satisfy the general continuity equation[L7]

$$-\frac{\partial f_{ij}}{\partial t} = \nabla_{\mathrm{Q}}\cdot(f_{ij}\mathbf{v}_{ij}) + \nabla_{\mathrm{P}}\cdot(f_{ji}\mathbf{v}_{ji}) = -\frac{\partial f_{ji}}{\partial t} \tag{5.2.4}$$

where $\mathbf{v}_{ij}$ stands for the velocity of a j ion at point Q when there is an i ion at P. In the case of low frequency (audio frequency) conductance measurements, a steady state is set up such that f_{ij} and f_{ji} are time independent. The velocities of the ions are given by

$$\mathbf{v}_{ij} = \mathbf{u}_{ij} + \omega_j(z_j e\mathbf{X} - z_j e\nabla_{\mathrm{Q}}\psi_{ij} - kT\nabla_{\mathrm{Q}}\ln f_{ij} - z_j e\nabla_{\mathrm{Q}}\psi_{ji}(a)) \tag{5.2.5}$$

where $\mathbf{u}_{ij}$ is the local velocity of solvent in the neighbourhood of ion j when it is near ion i; ω_j is the absolute mobility of ion j. The third term in eqn. 5.2.5 represents a diffusional contribution due to local variations in the chemical potential of ion j; this diffusional force opposes the electrostatic forces so that the ionic distribution reaches a steady state. The second and fourth terms inside the parentheses in eqn. 5.2.5 involve forces arising from ionic interactions. $\nabla_Q \psi_{ij}$ is the gradient of electrostatic potential of ion i at P and of its ionic atmosphere over point Q where ion j is situated; $\nabla_Q \psi_{ji}(a)$ is the gradient of electrostatic potential on the surface of ion j ($r = a$) due to the ionic atmosphere of ion j. This last term is considered only by Fuoss and Onsager[L5] and is due to the asymmetry of the ionic atmosphere around a moving ion. Pitts considers that this term is superfluous[L8] since all the ionic interactions should be included in $\nabla_Q \psi_{ij}$, ion j being part of the ionic atmosphere of i. The first term in the parentheses in eqn. 5.2.5 represents the force due to the external field and is the only term which contributes in the limit of infinite dilution.

In order to evaluate $\mathbf{u}_{ij}$, the local velocity of solvent near the ion j, the hydrodynamic equation

$$\eta \nabla \times \nabla \times {}_P \mathbf{s}_{ji} + \nabla_P p_{ji} = \mathbf{F}_{ji} = -\frac{\varepsilon_r}{4\pi}(\mathbf{X} - \nabla \psi_{ji}) \Delta \psi_{ji} \quad (5.2.6)^*$$

is used. From $\mathbf{u}_{ij}$ the solvent stress $\mathbf{F}'_j$ on the ion may be calculated. In eqn. 5.2.6 p_{ji} is the pressure of the solvent at P and $\mathbf{F}_{ji}$ is the electrostatic force per unit volume acting upon the ionic charge density at P considered uniform. The use of eqn. 5.2.6 implies that the solution is considered as a continuous fluid having a velocity $\mathbf{s}_{ji}$ relative to ion j. This is in keeping with the model chosen to represent the electrolytic solution. Further, the condition $\nabla_P \cdot \mathbf{s}_{ji} = 0$ specifies that the solution is incompressible.

The external electric field perturbs the ionic distribution and the electrostatic potential around the central ion. The distribution functions f_{ij} and the electrostatic potential ψ_{ij} which are related by the Poisson–Boltzmann equation

$$\Delta_Q \psi_{ij} = -\frac{4\pi}{\varepsilon_r n_i} \sum_{j=1}^{s} z_j e f_{ij} \quad (5.2.7a)^*$$

will be expressed as the sum of an equilibrium contribution (superscript ^{0}) and a small perturbation term which is proportional to the field $\mathbf{X}$.

$$f_{ij} = f_{ij}^0 + f'_{ij} = f_{ji}$$
$$\psi_{ij} = \psi_i^0 + \psi'_{ij}$$
$$\psi_{ji} = \psi_j^0 + \psi'_{ji} \quad (5.2.7b)$$

Since the electric fields employed in conductance measurements are small, such that the electrolyte solution follows Ohm's Law, the only

* $\Delta \equiv \nabla^2$ in these equations.

perturbation terms of importance are those depending linearly on the external electric field. The equilibrium contributions in (5.2.7b) have spherical symmetry depending only on the absolute magnitude of vector **r**. The Poisson–Boltzmann eqn. 5.2.7a is valid at equilibrium, so it follows from 5.2.7b that the perturbations are also related by Poisson–Boltzmann concepts. According to Debye and Hückel[L1] we have for the equilibrium contributions

$$f_{ij}^0 = n_i n_j \exp\left[-\frac{z_j e \psi_j^0}{kT}\right] = n_i n_j \exp(-\varphi_{ij}^0)$$

$$\psi_i^0 = \frac{z_i e}{\varepsilon_r r(1+\kappa a)} \exp[-\kappa(r-a)] \tag{5.2.8}$$

where a is the distance of closest approach of ions and

$$\kappa = \left(\frac{4\pi}{\varepsilon_r kT} \sum_{j=1}^{s} z_j^2 n_j\right)^{\frac{1}{2}}$$

is the reciprocal of the radius of the ionic atmosphere. The exponential in 5.2.8 is expanded and the series truncated after the quadratic term, thus

$$f_{ij}^0 = n_i n_j \left[1 - \varphi_{ij}^0 + \frac{(\varphi_{ij}^0)^2}{2}\right] \tag{5.2.9}$$

Equation 5.2.9 differs from that employed by Debye and Hückel in the presence of the quadratic term which they omitted. Conductance theory has dealt so far only with symmetrical electrolytes.* For these salts the even powers in the expansion of 5.2.8 do not contribute to the equilibrium properties of their solutions, affecting only their transport properties.

Replacing the ionic velocities in 5.2.4 by the expression 5.2.5, we have

$$\begin{aligned}\nabla_Q\cdot(f_{ij}\mathbf{v}_{ij}) = \mathbf{u}_{ij}\cdot\nabla_Q f_{ij}^0 + \omega_j z_j e\mathbf{X}\cdot\nabla_Q f_{ij}^0 - \omega_j z_j e f_{ij}^0 \Delta_Q \psi'_{ij} \\ - \omega_j z_j e f'_{ij}\Delta_Q \psi_i^0 - \omega_j z_j e \nabla_Q f'_{ij}\cdot\nabla_Q\psi_i^0 \\ - \omega_j z_j e \nabla_Q f_{ij}^0\cdot\nabla_Q \psi'_{ij} - \omega_j kT \Delta_Q f'_{ij} \\ - \omega_j z_j e\nabla_Q f_{ij}^0\cdot\nabla_Q\psi'_{ij}(a).\end{aligned} \tag{5.2.10}$$

In the derivation of 5.2.10 the following relations have been used

$$\begin{aligned}\nabla_Q \ln f_{ij} &= (\nabla_Q f_{ij})/f_{ij} \\ \nabla_Q\cdot\mathbf{X} &= 0 \\ \nabla\cdot(\nabla\psi_{ji}(a)) &= 0 \\ \nabla\cdot(b\mathbf{V}) &= b\nabla\cdot\mathbf{V} + \mathbf{V}\cdot\nabla b\end{aligned} \tag{5.2.11}$$

* Murphy and Cohen have recently derived a conductance equation for unsymmetrical electrolytes with a term linear in concentration (*J. chem. Phys.*, **53**, 2173 (1970)).

(b is a scalar, $\mathbf{V}$ a vector). The products of two perturbations have been omitted from 5.2.10 because they depend on $\mathbf{X}^2$. In the last term of 5.2.10, appearing in the Fuoss and Onsager treatment, only the perturbation remains since the equilibrium part produces a central force which will not result in any displacement of ion j. When the field $\mathbf{X}$ is switched off, all perturbations vanish, the equilibrium functions f_{ij}^0 and ψ_i^0 nevertheless satisfy 5.2.4, thus providing a relation which was used to eliminate the purely spherically symmetric terms from 5.2.10.

Equation 5.2.10 and the analogous relation for $\nabla_{\mathrm{P}}\cdot(f_{ij}\mathbf{v}_{ij})$ are substituted into 5.2.4. Remembering that $\mathbf{r}_{ij} = -\mathbf{r}_{ji} = \mathbf{r}$ and that the scalar products between perturbations and equilibrium functions may be written $\nabla f_{ij}^0\cdot\nabla\psi'_{ij} = (\mathrm{d}f_{ij}^0/\mathrm{d}r)(\partial\psi'_{ij}/\partial r)$, eqn. 5.2.4 becomes

$$kT(\omega_j + \omega_i)\Delta f'_{ij} + \omega_j z_j e\left[f_{ij}^0\Delta\psi'_{ij} + \frac{\partial\psi'_{ij}}{\partial r}\frac{\mathrm{d}f_{ij}^0}{\mathrm{d}r} + f'_{ij}\Delta\psi_i^0 + \frac{\mathrm{d}\psi_i^0}{\mathrm{d}r}\frac{\partial f'_{ij}}{\partial r} + \frac{\mathrm{d}f_{ij}^0}{\mathrm{d}r}\left(\frac{\partial\psi'_{ji}}{\partial r}\right)_a\right]$$
$$+ \omega_i z_i e\left[f_{ij}^0\Delta\psi'_{ji} + \frac{\partial\psi'_{ji}}{\partial r}\frac{\mathrm{d}f_{ij}^0}{\mathrm{d}r} + f'_{ij}\Delta\psi_j^0 + \frac{\mathrm{d}\psi_j^0}{\mathrm{d}r}\frac{\partial f'_{ij}}{\partial r} + \frac{\mathrm{d}f_{ij}^0}{\mathrm{d}r}\left(\frac{\partial\psi'_{ij}}{\partial r}\right)_a\right]$$
$$= (\omega_j z_j e - \omega_i z_i e)\mathbf{X}\frac{\mathrm{d}f_{ij}^0}{\mathrm{d}r}\cos\theta + (\mathbf{u}_{ij} - \mathbf{u}_{ji})\cdot\nabla f_{ij}^0 \quad (5.2.12)$$

where the terms containing $(\partial\psi'_{ji}/\partial r)_a$ are maintained in the Fuoss and Onsager treatment[L5] but omitted altogether by Pitts.[L4,L8]

Fuoss and Onsager[L5] rearranged eqn. 5.2.12 to the form

$$kT(\omega_j + \omega_i)\Delta f'_{ij} - \frac{4\pi}{\varepsilon_r}f_{ij}^0\left[\frac{\omega_j(z_je)^2}{n_i} - \frac{\omega_i(z_ie)^2}{n_j}\right]f'_{ij}$$
$$= (\omega_j z_j e - \omega_i z_i e)\mathbf{X}\frac{\mathrm{d}f_{ij}^0}{\mathrm{d}r}\cos\theta$$
$$+ T(g_1) + T(g_2) + T(g_3) + T(a) + T(v) \quad (5.2.13)$$

obtained from 5.2.12 using the Poisson–Boltzmann relation to replace $\Delta\psi'_{ij}$ in the term $f_{ij}^0\Delta\psi'_{ij}$. Expression 5.2.13 would give the limiting law 5.2.1 plus a correction for the fact that ions are rigid spheres, if the T terms are neglected. The latter will contribute higher order terms in the concentration to the final conductance equation. The $T(g_i)$ terms correspond respectively to the second, third and fourth terms inside the brackets of eqn. 5.2.12. $T(a)$ involves the terms containing $(\partial\psi'_{ji}/\partial r)_a$ and $T(v) = (\mathbf{u}_{ij} - \mathbf{u}_{ji})\cdot\nabla f_{ij}^0$.

The approach employed by Pitts[L4] to solve eqn. 5.2.12 consists in expressing all the perturbations (f'_{ij}, ψ'_{ij}, ψ'_{ji}) and the local velocities $\mathbf{u}_{ij}$ as power series of the dimensionless variable $q = (ze)^2\kappa/\varepsilon_r kT$, which is the ratio of the average electrostatic energy of the ions in the atmosphere surrounding the central ion to the average thermal energy. The perturbations are expressed by

$$\begin{aligned}
\psi'_{ij} &= \sum_{k=1}^{\infty}\sum_{\nu=0}^{\infty} \lambda_{ij}(k, \nu, x) S_\nu(\theta, \varphi) q^k \mathrm{X} \\
\psi'_{ji} &= \sum_{k=1}^{\infty}\sum_{\nu=0}^{\infty} \lambda_{ji}(k, \nu, x) S_\nu(\theta, \varphi) q^k \mathrm{X} \\
f'_{ij} &= \sum_{k=1}^{\infty}\sum_{\nu=0}^{\infty} \pi_{ij}(k, \nu, x) S_\nu(\theta, \varphi) q^k \mathrm{X} \\
\mathbf{u}_{ij} &= \sum_{k=1}^{\infty} \zeta_j(k) q^k \mathrm{X}
\end{aligned} \qquad (5.2.14)$$

and

$$\begin{aligned}
f &= e^y/(1 + y) \\
\varphi^0_{ij} &= \varepsilon_{ij} q f \lambda \\
\varepsilon_{ij} &= -1\ (i \neq j) \\
\varepsilon_{ij} &= 1\ (i = j) \\
\lambda &= e^{-x}/x \\
y &= \kappa a
\end{aligned}$$

In eqn. 5.2.14 the coefficients of the series expansions depend on the variable $x = \kappa r$, and $S_\nu(\theta, \varphi)$ are surface spherical harmonics depending on the polar angles θ and φ. The vector coefficients $\zeta_j(k)$ are independent of the position of ion j which may be considered stationary according to eqn. 5.2.6. At infinity ψ' and f' vanish. Besides, Pitts assumed that for $r = a$ the perturbations may be neglected; thus the ionic potentials and distribution functions on the surface of the central ion are not affected by the external field.

The solution of eqn. 5.2.12 is achieved by employing eqns. 5.2.8, 5.2.9 and 5.2.14 and equating the coefficients of equal powers of q. The coefficients of the terms linear in q give the first order contributions to the conductance equation

$$\begin{aligned}
(\omega_i + \omega_j)\Delta_\nu \pi_{ij}(1, \nu) &+ \frac{n^2}{kT}(\omega_j z_j e \Delta_\nu \lambda_{ij}(1, \nu) + \omega_i z_i e \Delta_\nu \lambda_{ji}(1, \nu)) \\
&= n^2 \varepsilon_{ij} \frac{(\omega_i z_i e - \omega_j z_j e)}{\kappa kT} f \frac{\mathrm{d}\lambda}{\mathrm{d}x}; \qquad \text{for} \quad \nu = 1 \\
&= 0; \qquad \text{for} \quad \nu \neq 1 \qquad (5.2.15)
\end{aligned}$$

The distinction between the case $\nu = 1$ and those where $\nu \neq 1$ is due to the fact that the angular dependence of the first term in the r.h.s. of eqn. 5.2.12 is that of the first spherical harmonic. Equation 5.2.15 is analogous to that of Fuoss and Onsager (eqn. 5.2.13) with the T terms set equal to zero. The terms grad·grad, $\mathbf{u}_{ij}\cdot\text{grad} f^0_{ij}$ and $f'_{ij}\Delta\psi^0_i$ which give contributions of order q^2 do not appear in eqn. 5.2.15. In what follows the equations are specialised for solutions of a single symmetrical electrolyte with ionic species 1 and 2.

Replacing eqn. 5.2.14 in the corresponding Poisson–Boltzmann relations

$$\Delta_\nu \lambda_{ij}(1, \nu) = -\frac{4\pi}{\varepsilon_r \kappa^2 n}[z_1 e\pi_{i1}(1, \nu) + z_2 e\pi_{i2}(1, \nu)]$$

$$\Delta_\nu \lambda_{ji}(1, \nu) = -\frac{4\pi}{\varepsilon_r \kappa^2 n}[z_1 e\pi_{1j}(1, \nu) + z_2 e\pi_{2j}(1, \nu)] \qquad (5.2.16)$$

with $$\Delta_\nu = \frac{\mathrm{d}^2}{\mathrm{d}x^2} + \frac{2}{x}\frac{\mathrm{d}}{\mathrm{d}x} - \frac{\nu(\nu+1)}{x^2}$$

then in the eqns. 5.2.15 and 5.2.16, i and j can be either ion 1 or 2, hence there are four equations of each of the three types. In this way the coefficients π and λ are determined. A general result is that they are zero for $\nu \neq 1$. For $\nu = 1$ we have

$$\pi_{22}(1, 1) = \pi_{11}(1, 1) = 0$$

$$\pi_{12}(1, 1) = -\pi_{21}(1, 1) = \frac{2n^2 ze}{\kappa kT}\left[f\left(\frac{1}{x} + \frac{1}{x^2}\right)\mathrm{e}^{-x} - g\right] \qquad (5.2.17)$$

and

$$\lambda_{1j}(1, 1) = \lambda_{2j}(1, 1) = -\lambda_{1i}(1, 1) = -\lambda_{2i}(1, 1)$$
$$= \frac{kT\pi_{12}(1, 1)}{2n^2 ze} + \frac{1}{\kappa}\left[\frac{1}{x^2} - g\right] \qquad (5.2.18)$$

where $$g = \frac{\mathrm{e}^{\delta y}}{1 + \delta y}\left(\frac{\delta}{x} + \frac{1}{x^2}\right)\mathrm{e}^{-\delta x};$$
$$\delta = 1/\sqrt{2}$$

The contribution $\mathbf{F}_j$ to the effective force on the ion, up to terms linear in q is now evaluated through eqn. 5.2.6. Following the procedure employed above, the velocity and pressure may be expanded in powers of q, e.g.,

$$\mathbf{s}_{ji} = \sum_{k=0}^{\infty} \mathbf{s}_j(k) q^k$$

and likewise for p_{ji}. The volume of the solution surrounding the central

j ion is divided into two regions: (1) an inner region extending from the surface of the ion ($r = R_j$) to the distance of closest approach of two ions. In this region the charge density is zero since there are no ions excepting the central one ($\mathbf{F}_{ji} = 0$ in eqn. 5.2.6); (2) an external region which extends from $r = a$ to infinity. It is in this region where interionic effects will modify the local velocity of solvent around ion j. In this way,

$$\eta\nabla \times \nabla \times \mathbf{s}_j(0) + \nabla p_j(0)$$

$$= 0, \qquad \text{for region (1)}$$

$$= -\frac{\varepsilon_r}{4\pi}\,\Delta\psi_j^0(\mathbf{X} - \nabla\psi_j^0), \qquad \text{for region (2)} \qquad (5.2.19)$$

Pitts determined the constants of integration appearing upon solution of eqn. 5.2.19 by setting $\mathbf{s}_j = 0$ at $r = R_j$ and making the velocity and pressure, as well as their first derivatives, continuous at $r = a$.

If $\mathbf{F}_j(0)_x$ is, to terms linear in q, the force $\mathbf{F}_j^{\mathrm{vis}} + \mathbf{F}_j'$ acting on the ion due to viscous flow and to solvent stress produced by the counterflow of the ionic atmosphere on the surface of particle j, Pitts' result gives

$$\mathbf{F}_j(0)_x = -6\pi\eta R_j\mathbf{V}_j - \frac{z_j e R_j \varepsilon_r kT}{(ze)^2(1 + y)}\,q\mathbf{X} \qquad (5.2.20)$$

At infinite dilution ($q = 0$) the effects of the ionic atmosphere vanish and only the resistive force characteristic of the displacement of a spherical particle through a fluid of viscosity η remains; hence $\omega_j = 1/6\pi\eta R_j$ which is the well known Stokes' law.

Finally from eqn. 5.2.20 the first coefficient to the series expansion of $\mathbf{u}_{ij}$ is

$$\zeta_j(1)_x = -\frac{z_j e \varepsilon_r kT}{6\pi\eta(ze)^2(1 + y)}$$

If the calculation of conductance were carried only up to this point, i.e., terms linear in q, the equation obtained for the molar conductance would be

$$\Lambda = \Lambda^\circ - \left[\frac{\alpha\Lambda^\circ}{1 + \delta\kappa a} + \beta\right]\frac{c^{\frac{1}{2}}}{1 + \kappa a} \qquad (5.2.21)$$

This expression was obtained by Leist[L9] and gives the first order correction to the limiting law 5.2.1 due to the finite size of the ions. Equation 5.2.21 appears as the first order correction in both the Pitts and Fuoss–Onsager treatments.

It is important to emphasise that the q^2 terms in eqn. 5.2.12 which will permit the calculation of the higher order contribution to the electrostatic force exerted by the ions on the central one, involve $\zeta_j(1)$ (electrophoretic effect). On the other hand, according to eqn. 5.2.6, the relaxation terms will contribute to the stress of the solvent on the ions. Thus, as soon as the calculation exceeds the terms linear in q, it becomes artificial to separate electrophoretic and relaxation terms. Using the approach explained above, the contribution of q^2 terms was obtained by Pitts from eqn. 5.2.12

$$\pi_{11}(2, 1) = \pi_{22}(2, 1) = 0$$

and

$$\pi_{21}(2, 1) = -\pi_{12}(2, 1) = \frac{2n^2zef}{\kappa kT}(H_1 + H_2) \tag{5.2.22}$$

where H_1 and H_2 are complex functions of x.

Using eqn. 5.2.6, $\zeta_j(2)$ is evaluated. In the second hydrodynamic region ($r \geqslant a$) only ψ'_{ji} appears, since $\Delta\psi_j^0$ does not contribute even powers of q for symmetrical electrolytes. The force $\mathbf{F}_j^{\text{vis}} + \mathbf{F}'_j$ is to a good approximation given by

$$\mathbf{F}_j^{\text{vis}} + \mathbf{F}'_j = -6\pi\eta R_j\mathbf{V}_j - \frac{z_je\varepsilon_rkT}{(ze)^2(1+y)}R_jq\left[1 - \frac{q}{3}T_1\right] \tag{5.2.23}$$

where

$$T_1 = \frac{3}{4(\sqrt{(2)}+y)} + \frac{3}{8}\left[\frac{3\sqrt{(2)}\,\mathrm{e}^{(\delta+1)y}}{\sqrt{(2)}+y}E_i[(1+\delta)y] - 2\,\mathrm{e}^{y}E_i(y)\right]$$

and

$$E_i(t) = \int_t^\infty \frac{\mathrm{e}^{-u}}{u}\,\mathrm{d}u$$

The mean velocity of the ion is determined by the three forces described in eqn. 5.2.2. The first two contributions are given in eqn. 5.2.23. The force due to electrostatic interactions of ions surrounding j is produced by the charge density ρ_{ji} around the central ion,

$$(\mathbf{F}''_j)_x = -\int_V \frac{z_je\rho_{ji}}{\varepsilon_rr^2}\cos\theta\,\mathrm{d}V \tag{5.2.24}$$

where the integrand is the force on the j ion, due to the charge existing in the element of volume $\mathrm{d}V$ at distance r from j in the direction of the external field. The asymmetric component ρ'_{ji} is substituted in eqn. 5.2.24 in terms of $\pi_{21}(1, 1)$ and $\pi_{21}(2, 1)$ according to eqns. 5.2.17 and 5.2.22, and the force arising from it becomes

$$(\mathbf{F}''_j)_x = -z_je\mathbf{X}\left[\frac{q}{3(1+\sqrt{2})(1+y)(\sqrt{(2)}+y)} + \frac{q^2}{3}(S_1 + S_2)\right] \tag{5.2.25}$$

We are now ready to obtain the final equation for the molar conductance. According to Ohm's law the current density due to j ions is $\mathbf{j}_j = \sigma_j \mathbf{X} = z_j F c \mathbf{V}_j$, where σ_j is the conductivity of the solution due to the j ionic species, and F is the Faraday. Besides, by definition

$$\Lambda_j = \sigma_j / c = z_j F \mathbf{V}_j / \mathbf{X}$$

and

$$\Lambda_j^\circ = z_j^2 e F \omega_j = z_j^2 e F / 6\pi\eta R_j.$$

Replacing in eqn. 5.2.2 the values for the three components of the total force acting on the ions, we get finally for the molar conductance of the electrolyte

$$\Lambda = \Lambda^\circ - \frac{z^2 e\kappa F}{3\pi\eta(1+y)} + \frac{z^4 e^3 \kappa^2 F}{3\pi\eta\varepsilon_r kT(1+y)} \frac{T_1}{3}$$
$$- \Lambda^\circ \left[\frac{(ze)^2\kappa}{3\varepsilon_r kT(1+\sqrt{2})(1+y)(\sqrt{(2)}+y)} + \frac{(ze)^4\kappa^2}{3(\varepsilon_r kT)^2} S_1 \right]$$
$$+ \frac{(\sqrt{(2)}-1) z^4 e^3 F \kappa^2}{9\pi\eta(1+y)^2(\sqrt{(2)}+y)} \qquad (5.2.26)$$

The first two terms plus the first one inside the square brackets give Leist's expression 5.2.21. The third term in eqn. 5.2.26 gives the high order effect of the local velocity which arose from the perturbation of the ionic electrostatic potential. The second term in the brackets represents the higher order contribution of the asymmetric potentials and distribution functions from the continuity equation. The last term in eqn. 5.2.26 arises from S_2 in eqn. 5.2.25, which is due to the introduction of the first order term in $\mathbf{u}_{ij}$ into the continuity eqn. 5.2.12 and is equal to $\alpha\beta c/(1+y)^2(\sqrt{(2)}+y)$. The functions S_1 and T_1 have been tabulated by Pitts[L4] for values of κa in the range 0.02–0.50.

The alternative approach to the conductance equation for electrolytic solutions followed by Fuoss and Onsager[L5] has many points in common to the one employed by Pitts. Some differences are, however, worth mentioning.

Fuoss and Onsager start their treatment assuming that

$$\Lambda = (\Lambda^\circ - \Lambda^e)(1 + \Delta\mathbf{X}/\mathbf{X}) \qquad (5.2.27)$$

where Λ^e is the electrophoretic term which they made equal to[L10] $\beta c^{\frac{1}{2}}/(1+y)$. In writing eqn. 5.2.27, the electrophoretic and relaxation effects are considered separable even beyond the first order contributions to Λ. Thus only $\Delta\mathbf{X}/\mathbf{X}$, the relaxation field, is calculated by Fuoss and Onsager, omitting altogether the effect of asymmetric ionic potentials and distributions on the local velocity of the medium where the ions are immersed (term T_1 in eqn. 5.2.26).

The expression for $\Delta\mathbf{X}/\mathbf{X}$ is obtained by a series of successive approximations yielding the terms of different order which contribute to the relaxation field. The first order term arises from eqn. 5.2.13 putting all the T terms equal to zero. Fuoss and Onsager obtain, in this way, an equation equivalent to 5.2.15. The first order expression for f'_{ij} is then replaced in the T terms and a further approximation to the perturbed distributions and ionic potentials is calculated.

The boundary conditions used by Fuoss and Onsager[L5] are:

(i)
$$(\partial\psi_{ij}/\partial r)_{r\to\infty} = 0 \tag{5.2.28a}$$

(ii) The perturbed potentials and electric fields are considered continuous, which may be expressed by[L5]

$$[r(\partial\psi'_{ij}/\partial r) - \psi'_{ij}]_{r=a} = 0 \tag{5.2.28b}$$

(iii) The radial component of the relative velocities of two ions vanishes when both ions collide,

$$[(\mathbf{V}_{ij} - \mathbf{V}_{ji})\cdot\mathbf{r}]_{r=a} = 0 \tag{5.2.28c}$$

The various terms obtained in the successive approximations to the perturbed ionic potential $(\psi'_{ij})^{(n)}$ contribute terms of the form

$$-\Delta\mathbf{X}^{(n)} = (\partial\psi'^{(n)}_{ij}/\partial r)_{r=a} \tag{5.2.29}$$

to the total relaxation field. This expression is analogous to eqn. 5.2.24 which was also used to obtain the interionic force which modifies the mobility of the ions.

The differences between Pitts (P) and Fuoss–Onsager (F–O) are: first, the above mentioned omission by F–O of the effect of asymmetric potential on the local velocities of the solvent near the ions; second, the use of the more usual boundary conditions 5.2.28b by F–O compared to the P assumption that perturbations cease to be important at $r = a$. Pitts, Tabor and Daly,[L8] who have analysed in detail both treatments, concluded that the discrepancy due to the different boundary conditions is small but has the effect of reducing ionic interactions in the P treatment with respect to the F–O. This is confirmed by the analysis of data with both theories. Usually P requires a smaller value of the a parameter than F–O.[L11] The third discrepancy between the theoretical treatments is in the expression of $\mathbf{v}_{ij}$ in eqn. 5.2.5, for which F–O add a term which involves the effect of the asymmetry of the ionic atmosphere upon the central ion surrounded by such atmosphere. The last difference lies in the hydrodynamic approaches and the corresponding boundary conditions. P imposes the condition that the velocity of the smoothed medium vanishes at $r = R_j$ while F–O consider that the radial component of the ionic relative velocities is zero at $r = a$. In this way Fuoss and Accascina[L6] equate the hydrodynamic radius to the contact distance a between two ions. It seems unlikely that the condition $R_j = a$ is valid. The hydrodynamic radius is a specific parameter for each ion forming the electrolyte, related by Stokes' law to their respective Λ_j°, while a is a collision distance which characterises the electrolyte as a whole. Experimental results do not show any simple relationship of the type $R_j = a$ to hold between both distance parameters.

Another contribution has been added by Onsager and Fuoss[L12] to their original expression for the relaxation field. Since the ionic atmosphere is deformed by the displacement of the central ion, there are more anions (cations) behind a central cation (anion) than in front of it. Hence collisions from behind a central cation (anion) are more frequent than in front of it. This will result in an increased velocity of the central ion. The calculated effect has a minor effect on the conductance, contributing only terms linear in the concentration. Valleau[L13] has cast doubts on the reality of this effect.

The original F–O equation has been further modified by Fuoss and Hsia,[L14] who recalculated the relaxation field, retaining terms which had previously been neglected.

The results of the conductance theories may be expressed in a general form by

$$\Lambda = \Lambda^\circ - \frac{\alpha\Lambda^\circ c^{\frac{1}{2}}}{(1 + \kappa a)(1 + \kappa a/\sqrt{2})} - \frac{\beta c^{\frac{1}{2}}}{1 + \kappa a} + G(\kappa a) \qquad (5.2.30)$$

where $G(\kappa a)$ is in general a complicated function of the variable. In order to simplify the analysis of the experimental results, often an equation of the form

$$\Lambda = \Lambda^\circ - S\sqrt{(c)} + Ec \ln c + J_1 c - J_2 c^{\frac{3}{2}} \qquad (5.2.31)$$

is employed. This is obtained by expanding eqn. 5.2.30 using the expressions

$$\begin{aligned} e^x &= 1 + x + x^2/2! + \ldots \\ 1/(1 + x) &= 1 - x + x^2 - \ldots \qquad (5.2.32) \\ E_i(x) &= -\Gamma - \ln x + x - \ldots; \end{aligned}$$

where Γ is Euler's constant, and neglecting those terms which depend on a power of concentration higher than $c^{\frac{3}{2}}$.

In eqn. 5.2.31, S is the limiting law coefficient given by eqn. 5.2.3. E originates in the expansion of the exponential integrals 5.2.32 contained in the function G. It depends on the solvent physical properties and on the charge of the electrolyte.

$$ba = \frac{(ze)^2}{\varepsilon_r kT} = 16.7099 \times 10^4 \frac{z^2}{\varepsilon_r T} \Big/ \text{Å};$$

$$\kappa = 50.2916 \frac{z\sqrt{c}}{(\varepsilon_r T)^{\frac{1}{2}}} \Big/ \text{Å}^{-1};$$

$$E = E_1 \Lambda^\circ - E_2;$$

$$E_1 = \frac{(\kappa ab)^2}{24c} = 2.94257 \times 10^{12} \frac{z^6}{(\varepsilon_r T)^3} \Big/ \text{mol}^{-1}\,\text{l}$$

$$E_2 = \frac{\kappa ab\beta}{16c^{\frac{1}{2}}} = 4.33244 \times 10^7 \frac{z^5}{(\varepsilon_r T)^2 \eta} \Big/ \Omega^{-1}\,\text{cm}^2\,\text{mol}^{-2}\,\text{l} \qquad (5.2.33)$$

The presence of this term in eqn. 5.2.31 has the effect that the limiting law is approached from below as concentration decreases, an effect which is usually masked by the J_1 term. J_1 and J_2 depend on the same parameters as S and E, but also on the closest distance of approach of two ions. For each theoretical treatment, different expressions for J_1 and J_2 are obtained; they may, however, be written in the general form

$$J_1 = 2E_1\Lambda^\circ \left[\ln\left(\frac{\kappa a}{c^{\frac{1}{2}}}\right) + \Delta_1\right] + 2E_2\left[\Delta_2 - \ln\left(\frac{\kappa a}{c^{\frac{1}{2}}}\right)\right]$$

$$J_2 = \frac{\kappa ab}{c^{\frac{1}{2}}}\left[4E_1\Lambda^\circ\Delta_3 + 2E_2\Delta_4\right] - \Delta_5 \qquad (5.2.34)$$

The expressions for the Δ_i terms are given in Table 5.2.1 according to the Pitts[L15] (P) and Fuoss and Hsia[L11] (F–H) treatments. Another theoretical treatment of conductances has been given by Kremp[L16] and by Kremp, Kraeft and Ebeling.[L17] Their result has been approximated by Kraeft[L18] to an equation of the form 5.2.31 with $J_2 = 0$; the expression for J_1 has been included in Table 5.2.1.

It is noteworthy that the majority of the conductance data have been analysed with the equations of Fuoss and Onsager[L5] and of Fuoss and Accascina.[L6] However, since it has been recently shown[L11,L14,L19] that both equations are incomplete and in some cases fail to fit experimental data, we quote here only the improved Fuoss and Hsia result.

Table 5.2.1
Expressions for the Δ_i Terms in Eqn. 5.2.34

Theory	Δ_1		Δ_2	
P	$\frac{2}{b} + 1.7718;$	(2.7718)	$\frac{8}{b} + 0.01387;$	(4.01387)
F–H	$\frac{1}{b^3}[2b^2 + 2b - 1] + 0.90735;$	(2.2824)	$\frac{22}{3b} + 0.01420;$	(3.6808)
Kraeft	$\frac{1}{b^3}\left[2.2125b^2 + \frac{3b}{4} - 1\right] + 1.1020;$	(2.2708)	$\frac{1}{b^2}[7.7500b + 1] - 0.7897;$	(3.3353)
	Δ_3		Δ_4	
P	$\frac{1.2929}{b^2} + \frac{1.5732}{b};$	(1.1098)	$\frac{8}{b^2} + \frac{1.4073}{b};$	(2.70365)
F–H	$\frac{0.9571}{b^3} + \frac{1.1187}{b^2} + \frac{0.1523}{b};$	(0.47546)	$\frac{1}{b^3}\left[0.5738b^2 + 7.0572b - \frac{2}{3}\right] - 0.6461;$	(1.3218)
	Δ_5			
P	0			
F–H	$\frac{E_2\beta}{\Lambda^\circ}\left[\frac{4}{3b} - 2.2194)\right];$	$\left(-1.5527\frac{E_2\beta}{\Lambda^\circ}\right)$		

Values in parentheses correspond to b equal to the Bjerrum distance (see p. 564).

In order to judge the effects of expanding the full equations 5.2.30 to the more approximate expression 5.2.31, values of Λ for 1–1 model electrolytes in *N*-methylacetamide (NMA) ($\Lambda^\circ = 19.00\ \Omega^{-1}\ cm^2\ mol^{-1}$, $a = 3.5$ Å, $t = 32°C$) and in dimethylformamide (DMF) ($\Lambda^\circ = 84.40\ \Omega^{-1}\ cm^2\ mol^{-1}$, $a = 4$ Å, $t = 25°C$) have been calculated for the full Pitts equation and also for eqn. 5.2.31 using both J_1 and J_2, and also putting $J_2 = 0$. This comparison may be seen in Fig. 5.2.4, where the

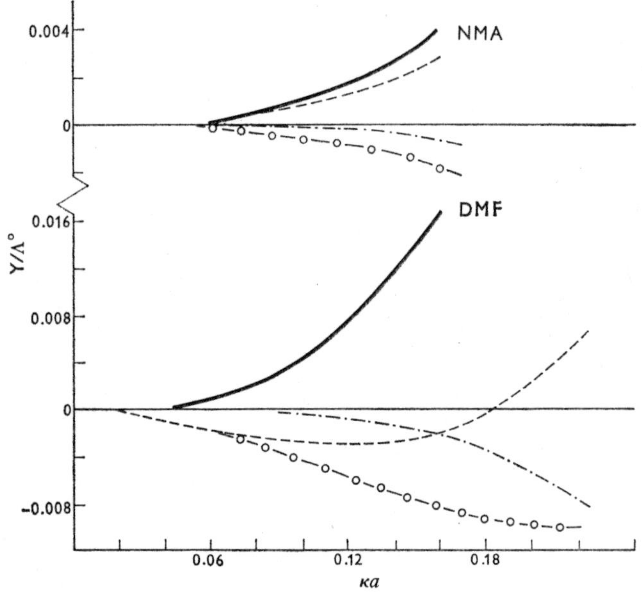

FIG. 5.2.4. Performance of various expanded conductance equations compared to the full (*P*) equation. $Y = \Lambda_{P\ full} - \Lambda$ (eqn. 5.2.31) against κa for (1:1) model electrolytes in NMA and DMF.

———— (P) eqn. 5.2.31 with $J_2 = 0$.
·—·—·—· (P) eqn. 5.2.31.
– – – – – (F–H) eqn. 5.2.31 with $J_2 = 0$.
–○–○–○– (F–H) eqn. 5.2.31.

relative deviation of an approximate eqn. 5.2.31 from the full Pitts equation, Y/Λ°, is plotted against κa. The Pitts and the Fuoss and Hsia expansions are shown; the plot of Kraeft's function is not included since it almost coincides with the Fuoss and Hsia curves for $J_2 = 0$. Figure 5.2.4 shows that the expanded equations 5.2.31 differ less from the full Pitts equation in NMA than in DMF. This is due to the fact that inter-ionic effects are much smaller in NMA, which has a dielectric constant nearly five times larger than DMF. Figure 5.2.4 illustrates the importance of the J_2 term in eqn. 5.2.31. The conductance equation with

$J_2 = 0$ is in some cases capable of representing the experimental data for $\kappa a < 0.2$, but obviously the value of the a parameter must be different. Comparing the Pitts and the Fuoss and Hsia expanded conductance equations, it is clear that for a given electrolyte Pitt's equation predicts a larger molar conductance, i.e., smaller interionic effects.

Table 5.2.2 illustrates the performance of the Pitts and the Fuoss and Hsia equations for some dissociated electrolytes in organic solvents at 25°C. The values of the fitting parameters for salts in DMF and in

Table 5.2.2

Comparison of Pitts (P) and Fuoss and Hsia (F–H) Expanded Conductance Equations

Solvent	Electrolyte	Ref.	$10^3 \cdot \Delta c$	Theory	Λ°	a	σ_Λ
DMF	KI	L3	0.4–4.0	P	83.00	4.54	0.05_6
				F–H	83.10	5.37	0.07_6
	$CsClO_4$	L3	0.1–5.0	P	87.87	3.52	0.02_4
				F–H	86.94	4.60	0.02_2
Methanol	KCl	L20	0.07–5.0	P	104.76	2.16	0.10_7
				F–H	104.85	3.40	0.06_7
	NaCl	L20	0.04–4.0	P	97.15	2.78	0.06_5
				F–H	97.23	4.04	0.03_6
PC	Et_4NCl	L21	1.4–10.3	P	31.55_9	4.81	0.01_9
				F–H	31.58_3	4.80	0.02_3

methanol are from refs. L11 and L15. Those for propylene carbonate (PC) have been calculated for the present contribution. It is quite obvious that both theoretical treatments account adequately for results in DMF and PC, the Pitts equation being only marginally more satisfactory as shown by the values of the standard deviations (σ_Λ) in Table 5.2.2. In methanol the Pitts equation does not fit the results so precisely and yields rather small values for a. This may be attributed to a small degree of ionic association in the methanol solutions.

Since the theoretical treatments which gave rise to the expanded eqns. 5.2.31 were only exact to terms linear in c, it may seem artificial to maintain the $J_2c^{\frac{3}{2}}$ term in 5.2.31. However, the coefficient of q^2 in the full eqn. 5.2.30 is concentration dependent, and it is in order to avoid large deviations from the full equation that $J_2c^{\frac{3}{2}}$ is kept. From the analysis of experimental data it has been observed[L14,L15,L22] that if J_2 is made zero, eqn. 5.2.31 shows a systematic deviation when compared to experimental data.

Leist's eqn. 5.2.21 and the simpler equation

$$\Lambda = \Lambda^\circ - S\sqrt{(c)}/(1 + \kappa a) \tag{5.2.35}$$

proposed by Robinson and Stokes,[L23] may be used to fit data within a few tenths of a per cent, but it must be realised that in this case a acts merely as an arbitrary parameter and its value may have little physical significance.

Considering that the viscosity of a fluid containing spherical particles dissolved in it, η_s, increases with the volume fraction φ according to Einstein's relation,

$$\eta_s = \eta\left(1 + \frac{5}{2}\varphi\right) \tag{5.2.36}$$

Fuoss and Accascina[L6] corrected the predicted theoretical conductances for this obstructive effect assuming the validity of Walden's rule, i.e., $\Lambda\eta_s$ = constant. Thus,

$$\Lambda_\eta = \Lambda(1 + \mathrm{F}c) \tag{5.2.37}$$

where $\mathrm{F}c = 5\varphi/2 = 4\pi R^3 N_\mathrm{A}/3$. The viscosity correction may be calculated using the value of F obtained from the hydrodynamic radii of ions, or by using the experimentally determined η_s/η values. Using eqns. 5.2.31 and 5.2.37,

$$\Lambda_\eta = \Lambda^\circ - S\sqrt{(c)} + Ec\ln c + J_1c - J_2c^{\frac{3}{2}} - \mathrm{F}\Lambda c \tag{5.2.38}$$

This correction has been frequently employed when dealing with the conductance of bulky ions. Its use modifies the value of a, but for most solvents the effect is small. It is not obvious that the viscosity correction is valid[L24] since it is known that when the viscosity of a solvent is increased by addition of a nonelectrolyte, the product $\Lambda^\circ\eta$ is not constant, that is the Walden rule is not obeyed. Treiner and Fuoss[L25] increased the viscosity of acetonitrile by addition of octacyanoethyl sucrose (CES), and measured the conductance of some tetraalkylammonium salts in these media. The viscosity increased up to a hundredfold from that in pure acetonitrile. They found that the extrapolated Λ° does not obey the Walden rule, rather $\Lambda^\circ\eta^{0.7}$ = constant for solvents with more than 40%

CES. In the case studied by Treiner and Fuoss it is to be expected that large variations of the specific interactions between ions and solvent molecules as the composition of the solvent is varied may be ruled out since both substances would interact with the ions through their —CN groups. Besides, all the solvent mixtures had very similar dielectric constants.

Stokes[L26] has also found a relationship, $\Lambda^\circ \eta^{0.7}$ = constant, to apply to aqueous solutions. Unfortunately, too few studies of this type have been carried out to enable the establishment of a general relationship between Λ and η. It may be concluded that eqn. 5.2.36 is an empirical correction which may be justified for more concentrated solutions where no exact treatment describes the variation of conductance with concentration. In dilute solutions, however, its use adds a new source of uncertainty to the interpretation of conductance data.

5.3 IONIC ASSOCIATION

The conductance eqn. 5.2.31 is capable of representing accurately the behaviour of completely dissociated electrolytes. For these electrolytes the experimental value of the conductance is found to be larger than that (Λ_{L-L}) predicted by Onsager's limiting law eqn. 5.2.1. This is called positive deviation and has been illustrated in Fig. 5.2.1 for $LiClO_4$ in dimethylformamide. The positive deviation may be attributed to short range hard-core repulsive interactions between ions. (A similar effect is responsible for the positive deviations from the Debye–Hückel limiting law found for the activity coefficients of electrolytes.[L27]) Leist's eqn. 5.2.21 shows that the first order effect of a finite collision diameter between ions on the conductance is to decrease the slope of the curve of Λ against $c^{\frac{1}{2}}$.

When the electrostatic interaction between ions is large, the conductance of the salt is smaller than predicted by eqn. 5.2.31 and usually smaller even than Λ_{L-L}. This type of deviation from the behaviour predicted by eqn. 5.2.31 is attributed to association of cations and anions. For symmetrical electrolytes this process leads to the formation of neutral ion pairs which do not contribute to the conductance of the solution. In Fig. 5.3.1 the values of $\Lambda - \Lambda_{L-L}$ are plotted against the square root of the concentration for lithium halide solutions in sulpholane;[L28] the halides show an increasing negative deviation going from LiI (slightly associated) to LiCl (typical weak electrolyte). It is worth mentioning that for electrolytes having association constants between that of LiI ($K_A = 5.6 \text{ mol}^{-1}\text{ l}$) and that of LiBr (278 $\text{mol}^{-1}\text{ l}$) it is very common to observe that the conductances vary almost linearly with $c^{\frac{1}{2}}$, the slope being, however, more negative than predicted by the limiting law. Therefore, a linear extrapolation of Λ against $c^{\frac{1}{2}}$ yields Λ°

values which differ from the real ones, the difference increasing with increasing K_A. In Fig. 5.3.1 linear extrapolation of the data for LiI in sulpholane would give a reasonable value for $\Lambda°$, but in the case of LiBr it would extrapolate to point A, about 5% larger than the real $\Lambda° = 13.25$ Ω^{-1} cm^2 mol^{-1}.

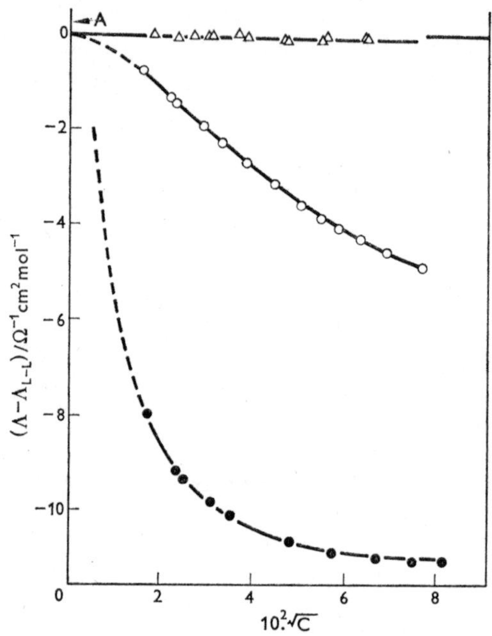

FIG. 5.3.1. Negative deviations from the limiting law of the conductance of lithium halide solutions in sulpholane at 30°C. △, LiI; ○, LiBr; and ●, LiCl.

In sect. 5.1 it was mentioned that some of the most important information obtained from conductance measurements are the values of the ion-pair association constants, K_A. For symmetrical electrolytes K_A refers to the equilibrium

$$M^{z+} + A^{z-} = MA \tag{5.3.1}$$

where one cation associates with an anion yielding a non-conducting species. If α is the degree of dissociation, we have

$$K_A = \frac{1-\alpha}{\alpha^2 c \gamma_\pm^2} \tag{5.3.2a}$$

$$\alpha = 1 - \alpha^2 c K_A \gamma_\pm^2 \tag{5.3.2b}$$

where $\gamma_{\pm}$ is the mean activity coefficient of the free ions at concentration αc. If Λ_f is the conductance of the free ions at concentration αc, the experimentally determined conductances are related to it by $\Lambda = \alpha\Lambda_f(\alpha c)$ and using eqn. 5.3.2b for α, we get

$$\Lambda = \Lambda_f(\alpha c) - \Lambda\gamma_{\pm}^2 K_A \alpha c \tag{5.3.3}$$

The conductance of free ions may be expressed by eqn. 5.2.31, so that for associated electrolytes we have

$$\Lambda = \Lambda^\circ - S\sqrt{(\alpha c)} + E(\alpha c)\ln(\mathrm{d}c) + J_1(\alpha c) - J_2(\alpha c)^{\frac{3}{2}} - K_A\Lambda\gamma_{\pm}^2(\alpha c) \tag{5.3.4}$$

The association of ions in a solution of typical electrolytes is essentially of electrostatic nature and called ion association. This electrostatic association has to be invoked because the electrostatic interactions between ions are underestimated by the linear approximation employed by Debye and Hückel to describe the distribution of ions. This approximation implies that $\Phi_{ij}^0 = z_j e\psi_i^0/kT \ll 1$.

The value of Φ_{ij}^0 at the distance $r = 1/\kappa$, where the ionic atmosphere is most densely populated, gives an estimate of the validity of the Debye–Hückel approximation. Neglecting factors of the order of unity, it turns out that $q = [(ze)^2/\varepsilon_r kT)]\kappa < 1$; hence at 25°C, $\sqrt{(c)} < 5.10^{-4}\varepsilon_r^{\frac{3}{2}}/z^3$. Thus the linearised Poisson–Boltzmann relation underestimates the electrostatic interactions in polyvalent electrolytes and even for (1–1) salts in solvents of low dielectric constant.

Bjerrum[L29] circumvented this problem in the following way. Let us divide the volume surrounding a central ion j into two regions; the first region extends from the surface of the ion ($r = a$) to $r = d$, the other region goes from d to infinity. The probability df that an i ion be at a distance between r and $r + \mathrm{d}r$ of a central j ion is given by

$$\mathrm{d}f = 4\pi n_i \exp(-\Phi_{ij}^0) r^2\,\mathrm{d}r \tag{5.3.5}$$

In the inner region both ions are relatively close and ion i interacts with the unshielded potential of the central ion; thus $\Phi_{ij}^0 = -(z^2e^2/\varepsilon_r r)$ for symmetrical electrolytes. The fraction of ions forming ion pairs separated by distances between a and d are obtained by integration of eqn. 5.3.5 in the inner region. This fraction was termed associated by Bjerrum. In sufficiently dilute solutions the expression (5.3.2a) for K_A becomes

$$K_A \cong \frac{1-\alpha}{c} = \frac{1-\alpha}{n} N_A$$

and it follows that

$$K_A = 4\pi N_A \int_a^d \exp\left(\frac{z^2e^2}{\varepsilon_r kTr}\right) r^2\,\mathrm{d}r \tag{5.3.6}$$

Setting $\xi = (z^2e^2/\varepsilon_r kTr)$ and rearranging eqn. 5.3.6

$$K_A = 4\pi N_A \left[\frac{(ze)^2}{\varepsilon_r kT}\right]^3 \int_{z^2e^2/\varepsilon_r kTd}^{b} (e^{\xi}/\xi^4)\,\mathrm{d}\xi \tag{5.3.7}$$

There is one parameter in Bjerrum's treatment which so far remains arbitrary. The upper limit of the integral in eqn. 5.3.6, d, was fixed by Bjerrum at $d = l = (ze)^2/2\varepsilon_r kT$ because at this distance the distribution function eqn. 5.3.5 presents a minimum. In this case

$$K_A = 4\pi N_A \left[\frac{(ze)^2}{\varepsilon_r kT}\right]^3 Q(b) \tag{5.3.8}$$

The function $Q(b)$ has been tabulated[L30] and thus K_A may be evaluated.

Bjerrum considered that the Debye–Hückel treatment may be applied to the ions which are present in the outer region, and activity coefficients for these 'free' ions are given by the Debye–Hückel equation

$$\ln \gamma_{\pm} = \frac{-A\sqrt{(\alpha c)}}{1 + d\kappa} \tag{5.3.9}$$

It should be pointed out that in dealing with ion association according to the treatment of Bjerrum, the distance d may be fixed at a value different from l,[L29,L31,L32] but once fixed in eqn. 5.3.6 the same value must consistently be used in eqn. 5.3.9.

It is important to digress briefly on the meaning of Bjerrum's approach, which has often been the subject of controversy.[L33] Onsager[L34] considered that 'one may say that Bjerrum applies different approximations to different regions of space in evaluating' the partition function of the electrolytic solution. In no case may Bjerrum association be taken to imply *necessarily* a 'chemical association' in the way protons and acetate ions associate to yield the new chemical species acetic acid.[L29] Bjerrum pointed out[L29] that for this chemical process intermediate species between the associated and dissociated states do not exist in measurable concentrations; but between free and associated ions intermediate forms exist in finite concentrations. We shall illustrate this point below. This approach deals with that part of the interaction which is underestimated by the approximation of Debye and Hückel and assumes that ions in the inner region may be considered to form a neutral pair, not affecting the rest of the solution.

This treatment of ion association may not seem entirely satisfactory on two accounts: (i) ion association ceases sharply at $r = d$, and (ii) it is difficult to accept that ions up to l may be considered associated to the central ion ($l = 14$ Å for a (1–1) electrolyte in acetone at 25°C). Fuoss[L35] has suggested that only those ions which are in actual contact should be considered associated. The distance between two ions forming a pair is then equal to the distance of closest approach for these ions a Fuoss considered that the probability of an ion becoming associated when added to a solution is proportional to the volume occupied by the free ions, which is taken to be $(\frac{4}{3})\pi a^3$ for each free ion. Thus, an anion is associated when its centre lies in the sphere $(\frac{4}{3})\pi a^3$ surrounding a cation. The probability of association is increased by the electrostatic attraction between oppositely charged ions, which is made equal to the coulombic interaction of an ion with the electrostatic potential of the central ion and its ionic atmosphere at $r = a$. In this way, Fuoss derived, for the association constant, the expression

$$K_A = \frac{4}{3}\pi N_A a^3 \exp b \tag{5.3.10}$$

Equation 5.3.10 predicts that there will always be some association in electrolyte solutions; furthermore K_A passes through a minimum at $a = (ze)^2/3\varepsilon_r kT$ and from there on, ionic association increases with the ionic size, which is not what would be expected for an electrostatic interaction. This effect is due to the volume pre-exponential factor in eqn. 5.3.10,[L36] which causes the probability of association to increase with a^3.

On the other hand, since Bjerrum's treatment is an extension of Debye and Hückel's, it is natural that when the electrostatic interactions are small enough to make the Debye–Hückel approximation valid $(r > d)$ there is no need to invoke ionic association.

To compare both treatments it is useful to calculate the distribution of associated ions around the central one as predicted by Bjerrum's distribution function (5.3.5). We may inquire at which distance d from the central ion does a given fraction of associated ions exist, on the average. This fraction will be $f(d)/f(l)$ and from eqn. 5.3.5

$$\frac{f(d)}{f(l)} = \frac{4\pi N_A \int_a^d r^2 \exp\ (\xi)\, dr}{4\pi N_A \int_a^l r^2 \exp\ (\xi)\, dr} \tag{5.3.11}$$

and making use of the function $Q(b)$

$$\frac{f(d)}{f(l)} = 1 - \frac{Q(ba/d)}{Q(b)} \tag{5.3.12}$$

Figure 5.3.2 gives the distance d, in which 3/4 and 9/10 of the total associated ions are contained, as a function of l. The calculation was done for a uni-univalent electrolyte with $a = 3$ Å at 25°C. The values of l for this electrolyte in some solvents of different dielectric constants have been indicated in Fig. 5.3.2; at the top of the graph the values of K_A predicted by eqn. 5.3.8 for different values of l are included. The maxima in the curves of Fig. 5.3.2 indicate the range of l values for which the larger dispersion of associated ions is predicted by Bjerrum's model. For these values of l there is a large number of intermediate species between free ions and contact pairs. As l increases the population of associated ions far from the central one decreases sharply, for $K_A \gtrsim 10^3$ mol^{-1} l, very few intermediate species exist. For large K_A we may conclude that the Bjerrum and the Fuoss models of ionic association are similar in that essentially only contact pairs are counted as associated.

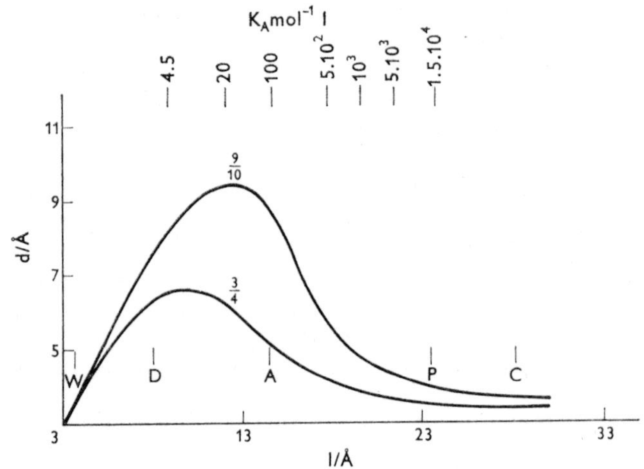

FIG. 5.3.2. Distance at which 9/10 and 3/4 of the associated ions lie from a central ion (a = 3 Å) against the Bjerrum distance l. The vertical bars denote the values of l for a (1:1) electrolyte in: W = water, D = dimethylformamide, A = acetone, P = pyridine, C = o-dichlorobenzene. In the top of the graph values of K_A corresponding to some l values are given.

In this case b is large and the function $Q(b)$ may be expanded in series and approximated to[L37]

$$Q(b) = \exp(b)/b^4 \tag{5.3.13}$$

using 5.3.13 in 5.3.8 the limiting expression of Bjerrum's K_A for large b is

$$K_A = 4\pi N_A a^3 \exp(b)/b \tag{5.3.14}$$

The same functional dependence of b appears in the corrections to the Debye–Hückel equation obtained by more exact statistical mechanical calculations.[L36] Equation 5.3.14 may be compared to eqn. 5.3.10, the difference between them being the factor $1/b$, which does not appear in eqn. 5.3.10. However, eqn. 5.3.14 is only valid for large b values where the dominant term is exp (b); no difference is found from experimental data between the Fuoss and the Bjerrum treatments for large b values.

Ionic association only depends on the parameter b which is the ratio of electrostatic energy to thermal energy of a pair of ions in contact. It might be expected that for a given electrolyte a would be a constant

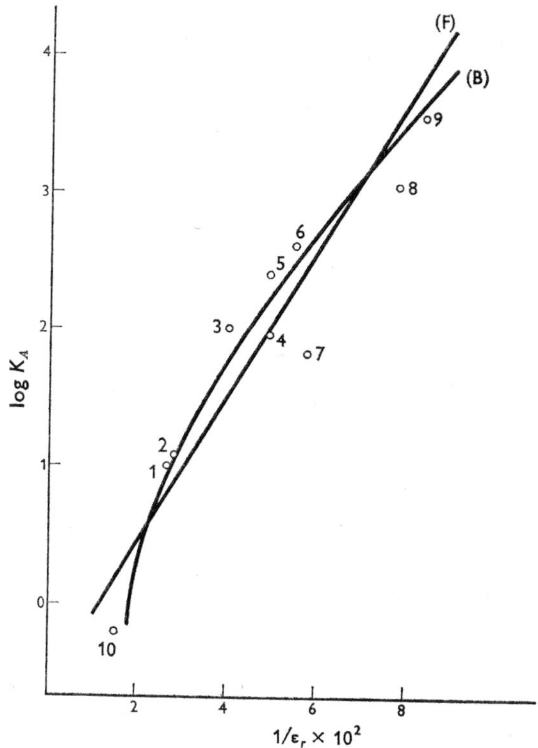

FIG. 5.3.3. Dependence of log K_A on $1/\varepsilon_r$ for KI solutions in various pure solvents at 25°C (after Janz *et al.*[L38,39]). (1) ethanolamine, (2) acetonitrile, (3) benzonitrile, (4) acetone, (5) propanol, (6) methylethyl ketone, (7) acetophenone, (8) ethylene diamine, (9) pyridine and (10) water.

independent of the solvent in which it is dissolved. If this were the case, ionic association for an electrolyte would only be dependent on $1/\varepsilon_r T$ and independent of the particular solvent. Figure 5.3.3 is a plot of log K_A against $1/\varepsilon_r$ for KI in a series of solvents at 25°C using the values of the association constants calculated by Janz *et al.*[L38,L39] Curve (B) corresponds to the predictions of eqn. 5.3.8 and the straight line (F) to that of the Fuoss eqn. 5.3.10, both for $a = 4.5$ Å, which was found to represent best the experimental values of the association constants.[L38,L39] It may be seen that the experimental association constants scatter from the predicted behaviour. This scatter may be attributed to a change in the distance between colliding solvated ions, which depends on other solvent properties besides dielectric constant. The process of ion pairing involves a competition between solvent molecules and ions for positions around the central ion. The oppositely charged ion will be able to approach more closely (smaller a) for a weaker ion-solvent interaction. Gilkerson[L40] extended eqn. 5.3.10, taking into account the difference between the solvation energy of free ions and that of the ion pair. If E_s denotes this energy difference, the association constant can be expressed by

$$K_A = K_A^0 \exp(b) \exp(-E_s/kT) \tag{5.3.15}$$

To judge the validity of eqns. 5.3.8 and 5.3.10, it is necessary to keep a constant while varying ε_r. This may be expected to be the case whenever the dielectric constant of the medium is varied by addition of a second solvent, and one of the solvents may be considered to solvate the ions strongly while the other does not, producing only a modification of dielectric constant of the mixture. The most favourable cases for which experimental results are available, are those solvent mixtures having *p*-dioxan or carbon tetrachloride together with a more strongly polar solvent.

Equation 5.3.10 predicts a linear relationship between $\log K_A$ and $1/\varepsilon_r$, while eqn. 5.3.8 gives a more complicated dependence on ε_r which becomes very similar to that of eqn. 5.3.10 only for large values of b. Figure 5.3.4 is a plot of $\log K_A$ against $1/\varepsilon_r$ for several solvent mixtures at 25°C. It will be observed that the data are best represented by curves concave downwards. This fact has been frequently observed.[L22,L42–45]

In order to appreciate the performance of eqn. 5.3.8, the following procedure is used. From the tables of the function $Q(b)$, the function $\log K_A^\star$

$$\begin{aligned} \log K_A^\star &\equiv \log K_A^{Bj}/a^3 \\ &= \log(4\pi N_A) + 3 \log b + \log Q(b) \end{aligned} \tag{5.3.16}$$

is plotted against $3 \log b$; K_A^{Bj} is the association constant predicted by expression 5.3.8. If the experimental association constant, $K_A(\text{exp})$, is described by Bjerrum's model, we have

$$\begin{aligned} \log K_A(\text{exp}) - 3 \log a = {} & \log(4\pi N_A) + \log Q(b) \\ & + 3 \log(ba) - 3 \log a \end{aligned} \tag{5.3.17}$$

A plot of $\log K_A(\text{exp})$ against $3 \log(ba)$ should coincide with the curve of log $K_A^\star$ against 3 log b, when its origin is displaced a distance 3 log a on the abscissa and the same distance on the ordinates, i.e., the origin is displaced along the line $y = x$. The displacement allows the evaluation of a for the electrolyte.

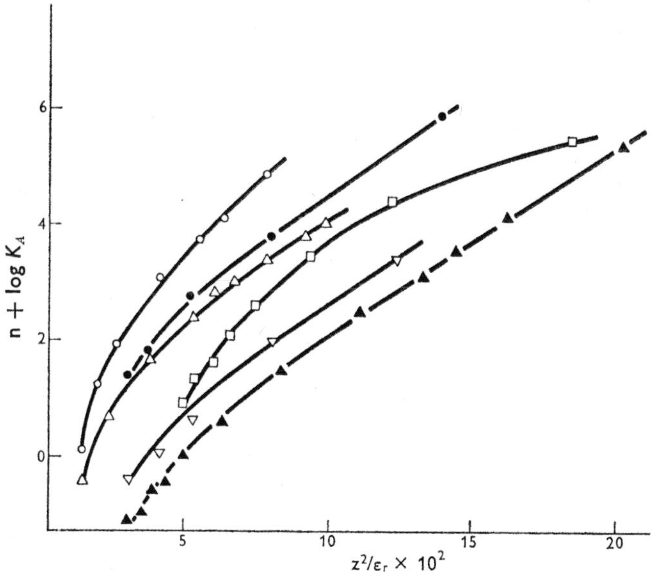

FIG. 5.3.4. Dependence of log K_A on z^2/ε_r for electrolytes in solvent mixtures. ○, CsBr in water–tetrahydrofuran ($n = 1$);[L43] △, CsBr in water–dioxan ($n = 0$);[L43] □, $MgSO_4$ in formamide–dioxan ($n = -1$);[L42] ▲, Bu_4NPi ($n = -2$); ▽, Bu_4N I($n = -1.75$); and ●, Bu_4NNO_3 ($n = 0$) in nitrobenzene–carbon tetrachloride.[L41]

Figure 5.3.5 is a plot of log K_A(exp) against 3 log (ba) for the same solvent mixtures illustrated in Fig. 5.3.4. The plots of log K_A(exp) against 3 log (ba) for each mixture have been displaced along the line $y = x$ so that the best agreement with eqn. 5.3.16 was obtained. The dashed curves in Fig. 5.3.5 correspond to a variation of $\pm 7\%$ in the a value for the electrolytes. It may be concluded that Bjerrum's eqn. 5.3.8 predicts satisfactorily the variation of ionic association with dielectric constant for the cases illustrated in Fig. 5.3.5.

For very low dielectric constants ($\varepsilon_r < 10$) it is sometimes observed that association of ions produces species more complex than ion pairs. Obviously in these circumstances the models of ion association and eqn. 5.3.4 break down. In the next section we shall consider the conductance of solutions in which triple ions are formed.

An interesting case of ion pairing where the Bjerrum and the Fuoss models fail to predict the association found experimentally, is that in which solvent-separated and contact ion pairs may be formed. Böhm and Schulz[L46] observed that the association of $NaBPh_4$ in tetrahydropyran decreases seven times when the temperature goes from 45°C to

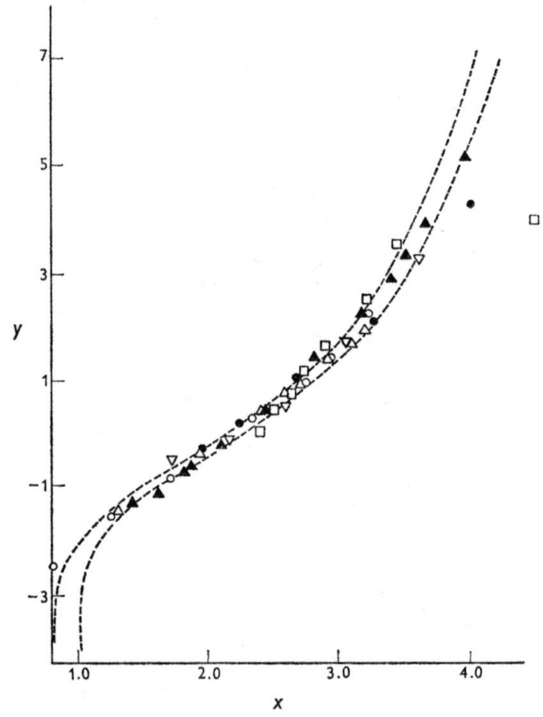

FIG. 5.3.5. Dependence of $y = \log K_A(\text{exp})$ on $x = \log (ba)^3$ for the same electrolytes and solvent mixtures of Fig. 5.3.5. The values found for a are: ○, 3.8; △, 4.7; □, 5.5; ▲, 5.4; ▽, 4.3 and ●, 3.6 Å.

−40°C; however, the product $\varepsilon_r T$ varies less than 1% in this temperature range. Böhm and Schulz assumed that ion pairs can be in contact or separated by solvent molecules according to

$$NaS_n^+ + BPh_4^- \underset{}{\overset{K_{ss}}{\rightleftarrows}} NaS_n \cdot BPh_4 \underset{}{\overset{K_{cont}}{\rightleftarrows}} Na \cdot BPh_4 + nS \quad (5.3.18)$$

Now the experimental association constant is, according to the equilibria 5.3.18, $K_A = K_{ss}(1 + K_{cont})$ and contact ion pairs would be more important at higher temperatures thus producing an increase in the total association. The models of ion association would require a temperature dependent collision distance to explain the behaviour of $NaBPh_4$ in tetrahydropyran.

It is important to point out that the association constants determined from conductance measurements according to eqn. 5.3.4 are independent of the nature of the association process (electrostatic or not), requiring only that association corresponds to pair formation.

Since ionic association is an electrostatic effect for equilibrium properties of electrolyte solutions, it may be included in the Debye–Hückel type of treatment by explicitly retaining further terms in the expansion of the Poisson–Boltzmann relation eqn. 5.2.8.[L47,L48] A similar calculation was attempted for conductance by Fuoss and Onsager.[L49] The mathematical approach and the model employed are similar to those used in their previous calculation,[L6] but they keep explicitly the exp $(-\Phi^0_{ij})$ term in the new calculation. The equation derived for Λ is

$$\Lambda = \Lambda^\circ - S\sqrt{(c)} + E'c \ln c + L(b)c - A(b)c\gamma^2_\pm \Lambda \quad (5.3.19)$$

The coefficient E' in eqn. 5.3.19 is different from E in eqn. 5.3.4 but still not dependent on d. $L(b)$ for low values of b is similar to the J_1 term in the Fuoss and Onsager version of eqn. 5.3.4 and $A(b)$ has the functional dependence on b typical of Bjerrum's association constant (eqn. 5.3.14).[L49b] In spite of the retention of the full Boltzmann exponential, ion association has to be invoked in order to replace the total concentration in eqn. 5.3.19 by αc.[L49b] The resulting equation is less capable of dealing with experimental results than eqn. 5.3.4.[L49b,L50]

5.4 EVALUATION AND INTERPRETATION OF EXPERIMENTAL RESULTS

5.4.1 Completely Dissociated Electrolytes

When $\Phi = (ze\psi)/kT$ is sufficiently small to ensure the applicability of the Debye–Hückel approximation to the electrolyte solution, eqn. 5.2.31 is capable of representing the experimental data very closely up to concentrations such that $\kappa a < 0.2$–0.5.[L51]

By fitting the experimental data to eqn. 5.2.31 it is possible to obtain the values of Λ° and a. Λ° gives information about the mobility of the ions, reflecting solute-solvent interactions. It is particularly useful when used together with transport numbers, enabling the separation of cation and anion contributions to the mobility of the electrolyte (sects. 5.12–5.15). On the other hand the a parameter, which is the distance separating two oppositely charged ions when they collide, furnishes information on the extent to which an ion is able to displace solvent molecules from the solvation sheath of the counterion when they approach each other; i.e., whether solvent-separated or contact ion pairs predominate. It is usual to compare a with the sum of the ionic crystallographic radii.

While this is probably the best reference that can be found, it seems important to remark that ionic radii depend on the environment surrounding the ions, that is, on the force field in which the ions lie. Studies of the ion pairs formed in alkali metal halide vapour[L52a] show that the distance between anion and cation may be 20% smaller than the interionic distance for the same pair in the crystal. Dimers in the vapour phase[L52b] have a distance between anion and cation which is intermediate between that in the monomer and that in the crystal. When an alkali salt melts, the first-neighbour distance may decrease up to 15%.[L52c]

An answer to the problem of what is the distance between colliding ions surrounded by solvent molecules has not yet been found. Extensive studies of the far infrared stretching vibration in electrolyte solutions, reported recently for the first time,[L53] may prove a valuable tool in clarifying this question.

Due to this uncertainty we may conclude that a is an effective collision distance in solution, reflecting the force field in the solution and the average success of an ion in displacing solvent molecules from the immediate surroundings of its ionic partner. Hence only relatively large deviations from crystallographic dimensions or trends in the a values under similar conditions (a series of similar salts in a solvent, or a given salt in various solvents) are meaningful.

In order to fit the experimental data for dissociated electrolytes to eqn. 5.2.31, either a graphical method or a numerical method may be employed. Both consist of a series of successive approximations to get the final values of Λ° and a. It is necessary to have an approximate value of the conductance at infinite dilution Λ_0° in order to calculate the constants S and E in eqn. 5.2.31. This is obtained from the extrapolation of Λ against $c^{\frac{1}{2}}$.

(a) *Graphical Method*

The function Λ' is calculated according to

$$\Lambda' \equiv \Lambda + S\sqrt{(c)} - Ec \ln c = \Lambda^\circ + J_1 c - J_2 c^{\frac{3}{2}} \qquad (5.4.1)$$

employing Λ_0°. A plot of Λ' as a function of c will be linear at low concentration when the J_2 term does not contribute significantly to Λ'. In this way Λ_1° may be obtained as the intercept of the curve. Λ_1° is then used to calculate new values for S and E and the procedure repeated until two successive iterations yield intercepts differing less than the experimental error. At this point eqn. 5.4.1 is rearranged

$$\Lambda'' \equiv (\Lambda' - \Lambda^\circ)/c = J_1 - J_2 c^{\frac{1}{2}} \qquad (5.4.2)$$

and if Λ'' is plotted against the square root of the concentration the intercept gives J_1 and the slope J_2.

It must be remembered that J_1 and J_2 are not independent of each other, both being functions of the solvent properties, of Λ° and of a. From a plot of these coefficients as a function of a according to the expressions in Table 5.2.1 for the various theoretical treatments of conductance, the value of a may be found.

It has been customary to analyse data using eqn. 5.2.31 with $J_2 = 0$ and in this case the function Λ' suffices to obtain both Λ° and a through J_1. However, it has been shown that the omission of the J_2 term leads to systematic deviations between calculated and experimental conductances.[L14,L15,L22]

(b) *Numerical Method*

Computers are now being commonly employed to analyse conductance data and, in this case, the omission of the J_2 term mentioned above is not at all justified since it does not involve additional work.

The computer programs, based on those proposed by Kay[L54], consist of a least squares fitting procedure. Differentiating eqn. 5.2.31 with respect to the adjustable parameters Λ° and a, and approximating the differentials by finite differences, we have

$$\Delta\Lambda = \Delta\Lambda^\circ + \left(\frac{\partial J_1}{\partial a}\, c + \frac{\partial J_2}{\partial a}\, c^{\frac{3}{2}}\right) \Delta a \tag{5.4.3}$$

It has been assumed that none of the coefficients in eqn. 5.2.31 depends on Λ°. This is a simplification justified by the iterative nature of the calculation. The expressions for the derivatives of the J terms with respect to a are calculated from the expressions in Table 5.2.1 according to the theory employed. As in the graphical method, a preliminary Λ_0° value is necessary and also an intelligent guess of the a_0 value must be made. These initial values of the parameters are used to evaluate

$$\Delta\Lambda_i = \Lambda_{\text{exp}} - \Lambda(\text{eqn. } 5.2.31) \tag{5.4.4}$$

and the term in parentheses on the r.h.s. of eqn. 5.4.3, for each experimental concentration (here designed with subindex i). Using the least squares method,[L55] the computer calculates $\Delta\Lambda^\circ$ and Δa; which in turn give a new value for the adjustable parameters $\Lambda_1^\circ = \Lambda_0^\circ + \Delta\Lambda^\circ$ and $a_1 = a_0 + \Delta a$. The calculation is repeated until $\Delta\Lambda^\circ$ and Δa are within the desired precision. The same program is used to evaluate the overall standard deviation and that of the adjusted parameters. Examples of this type of program may be found in several books, for example, ref. L56.

5.4.2 Associated Electrolytes

In this case eqn. 5.3.4 together with 5.3.9 have to be employed. Now the characteristic distance on which the J coefficients in eqn. 5.3.4 depend is d, the closest distance of approach of free ions, in place of the contact distance a. When pairwise association is of electrostatic nature and Bjerrum's model of ionic association is adopted, the value of d is fixed at l and eqn. 5.3.8 may be employed to calculate the contact distance from the experimental value of the association constant. However, the most frequently employed method for evaluating conductance data of associated electrolytes makes use of the Fuoss eqn. 5.3.10 for the association constant, where only ionic species in contact are considered associated and d becomes equal to a. According to eqn. 5.3.10, log K_A should be linear in $1/\varepsilon_r$, while as seen in Fig. 5.3.5, log K_A is usually curved according to the behaviour predicted by eqn. 5.3.8. When conductance data in solvent mixtures are analysed putting $d = a$, it is found that a increases with $1/\varepsilon_r$.[L49b]

Justice[L22] has recently shown very clearly that these inconsistencies largely disappear if Bjerrum's definition of ion pairs is accepted and d is made equal to $l = (ze)^2/2\varepsilon_r kT$ in eqns. 5.3.4 and 5.3.9. Some of the results of Justice[L43] are presented in Table 5.4.1. He has analysed conductance data in solvent mixtures by treating $\Lambda°$, J_1, J_2 and K_A as adjustable parameters in eqn. 5.3.4 and calculated their values for different $d_{\gamma_\pm}$ values employed in the activity coefficient eqn. 5.3.9. The values

Table 5.4.1

Solvent Mixture	ε_r	Salt	l	d_{J_1}*	d_{J_2}†	$d_{\lambda\pm}$‡	Ref.
Tetrahydrofuran-water				23.2	26.9	l	
	12.6	CsBr	22.23	9.9	17.5	4.0	L43
				7.0	14.4	0	
				18.0	19.0	l	
	18.3	CsBr	15.31	11.0	16.0	4.0	L43
				9.0	15.0	0	
Dioxan-water				23.4	26.3	l	
	12.81	CsBr	21.87	11.6	21.0	4.0	L43
				8.7	19.0	0	
				8.9	10.3	l	
	27.3	CsBr	10.26	7.9	10.8	4.0	L43
				7.4	11.3	0	
				21.0	23.5	l	
	12.74	KCl	22.0	17.0	22.0	15.0	
				10.0	18.5	3.14	L22
				8.0	17.0	0	

* Value of d which makes the theoretical J_1 equal to the experimental one.
† Value of d which makes the theoretical J_2 equal to the experimental one.
‡ Value of d used in the activity coefficient equation; the value 0 corresponds to the limiting law.

of J_1 and J_2 which best fitted the experimental data correspond, according to the theoretical treatments, to d_{J_1} and d_{J_2}. These three distances are reasonably close when $d_{\gamma_\pm}$ is made equal to l. The best agreement was found by Justice[L57] when the Fuoss and Hsia equation was used for the J coefficients, e.g., for CsBr in tetrahydrofuran-water ($\varepsilon_r = 12.6$) the best fit of experimental results is obtained with $J_1 = 35{,}390 \pm 629$ and $J_2 = 305{,}000 \pm 16{,}000$ while $J_1(l) = 34{,}760$ and $J_2(l) = 292{,}000$ according to Fuoss and Hsia's equation.[L11]

Setting $d = l$, it is obvious that the value of the distance parameter increases with $1/\varepsilon_r$, as observed. This also explains the finding that conductance data for tetraalkylammonium salts with alkyl groups varying very much in size, give the same d value in a given solvent, e.g., 3.5 ± 0.2 Å in methanol[L58] and acetonitrile,[L59,L60] 4.2 ± 0.2 Å in ethanol,[L61] 5.0 ± 0.5 Å in propanol,[L61] 5.3 ± 0.3 Å in methylethyl ketone,[L62] 6.0 ± 0.5 Å in butanol,[L63] and 6.7 ± 0.2 Å in pentanol;[L63] the values increasing with $1/\varepsilon_r$. This confirms that J_1 and J_2 depend on a distance parameter which varies with the solvent dielectric constant, and not with the size of the ions as a contact distance like a should. Data for associated electrolytes should therefore be treated taking account of the fact that J_1 and J_2 depend on the dielectric constant of the solvent. In Table 5.2.1 the numerical values of the terms Δ_i in eqn. 5.2.34 for d equal to the Bjerrum distance have been included in parentheses following each theoretical expression for the Δ_i. Information about the short-range interaction parameter a can only be obtained from an analysis of the association constant.[L22,L28]

Most of the data for associated electrolytes have been analysed with eqn. 5.3.4, putting $J_2 = 0$ and using the Debye–Hückel limiting law for the activity coefficients. Both these factors are responsible for the small values of d obtained from J_1, and thus it has been interpreted as if it were a contact distance between ions.

Graphical methods have been sometimes employed[L6,L64] to calculate the coefficients of eqn. 5.3.4 but are usually complicated and have to sacrifice the J_2 term. We shall here refer only to the numerical calculation of these coefficients.

From what has been said, eqn. 5.3.4 is clearly a two-parameter equation (Λ° and K_A), since J_1 and J_2 do not depend on the electrolyte. If association is purely electrostatic, the two parameters are finally Λ° and a, the latter being obtained from K_A by means of eqn. 5.3.8. The fact that only two parameters are adjustable in eqn. 5.3.4 is especially important when dealing with slightly associated electrolytes.

For the sake of completeness we describe the numerical method of evaluating conductance data using eqn. 5.3.4 as if it were a three-parameter equation. When d is fixed as suggested here, the procedure is simplified and becomes a special case of the three-parameter method.

Variation of the three parameters, Λ°, d and K_A in eqn. 5.3.4 produces a change $\Delta\Lambda$ in the conductance which, to a first approximation, is

$$\Delta\Lambda = \Delta\Lambda^\circ + \left(\frac{\partial J_1}{\partial d}\, c + \frac{\partial J_2}{\partial d}\, c^{\frac{3}{2}}\right) \Delta d - \gamma_\pm^2 \Lambda \alpha c \Delta K_A \qquad (5.4.5)$$

The coefficients of Δd and ΔK_A depend on all three parameters; however, as explained for dissociated electrolytes it is not necessary to consider explicitly this dependence in eqn. 5.4.5. To solve eqn. 5.4.5 by the least squares method it is necessary to have preliminary values of the three parameters (subscript $_0$). Λ_0° and $K_{A,0}$ may be obtained from a preliminary Shedlovsky plot of the data (see below) and d_0 is guessed or made equal to l. These approximate values of the adjustable parameters are employed to calculate the coefficients in eqn. 5.3.4.

A first approximation to the degree of dissociation at each concentration, $\alpha_{i,0}$ is obtained using the limiting law

$$\alpha_{i,0} = \Lambda_{\text{exp}} \bigg/ \left(\Lambda_0^\circ - S \bigg/ \sqrt{\left(\frac{c_i \Lambda_{\text{exp}}}{\Lambda_0^\circ}\right)}\right) \qquad (5.4.6)$$

(subscript i denotes values corresponding to concentration c_i). Then a new approximation is obtained from

$$\alpha_{i,1} = \Lambda_{\text{exp}}/(\Lambda_0^\circ - S\sqrt{(\alpha_{i,0} c_i)} + E(\alpha_{i,0} c_i) \ln (\alpha_{i,0} c_i) + J_1(\alpha_{i,0} c_i) - J_2(\alpha_{i,0} c_i)^{\frac{3}{2}}) \qquad (5.4.7)$$

and this last procedure is repeated until $|\alpha_{i,k} - \alpha_{i,k-1}|$ is smaller than a certain value, for example 0.0001. Equation 5.3.4 is now employed to obtain $\gamma_\pm$ using d_0 in its denominator; and $\Delta\Lambda_i$ becomes

$$\Delta\Lambda_i = \Lambda_{\text{exp}} - \Lambda(\text{eqn. } 5.3.4) \qquad (5.4.8)$$

Equation 5.4.5 is solved by least squares obtaining $\Delta\Lambda^\circ$, Δd, and ΔK_A. A second step starts by putting

$$\Lambda_1^\circ = \Lambda_0^\circ + \Delta\Lambda^\circ$$
$$d_1 = d_0 + \Delta d$$
$$K_{A,1} = K_{A,0} + \Delta K_A$$

New values of S, E, and of J_1, J_2 and their derivatives are obtained, and the calculation is repeated starting from eqn. 5.4.7 with $\alpha_{i,0}$ equal to $\alpha_i, k - 1$ as found in the previous step. The iteration is continued until $\Delta\Lambda^\circ$ and Δd are smaller than the precision desired.

Often the data at the lowest concentrations involve larger experimental errors because solutions are prepared by dilution of a stock solution or by dissolving previously weighted amounts of solid. Kay suggested[L54] that in this case the data should be given a weight proportional to $c^{\frac{1}{2}}$.

Since each of the theoretical treatments described in sect. 5.2.2 gives different values for J_1 and J_2, it is obvious that the numerical value obtained for K_A will depend to some extent on the particular theory employed. This effect is larger with smaller association constants because then the J terms in eqn. 5.3.4 are large compared to the last term. On the other hand, when K_A is about 10^5 mol^{-1} l, the number of free ions per unit volume is small (for $c = 0.01$ molar, the concentration of free ions is about 4.10^{-4} molar) so that the last term in eqn. 5.3.4 is the dominant one. In this case the theoretical form of the concentration dependence of the conductance is far less important and K_A is little dependent on it. Hence simpler equations are employed to deal with these strongly associated electrolytes.

5.4.3 Strongly Associated Electrolytes

In order to calculate association constants from conductance measurements, Ostwald considered that all the variations of Λ were due to association, making $\Lambda = \alpha\Lambda^\circ$ and neglecting interionic effects which were not known at that date. Thus from eqn. 5.3.2a

$$\frac{1}{\Lambda} = \frac{1}{\Lambda^\circ} + \frac{c\Lambda K_A}{\Lambda^\circ} \qquad (5.4.9)$$

This equation was improved by Fuoss and Kraus[L65] taking account of the interionic effects on conductance through the limiting law. On the other hand, Shedlovsky[L66] used the semiempirical equation,

$$\Lambda = \alpha\Lambda^\circ - \frac{\Lambda}{\Lambda^\circ} S\sqrt{(\alpha c)},$$

to correct eqn. 5.4.9 for interionic effects. Both these methods make use of the variable $z = S\sqrt{(\Lambda c)}\Lambda^{\circ -\frac{3}{2}}$ and of eqn. 5.3.2a for K_A. Equation 5.4.9 becomes

$$\frac{T(z)}{\Lambda} = \frac{1}{\Lambda^\circ} + \frac{K_A}{(\Lambda^\circ)^2}\frac{c\gamma_\pm^2\Lambda}{T(z)} \qquad (5.4.10)$$

where $T(z) \equiv F(z) = 1 - z(1 - z(1 - \ldots)^{-\frac{1}{2}})^{-\frac{1}{2}}$ for the Fuoss and Kraus' method and $1/T(z) \equiv S(z) = 1 + z + z^2/2 + z^3/8 + \ldots$ for the Shedlovsky method. Tables for $F(z)$[L67] and $S(z)$[L68] have been compiled.

A plot of $T(z)/\Lambda$ against $c\gamma_\pm^2\Lambda/T(z)$ should be a straight line having $1/\Lambda^\circ$ for its intercept and $K_A/(\Lambda^\circ)^2$ for its slope. To a first approximation $T(z) = 1 - z$ for both methods. This approximation usually suffices when a first approximate value of Λ° and K_A are desired in order to start the complete numerical calculation described in sect. 5.4.2. When K_A is very large the value of the conductance at infinite dilution obtained by these methods has a rather large uncertainty; hence K_A will be even more uncertain. In these cases Λ° is estimated by a different method, employing Walden's rule or using known values for the ionic conductances at infinite dilution.

Up to this point we have assumed that only pairwise association was present in the solution; however, when $\varepsilon_r < 10$, the equilibria

$$MA + M^+ = M_2A^+$$
$$MA + A^- = MA_2^-$$

have to be considered.[L41,L69] Usually the dissociation constants for both of these equilibria are assumed equal and denoted by K_3. Since in these cases the concentration of free ions is very small, we may start by using Ostwald's dilution law for the binary and ternary equilibria, that is, $\Lambda = \alpha\Lambda^\circ + \alpha_3\lambda_3^\circ$. Hence

$$\frac{\Lambda\sqrt{c}}{(1-\Lambda/\Lambda^\circ)^{\frac{1}{2}}} = \frac{\Lambda^\circ}{\sqrt{K_A}} + \frac{\lambda_3^\circ c}{K_3\sqrt{K_A}}(1-\Lambda/\Lambda^\circ) \qquad (5.4.11)$$

where α_3 is the fraction of triple ions and λ_3° their conductance at infinite dilution. If interionic effects are taken into account, eqn. 5.4.11 becomes[L70]

$$\Lambda\sqrt{(c)}g = \frac{\Lambda^\circ}{\sqrt{K_A}} + \frac{\lambda_3^\circ c}{K_3\sqrt{(K_A)}}(1-\Lambda/\Lambda^\circ) \qquad (5.4.12)$$

with $g = \gamma_\pm/(1-z)(1-\Lambda/\Lambda^\circ)^{\frac{1}{2}}$.

A plot of $\Lambda\sqrt{(c)}g$ against $(1-\Lambda/\Lambda^\circ)c$ should be a straight line having $\Lambda^\circ/\sqrt{K_A}$ for its intercept and a slope equal to $\lambda_3^\circ/K_3\sqrt{K_A}$. In order to obtain K_A and the triple ion dissociation constant K_3, a value of Λ° is calculated by means of Walden's rule and λ_3° is found using the approximation $\Lambda^\circ = 3\lambda_3^\circ$.

Since the effect of ternary association is to remove some nonconducting species MA from the solution and replace them by triple ions which contribute to the conductance, it is often observed that Λ passes through a minimum as c increases.[L69–71] The concentration corresponding to the minimum may be evaluated from eqn. 5.4.11.

5.4.4 Slightly Associated Electrolytes

For electrolytes having $K_A \gtrsim 20\ \text{mol}^{-1}\ \text{l}$, the precise value found from conductance measurements for the association constant is to a certain extent dependent upon the theory used to express J_1 and J_2. A more important difficulty arises many times when eqn. 5.3.4 is employed for slightly associated electrolytes as a three-parameter equation. It is usual to get negative values for K_A and/or d. This is so because for this type of electrolyte, it is difficult to separate the J term from that containing K_A in eqn. 5.3.4. When the association constant is small ($\alpha \simeq 1$) the degree of dissociation, as well as $\gamma_{\pm}$, can be expanded in a power series of the concentration and eqn. 5.3.4 becomes

$$\begin{aligned} \Lambda &\simeq \Lambda^\circ - S\surd(c) + Ec \ln c + J_1'c - J_2'c^{\frac{3}{2}} \\ J_1' &= J_1 - K_A\Lambda^\circ \\ J_2' &= J_2 - 3K_AS/2 + 2AK_A\Lambda^\circ \end{aligned} \tag{5.4.13}$$

where A is the coefficient of the Debye–Hückel limiting law. Equation 5.4.13 immediately shows why it is difficult to separate K_A from J_1 and J_2.

The best results for slightly associated electrolytes are obtained by fixing d, and letting only the association constant and Λ° be adjusted by least squares. When association is assumed to be of electrostatic nature, d is set equal to l.[L72] Sometimes association may be present in a solution where the sum of crystallographic radii of the dissolved ions is larger than l. In this case d can be set equal to the sum of crystal radii.[L24] In both cases this procedure has been shown to yield association constants which can be used in turn to predict thermodynamic properties for these solutions within experimental error.[L24,L71] Slightly associated electrolytes may be treated sometimes as dissociated salts, provided that the resulting a is reasonable despite being smaller than l (cf. KCl in methanol, Table 5.2.2, where $l = 8.75$ Å).

For very precise data it is possible to obtain the three parameters in eqn. 5.4.2.[L14,L51] A series of two-parameter computations are performed for different d values and the deviations, σ_Λ, of the calculated conductances from the experimental ones, are plotted against d. The resulting curve should pass through a minimum yielding the value of d which gives the best agreement with experiment. Figure 5.4.1 illustrates this procedure for the very precise conductance data of CsCl in pure ethanol and two water-ethanol mixtures[L73] using the Fuoss and Hsia expressions for J_1 and J_2. The minima for σ_Λ against d is at 5.3 Å, 9.2 Å and 10.4 Å for the three solvents, which have a Bjerrum distance 1 Å larger than the corresponding value found for d.

For unsymmetrical electrolytes there is no complete treatment of the conductance.* In some cases eqn. 5.2.31 has been employed to evaluate the conductance of unassociated unsymmetrical electrolytes[L74] and the a values obtained are reasonable. When unsymmetrical electrolytes undergo association the ion pairs formed are charged and will contribute to the conductance of the solution. Fuoss and Edelson[L75] derived a semi-empirical equation which may be employed in this case.

* See footnote on p. 533.

When $\kappa a < 0.5$ the treatments of conductance which have been described break down. Conductance of concentrated electrolyte solutions in non-aqueous solvents has been reported[L76,L77] to follow the Wishaw and Stokes equation,[L78] which is essentially the Robinson and Stokes equation 5.2.35 with the conductances being corrected for the increase of viscosity of the solution. Association and conductance in concentrated solutions in various solvents have also been considered by several investigators.[L79–81]

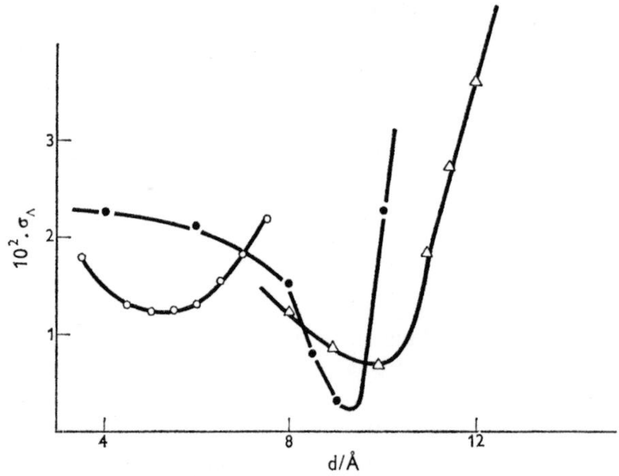

FIG. 5.4.1. Dependence of σ_Λ on the value assumed for *d*. CsCl in ethanol–water mixtures.[L73] ○, 60% ethanol; ●, 93% ethanol, and △, 100% ethanol. The values of K_A corresponding to the minima are respectively 10, 115 and 265 mol^{-1} l.

5.5 SOME EXPERIMENTAL ASPECTS

The method most widely employed to measure the resistance of electrolyte solutions is the a.c. Wheatstone bridge together with subsidiary equipment to amplify and detect the output signal of the bridge. Shedlovsky[L82] has reviewed the experimental techniques and the equipment employed for the measurements.

Stokes[L83] has observed that after a solution has attained thermal equilibrium it is necessary to stir it before a measurement of resistance is made; stirring eliminates the concentration gradient of electrolyte produced by the temperature difference when the solution is first introduced into the thermostat. Prue has reported[L84] variations of the measured resistance with time, probably due to surface conduction on the glass walls of the cell. This 'shaking effect' can be minimised by proper design of the conductance cells and by shaking the solution before each measurement is carried out. Both these effects are avoided by using cells of the type described by Marsh and Stokes,[L85] which allow operation under an adequately controlled inert-gas atmosphere which is also used to stir the solution. The cell is especially well suited for measurements in which the concentration of the solution is varied by addition of a concentrated stock solution under a countercurrent of inert gas.

Hawes and Kay[L73,L86] have designed a salt-dispensing cup for conductance cells which enabled them to carry out measurements in a completely closed system. The concentration of the solution was varied by successive additions of weighted amounts of solid salt into the cell.

In conductance studies of non-aqueous solutions it is important to consider how some impurities in the solvent may affect the conductance of the solution. The all important impurity in non-aqueous work is water. Several authors have confirmed the observations of Hughes and Hartley[L87] that at fixed electrolyte concentration the percentage variation in the conductance upon addition of small amounts of water to the solution is proportional to the water content of the solvents.

Butler, Schiff and Gordon[L88] verified this and also found that for NaCl solutions in methanol the relative decrease of Λ for 1% of added water varies with the concentration of the salt, extrapolating precisely to the percentage decrease of the fluidity of methanol containing 1% of water when the concentration of salt is zero. Barthel and Schwitzgebel[L89] observed the same behaviour when water was added to $NaOCH_3$ in methanol. Thus the reaction $CH_3O^- + H_2O = CH_3OH + OH^-$ seems not to occur in methanol to any appreciable degree. On the other hand CO_2 does modify the conductance of CH_3O^- salts.

Similar effects have been found for nitromethane,[L90] dimethylformamide,[L3] and sulpholane[L91] solutions upon addition of water. The observed conductance change can be essentially explained in terms of the viscosity change introduced in the media by the addition of water and may be considered of non-specific nature. This justifies the use by some workers of hydrates instead of the anhydrous salts, the first being more readily soluble in some non-aqueous solvents.[L92,L93]

Notwithstanding the non-specific role of water in many non-aqueous solutions, in some cases it may have specific effects due to ion-water interactions which depend on the particular nature of the ions and solvent. Thus the conductance of associated $AgNO_3$ and $(CH_3)_2TlI$ in dimethylformamide increase with the water content[L3] in contrast to the decrease found for dissociated salts In methanol and ethanol the conductance of perchloric acid (and presumably of other inorganic acids) decreases significantly upon addition of 0.3% water.[L94] This was attributed to a change in the proton transport mechanism.

D'Aprano and Fuoss[L95] found a very large increase in the conductance of picric acid dissolved in acetonitrile as water was added (Λ

increases ten times for 0.02% of water). Ammonia, which is a likely impurity in acetonitrile, also increases the conductance of picric acid probably due to formation of the ammonium salt which is highly conducting. The abnormal variation of conductance with concentration for transition metal and aluminium salts in *N*-methylacetamide was attributed to the presence of acetates,[L92,L96] a common impurity in this solvent.

It may be concluded that for precise conductance measurements it is important always to verify the effect of water and other impurities likely to be present in the solvent being used, particularly when specific effects may occur as in solutions of acids, bases or transition metal salts.

5.6 REVIEW OF RECENT CONDUCTANCE DATA IN NON-AQUEOUS SOLVENTS

The solvents have been divided in this section into protic and aprotic, according to whether they can form hydrogen bonds with the dissolved ions. In protic solvents hydrogen bonding, especially for small anions, contributes appreciably to the total solvent-ion interaction. In the case of aprotic solvents only ion-dipole and mutual polarisability are of importance in determining the interactions of the solvent with the ions.[L97] The aprotic solvents having $\varepsilon_r \gtrsim 12$ have been considered in a separate group. This way of ordering the solvents will help to make clearer some anomalies observed by conductance measurements in the behaviour of electrolytes in non-aqueous solvents.

5.6.1 Protic Solvents

The conductance of alkali metal (M) halides in methanol,[L20,L88,L98,L99] ethanol,[L100] and propanol;[L101] of $M\text{ClO}_4$ and $M\text{NO}_3$ in methanol,[L102] and of MSCN[L101] in propanol were determined before the extension of the conductance theory. Many of these results have been reanalysed[L3,L54] using the new conductance equations. For these electrolytic solutions, $\Lambda°$ increases with the size of the ions; that is, the solvodynamic units are larger for the smaller ions In methanol, alkali metal salts are essentially dissociated except the nitrates. Kay and Hawes have measured the conductance of CsCl in methanol[L103] and ethanol.[L73] Analysis of the data in the former solvent using Fuoss and Hsia's eqn. 5.3.4, yields for $d = 7.00$ Å, $K_A = 20.8\ \text{mol}^{-1}\,\text{l}$ corresponding to $a = 3.6$ Å. Recent data[L104] for $KClO_4$ and $CsClO_4$ in methanol are at variance with the behaviour described above and indicate both salts are associated. $AgNO_3$ in methanol[L105] and ethanol[L106,L107] is more associated than $M\text{NO}_3$. MPic[L108] and $M\text{BPh}_4$[L109] are dissociated in methanol except when $M = \text{K}$. However, $KBPh_4$ can be considered a dissociated electrolyte with $a = 5.6$ Å, if Fuoss and Hsia's eqn. 5.2.31 is used to evaluate the data. Lithium and caesium dinonylnaphthalene-sulphonates[L110] are probably associated in methanol but definitely so in ethanol and butanol. The association of lithium and ammonium halides in butanol[L111] in the range 0–50°C agrees with Bjerrum's model if $a = 3.7$ Å. The Walden product is almost constant in this range of temperatures.

Alkali metal salts in alcohols show in general a behaviour which may be accounted for by the electrostatic model if the ions are considered solvated. In addition, association is larger for large M^+ indicating that ion pairs are solvent separated. Interesting exceptions are the MRO bases in ROH[L89,L112a] (R = Me or Et). For R = Me the salts were considered dissociated but the contact distances increase from 1.74 Å for Li^+ to 3.13 Å for Cs^+. When R = Et the bases are associated and K_A goes through a minimum at the potassium salt ($K_A = 176$ for Li^+, 36 for K^+, and 65 for Cs^+). Barthel *et al.*[L89,L112a] assumed that large cations form solvent-separated pairs but the small ones polarise the solvent molecules around them such that contact pairs are formed.

$$M^+\cdots\overset{\displaystyle R}{\overset{|}{O}}\!-\!H\cdots\overset{\displaystyle R}{\overset{|}{O^-}} \rightleftarrows M^+\cdots\overset{\displaystyle R}{\overset{|}{O^-}}\cdots\overset{\displaystyle R}{\overset{|}{O}}\!-\!H$$

This assumption is akin to the hypothesis of localised hydrolysis invoked by Robinson and Harned[L112b] to explain the reversal of the order of activity coefficients of alkali metal hydroxides, fluorides, acetates and formates in water.

A large amount of conductance data for substituted ammonium salts in alcohols is avilable.[L20,L58,L60,L61,L63,L80,L108,L113–116] The mobilities and association constants of tetraalkylammonium (R_4N) halides are larger for iodides suggesting solvation of the anions through hydrogen bonds. In general R_4NI are more associated in ROH than in dipolar aprotic solvents of similar ε_r. Furthermore, in alcohols their K_A values go through a minimum for R = Pr. Both these effects are illustrated in Table 5.6.1 which also shows that the effect is especially marked for bulky cations. Kay, Evans and coworkers[L58,L60,L63] have invoked a

Table 5.6.1

Solvent	ε_r	K_A/mol^{-1} l						Ref.
		Me_4NBr	Bu_4NBr	Me_4NI	Bu_4NI	Bu_4NBPh_4	Bu_4NPic	
Methanol	32.7	14	3	18	16	37	7	L58
Acetonitrile	36.0	46	2	19	3	0	4	L59, L117
		Bu_4NCl	Bu_4NBr	Et_4NI	Pr_4NI	Bu_4NI	i-Am_4NI	
Propanol	20.3	149	266	466	391	415	462	L61
Acetone	20.7	602		162	143			L59, L118
		Et_4NBr	Bu_4NBr	Et_4NI	Bu_4NI	$Hept_4NI$		
Butanol	17.5	1330	860	1410	1180	1260		L63
Methylethyl ketone	18.0	958	787	434	382	309		L62

two-step association process similar to the equilibria 5.3.18 to explain these results. However, it is not clear how the large R_4N^+ are able to displace the ROH molecules solvating the halide ions to form contact pairs. The large association of R_4NI in alcohols resembles that found for the same salts in water[L119] attributed to noncoulombic structural effects which may be also present in ROH solutions being superimposed on electrostatic ionic association.

Similar behaviour was found for R_3SI in methanol.[L120] For $Me_nPh_{4-n}PI$ ($n = 1$ to 4) in the same solvent, association increases with n.[L121] The conductance of tetraethanolammonium salts in methanol shows evidence of cation solvation.[L116]

The conductance of acids dissolved in alcohols indicates that the proton has excess mobility[L122–124] due to a proton-jump mechanism of transport. HCl is associated in methanol[L123] ($K_A = 17$ mol^{-1} l) and in ethanol[L124,L125] ($K_A = 48$ mol^{-1} l). Picric acid is strongly associated in both solvents[L126] ($K_A = 3960$ and 5870 mol^{-1} l respectively).

The amides and N-monoalkylamides have very large dielectric constants such that no association would be expected for (1:1) salts ($l = 1.5$ Å in N-monomethylacetamide (NMA) at 32°C and 2.5 Å in formamide (F) at 25°C). However, a large number of uni-univalent salts show negative deviations from the Onsager limiting law in these solvents and also produce a considerable increase in the viscosity of the solution with increasing concentration. Positive deviations were observed for some uni-univalent salts in acetamide[L127] at 94°C. In F (1:1) salts give small positive deviations[L128,L129] with the exceptions of $LiNO_3$, sodium acetate, and Tl^+ salts. Notley and Spiro[L130] reassessed the conductance of salts in F with eqn. 5.2.35 finding an average $a = 1.8$ Å. If the conductances are corrected for the increased viscosity of the solution a becomes 3.8 Å. $\Lambda°$ in F increases with the size of the alkali metal and halide ions, while the opposite holds for R_4N^+. In NMA the conductance of alkali metal halides[L131,L132] varies linearly with $c^{\frac{1}{2}}$ but the slopes are 5–10% more negative than predicted by eqn. 5.2.1. Even when viscosity corrections were applied to Λ no satisfactory interpretation was found for KCl in NMA[L133] and in N-methylpropionamide (NMP).[L134] R_4N salts in NMA give larger negative deviations than alkali metal halides[L132,L135] and there is evidence[L136] that these salts disrupt the structure of the solvent. A similar behaviour was found for (1:1) salts in N-methylformamide (NMF).[L137] The suggestion[L131,L137] that the continuum model fails in these solvents because ions would depolymerise the solvent chains producing a local region of lower dielectric constant, seems unlikely in view of self-diffusion studies in NMA,[L138] of the increase in viscosity produced by increasing electrolyte concentration, and of the fact that polyvalent salts behave according to the continuum model even if conductances are not corrected for viscosity effects.

Alkaline earth halides and perchlorates in NMA[L139] show positive deviations which increase from Ba^{2+} to Mg^{2+} as expected for solvated ions. Johari and Tewari found bivalent sulphates to be dissociated in F[L140] but having small a values, $MgSO_4$[L42] may also be considered slightly associated in F but is completely dissociated in NMF ($a = 3.38$ Å).[L141] Copper *m*-benzenedisulphonate shows positive deviations in NMP.[L93] The conductance of (3:3) salts in F[L142] is in good agreement with the simple electrostatic model, e.g., $LaFe(CN)_6$ and $Co(en)_3Co(CN)_6$ have $K_A = 243$ and 605 respectively, corresponding to $a = 10.08$ and 7.0 Å. The values of $\Lambda°$ and K_A indicate that La^{3+} is a larger species in F than the complex ions $Co(en)^{3+}$ or $Co(NH_3)_6^{3+}$.[L142]

The behaviour found for the conductance of (1:1) electrolytes in these solvents agrees with some evidence of abnormal equilibrium properties.[L143,L144] As was the case for alcohols, it is probable that hydrogen-bonded solvents affect the behaviour of electrolytes in a way not predicted by the continuum model, this effect being more noticeable with smaller electrostatic interaction between ions.

Conductance measurements of acids in NMA[L96,L135] and F[L129] indicate that the protons have no excess mobility. Partially substituted ammonium salts show the same behaviour as R_4N salts in NMA[L145].

In formic acid (1:1) salts were found to be dissociated.[L146] HCl is associated ($K_A = 90$ mol^{-1} l) due to the acidic nature of the solvent, and formate and H^+ ions have excess mobility suggesting a proton-jump transport mechanism for these ions. Association of weak bases in acetic acid has been studied by conductance.[L147]

Brewster, Schmid and Schaap[L148] found that the anomalous conductances reported in ethanolamine[L149] are due to solvent impurities. In this solvent small ions are more solvated as indicated by $\Lambda°$, halides are slightly associated and the nitrates, nitrites and thiocyanates more associated. In ethylene and propylene diamines, Br^- has a lower conductance than I^-[L150] and R_4NI are more associated than KI.[L150,L151] The difference of K_A in both solvents can be accounted for by the change in ε_r.

5.6.2 Dipolar Aprotic Solvents

Dipolar aprotic solvents behave as differentiating solvents with the polarisability of the anions largely determining their interactions with the solvent molecules.[L97,L152] Ionic conductances at infinite dilution show[L3,L97] that the mobility of Br^- is roughly double that of K^+ supporting the idea that small halide ions are poorly solvated in these solvents.

The large number of conductance measurements for R_4N salts in dipolar aprotic solvents are in fair agreement with the continuum model, with their association being larger for small R. R_4N salts are essentially dissociated in nitromethane (NM)[L90,L117,L153,L154] with the exception of the bromides and chlorides of R = Me and Et. The Stokes radius of i-Am_4NBPh_4 is 8.3 Å in NM,[L117] almost the same as in acetonitrile (AN). In nitrobenzene (NB) tetraalkylammonium salts are associated[L41,L81,L155-162] with the association being larger for the smaller cations. They are, however, more associated than in other dipolar aprotic solvents of similar ε_r,[L59,L159] suggesting NB is a poor solvating medium.

Alkene carbonates are solvents with large dielectric constants. In ethylene carbonate,[L163] R_4N salts show small negative deviations but the restricted concentration range employed for the conductance measurements does not allow a more detailed analysis of the data. In propylene carbonate (PC) they are dissociated.[L21,L158]

In sulpholane ($TMSO_2$), R_4NClO_4[L164] and Et_4NI[L28] are slightly associated (K_A = 3–6 mol^{-1} l according to the Pitts eqn. 5.3.4 with d = 5.0 Å). In dimethylsulpholane[L165], $PhMe_3NI$ is associated but Ph_4AsI is not. Recent measurements[L166] of conductance of R_4N salts in dimethylsulphoxide (DMSO) show the salts to be essentially dissociated. Unfortunately the study was restricted to very dilute solutions so that not much weight can be put on the a values obtained.

Evans, Zawoyski and Kay[L59] analysed data for R_4N salts in acetone (AC)[L167-169] with the Fuoss–Onsager equation. They found K_A decreases with cation size, and for the anions, association decreases in the order $Bu_4NBr(K_A = 264) > I^-(143) \simeq NO_3^- > ClO_4^-(80) > Pic^-(17)$. This agrees with data for methylethylketone.[L62,L170] The fact that association of Bu_4NClO_4[L171] in AC, benzonitrile, and methylethylketone corresponds to a = 4.85 Å for the three solvents, indicates formation of contact ion pairs. Tetraalkylammonium halides in dimethylformamide (DMF)[L172] have small association constants when the data are evaluated with Shedlovsky's eqn. 5.4.10. When the data for Me_4NPic in DMF[L173] is assessed with Fuoss and Hsia's eqn. 5.2.31, a is 6.0 Å.

AN is probably the most commonly employed dipolar aprotic solvent medium for studies of electrolytic conductance. Picrates and halides,[L50,L59,L60,L174] hexafluorophosphates,[L175] perchlorates,[L117,L176] nitrates[L177] and substituted borates[L117,L178,L179] of R_4N^+ cations have been studied conductimetrically in AN. Older conductance data for R_4N salts in AN[L180-183] have been re-evaluated.[L59] These results show that the association constants for salts of a given R_4N^+ cation vary with change of anion as in AC, hexafluorophosphates and substituted borates being practically dissociated in AN. No association is found for solutions of Am_4N^+ salts and some Bu_4N^+ salts in AN. The behaviour of tetraalkylammonium salts in adiponitrile[L184] is similar to that in AN. Et_4NClO_4 in valeronitrile[L185] has an association corresponding to a = 3.3 Å, and for Bu_4NBF_4 in phenylacetonitrile,[L186] a = 5 Å independent of the temperature.

While $Me_nPh_{4-n}PI$ in AN[L121] are associated to a similar extent as Me_4NI, R_3SI[L120] are more associated in this solvent due to the fact that the positive charge in R_3S^+ is off-centre and more readily available to the approaching I^-.

In dipolar aprotic solvents partially substituted ammonium cations, $R_nH_{4-n}N^+$, can form internal hydrogen bonds with small anions,[L97] thus increasing their association. PhH_3NPic is more associated than R_4NPic in AN[L182] and its low Λ° suggests that two dissociation equilibria are coupled[L187]

$$\begin{aligned} PhH_3NPic &\rightleftarrows PhH_3N^+ + Pic^- \\ PhH_3NPic &\rightleftarrows PhH_2N + HPic \end{aligned} \qquad (5.6.1)$$

In the same solvent, $[Ph_4P_2N(NH_2)_2]Cl$ has $K_A = 3100$,[L188] internal hydrogen bonding being favoured in this case by the poor solvation of Cl^- by AN molecules. A good example of internal hydrogen-bonding enhanced association is provided by $(EtOH)_4N^+$ salts in AN.[L116] The I^- has $K_A = 142\ mol^{-1}\ l$ compared to 12.4 in methanol, the BPh_4^- is dissociated in AN. Evidence exists that the Br^- has $K_A = 10^3$ in acetonitrile,[L116] a fact explained by the stronger hydrogen bonding between Br^- and the cation. The conductance of Et_3HNBr, EtH_3NBr and Et_3HNPic in DMF[L173] when evaluated by means of the Fuoss–Hsia equation yields $K_A = 301$ and $134\ mol^{-1}\ l$ for the first two while the third is completely dissociated with $a = 7.3$ Å.

The conductance of acids in dipolar aprotic solvents indicates that H^+ has normal mobility in these media. Janz and Danyluk[L189] have reviewed the conductance data for acids in dipolar aprotic solvents up to 1959. Variations of the conductance of acids with time have been reported for some solvents. The conductances of aged solutions of HBr[L190] in benzonitrile and AN and of HCl in the former solvent[L191] show the acids are very associated. However, the Λ° values obtained are much smaller than expected, e.g., $\Lambda^\circ_{HBr} = 11.8$,[L190] $\Lambda^\circ_{HCl} = 1.5$[L191] compared to $\Lambda^\circ_{HClO_4} = 51.5\ \Omega^{-1}\ cm^2\ mol^{-1}$[L192] in benzonitrile. In AC and nitriles, $HClO_4$ is more associated than Bu_4NClO_4,[L192] the addition of small amounts of water to the acid in AN reduces its association to half of its value due to the efficiency of H_2O in solvating the ions CH_3CNH^+ and ClO_4^-. In DMSO, HCl has $\Lambda^\circ = 38._7\ \Omega^{-1}\ cm^2\ mol^{-1}$ and $K_A = 1.1_5 \times 10^2\ mol^{-1}\ l$.[L193] In DMF, HPic[L173] is more dissociated than HBr, HCl having an even larger association ($K_A(HCl) = 3.5 \times 10^3$ at 20°C[L194]). In AN, HPic is strongly associated[L95] suggesting that AN is a poorer solvating medium than DMF and DMSO.

Yeager and Kratochvil[L195,L196] have recently reported conductance data for Cu(I), Ag(I) and Tl(I) salts in AN which shows that the salts of the two first cations are essentially dissociated, while Tl(I) salts are associated similarly to KSCN. The Λ° values indicate that Cu(I) is a very large solvodynamic unit, probably in the form $Cu(AN)_4^+$. The association of $AgNO_3$ and $AgClO_4$ in AC and cyclohexanone[L106] can be reasonably accounted for by electrostatic interactions, but $AgNO_3$ in benzonitrile[L106] and $AgNO_3$ and Me_2TlI in DMF[L3] require too small a values (1.6, 1.3 and 1.1 Å) if association is considered purely electrostatic.

Prue and Sherrington[L3] measured the conductance of alkali metal (M) salts in DMF and reassessed existing data for some of these salts in DMSO,[L197] dimethylacetamide (DMA),[L198] and dimethylpropionamide (DMP).[L199] Alkali metal salts are dissociated in these solvents[L3,L77,L197] with the exception of nitrates and LiCl in DMF. The sum of the Stokes' radii for alkali metal salts corresponds to large solvodynamic units, e.g., for KI it is 5.3 Å in DMF, 5.6 Å in DMA, 5.7 Å in DMP, and 5.1 Å in DMSO. MClO$_4$ in $TMSO_2$[L28,L200] and AN[L86,L104] are slightly associated while MBPh$_4$ in AN are dissociated[L86] except for $CsBPh_4$. For all these electrolytic solutions the mobilities (excepting Li^+ in $TMSO_2$) and K_A increase with the size of M^+. However, it has been reported[L201] that for MSCN in $TMSO_2$ association is larger for M = Na.

Irregularities occur in the behaviour of salts of small M^+ ions, particularly Li^+, in dipolar aprotic solvents.[L28] The association of lithium halides in $TMSO_2$[L28] and AC[L118] is 2000 times greater for the Cl^- than for I^-. LiCl is 30 times more associated than LiBr in PC.[L21] The association constants for these solutions are given in Table 5.6.1 where a_{Bj} is the contact distance according to Bjerrum's expression (eqn. 5.3.8) for K_A. Abnormally large K_A values have been reported also for Li p-toluenesulphonate in AC,[L118] for $LiClO_3$ in AN ($K_A = 401$ compared to 5 mol^{-1} l in methanol),[L202] for MPic in AC and NB[L59] (the effect being larger in NB which is a poorer electron donor). On the other hand NaPic and KPic in DMF[L173] are dissociated ($a = 4.6$ and 5.0 Å) according to the Fuoss–Hsia conductance equation. For these picrate solutions, association increases from K^+ to Li^+. The same trend was reported for K_A of alkali metal salicylates[L203] in acetophenone and methylethylketone, opposite to the trend for iodides in these solvents. If Bjerrum's model for ion association is applied, the resulting a values for the salicylates are in the range 1.4–1.9 Å in methylethylketone and 1.5–2.5 Å in acetophenone, compared to a = 3.6–3.9 and 6.4–7.6 Å for the iodides. The formation of contact ion pairs by the salicylates is supported by the distinct decrease in association when 1% of water is added to the solvent.[L203]

Stabilisation of ion pairs by the solvent dipoles, invoked by Hyne[L204] to explain the behaviour of Bu_4NBr in NB, does not explain the anomalies observed for some alkali metal salts in dipolar aprotic solvents. This hypothesis cannot account for the large increase in K_A when going from

LiI to LiCl. Parker[L97] has shown that if the ability of dipolar aprotic solvents to solvate cations varies in the order DMSO = DMA > DMF > AC = $TMSO_2$ > AN = NM > benzonitrile = NB, many properties of electrolytes in these solvents can be explained. This ordering coincides with that found by Drago and co-workers[L205] for the electron donating power of dipolar aprotic solvents. It is interesting to note that the anomalies in conductance behaviour have been encountered most frequently in the solvents having lower cation solvating power.

The simple electrostatic model of charges in a continuum was extended[L28] to take account of dielectric saturation, a phenomenon which is only significant for the association of small ions. To correct the predicted association for dielectric saturation, a distance-dependent dielectric coefficient ε_r' must replace the static dielectric constant of the solvent. Eqn. 5.3.8 for K_A then becomes

$$K_A = 4\pi N_A \int_a^d \exp\left(-U(r)/kT\right) r^2 \,\mathrm{d}r \tag{5.6.2}$$

and

$$U(r) = \int_\infty^r \frac{e^2}{\varepsilon_r' r^2}\,\mathrm{d}r$$

The expression derived by Booth[L206] for the variation of ε_r' with the intensity of the electric field is easy to handle when the dielectric medium obeys Onsager's model[L207] for a dipolar liquid. $TMSO_2$, AC, NM and PC follow the behaviour predicted by Onsager's model,[L28,L208] hence for these solvents eqn. 5.6.2 can be employed. The variation of $\varepsilon_r'/\varepsilon_r$ with the distance separating two oppositely charged ions has been plotted in Fig. 5.6.1 for $TMSO_2$ at 30°C according to Booth's equation. Figure 5.6.1 also illustrates the sharp variation of $U(r)/U_0(r)$ with r, where $U_0(r)$ is the electrostatic energy of two ions separated by the distance r when there is no dielectric saturation. It may be seen that $U(r)/U_0(r)$ increases sharply for small variations of r, only when the two ions are near each other and thus reflects qualitatively the experimentally observed variations of association. Putting $d = l$ in eqn. 5.6.2 and using the experimental values of K_A, the contact distance a_{corr} is found by numerical integration. Its values are given in Table 5.6.1 for some electrolytic solutions showing abnormally large association. In the last column of Table 5.6.1, a_{crys} the sum of the crystallographic radii is given. A considerable improvement results when dielectric saturation is taken into account. It is noteworthy that for solutions where no anomalies are reported, e.g., LiI in $TMSO_2$ and in AC, Table 5.6.1 shows that the correction for dielectric saturation is negligible.

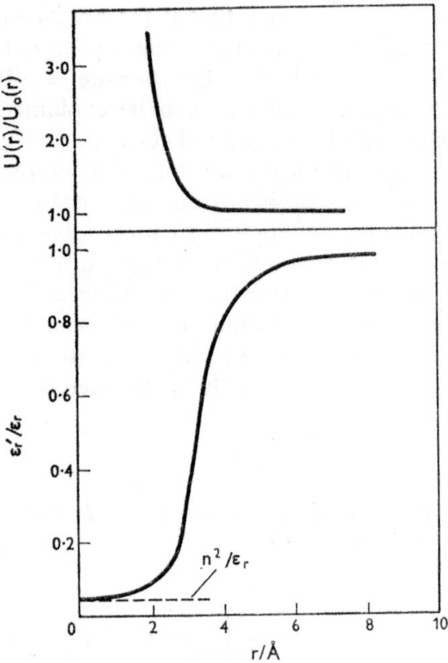

FIG. 5.6.1. Effect of dielectric saturation on the dielectric coefficient and on the electrostatic energy at distance r from a central ion in sulpholane at 30°C; n denotes the refractive index of the solvent.

Table 5.6.1

Solvent	Electrolyte	K_A/mol^{-1} l	a_{Bj}/Å	a_{corr}/Å	a_{crys}/Å	Ref.
$TMSO_2$	LiCl	13,860	0.70	2.2	2.46	L28
	LiBr	278	1.02	2.4	2.55	L28
	LiI	5.6	4.01	4.0	2.76	L28
NM	LiI	240	1.35	2.4	2.46	L209
	KI	40	2.10	2.6	3.49	L209
	$NaClO_4$	120	1.52	2.4	3.3	L209
AC	LiCl	3×10^5	1.4	2.5	2.46	L118
	LiBr	4.6×10^3	2.0	2.8	2.55	L118
	LiI	145	4.6	4.6	2.76	L118
	$LiClO_4$	5.3×10^3	1.9	2.8	3.0	L210
	$NaClO_4$	4.3×10^3	2.1	2.9	3.3	L210
PC	LiCl	557	0.75	1.9	2.46	L21
	LiBr	19	1.03	2.1	2.55	L21

5.6.3 Aprotic Solvents of Low Dielectric Constant

We shall consider in this section those solvents having $\varepsilon_r < 12$. At this low dielectric constant, electrostatic ionic interactions are very large and often triple ion formation is encountered. The conductance has been observed to pass through a minimum with concentration, indicating the formation of triple ions, for R_4N-*p*-toluenesulphonates in triaryl phosphides,[L211] for LiCl and bivalent chlorides in tetrahydrofuran (THF),[L212] and for $MeEt_3NI$ in CH_2Cl_2.[L71,L213]

In *o*-dichlorobenzene (*o*DCB), Bu_3NHPic[L214] has a very low $\Lambda°$ probably because it dissociates according to the equilibria 5.6.1. The bromide and iodide in *o*DCB and the picrate in chlorobenzene form triple ions.[L214]

When only pair formation is observed in these solvents it is usually not necessary to evaluate the data with eqn. 5.3.4. The simpler Shedlovsky or Fuoss expressions (eqn. 5.4.10) may be employed since the concentration of free ions is very small. However, Treiner and Justice[L215] have recently used the complete eqn. 5.3.4 with $d = l$ to evaluate the conductance of Bu_4NClO_4 in THF ($\varepsilon_r = 7.39$) finding the data can be fitted with extreme precision up to $c = 1.12 \times 10^{-4}$ molar; the value of K_A corresponds to $a = 4.4$ Å according to eqn. 5.3.8.

The conductance of Bu_4NPic[L216] in chlorobenzene, *m*DCB, and *o*DCB indicates that the association varies according to the values of the dielectric constant. The Bjerrum a values of R_4NPic in *o*DCB[L217] are lower than expected. Inami, Bodenseh and Ramsey[L171] observed that the association of Bu_4NClO_4 in ethylene chloride and 1,2 dichloropropane is smaller than that predicted from their dielectric constants and the association found for this salt in other solvents. They pointed out that spectroscopic and dielectric constant measurements give evidence that, in the presence of an electric field, both solvents (which are in isomeric equilibria between a *gauche*(polar) form and *trans*(non-polar) form) show an increase in the amount of polar isomer. They suggested that the microscopic ε_r is larger in the neighbourhood of the ions than in the pure solvent, thus decreasing the electrostatic interaction between ions. $MeBu_3NClO_4$ is also less associated in ethylene chloride than in *o*DCB[L218] having a similar ε_r. The addition of triphenylphosphine oxide produces a decrease in $\Lambda°$ and K_A, due to specific solvation of the ions. An association smaller than expected in ethylene chloride has been also reported for other substituted ammonium salts.[L219,L220] The association and $\Lambda°$ values of R_4NPic in *o*DCB[L219] decrease with the size of the cations. The contact distance of Bu_4NBr in ethylidene chloride[L221] agrees with that found for this salt in other solvents.

Evidence for the existence of solvent-separated and contact ion pairs in solutions of alkali metal salts in cyclic ethers has been reported. The conductances of $MBPh_4$ and R_4NBPh_4[L222,L223] in THF and dimethoxyethane (DME) indicate that the latter is a better solvating medium. In THF the association of $MBPh_4$ is similar to that of R_4NBPh_4 except for the caesium salt which is much more associated. Stokes' radii are 4.3 and 2.4 Å for M = Na and Cs respectively suggesting $CsBPh_4$ forms contact ionic pairs while the other M^+ only form solvent-separated pairs. In DME the Stokes' radii for $CsBPh_4$ and $NaBPh_4$ are very similar and the former is only marginally more associated than $NaBPh_4$. The contact distance for solvent-separated pairs is large, e.g., a = 7.5 and 7.3 Å for $NaBPh_4$ in THF and DME.[L222] Spectroscopic and conductance measurements of alkali metal salts of fluorenyl radicals[L224,L225] also indicate that DME is a better solvating medium than THF and that solvent-separated and contact pairs are present in the solutions.

Conductance and NMR measurements agree in suggesting that THF solvates the sodium ion specifically.[L226] The conductance of $NaAlBu_4$ in cyclohexane increases abruptly upon addition of THF to the solution until the ratio salt/THF is 1. This is due to a decrease in association because $NaTHF^+$ is formed. As the concentration of THF increases, Λ decreases slightly up to a value of 1/4 for the ratio salt/THF due to formation of $Na(THF)_4^+$. From this point onwards, Λ increases according to the variation of ε_r of the solvent mixture. No evidence of solvation was found when benzene or toluene was added to $NaAlBu_4$ in cyclohexane.[L227] For $NaAlEt_4$ in toluene, Λ increases upon addition of diethylether (Et_2O) until $NaEt_2O^+$ is formed.[L228]

5.6.4 Solvent Mixtures

In binary mixtures of MeOH, AN and NM having an almost constant ε_r, i-Am_3BuN^+ salts[L154] are dissociated; but the variation of the Walden product with solvent composition suggests specific ion-solvent interactions. D'Aprano and Fuoss[L229] observed specific effects on the conductance of the Pic^-, Br^- and BPh_4^- of Bu_4N^+ in isodielectric mixtures of dioxan with water, MeOH, AN and p-nitroaniline (PNA). K_A and the Walden products of the salts are reduced when PNA is present in the solvent mixture, the effect being largest for Br^- and least for BPh_4^-, due to anion-PNA interactions which are stronger for the bromide. HPic would interact with anions even more strongly than PNA.[L95] These interactions seem to depend not on the dipole moment of the added substance but primarily on the facility with which it hydrogen bonds to the anions.[L230]

In MeOH-AN mixtures, $CsClO_4$ and $KClO_4$[L104] become fully dissociated when the medium has 60% of AN, in spite of the small and monotonous change in ε_r with solvent composition. For Bu_4NBr in binary mixtures of NB, EtOH and MeOH with CCl_4[L161], association is somewhat larger than predicted by the electrostatic model. The excess association decreases in the order NB > EtOH > MeOH which agrees with the solvating power of these solvents. In NB-MeOH mixtures with practically constant ε_r, anomalies are observed in the association of Bu_4NBr.[L161,L162,L204] Specific interactions have been reported between Ag^+ in AC and cycloalkenes,[L231] and for the same ion in MeOH when nitroalkanes are added.[L105] The association constants of Me_4NNO_3 and Bu_4NNO_3 in AN-CCl_4,[L177] vary according to the predictions of the electrostatic model with contact distances of 4.01 and 5.41 Å respectively. In AN-dioxan and MeOH-dioxan mixtures, $LiClO_3$[L202] has $a = 3.70$ and 5.04 Å compared to 6.8 Å in the water-dioxan mixture, showing that contact pairs are favoured in dipolar aprotic solvents and solvent-separated pairs in the protic solvents. Agreement with the electrostatic model has been reported[L215] for the association of Bu_4NClO_4 and Bu_4NBr in THF with less than 15% of water ($a = 4.4$ and 4.9 Å respectively), for CsCl and KCl in EtOH-H_2O,[L73] for Bu_4NBPh_4[L158] ($a = 7$ Å) in NB-CCl_4, for R_4NBPh_4 in AN-CCl_4[L178,L232] ($a = 6.96$ Å for R = Me and 8.36 Å for R = Bu), and for polycyanocarbon salts[L159] in NB-CCl_4 and in AC-CCl_4. Alkali metal halides in mixtures of NMA-DMF[L233] provide no evidence of specific interactions. Et_4NPic in MeOH-EtOH[L234] does not show the anomalies encountered in ROH-H_2O mixtures.

As mentioned when discussing Fig. 5.3.4, it is common to find a curvature in the plots of log K_A against $1/\varepsilon_r$. However, this is not sufficient evidence for the breakdown of the simple electrostatic model of ionic association, since the curvature agrees qualitatively with that predicted by Bjerrum's eqn. 5.3.8 (cf. Fig. 5.3.5). Et_4NPic and Bu_4NPic in NB-anisole[L157] have Bjerrum contact distances which change with ε_r and which were attributed to specific solvation. For $MgSO_4$, log K_A varies in the same way in F-dioxan[L42] as in H_2O-dioxan mixtures,[L235] but this is not the case in F-AC mixtures.[L236] Johari[L237] has recently reported the conductances of $LaFe(CN)_6 \cdot 4H_2O$ in F-AC and F-dioxan mixtures; association in the second mixture resembles that in water-dioxan and AC-water mixtures.[L238] For F-AC the observed association corresponds to $a = 10.5$ Å; no evidence of La^{3+}-solvent interactions is found in this solvent mixture.

5.6.5 Effects of Pressure on the Electrolytic Conductance

Measurements of conductance at different pressures and temperatures may be used to test the predictions of the continuum model, especially as regards association, and to study the mechanism of ionic transport.

Skinner and Fuoss[L239] measured the conductance of i-Am_4NPic in diethylether and benzene up to 5×10^6 Nm^{-2}. Ion pairs and triplets are present because ε_r is small. Since ε_r increases with pressure, the variation of Λ may be expected to depend on the ratio of both associated species.

If pairs dominate Λ will increase with pressure. On the other hand, if triple ions predominate Λ decreases, the actual variation depending on whether the concentration of the solution is smaller or bigger than that corresponding to the minimum conductance (cf. sect. 5.4.3 end). According to the predictions, Λ is found to increase in benzene and to decrease in diethylether where $a = 4.9$ Å for the ion pairs. This behaviour was also encountered for Bu_4NPic in toluene[L240] where the variation of Λ with pressure depends on the concentration of the solution.

The effect of pressure on the conductance of salts and HBr in MeOH has been studied.[L241] The mobility of H^+ decreases less with pressure than that of the other ions. For HPic solutions in MeOH,[L241] Λ increases with the pressure because the enhanced dissociation of the acid dominates the variation of mobilities. In MeOH,[L242] Bu_4NBPh_4 has $\Lambda^\circ\eta$ constant up to 5×10^6 Nm^{-2}, NaBr shows an increase in the Walden product with pressure, and the R_4N bromides are intermediate cases. Variations of K_A values with pressure were also observed to depend on the particular salt.

Brummer and Hills derived[L243] relations between the activation parameters and the variations of conductance with temperature and pressure according to an activated transport mechanism. They have emphasised the importance of the isochoric activation energy (E_V) in characterising the energetics of activated transport. Barreira and Hills[L244] found that for R_4NPic in NB, E_V is similar for all R and also to the isochoric activation energy for viscous flow of the pure solvent. The activation volume ($\Delta V^{\ddagger}$) is also practically the same for all the salts. If the rate determining step in ionic transport is displacement of solvent molecules in the neighbourhood of the ions as suggested by the evidence mentioned above, the different ionic mobilities are due to the fact that the effective movement of ions caused by the solvent displacement depends on their size.[L244]

Similar behaviour was found for salts in MeOH and in NB.[L245] In DMF at various pressures the conductance of picrates[L246] shows that only for large cations is $\Delta V^{\ddagger}$ independent of cationic size. For them solvent displacement determines the transport mechanism of ions.

The variations of conductance with temperature and pressure for R_4NI in AC[L247] show $\Delta V^{\ddagger}$ for transport is almost independent of ionic size. It is found to be similar for conductance of dilute solutions, solvent self-diffusion and viscosity for a number of solvents. There is evidence[L247] that ions are not followed by their solvation sheaths when they move. From these studies it is suggested that the free volume of the solvent is an important parameter in determining the mobility of ions.

Acknowledgments

The author is grateful to Prof. Gordon Atkinson for many fruitful discussions and also for his hospitality at the time this manuscript was prepared.

Support of this work by the Office of Saline Water, U.S. Department of the Interior, under Grant 14–01–0001–1656 and by the National Aeronautics and Space Administration under Grant NsG–398 to the Computer Science Center of the University of Maryland is well appreciated.

REFERENCES

L1 P. Debye and E. Hückel, *Physik. Z.*, **24**, 185 (1923)
L2 L. Onsager, *Physik. Z.*, **28**, 277 (1927)
L3 J. E. Prue and P. J. Sherrington, *Trans. Faraday Soc.*, **57**, 1795 (1961)
L4 E. Pitts, *Proc. Roy. Soc.*, **217A**, 43 (1953)
L5 R. M. Fuoss and L. Onsager, *J. Phys. Chem.*, **61**, 668 (1957)
L6 R. M. Fuoss and F. Accascina, Electrolytic Conductance, Interscience, New York (1959)
L7 H. S. Harned and B. B. Owen, The Physical Chemistry of Electrolytic Solutions, p. 48, Reinhold, New York (1958)
L8 E. Pitts, B. E. Tabor and J. Daly, *Trans. Faraday Soc.*, **65**, 849 (1969)
L9 M. Leist, *Z. Phys. Chem. (Leipzig)*, **205**, 16 (1955)
L10 L. Onsager and R. M. Fuoss, *J. Phys. Chem.*, **36**, 2689 (1932)
L11 R. Fernández-Prini, *Trans. Faraday Soc.*, **65**, 3311 (1969)
L12 L. Onsager and R. M. Fuoss, *J. Phys. Chem.*, **62**, 1339 (1958)
L13 J. P. Valleau, *J. Phys. Chem.*, **69**, 1745 (1965)
L14 R. M. Fuoss and K-L. Hsia, *Proc. Nat. Acad. Sci.*, **57**, 1550 (1967)
L15 R. Fernández-Prini and J. E. Prue, *Z. Phys. Chem. (Leipzig)*, **228**, 373 (1965)
L16 D. Kremp, *Ann. Physik.*, **18**, 237, 278 (1966)
L17 D. Kremp, W. D. Kraeft and W. Ebeling, *Ann. Physik.*, **18**, 246 (1966)
L18 W. D. Kraeft, *Z. Phys. Chem. (Leipzig)*, **237**, 289 (1968)
L19 R. M. Fuoss, *J. Chim. Phys.*, **66**, 1191 (1969)
L20 E. C. Evers and A. G. Knox, *J. Amer. Chem. Soc.*, **73**, 1739 (1951)
L21 L. M. Mukherjee and D. P. Boden, *J. Phys. Chem.*, **73**, 3965 (1969)
L22 J-C. Justice, *J. Chim. Phys.*, **65**, 353 (1968)
L23 R. A. Robinson and R. H. Stokes, *J. Amer. Chem. Soc.*, **76**, 1991 (1954)
L24 R. Fernández-Prini, *Trans. Faraday Soc.*, **64**, 2146 (1968)
L25 C. Treiner and R. M. Fuoss, *J. Phys. Chem.*, **69**, 2576 (1965)
L26 R. H. Stokes, *in* The Structure of Electrolytic Solutions, p. 298 (W. J. Hamer, ed.), Wiley, New York (1959)
L27 R. A. Robinson and R. H. Stokes, Electrolyte Solutions, Revised Second Edition, p. 231, Butterworths (1965)
L28 R. Fernández-Prini and J. E. Prue, *Trans. Faraday Soc.*, **62**, 1257 (1966)
L29 N. Bjerrum, *Kgl. Danske. Videnskab. Selskab. Math-fys. Medd.*, **7**, 9 (1926)
L30 See ref. L7, p. 171; and ref. L27, p. 549
L31 E. A. Guggenheim, *Disc. Faraday Soc.*, **24**, 53 (1957)
L32 W. G. Davies, R. J. Otter and J. E. Prue, *Disc. Faraday Soc.*, **24**, 103 (1957)
L33 See Chemical Physics of Ionic Solutions, p. 282 ff. (B. E. Conway and R. G. Barradas, eds.), Wiley, New York (1966)
L34 L. Onsager, *Chem. Rev.*, **13**, 73 (1933)
L35 R. M. Fuoss, *J. Amer. Chem. Soc.*, **80**, 5059 (1958)
L36 G. Scatchard, B. Vonnegut and D. W. Beaumont, *J. Chem. Phys.*, **33**, 1292 (1960)
L37 R. M. Fuoss and C. A. Kraus, *J. Amer. Chem. Soc.*, **55**, 1019 (1933)
L38 G. J. Janz and M. J. Tait, *Canad. J. Chem.*, **45**, 1101 (1967)
L39 E. Andalaft, R. P. T. Tomkins and G. J. Janz, *Canad. J. Chem.*, **46**, 2959 (1968)

L40 W. R. Gilkerson, *J. Chem. Phys.*, **25**, 1199 (1956)
L41 E. Hirsch and R. M. Fuoss, *J. Amer. Chem. Soc.*, **82**, 1018 (1960)
L42 P. H. Tewari and G. P. Johari, *J. Phys. Chem.*, **69**, 2857 (1965)
L43 J-C. Justice, R. Bury and C. Treiner, *J. Chim. Phys.*, **65**, 1708 (1968)
L44 J. Hallada and G. Atkinson, *J. Amer. Chem. Soc.*, **83**, 3759 (1961)
L45 G. Atkinson and H. Tsubota, *J. Amer. Chem. Soc.*, **88**, 3901 (1966)
L46 L. L. Böhm and G. V. Schulz, *Ber. Bunsenges. Phys. Chem.*, **73**, 260 (1969)
L47 T. H. Gronwall, V. K. La Mer and K. Sandved, *Physik. Z.*, **29**, 358 (1927)
L48 E. A. Guggenheim, *Trans. Faraday Soc.*, **56**, 1152 (1960)
L49 (*a*) R. M. Fuoss and L. Onsager, *J. Phys. Chem.*, **66**, 1722 (1962); **67**, 621 (1963); **67**, 628 (1963); **68**, 1 (1964)
(*b*) R. M. Fuoss, L. Onsager and J. E. Skinner, *J. Phys. Chem.*, **69**, 2581 (1965)
L50 C. Treiner and R. M. Fuoss, *Z. Phys. Chem.* (*Leipzig*), **228**, 343 (1965)
L51 F. M. Hanna, A. D. Pethybridge and J. E. Prue, *Electrochim. Acta*, **16**, 677 (1971)
L52 (*a*) A. Honig, M. Mandel, M. L. Stitch and C. H. Townes, *Phys. Rev.*, **96**, 629 (1954)
(*b*) T. A. Milne and D. Cubicciotti, *J. Chem. Phys.*, **29**, 846 (1958)
(*c*) E. A. Ukshe, *Russ. Chem. Rev.*, **34**, 141 (1965)
L53 M. J. French and J. L. Wood, *J. Chem. Phys.*, **49**, 2358 (1968)
L54 R. L. Kay, *J. Amer. Chem. Soc.*, **82**, 2099 (1960)
L55 E. Whittaker and G. Robinson, The Calculus of Observation, p. 209, Blackie, London (1944)
L56 P. A. D. de Maine and R. D. Seawright, Digital Computer Programs for Physical Chemistry, Vols. I and II, Macmillan, New York (1963 and 1965)
L57 J-C. Justice, private communication
L58 R. L. Kay, C. Zawoyski and D. F. Evans, *J. Phys. Chem.*, **69**, 4208 (1965)
L59 D. F. Evans, C. Zawoyski and R. L. Kay, *J. Phys. Chem.*, **69**, 3878 (1965)
L60 R. L. Kay, D. F. Evans and G. P. Cunningham, *J. Phys. Chem.*, **73**, 3322 (1969)
L61 D. F. Evans and P. Gardam, *J. Phys. Chem.*, **72**, 3281 (1968)
L62 S. R. C. Hughes and D. H. Price, *J. Chem. Soc.*, **A**, 1093 (1967)
L63 D. F. Evans and P. Gardam, *J. Phys. Chem.*, **73**, 158 (1969)
L64 R. M. Fuoss, *J. Amer. Chem. Soc.*, **79**, 3301 (1957)
L65 R. M. Fuoss and C. A. Kraus, *J. Amer. Chem. Soc.*, **55**, 476 (1933)
L66 T. Shedlovsky, *J. Franklin Inst.*, **225**, 739 (1938)
L67 R. M. Fuoss, *J. Amer. Chem. Soc.*, **57**, 488 (1935)
L68 H. M. Daggett, *J. Amer. Chem. Soc.*, **73**, 4977 (1951)
L69 See ref. L6, p. 249
L70 R. M. Fuoss and C. A. Kraus, *J. Amer. Chem. Soc.*, **55**, 2387 (1933)
L71 J. H. Beard and P. H. Plesch, *J. Chem. Soc.*, 4075 (1964)
L72 R. Bury, M-C. Justice and J-C. Justice, *Compt. Rend.*, **268**, 670 (1969)
L73 J. L. Hawes and R. L. Kay, *J. Phys. Chem.*, **69**, 2420 (1965)
L74 B. R. Staples and G. Atkinson, *J. Phys. Chem.*, **71**, 667 (1967)
L75 R. M. Fuoss and D. Edelson, *J. Amer. Chem. Soc.*, **73**, 269 (1951)
L76 G. J. Janz, A. E. Marcinkowsky and I. Ahmad, *J. Electrochem. Soc.*, **112**, 104 (1965); R. P. T. Tomkins, E. Andalaft and G. J. Janz, *Trans. Faraday Soc.*, **65**, 1906 (1969)
L77 J. S. Dunnett and R. P. H. Gasser, *Trans. Faraday Soc.*, **61**, 922 (1965); M. D. Archer and R. P. H. Gasser, *Trans. Faraday Soc.*, **62**, 3451 (1966)
L78 B. F. Wishaw and R. H. Stokes, *J. Amer. Chem. Soc.*, **76**, 2065 (1954)
L79 L. Kenausis, E. C. Evers and C. A. Kraus, *Proc. Nat. Acad. Sci.*, **49**, 141 (1963); L. E. Strong and C. A. Kraus, *J. Amer. Chem. Soc.*, **72**, 166 (1950); C. A. Kraus, *J. Phys. Chem.*, **60**, 129 (1956)
L80 R. P. Seward, *J. Amer. Chem. Soc.*, **73**, 515 (1951)

L81 F. R. Longo, J. D. Kerstetter, T. F. Kumonsinski and E. C. Evers, *J. Phys. Chem.*, **70,** 431 (1966)
L82 T. Shedlovsky *in* Physical Methods of Organic Chemistry, Part IV *in* Technique of Organic Chemistry, p. 3011 (A. Weissberger, ed.), Interscience, New York (1960)
L83 R. H. Stokes, *J. Phys. Chem.*, **65,** 1277 (1961)
L84 J. E. Prue, *J. Phys. Chem.*, **67,** 1152 (1963)
L85 K. N. Marsh and R. H. Stokes, *Aust. J. Chem.*, **17,** 740 (1964)
L86 R. L. Kay, B. J. Hales and G. P. Cunningham, *J. Phys. Chem.*, **71,** 3925 (1967)
L87 O. L. Hughes and H. Hartley, *Phil. Mag.*, **15,** 610 (1933)
L88 J. P. Butler, H. I. Shiff and A. R. Gordon, *J. Chem. Phys.*, **19,** 752 (1951)
L89 (*a*) J. Barthel and G. Schwitzgebel, *Z. Phys. Chem. (Frankfurt)*, **54,** 181 (1967)
(*b*) J. Barthel, *Angew. Chem. Int. Ed.*, **7,** 260 (1968)
L90 A. K. R. Unni, L. Elias and H. I. Schiff, *J. Phys. Chem.*, **67,** 1216 (1963)
L91 R. Fernández-Prini and J. E. Prue, unpublished results
L92 L. R. Dawson, J. W. Vaughn, G. R. Lester, M. E. Pruitt and P. G. Sears, *J. Phys. Chem.*, **67,** 278 (1963)
L93 T. B. Hoover, *J. Phys. Chem.*, **68,** 3003 (1964)
L94 N. Goldenberg and E. S. Amis, *Z. Phys. Chem. (Frankfurt)*, **31,** 145 (1962)
L95 A. D'Aprano and R. M. Fuoss, *J. Phys. Chem.*, **73,** 223 (1969)
L96 L. R. Dawson, J. W. Vaughn, M. E. Pruitt and H. C. Eckstrom, *J. Phys. Chem.*, **66,** 2684 (1962)
L97 A. J. Parker, *Quart. Rev.*, **16,** 162 (1962)
L98 R. E. Jarvis, D. R. Muir and A. R. Gordon, *J. Amer. Chem. Soc.*, **75,** 2855 (1953)
L99 J. E. Frazer and H. Hartley, *Proc. Roy. Soc.*, **109A,** 351 (1925)
L100 J. R. Graham, G. S. Kell and A. R. Gordon, *J. Amer. Chem. Soc.*, **79,** 2352 (1957)
L101 G. A. Gover and P. G. Sears, *J. Phys. Chem.*, **60,** 330 (1956)
L102 E. D. Copley and H. Hartley, *J. Chem. Soc.*, 2488 (1930)
L103 R. L. Kay and J. L. Hawes, *J. Phys. Chem.*, **69,** 2787 (1965)
L104 F. Conti and G. Pistoia, *J. Phys. Chem.*, **72,** 2245 (1968)
L105 R. E. Busby and V. S. Griffiths, *J. Chem. Soc.*, 902 (1963)
L106 V. S. Griffiths, K. S. Lawrence and M. L. Pearce, *J. Chem. Soc.*, 3998 (1958)
L107 G. D. Parfitt and A. L. Smith, *Trans. Faraday Soc.*, **59,** 257 (1963)
L108 M. A. Coplan and R. M. Fuoss, *J. Phys. Chem.*, **68,** 1177 (1964)
L109 R. W. Kunze and R. M. Fuoss, *J. Phys. Chem.*, **67,** 385 (1963)
L110 R. C. Little and C. R. Singleterry, *J. Phys. Chem.*, **68,** 2709 (1964)
L111 H. K. Venkatasetty and G. H. Brown, *J. Phys. Chem.*, **66,** 2075 (1962)
L112 (*a*) J. Barthel, G. Schwitzgebel and R. Wachter, *Z. Phys. Chem. (Frankfurt)*, **55,** 33 (1967)
(*b*) R. A. Robinson and H. S. Harned, *Chem. Rev.*, **28,** 419 (1941)
L113 F. Accascina and S. Petrucci, *Ric. Sci. Suppl.*, **29,** 1383 (1959); F. Accascina and L. Antonucci, *Ric. Sci. Suppl.*, **29,** 1391 (1959)
L114 R. Whorton and E. S. Amis, *Z. Phys. Chem. (Frankfurt)*, **17,** 300 (1958)
L115 H. K. Venkatasetty and G. H. Brown, *J. Phys. Chem.*, **67,** 954 (1963)
L116 G. P. Cunningham, D. F. Evans and R. L. Kay, *J. Phys. Chem.*, **70,** 3998 (1966)
L117 J. F. Coetzee and G. P. Cunningham, *J. Amer. Chem. Soc.*, **87,** 2529 (1965)
L118 L. G. Savedoff, *J. Amer. Chem. Soc.*, **88,** 664 (1966)
L119 R. L. Kay and D. F. Evans, *J. Phys. Chem.*, **70,** 366 (1966)
L120 D. F. Evans and T. L. Broadwater, *J. Phys. Chem.*, **72,** 1037 (1968)
L121 A. Höniger and H. Schindlbauer, *Ber. Bunsenges. Phys. Chem.*, **69,** 138 (1965)
L122 I. G. Murgulescu, F. Barbulescu and A. Greff, *Rev. Roum. Chim.*, **10,** 387 (1965)

L123 T. Shedlovsky *in* The Structure of Electrolytic Solutions, p. 268 (W. J. Hamer, ed.), Wiley, New York (1959)
L124 H. O. Spivey and T. Shedlovsky, *J. Phys. Chem.*, **71,** 2165 (1967)
L125 I. I. Bezman and F. H. Verhoek, *J. Amer. Chem. Soc.*, **67,** 1330 (1945)
L126 A. D'Aprano and R. M. Fuoss, *J. Phys. Chem.*, **73,** 400 (1969)
L127 G. Jander and G. Winkler, *J. Inorg. Nucl. Chem.*, **9,** 39 (1959)
L128 P. H. Tewari and G. P. Johari, *J. Phys. Chem.*, **67,** 512 (1963)
L129 L. R. Dawson, T. M. Newell and W. J. McCreary, *J. Amer. Chem. Soc.*, **76,** 6024 (1954)
L130 J. M. Notley and M. Spiro, *J. Phys. Chem.*, **70,** 1502 (1966)
L131 L. R. Dawson, P. G. Sears and R. H. Graves, *J. Amer. Chem. Soc.*, **77,** 1986 (1955)
L132 C. M. French and K. H. Glover, *Trans. Faraday Soc.*, **51,** 1427 (1955)
L133 G. Kortüm and H. Quabeck, *Ber. Bunsenges. Phys. Chem.*, **72,** 53 (1968)
L134 T. B. Hoover, *J. Phys. Chem.*, **68,** 876 (1964)
L135 L. R. Dawson, E. D. Wilhoit, R. R. Holmes and P. G. Sears, *J. Amer. Chem. Soc.*, **79,** 3004 (1957)
L136 R. D. Singh, R. P. Rastogi and R. Gopal, *Canad. J. Chem.*, **46,** 3525 (1968)
L137 C. M. French and K. H. Glover, *Trans. Faraday Soc.*, **51,** 1418 (1955)
L138 W. D. Williams, J. A. Ellard and L. R. Dawson, *J. Amer. Chem. Soc.*, **79,** 4652 (1957)
L139 L. R. Dawson, G. R. Lester and P. G. Sears, *J. Amer. Chem. Soc.*, **80,** 4233 (1958)
L140 G. P. Johari and P. H. Tewari, *J. Phys. Chem.*, **69,** 696 (1965)
L141 G. P. Johari and P. H. Tewari, *J. Phys. Chem.*, **69,** 3167 (1965)
L142 G. P. Johari and P. H. Tewari, *J. Phys. Chem.*, **69,** 2862 (1965)
L143 R. P. Held and C. M. Criss, *J. Phys. Chem.*, **69,** 2611 (1965)
L144 O. D. Bonner, S. J. Kim and A. L. Torres, *J. Phys. Chem.*, **73,** 1968 (1969)
L145 L. R. Dawson, E. D. Wilhoit and P. G. Sears, *J. Amer. Chem. Soc.*, **78,** 1569 (1956)
L146 T. C. Wehman and A. I. Popov, *J. Phys. Chem.*, **72,** 4031 (1968)
L147 A. M. Shkodin and L. P. Sadovnichaya, *Russ. J. Phys. Chem.*, **36,** 990 (1962)
L148 P. W. Brewster, F. C. Schmidt and W. B. Schaap, *J. Amer. Chem. Soc.*, **81,** 5532 (1959)
L149 W. T. Briscoe and T. P. Dirkse, *J. Phys. Chem.*, **41,** 388 (1940)
L150 A. M. Harstein and S. Windwer, *J. Phys. Chem.*, **73,** 1549 (1969)
L151 G. W. A. Fowles and W. R. McGregor, *J. Phys. Chem.*, **68,** 1342 (1964)
L152 G. Choux and R. L. Benoit, *J. Amer. Chem. Soc.*, **91,** 6221 (1969)
L153 R. L. Kay, S. C. Blum and H. I. Schiff, *J. Phys. Chem.*, **67,** 1223 (1963)
L154 M. A. Coplan and R. M. Fuoss, *J. Phys. Chem.*, **68,** 1181 (1964)
L155 E. G. Taylor and C. A. Kraus, *J. Amer. Chem. Soc.*, **69,** 1731 (1947)
L156 C. R. Witschonke and C. A. Kraus, *J. Amer. Chem. Soc.*, **69,** 2472 (1947)
L157 A. L. Powell and A. E. Martell, *J. Amer. Chem. Soc.*, **79,** 2118 (1957)
L158 R. M. Fuoss and E. Hirsch, *J. Amer. Chem. Soc.*, **82,** 1018 (1960)
L159 R. H. Boyd, *J. Phys. Chem.*. **65,** 1834 (1961)
L160 W. Reed D. W. Secret, A. C. Thompson and P. A. Yeats, *Canad. J. Chem.*, **47,** 1725 (1969)
L161 H. Sadek and R. M. Fuoss, *J. Amer. Chem. Soc.*, **81,** 4507 (1959)
L162 R. L. Kay and D. F. Evans, *J. Amer. Chem. Soc.*, **86,** 2748 (1964)
L163 R. F. Kempa and W. H. Lee, *J. Chem. Soc.*, 100 (1961)
L164 M. Della Monica and U. Lamanna, *J. Phys. Chem.*, **72,** 4329 (1968)
L165 J. Eliassaf, R. M. Fuoss and J. E. Lind, *J. Phys. Chem.*, **67,** 1724 (1963)
L166 D. E. Arrington and E. Griswold, *J. Phys. Chem.*, **74,** 123 (1970)
L167 M. B. Reynolds and C. A. Kraus, *J. Amer. Chem. Soc.*, **70,** 1701 (1948)

L168 M. J. McDowell and C. A. Kraus, *J. Amer. Chem. Soc.*, **73,** 3293 (1951)
L169 P. Walden, H. Ulich and G. Busch, *Z. Phys. Chem.*, **121A,** 429 (1926)
L170 S. C. R. Hughes and S. H. White, *J. Chem. Soc.*, **A,** 1216 (1966)
L171 Y. H. Inami, H. K. Bodenseh and J. B. Ramsey, *J. Amer. Chem. Soc.*, **83,** 4745 (1961)
L172 P. G. Sears, E. D. Wilhoit and L. R. Dawson, *J. Phys. Chem.*, **59,** 373 (1955)
L173 P. G. Sears, R. K. Wolford and L. R. Dawson, *J. Electrochem. Soc.*, **103,** 633 (1956)
L174 A. C. Harkness and H. M. Daggett, *Canad. J. Chem.*, **43,** 1215 (1965)
L175 J. Eliassaf, R. M. Fuoss and J. E. Lind, *J. Phys. Chem.*, **67,** 1941 (1963)
L176 C. H. Springer, J. F. Coetzee and R. L. Kay, *J. Phys. Chem.*, **73,** 471 (1969)
L177 D. S. Berns and R. M. Fuoss, *J. Amer. Chem. Soc.*, **83,** 1321 (1961)
L178 D. S. Berns and R. M. Fuoss, *J. Amer. Chem. Soc.*, **82,** 5585 (1960)
L179 A. M. Brown and R. M. Fuoss, *J. Phys. Chem.*, **64,** 1341 (1960)
L180 P. Walden and E. J. Birr, *Z. Phys. Chem.*, **144A,** 269 (1929)
L181 G. Kortüm, S. D. Gokhale and H. Wilski, *Z. Phys. Chem. (Frankfurt)*, **4,** 86 (1955)
L182 C. M. French and D. F. Muggleton, *J. Chem. Soc.*, 2131 (1957)
L183 A. I. Popov and N. E. Skelly, *J. Amer. Chem. Soc.*, **76,** 5309 (1954); A. I. Popov, R. H. Rygg and N. E. Skelly, *J. Amer. Chem. Soc.*, **78,** 5740 (1956)
L184 P. G. Sears, J. A. Caruso and I. A. Popov, *J. Phys. Chem.*, **71,** 905 (1967)
L185 J. J. Banewicz, J. A. Maguire and P. S. Shih, *J. Phys. Chem.*, **72,** 1960 (1968)
L186 E. J. Del Rosario and J. E. Lind, *J. Phys. Chem.*, **70,** 2876 (1966)
L187 M. A. Elliott and R. M. Fuoss, *J. Amer. Chem. Soc.*, **61,** 294 (1939)
L188 I. Y. Ahmed and C. D. Schmulbach, *J. Phys. Chem.*, **71,** 2358 (1967)
L189 G. J. Janz and S. S. Danyluk, *Chem. Rev.*, **60,** 209 (1960)
L190 G. J. Janz and I. Ahmad, *Electrochim. Acta*, **9,** 1539 (1964)
L191 G. J. Janz, I. Ahmad and H. V. Venkatasetty, *J. Phys. Chem.*, **68,** 889 (1964)
L192 J. F. Coetzee and D. K. McGuire, *J. Phys. Chem.*, **67,** 1810 (1963)
L193 J. A. Bolzán and A. J. Arvia, *Electrochim. Acta*, **15,** 39 (1970)
L194 A. B. Thomas and E. G. Rochow, *J. Amer. Chem. Soc.*, **79,** 1843 (1957)
L195 H. L. Yeager and B. Kratochvil, *J. Phys. Chem.*, **73,** 1963 (1969)
L196 H. L. Yeager and B. Kratochvil, *J. Phys. Chem.*, **74,** 963 (1970)
L197 P. G. Sears, G. R. Lester and L. R. Dawson, *J. Phys. Chem.*, **60,** 1433 (1956)
L198 G. R. Lester, C. P. Groves and P. G. Sears, *J. Phys. Chem.*, **60,** 1076 (1956)
L199 E. D. Wilhoit and P. G. Sears, *Trans. Kentucky Acad. Sci.*, **17,** 123 (1956)
L200 M. Della Monica, U. Lamanna and L. Jannelli, *Gazz. Chim.*, **97,** 367 (1967)
L201 M. Della Monica and U. Lamanna, *Gazz. Chim.*, **98,** 256 (1968)
L202 A. D'Aprano and R. Triolo, *J. Phys. Chem.*, **71,** 3474 (1967)
L203 S. R. C. Hughes, *J. Chem. Soc.*, 634 (1957)
L204 J. B. Hyne, *J. Amer. Chem. Soc.*, **85,** 304 (1963)
L205 R. S. Drago, B. Wayland and R. L. Carson, *J. Amer. Chem. Soc.*, **85,** 3125 (1963); R. S. Drago and B. Wayland, *J. Amer. Chem. Soc.*, **87,** 3571 (1965)
L206 F. Booth, *J. Chem. Phys.*, **19,** 391 (1951)
L207 L. Onsager, *J. Amer. Chem. Soc.*, **58,** 1486 (1936)
L208 L. Simeral and R. L. Amey, *J. Phys. Chem.*, **74,** 1443 (1970)
L209 C. P. Wright, D. M. Murray-Rust and H. Hartley, *J. Chem. Soc.*, 199 (1931)
L210 F. Accascina and S. Schiavo, *Ann. Chim. (Italy)*, **43,** 695 (1953)
L211 C. M. French and R. C. B. Tomlinson, *J. Chem. Soc.*, 311 (1961)
L212 W. Strohmeier, A. E-S. Mahgoub and F. Gernert, *Z. Elektrochem.*, **65,** 85 (1961)
L213 J. H. Beard and P. H. Plesch, *J. Chem. Soc.*, 4879 (1964)
L214 E. K. Ralph and W. R. Gilkerson, *J. Amer. Chem. Soc.*, **86,** 4783 (1964)
L215 C. Treiner and J-C. Justice, *Compt. Rend.*, **269,** 1364 (1969)

L216 P. H. Flaherty and K. H. Stern, *J. Amer. Chem. Soc.*, **80,** 1034 (1958)
L217 H. L. Curry and W. R. Gilkerson, *J. Amer. Chem. Soc.*, **79,** 4021 (1957)
L218 W. R. Gilkerson and J. B. Ezell, *J. Amer. Chem. Soc.*, **87,** 3812 (1965)
L219 F. Accascina, E. L. Swarts, P. L. Mercer and C. A. Kraus, *Proc. Nat. Acad. Sci.*, **39,** 917 (1953)
L220 J. J. Zwolenik and R. M. Fuoss, *J. Phys. Chem.*, **68,** 434, 903 (1964); K. H. Stern and A. E. Martell, *J. Amer. Chem. Soc.*, **77,** 1983 (1955)
L221 H. K. Bosenseh and J. B. Ramsey, *J. Phys. Chem.*, **69,** 543 (1965)
L222 C. Carvajal, K. J. Tölle, J. Smid and M. Szwarc, *J. Amer. Chem. Soc.*, **87,** 5548 (1965)
L223 D. N. Bhattacharyya, C. L. Lee, J. Smid and M. Szwarc, *J. Phys. Chem.*, **69,** 608 (1965)
L224 T. Ellingsen and J. Smid, *J. Phys. Chem.*, **73,** 2712 (1969)
L225 T. E. Hogen-Esch and J. Smid, *J. Amer. Chem. Soc.*, **88,** 318 (1966)
L226 C. N. Hammonds and M. C. Day, *J. Phys. Chem.*, **73,** 1151 (1969)
L227 C. N. Hammonds, T. D. Westmoreland and M. C. Day, *J. Phys. Chem.*, **73,** 4347 (1969)
L228 M. C. Day, H. M. Barnes and A. J. Cox, *J. Phys. Chem.*, **68,** 2595 (1964)
L229 A. D'Aprano and R. M. Fuoss, *J. Phys. Chem.*, **67,** 1704, 1722, 1871 (1963)
L230 C. Treiner, M. Quintin and R. M. Fuoss, *J. Chim. Phys.*, **63,** 320 (1966)
L231 V. S. Griffiths and M. L. Pearce, *J. Chem. Soc.*, 1557 (1958)
L232 F. Accascina, S. Petrucci and R. M. Fuoss, *J. Amer. Chem. Soc.*, **81,** 1301 (1959)
L233 L. R. Dawson and W. W. Wharton, *J. Electrochem. Soc.*, **107,** 710 (1960)
L234 N. G. Foster and E. S. Amis, *Z. Phys. Chem. (Frankfurt)*, **7,** 360 (1956); R. Whorton and E. S. Amis, *Z. Phys. Chem. (Frankfurt)*, **8,** 10 (1956)
L235 J. C. James, *J. Chem. Soc.*, 2925 (1951)
L236 G. P. Johari and P. H. Tewari, *J. Amer. Chem. Soc.*, **87,** 4691 (1965)
L237 G. P. Johari, *J. Phys. Chem.*, **74,** 934 (1970)
L238 J. C. James, *J. Chem. Soc.*, 1094 (1950)
L239 J. F. Skinner and R. M. Fuoss, *J. Phys. Chem.*, **69,** 1437 (1965)
L240 C. M. Apt, F. F. Margosian, I. Simon, J. H. Vreeland and R. M. Fuoss, *J. Phys. Chem.*, **66,** 1210 (1962)
L241 W. Strauss, *Aust. J. Chem.*, **10,** 277 (1957)
L242 J. F. Skinner and R. M. Fuoss, *J. Phys. Chem.*, **70,** 1426 (1966)
L243 S. B. Brummer and G. J. Hills, *Trans. Faraday Soc.*, **57,** 1816 (1961)
L244 F. Barreira and G. J. Hills, *Trans. Faraday Soc.*, **64,** 1359 (1968)
L245 S. B. Brummer and G. J. Hills, *Trans. Faraday Soc.*, **57,** 1823 (1961)
L246 S. B. Brummer, *J. Chem. Phys.*, **42,** 1636 (1965)
L247 W. A. Adams and K. J. Laidler, *Canad. J. Chem.*, **46,** 1977, 1989, 2005 (1968).

APPENDIX 5.1

The tables in this Appendix contain a critical selection of the available conductance data in non-aqueous solvents. When more than one set of precise data existed for a given electrolyte solution, that believed to be the best was included in the tables.

The symbol Δc stands for the concentration range employed in the analysis of conductance data in pure solvents; in some cases, especially when the conductance results have been re-evaluated, this information is lacking. The values of the parameters which appear in parenthesis have been fixed before the data was fitted. In the column headed *a*, *d* values in square brackets correspond to the J_1 coefficient. The triple-ion dissociation constant K_3 is reported in the same column as K_A.

The methods of analysis of the data are indicated in the following fashion. (a) Equations of the form 5.2.31 or 5.3.4 used with $J_2 = 0$: Fuoss–Accascina (FA)[1], Fuoss–Onsager (FO)[2]. (b) Same type of eqns. but including the J_2 term: Fuoss–Accascina–Berns (FAB)[3], Pitts (P)[4], Fuoss–Hsia (FH)[5]; the latter two correspond to the *J* coefficients in Table 5.2.1. (c) Equation 5.3.19: Fuoss–Onsager–Skinner (FOS)[6]. (d) Full eqn. of type 5.2.30: Fuoss–Onsager (FO-full)[2]. (e) Simpler eqn. 5.4.10 for strongly associated electrolytes: Shedlovsky (S)[7], Fuoss (F)[8]. (f) Triple-ion association, eqn. 5.4.12: Fuoss–Kraus (FK)[9]. (g) When any of these procedures has been modified in a particular manner by the authors the symbol (*) appears. (h) Whenever various runs for the same electrolyte solution were evaluated separately, only parameters for one run are reported in the tables and this is indicated by the superscript #. (i) The symbol η implies that a viscosity correction has been applied to the data in the form given by eqn. 5.2.37.

The references quoted in the last column correspond to the authors who evaluated the parameters here reported. An exception being those data analysed for this contribution, this is indicated by the symbol (+) following the reference number; references are to be found at the end of this Appendix.

The following abbreviations have been employed: Pic, picrate; Ac, acetate; Ph, phenyl; EtOH, C_2H_5OH; en, ethylene diamine; Tolsul, *p*-toluenesulphonate; Sal, salicylate; TCVal, tricyanovinyl alcoholate; PCP, pentacyanopropenide; *bis*TCVam, *bis*(tricyanovinyl)amine; HCHT, hexacyanoheptatrienide. Me, Et, Pr, Bu, Am, Hex, Hept and Oct denote normal alkyl radicals from C_1 to C_8; with a preceding *i* it denotes the corresponding *iso* alkyl radical.

Table 5.1.1

Conductance Parameters for Electrolytes in Protic Solvents at 25°C

Solvent	Electrolyte	$10^4 \cdot \Delta c$/ mol l^{-1}	$\Lambda°$/ Ω^{-1} cm^2 mol^{-1}	a, d/Å	K_A/mol^{-1} l	Method of Analysis	Ref.
Methanol	LiCl	$\kappa a < 0.2$	92.0_5	3.7_3		FA	10
	NaCl	0.4–40.0	97.23	4.0_4		FH	5
	KCl	0.7–50.0	104.8_5	3.4_0		FH	5
	RbCl	$\kappa a < 0.2$	108.2_8	2.4		FA	10
	CsCl	7.0–49.0	113.26_4	(7.0)	20_8	FH	11+
	NaBr	$\kappa a < 0.2$	101.6_4	3.7_9		FA	10
	KBr	$\kappa a < 0.2$	108.88	3.45		FA	10
	KI	$\kappa a < 0.2$	115.22	3.78		FA	10
	$LiClO_4$		110.6	3.3		FO	12
	$NaClO_4$		116.2	2.2		FO	12
	$KClO_4$	4.0–13.0	123.0_6	2.5	11	FA	13
	$CsClO_4$	4.0–24.0	131.5	2.9	33	FA	13
	$AgClO_4$		121.1_5	3.6		FO	12
	$LiClO_3$	4.0–22.0	100.9_7	[1410]	5	FA	14
	$LiNO_3$	$\kappa a < 0.2$	100.21	5_2	10	FA	10
	$NaNO_3$	$\kappa a < 0.2$	106.2_9	4_9	19	FA	10
	KNO_3	$\kappa a < 0.2$	113.8_0	7	39	FA	10
	$RbNO_3$	$\kappa a < 0.2$	117.21	2_2	18	FA	10
	$AgNO_3$	0.3–17.0	111.00	5.2	67_8	FO	15
	NaPic	4.0–22.0	92.0_6	[1101]		FA#	16
	KPic	4.0–43.0	99.2_7	[1238]	12	FA#	16
	HPic	5.0–26.0	156.3		3.96×10^3	F	17
	HAc		(185.5)		4.22×10^9	S*	18
	$NaBPh_4$	1.0–50.0	81.6_3	7.4		FH	19+
	$KBPh_4$	3.0–16.0	88.6_4	5.6		FH	19+

Methanol	$LiOCH_3$	3.0–12.0	92.23	1.7_4		FA	20
	$NaOCH_3$	3.0–14.0	98.2_7	2.2_3		FA	20
	$KOCH_3$	4.0–11.0	105.48	3.1_2		FA	20
	$RbOCH_3$	5.0–15.0	109.34	3.2_0		FA	20
	$CsOCH_3$	1.0–11.0	114.11	3.1_3		FA	20
	Me_4NCl	10.0–90.0	120.8_2	3.3	7	FA, η	21
	Bu_4NCl	9.0–65.0	91.3_8	3.9		FA, η	21
	Me_4NBr	5.0–80.0	125.1_6	3.5	14	FA, η	21
	Et_4NBr	3.0–56.0	116.95	3.8	10	FA, η	21
	Pr_4NBr	3.0–50.0	102.55	3.7	6	FA, η	21
	Bu_4NBr	3.0–50.0	95.39	3.6	3	FA, η	21
	i-Am_3BuNBr	5.0–54.0	92.66	3.38		FA, η	22
	Am_4NBr	3.0–50.0	91.36	3.0_5	3_0	FA, η	22
	$(EtOH)_4NBr$	7.0–61.0	94.0_8	4.0_5	16	FA, η #	23
	Me_4NI	6.0–70.0	131.3_5	3.5	18	FA, η	21
	Pr_4NI	3.0–50.0	108.85	3.8	17	FA, η	21
	Bu_4NI	4.0–50.0	101.72	3.8	16	FA, η	21
	i-Am_3BuNI	6.0–47.0	99.16	3.6_9	14_4	FA, η	22
	i-Am_4NI	10.0–47.0	98.04	3.5	12_8	FA, η	22
	Am_4NI	4.0–110.0	97.42	3.7	16	FA, η	21
	$(EtOH)_4NI$	5.0–40.0	100.0_0	4.1	12_4	FA, η #	23
	Me_4NPic	3.0–50.0	115.80	3.8	11	FA, η	21
	Et_4NPic		107.63	4	13	FA, η	21
	Pr_4NPic		93.12	6_4	21	FA, η	21
	Bu_4NPic	3.0–50.0	86.14	3.4	7	FA, η	21
	i-$Am_3BuNPic$	3.0–25.0	83.68	4.2	12	FA, η	21
	i-Am_4NPic		82.5_5	3_9	15	FA, η	22
	i-Am_4NClO_4	7.0–46.0	106.20	3.1_2	32_6	FA, η	22
	Bu_4NBPh_4	2.0–15.0	75.99	5_1	37	FA, η	21
	i-$Am_3BuNBPh_4$	2.0–10.0	73.3	5_3	32	FA, η	21
	Me_3SI	6.0–57.0	130.3_2	3.6_9	23_4	FA	24
	Et_3SI	6.0–61.0	124.6_2	3.6	23_9	FA	24

Table 5.1.1 (*contd*)

Solvent	Electrolyte	$10^4 \cdot \Delta c$/ mol l^{-1}	$\Lambda°$/ Ω^{-1} cm^2 mol^{-1}	a, d/Å	K_A/mol^{-1} l	Method of Analysis	Ref.
Methanol	Pr_3SI	5.0–56.0	113.09	3.5_2	24_2	FA	24
	Bu_3SI	5.0–34.0	106.6_7	4_3	31	FA	24
Ethanol	LiCl	$\kappa a < 0.2$	38.94	4.4	27	FA	10
	NaCl	$\kappa a < 0.2$	42.17	4.0	44	FA	10
	KCl	$\kappa a < 0.2$	45.42	4.6	95	FA	10
	CsCl	9.0–38.0	48.43	(11.0)	207	FH	25+
	HCl		84.65	3.4	48	FOS	26
	$LiOC_2H_5$	3.0–11.0	39.99	[1853]	176	FA	27
	$NaOC_2H_5$	4.0–17.0	43.35	[1397]	49_2	FA	27
	KOC_2H_5	3.0–12.0	46.62	[1359]	36_0	FA	27
	$RbOC_2H_5$	3.0–11.0	47.73	[1315]	39_1	FA	27
	$CsOC_2H_5$	2.0–10.0	49.47	[1241]	65_1	FA	27
	$LiNO_3$	7.0–50.0	42.74	3.7	19_2	FA	28
	$AgNO_3$	10.0–167.0	44.88	3.8_5	210	FA	28
	HPic	6.0–38.0	74.1		5.83×10^3	F	17
	Me_4NCl	3.0–56.0	51.6_7	4.2	122	FA	29
	Bu_4NCl	2.0–42.0	41.5_4	4.4	39	FA	29
	Me_4NBr	5.0–36.0	53.5_6	4.1	146	FA	29
	Et_4NBr	4.0–39.0	53.15	4.5	99	FA	29
	Pr_4NBr	4.0–40.0	46.86	4.3	78	FA	29
	Bu_4NBr	5.0–46.0	43.5_1	4.3	75	FA	29
	Et_4NI	4.0–40.0	56.34	4.6	133	FA	29
	Pr_4NI	5.0–48.0	49.9_4	4.1	120	FA	29
	Bu_4NI	5.0–44.0	46.65	4.0	123	FA	29
	i-Am_3BuNI	4.0–38.0	45.31	4.0	130	FA	29
	$Hept_4NI$	3.0–31.0	41.93	4.3	139	FA	29
	Me_4NPic		55.0_3	6	$1.1_0 \times 10^2$	FA, η	21
	Et_4NPic		54.2	3.7	69	FA, η	21

Propanol	HCl	2.0–44.0	29.8_1	3.9_0	$1.7_1 \times 10^2$	FOS	29-A
	NaI	$\kappa a < 0.2$	23.8_7	4_2	1.0×10^2	FA	10
	KI	$\kappa a < 0.2$	25.69	3_6	2.3×10^2	FA	10
	NaSCN	1.0–16.0	24.40		2.4×10^2	S	30
	KSCN	1.0–26.0	26.1_2		3.2×10^2	S	30
	Me_4NCl	6.0–43.0	25.05	4.2	456	FA	29
	Bu_4NCl	2.0–45.0	21.16	4.4	149	FA	29
	Me_4NBr	2.0–27.0	26.91	6.4	638	FA	29
	Et_4NBr	4.0–40.0	27.19	5.0	373	FA	29
	Pr_4NBr	6.0–54.0	24.4_2	4.4	270	FA	29
	Bu_4NBr	4.0–42.0	22.92	4.6	266	FA	29
	Et_4NI	3.0–40.0	29.0_1	5.5	466	FA	29
	Pr_4NI	5.0–50.0	26.0_8	4.5	391	FA	29
	Bu_4NI	4.0–43.0	24.6_0	4.7	415	FA	29
	i-Am_3BuNI	3.0–36.0	24.02	4.9	462	FA	29
	$Hept_4NI$	3.0–32.0	22.18	4.8	442	FA	29
	Bu_4NClO_4	3.0–36.0	27.13	4.2	769	FA	29
Butanol	LiI	8.0–160.0	17.42		$7.2_7 \times 10^2$	S	31
	NH_4I	7.0–130.0	16.00		$5.1_2 \times 10^2$	S	31
	Me_4NCl	2.0–48.0	17.49	7.0_9	2.20×10^3	FA#	32
	Bu_4NCl	1.0–40.0	15.55	5.5	$6.4_0 \times 10^2$	FA#	32
	Me_4NBr	1.0–18.0	17.8_8	6_3	$2.1_1 \times 10^3$	FA	32
	Et_4NBr	4.0–39.0	18.7_0	6.3	1.33×10^3	FA	32
	Pr_4NBr	4.0–40.0	17.01	5.4	9.2×10^2	FA	32
	Bu_4NBr	3.0–35.0	16.07	5.4	$8.6_0 \times 10^2$	FA	32
	Et_4NI	4.0–40.0	19.7_2	6_4	1.41×10^3	FA	32
	Pr_4NI	5.0–46.0	18.1_2	5.5	1.16×10^3	FA	32
	Bu_4NI	4.0–37.0	17.16	5.5	1.18×10^3	FA	32
	i-Am_3BuNI	3.0–40.0	16.99	5.9	1.36×10^3	FA	32
	$Hept_4NI$	3.0–35.0	15.57	5.5	1.26×10^3	FA	32
	Bu_4NClO_4	3.0–49.0	19.0_6	5.7	2.20×10^3	FA	32
i-Butanol	$AgClO_4$		12.2		6.8×10^2	S	35

Table 5.1.1 (*contd*)

Solvent	Electrolyte	$10^4 \cdot \Delta c$/ mol l^{-1}	$\Lambda°$/ Ω^{-1} cm^2 mol^{-1}	a, d/Å	K_A/mol^{-1} l	Method of Analysis	Ref.
Pentanol	Bu_4NBr	3.0–41.0	11.3_1	6.8_2	$2.5_2 \times 10^3$	FA	32
	Bu_4NI	3.0–40.0	12.0_0	6.5	$3.2_2 \times 10^3$	FA	32
	i-Am_3BuNI	2.0–35.0	11.6_2	6.8	$3.2_9 \times 10^3$	FA	32
	$Hept_4NI$	2.0–30.0	10.7_6	6.6	$3.2_2 \times 10^3$	FA	32
Ethanolamine	NaI		6.23	4_7	10	FA	33
	KI		7.11	4_2	9	FA	33
	Bu_4NI	3.0–63.0	4.846	(7.4)	20_0	FH	34+
Ethylene diamine	NaI		61.4		1.6×10^3	F	33
	KI		$70._9$	16	2.0×10^3	FA	33
	AgI	3.0–35.0	45.5		$8.9_3 \times 10^4$	F	36
	$AgNO_3$	1.0–22.0	65.8		$2.1_7 \times 10^3$	F	36
	(en)HCl	4.0–80.0	61.5		$8.6_2 \times 10^3$	F	36

Propylenediamine	KI	1.0–40.0	50.0		$1.2_2 \times 10^4$	F	36
	$AgNO_3$	3.0–40.0	53.1		$6.6_2 \times 10^3$	F	36
	Me_4NBr		44.21		8.30×10^4	S	37
	Et_4NBr		40.41		2.06×10^4	S	37
	Pr_4NBr		25.79		6.19×10^3	S	37
	Bu_4NBr		23.69		5.20×10^3	S	37
	Me_4NI		55.84		8.95×10^4	S	37
	Pr_4NI		37.47		9.66×10^3	S	37
	Bu_4NI		35.10		8.78×10^3	S	37
Formamide	$MgSO_4$	1.0–20.0	56.76	(3.0)	9.3	FA	38
	$NiSO_4$	1.0–20.0	63.36	2.5		FA	39
	$Cu(en)_2SO_4$	1.0–20.0	63.90	2.3		FA	39
	$LaCo(CN)_6$	1.0–17.0	103.9_2	8.1	$2_4 \times 10^2$	FA	40
	$LaFe(CN)_6$	1.0–14.0	103.6_5	8.4	$2_5 \times 10^2$	FA	40
	$Co(en)_3Co(CN)_6$	1.0–19.0	109.8_6	7.9	$6_0 \times 10^2$	FA	40
	$Co(NH_3)_6Fe(CN)_6$	1.0–14.0	119.1_9	7.0	$4_3 \times 10^2$	FA	40
	$Co(en)_3Fe(CN)_6$	1.0–18.0	109.4_1	6.8	$4_6 \times 10^2$	FA	40
N-methyl-formamide	$MgSO_4$	3.0–27.0	90.62	3.4		FA	41

Table 5.1.2
Conductance Parameters for Electrolytes in Dipolar Aprotic Solvents at 25°C

Solvent	Electrolyte	$10^4 \cdot \Delta c$/ mol l^{-1}	$\Lambda°$/ Ω^{-1} cm^2 mol^{-1}	a, d/Å	K_A/mol^{-1} l	Method of Analysis	Ref.
Dimethyl sulphoxide	KBr		38.6	3.8		FO	12
	NaI		37.65	4.7_2		FA	33
	KI		38.32	4.4		FA	33
	CsI	9.0–200.0	40.0	3.4		FO	42
	$NaClO_4$		38.3	4.8		FO	12
	$KClO_4$		39.1	4.4		FO	12
Dimethyl-							12
formamide	LiCl	1.0–55.0	80.1_5	(5.0)	42	FO-full	12
	KBr	4.0–46.0	84.3_8	3.3		FO-full	12
	HBr	8.0–15.0	88.3	(7.5)	54	FH	43+
	NaI		82.36	4.7		FA	33
	KI	4.0–40.0	83.00	4.54		P	4
	$LiClO_4$	1.0–43.0	77.4_2	5.0		FO-full	12
	$NaClO_4$		82.3	5.3		FO-full	12
	$KClO_4$		83.2	5.0		FO-full	12
	$RbClO_4$	2.0–48.0	84.8_0	4.7		FO-full	12
	$CsClO_4$	1.0–50.0	86.87	3.52		P	4
	NH_4ClO_4	1.0–44.0	91.0_0	4.8		FO-full	12
	$AgClO_4$	1.0–45.0	87.6_2	4.9		FO-full	12
	$TlClO_4$	0.8–45.0	91.0_7	4.6		FO-full	12
	$AgNO_3$	2.0–46.0	92.5	(5.0)	401	FO-full	12
	Me_2TlI	1.0–45.0	79.3	(5.0)	787	FO-full	12
	NaPic	1.0–24.0	67.3_7	4.6_1		FH	43+

Dimethyl-formamide	KPic	1.0–35.0	68.7_9	5.0_5		FH	43+
	HPic	2.0–18.0	71.7_2	(7.5)	16	FH	43+
	EtH_3NBr	2.0–38.0	92.6	(7.5)	$1.3_4 \times 10^2$	FH	43+
	Et_3HNBr	0.9–17.0	89.4_1	(7.5)	301	FH	43+
	Me_4NPic	0.6–15.0	76.5_2	6.0		FH	43+
	Et_3HNPic	0.8–23.0	72.9_8	7.3		FH	43+
Dimethyl-acetamide	KBr		68.4	3.3		FO	12
	NaI		67.7_9	4.8		FA	33
	KI		67.0	5.4		FO	12
	$NaClO_4$		68.6	5.3		FO	12
	$KClO_4$		68.2	4.8		FO	12
Dimethylpropion-amide	KI		58.8	5.7		FO	12
	$NaClO_4$		61.4	5.0		FO	12
	$KClO_4$		60.9	5.0		FO	12
	NH_4ClO_4		62.7	5.2		FO	12
Propylene carbonate	LiCl	9.0–178.0	27.2_6	(4.35)	536	FH	44+
	LiBr	6.0–180.0	27.3_9	(4.35)	22_7	FH	44+
	$LiClO_4$	20.0–200.0	26.0_3	2.75		FA, η	44
	Et_4NCl	14.0–103.0	31.58	4.8_0		FH	44+
	Bu_4NBr	20.0–128.0	28.65	3.52		FA, η	44
	Et_4NClO_4	9.0–136.0	32.61	2.78		FH	44+
	Bu_4NClO_4	15.0–75.0	28.20	3.34		FH	44+
	Bu_4NBPh_4	2.0–39.0	17.14	5_1		FA, η	45
Acetone	LiCl	0.1–8.0	214		3.0×10^5	S	46
	LiBr	0.03–16.0	194		4.6×10^3	S	46
	LiI	0.08–10.0	195.0		145	S	46
	NaI		184_4	9.0	$1_6 \times 10^2$	FA	33
	KI	0.2–17.0	197.5_2		180	S	46
	$AgClO_4$		181		$1.8_6 \times 10^3$	S	35

Table 5.1.2 (*contd*)

Solvent	Electrolyte	$10^4 \cdot \Delta c$/ mol l^{-1}	$\Lambda°$/ Ω^{-1} cm^2 mol^{-1}	a, d/Å	K_A/mol^{-1} l	Method of Analysis	Ref.
Acetone	$HClO_4$	0.5–40.0	205		$4_3 \times 10^2$	S	48
	LiPic		157.7	1.5	819	FA	47
	NaPic		163.5	5.4	680	FA	47
	KPic		166.0	5.0	244	FA	47
	LiTolsul	0.1–2.0	172		$1.0_4 \times 10^5$	S	46
	Me_4NF		182.7	0.8	1.14×10^3	FA	47
	Et_4NCl		194.2	16	370	FA	47
	Bu_4NCl	0.1–17.0	188		602	S	46
	Bu_4NBr		183.2	5.4	264	FA	47
	Am_4NBr		174.7	7	220	FA	47
	Pr_4NI		190.72	5_2	162	FA	47
	Bu_4NI		180.3	6_1	143	FA	47
	Bu_4NClO_4		182.8	5.8	80	FA	47
	Bu_4NNO_3		187.11	4.96	143.1	FA	47
	Me_4NPic		183.4	5.8	67	FA	47
	Et_4NPic		176.65	6.3	45	FA	47
	Pr_4NPic		156.2	5_2	27	FA	47
	Bu_4NPic		152.34	4.6	17	FA	47
	$Bu_4NTolsul$	0.08–10.0	151.6		407	S	46
	$Me_4NTCVal$	2.0–11.0	209.8	(6.5)	47_2	FA	49
	Me_4NPCP	1.0–10.0	191.0	6.5		FA	49
	Me_4N*bis*TCVam	1.0–13.0	176.7	6.5		FA	49
	Me_4NHCHT	1.0–10.0	166.9	6.0		FA	49
Methylethyl ketone	LiI	0.8–20.0	147.2		367	F	50
	NaI	0.7–5.0	147.7		405	F	50
	KI		150.8		469	F	50
	LiSal	1.0–343.0	(126.3)		$2.0_1 \times 10^6$ $K_3 = 3.6 \times 10^{-4}$	FK	50

Methylethyl ketone	NaSal	0.6–7.0	(126.8)		$2.3_0 \times 10^5$	F	50
	KSal	0.3–10.0	129.9		$2.8_6 \times 10^4$	F	50
	Et_4NBr	2.0–10.0	158.3_2	7_4	958	FA, η	51
	Pr_4NBr	2.0–15.0	146.4_2	5.4	940	FA, η	51
	Bu_4NBr	2.0–15.0	139.5_0	5.6	787	FA, η	51
	Am_4NBr	2.0–15.0	134.7_7	5.8	761	FA, η	51
	Hex_4NBr	2.0–15.0	131.0_7	5.0	662	FA, η	51
	$Hept_4NBr$	2.0–15.0	128.2_9	5.2	655	FA, η	51
	Oct_4NBr	2.0–15.0	126.08	5.3	648	FA, η	51
	Et_4NI	2.0–15.0	160.3_4	6.0	434	FA, η	51
	Pr_4NI	2.0–15.0	148.66	5.5	442	FA, η	51
	Bu_4NI	2.0–13.0	141.66	5.3	382	FA, η	51
	Am_4NI	2.0–16.0	136.93	4.9	351	FA, η	51
	Hex_4NI	2.0–15.0	133.28	5.0	326	FA, η	51
	$Hept_4NI$	5.0–15.0	130.50	4.9	309	FA, η	51
	Oct_4NI	2.0–13.0	128.22	5.0	308	FA, η	51
	$MePr_3NI$	2.0–16.0	152.29	5.4	678	FA, η	51
	$MeBu_3NI$	2.0–16.0	145.69	5.6	664	FA, η	51
	$MeHept_3NI$	2.0–16.0	134.93	5.4	588	FA, η	51
	i-Am_3BuNI	3.0–14.0	139.19	5.3	386	FA, η	51
Acetophenone	LiI	1.0–19.0	35.27		145	F	50
	NaI		37.85	5_4	150	FA	33
	KI	0.8–8.0	38.13		191	F	50
	LiSal	2.0–29.0	(24.76)		1.43×10^6 $K_3 = 9.9 \times 10^{-4}$	FK	50
	NaSal	0.6–7.0	(27.04)		9.9×10^4	F	50
	KSal	0.4–6.0	27.62		5.26×10^3	F	50

Table 5.1.2 (*contd*)

Solvent	Electrolyte	$10^4 \cdot \Delta c$/ mol l^{-1}	Λ^0/ Ω^{-1} cm^2 mol^{-1}	a, d/Å	K_A/mol^1 − l	Method of Analysis	Ref.
Cyclohexanone	$AgClO_4$		27.5		1.14×10^3	S	35
Acetonitrile	NaI		179.96	2.80		FA	33
	KI		186.50	2.14		FA	33
	$NaClO_4$	4.0–40.0	180.63	3_7	11	FA#	52
	$KClO_4$	7.0–60.0	187.41	3.0	13	FA#	52
	$RbClO_4$	6.0–50.0	189.49	3.4	19	FA#	52
	$CsClO_4$	5.0–50.0	191.08	3.5	23	FA#	52
	$AgClO_4$	4.0–24.0	186.69	4.00		FA#	53
	$TlClO_4$	6.0–30.0	195.21	3.1	32_4	FA#	54
	$CuClO_4$	5.0–17.0	168.38	4.0_3		FA#	53
	$LiClO_3$	5.0–32.0	170.0	[1078]	$4_0 \times 10^2$	FA	14
	KSCN	7.0–30.0	197.00	3.1	26_3	FA#	54
	$AgNO_3$	6.0–41.0	192.34	3.8_3	72_3	FA#	53
	$AgBF_4$	7.0–26.0	194.56	4.16		FA#	53
	$TlBF_4$	4.0–32.0	199.08	3.3	15_4	FA#	54
	$CuBF_4$	6.0–22.0	173.16	7_1	12	FA#	53
	$AgPF_6$	4.0–20.0	189.9_9	4.46		FA#	53
	$CuPF_6$	4.0–20.0	168.92	7_6	14	FA#	53
	$NaBPh_4$	5.0–50.0	135.4	5.2		FA, η#	52
	$KBPh_4$	5.0–40.0	141.79	5.5		FA, η#	52
	$RbBPh_4$	3.0–30.0	143.86	5.3		FA, η#	52
	$CsBPh_4$	4.0–40.0	145.47	3.3	2_6	FA, η#	52
	$((Ph_2NH_2)P)_2NCl$	2.0–13.0	148.9		3.1×10^3	F	55
	Me_4NBr	8.0–57.0	195.2	4.4	46	FA, η	47
	Et_4NBr		185.3	4	10	FA, η	47

Acetonitrile	Pr_4NBr	3.0–65.0	171.1	3.4	4	FA, η	47
	Bu_4NBr		162.1	3.5	2	FA, η	47
	i-Am_3BuNBr	5.0–37.0	158.36	3.6	$1._2$	FA, η	22
	i-Am_4NBr	7.0–54.0	157.49	3.35		FA, η	22
	Am_4NBr	4.0–50.0	156.8	2.9		FA, η	22
	Me_4NI	6.0–70.0	196.7	3.5	19	FA, η	47
	Et_4NI		187.2_9	3.3	5	FA, η	47
	Pr_4NI	4.0–60.0	172.9	3.8	5	FA, η	47
	Pr_3MeNI	5.0–21.0	178.2_2	2.3	4	FOS	56
	Pr_3BuNI	3.0–17.0	170.4_2	4.5	8	FOS	56
	Bu_4NI	4.0–49.0	164.0	3.6	3	FA, η	47
	i-Am_3BuNI	6.0–49.0	160.39	3.5	$1._6$	FA, η	22
	Am_4NI		158.1	3.2		FA, η	22
	$(EtOH)_4NI$	5.0–40.0	165.9_1	1.6	143	FA, η#	23
	Me_3SI	2.0–34.0	199.23	3.6	$35._7$	FA	24
	Et_3SI	2.0–62.0	190.25	3.5	$19._8$	FA	24
	Pr_3SI	2.0–56.0	178.64	3.4	$18._7$	FA	24
	Me_4NClO_4	8.0–61.0	198.16	3.13	7	FA#	57
	Bu_4NClO_4	4.0–41.0	165.06	3.57		FA	58
	i-Am_4NClO_4	4.0–34.0	160.62	3.79		FA	58
	Ph_4AsClO_4	5.0–40.0	159.5_8	4.5_5		FA#	57
	Me_4NNO_3	4.0–20.0	200.5	4.64	23	FAB, η	59
	Bu_4NNO_3	4.0–20.0	168.2	3.73	7	FAB, η	59
	Me_4NPF_6	2.0–14.0	196.75	(5.0)	5	FA	60
	Bu_4NPF_6	2.0–10.0	164.75	5.44		FA	60
	Me_4NPic	4.0–48.0	171.8	3.7	1	FA, η	47
	Et_4NPic		164.6_2	$5._0$	10	FA, η	47
	Pr_4NPic		147.2	5.1		FA, η	47
	Bu_4NPic	2.0–42.0	139.4	4.0		FA, η	47
	i-$Am_3BuNPic$		135.70	[1550]		FA	61
	Am_4NPic		135.1	8	39	FA, η	47
	Me_4NBPh_4	1.0–10.0	152.27	5.1	3	FAB, η	3

Table 5.1.2 (*contd.*)

Solvent	Electrolyte	$10^4.\Delta c$/ mol l^{-1}	$\Lambda°$/ Ω^{-1} cm^2 mol^{-1}	$a \cdot d$/Å	K_A/mol^{-1} l	Method of Analysis	Ref.
Acetonitrile	Et_4NBPh_4	2.0–14.0	142.77	5.2	4	FAB, η	3
	Pr_4NBPh_4	2.0–11.0	128.36	5.8	4	FAB, η	3
	Bu_4NBPh_4	4.0–33.0	119.65	4.5_5		FA	58
	i-$Am_3BuNBPh_4$		116.26	[1678]		FA	61
	i-Am_4NBPh_4	4.0–36.0	114.96	4.7_4		FA	58
	$(EtOH)_4NBPh_4$	5.0–35.0	122.3_3	5.2_6		FA, η#	23
	i-$Am_4NB(i$-$Am_4)$	4.0–29.0	114.48	4.8_3		FA	58
	$Me_4NTCVal$	1.0–12.0	191.7	4.1		FA	49
	Me_4NPCP	1.0–12.0	179.3	3.9		FA	49
	Me_4N*bis*TCVam	1.0–13.0	167.0	4.4		FA	49
	Me_4NHCHT	0.8–9.0	156.3	6.8		FA	49
i-Butyronitrile	$HClO_4$	0.5–40.0	126		$5.0_0 \times 10^2$	S	48
	Bu_4NBPh_4	13.0–31.0	81.66	5.0_5	$18._4$	FA, η	62
Valeronitrile	Et_4NClO_4	1.0–63.0	88.3	$3._8$	194	FA, η	63
Adiponitrile	NaI	2.0–40.0	12.52	3.9		FA	64
	KI	2.0–40.0	13.26	3.6		FA	64
	$NaClO_4$	2.0–40.0	13.16	2.9		FA	64
	KSCN	2.0–60.0	15.91	3.1	20	FA	64
	$NaBPh_4$	2.0–40.0	9.16	5.2		FA	64
	Me_3PhNBr	2.0–40.0	12.93	4.2	28	FA	64
	Et_4NBr	2.0–40.0	13.06	3.3		FA	64
	Pr_4NBr	2.0–40.0	11.71	3.9		FA	64
	Bu_4NBr	2.0–40.0	10.94	4.2		FA	64
	Hex_4NBr	2.0–40.0	10.06	4.6		FA	64
	Me_3PhNI	2.0–40.0	13.27	2.7		FA	64
	Pr_4NI	2.0–40.0	12.08	4.0		FA	64
	Bu_4NI	2.0–40.0	11.30	4.2		FA	64
	i-Am_3BuNI	2.0–40.0	10.91	4.1		FA	64

Adiponitrile	Hex_4NI	2.0–40.0	10.40	4.4		FA	64
	i-$Am_3BuNBPh_4$	2.0–40.0	7.58	4.9		FA	64
	$Me_3PhNTolsul$	2.0–40.0	11.80	2.8	14	FA	64
Benzonitrile	NaI		47.6	(4.50)	76	FA	33
	KI		51.5	(4.50)	105	FA	33
	$HClO_4$	0.5–40.0	50.6		$6.0_4 \times 10^2$	S	48
	$AgNO_3$		55.2		4.0×10^3	S	35
Phenylaceto-nitrile	$HClO_4$	0.5–40.0	27.8		$6.0_4 \times 10^2$	S	48
	Bu_4NBF_4	6.0–31.0	31.444	7.75	$175._7$	FOS*	65
Nitromethane	Me_4NCl	<55	116.44	4.0	45	FA	66
	Et_4NCl	<55	110.06	3.4	2.0	FA	66
	Pr_4NCl	<55	101.61	3.64		FA	66
	Bu_4NCl	<55	96.58	4.12		FA	66
	Me_4NBr	<55	116.89	4.1	31	FA	66
	Et_4NBr	<55	110.45	3.3	1.7	FA	66
	Pr_4NBr	<55	102.13	3.44		FA	66
	Bu_4NBr	<55	96.97	3.19		FA	66
	Bu_4NBPh_4	3.0–29.0	67.26	4.85		FA	58
	i-Am_4NBPh_4	5.0–18.0	64.65	5.22		FA	58
	i-$Am_4NB(i$-$Am_4)$	4.0–24.0	64.58	4.72		FA	58
Nitrobenzene	Et_4NCl		38.54	7	81	FA	47
	Bu_4NBr		33.29	6.6	57	FA	67
	Me_4NPic		33.34	3.6	$23._6$	FA	47
	Et_4NPic		32.43	3.8_9	$7._4$	FA	47
	Pr_4NPic		29.48	3.1	3	FA	47
	Bu_4NPic		27.86	3.3	$2._6$	FA	47
	Am_4NSCN	0.7–57.0	33.26	6.1	37.0	FA	68
	Bu_4NBPh_4	3.0–14.0	22.31	(8.0)	$5._9$	FH	45+
	Me_4NPCP	1.0–12.0	35.43	(6.5)	$10._7$	FA	49

Table 5.1.3

Conductance Parameters for Electrolytes in Low Dielectric Constant Aprotic Solvents at 25°C

Solvent	Electrolyte	$10^4 \cdot \Delta c$/ mol l^{-1}	$\Lambda°$/ Ω^{-1} cm^2 mol^{-1}	a, d/Å	K_A/mol^{-1} l	Method of Analysis	Ref.
Pyridine	NaI		75.0_5	$7._1$	$2.2_2 \times 10^3$	FA	33
	KI		80.0	$7._9$	$3.9_8 \times 10^3$	FA	33
Tetrahydro-furan	$LiBPh_4$	0.04–1.0	$76._9$		$1.2_6 \times 10^4$	F	69
	$NaBPh_4$	0.02–0.8	$88._5$		$1.1_7 \times 10^4$	F	69
	$KBPh_4$	0.01–0.8	$90._1$		$3.1_1 \times 10^4$	F	69
	$CsBPh_4$	0.01–2.0	$108._7$		$5.3_5 \times 10^5$	F	69
	Bu_4NClO_4	<1.1	162.6	(37.9)	$1.58_0 \times 10^6$	FA*	70
	Bu_4NBPh_4	0.01–2.0	84.8		$2.3_1 \times 10^4$	F	69
	i-$Am_3BuNBPh_4$	0.03–1.0	80.6		$1.6_6 \times 10^4$	F	69
Tetrahydro-pyran	$NaBPh_4$		56.3		$4.3_7 \times 10^5$	F	71
Dimethoxyethane	$NaBPh_4$		$102._0$		1.8×10^4	F	72
	$CsBPh_4$		$100._0$		3.5×10^4	F	72
Ethylene chloride	$MeBu_3NClO_4$	2.0–13.0	67.5_5		1.17×10^4	S	73
	Bu_4NClO_4	0.1–1.0	65.2		6.41×10^3	S	74
	Me_4NPic		73.8		3.1×10^4	S	75
	Et_4NPic		69.4		$6.2_2 \times 10^3$	S	75
	Pr_4NPic		62.7		$5.1_6 \times 10^3$	S	75
	Bu_4NPic		57.4		$4.3_8 \times 10^3$	S	75
	Am_4NPic		54.5		$4.1_6 \times 10^3$	S	75
	Bu_4NBPh_4	1.0–7.0	49.8	[29100]	$9._5 \times 10^2$	FA	76

Ethylidene-chloride	Bu_4NClO_4		109.7	4.67×10^4	S	74
	Bu_4NPic		96.9	$2.2_0 \times 10^4$	S	75
1,2 Dichloro-propane	Bu_4NClO_4		56.5	9.16×10^4	S	74
	Bu_4NPic	0.6–4.0	51.3	3.75×10^4	S	75
2,2 Dichloro-propane	Bu_4NClO_4		84.3	$8.0_6 \times 10^4$	S	74
o-Dichloro-benzene	$MeBu_3NI$	0.1–1.7	42.2_5	$3.8_6 \times 10^5$	S	73
	Bu_4NI		39.3	$1.5_6 \times 10^5$	F	78
	$OctBu_3NI$		35.0	1.59×10^5	F	78
	Bu_4NNO_3		42.0	$2.1_8 \times 10^5$	F	78
	$MeBu_3NClO_4$	0.09–0.8	43.8_7	$1.8_7 \times 10^5$	S	73
	Bu_4NClO_4		42.3	9.34×10^4	S	74
	Bu_4NBPh_4	0.2–36.0	29.6	8.3×10^3	FA	76
	Et_4NPic		47.8	8.1×10^4	S	77
	Pr_4NPic		40.3	6.1×10^4	S	77
	Bu_4NPic		36.8	$5.2_1 \times 10^4$	S	77
	Am_4NPic		35.3	$5.5_2 \times 10^4$	F	78
	$OctMe_3NPic$		34.8	5.3×10^5	F	78

Table 5.1.4

Conductance Parameters for Electrolytes in Non-aqueous Solvents at Temperature t

Solvent	Electrolyte	t °C	$10^4 \cdot \Delta c$/ mol l^{-1}	$\Lambda°$/ Ω^{-1} cm^2 mol^{-1}	a, d/Å	K_A/mol^{-1} l	Method of Analysis	Ref.
Methanol	Me_4NBr	10	7.0–100.0	101.7_7	3.2	12	FA, η	21
	Et_4NBr	10	5.0–70.0	94.8_2	3.4	7	FA, η	21
	Pr_4NBr	10	5.0–70.0	83.0_7	3.3	5	FA, η	21
	Bu_4NBr	10	3.0–57.0	76.9_0	3.0		FA, η	21
	Me_4NI	10	7.0–70.0	106.95	3.6	19	FA, η	21
	Pr_4NI	10	4.0–60.0	88.1_1	3.8	17	FA, η	21
	Bu_4NI	10	4.0–60.0	82.12	3.7	17	FA, η	21
	$(EtOH)_4NI$	10	4.0–35.0	80.88	$4._9$	16	FA, η#	23
	Me_3SI	10	5.0–63.0	106.2_3	3.6	23	FA	24
	Et_3SI	10	3.0–74.0	101.2_8	3.5	24	FA	24
	Pr_3SI	10	4.0–77.0	91.97	3.6	$24._2$	FA	24
	Me_4PI	18	2.0–42.0	115.5	(8.44)	33	P	79+
	Me_3PhPI	18	1.0–48.0	104.0_6	(8.44)	25	P	79+
	Me_2Ph_2PI	18	1.0–49.0	97.5_7	(8.44)	20	P	79+
	$MePh_3PI$	18	1.0–50.0	94.0	(8.44)	20	P	79+
	Ph_4PI	18	1.0–20.0	92.1_8	(8.44)	18	P	79+
Propanol	HCl	15	9.0–45.0	22.0	3.8	1.2×10^2	FOS	29-A
	HCl	35	5.0–34.0	39.7_6	3.93	$2.5_5 \times 10^2$	FOS	29-A
Acetonitrile	$(EtOH)_4NI$	10	7.0–40.0	142.2_0	1.6	136	FA, η#	23
	Me_3SI	10	3.0–85.0	171.6_7	3.0	$31._7$	FA	24
	Et_3SI	10	2.0–64.0	163.5_7	3.3	$17._9$	FA	24
	Pr_3SI	10	8.0–70.0	153.27	2.76	13.5	FA	24
	Me_4PI	18	1.0–39.0	180.4	(7.59)	42	P	79+
	Me_3PhPI	18	1.0–46.0	168.5	(7.59)	37	P	79+
	Me_2Ph_2PI	18	0.7–45.0	159.6	(7.59)	29	P	79+
	$MePh_3PI$	18	1.0–9.0	150.5_4	(7.59)	$23._7$	P	79+
	Ph_4PI	18	1.0–28.0	143.8	(7.59)	24	P	79+

Sulpholane	LiCl	30	3.0–70.0	13.63	(5.0)	13860	P	80
	LiBr	30	2.0–60.0	13.25	(5.0)	278	P	80
	LiI	30	3.0–41.0	11.52_8	(5.0)	$5._6$	P	80
	NaI	30	2.0–52.0	10.865	(5.0)	$4._7$	P	80
	KI	30	3.0–50.0	11.25_3	(5.0)	$6._5$	P	80
	$LiClO_4$	30	2.0–34.0	11.05_3	(5.0)	$6._5$	P	80
	$NaClO_4$	30	2.0–56.0	10.32_5	(5.0)	$7._4$	P	80
	$KClO_4$	30	2.0–56.0	10.76_6	(5.0)	$8._5$	P	80
	$RbClO_4$	30	3.0–58.0	10.83_8	(5.0)	$8._6$	P	80
	$CsClO_4$	30	4.0–80.0	11.03	(5.0)	$9._4$	P	80
	NH_4ClO_4	30	3.0–44.0	11.66_4	(5.0)	9	P	81+
	KPF_6	30	2.0–60.0	9.99_5	(5.0)	4.6	P	80
	NaSCN	30	3.0–44.0	13.18_6	(5.0)	$47._2$	P	82+
	KSCN	30	12.0–52.0	13.67	(5.0)	13	P	82+
	RbSCN	30	6.0–71.0	13.87_1	(5.0)	8.9	P	82+
	CsSCN	30	6.0–133.0	13.96_7	(5.0)	$8._6$	P	82+
	Et_4NI	30	2.0–40.0	11.20_1	(5.0)	4.6	P	80
	Me_4NClO_4	30	8.0–99.0	10.97	(5.0)	$5._9$	P	83+
	Et_4NClO_4	30	13.0–93.0	10.58_9	(5.0)	$3._0$	P	83+
	Pr_4NClO_4	30	6.0–65.0	9.92	(5.0)	5	P	83+
	Bu_4NClO_4	30	8.0–93.0	9.45	(5.0)	3	P	83+
Butanol	LiI	0	8.0–150.0	8.58		$4.7_7 \times 10^2$	S	31
	NH_4I	0	8.0–130.0	9.09		$5.7_5 \times 10^2$	S	31
	Bu_4NI	0	7.0–150.0	8.85		$1.2_9 \times 10^3$	S	84
	LiI	50	8.0–150.0	30.30		1.35×10^3	S	31
	NH_4I	50	7.0–130.0	29.85		1.49×10^3	S	31
	Bu_4NI	50	7.0–150.0	37.73		$3.7_0 \times 10^3$	S	84
Valeronitrile	Et_4NClO_4	30	3.0–64.0	93.9_3	4.0	$2.0_2 \times 10^2$	FA, η	63
		40	1.0–63.0	106.3	4.2	$2.2_1 \times 10^2$	FA, η	63
		50	1.0–63.0	119.1	4.1	$2.3_7 \times 10^2$	FA, η	63
Phenylaceto-nitrile	Bu_4NBF_4	75	4.0–20.0	65.44	7.8_6	207	FOS*	65
		100	4.0–30.0	84.7	7.6	2.1×10^2	FOS*	65

Table 5.1.5

Conductance Parameters for Electrolytes in Solvent Mixtures at 25°C

Per cent of solvent (2)	ε_r	$\Lambda°/\Omega^{-1}$ cm^2 mol^{-1}	a, d/Å	K_A/mol^{-1} l	Method of Analysis
(1)Methanol-(2)Dioxan		Electrolyte: $LiClO_3$			Ref. 14
0.00	32.66	100.9_7	[1410]	5	FA
40.15	19.05	95.2	[7222]	46	FA
50.93	15.40	87.9	[13134]	2.2×10^2	FA
60.07	12.32	86.6_8	[24948]	1344	FA
65.71	10.50	78.40		6625	F
(1)Acetonitrile-(2)Dioxan		Electrolyte: $LiClO_3$			Ref. 14
0.00	36.01	170.0	[1078]	$4._1 \times 10^2$	FA
30.31	26.85	145.20		1760	F
39.68	23.80	133.15		3520	F
49.49	20.45	125.20		6300	F
59.92	17.03	113.00		3.93×10^4	F
(1)Methanol-(2)Acetonitrile		Electrolyte: $KClO_4$			Ref. 13
0.00	32.63	123.0_6	2.5	11	FA
5.85	32.80	130.4	4.3	18	FA
19.87	33.30	148.2	4.2	13	FA
40.82	34.00	166.4	4.2	11	FA
59.31	34.42	181.3	2.9	3	FA
62.79	34.55	185.38	2.6		FA
81.40	35.35	193.4_2	3.9	11	FA
100.00	36.01	187.6	3.4	17	FA

(1)Methanol-(2)Acetonitrile		Electrolyte: $CsClO_4$				Ref. 13
0.00	32.63	131.5	2.9	33	FA	
39.35	33.95	177.3	4.5	19	FA	
62.14	34.50	194.4_4	2.1	4	FA	
74.44	35.13	200.2_8	4.5	18	FA	
100.00	36.01	191.2	3.2	23	FA	
(1)Ethanol-(2)Acetone		Electrolyte: $CsClO_4$				Ref. 85
30.24	21.90	98.6	10	2.6×10^2	FA	
54.70	20.75	136.4	4.3	177	FA	
61.40	20.60	145.7	4.3	173	FA	
74.10	20.35	162.2	3.9	154	FA	
100.0	20.70	199.9	3.7	223	FA	
(1)Ethanol-(2)Acetone		Electrolyte: Et_4NClO_4				Ref. 85
5.90	23.70	68.0	4_5	164	FA, η	
36.30	21.55	110.8	3.9	74	FA, η	
51.80	20.85	135.5	4.2	74	FA, η	
64.08	20.50	153.1	4.0	62	FA, η	
93.70	20.50	198.8	4.5	84	FA, η	
(1)Nitrobenzene-(2)Carbon Tetrachloride		Electrolyte: Bu_4NBr				Ref. 67
0.00	34.69	33.29	6.6	57	FA	
8.13	31.13	33.96	6.6	65	FA	
18.13	27.20	35.27	6.35	130	FA	
26.56	24.23	36.14	6.10	210	FA	
40.30	19.07	37.78	6.30	655	FA	
50.74	15.69	38.80	(7.50)	2090	FA	
(1)Ethanol-(2)Carbon Tetrachloride		Electrolyte: Bu_4NBr				Ref. 67
0.00	24.91	43.35	(6.0)	90	FA	
1.41	24.40	43.20	(6.0)	115	FA	
3.37	23.68	43.95	(6.0)	145	FA	
5.10	23.05	42.75	(6.0)	180	FA	
8.80	21.77	41.90	(6.0)	240	FA	

Table 5.1.5 (*contd.*)

Per cent of Solvent (2)	ε_r	$\Lambda^\circ/\Omega^{-1}\,cm^2\,mol^{-1}$	$a, d/Å$	$K_A/mol^{-1}\,l$	Method of Analysis
(1)Methanol-(2)Carbon Tetrachloride		Electrolyte: Bu_4NBr			Ref. 67
0.00	32.63	96.30	(6.0)	12.4	FA
4.38	29.74	88.44	(6.0)	16.4	FA
9.40	26.72	81.42	(6.0)	27	FA
14.66	23.80	74.95	(6.0)	53	FA
17.69	22.22	72.75	(6.0)	85	FA
22.00	20.13	68.75	5.7	140	FA
33.11	15.31	58.55	6.0	655	FA
(1)Ethanol-(2)Water		Electrolyte: KCl			Ref. 25
0.00	24.3	45.42	4.6	95	FA
12.08	29.0	44.59	3.2_5	$23._5$	FA
20.71	33.1	44.05	3.0_3	$11._6$	FA
39.75	43.3	46.77	2.9_9	$3._0$	FA
(1)Ethanol-(2)Water		Electrolyte: CsCl			Ref. 25
0.00	24.3	48.33	4.20	$158._1$	FA
6.76	26.8	46.39	3.7_6	80	FA
8.75	27.5	45.99	3.7_2	68	FA
15.67	30.7	44.90	3.5_8	$38._5$	FA
26.10	35.7	44.85	3.5_3	$18._0$	FA
39.87	43.3	47.72_5	3.8	8.4	FA
(1)Nitrobenzene-(2)Anisole		Electrolyte: Et_4NPic			Ref. 86
0.00	34.9	32.15		33	S
12.40	29.6	35.34		63	S
33.09	21.8	40.78		$1.7_0 \times 10^2$	S
53.65	15.3	45.83		8.1×10^2	S

69.48	11.0	50.71		7.0×10^3	S	
82.37	8.0	(54.54)		8.9×10^4	S	
92.64	5.8	(57.70)		5.6×10^6	S	
100.00	4.3	(59.62)		1.05×10^9	S	
(1)Nitrobenzene-(2)Anisole		Electrolyte: Bu_4NPic				Ref. 86
0.00	34.9	27.97		13	S	
17.75	27.4	31.42		54	S	
32.19	22.1	34.26		$1.2_1 \times 10^2$	S	
53.79	15.3	38.37		7.3×10^2	S	
75.13	9.6	44.20		2.0×10^4	S	
86.51	7.0	(46.52)		3.8×10^5	S	
100.00	4.3	(49.60)		9.2×10^8	S	
(1)Acetonitrile-(2)Carbon Tetrachloride		Electrolyte: Me_4NBPh_4				Ref. 3
0.00	36.01	152.27	5.1	3	FAB, η	
54.84	22.32	114.85	7.1	20	FAB, η	
67.74	17.45	99.31	6.1	62	FAB, η	
77.73	13.06	83.60	6.6	328	FAB, η	
82.72	10.68	74.39	7.6	1110	FAB, η	
(1)Acetonitrile-(2)Carbon Tetrachloride		Electrolyte: Et_4NBPh_4				Ref. 3
0.00	36.01	142.77	5.2	4	FAB, η	
28.66	30.62	128.08	4.9	8	FAB, η	
60.15	20.60	102.95	5.4	21	FAB, η	
68.69	17.05	93.18	5.8	60	FAB, η	
75.22	14.21	84.91	6.9	184	FAB, η	
79.65	12.14	76.72	5.9	408	FAB, η	
(1)Acetonitrile-(2) Carbon Tetrachloride		Electrolyte: Pr_4NBPh_4				Ref. 3
0.00	36.01	128.36	5.8	4	FAB, η	
62.47	19.46	91.29	5.5	31	FAB, η	
70.81	16.13	81.55	5.5	70	FAB, η	
78.87	12.31	71.62	6.0	235	FAB, η	
84.27	9.80	63.10	7.0	880	FAB, η	

Table 5.1.5 (*contd.*)

Per cent of Solvent (2)	ε_r	$\Lambda°/\Omega^{-1}\ cm^2\ mol^{-1}$	$a, d/Å$	$K_A/mol^{-1}\ l$	Method of Analysis
(1)Acetonitrile-(2)Carbon Tetrachloride		Electrolyte: Bu_4NBPh_4			Ref. 3
0.00	36.01	119.48	5.8	6	FAB, η
60.61	19.87	85.03	6.4	21	FAB, η
68.06	17.18	78.37	6.0	45	FAB, η
75.97	13.92	70.04	6.4	133	FAB, η
80.68	11.16	64.18	7.1	365	FAB, η
(1)Acetonitrile-(2)Carbon Tetrachloride		Electrolyte: Me_4NNO_3			Ref. 59
0.00	36.01	200.5	4.64	23	FAB, η
63.20	18.91	137.6	5.93	465	FAB, η
68.46	16.99	127.2	5.39	840	FAB, η
75.81	13.93	111.4	7.50	2950	FAB, η
80.80	11.35	98.0	8.31	9600	FAB, η
(1)Acetonitrile-(2)Carbon Tetrachloride		Electrolyte: Bu_4NNO_3			Ref. 59
0.00	36.01	168.2	3.73	7	FAB, η
63.85	18.45	114.1	5.85	130	FAB, η
68.71	17.02	106.6	5.89	250	FAB, η
74.54	14.65	96.5	6.77	665	FAB, η
78.84	12.29	88.4	7.14	1670	FAB, η
(1)Tetrahydrofuran-(2)Water		Electrolyte: Bu_4NClO_4			Ref. 70
0.00	7.39	162.6	l†	1.580×10^6	FA*
2.00	8.47	149.3	l	2.219×10^5	FA*
3.00	9.00	140.8	l	97600	FA*
5.00	10.00	123.2	l	24000	FA*
10.00	12.60	83.0	l	3010	FA*
15.00	15.60	71.74	l	872	FA*

† l = Bjerrum distance parameter.

(1)Tetrahydrofuran-(2)Water		Electrolyte: Bu_4NBr				Ref. 70
6.55	10.85	75.6	*l*	10400	FA*	
10.00	12.60	59.87	*l*	1800	FA*	
15.00	15.60	52.74	*l*	539	FA*	
(1)Formamide-(2)Dioxan		Electrolyte: $MgSO_4$				Ref. 38
0.0	109.5	56.76	(3.0)	9.3	FA	
20.0	81.42	49.00	3.8	88	FA	
25.0	74.67	49.0	3.8	2.2×10^2	FA	
30.0	68.17	49.2	4.0	4.5×10^2	FA	
35.0	61.54	49.4	5	$1.2_4 \times 10^3$	FA	
40.0	55.38	51.1	6	3.8×10^3	FA	
50.0	43.17	53.0		2.47×10^4	F	
60.0	33.02	55.0		2.40×10^5	F	
70.0	21.77	60.0		2.64×10^6	F	
(1)Formamide-(2)Acetone		Electrolyte: $MgSO_4$				Ref. 87
0.0	109.5	56.76	(3.0)	9.3	FA	
10.0	97.61	60.32	2.37	43.4	FA	
20.0	85.20	65.92	2.47	79.7	FA	
25.0	79.12	68.62	3.53	263	FA	
30.0	73.47	78.56	4.53	693	FA	
35.0	67.88	82.34	6.82	1214	FA	
40.0	62.79	89.8	7.91	4480	FA	
50.0	53.48	102	8.93	31620	FA	
(1)Formamide-(2)Dioxan		Electrolyte: $LaFe(CN)_6$				Ref. 88
0.0	109.5	103.65		249	S	
20.0	81.4	104.94		4754	S	
30.0	68.2	106.53		13820	S	
40.0	55.4	106.95		60690	S	
50.0	43.2	108.78		3.6×10^5	S	
60.0	33.0	114.00		1.6×10^6	S	

Table 5.1.5 (*contd.*)

Per cent of solvent (2)	ε_r	$\Lambda°/\Omega^{-1}$ cm^2 mol^{-1}	a, d/Å	K_A/mol^{-1} l	Method of Analysis
(1)Formamide-(2)Acetone		Electrolyte: $LaFe(CN)_6$			Ref. 88
0.0	109.5	103.65		249	S
10.0	97.6	115.65		504	S
20.0	85.2	134.28		1041	S
30.0	73.5	153.18		2450	S
40.0	62.8	162.39		8220	S
50.0	53.5	174.3		2.65×10^4	S
(1)Methanol-(2)Carbon Tetrachloride		Electrolyte: Et_4NClO_4			Ref. 89
0.00	32.63	131.3_9	$4._6$	41	FA, η
10.05	31.05	124.70	3.8	47	FA, η
17.05	29.98	120.10	4.1	59	FA, η
25.04	28.52	114.7_3	$5._5$	89	FA, η
37.74	25.80	103.8_8	$3._4$	111	FA, η
56.40	20.46	86.2	2.4	285	FA, η
61.78	18.60	81.4	$3._7$	560	FA, η
(1)Methanol-(2)Pyridine		Electrolyte: Et_4NClO_4			Ref. 89
4.73	31.8	122.57	4.6	41	FA, η
16.56	30.9	110.45	3.7	36	FA, η
30.23	29.2	101.48	4.3	$45._6$	FA, η
53.49	25.7	93.4_0	4.8	99	FA, η
78.86	19.8	90.8	$5._1$	156	FA, η

REFERENCES TO APPENDIX 5.1.

1. R. M. Fuoss and F. Accascina, Electrolytic Conductance, Interscience, New York (1959)
2. R. M. Fuoss and L. Onsager, *J. Phys. Chem.*, **61**, 668 (1957)
3. D. S. Berns and R. M. Fuoss, *J. Amer. Chem. Soc.*, **82**, 5585 (1960)
4. R. Fernández-Prini and J. E. Prue, *Z. Phys. Chem.* (*Leipzig*), **228**, 373 (1965)
5. R. Fernández-Prini, *Trans. Faraday Soc.*, **65**, 3311 (1969)
6. R. M. Fuoss, L. Onsager and J. E. Skinner, *J. Phys. Chem.*, **69**, 2581 (1965)
7. T. Shedlovsky, *J. Franklin Inst.*, **225**, 739 (1938)
8. R. M. Fuoss and C. A. Kraus, *J. Amer. Chem. Soc.*, **55**, 476 (1933)
9. R. M. Fuoss and C. A. Kraus, *J. Amer. Chem. Soc.*, **55**, 2387 (1933)
10. R. L. Kay, *J. Amer. Chem. Soc.*, **82**, 2099 (1960)
11. R. L. Kay and J. L. Hawes, *J. Phys. Chem.*, **69**, 2787 (1965)
12. J. E. Prue and P. J. Sherrington, *Trans. Faraday Soc.*, **57**, 1795 (1961)
13. F. Conti and G. Pistoia, *J. Phys. Chem.*, **72**, 2245 (1968)
14. A. D'Aprano and R. Triolo, *J. Phys. Chem.*, **71**, 3474 (1967)
15. R. E. Busby and V. S. Griffiths, *J. Chem. Soc.*, 902 (1963)
16. M. A. Coplan and R. M. Fuoss, *J. Phys. Chem.*, **68**, 1177 (1964)
17. A. D'Aprano and R. M. Fuoss, *J. Phys. Chem.*, **73**, 400 (1969)
18. T. Shedlovsky, *in* The Structure of Electrolytic Solutions, p. 268 (W. J. Hamer, ed.), Wiley, New York (1959)
19. R. W. Kunze and R. M. Fuoss, *J. Phys. Chem.*, **67**, 385 (1963)
20. J. Barthel and G. Schwitzgebel, *Z. Phys. Chem.* (*Frankfurt*), **54**, 181 (1967)
21. R. L. Kay, C. Zawoyski and D. F. Evans, *J. Phys. Chem.*, **69**, 4208 (1965)
22. R. L. Kay, D. F. Evans and G. P. Cunningham, *J. Phys. Chem.*, **73**, 3322 (1969)
23. G. P. Cunningham, D. F. Evans and R. L. Kay, *J. Phys. Chem.*, **70**, 3998 (1966)
24. D. F. Evans and T. L. Broadwater, *J. Phys. Chem.*, **72**, 1037 (1968)
25. J. L. Hawes and R. L. Kay, *J. Phys. Chem.*, **69**, 2420 (1965)
26. H. O. Spivey and T. Shedlovsky, *J. Phys. Chem.*, **71**, 2165 (1967)
27. J. Barthel, *Angew. Chem.* (*Int. Ed.*), **7**, 260 (1968)
28. G. D. Parfitt and A. L. Smith, *Trans. Faraday Soc.*, **59**, 257 (1963)
29. D. F. Evans and P. Gardam, *J. Phys. Chem.*, **72**, 3281 (1968)

29A M. Goffredi and T. Shedlovsky, *J. Phys. Chem.*, **71**, 2182 (1967)

30. T. A. Gover and P. G. Sears, *J. Phys. Chem.*, **60**, 331 (1956)
31. H. K. Venkatasetty and G. H. Brown, *J. Phys. Chem.*, **66**, 2075 (1962)
32. D. F. Evans and P. Gardam, *J. Phys. Chem.*, **73**, 158 (1969)
33. G. J. Janz and M. J. Tait, *Canad. J. Chem.*, **45**, 1101 (1967)
34. P. W. Brewster, F. C. Schmidt and W. B. Schaap, *J. Amer. Chem. Soc.*, **81**, 5532 (1959)
35. V. S. Griffiths, K. S. Lawrence and M. L. Pearce, *J. Chem. Soc.*, 3998 (1958)
36. G. W. A. Fowles and W. R. McGregor, *J. Phys. Chem.*, **68**, 1342 (1964)
37. A. M. Harstein and S. Windwer, *J. Phys. Chem.*, **73**, 1549 (1969)
38. P. H. Tewari and G. P. Johari, *J. Phys. Chem.*, **69**, 2857 (1965)
39. G. P. Johari and P. H. Tewari, *J. Phys. Chem.*, **69**, 696 (1965)
40. G. P. Johari and P. H. Tewari, *J. Phys. Chem.*, **69**, 2862 (1965)
41. G. P. Johari and P. H. Tewari, *J. Phys. Chem.*, **69**, 3167 (1965)
42. M. D. Archer and R. P. H. Gasser, *Trans. Faraday Soc.*, **62**, 3451 (1966)
43. P. G. Sears, R. K. Wolford and L. R. Dawson, *J. Electrochem. Soc.*, **103**, 633 (1956)
44. L. M. Mukherjee and D. P. Boden, *J. Phys. Chem.*, **73**, 3965 (1969)
45. R. M. Fuoss and E. Hirsch, *J. Amer. Chem. Soc.*, **82**, 1013 (1960)
46. L. G. Savedoff, *J. Amer. Chem. Soc.*, **88**, 664 (1966)

47. D. F. Evans, C. Zawoyski and R. L. Kay, *J. Phys. Chem.*, **69,** 3878 (1965)
48. J. F. Coetzee and D. K. McGuire, *J. Phys. Chem.*, **67,** 1810 (1963)
49. R. H. Boyd, *J. Phys. Chem.*, **65,** 1834 (1961)
50. S. R. C. Hughes, *J. Chem. Soc.*, 634 (1957)
51. S. R. C. Hughes and D. H. Price, *J. Chem. Soc.*, **A,** 1093 (1967)
52. R. L. Kay, B. J. Hales and G. P. Cunningham, *J. Phys. Chem.*, **71,** 3925 (1967)
53. H. L. Yeager and B. Kratochvil, *J. Phys. Chem.*, **73,** 1963 (1969)
54. H. L. Yeager and B. Kratochvil, *J. Phys. Chem.*, **74,** 963 (1970)
55. I. Y. Amhed and C. D. Schmulbach, *J. Phys. Chem.*, **71,** 2358 (1967)
56. C. Treiner and R. M. Fuoss, *Z. Phys. Chem.* (*Leipzig*), **228,** 343 (1965)
57. C. H. Springer, J. F. Coetzee and R. L. Kay, *J. Phys. Chem.*, **73,** 471 (1969)
58. J. F. Coetzee and G. P. Cunningham, *J. Amer. Chem. Soc.*, **87,** 2529 (1965)
59. D. S. Berns and R. M. Fuoss, *J. Amer. Chem. Soc.*, **83,** 1321 (1961)
60. J. Eliassaf, R. M. Fuoss and J. E. Lind, *J. Phys. Chem.*, **67,** 1941 (1963)
61. T. L. Fabry and R. M. Fuoss, *J. Phys. Chem.*, **68,** 907 (1964)
62. A. M. Brown and R. M. Fuoss, *J. Phys. Chem.*, **64,** 1341 (1960)
63. J. J. Banewicz, J. A. Maguire and P. S. Shih, *J. Phys. Chem.*, **72,** 1960 (1968)
64. P. G. Sears, J. A. Caruso and A. I. Popov, *J. Phys. Chem.*, **71,** 905 (1967)
65. E. J. del Rosario and J. E. Lind, *J. Phys. Chem.*, **70,** 2876 (1966)
66. R. L. Kay, S. C. Blum and H. I. Schiff, *J. Phys. Chem.*, **67,** 1223 (1963)
67. H. Sadek and R. M. Fuoss, *J. Amer. Chem. Soc.*, **81,** 4507 (1959)
68. F. R. Longo, J. D. Kerstetter, T. F. Kumonsinski and E. C. Evers, *J. Phys. Chem.*, **70,** 431 (1966)
69. D. N. Bhattacharyya, C. L. Lee, J. Smid and M. Szwarc, *J. Phys. Chem.*, **69,** 608 (1965)
70. C. Treiner and J-C. Justice, *Compt. Rend.*, **269,** 1364 (1969)
71. L. L. Böhm and G. V. Schulz, *Ber. Bunsenges. Phys. Chem.*, **73,** 260 (1969)
72. C. Carvajal, K. J. Tölle, J. Smid and M. Szwarc, *J. Amer. Chem. Soc.*, **87,** 5548 (1965)
73. W. R. Gilkerson and J. B. Ezell, *J. Amer. Chem. Soc.*, **87,** 3812 (1965)
74. Y. H. Inami, H. K. Bodenseh and J. B. Ramsey, *J. Amer. Chem. Soc.*, **83,** 4745 (1961)
75. K. H. Stern and A. E. Martell, *J. Amer. Chem. Soc.*, **77,** 1983 (1955)
76. J. J. Zwolenik and R. M. Fuoss, *J. Phys. Chem.*, **68,** 903 (1964)
77. H. L. Curry and W. R. Gilkerson, *J. Amer. Chem. Soc.*, **79,** 4021 (1957)
78. F. Accascina, E. L. Swarts, P. L. Mercier and C. A. Kraus, *Proc. Nat. Acad. Sci.*, **39,** 917 (1953)
79. A. Höniger and H. Schindlbauer, *Ber. Bunsenges. Phys. Chem.*, **69,** 138 (1965)
80. R. Fernández-Prini and J. E. Prue, *Trans. Faraday Soc.*, **62,** 1257 (1966)
81. M. Della Monica, U. Lamanna and L. Jannelli, *Gazz. Chim.*, **97,** 367 (1967)
82. M. Della Monica and U. Lamanna, *Gazz. Chim.*, **98,** 256 (1968)
83. M. Della Monica and U. Lamanna, *J. Phys. Chem.*, **72,** 4329 (1968)
84. H. V. Venkatasetty and G. H. Brown, *J. Phys. Chem.*, **67,** 954 (1963)
85. G. Pistoia and G. Pecci, *J. Phys. Chem.*, **74,** 1450 (1970)
86. A. L. Powell and A. E. Martell, *J. Amer. Chem. Soc.*, **79,** 2118 (1957)
87. G. P. Johari and P. H. Tewari, *J. Amer. Chem. Soc.*, **87,** 4691 (1965)
88. G. P. Johari, *J. Phys. Chem.*, **74,** 934 (1970)
89. F. Conti, P. Delogu and G. Pistoia, *J. Phys. Chem.*, **72,** 1396 (1968)

Chapter 5

Conductance and Transference Numbers

Part 2

Transference Numbers

M. Spiro*
Department of Chemistry,
Imperial College of Science and Technology,
London, S.W.7, England

5.7 INTRODUCTION

Conductance and transference number are complementary properties of an electrolyte solution: the former depends on the sum of ionic conductances, the latter on their ratio. A knowledge of both properties is therefore essential for a proper understanding of ionic behaviour in any solvent medium. It is surprising that until a few years ago the abundance of non-aqueous conductance figures in the literature was matched only by a dearth of reliable transference numbers, and even today the disparity between the two sets of data is remarkable. The reason lies in the

* This chapter was written while the author was on leave in the Chemistry Department, University of Otago, Dunedin, New Zealand, July–Dec. 1970.

greater experimental difficulties met with in measuring transference numbers, and these difficulties, and ways of overcoming them, are discussed in sect. 5.9. Later in this chapter we shall also describe, and evaluate, two approximate procedures for dividing up limiting conductances between the constituent ions in the absence of transference numbers.

5.8 DEFINITIONS AND RELATIONSHIPS

Let us consider first a solution containing a completely dissociated electrolyte of molarity c (mol l^{-1} or mol dm^{-3}) made up of cations A^{z+} and anions X^{z-}. The transference number t of either ion is then defined as the number of faradays of electricity carried by the ion concerned across a reference plane, fixed with respect to the solvent, when one faraday of electricity passes across the plane. Since the number of faradays transported by any ionic species i depends directly upon the magnitude of its algebraic charge number z_i, its molarity c_i, and its mobility u_i (i.e. its velocity in unit electric field), it follows that[M1]

$$\frac{t_+}{t_-} = \frac{|z_+|c_+u_+}{|z_-|c_-u_-} = \frac{c_+\Lambda_+}{c_-\Lambda_-} \tag{5.8.1}$$

where Λ_i is the ionic conductance given by

$$\Lambda_i = |z_i|u_iF \tag{5.8.2}$$

In this chapter Λ_i is quoted in units of $cm^2\ \Omega^{-1}\ mol^{-1}$. F is the Faraday constant (96487 C mol^{-1}). Equation 5.8.1 leads directly to

$$t_\pm = \frac{c_\pm\Lambda_\pm}{c_+\Lambda_+ + c_-\Lambda_-} = \frac{c_\pm\Lambda_\pm}{c\Lambda} = \frac{\nu_\pm\Lambda_\pm}{\Lambda} \tag{5.8.3}$$

$$t_+ + t_- = 1 \tag{5.8.4}$$

where Λ is the molar conductance of the electrolyte $A_{\nu_+}X_{\nu_-}$. The subscript $\pm$ means that the equation applies to the cation when all the upper subscripts are taken, and to the anion for the lower subscripts. In the case of electrolytes of symmetrical valence type $z_+ = |z_-|$ and, almost invariably, $\nu_+ = \nu_- = 1$. It is electrolytes of this kind, and chiefly 1:1 electrolytes at that, which have been employed for accurate transference measurements in organic solvents (Appendix 5.8.1.)

The simple treatment above is inadequate whenever the electrolyte in solution produces more than two kinds of ion. Suppose there is present a variety of species such as A, AX, AX_2, . . ., X. The transference number of each one—the number of faradays each carries across a reference plane—cannot be measured because of the normally rapid dynamic equilibria between the species. What *can* be measured, however, is the *net* number of faradays carried by the ion-constituent or radical A, and by the ion-constituent or radical X, as the result of the migration of all the various ionic and molecular species. This is easy to do, at least in principle, for the net amount of A (or X) transported across a plane can be analysed and the number of faradays involved calculated from the known charge number. Thus, quite generally, we define the transference number t_R, of a cation- or anion-constituent R, as the net number of faradays carried by that constituent in the direction of the cathode or anode, respectively, across a reference plane fixed with respect to the solvent, when one faraday of electricity passes across the plane. Numerically, t_R equals the net number of moles of R transported for every z_R faradays. It has been shown in detail elsewhere[M1] that

$$t_R = \frac{\sum_i (z_R/z_i) N_{R/i} c_i \Lambda_i}{\sum_i c_i \Lambda_i} \tag{5.8.5}$$

$$\sum_R t_R = 1 \tag{5.8.6}$$

where $N_{R/i}$ is the number of moles of ion-constituent R in one mole of ion i.

An example is provided by the formation of triple ions in solvents of low dielectric constant.[M2] If the ionic species present in such a solution of, say, LiBr are Li^+, Br^-, $LiBrLi^+$ and $BrLiBr^-$, then

$$t_{Li^+} = \frac{c_{Li^+}\Lambda_{Li^+} + 2c_{LiBrLi^+}\Lambda_{LiBrLi^+} - c_{BrLiBr^-}\Lambda_{BrLiBr^-}}{c_{Li^+}\Lambda_{Li^+} + c_{Br^-}\Lambda_{Br^-} + c_{LiBrLi^+}\Lambda_{LiBrLi^+} + c_{BrLiBr^-}\Lambda_{BrLiBr^-}} \tag{5.8.7}$$

Application of this equation will be deferred to sect. 5.10.2.

5.9 SOME EXPERIMENTAL ASPECTS

The detailed theory and mode of operation of the main experimental methods of obtaining transference numbers—Hittorf, direct and indirect moving boundary, analytical boundary, e.m.f. of cells with transference or of cells in centrifugal fields—have been published elsewhere.[M1] Only the features particularly pertinent to work with electrolytes in organic solvents will be dealt with here.

Joule heating can be a major problem. Both Hittorf and moving boundary methods require the passage of current, and the solutions are frequently of high electrical resistance. This arises in part because the concentrations of electrolyte are low, for reasons of solubility or boundary stability, and partly as a result of the nature of the solvent itself. Many organic solvents have low dielectric constants which cause ion association, and others possess a high viscosity and the ionic mobilities are small in consequence. Moreover, the thermal conductivities of the organic liquids of interest[M3] are smaller than the thermal conductivity of water by a factor of roughly 3, and the Joule heat generated causes correspondingly greater convection currents.[M4]

Excessive Joule heating in Hittorf experiments might be obviated by selecting appropriate cell design and operating conditions, such as short distance between electrodes, wide tubing, low currents and long times of electrolysis. These, however, are the very factors that cause diffusive mixing between the cell compartments. In practice compromise decisions must be made and the precision of the results suffers. One suggestion for cutting down interdiffusion is to employ several narrow and curved tubes in the middle section of the cell instead of a wide tube.[M5] The middle compartments, incidentally, are best placed off-centre when ionic mobilities differ widely[M6] as, for example, when one ion moves by proton-jumping and the other is hindered by high solvent viscosity.

In moving boundary experiments Joule heating causes the solution along the axis of the tube to rise. This produces a curved boundary at moderate currents and breaks the boundary up completely at high ones. Very low currents must therefore be employed together with appropriate means of boundary detection, electrical probes coupled with either resistance or potential[M7] measurements being particularly suitable. Isotopic boundary systems with a radioactive ion also seem promising.[M8] Joule heating problems as such do not arise in the e.m.f methods of determining transference numbers. The difficulties that were at one time encountered[M9,M10] in measuring the e.m.f. of a cell of high internal resistance with a potentiometer are now easily overcome with electrometer devices.

An irritating problem is that of leakage at taps and joints. Mobile liquids like methanol that are also good solvents for greases are especially prone to be troublesome. One remedy is to use other lubricants (e.g. lithium stearate + oil,[M11] a fluorocarbon lubricant,[M12] or silicone grease[M13]), another is to employ teflon sleeves, barrels or joints that do not require greasing. Modification of cell design provides a third solution, and boundary formation by an 'air bubble'[M14] or better still, by the recently developed 'flowing junction',[M7] are excellent examples.

Undoubtedly the most accurate transference number method is that of moving boundaries. It sometimes happens that several following electrolytes, chosen to meet certain theoretical conditions,[M15] form detectable and at first apparently promising boundaries with a given leading electrolyte, only for later investigations to reveal instability, progression, irreproducibility or some other deficiency in the systems selected. The most glaring example is acetonitrile where only one electrolyte out of 17 tried gave reproducible results.[M16] In non-aqueous solvents it is particularly important to test the correctness of the results obtained by varying the current and the concentration and type of following electrolyte, and to measure, wherever possible, the velocities of both cation and anion boundaries. Boundaries in organic solvents tend to become distorted at higher concentrations[M11] and it is often necessary to set an upper concentration limit of ca. 0.01 molar.

The success of all transference methods depends in large measure on the availability of suitable electrodes. These can be difficult to find in certain organic media[M17] and rarely is there enough information in the literature to enable selection to be made on paper. Two different kinds of electrode are involved in transference measurements. Hittorf and moving boundary work demand electrodes capable of passing several coulombs for one or more hours with known or ascertainable electrode reactions. These may differ considerably from their aqueous counterparts and solvent oxidation or reduction is frequently encountered. A special requirement of moving boundary experiments is that one of the two working electrodes be completely non-gassing. The e.m.f. methods, on the other hand, call for electrochemically reversible electrodes which, however, need not sustain any prolonged current flow. For e.m.f. work it is essential that the potentials be constant and reproducible to 0.1 mV or better; in organic solvents few electrodes can satisfy this condition.

Finally, the quality of the solvent itself is clearly of importance. The great care normally lavished on elaborate purification must not be wasted by careless cell filling, and solutions are generally prepared under dry nitrogen and transferred by nitrogen pressure. The effect of water as an impurity has seldom been tested. In both acetonitrile[M16] and formamide[M18] addition of small amounts of water caused no measurable change in the transference number and only in ethylene glycol,[M13] where large amounts of water were deliberately added, did the transference number vary significantly. A change of 0.002 was produced by $0.5M$ (0.8 w/w %) water.

5.10 VARIATION WITH CONCENTRATION

First, a general comment: transference numbers, being essentially conductance ratios, are far less sensitive than conductances to changes in concentration, or, for that matter, to changes in solvent composition, temperature or pressure. A typical example is provided by $0.01M$ NaCl in methanol at 25°C: the cation transference number is 1.1% less than the limiting value,[M11] the conductance is 21% lower.[M19] Again, the cation transference number of $0.1M$ KCl in formamide at 25°C is 1.6% below the value at zero concentration[M18] whereas the conductance is 16% less.[M20] Extrapolations or interpolations of transference data are therefore subject to much less uncertainty than those of conductances. On the other hand, the larger conductance concentration shifts give more precise information about the distances of closest approach.

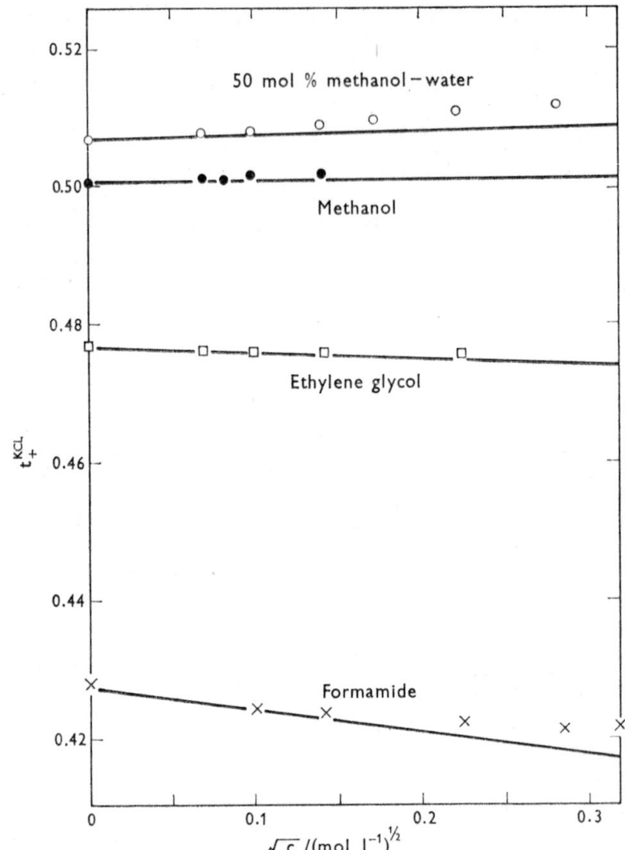

FIG. 5.10.1. Dependence on concentration of the cation-constituent transference number of KCl at 25°C in 50 mol % methanol–water[M22] (○), methanol[M11] (●), ethylene glycol[M13] (□) and formamide[M18] (×). The straight lines are the theoretical limiting slopes according to eqn 5.10.3.

5.10.1 Solvents of Moderate and High Dielectric Constant

For reasons given later the discussion in this section will be written in terms of symmetrical and completely dissociated electrolytes. In the event of ion association the transference number is not affected directly since c_+ always equals c_- and these cancel out in eqn. 5.8.3. There remains the indirect effect, in that the concentration c in the interionic terms must be replaced by the ionic concentration αc. The variation of the transference number with concentration is then even smaller than is the case with strong electrolyte solutions. By contrast, ion association affects conductances directly through the degree of dissociation as well as in the interionic terms (eqns. 5.3.3 and 5.3.4), and the concentration dependence is greatly accentuated.

The variation of transference numbers with concentration in dilute solutions can be explained by the now classic interionic attraction theory of Debye and Hückel.[M21a,b] Using a model of hard-sphere ions in a dielectric continuum, they found that the interplay of order-producing coulombic forces and the randomising effect of thermal motion forms around each ion a diffuse, radially symmetric, ionic atmosphere. How this influences the motion of the ion in an electric field has been described in sect. 5.2.1 and Fig. 5.2.1. In brief, the ion is slowed down both by an electrophoretic effect, caused by the counterflow of solvent in the atmosphere, and by an asymmetry or relaxation effect. This arises because the ion and its atmosphere travel in opposite directions; the latter cannot instantaneously rearrange itself symmetrically around the ion and so exerts a net electrostatic pull backwards. The resulting eqn. 5.2.1, with 5.2.3, for the molar conductance of a symmetrical electrolyte is often known as the limiting Debye–Hückel–Onsager (D–H–O) equation or more simply as the Onsager equation, and is valid in extremely dilute solutions. The corresponding equation for the molar conductance Λ_i of any given ion i at molarity c is

$$\Lambda_i = \Lambda_i^\circ - (\alpha\Lambda_i^\circ + \tfrac{1}{2}\beta)\sqrt{c} \tag{5.10.1}$$

Λ_i° is the limiting molar conductance at infinitesimal ionic strength, and the relaxation and electrophoretic parameters α and β, respectively, are defined in eqn. 5.2.3. From eqns. 5.8.3 and 5.10.1 the transference number of a completely dissociated symmetrical electrolyte is given by

$$\begin{aligned} t_\pm &= \frac{\Lambda_\pm^\circ - (\alpha\Lambda_\pm^\circ + \tfrac{1}{2}\beta)\sqrt{c}}{\Lambda^\circ - (\alpha\Lambda^\circ + \beta)\sqrt{c}} \\ &= t_\pm^\circ + \frac{(t_\pm^\circ - 0.5)\beta\sqrt{c}}{\Lambda'} \end{aligned} \tag{5.10.2}$$

where

$$\Lambda' = \Lambda^\circ - (\alpha\Lambda^\circ + \beta)\sqrt{c}$$

$t_\pm^\circ$ is the limiting transference number at infinitesimal ionic strength and equals $\Lambda_\pm^\circ/\Lambda^\circ$. It follows that at very low concentrations the transference number varies linearly with $\sqrt{c}$, the limiting slope being

$$\left(\frac{\mathrm{d}t_\pm}{\mathrm{d}\sqrt{c}}\right)_{c\to 0} = \frac{(t_\pm^\circ - 0.5)\beta}{\Lambda^\circ} \tag{5.10.3}$$

Clearly transference numbers greater than 0.5 should increase with increasing concentration, those smaller than 0.5 should decrease, and transference numbers close to 0.5 should be virtually concentration independent. On the whole these qualitative conclusions are borne out by the facts, as Fig. 5.10.1 demonstrates, but several instances are known where the trend is different. The transference numbers of HCl in 50 mol % H_2O—EtOH[M23] and of KBr in MeOH[M24] vary in the opposite direction to that predicted, and those of KSCN in MeOH,[M24] EtOH,[M24] and dimethyl formamide,[M25] though not equal to 0.5, are nearly constant over a range of concentrations. Confirmation of some of these observations would be of interest.

In the diagram the experimental points approach the limiting D–H–O slopes as the concentration falls. In comparing data obtained with different solvents one notes that the magnitude of the limiting slope depends not only upon ($t^\circ_\pm$ − 0.5) but also on β/Λ° and therefore, since

$$\beta = \frac{82.487|z|^3}{\eta(\varepsilon_r T)^{\frac{1}{2}}} \Big/ \Omega^{-1}\,\mathrm{cm}^2\,\mathrm{mol}^{-\frac{3}{2}}\,\mathrm{l}^{\frac{1}{2}}, \tag{5.2.3}$$

should be greater the lower the dielectric constant ε_r and the viscosity η (in poise) of the medium. This last point deserves comment. Were Stokes' law valid, or even Walden's rule (eqns. 5.13.1 and 5.13.8), the limiting slope would be independent of η since Λ° appears in the denominator. In fact, however, limiting conductances vary less rapidly with viscosity than Stokes' law demands,[M26–28] and so the viscosity of the medium still inversely affects the limiting D–H–O transference slope.

Equation 5.10.2 holds well in extremely dilute solutions but is no longer quantitatively successful in the concentration range in which measurements are normally made (cf. Fig. 5.10.1). To overcome this problem the equation may be rearranged to read

$$t^\circ_\pm = \frac{t_\pm \Lambda' + \frac{1}{2}\beta\sqrt{c}}{\Lambda' + \beta\sqrt{c}} \tag{5.10.4}$$

At moderate concentrations $t^\circ_\pm$ so calculated is not constant—call it $t^{\circ\prime}_\pm$—and varies linearly with c:

$$t^{\circ\prime}_\pm = t^\circ_\pm + bc \tag{5.10.5}$$

This useful empirical relation is due to Longworth[M29] and seems entirely reasonable in form, since theory shows that terms in c are the next to appear in extensions of eqn. 5.10.1 to higher concentrations. Equation 5.10.5 has proved very popular and most workers in the field have employed it to extrapolate transference numbers to zero concentration (see Appendix 5.8.1).

The limiting slope in eqn. 5.10.3 depends only on the electrophoretic parameter β, the relaxation factor α having disappeared. A later treatment by Stokes,[M30] generalised by Kay and Dye[M31] and based on the D–H–O theory as modified by Fuoss and Onsager,[M32,M33] showed transference numbers to be entirely free of the relaxation effect at *all* (low) concentrations, a point originally made by Onsager.[M34] The essential relation is the Fuoss–Onsager conductance eqn. 5.2.27 written for an individual ionic species i

$$\Lambda_i = (\Lambda_i^\circ - \Lambda_i^e)\left(1 + \frac{\Delta \mathbf{X}_i}{\mathbf{X}}\right) \tag{5.10.6}$$

where Λ_i^e is the electrophoretic term and $\Delta\mathbf{X}_i/\mathbf{X}$ the so-called relaxation field. Here $\mathbf{X}$ is the applied field and $\Delta\mathbf{X}_i$ the additional field at the ion i caused by the ion-atmosphere asymmetry; since $\Delta\mathbf{X}_i$ is in the opposite direction to $\mathbf{X}$ the sign of $\Delta\mathbf{X}_i/\mathbf{X}$ is always negative. The basic assumption made in writing eqn. 5.10.6 is to regard as negligible certain interactions between the electrophoretic and relaxation effects and, in particular, to omit the way in which the asymmetry potential affects the local velocity of the solvent near the ions. The applicability of eqn. 5.10.6 to proton-jumping ions is discussed in sect. 5.14.

Two other consequences emerge from the theory.[M30–33,M21b] For an electrolyte dissociating into only two kinds of ions

$$\Delta\mathbf{X}_+ = \Delta\mathbf{X}_- \tag{5.10.7}$$

and if the electrolyte is also of symmetrical charge type

$$\Lambda_+^e = \Lambda_-^e = \Lambda_\pm^e = \tfrac{1}{2}\Lambda^e \tag{5.10.8}$$

If it is not of this type the degree of self-consistency of the interionic attraction theory is relatively low.[M30,M21a] Since, moreover, virtually no transference numbers have been accurately measured in organic solvents for other than 1:1 electrolytes, we are restricting the present treatment to symmetrical electrolytes. Then combination of eqns 5.8.3, 5.10.6, 5.10.7 and 5.10.8 gives

$$t_\pm = \frac{\Lambda_\pm^\circ - \frac{1}{2}\Lambda^e}{\Lambda^\circ - \Lambda^e} \tag{5.10.9}$$

The relaxation field, it should be noted, has cancelled out completely. Rearranging,

$$t_\pm = t_\pm^\circ + \frac{(t_\pm^\circ - 0.5)\Lambda^e}{\Lambda^\circ - \Lambda^e} \tag{5.10.10}$$

$$= t_\pm^\circ + \frac{(t_\pm - 0.5)\Lambda^e}{\Lambda^\circ} \tag{5.10.11}$$

The form of the electrophoretic function Λ^e depends upon the particular version of the interionic attraction theory chosen. According to the limiting D–H–O theory

$$\Lambda^e = \beta\sqrt{c} \tag{5.10.12}$$

This transforms 5.10.10 into an expression that bears a striking resemblance to eqn. 5.10.2 and, like the latter, leads to the limiting slope 5.10.3. When allowance is made for the finite sizes of the ions and a closest distance of approach between cation and anion, *a*, is introduced, the Fuoss–Onsager prediction[M32,M33] is (cf. eqn. 5.2.21):

$$\Lambda^e = \frac{\beta\sqrt{c}}{1 + \kappa a} \tag{5.10.13}$$

with the Debye–Hückel distance κ defined in sect. 5.2. However, the right-hand side of eqn. 5.10.13 is only the first term of a series.[M30] The second and other even terms are zero for symmetrical electrolytes and for these the ratio of the third term to the first is of the order z^2/a^2. The series therefore converges rapidly only if z is unity, and 1:1 electrolytes alone can provide a meaningful test of the equation

$$t_\pm = t^\circ_\pm + \frac{(t_\pm - 0.5)\beta\sqrt{c}}{\Lambda^\circ(1 + \kappa a)} \tag{5.10.14}$$

This equation promised to explain the observed deviations from the limiting D–H–O slope and to follow the experimental concentration dependence of $t_\pm$ up to $\kappa a = 0.2$ or even, since the electrophoretic effect alone is involved, up to higher concentrations.[M35] Its fit to aqueous data[M30,M21b] seemed fairly encouraging but a scrutiny based on the most accurate (moving boundary) data available for organic solvents exposes the theory as unsatisfactory. Table 5.10.1 summarises the evidence. For each solution a value of *a* has been sought that would lead to concentration-invariant values of $t^\circ_\pm$ in eqn. 5.10.14. In five cases only out of 18, those of NaCl in methanol and the four salts in nitromethane, are the distances of closest approach numerically reasonable and in fair agreement with the figures derived from conductances (Appendix 5.1). Negative values of *a* fit the data for four salts in alcoholic solvents and one in acetone, which means that the experimental points approach the limiting slope from the wrong side. Frequently *a* appears to be too large, in some instances ludicrously so. No physical significance in terms of ionic sizes can be ascribed to such results, and we are forced to conclude that eqn. 5.10.14 fails to describe adequately the concentration dependence of transference numbers in organic solvents.

This failure must arise out of deficiencies in the Onsager–Fuoss treatment of the electrophoretic and relaxation effects. Thus eqn. 5.10.13, based as it is on Stokes' law, may not provide a sufficiently good description of the electrophoretic correction, and extended electrophoretic expressions have recently been published.[M39,M40] More elaborate treatments (sect. 5.2) have also indicated that the relaxation effect would not completely cancel out in the transference number and the neglected

Table 5.10.1

Test of Eqn. 5.10.14 for the Variation of Transference Numbers with Concentration at 25°C

Solvent	Salt	$10^3.\Delta c/$ mol l^{-1}	Ref.	Approx. a/Å needed for const. $t_{\pm}^{0}$
50 mol% methanol-water	NaCl	5–80	M22	12
	KCl	5–80	M22	−8
Methanol	NaCl	3–10	M11	$4._5$
	KCl	5–20	M11	−10*
	KBr	4–10	M24	>20
	KSCN	2.3–10	M24	>20
Ethanol	LiCl	0.9–2.5	M36	−4†
	NaCl	1–2.5	M36	−9†
	KSCN	1.5–7.5	M24	ca. 20‡
Ethylene glycol	KCl	5–50	M13	9
Acetone	KSCN	1.1–2.4	M38	−12§
Formamide	KCl	10–100	M18	8
Dimethylformamide	KSCN	6–13	M25	15
Acetonitrile	Me_4NClO_4	0.6–12.5	M16	8†
Nitromethane	Me_4NCl	0.2–10	M37	4‖
	Me_4NBr	0.2–10	M37	$3._5$‖
	Et_4NCl	0.2–10	M37	$4._5$
	Et_4NBr	0.2–10	M37	$3._5$

* t_{+}^{0} is very close to 0.5 and the transference numbers are consistent with a wide range of a values.

† The results are only marginally affected by allowing for association to the extent given in Appendix 5.1.

‡ No reliable association constant available. The transference numbers show some concentration scattering.

§ After allowing for association to the extent given in ref. M38.

‖ After allowing for association to the extent given in Appendix 5.1.

interaction between the two effects could become significant.[M41] However, a recently attempted fit of transference data by the Pitts[M42] equation in the author's laboratory has fared little better.[M43] In any case, improvements in theory of the kind just mentioned hardly seem enough to account for the breakdown of eqn. 5.10.14, an equation that proceeds no more than one step beyond the limiting law. It must be remembered that the similarly based Robinson–Stokes[M44,M21b] and Fuoss–Onsager[M45] equations describe moderately well the conductances of electrolyte solutions. One is indeed forced to wonder whether the conductance sum cancels out some aspect of theory that is revealed only by testing the conductance ratio.

One aspect which affects the conductance sum not at all[M46] while strongly influencing the individual ionic mobility and the transference number, is the frame of reference with respect to which the ionic motions are considered. The basic equation is

$$\mathbf{v}_i = u_i\mathbf{X} \tag{5.10.15}$$

which relates the velocity of a charged species i to the applied field $\mathbf{X}$. The velocity is generally regarded as being with respect to the solvent as a whole, as is the transference number in the definition in sect. 5.8. In one of the few explicit statements on this point,[M47] the ion in eqn. 5.10.15 was said to 'move relatively to the surrounding solvent'. It could be argued from this that the frame of reference appropriate for the motion of a (solvated) ion is provided by the 'free' unsolvated solvent molecules,[M48] those that are not bound to the ions and travelling with them. If so, it follows that the Onsager–Fuoss equation should apply to the so-called 'true' transference numbers 't' based on a free-solvent reference plane (see sect. 5.11) and not to the observed values which are determined relative to the solvent as a whole. This is the interpretation which Carman[M35] has recently favoured. He believes that eqn. 5.10.14 holds for transference numbers on a Darken frame of reference, one which depends on the (real or imagined) presence in the system of inert markers[M49,M50] which in this case take the shape of the 'free' molecules of solvent. Substituting (from eqn. 5.11.3) '$t_\pm$' for $t_\pm$ in eqn. 5.10.9 leads to Carman's equation (28):

$$t_\pm = \left(\frac{\Lambda_\pm^\circ - \frac{1}{2}\Lambda^e}{\Lambda^\circ - \Lambda^e}\right)\left(1 - \frac{c}{c_1}\left[\omega_+ + \omega_-\right]\right) + \frac{c\omega_\mp}{c_1} \tag{5.10.16}$$

where c_1 is the molarity of the solvent and ω_+, ω_- the solvation numbers of cations and anions.

To test Carman's hypothesis we can rearrange eqn. 5.10.16 and incorporate the Fuoss–Onsager electrophoretic term 5.10.13 to give the convenient expression

$$t_\pm = t_\pm^\circ + \frac{(t_\pm - 0.5)\beta\sqrt{c}}{\Lambda^\circ(1 + \kappa a)} + \frac{cW}{c_1} \tag{5.10.17}$$

$$W = \omega_\mp - t_\pm^\circ(\omega_+ + \omega_-) - \frac{\beta\sqrt{c}}{\Lambda^\circ(1 + \kappa a)}\left[\omega_\mp - 0.5(\omega_+ + \omega_-)\right] \tag{5.10.18}$$

$$= \omega_\mp - t_\pm^\circ(\omega_+ + \omega_-) \quad \text{[at low concentrations]} \tag{5.10.19}$$

It is the term in W that distinguishes 5.10.17 from 5.10.14. In principle the presence of W must permit a better representation of transference behaviour because it introduces a new linear concentration term and two new parameters, ω_+ and ω_-. In practice, however, the real test of eqn. 5.10.17 lies in the physical reasonableness of the numerical values of these parameters.

The procedure adopted has been to assign to the distances of closest approach a constant value of 4 Å for all solutes, and then to calculate the values of W that lead to constant limiting transference numbers in eqn. 5.10.17. The results are summarised in Table 5.10.2, the experimental moving boundary data being those quoted in Table 5.10.1. As

Table 5.10.2

Test of Eqns. 5.10.17 and 5.10.19 for the Variation of Transference Numbers with Concentration at 25°C

Solvent	c_1 mol l^{-1}	Salt	t_+^0	W/c_1 l mol^{-1}	W*	ω_- if $\omega_+ = 0$	ω_- if $\omega_+ = 4$
50 mol% methanol-water	35.21	NaCl	0.4437	+0.13	+4.6	8.2	11.5
		KCl	0.5068	+0.05	+1.8	3.6	7.7
Methanol	24.55	NaCl	0.4634	+0.02	+0.5	0.9	4.4
		KCl	0.5001†	+0.08†	+2.0†	4.0	8.0
		KBr	0.4795‡	+0.24	+5.9	11.3	15.0
		KSCN	0.4555	+0.28	+6.9	12.7	16.0
Ethanol	17.04	KSCN	0.4612	+0.40	+6.8	12.6	16.0
Ethylene glycol	17.88	KCl	0.4765	+0.01	+0.2	0.4	4.0
Formamide	25.07	KCl	0.4270	+0.02	+0.5	0.9	3.8
Dimethylformamide	12.91	KSCN	0.340	+0.5	+6.6	10.0	12.1
Acetonitrile	18.92	Me_4NClO_4	0.4768	+0.04	+0.8	1.5	5.2
Nitromethane	18.53	Me_4NCl	0.4674	+0.02	+0.4	0.8	4.3
		Et_4NCl	0.4321	0.00	0.0	0.0	3.0
						ω_+ if $\omega_- = 0$	ω_+ if $\omega_- = 4$
Nitromethane	18.53	Me_4NBr	0.4663	−0.01	−0.2	0.4	5.0
		Et_4NBr	0.4314	−0.01	−0.2	0.5	5.7
Ethanol	17.04	LiCl	0.4392	−0.90	−15.3	34.8	39.8
		NaCl	0.4813	−0.46	−7.8	16.2	20.6
Acetone	13.50	KSCN	0.376_3	−3.0	−40.5	108	114

* W, ω_+ and ω_- are usually ±0.5.
† If $t_+^0 = 0.5006$, $(W/c_1) = +0.02$ and $W = +0.5$.
‡ The value calculated from the ionic conductances in Appendix 5.12.2 is 0.4816.

expected, W is close to zero in those cases in which the transference numbers in Table 5.10.1 fitted eqn. 5.10.14 with a values close to 4 Å. Significantly negative W values occur only for LiCl and NaCl in ethanol, and KSCN in acetone, the salts whose cation-constituent transference numbers are less than 0.5 and which, in Table 5.10.1, required negative a distances. On the basis of W alone, however, judgement cannot be passed as to its fitness: it is the magnitudes of the derived solvation numbers that will prove decisive. They are listed in the last two columns of Table 5.10.2, calculated from eqn. 5.10.19. In the majority of solutions the anion solvation number is seen to exceed, and sometimes to exceed greatly, the cation solvation number, a result that surprises. The mere fact that almost all the limiting cation transference numbers are less than 0.5 shows that the migrating cations are slower than their anionic partners; slower, and hence larger, and therefore more strongly solvated, for most of the bare cations are smaller than their respective bare anions (Table 5.12.1). Certain counter-arguments could be produced—the non-applicability of even the qualitative Stokes' law approach employed, the inclusion in the solvation numbers of solvent molecules swept along by momentum transfer. But on balance the evidence remains unfavourable. In three instances only, LiCl and NaCl in ethanol and KSCN in acetone, is ω_+ appreciably greater than ω_-, and here the cation solvation numbers are of a magnitude so enormous as to make them impossible. It appears, then, that eqn. 5.10.17 falls short in just those cases in which eqn. 5.10.14 failed, and Carman's hypothesis[M35] has not been substantiated. Whatever the truth of the model used, however, eqn. 5.10.17 does provide a convenient semi-empirical relationship on the lines of the Longsworth eqn. 5.10.5.

5.10.2 Solvents of Low Dielectric Constant

The conductance-concentration curves of electrolytes in solvents of dielectric constant below about 15 contain minima which appear at lower and lower concentrations as the dielectric constant falls.[M51] In a classic paper,[M52] Fuoss and Kraus proposed as an explanation the formation of triple ions $+-+$ and $-+-$ and even of higher aggregates. In very dilute solutions, simple cations and anions are present and the molar conductance decreases with rising concentration as a result of ion-pairing. At still higher concentrations triplet ions are produced and the molar conductance rises accordingly. Dole[M2] suggested that transference numbers could shed further light on this phenomenon, and Sukhotin[M12,M53] has now reported several appropriate Hittorf determinations which are summarised in Table 5.10.3.

The equation necessary for their interpretation is 5.8.7. At concentrations well below the conductance minimum, where the presence of ionic triplets may be disregarded, the cation-constituent transference number (with LiBr as the example) becomes:

$$t_{Li^+} = \frac{\Lambda_{Li^+}}{\Lambda_{Li^+} + \Lambda_{Br^-}} \tag{5.10.20}$$

If at much higher molarities, above the conductance minimum, the content of single ions is insignificant in comparison with the concentration of triple ions, then $c_{LiBrLi^+} = c_{BrLiBr^-}$ and eqn. 5.8.7 approximates to

$$t_{Li^+} = \frac{2\Lambda_{LiBrLi^+} - \Lambda_{BrLiBr^-}}{\Lambda_{LiBrLi^+} + \Lambda_{BrLiBr^-}} \tag{5.10.21}$$

Table 5.10.3

Cation-Constituent Transference Numbers and Conductance Ratios in Media of Low Dielectric Constant at 25°C

Solvent	ε_r	Salt	Molarity	t_{Li^+}	Ref.	$\frac{\Lambda_{Li^+}}{\Lambda_{X^-}}$	$\frac{\Lambda_{LiXLi^+}}{\Lambda_{XLiX^-}}$
70 wt% butanol + 30 wt% hexane	10.1	LiCl	0.01 (below min.)	0.450 ± 0.011	M53	0.82 ± 0.04	
			0.15 (above min.)	0.443 ± 0.015			0.93 ± 0.02
40 wt% butanol + 60 wt% hexane	4.77	LiBr	0.002 (below min.)	0.453 ± 0.002	M12	0.83 ± 0.01	
			0.1 (above min.)	0.461 ± 0.014			0.95 ± 0.02
10 wt% water + 90 wt% dioxan	5.75	LiI	0.008 (below min.)	0.481 ± 0.027	M53	0.93 ± 0.10	
			0.1 (above min.)	0.455 ± 0.022			0.94 ± 0.03

Sukhotin pointed out[M53] that the transference numbers of any given salt in Table 5.10.3 are, within experimental error, the same in dilute as in concentrated solution. From this he concluded[M53] that triple ions are absent altogether, the only charged species present being lithium cations and halide anions. However, the dielectric constants are so low that the ionic conductances will differ quite appreciably from their limiting values even though the ionic strengths are extremely small ($10^{-5} - 10^{-4}M$)[M12] and the transference numbers should change more with concentration than they do. Indeed, this argument is an essential corollary to his own explanation of the conductance minima which he ascribes to rapid decreases in ionic activity coefficients.[M12]

We shall now show that the results are in fact quite consistent with the existence of ionic triplets. Interpretation of the transference numbers themselves is difficult, and a better prospect is offered by the conductance ratios where such effects as the changing viscosity of the medium will largely cancel out. From eqns. 5.10.20 and 5.10.21, respectively:

$$\frac{\Lambda_{Li^+}}{\Lambda_{Br^-}} = \frac{t_{Li^+}}{1 - t_{Li^+}} \quad \text{(dilute solutions)} \qquad (5.10.22)$$

$$\frac{\Lambda_{LiBrLi^+}}{\Lambda_{BrLiBr^-}} = \frac{1 + t_{Li_+}}{2 - t_{Li_+}} \quad \text{(concd. solutions)} \qquad (5.10.23)$$

The ratios are listed in the last two columns of Table 5.10.3. Inspection shows that the conductance ratio of the triple ions is always closer to 1.0 than is that of the single ions. This is just what would be expected, for 'solvation' of both lithium cation and halide anion by the same lithium halide ion-pair should make the conductances of the new species more equal. Further than this we cannot at present go, since there are insufficient data available to enable the D–H–O terms to be estimated. Accurate measurements of several properties over a wide concentration range are needed before we can hope to understand better the nature of these most interesting solutions.

5.11 WASHBURN NUMBERS

The frame of reference used in the definition of transference numbers in solution, and therefore in their determination, is an imaginary plane fixed with respect to the solvent as a whole (sect. 5.8). Yet some solvent molecules must be solvating and travelling along with the ions while other molecules of solvent remain relatively unaffected by the passage of current and could be regarded as 'free'. This concept has already been referred to in sect. 5.10.1. Were it possible to set up a plane stationary with respect to the 'free' part of the solvent only,[M48] one would be able to determine the 'true' or 'absolute' transference number, 't', as well as the Washburn number[M54] ω. The latter is simply the net number of moles of solvent carried by the electrolyte from anode to cathode per faraday of electricity, and is related to the solvation numbers ω_+ and ω_- of the cations and anions by

$$\omega = \frac{`t\text{'}_+\omega_+}{z_+} + \frac{`t\text{'}_-\omega_-}{z_-} \tag{5.11.1}$$

in the case of a completely dissociated electrolyte. The algebraic charge number z_- is negative, and so ω can be positive, negative or zero. The difference between the 'true' transference number 't' and the ordinary or Hittorf transference number t depends directly on ω, as can easily be shown. If before passage of current the two imaginary reference planes coincide, after it the all-solvent reference plane will be displaced relative to the free-solvent plane in the direction of the cathode by a volume of solution containing ω moles of solvent. Hence

$$`t\text{'}_\pm - t_\pm = z_\pm \nu_\pm c\omega/c_1 \tag{5.11.2}$$

where one litre of the solution investigated contains c_1 moles of solvent as well as ν_+c moles of cation and ν_-c moles of anion. The two transference numbers clearly become identical at zero ionic strength. At any finite concentration they are connected by the relation

$$t_\pm = `t\text{'}_\pm \left\{1 - \frac{c(\omega_+ + \omega_-)}{c_1}\right\} + \frac{c\omega_\mp}{c_1} \tag{5.11.3}$$

obtained by combining 5.11.1 and 5.11.2, and simplified by taking the special case of a 1:1 electrolyte.

Nernst[M55] suggested in 1900 that 't' could be found by adding to the solution a small known amount of a nonelectrolyte which was assumed to remain stationary in the electric field. The movement of solvent and of the various ion-constituents, relative to the inert nonelectrolyte, would then give ω and 't' respectively. Early work on this principle was confined to aqueous systems but within the last decade it has also been applied to water-ethanol[M56] and water-dioxan[M57] mixtures. Unfortunately, however, the results are devoid of any simple physical meaning[M58] because experimental evidence has accumulated[M59] to show that the basic assumption of the method is invalid. On reflection we can see why. The reference substances employed—raffinose, fructose and the like—are polar, partially solvate the ions, and so are no more stationary or inert than the solvent itself. The Washburn numbers obtained are therefore not solvent transference numbers but simply convenient parameters for expressing certain experimental results.

Washburn numbers or a quantity involving them can, however, be determined in mixed solvents without the use of a reference substance because the composition of the solvent itself changes in the electrode compartments. Take, for example, a 1:1 electrolyte in a medium of mole fraction x_1 in component 1 and x_2 in component 2. Equation 5.11.1 tells us that, when f faradays have passed, Δn_1 moles of solvent 1 and Δn_2 moles of solvent 2 are carried out of the anode and into the cathode compartment, where

$$\Delta n_1/f = \omega_1 = `t'_+\omega_{1+} - `t'_-\omega_{1-} \tag{5.11.4a}$$

$$\Delta n_2/f = \omega_2 = `t'_+\omega_{2+} - `t'_-\omega_{2-} \tag{5.11.4b}$$

This alters the composition of the solvent in the cathode compartment (say) from x_2 to x_2^t, and it now contains n_1^t moles of solvent 1 and n_2^t of solvent 2 instead of n_1 and n_2 at the start of the experiment. This composition change is measurable (by density[M60] or refractive index,[M61,M62] for example) and can be related as follows[M60] to the solvation numbers $\omega_{1\pm}$ and $\omega_{2\pm}$:

$$\begin{aligned}\Delta &= (n_1^t + n_2^t)(x_2^t - x_2)/f \\ &= (n_1^t + n_2^t)(x_1x_2^t - x_2x_1^t)/f \\ &= (n_1n_2^t - n_2n_1^t)/(n_1 + n_2)f \\ &= [n_1(n_2^t - n_2) - n_2(n_1^t - n_1)]/(n_1 + n_2)f \\ &= (x_1\Delta n_2 - x_2\Delta n_1)/f \\ &= x_1\omega_2 - x_2\omega_1 \\ &= x_1`t'_+\omega_{2+} - x_2`t'_+\omega_{1+} \\ &\quad - (x_1`t'_-\omega_{2-} - x_2`t'_-\omega_{1-})\end{aligned} \tag{5.11.5}$$

The experiment thus yields one quantity, Δ, which involves the 4 solvation numbers ω_{1+}, ω_{1-}, ω_{2+} and ω_{2-}. A figure for each can be derived only by making further assumptions. Strehlow and his co-workers have measured Δ by the Hittorf method for $CaCl_2$ in H_2O—MeOH mixtures[M61] and for $AgNO_3$[M60] and $ZnCl_2$[M62] in H_2O—CH_3CN mixtures, and it must be pointed out that the changes in solvent composition were at most 1%.[M63] The individual solvation numbers were then estimated by assuming that each ion has a reasonable maximum solvation number, that solvation numbers vary monotonically with solvent composition, and by using Hittorf instead of 'true' transference numbers. The last is a fair approximation in dilute salt solutions as eqn. 5.11.2 shows. The qualitative conclusions drawn from the results, for example that Ag^+ ions are preferentially solvated by CH_3CN molecules whereas NO_3^- ions prefer H_2O, are in agreement with evidence from other sources.[M63]

The above treatment has glossed over the reference state on which the experimental Washburn numbers are based. The data of Strehlow *et al.* seem to refer to the final and initial contents of the cathode compartment of their cell so that their Washburn numbers have been determined relative to the glass walls of the apparatus, with no correction for any increase or decrease of electrode volume. Composition changes should have been calculated with respect to the total mass of solvent in the cathode section. Two extreme but interesting cases arise when one or other of the two solvent components is regarded as fixed. If this is component 1,

$$\Delta n_1 = 0, \qquad \Delta = x_1{}^1 \Delta n_2/f = x_1{}^1 \omega_2$$

If component 2 is considered to be stationary,

$$\Delta n_2 = 0, \qquad \Delta = -x_2{}^2 \Delta n_1/f = -x_2{}^2 \omega_1$$

The connection between these two Washburn numbers is thus:

$$^1\omega_2/^2\omega_1 = -x_2/x_1 \tag{5.11.6}$$

a relation first deduced by Feakins[M64] from eqn. 5.11.2.

Chapter 5

Conductance and Transference Numbers

Part 3

Ionic Conductances

M. Spiro
Department of Chemistry,
Imperial College of Science and Technology,
London, S.W.7, England

5.12 INTRODUCTION

The properties of an electrolyte solution differ from those expected from a mixture of the pure solvent and the bare ions by virtue of ion-solvent and ion–ion interactions. The latter become negligible at infinitesimal ionic strength (a condition often incorrectly termed 'infinite dilution')[M65] and in this state only cation-solvent and anion-solvent interactions remain. Numerous attempts have been made to separate these two, although arguments still rage as to the best manner of doing so.[M66] For one property alone argument is superfluous: the limiting ionic conductance. There is no ambiguity here because not only is the sum of the ionic conductances measurable but also, via the transference number, their ratio. The product of the electrolyte conductance and the transference number, extrapolated to infinitesimal ionic strength as described in sect. 5.4 and 5.10, leads directly through eqn. 5.8.3 to the limiting molar ionic conductance $\Lambda^\circ_\pm$. Its value depends solely on the properties of the solvent, the ion in question, and their interaction with each other, and in this lies its importance.

In Appendices 5.12.1 to 5.12.12 are given tables of limiting molar ionic conductances in those organic solvents for which both accurate conductances and at least reasonably accurate transference numbers are at present available. Inspection of the figures in the appendices yields the following broad generalisations:

1. In every solvent except sulpholane (and HCN[M67]) the conductances of the alkali metal ions increase as the crystal radii (Table 5.12.1) increase, i.e.

$$Li^+ < Na^+ < K^+ < Rb^+ < Cs^+.$$

2. Iodide is the fastest halide ion in all protic solvents with the exception of formamide, and the slowest one in the aprotic solvents apart from acetonitrile.

3. Cs^+ and Cl^- seem suitable ions for testing the effect of charge type: their outer electronic configurations are both inert-gas ones and their radii are approximately equal (Table 5.12.1). Cs^+ is the faster ion in protic solvents other than formamide and HCN[M67] while Cl^- is faster in aprotic media.

4. The conductances of R_4N^+ cations decrease as the size of R increases, in contradistinction to the behaviour of the alkali cations. In propanol,[M68a] iso-propanol[M68b] and butanol,[M68c] however, Me_4N^+ is slower than Et_4N^+.

5. On the whole, organic anions migrate more slowly the larger the anion.

6. The H^+ ion is exceptionally fast in hydroxylic media such as the alcohols and formic acid.[M69]

7. The rule that a given ion is slower the more viscous the solvent holds fairly well for the large (usually organic) ions but frequently breaks down when applied to small inorganic ions (Table 5.12.2).

5.13 THEORIES OF IONIC MOTION

5.13.1 Stokes' Law and its Modifications

Most theoretical treatments of ionic conductance have been based on an equation published well over a century ago by Sir George Stokes[M70] although it was not applied to ions until much later.[M71] Stokes derived hydrodynamically the steady-state velocity v_i with which a large sphere

Table 5.12.1
Ionic Radii (in Ångstroms)

Ion	(a)	(b)	(c)	(d)	(e)
Li^+	0.60	0.94			
Na^+	0.95	1.17			
K^+	1.33	1.49			
Rb^+	1.48	1.63			
Cs^+	1.69	1.82			
Ag^+	1.26				
H_4N^+	1.48				
Me_4N^+			2.83	3.47	3.2
Et_4N^+			3.39	4.00	4.0
Pr_4N^+			3.81	4.52	4.6
Bu_4N^+			(4.15)	4.94	5.0
Am_4N^+			(4.44)	5.29	
$(i\text{-Am})_4N^+$					5.4
Hex_4N^+				5.60	
$Hept_4N^+$				5.88	
Cl^-	1.81	1.64			
Br^-	1.95	1.80			
I^-	2.16	2.05			
$B(i\text{-Am})_4^-$					5.4
BPh_4^-				4.8	

(a) Crystal radii given by L. Pauling, The Nature of the Chemical Bond, 2nd edn. pp. 346, 350, Cornell University Press, Ithaca, N.Y. (1940).

(b) Calculated from X-ray electron densities in crystalline NaCl by B. S. Gourary and F. J. Adrian, *Solid State Phys.*, **10**, 127 (1960), the value for Cs^+ being that given by M. H. Panckhurst, *Rev. Pure Appl. Chem.*, **19**, 45 (1969). More recent electron density determinations by V. Meisalo and O. Inkinen, *Acta Cryst.*, **22**, 58 (1967), for KBr crystals indicate that the radius of K^+ may be higher and that for Br^- lower by about 0.07 Å, and this would affect the values for several other ions.

(c) Calculated from van der Waals' volumes by the method of J. T. Edward, *J. Chem. Educ.*, **47**, 261 (1970), based on the data of A. Bondi, *J. Phys. Chem.*, **68**, 441 (1964). The values for the larger ions are less meaningful. The Me_4N^+ radius is in fair agreement with a figure of 2.67 Å deduced from crystal structures by L. G. Hepler, J. M. Stokes and R. H. Stokes, *Trans. Faraday Soc.*, **61**, 20 (1965).

(d) Estimated from bond lengths and angles, and for the larger ions from molar volumes, by R. A. Robinson and R. H. Stokes, Electrolyte Solutions, 1st edn., Chapter 6, Butterworths, London (1955), and in the case of BPh_4^-, by E. Grunwald, *in* Electrolytes, pp. 74–75 (B. Pesce, ed.), Pergamon, Oxford (1962). The figures for Hex_4N^+ and $Hept_4N^+$ were calculated by Robinson and Stokes' volume formula.

(e) Estimated from Fisher–Taylor–Hirschfelder models by J. F. Coetzee and G. P. Cunningham, *J. Amer. Chem. Soc.*, **87**, 2529 (1965).

Table 5.12.2

Walden Products at 25°C

Solvent	η/cpoise	ε_r	$\Lambda_i^0\eta$/cm² (int. Ω)⁻¹ mol⁻¹ poise						
			K^+	Et_4N^+	Bu_4N^+	$(i\text{-}Am)_4N^+$	Cl^-	Pic^-	BPh_4^-
Acetonitrile	0.3409	36.0	0.285	0.289	0.209	0.194	0.337	0.265	0.199
Methanol	0.5445	32.7	0.286	0.329	0.213	0.193	0.285	0.256	0.199
Nitromethane	0.627	36.7		0.299	0.214	0.198	0.392		0.198
Dimethylformamide	0.796	36.7	0.245	0.282	0.202		0.439	0.298	
Water[M21c]	0.8903	78.3	0.654	0.291	0.173		0.680	0.271	0.18_7[M76]
Ethanol	1.078	24.6	0.254	0.317	0.213		0.236	0.270	
50 mol% water-methanol	1.319	49.8	0.502				0.489		
N-methylformamide	1.65	182.4	0.366	0.432			0.326	0.216	
Dimethylsulphoxide	1.99	46.7	0.299	0.348	0.235	0.219	0.478	0.334	0.203
N-methylacetamide (40°C)	3.019	165.5	0.251	0.347	0.229		0.353	0.361	
Formamide	3.302	109.5	0.421	0.364	0.226		0.565	0.301	0.199
Sulpholane (30°C)	10.29	43.3	0.416	0.405	0.284		0.956	0.547	
Ethylene glycol	16.84	40.7	0.778	0.370	0.255		0.854	0.373	

of radius r_i would move through a continuous incompressible fluid of viscosity η when subjected to a force **F**, and obtained

$$\mathbf{F} = 6\pi\eta r_i \mathbf{v}_i \tag{5.13.1}$$

provided the liquid immediately adjacent to the interface 'wets' the sphere and so moves along with it. If there is any slipping between the surface of the sphere and the fluid[M72]

$$\mathbf{F} = 6\pi\eta r_i \mathbf{v}_i \left(\frac{\beta r_i + 2\eta}{\beta r_i + 3\eta}\right) \tag{5.13.2}$$

β being a coefficient of sliding friction. In the extreme case of a completely slippery sphere that moves through the liquid with no drag, needing only to push the medium out of its path,

$$\mathbf{F} = 4\pi\eta r_i \mathbf{v}_i \tag{5.13.3}$$

It is eqn. 5.13.1 that is normally called Stokes' law. In this chapter we shall be concerned entirely with spherical particles although modifications of 5.13.1 exist making it applicable to ellipsoids of various kinds[M73] and to macromolecules that can be split into spherical subunits.[M74]

If the force arises from the application of a potential gradient **X** and the sphere possesses a charge $|z_i|e$,

$$\mathbf{F} = |z_i|e\mathbf{X} \tag{5.13.4}$$

Then, assuming no slipping between sphere and liquid,

$$u_i^\circ = \mathbf{v}_i/\mathbf{X} = |z_i|e/6\pi\eta r_i \tag{5.13.5}$$

$$\Lambda_i^\circ = |z_i|u_i^\circ F = z_i^2 F^2/6\pi N_A \eta r_i \tag{5.13.6}$$

where $-e$ is the charge on the electron, F is the Faraday constant, and N_A the Avogadro number. The relations between the velocity at infinitesimal ionic strength, the limiting ionic mobility u_i°, and the limiting molar ionic conductance Λ_i° are taken from eqns. 5.101.5 and 5.8.2. In terms of the units commonly employed by workers in the field:

$$\left(\frac{\Lambda_i^\circ}{\text{cm}^2\,(\text{int. }\Omega)^{-1}\,\text{mol}^{-1}}\right)\left(\frac{\eta}{\text{poise}}\right) = \frac{0.820_4 z_i^2}{(r_S/\text{Å})} \tag{5.13.7}$$

The radii r_S calculated from this equation are known as Stokes radii. Should r_S for a given ion i remain constant when the temperature, pressure, or solvent is varied, it follows that

$$\Lambda_i^\circ \eta = \text{constant} \tag{5.13.8}$$

a relation called Walden's rule after the man who empirically discovered it.[M75]

A test of Stokes' equation is easily arranged. Table 5.12.1 lists sets of ionic radii, and Table 5.12.2 the so-called Walden products $\Lambda_i^\circ\eta$ for several cations and anions using data drawn from Appendices 5.12.1 to 5.12.12. The first point to notice is that the Walden product for a given ion varies from solvent to solvent and attains approximate constancy only when the ion is very large. In most cases, therefore, the Stokes radius is a function of the solvent as well as of the ion. For K^+, r_S ranges from 3.35 Å (in DMF) to 1.05 Å (in ethylene glycol), for Cl^- from 3.48 Å (in ethanol) to 0.86 Å (in sulpholane). Stokes radii greater than crystal radii are customarily accounted for by postulating solvation but smaller values suggest a breakdown of the theory. It may be noted that even for the relatively large ion Bu_4N^+, whose Walden product is roughly constant at 0.22, the average Stokes radius of 3.7_3 Å is less than the bare radius in Table 5.12.1. In aqueous solution the radius of a solute particle must exceed ca. 5.5 Å before hydrodynamic equations of the Stokes' law type become valid[M77] and this seems quite reasonable in view of the model employed. The medium can hardly be regarded as a continuum when solvent molecules and solute ions are of comparable size.

Several attempts have been made to modify Stokes' equation so as to render it applicable to ionic solutions. Whether theoretical or semi-empirical, these efforts have not had the success at first hoped for and it has become fashionable to deride Stokes' law as having reached a state of sterility.[M78] Such a view is overly pessimistic, and considerable insight into the problem will be gained by a careful examination of the different approaches that have been employed so far.

An easy method to visualise is the geometrical one of Robinson and Stokes[M79a] in which the Stokes radius r_S was corrected to give the real or effective radius in solution, r_{eff}, by means of a calibration graph. It was constructed on the assumption that tetraalkylammonium cations in solution have such a low surface charge density that they are not solvated, and their effective radii were therefore taken as equal to their crystal radii r_x (column *d* in Table 5.12.1). From a plot of the correction factor r_{eff}/r_S versus r_S for the ions Me_4N^+, Et_4N^+, Pr_4N^+, Bu_4N^+ and Am_4N^+, the effective radii of other ions of known Stokes radius (i.e. of known ionic conductance) could be evaluated. Figure 5.13.1 shows that, as expected, the correction factor for aqueous solutions tends to unity for $r_S > 5$ Å. The curve for water rises fairly steeply at the other end and the determination of the effective radii of ions of $r_S < 2$ Å is thus very uncertain. To overcome this limitation Nightingale[M80] put forward a modified procedure. He plotted r_{eff} versus r_S for the tetraalkylammonium ions in water (omitting Me_4N^+ because its Walden product $\Lambda_i^\circ\eta$ varied with temperature) and extrapolated the curve to the point $r_S = 0$, $r_{eff} = 2.7$ Å, on the grounds that the effective radius of an ion of infinitesimally small Stokes radius should approach the diameter of a water molecule. From this extended graph the effective radii of quite small ions could be read off. An r_x versus r_S plot has recently been published for methanol solutions.[M81]

It is ironic that water, the solvent to which these ideas were originally applied,[M79a,M80] is the very one where the basic assumption appears to have least validity. Me_4N^+ is probably a net structure-breaker in aqueous solutions and Pr_4N^+ and higher alkyl homologues are hydrophobic structure-makers,[M82,M83] being surrounded by water molecules possessing a greater degree of hydrogen bonding than bulk water. Their effective radii are therefore greater than the crystal radii by some unknown amount. Robinson and Stokes' plots were later drawn for other protic solvents such as NMA[M84] and formamide[M18] although there are indications[M18,M85,M86] that in these media tetraalkylammonium ions exert a structure-breaking effect. It seems reasonable to suppose that in aprotic solvents (in which any hydrogen present is bonded only to carbon) the interaction between large R_4N^+ ions and the solvent would be less, a hypothesis that could be tested by determining the viscosity B_η coefficients of the ions and the temperature dependence of their Walden products. Figure 5.13.1 has been constructed on the assumption that the hypothesis is valid. The Stokes radii were calculated from the ionic conductances in Appendices 5.12.8–5.12.12, and the crystal radii of the alkylammonium ions were again taken from column *d* of Table 5.12.1.

In Fig. 5.13.1 the curves of r_{eff}/r_S versus r_S for sulpholane, DMSO, nitromethane, acetonitrile and DMF are roughly parallel and are crossed by the water curve. All the points for any given tetraalkylammonium ion must of course lie on a hyperbola, and the overall picture of an interlaced network is very similar to that exhibited by the amide solvents.[M86] Two features call for special comment. All the aprotic solvent plots show a maximum at Et_4N^+, a fact previously noted[M84] for several other aprotic media (acetone, nitrobenzene, pyridine, ethylene chloride) as well as for NMA and methanol. The implication is that Me_4N^+ is partially solvated, and the minimum would disappear if an effective solvated radius were substituted for the bare crystal radius of Me_4N^+. In water the ion's structure-breaking ability obscures the influence of hydration. The second striking aspect of the aprotic solvent curves is that they would appear to reach the limiting value of $r_{\text{eff}}/r_S = 1$ at very much higher values of the Stokes radius than in water. The DMSO curve looks as if it is not approaching the unity axis at all but levelling off towards $r_{\text{eff}}/r_S = 1.2$. This phenomenon may be illusory, residing in the choice of the crystal radii in column *d* of Table 5.12.1 instead of the smaller van der Waals radii in column *c*, which might be more appropriate for aprotic solvents. The more fundamental reason, however, should be sought in the relative sizes of the solvent species. One molecule of acetonitrile or

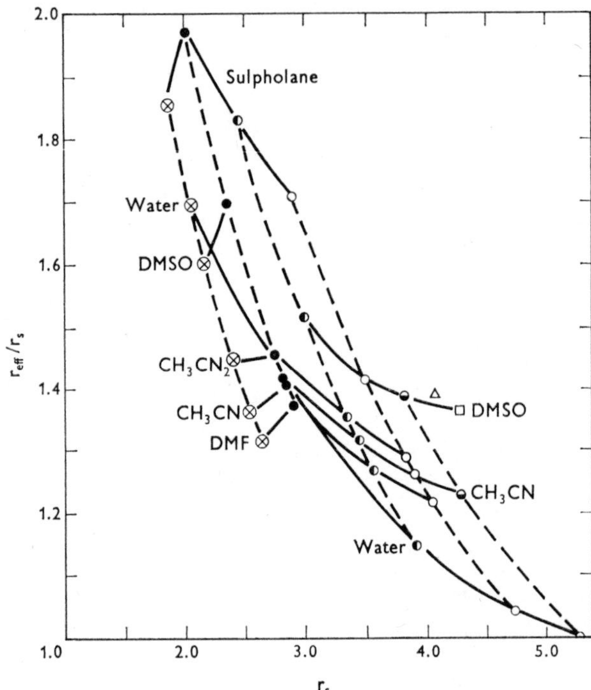

FIG. 5.13.1. Correction plots of the Robinson and Stokes[M79a] type for Me_4N^+(⊗), Et_4N^+(●), Pr_4N^+(◐), Bu_4N^+(○), Am_4N^+(◒), Hex_4N^+(△) and $Hept_4N^+$(□), in sulpholane, dimethylsulphoxide, nitromethane, acetonitrile, dimethylformamide and water.

nitromethane occupies approximately three times the volume of a water molecule, and the factor for sulpholane is more than five. A solute ion at least 10 Å in radius would be required before the sulpholane curve could approach the Stokes' law theoretical limit of unity, and experiments designed to test this point would be welcome. Further work with large ions in DMSO solutions, in particular, is called for.

Following Robinson and Stokes[M79a] we can use the correction plots in Fig. 5.13.1 to estimate effective solution radii and hence to evaluate tentative solvation numbers n_i by means of the equation

$$n_i = \frac{4}{3}\pi(r_{\text{eff}}^3 - r_x^3)/V_1 \tag{5.13.9}$$

V_1 is the average volume of a solvent molecule and electrostriction effects will be ignored. A summary of the results appears in Table 5.13.1, the calculations being restricted to those ions whose Stokes radii exceed that of Et_4N^+ in the solvent concerned. The ion conductances were taken from Appendices 5.12.8–5.12.12. The correction in eqn. 5.13.9 for the volumes of the bare ions is very small; for the purpose of the table the crystal radii from column *a* of Table 5.12.1 were employed although there is no material difference if those in column *b* are chosen instead.

Table 5.13.1

Stokes Radii and Solvation Numbers for Ions in Four Aprotic Solvents

Ion	DMF, 25°C		Acetonitrile, 25°C		DMSO, 25°C		Sulpholane, 30°C	
	r_S/Å	n_i	r_S/Å	n_i	r_S/Å	n_i	r_S/Å	n_i
H^+	2.97	2.2			2.7	3		
Li^+	4.12	4.0	3.47	4.4				
Na^+	3.45	2.8	3.13	3.7	2.90	3.1	2.21	2.0
K^+	3.35	2.7	2.88	3.0	2.75	2.8	1.97	1.6
Rb^+	3.18	2.4						
Cs^+	2.99	2.1			2.48	2.4		
Cl^-	1.87		2.44		1.72		0.86	
Br^-	1.92		2.39		1.75		0.89	
I^-	1.97		2.35		1.76		1.10	
V_1/Å^3	128.6		87.8		118.4		158.1	

Inspection of Table 5.13.1 shows that the magnitudes of the solvation numbers of the alkali metal ions are eminently reasonable. Whether they are meaningful, or correct, must be left to the future. Turning to the halide ions, the Stokes radii in DMF and DMSO are seen to be remarkably close to the crystal radii and this has been interpreted[M25] to mean that the anions are unsolvated. This is consistent with the structures of the DMF and DMSO molecules in which the positive ends of the dipoles are shielded by two methyl groups. On the other hand, it is unlikely that the motion of ions smaller in size than the solvent molecules themselves should obey Stokes' law so well, and the agreement between r_S and r_x must surely be regarded as fortuitous. A cautionary attitude is strengthened by the sulpholane results. Here the positive end of the solvent dipole is shielded to an even greater extent, yet the Stokes and crystal radii are far from equal. The small magnitudes of r_S in sulpholane can have no possible physical significance except to underline the breakdown of Stokes' law when the solvent entities are large compared with the solute ions. Nevertheless, an understanding of the way in which the structures of solvent molecules influence their solvation behaviour does suggest that a more suitable correction plot of the Robinson and Stokes type could be provided in these aprotic solvents by utilising a series of anions of graded sizes rather than cations. In this connection the determination and analysis of the conductances of the BR_4^- ions should prove instructive.

Several attempts have been made to attribute the failures of Stokes' law to the occurrence of slip between the ion and the contiguous solvent molecules. Equation 5.13.2 predicts a hydrodynamic radius up to 50% greater than does eqn. 5.13.1 and can account for r_{eff}/r_S values in Robinson and Stokes' plots between 1 and 1.5. The next step forward should be to relate slip to other physical properties such as the rate of exchange of solvent.[M88] Edward[M89] went so far as to fit diffusion coefficients, of small molecular solutes in carbon tetrachloride and in water, to Stokes' law by treating the numerical factor relating kT/D_i^0 to $\pi\eta r_i$ as an adjustable parameter and found that it ranged from just over 6 for large solutes to 1.1 for H_2 in CCl_4. A value so small—indeed, any value less than 4—cannot rightly be laid at the door of eqn. 5.13.2. The word slip, if used in such a context, means something quite different, the sense of cunningly infiltrating in between large solvent species without having to push them bodily out of the way, rather like a cat dodging through a crowd of people without physically disturbing them. Slipping of this sort will be more pronounced the smaller the solute (very small ions, of course, grow again on the accretion of a solvation shell) and the larger the entities of solvent, be they individual molecules or hydrogen-bonded agglomerates. The plots in Fig. 5.13.1 are in fair agreement with these predictions, as are those of the amide solvents[M86] where the curves of the highly structured and viscous NMA and formamide lie above those of the aprotic DMA and DMF. Nevertheless, before slipping can be accepted as a reality the geometries of the species concerned must be examined more closely. A Bu_4N^+ ion, for example, is twice as wide and appreciably bulkier than a nitromethane molecule, and it is hard to visualise the ion dodging through the gaps in a milling nitromethane throng. An Et_4N^+ ion in sulpholane is quite a different proposition since here the sizes of solute and solvent are very similar and slipping could be a real possibility. How deceptive the slipping hypothesis[M90] can be is illustrated by conductance experiments which were carried out specifically to test it, using Et_4NPic[M91] and Am_4NPic[M92] in a series of phthalates. The results were completely contrary to expectation, for the larger the phthalate, the lower was the Walden product. It is clear that significant progress in this field must be preceded by further theoretical developments, and computer-simulated random walk experiments, with particles of graded sizes and shapes, would seem to offer the best hope.

The idea of slip, and the even vaguer one of microscopic viscosity, are frequently invoked when deviations from Stokes' law occur. Before concepts of this kind can become useful they must be cast into a more quantitative mould and related to measurable properties of solute and solvent. A beginning has been made by Broersma.[M93] There is much evidence from partial molar entropies NMR residence times, and other properties[M94] that, very often, solvent in the immediate vicinity of an ion is 'stickier' than in the bulk of the solvent, and Broersma expressed this in the form

$$\frac{1}{\eta(s)} = \frac{1}{\eta}\left[1 - \varepsilon_m\left(\frac{a_m}{s}\right)^m\right] \tag{5.13.10}$$

where s is the distance from the centre of the solute particle and the pure solvent viscosity $\eta(\infty)$ is written η for short. The viscosity gradient is steeper the larger are the parameters m, ε_m and a_m. If the local non-uniformity is relatively small ($|\varepsilon_m| < 0.2$, $m \geqslant 0$), Stokes' law is modified to

$$\frac{\mathbf{F}}{6\pi\eta r\mathbf{v}} = 1 + \left(\frac{\varepsilon_m}{m+1}\right)\left(\frac{m^2+4m+15}{m^2+8m+15}\right)$$

$$\approx 1 + \frac{(0.80 \pm 0.05)\varepsilon_m}{m+1} \quad [m = 12 \pm 6] \tag{5.13.11}$$

Evidence adduced for solvation from a large Stokes radius (eqn. 5.13.7) could therefore be alternatively regarded as evidence of an extremely high viscosity gradient. A large reverse viscosity gradient ($\varepsilon_m < 0$), occasioned by structure-breaking along the ion's periphery, would explain a Stokes radius smaller than that expected from crystal sizes.

Putting microscopic viscosity on a more quantitative basis is a small but necessary step forward, particularly if we heed Miller's warning,[M95] on mathematical grounds, not to expect a universally valid connection between ionic conductance and macroscopic viscosity. It is now generally acknowledged that no causal relationship exists.[M96] One way of overcoming this restriction is to examine every case on its own merits. A successful example (and one, incidentally, that illustrates Miller's thesis) is the study by R. H. Stokes and his group of the conductances of aqueous electrolyte solutions to which an organic component—sucrose, glycerol, mannitol,[M27] and lately Ficoll[M97] (*vide infra*)—had been added. In each case the limiting molar conductance decreased less on the addition of the organic nonelectrolyte than did the fluidity of the medium. The experimental behaviour was well described by the equation

$$\Lambda_i^\circ \eta^p = \text{constant} \tag{5.13.12}$$

with p equal to unity only for the large Am_4N^+ ion and less than one (often around 0.7) for alkali metal and halide ions. A theoretical analysis[M98,M99a] led to the conclusion that the big nonelectrolyte molecules played, at least in part, two different roles: towards ionic migration they acted largely as inert obstructions and made the ion's pathway longer, whereas they affected the solvent viscosity by distorting the streamlines during viscous flow and introducing a rotational factor. Since the effect on viscosity was much larger than that on conductance the Walden product increased with increasing viscosity and eqn. 5.13.12 resulted. A most interesting organic additive was Ficoll,[M97] a synthetic sucrose polymer consisting of approximately spherical molecules of molecular weight *ca.* 10^5. Ficoll present in water to the extent of 9.85 wt%, increased the viscosity by 341% but decreased the limiting conductance of Et_4NI by only 26%. The picture of the polymer molecules that emerged from these and other results was of sponge-like networks which immobilised the water within them while allowing a relatively easy passage to small ions and a more difficult one to larger ions such as Et_4N^+. Systems even more extreme are rigid gelatin and agar solutions whose viscosities are enormous yet whose conductances are high.[M100]

In the last decade another aspect of ionic movement, that of dielectric relaxation drag, has received attention. A moving ion orients the solvent dipoles around it, and these can relax again into a random distribution only after the ion has passed. Such a re-orientation requires a finite relaxation time (τ) of the order of 10^{-11} s during the course of which the attendant electrostatic field in the medium opposes the ion's movement (compare the D–H–O relaxation effect). This phenomenon was originally described by Born,[M101a] independently rediscovered by Fuoss[M101b] and treated explicitly by Boyd[M101c] and then by Zwanzig.[M102] As a result of various theoretical criticisms,[M88,M103,M104,M108] and because the fit to experimental data was very poor,[M88,M105] Zwanzig[M106] later revised his theory, taking into account the fact that some of the solvent near the ion is dragged along by viscous forces and does not need to be reoriented. As before, the ion was treated as a rigid sphere of radius r_i moving with a steady state velocity v_i through a viscous incompressible dielectric continuum. The dielectric retarding force so calculated must be added to that produced by viscous drag, changing eqn. 5.13.1 to

$$\mathbf{F} = 6\pi\eta r_i \mathbf{v}_i + \frac{3z_i^2e^2}{8} \cdot \frac{\varepsilon_r^\circ - \varepsilon_r^\infty}{\varepsilon_r^\circ(2\varepsilon_r^\circ + 1)} \cdot \frac{\tau \mathbf{v}_i}{r_i^3} \tag{5.13.13}$$

with perfect sticking at the surface of the ion, and eqn. 5.13.3 to

$$\mathbf{F} = 4\pi\eta r_i \mathbf{v}_i + \frac{3z_i^2e^2}{4} \cdot \frac{\varepsilon_r^\circ - \varepsilon_r^\infty}{\varepsilon_r^\circ(2\varepsilon_r^\circ + 1)} \cdot \frac{\tau \mathbf{v}_i}{r_i^3} \tag{5.13.14}$$

with perfect slipping. The latter boundary condition allows greater relative motion between the ion and the surrounding polar solvent molecules and thus increases the extent of dielectric friction. In both equations τ is the dielectric relaxation time and ε_r°, ε_r^∞ are the static and the limiting high-frequency (optical) dielectric constants, respectively. Equation 5.13.6 then becomes[M107]

$$\Lambda_i^\circ = \frac{z_i^2 eF}{A_v \pi \eta r_i + A_D[z_i^2 e^2(\varepsilon_r^\circ - \varepsilon_r^\infty)\tau/\varepsilon_r^\circ(2\varepsilon_r^\circ + 1)r_i^3]} \qquad (5.13.15)$$

with $A_v = 6$ and $A_D = \frac{3}{8}$ for perfect sticking, and $A_v = 4$ and $A_D = \frac{3}{4}$ for perfect slipping.

In any given medium eqn. 5.13.15 takes the form

$$\Lambda_i^\circ = Ar_i^3/(r_i^4 + B), \qquad (5.13.16)$$

and the theory accordingly predicts[M104] that Λ_i° passes through a maximum of $27^{\frac{1}{4}}A/4B^{\frac{1}{4}}$ at $r_i = (3B)^{\frac{1}{4}}$. The phenomenon of maximum conductance is indeed well known; with increase in the ionic radius, the conductances of univalent cations generally rise through Li^+, Na^+ and K^+ only to fall again as the larger tetraalkylammonium ions are reached. Anions, particularly in aprotic solvents, exhibit a flatter and often higher maximum conductance. Quantitative results are presented in Table 5.13.2. It is evident that the experimental maximum conductances are considerably greater than the theoretical ones, on average by 62% (37% for cations only) in the case of the slipping model and by 85% (67% for cations only) for sticking. The agreement between experiment and the slipping version of the theory is good for three out of the four aprotic solvents provided cations alone are considered.

The maximum apart, for the dipolar aprotic solvents plots of Λ_i°(exp.) *vs.* r_i(cryst.) show[M107] that the tetraalkylammonium ions fall fairly close to the theoretical (slipping) curves, though the conductance does fall off rather more rapidly than predicted with increasing radius. Another convenient way of demonstrating this is by rearranging eqn. 5.13.15 to

$$\frac{z_i^2 eF}{\Lambda_i^\circ \eta} = A_v \pi r_i + \frac{A_D z_i^2}{r_i^3} \cdot \frac{e^2(\varepsilon_r^\circ - \varepsilon_r^\infty)}{\varepsilon_r^\circ(2\varepsilon_r^\circ + 1)} \cdot \frac{\tau}{\eta} \qquad (5.13.17)$$

and re-writing as

$$L^* = A_v \pi r_i + \frac{A_D z_i^2}{r_i^3} \cdot P^* \qquad (5.13.18)$$

Plots of L^* against the solvent function P^* are linear[M107] for Me_4N^+ and for Et_4N^+ in the four pure aprotic solvents in Table 5.13.2 and in acetonitrile-carbon tetrachloride mixtures (P^* values for the latter were estimated[M107] by taking τ/η as constant over the composition range). The radii calculated from the intercepts and slopes agree moderately

Table 5.13.2

Maximum Conductances of Spherical Univalent Ions at 25°C Predicted from the Zwanzig (1970) Theory (after Ref. M107)

Solvent	$\Lambda_i^\circ(\max)/\text{cm}^2(\text{int. }\Omega)^{-1}\text{ mol}^{-1}$			
	Stick	Slip	Experimental: Cations	Experimental: Anions
Water	42.2	47.5	77.2† (Rb^+)	78.1‡ (Br^-)
Methanol	31.2	36.7	68.7* (Me_4N^+)	70.8* (ClO_4^-)
Ethanol	13.9	16.0	29.8* (Me_4N^+)	26.9* (I^-)
Propanol	6.56	7.48	15.1§ (Et_4N^+)	16.4§ (ClO_4^-)
Butanol	4.71	5.41	10.4‖ (Et_4N^+)	11.2‖ (ClO_4^-)
Ethylene glycol	2.08	2.37	5.3* (Tl^+)+	5.07* (Cl^-)
Formamide	11.4	13.0	15.8* (Tl^+)	17.2* (Br^-)
Dimethylformamide	36.2	41.1	39.1* (Me_4N^+)	55.1* (Cl^-)
Acetonitrile	85.1	96.8	94.5* (Me_4N^+)	108.2* (BF_4^-)
Nitrobenzene	13.4	15.2	18.1¶ (NH_4^+)	22.7¶ (Cl^-)
Acetone	86.5	98.7	96.6** (Me_4N^+)	121.2** (Cl^-)

* From Appendices 5.12.2–5.12.9.
† T. L. Fabry and R. M. Fuoss, *J. Phys. Chem.*, **68**, 974 (1964).
‡ Ref. M21*c*
§ Ref. M68*a*
‖ Ref. M68*c*
¶ Ref. M184
** Ref. M183
(§, ‖, ¶, **: Individual ionic assignments based on indirect methods; see sect. 5.15.)

well with each other, and with the crystal radii in column *d* of Table 5.12.1, if the A_v and A_D values corresponding to perfect slipping are employed. Numerically,

radius from:	intercept	slope	crystal
Me_4N^+	3.6 Å	3.9 Å	3.5 Å
Et_4N^+	4.2 Å	4.7 Å	4.0 Å

In protic solvents, the agreement between the experimental and the calculated limiting conductances of R_4N^+ ions is only fair in those cases in which P^* is small (water and formamide) and becomes much worse for the higher alcohols of large P^* (large τ),[M107] the ions moving more rapidly than expected. Plots of L^* versus P^* for protic media are linear but with quite different slopes and intercepts to those found for the aprotic solvents; the values of r_i calculated from the intercepts are too small while those from the slopes are enormous and physically meaningless. Nevertheless, certain qualitative trends noted in the introduction (sect. 5.12) are explained by the theory. In particular, it accounts for the greater mobility of Me_4N^+ over Et_4N^+ in the lower alcohols (MeOH, EtOH) where P^* is relatively low and for the reversal of this order, Et_4N^+ becoming the fastest cation, in PrOH and BuOH where P^* is large.

The alkali metal cations, in all the solvents examined, possess higher conductances than are predicted by the Zwanzig theory. The discrepancies become more striking the smaller the ionic radius, and it has been suggested[M109,M107] that the reason may be dielectric saturation of solvent close to the ion. Some simplified calculations of this effect show promise,[M107] at least for dipolar aprotic media.

It is noticeable that the Zwanzig theory is most successful for large organic cations in aprotic media where solvation is likely to be minimal and where viscous friction predominates over that caused by dielectric relaxation. The theory breaks down whenever the dielectric relaxation term becomes large, i.e. for solvents of high P^* and for ions of small r_i. Further developments of the Zwanzig equation are clearly desirable, and a more sophisticated treatment of dielectric saturation is an obvious next step. It must be recognised, however, that an inevitable weakness of any continuum theory is its inability to account for structural features. Two such will be mentioned. First, the Zwanzig model does not allow for any correlation in the reorientation of the solvent molecules as the ion passes by, and this may be the reason why the equations do not apply to hydrogen-bonded solvents.[M107] Second, the theory does not distinguish between positively and negatively charged ions and therefore cannot explain why certain anions in dipolar aprotic media possess considerably higher molar conductances than the fastest cations, the average difference for the four last solvents in Table 5.13.2 being 26%.

Limitations in continuum theories of ionic motion have led one school of thought[M105] to suggest that the key to understanding is to be sought in specific structural effects. Certainly if we wish to understand, for instance, the differences in conductance behaviour between noble-gas type anions and cations,[M109,M110] a detailed picture of ion-solvent interaction is an essential prerequisite. Much work has been done in this field[M94] and more will be required, in view of the growing realisation that solvation in both organic[M111] and inorganic[M112] media depends on more than electrostatic factors. So does the conductance of ions, as Gordon emphasised several years ago.[M113]

5.13.2 Transition State Theory

The hydrodynamic treatment envisages an ion as a charged billiard ball slowly and relentlessly ploughing its way through the medium in response to an impressed electric field. That the theory has met with the success it has is quite remarkable[M99b] for the real situation is utterly different.[M114] Every particle in the solution executes vigorous random

Brownian movements upon which, if the particle is an ion, is superimposed a feeble perturbation in the direction of the field, smaller by a factor of around 10^7. Moreover, the ionic motion is not a steady drift. An ion may be pictured[M114] as oscillating inside a solvent cage; only after numerous oscillations does it manage to change its position and its environment by one of several possible manoeuvres—perhaps a cooperative shift such as interchange with a neighbour, or by means of squeezing between several of the cage molecules. The ion must therefore acquire a certain minimum energy before it can escape. These features are essentially those of the transition-state theory of conductance[M114–116,M78] in which the ion is assumed to migrate by jumping (over a mean jump distance L) from one quasi-equilibrium state to another, and in so doing requires a mean partial molar standard Gibbs free energy of activation, $\Delta G^{0\ddagger}$. The word 'jump', incidentally, does not imply any specific mechanism of movement. In the absence of an applied field, the ion's mean velocity in any direction is thus

$$v = Lk' = L(kT/h) \exp(-\Delta G^{0\ddagger}/RT) \qquad (5.13.19)$$

where k' is a mean first-order rate constant, and k is the Boltzmann constant, R the gas constant, T the absolute temperature, and h the Planck constant. The transmission coefficient has been taken as unity. Let us now make the assumption, common in elementary kinetic theory and fully justified by a more sophisticated mathematical treatment, that effectively 1/3 of the ions jump in the x-direction, half of them one way and half the other. When the electric field (X) is turned on, in the x-direction, the velocity of the ions either downfield or upfield would be $Lk'/6$—with no *net* migration at all—were it not for the small bias introduced into the activation energy by the field's electrical energy, $z_i eXN_A(L/2)$. Here $\frac{1}{2}L$ is the mean distance between the ion's starting position and the transition state which should lie half-way between the virtually identical initial and final states. Thus the downfield velocity is increased to

$$\vec{v} = (LkT/6h) \exp(-[\Delta G^{0\ddagger} - \tfrac{1}{2}|z_i|eN_A XL]/RT)$$

while the velocity upfield is reduced to

$$\overleftarrow{v} = (LkT/6h) \exp(-[\Delta G^{0\ddagger} + \tfrac{1}{2}|z_i|eN_A XL]/RT).$$

The net velocity of the ion i in the direction of the field is accordingly

$$v_i = \vec{v} - \overleftarrow{v} = \frac{LkT}{6h} \cdot e^{\frac{-\Delta G^{\circ\ddagger}}{RT}} \left\{ e^{\frac{|z_i|eN_A XL}{2RT}} - e^{\frac{-|z_i|eN_A XL}{2RT}} \right\}$$

$$= \frac{|z_i|eXL^2}{6h} e^{\frac{-\Delta G^{\circ\ddagger}}{RT}} \qquad (5.13.20)$$

for, since the electrical energy is normally tiny compared with the thermal energy, exponentials in the curly brackets can be expanded and all but the first terms dropped. No inter-ionic effects have been allowed for and the velocity therefore refers to a condition of infinitesimal ionic strength. It follows from eqns. 5.10.15 and 5.8.2 that the limiting molar conductance of i is

$$\Lambda_i^\circ = |z_i|Fu_i^\circ = \frac{|z_i|Fv_i^\circ}{X} = \frac{z_i^2 eFL^2}{6h}\, e^{\frac{-\Delta G^{\circ \ddagger}}{RT}} \tag{5.13.21}$$

or

$$\Lambda_i^\circ = \frac{z_i^2 eFL^2}{6h}\, e^{\frac{\Delta S^{\circ \ddagger}}{R}}\, e^{-\frac{\Delta H^{\circ \ddagger}}{RT}} \tag{5.13.22}$$

The pre-exponential factor equals 3500 $cm^2\,\Omega^{-1}\,mol^{-1}$ if $L = 3$ Å. The actual value of L is unknown and there has been speculation[M105] whether the ion proceeds by many small displacements or few longer ones. But far more important than any uncertainty in L is our present ignorance as to the magnitudes of $\Delta G^{0\ddagger}$, $\Delta H^{0\ddagger}$ and $\Delta S^{0\ddagger}$. It must be emphasised that eqns. 5.13.21 and 5.13.22 do not enable us to predict a priori even the order of magnitude of Λ_i°, and in this respect they are greatly inferior to Stokes' law which, small ions apart, can usually give an answer correct to within a factor of 2. The usefulness of transition state theory lies rather in establishing certain relationships leading to thermodynamic properties of activation, whose experimental values will provide further insight into the phenomenon of ionic transport. The most significant of these properties are obtained from the temperature and pressure derivatives of Λ_i°:

$$\left(\frac{\partial \ln \Lambda_i^\circ}{\partial T}\right)_P = \frac{\Delta H^{0\ddagger}}{RT^2} + \frac{2\alpha}{3} \equiv \frac{E_p}{RT^2} \tag{5.13.23}$$

$$\left(\frac{\partial \ln \Lambda_i^\circ}{\partial T}\right)_V = \frac{\Delta U^{0\ddagger}}{RT^2} \equiv \frac{E_V}{RT^2} \tag{5.13.24}$$

$$\left(\frac{\partial \ln \Lambda_i^\circ}{\partial P}\right)_T = -\frac{\Delta V^{0\ddagger}}{RT} - \frac{2\beta}{3} \tag{5.13.25}$$

α and β are, respectively, the coefficients of cubical expansion and of compressibility of the system and arise from the variation of L with temperature and pressure:

$$\left(\frac{\partial \ln L}{\partial T}\right)_P = \frac{1}{3}\left(\frac{\partial \ln V}{\partial T}\right) = \frac{\alpha}{3}$$

$$\left(\frac{\partial \ln L}{\partial P}\right)_T = \frac{1}{3}\left(\frac{\partial \ln V}{\partial P}\right)_T = -\frac{\beta}{3}$$

A refined analysis[M117,M118] shows that if α and β are taken as the properties of the pure solvent, the differentiation in eqn. 5.13.24 should properly be carried out at constant volume of the activated state rather than at constant total volume.

Non-aqueous measurements of conductances at different temperatures and pressures have so far been reported for several electrolytes, notably tetraalkylammonium salts, in methanol,[M119,M117] dimethylformamide,[M120] nitrobenzene,[M119,M118] and acetone.[M121] It has been shown[M117] that E_p, E_V and $\Delta V^{0\ddagger}$, but not $\Delta S^{0\ddagger}$, are additive functions of the corresponding ionic properties of the type

$$E_V = t_+^\circ (E_+)_V + t_-^\circ (E_-)_V \tag{5.13.26}$$

providing the limiting transference numbers are independent of temperature and pressure. This is only approximately true for aqueous solutions,[M122,M123] and non-aqueous transference numbers over a range of physical conditions are awaited with interest so that the properties of the individual ions can be determined.

In most solutions E_V is both much smaller that E_p and displays a greater constancy with temperature. In solvents other than water[M119] and acetone[M121] E_V declines markedly with increasing specific volume of solvent. Plots of $1/E_V$ versus V are frequently linear and, on extrapolation to $1/E_V = 0$, meet the volume axis at values near those of the close-packed solvent.[M120,M118] The activation energy is therefore infinite, and conductance ceases, when the solvent no longer contains any 'free volume'. In a given medium E_V is almost independent of the electrolyte when the latter is composed of large ions like R_4N^+ and Pic^-, and this is only to be expected under circumstances in which Walden's rule holds. It would be interesting to examine E_V for the ions Li^+, Na^+ and K^+ in which the cation-solvent interaction becomes progressively weaker.

Conductances in organic solvents (but not in water) invariably fall as pressure is applied, and $\Delta V^{0\ddagger}$ is accordingly positive. The transition state therefore occupies a greater volume than does the initial state, suggestive of a loosening of structure—the role of electrostriction would be made clearer by a study of Li^+, Na^+, K^+ and multiply charged ions—or the participation of more solvent molecules. The second interpretation is favoured by the rise of $\Delta V^{0\ddagger}$ with increasing specific volume of solvent. (In either case $\Delta S^{0\ddagger}$ should be positive, a conclusion that is untested because the parameters L and $\Delta S^{0\ddagger}$ cannot be separated. Certain inferences, drawn from results in DMF[M120] and acetone,[M121] hint that $\Delta S^{0\ddagger}$ may be negative!). $\Delta V^{0\ddagger}$ also increases as the temperature falls and as the ions become bulkier. In acetone,[M121] however, $\Delta V^{0\ddagger}$ is independent both of temperature and ionic size and the results have been discussed in terms of the hole theory of liquids. With sufficiently large ions that obey Walden's rule, $\Delta V^{0\ddagger}$ becomes equal to the volume of activation of viscous flow.

5.14 PROTON-JUMP AND ELECTRON-JUMP CONDUCTANCES

In many hydroxylic solvents, as Table 5.14.1 shows, the hydrogen ion is unusually mobile in comparison with K^+ or other alkali metal ions. Size is not the explanation, for both experimental evidence and theoretical calculations affirm that the proton is attached to at least one molecule of solvent[M126,M78] whose radius is equal to or greater than that of K^+. No significantly high H^+ conductance is found in protic solvents in which hydrogen is bonded to nitrogen (formamide, NMA, or ethylene-diamine[M127]) or in aprotic media such as DMF, acetone[M128] and nitriles.[M128] Here the conductance can safely be attributed to bodily transport of the protonated solvent species whereas in certain hydroxylic liquids a special additional mechanism must be present. It is generally accepted[M78] that this mechanism involves proton-jumping between favourably orientated solvent molecules.

While theoretical work on this topic has concentrated almost exclusively on aqueous solutions, the essential features apply to other media also. The problem is the kinetic one of discovering and describing the rate-determining step. Calculations of individual molecular processes in highly structured media are of necessity difficult and involve numerous assumptions. Undaunted, Conway, Bockris and Linton[M129] have examined in detail various possible stages in the overall process, and reached the following conclusions:

1. The transfer of the proton from H_3O^+ to H_2O by quantum-mechanical tunnelling is too fast to account for the observed conductance, although it is true that the calculation is very sensitive to the assumed shape and width (the latter was taken as 0.35 Å) of the proton-transfer barrier.[M129,M130] A more crucial test is the ratio of the conductance of H_3O^+ in H_2O to that of D_3O^+ in D_2O; quantum-mechanical leakage demands a value of at least 6 whereas the experimental result is 1.4.

2. Transfer of a proton from H_3O^+ to H_2O by a classical path can be treated by the transition state theory.[M129,M131] The symmetrical nature of the transition state $H_2O \ldots H^+ \ldots OH_2$, and of the reactants and products, makes calculation of the potential energy surface relatively easy. The velocity of the classical mechanism is lower than that of tunnelling yet not rate-controlling, for it leads to too high (>2) an isotopic conductance ratio.[M131]

Table 5.14.1

Comparisons of the Limiting Conductances ($cm^2(int.\ \Omega)^{-1}\ mol^{-1}$) of Hydrogen and Solvate Ions and Electrons with those of Potassium and Chloride Ions

	Water* 25°	Methanol† 25°	Ethanol† 25°	Ethylene glycol† 25°	Formic acid‡ 25°	Form-amide† 25°	*N*-methyl acetamide† 40°	Dimethyl formamide† 25°	Aceto-nitrile† 25°
H^+	349.8	146.2	62.80	27.7	79.6	10.8	8.9_5	34.7	99
K^+	73.50	52.4_5	23.57	4.62	24.0	12.75	8.3	30.8	83.6
Solvate$^-$	199.2	53.1	22.97		50.0_5	§			
Cl^-	76.35	52.3_5	21.85	5.07	26.5	17.12	11.7	55.1	98.7
e^-	185‖								

* Ref. M21c.
† From Appendices 5.12.2–5.12.9.
‡ Ref. M69. The electrolyte conductances were split into ionic contributions by assuming $(i\text{-Am})_3BuN^+$ and BPh_4^- to be equi-mobile (see sect. 5.15).
§ The solvate ion $HCONH^-$ appears to migrate normally.[M124]
‖ Ref. M125.

3. Prior to any jump, the initial unfavourable orientation

```
      H         H        H
      |         |        |
H—O—HH—O·····H—O·····
      ⊕        ↺
```

must be changed. Ordinary thermal rotation of the neighbouring water molecule is not rapid enough to explain the observed conductance, but the rotation is accelerated by the effect of the H_3O^+ field on the H_2O dipole and by the repulsive O—HH—O dipole interaction. The rate of this field-assisted orientation is of the right magnitude, and the observed isotopic conductance ratio is also reasonably accounted for.[M129] (Eigen and De Maeyer[M130] prefer as the slow step a more co-operative process of structural diffusion, in which a large $H_9O_4^+$ complex shifts by breaking and forming several hydrogen bonds and re-orienting water molecules around its periphery. Few detailed calculations have been carried out on the basis of this model, which has been further discussed by Conway.[M132])

One point that had long puzzled physical chemists is why the concentration dependence of acids obeys the Debye–Hückel–Onsager theory. Pitts *et al.*[M133] have now calculated from Conway's theory that for over 99 % of the time the proton is attached to a water molecule. The H_3O^+ species moves through the solvent as do normal ions, and its conductance is then subject to the usual hydrodynamic electrophoretic correction term Λ^e (eqn. 5.10.6) whose magnitude is independent of the ion's mobility. The relaxation term $\Delta\mathbf{X}/\mathbf{X}$ is a purely electrostatic one and applies whatever the mechanism of movement. Thus[M99c,M133]

$$\Lambda_{H^+} = (\Lambda^\circ_{H_3O^+} + \Lambda^\circ_{JUMP} - \Lambda^e_{H^+})\left(1 + \frac{\Delta\mathbf{X}_{H^+}}{\mathbf{X}}\right)$$

The proton-jump contribution to the overall conductances of acids and alkalies decreases markedly in concentrated solutions.[M134a]

It seems highly probable that the acid conductance mechanism in alcoholic solvents is similar to that in water. The rate-determining step is then the orientation of *R*OH molecules in the field of the $RO{H_2}^+$ ion, followed by fast quantum-mechanical or classical proton transfer. The larger the *R*OH molecule the more slowly will it be able to rotate, and this explains why the enhancement of H^+ conductance over that of (say) K^+ decreases along the series[M129]

$$\text{HOH} > \text{MeOH} > \text{EtOH} > \text{PrOH}$$

The conductance behaviour of acids in water-alcohol mixtures has attracted the attention of several research groups, and in the last 15 years results have been published for 0.1*M* HCl,[M129] 0.01*M* HCl[M134b] and HCl at infinitesimal ionic strength.[M135–137] Unlike the gently sloping curves of salt conductances *vs.* solvent composition, those for acids pass through a pronounced minimum which appears at *ca.* 90 wt% methanol in H_2O—MeOH mixtures and shifts ever closer to 100% alcohol as the chain length increases. In all these solutions the equilibrium

$$ROH_2^+ + H_2O \rightleftarrows ROH + H_3O^+ \tag{5.14.1}$$

is set up, and lies far to the right since H_2O is a relatively strong base. Thus, when small amounts of water are added to alcoholic acid solutions, the water molecules steal a disproportionate number of protons from ROH_2^+ ions and thereby diminish the extent of ROH_2^+—ROH proton jumping. In the newly formed H_3O^+ ions, however, the proton is effectively trapped, for jumping to alcohol molecules is energetically unfavourable and there are too few water molecules present to enable many H_3O^+–H_2O transfers to take place. The H_3O^+ ions can only move bodily through the solution and the overall conductance is therefore less than in pure alcohol as solvent. Addition of more and more water progressively increases the chances of H_3O^+–H_2O proton transfers and the molar conductance rises steadily as the solvent becomes more aqueous.

A simple quantitative treatment is possible at the alcohol-rich end of the range.[M138,M139] If K is the equilibrium constant of reaction 5.14.1, then at infinitesimal ionic strength

$$\frac{[H_3O^+]}{[H_3O^+] + [ROH_2^+]} = \frac{c_w}{c_w + r}; \qquad r = \frac{c_{alc}}{K} \tag{5.14.2}$$

where c_w and c_{alc} are the molarities of water and of alcohol, respectively. As long as c_w is small, c_{alc} and r will be virtually constant. The limiting conductance of HCl at any given water concentration is given by

$$\Lambda_m^\circ = \left(\frac{r}{c_w + r}\right) \Lambda^\circ_{ROH_2^+} + \left(\frac{c_w}{c_w + r}\right) \Lambda^\circ_{H_3O^+} + \Lambda^\circ_{Cl^-} \tag{5.14.3}$$

When traces of water are added to the lower alcohols, the percentage increase in solvent viscosity is much less than the percentage decrease in the conductance of HCl.[M135–137] One may therefore assume, as a first approximation, that the limiting ionic conductances are essentially the same as in the pure alcohol. Then, on rearrangement,

$$\frac{c_w}{\Lambda^\circ_{ROH_2^+} + \Lambda^\circ_{Cl^-} - \Lambda^\circ_m} = \frac{c_w + r}{\Lambda^\circ_{ROH_2^+} - \Lambda^\circ_{H_3O^+}} \tag{5.14.4}$$

Plots of the left-hand side of eqn. 5.14.4 against c_w were found to be linear up to $c_w = 1M$ in methanol and ethanol.[M138,139a] The values of K calculated from slopes and intercepts agree well with colorimetrically determined constants,[M139b] so vindicating the above analysis, and the values of $\Lambda^{\circ}_{H_3O^+}$ in the two alcohols are close to the corresponding limiting conductances of the Na^+ ion.[M139a] More recently,[M140] the linearity of the left-hand side of 5.14.4 against c_w up to $0.25M$ water has been confirmed for HCl solutions in *n*-butanol and *iso*-butanol. The K value in *n*-butanol calculated from the small intercept is twice as large as the colorimetric result,[M139b] but the latter may be influenced by deviations from ideality.

Of the solvate species listed in Table 5.14.1, OH^- in water and $HCOO^-$ in formic acid possess considerable excess mobility when compared with more pedestrian anions like Cl^-. In these cases the normal ionic motion is once again supplemented by a proton-jumping mechanism, with the slow step the prior re-orientation of the solvent molecule in the field of the anion.[M129] Pictorially, in water

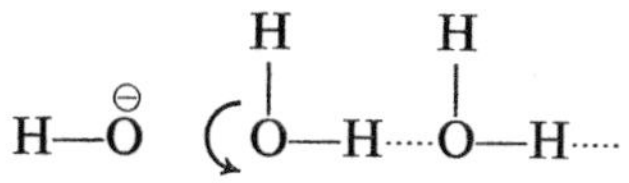

```
                 H       H
       ⊖         |       |
    H—O      ( O—H·····O—H·····
```

The proton deficiency between the anion and the neighbouring water molecule means that there can be no repulsion between opposed O—H bonds, as there is before H_3O^+–H_2O proton transfers, although lone-pair repulsion should play a part. If the only driving force for rotation is taken to be the derivative of the energy of ion-dipole interaction with respect to the angular displacement, then numerical calculations suggest[M129] that the proton-jumping contribution of OH^- will be approximately half that of H_3O^+, as is indeed the case. No estimates have been made for other solvate species, and it is still not clear why the alcoholate ions display no signs at all of higher conductivity.

An abnormally high mobility is displayed by another species: the electron. Liquid ammonia[M141] and many amines (alkylamines, ethylenediamine, hexamethylphosphotriamide)[M142,M143] dissolve alkali metals to produce solutions of a characteristic blue colour at low concentrations and with a bronze metallic sheen at high concentrations. Alternative methods of preparation include electrolysis and the use of ionising radiation.[M142] The many unusual properties—extremely high conductance, ready reducing power, paramagnetism, photoelectron and even thermionic[M144] emission—all testify to the presence of solvated electrons. The amine systems are of considerable importance in providing a bridge between electronic conduction (in concentrated solutions) and electrolytic conduction (in dilute solutions). In this chapter we consider only the latter.

Less stable blue solutions form when alkali metals are added to tetrahydrofuran, 1,2-dimethoxyethane, and other polyethers,[M145] and transient blue colours appear in alcohols and water.[M146,M147a] The ether solutions are mainly diamagnetic[M148] and the existence of a diamagnetic negative charge carrier seems necessary to account for their conductances. Both spin-compensated electron pairs, e_2^{2-}, and triple ions, $e^-M^+e^-$, have been suggested, and there is recent photolytic evidence for the Na^- ion in amine solvents.[M149] In amines a variety of ionic (M^+, e^-, M^-), ion-pair (M^+e^-), and dimeric (M_2, e_2^{2-}) species appears to be present.[M143]

Experimental difficulties, theoretical uncertainties, and poor planning have so conspired together as to frustrate most attempts to determine the conductances or excess conductances of the electrons in amine solvents. One of the main problems in the laboratory has been the low chemical stability of the alkali metal solutions. Their blue colour gradually fades as the solutions decompose with the formation of hydrogen, a process catalysed by impurities and especially by the platinum electrodes of the cell itself. Pyrex vessels, it was recently discovered,[M150] cause sodium contamination, and for this reason much of the early research is now of doubtful worth. The experimental problems are exacerbated in the case of methylamine, whose volatility demands the use of low temperatures at which the metals dissolve but slowly. A further problem arises in the extrapolation of the data to infinitesimal ionic strength, for the appropriate conductance function to be applied depends upon the kind of species which the solution contains.[M151] And when, after all these hazards, the limiting conductance of an alkali metal solution has finally been obtained, it turns out as often as not that it can neither be compared with values for other metals because each experimenter has worked at a different temperature, nor with the conductances of normal salts because in the excitement their measurement has been overlooked.

In ethylenediamine at 25°C the excess conductance of the electron can be estimated from[M152,M153]

$$\Lambda°(\mathrm{K}) - \Lambda°(\mathrm{KI}) = 139 - 71 = 68\ \mathrm{cm^2\ \Omega^{-1}\ mol^{-1}}$$

The analogous calculation for sodium solutions[M152] yields a negative value; this fact, and the zero paramagnetism of the solutions,[M154] suggest that the anion is not e^- but e_2^{2-} or Na^-. A further peculiarity of the sodium conductances is their small variation with increasing concentration. Limiting conductances of rubidium and caesium solutions, 117 and 204 respectively, are available[M152] but no data exist for the corresponding metal salts. In methylamine, too, the limiting conductance of sodium (54 at −70°C)[M155] is considerably lower than that of lithium (228 at −78.3°C)[M156] or of caesium (167 at −70°C),[M155] probably for the same reason. The only possibility of estimating the excess conductance of the electron in methylamine is at $-50°C$, where[M155,M157]

$$\Lambda°(Cs) - \Lambda°(CsI) = 249 - 56 = 193 \text{ cm}^2\,\Omega^{-1}\,\text{mol}^{-1}$$

The CsI value comes from a single run at $-48.5°C$, and $\Lambda°(Cs)$ at $-48.5°C$ obtained by the same workers[M157] (the purity of whose solvent has been criticised)[M158] is only 174.

It would be of great interest to know not only the relative but also the absolute electron conductances. An approximate value of the latter could be deduced (see sect. 5.15) by estimating from Walden's rule, say, $\Lambda°(Bu_4N^+)$ and measuring $\Lambda°(Bu_4NI)$ and the limiting conductances of the appropriate metal iodides. For ethylenediamine $\Lambda°(Bu_4NI)$ has in fact been published,[M159] but the limiting conductances of the various salts reported in the paper in question[M159] display such appalling deviations from additivity that the results must be discarded. It is to be hoped that the hard-won alkali metal conductances in amines will eventually be supplemented by transference number determinations: a formidable experimental challenge, but one already successfully met in the case of liquid ammonia as a solvent.[M160]

Early theories of the state of the solvated electron suggested that it was located in a cavity in the liquid where it was trapped by its polarisation of the surrounding medium.[M161] In the latest theory,[M162] the electron cavity is characterised by a loosely packed first coordination layer containing an appreciable amount of empty space. The high electron mobility in an electric field cannot be reconciled with hydrodynamic motion of the whole cavity, and instead it is proposed that the loosely packed structure allows the electron to jump or leak away, by quantum-mechanical tunnelling,[M142,M147b] No numerical estimates of electron conductances have yet been made on this model.

5.15 INDIRECT ESTIMATION OF IONIC CONDUCTANCES

A. R. Gordon's data on solutions in methanol, wrote Robinson and Stokes in 1955,[M79b] 'form the sole oasis of exact knowledge in a desert of ignorance'. In this vivid and now much quoted phrase they were expressing their concern at the lack of accurate transference numbers in other non-aqueous solvents, a deficiency which made it impossible to assign with certainty conductance values to individual ions. Appendix 5.8.1 shows that the situation has significantly improved, although a comparison of tnose solvents for which reliable transference numbers are known with the much greater number of solvents for which we possess accurate conductances (Appendix 5.1) highlights how much there is still to be done. The paucity of transference information is particularly striking for temperatures other than 25°C, in mixed solvents, and in solvents of low dielectric constant.

In the absence of transference data, certain indirect methods have been resorted to over the years in order to provide at least approximate values of single ionic conductances. The first method is due to Walden, who had discovered[M75] that the product of the limiting molar conductance of a salt, comprising large ions, such as Et_4NPic, and the solvent viscosity, was constant in a variety of solvents and at different temperatures. It was a logical next step to postulate that $\Lambda_i^\circ\eta$ was constant for each of the ions involved, both when the temperature[M163] and the solvent[M164] were changed. Since limiting ionic conductances were well known in water (the only medium in which transference numbers were plentiful), he was able to calculate the limiting conductance of Et_4N^+ (or Pic^-) in any chosen solvent from

$$\Lambda^\circ_{\text{solvent}} = \Lambda^\circ_{\text{water}}\, \eta_{\text{water}}/\eta_{\text{solvent}} \tag{5.15.1}$$

Λ° values for other ions in the given solvent followed readily from the conductances of tetraethylammonium or picrate salts by application of Kohlrausch's rule of independent ionic migration. Ulich[M165] has summarised many of the earlier results; some later examples of solvents to which this approach has been applied are ethylenedichloride (Pic^-),[M166] HCN (Pic^-[M67] and Bu_4N^+[M167]), dimethylacetamide (Bu_4N^+),[M168] and propanol,[M68a] *iso*propanol[M68b] and butanol[M68c] ($Hept_4N^+$ and $(i\text{-Am})_3BuN^+$).

The validity of Walden's approach can now be tested by data for several solvents in which transference numbers are known. Table 5.12.2 lists the relevant Walden products in order of increasing solvent viscosity, the ionic conductances being taken from Appendices 5.12.1–5.12.12. It is at once apparent that the constancy of $\Lambda_i^\circ\eta$ is poor for Et_4N^+, not much better for Pic^-, and fair for the bigger Bu_4N^+ (with water, sulpholane and glycol notable exceptions). As expected, the still larger ions $(i\text{-Am})_4N^+$ and BPh_4^- fare better, though here the number of solvents available for comparison is limited. A pictorial test is provided by Fig. 5.13.1: if Walden's rule held, all the curves would coincide. Clearly Walden's method carries no guarantee of reliability and ionic conductances so estimated may occasionally be out by 20 or 30%, especially in highly structured solvents or those whose molecules are intrinsically bulky. On the whole, however, if large ions are employed, the results are frequently correct to within a few per cent.

The second method of estimating single ion conductances originated with Kraus and his group.[M169–174] They suggested the use of a reference electrolyte, initially $Bu_4N^+BFPh_3^-$, whose ions were both large and contained approximately the same number of carbon atoms. These ions, it was postulated, would be almost equi-mobile and possess a conductance equal to half the limiting conductance of the reference electrolyte. On the basis of this assumption other ionic conductances could then be evaluated from the conductances of electrolytes containing either of these ions, and of other electrolytes with common cations and anions. A major advantage of this method is that it depends solely on conductance measurements in the solvent of interest.

Within the last three decades 6 different equi-transferent electrolytes have been proposed by different workers, and their main applications in pure solvents are summarised in Table 5.15.1. Some salts were picked for practical reasons such as ready availability or ease of synthesis, and adequate solubility. Others were chosen because models showed the ions to be symmetrical and of roughly equal size. In time, finer structural considerations began to play a role. It was realised that the presence in one of the ions of a polar group that exposed it to solvation influences was a marked disadvantage. Even phenyl groups must be disqualified on this count[M186] since there is now evidence,[M187,M188] from proton magnetic resonance and other sources, of interaction between aromatic groups and alcohols or other hydrogen donors. The ultimate criterion of selection, of course, remains the actual performance of the salts in question, and the last column of Table 5.15.1 lists the relative limiting conductances of the constituent ions in all solvents of known transference number. By a quirk of fate the original reference electrolyte cannot be tested because there is no information on the conductances of the $BFPh_3^-$ ion. For the next 3 salts in the table, $\Lambda_+^\circ/\Lambda_-^\circ$ deviates from unity by an average of 9.1% (6.6% if the disastrous result for $Me_3PhN^+PhSO_3^-$ in DMSO is ignored). The mean deviation is only 2.4% for the last two electrolytes, a greatly superior result and one due partly to the fact that $(i\text{-Am})_3BuN^+BPh_4^-$ was deliberately chosen because its two ions were equi-mobile in methanol.[M180] Further tests for $(i\text{-Am})_4N^+B(i\text{-Am})_4^-$, while highly desirable, particularly in water where both ions should be hydrophobic structure-makers, are restricted by the limited solubility of this electrolyte in hydroxylic solvents and by the relative instability of alkali metal tetra*iso*amylborides.[M183]

The two indirect methods of estimating ionic conductances are closely related. Equally mobile ions must have equal Walden products. Table 5.12.2 has shown that Walden products even of fairly large ions vary somewhat from solvent to solvent, and it is only a certain parallelism in these variations that gives a reference electrolyte a better chance of succeeding. For example,[M176] the Walden products of Bu_4N^+ and of Pic^- are both ca. 33% greater in NMA than in water, but the ratios of their Walden products, and of their conductances, differ by only 1% in these two media. The correlation in other instances is far from perfect, however, and Table 5.12.2 makes it plain that it cannot always be relied

Table 5.15.1

Applications and Tests of Reference Electrolytes in Pure Solvents

Electrolyte	Applications	Λ_+^0/Λ_-^0 (Appendices)
$Bu_4N^+BFPh_3^-$	Fowler and Kraus (1940), ethylenedichloride[M169]	
	Burgess and Kraus (1948), pyridine[M170]	
	Reynolds and Kraus (1948), acetone[M171]	
	Pickering and Kraus (1949), nitrobenzene[M172]	
$Me_3OctdN^+OctdSO_4^-$	Thompson and Kraus (1947), ethylenedichloride[M173]	0.972, NMA
	Sears, Lester and Dawson (1956), DMSO[M175]	1.095, DMSO
	Dawson, Wilhoit, Holmes and Sears (1957), NMA[M176]	
$Me_3PhN^+PhSO_3^-$	Dawson, Wilhoit, Holmes and Sears (1957), NMA[M176]	0.981, NMA
	Dawson, Wilhoit and Sears (1957), formamide[M20]	1.029, formamide
		0.695, DMSO
		0.936, water[M21c,M177]
$Bu_4N^+BPh_4^-$	Fuoss and Hirsch (1960), propylene carbonate, nitrobenzene[M178]	1.067, methanol
	Brown and Fuoss (1960), acetonitrile, *i*-butyronitrile[M179]	1.131 formamide
	Coetzee and McGuire (1963), acetone, several nitriles[M128]	1.053, acetonitrile
		1.079, nitromethane
		1.157, DMSO
		0.93, water[M21c,M76]
$(i\text{-Am})_3BuN^+BPh_4^-$	Coplan and Fuoss (1964), methanol[M180]	0.996, methanol
	Hughes and Price (1967), ethylmethylketone[M181]	1.043, formamide
	Sears, Caruso and Popov (1967), adiponitrile[M182]	0.991, acetonitrile
	Wehman and Popov (1968), formic acid[M69]	
	Evans, Thomas, Nadas and Matesich (1971), acetone[M183]	
$(i\text{-Am})_4N^+B(i\text{-Am})_4^-$	Coetzee and Cunningham (1965), nitrobenzene[M184]	0.988, acetonitrile
	Mukherjee, Boden and Lindauer (1970), propylene carbonate[M185]	1.000, nitromethane
		1.078, DMSO

Octd = Octadecyl

upon. In fact, the essential strategy in devising a reference electrolyte is founded on Stokes's law—the use of very large spherical ions of equal radius—and solvation considerations. In sufficiently large anions and cations, the charge is spread very thinly over the surface, and the solvent might be deceived into not recognising them as ions at all. The flaw in this argument is that the ability of the solvent to differentiate between positively, negatively and uncharged solutes is one of the very factors on which modern researchers seek more enlightenment. No doubt $(i\text{-Am})_3BuN^+BPh_4^-$ and $(i\text{-Am})_4N^+B(i\text{-Am})_4^-$ will prove extremely useful for approximate and tentative assignments of individual ionic conductances, but final judgement in each and every solvent must be held in abeyance until accurate transference numbers have been determined in it.

REFERENCES

M1 M. Spiro, Transference Numbers, *in* Physical Methods of Chemistry, Part IIA; Electrochemical Methods, Chapter 4 (A. Weissberger and B. W. Rossiter, ed.), Interscience, New York (1971)

M2 M. Dole, *Trans. Electrochem. Soc.*, **77**, 385 (1940)

M3 R. C. Weast (Ed.), Handbook of Chemistry and Physics, 50th edition, Section E-4, Chemical Rubber Co., Cleveland, Ohio (1969)

M4 M. Mooney, Minimizing Convection Currents in Electrophoresis Measurements, *in* Temperature: Its Measurement and Control in Science and Industry p. 428, Reinhold, New York (1941)

M5 L. P. Hammett and F. A. Lowenheim, *J. Amer. Chem. Soc.*, **56**, 2620 (1934)

M6 W. C. Lanning and A. W. Davidson, *J. Amer. Chem. Soc.*, **61**, 147 (1939)

M7 R. L. Kay, G. A. Vidulich and A. Fratiello, *Chem. Instrum.*, **1**, 361 (1969); K. S. Pridabi, *J. Solution Chem.*, **1**, 455 (1972)

M8 G. Marx and D. Hentschel, *Talanta*, **16**, 1159 (1969); J. Vehlow and G. Marx, *Naturwissensch.*, **58**, 320 (1971)

M9 H. S. Harned and E. C. Dreby, *J. Amer. Chem. Soc.*, **61**, 3113 (1939)

M10 A. Gemant, *J. Chem. Phys.*, **10**, 723 (1942)

M11 J. A. Davies, R. L. Kay and A. R. Gordon, *J. Chem. Phys.*, **19**, 749 (1951)

M12 A. M. Sukhotin, *Russ. J. Phys. Chem.*, **34**, 29 (1960)

M13 M. Carmo Santos and M. Spiro, *J. Phys. Chem.*, **76**, 712 (1972)

M14 G. S. Hartley and G. W. Donaldson, *Trans. Faraday Soc.*, **33**, 457 (1937)

M15 M. Spiro, *Trans. Faraday Soc.*, **61**, 350 (1965)

M16 C. H. Springer, J. F. Coetzee and R. L. Kay, *J. Phys. Chem.*, **73**, 471 (1969)

M17 (*a*) H. Strehlow, Electrode Potentials in Non-aqueous Solvents, *in* The Chemistry of Non-aqueous Solvents, Vol. 1, Chapter 4 (J. J. Lagowski, ed.), Academic Press, New York (1966)

M17 (*b*) J. N. Butler, Reference Electrodes in Aprotic Organic Solvents, *in* Advances in Electrochemistry and Electrochemical Engineering, Vol. 7, Wiley, New York (1970)

M18 J. M. Notley and M. Spiro, *J. Phys. Chem.*, **70**, 1502 (1966)

M19 J. P. Butler, H. I. Schiff and A. R. Gordon, *J. Chem. Phys.*, **19**, 752 (1951)

M20 L. R. Dawson, E. D. Wilhoit and P. G. Sears, *J. Amer. Chem. Soc.*, **79**, 5906 (1957)

M21 R. A. Robinson and R. H. Stokes, Electrolyte Solutions, 2nd edition revised, (a) Chapter 4, (b) Chapter 7, (c) Appendix 6.1, Butterworths, London (1965)

M22 L. W. Shemilt, J. A. Davies and A. R. Gordon, *J. Chem. Phys.*, **16,** 340 (1948)
M23 H. S. Harned and M. H. Fleysher, *J. Amer. Chem. Soc.*, **47,** 92 (1925)
M24 J. Smisko and L. R. Dawson, *J. Phys. Chem.*, **59,** 84 (1955)
M25 J. E. Prue and P. J. Sherrington, *Trans. Faraday Soc.*, **57,** 1795 (1961)
M26 J. M. Stokes and R. H. Stokes, *J. Phys. Chem.*, **60,** 217 (1956); **62,** 497 (1958); **63,** 2089 (1959)
M27 B. J. Steel, J. M. Stokes and R. H. Stokes, *J. Phys. Chem.*, **62,** 1514 (1958); **63,** 2089 (1959)
M28 C. Treiner and R. M. Fuoss, *J. Phys. Chem.*, **69,** 2576 (1965)
M29 L. G. Longsworth, *J. Amer. Chem. Soc.*, **54,** 2741 (1932)
M30 R. H. Stokes, *J. Amer. Chem. Soc.*, **76,** 1988 (1954)
M31 R. L. Kay and J. L. Dye, *Proc. Nat. Acad. Sci.*, **49,** 5 (1963)
M32 L. Onsager and R. M. Fuoss, *J. Phys. Chem.*, **36,** 2689 (1932)
M33 R. M. Fuoss and L. Onsager, *J. Phys. Chem.*, **61,** 668 (1957)
M34 L. Onsager, *Physik. Z.*, **28,** 277 (1927)
M35 P. C. Carman, *J. Phys. Chem.*, **73,** 1095 (1969)
M36 J. R. Graham and A. R. Gordon, *J. Amer. Chem. Soc.*, **79,** 2350 (1957)
M37 S. Blum and H. I. Schiff, *J. Phys. Chem.*, **67,** 1220 (1963)
M38 H. C. Brookes, M. C. B. Hotz and A. H. Spong, *J. Chem. Soc.* **A,** 2415 (1971)
M39 R. M. Fuoss and L. Onsager, *J. Phys. Chem.*, **67,** 628 (1963)
M40 T. M. Murphy and E. G. D. Cohen, *J. Chem. Phys.*, **53,** 2173 (1970); **56,** 4091 (1972)
M41 P. C. Carman, *J. Phys. Chem.*, **74,** 1653 (1970)
M42 E. Pitts, *Proc. Roy. Soc.*, **A217,** 43 (1953); E. Pitts, B. E. Tabor and J. Daly, *Trans. Faraday Soc.*, **65,** 849 (1969); **66,** 693 (1970)
M43 D. P. Sidebottom and M. Spiro, *J. Chem. Soc. Faraday Trans.*, **69,** 1287 (1973)
M44 R. A. Robinson and R. H. Stokes, *J. Amer. Chem. Soc.*, **76,** 1991 (1954)
M45 R. M. Fuoss and F. A. Accascina, Electrolytic Conductance, Chapters 15–17, Interscience, New York (1959)
M46 D. G. Miller, *J. Phys. Chem.*, **70,** 2639 (1966)
M47 Ref. 32, p. 2736
M48 R. Haase, *Z. Phys. Chem. Frankfurt*, **14,** 292 (1958); *Z. Elektrochem.*, **62,** 279 (1958)
M49 L. S. Darken, *Trans. Amer. Inst. Min. Met. Eng.*, **175,** 184 (1948)
M50 G. S. Hartley and J. Crank, *Trans. Faraday Soc.*, **45,** 801 (see p. 815) (1949)
M51 C. A. Kraus and R. M. Fuoss, *J. Amer. Chem. Soc.*, **55,** 21 (1933)
M52 R. M. Fuoss and C. A. Kraus, *J. Amer. Chem. Soc.*, **55,** 2387 (1933)
M53 A. M. Sukhotin, D. N. Saburova and G. V. Smirnova, *Russ. J. Phys. Chem.*, **35,** 347 (1961)
M54 J. N. Agar, *in* The Structure of Electrolytic Solutions, Chapter 13 Appendix (W. J. Hamer, ed.), Wiley, New York (1959)
M55 W. Nernst, *Nachr. kgl. Ges. Wiss. Göttingen, Math.-phys. Klasse*, 68 (1900)
M56 D. M. Mathews, J. O. Wear and E. S. Amis, *J. Inorg. Nucl. Chem.*, **13,** 298 (1960); W. V. Childs and E. S. Amis, *J. Inorg. Nucl. Chem.*, **16,** 114 (1960); J. O. Wear, C. V. McNully and E. S. Amis, *J. Inorg. Nucl. Chem.*, **18,** 48 (1961); **19,** 278 (1961); **20,** 100 (1961); J. O. Wear, J. T. Curtis Jr. and E. S. Amis, *J. Inorg. Nucl. Chem.*, **24,** 93 (1962); J. O. Wear and E. S. Amis, *J. Inorg. Nucl. Chem.*, **24,** 903 (1962); R. G. Griffin, E. S. Amis and J. O. Wear, *J. Inorg. Nucl. Chem.*, **28,** 543 (1966)
M57 J. R. Bard, J. O. Wear, R. G. Griffin and E. S. Amis, *J. Electroanal. Chem.*, **8,** 419 (1964); J. R. Bard, E. S. Amis and J. O. Wear, *J. Electroanal. Chem.*, **11,** 296 (1966)
M58 M. Spiro, *J. Inorg. Nucl. Chem.*, **27,** 902 (1963)

M59 P. Z. Fischer and T. E. Koval, *Bull. Sci. Univ. Kiev*, No. 4, 137 (1939); L. G. Longsworth, *J. Amer. Chem. Soc.*, **69,** 1288 (1947); C. H. Hale and T. de Vries, *J. Amer. Chem. Soc.*, **70,** 2473 (1948); A. Hunyar, *J. Amer. Chem. Soc.*, **71,** 3552 (1949); F. J. Kelly and R. H. Stokes, *in* Electrolytes, p. 96 (B. Pesce, ed.), Pergamon, Oxford (1962)

M60 H. Strehlow and H.-M. Koepp, *Z. Elektrochem.*, **62,** 372 (1958)

M61 H. Schneider and H. Strehlow, *Z. Elektrochem.*, **66,** 309 (1962)

M62 H. Schneider and H. Strehlow, *Ber. Bunsenges. Phys. Chem.*, **69,** 674 (1965)

M63 H. Strehlow and H. Schneider, *J. Chim. Phys.* (Special October 1969 Issue), 118 (1969)

M64 D. Feakins, *J. Chem. Soc.*, 5308 (1961)

M65 M. Spiro, *Educ. Chem.*, **3,** 139 (1966)

M66 M. H. Panckhurst, *Rev. Pure Appl. Chem.*, **19,** 45 (1969)

M67 J. E. Coates and E. G. Taylor, *J. Chem. Soc.*, 1245 (1936)

M68 (*a*) D. F. Evans and P. Gardam, *J. Phys. Chem.*, **72,** 3281 (1968)
(*b*) M. A. Matesich, J. A. Nadas and D. F. Evans, *J. Phys. Chem.*, **74,** 4568 (1970)
(*c*) D. F. Evans and P. Gardam, *J. Phys. Chem.*, **73,** 158 (1969)

M69 T. C. Wehman and A. I. Popov, *J. Phys. Chem.*, **72,** 4031 (1968)

M70 G. G. Stokes, *Trans. Cambr. Phil. Soc.*, **9,** Part II, Paper X (see p. 51) (1856) (read in 1850)

M71 W. Sutherland, *Phil. Mag.*, **3,** 161 (1902); H. Hartley, N. G. Thomas and M. P. Applebey, *J. Chem. Soc.*, **93,** 538 (see p. 555) (1908)

M72 H. Lamb, Hydrodynamics, 6th edition, p. 602, Cambridge University Press (1932)

M73 R. O. Herzog, R. Illig and H. Kudar, *Z. Phys. Chem.*, **A167,** 329 (1933); F. Perrin, *J. Phys. Radium*, **7,** 1 (1936); B. E. Conway, Electrochemical Data, p. 254, Elsevier, Amsterdam (1952); P. H. Elworthy, *J. Chem. Soc.*, 3718 (1962); S. A. Rice, *J. Amer. Chem. Soc.*, **80,** 3207 (1958)

M74 V. Bloomfield, W. D. Dalton and K. E. van Holde, *Biopolymers*, **5,** 135 (1967)

M75 P. Walden, *Z. Phys. Chem.*, **55,** 207 (1906); **78,** 257 (1912)

M76 W. Rüdorff and H. Zannier, *Z. Naturforsch.*, **8*b*,** 611 (1953)

M77 M. C. Baker, P. A. Lyons and S. J. Singer, *J. Amer. Chem. Soc.*, **77,** 2011 (1955); T. Kurucsev, A. M. Sargeson and B. O. West, *J. Phys. Chem.*, **61,** 1567 (1957)

M78 B. E. Conway, Some Aspects of the Thermodynamic and Transport Behavior of Electrolytes, *in* Physical Chemistry: An Advanced Treatise, Vol. IXA/ Electrochemistry, Chapter 1 (H. Eyring, ed.), Academic Press, New York (1970)

M79 R. A. Robinson and R. H. Stokes, Electrolyte Solutions, 1st edition, (a) Chapter 6, (b) p. 125, Butterworths, London (1955)

M80 E. R. Nightingale, Jr., *J. Phys. Chem.*, **63,** 1381 (1959)

M81 M. Della Monica and L. Senatore, *J. Phys. Chem.*, **74,** 205 (1970)

M82 R. L. Kay and D. F. Evans, *J. Phys. Chem.*, **69,** 4216 (1965); **70,** 2325 (1966)

M83 R. L. Kay, *Advan. Chem. Ser.*, **1,** 1 (1968)

M84 J. M. Hale and R. Parsons, Polarography in the Formamides, *in* Advances in Polarography, Vol. 3, p. 829 (I. S. Longmuir, ed.), Pergamon, Oxford (1960)

M85 P. P. Rastogi, *Bull. Chem. Soc. Japan*, **43,** 2442 (1970)

M86 D. S. Reid and C. A. Vincent, *J. Electroanal. Chem.*, **18,** 427 (1968)

M87 M. Della Monica and U. Lamanna, *J. Phys. Chem.*, **72,** 4329 (1968)

M88 G. Atkinson and Y. Mori, *J. Phys. Chem.*, **71,** 3523 (1967)

M89 J. T. Edward, *J. Chem. Educ.*, **47,** 261 (1970)

M90 M. A. Elliott with R. M. Fuoss, *J. Amer. Chem. Soc.*, **61,** 294 (1939)

M91 C. M. French and N. Singer, *J. Chem. Soc.*, 1424 (1956)

M92 C. M. French and N. Singer, *J. Chem. Soc.*, 2428 (1956)
M93 S. Broersma, *J. Chem. Phys.*, **28,** 1158 (1958)
M94 H. G. Hertz, *Angew. Chem. Int. Ed.*, **9,** 124 (1970)
M95 D. G. Miller, *J. Phys. Chem.*, **64,** 1598 (1960)
M96 G. J. Hills, *in* Chemical Physics of Ionic Solutions, pp. 571–572 (B. E. Conway and R. G. Barradas, eds.), Wiley, New York (1966)
M97 R. H. Stokes and I. A. Weeks, *Austral. J. Chem.*, **17,** 304 (1964)
M98 R. H. Stokes, Mobilities of Ions and Uncharged Molecules in Relation to Viscosity, *in* The Structure of Electrolytic Solutions, Chapter 20 (W. J. Hamer, ed.), Wiley, New York (1959)
M99 R. A. Robinson and R. H. Stokes, Electrolyte Solutions, 2nd edition revised, (a) Chapter 11, (b) Chapter 6, (c) Chapter 13, Butterworths, London (1965)
M100 A. W. Porter, *Trans. Faraday Soc.*, **23,** 413 (1927); A. G. Langdon and H. C. Thomas, *J. Phys. Chem.*, **75,** 1821 (1971)
M101 (*a*) M. Born, *Z. Physik*, **1,** 221 (1920)
(*b*) R. M. Fuoss, *Proc. Nat. Acad. Sci.*, **45,** 807 (1959)
(*c*) R. H. Boyd, *J. Chem. Phys.*, **35,** 1281 (1961); **39,** 2376 (1963)
M102 R. Zwanzig, *J. Chem. Phys.*, **38,** 1603 (1963)
M103 J. N. Agar, *Ann. Rev. Phys. Chem.*, **15,** 469 (see p. 483) (1964)
M104 H. S. Frank, Solvent Models and the Interpretation of Ionization and Solvation Phenomena, *in* Chemical Physics of Ionic Solutions, Chapter 4 (B. E. Conway and R. G. Barradas, eds.), Wiley, New York (1966)
M105 R. L. Kay, G. P. Cunningham and D. F. Evans, The Effect of Solvent Structure on Ionic Mobilities in Aqueous Solvent Mixtures, *in* Hydrogen-Bonded Solvent Systems, p. 249 (A. K. Covington and P. Jones, eds.), Taylor & Francis, London (1968)
M106 R. Zwanzig, *J. Chem. Phys.*, **52,** 3625 (1970)
M107 R. Fernández-Prini and G. Atkinson, *J. Phys. Chem.*, **75,** 239 (1971)
M108 E. Glueckauf, *in* Chemical Physics of Ionic Solutions, p. 112 (B. E. Conway and R. G. Barradas, eds.), Wiley, New York (1966)
M109 E. J. Passeron, *J. Phys. Chem.*, **68,** 2728 (1964)
M110 R. L. Kay, B. J. Hales and G. P. Cunningham, *J. Phys. Chem.*, **71,** 3925 (1967); H. L. Yeager and B. Kratochvil, *J. Phys. Chem.*, **73,** 1963 (1969); **74,** 963 (1970)
M111 A. J. Parker, *Quart. Rev.*, **16,** 163 (1962)
M112 R. S. Drago and B. B. Wayland, *J. Amer. Chem. Soc.*, **87,** 3571 (1965); V. Gutmann, *Coord. Chem. Rev.*, **2,** 239 (1967); Coordination Chemistry in Non-aqueous Solutions, Springer, Vienna (1968)
M113 J. R. Graham, G. S. Kell and A. R. Gordon, *J. Amer. Chem. Soc.*, **79,** 2352 (1957)
M114 M. J. Polissar, *J. Chem. Phys.*, **6,** 833 (1938)
M115 A. E. Stearn and H. Eyring, *J. Phys. Chem.*, **44,** 955 (see p. 976) (1940)
M116 S. B. Brummer and G. J. Hills, *Trans. Faraday Soc.*, **57,** 1816 (1961)
M117 G. J. Hills, Kinetic Studies of Ionic Migration, *in* Chemical Physics of Ionic Solutions, Chapter 23 (B. E. Conway and R. G. Barradas, eds.), Wiley, New York (1966)
M118 F. Barreira and G. J. Hills, *Trans. Faraday Soc.*, **64,** 1359 (1968)
M119 S. B. Brummer and G. J. Hills, *Trans. Faraday Soc.*, **57,** 1823 (1961)
M120 S. B. Brummer, *J. Chem. Phys.*, **42,** 1636 (1965)
M121 W. A. Adams and K. J. Laidler, *Canad. J. Chem.*, **46,** 1977, 1989 (1968)
M122 J. E. Smith, Jr. and E. B. Dismukes, *J. Phys. Chem.*, **67,** 1160 (1963); **68,** 1603 (1964); R. L. Kay and G. A. Vidulich, *J. Phys. Chem.*, **74,** 2718 (1970)
M123 R. L. Kay, K. S. Pribadi and B. Watson, *J. Phys. Chem.*, **74,** 2724 (1970)
M124 H. Röhler, *Z. Elektrochem.*, **16,** 419 (1910)

M125 K. H. Schmidt and S. M. Ander, *J. Phys. Chem.*, **73,** 2846 (1969); G. C. Barker, P. Fowles, D. C. Sammon and B. Stringer, *Trans. Faraday Soc.*, **66,** 1498 (1970)
M126 R. P. Bell, The Proton in Chemistry, Chapter III, Cornell University Press, Ithaca, N.Y. (1959)
M127 G. W. A. Fowles and W. R. McGregor, *J. Phys. Chem.*, **68,** 1342 (1964)
M128 J. F. Coetzee and D. K. McGuire, *J. Phys. Chem.*, **67,** 1810 (1963)
M129 B. E. Conway, J. O'M. Bockris and H. Linton, *J. Chem. Phys.*, **24,** 834 (1956)
M130 M. Eigen and L. De Maeyer, *Proc. Roy. Soc.*, **A247,** 505 (1958)
M131 B. E. Conway and M. Salomon, Classical H/D Isotope Effect in Proton Conductance, and the Proton Solvation Energy, *in* Chemical Physics of Ionic Solutions, Chapter 24 (B. E. Conway and R. G. Barradas, eds.), Wiley, New York (1966)
M132 B. E. Conway, Proton Solvation and Proton Transfer Processes in Solution, *in* Modern Aspects of Electrochemistry, Vol. 3, Chapter 2 (J. O'M. Bockris and B. E. Conway, eds.), Butterworths, London (1964)
M133 E. Pitts, B. E. Tabor and J. Daly, *Trans. Faraday Soc.*, **66,** 693 (1970)
M134 (*a*) S. Lengyel, J. Giber and J. Tamas, *Acta Chim. Acad. Sci. Hung.*, **32,** 429 (1962); D. A. Lown and H. R. Thirsk, *Trans. Faraday Soc.*, **67,** 132 (1971)
M134 (*b*) T. Erdey-Grúz, E. Kugler and L. Majthényi, *Electrochim. Acta*, **13,** 947 (1968)
M135 T. Shedlovsky and R. L. Kay, *J. Phys. Chem.*, **60,** 151 (1956)
M136 H. O. Spivey and T. Shedlovsky, *J. Phys. Chem.*, **71,** 2165 (1967)
M137 M. Goffredi and T. Shedlovsky, *J. Phys. Chem.*, **71,** 2182 (1967)
M138 L. Thomas and E. Marum, *Z. Phys. Chem.*, **143,** 213 (1929)
M139 (*a*) R. W. Gurney, Ionic Processes in Solution, pp. 226–227, McGraw-Hill, New York (1953)
M139 (*b*) L. S. Guss and I. M. Kolthoff, *J. Amer. Chem. Soc.*, **62,** 1494 (1940)
M140 R. de Lisi and M. Goffredi, *Electrochim. Acta*, **16,** 2181 (1971)
M141 G. Lepoutre and M. J. Sienko (eds.), Metal-Ammonia Solutions, Benjamin, New York (1964); J. J. Lagowski and M. J. Sienko (eds.), Metal-Ammonia Solutions, Butterworths, London (1970)
M142 U. Schindewolf, *Angew. Chem. Int. Ed.*, **7,** 190 (1968)
M143 J. L. Dye, *Accounts Chem. Res.*, **1,** 306 (1968)
M144 B. Baron, P. Delahay and R. Lugo, *J. Chem. Phys.*, **53,** 1399 (1970)
M145 J. L. Down, J. Lewis, B. Moore and G. Wilkinson, *Proc. Chem. Soc.*, 209 (1957); *J. Chem. Soc.*, 3767 (1959)
M146 J. Jortner and G. Stein, *Nature*, **175,** 893 (1955)
M147 E. J. Hart and M. Anbar, The Hydrated Electron, (a) pp. 30 *et seq.*, (b) pp. 185–186, Wiley–Interscience, New York (1970)
M148 F. Cafasso and B. R. Sundheim, *J. Chem. Phys.*, **31,** 809 (1959); F. S. Dainton, D. M. Wiles and A. N. Wright, *J. Chem. Soc.*, 4283 (1960)
M149 D. Huppert and K. H. Bar-Eli, *J. Phys. Chem.*, **74,** 3285 (1970)
M150 I. Hurley, T. R. Tuttle Jr. and S. Golden, *J. Chem. Phys.*, **48,** 2818 (1968)
M151 E. C. Evers and P. W. Frank Jr., *J. Chem. Phys.*, **30,** 61 (1959)
M152 R. R. Dewald and J. L. Dye, *J. Phys. Chem.*, **68,** 128 (1964). See also ref. 155.
M153 W. H. Bromley Jr. and W. F. Luder, *J. Amer. Chem. Soc.*, **66,** 107 (1944); G. W. A. Fowles and W. R. McGregor, *J. Phys. Chem.*, **68,** 1342 (1964); G. J. Janz and M. J. Tait, *Canad. J. Chem.*, **45,** 1101 (1967)
M154 G. W. A. Fowles, W. R. McGregor and M. C. R. Symons, *J. Chem. Soc.*, 3329 (1957)
M155 R. R. Dewald and K. W. Browall, *J. Phys. Chem.*, **74,** 129 (1970)
M156 D. S. Berns, E. C. Evers and P. W. Frank Jr., *J. Amer. Chem. Soc.*, **82,** 310 (1960)

M157 G. E. Gibson and T. E. Phipps, *J. Amer. Chem. Soc.*, **48,** 312 (1926)
M158 E. C. Evers, A. E. Young and A. J. Panson, *J. Amer. Chem. Soc.*, **79,** 5118 (1957)
M159 B. B. Hibbard with F. C. Schmidt, *J. Amer. Chem. Soc.*, **77,** 225 (1955)
M160 J. L. Dye, R. F. Sankuer and G. E. Smith, *J. Amer. Chem. Soc.*, **82,** 4797 (1960); **83,** 5047 (1961)
M161 J. Jortner, *J. Chem. Phys.*, **30,** 839 (1959), and references quoted therein
M162 D. A. Copeland, N. R. Kestner and J. Jortner, *J. Chem. Phys.*, **53,** 1189 (1970); J. Jortner, *Ber. Bunsenges. Phys. Chem.*, **75,** 696 (1971)
M163 P. Walden and H. Ulich, *Z. Phys. Chem.*, **107,** 219 (1923)
M164 P. Walden and H. Ulich, *Z. Phys. Chem.*, **114,** 297 (1924); P. Walden, H. Ulich and G. Busch, *Z. Phys. Chem.*, **123,** 429 (1926); P. Walden and E. J. Birr, *Z. Phys. Chem.*, **A153,** 1 (1931)
M165 H. Ulich, *Trans. Faraday Soc.*, **23,** 388 (1927)
M166 D. J. Mead with R. M. Fuoss and C. A. Kraus, *Trans. Faraday Soc.*, **32,** 594 (1936)
M167 R. H. Davies and E. G. Taylor, *J. Phys. Chem.*, **68,** 3901 (1964)
M168 G. R. Lester, T. A. Glover and P. G. Sears, *J. Phys. Chem.*, **60,** 1076 (1956)
M169 D. L. Fowler and C. A. Kraus, *J. Amer. Chem. Soc.*, **62,** 2237 (1940)
M170 D. S. Burgess and C. A. Kraus, *J. Amer. Chem. Soc.*, **70,** 706 (1948)
M171 M. B. Reynolds and C. A. Kraus, *J. Amer. Chem. Soc.*, **70,** 1709 (1948)
M172 H. L. Pickering and C. A. Kraus, *J. Amer. Chem. Soc.*, **71,** 3288 (1949)
M173 W. E. Thompson and C. A. Kraus, *J. Amer. Chem. Soc.*, **69,** 1016 (1947)
M174 C. A. Kraus, *Ann. N.Y. Acad. Sci.*, **51,** 789 (1949)
M175 P. G. Sears, G. R. Lester and L. R. Dawson, *J. Phys. Chem.*, **60,** 1433 (1956)
M176 L. R. Dawson, E. D. Wilhoit, R. R. Holmes and P. G. Sears, *J. Amer. Chem. Soc.*, **79,** 3004 (1957)
M177 G. H. Jeffery and A. I. Vogel, *J. Chem. Soc.*, 400 (1932)
M178 R. M. Fuoss and E. Hirsch, *J. Amer. Chem. Soc.*, **82,** 1013 (1960)
M179 A. M. Brown and R. M. Fuoss, *J. Phys. Chem.*, **64,** 1341 (1960)
M180 M. A. Coplan and R. M. Fuoss, *J. Phys. Chem.*, **68,** 1177 (1964)
M181 S. R. C. Hughes and D. H. Price, *J. Chem. Soc.* **A,** 1093 (1967)
M182 P. G. Sears, J. A. Caruso and A. I. Popov, *J. Phys. Chem.*, **71,** 905 (1967)
M183 D. F. Evans, J. Thomas, J. A. Nadas and M. A. Matesich, *J. Phys. Chem.*, **75,** 1714 (1971)
M184 J. F. Coetzee and G. P. Cunningham, *J. Amer. Chem. Soc.*, **87,** 2529 (1965)
M185 L. M. Mukherjee, D. P. Boden and R. Lindauer, *J. Phys. Chem.*, **74,** 1942 (1970)
M186 J. F. Coetzee and G. P. Cunningham, *J. Amer. Chem. Soc.*, **86,** 3403 (1964)
M187 P. von R. Schleyer, D. S. Trifan and R. Bacskai, *J. Amer. Chem. Soc.*, **80,** 6691 (see footnote 1) (1958)
M188 J. F. Coetzee and W. R. Sharpe, *J. Phys. Chem.*, **75,** 3141 (1971)

APPENDIX 5.8.1

Selected Transference Number Data for Electrolytes in Organic Solvents*

Solvent	Temp/°C	Solute	$10^3 . \Delta c/\text{mol l}^{-1}$	t^0_+ ‡	Method	Ref.
50 mol% Methanol-Water	25	NaCl	5–80	0.4437	m.b.	1
		KCl	5–80	0.5068	m.b.	1
Methanol	25	NaCl	3–10	0.4633	m.b.	2
		KCl	5–20	0.5001§	m.b.	2
		KBr	4–10	0.4795‖	m.b.	3
		KSCN	2.3–10	0.4555	m.b.	3
50 mol% Ethanol-Water	25	HCl	1–2000†		e.m.f.	4
Ethanol	25	LiCl	0.9–2.5	0.4393	m.b.	5
		NaCl	1–2.5	0.4813	m.b.	5
		KSCN	1.5–7.5	0.4612	m.b.	3
20 mol% *t*-Butanol-Water	25	KBr	20	0.480	m.b.	20
20 wt% Dioxan-Water	0–50	HCl	2–3000†	0.83_1 (25°)	e.m.f.	6
45 wt% Dioxan-Water	0–50	HCl	2–3000†	0.80_6 (25°)	e.m.f.	6
70 wt% Dioxan-Water	5–50	HCl	4–1300†	0.75_5 (25°)	e.m.f.	6
82 wt% Dioxan-Water	5–45	HCl	5–460†	0.67 (25°)	e.m.f.	6
Ethylene glycol	25	KCl	5–50	0.4765	m.b.	7
Glycerol	5, 25	KCl, KF, KOH	10		m.b.	8
	5, 25	HCl	25		m.b.	8
Acetone	25	KSCN	1.1–2.4	0.376	m.b.	21
Acetonitrile	25	$AgNO_3$	10–1000		Hittorf	9
		Me_4NClO_4	0.6–12.5	0.4768	m.b.	10

APPENDIX 5.8.1 (*contd*)

Solvent	Temp/°C	Solute	$10^3 . \Delta c$/mol l^{-1}	t_+^0 ‡	Method	Ref.
Nitromethane	25	Me_4NCl	0.2–10	0.4674	m.b.	11
		Me_4NBr	0.2–10	0.4663	m.b.	11
		Et_4NCl	0.2–10	0.4320	m.b.	11
		Et_4NBr	0.2–10	0.4314	m.b.	11
		Pr_4NCl	0.2–10	0.3843	m.b.	11
		Pr_4NBr	0.2–10	0.3835	m.b.	11
		Bu_4NCl	0.2–10	0.3526	m.b.	11
		Bu_4NBr	0.2–10	0.3513	m.b.	11
Formamide	25	KCl	10–100	0.4270	m.b.	12
Methylformamide	15–45	KBr	50–300	0.508 (25°C)	Hittorf	13
Dimethylformamide	25	KSCN	6–13	0.340	m.b.	14
		LiCl	86–550	0.31_5¶	Hittorf	15
N-Methylacetamide	35–50	KBr	100–300	0.390 (40°C)	Hittorf	16
	40	KCl	20–63	0.429	Hittorf	17
Dimethylsulphoxide	25	$AgClO_4$	20–100	0.404	Hittorf	18
Sulpholane	30	$AgClO_4$	20–100	0.418**	Hittorf	19

* Papers in which one concentration only was used are mentioned under the appropriate reference numbers but as a rule the data are not tabulated.

† Concentrations in mol (kg solvent)$^{-1}$.

‡ Most extrapolations were carried out with the Longsworth equation or some variant of it. The limiting transference number is generally not very sensitive to the extrapolation method employed.

§ Alternative extrapolations can give 0.5006.

|| The value calculated from the ionic conductances in Appendix 5.12.2 is 0.4816.

¶ Extrapolated by eqn. 5.10.17 using $a = 3$ Å. The Longsworth extrapolation is not applicable because Λ' is negative at the higher concentrations, and the naive t_+ vs. $\sqrt{c}$ plot[15] gave a result 0.02 lower.

** From separate extrapolations of Λ_+ and Λ_-.

APPENDIX 5.8.1 REFERENCES

1. L. W. Shemilt, J. A. Davies and A. R. Gordon, *J. Chem. Phys.*, **16,** 340 (1948). M.b. transference numbers of 0.05 *M* LiCl and NaCl at 25°C were measured by L. G. Longsworth and D. A. MacInnes, *J. Phys. Chem.*, **43,** 239 (1939) in 10, 20, 40, 60 and 80 mol % MeOH-H_2O, and those of 0.025 *M* KCl and HCl, 0.023 *M* KOH and 0.01 *M* KF in various MeOH-H_2O mixtures at 5° and 25°C by T. Erdey-Grúz and L. Majthényi, *Acta Chim. Acad. Sci. Hung.*, **16,** 417 (1958). The Hittorf transference numbers of 0.53 *m* $CaCl_2$ in 30 and 60 mol % MeOH-H_2O at 25°C were determined by H. Schneider and H. Strehlow, *Z. Elektrochem.*, **66,** 309 (1962)
2. J. A. Davies, R. L. Kay and A. R. Gordon, *J. Chem. Phys.*, **19,** 749 (1951). The value for 0.01 *M* NaCl in MeOH has been confirmed by G. Marx and D. Hentschel, *Talanta*, **16,** 1159 (1969)
3. J. Smisko and L. R. Dawson, *J. Phys. Chem.*, **59,** 84 (1955). The limiting potassium transference number for KBr in MeOH is 0.0021 smaller than the value calculated from the transference numbers in ref. 2 and conductance data (Appendix 5.12.2)
4. H. S. Harned and M. H. Fleysher, *J. Amer. Chem. Soc.*, **47,** 92 (1925). These authors also reported transference numbers in pure EtOH, but J. W. Woolcock and H. Hartley, *Phil. Mag.*, [7], **5,** 1133 (1928), later showed that their solvent had probably contained some water. M.b. transference numbers at 5° and 25°C in various EtOH-H_2O mixtures of 0.01–0.02 *M* KCl and HCl, 0.016 *M* KOH and 0.01 *M* KF were measured by T. Erdey-Grúz and L. Majthényi, *Acta Chim. Acad. Sci. Hung.*, **20,** 73 (1959), and those of 0.01 *m* HCl at 25°C by J. Lin and J. J. de Haven, *J. Electrochem. Soc.*, **116,** 805 (1969). G. Kortüm and A. Weller, *Z. Naturforsch.* **5a,** 590 (1950), determined the transference numbers of dilute lithium picrate in various EtOH-H_2O mixtures at 25°C by the Hittorf method
5. J. R. Graham and A. R. Gordon, *J. Amer. Chem. Soc.*, **79,** 2350 (1957)
6. H. S. Harned and E. C. Dreby, *J. Amer. Chem. Soc.*, **61,** 3113 (1939)
7. M. Carmo Santos and M. Spiro, *J. Phys. Chem.*, **76,** 712 (1972). T. Erdey-Grúz and L. Majthényi, *Acta Chim. Acad. Sci. Hung.*, **20,** 175 (1959), reported m.b. transference numbers of 0.02 *M* KCl, 0.01–0.02 *M* HCl, 0.01 *M* KOH and KF in several different glycol-water mixtures at 5° and 25°C; their value in slightly wet glycol differed by 0.02 from that of M. Carmo Santos and M. Spiro
8. T. Erdey-Grúz, L. Majthényi and I. Nagy-Czako, *Acta Chim. Acad. Sci. Hung.*, **53,** 29 (1967). Numerous water-glycerol mixtures, ranging in composition from pure water to almost pure glycerol, were used as solvent
9. H. Strehlow and H.-M. Koepp, *Z. Elektrochem.*, **62,** 373 (1958). Experiments were also carried out with various acetonitrile-water mixtures. Hittorf transference numbers of 0.25 *m* $ZnCl_2$ in 25, 50 and 75 mol % CH_3CN-H_2O at 25°C were determined by H. Schneider and H. Strehlow, *Ber. Bunsenges. Phys. Chem.*, **69,** 674 (1965)
10. C. H. Springer, J. F. Coetzee and R. L. Kay, *J. Phys. Chem.*, **73,** 471 (1969)
11. S. Blum and H. I. Schiff, *J. Phys. Chem.*, **67,** 1220 (1963). It is possible that the volume correction was affected by solvent diffusion between the aqueous electrode compartments and the main cell
12. J. M. Notley and M. Spiro, *J. Phys. Chem.*, **70,** 1502 (1966)
13. R. Gopal and O. N. Bhatnagar, *J. Phys. Chem.*, **70,** 3007 (1966). These workers have also reported Hittorf transference numbers for KBr in *N*-methylpropionamide (*J. Phys. Chem.*, **70,** 4070 (1966); their values for KCl in formamide (*J. Phys. Chem.*, **68,** 3892 (1964) differ by ca. 0.01 from those in ref. 12
14. J. E. Prue and P. J. Sherrington, *Trans. Faraday Soc.*, **57,** 1795 (1961)

15. R. C. Paul, J. P. Singla and S. P. Narula, *J. Phys. Chem.*, **73**, 741 (1969). The limiting LiCl transference number, obtained by a modified Fuoss-Onsager extrapolation, is in good accord with the limiting ionic conductances in ref. 14
16. R. Gopal and O. N. Bhatnagar, *J. Phys. Chem.*, **69**, 2382 (1965)
17. G. P. Johari and P. H. Tewari, *J. Phys. Chem.*, **70**, 197 (1966). The results are, however, not consistent with those of ref. 16 by ca. 0.01, and their Hittorf data for KCl in formamide differ by 0.02 from those of ref. 12
18. M. Della Monica, D. Masciopinto and G. Tessari, *Trans. Faraday Soc.*, **66**, 2872 (1970)
19. M. Della Monica, U. Lamanna and L. Senatore, *J. Phys. Chem.*, **72**, 2124 (1968). However, the resulting ionic conductances are completely at variance with a m.b. value for $NaClO_4$ reported by J. Lawrence and R. Parsons, *Trans. Faraday Soc.*, **64**, 751 (see p. 761) (1968)
20. T. L. Broadwater and R. L. Kay, *J. Phys. Chem.*, **74**, 3802 (1970). Transference numbers of 0.02 *M* KBr in several aqueous solutions containing less than 20 mol % *t*-BuOH were also reported
21. H. C. Brookes, M. C. B. Hotz and A. H. Spong, *J. Chem. Soc.* **A**, 2415 (1971)

APPENDIX 5.12.1

Limiting Single Ion Conductances* in 50 mol % Methanol-Water at 25°C

Cation	Λ°_+	Ref.	Anion	Λ°_-	Ref.
Na^+	29.57	1	Cl^-	37.05	1
K^+	38.05	1			

* In this and the following tables, $\Lambda^\circ_\pm$ is cited in units of cm^2 (int. $\Omega)^{-1}$ mol^{-1}.

1. H. I. Schiff and A. R. Gordon, *J. Chem. Phys.*, **16**, 336 (1948)

APPENDIX 5.12.2

Limiting Single Ion Conductances in Methanol at 25°C

Cation	Λ_+°	Ref.	Anion	Λ_-°	Ref.
H^+	146.2	1	CH_3O^-	53.1	2
Li^+	39.6	2	Cl^-	52.3_5	3
Na^+	45.2	2	Br^-	56.4_5	2
K^+	52.4_5	2	I^-	62.7_5	2
Rb^+	55.9	2	SCN^-	62.3	4
Cs^+	60.9	2	NO_3^-	61.1	2
Ag^+	50.1	2	ClO_3^-	61.4	2
Me_4N^+	68.7	2	ClO_4^-	70.8	2
Et_4N^+	60.5	2	CH_3COO^-	39.3	1
Pr_4N^+	46.1	2	Pic^-	47.0_5	2
Bu_4N^+	39.0_5	2	BPh_4^-	36.6	2
Am_4N^+	34.8_5	2			
$(i\text{-}Am)_4N^+$	35.4	2			
$(i\text{-}Am)_3BuN^+$	36.4_5	2			
$(C_2H_4OH)_4N^+$	37.5	2			
$Hept_4N^+$	29.3	5			
Me_3S^+	67.6	2			
Et_3S^+	61.9	2			
Pr_3S^+	50.3	2			
Bu_3S^+	43.9	2			

1. T. Shedlovsky and R. L. Kay, *J. Phys. Chem.*, **60,** 151 (1956). The very much lower hydrogen ion conductance of A. D'Aprano and R. M. Fuoss, *J. Phys. Chem.*, **73,** 400 (1969), is unlikely to be correct (see D. A. MacInnes, The Principles of Electrochemistry, p. 362, Reinhold, New York (1939))
2. Calculated from the limiting conductances in Appendix 5.1 and from ref. 3
3. The limiting conductance of Cl^- was obtained by combining the conductances of NaCl and KCl, determined by J. P. Butler, H. I. Schiff and A. R. Gordon, *J. Chem. Phys.*, **19,** 752 (1951) and re-extrapolated by R. L. Kay, *J. Amer. Chem. Soc.*, **82,** 2099 (1960), with the transference numbers of NaCl and KCl of J. A. Davies, R. L. Kay and A. R. Gordon, *J. Chem. Phys.*, **19,** 749 (1951) extrapolated according to Appendix 5.8.1
4. Calculated from the data given by J. Smisko and L. R. Dawson, *J. Phys. Chem.*, **59,** 84 (1955)
5. D. F. Evans and P. Gardam, *J. Phys. Chem.*, **72,** 3281 (1968)

APPENDIX 5.12.3

Limiting Single Ion Conductances in Ethanol at 25°C

Cation	Λ°_+	Ref.	Anion	Λ°_-	Ref.
H^+	62.80	1	$CH_3CH_2O^-$	22.97	1
Li^+	17.09	1	Cl^-	21.85	1
Na^+	20.32	1	Br^-	23.78	1
K^+	23.57	1	I^-	26.92	1
Rb^+	24.76	1	SCN^-	27.4	2
Cs^+	26.58	1	NO_3^-	25.65	1
Ag^+	19.23	1	Pic^-	25.0	1
Me_4N^+	29.80	1			
Et_4N^+	29.40	1			
Pr_4N^+	23.05	1			
Bu_4N^+	19.72	1			
$(i\text{-Am})_3BuN^+$	18.39	1			
$Hept_4N^+$	15.01	1			

1. Calculated from the limiting conductances in Appendix 5.1 and the limiting transference numbers of J. R. Graham and A. R. Gordon, *J. Amer. Chem. Soc.*, **79**, 2350 (1957)
2. J. Smisko and L. R. Dawson, *J. Phys. Chem.*, **59**, 84 (1955)

APPENDIX 5.12.4

Limiting Single Ion Conductances in Ethylene Glycol at 25°C

Cation	Λ°_+	Ref.	Anion	Λ°_-	Ref.
H^+	27.7	1	F^-	3.26	1
Li^+	2.112	1	Cl^-	5.073	1
Na^+	3.10_7	1	Br^-	4.98_0	1
K^+	4.620	1	I^-	4.60_8	1
Tl^+	5.31	1	Pic^-	2.21_3	1
Me_4N^+	2.97_3	1	NO_3^-	4.8_1	2
Et_4N^+	2.19_7	1			
Pr_4N^+	1.73_7	1			
Bu_4N^+	1.51_3	1			
Pb^{++}	2.66	1			

1. M. Carmo Santos and M. Spiro, *J. Phys. Chem.*, **76**, 712 (1972)
2. B. Sesta and M. L. Berardelli, *Electrochim. Acta*, **17**, 915 (1972)

APPENDIX 5.12.5

Limiting Single Ion Conductances in Formamide at 25°C

Cation	Λ_+°	Ref.	Anion	Λ_-°	Ref.
H^+	10.8	1	Cl^-	17.12	2
Li^+	9.03	2	Br^-	17.17	2
Na^+	10.10	2	I^-	16.73	2
K^+	12.75	2	SCN^-	17.2	1
Rb^+	12.8	1	NO_3^-	17.4	1
Cs^+	13.9	2	CH_3COO^-	11.9	1
Tl^+	15.8	1	$PhCOO^-$	9.8	1
H_4N^+	$15._6$	1	$PhSO_3^-$	10.4	1
Me_4N^+	13.38	2	BPh_4^-	6.04	2
Et_4N^+	11.02	2	Pic^-	9.1_3	3
Pr_4N^+	8.12	2			
Bu_4N^+	6.83	2			
$(i\text{-Am})_3BuN^+$	6.30	2			
Am_4N^+	5.8	3			
Hex_4N^+	4.89	2			
Me_3PhN^+	10.7	1			

1. J. M. Notley and M. Spiro, *J. Phys. Chem.*, **70,** 1502 (1966)
2. J. Thomas and D. F. Evans, *J. Phys. Chem.*, **74,** 3812 (1970)
3. P. Bruno and M. Della Monica, *J. Phys. Chem.*, **76,** 1049 (1972)

APPENDIX 5.12.6

Limiting Single Ion Conductances in *N*-Methylformamide (NMF) at 25°C

Cation	Λ_+°	Ref.	Anion	Λ_-°	Ref.
Na^+	21.6_1	1	Cl^-	19.7_5	1
K^+	22.1_9	1	Br^-	21.5_0	1
Cs^+	24.3_8	1	I^-	22.8_1	1
Et_4N^+	26.1_9	1	Pic^-	13.0_9	1

1. Calculated from the limiting transference number of R. Gopal and O. N. Bhatnagar, *J. Phys. Chem.*, **70,** 3007 (1966), and the limiting conductances of C. M. French and K. H. Glover, *Trans. Faraday Soc.*, **51,** 1418 (1955)

APPENDIX 5.12.7

Limiting Single Ion Conductances in *N*-Methylacetamide (NMA) at 40°C

Cation	Λ_+°	Ref.	Anion	Λ_-°	Ref.
H^+	8.9_5	1	Cl^-	11.7	2, 3
Li^+	6.4_5	2	Br^-	12.9_5	4
Na^+	8.1	1, 2	I^-	14.7_5	1, 2, 3
K^+	8.3	4	SCN^-	16.1	1
Ba^{2+}	19.9	5	NO_3^-	14.6_5	1
H_4N^+	9.5_5	1, 3	ClO_4^-	16.9	1
MeH_3N^+	11.7_5	3	BrO_3^-	13.7	1
$Me_2H_2N^+$	13.5	3	Pic^-	11.9_5	1
Me_3HN^+	12.7_5	3	$PhSO_3^-$	10.3	1
Me_4N^+	11.9	3, 7	$OctdSO_4^-$	7.2	1
Et_4N^+	11.5	3, 7	$Fe(CN)_6^{3-}$	51.1	6
Pr_4N^+	9.0	3, 7			
Bu_4N^+	7.7	3, 7			
Am_4N^+	7.2_5	7			
Hex_4N^+	7.0_5	7			
$Hept_4N^+$	6.7	7			
Me_3PhN^+	10.1	1			
$OctdMe_3N^+$	7.0	1			
Octd = octadecyl					

1. L. R. Dawson, E. D. Wilhoit, R. R. Holmes and P. G. Sears, *J. Amer. Chem. Soc.*, **79,** 3004 (1957), and ref. 4
2. L. R. Dawson, P. G. Sears and R. H. Graves, *J. Amer. Chem. Soc.*, **77,** 1986 (1955), and ref. 4
3. L. R. Dawson, E. D. Wilhoit and P. G. Sears, *J. Amer. Chem. Soc.*, **78,** 1569 (1956), and ref. 4. This paper contains conductance data for several other substituted ammonium ions
4. The limiting conductances of K^+ and Br^- were obtained by multiplying the limiting transference number of KBr reported by R. Gopal and O. N. Bhatnagar, *J. Phys. Chem.*, **69,** 2382 (1965), by the weighted mean limiting conductance of KBr. The latter was calculated from the directly measured conductance of KBr in ref. 2 and from the values derived indirectly by use of Kohlrausch's law of independent ionic migration
5. L. R. Dawson, G. R. Lester and P. G. Sears, *J. Amer. Chem. Soc.*, **80,** 4233 (1958) Data for other alkaline earth ions are also given
6. L. R. Dawson, J. W. Vaughn, G. R. Lester, M. E. Pruitt and P. G. Sears, *J. Phys. Chem.*, **67,** 278 (1963). Conductances are also given for several other large anions
7. R. D. Singh, P. P. Rastogi and R. Gopal, *Canad. J. Chem.*, **46,** 3525 (1968)

APPENDIX 5.12.8

Limiting Single Ion Conductances in Dimethylformamide (DMF) at 25°C

Cation	Λ_+°	Ref.	Anion	Λ_-°	Ref.
H^+	34.7	1	Cl^-	55.1	2
Li^+	25.0	2	Br^-	53.6	2
Na^+	29.9	2	I^-	52.3	2
K^+	30.8	2	SCN^-	59.7	2
Rb^+	32.4	2	NO_3^-	57.3	2
Cs^+	34.5	2	ClO_4^-	52.4	2
Ag^+	35.2	2	Pic^-	37.4	1
Tl^+	38.7	2			
H_4N^+	38.6	2			
Me_4N^+	39.1	1			
Et_4N^+	35.4	3			
Pr_4N^+	29.0	3			
Bu_4N^+	25.4	3			
EtH_3N^+	39.0	1			
Et_3HN^+	35.7	1			
Me_2Tl^+	27.0	2			
Am_4N^+	22.9	4			

1. Calculated from the limiting conductances in Appendix 5.1, Tables 5.1.2, and the limiting single ion conductances in ref. 2
2. J. E. Prue and P. J. Sherrington, *Trans. Faraday Soc.*, **57**, 1795 (1961)
3. P. G. Sears, E. D. Wilhoit and L. R. Dawson, *J. Phys. Chem.*, **59**, 373 (1955)
4. P. Bruno and M. Della Monica, *J. Phys. Chem.*, **76**, 1049 (1972)

APPENDIX 5.12.9

Limiting Single Ion Conductances in Acetonitrile at 25°C

Cation	Λ_+°	Ref.	Anion	Λ_-°	Ref.
H^+	99	4	Cl^-	98.7	3
Li^+	69.3	1	Br^-	100.7	1
Na^+	76.9	1	I^-	102.4	1
K^+	83.6	1	SCN^-	113.4	2
Rb^+	85.6	1	NO_3^-	106.4	2
Cs^+	87.3	1	ClO_3^-	100.7	2
Cu^+	65.0	2	ClO_4^-	103.7	1
Ag^+	86.2	2	BF_4^-	108.2	2
Tl^+	91.2	2	PF_6^-	103.5	2
Me_4N^+	94.5	1	Pic^-	77.7	1
Et_4N^+	84.8	1	$B(i\text{-}Am)_4^-$	57.6	1
Pr_4N^+	70.3	1	BPh_4^-	58.3	1
Bu_4N^+	61.4	1	$TCVal^-$	97.2	2
Am_4N^+	56.0	2	PCP^-	84.8	2
$(i\text{-}Am)_4N^+$	56.9	1	$bisTCVam^-$	72.5	2
$(i\text{-}Am)_3BuN^+$	57.8	1	HCHT	61.8	2
$(C_2H_4OH)_4N^+$	64.0	1			
Me_3S^+	96.8	2			
Et_3S^+	87.9	2			
Pr_3S^+	76.2	2			
Ph_4As^+	55.8	1			
$((Ph_2NH_2)P)_2N^+$	50.2	2			

1. C. H. Springer, J. F. Coetzee and R. L. Kay, *J. Phys. Chem.*, **73**, 471 (1969)
2. Calculated from the limiting single ion conductances in ref. 1 and the limiting conductances in Appendix 5.1, Table 5.1.2. See this Appendix for the abbreviations employed
3. R. L. Kay and D. F. Evans, *J. Phys. Chem.*, **70**, 2325 (1966)
4. J. F. Coetzee and D. K. McGuire, *J. Phys. Chem.*, **67**, 1810 (1963)

APPENDIX 5.12.10
Limiting Single Ion Conductances in Nitromethane at 25°C

Cation	Λ°_+	Ref.	Anion	Λ°_-	Ref.
Me_4N^+	54.5	1, 2	Cl^-	62.5	1
Et_4N^+	47.7	1	Br^-	62.9	1
Pr_4N^+	39.2	1	$B(i\text{-Am})_4^-$	31.5	1
Bu_4N^+	34.1	1	BPh_4^-	31.6	1
$(i\text{-Am})_4N^+$	31.5	1			

1. J. F. Coetzee and G. P. Cunningham, *J. Amer. Chem. Soc.*, **87,** 2529 (1965). Most of their values are based on those given by R. L. Kay, S. C. Blum and H. I. Schiff, *J. Phys. Chem.*, **67,** 1223 (1963), and refer to a solvent of viscosity 0.00627 poise
2. If this value for the tetramethylammonium ion is correct, the limiting molar conductances of Me_4NCl and Me_4NBr are too low by 0.6 $cm^2\ \Omega^{-1}\ mol^{-1}$

APPENDIX 5.12.11
Limiting Single Ion Conductances in Dimethylsulphoxide (DMSO) at 25°C

Cation	Λ°_+	Ref.	Anion	Λ°_-	Ref.
H^+	15.0	4	Cl^-	24.0	3
Na^+	14.2	1	Br^-	23.6	1
K^+	15.0	1	I^-	23.4	1, 2
Cs^+	16.6	2	SCN^-	28.6	1
Ag^+	16.4	1	NO_3^-	26.5	1
Me_4N^+	19.0	3	ClO_4^-	24.1	1
Et_4N^+	17.5	3	Pic^-	16.8	1
Pr_4N^+	13.8	3	$PhSO_3^-$	16.4	1
Bu_4N^+	11.8	1, 3	$OctdSO_4^-$	9.5	1
Am_4N^+	10.8	3	$B(i\text{-Am})_4^-$	10.2	3
$(i\text{-Am})_4N^+$	11.0	3	BPh_4^-	10.2	3
Hex_4N^+	10.2	3			
$Hept_4N^+$	9.6	3			
Me_3PhN^+	11.4	1			
Me_3OctdN^+	10.4	1			
Octd = octadecyl					

1. M. Della Monica, D. Masciopinto and G. Tessari, *Trans. Faraday Soc.*, **66,** 2872 (1970). Somewhat lower conductances for K^+, ClO_4^-, and Pic^- are indicated by P. Bruno and M. Della Monica, *J. Phys. Chem.*, **76,** 1049 (1972)
2. Calculated from the limiting conductances in Appendix 5.1, Table 5.1.2 and ref. 1
3. Calculated from the limiting conductances of D. E. Arrington and E. Griswold, *J. Phys. Chem.*, **74,** 123 (1970)
4. J. A. Bolzan and A. J. Arvia, *Electrochim. Acta*, **15,** 827 (1970); **16,** 531 (1971)

APPENDIX 5.12.12

Limiting Single Ion Conductances in Sulpholane (Tetramethylenesulphone) at 30°C

Cation	Λ°_{+}	Ref.	Anion	Λ°_{-}	Ref.
Li^+	4.34	1	Cl^-	9.29	1
Na^+	3.61	1	Br^-	8.91	1
K^+	4.04	1	I^-	7.22	1
Rb^+	4.20	1	SCN^-	9.63	1
Cs^+	4.34	1	ClO_4^-	6.69	2
Ag^+	4.81	3	PF_6^-	5.95	1
H_4N^+	4.98	1	Pic^-	5.3_2	4
Me_4N^+	4.28	1			
Et_4N^+	3.94	1			
Pr_4N^+	3.23	1			
Bu_4N^+	2.76	1			
Am_4N^+	2.51	4			

1. Calculated from the limiting conductances in Appendix 5.1, Table 5.1.4, and the limiting ionic conductance of ClO_4^- in ref. 2
2. M. Della Monica, U. Lamanna and L. Senatore, *J. Phys. Chem.*, **72**, 2124 (1968)
3. Calculated from the conductance and transference data in ref. 2
4. P. Bruno and M. Della Monica, *J. Phys. Chem.*, **76**, 1049 (1972)

Chapter 6

Reaction Kinetics and Mechanism

Donald W. Watts
School of Chemistry, University of Western Australia
Nedlands W.A. 6009, Australia

6.1 INTRODUCTION

In discussing reactions in organic non-aqueous solvents much that is interesting comes from comparing and contrasting the results in water, especially in that it is one of an important group of solvents, the hydroxylic solvents, the majority of which are organic. Thus although the emphasis here will be on organic solvents, results in water will occasionally become relevant to the discussion, as will some results in aqueous-organic solvent mixtures.

Amis,[N1] in his text 'Solvent Effects on Reaction Rates and Mechanisms', clearly states the varied interests of chemists in this field. 'The physical chemist will envision the effect on rates of reactions of dielectric constant, viscosity, internal cohesion and external pressure as these are influenced by solvent. The physical organic chemist will perhaps call to mind acidity, basicity, hydrogen bonding, structure effects, electronegativity and solvating ability as related to solvent. The strictly organic chemist may simply think in terms of a medium in which reactants can be made to form products merely because of solubility relations etc.' Since these statements were made, considerable work has accumulated concerned with the kinetics of inorganic reactions in nonaqueous media and particularly with the substitution reactions of metal coordination and organometallic molecules. The philosophy of the inorganic-reaction kineticist like that of his physical-organic counterpart has been to seek empirical correlations.

Because this text by Amis[N1] and other similar treatises[N2-9] give detailed accounts of the physiochemical theories, the emphasis here will be on the attempts of physical-organic and physical-inorganic chemists to correlate the rapidly increasing volume of kinetic data. Recent reviews[N9-19] of the organic field make extensive coverage of this area unnecessary. The emphasis here will be on inorganic systems with reference to the principles established in organic chemistry where these are applicable and useful. In particular there is now a considerable literature on the kinetics, in organic solvents, of inorganic substitution reactions involving non-labile octahedral centres, in particular cobalt(III) and chromium(III), square-planar complexes of platinum(II), organometallic complexes, particularly metal-carbonyl and phosphine complexes, and solvent-exchange reactions between the bulk solvent and the first solvation sphere of a great range of metal ions. This last set of exchange data is not only interesting in itself but forms the basis of a systematic study of other substitution reactions at these metal centres where knowledge of the solvent-exchange rate aids the interpretation of mechanism (see p. 693).

6.1.1 Solvents

The physical-organic chemist has classified solvents according to their polarity into three broad categories which are distinguishable in their effects on reaction rates and mechanisms. These are non-polar and weakly polar, dipolar aprotic solvents and protic solvents. The first group of solvents is clearly distinguished by their properties but the dipolar aprotic and protic solvents show many similarities in their physical properties. In addition to the obvious differentiation on the basis of the acidic proton, these last two groups also differ markedly in their ability to solvate anions, particularly anions of high charge and small radius. This difference is manifest in profound rate effects particularly in bimolecular reactions involving anionic nucleophiles.[N10-13,N15,N17]

It is the difference between these last two classes of solvents which has attracted most interest in the activities of physical organic chemists in the last ten years in systematising solvent effects and reaction rates. Because of their greater solubility for ionic species, these solvents have proved of greatest value to inorganic chemists. Non-polar and weakly polar solvents are extensively used, however, in the studies of organometallic compounds.

Even when the nucleophile is an anion, the rates of substitution reactions at metal centres are not greatly different in protic and dipolar aprotic solvents. This is in marked contrast to bimolecular substitution reactions at carbon and arises because most metal-ion substitution reactions are of a dissociative type in which entering-anion activity is not directly reflected in the rate. The increased anion activity in aprotic solvents does lead, however, to many interesting effects related to fast equilibria prior to the rate-determining step, such as ion-pair formation, which influence product ratios and steric course.[N23] In both protic and dipolar aprotic solvents, solvent coordination is common. Coordination species containing dipolar aprotic solvents are usually less stable to anion substitution than the water analogues,[N24–26] not because of the strength of the metal-ligand bond in the aquo-complexes, as is often stated, but rather because of the increased anion activity in aprotic media.[N23,N26]

It should be emphasised that bulk-solvent properties (Table 1.3.1) have been used in the past and will be used here in both the physicochemical theories and in empirical correlations, although dielectric saturation around ions and strongly polar species could make the procedure invalid. To date, allowance has been made for dielectric saturation in the calculation of ionic free energies but as yet there is no successful application to reaction rates. Unfortunately, with strongly polar solvents where the solvent effects on rate are largest, and thus hopefully more amenable to theoretical treatment, solvent orientation is strongest, and as a result, the bulk solvent properties are less applicable. The limited success of theories and correlations based on the use of bulk dielectric constant must arise fortuitously because forces, such as hydrogen-bonding and ion-dipole interactions, respond similarly to factors that determine the bulk dielectric properties.

6.1.2 Present State of Theory

In a number of recent publications[N1–8] solvent effects in relation to substitution kinetics have been treated under the headings of reactions between ions, ions and dipoles and between dipoles. All the current theories depend on the application of transition-state theory to the solution kinetics. However, there is little doubt that identical expressions can be rationalised without the acceptance of transition state theory. The existence of an 'encounter complex'[N34] must be postulated in a modification of the collision theory, in which activation is accounted for by favourable phonon concentration achieved by concerted collision within the solvent cage.

All theories depend upon the development of expressions for the variation in the relative free energies of reactants and the transition state as a function of the solvation by the medium. At this stage the fully developed theories consider the solvent as a continuous medium and use electrostatic theories to calculate the work required to produce a certain charge in the dielectric and hence changes in solute free energy resulting from changes in the solvent.

All these theories arrive at expressions typified by the Brönsted–Christiansen–Scatchard equation, for the dependence of the rate of the reaction on the dielectric constant of the medium.[N1,N35–38] Thus for the reaction

$$A + B \rightleftarrows X \rightarrow \text{PRODUCTS}$$

where A and B are ions of charge z_A and z_B and X the transition state complex. This equation gives:

$$\ln k'_{x=0} = \ln k'_{\substack{x=0\\ \varepsilon_r=\infty}} - \frac{z_A z_B e^2}{kTr\varepsilon_r} \tag{6.1.1}$$

In this expression $k'_{x=0}$ is the rate constant for the reaction at absolute temperature T and zero ionic strength in a medium of dielectric constant ε_r and $k'_{\substack{x=0\\ \varepsilon_r=\infty}}$ is the rate constant in a standard reference state of infinite dielectric constant. k is the Boltzmann constant, e the electronic charge and r is the radius of the transition complex which is normally expected to be $r_A + r_B$, the distance two ionic reactants must approach in order to react. Equations such as this have been differentiated with respect to T to produce theoretical expressions for the dependence of the activation energy upon the dielectric constant of the medium.[N39]

This theory is tested by plotting $\ln k'_{x=0}$ against $1/\varepsilon_r$, a linear relationship of appropriate slope yields the parameter r and success is judged largely on the acceptability of these dimensions.

In reactions between dipolar molecules, Laidler and Eyring[N40] have used the Kirkwood[N41] equation to derive an expression for $\log k$ which suggests a linear relationship with $(\varepsilon_r - 1)/(2\varepsilon_r + 1)$ where k is the second order rate constant. The slope of a plot of $\log k$ against $(\varepsilon_r - 1)/(2\varepsilon_r + 1)$ depends on the dipole moments and radii of the reactants and the transition state. Although linear relationships have been found in many cases the results can hardly be classed as successful in that, again, there is no reasonable basis for estimating the dimensions or dipole moment of the transition state.

The validity of all these formulae has been successfully tested only for mixtures of weakly solvating solvents such as dioxan and nitrobenzene with protic solvents such as water and methanol. In aqueous mixtures this treatment is seldom valid when the dielectric constant is below 60, and in all cases the Kirkwood function, $(\varepsilon_r - 1)/(2\varepsilon_r + 1)$, varies over such a small range that the validity of the test remains in doubt. In any case where one mixes such vastly different solvents as dioxan and water the assumption of solvent homogeneity in the bulk and inner solvation layers must always be questioned (see p. 715).[N42] It is likely that in such mixtures water is fractionated either into, or out of, the inner solvation spheres of ionic species and, small dipolar molecules and the surface dipolar sites of large molecules. If water is fractionated into the solvation sphere, then not only is the assumption of homogeneity in error, but in addition this strongly bonded water of solvation will produce a dielectrically saturated envelope, with a dielectric constant considerably different from that of bulk solvent mixture.

The electrostatic theories are also unsatisfactory in that they assume that the solvation effects will depend only on the magnitude of the charge and not on its nature. This assumption completely neglects specific short range interactions, such as hydrogen-bonding, which are appreciable in the case of small anions in water.[N10–15] Hydrogen bonding in these cases must increase the solvation free energy and will contribute to solvent orientation, and thus dielectric saturation, beyond that expected for cations of the same size and charge where only charge-dipole orientation occurs.[N33] Clear evidence for this comes from attempts to introduce allowance for dielectric saturation into calculation of solvation energies of ions through modifications to the Born equation using field dependent dielectric constants.[N28–33] A number of these models are successful for cations in water where the forces are mainly of the charge-dipole type, but fail for simple anions, such a shalide ions, because the models do not allow for the additional hydrogen-bonding terms in the solvation energy.

The inadequacies of the present electrostatic models for solvent effects on reaction rates has been emphasised in the last ten years by the availability of a range of strongly dipolar aprotic solvents, typified by dimethylsulphoxide (DMSO), *N,N*-dimethylacetamide (DMA), *N,N*-dimethylformamide (DMF) and sulpholane ($TMSO_2$). The magnitude of the effect that remains unaccounted for by the present electrostatic models can be demonstrated from measurements of the rate of the Finkelstein reaction:

$$CH_3I + Cl^- \rightarrow CH_3Cl + I^-$$

measured in methanol ($\varepsilon_r = 32.7$), DMF ($\varepsilon_r = 36.7$) and DMA ($\varepsilon_r = 37.8$).[N17] The ratios of the second order rate constants

$$\frac{k_2\,(\text{DMF})}{k_2\,(\text{MeOH})} \quad \text{and} \quad \frac{k_2\,(\text{DMA})}{k_2\,(\text{MeOH})}$$

at 0°C are 1.6×10^6 and 7.4×10^6 and electrostatic models cannot account for this large difference. The extraordinary enhancement of the rate of bimolecular anion attack in dipolar aprotic solvents, compared with the rates of attack in protic media, is due to solvation stabilisation of the anionic nucleophile in protic solvents, which is not compensated by stabilisation of the large anionic transition state.[N13] This stabilisation of small anions is clearly a function of hydrogen-bond solvation and not of a solvent property reflected by the bulk dielectric properties. It now seems, as suggested by Amis[N1] in quoting Laidler,[N43] that the time has come for a new theoretical approach in which the immediate solvation layers are treated separately, using properties assigned to fixed dipoles, and that only outside these layers should the bulk-solvent properties be used.

The success of the Born treatment, in accounting for cation solvation in water and dipolar aprotic solvents, when modified with a field-dependent dielectric constant, but its failure for anion solvation because of hydrogen bonding, suggests that the Born equation with allowance for dielectric saturation, should account for anion solvation in dipolar aprotic media where the hydrogen-bonding complication is absent.

6.1.3 Solvent Activity Coefficients (Medium Effects)

In the absence of acceptable theories to explain the effects of the solvent on the rates of reactions, the established effects of a number of solvent properties on thermodynamic properties have been used to infer the effect of the solvent on the properties of the transition state. An alternative approach has been to assume the nature of the transition state and infer the detailed role of the solvent in the mechanism.

To be of value the treatment must explain the enormous enhancement of rate in the Finkelstein reaction above, and it is clear that no single bulk-solvent property is adequate. If the transition state is in chemical equilibrium with reactants or, in collision theory parlance, if the stability of the encounter complex is directly reflected in the rate expression, then the change in rate with medium must reflect the free energy difference between the reactants and the transition state or encounter complex. This free energy difference must itself reflect the solvation free energies of the reactants and the transition state or encounter complex, as long as these latter two can be considered to retain their identity upon solvent change.

(i) *Measurement of Solvent Activity Coefficients*

The change in solvation free energy of a species on transfer from one solvent to another is most conveniently expressed in terms of the medium effect or solvent activity coefficient[N10] ${}^{w}_{s}\gamma_i$ of a solute (i), (see sect. 2.11.3) defined by:

$$ {}^{s}\mu_i^0 = {}^{w}\mu_i^0 + RT \ln {}^{w}_{s}\gamma_i = {}^{w}\mu_i^0 + RT \ln {}_{m}\gamma_i \tag{6.1.2} $$

cf. 2.11.13 and 3.3.9, where ${}^{s}\mu_i^0$ and ${}^{w}\mu_i^0$ are the standard chemical potentials of the solute (hypothetically ideal, with respect to Henry's Law, in unimolar solution) in a solvent, s, and an arbitrary standard reference solvent, often water, w, at a temperature T. The term $RT \ln {}^{w}_{s}\gamma_i$ then expresses the difference in the chemical potential between the hypothetical ideal unimolar solutions of a solute in the two solvents. The term 'solvent activity coefficient'[N10] is not entirely satisfactory because ${}^{w}_{s}\gamma_i$ is a property of a solute and a pair of solvents, however, it seems better than some alternatives used[N45–47] although a term such as 'solvent-transfer activity coefficient' would perhaps be better.

It should be emphasised that the solvent activity coefficients recorded to date are of greatest value in reactions involving at least one ionic species as a reactant or product, in particular bimolecular nucleophilic attack by anions and molecular solvolyses producing ionic products. These reactions show the greatest sensitivity to solvent change and thus the relatively imprecise values of these activity coefficients do not endanger the validity of the interpretation. Allowance for solute non-ideality in terms of Debye–Hückel activity coefficients and on association can be made in principle, but is time consuming and requires both additional care and more experiments than have been done, indeed warranted, to date.

The methods used to establish values of ${}^{w}_{s}\gamma_i$ have been considered elsewhere in this book (sect. 2.11.3). It is appropriate here only to mention the most widely applicable methods, where the solute is equilibrated separately with the solvents under consideration. The solvent activity coefficient or medium effect is then obtained from eqn. 2.11.13b (neglecting solute activity coefficients):

$$ \log {}^{w}S_i - \log {}^{s}S_i = \log {}^{w}_{s}\gamma_i \tag{6.1.3} $$

where ${}^{w}S_i$ and ${}^{s}S_i$ are the solubilities (mol l^{-1}) of the solute (i) in the solvents ('w' and 's') equilibrated with the solute, both measurements being at the same specified temperature. If solvents 's' and 'w' are immiscible then ${}^{w}_{s}\gamma_i$ can be established by direct distribution of the solute. Alternatively if a third solvent is immiscible with two miscible solvents then ${}^{w}S_i/{}^{s}S_i$ can be established in two distribution experiments in which ${}^{w}S_i$ and ${}^{s}S_i$ are compared through equilibration with equal solute concentrations in the third solvent.

For gaseous solutes one can, by choice of pressure, establish ${}^{w}_{s}\gamma_i$ at a concentration, or over a range of concentrations, chosen to cover those used in rate experiments, or alternatively extrapolation of data can lead to an 'ideal' ${}^{w}_{s}\gamma_i$ value.

For a 1:1 electrolyte (MX), the relation eqn. 2.11.13c becomes

$$\log \frac{{}^{w}K_{MX}}{{}^{s}K_{MX}} = \log {}^{w}_{s}\gamma_{MX} = \log {}^{w}_{s}\gamma_M {}^{w}_{s}\gamma_X \qquad (6.1.4)$$

where K_{MX} is the solubility product. ${}^{w}_{s}\gamma_M$ and ${}^{w}_{s}\gamma_X$ are properties of each ion, and although they can be defined they cannot be independently measured. To achieve division into ionic contributions to the solvent activity coefficient, use must be made of one of a number of extra-thermodynamic assumptions,[N13,N45,N48–51] which have been reviewed in sect. 2.11.4.

These assumptions have been critically evaluated[N13,N48] and perhaps the most intuitively satisfying is that of Grunwald, Baughman and Kohnstam[N52] in equating the anion and cation contributions of two large and equal sized ions. In practice the solubility product of tetraphenylarsonium tetraphenylboride is determined in a range of solvents and using this assumption and eqn. 6.1.4 the values of ${}^{w}_{s}\gamma_{Ph_4As^+} = {}^{w}_{s}\gamma_{Ph_4B^-}$ are established. By determining the solubility products of a range of salts of the types $Ph_4As^+X^-$ and $M^+Ph_4B^-$, a self-consistent set of values of ${}^{w}_{s}\gamma_{X^-}$ and ${}^{w}_{s}\gamma_{M^+}$ can then be established.

It should be emphasised that there is always some doubt about our ability to measure solubility products. It appears that in many cases different crystalline forms of solids equilibrate with saturated solutions in different solvents, even when solvates are not formed and eqn. 6.1.4 is invalid if this is so. This difficulty is recognised for $Ag^+Ph_4B^-$ in water[N53,N54] and for *cis*-$[CoCl_2(en)_2]ClO_4$ in DMSO.[N27]

Finally mention should be made of the method of Parker *et al.*[N48,N55] based on electrode systems in different solvents, where a salt bridge is assumed to remove the effects of liquid-junction potentials.

We will use data based on the Ph_4As^+ Ph_4B^- assumption, although the most popular assumption is that of Strehlow,[N49] based on the ferrocene (F)—ferricinium cation (F^+) pair, a choice which has been supported by others.[N50,N56]

Table 6.1.1 collects together some values of $\log {}^{M}_{s}\gamma_i$ which have been drawn from the work of Parker *et al.*[N13] These values have all been expressed relative to methanol as reference solvent. A negative value of $\log {}^{M}_{s}\gamma_i$ indicates that the species '*i*' is more strongly solvated (that is its free energy is decreased) by transfer into the solvent '*s*'. A positive value indicates poorer solvation of the species in the solvent compared to the reference methanol. A difference in $\log {}^{M}_{s}\gamma_i$ of unity corresponds to a difference in free energy of $2.303RT \log {}^{M}_{s}\gamma_i$ or 5.8 kJ mol^{-1} at 25°C.

Table 6.1.1

Some Representative Values of Solvent Activity Coefficients at 25°C

(Reference Solvent: Methanol)*

Solute	$\log {}^{MeOH}_{s}\gamma_i$					
	$s = H_2O$	s = DMF	s = DMA	s = DMSO	$s = CH_3CN$	s = TMS
Xenon	1.7†			1.1‡		
CH_3I	1.4	−0.5		−0.5	−0.4	
n-BuBr		−0.1		0.1	−0.2	
$(C_5H_5)_2Fe$	3.6				−0.3	
I_2	2.3	−1.8		−4.1	−0.2	
Cl^-	−2.5	6.5	7.8	5.5	6.3	5.8
Br^-	−2.1	4.9	5.9	3.6	4.2	
I^-	−1.5	2.6	3.0	1.3	2.4	2.4
N_3^-	−1.8	4.9	6.2	3.5	4.7	5.4
SCN^-	−1.2	2.7	3.2	1.4	2.6	2.6
Ph_4B^-	4.1	−2.7	−2.7	−2.6	−1.6	−2.0
ClO_4^-	−1.9	−0.4		−0.3		
$AgCl_2^-$	−3.3	−0.5	−0.3	−1.3	−0.2	−1.9
I_3^-	2.2	−2.0	−3.0	−3.6	−0.4	
Ag^+	−0.8	−5.1	−6.6	−8.2	−6.3	
Na^+		−3.9		−3.6	1.4	
K^+	−1.5	−3.7		−4.5	−0.8§	
Me_3S^+		−3.1	−3.6		−1.6	
Ph_4As^+	4.1	−2.7	−2.7	−2.6	−1.6	
cis-$[CoCl_2(en)_2]^+$	−2.2‖	−6.0‖	−6.2‖	−6.8‖		−4.3‖
trans-$[CoCl_2(en)_2]^+$	−0.9‖	−4.2‖	−4.2‖	−4.7‖		−2.9‖

* Data from ref. N17 except where acknowledged.
† Ref. N57.
‡ Ref. N58.
§ Ref. N13.
‖ Ref. N27.

The data presented in Table 6.1.1 will be used later where relevant. No detailed discussion of these data is necessary here, but it is worth emphasising the contrasting behaviour of anions and cations to transfer from protic solvents, such as water and methanol, to dipolar aprotic solvents, such as DMSO, DMF and DMA. It is clear that cations, particularly small cations such as Ag^+ and K^+, and even large ones such as *trans*- and *cis*-$[CoCl_2(en)_2]^+$ which possess hydrogen-bond donor sites, are substantially better solvated in the strongly dipolar aprotic solvents. Protic solvents such as MeOH and water on the other hand strongly interact with simple anions through hydrogen-bond donor solvation. Large polarisable anions such as picrate, perchlorate and $AgCl_2^-$ are better solvated by the strongly dipolar solvents, presumably through dipole—induced dipole interactions.

It is this contrasting response of anions and cations that makes difficult the direct, simple interpretation of solvent changes in terms of mechanism in the reactions of cationic coordination compounds with anions. It is the counteraction of increased anion activity by decreased cation activity on transfer from protic to dipolar aprotic solvents that causes such reactions to be insensitive to solvent transfer.

(ii) *Application of Solvent Activity Coefficients to the Discussion of Mechanism*

Before discussing individual metal coordination systems to which solvent activity coefficients can be applied, it is constructive to see how they have been applied with success to an organic system, leading to a considerable insight into the nature of the transition state. Here we will develop their application to anionic S_N2 attack at tetrahedral carbon centres.

Reactions of the type

$$XCR_3 + Y^- = YCR_3 + X^- \tag{6.1.5}$$

have been well studied and the usefulness of solvent activity coefficients in understanding the solvent dependence of the nucleophilic tendancies of X^- and Y^- is clearly established. The transition state is the same for both the forward and reverse processes and thus potentially we can get to the solvent activity coefficient of the transition state (${}^M_s\gamma_{\ddagger}$), from a knowledge of the values of the ${}^M_s\gamma_i$ of the reactants or 'products', and a knowledge of the rate constant for either the forward or the reverse reaction. In this case:

$$\log {}^M_s\gamma_{\ddagger} = \log (k^s_F/k^M_F) - \log {}^M_s\gamma_{Y^-} - \log {}^M_s\gamma_{XCR_3} \tag{6.1.6}$$

$$= \log (k^s_B/k^M_B) - \log {}^M_s\gamma_{X^-} - \log {}^M_s\gamma_{YCR_3} \tag{6.1.7}$$

where k^s_F and k^M_F are the forward second-order rate constants and k^s_B and k^M_B are the reverse second-order rate constants in the solvents (s) and (M). Clearly ${}^M_s\gamma_{\ddagger}$ can be established, even for reactions whose equilibrium is strongly solvent dependent, because the Law of Microscopic Reversibility allows us to establish the value of ${}^M_s\gamma_{\ddagger}$ by a study of the rate of either the formalised forward or reverse reaction.

Once ${}^M_s\gamma_{\ddagger}$ is established from rate data and the ${}^M_s\gamma$ values of the reactants, one can then compare this calculated ${}^M_s\gamma_{\ddagger}$ with values for stable model species, or with values for comparable transition states and thus gain insight into the detailed structure of the transition state.

Table 6.1.2 collects data for the reaction:

$$N_3^- + CH_3X = (N_3CH_3X^-)^{\ddagger} = N_3CH_3 + X^- \tag{6.1.8}$$

and compares the rates and thermodynamic functions on transfer from reference methanol (abbreviated further to M) to DMF (D) at 25°C; X is a halogen.

Table 6.1.2[a]

	$\log {}^{M}_{D}\gamma_{N_3}$	$\log {}^{M}_{D}\gamma_{CH_3X}$	$\log k^D/k^M$	$\log {}^{M}_{D}\gamma_{X^-}$	$\log {}^{M}_{D}\gamma_{\ddagger}$
$X = \mathrm{Cl}$	4.9	−0.4	3.3	6.5	1.2
$X = \mathrm{Br}$	4.9	−0.4	3.9	4.9	0.6
$X = \mathrm{I}$	4.9	−0.5	4.6	2.6	−0.2

[a] All data taken from ref. N13. D = DMF.

The increase in the rate of these three analogous bimolecular nucleophilic displacement reactions on transfer from methanol (a typical protic, hydrogen-bond donor type solvent) to DMF (a typical dipolar aprotic solvent) varies from a factor of $10^{3.3}$ to $10^{4.6}$, being faster in the aprotic media. The medium has little effect on the CH_3X free energy, these molecules being slightly better solvated in DMF. This stabilisation is almost independent of X. The increase in rate clearly reflects the substantial destabilisation of the N_3^- on the transfer to DMF, modified by a small destabilisation of the transition state (${}^{M}_{D}\gamma_{\ddagger}$, positive for $X = \mathrm{Cl}$ and Br, showing the anionic nature of the transition state) but with the superposition of a relative increase in solvation of the transition state in DMF from CI $<$ Br $<$ I. This increase in transition state stability correlates with increasing polarisability ($N_3CH_3I^- > N_3CH_3Br^- > N_3CH_3Cl^-$). All of these trends can be observed on a whole range of species and reactions of this type. These results clearly emphasise the bimolecular nature of the reaction, in addition to inferring the electrical properties of the transition state.

The diagnostic value of solvent activity coefficients is well illustrated by considering the consequences of these reactions being controlled by a simple dissociative process (S_N1 lim). In each case the rate determining step would now result in the formation of an intermediate CH_3^+ cation and the anion X^-. The relevant values of ${}^{M}_{D}\gamma_i$ are accumulated in Table 6.1.3.

Although the assumption used to produce a value for $\log {}^{M}_{D}\gamma_{CH_3^-}$ may be in error it does not affect the principle, as this term is constant for each

Table 6.1.3*

	$\log {}^{M}_{D}\gamma_{CH_3X}$	$\log {}^{M}_{D}\gamma_{X^-}$	$\log {}^{M}_{D}\gamma_{CH_3^+}$†	$(\log k^D/k^M)_{calc}$‡	$\log k^D/k^M$
$X = \mathrm{Cl}$	−0.4	6.5	−4.6	−2.3	3.3
$X = \mathrm{Br}$	−0.4	4.9	−4.6	−0.7	3.9
$X = \mathrm{I}$	−0.5	2.6	−4.6	1.5	4.6

* All data taken from ref. N13. D = DMF.
† $\log {}^{M}_{D}\gamma_{CH_3^+}$ taken as equal to $\log {}^{M}_{D}\gamma_{SMe_3^+}$.
‡ $(\log k^D/k^M)_{calc}$ is calculated from the data in the first three columns of the table, assuming that the transition state is the fully developed intermediate.

reaction. Clearly such a dissociative reaction producing ionic intermediates has its transfer properties dominated by the solvation of the developing anion, which is greater in protic solvents and varies directly with the ionic potential of the ion. If the reaction is thus rate determined then it is expected to be much faster in methanol than in DMF (witness $(\log k^{D}/k^{M})_{calc}$) for $X = Cl$, slightly faster in methanol for $X = Br$ and faster in DMF for $X = I$. These predictions are clearly not in agreement with the experimental values of $\log k^{D}/k^{M}$ which we have already shown fit a bimolecular, S_N2(lim) mechanism.

Here we have taken the extreme possible mechanisms of S_N1 lim and S_N2 lim. The application of solvent activity coefficients to the investigation of the transition state has proved valuable not only in these cases but in providing information for reactions where the mechanism is less well defined. We will not pursue this further here since the subject is elegantly presented by Parker[N13] in the description of 'tight and loose S_N2 transition states'. In addition solvent activity coefficients have given useful information on the degree of bond breaking in the dissociative reactions of cobalt(III) complexes.[N27]

It should be emphasised that the use of solvent activity coefficients in investigating the transition state is not limited to one component solvent systems. Provided the thermodynamics of solvent transfer can be established for the reactant solutes in the mixed solvent system then the transfer properties of the transition state can be established.

There is increasing interest in the degree of fractionation which occurs from the bulk solvent to the solvation sphere of solutes[N42,N59–63] but at the moment even in fields of strong interaction (e.g. Ni^{2+} in DMSO-H_2O mixtures) no technique appears sufficiently definitive, although high precision NMR studies have produced promising results.[N59,N63] The advantage of using solvent activity coefficients in a discussion of the transition state for non-solvolytic reactions in mixed solvent media, is that the solvent activity coefficient is a function of state of the system and can be used to infer transition state properties without cognisance of, or concern for, the structure or detailed identity of the solvation sphere. If on the other hand the mechanism of a bimolecular reaction is clearly defined on other grounds, then the solvent activity coefficients should lead to understanding of the solvent reorganisation upon activation. At this stage very few results have been accumulated in binary solvent mixtures.[N13]

6.2 REACTIONS OF OCTAHEDRAL COMPLEXES

6.2.1 Introduction

It is difficult rigorously to divide kinetic studies into exclusive compartments. This is particularly true of the kinetics of octahedral metal coordination complexes for which, irrespective of rate and the nature of the metal, the rate of substitution is determined by a dissociative mechanism.

Here we will discuss a few broad principles in relation to the rather special type of substitution process represented by the equation:

$$[M(\mathrm{SOL})_6]^{n+} + L^{m-} \rightleftarrows [ML(\mathrm{SOL})_5]^{(n-m)+} + \mathrm{SOL} \quad (6.2.1)$$

Such reactions have been studied over a large range of rates, from the slow reactions[N64] of Cr^{3+} to the fast reactions of Ni^{2+}[N65,N66] and over a range of solvents with both conventional kinetic techniques as well as T-jump, ultrasonic absorption, flow techniques, and NMR relaxation techniques.[N66–70]

Following a discussion of these conceptually simple reactions we will discuss the non-aqueous chemistry of the cobalt(III) and chromium(III) 'amine' complexes and finally a few other examples of importance that do not strictly fit either of these categories.

These reactions have been classified as dissociative as a result of three types of information.[N23,N67,N68,N70–81] Firstly the reactions of one metal in one solvent are found to be largely independent of the nature of L^{m-}, secondly these rates are found to approach a limiting rate which is independent of further increases in the concentration of L^{m-}, and finally this limiting rate is, within a power of ten, the same as the rate of the independently measured solvent exchange process:

$$[M(\mathrm{SOL})_6]^{n+} + \mathrm{SOL}^* \rightleftarrows [M(\mathrm{SOL})_5(\mathrm{SOL}^*)]^{n+} + \mathrm{SOL} \quad (6.2.2)$$

The relationship between this solvent exchange rate and the rate of the anation reaction 6.2.1 has been explained by a number of workers but is perhaps most clearly expressed by Langford and Stengle[N67,N68] who carefully follow the descriptive parlance of Langford and Gray.[N75] Two paths are recognised by Langford and Stengle:

$$\begin{array}{ccc}
\text{(I)} & & \text{(II)} \\
[M(\mathrm{SOL})_6]^{n+} + L^{m-} & \underset{-L}{\overset{+L(K_a)}{\rightleftarrows}} & [M(\mathrm{SOL})_6]^{n+}\ldots L^{m-} \\
k_{+\mathrm{SOL}} \uparrow\downarrow k_{-\mathrm{SOL}} & & k'_{+\mathrm{SOL}} \uparrow\downarrow k'_{-\mathrm{SOL}} \\
[M(\mathrm{SOL})_5]^{n+} + L^{m-} + \mathrm{SOL} & \underset{k_{-L}}{\overset{k_{+L}}{\rightleftarrows}} & [ML(\mathrm{SOL})_5]^{(n-m)+} + \mathrm{SOL} \\
\text{(III)} & & \text{(IV)}
\end{array}$$

where K_a is the equilibrium constant for the formation of the one-to-one adduct (in this case an ion pair) and $k_{-\mathrm{SOL}}$ is the first-order rate constant for the formation of the five-coordinate intermediate which is captured by either solvent (k_{+SOL}) or ligand (k_{+L}), both second-order processes. k'_{-SOL} is the first-order rate constant for the incorporation of L^{m-} into the co-ordination sphere from the solvation sphere by a dissociative interchange.[N82] The pseudo-first order rate constant k for the two paths is given by:

$$k = \frac{k'_{-SOL}K_a[L]}{1 + K_a[L]} \tag{6.2.3}$$

for path (I → II → IV) and

$$k = \frac{k_{-SOL}k_{+L}[L]}{k_{+SOL} + k_{+L}[L]} \tag{6.2.4}$$

for path (I → III → IV). The limiting rate, at high $[L^{m-}]$, by pathway (I → II → IV) is k'_{-SOL} and through the (I → III → IV) path is k_{-SOL} where the concentration of the ligand is such that it captures all the intermediates ($M(SOL)_5^{n+}$). Through path (I → III → IV) the limiting rate k_{-SOL} is, of course, identical with the solvent exchange rate which, in general, can be separately measured.

It is argued by Langford and Stengle[N67,N68] that the limiting rate through (I → II → IV), that is k'_{-SOL}, will be $(1/s)$ of the rate of the separately measured solvent exchange rate (k_{-SOL}), where one of the s outer sphere ligand sites (in our parlance, solvation sites) is occupied by L^{m-}, leaving only $(s - 1)$ sites for the solvent. It is further argued that k'_{-SOL} can be evaluated from rates measured at other than the limiting condition and then similarly compared to k_{-SOL} if K_a is separately measured.[N18] This has been done in some studies.[N66,N83]

These simple statistical arguments, by which k'_{-SOL} and k_{-SOL} are inter-related, are naïve and already have been known not to stand the test of experiment, as is emphasised by the studies in cobalt(III) chemistry (p. 710). Firstly the ion pair is a new species and in general its intrinsic dissociative rate will differ from that of the free ion, not only because the ground state energy is different but also because the transition state is different. Secondly there is no reason to assume, and indeed there is good experimental evidence to doubt, that the occupancy of a solvation site gives one ligand an equal chance of capturing the developing dissociative intermediate as a different ligand in another solvation site. Evidence, again from cobalt(III) chemistry (p. 710) suggests that ligands have individual characteristics, as must be expected, for the capture of any dissociative intermediate and in this case the entry process must reflect some of the ligand's nucleophilic character. Further, in the examples often quoted by these workers, the dissociating species has not six identical ligands[N67,N85,N87] and such dipolar species have been shown to infer special properties on some of the solvation sites.[N23,N84]

The solvent exchange which accompanies ligand entry through the path (I → II → IV) is of interest in this connection. Evidence from cobalt(III) chemistry[N23,N84] suggests that the entering anion can capture up to 100% of the developing dissociative species. It must be emphasised that although we can expect k'_{-SOL} to remain, as has been found, within a factor of ten of k_{-SOL}, it will differ from it not by a simple statistical factor, but could indeed be greater than or less than k_{-SOL} and still fit a limiting dissociative mechanism. The effect of the ion association on the free energy of the reactant and the transition state must be considered, along with inherent nucleophilicity of the ligand, and asymmetric electrostatic influences in the reactants and intermediates.

Solvent activity coefficients could prove of immense value in establishing mechanisms in this field as has been shown recently in eliminating the bimolecular mechanism for the reaction:[N88]

$$[Fe(DMSO)_6]^{3+} + SCN^- \rightleftharpoons [Fe(SCN)(DMSO)_5]^{2+} + DMSO$$

in DMSO, by comparing the rate and equilibrium data for the aqueous reaction:

$$[Fe(OH_2)_6]^{3+} + SCN^- \rightleftharpoons [Fe(SCN)(OH_2)_5]^{2+} + H_2O$$

Here the situation is a little more complicated than one would like, in that changing the solvent changes the nature of the reactant from $[Fe(DMSO)_6]^{3+}$ to $[Fe(H_2O)_6]^{3+}$. The changes introduced in rate data, however, will be parallelled by free energy changes in the equilibrium data. So if we for the moment neglect this and symbolise the reaction:

$$[Fe(SOL)_6]^{3+} + SCN^- \rightleftharpoons [Fe(SCN)(SOL)_5]^{2+} + SOL \quad (6.2.5)$$

as

$$\text{'}A\text{'} + SCN^- \rightleftharpoons \text{'}B\text{'} + SOL \quad (6.2.6)$$

then for the equilibrium constants (K) in the two solvents, DMSO and $H_2O(w)$ we have

$$\log \frac{K_{DMSO}}{K_{H_2O}} = (\log {}_{DMSO}^{\ \ w}\gamma_A - \log {}_{DMSO}^{\ \ w}\gamma_B) + \log {}_{DMSO}^{\ \ w}\gamma_{SCN^-}$$

and for the rate constants (k) if a bimolecular transition state is involved:

$$\log \frac{k_{DMSO}}{k_{H_2O}} = (\log {}_{DMSO}^{\ \ w}\gamma_A - \log {}_{DMSO}^{\ \ w}\gamma_{\ddagger}) + \log {}_{DMSO}^{\ \ w}\gamma_{SCN^-}$$

We would expect $\log {}_{DMSO}^{\ \ w}\gamma_A$ and $\log {}_{DMSO}^{\ \ w}\gamma_B$ to be similar and negative and for $\log (K_{DMSO}/K_{H_2O})$ to mainly reflect $\log {}_{DMSO}^{\ \ w}\gamma_{SCN^-}$ which[N13] is 2.6. In practice at 25°C, $K_{DMSO} = 5.4 \times 10^4$ (ionic strength (I) = 0.036 M) and $K_{H_2O} = 1.72 \times 10^2$ at $I = 0.286M$,[N89] the ionic strength in water which gives a Debye–Hückel activity coefficient equal to that in a solution of $I = 0.036M$ in DMSO. Thus

$$\log \frac{K_{\mathrm{DMSO}}}{K_{\mathrm{H_2O}}} = 2.5$$

reflecting $\log {}_{\mathrm{DMSO}}^{w}\gamma_{\mathrm{SCN^-}}$ modified by the small difference in the solvation of cations A and B.

If the mechanism is bimolecular then $\log (k_{\mathrm{DMSO}}/k_{\mathrm{H_2O}})$ should also be dominated by $\log {}_{\mathrm{DMSO}}^{w}\gamma_{\mathrm{SCN^-}}$ with modification because of the difference between the solvation of the cationic transition state, which must behave much as the product, and the cationic reactant. Thus an increase in rate of one hundred to three hundred-fold must be predicted for such a mechanism on transfer from water to DMSO. In practice there is a decrease in rate from 1.0×10^4 mol l^{-1} s^{-1} ($I \sim 0.01$ in water at 25°C)[N90] to 8.70×10^2 mol l^{-1} s^{-1} ($I = 0.006$ in DMSO at 25°C), which is incompatible with a bimolecular mechanism. The decrease in rate, however, is consistent with other dissociative rates for breaking metal-DMSO and metal-water bonds, where the rate of such a dissociative process was found to decrease in the order $H_2O \simeq$ DMA $>$ DMF $>$ DMSO.[N24,N85] This order is not the same as that established by Drago, Hart and Carlson[N91] using criteria based on the value of the ligand field or spectrochemical parameter Dq for Cr^{3+}. The solvent exchange data available for this system[N92] are not precise but as emphasised by Langford and Chung[N92] are compatible with a dissociative mechanism.

An extension of these anation studies in non-aqueous media will lead to a greater understanding of the role of ion association pre-equilibria in mechanism, especially as there is already accumulated a good volume of solvent exchange data, mostly from NMR measurements. This field has been reviewed;[N68] however, quite a lot of data has been published recently and much of it is collected in Table 6.2.1.

6.2.2 Non-labile Cobalt(III) and Chromium(III) 'Amine' Complexes

Although there have been recent kinetic studies in this field involving complexes of ions other than cobalt(III) and chromium(III), little of this work involves non-aqueous media. At this point of time the relevant work on cobalt compounds out-numbers the chromium studies by ten to one. The extension of kinetic studies beyond water in these two cases has been important in relation to two aspects of solution kinetics. Firstly, it has led to a greater understanding of the ligand-for-ligand substitution mechanism, and secondly, it has led to a greatly increased understanding of the role of ion-association processes in mechanism. The role of the dipolar aprotic solvent should also be recognised in that it allows the study of a wider range of substrates because of the greater solubility of ionic reactants. These solvents also allow the use of more basic ligands without interference from solvolysis reactions which, in protic media,

Table 6.2.1

Rates of Solvent Exchange with Solvation Spheres of Metal Ions

Complex	Solvent	log (k/s^{-1}) (25°C)	Activation Parameters		Ref.
			$\Delta H\ddagger$ (kJ mol^{-1})	$\Delta S\ddagger$ (J K^{-1} mol^{-1})	
$Al(DMSO)_6^{3+}$	DMSO	−0.3	84	9	N93
$Al(DMF)_6^{3+}$	DMF*,§	1.15	63		N94
	DMF†,§	0.00	71		N94
	DMF‡,§	0.08	75	4	N94
	DMF‖	0.08	75	4	N96
$Be(DMF)_6^{2+}$	DMF	2.49	63	13	N97
$Co(DMF)_6^{2+}$	DMF	5.3	29	−42	N98
$Co(DMF)_6^{2+}$	DMF	5.5	59	54	N99
$Co(MeOH)_6^{2+}$	MeOH	5.0	54		N100
$Co(MeOH)_6^{2+}$	MeOH	4.25	59	29	N100
$[Co(MeOH)_4(OH_2)_2]^{2+}$	MeOH	6.3			N100
$[Co(MeOH)_5(OH_2)]^{2+}$	MeOH	5.8			N100
$[CoCl(MeOH)_5]^+$	MeOH	3.2¶			N101
$[CoCl(MeOH)_5]^+$	MeOH	2.9**			N101
$[Co(CH_3CN)_6]^{2+}$	CH_3CN	5.15	33	−29	N102
$Fe(DMF)_6^{2+}$	DMF	5.7			N103
$Fe(DMSO)_6^{3+}$	DMSO	1.7	42	−46	N92
$Fe(EtOH)_6^{3+}$	EtOH	4.3	25	−75	N104
$Fe(DMF)_6^{3+}$	DMF	1.5	52	−42	N104
$Fe(CH_3CN)_6^{3+}$	CH_3CN	<1.6			N104
$Mg(MeOH)_6^{2+}$	MeOH	3.7	71	59	N105
$Mg(EtOH)_6^{2+}$	EtOH	6.45	75	125	N106
$Mn(DMF)_6^{2+}$	DMF	6.6			N101
$Mn(MeOH)_6^{2+}$	MeOH	6.0	33	−21	N107
$Ni(DMF)_6^{2+}$	DMF	3.8	37	−38	N98
$Ni(DMF)_6^{2+}$	DMF	3.5	63	33	N99
$Ni(DMSO)_6^{2+}$	DMSO	3.9	33	−71	N108
$Ni(DMSO)_6^{2+}$	DMSO	3.6	50	4	N113
$Ni(MeOH)_6^{2+}$	MeOH	3.0	67	33	N109
$Ni(CH_3CN)_6^{2+}$	CH_3CN	3.4	50	−17	N110
$Ni(CH_3CN)_6^{2+}$	CH_3CN	3.6	46	−39	N102
$Ni(tpp)_2Br_2$	tpp††	3.8	21	105	N111
$Ni(tpp)_2I_2$	tpp††	3.8	29	92	N111
$Ni(EtOH)_6^{2+}$	EtOH	4.04	46	−17	N105
$Np(MeOH)_n^{5+}$	MeOH	4.9‡‡	33	−33	N112
VO^+	DMSO	>3.2			N113
	TMPI§§	<2.5			N113
	TMPA‖‖	>2.9			N113
	CH_3CN	3.45	29	−84	N113
	DMF	2.75	29	−69	N114
	MeOH	2.52	50	−4	N113

* As $AlCl_3$.
† As $AlBr_3$.
‡ As AlI_3.
§ All these results discussed in terms of role of ion association (ref. N95).
‖ As $Al(ClO_4)_3$, k at 80°C.
¶ MeOH *cis* to chloro-ligand.
** MeOH *trans* to chloro-ligand.
†† tpp ≡ triphenylphosphine
‡‡ k at 60°C.
§§ TMPI ≡ trimethyl phosphite.
‖‖ TMPA ≡ trimethyl phosphate.

lead to uncertainty due to the reactions of the lyate ion (i.e. OH^- in water and OMe^- in MeOH). Many of the reaction schemes in dipolar aprotic media are also simplified because the solvent-coordinated species do not undergo solvolysis reactions to produce complexes of the coordinated lyate ion, which in many systems have shown strong tendencies to form bridged polynuclear complexes.

Since the early 1950's when systematic studies of the substitution reactions of non-labile octahedral complexes were begun by Taube,[N115] Brown, Ingold and Nyholm[N116] and Basolo,[N117] there has been a preoccupation with features of aquation and base hydrolysis reactions. This work has been excellently summarised in all its features in a progressive series of reviews,[N71,N72,N74–80,N117–119] and at least to this contributor, the fundamental arguments seem to be settled, following the elegant experiments of Green and Taube[N120] and subsequently of Buckingham, Olsen and Sargeson.[N121,N122] In summary it can be said that octahedral complexes of the 'cobalt(III)-amine' type react by a dissociative mechanism (S_N1 or I_d) and that hydroxide ion enjoys a special position as a nucleophile because of its ability to extract a proton from the 'amine' ligand. The enhanced rate is a function of amine conjugate base complex—this mechanism is often classified as S_N1CB. The persistence of Chan[N123–125] upon an ion-pair mechanism, which differs in degree of proton extraction rather than in principle, seems misguided in that genuine ion-pair mechanisms in non-aqueous media have shown only small rate enhancement above the free ion rate.[N23,N67,N70,N73,N92] Some of the experimental results of Chan[N125,N126] seem to be in doubt. The alternative redox mechanism of Gillard[N127] does not appear to stand the test of experiment. Certainly experience in non-aqueous solvents, both hydroxylic and dipolar aprotic, suggests that if reduction occurs in cobalt(III) substitution, then it is totally irreversible.[N25,N128]

These results do not suggest that a dissociative, five coordinate intermediate is ever achieved in these reactions, although it is well established by the work of Loeliger and Taube[N129,N130] and Buckingham, Olsen and Sargeson[N131,N132] for a series of induced dissociative reactions. In addition certain anionic complexes (e.g. $[Co(CN)_5OH_2]^{2-}$ and several anation products) have been shown[N133,N134] to develop a five coordinate intermediate, although the $[Co(CN)_5SO_3]^{4-}$ ion may be in a special category since even reactions of the cationic *trans*-$[Co(SO_3)(OH)_2(en)_2]^+$ complex exhibit unusual features.[N135] The conclusions above are based on competition reactions between the solvent and the anion, or between the different isotopic forms of the solvent. While the conclusions based upon anion competitions seem sound, the warning given by Langford and Stengle[N130] should not be forgotten, namely that it is quite possible for fractionation into the first solvation shell to lead to a standard competition ratio in solvolysis reactions irrespective of mechanism.

The degree to which dissociation has developed at the rate-determining transition state still remains a major point of interest, and the arguments are discussed with detachment by Langford and Stengle in their recent review.[N67] These authors have pursued further the logic of Langford and Gray[N75,N81] and use the symbolism D, I_d, I_a, A to describe the range of mechanisms which Basolo and Pearson,[N72] following Ingold,[N137] describe as S_N1(lim), S_N1, S_N2 and S_N2(lim). There is no real advantage of one set of these symbols over the other, indeed, there seems little justification for the continued use of either, in that this involves classifying mechanisms which no doubt cover the full range from $S_N1(D)$ to $S_N2(A)$, in terms of a four-point scale.

Three aspects of the dynamics of ligand for ligand substitution in which the non-aqueous results are illuminating will be discussed, namely, the degree of dissociation at the transition state, the steric course, and the role of ion association in mechanism.

In addition to the already stated advantages of dipolar aprotic media over water in studying the modes of reaction of octahedral cobalt(III) complexes, there is the further advantage that for solvolysis reactions, being often reversible, the transition state can be investigated both through solvolysis reactions and the reverse anation reaction.[N23] Since solvent coordinated species can be prepared and yet are unstable to anation, such studies lead to a clearer understanding of the role of solvents in anion-for-anion substitution in weakly coordinating solvents. In contrast, in methanol the great relative instability of the solvent-containing species led to some confusion in the interpretation of the role of these species in the mechanism of anion-for-anion substitution (e.g. in the interpretation of the mechanism of simple reactions such as the *cis-trans*-$[CoCl_2(en)_2]^+$ isomerisation in methanol).[N116,N138–141]

This promise has been only partially fulfilled because of the difficulty of interpreting anation mechanisms where second order kinetics, first order in entering anion and first order in complex, are often found because of ion association which contributes a term in anion concentration to the rate law. A further difficulty, emphasised by Archer in his recent review on the stereochemistry of octahedral substitution reactions,[N78] is found in cobalt(III) chemistry because of the difficulty in isolating *trans* solvent-containing species. This results in continued doubt in the study of such systems as:

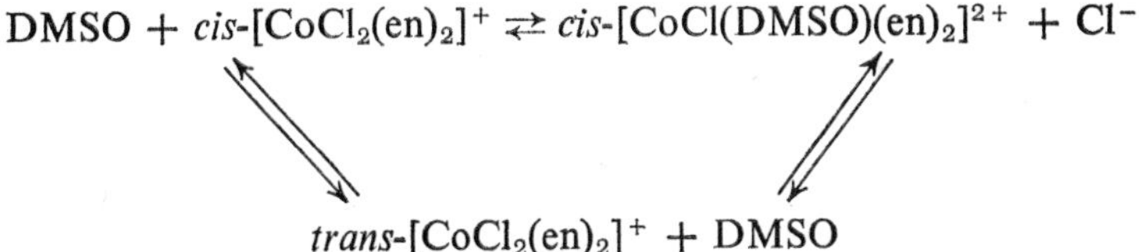

concerning the possible role of *trans*-[CoCl(DMSO)(en)$_2$]$^{2+}$ as an intermediate. It has always seemed likely that this species was the immediate product of the solvolysis of the *trans*-[CoCl$_2$(en)$_2$]$^+$ species and that its combined kinetic and thermodynamic instability led to the formation of *cis*-[CoCl(DMSO)(en)$_2$]$^{2+}$ by isomerisation and *cis*-[CoCl$_2$(en)$_2$]$^+$ by anation, either directly or through *cis*-[CoCl(DMSO)(en)$_2$]$^{2+}$.

A recent study of the [CrCl$_2$(en)$_2$]$^+$ and [CrBr$_2$(en)$_2$]$^+$ systems in DMSO[N142,N143] where both *trans*-[Cr*X*(DMSO)(en)$_2$]$^{2+}$ and *trans*-[Cr(DMSO)$_2$(en)$_2$]$^{3+}$ species have been isolated may point to missing details in our knowledge of the cobalt(III) systems.

The total reaction sequence is represented on the facing page.

In addition of course one must recognise that (1) and (4) have the possibility of exchange retention paths with free chloride ion, (2) and (5) have possible exchange retention paths with solvent and chloride ion, and (3) and (6) have the possibility of solvent exchange retention paths. These reactions are referred to as k_{11}, k_{44}, k_{22}^{DMSO}, $k_{22}^{Cl^-}$, k_{55}^{DMSO}, $k_{55}^{Cl^-}$, k_{33} and k_{66} respectively, all other paths are labelled on the diagram. All constants are initial, first order rate constants.

The relative magnitudes of all these rate constants are a function of the ambient, uncoordinated chloride concentration and, to a lesser degree, temperature. The activation energies are found to vary from 84 kJ mol^{-1} for k_{32} to 122 kJ mol^{-1} for k_{21}. The following processes are found to be insignificant, irrespective of starting materials at zero initial uncoordinated chloride concentration (that is the rate constants may be taken as zero); k_{11}, k_{44}, $k_{22}^{Cl^-}$, k_{23}, $k_{55}^{Cl^-}$, k_{14}, k_{41}, k_{24}, k_{42}, k_{15}, k_{51}, k_{35}, k_{53}, k_{26}, k_{62}, k_{21}, k_{54}, k_{32}, k_{65} and k_{56}.

Thus the solvolytic and isomerisation paths in this system in the absence of uncoordinated chloride are k_{12} for *cis*-[CrCl$_2$(en)$_2$]$^+$, k_{45} for *trans*-[CrCl$_2$(en)$_2$]$^+$, k_{25} and k_{52} (by solvent exchange, accompanied almost certainly by exchange k_{22}^{DMSO} and k_{55}^{DMSO}, with $k_{52} > k_{25}$) for *cis* and *trans*-[CrCl(DMSO)(en)$_2$]$^{2+}$, and k_{36} and k_{63} for isomerisation of *cis* and *trans*-[Cr(DMSO)$_2$(en)$_2$]$^{3+}$ (accompanied again by k_{33} and k_{66}) with $k_{33} \gg k_{36}$. In summary all paths involving the solvolytic loss of chloride ion involve retention with isomerisation accomplished in solvent-for-solvent exchange paths.

At high uncoordinated chloride ion concentrations, that is more than is necessary to form 1 : 1 aggregates with the reactant, the picture changes. Now *cis*-[CrCl$_2$(en)$_2$]$^+$ is still removed only through k_{12} (that is k_{14} and k_{15} remain insignificant) however, *trans*-[CrCl$_2$(en)$_2$]$^+$ now reacts through k_{41} (or a path k_{42}, k_{21}) and k_{42} (or a path k_{45}, k_{52}) and a chloride-for-chloride exchange by retention ($k_{44}^{Cl^-}$) becomes significant. For *cis*-[CrCl(DMSO)(en)$_2$]$^{2+}$ only the retention anation (k_{21}) is significant. Regrettably the reactions of *trans*-[CrCl(DMSO)(en)$_2$]$^{2+}$ k_{51}, k_{54}, k_{52} cannot be studied precisely because of the loss of

The total reaction sequence is represented as follows:

$$\begin{array}{ccccc}
\text{(1)} & & \text{(2)} & & \text{(3)} \\
cis\text{-}[CrCl_2(en)_2]^+ + 2DMSO & \underset{k_{21}}{\overset{k_{12}}{\rightleftharpoons}} & cis\text{-}[CrCl(DMSO)(en)_2]^{2+} + Cl^- + DMSO & \underset{k_{32}}{\overset{k_{23}}{\rightleftharpoons}} & cis\text{-}[Cr(DMSO)_2(en)_2]^{3+} + 2Cl^- \\
k_{41} \upharpoonleft\downharpoonright k_{14} & \times \; (k_{24}, k_{42}, k_{15}, k_{51}) & k_{52} \upharpoonleft\downharpoonright k_{25} & \times \; (k_{26}, k_{62}, k_{35}, k_{53}) & k_{63} \upharpoonleft\downharpoonright k_{36} \\
trans\text{-}[CrCl_2(en)_2]^+ + 2DMSO & \underset{k_{54}}{\overset{k_{45}}{\rightleftharpoons}} & trans\text{-}[CrCl(DMSO)(en)_2]^{2+} + Cl^- + DMSO & \underset{k_{65}}{\overset{k_{56}}{\rightleftharpoons}} & trans\text{-}[Cr(DMSO)_2(en)_2]^{3+} + 2Cl^- \\
\text{(4)} & & \text{(5)} & & \text{(6)}
\end{array}$$

ethylene-diamine. For *cis*-$[Cr(DMSO)_2(en)_2]^{3+}$ only the retention path k_{32} is significant, while for *trans*-$[Cr(DMSO)_2(en)_2]^{3+}$ k_{62} becomes much greater than k_{64} (possibly due to the fact that $k_{32} > k_{62}$ or k_{63}) with no evidence for k_{65}. In the real situation, the reaction sequence becomes that shown on the facing page where only the significantly important paths are included.

In some cases only forward paths are given. This is not contrary to the principle of microscopic reversibility but reflects the relative thermodynamic stability of various other intermediates and their subsequent reaction products. This sequence of reactions, with the same return paths omitted, can account for all the observations so far made on the analogous cobalt(III) system, with minor modifications consistent with the greater stability of chromium(III) solvent complexes[N142,N144] and with the general greater relative stability of *cis* species for chromium(III) complexes of this type.[N142,N144] In fact the extraordinary insignificance of the metal in determining the geometry and ligand specificity in these complexes, inevitably leads one to speculate on the influence of structure in the solvation sphere, produced by dipolar and hydrogen bonding sites at the 'amine' nitrogen protons.[N145,N146] This structure, to a very reasonable approximation, must be independent of the metal involved, except for minor differences reflected in the greater acidity of these nitrogen protons in the cobalt(III) complexes.[N146,N147] This aspect of solvation sphere structure is discussed later. It is regrettable that the steric course of solvent exchange for the chromium(III) system is not yet available; the paramagnetic nature of the complexes prevents the use of NMR techniques, as used for the cobalt(III) analogues.[N148]

The degree of bond fission (D or I_d character, S_N1(lim) or S_N1 character) in the transition state of these substitution reactions involving complexes of the $[MX(en)_2]^+$ type remains a major question of interest, and there seems little doubt that the role of the solvent in this dissociation act, and the degree of free-ion character developed in the departing anion will come from a careful systematic study of solvent activity coefficients. This has been attempted for the isomerisation system:

$$cis\text{-}[CoCl_2(en)_2]^+ = trans\text{-}[CoCl_2(en)_2]^+$$

by Fitzgerald, Parker and Watts.[N27] This study not only led to an improved understanding of the mechanism but also, for the first time, gave a clear picture of the dependence of the free energies of reactants and products on the solvent. Figure 6.2.1 shows the free energies of these two species and the transition states for their removal by solvolysis or direct isomerisation (which could be solvolysis controlled through a very unstable solvent containing species). The equilibrium studies together with the solvent activity coefficients for both the *cis* and *trans*-$[CoCl_2(en)_2]^+$ species, determined from solubility product measurements, clearly

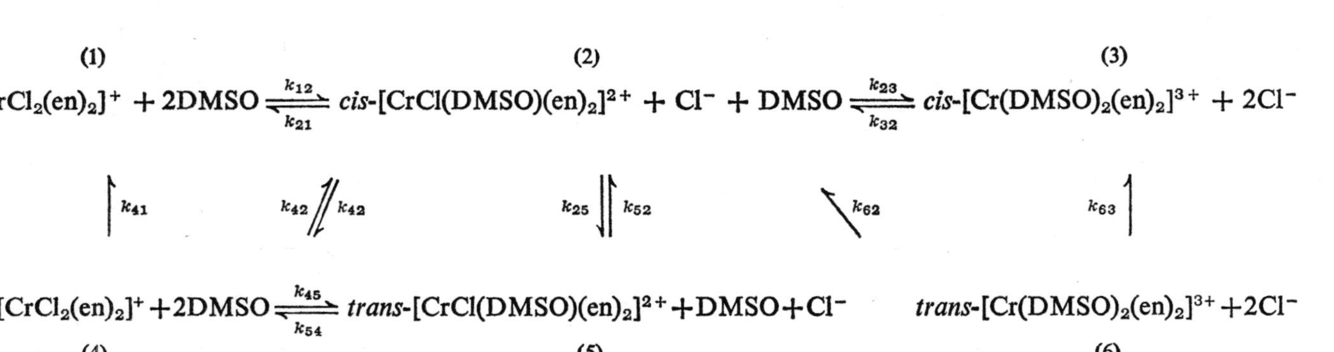
(1)
cis-[CrCl2(en)2]+ + 2DMSO
k12
k21
(2)
cis-[CrCl(DMSO)(en)2]2+ + Cl− + DMSO
k23
k32
(3)
cis-[Cr(DMSO)2(en)2]3+ + 2Cl−
k41
k42
k42
k25
k52
k62
k63
trans-[CrCl2(en)2]+ +2DMSO
(4)
k45
k54
trans-[CrCl(DMSO)(en)2]2+ +DMSO+Cl−
(5)
trans-[Cr(DMSO)2(en)2]3+ +2Cl−
(6)

reinforce the idea established for simple cations,[N13] that dipolar aprotic solvents like DMF, DMSO, DMA and $TMSO_2$ (sulpholane) solvate cations better than the less dipolar hydroxylic solvents (water and MeOH). In addition the results show that the dipolar *cis*-$[CoCl_2(en)_2]^+$

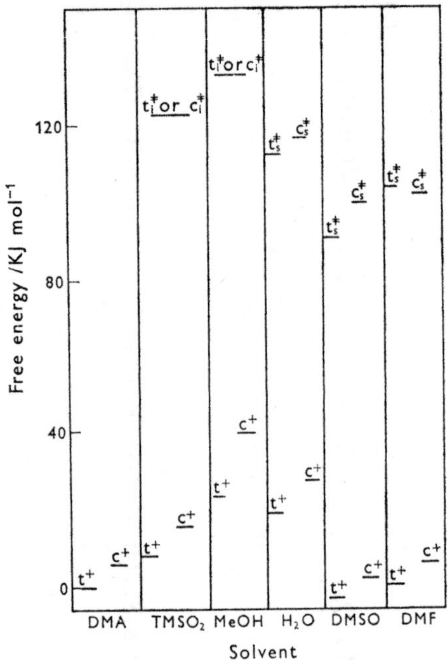

FIG. 6.2.1 Relative molar free energies of reactants and transition states for an assumed S_N1 process of isomerisation and solvolysis at 25°C. Reference point is the molar free energy of *trans*-$[CoCl_2(en)_2]^+$ in DMA. Abbreviations: c^+ is *cis*-$[CoCl_2(en)_2]^+$; t^+ is *trans*-$[CoCl_2(en)_2]^+$; $t_s^{\ddagger}$ and $c_s^{\ddagger}$ are transition states for solvolysis of the *trans* and *cis*-cations, respectively; and $t_i^{\ddagger}$ and $c_i^{\ddagger}$ are transition states for isomerisation.

gains in solvation energy in dipolar aprotic solvents relative to the symmetrical *trans*-$[CoCl_2(en)_2]^+$ ion, through either additional dipole-dipole terms in the solvation energy or through specific site solvation of a hydrogen-bonding type between these strongly dipolar hydrogen bond acceptor type solvents and more acidic 'amine' protons *trans* to the *cis* chloro-ligands.

Table 6.2.2 shows the determined values of ${}^{DMF}_{s}\gamma$ for the reactant and product ions, and for the transition state for their removal in the solvents DMF, DMSO, H_2O, MeOH and $TMSO_2$. Some of these reactions appear to be straight solvolyses and others isomerisations although the mechanism, even in the cases of isomerisation, could

Table 6.2.2‡

Solvation of Transition States for Isomerisation and Solvolysis in the *cis* and *trans*-$[CoCl_2(en)_2]^+$ System at 25°C

(Reference Solvent DMF, k/min^{-1})

<table>
<tr><th>Solvent (S)</th><th>Reaction</th><th>*log k_c</th><th>*log k_t</th><th>log ${}^{DMF}_{s}\gamma_c$</th><th>log ${}^{DMF}_{s}\gamma_t$</th><th>†log ${}^{DMF}_{s}\gamma_{c\ddagger}$</th><th>†log ${}^{DMF}_{s}\gamma_{t\ddagger}$</th></tr>
<tr><td>DMF</td><td>Solvolysis</td><td>−2.8</td><td>−4.3</td><td>0.0</td><td>0.0</td><td>0.0</td><td>0.0</td></tr>
<tr><td>DMSO</td><td>Solvolysis</td><td>−3.6</td><td>−3.8</td><td>−0.8</td><td>−0.5</td><td>0.0</td><td>−1.0</td></tr>
<tr><td>H_2O</td><td>Solvolysis</td><td>−1.8</td><td>−2.7</td><td>3.8</td><td>3.3</td><td>2.8</td><td>1.6</td></tr>
<tr><td>MeOH</td><td>Isomerisation</td><td>−3.0</td><td>−5.9</td><td>6.0</td><td>4.2</td><td colspan="2">6.0</td></tr>
<tr><td>$TMSO_2$</td><td>Isomerisation</td><td>−5.4</td><td>−7.0</td><td>1.9</td><td>1.3</td><td colspan="2">4.2</td></tr>
</table>

* k_c and k_t are the initial first order rate constants for removal of *cis*-$[CoCl_2(en)_2]^+$ and *trans*-$[CoCl_2(en)_2]^+$ (that is 'c' and 't') respectively.

† $c\ddagger$ and $t\ddagger$ are the transition states for the removal of *cis* and *trans*-$[CoCl_2(en)_2]^+$ respectively.

‡ This table is taken from W. R. Fitzgerald, A. J. Parker and D. W. Watts[N27] in which paper the source of all the rate data is acknowledged.

Table 6.2.3†

Data for Isomerisation in the *cis* and *trans*-$[CoCl_2(en)_2]^+$ System at 25°C

(Reference Solvent MeOH, k/min^{-1})

Solvent (S)	$\log k_c$	$\log k_t$	$\log {}^{MeOH}_{S}\gamma_c$	$\log {}^{MeOH}_{S}\gamma_t$	$\log {}^{MeOH}_{S}\gamma_{Cl^-}$	$^*\log {}^{MeOH}_{S}\gamma_{c^{\ddagger}(t^{\ddagger})}$
MeOH	−3.03	−5.9	0.0	0.0	0.0	0.0
$TMSO_2$	−5.37	−7.0	−4.3	−2.9	5.8	−1.9

* Here where no intermediate is observed, the transition states for the removal of *cis*-$[CoCl_2(en)_2]^+$ and *trans*-$[CoCl_2(en)_2]^+$ are indistinguishable.

† This table is taken from W. R. Fitzgerald, A. J. Parker and D. W. Watts[N27] in which paper the source of all data is acknowledged.

involve the formation of a thermodynamically and kinetically unstable solvent containing an octahedral intermediate.

Table 6.2.3 compares the relevant solvent activity coefficients for reactions in the solvents MeOH and $TMSO_2$, where only isomerisation is observed, expressed using MeOH as reference.

It must be emphasised that in the discussion that follows it is assumed that the basic mechanism is one of dissociation, and that the solvent is not involved in covalent interaction with the metal in the transition state. If this were so the calculations of ${}^{M}_{s}\gamma_{c\ddagger}$ and ${}^{M}_{s}\gamma_{t\ddagger}$ would be invalid. If such data are to be useful they must lead to information on the degree of charge separation in the transition state, which we can formulate as $[CoCl(en)_2]^{+\delta+}$ . . . $Cl^{\delta-}$. Firstly take the two systems in which isomerisation alone is observed. The isomerisations are faster in MeOH than in the dipolar aprotic solvent $TMSO_2$, so that on these grounds alone, a bimolecular reaction between a cation and chloride ion is unlikely (see p. 690), because chloride ion will have an overwhelmingly increased activity in $TMSO_2$, as reflected in ${}^{MeOH}_{TMSO_2}\gamma_{Cl^-}$. The transition state cation is more solvated by $TMSO_2$ than by MeOH (Fig. 6.2.1), as are most cations, but is less influenced by solvent transfer than either of the reactant cations. This is consistent with the model transition state, in which the chloride ion, poorly solvated by $TMSO_2$, has made significant progress towards dissociation from a doubly-charged cation. The unfavourable solvation of the developing chloride ion counteracts the favourable solvation of the doubly-charged cation in $TMSO_2$ compared with MeOH. This response to solvent transfer is not in accord with a transition state in which there is little charge separation, as would be the case in the intramolecular rearrangement suggested by Springer and Sievers.[N149] It should be emphasised again that bimolecular solvolysis followed by chloride re-entry cannot be excluded by these considerations, but it has surely been excluded by other studies on these systems particularly those on the anation reactions.

A comparison of the solvolysis in water and the isomerisation in $TMSO_2$, Table 6.2.2 and Fig. 6.2.1, fits well within the concept of this dissociative transition state. This solvolytic transition state is more solvated by water than is the isomerisation transition state by $TMSO_2$. Normally cations, particularly doubly-charged ones, are expected to be better solvated by solvents such as $TMSO_2$ and the present observation can only be accounted for by an overwhelming destabilisation of the transition state in $TMSO_2$ due to the relatively poor solvation of the well-developed chloride ion. Alternatively expressed, the developing chloride ion is well solvated by water ($\log {}^{TMSO_2}_{w}\gamma_{Cl^-} = -8.3$).

The results for DMSO and particularly DMF do not fit the present model so well. The transition states for solvolysis in these solvents are clearly stabilised relative to the transition state for hydrolysis and this is inconsistent with a well-developed dissociation of chloride ion. It seems most unlikely, at least in the case of DMF, that the great activity of chloride ion ($\log {}^{\mathrm{DMF}}_{w}\gamma_{\mathrm{Cl}^-} = -9.0$) could be compensated by the favourable solvation energy of the doubly-charged cationic moiety. To be sure of this statement, solvent activity coefficients must be determined for doubly-charged cations, but it seems likely that the transition states for the reactions of these ions in DMF are of a different character to those in the other media, and probably involve an advance of solvent entry upon the developing dissociation (some I_a or S_N2 character). It is consistent with the present picture to say that this interaction is equivalent to an especially high solvation energy term, which counteracts the relative destabilisation caused by the development of the especially poorly-solvated chloride in DMF. There has been other evidence to suggest that the transition state in this system may be better described as a developing seven-coordinate intermediate rather than a developing five-coordinate species.[N150]

It is logical at this stage to discuss the reverse of these solvolysis reactions, namely the anation reactions of the type,

$$cis\text{-}[\mathrm{Co}X(\mathrm{SOL})(\mathrm{en})_2]^{2+} + Y^- \rightarrow cis\text{- and } trans\text{-}[\mathrm{Co}XY(\mathrm{en})_2]^+ + \mathrm{SOL}$$

particularly in relation to the role of ion association. Although ion association was recognised as important in very early work on octahedral substitution in water,[N151] a better understanding has come from studies in non-aqueous media[N23,N67,N74] and in particular from recent studies in dipolar aprotic solvents.[N23,N25,N27,N67,N84,N85,N142,N144] The role of ion association has been a particular interest of Tobe,[N74] Langford,[N67,N68,N75,N85] Eigen and Wilkins[N70,N73] and others.[N23,N152]

Ion association is more significant in dipolar aprotic solvents because of the poor solvation of the small anionic nucleophiles of interest (e.g. $\log {}^{\mathrm{MeOH}}_{\mathrm{DMF}}\gamma_{\mathrm{Cl}^-} = 6.5$ and $\log {}_{\mathrm{DMF}}^{w}\gamma_{\mathrm{Cl}^-} = 9.0$).[N13] This increased anion activity in the dipolar aprotic media is not compensated by the decrease in cation activity which accompanies transfer to the dipolar aprotic solvent (e.g.

$$\log {}^{\mathrm{MeOH}}_{\mathrm{DMF}}\gamma_{cis\text{-}[\mathrm{CoCl_2(en)_2}]^+} = -6.0$$

and

$$\log {}_{\mathrm{DMF}}^{w}\gamma_{trans\text{-}[\mathrm{CoCl_2(en)_2}]^+} = -3.8).$$ [N27]

One cannot neglect of course the changes in solvation of the ion aggregate itself but these changes are small and variations in the ion-association constant for this type of aggregate are largely a reflection of changes in activity of the parent ions and in particular the anion.[N27]

Ion association has a particular significance in the substitution reactions of the 'cobalt(III)-amine' type complexes in aprotic media because the poorly solvated hydrogen-bond acceptor anions find refuge in the acidic protons on the 'amine'-nitrogens of the complex cations. The role of these protons in ion association in non-aqueous media was first recognised by Basolo, Henry and Pearson[N138,N139,N153] following their investigation of anomalies in the rate of buffered azide attack on *cis*-$[CoCl_2(en)_2]^+$ in methanol. Since then, the existence of specific sites in the solvation sphere of these complexes has been well established.[N23,N145–7] It is also clear that the *cis*-complexes (and to a lesser extent *trans*-complexes of the *trans*-$[MXY(en)_2]^+$ type[N146]) possess dipole moments, the charge separation providing favourable sites for anion occupation within the solvation sphere, and that this dipole interaction is often complemented by H-bonding interaction.[N145–7,N153]

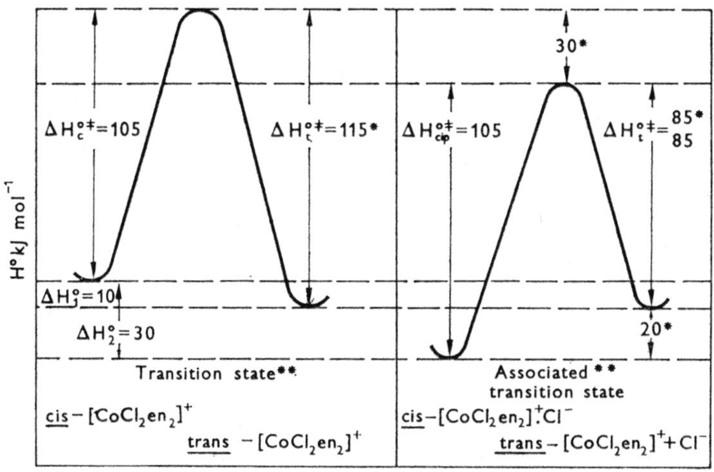

FIG. 6.2.2 Reaction profile for the reactions (a) *cis*-$[CoCl_2(en)_2]^+ \leftrightarrows$ *trans*-$[CoCl_2(en)_2]^+$ and (b) *cis*-$[CoCl_2(en)_2]^+ \cdot Cl^- \leftrightarrows$ *trans*-$[CoCl_2(en)_2]^+ + Cl^-$. * These values have been calculated from the other energies which have been measured independently. ** In the region of these transition states there must be at least one dissociative intermediate and one other transition state. A complex $[CoCl(TMSO_2)(en)_2]^{2+}$ could also have transient existence.

The mechanistic roles of the pre-association equilibria depend on the detailed nature of the reaction under consideration. These roles cannot be deduced from consideration of the reactant free energies alone since changes in reactant free energy due to ion association are often compensated by changes in transition state free energy.[N23,N154–6]

Figure 6.2.2 demonstrates this point for the isomerisation reaction:

$$cis\text{-}[CoCl_2(en)_2]^+ = trans\text{-}[CoCl_2(en)_2]^+$$

in $TMSO_2$. Here as in all reactions of this type the position of equilibrium is strongly dependent on the accompanying chloride ion concentration, because the dipolar *cis*-isomer is relatively stabilised as the ion pair ($K_{IP,cis} \sim 5000$ and $K_{IP,trans} \sim 4$).[N23,N154] Clearly the transition state has comparable sensitivity to this ion association, the result being that the rate of *cis* removal to form *trans* is much less sensitive to the ambient chloride ion concentration than is the rate of *trans* removal to form the *cis* isomer. The same effect was observed in this isomerisation reaction is DMF.[N128]

The occupation of a solvation site by a nucleophilic anion does not necessarily lead to entry following the commencement of a dissociative act. As a consequence, an associated anion can remain in a

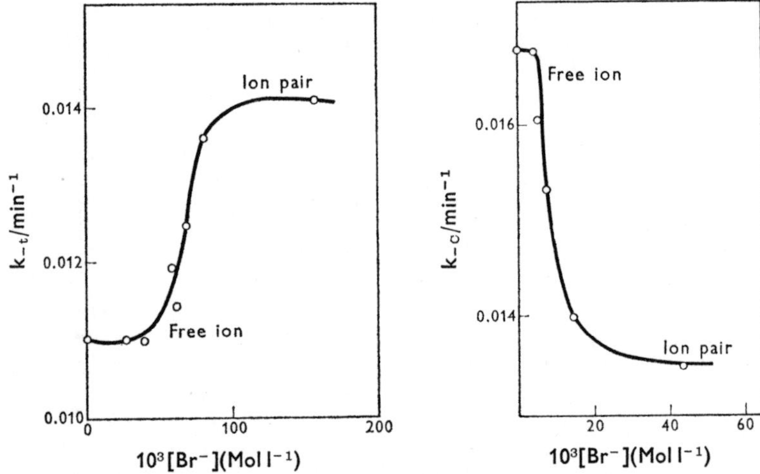

FIG. 6.2.3 Dependence of k_{-t} (at 31°C) and k_{-c} (at 45°C) on Br^- concentration for the solvolysis of *trans*-$[CoBr_2(en)_2]^+$ and *cis*-$[CoBr_2(en)_2]^+$ in DMSO: total complex concentration $\sim 8 \times 10^{-3}$ mol l^{-1}.

solvation site through the solvolytic act, its presence resulting in a dependence of the rate on anion concentration but not affecting the product composition.[N157] This is shown by the results in Fig. 6.2.3 for the solvolysis of *cis* and *trans*-$[CoBr_2(en)_2]^+$ by DMSO to give *cis*-$[CoBr(DMSO)(en)_2]^{2+}$. The rate dependences correlate well with the known ion-association constants of the reactant species. In the case of the *trans*-isomer, ion association narrows the energy gap between the reactant and the transition state, while in the *cis* case the gap is increased.[N157]

It should be emphasised that while ion association can be expected to affect reaction rates, it is not always possible to evaluate the ion-association properties of the reactant thereby, because the effect of ion association on the free energy of the reactant is often compensated or outweighed by corresponding changes in the free energy of the transition state.

It is indeed only in the anation reactions of the type:

$$cis\text{-}[\mathrm{Co}X(\mathrm{SOL})(\mathrm{en})_2]^{2+} + Y^- = cis + trans\text{-}[\mathrm{Co}XY(\mathrm{en})_2]^+ + \mathrm{SOL}$$

that the kinetic features appear to be thoroughly understood. These features have been well described by Tobe[N80,N141,N158] Langford[N67,N68,N75] and others.[N23,N159] Here the rate law includes a term in both cation-complex and anion, until the anion concentration is sufficient to produce complete ion association of the complex. For reactions such as the anation of *trans*-$[Co(NO_2)(OH_2)(en)_2]^{2+}$ by halide ions in acetone, association of anions about the cation continues until a negatively charged aggregate is achieved, while in more strongly cation-solvating solvents such as DMF, DMSO and DMA, aggregation seems to continue until the aggregate carries no charge.[N23]

Because of the high anion activity in dipolar aprotic solvents, there seems to be great variety in the stoichiometry of ion association aggregates, especially when highly charged cations are involved. Here even the aggregates carry substantial positive charge and are thus well solvated by these solvents. Lo[N160] in a recent study of the anation reaction:

$$[\mathrm{Cr(DMF)_6}]^{3+} + \mathrm{Cl^-} \rightarrow [\mathrm{CrCl(DMF)_5}]^{2+} + \mathrm{DMF}$$

in dimethylformamide found kinetic features consistent with the existence of the ion-triplet species

$$[\mathrm{Cr(DMF)_6}]^{3+} \ldots \mathrm{Cl^-} \ldots [\mathrm{Cr(DMF)_6}]^{3+}$$

when the chloride concentration was less than half that of the complex. At greater chloride concentrations a more normal behaviour consistent with the existence of the ion pair $[Cr(DMF)_6]^{3+} \ldots Cl^-$ was observed. The charge-transfer spectra of the system at 290 nm also correlate with the presence of this ion triplet.

Langford and others[N23,N67,N68,N70–75,N80] have clearly described the role of pre-association in anation reactions of this type and the principles we have discussed already (p. 693). That the anion concentration term in the rate law arises as a consequence of a pre-equilibrium and not as a result of a bimolecular nucleophilic attack, has been established by a comparison of the solvent exchange rate with the saturation rate of anion entry. Such results had been accumulated previously in water, for example by Schmidt and Taube,[N161] and more recently by Duffy and Earley[N86] and Murray and Barraclough.[N87] Muir and Langford[N85] have recently measured anation rates of *cis*-$[CoNO_2(DMSO)(en)_2]^{2+}$ in DMSO and compared these with the rate of DMSO exchange obtained from deuterium NMR experiments.[N148]

All these experiments show the anation rate to range from about 10% of the exchange rate up to values slightly in excess of it, and thus add support to the assignment of the basically dissociative mechanism (I_d or S_N1). Langford and Stengle[N67] state that 'the difference between I_d and D pathways is revealed by the observation that on a D path a preferred ligand would enter from the ion pair at a rate equal to the solvent exchange rate, instead of a statistical factor below it'. Like many statements in this field, this one is not generally true. A '*D*' pathway for an anation reaction is easily envisaged in which a full five-coordinate intermediate is established with the anion retaining its occupation of a site in the solvation shell. The rate of formation of this intermediate from the

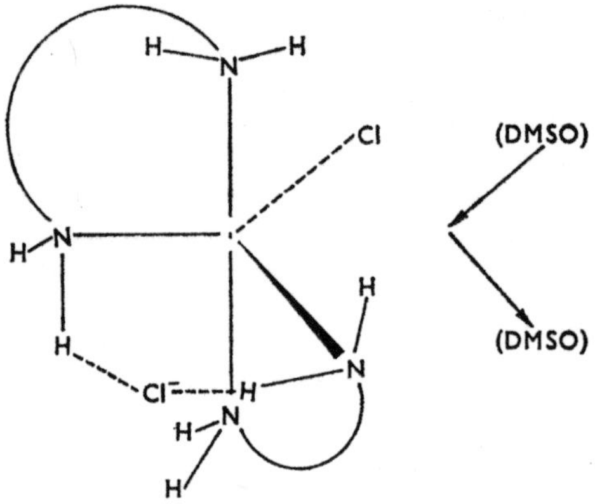

Fig. 6.2.4 A retention path for solvent exchange in the reaction of an ion-paired cobalt(III) complex by a dissociative mechanism developing a five-coordinate intermediate. H_2N——NH_2 represents ethylenediamine.

ion-paired complex (see the discussion above for the DMSO solvolysis reactions of the $[CoBr_2(en)_2]^+$ complexes), is not necessarily the same as the rate shown by the free complex ion in solvent exchange, nor is it reasonable to assume that the capture of the intermediate will always be by the anion and not a solvent molecule. While it must be admitted that ion association could leave the anion favourably placed for entry, it is not difficult to see structures, such as shown in Fig. 6.2.4, permitting solvent exchange with retention while retaining the anion in the solvation sphere even if the five-coordinate intermediate is manifest.

There is no doubt that such interactions produce notable effects on the steric course of the anation reaction and it has been shown in one system, where product isomerisation is slow enough not to interfere with the definition of the steric course, that anation through the ion pair and the ion triplet species have different stereochemical consequences. The results in Fig. 6.2.5 show that in the anation reaction:

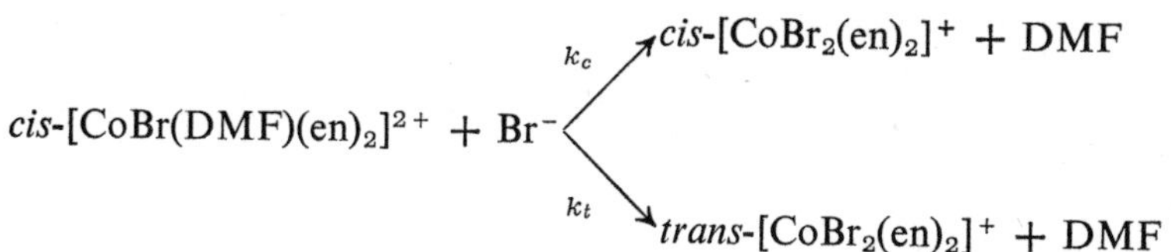

$$cis\text{-}[CoBr(DMF)(en)_2]^{2+} + Br^- \xrightarrow{k_c} cis\text{-}[CoBr_2(en)_2]^+ + DMF$$
$$cis\text{-}[CoBr(DMF)(en)_2]^{2+} + Br^- \xrightarrow{k_t} trans\text{-}[CoBr_2(en)_2]^+ + DMF$$

in DMF,[N84] anation through the ion pair (*cis*-$[CoBr(DMF)(en)_2]^{2+}$ . . . Br^-) favours the *trans*-product by a ratio of 3:1 while the ion triplet (*cis*-$[CoBr(DMF)(en)_2]^{2+}$ . . . $2Br^-$) favours *cis*-isomer by 2:1. These results tempt one to speculate that the predisposition of the bromide ion

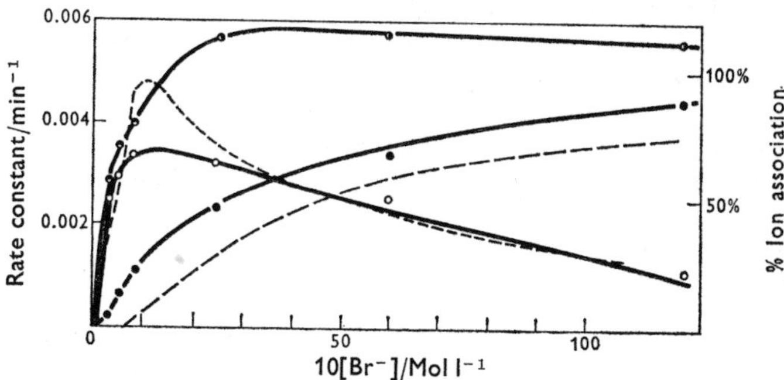

FIG. 6.2.5 Dependence of k_c, k_t and $(k_c + k_t)$ on bromide concentration for the reaction: *cis*-$[CoBr(DMF)(en)_2]^{2+} + Br^- \rightleftarrows$ *cis*- and *trans*-$[CoBr_2(en)_2]^+$ + DMF at 45°C, and the percentage of complex initially as *cis*-$[CoBr(DMF)(en)_2]^{2+} \cdot Br^-$ and *cis*-$[CoBr(DMF)(en)_2]^{2+} \cdot 2Br^-$: total complex concentration $\sim 8 \times 10^{-3}$ mol l^{-1}, ◑, $k_c + k_t$; ○, k_t; ●, k_c; - - - -, % ion pair; – – –, % ion triplet.

in a *trans*-position by the dipole of the reactant complex leads to *trans*-product from the ion pair, and that the second anion is positioned near the coordinated solvent molecule where interchange favours *cis*-entry and produces the minimum reorganisation within the complex. An anion in the second site is not held in the symmetrical hydrogen-bonded site pictured in Fig. 6.2.4, and thus presumably is able to participate more readily in the interchange process. In the ion pair the *trans* anation path must be accompanied by the unobserved *cis* solvent-for-solvent exchange path pictured in Fig. 6.2.4.

These results convincingly support an S_N1 (I_d) mechanism for anation in these systems, as do considerations based upon solvent activity coefficients as discussed earlier for the anation of $[Fe(DMSO)_6]^{3+}$.

In addition these results indicate the existence of a first solvation sphere with a structure determined by the geometry of the complex, and which is as important in determining the fate of the reacting complex as is the complex itself. Further in the comparison of analogous cobalt(III) and chromium(III) complexes this structure may be more important than the identity of the metal itself. It is clear that some sites in the outer solvation sphere have sufficient stability to maintain the geometry during a substitution act in another part of the assembly. It is also clear that other sites for the preassembly of nucleophiles, be they anions or solvent molecules, are compatible with the developing dissociation of the complex and thus lead to a strongly preferential incorporation of the nucleophile into the complex. The geometry of this entry is thus preselected by the structure of the solvation sphere as much as it is by the metal and its immediate coordinated ligands.

There seems no doubt, that a greater insight into the processes which follow the commencement of the dissociative act in these complexes will come from studies of competitive anation reactions in a range of organic solvents. It is imperative that these studies are backed with a full knowledge of the thermodynamics of the ion-association equilibria.[N23]

Such experiments will require greater attention to analytical accuracy if their full potential is to be realised, but some preliminary results[N162] on which work is now continuing[N163], on the competition of N_3^-, Cl^- and SCN^- in methanol for entry into *cis*-$[CoCl_2(en)_2]^+$, are interesting. The effectiveness of these three ions at entering the dissociating complex is in the ratio 7:3:1. The steric course of this entry is such as to give 100% *trans*-$[CoClN_3(en)_2]^+$, 80% *trans*-$[CoCl_2(en)_2]^+$ and 30% *trans*-$[CoClSCN(en)_2]^+$. Azide ion, consistent with the interpretation of Basolo, Henry and Pearson,[N138,N139,N164] appears to interact strongly with the complex and is in a position to gain entry into all the dissociative intermediates indicated in separate chloride exchange experiments.[N138-140] This interaction favours the *trans* substitution path. Only 80% of the acts of chloride exchange, however, give the *trans* path suggesting a less favourable preorganisation. Thiocyanate is still less effective either because of a weaker interaction (expected on size and hydrogen-bonding potential) or because its orientation leads to unfavourable positioning of the coordinating nitrogen atom.

It is not infrequently stated that 'these results in non-aqueous media are interesting but bear no relevance to the discussion of aqueous cobalt(III) chemistry'. It should be remembered that the very observation that ion association is so much weaker in water means that these special sites for preferential ion association are occupied by water molecules. The water molecule is of course dipolar and both a hydrogen-bond donor and acceptor and will be structured into favourable sites about these cationic complexes. Thus, to a first approximation, even around these large cations the solvation sphere is more part of the complex than part of the solvent and thus, although the reaction is overall of dissociative type, there is strong and early commitment of the entering group. These considerations discredit the interpretation of the steric course of aquation and hydrolysis in terms of the development of a five-coordinated intermediate with statistical entry into the various faces of this trigonal bipyramidal complex.[N72]

It may be significant that, except for the cases of catalytic inducement, the best authenticated examples of the fully developed dissociative intermediate are for anionic complexes, such as $[Co(CN)_5(H_2O)]^{2-}$. Here the water molecule dipole will normally be directed unfavourably for coordination and becomes more part of the solvent continuum than an integral part of the complex. Thus the retardation of entry, leading to the full formation of the five-coordinated intermediate, is probably the result of the strong solvent coherence even in the solvation sphere of the complex rather than something inherently different in the electronic properties of the complex itself.

6.2.3 Other Octahedral Reactions

Two fields of work which have been the subject of increasing interest in the last few years warrant special mention; both are likely to lead to a special understanding of the role of the solvent in substitution processes.

(i) *Substitution in Mixed Solvent Systems*

The major contributions falling within the scope of this section are those of Frankel, Langford, King and others,[N165–171] who have concentrated on the thermal and photochemical reactions of chromium(III) complexes. Additional interesting results relevant to cobalt(III) aquation reactions have been produced by Barraclough and McTigue[N172] while a recent review by Amis[N42] on the solvation of a range of 'main block' elements is relevant.

The use of mixed solvents for distinguishing associative from dissociative mechanisms has not been successful, since a solvolytic rate proportional to the mole fraction of coordinating solvent in the presence of a non-coordinating solvent can be correlated with either mechanism.[N166,N173] It seems likely that delineation could be achieved through solvent activity coefficients in much the same way as Parker has succeeded in the treatment of the solvolysis reactions at carbon centres.[N13]

What is impressive, however, is the increasing weight of evidence for important solvent fractionation into the solvation sphere of complex ions. These results support the already strong evidence that specific short range ion-solvent interactions lead to immediate reactant environments which significantly differ in composition from the bulk medium.

The results represented in Fig. 6.2.6 (taken from Behrendt, Langford and Frankel)[N166] show the fractionation of CH_3CN from H_2O–CH_3CN mixtures, into the solvation sphere of $[Cr(NCS)_6]^{3-}$. These results, like those for the solvation of the Reineckate ion (*trans*-$[Cr(NH_3)_2(NCS)_4]^-$) in acetone-water[N168] mixtures, confirm that there is a preferential incorporation of the organic solvent molecule into the solvation sphere. Hinton, McDowell and Amis[N174] found also that dioxan selectively solvated Al^{3+} ions in 0.85*M* $Al(ClO_4)_3$ solution in 3:2 by weight mixtures of dioxan and water. Other studies[N175,N176] show that small anions such as Cl^-, Br^- and NO_3^- are selectively hydrated in dioxan-water mixtures, and in contrast Kobayasi, Toka and Muiri[N177] have shown that lithium ion is preferentially solvated by water in ethanol-water mixtures.

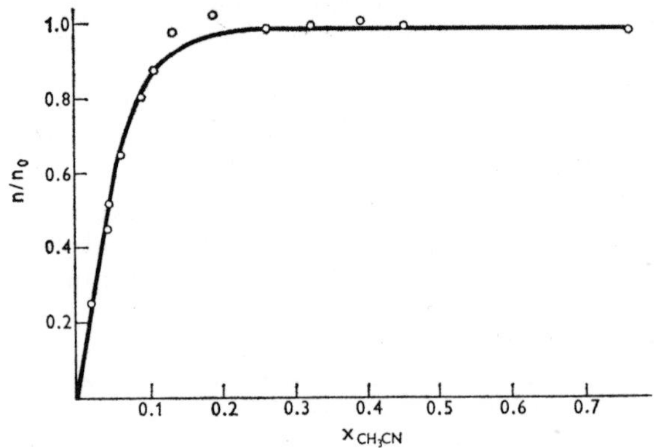

FIG. 6.2.6 Values of the solvation parameter n/n_0 for CH_3CN as a function of the mol fraction of CH_3CN. n/n_0 is the number of solvent molecules in the outer coordination sphere (solvation sphere) in a mixed solvent compared to that in the pure solvent (from Ref. N62).

Solvent activity coefficients allow the construction of diagrams such as Fig. 6.2.1 which show the relative free energies of ions in a range of solvents. However, they do not permit us to forecast which solvent is fractionated into the solvation sphere in a mixed solvent environment, since in addition to the ion-solvent interaction we must also consider the solvent-solvent interactions. The preferential solvation of ions such as $[Cr(NCS)_6]^{3-}$ by CH_3CN in CH_3CN–H_2O solvent mixtures is perhaps best regarded as a rejection of CH_3CN by the water structure, rather than a predominance of any ion-solvent interactions.

In general then in addition to ion-solvent interactions one must consider the self-coherence of each solvent as well as the complementary interactions between the solvents. The balance between these determines whether fractionation occurs. The tendency of even totally miscible solvents, particularly if one has strong structure (e.g. water or DMSO), to show a number of discontinuities in chemical properties with solvent composition,[N42] is strong evidence not only of such complementary interactions leading to greater stability of certain composition mixtures, but also of a tendency for the strongly structured solvent to persist in its structure with a surprising tolerance to contamination. Amis[N42] concludes that in mixtures of organic solvents and water, 10–30 and 75–90 weight percent of organic component are critical with respect to replacement of water by organic components in the solvation sphere of ions, and that the marked effects of solvent composition on chemical phenomena affecting dissolved solutes are a combination of critical solvation and solvent structure effects.

It must be remembered that irrespective of the composition of the solvation sphere, the chemical potential of a solvent molecule in this sphere, is the same as the bulk chemical potential, since the system is in equilibrium. However, following the commencement of the dissociative act in an octahedral complex, equilibrium may not exist at a particular site and entry may depend on the immediate availability of a potential ligand rather than on the statistical population of the solvation sphere or the bulk solvent composition.

If this is true then the assignment of mechanism of aquation based upon the order with respect to water activity or mole fraction in mixed aqueous organic solvent media may be a very dangerous procedure.[N172]

(ii) *Substitution in Optically Active Solvents*

One of the interesting fields of kinetics, opened up by the availability of accurate optical rotatory dispersion and circular dichoism recording spectrophotometers, is that of stereospecific reactions in optically-active organic solvents. Both Bosnich *et al.*[N178–180] and Haines and Smith[N180–182] have demonstrated asymmetric interactions in stereospecific substitution reactions. An estimate of the magnitude of the asymmetric interaction in the case of the *d* and *l* form of *cis*-$[CoCl_2(en)_2]^+$ with the solvent (—)-2,3-butanediol[N180], from solubility product measurements, suggests that antiracemisation in systems where solvolytic interference is less important (perhaps with *cis*-$[Co(NO_2)_2(en)_2]^+$), will lead to interesting kinetics. In this solvent, as well as 1,2-propanediol, such studies, when compared with the racemisation rates of the resolved complexes in the racemic or meso forms of the solvents, must lead to a greater understanding of the role of the solvent in a class of reactions whose precise mechanism has proved difficult to define.

6.3 REACTIONS OF SQUARE PLANAR COMPLEXES

6.3.1 Introduction

There has been a great deal of work on the kinetics of substitution in the square planar complexes of platinum(II) and, to a much less extent, on the analogous complexes of gold(III), low spin nickel(II), palladium(II) rhodium(I) and iridium(I). This work has been excellently reviewed by Basolo and Pearson[N72] and Langford and Gray.[N75] Various aspects of the role of the solvent are described elsewhere.[N184–201]

The mechanism of substitution in all these complexes involves an increase of coordination number to include the entering ligand. The reaction then is truly bimolecular. In addition to the form of the rate law which is second order the following evidence strongly supports the assignment of a bimolecular mechanism:

(i) The known existence of stable five-coordinate species of d^8 metals.[N202–205] Such compounds have been isolated for nickel(II),[N206,N207] platinum(II),[N208] iridium(I)[N209,N210] and rhodium(I),[N211] while there is also strong evidence for the existence of five-coordinate complexes in nitrobenzene solutions of the nickel(II), palladium(II) and platinum(II) analogues of $[M(\text{diars})_2X_2]$.[N212]

(ii) The strong steric hindrance to nucleophilic attack on platinum(II) complexes where the bound groups are bulky.[N213]

The general form of the rate law for substitution by Y into a square planar complex $[MA_3X]$ to give $[MA_3Y]$ in a solvent (s) is:

$$k_{\text{obs}} = k_s + k_Y[Y] \qquad (6.3.1)$$

k_{obs} is the observed pseudo-first order rate constant, k_s expresses a reagent independent path, which is rate determined by bimolecular solvolysis, and k_Y expresses the path through direct bimolecular nucleophilic attack of reagent Y. The two-way mechanism is shown in Fig. 6.2.7.

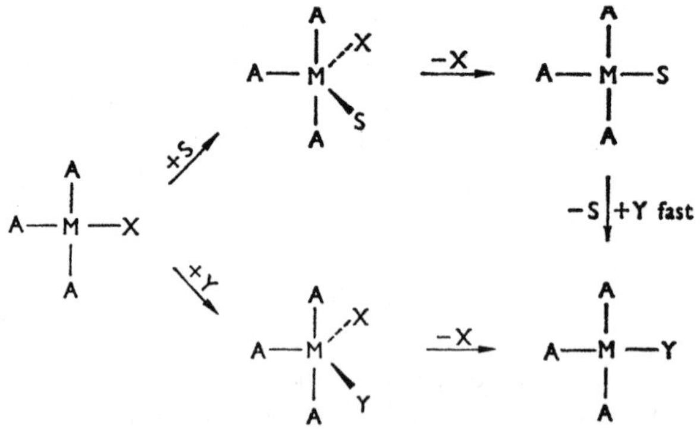

FIG. 6.2.7 Two-path mechanism for the reaction of a square planar complex, MA_3X, with Y to yield MA_3Y. The upper path is the solvent path (k_s) and the lower one is the direct path (k_Y) (from Ref. N72).

6.3.2 Square Planar Platinum(II) Complexes

The solvent dependence of the substitution reactions of anionic nucleophiles into both

trans-$[PtCl_2(py)_2]$[N184,192] and *trans*-$[PtCl_2(PEt_3)_2]$[N184,N185] have shown that the solvents fall into two classes. The solvents which coordinate poorly, such as carbon tetrachloride and benzene, show a predominance of the k_Y term while strong coordinators, such as dimethylsulphoxide and water, show the k_s term to be dominant. The direct dependence of the solvent path on coordinating ability is strong evidence that the solvent participates in a bimolecular mechanism and is thus coordinated to the platinum in an intermediate. Dimethylsulphoxide is known to coordinate strongly to platinum through the sulphur atom[N24,N25] and leads to a substantially faster rate of chloride ion exchange with $[PtCl_2(py)_2]$[N192] than the solvents water and nitromethane.

Belluco, Palazzi and Parker[N185] have attempted to correlate the difference in the k_Y term, for chloride ion exchange with $[PtCl_2(PEt_3)_2]$, with solvent activity coefficients in methanol and dimethylsulphoxide, and have contrasted the behaviour with chloride ion exchanges at tetrahedral carbon centres. They are contrasting the response of the following two reactions to solvent transfer:

$$\textit{trans}\text{-}[PtCl_2(PEt_3)_2] + Cl^{-*} \rightleftarrows \underset{\text{Transition State}}{[PtCl^*Cl_2(PEt_3)_2]^-} \rightleftarrows [PtClCl^*(PEt_3)_2] + Cl^- \quad (6.3.2)$$

$$RCH_2Cl + Cl^{-*} \rightleftarrows \underset{\text{Transition State}}{[Cl^*RCH_2Cl]^-} \rightleftarrows RCH_2Cl^* + Cl^- \quad (6.3.3)$$

As explained previously, reactions of the type 6.3.3 show a great increase in rate on transfer from methanol to dimethylsulphoxide (in this case $\sim 10^4$–10^5):[N13] this increase is due to the great destabilisation of the chloride ion in dimethylsulphoxide which is not compensated by destabilisation of the large but 'tight'[N13,N185] trigonal-bipyramidal transition state. This great increase in rate is not shown in the transfer of the platinum reaction 6.3.2 into the dipolar aprotic solvent. Thus in the platinum case there must be stabilisation of the reactant complex or substantial destabilisation of the transition state, or a combination of both, to counteract the great increase of chloride ion activity on transfer to dimethylsulphoxide.

Solubility measurements[N185] on the platinum complex show that the difference in *trans*-$[PtCl_2(PEt_3)_2]$ solvation on transfer is very small and certainly could not account for a change in rate of more than one power of ten. This conclusion is supported by the work of Drago, Mode and Kay.[N190] Belluco, Palazzi and Parker[N185] conclude that the platinum transition state, unlike the carbon transition state, is destabilised substantially in dimethylsulphoxide because the transition states are 'loose' and the chloride ion retains much of its free anion characteristic, which allows it to contribute substantial interaction with the hydroxylic solvent despite its incorporation into a coordination position on the platinum. This is equivalent to saying that the negative charge on the transition state in the carbon case is distributed equally between the entering and leaving anions, whereas in the platinum case the charge remains localised on the entering group.

These results, if not the explanations, are indisputable. It does, however, seem unlikely that an entering anion, even if it maintains its localised charge characteristics, could retain almost all of its solvation characteristics when coordinated. The explanations can be disputed only on the grounds that the platinum reaction in the two solvents may not involve an increase of coordination from four to five. Conductance and ultra-violet spectral results[N185] suggest that the reactant species has only very weak interaction in the fifth and sixth coordination positions. This does not prove, however, that an increasing involvement of a chloride ion with the platinum will not encourage increased interaction of the sulphur atom of the dimethylsulphoxide ligand in the sixth position of the transition state. This type of synchronous involvement of the fifth and sixth positions can be justified in this case on steric and electrostatic grounds, and six-coordinate intermediates have been postulated before for platinum(II) reactions.[N26]

If the transition state does involve this sort of interaction then the simple considerations based on solvent transfer activity coefficients are invalidated. The mechanism of substitution in square planar platinum(II) complexes, and in particular the role of the fifth and sixth positions in relation to reactant and transition state stability, is one of the most interesting and challenging mechanistic problems in transition-metal substitution kinetics, and there is no doubt that a systematic application of solvent activity coefficients to a range of neutral and charged reactant complexes will lead to a better insight into these problems.

The relative insensitivity of platinum(II) reactions with anionic nucleophiles to solvent transfer is paralleled by an almost insensitivity of anionic nucleophilic strength to changes in solvent.[N188,N219,N220] Parker[N12] and others[N17,N221] have demonstrated the order of nucleophilic strength to carbon for the halide ions in water or alcohols to be $Cl < Br < I$ and to be reversed $(Cl > Br > I)$ in aprotic solvents. This change is attributed to hydrogen-bonding solvation in protic solvents which is stronger for small 'hard' nucleophiles. The order in aprotic solvents is thought to parallel the inherent gas-phase nucleophilic strength.

In platinum(II) substitution reactions, however, the order of nucleophilic strength is $I^- > Br^- > Cl^-$ and is independent of solvent. This is clearly shown by the results in Table 6.3.1 which compares values of k_Y for the reaction:

$$trans\text{-}[PtCl_2(pip)_2] + Y^- \rightarrow trans\text{-}[PtClY(pip)_2] + Cl^-$$

in a range of solvents (pip $\equiv$ piperidine).

Table 6.3.1

Effect of Solvent on Nucleophilic Strengths of Anionic Ligands* (expressed as $10^3k/\text{mol}^{-1}\,\text{l s}^{-1}$) for Reaction *trans*-[$PtCl_2(pip)_2$] + $Y^- \rightarrow$ *trans*-[$PtClY(pip)_2$] + Cl^-

Y	MeOH	$(CH_3)_2CO$	DMSO	DMF	CH_3CN	CH_3NO_2
Cl^-	0.9	3.5	1.0	0.56	1.1	1.5
Br^-	6.2	80	5.2	6.7	13	15
I^-	300	165	18	27	86	78

* Results taken from refs N72, N188, N219, N220.

The results do not reflect the great changes in anion stability which are demonstrated by solvent activity coefficients (e.g. ${}^{\text{MeOH}}_{\text{DMSO}}\gamma$ for Cl^-, Br^- and I^- of $10^{5\cdot5}$, $10^{3\cdot6}$ and $10^{1\cdot3}$). It seems unlikely that the transition states involving anions with this range of polarisability will reflect the response of the anion to change in the solvent. The only firm generalisation is that the nucleophilic strength for platinum correlates with the polarisability of the anionic nucleophile and is almost independent of solvent. It is possible that the solvent has a more particular involvement in these reactions, which counteracts the solvent dependence of anion activity, and invalidates the application of the solvent activity coefficient made by Belluco, Palazzi and Parker.[N185] It should be stressed that specific solvation interaction in the fifth and sixth positions is not a new concept and has been emphasised by others.[N72,N178,N195,N199,N222,N223]

There is an interesting contrast in the substitution kinetics of platinum(II) complexes and the complexes of cobalt(III) and chromium(III), in relation to the two solvents DMF and DMSO. As emphasised by solvolysis and isomerisation studies with both cobalt(III) and chromium(III),[N25,N128,N142,N144] these two solvents differ in only minor ways. DMSO is a slightly stronger ligand, based on its resistance to substitution replacement by anions, but this difference is small as are the differences in their mutual interchange rates.[N24,N217]

In square planar platinum(II) chemistry, however, the behaviour of the complexes, particularly in response to anion attack, differs greatly in the two solvents. In the reactions of *trans*-[$PtCl_2(py)_2$] with radioactive chloride ion in DMSO, the exchange is only through the anion independent path (k_s) while in DMF, exchange involves the anion dependent path (k_Y).[N192,N218]

The predominance of the exchange through the solvolysis path in

DMSO reflects the greater strength of 'soft' polarisable nucleophiles in platinum(II) chemistry, in that DMSO coordinates through the lone pair of electrons on the polarisable sulphur atom.[N214,N215] DMF on the other hand is a 'harder' nucleophile and, for platinum(II), behaves much as acetone and acetonitrile. In contrast in cobalt(III) and chromium(III), where DMSO coordinates through the 'hard' oxygen atom, DMF behaves much like DMSO.

6.3.3 Other Square Planar Complexes

Little systematic work exists on the role of the solvent in substitution reactions of other square planar complexes. The two-term rate law is found to characterise all the reactions so far studied.

In the case of gold(III), the solvent path (k_s) is much less important than in platinum(II) and the reactions are approximately 10^4 times faster. There is evidence that bond making is more important in gold(III) complexes, which correlates with the greater charge on the metal as does the more favourable activation entropies, which correspond to greater loss of solvation order around the metal ion of greater charge.[N224,N225]

Cattalini, Ricevuto, Orio and Tobe[N226] have studied the reactions of $[AuCl_4]^-$ and $[AuCl_3py]$ in a series of hydroxylic solvents and conclude that 'specific solvation effects are less important than the general electrostatic effects'. It must be remembered that the solvents used here are not only very similar in properties, but are also 'hard' and thus less likely to be involved in strong interactions in the fifth and sixth coordination positions.

In the reactions of palladium(II), the solvent path (k_s) seems to be more important than with either platinum(II) or gold(III), and conversely the reactions involve less bond making at the transition state. A most interesting reaction in rhodium(I) chemistry from the point of view of the solvent, was recently studied by Cattalini, Ugo and Orio.[N227] They found the rate of replacement of stibine (Sb(p-tolyl)$_3$) by amines from [RhCl(1,5-cyclooctadiene)(SbR_3)] in a range of solvents, to be virtually independent of the nature of the solvent, independent of the concentration of the amine, but dependent on the nature of the amine. These results have been interpreted in terms of a relatively inert five-coordinate intermediate.

6.4 REACTIONS OF ORGANOMETALLIC COMPOUNDS

There are two recent reviews on mechanism in this field, one on metal carbonyl complexes[N228] and one on organometallics in general.[N229] Both reviews emphasise that the rates of all classes of reaction are virtually insensitive to changes in solvent. This insensitivity to solvent results because the reactions involve uncharged nucleophiles reacting with uncharged reactant complexes of metals in low oxidation states. In addition, the central metal atom is usually saturated with eighteen valence electrons in the *ns*, *np* and $(n - 1)d$ orbitals and is thus not susceptible to bimolecular solvent attack. It should be emphasised, however, that not much work has been attempted in strongly polar solvents whose participation by coordination would be anticipated.

Some small changes in rate with changes in the dielectric constant of the solvent have been used to infer the changes in polarity of the reactant complex upon activation. The substitution reactions of metal carbonyl halides[N230,N232] show a decrease in rate with increasing dielectric constant, which correlates with an expected decrease in polarity on going to the transition state.[N233] On the other hand, the insertion reactions of carbon monoxide into methyl manganese pentacarbonyl show an increase in rate,[N234] which suggests a transition state more polar than the reactant.

Although these rate changes are small they are often the result of quite large changes in enthalpy of activation, compensated by changes in the activation entropy.[N230,N235] The reaction

$$Ni(CO)_4 + P(C_6H_5)_3 \rightarrow [Ni(CO)_3P(C_6H_5)_3] + CO$$

at 25°C, has first order rate constants of 13.7×10^{-3} and $19.4 \times 10^{-3}\ s^{-1}$ in cyclohexane and toluene, respectively.[N236] The activation parameters are:

Cyclohexane	$\Delta H^{\ddagger}$ 111 kJ mol^{-1}	$\Delta S^{\ddagger}$ 87 J K^{-1} mol^{-1}
Toluene	$\Delta H^{\ddagger}$ 85 kJ mol^{-1}	$\Delta S^{\ddagger}$ 8.4 J K^{-1} mol^{-1}

Aromatic solvents give much lower activation enthalpies for reactions of nickel carbonyl.[N228]

Insertion or ligand migration reactions[N237] show a much greater sensitivity to solvent. The reaction:

$$L + CH_3Mn(CO_5) \rightarrow CH_3COMn(CO)_4L$$

has been studied in a range of solvents.[N234,N238] Here weakly polar solvents give second order kinetics while strongly polar solvents, such as dimethylformamide, give first order kinetics. In these latter cases it seems that solvent coordination produces an intermediate prior to the incorporation of the ligand (L). There are examples of solvents of intermediate character that give a two-term rate law, familiar in the substitution reactions of square planar complexes (p. 718).

A two-term rate law has also been found for the substitution of bases into dinitrosyldicarbonyliron(0)[N239] in a range of solvents. Here, although the coordination of solvents, such as tetrahedrofuran, is recognised as a possible cause of the first order term, a mechanism involving the attack of the 'hard' solvent on the electrophilic carbon atom of a carbonyl ligand is preferred.[N239]

The solvent effects on oxidative addition reactions to square planar iridium(I) complexes have drawn some comment. The general acceleration of the addition reactions of hydrogen, oxygen and methyl iodide to *trans*-$[IrX(CO)(PPh_3)_2]$ (where X is a halogen) by polar solvents, such as dimethylformamide, is taken by Chock and Halpern to be evidence for a polar transition state.[N240] Perhaps more interesting is the stereochemical result of the addition of alkyl and hydrogen halides to these iridium(I) complexes.[N241–247]

Blake and Kubota[N247] have convincingly shown that in anhydrous benzene, chloroform and ether, as in the gaseous reaction on the solid complex, hydrogen halides (HX) and *trans*-$[IrY(CO)(PPh_3)_2]$ (Y = halogen) react to give octahedral *cis*-addition products (i.e. H and X *cis*). In contrast in wet solvents, and in such solvents as dimethylformamide, acetonitrile and ethanol, mixtures of *cis* and *trans* products are formed. Whether these solvents cause rapid halide-ion exchange before the addition reaction or in the products is yet to be demonstrated.[N247]

Polar solvents have been found to increase the efficiency of hydrogenation of square planar rhodium complexes. Osborn, Jardine, Young and Wilkinson[N248] suggested that the activated complex must have a greater dipole moment than the reacting species, which could result in one of the added hydrogen atoms being positive with respect to the other. They also recognised that the solvent could accelerate the reaction by coordination.[N249]

More recent work[N250–252] shows a very specific coordination role for such solvents as dimethylformamide, dimethylacetamide and dimethylsulphoxide in the catalysis of hydrogenation reactions by rhodium(III) complexes such as $[RhCl_3(py)_3]$. Particularly interesting are the results in (+) or (−)-1-phenylethylformamide,[N252] where through simultaneous coordination to the rhodium with the substrate methyl-3-phenylbut-2-en-oate, the optically active solvent encourages stereospecific hydrogenation to yield (+) or (−)-methyl 3-phenylbutanoate. These results emphasise that further kinetic studies in optically active solvents would be interesting (see p. 717).

Results such as these suggest that as preoccupation with the preparative chemistry of organometallics decreases and the interest in kinetics and mechanism increases, many important solvent effects will be found even with solvents whose normal kinetic roles are similar. This is emphasised by the change in the mechanism of exchange reactions involving $[(COD)RhCl(AsPh_3)]$ (where COD represents 1,5-cyclooctadiene), in the solvents CD_2Cl_2 and $CDCl_3$.[N253] In CD_2Cl_2, reactions of the free ligand ($AsPh_3$) are important while in $CDCl_3$, exchange is accomplished through reactions between activated monomeric complexes and the chloro-bridged dimer.

REFERENCES

N1 E. S. Amis, Solvent Effects on Reaction Rates and Mechanisms, Academic Press, New York, N.Y. (1966)

N2 E. S. Amis, Kinetics of Chemical Change in Solution, MacMillan, New York, N.Y. (1949).

N3 E. A. Moelwyn-Hughes, The Kinetics of Reactions in Solution, 2nd edition, Oxford University Press, London (1947)

N4 A. A. Frost and R. G. Pearson, Kinetics and Mechanism, 2nd edition, John Wiley and Sons, Inc., New York, N.Y. (1961)

N5 K. J. Laidler, Chemical Kinetics, 2nd edition, McGraw-Hill Book Co. Inc., New York, N.Y. (1965)

N6 S. W. Benson, The Foundations of Chemical Kinetics, McGraw-Hill Book Co. Inc., New York, N.Y. (1960)

N7 K. B. Wiberg, Physical Organic Chemistry, John Wiley and Sons, Inc., New York, N.Y. (1964)

N8 I. Amdur and G. G. Hammes, Chemical Kinetics, McGraw-Hill, Inc., New York (1966)

N9 J. E. Leffler and E. Grunwald, Rates and Equilibria of Organic Reactions, John Wiley and Sons, Inc., New York, N.Y. (1963)

N10 A. J. Parker, *Advan. Phys. Org. Chem.*, **5**, 173 (1967)

N11 A. J. Parker, *Advan. Org. Chem. Methods Results*, **5**, 1 (1965)

N12 A. J. Parker, *Quart. Rev. (London)*, **16**, 163 (1962)

N13 A. J. Parker, *Chem. Rev.*, **69**, 1 (1969)

N14 H. Normant, *Angew. Chem.* **6**, 1046 (1967)

N15 C. D. Ritchie *in* Solute-Solvent Interactions (J. F. Coetzee and C. D. Ritchie eds.), Marcel Dekker Ltd., London (1969)

N16 E. M. Arnett and D. R. McKelvey, *Record Chem. Progr.*, **26**, 185 (1965)

N17 R. Alexander, E. C. F. Ko, A. J. Parker and T. J. Broxton, *J. Amer. Chem. Soc.*, **90**, 5049 (1968)

N18 E. M. Kosower, An Introduction to Physical Organic Chemistry, John Wiley and Sons, Inc., New York, N.Y. (1968)

N19 J. Miller *in* Reaction Mechanisms in Organic Chemistry, Monograph 8 (C. Eaborn and N. B. Chapman, eds.), Elsevier Publishing Co., New York, N.Y. (1968)

N20 L. R. Dawson, Chemistry in Non-Aqueous Ionizing Solvents, Vol. IV, Interscience Publishers, New York, N.Y. (1963)

N21 D. Martin, A. Weise and N. J. Niclas, *Angew. Chem.*, **6**, 318 (1967)

N22 G. R. Leader and J. F. Gormley, *J. Amer. Chem. Soc.*, **73**, 5731 (1951)

N23 D. W. Watts, *Record Chem. Progr.*, **29**, 131 (1968)

N24 I. R. Lantzke and D. W. Watts, *J. Amer. Chem. Soc.*, **89**, 815 (1967)

N25 M. L. Tobe and D. W. Watts, *J. Chem. Soc.*, 2991 (1964)

N26 C. H. Langford, *Inorg. Chem.*, **3**, 228 (1964)

N27 W. R. Fitzgerald, A. J. Parker and D. W. Watts, *J. Amer. Chem. Soc.*, **90**, 5744 (1968)

N28 T. J. Webb, *J. Amer. Chem. Soc.*, **48**, 2589 (1926)

N29 C. K. Ingold, *J. Chem. Soc.*, 2179 (1931)

N30 K. J. Laidler and C. Pegis, *Proc. Roy. Soc.*, **A241,** 80 (1957)
N31 E. Glueckauf, *Trans. Faraday Soc.*, **60,** 572 (1964)
N32 R. H. Stokes, *J. Amer. Chem. Soc.*, **86,** 979 (1964)
N33 W. A. Millen and D. W. Watts, *J. Amer. Chem. Soc.*, **89,** 6051 (1967)
N34 A. W. Adamson, A Textbook of Physical Chemistry, Harper and Row, New York, N.Y. (1971)
N35 G. Scatchard, *Chem. Rev.*, **10,** 229 (1932)
N36 J. N. Brönsted, *Z. Phys. Chem.*, **102,** 169 (1922)
N37 J. N. Brönsted, *Z. Phys. Chem.*, **115,** 337 (1925)
N38 J. A. Christiansen, *Z. Phys. Chem.*, **113,** 35 (1924)
N39 E. S. Amis and F. C. Holmes, *J. Amer. Chem. Soc.*, **63,** 2231 (1941)
N40 K. J. Laidler and H. Eyring, *Ann. N. Y. Acad. Sci.*, **39,** 303 (1940)
N41 J. G. Kirkwood, *J. Chem. Phys.*, **2,** 351 (1934)
N42 E. S. Amis, *Inorg. Chim. Acta Rev.*, **3,** 7 (1969)
N43 K. J. Laidler, Symposium on Solvation Phenomena, p. 16, The Chemical Institute of Canada, Calgary Section (July 1963)
N44 W. A. Millen, Ph.D. Thesis, The University of Western Australia (1967)
N45 R. G. Bates *in* The Chemistry of Non-Aqueous Solvents, Vol. I, p. 97 (J. J. Lagowski, ed.), Academic Press, New York, N.Y. (1966)
N46 I. M. Kolthoff, J. J. Lingane and W. D. Larson, *J. Amer. Chem. Soc.*, **60,** 2512 (1938)
N47 E. Grunwald and B. J. Berkowitz, *J. Amer. Chem. Soc.*, **73,** 4939 (1951)
N48 R. Alexander and A. J. Parker, *J. Amer. Chem. Soc.*, **90,** 3313 (1968)
N49 H. Strehlow *in* The Chemistry of Non-Aqueous Solvents, Vol. I, p. 129 (J. J. Lagowski, ed.), Academic Press, New York, N.Y. (1966)
N50 I. M. Kolthoff and F. G. Thomas, *J. Phys. Chem.*, **69,** 3049 (1965)
N51 A. J. Parker, *J. Amer. Chem. Soc.*, in press
N52 E. Grunwald, G. Baughman and G. Kohnstam, *J. Amer. Chem. Soc.*, **82,** 5801 (1960)
N53 R. Alexander, E. C. F. Ko, Y. C. Mac and A. J. Parker, *J. Amer. Chem. Soc.*, **89,** 3703 (1967)
N54 R. Alexander and A. J. Parker, *J. Amer. Chem. Soc.*, **89,** 5549 (1967)
N55 A. J. Parker, *J. Chem. Soc.*, **A,** 220 (1966)
N56 R. T. Iwamoto and I. V. Nelson, *Anal. Chem.*, **35,** 867 (1963)
N57 H. Stephen and T. Stephen, Solubilities of Inorganic and Organic Compounds, Pergamon Press, London (1963)
N58 J. H. Dymond, *J. Phys. Chem.*, **71,** 1829 (1967)
N59 C. H. Langford and J. F. White, *Canad. J. Chem.*, **45,** 3049 (1967)
N60 L. P. Scott, T. J. Weeks, D. E. Bracken and E. L. King, *J. Amer. Chem. Soc.*, **91,** 5219 (1969)
N61 D. W. Kemp and E. L. King, *J. Amer. Chem. Soc.*, **89,** 3433 (1967)
N62 S. Behrendt, C. H. Langford and L. S. Frankel, *J. Amer. Chem. Soc.*, **91,** 2236 (1969)
N63 L. S. Frankel, T. R. Stengle and C. H. Langford, *Chem. Comm.*, 393 (1965)
N64 J. C. Jayne and E. L. King, *J. Amer. Chem. Soc.*, **86,** 3989 (1964
N65 R. G. Pearson and P. Ellgen, *Inorg. Chem.*, **6,** 1379 (1967)
N66 H. Brintzinger and G. G. Hammes, *Inorg. Chem.*, **5,** 1286 (1966)
N67 C. H. Langford and T. R. Stengle, *Ann. Rev. Phys. Chem.*, **19,** 193 (1968)
N68 T. R. Stengle and C. H. Langford, *Coord. Chem. Rev.*, **2,** 349 (1967)
N70 M. Eigen *in* Advances in the Chemistry of Coordination Compounds, pp. 371–(1958)
N70 M. Eigen, *in* Advances in the Chemistry of Coordination Compounds, pp. 371–378 (S. Kirschner, ed.), MacMillan, New York, N.Y. (1961)

N71 J. O. Edwards Inorganic Reaction Mechanisms, W. A. Benjamin, New York (1964)

N72 F. Basolo and R. G. Pearson, Mechanism of Inorganic Reactions, First Edition, 1958 and Second Edition, 1967, John Wiley and Sons, Inc., New York, N.Y.

N73 M. Eigen and R. G. Wilkins, Mechanism of Inorganic Reactions, pp. 55–80, Advances in Chemistry Series, No. 49, American Chemical Society, Washington, D.C. (1965)

N74 M. L. Tobe, Mechanisms of Inorganic Reactions, pp. 7–30, Advances in Chemistry Series, No. 49, American Chemical Society, Washington, D.C. (1965)

N75 C. H. Langford and H. B. Gray, Ligand Substitution Process, W. A. Benjamin, Inc., New York, N.Y. (1966)

N76 D. R. Stranks *in* Modern Coordination Chemistry, pp. 78–173 (J. Lewis and R. G. Wilkins, eds.), Interscience Publishers, Inc., New York, N.Y. (1960)

N77 M. L. Tobe, *Inorg. Chem.*, **7**, 1260 (1968)

N78 R. D. Archer, *Coord. Chem. Rev.*, **4**, 243 (1969)

N79 S. C. Chan and J. Miller, *Rev. Pure and Appl. Chem.*, **15**, 11 (1965)

N80 M. L. Tobe, *Record Chem. Progr.*, **27**, 79 (1966)

N81 C. H. Langford and H. B. Gray, *Chem. Eng. News*, 68 (April 1, 1968)

N82 R. G. Pearson, Mechanism of Inorganic Reactions, pp. 21–28 Adv. in Chem. Series, No. 49, American Chemical Society (1965)

N83 G. Atkinson and S. K. Kor, *J. Phys. Chem.*, **69**, 128 (1965)

N84 W. R. Fitzgerald and D. W. Watts, *J. Amer. Chem. Soc.*, **90**, 1743 (1968)

N85 W. R. Muir and C. H. Langford, *Inorg. Chem.*, **7**, 1032 (1968)

N86 N. V. Duffy and J. E. Earley, *J. Amer. Chem. Soc.*, **89**, 816 (1967)

N87 R. Murray and C. Barraclough, *J. Chem. Soc.*, 7074 (1965)

N88 D. H. Devia, Honours Thesis, University of Western Australia (1969)

N89 M. W. Lister and D. E. Rivington, *Canad. J. Chem.*, **33**, 1572 (1955)

N90 J. F. Below, R. E. Connick and C. P. Coppel, *J. Amer. Chem. Soc.*, **80**, 2961 (1958)

N91 R. S. Drago, D. M. Hart and R. L. Carlson, *J. Amer. Chem. Soc.*, **87**, 1900 (1965)

N92 C. H. Langford and F. M. Chung, *J. Amer. Chem. Soc.*, **90**, 4485 (1970)

N93 S. Thomas and W. L. Reynolds, *J. Chem. Phys.*, **44**, 3148 (1966)

N94 A. Fratiello and R. Schuster, *J. Phys. Chem.*, **71**, 1948 (1967)

N95 W. G. Movius and N. A. Matwiyoff, *J. Phys. Chem.*, **72**, 3063 (1968)

N96 W. G. Movius and N. A. Matwiyoff, *Inorg. Chem.*, **6**, 847 (1967)

N97 N. A. Matwiyoff and W. G. Movius, *J. Amer. Chem. Soc.*, **89**, 6077 (1967)

N98 J. S. Babiec, C. H. Langford and T. R. Stengle, *Inorg. Chem.*, **5**, 1363 (1966)

N99 N. A. Matwiyoff, *Inorg. Chem.*, **5**, 788 (1966)

N100 Z. Luz and S. Meiboom, *J. Chem. Phys.*, **40**, 1058, 1066, 2686 (1964)

N101 Z. Luz, *J. Chem. Phys.*, **41**, 1748 (1964)

N102 N. A. Matwiyoff and S. V. Hooker, *Inorg. Chem.*, **6**, 1127 (1967)

N103 A. C. Adams and E. M. Larsen, *Inorg. Chem.*, **5**, 814 (1966)

N104 F. W. Breivogal, *J. Phys. Chem.*, **73**, 4203 (1969)

N105 S. Namakura and S. Meiboom, *J. Amer. Chem. Soc.*, **89**, 1765 (1967)

N106 T. D. Alger, *J. Amer. Chem. Soc.*, **91**, 2220 (1969)

N107 H. Levanon and Z. Luz, *J. Chem. Phys.*, **49**, 2031 (1968)

N108 S. Thomas and W. L. Reynolds, *J. Chem. Phys.*, **46**, 4164 (1967)

N109 Z. Luz and S. Meiboom, *J. Chem. Phys.*, **40**, 2686 (1964)

N110 D. K. Ravage, T. R. Stengle and C. H. Langford, *Inorg. Chem.*, **6**, 1252 (1967)

N111 W. D. Horrocks and L. H. Pignolet, *J. Amer. Chem. Soc.*, **88**, 5929 (1966)

N112 J. C. Sheppard and J. L. Burdett, *Inorg. Chem.*, **5**, 921 (1966)

N113 N. S. Angerman and R. B. Jordon, *Inorg. Chem.*, **8,** 65, 2579 (1969)
N114 N. S. Angerman and R. B. Jordon, *J. Chem. Phys.*, **48,** 3983 (1968)
N115 H. Taube, *Chem. Rev.*, **50,** 69 (1952)
N116 D. D. Brown, C. K. Ingold and R. S. Nyholm, *J. Chem. Soc.*, 2678 (1953)
N117 F. Basolo, *Chem. Rev.*, **52,** 459 (1953)
N118 F. Basolo *in* The Chemistry of the Coordination Compounds, Chapter 8 (J. C. Bailar, ed.), Reinhold Publishing Co., New York, N.Y. (1956)
N119 F. Basolo and R. G. Pearson, *Adv. Inorg. Chem. Radiochem.*, **3,** 1 (1961)
N120 M. Green and H. Taube, *Inorg. Chem.*, **2,** 948 (1963)
N121 D. A. Buckingham, I. I. Olsen and A. M. Sargeson, *J. Amer. Chem. Soc.*, **89,** 5129 (1967)
N122 D. A. Buckingham, I. I. Olsen and A. M. Sargeson, *J. Amer. Chem. Soc.*, **90,** 6539 (1968)
N123 S. C. Chan and F. Leh, *J. Chem. Soc.*, **A,** 126 (1966)
N124 S. C. Chan and F. Leh, Proceedings of the VIIIth Int. Conf. on Coord. Chem., p. 298, Vienna (1964)
N125 S. C. Chan, *J. Chem. Soc.*, **A,** 1124 (1966)
N126 D. A. Buckingham, I. I. Olsen and A. M. Sargeson, *Inorg. Chem.*, **7,** 174 (1968)
N127 R. D. Gillard, *J. Chem. Soc.*, **A,** 917 (1967)
N128 M. L. Tobe and D. W. Watts, *J. Chem. Soc.*, 4614 (1962)
N129 D. Loeliger and H. Taube, *Inorg. Chem.*, **4,** 1032 (1965)
N130 D. Loeliger and H. Taube, *Inorg. Chem.*, **5,** 1376 (1966)
N131 A. M. Sargeson, *Australian J. Chem.*, **17,** 385 (1964)
N132 D. A. Buckingham, I. I. Olsen and A. M. Sargeson, *Inorg. Chem.*, **6,** 1807 (1967)
N133 A. Haim and W. K. Wilmarth, *Inorg. Chem.*, **1,** 573 (1962)
N134 A. Haim, R. J. Grassie and W. K. Wilmarth, *Adv. Chem. Ser.*, **49,** 31 (1965)
N135 I. R. Jonasson, R. S. Murray, D. R. Stranks and J. K. Yandell, Proceedings of the XIIth Intern. Conf. on Coord. Chem., Sydney (1969)
N136 Ref. 67, p. 207.
N137 C. K. Ingold, Structure and Mechanism in Organic Chemistry, p. 141, Cornell University Press, Ithaca (1953)
N138 F. Basolo, P. M. Henry and R. G. Pearson, *J. Amer. Chem. Soc.*, **79,** 5379 (1957)
N139 F. Basolo, P. M. Henry and R. G. Pearson, *J. Amer. Chem. Soc.*, **79,** 5582 (1957)
N140 B. Bosnich, C. K. Ingold and M. L. Tobe, *J. Chem. Soc.*, 4074 (1965)
N141 B. Bosnich, J. Ferguson and M. L. Tobe, *J. Chem. Soc.*, **A,** 1636 (1966)
N142 D. A. Palmer and D. W. Watts, *Australian J. Chem.*, **21,** 2895 (1968)
N143 D. A. Palmer, Ph.D. Thesis, University of Western Australia (1970)
N144 D. A. Palmer and D. W. Watts, *Australian J. Chem.*, **20,** 53 (1967)
N145 W. A. Millen and D. W. Watts, *J. Amer. Chem. Soc.*, **89,** 6858 (1967)
N146 D. A. Palmer and D. W. Watts, in press
N147 D. A. Palmer and D. W. Watts, Proceedings of the XIIth Intern. Conf. on Coord. Chem., Sydney (1969)
N148 I. R. Lantzke and D. W. Watts, *J. Amer. Chem. Soc.*, **89,** 815 (1967)
N149 C. S. Springer and R. E. Sievers, *Inorg. Chem.*, **6,** 852 (1967)
N150 I. R. Lantzke and D. W. Watts, *Australian J. Chem.*, **19,** 949 (1967)
N151 H. Taube and F. A. Posey, *J. Amer. Chem. Soc.*, **75,** 1463 (1953)
N152 Ref. N72, Second Edition, p. 178, p. 183, pp. 195–197 and p. 213
N153 Ref. N72, First Edition, p. 150
N154 W. R. Fitzgerald and D. W. Watts, *J. Amer. Chem. Soc.*, **89,** 821 (1967)
N155 W. R. Fitzgerald and D. W. Watts, *Australian J. Chem.*, **21,** 595 (1968)
N156 J. L. Kurz, *J. Amer. Chem. Soc.*, **85,** 987 (1963)

N157 W. R. Fitzgerald and D. W. Watts, *Australian J. Chem.*, **19,** 1411 (1966)
N158 M. N. Hughes and M. L. Tobe, *J. Chem. Soc.*, **A,** 1204 (1965)
N159 Ref. N72, Second Edition, p. 213
N160 D. Lo and D. W. Watts, unpublished results
N161 W. Schmidt and H. Taube, *Inorg. Chem.*, **2,** 698 (1963)
N162 Ref. N23, p. 151
N163 D. Druskovich and D. W. Watts, unpublished results
N164 Ref. N72, First Edition, p. 150
N165 L. S. Frankel, *J. Phys. Chem.*, **73,** 3897 (1969)
N166 S. Behrendt, C. H. Langford and L. S. Frankel, *J. Amer. Chem. Soc.*, **91,** 2236 (1969)
N167 L. P. Scott, T. J. Weeks, D. E. Bracken and E. L. King, *J. Amer. Chem. Soc.*, **91,** 5219 (1969)
N168 C. H. Langford and J. F. White, *Canad. J. Chem.*, **45,** 3049 (1967)
N169 J. C. Jayne and E. L. King, *J. Amer. Chem. Soc.*, **86,** 3989 (1964)
N170 J. Zsakó, C. Várhelyi, J. Gănescu and J. Turós, *Acta Chim. Acad. Sci. Hung.*, **61,** 167 (1969)
N171 J. Zsakó, C. Várhelyi, J. Gănescu, *Rev. Roum. Chim.*, **13,** 581 (1968), *Chem. Abstr.*, **69,** 9388 (100090x) (1968)
N172 C. G. Barraclough and P. T. McTigue, Proceedings of the XIIth Intern. Conf. on Coord. Chem., Sydney (1969)
N173 A. W. Adamson, *J. Inorg. Nucl. Chem.*, **28,** 1955 (1964)
N174 J. F. Hinton, L. S. McDowell and E. S. Amis, *Chem. Comm.*, 776 (1966)
N175 J. L. Walker and M. Rosalie, *Anal. Chem.*, **37,** 45 (1965)
N176 E. Grunwald, G. Baughman and G. Kohnstam, *J. Amer. Chem. Soc.*, **82,** 580 (1960)
N177 Y. Kobayasi, K. Toka and U. Muiri, *J. Sci. Hiroshima University*, **9A,** 33 (1939)
N178 B. Bosnich, *J. Amer. Chem. Soc.*, **88,** 2606 (1966)
N179 B. Bosnich, *J. Amer. Chem. Soc.*, **89,** 6143 (1967)
N180 B. Bosnich and D. W. Watts, *J. Amer. Chem. Soc.*, **90,** 6228 (1968)
N181 R. A. Haines and A. A. Smith, *Canad. J. Chem.*, **46,** 1444 (1968)
N182 A. A. Smith and R. A. Haines, *Canad. J. Chem.*, **47,** 2727 (1969)
N183 A. A. Smith and R. A. Haines, *J. Amer. Chem. Soc.*, **91,** 6280 (1969)
N184 U. Belluco, *Coord. Chem. Rev.*, **1,** 111 (1966)
N185 U. Belluco, A. Palazzi and A. J. Parker, *Inorg. Chem.*, in press
N186 U. Belluco, P. Rigo, M. Graziani and R. Ettorre, *Inorg. Chem.*, **5,** 1125 (1966)
N187 A. A. Grinberg, L. E. Nikol'skaya and N. M. Sprikova, *Radiokhimiya*, **9,** 589 (1967), *Chem. Abstr.*, **68,** 63091 (1968)
N188 U. Belluco, M. Martelli and A. Orio, *Inorg. Chem.*, **5,** 582 (1966)
N189 D. G. McMane and D. S. Martin, *Inorg. Chem.*, 1169 (1968)
N190 R. S. Drago, V. A. Mode and J. G. Kay, *Inorg. Chem.*, **5,** 2050 (1966)
N191 R. Ettorre, M. Graziani and P. Rigo, *Gazz. Chim. Ital.*, **97,** 58 (1967)
N192 R. G. Pearson, H. B. Gray and F. Basolo, *J. Amer. Chem. Soc.*, **82,** 787 (1960)
N193 H. Gray and R. J. Olcott, *Inorg. Chem.*, **1,** 481 (1962)
N194 P. B. Chock and J. Halpern, *J. Amer. Chem. Soc.*, **88,** 3511 (1966)
N195 L. Cattalini, M. Martelli and A. Orio, *Chim. Ind. (Milan)*, **49,** 623 (1967); *Chem. Abstr.*, **68,** 53828 (1968)
N196 L. Cattalini, V. Ricevuto, A. Orio and M. L. Tobe, *Inorg. Chem.*, **7,** 51 (1968)
N197 V. D. Panasyuk and N. F. Malashok, *Russ. J. Inorg. Chem.*, **14,** 661 (1969)
N198 G. Faraone, U. Belluco, V. Ricevuto and R. Ettorre, *J. Inorg. Nucl. Chem.*, **28,** 863 (1966)
N199 U. Belluco, M. Graziani, M. Nicolini and P. Rigo, *Inorg. Chem.*, **6,** 721 (1967)
N200 J. V. Rund and F. A. Palocsay, *Inorg. Chem.*, **8,** 2242 (1969)

N201 J. B. Goddard and F. Basolo, *Inorg. Chem.*, **7,** 2456 (1968)
N202 C. M. Harris and S. E. Livingstone, *Rev. Pure Appl. Chem.*, **12,** 16 (1962)
N203 L. M. Venanzi, *Angew. Chem.* **3,** 453 (1964)
N204 J. A. Ibers, *Ann. Rev. Phys. Chem.*, **16,** 380 (1965)
N205 E. L. Muetterties and R. A. Schunn, *Quart. Rev.* (*London*), **20,** 245 (1966)
N206 J. S. Coleman, H. Peterson and R. A. Penneman, *Inorg. Chem.*, **4,** 135 (1965)
N207 P. Rigo, C. Pecile and A. Turco, *Inorg. Chem.*, **6,** 1636 (1967)
N208 R. D. Cramer, R. V. Lindsey, C. T. Prewitt and U. G. Stollberg, *J. Amer. Chem. Soc.*, **87,** 658 (1965)
N209 M. Angaletta, *Gazz. Chim. Ital.*, **89,** 2359 (1959)
N210 M. Angaletta, *Gazz. Chim. Ital.*, **90,** 1021 (1960)
N211 S. S. Bath and L. Vaska, *J. Amer. Chem. Soc.*, **85,** 3500 (1963)
N212 C. M. Harris, R. S. Nyholm and D. J. Phillips, *J. Chem. Soc.*, 4379 (1960)
N213 F. Basolo, J. Chatt, H. B. Gray, R. G. Pearson and B. L. Shaw, *J. Chem. Soc.*, 2207 (1961)
N214 F. A. Cotton and R. Francis, *J. Amer. Chem. Soc.*, **82,** 2986 (1960)
N215 D. W. Meek, D. K. Straub and R. S. Drago, *J. Amer. Chem. Soc.*, **82,** 6013 (1960)
N216 Ref. N72, Second Edition, p. 395
N217 I. R. Lantzke and D. W. Watts, *Australian J. Chem.*, **19,** 1821 (1966)
N218 Ref. N72, Second Edition, p. 390
N219 U. Belluco, M. Martelli and A. Orio, *Inorg. Chem.*, **5,** 1125 (1966)
N220 U. Belluco, M. Martelli and A. Orio, *Inorg. Chem.*, **5,** 1370 (1966)
N221 J. F. Bunnett, *Ann. Rev. Phys. Chem.*, 271 (1963)
N222 C. M. Harris, S. E. Livingstone and I. H. Reece, *J. Chem. Soc.*, 1505 (1959)
N223 L. Cattalini, V. Ricevuto, A. Orio and M. L. Tobe, *Inorg. Chem.*, **7,** 51 (1968)
N224 W. H. Baddley, F. Basolo, H. B. Gray, C. Nolting and A. J. Poë, *Inorg. Chem.*, **2,** 921 (1963)
N225 W. H. Baddley and F. Basolo, *Inorg. Chem.*, **3,** 1087 (1964)
N226 L. Cattalini, V. Ricevuto, A. Orio and M. L. Tobe, *Inorg. Chem.*, **7,** 51 (1968)
N227 L. Cattalini, R. Ugo and A. Orio, *J. Amer. Chem. Soc.*, **90,** 4800 (1968)
N228 R. J. Angelici, *Organometal. Chem. Rev.*, **3,** 173 (1968)
N229 Ref. N72, Second Edition, p. 526
N230 R. J. Angelici and F. Basolo, *J. Amer. Chem. Soc.*, **84,** 2495 (1962)
N231 R. J. Angelici, F. Basolo and A. J. Poë, *J. Amer. Chem. Soc.*, **85,** 2215 (1963)
N232 R. J. Angelici, *Inorg. Chem.*, **3,** 1099 (1964)
N233 Ref. N72, Second Edition, p. 565
N234 F. Calderazzo and F. A. Cotton, *Inorg. Chem.*, **1,** 30 (1962)
N235 R. J. Angelici and J. R. Graham, *J. Amer. Chem. Soc.*, **87,** 5586 (1965)
N236 R. J. Angelici and B. E. Leach, *J. Organometal. Chem.*, **11,** 203 (1968)
N237 R. F. Heck, *Adv. Chem. Ser.*, **49,** 181 (1965)
N238 R. J. Mawby, F. Basolo and R. G. Pearson, *J. Amer. Chem. Soc.*, **86,** 3994 (1964)
N239 D. E. Morris and F. Basolo, *J. Amer. Chem. Soc.*, **90,** 2536 (1968)
N240 P. B. Chock and J. Halpern, *J. Amer. Chem. Soc.*, **88,** 3511 (1966)
N241 J. P. Collman and C. T. Sears, *Inorg. Chem.*, **7,** 27 (1968)
N242 A. J. Deeming and B. L. Shaw, *Chem. Comm.*, 751 (1968)
N243 L. Vaska, *Accounts Chem. Res.*, **1,** 335 (1968)
N244 M. A. Bennett, R. J. H. Clark and D. L. Milner, *Inorg. Chem.*, **6,** 1647 (1967)
N245 L. Vaska, *J. Amer. Chem. Soc.*, **88,** 5325 (1966)
N246 M. C. Baird, J. T. Mague, J. A. Osborn and G. Wilkinson, *J. Chem. Soc.*, **A,** 1347 (1967)
N247 D. M. Blake and M. Kubota, *Inorg. Chem.*, **9,** 989 (1970)

N248 J. A. Osborn, F. H. Jardine, J. F. Young and G. Wilkinson, *J. Chem. Soc.*, **A,** 1711 (1966)
N249 F. H. Jardine, J. A. Osborn and G. Wilkinson, *J. Chem. Soc.*, **A,** 1574 (1967)
N250 B. R. James and G. L. Rempel, *Disc. Faraday Soc.*, **46,** 48 (1968)
N251 I. Jardine and F. J. McQuillin, *Chem. Comm.*, 477 (1969)
N252 P. Abley and F. J. McQuillin, *Chem. Comm.*, 477 (1969)
N253 K. Vrieze, H. C. Volger and A. P. Praat, *J. Organometal. Chem.*, **15,** 447 (1968)

Chapter 7

Electrode Processes

Part 1

The Electrical Double Layer

Richard Payne
Air Force Cambridge Research Laboratories
L. G. Hanscom Field, Bedford, Massachusetts, U.S.A.

7.1 INTRODUCTION

The object here will be to review in outline the current status of the electrical double layer in organic solvent systems. The emphasis will be on broad conclusions rather than experimental methods and numerical data, details of which can be found in the original papers. In contrast to the more general field of electrode processes in organic media on which a substantial literature exists, the double layer has been neglected by comparison. Less than 40 papers have been published, the majority in recent years. Consequently the field is still in a relatively exploratory stage and many of the conclusions are tentative. As is the case in aqueous solutions most of the work has been concerned with the mercury electrode to which the following discussion refers exclusively.

The first extensive electrocapillary measurements in organic solvents were published by Gouy[P1] in 1906 although references to earlier work are given by Frumkin.[P2] Gouy measured the interfacial tension of mercury in 46 organic solvents and a number of mixtures. The interfacial tension was shown to be lower than the water value by 3–8 μ J cm^{-2} implying that organic solvents interact more strongly with mercury. However, this conclusion is an oversimplification since it neglects important differences in the internal cohesion of different solvents. A more reliable indicator of the strength of the metal-solvent interaction is the work of adhesion defined by the Dupré equation,

$$W_a = \gamma_{\mathrm{Hg}} + \gamma_s - \gamma_i \tag{7.1.1}$$

where γ_{Hg} and γ_s are the surface tensions of mercury and the solvent respectively, and γ_i is the interfacial tension between mercury and solvent. Comparison of the work of adhesion for different solvents indicates some surprising variations in the mercury-solvent interaction which are not always obvious from the interfacial tension. For example, water is evidently more strongly adsorbed than methanol although the interfacial tension for water is considerably higher. Data for a few solvents are shown in Table 7.1.1; more are given elsewhere.[P3] Solvents like aniline, which can undergo a specific interaction with mercury through the aromatic π-electron structure[P4] and others for which the interaction is not so obvious, notably formamide, have large values for the work of adhesion. Interaction of the solvent with the electrode is obviously an important factor to be considered in relation to adsorption from solution.

The capacity of the double layer depends strongly on the solvent but there is a qualitative similarity in the shape of capacity-potential curves in all solvents including water. Early measurements seemed to indicate qualitative differences between water and such non-aqueous solvents as methanol, ethanol and ammonia which have featureless capacity curves in contrast to the characteristic humped curves found in water. However, more recent studies have shown that capacity humps occur commonly in solvents of widely differing types. They are found in solvents of high and medium dielectric constant and probably have a common origin in field reorientation of solvent dipoles.

The absence of a proportionality between the capacity and the macroscopic dielectric constant was demonstrated many years ago by Frumkin[P2] and has since been fully confirmed. However, the validity of the use of the macroscopic dielectric constant in calculations of ionic interaction in the diffuse part of the double layer has been confirmed for a number of solvents. The major influence of the solvent on the capacity seems to be its effect on the average thickness of the compact layer. This

Table 7.1.1

Work of Adhesion of Organic Liquids on Mercury

Calculated from eqn. 7.1.1 using surface tension values at 20°C from *Handbook of Chemistry and Physics* (The Chemical Rubber Company, Cleveland, 1965) and interfacial tension values at 18°C from Ref. P1 except as otherwise stated. Surface tension of mercury taken as 48.4μJ cm^{-2} (C. Kemball, *Trans. Faraday Soc.*, **42,** 526 (1946).

Solvent	Surface Tension μJ cm^{-2}	Interfacial Tension μJ cm^{-2}	Work of Adhesion μJ cm^{-2}
Water	7.2 (25°C)	42.5 (25°C)	13.1
Methanol	2.2 (25°C)	39.3	11.4
n-Propanol	2.4	38.4	12.4
Formic acid	2.8	39.9	12.3
Glycerol	6.3	33.5	16.3
Formamide	5.8	38.9 (25°C)*	15.4
Dimethylformamide	3.6†	37.6 (20°C)†	14.4
Dimethylsulphoxide	4.5 (25°C)‡	37.1 (25°C)§	15.8
Sulpholane	5.3 (30°C)‖	37.4 (30°C)‖	16.4
Aniline	4.3	35.7	17.0
Pyridine	3.8	36.3	15.9

* R. Payne.[P28]
† V. D. Bezuglyi and L. A. Korshikov.[P29]
‡ *Technical Information on Dimethylsulphoxide* (Crown Zellerbach Corporation).
§ R. Payne.[P38]
‖ J. Lawrence and R. Parsons.[P37]

is shown for groups of related solvents where the capacity is roughly inversely proportional to the molecular weight of the solvent.

Specific adsorption of ions from organic solvents is broadly similar to that from aqueous solutions. Anions are more or less strongly adsorbed in the same general order as in water. Variations in the strength of adsorption for a given anion can usually be attributed to solvation effects. Cation adsorption, which cannot normally be detected in aqueous solutions by thermodynamic methods, is more evident in organic solvents. Adsorption of organic molecules from organic solvents is much weaker than from aqueous solutions as would be expected. Adsorption phenomena are discussed further below.

7.2 SOLVENT TYPES

7.2.1 Alcohols

Values of the interfacial tension at the electrocapillary maximum for 13 alcohols are given by Gouy[P1] but systematic measurements have been made only in methanol,[P2,P3,P5–17] ethanol,[P2,P9,P13,P18–20] *n*-propanol[P19] and *n*-butanol.[P21] Extensive electrocapillary and capacity measurements have been reported for methanol.

The work of adhesion of aliphatic alcohols on mercury is generally lower than that of water and most other solvents, and increases with molecular weight (Table 7.1.1). The interfacial tension decreases in the same order. Alcohols are strongly preferentially adsorbed from aqueous solutions close to the electrocapillary maximum.[P2,P14–19,P21] This is due to the tendency of water to 'squeeze-out' the alcohol rather than specific interaction with the metal, as shown by the close similarity of the surface excesses at the mercury-solution and air-solution interfaces.[P14] The interfacial tension in alcohol solutions is lowered by addition of water at sufficiently negative potentials where preferential adsorption of the water molecule occurs due to its small size and large dipole moment.[P2,P18,P19] The displacement of the alcohol by water is more marked the higher the molecular weight of the alcohol, as would be expected, and gives rise to large desorption peaks on the capacity-potential curve in *n*-butanol.[P21]

Anions are strongly adsorbed from alcohol solutions in the same general order as from water. Frumkin's early results for methanol shown in Fig. 7.2.1 are typical. Anion adsorption appears to be somewhat weaker in methanol than in water according to the relative lowering of the interfacial tension. As in water, no significant specific adsorption of fluoride ions can be detected from capacity measurements.[P6] On the other hand specific adsorption of certain cations, e.g. Cs^+ and NH_4^+, which is difficult to detect in water, is evident in methanol.[P6,P10] A few studies of nonionic adsorption from methanol[P7,P11] and ethanol[P20] have been reported. Adsorption of organics is weaker from alcoholic solution than from water as would be expected since the interaction of the solute with an organic solvent should be stronger than with water.

The double layer capacity decreases in the primary alcohol series with increasing molecular weight.[P3] The lower capacity in methanol compared with water has been attributed to the difference in dielectric constant[P6] but this is probably incorrect.[P2,P3,P22] The temperature-dependent capacity hump, characteristic of aqueous solutions, is not observed in methanol even at temperatures as low as -30°C, although effects which may be related have been recorded.[P3,P15] The preferred orientation of alcohol dipoles on mercury is uncertain but the available experimental evidence is generally consistent with the negative (toward the metal) orientation.[P3,P7]

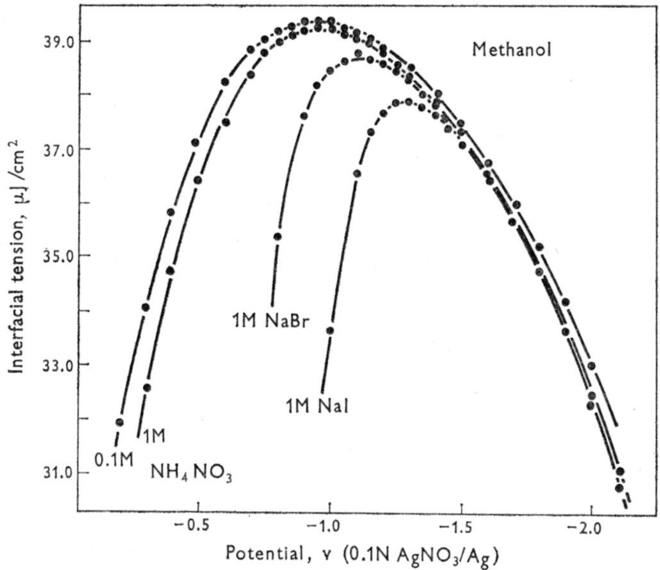

FIG. 7.2.1 Electrocapillary curves for methanol at 8°C (after Ref. P2).

7.2.2 Amides

Capacity measurements in formamide,[P23] *N*-methylformamide (NMF)[P24] and dimethylformamide (DMF)[P23] were first described in 1961. The results differed from earlier non-aqueous solution measurements insofar as capacity humps were found in formamide and NMF (Fig. 7.2.2). Humps had previously been found only in aqueous solution and had been attributed to some special property of water.[P25] The amide hump occurs on the negative side of the potential of zero charge (pzc) in contrast to the water hump which is on the positive side. Recent measurements[P22] in a range of *N*-alkyl formamides, acetamides and propionamides show that the cathodic hump occurs generally in amides having at least one unsubstituted *N*-hydrogen atom. In fully substituted solvents like dimethylformamide and dimethylacetamide (DMA) the cathodic hump is absent but a different hump occurs on the anodic side of the pzc. In the *N*-alkyl acetamides and *N*-methylpropionamide both humps appear on the same capacity curve (Fig. 7.2.3). The two humps have been tentatively attributed to solvent dipole reorientation effects but the evidence for this is inconclusive.[P22]

The dielectric constants of these solvents vary over a wide range but there is no correlation with the double layer capacity as shown by the

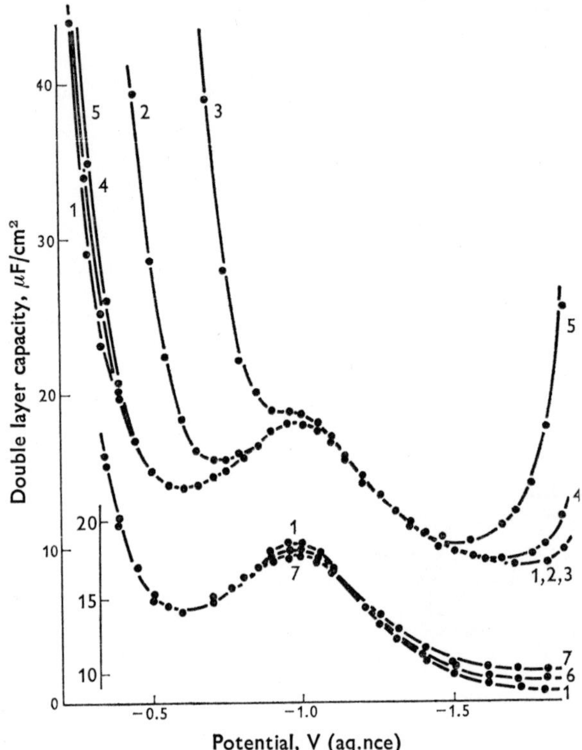

FIG. 7.2.2 Differential capacity curves for 0.1 mol l^{-1} solutions in NMF at 25°C; (1) KCl, (2) KBr, (3) KI, (4) RbCl, (5) CsCl, (6) NaCl, (7) LiCl.[P24]

results given in Table 7.2.1. On the other hand the capacity decreases with molecular weight of the solvent suggesting a proportionality with the thickness of the monolayer of solvent molecules constituting the inner region of the double layer.

Specific adsorption of anions occurs in the same general order as in water.[P12,P23,P24,P26–31] Wide variations in the strength of adsorption among different solvents have been attributed to differences in the solvation energy.[P22,P26] Solvents with unsubstituted *N*-hydrogen (e.g. formamide) are able to solvate anions through a hydrogen-bonding mechanism whereas fully substituted amides such as DMF are not.[P32] Consequently simple anions are considerably more strongly adsorbed from DMF than from formamide.[P3,P22] Specific solvation effects are also apparent in a comparison of water with DMF. Thus halide anions are more strongly adsorbed from DMF than from water whereas the reverse is true for polynuclear anions such as SCN^- and ClO_4^-.[P3,P22]

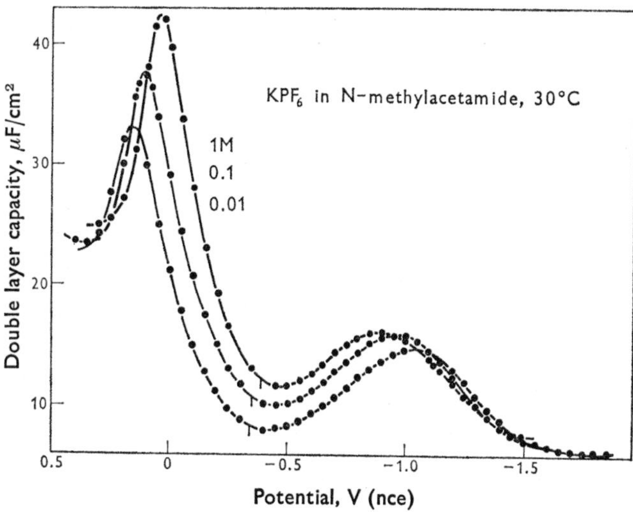

FIG. 7.2.3 Capacity Curves for KPF_6 in NMA at 30°C.[P22]

There is no significant specific adsorption of fluoride ions from either formamide[P33,P34] or NMF.[P33] However, specific adsorption of alkali metal cations, especially Cs^+, is evident both from electrocapillary and capacity measurements in most of these solvents.[P22–24,P26,P29,P35] A study of specific adsorption of Cs^+ and I^- ions in DMF solutions was reported in a recent paper.[P36]

Table 7.2.1

Physical Properties and Double-Layer Capacity in Amide Solvents at 25°C

Solvent	C_{min}* μF/cm²	Ref.	Mol. wt.	Length of† molecule, Å	Dielectric constant
Formamide	12.5 (0.1*M* KCl)	P26	45	5.6	109.5
N-Methylformamide	8.5 (0.1*M* KCl)	P24	69	6.4	182.4
Dimethylformamide	6.8	P22	73 } isomers	6.7	36.7
N-Methylacetamide	6.7 (30°)	P22	73 } isomers	6.8	178.9
N-Methylformamide	9.5	P22	73 } isomers	7.7	
Diethylacetamide	6.0	P22	87 } isomers	6.9	37.8
N-Methylpropionamide	5.6	P22	87 } isomers	7.8	176
N-Ethylacetamide	5.6	P22	87 } isomers	7.7	129.9
N-*t*-Butylformamide	8.4	P22	101	7.5	
N-*n*-Butylacetamide	5.2	P22	115	9.6 or 7.0‡	101.7

* C_{min} is measured at the minimum on the cathodic branch of the capacity-potential curve for 0.1*M* KPF solutions except where otherwise stated.

† Measured along axis of the dipole assuming a planar molecule.

‡ *n*-Butyl group and oxygen atom in *cis* (7.0 Å) or *trans* (9.6 Å) configuration.

Amides are preferentially adsorbed from aqueous solutions.[P22,P24,P26,P28] Maximum adsorption occurs close to the pzc. The controlling factor appears to be the 'squeezing-out' effect of the water rather than specific interaction of the amide molecule with the electrode even though this apparently occurs according to the high work of adhesion of formamide and DMF (Table 7.1.1).

7.2.3 Dimethylsulphoxide and Sulpholane

Dimethylsulphoxide (DMSO) and sulpholane are related solvents with similar physical and double layer properties. They are normal polar liquids with dielectric constants of 46.7 and 43.3 (30°C) respectively. According to the work of adhesion (Table 7.1.1) both solvents appear to undergo some kind of specific interaction with mercury the nature of which is not clear. This is confirmed for sulpholane which was shown to be more strongly adsorbed at the mercury-solution interface than at the air-solution interface.[P37] A capacity hump occurs in both solvents on the anodic side of the pzc.[P37,P38] The origin of the hump has not been fully investigated but, as in the case of the water and formamide systems, it appears to be a property of the solvent rather than of the electrolyte. The location of the hump is consistent with a dipole reorientation interpretation since both the DMSO and sulpholane dipoles appear to be preferentially oriented at the pzc with the positive end facing the electrode. Anions are strongly adsorbed from DMSO in the usual order $I^- > Br^- > Cl^- > ClO_4^- > PF_6^-$ (Fig. 7.2.4). This is consistent with weak solvation of anions as is also indicated by conductance studies. Strong specific adsorption of ClO_4^- ions in sulpholane has been reported.[P37] There is no significant specific adsorption of inorganic cations from DMSO.[P38]

7.2.4 Cyclic Esters and Lactones

A few measurements have been reported for ethylene carbonate, propylene carbonate, 4-butyrolactone and 4-valerolactone.[P39,P40] These compounds are aprotic 5-membered heterocylic ring structures with unusually large dipole moments. Dielectric constants range from 36 for valerolactone to 89.6 for ethylene carbonate. All appear to be normal polar liquids. In all four solvents a capacity hump is observed on the positive side of the pzc (Fig. 7.2.5). The temperature coefficient of capacity at the hump in the butyrolactone solution is only one third of its value in DMSO and sulpholane, and less than one tenth of the water value.[P3] Anions appear to be specifically adsorbed as in other solvents. Strong specific adsorption of ClO_4^- anions from propylene carbonate has been reported.[P40] Butyrolactone is strongly adsorbed from aqueous solutions.[P28]

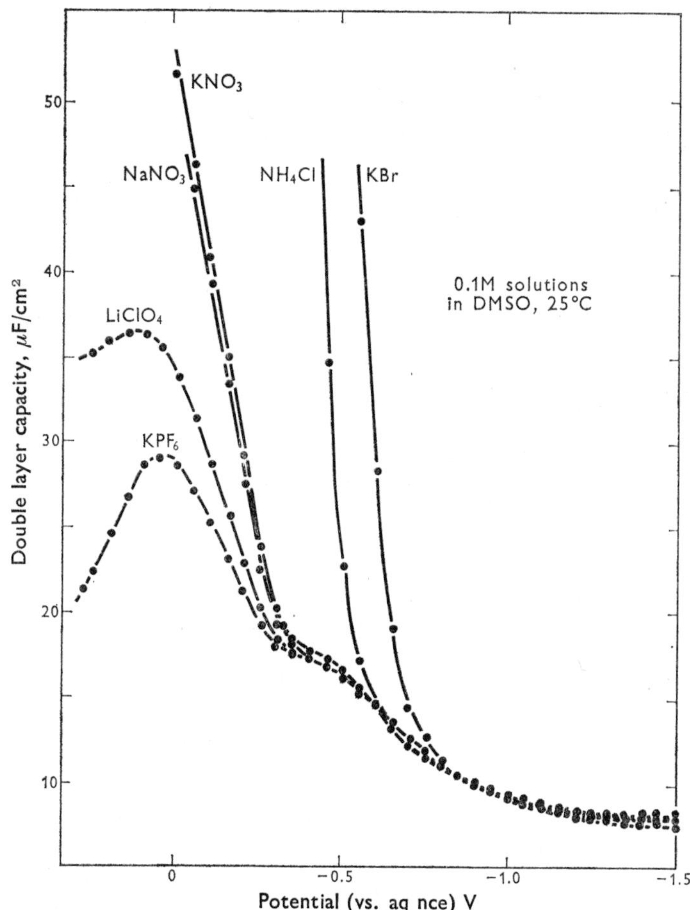

FIG. 7.2.4 Differential Capacity Curves for some electrolytes in DMSO at 25°C.[P38]

7.2.5 Formic Acid

Measurements in the formic acid system were reported by Lawrence and Parsons.[P41] The interfacial tension in formic acid (39.90 μJ cm^{-2} at the electrocapillary maximum) is the highest of any non-aqueous solvent so far investigated. The work of adhesion is correspondingly low in contrast to the high value in formamide (Table 7.1.1). Preferential adsorption of formic acid from aqueous solutions occurs to a similar extent at both the (uncharged) mercury-solution interface and the air-solution interface. The formic acid dipole appears to be preferentially oriented with the negative end facing the electrode. The only anions to be investigated were formate and dihydrogen-phosphate neither of which are appreciably specifically adsorbed. However, there is some evidence of specific adsorption of Cs^+ ions. Capacity curves in formic acid resemble the formamide curves but no hump is observed and the capacity is generally lower. The adsorption of *n*-butyl ether from formic acid was studied in some detail.

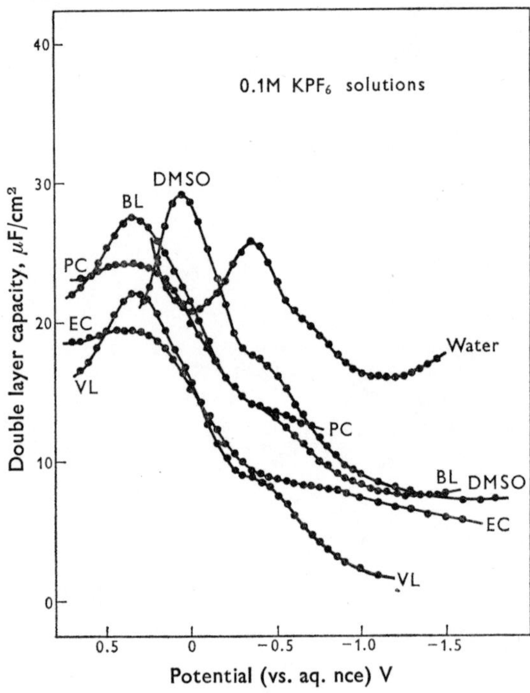

FIG. 7.2.5 Differential Capacity Curves for 0.1 mol l^{-1} KPF_6 in some solvents.[P39]

7.2.6 Other Solvents

A few measurements have been reported for acetonitrile which resembles dimethylformamide in its double layer properties.[P23,P39,P42] Somewhat higher capacities found in acetonitrile are probably due to the smaller size of the molecule compared with DMF.

The acetone solvent system was studied by Frumkin.[P2] Electrocapillary measurements showed that NO_3^- ions are much less strongly adsorbed from acetone than from water whereas for Cl^- ions the opposite is true. This again illustrates the importance of specific solvation effects in ionic adsorption.

Frumkin[P2] has also reported electrocapillary measurements for NH_4CNS and NH_4I solutions in pyridine which show that CNS^- and I^- are less strongly adsorbed from pyridine than from water. Pyridine itself is evidently strongly adsorbed on mercury due to interaction of the aromatic nucleus with the metal.[P4] However, the relative importance of ionic solvation and solvent co-adsorption to the anion adsorption process is unknown.

A large number of electrocapillary measurements in amyl alcohol, phenol, furfural, ethyl acetate, aniline, chloroform, propanol, *iso*butanol and diethylether, and in binary and tertiary mixtures of these solvents with each other and with water are given in an early paper by Wild.[P43]

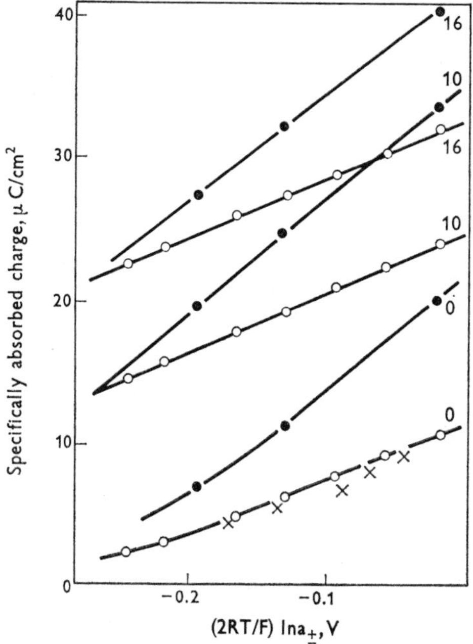

FIG. 7.3.1 Adsorption Isotherm for I^- adsorbed from KI solutions in water (●)[P44], F (○)[P27], and MeOH (×)[P7] at 25°C. Numbers at the end of each curve indicate electrode charge in $\mu C\ cm^{-2}$.

7.3 ADSORPTION FROM ORGANIC SOLVENTS

Detailed adsorption studies in organic solvents have been made for only a few systems. The adsorption of I^- ions from KI solutions has been studied in formamide[P27] and methanol[P7] and can be compared with Grahame's results for aqueous KI.[P44] In all three solvents the adsorption at constant electrode charge is represented by a logarithmic form of isotherm over a wide range of charge and concentration (Fig. 7.3.1). The logarithmic isotherm can be regarded as a limiting form of Frumkin's isotherm,[P45]

$$\ln\frac{\theta}{1-\theta} + A\theta = \ln \beta a \qquad (7.3.1)$$

where θ is the fractional coverage of the electrode with the adsorbed species, a is the activity of the adsorbate in the solution, β is the adsorption equilibrium constant and A is the lateral interaction parameter. When A is large and positive (repulsive interaction) or at intermediate values of θ, eqn. 7.3.1 reduces to the logarithmic form

$$\theta = \frac{1}{A}\ln \beta a \qquad (7.3.2)$$

According to Fig. 7.3.1 the isotherm slopes are approximately equal in formamide and methanol whereas the slope for the aqueous system is considerably larger. The mutual repulsion of adsorbed anions is therefore evidently stronger in methanol and formamide than in water. The interaction parameter is also found to depend strongly on the anion for a given solvent. For example in the formamide system the second virial coefficient (which is directly related to the interaction parameter) for adsorption of I^- ions is 310 Å^2/ion[P27] compared with ~2000 Å^2/ion for Cl^- ions.[P28] Thus the simple adsorption model of point charges undergoing lateral coulombic repulsion represents a considerable oversimplification in non-aqueous solutions as in aqueous solutions. Studies of adsorption of halide anions from mixed electrolyte solutions in formamide[P31] and methanol[P7] reveal complex behaviour which cannot be explained in terms of a simple model.

Parsons and coworkers have studied the adsorption of thiourea from solutions in formamide[P34] and methanol.[P7] The adsorption is rather weaker in both solvents than in water.[P46] For example the amounts adsorbed at the pzc for a 0.5*M* solution are 1.2, 1.0 and 2.0 × 10^{14} molecule cm^{-2} for methanol, formamide and water respectively. The adsorption of thiourea in all solvents is strongest on a positively charged electrode. Desorption occurs at sufficiently negative potentials, virtually complete in water and methanol but incomplete in formamide at the limit of polarisation. The adsorption of thiourea at constant charge produces a strong negative shift of potential, consistent with a permanent negative (toward the metal) orientation of the dipole. Interaction of the sulphur atom with the electrode appears to be the major contribution to the adsorption energy in all solvents.

REFERENCES

P1 G. Gouy, *Ann. Chim. Phys.*, **9(8),** 75 (1906)
P2 A. N. Frumkin, *Z. Elektrochem.*, **103,** 43 (1923)
P3 R. Payne *in* Advances in Electrochemistry and Electrochemical Engineering, Vol. 7, p. 1 (P. Delahay, ed.), Interscience, New York (1970)
P4 A. N. Frumkin and B. B. Damaskin *in* Modern Aspects of Electrochemistry No. 3, p. 149 (J. O'M. Bockris, ed.), Butterworths, London (1964)
P5 A. A. Moussa, H. M. Sammour and H. A. Ghaly, *J. Chem. Soc.*, 1269 (1958)
P6 D. C. Grahame, *Z. Elektrochem*, **59,** 740 (1955)
P7 J. D. Garnish and R. Parsons, *Trans. Faraday Soc.*, **63,** 1754 (1967)
P8 S. Minc and J. Jastrzebska, *Rocz. Chem.*, **31,** 1339 (1957); *Dokl. Akad. Nauk. SSSR*, **120,** 114 (1958)
P9 S. Minc and J. Jastrzebska, *J. Electrochem. Soc.*, **107,** 135 (1960)
P10 S. Minc and M. Brzostowska, *Rocz. Chem.*, **34,** 1109 (1960)
P11 A. M. Murtazaev and M. Abramov, *Zh. Fiz. Khim.*, **13,** 350 (1939)
P12 S. Minc and J. Jastrzebska, *Rocz. Chem.*, **36,** 1901 (1962)
P13 V. F. Ivanov, B. B. Damaskin, N. I. Peshkova, A. A. Ivashchenko and V. F. Balashov, *Elektrokhimiya*, **4,** 851 (1968)
P14 R. Parsons and M. A. V. Devanathan, *Trans. Faraday Soc.*, **49,** 673 (1953)
P15 R. Payne, Thesis, University of London (1962)
P16 A. A. Moussa, H. M. Sammour and H. A. Ghaly, *Egypt. J. Chem.*, **1,** 165 (1958)
P17 S. Minc and J. Jastrzebska, *Rocz. Chem.*, **31,** 735 (1957)
P18 C. Ockrent, *J. Phys. Chem.*, **35,** 3354 (1931)
P19 R. S. Maizlish, I. P. Tverdovsky and A. N. Frumkin, *Zh. Fiz. Khim.*, **28,** 87 (1954)
P20 G. A. Korchinskii, *Ukr. Khim. Zh.*, **28,** 473 (1962)
P21 P. A. Kirkov, *Zh. Fiz. Khim.*, **34,** 2375 (1960)
P22 R. Payne, *J. Phys. Chem.*, **73,** 3598 (1969)
P23 S. Minc, J. Jastrzebska and M. Brzostowska, *J. Electrochem. Soc.*, **108,** 1160 (1961)
P24 B. B. Damaskin and Yu. M. Povarov, *Dokl. Akad. Nauk SSSR*, **140,** 394 (1961)
P25 D. C. Grahame, *J. Amer. Chem. Soc.*, **79,** 2093 (1957)
P26 B. B. Damaskin, R. V. Ivanova and A. A. Survila, *Elektrokhimiya*, **1,** 767 (1965)
P27 R. Payne, *J. Chem. Phys.*, **42,** 3371 (1965)
P28 R. Payne, unpublished data
P29 V. D. Bezuglyi and L. A. Korshikov, *Elektrokhimiya*, **1,** 1422 (1965)
P30 V. D. Bezuglyi and L. A. Korshikov, *Elektrokhimiya*, **3,** 390 (1967)
P31 E. Dutkiewics, *Rocz. Chem.*, **41,** 1965 (1967)
P32 A. J. Parker, *Quart. Rev.*, **16,** 163 (1962)
P33 S. Minc and M. Brzostowska, *Rocz. Chem.*, **38,** 301 (1964)
P34 E. Dutkiewicz and R. Parsons, *J. Electroanal. Chem.*, **11,** 196 (1966)
P35 G. H. Nancollas, D. S. Reid and C. A. Vincent, *J. Phys. Chem.*, **70,** 3300 (1966)
P36 Ya. Doilido, R. V. Ivanova and B. B. Damaskin, *Elektrokhimiya*, **6,** 3 (1970)
P37 J. Lawrence and R. Parsons, *Trans. Faraday Soc.*, **64,** 751 (1968)
P38 R. Payne, *J. Amer. Chem. Soc.*, **89,** 489 (1967)
P39 R. Payne, *J. Phys. Chem.*, **71,** 1548 (1967)
P40 T. Biegler and R. Parsons, *J. Electroanal. Chem.*, **21,** App. 4–6 (1969)
P41 J. Lawrence and R. Parsons, *Trans. Faraday Soc.*, **64,** 1656 (1968)
P42 G. A. Korchinskii, *Ukr. Khim. Zh.*, **28,** 693 (1962)
P43 H. Wild, *Z. Elektrochem*, **103,** 1 (1923)
P44 D. C. Grahame, *J. Amer. Chem. Soc.*, **80,** 4201 (1958)
P45 A. N. Frumkin, *Z. Phys. Chem.* (*Leipzig*), **116,** 466 (1925)
P46 R. Parsons, *Proc. Roy. Soc.* (*London*), **261A,** 79 (1961)

Chapter 7

Electrode Processes

Part 2

Electrode Reactions

O. R. Brown
Electrochemistry Research Laboratory,
Department of Physical Chemistry,
University of Newcastle upon Tyne,
Newcastle upon Tyne, NE1 7RU, England

7.4 INTRODUCTION

Since about 1955, organic solvents have been increasingly used as media for the operation and study of electrode processes. They now find application in many areas of electrochemistry, e.g. power sources, analysis and synthesis. The principal reasons for choosing organic solvents are usually one or more of the following:

1. Good solvent properties, particularly towards organic reactants and products.
2. The stability of species which would decompose by chemical reaction with other solvents.
3. The desired participation of the solvent in the overall reaction.
4. The stability of the solvent towards electrolytic oxidation or reduction so that extreme values of electrode potential may be obtained, permitting the occurrence of charge transfer processes which would be precluded in aqueous media by the more facile evolution of oxygen or hydrogen.

A consequence of these factors is that quite different electrode processes are possible in non-aqueous media from those which occur in aqueous solutions. It is with such processes that these sections are chiefly concerned.

Several other factors are also responsible for important differences in electrochemical behaviour in different solvent systems. More viscous solvents impede the mass transfer processes, by which reactants are brought to the electrode surface and products are swept away. In such solvents the solution resistivities are high with consequently large heating effects, power wastage and the exacerbation of experimental problems associated with the maintenance of uniform electrical conditions across the electrode surface. However, the choice of viscous solvents such as glycerol or sulpholane is advantageous when diffusion phenomena are under study. Solution resistivity problems also arise when the solvent fails to promote complete dissociation of the electrolyte or permits extensive ion association.

Other questions which arise when the use of organic solvents is contemplated concern the temperature range of the liquid phase, chemical stability, toxicity, vapour pressure, and the elimination of impurities which are either electroactive or interact chemically with components of the system (see sect. 1.3). The solvent must also be capable of dissolving appreciable amounts (0.1M to 10M depending upon the application) of a suitable supporting electrolyte, which is often required to remain chemically and electrochemically inactive in the system.

Of considerable interest is the effect of the solvent upon the rate constants of both the charge-transfer steps and any coupled chemical steps in the overall electrode process. In addition the dependence of the double layer structure upon the nature of the solvent determines both the surface concentration and orientation of species involved in the electrode process. Whether the reactant is concentrated at, or expelled from, the interface depends upon both the electrostatic and specific chemical interactions. The latter involve solvent molecules, either directly or indirectly, through interactions of both solvent and reactant with other components in the system. Adsorption is important in affecting reaction rates, not only in cases where the reaction mechanism involves overlap between the orbitals of the reactant and the electrode material (heterogeneous catalysis), but also in simple electron transfer reactions where the probability of transfer decreases rapidly with their increasing separation.

An attempt will be made to review electrode processes in organic solvents with minimal reference to electrochemical techniques; conclusions are reported without the inclusion of detailed electrochemical arguments which are not immediately understood by the non-electrochemist. However, the section on electrode kinetics requires a grasp of electrochemical concepts; excellent standard textbooks[Q1,Q2] are recommended to the reader unfamiliar with this field.

The abundance of work in this area of electrochemistry precludes any attempt at comprehensive coverage in the space presently available. Indeed fully representative coverage is not claimed; instead some of the factors already mentioned will be illustrated by examples which particularly interest the author. Some topics such as electrochemiluminescence, which is reviewed elsewhere,[Q3] will not be treated. A fairly comprehensive source of literature references to electroanalytical and mechanistic aspects is provided by the recent book of Mann and Barnes.[Q4]

7.5 RANGES OF ELECTRO-INACTIVITY OF ORGANIC SOLVENTS

Compared with water, most of the commonly used organic solvents are stable toward decomposition at the electrode under more severe anodic or cathodic conditions or both. Mann's review[Q5] provides a valuable compilation of electro-inactive ranges, physical properties, purification, toxicity, solubilities of salts, etc. In general an essential requirement of an electrolytic solvent is a high (>20) dielectric constant in order that supporting electrolytes should dissolve. There are, however, exceptions, e.g. monoglyme has the ability to solubilise metal ions by chelation,[Q6] whereas nitrobenzene is unable to use fully its high dipole moment for solvation purposes.[Q7] Here we shall consider only the decomposition potentials of organic solvents. In general the anodic and cathodic 'limits' depend also upon the background electrolyte and the electrode material. Comparison of potentials in different solvents can not be made accurately if liquid-junction potentials are involved. Instead of using the aqueous calomel electrode as reference, either the ferrocene-ferrocinium couple or the rubidium ion-rubidium amalgam electrode can be used in the solvent being examined.[Q8] This subject is discussed in detail in Chapter 2. The decomposition potential is commonly taken as that at which a current density of 1 mA cm^{-2} is recorded.

In acetonitrile, probably the most commonly used organic solvent, both limits are determined by the decomposition of the support electrolyte. The largest cathodic range (−3.5 V *vs.* Ag|0.01*M* $AgClO_4$, 0.1*M* $LiClO_4$) is given by a $LiClO_4$ supporting electrolyte[Q9] and the largest anodic range by hexafluorophosphates or tetrafluorborates.[Q10–12] In each case platinum electrodes are used; mercury electrodes reduce the cathodic limit owing to amalgam formation and the anodic limit is greatly restricted by dissolution of the metal.

Propylene carbonate also possesses a wide potential range;[Q13] with respect to the Ag|$AgClO_4$ 0.01*M* reference electrode, the anodic limit on platinum is 3.2 V when KPF_6 is used and +2.2 V when perchlorates are employed. Lithium can be deposited at −3.5 V whereas the use of Bu_4NBr extends the limit to −4.0 V. At concentrations below 0.1*M*, water does not affect the anodic range but it causes a small current peak to appear at −2.6 to −2.7 V.

N,N-Dimethylformamide, DMF, a popular, inexpensive aprotic solvent, has a limited anodic range, being itself oxidised, but its cathodic range equals that of acetonitrile. Its decomposition potential is remarkably unaffected by substantial concentrations of water in the solvent[Q5] probably due to extensive hydrogen-bonded incorporation into the DMF structure. Dimethylsulphoxide, DMSO, also has a restricted anodic range and a cathodic limit determined by discharge of the cation. Sulpholane[Q14,Q15] gives a wide cathodic range (−3.5 V *vs.* Ag|$AgClO_4$ 1.0*M* when 0.1*M* tetraethylammonium perchlorate electrolyte is used). Pyridine is a useful solvent for non-polar materials although it is reducible at very negative potentials.[Q16] The acids and alcohols have limited cathodic ranges owing to hydrogen evolution.

7.6 CATHODIC GENERATION OF SOLVATED ELECTRONS

Several organic solvents, notably aliphatic amines and ethers are stable over prolonged periods towards solvated electrons, which can be generated cathodically by the discharge of alkali metal cations. The characteristic blue colour of the solvated electron is readily formed near cathodes of platinum, graphite or carbon in solutions of LiCl in ethylene diamine[Q17,Q18] or methylamine.[Q19] The colour is much less obvious when aromatic species are present; they are reduced to their radical anions which are then protonated by the amine solvent to yield a radical which is readily reduced further. The product, a dihydrocompound, may in the case of polycyclic aromatics be reduced still further. Benzene derivatives are only reduced further if the 1,4-dihydrocompound rearranges to the 1,2-dihydroisomer, a transition which is facilitated by the strongly basic amide ion remaining from the protonation reactions. In an undivided cell, the amide ion is neutralised rapidly by acid generated at the anode. The replacement of LiCl by NH_4Cl eliminates the reduction of benzene in favour of hydrogen evolution, but the use of LiCl + NH_4Cl or alternatively of Bu_4NI as electrolyte does permit some aromatic hydrogenation to proceed. Clearly in these solvents, as in ammonia, the solvation energy of the electron when accompanied by a suitable cation is considerable. Thus the influence of the electrode becomes extended into the bulk solution as electrons are transported away from the surface. Consequently the total rate of any reaction involving them is greatly increased and processes which would proceed only imperceptibly if confined to the interface occur in solution at considerable rates.

Misono *et al.* have reported[Q20] similar reactions in diglyme (1,2-dimethoxyethane) at a mercury cathode with Bu_4NBr electrolyte and added water. In a divided cell, substantial yields of 1,4-dihydrobenzene and 2,5-dihydrotoluene are produced from benzene and toluene respectively. No 1,4-dihydrotoluene or 1,2-dihydrobenzene were found although a little cyclohexene was detected. The current-voltage curves are essentially unaffected by the presence of the aromatic substance in the system, indicating the intermediacy of the solvated electron and the relatively slow nature of the reduction process.

Hexamethylphosphotriamide, HMPT, containing LiCl stabilises solvated electrons permitting their cathodic formation on aluminium at quite positive potentials. Surprisingly large amounts (66 mole %) of ethanol can be added without causing excessively rapid hydrogen evolution.[Q21] In this medium not only benzene but also olefinic bonds[Q22] can be reduced, ethanol acting as proton donor. In the absence of HMPT, the ethanolic solution evolved hydrogen at potentials 800 mV more positive than those previously required to form the solvated electron. The mechanism of the suppression of the hydrogen evolution reaction by HMPT is not fully understood. The presence of ethanol is essential for the reduction of aromatics in HMPT. The electroreduction of benzene to 1,4-dihydrobenzene in this medium has been patented.[Q23]

Naphthalene can be reduced in HMPT *via* solvated electrons formed at potentials more positive than permit the cathodic reduction of naphthalene in other solvent systems.[Q24]

7.7 DIRECT CATHODIC REDUCTIONS OF UNSATURATED ORGANIC MATERIALS

In aprotic organic solvents, e.g. AN, DMF, PC, HMPT and DMSO, the available cathodic potential range is so large that all unsaturated hydrocarbon groups, with the exception of unconjugated olefinic bonds and isolated benzene rings, can be made to receive an electron from the cathode into their lowest vacant molecular orbital.[Q25] The resulting radical anions can undergo several following chemical reactions of which the commonest is protonation. The rates of these reactions depend upon (a) the nature of the counter ion, alkali metal ions undergoing strong interactions with the anions, (b) the nature of the radical anion itself, in particular whether or not it contains centres of high charge density, and (c) the nature of the solvent.

Anthracene and naphthalene radical anions, in the presence of quaternary ammonium counter ions, are moderately stable when proton donors are absent. If phenols are added, the radical ion $R^{\cdot-}$ decomposes according to the scheme

$$R^{\cdot-} + X\mathrm{H} \rightarrow R\mathrm{H}^{\cdot} + X^-$$
$$R\mathrm{H}^{\cdot} + R^{\cdot-} \rightarrow R\mathrm{H}^- + R$$
$$R\mathrm{H}^- + X\mathrm{H} \rightarrow R\mathrm{H}_2 + X^-$$

Water also is an effective proton donor in acetonitrile solution. In DMF or DMSO, however, small amounts of water interact with the solvent in such a manner that it is not available for protonation which consequently occurs only slowly.[Q26]

Another chemical reaction of radical anions is the addition of carbon dioxide.[Q27,Q28]

$$R^{\cdot-} + \mathrm{CO}_2 \rightarrow R\mathrm{CO}_2^{\cdot-}$$
$$R\mathrm{CO}_2^{\cdot-} + R^{\cdot-} \rightarrow R + R\mathrm{CO}_2^{=}$$
$$R\mathrm{CO}_2^{=} + \mathrm{CO}_2 \rightarrow R(\mathrm{CO}_2^-)_2$$

If either carbon dioxide or proton donors are present in the catholyte during the formation of the radical anion and if the first chemical step is sufficiently rapid then the succeeding reduction is electrochemical instead of the chemical disproportionation.

Those radical anions in which the charge is highly concentrated, and also dianions which can be produced from radical anions at more negative potentials, abstract hydrogen atoms rather indiscriminately from the solvent or quaternary ammonium cations. Tetraphenylallene and tetraphenylpropene radical anions protonate rapidly under conditions (0.1 M $\mathrm{Bu_4NClO_4}$ in DMF) where those of planar aromatic hydrocarbons are stable.[Q29] Radical anion dimerisation reactions are not general but have been reported for both stilbene and phenanthrene.[Q28] In both cases, in the presence of CO_2, the addition reaction occurs more rapidly than the dimerisation. In a few cases radical anions reduce to dianions more readily than they are formed, e.g. the tetraphenylethylene radical anion cannot assume a planar configuration and is relatively unstable. Some of the resulting dianions dissociate to form diphenylmethane. The remainder are reduced to tetraphenylethane.[Q27]

The extent of radical anion-solvent interactions has been considered by Peover by comparing $E_{\frac{1}{2}}$ potentials in different organic solvents.[Q30] Amongst aprotic solvents the variations can, with few exceptions, be explained purely in terms of solvation energy as predicted by the Born equation. However, when $E_{\frac{1}{2}}$ potentials of quinones in acetonitrile and in alkaline aqueous media are compared, the semiquinone radical anions are seen to be vastly more stable in water, indicating the specific nature of the solvent-radical anion interactions in that system.

The addition of potassium ions to a solution of biphenyl radical anions in wet DMF containing tetrabutylammonium ions increases the rate of protonation.[Q31]

Carbanions, formed by the protonation of electrogenerated dianions of polycyclic hydrocarbons in nominally aprotic media, can be oxidised to radicals when the polarity of the electrode is reversed.[Q32]

The presence of electronegative groups in unsaturated hydrocarbons facilitates their reduction. For example, the presence of a nitrile group adajcent to a single olefinic bond or benzenoid ring enables those systems to accept an electron within the potential range available in the usual solvents. In DMF, benzonitrile forms a radical anion.[Q33] Phthalonitrile behaves similarly but can be reduced further with the elimination of one cyanide ion

$$(C_6H_4(CN)_2)^{\cdot -} + e = (C_6H_4(CN)_2)^{=} \xrightarrow{H^+} CN^- + (C_6H_5CN)^{\cdot -}$$

Acrylonitrile has been reduced in DMF containing tetraalkylammonium iodides; in dilute solutions of the substrate conversion to propionitrile occurs but at higher concentrations dimerisation to adiponitrile prevails.[Q34] β-phenylacrylonitrile forms the hydrodimer, 2,3-diphenyl adiponitrile.[Q35]

When nitro, nitroso or carbonyl groups are present in the organic substrate, the charge of the corresponding radical anion is concentrated in those groups, a fact which influences subsequent chemical reactions. In aprotic media the radical anions may be stable for considerable periods but in the presence of proton donors, reactions are similar to those obtained in aqueous media.

Dianions of aryl ketones in aprotic solutions protonate to become secondary alcohols whereas those of α–β unsaturated ketones form the corresponding saturated ketone.[Q36] Radical anions derived from α–β unsaturated carbonyl compounds cause extensive polymerisation of the substrate.[Q37] Dimerisation at the β-positions occurs in some systems with the formation of diketones.[Q38–40]

An interesting effect occurs with 1,3-diketone radical anions in DMSO in which a second substrate molecule, being the best proton donor available, permits formation of a dimeric product.[Q41]

$$(PhCOCH_2COPh)^{\cdot -} + PhCOCH_2COPh \rightarrow PhCO\bar{C}HCOPh + Ph\dot{C}OHCH_2COPh$$

$$Ph\dot{C}OHCH_2COPh \xrightarrow{\times 2} \begin{array}{c} PhCOHCH_2COPh \\ | \\ PhCOHCH_2COPh \end{array}$$

In the strongly basic environment the product decomposes to acetophenone and benzil which is further reducible at the reduction potential of the starting material. A similar auto-protonation effect occurs in the reduction at mercury of benzyldimethylphenacylammonium bromide in DMF containg LiCl.[Q42]

$$BzMe_2\overset{+}{N}CH_2COPh + e \rightarrow BzMe_2N + PhCO\dot{C}H_2 \xrightarrow{e} PhCO\bar{C}H_2$$

$$PhCO\bar{C}H_2 \xrightarrow{BzMe_2\overset{+}{N}CH_2COPh} CH_3COPh + BzMe_2\overset{+}{N}\bar{C}HCOPh \rightarrow Me_2NCHBzCOPh$$

Phenacyltriphenylphosphonium salts also undergo auto-protonation reactions when reduced in acetonitrile.[Q43]

The reduction of benzil itself is of interest as the presence of lithium ions can cause the initially formed radical anion to be reduced more readily than it is formed.[Q44] Anion pairing with lithium or protonation produce similar effects.

$$PhCOCOPh + e \rightleftharpoons PhCO\dot{C}\bar{O}Ph \overset{Li^+}{\rightleftharpoons} PhCO\dot{C}(\overset{-+}{OLi})Ph \xrightarrow{e} PhC(OLi){=}C(O^-)Ph$$

Heterocyclic aromatic compounds can be reduced to radical anions in organic solvents. Those derived from azines are in general short-lived reacting with further substrate molecules to form polymeric products.[Q45]

Stable electrogenerated nitro aromatic radical anions have been extensively studied by ESR Electrochemical $E_{\frac{1}{2}}$ data have been correlated with variations in coupling constants and interpreted in terms of steric effects, ion-pairing and induction effects.[Q46] ESR provides a particularly sensitive tool for the detection of interactions between radical anions and the medium. Aliphatic nitro radical anions decay by several possible routes. The *t*-nitrobutane radical anion produces some di-*t*-butyl nitroxide[Q48] whereas that derived from *t*-nitrooctane loses nitrite ion to form the octyl radical which gives dimer and disproportionation products.[Q49] In the reductions of compounds containing more than one functional group, particularly if a material which normally gives rise to a radical anion is substituted by a halogen atom, then a common follow-up reaction is the elimination of halide ion to form a neutral radical. As the chemical step has a finite rate constant and radical anions are desorbed from the cathodic surface, the resulting radical is formed in solution and undergoes chemical decomposition, e.g. by dimerisation, hydrogen abstraction and disproportionation routes. Examples include the reductions of halogenated benzophenones,[Q50] nitrobenzyl halides[Q51] and halo-nitrobenzenes.[Q47]

7.8 REDUCTIONS OF ORGANIC HALIDES

The basic mechanism for the reduction of alkyl halides on mercury cathodes is essentially the same in aqueous and in organic media. In general the mechanism can be written

$$RX + e \rightarrow X^- + R^\cdot \quad \begin{cases} \xrightarrow{M} \text{Metal alkyl} \\ \xrightarrow{e} R^- \end{cases}$$

However in aprotic media the reactions of R^- are more varied. Thus in the presence of carbon dioxide, benzyl halide reduction in aprotic solvents produces some phenyl acetate.[Q52] In aqueous solutions only toluene is formed at negative potentials. Rifi has reported[Q53] the synthesis of bicyclobutane derivatives from the corresponding 1,3-dihalo-cyclo butanes at a mercury pool cathode in DMF containing LiBr e.g.

$$\text{1,3-dibromo-1,3-dimethylcyclobutane (Me, Br)} \xrightarrow{2e} 2Br^- + \text{Me–bicyclobutane–Me}$$

Several other cyclopropane rings were synthesised in analogous reactions, e.g. spiropentane formation.

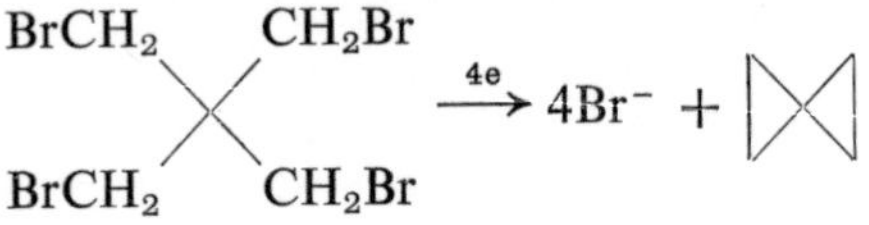

$$C(CH_2Br)_4 \xrightarrow{4e} 4Br^- + \text{spiropentane}$$

Four-membered rings could also be created,

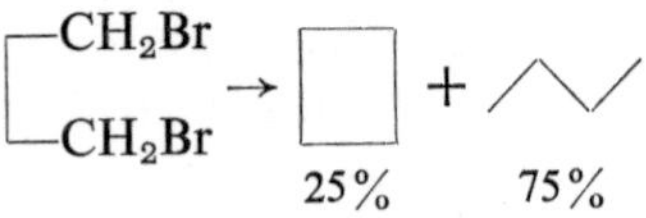

$$BrCH_2CH_2CH_2CH_2Br \rightarrow \text{cyclobutane } (25\%) + \text{butene } (75\%)$$

The mechanism involved is believed to be an internal nucleophilic displacement by a carbanion. In some cases the process is concerted; $E_{\frac{1}{2}}$ measurements show that the transition state for the initial electron transfer to α-ω-dibromopropane is stabilised by interaction between the two ends of the molecule to a degree which cannot be explained as an inductive effect. By contrast, 1,3-dimethyl-1,3-dibromocyclobutane shows no evidence of a concerted mechanism and the carbanionic intermediate can undergo either internal nucleophilic displacement or protonation by a second substrate molecule. The second alternative is favoured in concentrated solutions.

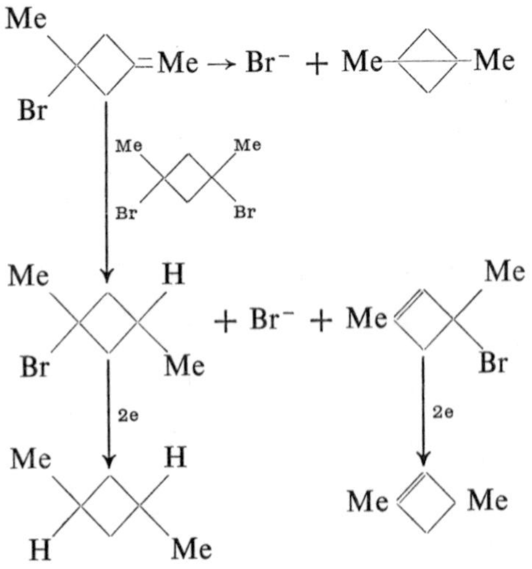

Me, Br (cyclobutane carbanion) =Me → Br⁻ + Me–(bicyclobutane)–Me

↓ (Me, Br, Me, Br cyclobutane)

Me, H, Br, Me cyclobutane + Br⁻ + Me(cyclobutene)Me, Br

↓ 2e ↓ 2e

Me, H, H, Me cyclobutane; Me(cyclobutene)Me

An interesting solvent effect is observed in the reduction of 1-bromomethyl-3 bromocyclobutane:

$$\text{◇Me and (diene)} \xleftarrow{\text{DMF}} \text{Br◇CH}_2\text{Br} \xrightarrow[\text{Bu}_4\text{NClO}_4]{\text{HMPT}} \text{(bicyclobutane)} + \text{(diene)}$$

The choice of organic solvents permits the occurrence of certain reduction reactions of alkyl halides which, if attempted in aqueous solutions, would rapidly be inhibited by the precipitation of insoluble products on the electrode. Under mildly cathodic conditions, the formation of metal alkyls occurs in high yields when alkyl halides are reduced at cathodes of low boiling metals in acetonitrile, DMF or propylene carbonate containing quaternary ammonium supporting electrolytes.[Q54–56] Whilst the formation of lead tetraethyl from ethyl bromide in DMF proceeds irrespective of the substitution of a sodium salt for the quaternary ammonium salt,[Q56] the same change of electrolyte in propylene carbonate no longer leads to metal alkyl products.[Q55]

7.9 ONIUM CATION REDUCTIONS

In DMF, tetraphenylphosphonium and triphenylbutylphosphonium halides are readily reduced polarographically.[Q57] The reactions are not properly characterised but it is believed that products include triphenyl phosphine; a further wave observed at more negative potentials corresponds to the reduction of that compound.

Cyano-substituted sulphonium salts have been reduced in DMSO solution in two successive steps[Q58]

$$R_2\overset{+}{S}(CH_2)_nCN \xrightarrow{e} R_2\dot{S}(CH_2)_nCN \xrightarrow{e} R_2S + (CH_2)_nCN^-$$

The second step gives rise to a polarographic wave only in the presence of proton donors; otherwise a further chemical reaction occurs to form an ylid.

$$(CH_2)_nCN^- + R_2\overset{+}{S}(CH_2)_nCH \rightarrow H(CH_2)_nCN + R_2^+S^-CH(CH_2)_{n-1}CN$$

When $n = 1$ the second reduction step occurs more readily than the first owing to the increased stability conferred on the carbanion by the close proximity of the nitrile group.

Quaternary ammonium salt reductions in DMF, DMA or acetonitrile are of two types. If the cation contains a group which forms a relatively stable radical such as benzyl, then the reaction scheme is

$$R_3NBz^+ + e \rightarrow R_3N + Bz^{\cdot} \begin{array}{l} \nearrow \tfrac{1}{2}Bz_2 \\ \xrightarrow{e} Bz^- \xrightarrow{XH} BzH + X^- \end{array}$$

The dimerisation reaction is favoured by aluminium cathodes at whose partially insulated surfaces high cathodic potentials exist.[Q59,Q60] Simple tetraalkylammonium cations reduce reversibly at mercury to form insoluble compounds of mercury with quaternary ammonium radicals. These compounds, stable only at sub-ambient temperatures. decompose on warming with the formation of hydrocarbon and amine products.[Q61]

7.10 INORGANIC CHEMISTRY

The literature concerned with the polarographic reduction of metal ions on mercury in non-aqueous solvents is vast and has been reviewed by Mann and Barnes.[Q4] Much less effort has been expended in the equally important field of the anodic behaviour of metal ions in such solvents. Redox couples, which may have very different standard potentials in non-aqueous media from those observed in aqueous solutions (e.g. $Fe^{2+}|Fe^{3+}$ ref. Q62), might form the basis of novel oxidative systems.

Several oxidations and reductions involving boron hydrides and their anions, particularly of decaborane, have been carried out electrochemically in aprotic organic solvents.[Q63–66] Decaborane reduces to a mixture of anions $B_{10}H_{13}^-$ and $B_{10}H_{15}^-$. *closo*-Decahydrodecaborate (2-) undergoes oxidative dimerisation.

$$2B_{10}H_{10}^{=} \rightarrow 2e + 2H^+ + B_{20}H_{13}^{=}$$

Sulphur tetranitride S_4N_4 can be reduced in THF[Q67] or in pyridine[Q68] to an anion radical which rapidly decomposes to a secondary anion radical and finally a mixture of anionic products.

The electrochemistry of a large number of transition metal organometallic compounds has been examined in various organic solvents, particularly diglyme.[Q69]

7.11 METAL DEPOSITION

A considerable number of metallic elements cannot be cathodically deposited from aqueous solutions because of prior discharge of water or chemical reaction between the metal and the solvent. The possible production of pure electrodeposits of these 'non-aqueous metals' from organic solvent systems is of technological importance. Brenner has critically reviewed[Q70] the literature in this area and considers doubtful almost all of the many claims of successful deposition of the 'non-aqueous metals'. Whilst it is true that cathodic deposits can be obtained from solutions of many metal salts in several organic solvents, the purity of the deposits has seldom been checked. When analysis has been conducted, the organic content of 'non-aqueous metal' deposits has been found usually to be substantial.

The lithium-propylene carbonate system is of particular interest in high energy density batteries and it has been claimed[Q70] that the metal can be deposited with 100% current efficiency. However, that electrodeposited lithium displays a much less stable open-circuit potential than lithium ribbon electrodes, can be taken as evidence that the former contains impurities.[Q71] Kinetic studies of the $Li|Li^+$ couple in PC have been made in order to understand the nature of the considerable polarisation of that electrode,[Q61]

Sodium, free from other metals, has been deposited from a solution of $NaB(C_2H_5)_4$ in THF at 70°C.[Q70] Of the alkaline earth metals only beryllium has been electrodeposited but the product was highly impure.[Q70] Aluminium has been deposited from several different baths[Q70] but in all cases the metal itself is required to prepare the baths. This process, therefore, may find application in electrorefining but is unsuitable for electrowinning. Attempts to deposit metals from groups IV, V and VI as well as the rare earths, have failed to produce pure deposits.[Q70]

Brenner has suggested[Q70] that organic solvents which provide the best media for the electrodeposition of 'non-aqueous metals' are those with weak co-ordinating centres, e.g. ethyl ether, which loosely co-ordinates with reducible metal alkyls. By contrast, polar solvents like water, form stable complexes which cannot satisfactorily be electrolysed.

7.12 CATHODIC GENERATION OF SUPEROXIDE ION

Molecular oxygen is reduced cathodically in aprotic solvents to the superoxide ion. The reaction in DMSO is quasi-reversible with $k_a \sim 10^{-3}$ cm s^{-1} on gold, mercury and platinum.[Q72]

$$O_2 + e \rightleftharpoons O_2^{\cdot -}$$

Although the disproportionation to peroxide and oxygen is thermodynamically important, it is slow when millimolar concentrations are involved.

$$2O_2^{\cdot -} \rightleftharpoons O_2 + O_2^{=}$$

Likewise the electrochemical formation and oxidation of peroxide ion in aprotic media require large overpotentials. Quaternary ammonium salt supporting electrolytes are chosen for superoxide generation because alkali metal superoxides are remarkably insoluble. The solubility of the tetraalkylammonium superoxides enables their electrochemical generation to provide a unique and useful reagent.[Q73] For example, the addition of an excess of *n*-butyl bromide to the catholyte causes the superoxide to react rapidly by nucleophilic displacement.

$$\mathrm{BuBr} + \mathrm{O_2^{\cdot -}} \rightarrow \mathrm{Br^-} + \mathrm{BuO_2^{\cdot}}$$

This reaction is followed rapidly by

$$\mathrm{BuO_2^{\cdot}} \xrightarrow[+\mathrm{O_2^{\cdot -}} - \mathrm{O_2}]{+e} \mathrm{BuO_2^-} \xrightarrow{\mathrm{BuBr}} \mathrm{Br^-} + \mathrm{BuO_2Bu}$$

At 0°C, in a DMF solution containing 0.2*M* BuBr and saturated with a 5% oxygen-nitrogen gas mixture, the second reduction step was essentially non-electrochemical. The method is general for the formation of dialkyl peroxides; diethyl peroxide can be conveniently prepared by using $\mathrm{Et_2SO_4}$ as both the aprotic solvent and the reactant. From the point of view of practical electrosynthesis, the system $\mathrm{O_2/BuBr/Bu_4NBr}$/DMF is convenient, as the anode reaction,

$$2\mathrm{Br^-} \rightarrow 2e + \mathrm{Br_2}$$

exactly keeps pace with the formation of bromide ion from butyl bromide.

7.13 ANODIC OXIDATION OF ORGANIC MATERIALS

The employment of suitable organic solvents, such as acetonitrile and acetic acid, with oxidation-resistant supporting electrolytes permits the anodic formation of reactive radical cations from many organic materials. Most aromatic compounds and olefins, as well as those alkanes which have particularly weak C—H bonds, are oxidised in acetonitrile containing fluoroborate or hexafluorophosphate electrolytes.[Q10–12] Some aromatic radical cations can be further oxidised to dications within the available potential range.[Q74] Radical cations in general either deprotonate or attack nucleophiles present in the medium; reactions with pyridine, methanol, water, cyanide ion, acetate ion or acetonitrile itself produce addition or substitution products. The complete reactions involve a second electron transfer and coupled chemical steps, e.g. deprotonation. In the case of polycyclic aromatic hydrocarbons, several successive electron transfers may occur. Some examples of anodic reactions of aromatic hydrocarbons are:[Q75,Q76]

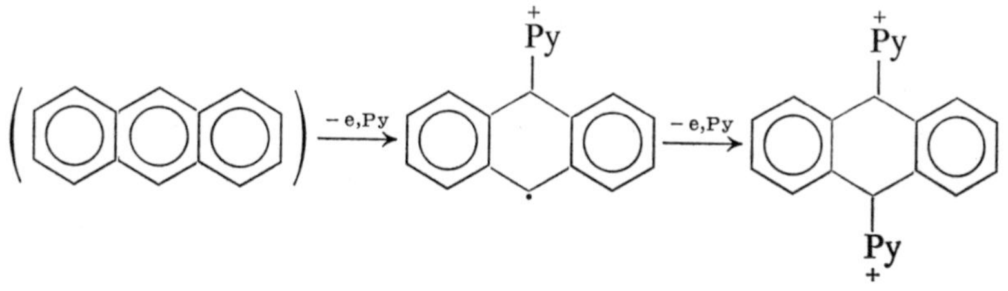

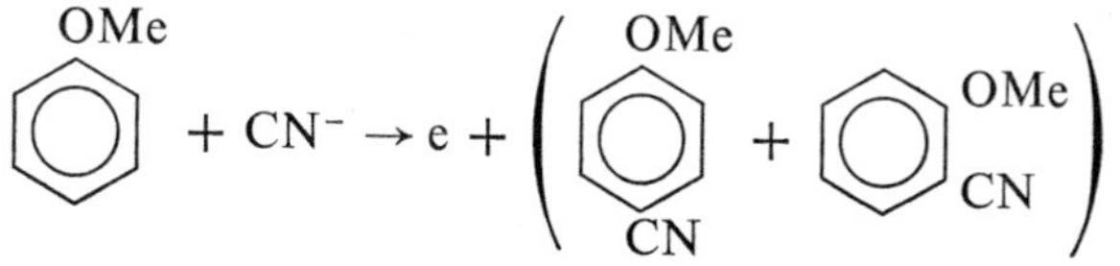

In general, radical cations are less stable than their anionic counterparts, but relatively stable cations can be obtained from reactants in which those positions which carry the highest charge density in the radical are blocked with respect to either proton loss or nucleophilic attack. For example the 9,10-diphenylanthracene and rubrene radical cations are sufficiently stable to give reversible cyclic voltammograms.[Q77] Radical cations appear to be more stable in nitrobenzene than in acetonitrile solutions.[Q78] Coating of the anode with insoluble insulating polymeric films is a common hazard in anodic oxidation systems but it can be alleviated by the use of scraped electrodes and periodic polarity-reversal techniques.

In non-nucleophilic solvents, free from added nucleophiles, the reactive intermediates attack other substrate molecules.[Q79] For example, in methylene chloride with Bu_4NBF_4 as electrolyte, mesitylene is oxidised at 1.8 V *vs.* SCE to form coupling products of the biphenyl type: bimesityl and some termesityl,

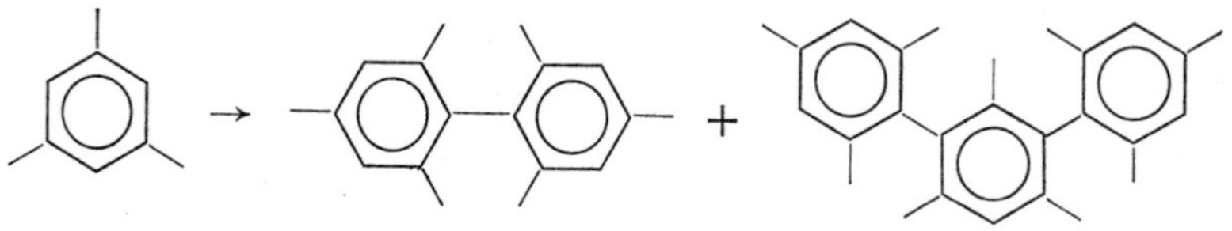

The yields improve markedly with increase in substrate concentration.

When side-chain groups are present in the organic molecule, side chain substitution may occur in competition with nuclear substitution.[Q80-83] In acetonitrile containing 0.1*M* Et_4NClO_4 and 0.6*M* water, pentamethylanisole is oxidised[Q84] according to

$$\text{pentamethylanisole} \underset{}{\overset{-e}{\rightleftharpoons}} (\text{pentamethylanisole})^{\cdot +} \longrightarrow H^+ + \text{substituted benzyl radical} \xrightarrow{-e} \text{benzyl carbonium ion}$$

Further examples include:[Q85,Q86]

$$AcO^- + \text{hexamethylbenzene} \rightarrow 2e + \text{pentamethylbenzyl acetate } (C_6(CH_3)_5CH_2OAc) + H^+$$

and

$$\text{durene} + MeCN \rightarrow 2e + \text{trimethylbenzyl-}CH_2\overset{+}{N}{:}CCH_3 + H^+$$

When hexamethylbenzene was oxidised in acetonitrile containing acetic acid, perchlorate electrolytes favoured the acetamidation reaction, whereas fluoroborate led to relatively higher acetoxylation yields.[Q86] This fact was interpreted in terms of preferential solvation of fluoroborate by acetic acid.

In a non-nucleophilic environment, coupling tends to occur mainly at the side chain if the unpaired-electron density is low at the free ring position, e.g. durene when oxidised at 1.35 V (*vs.* SCE) in methylene chloride containing Bu_4NBF_4 gave appreciable yields of the biphenylmethane derivative.[Q79]

The analogies which exist between anodic oxidation and cathodic reduction reactions of aromatic compounds are extensive; radicals formed from radical cations by chemical reaction can be further oxidised by other radical cations, i.e. by an overall disproportionation mechanism. Thus the formation of *trans* 9,10-dihydroxy 9,10-diphenyl anthracene on the addition of water to a solution of 9,10-diphenyl anthracene radical cations[Q87,Q88] is analogous to the production of 9,10-dihydro-anthracene when phenol is added to anthracene radical anions. If nucleophiles are present in solution during the anodic process the reaction proceeds through a disproportionation path when the chemical reaction with the radical cation is slow. If it is rapid, however, the resulting radical is more likely to diffuse back to the anode for further oxidation (ECE mechanism) than to accomplish the same transition through a disproportionation route. No cases appear to have been recorded of the disproportionation of radical cations preceding chemical reaction analogous to the decay of benzophenone radical anions[Q89,90] in the presence of water:

$$2R^{\cdot -} \rightleftharpoons R + R^{=} \xrightarrow{2H_2O} RH_2 + 2OH^-$$

According to the nature of the substituent group anodically introduced into the aromatic hydrocarbon, various deviations from the basic behaviour can occur. For example oxidation of anthracene in acetonitrile containing a trace of water leads to the reaction scheme[Q91,Q92]

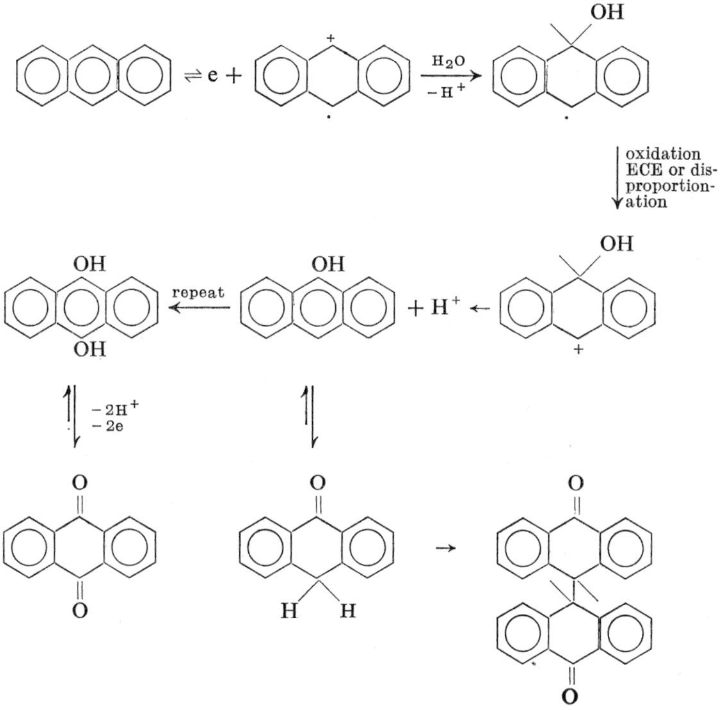

When the substrate molecule contains functional groups, quite different chemical reactions can occur following the initial radical cation formation. For example, benzyl alcohol derivatives can deprotonate at the α-hydrogen and then, after the second charge transfer, deprotonate once more; substantial yields of the benzaldehyde derivatives are obtained.[Q93,Q94] Aryl iodide radical cations attack a second substrate molecule;[Q95]

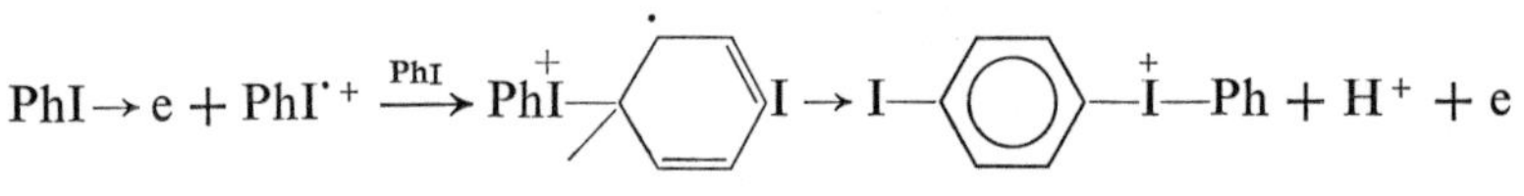

$$PhI \rightarrow e + PhI^{\cdot +} \xrightarrow{PhI} Ph\overset{+}{I}\text{—}(C_6H_5I)^{\cdot} \rightarrow I\text{—}C_6H_4\text{—}\overset{+}{I}\text{—}Ph + H^+ + e$$

Displacement of bromine occurs when 9,10-dibromoanthracene is oxidised in acetontrile containing lutidines.[Q96]

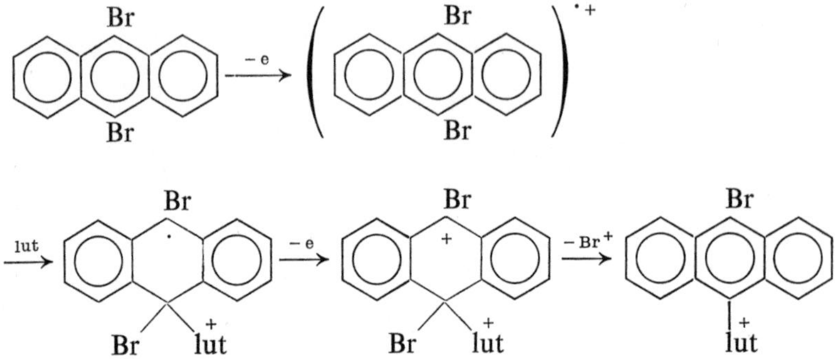

$$C_{14}H_8Br_2 \xrightarrow{-e} (C_{14}H_8Br_2)^{\cdot +} \xrightarrow{lut} \cdot C_{14}H_8Br_2(\overset{+}{lut}) \xrightarrow{-e} {}^{+}C_{14}H_8Br_2(\overset{+}{lut}) \xrightarrow{-Br^+} C_{14}H_8Br(\overset{+}{lut})$$

The cation radicals formed during the oxidation of phenols are active in initiating polymerisation processes through reaction at the ortho or para positions If those positions are blocked, as in 2,4,6-tri-*t*-butyl phenol, the expected deprotonation and further oxidation occurs:[Q97]

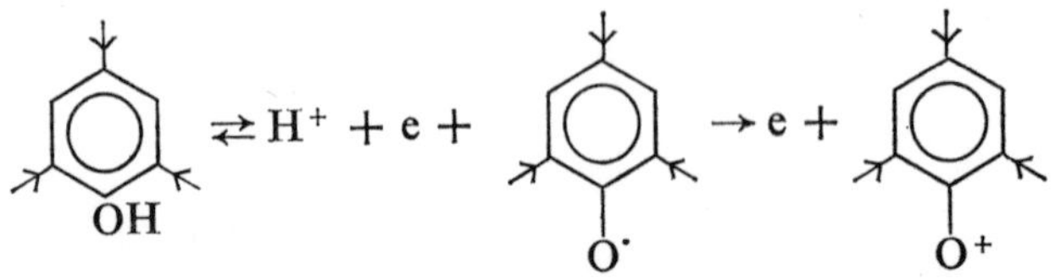

$$ArOH \rightleftarrows H^+ + e + ArO^{\cdot} \rightarrow e + ArO^+$$

The resulting carbonium ion is susceptible to nucleophilic attack at the *o* and *p* positions where it can be acetoxylated or methoxylated. In the presence of water (1*M*) 2,6-di-*t*-butyl-1,4-benzoquinone was formed and in dry acetontrile the product appeared to be 3-methyl-5,7-di-*t*-butyl-1,2-benzisoxazole

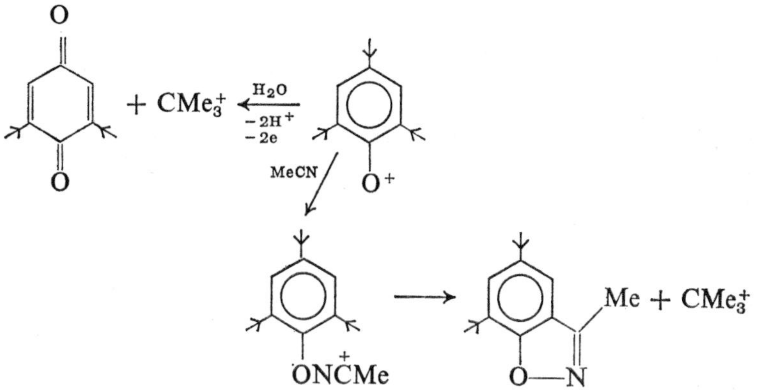

$$\text{2,6-di-}t\text{-butyl-1,4-benzoquinone} + CMe_3^+ \xleftarrow[-2H^+,\ -2e]{H_2O} \text{ArO}^+ \xrightarrow{MeCN} \text{ArONCMe}^+ \longrightarrow \text{benzisoxazole (Me)} + CMe_3^+$$

Phenoxide ions oxidise at lower potentials than either the corresponding phenols or phenoxy radicals. In consequence phenoxy radicals formed by the oxidation of hindered phenoxides are relatively stable and are only transformed to the carbonium ion at more anodic potentials. Phenoxy radicals from less hindered phenoxides undergo dimerisation.[Q98]

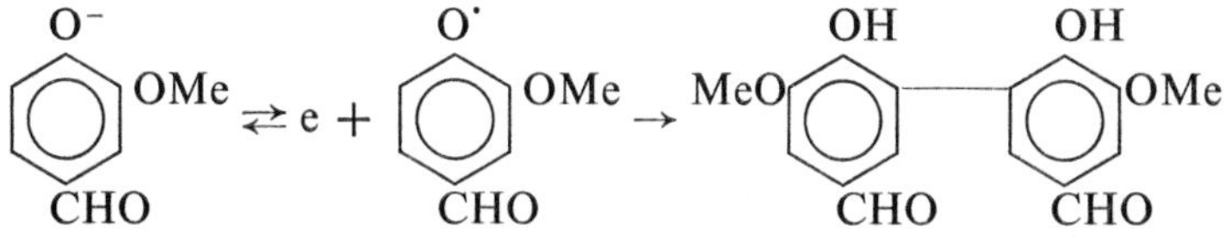

$$\text{O}^- (\text{OMe}, \text{CHO}) \rightleftarrows e + \text{O}^{\cdot} (\text{OMe}, \text{CHO}) \rightarrow \text{dimer: MeO, OH, CHO / OH, OMe, CHO}$$

p-Dimethoxybenzene and *p*-diacetoxybenzene undergo respectively methyl and acyl transfers when oxidised in acetonitrile solutions of $LiClO_4$ at +1.9 V *vs.* SCE:[Q99,Q100]

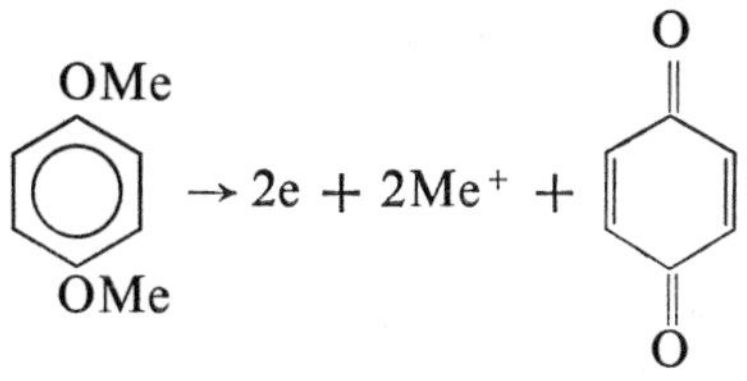

$$\text{p-}C_6H_4(OMe)_2 \rightarrow 2e + 2Me^+ + \text{1,4-benzoquinone}$$

Another interesting case involving a symmetrical molecule results in dehydrogenation:[Q101]

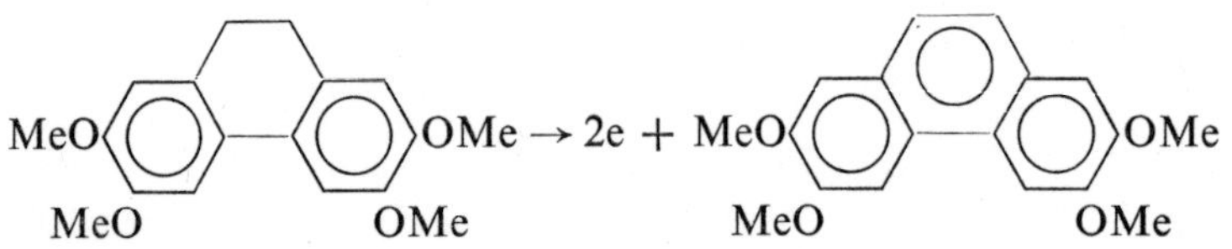

$$\text{MeO, MeO-substituted (CH}_2\text{CH}_2\text{-bridged) biphenyl (OMe, OMe)} \rightarrow 2e + \text{MeO, MeO, OMe, OMe-phenanthrene}$$

This last compound can be prepared also by anodic oxidation of 3,4,3′,4′-tetramethoxybibenzyl:[Q101]

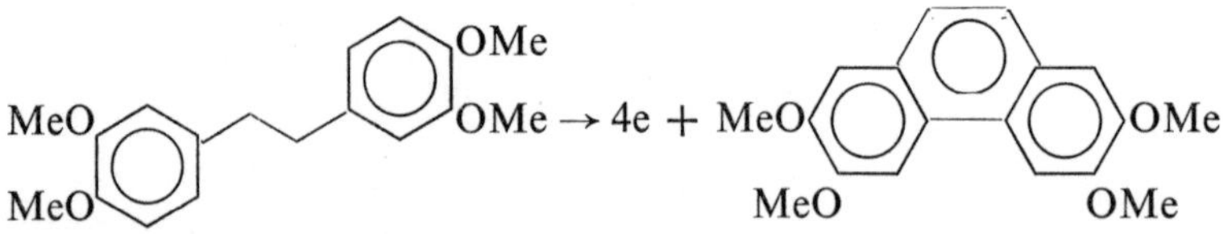

$$\text{(MeO)}_2\text{C}_6\text{H}_3\text{CH}_2\text{CH}_2\text{C}_6\text{H}_3\text{(OMe)}_2 \rightarrow 4e + \text{MeO, MeO, OMe, OMe-phenanthrene}$$

Aromatic ethers can undergo displacement of the ether group by cyanide. Thus *p*-dimethoxybenzene when oxidised in acetonitrile containing Et_4NCN electrolyte at 2.0 V *vs*. SCE, yielded 1-methoxy-4-cyanobenzene,[Q102]

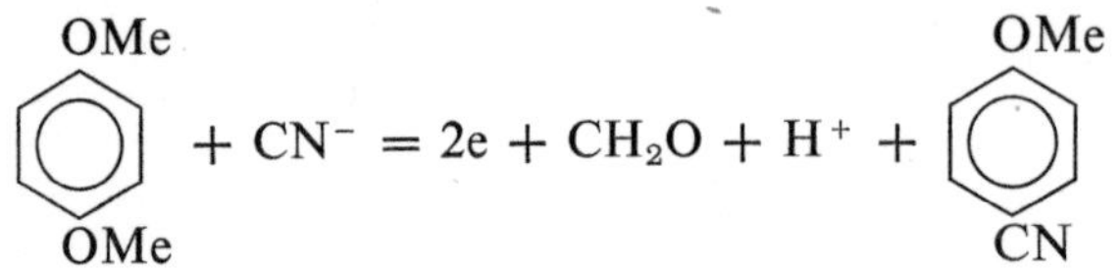

$$p\text{-(MeO)}_2\text{C}_6\text{H}_4 + \text{CN}^- = 2e + \text{CH}_2\text{O} + \text{H}^+ + p\text{-MeOC}_6\text{H}_4\text{CN}$$

o-dimethoxybenzene reacted similarly.

Aromatic amines oxidise to radical cations in which the charge is concentrated on the nitrogen, leaving the para position able to function as a radical centre,[Q103] e.g.

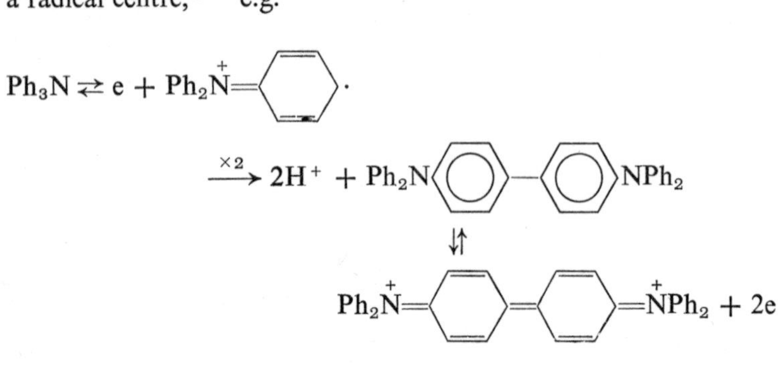

$$\text{Ph}_3\text{N} \rightleftarrows e + \text{Ph}_2\overset{+}{\text{N}}{=}\text{C}_6\text{H}_5\cdot$$

$$\xrightarrow{\times 2} 2\text{H}^+ + \text{Ph}_2\text{N}\text{C}_6\text{H}_4\text{-C}_6\text{H}_4\text{NPh}_2$$

$$\downarrow\uparrow$$

$$\text{Ph}_2\overset{+}{\text{N}}{=}\text{C}_6\text{H}_4{=}\text{C}_6\text{H}_4{=}\overset{+}{\text{N}}\text{Ph}_2 + 2e$$

If the *p*-position of the amine is blocked by methoxy, methyl, chlorine or bromine groups, the species is stable with respect to dimerisation[Q104] and higher anodic potentials are required for its further oxidation. However, further decomposition of the tri *p*-anisylamine radical cation occurs in the presence of any traces of cyanide ion present in the acetonitrile solution.[Q105] Primary aromatic amines, like phenols, tend to polymerise upon oxidation unless the *o* and *p* positions are blocked. 2,4,6-tri-t-butylaniline in acetonitrile solution yields a fairly stable radical cation which in the presence of water forms 3,5-di-*t*-butyl-4-amino-2,5-cyclohexadienone.[Q106]

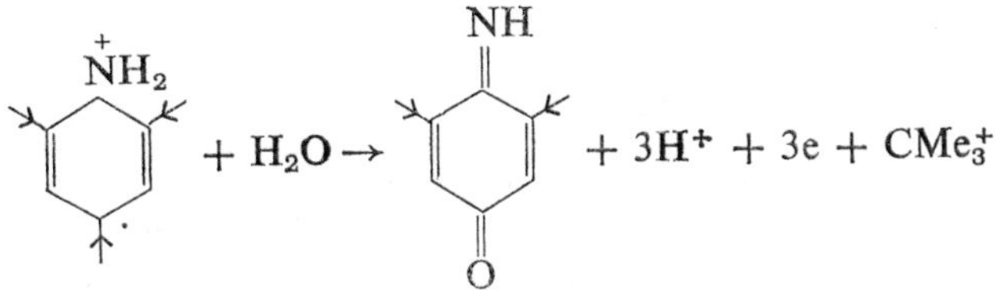

$$\overset{+}{N}H_2\text{-}(2,4,6\text{-tri-}t\text{-butylphenyl})^{\cdot} + H_2O \rightarrow (NH{=}C_6H_2(t\text{-Bu})_2{=}O) + 3H^+ + 3e + CMe_3^+$$

The use of dry methanol instead of wet acetontrile for this reaction permits the synthesis of the anisidine derivative.

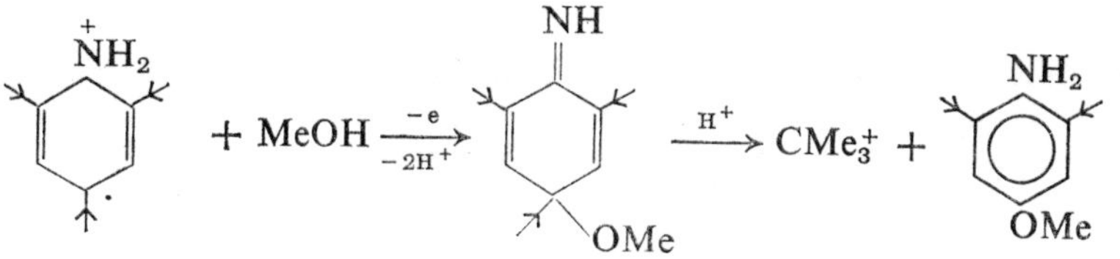

$$\overset{+}{N}H_2\text{-}(2,4,6\text{-tri-}t\text{-butylphenyl})^{\cdot} + MeOH \xrightarrow[-2H^+]{-e} (NH{=}C_6H_2(t\text{-Bu})_2(t\text{-Bu})(OMe)) \xrightarrow{H^+} CMe_3^+ + NH_2\text{-}C_6H_2(t\text{-Bu})_2\text{-}OMe$$

Deprotonation of the primary amine group by the addition of a base promotes head-to-tail coupling of aromatic amine radical cations. When the base diphenylguanidine is added to solutions of the 9-amino-10-phenylanthracene radical cation one obtains the dimer[Q107]

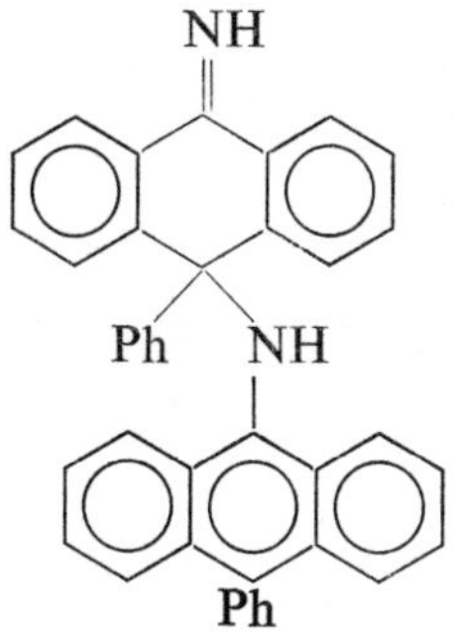

NH = (9,10-dihydroanthracene), 10-Ph, 10-NH–(10-Ph-anthracen-9-yl)

Another radical reaction shown by the same cation is the addition across an oxygen molecule to yield the peroxide

Not all aromatic oxidations in organic solvents proceed through the initial ionisation of the substrate molecule. In particular, several methoxylations appear to occur by primary oxidation of the solvent, methanol, to form a radical. Parker showed the methoxylation of anisole, in methanol containing 1% H_2SO_4, to proceed at electrode potentials less positive than those required for the ionisation of anisole.[Q108] The proposed mechanism is:

$$CH_3OH \rightarrow 2e + H^+ + \overset{+}{C}H_2OH$$

$$\overset{+}{C}H_2OH + PhOMe \rightarrow MeO\text{-}C_6H_4\text{-}CH_2OH + H^+$$

When alkoxide ion is present, the alkoxylation reactions proceed through its initial discharge:[Q109]

$$PhCH{=}CH_2 + 2MeO^- \rightarrow 2e + PhCH(OMe)CH_2OMe$$

Additional complicating factors in anodic substitution arise when an oxidisable supporting electrolyte is used. For example, the oxidation of mesitylene in an acetic acid solution of $R_4N\ NO_3$ produces substituted benzyl acetates and nitrates at potentials below those required for the ionisation of mesitylene:[Q110]

$$NO_3^- \rightarrow e + NO_3^{\cdot} \xrightarrow{\text{mesitylene}} HNO_3 + \dot{C}H_2\text{-}C_6H_3(CH_3)_2$$

$$\xrightarrow{-e} \overset{+}{C}H_2\text{-}C_6H_3(CH_3)_2 \xrightarrow{OAc^-} CH_2OAc\text{-}C_6H_3(CH_3)_2$$

$$\overset{+}{C}H_2\text{-}C_6H_3(CH_3)_2 \xrightarrow{NO_3^-} CH_2^+NO_3^-\text{-}C_6H_3(CH_3)_2$$

However, the presence in the anolyte of an anion, which oxidises more readily than the organic substrate and which appears in the substituted product, does not preclude ionisation of the substrate as the operative mechanism. Thus Parker and Burget showed that cyanation of anisole does not proceed at potentials where only cyanide is electroactive; ionisation of the anisole is a prerequisite for substitution.[Q76] A further example is the chlorination of cyclohexene where, by operating at high electrode potentials at which the olefin is electroactive, chloride ion can be intercepted *en route* to the anode surface by cyclohexene radical cations diffusing away from it.[Q111]

In addition to the aromatic and unsaturated species already considered, several saturated materials may be oxidised anodically in organic solvent systems. Organic anions are invariably oxidisable. Carboxylate anions undergo decarboxylation reactions to form radicals, which may lead to Kolbe dimers, and carbonium ions which usually attack the solvent, e.g. reaction with acetonitrile results in acetamidation.[Q112,Q113]

Many organic ions are stable only in organic solvents, in particular aprotic media. Thus the oxidation of Grignard reagents and of the alkyls and aryls of alkali metals can be conducted only in ethers. The field has been reviewed by several authors.[Q114–116] The products are characteristic of radical intermediates, hydrocarbons being formed by disproportionation and dimerisation reactions and metal alkyls by reaction of the radicals with the anode material when anodes of lead, cadmium, zinc or aluminium are employed.

Aliphatic amines are readily oxidised in organic solvents but the apparent absence of any synthetic applications has resulted in relatively few studies of these systems. Mann and co-workers[Q11] have proposed oxidation mechanisms involving the initial formation of radical cations, which undergo fission either to lose a proton or to break a C—N bond, forming respectively either a radical or a carbonium ion. The overall reactions are complex and are affected by quite small amounts of water. Protons liberated in the reaction acidify most of the starting amine, preventing its oxidation.

Recent reviews[Q4,Q116] covering thoroughly the topics of this section are recommended.

7.14 ELECTRODE KINETICS

Published data on the rates of charge-transfer processes in organic solvent systems are remarkably scarce. This is particularly true of comparative data pertaining to different solvents or mixtures and contrasts noticeably with the extensive literature on polarographic $E_{\frac{1}{2}}$ data in organic solvents. Nevertheless the role of solvent in electrode kinetics is of considerable importance, particularly when the charge transfer event occurs without the formation or rupture of any chemical bond in the substrate and when the orbital overlap between the redox particle and the electrode is small. This behaviour will be approached most closely when the reactant species has the same charge as the electrode, although solvents which permit ion-pair formation might allow close contact even in such circumstances. The theory of this class of charge-transfer reactions has developed alongside that for homogeneous electron-transfer processes in solution. The major recent contributions have been made by Marcus[Q117,Q118] and by Levich and co-workers.[Q119] Here we briefly review their work to anticipate solvent effects in electrode kinetics. A more comprehensive review was published by Marcus in 1964.[Q118] The activation free energy for simple electron-transfer reactions involving solvated species arises principally because the equilibrium arrangement of solvent molecules around the ions in the two valence states is not the same. Thus the transition state during which the electron-transfer event occurs has a solvation structure intermediate between those of reactant and product particles. Broadly speaking we might anticipate that solvents which strongly solvate ions will promote large differences between the solvation energies of different valence states and will also show large heats of activation for the charge-transfer reaction.

The contributions to the activation process considered by Marcus[Q118] were fluctuations in distance between the reactant and the electrode, in bond lengths within the inner co-ordination shell and in bond lengths, bond angles and molecular orientations of adjacent solvent molecules. Levich *et al.* emphasise the last contribution. Marcus essentially ignored the distribution and occupancy of electron levels in the metal electrode considered by the Soviet workers. Marcus' method of treating the fluctuations was through statistical mechanics whereas Levich *et al.* showed that a classical treatment gives in fact only one limiting result of the more general solution obtained when the motions of both the electron and the heavy particles are treated quantum mechanically. Two other limiting solutions are also considered important by the Soviet school (*vide infra*). Both theories assume that reaction is insufficiently rapid to disturb the thermal equilibrium distribution of reactant energies. Both groups of authors discuss the reaction in terms of energy surfaces for the reactant and product although the reaction co-ordinate, which is a transform of a multidimensional co-ordinate, clearly depends upon the contributions considered as making up the potential energy of the system. The reaction path is considered to be that which corresponds to the minimum value of the activation energy. Vibrations are essentially regarded as simple harmonic.

Both schools, unlike several earlier contributors in the field, recognise the importance of the dynamic nature of the nuclear configuration so that electron tunnelling relations developed for external static potential-energy barriers are not applicable and instead the Zener–Landau theory is used. Thus Marcus regards the probability of the transition from reactants to products as the product of the individual probabilities of (a) the system (i.e. the reactant particle, the electrode and neighbouring solvent molecules) attaining certain nuclear configurations and momentum values and (b) electron transfer occurring in those states. The reaction rate is determined by a suitable summation over all states. Levich *et al.* regard the separation of the total probability in this way as permissible only in the limiting case of adiabatic transitions where the electron-transfer probability is unity. For the non-adiabatic case, even using the semi-classical treatment, the probabilities are separable only before the final summation over states whereas in the quantum mechanical treatment, the division of the system into slow (nuclear) and fast (electronic) components is not meaningful.

The result of the quantum treatment gives two limiting cases:[Q119]

1. *High temperature case*, i.e. when

$$kT \gg \hbar\omega_0 \text{ and } kT \gg \hbar\omega_0(J_2 - J_1)/E_s$$

ω_0 is the limiting frequency of the polarisation wave in the solvent, treated as a dielectric continuum, and J_1 and J_2 are the equilibrium term energies of the reactant and product states at the electrode surface. The term energies used to construct the intersecting curves are the total energies of the reacting subsystem less the kinetic energies of the nuclei. E_s, the solvent reorganisation energy, is the sum of terms, each of which is the square of differences between parameters (e.g. dielectric induction in the continuum model) which measure ionic solvation in the initial and final states.

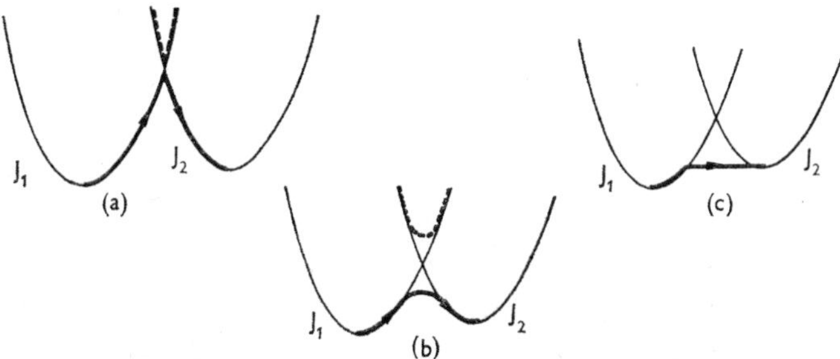

At high temperatures the transition probability for the non-adiabatic case (Fig. 7.14.1a) is given by[Q119] eqn. 7.14.1:

$$W_{12} \sim \left(\frac{\pi}{\hbar^2 E_s}\right)^{\frac{1}{2}} (L_{21}^{(2)})^2 \exp\left\{\frac{-(J_2 - J_1 + E_s)^2}{4E_s kT}\right\} \qquad (7.14.1)$$

which is similar to the result obtained by the semi-classical method. If the exchange integral, $L_{21}^{(2)}$, is sufficiently large, there is considerable splitting in the term curves and the transfer becomes adiabatic, with unit probability (Fig. 7.14.1b). Then

$$W_{12} = (\omega_0/2\pi) \exp\left\{\frac{-(J_2 - J_1 + E_s)^2}{4E_s kT}\right\}$$

2. *Low temperature case*, i.e. when

$$kT \ll \hbar\omega_0 \text{ and } kT \ll \hbar\omega_0(J_2 - J_1)/E_s$$

Under these circumstances (Fig. 7.14.1c), where the normal solvent fluctuations are insufficient to promote appreciable reaction rates,

$$W_{12} = \frac{2\pi|L_{12}^{(2)}|^2}{\hbar^2\omega_0[(J_2 - J_1)/\hbar\omega_0]!}\left(\frac{E_s}{\hbar\omega_0}\right)^{\frac{J_2-J_1}{\hbar\omega_0}} \exp\left(\frac{-E_s}{\hbar\omega_0}\right) \exp\left(\frac{J_1 - J_2}{kT}\right) \tag{7.14.2}$$

As the probabilities W_{12} are proportional to the heterogeneous rate constants, we see that these expressions have the familiar Arrhenius form where at high temperatures the activation energy is

$$(J_2 - J_1 + E_s)^2/4E_s$$

and at low temperatures $(J_2 - J_1)$. The second case is referred to as a tunnel transition of the system, as the activation energy has the lowest possible value. Both the rates and the temperatures at which the two mechanisms are of equal importance will in general be functions of the solvent.

Marcus gives a result similar to Levich's high temperature solution.[Q117]

$$k_{\text{rate}} = \kappa\rho Z \exp(-\Delta G^{\ddagger}/kT)$$

where

$$Z = (kT/2\pi m)^{\frac{1}{2}} \sim 3 \times 10^4 \text{ cm s}^{-1}$$

m is the mass of the reactant particle, ρ is a ratio of the root mean square fluctuations with a value approximately unity and κ is the nuclear velocity weighted average of the electron-transfer probability. The activation free energy is given as

$$\Delta G^{\ddagger} = \frac{\omega^r + \omega^p}{2} + \frac{E_s}{4} + \frac{ne(E - E^0)}{2} + \frac{[ne(E - E^0) + \omega^p - \omega^r]^2}{4E_s} \tag{7.14.3}$$

where ω^r and ω^p are respectively the amounts of work which are required to bring reactant and product particles to the electrode surface from infinity. The similarity to eqn. 7.14.1 is evident if we identify

$(J_2 - J_1)$ with $\omega^p - \omega^r + ne(E - E^0)$. α, the transfer coefficient, is obtained by differentiation:

$$\alpha = (\partial \Delta G^{\neq} / \partial neE)_\omega$$
$$= \tfrac{1}{2} + neE - neE^0 + \omega^p - \omega^r)/2E_s \quad (7.14.4)$$

In fact this is an approximation, based on neglect of antisymmetrical contributions to the potential energy curves. When these terms are considered:[Q117]

$$\alpha = \tfrac{1}{2} + (neE - neE^0 + \omega^p - \omega^r + \tfrac{1}{2}\lambda_i \langle l_s \rangle)/2E_s \quad (7.14.5)$$

where λ_i is the reorganisation energy minus the contribution made to it by the outer solvation shell. $\langle l_s \rangle$ is the mean value of a term which measures the difference between the vibrational force constants for reactants and products.

If it is assumed that the particle is spherical, the dielectric region near it is unsaturated, dielectric images and differences in size between reactants and products are unimportant and the solution is infinitely dilute, then the outer layer contribution to E_s can be written:

$$\lambda_0 = (ne)^2(a^{-1} - r^{-1})(\varepsilon^{-1} - \varepsilon_0^{-1}) \quad (7.14.6)$$

Here r is the distance between the ion and its image in the metal in the transition state and a is the radius of the reactant ion including its primary solvation sheath. ε is the optical permitivity and ε_0 the static value.

From this picture the rate constants for weak-overlap charge-transfer reactions would be expected to vary markedly with solvent through the dependence of E_s upon ε, ε_0, a and r.

Levich *et al.* have extended their low temperature case to include reactions involving bond breaking in which the activation energy arises from internal vibrations as well as solvent fluctuation. In this case a reactant species in an excited vibrational state is supposed to undergo an electron-transfer reaction to form a product species in the ground vibrational state.

Biegler, Gonzalez and Parsons[Q120] have made one of the few comparative studies of the kinetics of a simple inorganic redox reaction in different solvents. The rate constants are comparable for the polarographic reduction of V^{3+} to V^{2+} in water (3.36×10^{-3} cm s^{-1}) and *N*-methylformamide (1.16×10^{-3} cm s^{-1}) but the transfer coefficients are respectively 0.60 and 0.20. The discharge of cadmium into cadmium amalgam was examined in greater detail. From the viewpoint of studying solvent effects this system was an unfortunate choice as the mechanism is not fully understood in any solvent. However, the kinetic results indicated comparable rate constants at 25°C in the pure solvents water (0.45 cm s^{-1}), acetonitrile (0.19), DMF (0.15) and NMF (0.12) whereas noticeably lower values were obtained for PC (~0.01) methanol (0.013) and surprisingly for equimolar solvent mixtures: water/MeCN (0.0145), MeCN/MeOH (0.028) and DMF/NMF (0.089). No double-layer correction was undertaken; this is one possible explanation of the lack of correlation between the rates and the free energies of solvation of Cd^{2+} in the solvents water (−1805 kJ mol^{-1}) methanol (−1850) and acetonitrile (−1790). However, the temperature dependence of the results indicates an alternative explanation: compensation of the changes in the exponential and pre-exponential terms. Consistently low values of α suggest a common mechanism in all solvents.

The lower rates observed in mixtures of aqueous and organic solvents are explained in terms of the solvation of the reactant ion and the electrode by different solvent species; the ion is hydrated whereas the organic component is concentrated in the interface. Thus lower rates may be attributed to the greater separation in the transition state caused by the preclusion of solvent-dipole sharing between the electrode and the ion or alternatively to the increase in the reorganisation energy term in eqn. 7.14.6 arising from the low value of a and the large value of r.

Jaenicke and Schweitzer had earlier[Q121] noticed the relatively low rate constants obtained with mixtures of aqueous and organic solvents during their study of zinc-ion discharge on a hanging zinc-amalgam drop. Minima in the values of the exchange current for 0.002*M* $Zn(ClO_4)_2$ solution were observed when the organic content was between 10 and 40 mol % in mixtures of water with ethanol, 1,4-dioxan, THF, MeCN and acetone. However ethanol + acetone mixtures showed no such minimum and the rate constants in the pure solvents differed by a factor below 30. Mixtures of water with the strongly complexing solvent ethylenediamine showed remarkable behaviour; whereas for more than 95 % of the solvent mixture range, including both pure solvents, the rate constant varied only by a factor of two, it dropped by an order of magnitude when the Zn : *en* ratio was 1 : 2. Under these conditions zinc is present almost exclusively as $Znen_2^{2+}$ but the free *en* content of the medium is essentially zero. Once more therefore the minimum corresponds to circumstances where the ion and the electrode are solvated by different species.

Another instance of a similar phenomenon is found in the work of Bockris and Parsons[Q122] where the exchange current for hydrogen evolution on mercury was found to reach a minimum in approximately equimolar H_2O + MeOH solutions, both pure solvents exhibiting values an order of magnitude higher. Simultaneously the transfer coefficient increased from 0.5 in water to 0.6 in MeOH. For this system equilibrium data are available to show that MeOH is concentrated at the interface whilst water is the predominant ion solvator.

Recently, stationary electrode polarography has been used to study the electrode kinetics of several metal-ion systems on mercury in DMSO solutions.[Q123] In cadmium discharge, the rate-determining step was believed to be a chemical reaction, the loss of molecules of solvation from the reactant ion, preceding a reversible charge transfer. The sensitivity of the rate towards the electrolyte cation suggests that the slow step is essentially confined to the double layer.

The double-layer effect is pronounced also in the reduction of zinc, a quasi-reversible process with $\alpha n_a \simeq 0.5$. Apparent standard rate constants, k_a, vary from 5.6×10^{-3} cm s^{-1} in $0.1M$ $KClO_4$ to 2.7×10^{-4} in $0.5M$ $KClO_4$ and 4.0×10^{-5} in $0.5M$ Bu_4NClO_4, indicating strong specific adsorption of Bu_4N^+ ions. Lead discharge was somewhat similar with k_a equal to 1.9×10^{-2} cm s^{-1} in $1M$ $KClO_4$ and 3.0×10^{-7} in $0.5M$ Bu_4NPF_6.

In general, k_a values for metal-ion discharge were found to be lower in DMSO than in aqueous media, apparent exceptions to this rule being of dubious validity owing to the methods used in their evaluation. However, in a $1M$ $KClO_4$ solution in DMSO, the redox couple $Cr^{2+}|Cr^{3+}$ exhibited a k_a value of 3.5×10^{-4} cm s^{-1}, which compares with the value 1×10^{-5} cm s^{-1} in aqueous $1M$ KCl. Values of k_a in propylene carbonate solution for sodium (10^{-1} cm s^{-1}) and lithium (10^{-2} cm s^{-1}) discharge were higher than in DMSO (1.7×10^{-2} and 10^{-4} cm s^{-1} respectively).

Ferrocene has been frequently recommended as a system by which to relate potential scales in different solvents because the solvation energies of ferricinium ions and ferrocene are believed to be smaller than for most other ions.[Q8] It follows that the solvent contribution to the activation free energy should be relatively small in that system. However, no kinetic studies appear to have been made there; ferrocene is known to be polarographically reversible[Q124] but osmocene and ruthenocene are sufficiently slow[Q125] for the rate parameters to be determined polarographically and compared for different solvents.

Polycyclic aromatic hydrocarbons are also systems in which the solvation energy and therefore the activation energy for electron-transfer reactions might be expected to be small; the charges in the radical anions and radical cations, which are readily formed electrochemically, are delocalised throughout the structures. Aten, using a.c. polarography showed the systems polycyclic hydrocarbon/radical anion to be fully reversible in aprotic solvents in which the radical anions are not protonated.[Q126] Later the a.c. bridge method was used to determine the rate constants for these redox systems.[Q127] Using millimolar solutions of the hydrocarbons in DMF containing Bu_4NI supporting electrolyte, approximate values of k_a were found for the one-electron reductions of anthracene, tetracene, perylene, and also for stilbene, 1-phenyl-4-

biphenylenebutadiene, dibiphenyleneethene and dibiphenylenebutadiene; in all cases $\log k_a \sim 0.7$. In addition, the second electron-transfer step (to form dianions) was accessible for the dibiphenylenepolyenes but for the other hydrocarbons the rapid protonation of the dianion precluded the acquisition of reliable kinetic data.

Dietz and Peover[Q128] made similar measurements in their investigation of the effect of sterically produced non-planarity of stilbenes upon the reduction rate constants. The values were corrected for diffuse double-layer effects in the DMF solutions using Gouy–Chapman theory and pzc data. Whereas all of the planar conjugated aromatic hydrocarbons studied, naphthalene, 2-methylnaphthalene, tetracene and *trans*-stilbene reduced to their radical anions with true rate constants within the range 20 to 27 cm s^{-1} at 30°C, sterically distorted stilbenes (*cis*-stilbene and α-methyl-2,4,6,2′,4′,6′-hexamethyl-*trans*-stilbene) gave values an order of magnitude lower. Measurements were made also at −20°C and Arrhenius parameters were presented which showed the lower rate constants for the hindered stilbenes to be due to the lower pre-exponential terms. The Arrhenius parameters obtained for charge-transfer reactions at electrodes are composite values for the forward and reverse processes. It would appear that entropy and electron-transfer probability terms are the main reasons for differences between hindered and other hydrocarbons. The cathodic transfer coefficients were noticeably below 0.5 for the hindered molecules but in excess of that value for the planar species. Such variations are explained in terms of Marcus' theory as a lack of symmetry between the PE curves of the reactant and of the radical anion near to the intersection point (eqn. 7.14.5) suggesting that strained reactants must undergo additional reorganisation in order to reach the transition state.

In a study of the cathodic reduction of cyclo-octatetraene in DMF and DMSO, Allendoerfer and Rieger[Q129] used cyclic voltammetry and a.c. polarography to show that the reduction to the radical anion is quasi-reversible with a rate constant, k_h, equal to 8.5×10^{-3} cm s^{-1} at 25°C. The heat of activation was found to be 7.7 kcal mol^{-1} and the small values of α which were observed were interpreted in terms of the transition state at the equilibrium potential resembling the product radical anion more closely than it resembles the reactant molecule. The transition state is therefore presumably planar, as is the radical anion. This conclusion is supported by the similarity of the experimental value for the free energy of activation to the literature derived values for the free energy of activation in the bond isomerisation reaction of cyclo-octatetraene.

In the same paper the electrode kinetics of the second electron-transfer step to form the dianion are presented. In contrast to the reported conclusions for aromatic hydrocarbons, the rate constant for this step is some 200 times higher than that for the first stage. This is not surprising in view of the planar nature of both ions.

Qualitatively similar conclusions can be reached for the oxidation of tropilidene where the first charge transfer is relatively slow with a small anodic transfer coefficient:

$$C_7H_8 = e + C_7H_8^{+}$$

The intermediate radical cation is rapidly further oxidised at the potentials of its formation:

$$C_7H_8^{+} = e + H^{+} + C_7H_7^{+}$$

The rates of electron-transfer processes in the oxidation of polycyclic aromatic hydrocarbons have been considered by Peover and White, who estimated from cyclic voltammograms the apparent standard rate constant for 9,10-diphenylanthracene radical cation formation to be of the order 1 cm s^{-1}.[Q130]

Peover and Powell[Q131] satisfactorily correlated the rate constants for the reductions of a series of nitro compounds with the degree of spin localisation on the resulting anion radical. They justified the neglect of specific interactions with the solvent, DMF, and the electrolyte, Bu_4NI, and considered the free energy of activation to arise from the reorientation of solvent dipoles according to the simplified Marcus theory (eqn. 7.14.6.). For the series of compounds, r, ε and ε_0 were supposed to be constant so that differences in free energy of activation arose solely through the variations in a. The term e^2/a was split into contributions from the nitro group and from the ring, regarding the molecule as two separate spheres with $a_{NO_2} = 2.75$ Å and $a_{ring} = 3.5$ Å. The fractional charges were ascribed to the respective spheres on the basis of spin density data obtained by ESR measurements. The experimental data was plotted in the form ln k_a *vs.* the quantity

$$[\Sigma(a_N Q_N)^2/a_{NO_2} + (1 - \Sigma(a_N Q_N))^2/a_{ring}]$$

where $a_N Q_N$ is the fraction of charge on the nitro group of the radical anion transition state. The relationship was linear with the predicted slope. Good agreement can also be obtained with the expected values of the frequency factors if either a term allowing for repulsion between the spheres is introduced or alternatively the image term (that in r) is neglected.

REFERENCES

Q1 P. Delahay, Double Layer and Electrode Kinetics, Interscience, New York (1965)

Q2 J. Koryta, J. Dvorak and V. Bohackova, Electrochemistry, Methuen, London (1970)

Q3 T. Kuwana *in* Electroanalytical Chemistry, Vol. I (A. J. Bard, ed.), Marcel Dekker, New York (1966)

Q4 C. K. Mann and K. K. Barnes, Electrochemical Reactions in Non-aqueous Systems, Marcel Dekker, New York (1970)

Q5 C. K. Mann *in* Electroanalytical Chemistry, Vol. III (A. J. Bard, ed.), Marcel Dekker, New York (1969)

Q6 C. Agami, *Bull. Soc. Chim. France*, 1205 (1968)

Q7 J. E. Dubois and A. Marie de Ficquelmont, *J. Chim. Phys.*, **64,** 904 (1967)

Q8 H. Strehlow *in* Chemistry of Non-aqueous Solvents, Vol. I (J. J. Lagowski, ed.), Academic Press, New York (1966)

Q9 J-P. Billon, *J. Electroanal. Chem.*, **1,** 486 (1960)

Q10 T. Osa, A. Yildiz and T. Kuwana, *J. Amer. Chem. Soc.*, **91,** 3994 (1969)

Q11 K. K. Barnes and C. K. Mann, *J. Org. Chem.*, **32,** 1474 (1967)

Q12 M. Fleischmann and D. Pletcher, *Tetrahedron Lett.*, 6255 (1968)

Q13 J. Courtot-Coupez and M. L'Her, *Bull. Soc. Chim. France*, 1631 (1970)

Q14 J. Martinmaa, *Suomen Kem.*, **42B,** 33 (1969)

Q15 J. F. Coetzee, J. M. Simon and R. J. Bertozzi, *Anal. Chem.*, **41,** 766 (1969)

Q16 P. H. Given, *J. Chem. Soc.*, 2684 (1958)

Q17 H. W. Sternberg, R. E. Markby and I. Wender, *J. Electrochem. Soc.*, **110,** 425 (1963)

Q18 H. W. Sternberg, R. E. Markby, I. Wender and D. M. Mohilner, *J. Electrochem. Soc.*, **113,** 1060 (1966)

Q19 R. A. Benkeser, E. M. Kaiser and R. F. Lambert, *J. Amer. Chem. Soc.*, **86,** 5272 (1964)

Q20 A. Misono, T. Osa, T. Yamagishi and T. Kodama, *J. Electrochem. Soc.*, **115,** 266 (1968)

Q21 H. W. Sternberg, R. E. Markby and I. Wender, *J. Amer. Chem. Soc.*, **89,** 186 (1967)

Q22 H. W. Sternberg, R. E. Markby, I. Wender and D. M. Mohilner, *J. Amer. Chem. Soc.*, **91,** 4191 (1969)

Q23 K. Yang, J. D. Reedy and R. L. Hartshorn, *Chem. Abstr.*, **72,** 74124b

Q24 T. Asahari, M. Seno and H. Kaneko, *Bull. Chem. Soc. Japan*, **41,** 2985 (1968)

Q25 M. E. Peover, *in* Electroanalytical Chemistry, Vol. II (A. J. Bard, ed.), Marcel Dekker, New York (1967)

Q26 J. R. Jezorek and H. B. Mark, *J. Phys. Chem.*, **74,** 1627 (1970)

Q27 S. Wawzonek, E. W. Blaha, R. Berkey and M. E. Runner, *J. Electrochem. Soc.* **102,** 235 (1955)

Q28 S. Wawzonek and D. Wearring, *J. Amer. Chem. Soc.*, **81,** 2067 (1959)

Q29 R. Dietz, M. E. Peover and R. Wilson, *J. Chem. Soc.*, **B,** 75 (1968)

Q30 M. E. Peover, *Electrochim. Acta*, **13,** 1083 (1968)

Q31 D. L. Maricle, *Anal. Chem.*, **35,** 683 (1963)

Q32 R. Dietz and M. E. Peover, *Trans. Faraday Soc.*, **62,** 3535 (1966)

Q33 P. H. Reiger, I. Bernal, W. H. Reinmuth and G. K. Fraenkel, *J. Amer. Chem. Soc.*, **85,** 683 (1963)

Q34 I. G. Sevastyanova and A. P. Tomilov, *Zh. Obshch. Khim.*, **33,** 2815 (1963)

Q35 I. G. Sevastyanova and A. P. Tomilov, *Electrokhimiya*, **3,** 563 (1967)

Q36 P. E. Given and M. E. Peover, *J. Chem. Soc.*, 385 (1960)

Q37 S. Wawzonek and A. Gunderson, *J. Electrochem. Soc.*, **111,** 324 (1964)

Q38 J. Simonet, *Compt. Rend.*, **C263,** 685, 1546 (1966)

Q39 J. Weimann and M. L. Bougerra, *Ann. Chem. (Paris)*, **35,** (1967)

Q40 K. W. Bowers, R. W. Giese, J. Grimshaw, H. O. House, N. H. Kolodny, K. Kronberber and D. K. Roe, *J. Amer. Chem. Soc.*, **92,** 2783 (1970)

Q41 R. C. Buchta and D. H. Evans, *Anal. Chem.*, **40,** 2181 (1968); *J. Electrochem. Soc.*, **117,** 1494 (1970)

Q42 P. E. Iverson, *Tetrahedron Lett.*, 55 (1971)
Q43 J. M. Saveant, *J. Electroanal. Chem.*, **29,** 87 (1971)
Q44 R. H. Philp, T. Laycoff and R. N. Adam, *J. Electrochem. Soc.*, **111,** 1189 (1964)
Q45 K. B. Wiberg and T. P. Lewis, *J. Amer. Chem. Soc.*, **92,** 7154 (1970)
Q46 A. H. Maki and D. H. Geske, *J. Amer. Chem. Soc.*, **83,** 1852 (1961)
Q47 J. G. Lawless and M. D. Hawley, *J. Electroanal. Chem.*, **21,** 365 (1969)
Q48 A. K. Hoffman and A. T. Henderson, *J. Amer. Chem. Soc.*, **83,** 4671 (1961)
Q49 A. K. Hoffmann, W. G. Hodgson, D. L. Maricle and W. H. Jura, *J. Amer. Chem. Soc.*, **86,** 631 (1964)
Q50 J. G. Lawless, D. E. Bartok and M. D. Hawley, *J. Amer. Chem. Soc.*, **91,** 7121 (1969)
Q51 P. Petersen, A. K. Carpenter and B. F. Nelson, *J. Electroanal. Chem.*, **27,** 1 (1971)
Q52 S. Wawzonek, R. C. Duty, J. H. Wagenknecht, *J. Electrochem. Soc.*, **111,** 74 (1964)
Q53 M. R. Rifi, *J. Amer. Chem. Soc.*, **89,** 4442 (1967); *Tetrahedron Lett.*, 1043 (1969)
Q54 H. E. Ulery, *J. Electrochem. Soc.*, **116,** 1201 (1969)
Q55 R. Galli and F. Olivani, *J. Electroanal. Chem.*, **25,** 331 (1970)
Q56 M. Fleischmann, D. Pletcher and C. J. Vance, *J. Electroanal. Chem.*, **29,** 325 (1971)
Q57 S. Wawzonek and J. H. Wagenknecht *in* Polarography, 1964, p. 1035 (G. J. Hills, ed.), Interscience, New York (1966)
Q58 M. Baizer and J. H. Wagenknecht, *J. Electrochem. Soc.*, **114,** 1095 (1967)
Q59 S. D. Ross, M. Finkelstein, R. C. Peterson, *J. Amer. Chem. Soc.*, **92,** 6003 (1970)
Q60 O. R. Brown and E. R. Gonzalez, *J. Electroanal. Chem.*, **35,** 13 (1972)
Q61 J. D. Littlehailes and B. J. Woodhall, *Dis. Faraday Soc.*, **45,** 187 (1968)
Q62 B. Kratochvil and R. Long, *Anal. Chem.*, **42,** 43 (1970)
Q63 R. J. Wiersma and R. L. Middaugh, *J. Amer. Chem. Soc.*, **89,** 5078 (1967); *Inorg. Chem.*, **8,** 2074 (1969)
Q64 R. L. Middaugh and F. Farha, *J. Amer. Chem. Soc.*, **88,** 4147 (1966)
Q65 D. E. Smith, E. B. Rupp and D. F. Schriver, *J. Amer. Chem. Soc.*, **89,** 5562 (1967)
Q66 J. Q. Chambers, A. D. Norman, M. R. Bichell and S. H. Cadle, *J. Amer. Chem. Soc.*, **90,** 6056 (1968)
Q67 R. A. Meinzer, D. W. Pratt and R. J. Myers, *J. Amer. Chem. Soc.*, **91,** 6623 (1969)
Q68 A. J. Banister, O. R. Brown, H. Clarke and B. Thornton, in preparation
Q69 R. E. Dessy and R. L. Pohl, *J. Amer. Chem. Soc.*, **90,** 1995, 2005 (1968)
Q70 A. Brenner *in* Advances in Electrochemistry and Electrochemical Engineering, Vol. V (C. W. Tobias, ed.), Interscience, New York (1967)
Q71 B. Burrows and R. Jasinski, *J. Electrochem. Soc.*, **115,** 365 (1965)
Q72 D. T. Sawyer and J. L. Roberts, *J. Electroanal. Chem.*, **12,** 90 (1966)
Q73 R. Dietz, A. E. J. Forno, B. E. Larcombe and M. E. Peover, *J. Chem. Soc.*, **B,** 816 (1970)
Q74 V. D. Parker, *Chem. Comm.*, 848 (1969)
Q75 G. Manning, V. D. Parker and R. N. Adams, *J. Amer. Chem. Soc.*, **91,** 4584 (1969)
Q76 V. D. Parker and T. L. Burgert, *Tetrahedron Lett.*, 4065 (1965)
Q77 J. Phelps, K. S. V. Santhanam and A. J. Bard, *J. Amer. Chem. Soc.*, **89,** 1752 (1967)
Q78 L. S. Marcoux, J. M. Fritsch and R. N. Adams, *J. Amer. Chem. Soc.*, **89,** 5766 (1967)

Q79 K. Nyberg, *Acta Chem. Scand.*, **24,** 1609 (1970)
Q80 L. Eberson and K. Nyberg, *Tetrahedron Lett.*, 2389 (1966)
Q81 L. Eberson, *J. Amer. Chem. Soc.*, **89,** 4669 (1967)
Q82 V. D. Parker, *Chem. Comm.*, 1164 (1968)
Q83 S. D. Ross, M. Finkelstein and R. C. Petersen, *J. Amer. Chem. Soc.*, **89,** 4088 (1967)
Q84 V. D. Parker and R. N. Adams, *Tetrahedron Lett.*, 1721 (1969)
Q85 L. Eberson and K. Nyberg, *J. Amer. Chem. Soc.*, **88,** 1686 (1966)
Q86 K. Nyberg, *Chem. Comm.*, 774 (1969)
Q87 R. E. Sioda, *J. Phys. Chem.*, **72,** 2322 (1968)
Q88 H. N. Blount and T. Kuwana, *J. Electroanal. Chem.*, **27,** 464 (1970)
Q89 K. Umemoto, *Bull. Chem. Soc., Japan*, **40,** 1058 (1967)
Q90 G. Mengoli and G. Vidotto, *Macromol. Chem.*, **129,** 73 (1969)
Q91 J. D. Stuart, E. J. Majeski and W. E. Ohnesorge, *Chem. Comm.*, 353 (1970)
Q92 V. D. Parker and L. Eberson, *Chem. Comm.*, 1290 (1970)
Q93 H. Lund, *Acta Chem. Scand.*, **11,** 491 (1957)
Q94 O. R. Brown, S. Chandra and J. A. Harrison, *J. Electroanal. Chem.*, **38,** 185 (1972)
Q95 L. A. Miller and A. K. Hoffman, *J. Amer. Chem. Soc.*, **89,** 593 (1967)
Q96 V. D. Parker and L. Eberson, *Chem. Comm.*, 973 (1969)
Q97 A. B. Suttie, *Tetrahedron Lett.*, 953 (1969)
Q98 T. J. Vermillion and I. A. Pearl, *J. Electrochem. Soc.*, **111,** 1392 (1964)
Q99 V. D. Parker, *Chem. Comm.*, 610 (1969); *J. Amer. Chem. Soc.*, **91,** 5380 (1969)
Q100 D. Hawley and R. N. Adams, *J. Electroanal. Chem.*, **8,** (1964) 163
Q101 V. D. Parker and A. Ronlen, *Chem. Comm.*, 1567 (1970)
Q102 S. Andreades and E. W. Zahnow, *J. Amer. Chem. Soc.*, **91,** 4181 (1969)
Q103 M. Melicharek and R. F. Nelson, *J. Electroanal. Chem.*, **26,** 201 (1970)
Q104 E. T. Seo, R. F. Nelson, J. M. Fritsch, L. S. Marcoux, O. W. Leedy and R. N. Adams, *J. Amer. Chem. Soc.*, **88,** 3498 (1966)
Q105 L. Papouchado, R. N. Adams and S. Feldberg, *J. Electroanal. Chem.*, **21,** 408 (1969)
Q106 G. Cauquis, G. Fauvelot and J. Rigaudy, *Compt. Rend.*, **264C,** 1758, 1958 (1967); *Bull. Soc. Chim. France*, 4928 (1968)
Q107 G. Cauquis, J. Badoz-Lambling and J-P. Billon, *Bull. Soc. Chim. France*, 1433 (1965)
Q108 V. D. Parker, *Chem. and Ind. (London)*, 1363 (1968)
Q109 T. Inoue and S. Tsutsumi, *Bull. Chem. Soc., Japan*, **38,** 661 (1965)
Q110 K. Nyberg, *Acta. Chem. Scand.*, **24,** 473 (1970)
Q111 G. Faita, M. Fleischmann and D. Pletcher, *J. Electroanal. Chem.*, **25,** 455 (1970)
Q112 J. M. Kornprobst and A. Laurent, *Bull. Soc. Chim. France*, 3653 (1968); 1490 (1970)
Q113 D. L. Muck and E. R. Wilson, *J. Electrochem. Soc.*, **117,** 1358 (1970)
Q114 E. M. Martlett, *Ann. N.Y. Acad. Sci.*, **125,** 12 (1965)
Q115 A. P. Tomilov and I. N. Brago *in* Progress in Electrochemistry of Organic Compounds (A. N. Frumkin and A. B. Ershler, eds.), Plenum, London (1971)
Q116 N. L. Weinberg and H. R. Weinberg, *Chem. Rev.*, **68,** 449 (1968)
Q117 R. A. Marcus, *Can. J. Chem.*, **37,** 155 (1959); *J. Chem. Phys.*, **43,** 679 (1965)
Q118 R. A. Marcus, *Ann. Rev. Phys. Chem.*, **15,** 155 (1964)
Q119 V. G. Levich, *in* Advances in Electrochemistry and Electrochemical Engineering, Vol. IV (P. Delahay, ed.) (1965); and in Physical Chemistry, Vol. IXA (H. Eyring, D. Henderson and W. Jost, eds.), Academic, New York (1970)
Q120 T. Biegler, E. R. Gonzalez and R. Parsons, *Coll. Czech. Chem. Comm.*, **36,** 414 (1971)

Q121 W. Jaenicke and P. H. Schweitzer, *Z. Phys. Chem. (Frankfurt)*, **52**, 104 (1967)
Q122 J. O'M. Bockris and R. Parsons, *Trans. Faraday Soc.*, **45**, 916 (1949)
Q123 L. Peter, Ph.D. Thesis, University of Southampton (1970)
Q124 T. Kuwana, D. E. Bublitz and G. Hoh, *J. Amer. Chem. Soc.*, **82**, 5811 (1960)
Q125 D. E. Walker, R. N. Adams and A. L. Juliard, *Anal. Chem.*, **32**, 1526 (1960)
Q126 A. C. Aten, C. Büthker and G. J. Hoytink, *Trans. Faraday Soc.*, **55**, 324 (1959)
Q127 A. C. Aten and G. J. Hoytink, Advances in Polarography, p. 777, Pergamon, Oxford (1961)
Q128 R. Dietz and M. E. Peover, *Disc. Faraday Soc.*, **45**, 154 (1968)
Q129 R. D. Allendoefer and P. H. Rieger, *J. Amer. Chem. Soc.*, **87**, 2336 (1965)
Q130 M. E. Peover and B. S. White, *J. Electroanal. Chem.*, **13**, 93 (1967)
Q131 M. E. Peover and J. S. Powell, *J. Electroanal. Chem.*, **20**, 427 (1969)

Subject Index

The numbers refer to pages; when followed by T or A they refer to data in a Table or Appendix.

Il n'était manifestement pas très heureux qu'on lui rappelle ses liens avec un social-démocrate. Il prit une tranche de gâteau, qu'il laissa tomber maladroitement sur le tapis, essaya vainement de ramasser les miettes et finit par renoncer.

Carla se demanda : De quoi a-t-il peur ?

Heinrich alla droit au but. « Tu as dû entendre parler d'Akelberg, j'imagine. »

Carla observait attentivement Gottfried. L'espace d'un instant, une expression fugace passa sur son visage, mais il se composa très vite un air indifférent. « Une petite ville de Bavière ?

— Il y a un hôpital là-bas, continua Heinrich. Pour déficients mentaux et physiques.

— Je ne savais pas.

— Nous pensons qu'il s'y passe des choses étranges et nous nous demandions si tu avais des informations à ce sujet.

— Certainement pas. Que s'y passe-t-il d'étrange ? »

Ce fut Werner qui répondit.

« Mon frère est mort dans cet hôpital. D'appendicite, nous a-t-on dit. Le fils de la domestique de Herr von Ulrich est mort en même temps, dans le même hôpital, de la même chose.

— C'est très triste. Une coïncidence, probablement.

— Le fils de notre domestique n'avait plus d'appendice, intervint Carla. Il a été opéré il y a deux ans.

— Je comprends votre inquiétude, dit Gottfried. Mais il s'agit très probablement d'une erreur administrative.

— Dans ce cas, nous aimerions en être certains, insista Werner.

— Bien sûr. Avez-vous écrit à l'hôpital ?

— J'avais écrit il y a quelque temps pour demander quand notre bonne pourrait aller voir son fils, dit Carla. On ne m'a jamais répondu.

— Mon père a téléphoné ce matin, ajouta Werner. Le médecin-chef lui a raccroché au nez.

— Mon Dieu, quelles manières déplorables ! Mais, vous savez, cela ne concerne pas vraiment les Affaires étrangères. »

Werner se pencha en avant. « Herr von Kessel, est-il possible que ces deux enfants aient fait l'objet d'une expérience secrète qui aurait mal tourné ? »

de Frieda, Ludwig Franck, lui lança : « Qu'est-ce qu'ils t'ont dit à Wannsee ? »

Carla n'aimait pas beaucoup Ludwig. C'était un homme de droite qui avait soutenu les nazis dans les premiers temps. Sans doute avait-il évolué, c'était le cas de nombreux industriels désormais, mais il ne manifestait pas l'humilité que l'on pouvait attendre de ceux qui, comme lui, s'étaient aussi lourdement trompés.

Elle ne répondit pas tout de suite. Elle s'assit à la table et regarda les membres de la famille : Ludwig, Monika, Werner et Frieda, et le majordome qui s'activait à l'arrière-plan. Elle rassembla ses idées.

« Allons, mon petit, réponds-moi ! » insista Ludwig d'une voix impérieuse.

Il tenait à la main une lettre semblable à celle qu'avait reçue Ada et l'agitait d'un geste rageur.

Monika posa une main apaisante sur le bras de son mari. « Du calme, Ludi.

— Je veux savoir ! »

Carla considéra son visage empourpré, barré d'une petite moustache noire. Il était éperdu de chagrin, cela se voyait. En d'autres circonstances, Carla aurait refusé de répondre à un interlocuteur aussi discourtois. Mais il avait des excuses. Elle décida de passer outre. « Le directeur – le professeur Willrich – nous a dit qu'on disposait d'un nouveau traitement pour la pathologie de Kurt.

— On nous a dit la même chose. Quel genre de traitement ?

— Je le lui ai demandé. Il m'a répondu que je ne comprendrais pas. J'ai insisté. Il a parlé de médicaments, mais n'a pas voulu m'en dire davantage. Je peux voir votre lettre, Herr Franck ? »

Ludwig lui signifia par son attitude que c'était à lui de poser les questions. Il lui tendit néanmoins la lettre.

Elle était absolument identique à celle d'Ada. Carla eut la pénible intuition que la secrétaire en avait tapé un certain nombre en se contentant de changer les noms.

« Comment deux garçons peuvent-ils mourir d'appendicite au même moment ? s'interrogea Herr Franck. Ce n'est pas une maladie contagieuse.

— Kurt n'est certainement pas mort d'appendicite, répondit Carla. Pour la bonne raison qu'il n'avait plus d'appendice. On le lui a enlevé il y a deux ans.

— Bien, lança Ludwig. Assez parlé. » Il arracha la lettre des mains de Carla. « Je vais en toucher un mot à quelqu'un du gouvernement. » Il quitta la pièce.

Monika le suivit, ainsi que le majordome.

Carla s'approcha de Frieda et lui prit la main. « Je suis de tout cœur avec toi, dit-elle.

— Merci », murmura Frieda.

Carla se dirigea vers Werner. Il se leva et la prit dans ses bras. Elle sentit une larme couler sur son front. Elle était en proie à une vive émotion, vaguement ambiguë. Elle avait le cœur lourd mais ne pouvait s'empêcher de savourer le plaisir de sentir son corps contre le sien et le doux contact de ses mains.

Au bout d'un long moment, Werner s'écarta et s'écria d'une voix vibrante de colère : « Mon père a appelé deux fois l'hôpital. La seconde fois, on lui a dit qu'il n'y avait rien à ajouter et on lui a raccroché au nez. Mais je compte bien savoir ce qui est arrivé à mon frère et on ne m'enverra pas balader, je ne me laisserai pas faire.

— Ça ne le fera pas revenir, objecta Frieda.

— Je veux quand même savoir. Au besoin j'irai à Akelberg.

— Je me demande s'il y a quelqu'un à Berlin qui pourrait nous aider, dit Carla.

— Il faudrait quelqu'un qui soit au gouvernement, précisa Werner.

— Le père d'Heinrich… », suggéra Frieda.

Werner claqua des doigts. « C'est notre homme. Il a appartenu au parti du Centre. Maintenant, il est nazi et occupe un poste important aux Affaires étrangères.

— Tu crois qu'Heinrich obtiendrait un rendez-vous pour nous ? demanda Carla.

— Oui, si Frieda le lui demande. Heinrich ferait n'importe quoi pour elle. »

Carla le croyait volontiers. Heinrich mettait toujours tant de passion dans tout ce qu'il faisait.

« Je l'appelle tout de suite », dit Frieda.

Elle se rendit dans le vestibule. Carla et Werner se retrouvèrent assis côte à côte. Il l'entoura de son bras et elle posa la tête sur son épaule tout en se demandant s'il fallait attribuer ces marques d'affection au contrecoup du drame qui les frappait ou à autre chose.

Frieda revint en annonçant : « Le père d'Heinrich peut nous recevoir immédiatement. »

Ils s'entassèrent sur le siège avant de la voiture de sport de Werner.

« Je ne sais pas comment tu te débrouilles pour rouler avec cette voiture, remarqua Frieda alors qu'il démarrait. Même notre père n'arrive pas à obtenir d'essence pour son usage personnel.

— Je dis à mon supérieur que c'est pour mes déplacements officiels. » Werner travaillait pour un général important. « Mais je ne sais pas combien de temps ça va durer. »

Les von Kessel habitaient le même quartier et ils arrivèrent chez eux en moins de cinq minutes.

Quoique plus petite que celle des Franck, la maison était luxueuse. Heinrich les accueillit à la porte et les fit entrer dans un salon, ornementé d'une sculpture sur bois ancienne représentant un aigle, où s'alignaient des étagères chargées de livres reliés en cuir.

Frieda l'embrassa. « Merci, lui dit-elle. Cela n'a pas dû être facile. Je sais que tu ne t'entends pas très bien avec ton père. »

Heinrich rougit de plaisir.

Sa mère leur apporta un gâteau et du café. C'était une personne simple et chaleureuse. Après les avoir servis, elle s'éclipsa, comme une domestique.

Le père d'Heinrich, Gottfried, entra à son tour. Il avait la même tignasse épaisse et raide que son fils, mais argentée au lieu d'être noire.

Heinrich lui dit : « Je te présente Werner et Frieda Franck. C'est leur père qui fabrique la *Volksradio*, la radio du peuple.

— Ah, oui, acquiesça Gottfried. Je l'ai rencontré au Herrenklub.

— Et voici Carla von Ulrich. Je crois que tu connais aussi son père.

— Nous avons travaillé ensemble à l'ambassade allemande à Londres. En 1914. »

Gottfried se cala dans son fauteuil. « C'est parfaitement exclu », assura-t-il. Carla eut l'impression qu'il disait la vérité. « Cela n'existe pas. » Il paraissait soulagé.

Werner n'avait apparemment plus de questions à poser. Carla, pour sa part, n'était pas satisfaite. Elle se demandait pourquoi Gottfried semblait si tranquillisé par la réponse qu'il venait de leur donner. Était-ce parce qu'il taisait une réalité bien plus affreuse encore ?

Une idée épouvantable lui traversa l'esprit, tellement effroyable qu'elle osait à peine l'envisager.

« Bon, eh bien, si nous en avons fini…, dit Gottfried.

— Vous êtes absolument certain qu'ils n'ont pas succombé à un traitement expérimental qui s'est révélé fatal ? l'interrompit Carla.

— Absolument.

— Pour pouvoir affirmer avec autant d'assurance ce qui *n'existe pas* à Akelberg, vous devez avoir une petite idée de ce qui s'y fait.

— Pas nécessairement, nia-t-il avec un regain de nervosité manifeste qui confirma à Carla qu'elle avait mis le doigt sur quelque chose.

— Je me souviens d'une affiche nazie », continua-t-elle. C'était ce souvenir qui avait fait surgir l'idée insupportable qui lui était venue. « On y voyait un infirmier à côté d'un homme atteint d'une infirmité congénitale. Le texte était approximativement le suivant : "Soixante mille marks, voilà ce que coûte à la communauté pendant la durée de sa vie cette personne atteinte d'une maladie héréditaire. Peuple allemand, c'est ton argent !" C'était une publicité pour une revue, je crois.

— J'ai vu ces annonces », lâcha Gottfried d'un ton dédaigneux, comme si elles ne le concernaient pas.

Carla se leva. « Vous êtes catholique, monsieur, et vous avez élevé Heinrich dans la foi catholique. »

Gottfried émit un petit bruit méprisant : « Heinrich se dit athée.

— Mais vous ne l'êtes pas. À vos yeux, la vie humaine est sacrée.

— Oui.

— Vous dites que les médecins d'Akelberg ne se livrent à aucune expérience thérapeutique risquée sur leurs patients et je vous crois.

— Merci.

— Mais ne font-ils pas autre chose ? Quelque chose de bien pire ?

— Mais non, voyons.

— Êtes-vous sûr qu'ils ne *tuent* pas délibérément les incurables ? »

Gottfried secoua la tête en silence.

Carla s'approcha et lui parla à voix basse, comme s'ils étaient seuls dans la pièce. « En tant que catholique convaincu du caractère sacré de la vie humaine, pouvez-vous m'assurer, la main sur le cœur, qu'on ne tue pas les petits malades d'Akelberg ? »

Gottfried sourit, esquissa un geste qui se voulait rassurant et ouvrit la bouche, mais il n'en sortit aucun son.

Carla s'agenouilla devant lui. « Pouvez-vous me le jurer, s'il vous plaît ? Maintenant. Ici, dans votre maison, en présence de quatre jeunes Allemands, votre fils et ses amis. Dites-nous simplement la vérité. Regardez-moi dans les yeux et dites-moi que notre gouvernement ne supprime pas les enfants déficients. »

Un silence absolu régnait dans la pièce. Gottfried parut sur le point de parler, mais il se ravisa. Il ferma les yeux, une grimace déforma sa bouche et il baissa la tête. Les quatre jeunes gens considéraient, ahuris, son visage crispé.

Il rouvrit enfin les yeux, posa le regard sur chacun d'eux tour à tour, en terminant par son fils.

Puis il se leva et sortit.

3.

Le lendemain, Werner dit à Carla : « Je n'en peux plus. Cela fait vingt-quatre heures que nous ne parlons que de ça. Nous allons devenir fous si nous ne nous changeons pas les idées. Je t'emmène au cinéma. »

Ils rejoignirent le Kurfürstendamm, une rue aux nombreuses salles de spectacle et boutiques que tout le monde appelait le Ku'damm. Les meilleurs réalisateurs allemands s'étaient exilés à Hollywood depuis plusieurs années et les films locaux étaient désormais de qualité médiocre. Ils virent *Trois soldats*, qui se passait pendant l'invasion de la France.

Les trois soldats en question étaient un solide sergent nazi, un mou pleurnichard à l'air vaguement juif et un jeune homme grave. Ce dernier posait des questions naïves – comme « Est-ce que les Juifs sont vraiment nuisibles ? » – auxquelles le sergent répondait par de longs sermons assommants. Au moment de prendre part aux combats, le pleurnichard avouait être communiste, désertait et se faisait finalement tuer dans un raid aérien. Le jeune homme grave se battait vaillamment, était promu sergent et devenait un admirateur du Führer. Malgré un scénario lamentable, les scènes de batailles étaient spectaculaires.

Werner tint la main de Carla pendant toute la durée du film. Elle espérait qu'il l'embrasserait dans l'obscurité, mais il n'en fit rien.

Quand les lumières se rallumèrent, il déclara : « Bon, c'était affreusement mauvais, mais au moins, j'ai pensé à autre chose pendant deux heures. »

Ils sortirent et retournèrent à sa voiture. « Et si on allait se balader ? proposa-t-il. C'est sans doute la dernière occasion. La voiture va être mise sur cales la semaine prochaine. »

Il prit le volant en direction de Grünewald. Pendant le trajet, la conversation de la veille avec Gottfried von Kessel revint hanter les pensées de Carla. Elle avait beau la tourner et la retourner dans sa tête, elle ne pouvait balayer l'horrible conclusion à laquelle ils étaient tous arrivés. Kurt et Axel n'avaient pas été les victimes accidentelles d'une expérimentation médicale dangereuse comme elle l'avait d'abord cru. Gottfried avait catégoriquement réfuté cette possibilité. Mais il n'avait pu se résoudre à leur mentir en niant que le gouvernement assassinait volontairement les handicapés et les malades mentaux. C'était difficile à croire, même de la part d'individus aussi barbares et dénués de scrupules que les nazis. Pourtant, la réaction de Gottfried dont Carla avait été témoin révélait clairement un sentiment de culpabilité.

Une fois dans la forêt, Werner quitta la route et s'engagea dans une allée qu'il suivit jusqu'à ce qu'ils se trouvent perdus au milieu de la végétation. Carla le soupçonnait d'avoir déjà amené là d'autres filles.

Il éteignit les phares. Ils furent plongés dans le noir complet. « Je vais parler au général Dorn », dit-il. Dorn était son supérieur, un important officier de l'armée de l'air. « Et toi ?

— Mon père prétend qu'il n'existe plus d'opposition politique mais que les Églises ont toujours de l'influence. Aucun vrai croyant ne peut excuser pareils agissements.

— Tu es pratiquante ?

— Pas vraiment. Mon père si. Pour lui, la foi protestante fait partie de l'héritage allemand auquel il est profondément attaché. Mutter l'accompagne au temple, pourtant je la soupçonne d'avoir des opinions un peu excentriques en matière de théologie. Pour ma part, je crois en Dieu, mais je suis persuadée qu'il se fiche pas mal que les gens soient catholiques ou protestants, musulmans ou bouddhistes. Et j'aime chanter des cantiques. »

Werner baissa la voix pour dire dans un murmure : « Il m'est impossible de croire en un Dieu qui permet aux nazis d'assassiner des enfants.

— Je ne peux pas t'en vouloir.

— Que va faire ton père ?

— Il va en parler au pasteur.

— Bien. »

Ils restèrent silencieux un moment puis il la prit dans ses bras. « Je peux ? » chuchota-t-il.

Tendue et impatiente, elle eut du mal à émettre un son. Sa réponse ne fut d'abord qu'un vague grognement. Elle se reprit pour répondre d'une voix plus claire : « Si ça peut apaiser ta tristesse…, oui. »

Il l'embrassa.

Elle lui rendit son baiser avec ferveur. Il lui caressa les cheveux, puis la poitrine. Elle eut un instant d'hésitation. Les autres filles disaient que si on laissait les choses aller plus loin, on perdait tout contrôle de soi.

Elle décida de courir ce risque.

Elle frôla sa joue en l'embrassant. Lui caressa le cou du bout des doigts en savourant la chaude douceur de sa peau. Elle glissa la main sous sa veste et la promena sur son corps, explorant son dos, ses côtes, ses omoplates.

Elle soupira au moment où elle sentit sa main sur sa cuisse, sous la jupe. Dès qu'il l'effleura entre les jambes, elle les écarta. Les autres disaient que les garçons vous prenaient pour une fille facile quand on faisait ça. Mais c'était plus fort qu'elle.

Il posa le doigt juste au bon endroit. Il ne chercha pas à glisser sa main dans sa culotte, se contentant d'une caresse légère à travers le coton. Elle entendit des plaintes monter de sa propre gorge, doucement d'abord, puis de plus en plus fortes. Elles finirent par éclater en un cri de plaisir qu'elle assourdit en enfouissant son visage au creux de son épaule. Elle dut écarter sa main car la sensation était devenue insupportable.

Elle était haletante. Quand elle eut un peu repris son souffle, elle l'embrassa dans le cou. Il passa affectueusement sa main sur sa joue. Elle lui demanda : « Je peux faire quelque chose pour toi ?

— Seulement si tu le souhaites. »

Elle en avait tellement envie qu'elle en était gênée. « Ce qu'il y a, c'est que je n'ai jamais…

— Je sais. Je vais te montrer. »

4.

Le pasteur Ochs était un homme tranquille et corpulent, qui avait une grande maison, une gentille femme et cinq enfants. Carla craignait qu'il ne refuse de s'impliquer. C'était le sous-estimer. Il avait déjà entendu des rumeurs qui troublaient sa conscience et accepta d'accompagner Walter au centre de soins pour enfants de Wannsee. Le professeur Willrich n'éconduirait pas un homme d'Église.

Ils décidèrent d'emmener Carla puisqu'elle avait assisté à l'entretien avec Ada. Le directeur hésiterait à changer de version en sa présence.

Pendant le trajet en train, Ochs leur suggéra de le laisser mener la conversation. « Le directeur est probablement nazi », expliqua-t-il. Les hommes qui occupaient des postes importants étaient pour la plupart membres du parti. « À ses yeux, un ancien député social-démocrate sera d'emblée un ennemi. Je jouerai le rôle d'arbitre neutre. Cela devrait nous permettre d'en apprendre davantage. »

Carla n'en était pas convaincue. Elle pensait que son père saurait mieux poser les bonnes questions. Mais Walter accepta la proposition du pasteur.

C'était le printemps et il faisait plus doux que lors de la dernière visite de Carla. Des bateaux sillonnaient le lac. Carla décida de proposer à Werner de venir prochainement y faire un pique-nique. Elle voulait profiter de lui au maximum avant qu'il ne la délaisse pour une autre.

Un feu brûlait dans la cheminée du professeur Willrich, mais une fenêtre de son bureau était ouverte, laissant entrer une brise fraîche venue du lac.

Le directeur serra la main du pasteur et de Walter. Ayant reconnu Carla, il l'ignora après lui avoir jeté un bref regard. Il les invita à s'asseoir, mais Carla sentit une hostilité agacée derrière sa courtoisie. Il n'avait manifestement pas envie d'être questionné. Il prit une de ses pipes et se mit à la tripoter nerveusement. Il était moins arrogant devant ces deux hommes mûrs qu'il ne l'avait été en présence de deux jeunes femmes.

Ochs entama la discussion. « Professeur Willrich, Herr von Ulrich et d'autres membres de ma congrégation s'inquiètent des morts mystérieuses de plusieurs jeunes malades dont ils ont eu connaissance.

— Aucun enfant n'est décédé ici de mort mystérieuse, rétorqua Willrich. En réalité, aucun enfant n'est mort ici depuis deux ans. »

Ochs se tourna vers Walter. « C'est très rassurant, Walter, vous ne trouvez pas ?

— En effet. »

Ce n'était pas l'avis de Carla, mais elle préférait garder le silence pour le moment.

Ochs poursuivit d'une voix onctueuse : « Je suis certain que vous accordez toute votre attention aux enfants dont vous avez la charge.

— C'est un fait. » Willrich semblait se détendre.

« Il vous arrive néanmoins d'envoyer dans d'autres établissements les enfants qui vous ont été confiés ?

— Bien sûr, lorsque d'autres institutions peuvent leur offrir des traitements dont nous ne disposons pas.

— Quand un enfant est transféré ailleurs, je suppose qu'on ne vous tient pas forcément informé des traitements qu'il suit, ni de son état de santé.

— Absolument !

— Sauf s'il revient. »

Willrich ne répondit pas.

« Certains sont-ils revenus ?

— Non. »

Ochs haussa les épaules. « Ainsi, vous ne savez pas ce qui leur est arrivé.

— C'est exact. »

Ochs s'appuya au dossier de son siège et écarta les mains en signe d'ouverture. « Vous n'avez donc rien à cacher.

— Rien du tout.

— Certains des enfants qui ont été transférés sont morts. »

Willrich garda le silence.

Ochs insista gentiment. « C'est le cas, n'est-ce pas ?

— Je ne peux pas vous répondre avec certitude, monsieur le pasteur.

— Ah ! Car même si l'un de ces enfants venait à mourir, vous n'en seriez pas informé.

— C'est ce que je vous ai dit.

— Pardonnez-moi si je me répète. Je tiens seulement à m'assurer, sans doute possible, que vous n'êtes pas en mesure de faire la lumière sur ces décès.

— C'est le cas. »

Ochs se tourna à nouveau vers Walter. « Je pense que nous commençons à y voir remarquablement clair. »

Walter acquiesça.

Carla avait envie de hurler : *Nous n'y voyons pas clair du tout !*

Ochs avait déjà repris la parole. « Combien d'enfants approximativement avez-vous transférés, disons, au cours des douze derniers mois ?

— Dix, répondit Willrich. Très exactement. » Il s'accorda un sourire condescendant. « Nous autres scientifiques ne pouvons nous contenter d'approximations.

— Dix patients, sur… ?

— À ce jour, nous avons cent sept enfants ici.

— C'est une très faible proportion ! » admit Ochs.

Carla sentait la moutarde lui monter au nez. Ochs jouait le jeu de Willrich, c'était évident ! Pourquoi son père acceptait-il cette mascarade ?

Ochs poursuivit : « Ces enfants souffraient-ils tous de la même affection ou étaient-ils atteints de différentes maladies ?

— Différentes. » Willrich ouvrit un dossier posé sur son bureau. « Crétinisme, trisomie, microcéphalie, hydrocéphalie, malformation des membres, de la boîte crânienne et de la colonne vertébrale, et paralysie.

— Ce sont les types de patients qu'on vous a demandé d'envoyer à Akelberg ? »

Il y avait du progrès. C'était la première fois qu'il était question d'Akelberg et de directives des autorités supérieures. Ochs était peut-être plus malin qu'il n'en avait l'air.

Willrich ouvrit la bouche pour parler, mais Ochs le devança en lui posant une autre question. « Étaient-ils tous censés recevoir le même traitement spécial ? »

Willrich retrouva son sourire. « Là encore, je n'ai pas été informé et je ne peux donc pas vous le dire.

— Vous n'avez fait que respecter…

— Les ordres que j'ai reçus, en effet. »

Ce fut au tour d'Ochs de sourire. « Vous êtes un homme avisé. Vous choisissez vos mots avec soin. Ces enfants avaient-ils tous le même âge ?

— Au départ, le programme ne concernait que les enfants de moins de trois ans. Il a été élargi par la suite à tous les âges, oui. »

Carla nota au passage l'allusion à un « programme ». Personne n'avait encore admis l'existence d'un programme quelconque. Carla commençait à s'apercevoir qu'Ochs était un homme très habile.

Le pasteur prononça la phrase suivante comme pour confirmer un fait déjà établi. « Tous les enfants juifs qui vous sont confiés sont concernés, quelle que soit leur infirmité. »

Il y eut un long silence. Willrich semblait paralysé. Carla se demanda comment Ochs était informé du sort des enfants juifs. Peut-être n'était-il sûr de rien et n'était-ce qu'une hypothèse.

Le silence se prolongeant, Ochs ajouta : « Les enfants juifs et les *Mischlinge*, les demi-juifs, devrais-je dire. »

Sans parler, Willrich confirma d'un bref hochement de tête.

Ochs continua : « Il est rare, de nos jours, d'accorder la préférence à des enfants juifs, n'est-ce pas ? »

Willrich détourna les yeux.

Le pasteur se leva. Quand il reprit la parole, sa voix vibrait de colère. « Vous me dites que dix enfants souffrant d'affections diverses, et qui ne pouvaient donc pas tous recevoir le même traitement, ont été envoyés dans un hôpital spécial dont ils ne sont jamais revenus ; et que les Juifs y ont été transférés en priorité. Que croyiez-vous qu'il leur arrivait, professeur Willrich ? Au nom du ciel, que croyiez-vous ? »

Willrich semblait prêt à fondre en larmes.

« Vous n'avez pas de réponse à me donner, bien sûr, reprit Ochs plus calmement. Mais un jour, cette question vous sera posée par une autorité supérieure, la plus haute de toutes, en fait. »

Il tendit le bras, pointant sur lui un index vengeur.

« Et ce jour-là, mon fils, vous répondrez. »

Sur ces mots, il tourna les talons et quitta la pièce.

Carla et Walter lui emboîtèrent le pas.

5.

Le commissaire Thomas Macke affichait un sourire satisfait. Parfois, les ennemis de l'État lui mâchaient le travail. Au lieu d'agir en secret et de se cacher là où on ne les trouverait pas, ils livraient d'eux-mêmes leur identité et lui fournissaient des preuves irréfutables de leurs crimes. Ils étaient comme des poissons pour lesquels on pouvait se passer d'amorce et d'hameçon : ils sautaient de la rivière directement dans le panier du pêcheur pour passer à la casserole.

Le pasteur Ochs était de ceux-là.

Macke relut sa lettre. Elle était adressée au ministre de la Justice, Franz Gürtner.

Monsieur le ministre,

Le gouvernement a-t-il entrepris de tuer les enfants déficients physiques et mentaux ? Je vous pose la question sans détour car je veux une réponse claire et nette.

Quel imbécile ! Si la réponse était négative, sa lettre tenait de la calomnie. Si elle était positive, il se rendait coupable de trahison de secret d'État. Fallait-il être bête pour ne pas s'en rendre compte !

Ne pouvant ignorer les rumeurs qui circulaient dans ma paroisse, je me suis rendu au centre pour enfants de Wannsee où je me suis entretenu avec le directeur, le professeur Willrich. Ses réponses évasives m'ont convaincu qu'un projet terrible était en cours, un projet qui constitue probablement un crime et incontestablement un péché.

Il avait le culot de parler de crimes ! Ne lui était-il pas venu à l'esprit qu'accuser le gouvernement d'actes illégaux était en soi un acte illégal ? Croyait-il vivre dans une démocratie libérale dégénérée ?

Macke savait à quoi Ochs faisait allusion. Le programme s'appelait Aktion T4, en raison de l'adresse de son siège, Tiergartenstrasse 4. Le nom officiel du service était la « Fondation caritative de traitements et soins institutionnels », mais il était supervisé par le cabinet personnel d'Hitler, la chancellerie du Führer. Il était chargé d'organiser la mise à mort en douceur des incurables qui nécessitaient des soins coûteux et vains. Il avait fait un travail fantastique au cours des deux dernières années en débarrassant le pays de dizaines de milliers de bouches inutiles.

Malheureusement, l'opinion publique allemande n'étant pas encore assez mûre pour comprendre le bien-fondé de cette mesure, le programme devait être tenu secret.

Macke était dans la confidence. Il avait été promu commissaire et récemment admis au sein de l'élite paramilitaire nazie, les Schutzstaffel, autrement dit les SS. On l'avait mis au courant du programme Aktion T4 quand on lui avait confié l'affaire Ochs. Il en était fier. Il faisait vraiment partie des initiés désormais.

Malheureusement, certains avaient manqué de prudence, et le secret de l'Aktion T4 risquait d'être éventé.

Macke était chargé de stopper les fuites.

Les premières enquêtes avaient rapidement montré qu'il y avait trois hommes à faire taire de toute urgence : Peter Ochs, Walter von Ulrich et Werner Franck.

Franck était le fils aîné d'un fabricant de radios qui avait été un important partisan nazi de la première heure. L'industriel lui-même, Ludwig Franck, avait peu de temps auparavant déposé des demandes péremptoires d'information sur la mort de son fils infirme, mais avait promptement mis fin à ses démarches quand on l'avait menacé de fermer ses usines. En revanche, le jeune Werner, brillant officier du ministère de l'Air, avait continué à poser des questions embarrassantes en cherchant à se prévaloir de l'influence de son supérieur, le général Dorn.

Le ministère de l'Air, le plus grand immeuble de bureaux d'Europe, disait-on, était un édifice ultramoderne qui occupait tout un pâté de maisons de la Wilhelmstrasse, à deux pas du siège de la Gestapo, Prinz-Albrecht-Strasse. Macke s'y rendit à pied.

Grâce à son uniforme de SS, il put se permettre d'ignorer les gardes. Au comptoir d'accueil, il aboya : « Je veux voir le lieutenant Werner Franck. Immédiatement. » Le réceptionniste le fit monter dans un ascenseur et le précéda le long d'un couloir jusqu'à une porte ouverte sur un petit bureau. Le jeune homme assis à la table ne leva pas tout de suite les yeux des papiers étalés devant lui. En l'observant, Macke jugea qu'il devait avoir vingt-deux ans. Pourquoi n'était-il pas dans une unité du front, en train de bombarder l'Angleterre ? Son père avait dû le pistonner, se dit Macke avec rancœur. Werner avait bien l'air d'un fils à papa : uniforme de bonne coupe, chevalière en or et des cheveux trop longs qui n'avaient rien de militaire. Macke le méprisa sur-le-champ.

Werner griffonna une note au crayon et leva les yeux. Son expression aimable s'effaça dès qu'il reconnut l'uniforme des SS. Macke releva, non sans satisfaction, un bref éclair de peur dans son regard. Le jeune homme tenta aussitôt de masquer sa réaction sous des dehors affables, en se levant poliment avec un sourire accueillant, mais Macke ne s'y laissa pas prendre.

« Bonjour, commissaire. Asseyez-vous, je vous en prie.

— *Heil Hitler*, dit Macke.

— *Heil Hitler*. En quoi puis-je vous être utile ?

— Asseyez-vous et taisez-vous, jeune crétin », siffla Macke.

Werner s'efforça de dissimuler sa frayeur. « Alors ça, qu'ai-je fait pour mériter tant de colère ?

— Évitez de poser des questions. Vous parlerez quand on vous le demandera.

— Comme vous voudrez.

— À partir de maintenant, vous allez cesser de vous intéresser à ce qui est arrivé à votre frère Axel. »

Macke fut étonné de voir passer sur le visage de Werner une expression qui ressemblait à du soulagement. Curieux. Avait-il craint autre chose, quelque chose de plus redoutable que l'interdiction de poser des questions sur son frère ? Werner serait-il mêlé à d'autres activités subversives ?

Probablement pas, se dit Macke après réflexion. Il était sans doute soulagé de ne pas être arrêté et conduit dans les sous-sols de la Prinz-Albrecht-Strasse.

Werner n'était pas encore complètement dompté et trouva le courage de répliquer : « Pourquoi ne devrais-je pas essayer de savoir comment mon frère est mort ?

— Je vous ai dit de ne pas me poser de questions. Mettez-vous bien ça dans la tête : vous avez droit à un traitement de faveur parce que votre père a été un précieux allié du parti nazi. Sans quoi, c'est vous qui seriez dans mon bureau, et non l'inverse. »

La menace était limpide.

« Je vous sais gré de votre indulgence, dit Werner en essayant de conserver un semblant de dignité. Mais je veux savoir qui a tué mon frère et pourquoi.

— Quoi que vous fassiez, vous n'en apprendrez pas davantage. En revanche, toute nouvelle recherche d'information sera considérée comme une trahison de votre part.

— Cet entretien m'en a suffisamment appris. Il est désormais évident que mes pires craintes étaient fondées.

— Vous avez intérêt à renoncer à ce comportement séditieux, croyez-moi. »

Werner lui jeta un regard de défi mais ne répondit pas.

Macke reprit : « Sinon, le général Dorn sera informé que votre loyauté est toute relative. »

Werner savait très bien ce que cela signifiait. Il perdrait son poste confortable à Berlin et serait envoyé dans une caserne, sur un terrain d'aviation du nord de la France.

Il avait maintenant l'air moins insolent.

Macke se leva. Il avait perdu assez de temps. « Il paraît que le général Dorn vous trouve capable et intelligent, dit-il encore. Si vous vous montrez raisonnable, vous pourrez conserver votre place. »

Il s'en alla, nerveux et insatisfait. Il n'était pas sûr d'avoir brisé la volonté de Werner. Il avait perçu un fond de résistance inébranlable.

Il reporta son attention sur le cas du pasteur Ochs. Avec lui, il faudrait s'y prendre différemment. Macke retourna au siège de la Gestapo et réunit une petite équipe : Reinhold Wagner, Klaus Richter et Günther Schneider. Ils prirent une Mercedes 260D noire, la voiture préférée des hommes de la Gestapo, parfaitement discrète car analogue, par son modèle et sa couleur, à la plupart des taxis de Berlin. Dans les premiers temps, on les avait incités à agir ostensiblement pour que le public n'ignore pas les méthodes brutales qu'employait la Gestapo pour écraser l'opposition. Cependant, la terreur avait depuis longtemps fait son œuvre et il n'était plus nécessaire d'user de violence ouvertement. La Gestapo intervenait dorénavant discrètement, et toujours sous couvert de la légalité.

Ils se rendirent chez le pasteur, dont la maison jouxtait le grand temple protestant du Mitte, le quartier central. Tout comme Werner s'imaginait être protégé par son père, Ochs croyait sans doute être à l'abri à l'ombre de son Église. Il allait déchanter.

Macke sonna à la porte : autrefois, ils l'auraient enfoncée, juste pour le plaisir.

Une domestique vint ouvrir et il pénétra dans une vaste entrée bien éclairée, au parquet ciré revêtu d'épais tapis. Ses trois compagnons le suivirent. « Où est le maître de maison ? » demanda aimablement Macke à la servante.

Il ne l'avait pas menacée, et pourtant, elle avait l'air terrorisée. « Dans son bureau, monsieur, dit-elle en désignant une porte.

— Fais descendre la femme et les enfants dans la pièce voisine », commanda Macke à Wagner.

Ochs ouvrit la porte de son bureau et passa la tête. « Que se passe-t-il ? » demanda-t-il d'un ton indigné.

Macke se dirigea vers lui d'un pas décidé qui l'obligea à reculer et à le laisser entrer. C'était un petit cabinet de travail bien aménagé, avec un bureau à dessus de cuir et une bibliothèque remplie d'ouvrages de théologie. « Fermez la porte », ordonna Macke.

Ochs s'exécuta à contrecœur en disant : « J'espère que vous avez un motif valable pour justifier cette intrusion.

— Asseyez-vous et taisez-vous. »

Ochs en resta ébahi. On ne lui avait sans doute pas parlé sur ce ton depuis sa petite enfance. Les ecclésiastiques ne se faisaient généralement pas insulter, même par la police. Les nazis se moquaient bien de ces conventions débilitantes.

« C'est un affront ! » parvint enfin à articuler Ochs avant de s'asseoir.

Dans la pièce voisine, une voix féminine s'éleva : probablement l'épouse qui protestait. Ochs pâlit et se leva.

Macke le fit brutalement retomber sur sa chaise. « Restez où vous êtes. »

Ochs était un homme robuste, plus grand que Macke, mais il ne tenta pas de résister.

Macke adorait voir ces poseurs prétentieux se déballonner.

« Qui êtes-vous ? » demanda Ochs.

Macke ne le leur disait jamais. Ils pouvaient deviner, mais le doute était bien plus effrayant. Dans le cas improbable où une enquête serait engagée plus tard, tous les membres de l'équipe jureraient leurs grands dieux qu'ils avaient commencé par se présenter comme des officiers de police et montré leurs insignes.

Il sortit. Ses hommes poussaient sans ménagement des enfants dans le salon. Macke demanda à Reinhold Wagner d'aller garder un œil sur Ochs et de l'empêcher de quitter son bureau. Puis il suivit les enfants dans le salon.

Il y avait des rideaux à fleurs, des photos de famille sur le manteau de la cheminée et des fauteuils confortables recouverts de tissu à carreaux. Une agréable maison, une gentille famille.

Pourquoi ne se conduisaient-ils pas en fidèles sujets du Reich et se mêlaient-ils de ce qui ne les regardait pas ?

La domestique se tenait près de la fenêtre, la main sur la bouche comme pour retenir un cri. Quatre enfants étaient regroupés autour de Frau Ochs, une femme sans beauté, à la poitrine imposante, âgée d'une trentaine d'années. Elle tenait un cinquième enfant dans ses bras, une petite blonde bouclée d'environ deux ans.

Macke tapota la tête de la fillette. « Comment s'appelle cette petite ? » demanda-t-il.

Frau Ochs était terrifiée. Elle murmura : « Liselotte. Qu'est-ce que vous nous voulez ?

— Viens voir oncle Thomas, Liselotte, dit Macke en tendant les bras.

— Non ! » cria Frau Ochs. Elle serra l'enfant contre elle et se détourna.

Liselotte se mit à pleurer.

Macke adressa un signe à Klaus Richter.

Richter attrapa Frau Ochs par-derrière, lui tirant les bras, la forçant à lâcher sa fille. Macke rattrapa Liselotte avant qu'elle ne tombe. La fillette se tortilla dans tous les sens. Il l'agrippa plus fermement, la maintenant comme un chat récalcitrant. La fillette hurla encore plus fort.

Un garçon d'une douzaine d'années se précipita sur Macke, qu'il bourra de coups de poing inoffensifs. Il était temps de lui apprendre à respecter l'autorité. Macke cala Liselotte sur sa hanche gauche ; de la main droite, il empoigna l'effronté par le devant de sa chemise et le projeta à l'autre bout de la pièce en veillant à ce qu'il atterrisse sur un fauteuil. Le garçon poussa un glapissement de frayeur. Frau Ochs cria elle aussi. Le fauteuil se renversa et l'enfant bascula avec lui. Il ne s'était pas fait très mal, mais il fondit en larmes.

Macke emmena la fillette dans l'entrée. Elle appelait sa mère en poussant des cris suraigus. Macke la posa par terre. Elle courut vers la porte du salon et tambourina dessus en poussant des hurlements de terreur. Elle n'avait pas encore appris à tourner les poignées, remarqua Macke.

Macke la laissa dans le vestibule et regagna le bureau. Wagner montait la garde près de la porte. Ochs était debout au milieu de la pièce, blanc de peur. « Qu'est-ce que vous faites à mes enfants ? Pourquoi Liselotte pleure-t-elle ?

— Vous allez écrire une lettre.

— Oui, tout ce que vous voudrez, dit Ochs en se dirigeant vers sa table recouverte de cuir.

— Pas maintenant. Plus tard.

— Très bien. »

Macke était aux anges. Ochs était complètement soumis, contrairement à Werner. « Une lettre au ministre de la Justice, précisa-t-il.

— C'était donc ça…

— Vous lui expliquerez que vous avez compris que les allégations de votre première lettre étaient totalement erronées. Vous avez été manipulé à votre insu par des communistes. Vous présenterez vos excuses au ministre pour les problèmes provoqués par vos initiatives irréfléchies et lui promettrez de ne plus jamais évoquer ce sujet avec qui que ce soit.

— Oui, oui, comptez sur moi. Que font-ils à ma femme ?

— Rien. Elle crie à l'idée de ce qui lui arrivera si vous n'écrivez pas cette lettre.

— Je veux la voir.

— Vous n'arrangerez rien avec des exigences ineptes.

— Oui, bien sûr. Je suis désolé. Je vous prie de m'excuser. »

Les ennemis du nazisme étaient d'une veulerie ! « Vous écrirez cette lettre ce soir et la posterez demain.

— C'est entendu. Dois-je vous en envoyer une copie ?

— Je l'aurai de toute façon, espèce d'idiot. Vous croyez vraiment que le ministre lit lui-même vos gribouillages débiles ?

— Non, bien sûr que non, c'est évident. »

Macke marcha vers la porte. « Et évitez les gens comme Walter von Ulrich.

— Oui, je vous le promets. »

Macke sortit en faisant signe à Wagner de le suivre. Assise par terre, Liselotte braillait à pleins poumons. Macke ouvrit la porte du salon et appela Richter et Schneider.

Ils quittèrent la maison.

« La violence est parfois superflue », observa Macke d'un air songeur en montant dans la voiture.

Wagner prit le volant. Macke lui indiqua l'adresse des von Ulrich.

« Mais dans bien des cas, c'est la méthode la plus efficace », ajouta-t-il.

Von Ulrich habitait à proximité du temple protestant. Il vivait dans un vaste hôtel particulier qu'il n'avait manifestement pas les moyens d'entretenir. La peinture s'écaillait, les balustrades étaient rouillées, une vitre cassée avait été bouchée avec du carton. Ce n'était pas rare : beaucoup de maisons étaient en mauvais état en raison des mesures d'austérité dues à la guerre.

Une domestique vint leur ouvrir. Macke supposa qu'il s'agissait de la mère de l'enfant déficient qui était à l'origine de toute cette affaire. Il ne prit pas la peine de s'en assurer. Il ne servait à rien d'arrêter les femmes.

Walter sortit d'une pièce qui donnait sur le vestibule.

Macke se souvenait de lui. C'était le cousin de ce Robert von Ulrich dont il avait racheté le restaurant avec son frère huit ans auparavant. À l'époque, c'était un homme fier et arrogant. Il portait maintenant un costume défraîchi, mais avait gardé ses manières hautaines. « Que voulez-vous ? » demanda-t-il du ton de celui qui croit encore pouvoir exiger des explications.

Macke n'avait pas l'intention de perdre son temps. « Passez-lui les menottes », ordonna-t-il.

Wagner s'avança.

Une grande femme séduisante surgit et vint se placer devant von Ulrich. « Dites-moi qui vous êtes et ce que vous voulez », lança-t-elle d'une voix autoritaire.

C'était sa femme, certainement. Elle avait une pointe d'accent étranger. Pas étonnant.

Macke la gifla brutalement. Elle recula en chancelant.

« Tournez-vous et croisez les mains dans le dos, dit Macke à von Ulrich. Sinon, je lui fais avaler ses dents. »

Von Ulrich obtempéra.

Une jolie jeune fille en tenue d'infirmière arriva en dévalant l'escalier. « Vati ! cria-t-elle. Que se passe-t-il ? »

Macke se demanda, avec un léger pincement d'angoisse, combien il y avait de personnes dans la maison. Une famille ordinaire ne pouvait entraver l'action d'officiers de police entraînés, mais s'ils étaient vraiment nombreux, ils pouvaient créer suffisamment de désordre pour permettre à von Ulrich de s'échapper.

Apparemment, celui-ci ne cherchait pas la bagarre. « Ne fais rien ! dit-il à sa fille d'un ton suppliant. Tiens-toi à l'écart ! »

Terrifiée, la jeune infirmière obéit.

« Embarquez-le », lança Macke.

Wagner poussa von Ulrich dehors.

La femme se mit à sangloter.

L'infirmière demanda : « Où l'emmenez-vous ? »

Sur le seuil, Macke se tourna vers les trois femmes : la domestique, l'épouse et la fille. « Tout ça à cause d'un débile de huit ans, déclara-t-il. Je ne vous comprendrai jamais, vous autres. »

Il sortit et monta dans la voiture.

Ils franchirent la courte distance qui les séparait de la Prinz-Albrecht-Strasse. Wagner rangea la voiture à l'arrière du bâtiment de la Gestapo, derrière une file de véhicules noirs identiques. Ils descendirent.

Ils firent entrer von Ulrich par-derrière et l'emmenèrent au sous-sol, où ils l'enfermèrent dans une pièce recouverte de carreaux blancs.

Macke ouvrit un placard et y prit trois longues et lourdes matraques qui ressemblaient à des battes de base-ball américain. Il en tendit une à chacun de ses acolytes.

« Rossez-le », leur dit-il. Et il le leur abandonna.

6.

Le capitaine Volodia Pechkov, chef de la section berlinoise des services de renseignement de l'armée Rouge, rencontra Werner Franck au cimetière des Invalides, près du canal Berlin-Spandau.

C'était un bon choix. En regardant attentivement autour de lui, Volodia eut la certitude que personne ne les avait suivis à l'intérieur du cimetière. Il n'aperçut qu'une vieille femme coiffée d'un foulard noir qui, d'ailleurs, était en train de repartir.

Ils s'étaient donné rendez-vous sur la tombe du général von Scharnhorst, un grand mausolée orné d'un lion assoupi, coulé dans des canons ennemis fondus. C'était un jour ensoleillé de printemps et les deux espions ôtèrent leurs vestes pendant qu'ils déambulaient entre les sépultures de héros allemands.

Malgré la signature du pacte germano-soviétique deux ans plus tôt, l'espionnage soviétique n'avait pas désarmé en Allemagne, pas plus que la surveillance du personnel de l'ambassade soviétique. Tout le monde savait que le pacte serait provisoire, mais personne ne savait combien de temps ce provisoire durerait. Des agents du contre-espionnage continuaient donc à filer Volodia dans tous ses déplacements.

Ils devaient savoir quand il partait pour une véritable mission secrète, se disait-il, car c'était dans ces cas-là qu'il les semait. S'il sortait s'acheter des saucisses de Francfort, il se laissait suivre. Il se demanda s'ils étaient assez malins pour s'en être rendu compte.

« As-tu vu Lili Markgraf dernièrement ? » demanda Werner.

Elle avait été leur petite amie à tous les deux à des périodes différentes du passé. Depuis, Volodia l'avait recrutée. Elle avait appris à chiffrer et déchiffrer les messages en code de l'armée Rouge. Mais Volodia ne voulait pas encore en parler à Werner. « Je ne l'ai pas vue depuis un moment, mentit-il. Et toi ? »

Werner secoua la tête. « Mon cœur appartient à quelqu'un d'autre. » Il avait l'air tout intimidé en disant cela. Il était peut-être ennuyé de démentir sa réputation de tombeur. « À part ça, tu voulais me voir pourquoi ?

— Nous avons reçu une information fracassante. Une nouvelle qui va changer le cours de l'histoire… si elle est vraie. »

Werner parut sceptique.

Volodia continua : « Une source nous a appris que l'Allemagne avait l'intention d'envahir l'Union soviétique en juin. » Rien que de le dire, il en frissonnait encore. Cela représentait une grande victoire des services de renseignement de l'armée Rouge et une terrible menace pour l'URSS.

Werner repoussa de son front une mèche de cheveux, un geste qui avait dû faire battre bien des cœurs féminins. « Une source fiable ? » demanda-t-il.

Il s'agissait d'un journaliste en poste à Tokyo, qui avait l'oreille de l'ambassadeur d'Allemagne tout en étant secrètement communiste. Tout ce qu'il avait rapporté jusqu'à ce jour s'était vérifié. Mais Volodia ne pouvait raconter tout cela à Werner. « Fiable, répondit-il simplement.

— Donc, tu y crois ? »

Volodia hésita. C'était tout le problème. Staline n'y croyait pas. Il y voyait une manœuvre de désinformation des Alliés destinée à semer la défiance entre Hitler et lui. Le scepticisme de Staline à l'égard de ce coup de maître du Renseignement avait consterné les supérieurs de Volodia, rendant leur joie amère. « Nous vérifions encore », dit-il.

Werner regarda les arbres du cimetière qui se couvraient de leurs premières feuilles. « J'espère de toute mon âme que c'est vrai, déclara-t-il avec une ferveur soudaine. Ce serait la fin de ces maudits nazis.

— Oui. Si l'armée Rouge est prête. »

Werner lui jeta un regard surpris. « Vous n'êtes pas prêts ? »

Cette fois encore, Volodia ne pouvait lui dévoiler toute la vérité. Staline était persuadé que les Allemands n'attaqueraient pas tant qu'ils ne seraient pas venus à bout des Britanniques, pour éviter de mener la guerre sur deux fronts. Aussi longtemps que la Grande-Bretagne continuerait à défier l'Allemagne, l'Union soviétique serait en sécurité, selon lui. En conséquence, l'armée Rouge n'était absolument pas en état d'affronter une invasion allemande.

« Nous *serons* prêts, assura Volodia, si tu peux m'apporter la confirmation de ce projet d'invasion. »

Il ne put s'empêcher de savourer l'importance du moment. Son espion pouvait jouer un rôle capital.

« Malheureusement, je ne peux pas t'aider », murmura Werner.

Volodia se rembrunit. « Comment ça ?

— Je ne peux t'apporter ni la confirmation de cette information ni quoi que ce soit d'autre. Je suis sur le point de me faire virer du ministère de l'Air. Je serai sans doute envoyé en France, ou, si tes informations sont exactes, sur le front soviétique. »

Volodia fut horrifié. Werner était son meilleur agent. C'était grâce à ses informations qu'il avait été promu capitaine. Le souffle coupé, il demanda : « Bon sang, mais qu'est-ce qui s'est passé ?

— Mon frère est mort dans un établissement médical, en même temps que le filleul de ma petite amie. Et nous posons trop de questions.

— Pourquoi serais-tu renvoyé à cause de ça ?

— Les nazis exterminent les infirmes mais c'est un programme secret. »

Volodia en oublia un instant sa mission. « Quoi ? Ils les tuent purement et simplement ?

— Ça m'en a tout l'air. Nous n'avons pas encore tous les détails. Mais s'ils n'avaient rien à cacher, ils ne s'en prendraient pas à nous, moi et d'autres, parce que nous cherchons à nous informer.

— Quel âge avait ton frère ?

— Quinze ans.

— Bon sang ! C'était encore un gosse !

— Ils ne s'en tireront pas comme ça. Je refuse de la fermer. »

Ils s'arrêtèrent devant la tombe de Manfred von Richthofen, l'as de l'aviation. C'était une immense dalle de deux mètres de haut sur deux fois autant de large. Un seul mot y était gravé en majuscules : RICHTHOFEN. Volodia trouvait cette simplicité particulièrement émouvante.

Il essaya de reprendre ses esprits. Après tout, la police secrète soviétique tuait des gens elle aussi, en particulier ceux qu'elle soupçonnait de déloyauté. Le chef du NKVD, Lavrenti Beria, était un tortionnaire dont le passe-temps préféré, selon la rumeur, consistait à envoyer ses hommes enlever deux ou trois jolies filles dans la rue pour qu'il puisse les violer en guise de distraction nocturne. Cependant, l'idée que les communistes puissent être aussi barbares que les nazis n'était pas une consolation. Un jour, les Soviétiques se débarrasseraient de Beria et de ses semblables pour bâtir le vrai communisme. Mais avant, il fallait vaincre les nazis.

Ils atteignirent le bord du canal et restèrent là, à regarder une péniche avancer lentement sur l'eau en crachant une fumée noire de suie. Volodia réfléchissait à l'inquiétant aveu de Werner. « Qu'arriverait-il si tu cessais d'enquêter sur ces morts d'enfants ? demanda-t-il.

— Je perdrais ma petite amie. Elle est aussi révoltée que moi par cette affaire. »

Volodia songea tout à coup avec effroi que Werner risquait de révéler la vérité à cette fille. « Tu ne pourrais évidemment pas lui avouer la vraie raison de ce revirement », dit-il catégoriquement.

Werner accusa le coup mais ne protesta pas.

Volodia n'ignorait pas qu'en persuadant Werner d'abandonner ce combat il aiderait les nazis à dissimuler leurs crimes. Il écarta cette idée embarrassante. « Tu pourrais conserver ton poste auprès du général Dorn si tu promettais de laisser tomber cette affaire ?

— Oui. C'est ce qu'ils veulent. Mais je ne vais quand même pas les laisser assassiner mon frère sans rien dire ! Qu'ils m'envoient au front si ça leur chante, je ne me tairai pas.

— Que feront-ils à ton avis quand ils auront compris que tu ne céderas pas ?

— Ils m'enverront dans un camp de détention.

— Et qu'est-ce que tu y gagneras ?

— Je ne peux pas me laisser bâillonner dans une affaire pareille ! »

Il fallait que Volodia rallie Werner à son point de vue, mais comment lui faire entendre raison ? Werner avait réponse à tout. Ce type était remarquablement intelligent. C'était bien ce qui en faisait un espion aussi précieux.

« Et les autres ? demanda Volodia.

— Quels autres ?

— Il doit bien y avoir des milliers d'autres infirmes, enfants et adultes. Les nazis vont les tuer, eux aussi ?

— Probablement.

— Et comment comptes-tu les en empêcher si tu es dans un camp de détention ? »

Pour la première fois, Werner resta coi.

Volodia se détourna du canal pour scruter le cimetière. Un jeune homme en costume était agenouillé devant une petite pierre tombale. Les surveillait-il ? Volodia l'observa attentivement. L'homme était secoué de sanglots. Il paraissait sincère. Les agents du contre-espionnage n'étaient pas d'aussi bons acteurs.

« Regarde-le, dit Volodia à Werner.
— Pourquoi ?
— Il est triste. Comme toi.
— Et alors ?
— Regarde. »

Au bout d'un moment, l'homme se leva, s'essuya les yeux avec un mouchoir et s'en alla.

Volodia commenta : « Maintenant, il est soulagé. Voilà à quoi ça sert de donner libre cours à sa peine. Ça ne change rien. Mais ça permet de se sentir mieux. »

Il se tourna vers Werner et le regarda droit dans les yeux. « Je ne te critique pas. Tu veux savoir la vérité et la crier haut et fort. Mais réfléchis, essaie d'être logique. La seule manière de mettre fin à cette horreur, c'est de faire tomber ce régime. Et la seule manière d'y parvenir, c'est de permettre à l'armée Rouge d'écraser les nazis.

— Peut-être. »

Werner commençait à fléchir et Volodia reprit espoir. « Peut-être ? reprit-il. Tu vois une autre solution, toi ? Les Anglais sont à genoux, bombardés jour et nuit par la Luftwaffe. Les Américains ne s'intéressent pas aux chamailleries européennes. Et tous les autres soutiennent les fascistes. » Il posa les deux mains sur les épaules de Werner. « L'armée Rouge est ton seul espoir, mon ami. Si nous perdons, les nazis continueront à tuer les enfants infirmes, et les Juifs, et les communistes, et les homosexuels. Leur sang continuera à couler pendant mille ans encore.

— Et merde, capitula Werner. Tu as raison. »

7.

Le dimanche, Carla se rendit au temple avec sa mère. Maud était folle d'angoisse depuis l'arrestation de Walter et cherchait désespérément à savoir où on l'avait emmené. Évidemment, la Gestapo refusait de donner la moindre information. Mais le temple du pasteur Ochs était un lieu bien fréquenté, où se pressaient les habitants des quartiers huppés et qui comptait parmi ses fidèles quelques personnages importants dont certains pourraient se renseigner.

Carla courba la tête et pria pour que son père ne soit ni battu ni torturé. Elle ne croyait pas vraiment aux vertus de la prière, mais du fond de son désespoir, elle était prête à tout essayer.

Elle remarqua avec plaisir la famille Franck, assise quelques rangs plus haut. Elle contempla le dos de Werner. Ses cheveux bouclaient légèrement sur sa nuque, contrairement à ceux de la plupart des hommes qui les portaient très courts. Elle avait effleuré ce cou, embrassé cette gorge. Il était adorable. C'était certainement le plus gentil de tous les garçons qui l'avaient embrassée. Tous les soirs, avant de s'endormir, elle revivait en pensée leur soirée dans la forêt de Grünewald.

Mais elle n'était pas amoureuse de lui, se dit-elle.

Pas encore.

Le pasteur Ochs fit son entrée ; c'était un homme brisé. Le changement était impressionnant. Il s'avança lentement vers le lutrin, la tête basse et les épaules voûtées, suscitant des murmures consternés dans l'assemblée. Il récita les prières d'une voix terne et lut son sermon dans un livre. Il y avait maintenant deux ans que Carla était infirmière. Elle reconnut immédiatement les symptômes de la dépression. Elle en déduisit qu'il avait, lui aussi, reçu la visite de la Gestapo.

Frau Ochs et leurs cinq enfants n'étaient pas à leur place habituelle, au premier rang.

En entonnant le chant final, Carla se promit de ne pas renoncer, quelle que soit sa peur. Elle avait encore des alliés : Frieda, Werner et Heinrich. Et pourtant, que pouvaient-ils faire ?

Elle aurait voulu avoir des preuves irréfutables des atrocités que commettaient les nazis. Elle était persuadée qu'ils avaient entrepris d'exterminer les personnes déficientes. Ces descentes de la Gestapo le confirmaient. Mais comment convaincre autrui sans preuves concrètes ?

Comment se procurer celles-ci ?

À la fin de l'office, elle sortit avec Frieda et Werner, les attira à l'écart et leur dit : « Il nous faut des preuves de ce qui se passe. »

Frieda comprit tout de suite. « Nous devrions aller à Akelberg, suggéra-t-elle. Visiter cet hôpital. »

Werner l'avait proposé dès le début, mais ils avaient décidé de commencer leurs recherches à Berlin. Carla considérait maintenant l'idée sous un nouveau jour. « Il nous faut des autorisations pour voyager.

— Comment les obtenir ? »

Carla claqua dans ses doigts. « Nous appartenons toutes les deux au club cycliste Mercury. Ils peuvent obtenir des laissez-passer pour des excursions à bicyclette. » C'était le genre d'activités qui plaisait aux nazis, convaincus qu'une jeunesse saine devait aimer le sport et la nature.

« Tu crois qu'on pourra entrer dans l'hôpital ?

— On peut toujours essayer.

— Vous feriez mieux de laisser tomber », intervint Werner.

Carla n'en revint pas. « Qu'est-ce que tu veux dire ?

— Le pasteur Ochs a manifestement été terrorisé. C'est une affaire très dangereuse. Vous pourriez être jetées en prison, torturées. Et ça ne fera revenir ni Axel ni Kurt. »

Elle le regarda, incrédule. « Tu veux qu'on renonce ?

— Vous n'avez pas le choix. Vous parlez comme si l'Allemagne était un pays libre ! Vous allez vous faire tuer, toutes les deux.

— Nous devons prendre le risque, répliqua Carla, furieuse.

— Alors laissez-moi en dehors de tout ça. J'ai eu la visite de la Gestapo, moi aussi. »

La colère de Carla se transforma aussitôt en inquiétude. « Oh, Werner… qu'est-ce qui s'est passé ?

— Rien que des menaces, pour le moment. Mais si je continue à poser des questions, je serai envoyé au front.

— Oh, Dieu merci. Si ce n'est que ça !

— Ce n'est pas tellement réjouissant, tu sais. »

Les filles se turent un moment, puis Frieda exprima tout haut ce que Carla pensait tout bas. « C'est tout de même plus important que ton poste, non ?

— Je ne te permets pas de me donner de leçons », rétorqua Werner. Carla prit conscience que sous sa colère apparente perçait la honte. « Ce n'est pas votre carrière qui est en jeu, reprit-il. Et vous n'avez pas encore eu affaire à la Gestapo. »

Carla était déconcertée. Elle croyait connaître Werner et n'avait pas douté un instant qu'il envisagerait les choses sous le même angle qu'elle. « En fait, si, j'ai eu affaire à eux, murmura-t-elle. Ils sont venus arrêter mon père. »

Frieda fut effarée. « Oh, Carla ! » Elle la prit dans ses bras.

« Nous n'arrivons pas à savoir où il est », ajouta Carla.

Werner ne manifesta aucune compassion. « Dans ce cas, tu devrais avoir compris ! C'est toi qu'ils auraient dû arrêter, mais le commissaire Macke s'imagine que les filles ne sont pas dangereuses. »

Carla avait envie de pleurer. Elle était sur le point de tomber amoureuse de Werner et voilà qu'elle découvrait un lâche.

« Tu veux dire que tu ne nous aideras pas, c'est ça ? demanda Frieda.

— Oui.

— Parce que tu veux garder ton poste ?

— Ça ne sert à rien. Vous ne pouvez rien contre eux ! »

Carla lui en voulait de sa veulerie et de son défaitisme. « Tu ne peux quand même pas laisser faire ça !

— C'est idiot d'essayer de s'opposer à eux de front. Il y a d'autres moyens de lutter contre eux.

— Et lesquels ? demanda Carla. En ralentissant notre rythme de travail, comme le proposent les tracts ? Ça ne les empêchera pas de continuer à tuer les enfants déficients !

— Ce que vous voulez faire est suicidaire !

— Parce qu'il vaut mieux être lâche ?

— Je refuse d'être jugé par deux gamines ! » Sur ces mots, il s'éloigna.

Carla retint ses larmes. Elle n'allait tout de même pas pleurer devant les deux cents fidèles rassemblés au soleil devant le temple. « Je n'aurais jamais cru qu'il était comme ça », murmura-t-elle.

Frieda était perplexe, elle aussi. « Il n'est *pas* comme ça, affirma-t-elle. Je le connais depuis toujours. Il y a autre chose, quelque chose qu'il ne nous dit pas. »

La mère de Carla s'approcha. Elle ne remarqua pas le désarroi de sa fille, ce qui ne lui ressemblait pas. « Personne ne sait rien ! soupira-t-elle d'un air découragé. Je n'arrive pas à avoir d'informations sur l'endroit où ils ont pu emmener ton père.

— On va continuer à chercher, dit Carla. Il n'avait pas des amis à l'ambassade américaine ?

— Des connaissances. Je leur ai déjà posé la question, mais personne ne sait rien.

— On leur redemandera demain.

— Oh, Seigneur, je suppose qu'il y a des millions de femmes en Allemagne dans la même situation que moi. »

Carla hocha la tête.

« Rentrons, Mutti. »

Elles repartirent en marchant lentement, silencieuses, plongées dans leurs pensées. Carla était furieuse contre Werner, et s'en voulait de s'être fait une fausse idée de lui. Comment avait-elle pu s'amouracher d'une telle lavette ?

Elles arrivèrent dans leur rue. « J'irai à l'ambassade américaine dans la matinée, déclara Maud en se dirigeant vers leur maison. J'attendrai dans le hall la journée entière s'il le faut. Je les supplierai de faire quelque chose. S'ils veulent, ils peuvent lancer une enquête semi-officielle pour rechercher le beau-frère d'un ministre du gouvernement britannique. Oh ! Mais pourquoi la porte est-elle ouverte ? »

Carla crut tout d'abord que la Gestapo était revenue. Mais il n'y avait pas de voiture noire rangée le long du trottoir. Et la clé de la porte dépassait de la serrure.

Maud entra dans le vestibule et poussa un cri.

Carla se précipita.

Un homme gisait sur le sol, baignant dans son sang.

Carla réussit à retenir un cri. « Qui est-ce ? »

Maud s'agenouilla près de l'homme. « Walter, murmura-t-elle. Oh, Walter, qu'est-ce qu'ils t'ont fait ? »

C'est alors que Carla reconnut son père, si gravement blessé qu'il en était méconnaissable. Il avait un œil fermé, sa bouche enflée n'était qu'un énorme hématome et ses cheveux étaient collés par le sang coagulé. Un de ses bras formait un angle bizarre. Le devant de sa veste était maculé de vomissures.

« Walter, parle-moi, parle-moi », gémissait Maud.

Il ouvrit la bouche et émit une sorte de grognement.

Carla réprima son angoisse en retrouvant ses réflexes professionnels. Elle alla chercher un coussin et le cala sous la tête de son père. Elle prit un verre d'eau dans la cuisine et lui en versa un peu sur les lèvres. Il avala et ouvrit la bouche pour en réclamer davantage. Quand elle jugea que c'était suffisant, elle alla dans son bureau chercher un flacon de schnaps et lui en donna quelques gouttes. Il déglutit et toussa.

« Je vais chercher le docteur Rothmann, dit-elle. Lave-lui le visage et donne-lui encore un peu à boire. Surtout, n'essaie pas de le déplacer.

— Oui, oui. Dépêche-toi ! » supplia Maud.

Carla sortit sa bicyclette et s'élança. Le docteur Rothmann n'avait plus le droit d'exercer mais il soignait encore clandestinement les plus pauvres.

Carla pédala comme une furie. Comment son père était-il revenu ? On avait dû le ramener en voiture et il avait réussi à se traîner du trottoir jusqu'à la maison pour s'effondrer dans l'entrée.

Elle arriva devant la maison du médecin. Comme la leur, elle était en piteux état. Les carreaux avaient été cassés par des brutes. Frau Rothmann lui ouvrit. « Mon père a été battu, haleta Carla. La Gestapo.

— Mon mari va y aller. » Frau Rothmann se tourna vers l'escalier et cria : « Isaac ! »

Le médecin descendit aussitôt.

« C'est Herr von Ulrich », lui dit sa femme.

Le médecin saisit un cabas en toile posé près de la porte. Comme il n'était pas autorisé à exercer, il ne pouvait pas se promener avec une sacoche ressemblant à celle d'un médecin.

Ils sortirent de la maison.

« Je vous précède à bicyclette », lui dit Carla.

Quand elle arriva, elle trouva sa mère assise en pleurs sur le pas de la porte.

« Le docteur sera là dans une minute, annonça-t-elle.

— C'est trop tard, murmura Maud. Ton père est mort. »

8.

Volodia se retrouva devant le grand magasin Wertheim, tout près de l'Alexanderplatz, à deux heures et demie de l'après-midi. Il arpenta plusieurs fois le trottoir, à l'affût d'individus susceptibles d'être des policiers en civil. Il était certain de ne pas avoir été suivi, pourtant il n'était pas impossible qu'un agent de la Gestapo l'ait reconnu en passant et se soit demandé ce qu'il faisait là. Les lieux très fréquentés permettaient de se fondre dans la foule, mais n'étaient jamais parfaitement sûrs.

Cette histoire d'invasion était-elle vraie ? Si c'était le cas, Volodia ne resterait plus longtemps à Berlin. Il dirait adieu à ses bonnes amies Gerda et Sabine et serait sans doute renvoyé au siège des services de renseignement de l'armée Rouge à Moscou. Il était impatient de retrouver les siens. Sa sœur Ania avait eu des jumeaux en son absence. Et un peu de repos ne serait pas pour lui déplaire. Ses activités clandestines le soumettaient à une tension permanente : semer les hommes de la Gestapo, organiser des rendez-vous secrets, recruter des agents et redouter la trahison. Une année ou deux au siège seraient les bienvenues, en admettant que l'Union soviétique survive jusque-là… Il pourrait aussi être affecté à l'étranger. Washington le tentait bien. Il avait toujours eu envie de voir l'Amérique.

Il sortit de sa poche une boulette de papier qu'il jeta dans une poubelle. À trois heures moins une, il alluma une cigarette, alors qu'il ne fumait pas. Il laissa soigneusement tomber l'allumette dans la poubelle de manière qu'elle atterrisse à proximité de la boulette de papier. Puis il s'éloigna.

Quelques secondes plus tard, une voix cria « Au feu ! ».

Alors que tout le monde regardait les flammes qui s'élevaient de la poubelle, un taxi, une Mercedes 260D noire, s'arrêta devant l'entrée du magasin. Un élégant jeune homme en uniforme de lieutenant de l'armée de l'air en sortit prestement. Pendant que le jeune officier payait le taxi, Volodia monta dans la voiture et claqua la porte.

Sur le sol du taxi, à un endroit invisible du chauffeur, gisait un exemplaire du *Neues Volk*, la revue de propagande raciste des nazis. Volodia le ramassa, mais ne le lut pas.

« Un imbécile a mis le feu à une poubelle, remarqua le chauffeur du taxi.

— Hôtel Adlon », dit Volodia, et le taxi démarra.

Il feuilleta le magazine et vérifia qu'une enveloppe brune y était bien dissimulée.

Malgré son impatience il ne l'ouvrit pas.

Il descendit du taxi devant l'hôtel sans y entrer. Au lieu de cela, il franchit la porte de Brandebourg et pénétra dans le parc.

Les arbres arboraient un feuillage tout neuf. Cet après-midi de printemps était doux et les promeneurs nombreux.

La revue lui brûlait les doigts. Il trouva un banc abrité des regards et s'assit.

En se cachant derrière le magazine déployé, il décacheta l'enveloppe. Il en tira un document, un carbone sur lequel le texte dactylographié était estompé, mais lisible. Il était intitulé :

Directive n° 21 : Opération Barbarossa

Frédéric Barberousse était l'empereur germanique qui avait mené la troisième croisade en 1189.

Le texte commençait ainsi : *La Wehrmacht allemande doit se préparer à écraser la Russie soviétique au cours d'une campagne rapide, avant même la fin de la guerre contre l'Angleterre.*

Volodia retint son souffle. C'était de la dynamite. L'espion de Tokyo avait eu raison et Staline tort. L'Union soviétique courait un grave danger.

Le cœur battant, Volodia se reporta à la fin du document. Il était signé *Adolf Hitler*.

Il balaya les pages du regard à la recherche d'une date. Il en trouva une. L'invasion était prévue pour le 15 mai 1941.

La date était accompagnée d'une note manuscrite de Werner Franck : *Repoussée au 22 juin.*

« Oh, bon sang, il l'a fait, dit Volodia tout haut. Il a eu confirmation de l'invasion. »

Il remit le document dans l'enveloppe et celle-ci à l'intérieur de la revue.

Voilà qui changeait tout.

Il se leva et repartit en direction de l'ambassade soviétique pour transmettre cette information capitale.

9.

Comme il n'y avait pas de gare à Akelberg, Carla et Frieda descendirent à l'arrêt le plus proche, à une quinzaine de kilomètres, et enfourchèrent leurs bicyclettes.

Elles étaient en short, pull et sandales plates et avaient natté leurs cheveux en tresses relevées sur la tête. Elles avaient l'air de membres du Bund Deutscher Mädel ou BDM, l'association des jeunes Allemandes, qui organisait de nombreuses sorties à vélo. Quant à savoir si ces jeunes filles faisaient autre chose que pédaler – en particulier au cours des soirées dans les hôtels inconfortables où elles séjournaient –, la question donnait lieu à toutes sortes de supputations. Les garçons prétendaient que BDM signifiait *Bubi drück mich*, chéri, pelote-moi.

Carla et Frieda consultèrent leur carte et prirent la direction d'Akelberg.

Carla pensait à son père à chaque heure du jour. Elle savait qu'elle n'oublierait jamais l'instant abominable où elle l'avait découvert battu à mort et agonisant. Elle avait pleuré pendant des jours. Mais sa peine s'accompagnait d'un autre sentiment : la colère. Elle n'allait certainement pas laisser la tristesse l'anéantir. Elle réagirait.

Éperdue de chagrin, Maud avait d'abord tenté de dissuader Carla d'aller à Akelberg. « Mon mari est mort, mon fils est à l'armée, je ne veux pas que ma fille mette sa vie en danger, elle aussi ! » avait-elle gémi.

Après l'enterrement de son père, quand le choc et l'horreur avaient fait place à une douleur plus sourde et plus profonde, Carla lui avait demandé ce qu'aurait souhaité Walter. Maud avait longuement réfléchi. Elle avait attendu le lendemain pour lui donner sa réponse : « Il aurait voulu que tu continues la lutte. »

Il n'était pas facile pour Maud de l'admettre, mais elles savaient toutes les deux que c'était vrai.

Frieda n'avait pas eu de discussion de ce genre avec ses parents. Sa mère, Monika, avait aimé Walter autrefois. Elle était bouleversée par sa mort. Mais elle aurait été épouvantée si elle avait su ce que projetait sa fille. Son père, Ludi, l'aurait enfermée à la cave. Ils pensaient qu'elle était partie en randonnée à bicyclette. Au pire, ils pouvaient soupçonner qu'elle était allée retrouver un petit ami peu recommandable.

La route traversait un paysage vallonné, mais elles étaient en forme et une heure plus tard, après une dernière descente, elles arrivèrent dans la petite ville d'Akelberg. Carla était angoissée à l'idée de s'engager en territoire ennemi.

Elles entrèrent dans un café. Il n'y avait pas de Coca. « On n'est pas à Berlin, ici ! » protesta la femme debout derrière le comptoir, avec la même indignation que si elles avaient demandé à ce qu'un orchestre leur donne la sérénade. Carla se demanda pourquoi quelqu'un qui détestait manifestement les étrangers tenait un café.

On leur servit des Fanta, une boisson allemande, et elles en profitèrent pour remplir leurs gourdes d'eau.

Elles ne connaissaient pas l'emplacement exact de l'hôpital. Elles devraient se renseigner, mais Carla craignait d'éveiller les soupçons. Les nazis du lieu pourraient s'intéresser à deux inconnues trop curieuses. Au moment de payer, Carla demanda : « Nous devons retrouver le reste du groupe au croisement, près de l'hôpital. C'est dans quelle direction ? »

La tenancière évita son regard. « Il n'y a pas d'hôpital ici.

— L'Institut médical d'Akelberg, insista Carla en citant l'en-tête de la lettre.

— C'est sans doute un autre Akelberg. »

Carla était sûre qu'elle mentait. « C'est bizarre, murmura-t-elle, continuant à jouer son rôle. J'espère qu'on ne s'est pas trompées d'endroit. »

Elles suivirent la rue principale. Il n'y avait pas d'autre solution, se dit Carla. Elles allaient devoir demander leur chemin.

Un vieil homme à l'air inoffensif prenait le soleil sur un banc, devant un bar. « Où se trouve l'hôpital ? lui demanda Carla en dissimulant son inquiétude sous une attitude joviale.

— De l'autre côté de la ville, au sommet de la colline, à gauche. Mais n'y entrez surtout pas… la plupart des gens n'en ressortent pas ! » Il gloussa comme s'il avait dit une bonne blague.

Ses indications étaient un peu vagues, mais sans doute suffisantes. Carla décida de ne pas attirer davantage l'attention en demandant des précisions.

Une femme coiffée d'un foulard prit le vieux par le bras. « Ne faites pas attention à lui. Il ne sait pas ce qu'il dit », marmonna-t-elle, l'air ennuyée. Elle le força à se lever et s'éloigna avec lui le long du trottoir en murmurant : « Tu ferais mieux de la fermer. Tu es complètement fou ! »

Apparemment, ces gens se doutaient de ce qui se passait à leur porte. Heureusement, ils préféraient se montrer revêches et dissuasifs, évitant soigneusement de se compromettre. Ils n'allaient sans doute pas courir prévenir la police ou le parti nazi.

Carla et Frieda poursuivirent leur route et tombèrent sur l'auberge de jeunesse. Il existait des milliers d'établissements de ce genre en Allemagne, destinés à héberger le type de clients dont elles prétendaient faire partie : des jeunes, sains, vigoureux et sportifs en vacances dans la nature. Elles s'inscrivirent pour la nuit. L'installation, constituée de lits superposés à trois étages, était rudimentaire mais ce n'était pas cher.

L'après-midi était déjà bien avancé quand elles sortirent de la ville. Au bout d'un kilomètre, elles aperçurent une route sur la gauche. Il n'y avait aucun panneau, mais comme elle gravissait la colline, elles la prirent.

L'inquiétude de Carla grandissait. Plus elles s'approchaient de l'hôpital, plus il deviendrait difficile de feindre l'innocence si on les interrogeait.

Un kilomètre plus loin, elles aperçurent une grande maison au milieu d'un parc. Il n'y avait ni mur d'enceinte ni clôture. La route s'achevait devant la porte. Là encore, aucun panneau.

Inconsciemment, Carla avait imaginé un château fort, perché sur une hauteur, en pierres grises sinistres, aux fenêtres défendues par des grilles et aux portes de chêne ornées de ferrures. Or elle avait devant elle une maison de campagne bavaroise aux toits pointus à double pente, avec des balcons en bois et un petit campanile. Comment croire qu'on y assassinait des enfants ? La bâtisse semblait petite pour un hôpital. Elle remarqua alors qu'on y avait ajouté une longue annexe moderne surmontée d'une haute cheminée.

Elles descendirent de leurs bicyclettes et les appuyèrent contre le mur du bâtiment. Carla avait le cœur qui battait la chamade en montant les marches menant à l'entrée. Pourquoi n'y avait-il pas de gardiens ? Parce que personne ne risquait d'avoir l'audace de tenter de visiter les lieux ?

En l'absence de sonnette ou de marteau, Carla poussa la porte, qui s'ouvrit. Elle pénétra à l'intérieur, suivie de Frieda. Elles se retrouvèrent dans une entrée dallée de pierre, aux murs nus. Plusieurs portes donnaient sur le vestibule, mais elles étaient toutes fermées. Une femme à lunettes d'un certain âge descendait un grand escalier. Elle portait une robe grise assez élégante. « Oui ? dit-elle.

— Bonjour, lança Frieda d'un ton désinvolte.

— Qu'est-ce que vous faites ici ? L'entrée est interdite. »

Frieda et Carla avaient préparé une histoire.

« Je voulais voir l'endroit où mon frère est mort, expliqua Frieda. Il avait quinze ans…

— Ce n'est pas un établissement public ! s'indigna la femme en gris.

— Mais si ! » Frieda, qui avait été élevée dans une famille aisée, n'était pas du genre à se laisser intimider par des employés subalternes.

Une infirmière de dix-neuf ans tout au plus surgit de l'une des pièces et les dévisagea.

La femme en gris s'adressa à elle : « König, allez chercher Herr Römer immédiatement. »

L'infirmière s'exécuta.

La femme reprit : « Vous auriez dû écrire pour nous avertir de votre visite.

— Vous n'avez pas eu ma lettre ? s'étonna Frieda. J'ai écrit au médecin-chef. » C'était parfaitement faux. Frieda improvisait.

« Nous n'avons rien reçu de ce genre ! » Pour elle, il était évident qu'une requête aussi extravagante ne serait pas passée inaperçue.

Carla tendait l'oreille. L'endroit était étrangement silencieux. Elle avait eu affaire à des patients, enfants ou adultes, atteints d'infirmités physiques et mentales. Ils étaient en général bruyants. Même à travers ces portes, elle aurait dû entendre des cris, des rires, des pleurs, des protestations, des discours délirants. Rien. Aucun bruit. On se serait cru dans une morgue.

Frieda tenta une nouvelle approche. « Peut-être pourriez-vous me dire où se trouve la tombe de mon frère ? J'aurais bien voulu m'y recueillir.

— Il n'y a pas de tombes. Nous avons un four… » Elle s'interrompit et se reprit immédiatement. « Un crématorium.

— J'ai remarqué la cheminée », acquiesça Carla.

Frieda continua : « Où sont les cendres de mon frère ?

— Elles vous seront envoyées en temps utile.

— Faites bien attention à ne pas les mélanger avec d'autres, n'est-ce pas ? »

La femme en gris rougit. Carla en déduisit que c'était ce qu'ils faisaient, convaincus que personne ne viendrait les réclamer.

L'infirmière König reparut, suivie d'un homme trapu portant une tenue blanche d'infirmier.

« Ah, Römer, dit la femme. Veuillez raccompagner ces demoiselles.

— Une minute, intervint Frieda. Êtes-vous sûre de réagir comme il convient ? Je voulais simplement voir l'endroit où mon frère est mort.

— Parfaitement sûre.

— Dans ce cas, vous ne verrez pas d'inconvénient à m'indiquer votre nom. »

Elle eut une seconde d'hésitation. « Frau Schmidt. Maintenant, veuillez nous laisser, je vous prie. »

Römer s'avança d'un air menaçant.

« Nous partons, lança Frieda d'un ton glacial. Nous n'avons pas l'intention d'offrir à Herr Römer une occasion de nous malmener. »

L'homme fit demi-tour pour aller leur ouvrir la porte.

Elles sortirent, remontèrent sur leurs bicyclettes et reprirent la route.

« Tu crois qu'elle a avalé notre histoire ? demanda Frieda.

— Mais oui. Elle ne nous a même pas demandé nos noms. Si elle avait soupçonné la vérité, elle aurait tout de suite appelé la police.

— Malheureusement, nous n'avons pas appris grand-chose. Nous avons vu la cheminée. Mais nous n'avons toujours pas l'ombre d'une preuve. »

Carla était elle aussi un peu dépitée par l'échec de leur enquête.

Elles retournèrent à l'auberge de jeunesse, firent leur toilette, se changèrent et ressortirent pour trouver un endroit où manger. Le seul café-restaurant était celui de la tenancière revêche. Elles prirent des saucisses accompagnées de galettes de pommes de terre avant de se rendre au bar de la ville. Elles commandèrent des bières et essayèrent d'engager la conversation avec les autres clients, mais personne ne voulait leur parler. Ce silence leur parut suspect. Bien sûr, tout le monde se méfiait des étrangers, qui pouvaient être des mouchards de la police. Tout de même, Carla se demandait s'il existait beaucoup de villes dans lesquelles deux jeunes filles pouvaient passer une heure dans un bar sans se faire aborder.

Ayant décidé de se coucher tôt, elles regagnèrent l'auberge. Carla ne voyait pas ce qu'elles auraient pu faire d'autre. Elles rentreraient le lendemain chez elles les mains vides. Comment accepter d'être au courant de ces horribles meurtres et de ne rien pouvoir faire pour les empêcher ? Elle était tellement frustrée qu'elle en aurait hurlé.

Elle s'avisa soudain que Frau Schmidt – si c'était vraiment son nom – risquait de s'interroger sur ses visiteuses après coup. Sur le moment, elle avait pris Carla et Frieda pour ce qu'elles disaient être, mais elle pouvait parfaitement avoir eu des doutes et avoir appelé la police par acquit de conscience. Dans ce cas, elles ne seraient pas difficiles à trouver. Il n'y avait que cinq personnes à l'auberge et elles étaient les seules filles. Elle attendit, l'oreille aux aguets, le coup fatal frappé à la porte.

Si on les interrogeait, elles diraient une partie de la vérité, que le frère de Frieda et le filleul de Carla étaient morts à Akelberg et qu'elles avaient voulu voir leur tombe, ou au moins l'endroit où ils étaient morts pour y passer quelques minutes, en guise de commémoration. La police s'en contenterait peut-être. En revanche, si elle prenait la peine de vérifier auprès de la police berlinoise, elle ne tarderait pas à faire le lien avec Walter von Ulrich et Werner Franck, deux hommes qui avaient eu maille à partir avec la Gestapo pour avoir posé trop de questions. Carla et Frieda seraient alors en mauvaise posture.

Comme elles s'apprêtaient à se coucher dans les lits superposés à l'air peu accueillant, on frappa à la porte.

Le cœur de Carla fit un bond dans sa poitrine. Elle songea à ce que la Gestapo avait fait à son père. Elle savait qu'elle ne résisterait pas à la torture. Elle ne mettrait pas deux minutes à dénoncer tous les swing kids qu'elle connaissait.

Frieda, qui avait moins d'imagination, lui dit : « Ne fais pas cette tête ! » Et elle ouvrit la porte.

Ce n'était pas la Gestapo, mais une jolie fille blonde et menue. Carla mit un moment à reconnaître l'infirmière König sans son uniforme.

« Il faut que je vous parle », murmura-t-elle. Elle était bouleversée, essoufflée et avait les larmes aux yeux.

Frieda l'invita à entrer. Elle s'assit sur un lit et essuya ses yeux avec sa manche. « Je ne peux plus garder ça pour moi », poursuivit-elle.

Carla croisa le regard de Frieda. Elles pensaient la même chose.

« Garder quoi pour vous, Fräulein König ? demanda Carla.

— Je m'appelle Ilse.

— Moi, c'est Carla et elle, c'est Frieda. De quoi veux-tu parler, Ilse ? »

Sa voix était si basse qu'elles l'entendirent à peine. « Nous les tuons. »

Carla avait le souffle court et les mots franchirent difficilement ses lèvres : « À l'hôpital ? »

Ilse hocha la tête. « Ces pauvres gens qui arrivent dans les autocars gris. Des enfants, parfois des bébés, et puis des vieux, des grands-mères. Ils sont tous plus ou moins infirmes. Certains sont affreux à regarder, ils bavent, ils font sous eux, mais ils n'y peuvent rien. D'autres sont adorables et complètement inoffensifs. Peu importe… nous les tuons, tous.

— Comment faites-vous ?

— Une injection d'un mélange de morphine et de scopolamine. »

Carla hocha la tête. C'était un anesthésique courant, mortel à trop fortes doses.

« Et le traitement spécial qu'ils sont censés recevoir ?

— Il n'y a pas de traitement spécial.

— Ilse, soyons bien claires. Est-ce qu'ils tuent tous les patients qui arrivent ?

— Tous.

— Dès leur arrivée ?

— Dans les vingt-quatre heures, quarante-huit tout au plus. »

C'était bien ce que soupçonnait Carla, mais la réalité brute n'en était pas moins terrifiante. Elle en avait la nausée.

Après un instant de silence, elle reprit : « Il y a des patients là-bas en ce moment ?

— Pas de vivants. Nous avons administré les injections cet après-midi. C'est pour ça que Frau Schmidt était dans tous ses états quand vous êtes arrivées.

— Pourquoi ne font-ils rien pour dissuader les curieux d'entrer ?

— Ils se disent que si l'hôpital est gardé et entouré de barbelés, tout le monde se doutera qu'il s'y passe des choses terribles. D'ailleurs, vous êtes les premières à avoir cherché à y entrer.

— Combien de personnes sont mortes aujourd'hui ?

— Cinquante-deux. »

Carla en eut la chair de poule. « L'hôpital a mis à mort cinquante-deux personnes cet après-midi, à peu près au moment où nous y étions ?

— Oui.

— Elles sont toutes mortes maintenant ? »

Ilse hocha la tête. Une idée avait germé dans l'esprit de Carla et elle était décidée à la mettre à exécution. « Je veux voir », déclara-t-elle.

Ilse eut l'air épouvantée.

« Comment ça ?

— Je veux m'introduire dans l'hôpital et voir les cadavres.

— Ils sont déjà en train de les brûler.

— Alors je veux les voir brûler. Tu peux nous faire entrer discrètement ?

— Cette nuit ?

— Tout de suite.

— Oh, mon Dieu.

— Tu n'es pas obligée. Tu as déjà été extrêmement courageuse de venir nous parler. Si tu ne veux pas en faire plus, ça ne fait rien. Mais si nous voulons empêcher que ça continue, il nous faut un témoignage irrécusable.

— Un témoignage ?

— Oui. Tu vois, le gouvernement est embarrassé : il sait que les Allemands n'accepteraient pas qu'on tue des enfants. C'est la raison pour laquelle il garde le secret sur ce programme. Il peut faire croire aux gens que ça n'existe pas. Les nazis n'ont aucun mal à écarter une rumeur, surtout si elle vient d'une jeune fille. Voilà pourquoi nous devons constater *de visu* ce qui se passe.

— Je vois. » Une sombre détermination durcit le joli visage d'Ilse. « D'accord. Je vous emmène. »

Carla se leva. « Comment tu y vas d'habitude ?

— À bicyclette. Je l'ai laissée dehors.

— Très bien, nous prendrons aussi les nôtres. »

Elles sortirent. Il faisait sombre. Le ciel était couvert de nuages qui masquaient les étoiles. Elles allumèrent les lampes de leurs vélos pour se diriger sur la route. Arrivées en vue de l'hôpital, au sommet de la colline, elles les éteignirent et continuèrent à pied, en poussant leurs bicyclettes. Ilse les entraîna sur un chemin boisé qui rejoignait l'arrière du bâtiment.

Une odeur désagréable, comparable à celle de gaz d'échappement, parvint aux narines de Carla. Elle renifla.

Ilse murmura : « Le four crématoire.

— Oh, non ! »

Elles dissimulèrent leurs bicyclettes dans des buissons et se dirigèrent vers la porte de derrière. Elle n'était pas verrouillée. Elles entrèrent.

Les couloirs étaient inondés de lumière. Pas de recoins obscurs : le lieu était aussi brillamment éclairé que l'hôpital qu'il prétendait être. Si elles croisaient quelqu'un, on ne pourrait pas ne pas les voir. Leurs vêtements les trahiraient comme des intruses. Que feraient-elles alors ? Elles fuiraient à toutes jambes, sans doute.

Ilse suivit un couloir d'un pas vif, tourna à un angle et ouvrit une porte. « Ici », chuchota-t-elle.

Elles passèrent à l'intérieur.

Frieda poussa un cri d'horreur et plaqua sa main sur sa bouche.

Carla murmura : « Oh, ce n'est pas possible ! »

Dans une grande salle où régnait une température glaciale, une trentaine de morts gisaient, allongés sur le dos, nus, sur des tables. Certains étaient gros, d'autres maigres, certains vieux et fripés, d'autres n'étaient que des enfants, et il y avait même un bébé d'environ un an. Quelques-uns étaient tordus et contrefaits, mais la plupart semblaient physiquement normaux.

Tous portaient un sparadrap à l'avant-bras gauche, là où l'aiguille avait été plantée.

Carla entendit Frieda pleurer tout bas.

Elle prit son courage à deux mains. « Où sont les autres ? demanda-t-elle doucement.

— Déjà brûlés », répondit Ilse.

Des voix s'élevèrent derrière la double porte, à l'autre extrémité de la pièce.

« Il faut filer », conseilla Ilse.

Elles s'éclipsèrent dans le couloir. Carla tira la porte derrière elles, en la laissant à peine entrouverte et regarda par l'interstice. Elle vit Herr Römer et un autre homme entrer en poussant un chariot d'hôpital.

Les hommes ne regardaient pas dans sa direction. Ils parlaient football. Römer disait : « Il n'y a que neuf ans que nous avons remporté le championnat national. On a battu l'Eintracht Francfort deux à zéro.

— Oui, mais la moitié de vos meilleurs joueurs étaient juifs et ils ne sont plus là. »

Carla comprit qu'ils parlaient de l'équipe du Bayern de Munich.

« Les beaux jours reviendront, reprit Römer, mais il faut avoir une bonne tactique. »

Tout en continuant à bavarder, les deux hommes s'approchèrent d'une table sur laquelle reposait le corps d'une femme corpulente. Ils la prirent par les genoux et les épaules et, grognant sous l'effort, la balancèrent sans cérémonie sur le chariot. Ils se dirigèrent vers une autre table et jetèrent un deuxième cadavre sur le premier. Quand ils en eurent ramassé trois, ils repartirent avec le chariot.

« Je vais les suivre », annonça Carla.

Elle traversa la morgue en direction de la double porte. Frieda et Ilse lui emboîtèrent le pas. Elles débouchèrent dans une aile du bâtiment qui paraissait plus industrielle que médicale. Les murs étaient peints en brun, le sol était en béton et on y voyait des placards de rangement et des râteliers à outils.

Au bout d'un couloir, elles découvrirent une vaste pièce qui ressemblait à un garage, éclairée d'une lumière vive projetant de grandes ombres. Il y faisait chaud et il flottait une légère odeur de cuisine. Un caisson d'acier assez grand pour contenir un moteur de voiture trônait au milieu de l'espace. Il était surmonté d'un conduit métallique qui rejoignait une ouverture percée dans le toit. Carla comprit que c'était un fourneau.

Les deux hommes soulevèrent un corps et le posèrent sur un tapis roulant. Römer se dirigea vers le mur et appuya sur un bouton. Le tapis se mit en marche, un battant du caisson s'ouvrit et le corps disparut dans le four.

Ils chargèrent le deuxième corps sur le tapis roulant.

Carla en avait assez vu.

Elle se retourna et fit signe aux autres de reculer. Frieda se heurta à Ilse qui laissa échapper un cri involontaire. Elles se figèrent.

La voix de Römer leur parvint : « Qu'est-ce que c'était ?

— Un fantôme, suggéra son acolyte.

— Ne plaisante pas avec ces choses-là ! répliqua Römer d'une voix tremblante.

— Alors, tu prends l'autre bout de ce macchabée, oui ou non ?

— Tout de suite, ça vient ! »

Les trois filles regagnèrent furtivement la morgue. Devant tous ces corps allongés, Carla sentit monter en elle une immense tristesse en songeant au petit Kurt d'Ada. Il avait été couché là, un pansement adhésif sur le bras, avant d'être jeté sur le tapis roulant et éliminé comme un sac d'ordures. Nous ne t'oublions pas, Kurt, lui dit-elle intérieurement.

Elles repartirent par le couloir. Au moment où elles allaient atteindre la porte ouvrant sur l'extérieur, elles entendirent un bruit de pas et la voix de Frau Schmidt : « Pourquoi mettent-ils tout ce temps ? »

Elles franchirent les derniers mètres au pas de course et se précipitèrent vers la sortie. La lune s'était dégagée et éclairait le parc. Carla distinguait les buissons derrière lesquels elles avaient dissimulé leurs bicyclettes, deux cents mètres plus loin, de l'autre côté de la pelouse.

Frieda sortit la dernière. Dans sa précipitation, elle laissa la porte se refermer en claquant.

Carla évalua aussitôt la situation. Frau Schmidt chercherait sûrement à savoir d'où venait ce bruit. Elles n'auraient sans doute pas le temps d'atteindre les buissons avant qu'elle n'ouvre la porte. Elles devaient se cacher. « Par ici ! » souffla-t-elle en contournant le bâtiment en courant. Les deux autres la suivirent.

Elles s'aplatirent contre le mur. Carla entendit la porte s'ouvrir. Elle retint son souffle.

Il y eut un long silence. Puis Frau Schmidt marmonna des paroles inintelligibles et la porte se referma en claquant.

Carla glissa un œil. Frau Schmidt n'était plus en vue.

Les trois filles s'élancèrent sur la pelouse et récupérèrent leurs vélos. Elles les poussèrent sur le chemin forestier et débouchèrent sur la route. Là, elles les enfourchèrent et s'éloignèrent en pédalant comme des forcenées, lumière allumée. Carla exultait : elles s'en étaient tirées.

Peu avant de regagner la ville, son sentiment de victoire laissa place à des considérations plus réalistes. Qu'avaient-elles obtenu exactement ? Qu'allaient-elles faire maintenant ?

Raconter ce qu'elles avaient vu. Mais à qui ? Il faudrait réussir à convaincre quelqu'un. Les croirait-on ? Plus elle y pensait, moins elle en était sûre.

Quand, arrivées à l'auberge, elles descendirent de vélo, Ilse soupira : « Dieu merci, c'est fini. Je n'ai jamais eu aussi peur de ma vie.

— Ce n'est pas fini, objecta Carla.

— Que veux-tu dire ?

— Ce ne sera pas fini tant que nous n'aurons pas fait fermer cet hôpital et tous ceux qui se livrent aux mêmes activités.

— Comment comptes-tu faire ?

— Nous avons besoin de toi. De ton témoignage.

— C'est bien ce que je craignais.

— Accepterais-tu de venir avec nous, demain, quand nous retournerons à Berlin ? »

Après un long silence, Ilse acquiesça : « Je viendrai. »

10.

Volodia Pechkov était content d'être de retour chez lui. Moscou était resplendissante sous le soleil et la chaleur de l'été. Le lundi 30 juin, il réintégra le siège du service de renseignement de l'armée Rouge près de l'aérodrome de Khodynka.

Werner Franck et l'espion de Tokyo ne s'étaient pas trompés : les troupes allemandes avaient envahi l'Union soviétique le 22 juin. Volodia, ainsi que tout le personnel de l'ambassade soviétique à Berlin étaient repartis pour Moscou, en bateau et en train. Volodia était prioritaire, ce qui lui avait permis de rentrer plus vite que les autres. Certains étaient encore en route.

Volodia se rendait compte maintenant à quel point la vie à Berlin avait été déprimante. Les nazis étaient insupportables d'autosatisfaction et de triomphalisme. On aurait dit les joueurs d'une équipe de foot victorieuse après le match, de plus en plus saouls et de plus en plus pénibles, qui refusent de rentrer chez eux. Il ne pouvait plus les voir en peinture.

Certains prétendaient que l'URSS ne valait guère mieux avec sa police secrète, son orthodoxie rigide, son puritanisme à l'égard de divertissements aussi inoffensifs que la mode ou les arts abstraits. Ils avaient tort. Le communisme était un système en cours d'édification, et quelques erreurs étaient inévitables sur la voie d'une société plus juste. Le NKVD et ses chambres de torture étaient une aberration, un cancer dont le communisme guérirait un jour. Mais sans doute pas en temps de guerre.

Anticipant le déclenchement des hostilités, Volodia avait depuis longtemps équipé ses espions berlinois de radios clandestines et de manuels de chiffrage. Il était désormais plus vital que jamais que les quelques courageux résistants au nazisme continuent à transmettre des informations aux Soviétiques. Avant de partir, il avait détruit tous les documents où apparaissaient leurs noms et leurs adresses, dont la liste n'existait plus que dans sa mémoire.

Il avait trouvé ses parents en bonne forme, bien que son père fût surchargé de travail : il avait pour mission de préparer Moscou aux raids aériens. Volodia était allé voir sa sœur, Ania, son mari, Ilia Dvorkine, et leurs jumeaux, qui avaient déjà dix-huit mois : Dimitri, surnommé Dimka, et Tatiana, surnommée

Tania. Malheureusement, leur père avait toujours la même tête de fouine et restait un sale type.

Après avoir passé une agréable journée chez lui et une bonne nuit de sommeil dans son ancienne chambre, il se sentait prêt à se remettre au travail.

Il passa sous le détecteur de métaux installé à l'entrée du siège des services de renseignement. Malgré le décor triste et fonctionnel, il retrouva les escaliers et les couloirs familiers avec une pointe de nostalgie. En traversant le bâtiment, il s'attendait plus ou moins à voir les gens sortir des bureaux pour le féliciter. La plupart d'entre eux devaient savoir que c'était lui qui avait apporté la confirmation du projet Barbarossa. Pourtant, il ne vit personne. Par discrétion, peut-être.

Il entra dans une vaste salle occupée par des dactylos et des employés de bureau et s'adressa à la réceptionniste, une femme d'âge mûr : « Bonjour, Nika… toujours fidèle au poste ?

— Bonjour, capitaine Pechkov, répondit-elle, d'un ton moins chaleureux qu'il ne l'avait espéré. Le colonel Lemitov va vous recevoir tout de suite. »

Comme le père de Volodia, Lemitov n'avait pas été un personnage assez important pour être victime des grandes purges des années 1930. Il avait obtenu de l'avancement depuis et occupait le poste d'un ancien supérieur moins chanceux. Volodia ne savait pas grand-chose des purges, mais il avait du mal à croire qu'autant de hauts responsables aient pu faire preuve d'une déloyauté assez grave pour mériter de telles sanctions. Il ignorait aussi en quoi consistaient celles-ci. Ils étaient peut-être exilés en Sibérie, emprisonnés quelque part, ou morts. Une seule chose était sûre : ils avaient disparu de la circulation.

Nika ajouta : « Il est maintenant dans le grand bureau, au fond du couloir principal. »

Volodia franchit la porte ouverte et suivit le couloir en adressant saluts et sourires aux têtes connues, sans arriver, là encore, à se défaire du sentiment de n'être pas reçu en héros, comme il aurait dû l'être. Il frappa à la porte de Lemitov, espérant que son supérieur lui fournirait une explication.

« Entrez. »

Volodia entra, salua et referma la porte derrière lui.

« Bienvenue au pays, camarade. » Lemitov fit le tour de son bureau. « Entre nous, tu as fait du bon travail à Berlin. Merci.

— Je suis très honoré, mon colonel. Mais pourquoi “entre nous” ?

— Parce que tu as donné tort à Staline. » Il leva la main pour prévenir sa réaction. « Staline ne sait pas que ça vient de toi. Malgré tout, après les purges, les gens d'ici n'ont pas envie d'être associés à quelqu'un qui s'est écarté de la ligne.

— Qu'est-ce que j'aurais dû faire ? protesta Volodia, incrédule. Envoyer de fausses informations ? »

Lemitov secoua énergiquement la tête. « Tu as fait exactement ce qu'il fallait, je suis parfaitement d'accord avec toi. Et je t'ai protégé. Mais ne t'attends pas à ce que tes collègues t'accueillent à bras ouverts.

— Compris », se résigna Volodia. La situation était pire qu'il ne l'avait imaginée.

« Maintenant, au moins, tu as un bureau pour toi tout seul. Trois portes plus loin. Il va te falloir une bonne journée pour te mettre au courant. »

Volodia comprit qu'il lui donnait congé. « Oui, mon colonel. » Il salua et s'en alla.

Son bureau, une petite pièce sans tapis, n'avait rien de luxueux, mais il y était chez lui. Soucieux de regagner son pays au plus vite, il n'avait pas eu le temps de se tenir au courant de la progression de l'invasion allemande. Ravalant son amertume, il s'attela à la lecture des rapports des commandants militaires sur le terrain pendant la première semaine de la guerre.

Chaque page qu'il lisait aggravait son désespoir.

L'invasion avait pris l'armée Rouge au dépourvu.

Cela paraissait impossible, et pourtant les preuves s'étalaient sur son bureau.

Le 22 juin, quand les Allemands avaient attaqué, un grand nombre d'unités de première ligne n'avaient pas de munitions.

Ce n'était pas tout. Les avions étaient soigneusement alignés sur les aérodromes sans camouflage, ce qui avait permis à la Luftwaffe d'en détruire mille deux cents dans les premières heures de l'offensive. Les troupes avaient été lancées contre les unités allemandes sans armement suffisant, sans couverture aérienne et sans information sur les positions ennemies. Elles avaient été anéanties.

Pire encore, Staline avait donné à l'armée Rouge un ordre formel : toute retraite était interdite. Chaque unité devait se battre jusqu'au dernier de ses hommes et les officiers étaient censés se tirer une balle dans la tête pour éviter d'être faits prisonniers. Les soldats n'étaient jamais autorisés à se regrouper sur de nouvelles positions défensives mieux protégées. Autrement dit, chaque défaite tournait au massacre.

L'armée Rouge subissait une véritable hémorragie en hommes et en matériel.

Staline n'avait pas tenu compte de l'avertissement de l'espion de Tokyo ni de la confirmation apportée par Werner Franck. Quand l'offensive avait débuté, il avait continué à affirmer que ce n'était qu'une opération de provocation isolée, menée par des officiers allemands à l'insu d'Hitler, qui y mettrait fin dès qu'il en serait informé.

Le temps pour Staline d'admettre qu'il ne s'agissait pas d'une provocation mais de la plus grande campagne d'invasion de l'histoire de la guerre, les Allemands avaient écrasé les avant-postes. Au bout d'une semaine, ils avaient fait une percée de cinq cents kilomètres en territoire soviétique.

C'était une catastrophe, et elle aurait pu être évitée, ce qui donnait à Volodia envie de hurler.

L'identité du coupable ne faisait aucun doute. L'Union soviétique était une autocratie. Toutes les décisions étaient prises par une seule personne : Joseph Staline. Il s'était obstinément, stupidement, tragiquement enferré dans ses propres erreurs. Et maintenant, son pays courait un danger mortel.

Jusqu'alors, Volodia avait cru que le communisme soviétique était la seule vraie idéologie, qui n'était entachée que par les excès de sa police secrète, le NKVD. Il se rendait compte maintenant que les failles se situaient tout en haut. Beria et le NKVD n'existaient que parce que Staline le voulait bien. C'était Staline qui entravait la marche vers le communisme authentique.

En fin d'après-midi, alors que Volodia contemplait par la fenêtre le terrain d'aviation en ruminant tout ce qu'il venait d'apprendre, il reçut la visite de Kamen. Ils avaient été lieutenants ensemble quatre ans plus tôt, frais émoulus de l'Académie militaire, section Renseignement, et avaient partagé une chambrée avec deux autres camarades. À cette époque, Kamen était le plaisantin de la troupe, se moquant des uns et des autres, osant tourner en dérision le dogme soviétique. Il avait pris du poids et un air plus sérieux. Il s'était laissé pousser une petite moustache noire comme celle du ministre des Affaires étrangères Molotov, peut-être pour se donner l'air plus mûr.

Kamen ferma la porte derrière lui et s'assit. Il sortit un jouet de sa poche, un petit soldat de fer-blanc muni d'une clé dans le dos. Il tourna la clé et posa le soldat sur le bureau de Volodia. Le soldat se mit à balancer les bras comme au défilé pendant que le mécanisme émettait un bruit d'horloge.

Kamen déclara tout bas : « Staline est introuvable depuis deux jours. »

Volodia comprit que le jouet était destiné à brouiller les éventuels appareils d'écoute qui pourraient être cachés dans son bureau.

« Qu'est-ce que tu entends par "introuvable" ?

— On ne l'a pas vu au Kremlin et il ne répond pas au téléphone. »

Volodia était stupéfait. Le chef d'une nation ne pouvait pas disparaître comme par enchantement. « Qu'est-ce qu'il fabrique ?

— Personne n'en sait rien. » Le soldat s'arrêta. Kamen le remonta et le remit en marche. « Samedi soir, quand il a appris que le groupe d'armées ouest avait été encerclé par les Allemands, il a dit : "Tout est perdu. Je renonce. Lénine a fondé notre État et nous l'avons fichu en l'air." Puis il est parti pour Kountsevo. » Staline avait une datcha à proximité de Kountsevo, une ville située à la périphérie de Moscou. « Hier, il n'est pas venu au Kremlin à midi comme à son habitude. Quand on a téléphoné à Kountsevo, personne n'a répondu. Pareil aujourd'hui. »

Volodia se pencha en avant. « Souffre-t-il… » Sa voix se perdit dans un murmure : « … de dépression nerveuse ? »

Kamen eut un geste d'impuissance. « Ce ne serait pas surprenant. Il a affirmé, contre toute évidence, que l'Allemagne ne nous attaquerait pas durant l'année 1941 et tu vois ce qui s'est passé. »

Volodia hocha la tête. Cela se tenait. Staline s'était fait appeler petit père des peuples, guide, dirigeant suprême, grand transformateur de la nature, grand timonier, génie de l'humanité, plus grand génie de tous les temps. Or les événements venaient de démontrer à tous, lui-même compris, qu'il s'était trompé quand tous les autres avaient raison. Beaucoup d'hommes se suicidaient dans de telles circonstances.

La crise était encore plus grave que Volodia ne l'avait pensé. Non seulement l'Union soviétique était attaquée et en train de perdre. En plus, elle n'était plus gouvernée. Depuis la révolution, la nation n'avait pas connu plus grand danger.

Mais n'était-ce pas en même temps une occasion à saisir? L'occasion de se débarrasser de Staline?

La dernière fois que Staline s'était trouvé en position de faiblesse, c'était en 1924, quand le testament de Lénine l'avait déclaré inapte au pouvoir. Depuis que Staline avait survécu à cette crise, son pouvoir semblait indestructible, même lorsque ses décisions frisaient la folie : les purges, les erreurs commises en Espagne, la nomination de Beria, un vrai sadique, à la tête de la police secrète, le pacte avec Hitler. Cette situation d'urgence offrait-elle enfin une possibilité de briser son emprise?

Volodia dissimula sa fébrilité à Kamen et aux autres. Dans le bus qui le ramenait chez lui dans la douce lumière du soir d'été, il continua à jubiler secrètement. Ils furent ralentis par un convoi de camions remorquant des canons antiaériens.

Pouvait-on chasser Staline?

Il se demandait combien de personnes au Kremlin se posaient la même question.

Il arriva à l'appartement de ses parents, dans la Maison du gouvernement, l'immeuble de neuf étages situé en face du Kremlin. Ils étaient sortis. En revanche, sa sœur était là avec ses deux jumeaux, Dimka et Tania. Le garçon, Dimka, avait les yeux et les cheveux noirs. Armé d'un crayon rouge, il gribouillait sur un vieux journal. Sa sœur avait le même regard bleu intense que Grigori. Et que Volodia, disaient les gens. Elle lui montra aussitôt sa poupée.

Zoïa Vorotsintseva était là, elle aussi. Volodia avait vu la ravissante physicienne pour la dernière fois quatre ans plus tôt, juste avant de partir pour l'Espagne. Ania et elle s'étaient découvert une passion commune pour la musique folklorique russe : elles allaient ensemble à des récitals et Zoïa jouait du goudok, une sorte de violon à trois cordes. N'ayant ni l'une ni l'autre les moyens de s'acheter un phonographe, elles écoutaient un enregistrement d'orchestre de balalaïkas sur celui de Grigori. Grigori n'était pas très mélomane, mais il aimait bien ce disque.

Zoïa portait une robe d'été à manches courtes du même bleu clair que ses yeux. Quand Volodia lui posa la question conventionnelle « Comment vas-tu ? » elle répondit : « Je suis furieuse. »

Les Russes avaient de nombreuses raisons d'être furieux par les temps qui couraient.

« Pourquoi ?

— Mes recherches en physique nucléaire ont été annulées. Tous les scientifiques avec lesquels je travaille ont été affectés à d'autres missions. Je suis moi-même chargée d'améliorer la conception des viseurs de bombardement. »

Volodia trouvait cela assez naturel. « Nous sommes en guerre, après tout.

— Tu ne comprends pas. Écoute. Quand l'uranium subit un processus appelé fission, le phénomène dégage d'énormes quantités d'énergie. Je dis bien *énormes*. Nous le savons, tout comme les chercheurs occidentaux d'ailleurs… nous avons lu leurs articles dans des revues scientifiques.

— La question des viseurs est sans doute plus urgente. »

Zoïa répondit d'un ton irrité : « Ce processus de fission pourrait être exploité pour fabriquer des bombes cent fois plus puissantes que tout ce qui existe actuellement. Une seule explosion nucléaire suffirait à effacer Moscou de la carte. Qu'arrivera-t-il si les Allemands fabriquent cette bombe et que nous ne l'avons pas ? Ce sera comme s'ils avaient des armes à feu et nous, des épées !

— Y a-t-il des raisons de penser que les scientifiques d'autres pays travaillent sur la bombe atomique ?

— Nous en sommes certains. Le concept de fission conduit tout droit à l'idée de bombe. Nous y avons pensé… pourquoi pas eux ? Mais il y a autre chose. Ils publiaient tous leurs résultats dans les revues scientifiques et tout à coup, il y a un an, ils ont arrêté. Il n'y a pas eu un seul article sur la fission nucléaire depuis.

— Tu crois que les responsables politiques et les généraux du monde occidental ont compris le potentiel militaire de ces recherches et ont décidé de les garder secrètes ?

— Ça tombe sous le sens. Or, ici, en Union soviétique, nous n'avons même pas commencé à essayer de nous procurer de l'uranium.

— Hum. » Volodia prit l'air sceptique alors même qu'il était persuadé qu'elle avait raison. Même les plus grands admirateurs de Staline, parmi lesquels le père de Volodia, Grigori, admettaient qu'il n'était pas très versé dans les sciences. Et les autocrates ont une fâcheuse tendance à négliger ce qu'ils ne maîtrisent pas.

« J'en ai parlé à ton père, continua Zoïa. Il m'écoute, mais lui, personne ne l'écoute.

— Qu'est-ce que tu vas faire alors ?

— Que veux-tu que je fasse ? Mettre au point un excellent viseur pour nos aviateurs en espérant que ça suffira. »

Volodia hocha la tête. Cette attitude lui plaisait. Cette fille lui plaisait. Elle était vive et intelligente… et si jolie ! Il se demanda si elle accepterait d'aller au cinéma avec lui.

Cette discussion à propos de physique lui rappela Willi Frunze, qu'il avait connu à l'Académie de garçons de Berlin. D'après Werner Franck, Willi était un brillant physicien qui poursuivait actuellement ses études à Londres. Il avait peut-être des informations sur la bombe atomique qui préoccupait tant Zoïa. S'il était resté communiste, il serait peut-être disposé à dire ce qu'il savait. Volodia décida d'envoyer un télégramme au bureau du Renseignement de l'armée Rouge à leur ambassade de Londres.

Ses parents rentrèrent enfin, son père en grand uniforme, sa mère en manteau et chapeau. Ils revenaient de l'une de ces interminables cérémonies qu'affectionnait l'armée. Staline tenait à conserver ce genre de manifestations, malgré l'invasion allemande, parce qu'elles étaient bonnes pour le moral.

Ils jouèrent un moment avec les jumeaux, mais Grigori semblait préoccupé. Il marmonna quelque chose au sujet d'un coup de téléphone urgent à donner et se retira dans son bureau. Katerina commença à préparer le dîner.

Volodia resta à bavarder avec les femmes dans la cuisine, mais il était impatient de parler à son père. Il croyait deviner l'objet de son coup de fil urgent : la question du renversement ou du maintien de Staline devait être en cours de discussion, probablement même dans ce bâtiment.

Au bout d'un moment, il décida de braver l'éventuelle colère de son père et de l'interrompre. Il demanda aux femmes de l'excuser et se dirigea vers le bureau. Son père en sortait justement. « Il faut que j'aille à Kountsevo », annonça-t-il.

Volodia mourait d'envie de savoir ce qui se passait. « Pourquoi ? »

Grigori ne répondit pas. « J'ai demandé la voiture, poursuivit-il, mais mon chauffeur est rentré chez lui. Tu peux me conduire ? »

Volodia était aux anges. Il n'était jamais allé à la datcha de Staline. Mieux encore, il allait s'y rendre en pleine crise.

« Allez, dépêche-toi ! » s'impatienta son père.

Ils lancèrent des au revoir depuis le couloir et sortirent.

Grigori avait une ZIS 101-A, la copie soviétique de la Packard américaine, avec transmission automatique à trois vitesses. Sa vitesse de pointe était d'environ cent vingt kilomètres à l'heure. Volodia s'installa au volant et démarra.

Il traversa l'Arbat, un quartier d'intellectuels et d'artisans, et rejoignit la voie rapide qui filait vers l'ouest. « Le camarade Staline t'a convoqué ? demanda-t-il à son père.

— Non. Staline est injoignable depuis deux jours.

— C'est ce que j'ai entendu dire.

— Ah bon ? C'est censé être un secret.

— Ce genre de choses ne peut pas rester secret bien longtemps. Qu'est-ce qui se passe actuellement ?

— Quelques-uns d'entre nous se rendent à Kountsevo pour le voir. »

Volodia posa la question clé. « Dans quel but ?

— Avant tout pour vérifier s'il est encore vivant. »

Se pouvait-il qu'il soit déjà mort et que personne ne soit au courant ? s'étonna Volodia. Cela paraissait peu vraisemblable. « Et s'il est encore en vie ?

— Je ne sais pas. Quoi qu'il advienne, je préfère être sur place qu'en être informé plus tard. »

Volodia savait que les appareils d'écoute ne fonctionnaient pas dans les voitures. Les micros ne captaient que le bruit du moteur. Il ne craignait donc pas d'être entendu. C'est néanmoins avec une certaine appréhension qu'il évoqua l'impensable. « Staline pourrait-il être renversé ? »

Son père répondit d'un ton agacé : « Je te l'ai dit, je n'en sais rien. »

Volodia frémit. Pareille question aurait dû susciter une réponse fermement négative. Toute hésitation équivalait à un oui. Son père venait d'admettre la possibilité d'une destitution de Staline.

Un espoir indicible fit battre le cœur de Volodia. « Tu imagines ? s'enthousiasma-t-il. Plus de purges, plus de camps de travail ! Plus de police secrète pour enlever les jeunes filles dans la rue et les violer ! » Il s'attendait plus ou moins à être interrompu par les protestations de son père. Grigori se contenta d'écouter, les yeux mi-clos. Volodia continua sur sa lancée. « La stupide appellation de trotsko-fasciste disparaîtrait du vocabulaire. Les troupes pourraient battre en retraite quand leurs effectifs et leur armement sont insuffisants au lieu de se sacrifier inutilement. Des décisions raisonnables seraient prises par des groupes d'hommes soucieux du bien public. Ce serait le communisme dont tu rêvais il y a trente ans !

— Foutaises ! lâcha son père avec mépris. Nous serions dans un sacré pétrin si nous perdions notre chef en ce moment. Nous sommes en guerre, et l'ennemi a envahi notre pays ! Notre seul objectif doit être de sauver la révolution, coûte que coûte. Nous avons besoin de Staline, aujourd'hui plus que jamais. »

Volodia eut l'impression de recevoir une gifle. Il y avait bien longtemps que son père ne lui avait plus parlé sur ce ton.

Et s'il avait raison ? Si l'Union soviétique avait besoin de Staline ? Tout de même, il avait pris tellement de décisions catastrophiques que Volodia voyait mal comment les choses pourraient être pires avec un autre.

Ils arrivèrent à destination. On parlait de « datcha » mais la résidence de Staline n'avait rien d'une modeste maison de campagne. C'était un long bâtiment bas percé de cinq grandes fenêtres de part et d'autre d'une entrée majestueuse, niché au milieu d'un bois de conifères et peint en vert, comme pour le dissimuler. Des centaines d'hommes en armes en gardaient les grilles et la double clôture de barbelés. Grigori désigna une batterie antiaérienne abritée sous un filet de camouflage. « C'est moi qui l'ai installée là », dit-il.

Le garde posté à l'entrée le reconnut, mais leur demanda quand même leurs papiers. Grigori avait beau être général et Volodia capitaine du Renseignement, ils n'en furent pas moins fouillés.

Volodia roula jusqu'à la porte de la maison. Ils étaient les premiers. « Nous allons attendre les autres », décida son père.

Quelques instants plus tard, trois autres limousines ZIS apparurent. Volodia n'ignorait pas que ZIS signifiait *Zavod Imeni Stalina*, « usine nommée Staline ». Les bourreaux arrivaient-ils dans des véhicules portant le nom de leur victime ?

Les occupants en sortirent, huit hommes d'âge mûr en costume et chapeau qui tenaient l'avenir du pays entre leurs mains. Volodia reconnut parmi eux le ministre des Affaires étrangères Molotov et Beria, le chef de la police secrète.

« Allons-y », dit Grigori.

Volodia fut surpris. « Je vous accompagne ? »

Grigori glissa la main sous son siège et lui tendit un pistolet Tokarev TT-33. « Fourre ça dans ta poche. Si ce con de Beria fait mine de m'arrêter, tu l'abats. »

Volodia s'en saisit prudemment. Le TT-33 n'avait pas de cran de sécurité. Il enfonça dans sa poche l'arme longue d'une bonne quinzaine de centimètres et descendit de voiture. Il se rappela qu'il y avait huit balles dans le chargeur.

Ils entrèrent tous dans la maison. Volodia craignait d'être soumis à une nouvelle fouille qui aurait révélé qu'il était armé mais il n'en fut rien.

Les peintures étaient de couleur sombre et la maison mal éclairée. Un officier les conduisit dans ce qui semblait être une petite salle à manger. Staline était assis dans un fauteuil.

L'homme le plus puissant de l'Est avait l'air hagard et abattu. En voyant le petit groupe, il demanda : « Pourquoi êtes-vous venus ? »

Volodia se raidit. Staline pensait manifestement que ces hommes étaient là pour l'arrêter ou l'exécuter.

Il y eut un long silence et Volodia comprit qu'ils n'avaient préparé aucun plan. Comment l'auraient-ils pu, ne sachant même pas si Staline était encore vivant ?

Qu'allaient-ils faire maintenant ? Le tuer ? L'occasion ne se présenterait sans doute plus jamais.

Molotov s'avança enfin. « Nous sommes venus vous demander de vous remettre au travail. »

Volodia réprima un cri de protestation.

Staline secoua la tête. « Puis-je être à la hauteur des attentes du peuple ? Puis-je conduire le pays à la victoire ? »

Volodia attendit, médusé. Allait-il refuser ?

« Il y a sans doute des hommes plus capables que moi », ajouta Staline.

Il leur offrait une deuxième chance de le démettre !

Un autre membre du groupe prit la parole. Volodia reconnut le maréchal Vorochilov. « Aucun n'est plus digne que vous, camarade Staline, de diriger le pays. »

Quel intérêt ? L'heure n'était pas aux flagorneries.

« C'est certain ! » renchérit alors son père.

Ils n'allaient donc pas laisser Staline démissionner ? Quelle bêtise !

Molotov fut le premier à prononcer des paroles raisonnables. « Nous proposons de former un cabinet de guerre qu'on appellera le Comité d'État à la Défense, une sorte de politburo spécial composé d'un très petit nombre de membres et investi de très larges pouvoirs. »

Staline réagit immédiatement. « Qui le dirigera ?

— Vous, camarade Staline ! »

Volodia faillit crier « Non ! »

Nouveau silence.

Staline reprit alors la parole. « Très bien. Et quels seront les autres membres de ce comité ? »

Beria lui présenta une liste de propositions.

C'est fini, songea Volodia, accablé de désespoir. Ils avaient laissé passer l'occasion. Ils auraient pu se débarrasser d'un tyran, mais n'en avaient pas eu le courage. Tels les enfants d'un père violent, ils avaient craint de ne pas pouvoir se passer de lui.

En réalité, c'était même pire, se dit-il, de plus en plus consterné. Peut-être Staline avait-il vraiment souffert de dépression. Son état d'abattement semblait bien réel. Et pourtant, il venait de mener avec succès une superbe manœuvre politique. Tous les hommes susceptibles de le remplacer étaient présents dans la pièce. Au moment où son manque de jugement s'était imposé aux yeux de tous en même temps que les conséquences dramatiques qu'il avait entraînées, Staline avait obligé ses rivaux à venir le supplier de reprendre sa place à leur tête. Il avait tiré un trait sur ses erreurs monumentales pour s'offrir un nouveau départ.

Staline n'était pas seulement de retour aux affaires.

Il était plus puissant que jamais.

11.

Qui aurait le courage de dénoncer publiquement ce qui se passait à Akelberg ? Carla et Frieda l'avaient vu de leurs yeux et avaient un témoin en la personne d'Ilse König. Mais elles avaient besoin de quelqu'un qui défende leur cause. Il n'y avait plus de représentants élus : tous les députés du Reichstag étaient nazis. Il n'y avait plus non plus de vrais journalistes ; il ne restait que des flagorneurs professionnels. Les juges étaient tous nommés par le gouvernement nazi qu'ils servaient docilement. Carla n'avait jamais pris la mesure de la protection que représentaient les politiciens, la presse et les avocats. Sans eux, le gouvernement était libre d'agir comme bon lui semblait, et même d'assassiner des gens.

Vers qui se tourner ? L'admirateur de Frieda, Heinrich von Kessel, avait un ami qui était prêtre catholique. « Peter était le meilleur de la classe, leur dit-il. Mais pas le plus apprécié. Un peu coincé et collet monté. Je pense qu'il nous écoutera. »

Carla se dit que cela valait la peine d'essayer. Son pasteur protestant les avait soutenus avec bienveillance jusqu'à ce que la Gestapo le réduise au silence par la terreur. On ne pouvait exclure qu'il se passe la même chose. Mais elle ne voyait pas d'autre solution.

Heinrich emmena Carla, Frieda et Ilse à l'église de Peter dans le quartier de Schöneberg, un dimanche matin de juillet. Heinrich était très séduisant dans son costume noir. Les trois jeunes filles avaient mis leurs tenues d'infirmières, gages de crédibilité. Une porte latérale les conduisit dans une petite pièce poussiéreuse meublée de quelques vieilles chaises et d'une grande armoire. Le père Peter était seul, en prière. Il dut les entendre arriver, mais resta néanmoins agenouillé encore un moment avant de se lever pour s'avancer à leur rencontre.

Peter était grand et mince. Il avait des traits réguliers et une coupe de cheveux soignée. Il devait avoir vingt-sept ans puisqu'il avait le même âge qu'Heinrich, estima Carla. « Je me prépare pour la messe, leur dit-il d'un ton sévère. Je suis heureux de t'accueillir dans mon église, Heinrich, mais il faut me laisser maintenant. Je vous verrai tout à l'heure.

— Il s'agit d'une urgence spirituelle, Peter, insista Heinrich. Asseyons-nous un instant. Nous avons quelque chose d'important à te confier.

— Rien ne saurait être plus important que la messe.

— Si, crois-moi, Peter. Dans cinq minutes, tu me donneras raison.

— Très bien.

— Je te présente ma petite amie, Frieda Franck. »

Parce que Frieda était devenue sa « petite amie » ? s'étonna Carla.

« J'avais un petit frère qui est né avec un spina-bifida, expliqua Frieda. Au début de l'année, il a été transféré dans un hôpital à Akelberg, en Bavière, pour suivre un traitement spécial. Peu après, nous avons reçu une lettre nous annonçant qu'il était mort d'appendicite. »

Elle se tourna vers Carla, qui prit la suite. « Notre domestique, Ada, avait un fils atteint de lésions cérébrales congénitales. Il a lui aussi été transféré à Akelberg. Ada a reçu la même lettre, le même jour. »

Peter écarta les mains d'un geste élusif. « J'ai déjà entendu des histoires de ce genre. C'est de la propagande antigouvernementale. L'Église ne se mêle pas de politique. »

N'importe quoi, se dit Carla. L'Église était plongée dans la politique jusqu'au cou. Mais elle ne releva pas. « Le fils d'Ada n'avait plus d'appendice, continua-t-elle. On le lui avait retiré deux ans plus tôt.

— Eh bien, dit Peter. Qu'est-ce que ça prouve ? »

Carla commençait à perdre courage. Peter avait de toute évidence des a priori contre eux.

Heinrich insista : « Attends, Peter. Tu n'as pas encore tout entendu. Ilse, que voici, a travaillé à l'hôpital d'Akelberg. »

Peter lui jeta un regard interrogateur.

« J'ai été élevée dans la foi catholique, mon père », précisa Ilse.

Carla ignorait ce détail.

« Je ne suis pas une bonne catholique, poursuivit Ilse.

— Dieu est bon, l'homme ne l'est pas, ma fille.

— Je savais pourtant que ce que je faisais était un péché. Je l'ai fait quand même, parce qu'on me l'a ordonné et que j'avais peur. » Elle se mit à pleurer.

« Qu'as-tu fait ?

— J'ai tué des gens. Oh, mon père, Dieu me pardonnera-t-il ? »

Le prêtre dévisagea la jeune infirmière. Devant cette âme tourmentée, il ne pouvait plus prétendre qu'il s'agissait de propagande. Il devint livide.

Les autres étaient muets. Carla retint son souffle.

Ilse reprit : « Les malades arrivent à l'hôpital dans des cars gris. Ils ne reçoivent pas de traitement spécial. Nous leur faisons une injection et ils meurent. Ensuite, nous les incinérons. » Elle leva les yeux vers Peter. « Dieu pourra-t-il un jour me pardonner ce que j'ai fait ? »

Il ouvrit la bouche pour parler, mais les mots restèrent coincés dans sa gorge. Il toussa et dit enfin, d'une voix sourde : « Combien ?

— D'habitude quatre. Quatre autocars, je veux dire. Chacun contient environ vingt-cinq passagers.

— Cent personnes ?

— Oui. Chaque semaine. »

Toute la morgue de Peter s'était évanouie. Il avait le teint gris, la bouche ouverte. « Cent personnes par semaine ?

— Oui, mon père.

— De quoi souffrent-ils ?

— De toutes sortes d'affections mentales ou physiques. Il y a des vieillards séniles, des bébés atteints de malformations, des hommes et des femmes, paralysés, retardés mentaux ou simplement incurables. »

Il avait besoin de se l'entendre répéter. « Le personnel de l'hôpital les tue tous ?

— Je regrette, je regrette, sanglota Ilse. Je savais que c'était mal. »

Carla observa Peter. La métamorphose était étonnante. Après avoir entendu pendant des années les chrétiens aisés de cette banlieue verdoyante confesser leurs menus péchés, il était soudain confronté au mal à l'état pur. Il en était ébranlé jusqu'à la moelle.

Mais que ferait-il ?

Peter se leva. Il prit Ilse par la main et l'aida à se remettre debout. « Reviens dans le sein de l'Église, lui dit-il. Va te confesser au prêtre de ta paroisse. Dieu te pardonnera. Ça, je le sais.

— Merci », murmura Ilse.

Il lui lâcha la main et s'adressa à Heinrich. « Ça risque d'être moins simple pour nous », murmura-t-il.

Il leur tourna le dos, s'agenouilla et se remit à prier.

Carla interrogea Heinrich du regard. Il haussa les épaules. Ils se levèrent et sortirent, Carla entourant les épaules d'Ilse en larmes.

« Restons pour la messe, proposa Carla. Il nous reparlera peut-être après. »

Ils pénétrèrent tous les quatre dans la nef. Ilse sécha ses larmes et se calma. Frieda serra le bras d'Heinrich. Ils s'assirent au milieu des fidèles qui prenaient place sur les bancs, des hommes à l'air prospère, des femmes replètes, des enfants agités, vêtus de leurs plus beaux habits. Ces gens-là ne tueraient jamais des infirmes, songea Carla. Pourtant le gouvernement le faisait en leur nom. Comment était-ce arrivé ?

Elle ne savait qu'attendre du père Peter. Il était clair qu'il avait fini par les croire. Il avait cherché à ignorer leur requête qu'il pensait entachée de motivations politiques, mais s'était laissé convaincre par la sincérité d'Ilse. Il avait été épouvanté.

Toutefois il ne leur avait rien promis, sinon le pardon de Dieu pour Ilse.

Carla regarda autour d'elle. Le décor était plus coloré que dans les temples protestants qu'elle connaissait. Il y avait davantage de tableaux et de statues, plus de marbre et de dorures, d'étendards et de cierges. Les catholiques et les protestants s'étaient entretués pour des détails aussi dérisoires. Comme il semblait futile, dans un monde où l'on assassinait les enfants, qu'on puisse se préoccuper de cierges…

La messe commença, les prêtres – le père Peter dépassant les autres de sa haute taille – entrèrent, vêtus de leurs aubes. Carla se demanda comment interpréter son expression grave, empreinte d'une austère piété.

Elle écouta les cantiques et les prières avec indifférence. Elle avait prié pour son père avant de le découvrir, deux heures plus tard, battu à mort et agonisant dans l'entrée de leur maison. Il lui manquait tous les jours, à chaque heure du jour parfois. Les prières ne l'avaient pas sauvé, pas plus qu'elles ne protégeraient ceux que le gouvernement jugeait inutiles. Il fallait des actes, pas des paroles.

L'évocation de son père lui fit penser à son frère Erik. Il se trouvait quelque part en Russie. Il avait envoyé une lettre dans laquelle il se réjouissait de l'avancée rapide de l'invasion et refusait en des termes acerbes de croire que Walter avait été tué par la Gestapo. Pour lui, leur père avait été relâché sain et sauf et s'était fait attaquer dans la rue par des voyous, des communistes ou des Juifs. Erik vivait dans un monde imaginaire où la raison n'avait pas sa place.

Était-ce également le cas du père Peter ?

Celui-ci monta en chaire. Carla ne savait pas qu'il devait faire un sermon. Elle se demanda ce qu'il allait dire. S'inspirerait-il de ce qu'il venait d'entendre ? Parlerait-il de tout autre chose ? Des vertus de la pudeur ou du péché d'envie ? Ou fermerait-il les yeux pour remercier avec dévotion le Seigneur des victoires de l'Allemagne en Russie ?

Se dressant en chaire de toute sa haute taille, il balaya l'assemblée d'un regard dont on ne savait s'il était arrogant, fier ou provocant.

« Le cinquième commandement dit : "Tu ne tueras point." »

Carla croisa le regard d'Heinrich.

La voix du prêtre résonnait dans la nef, et les pierres en renvoyaient l'écho. « Il y a un endroit à Akelberg, en Bavière, où notre gouvernement transgresse ce commandement cent fois par semaine ! »

Carla étouffa une exclamation. Il les avait entendus ! Il prononçait un sermon contre le programme du gouvernement ! Cela pouvait tout changer !

« Peu importe que les victimes soient infirmes, déficientes mentales, incapables de se nourrir seules ou paralysées. » Peter ne contenait pas sa colère. « Les bébés innocents et les vieillards séniles sont tous des enfants de Dieu et leur vie est aussi sacrée que la vôtre ou la mienne. » Sa voix enfla. « Tuer est un péché mortel ! » Il leva le bras et brandit son poing fermé. Sa voix vibrait d'émotion. « Je vous le dis, si nous ne faisons rien pour nous opposer à de tels agissements, notre péché est aussi grand que celui des médecins et des infirmières qui administrent les injections mortelles. Si nous nous taisons… » Il s'interrompit avant de conclure. « Si nous nous taisons, nous sommes, nous aussi, des assassins ! »

12.

Le commissaire Thomas Macke était furieux. Il passait pour un imbécile aux yeux du commissaire principal Kringelein et de tous ses supérieurs. Il leur avait juré qu'il n'y aurait pas d'autres fuites. Que le secret d'Akelberg et de tous les hôpitaux du même genre dans le reste du pays était désormais bien gardé. Il avait retrouvé les trois fauteurs de troubles, Werner Franck, le pasteur Ochs et Walter von Ulrich, et les avait réduits d'une manière ou d'une autre au silence.

Pourtant le secret avait transpiré.

À cause d'un jeune prêtre insolent appelé Peter.

Le père Peter se trouvait devant Macke, nu, lié par les mains et les chevilles à un siège conçu spécialement à cet effet. Du sang lui coulait des oreilles, du nez et de la bouche. Sa poitrine était couverte de vomissures. Des électrodes étaient reliées à ses lèvres, ses mamelons et son sexe. Une sangle entourant son front immobilisait sa tête pour empêcher son cou de se briser sous la violence des convulsions.

Un médecin assis près de lui ausculta son cœur au stéthoscope, l'air dubitatif. « Il ne tiendra plus très longtemps », déclara-t-il d'un ton neutre.

Le prêche séditieux du père Peter avait fait boule de neige. L'évêque de Münster, monseigneur von Galen, un ecclésiastique bien plus influent, avait prononcé un sermon de la même veine, dans lequel il dénonçait le programme T4. L'évêque en avait appelé à Hitler pour qu'il protège le peuple contre la Gestapo, laissant habilement entendre que le Führer ne pouvait pas avoir connaissance de ce programme et lui offrant par là même un alibi de choix.

Son sermon avait été dactylographié, recopié et distribué dans toute l'Allemagne.

La Gestapo avait arrêté toutes les personnes surprises en possession d'un exemplaire de ce texte, en vain. Pour la première et dernière fois de l'histoire du IIIe Reich, l'opinion manifesta sa réprobation contre une action du gouvernement.

La répression fut brutale, mais ne servit à rien : les copies du sermon continuèrent à circuler, les hommes d'Église furent de plus en plus nombreux à prier pour les malheureux assassinés. Il y eut même une marche de protestation à Akelberg. La situation échappait à tout contrôle.

Et la faute en revenait à Macke.

Il se pencha sur Peter. Le prêtre avait les yeux fermés et respirait faiblement, mais il était conscient. Macke lui hurla à l'oreille : « Qui t'a parlé d'Akelberg ? »

Pas de réponse.

Peter était sa seule piste. Les enquêtes menées à Akelberg n'avaient rien donné. On avait bien parlé à Reinhold Wagner de deux filles qui étaient passées à l'hôpital, mais personne ne savait qui elles étaient ; on lui avait aussi signalé une infirmière qui avait quitté sa place soudainement, en expliquant dans sa lettre de démission qu'elle devait se marier précipitamment, mais sans donner le nom du prétendu mari. Ces deux indices n'avaient mené à rien. De toute façon, Macke était persuadé qu'un tel désastre ne pouvait pas être l'œuvre de quelques gamines.

Macke fit signe au technicien qui manipulait l'appareil. Il tourna un bouton.

Peter poussa un cri de souffrance quand le courant électrique parcourut son corps, mettant ses nerfs au supplice. Il se contorsionna comme s'il souffrait d'une crise d'épilepsie. Ses cheveux se dressèrent sur sa tête.

Le technicien coupa le courant.

Macke brailla : « Dis-moi le nom de cet homme ! »

Peter ouvrit enfin la bouche.

Macke se pencha encore.

Peter murmura : « Pas un homme.

— Une femme alors ! Son nom !

— C'était un ange.

— Va chier en enfer ! » Macke saisit le bouton et le tourna. « Ça ne s'arrêtera que quand tu parleras ! » glapit-il à l'adresse de Peter qui hurlait de douleur.

La porte s'ouvrit. Un jeune policier en civil passa la tête, pâlit et adressa un signe à Macke.

Le technicien coupa le courant. Les hurlements cessèrent. Le médecin se pencha pour examiner Peter.

« Excusez-moi, commissaire Macke, dit le policier. Le commissaire principal Kringelein veut vous voir.

— Maintenant ?

— C'est ce qu'il a dit, commissaire. »

Macke se tourna vers le médecin qui haussa les épaules. « Il est jeune. Il sera encore vivant quand vous reviendrez », assura-t-il.

Macke sortit et suivit le policier dans l'escalier. Le bureau de Kringelein se trouvait au rez-de-chaussée. Macke frappa et entra. « Ce foutu prêtre n'a toujours pas parlé, dit-il sans préambule. Il me faut plus de temps. »

Kringelein était un petit homme mince à lunettes, intelligent mais velléitaire. Converti tardivement au nazisme, il n'était pas membre de l'élite SS et n'avait pas la ferveur des inconditionnels comme Macke. « Laissez tomber le prêtre, ordonna-t-il. Les membres du clergé ne nous intéressent plus. Envoyez-les dans des camps et qu'on n'en parle plus. »

Macke n'en croyait pas ses oreilles. « Mais ils ont comploté contre le Führer !

— Et ils ont réussi. Alors que vous, vous avez échoué. »

Macke soupçonnait Kringelein de s'en réjouir intérieurement.

« Une décision a été prise au sommet de l'État, continua le commissaire principal. L'Aktion T4 est annulée. »

Macke n'en revenait pas. Les nazis ne se laissaient jamais dicter leurs décisions par les scrupules des ignorants. « Nous n'en sommes pas arrivés où nous en sommes en cédant à l'opinion publique ! s'indigna-t-il.

— Cette fois-ci pourtant, c'est le cas.

— Pourquoi ?

— Le Führer n'a pas pris la peine de m'expliquer ses raisons personnellement, répliqua Kringelein d'un ton sarcastique. Mais je peux les deviner. Ce programme a suscité des protestations particulièrement vives de la part d'une opinion ordinairement passive. Le maintenir, c'est risquer de heurter de front les Églises de toutes confessions. Ce n'est pas souhaitable. Nous ne devons pas entamer l'unité et la détermination du peuple allemand, surtout maintenant que nous sommes en guerre contre l'Union soviétique, notre plus puissant ennemi jusqu'à présent. Le programme est donc annulé.

— Bien, commissaire principal, capitula Macke en ravalant sa colère. Autre chose ?

— Vous pouvez disposer. »

Macke se dirigea vers la porte.

« Macke.

— Oui, commissaire principal.

— Changez de chemise.

— De chemise ?

— Elle est tachée de sang.

— Oui, commissaire principal. Excusez-moi, commissaire principal. »

Macke reprit l'escalier en bouillant intérieurement. Il regagna la pièce du sous-sol. Le père Peter était toujours en vie.

Furieux, il lui cria encore : « Qui t'a parlé d'Akelberg ? »

Pas de réponse.

Il tourna le bouton à fond.

Le père Peter hurla longtemps avant de retomber dans un silence définitif.

13.

La villa des Franck se dressait au milieu d'un parc. Sur un talus, à deux cents mètres de la maison, se trouvait un petit pavillon ouvert aux quatre vents, équipé de sièges. Quand elles étaient petites, Carla et Frieda disaient que c'était leur maison de campagne et y jouaient pendant des heures à donner de grandes réceptions fictives, servies par des dizaines de domestiques empressés auprès de leurs invités de marque. Plus tard, elles s'y étaient retrouvées pour bavarder à l'abri des oreilles indiscrètes.

« La première fois que je me suis assise sur ce banc, mes pieds ne touchaient pas terre, remarqua Carla.

— Je regrette ce temps-là », soupira Frieda.

Elles étaient vêtues de robes sans manches pour supporter la chaleur humide et étouffante, sous le ciel couvert de l'après-midi. Elles étaient d'humeur sombre. Le père Peter était mort. D'après la police, il s'était suicidé en prison, accablé par le poids de ses crimes. Carla se demandait s'il n'avait pas plutôt été battu à mort comme son père. C'était affreusement probable.

Des dizaines d'autres prêtres croupissaient dans des cellules partout en Allemagne. Certains avaient protesté publiquement contre le meurtre des handicapés, d'autres n'avaient fait que diffuser des exemplaires du sermon de monseigneur von Galen. Carla se demandait s'ils seraient tous torturés. Et pendant combien de temps elle échapperait elle-même à ce sort.

Werner sortit de la maison avec un plateau. Il traversa la pelouse pour l'apporter au pavillon. « Une limonade, les filles ? » lança-t-il joyeusement.

Carla se détourna. « Non merci », répondit-elle sèchement. Elle ne comprenait pas comment il pouvait encore prétendre être son ami après s'être lâchement défilé.

« Moi non plus, refusa Frieda.

— J'espère que nous ne sommes pas fâchés », dit-il en regardant Carla.

Quelle question ! Bien sûr que si.

« Le père Peter est mort, Werner », lui annonça Frieda.

Carla ajouta : « Probablement torturé à mort par la Gestapo parce qu'il n'admettait pas qu'on assassine des gens comme ton frère. Mon père est mort, lui aussi, pour la même raison. Quantité d'autres sont en prison ou dans des camps. Mais toi, tu as gardé ton petit travail de bureau pépère, donc tout va bien. »

Werner parut blessé, ce qui étonna Carla. Elle s'attendait à une protestation arrogante ou au moins à une feinte indifférence. Il avait l'air sincèrement malheureux. « Tu ne crois pas que c'est à chacun de nous de trouver comment agir de son mieux ? » demanda-t-il.

L'argument manquait de poids. « Tu n'as rien fait !

— Peut-être, acquiesça-t-il tristement. Pas de limonade alors ? »

Devant le silence des deux filles, il repartit vers la maison.

Carla était furieuse et indignée, mais ne pouvait s'empêcher d'éprouver des regrets. C'était plus fort qu'elle. Avant de découvrir que Werner était un lâche, elle s'était engagée dans une histoire d'amour avec lui. Elle l'aimait beaucoup, dix fois plus que tous les garçons qu'elle avait embrassés avant lui. Elle n'avait pas à proprement parler le cœur brisé, mais elle était profondément déçue.

Frieda avait plus de chance, se dit-elle en voyant Heinrich sortir de la maison. Frieda était expansive et aimait s'amuser, Heinrich était tourmenté et sérieux, mais ils s'entendaient bien. « Tu es amoureuse de lui ? lui demanda Carla alors qu'il était encore hors de portée de voix.

— Je ne sais pas encore. Il est tellement gentil ! Je l'adore. »

Si ce n'était pas de l'amour, ça n'en était pas loin, songea Carla.

Heinrich arrivait avec des nouvelles. « Il fallait que je vienne vous l'annoncer tout de suite, dit-il. Mon père me l'a appris juste après le déjeuner.

— Quoi ? demanda Frieda.

— Le gouvernement a annulé l'opération. Elle s'appelait Aktion T4. Le meurtre des incurables. Ils arrêtent tout.

— Tu veux dire qu'on a gagné ? » demanda Carla.

Heinrich hocha la tête avec vigueur. « Mon père n'en revient pas. Il dit qu'il n'a jamais vu le Führer céder à l'opinion publique.

— Et c'est nous qui l'y avons obligé ! se réjouit Frieda.

— Heureusement que personne ne le sait ! » remarqua Heinrich gravement.

Carla insista : « Ils vont fermer les hôpitaux et mettre fin à tout le programme ?

— Pas exactement.

— Que veux-tu dire ?

— D'après mon père, les médecins et les infirmières vont tous être transférés.

— Où ça ? s'inquiéta Carla.

— En Russie. »

IX

1941 (II)

1.

Le téléphone posé sur le bureau de Greg Pechkov se mit à sonner. Il décrocha. C'était une chaude matinée de juillet. Il venait de finir son avant-dernière année à Harvard et, pour la seconde fois, effectuait un stage au Département d'État, au bureau de l'Information. Doué en physique et en maths, il avait réussi ses examens brillamment. Cependant, il n'avait pas envie de se consacrer à la science. La politique, voilà ce qui le passionnait.

« Greg Pechkov

— Bonjour, monsieur Pechkov. Tom Cranmer au téléphone. »

Le cœur de Greg se mit à battre plus vite. « Merci de me rappeler. Je vois que vous vous souvenez de moi.

— Hôtel Ritz-Carlton, 1935. La seule fois de ma vie où j'ai eu ma photo dans le journal.

— Vous êtes toujours détective dans cet hôtel ?

— J'ai quitté l'hôtellerie pour le commerce. Je travaille maintenant dans un grand magasin.

— Vous arrive-t-il d'effectuer des enquêtes pour votre propre compte ?

— Oui, bien sûr. Qu'avez-vous en tête ?

— Je suis au bureau actuellement, et j'aimerais vous parler seul à seul.

— Vous travaillez juste en face de la Maison Blanche, dans l'Old Executive Office Building, c'est bien ça ?

— Comment le savez-vous ?

— Je suis détective.

— Évidemment.

— Je suis juste à côté, au café Aroma, à l'angle de F Street et de la 19e.

— Je ne suis pas libre pour le moment, dit Greg en regardant sa montre. D'ailleurs, je devrais déjà être parti.

— Je peux vous y attendre.

— Dans une heure, alors. »

Greg descendit l'escalier au pas de course. Il arriva à l'entrée principale juste au moment où une Rolls-Royce s'arrêtait silencieusement devant le perron. Un chauffeur adipeux s'en extirpa et alla ouvrir la portière arrière, d'où sortit un bel homme élancé, à l'abondante chevelure argentée. Son costume croisé en flanelle gris perle, de coupe parfaite, l'habillait avec une élégance que seuls les tailleurs de Londres sont capables d'atteindre. Il grimpa les marches de granit menant à l'immense bâtiment, son gros chauffeur courant derrière lui, chargé de sa serviette.

C'était le sous-secrétaire d'État Sumner Welles, numéro deux du Département d'État et ami personnel du président Roosevelt.

Le chauffeur était sur le point de remettre la mallette à un huissier qui attendait là quand Greg s'avança. « Bonjour, monsieur », dit-il, et, d'un même geste, il s'empara du porte-documents et tint la porte à Welles, avant de pénétrer à sa suite dans le bâtiment.

Greg avait été embauché au bureau de l'Information parce qu'il avait pu présenter plusieurs articles bien écrits et solidement documentés qu'il avait rédigés pour le *Harvard Crimson*. Pour autant, il ne souhaitait pas finir dans la peau d'un journaliste officiel. Il avait de plus hautes ambitions.

Il admirait Sumner Welles, parce qu'il lui rappelait son père : la même prestance, la même élégance et le même charme dissimulant la même nature impitoyable. Welles était déterminé à prendre la succession de son patron, le secrétaire d'État Cordell Hull, et n'hésitait pas à aller parler au Président directement, derrière son dos, ce qui mettait Hull en rage. Quant à Greg, il trouvait passionnant d'être proche de quelqu'un qui avait du pouvoir et ne craignait pas de l'utiliser. C'était exactement la vie qu'il voulait mener.

Welles s'était entiché de lui, comme bien des gens, surtout quand Greg avait décidé de tout faire pour ça. Dans le cas présent, cependant, un facteur supplémentaire jouait en sa faveur : en fait, Welles avait un faible pour les jeunes gens séduisants, bien qu'il fût marié à une riche héritière et, apparemment, heureux en ménage.

Greg était hétérosexuel jusqu'au bout des ongles. À Harvard, il entretenait une relation suivie avec une étudiante de Radcliffe du nom d'Emily Hardcastle, qui lui avait promis de mettre la main sur un moyen de contraception avant le mois de septembre, et ici, à Washington, il sortait avec Rita Lawrence, une demoiselle pulpeuse, fille du représentant du Texas au Congrès. Avec Welles, Greg marchait sur des œufs, évitant tout contact physique mais se montrant assez aimable pour ne pas perdre les bonnes grâces de son patron. En conséquence, après les cocktails, quand l'alcool avait raison des inhibitions de son aîné et qu'il commençait à avoir les mains baladeuses, Greg s'efforçait de ne pas croiser son chemin.

À dix heures, lorsque les cadres se rassemblèrent dans son bureau pour le point quotidien, Welles demanda à Greg de rester. « C'est bon pour votre instruction, mon garçon. » Le jeune homme en fut ravi. Cette réunion lui offrirait-elle l'occasion de briller ? C'était son désir le plus vif : être remarqué et faire impression.

Quelques minutes plus tard, le sénateur Dewar fit son entrée, accompagné de son fils Woody. Grands et maigres tous les deux, ils avaient une grosse tête et étaient vêtus à l'identique d'un costume d'été en lin bleu marine à veston droit. Le talent artistique différenciait toutefois Woody de son père : ses photographies publiées dans le *Harvard Crimson* lui avaient valu des prix. Woody adressa un signe de tête à Bexforth Ross, le bras droit de Welles, qu'on surnommait Greg « Russki », à cause de son nom de famille.

Welles ouvrit la réunion par ces mots : « Je dois vous annoncer une nouvelle hautement confidentielle, qui ne doit en aucun cas sortir de cette pièce : le Président va rencontrer le Premier ministre britannique au début du mois prochain. »

Greg faillit pousser une exclamation et se retint juste à temps.

« Bien ! déclara Gus Dewar. Où ça ?

— La rencontre est prévue à bord d'un navire, quelque part au milieu de l'Atlantique, pour des raisons de sécurité évidentes, et pour éviter à Churchill un trop long voyage. Le Président tient à ce que je sois présent. Le secrétaire d'État restera ici, à Washington, pour veiller au grain. Il veut que vous soyez également du voyage, Gus.

— J'en suis très honoré, répondit celui-ci. Quel est l'ordre du jour ?

— Pour le moment, les Britanniques semblent avoir écarté toute menace d'invasion, mais ils sont trop faibles pour attaquer les Allemands sur le continent – sans notre aide du moins. Churchill va donc nous demander de déclarer la guerre à l'Allemagne, ce que nous refuserons bien entendu. Cette question réglée, le Président souhaite que nos deux pays exposent leurs objectifs dans une déclaration commune.

— Des objectifs sans rapport avec la guerre, je suppose ? demanda Gus.

— En effet. Pour la bonne raison que les États-Unis ne sont pas en guerre et n'ont pas l'intention d'y entrer. Ce qui ne nous empêche pas d'être – en tant que puissance non belligérante – les alliés des Britanniques : nous leur fournissons à peu près tout ce dont ils ont besoin et leur assurons un crédit illimité. Une fois la paix revenue, nous comptons bien avoir notre mot à dire sur la marche du monde.

— Cela comprendra-t-il le renforcement de la Société des nations ? » demanda Gus. C'était une idée à laquelle il tenait beaucoup, comme Welles, Greg ne l'ignorait pas.

« C'est exactement ce dont je souhaitais m'entretenir avec vous, Gus. Si nous voulons mener ce projet à bien, nous devons être soigneusement préparés. Et obtenir de FDR et de Churchill un engagement en ce sens dans leur déclaration.

— En théorie, le Président est favorable à cette idée, nous le savons tous les deux, observa Gus. Ce qui l'inquiète, c'est l'opinion publique. »

Un assistant, entré discrètement, remit une note à Bexforth qui la lut et s'écria : « Oh ! Mon Dieu !

— Qu'y a-t-il ? lança Welles avec humeur.

— Le Conseil impérial japonais s'est réuni la semaine dernière, vous le savez, expliqua Bexforth. Nous avons obtenu certains renseignements sur le contenu de ses délibérations. »

Bexforth restait vague sur la provenance de ces renseignements, mais Greg avait compris de quoi il parlait. La Signal Intelligence Unit de l'armée américaine était capable d'intercepter et de décrypter les messages radio que le ministère des Affaires étrangères japonais adressait à ses ambassades à l'étranger. Ces décryptages avaient pour nom de code MAGIC. Greg savait bien des choses qu'il n'était pas censé savoir – si l'armée avait appris qu'il était dans le secret, il aurait même eu de sacrés ennuis !

« Les Japonais envisagent d'étendre encore leur empire », continua Bexforth. Ils avaient déjà annexé la vaste région de Mandchourie et fait pénétrer des troupes dans une grande partie du reste de la Chine, Greg en était informé. « Ils ne sont pas favorables à une poursuite de l'expansion vers l'ouest en direction de la Sibérie, car cela les entraînerait dans une guerre avec l'Union soviétique.

— Heureusement ! s'exclama Welles. Ainsi, les Russes garderont comme priorité de combattre les Allemands.

— C'est exact, monsieur. Mais les Japonais envisagent en revanche de s'étendre vers le sud, de prendre le contrôle de toute l'Indochine et de s'emparer ensuite des Indes orientales néerlandaises. »

Greg en resta abasourdi. Il s'agissait là de renseignements d'une actualité brûlante, et il était parmi les premiers à en être informé.

« Comment ? s'indigna Welles. Mais c'est une guerre impérialiste, ni plus ni moins ! »

Gus s'interposa : « Pas à proprement parler, Sumner. Les Japonais ont déjà des troupes en Indochine avec l'accord en bonne et due forme de la puissance coloniale en place, à savoir la France, représentée par le gouvernement de Vichy.

— Des marionnettes entre les mains des nazis !

— J'ai bien dit "à proprement parler". Quant aux Indes néerlandaises, elles sont théoriquement sous contrôle des Pays-Bas, lesquels sont actuellement occupés par les Allemands.

L'Allemagne ne peut que se réjouir de voir le Japon, son allié, étendre son contrôle sur une colonie hollandaise.

— C'est ce qu'on appelle finasser.

— Mais c'est une finasserie qui nous sera opposée. À commencer par l'ambassadeur du Japon.

— Vous avez raison, Gus, merci de me mettre en garde. »

Greg guettait l'occasion d'apporter sa contribution au débat. Impressionner ces hommes d'expérience, voilà ce qu'il voulait par-dessus tout. Mais ils en savaient tous tellement plus que lui !

« Qu'est-ce qu'ils veulent à la fin, ces Japonais ? s'exclama Welles.

— Du pétrole, du caoutchouc, de l'étain, répondit Gus. Ils veulent s'assurer un accès permanent aux ressources naturelles. Ce n'est pas étonnant puisque nous continuons à les empêcher de s'approvisionner. » Les États-Unis avaient placé sous embargo l'exportation de matières premières telles que le pétrole et le fer en direction du Japon, dans l'espoir de freiner les visées expansionnistes de ce pays en Asie.

« Pour l'efficacité avec laquelle nous appliquons nos embargos ! s'écria Welles avec irritation.

— C'est vrai, mais cette menace suffit à terrifier les Japonais. Leur sol ne renferme quasiment aucune matière première.

— Il est clair que nous devons prendre des mesures plus sévères, lança Welles sèchement. Les Japonais ont beaucoup d'argent dans les banques américaines. Ne pourrait-on pas geler leurs avoirs ? »

Les fonctionnaires rassemblés dans la salle prirent un air désapprobateur. C'était une proposition franchement radicale. Au bout d'un moment, Bexforth déclara : « Ce serait sans doute possible. Et plus efficace que tous les embargos du monde. Au moins, ils ne pourraient plus acheter de pétrole ni d'autres matières premières ici, aux États-Unis, faute de moyens financiers. »

Gus Dewar déclara : « Le secrétaire d'État veillera évidemment, comme d'habitude, à éviter toute action qui risquerait de déboucher sur la guerre. »

Il disait vrai. La prudence de Cordell Hull frôlait la pusillanimité, et il était souvent en conflit avec Welles, beaucoup plus agressif.

« Mr. Hull a toujours suivi cette ligne, et il a eu bien raison », déclara Welles. Tout le monde savait qu'il n'en pensait pas un mot, mais que l'étiquette exigeait ce semblant d'approbation. « Toutefois, les États-Unis doivent garder la tête haute sur la scène internationale. Prudence n'est pas synonyme de lâcheté. Cette idée du gel des avoirs mérite que je la soumette au Président. »

Greg était émerveillé. Voilà donc ce qu'était le pouvoir : en l'espace d'un instant, Welles proposait une initiative susceptible d'ébranler une nation tout entière !

Gus Dewar fronça les sourcils. « Si le Japon ne peut plus importer de pétrole, toute l'économie du pays sera paralysée et son armée réduite à l'impuissance.

— Que demander de mieux ! déclara Welles.

— Vous croyez ? À votre avis, comment le gouvernement militaire japonais réagira-t-il à une catastrophe de cette ampleur ?

— Eh bien dites-le-moi, monsieur le sénateur ! répliqua Welles, qui n'appréciait guère la contestation.

— Je ne sais pas. Mais il me semble préférable d'avoir la réponse à cette question avant d'entreprendre quoi que ce soit. Les gens aux abois sont dangereux. Et les États-Unis ne sont pas prêts à entrer en guerre contre le Japon, je le sais. Notre marine n'est pas prête, et notre armée de l'air pas davantage. »

Greg vit là une occasion de prendre la parole et n'hésita pas un instant. « Excusez-moi, monsieur le sous-secrétaire. Savez-vous que l'opinion publique américaine est plus favorable à la guerre avec le Japon qu'à l'apaisement ? À deux contre un.

— Bonne remarque, Greg, merci. Les Américains ne veulent pas que le Japon puisse assassiner des gens impunément.

— Ils ne veulent pas non plus la guerre, pas vraiment, rétorqua Gus. N'en déplaise aux sondages. »

Welles referma le dossier posé devant lui. « Eh bien, monsieur le sénateur, nous sommes d'accord en ce qui concerne la Société des nations et nous ne sommes pas d'accord en ce qui concerne le Japon.

— Et dans les deux cas, il revient au Président d'en décider, dit Gus en se levant.

— Merci à vous d'être venus me voir. »

La réunion prit fin.

Greg était euphorique lorsqu'il sortit de la pièce. Il avait été invité à assister au débat, il avait appris des nouvelles étonnantes, et son intervention lui avait valu les remerciements de Welles. La journée commençait bien !

Il quitta discrètement le bâtiment et se dirigea vers le café Aroma.

Il n'avait jamais engagé de détective privé et avait un peu l'impression de frôler l'illégalité. Mais Cranmer était un homme respectable, et il n'y avait rien de délictueux à vouloir renouer avec une ancienne petite amie.

À l'Aroma, la clientèle se limitait à deux filles qui avaient l'air de secrétaires prenant une pause, à un couple âgé sorti faire des courses et à un homme costaud en costume de seersucker tout froissé qui tirait sur une cigarette : Cranmer. Greg se glissa dans le box qu'il occupait et commanda un café.

« Je voudrais retrouver Jacky Jakes.

— La jeune Noire ? »

Eh oui, songea Greg avec nostalgie. Elle était très jeune à l'époque. Seize ans, bien qu'elle se prétendît plus âgée. « Six ans ont passé, fit-il remarquer à Cranmer. Ce n'est plus une toute jeune fille aujourd'hui.

— Vous savez, ce n'est pas moi qui l'avais embauchée pour cette petite comédie, c'était votre père.

— Je ne veux pas m'adresser à lui. Mais vous pourrez la retrouver, n'est-ce pas ?

— Probablement, répondit Cranmer en sortant calepin et stylo. Jacky Jakes, ce n'était pas son vrai nom, si ?

— En réalité, elle s'appelle Mabel Jakes.

— Profession : actrice, c'est bien ça ?

— C'était son rêve. Je ne sais pas si elle l'a réalisé. » Elle était jolie, avec du charme à revendre, mais il n'y avait pas beaucoup de rôles pour les acteurs noirs.

« À l'évidence, elle n'est pas dans l'annuaire, sinon vous n'auriez pas besoin de moi.

— Elle est peut-être sur liste rouge, mais je pense plutôt qu'elle ne peut pas s'offrir le téléphone.

— Vous l'avez revue depuis 1935 ?

— Deux fois. La première, il y a deux ans, tout près d'ici, dans E Street. La deuxième, il y a deux semaines, à deux rues d'ici.

— Comme elle n'habite certainement pas dans ce quartier huppé, elle doit travailler dans le coin. Vous avez sa photo?

— Non.

— Je me souviens vaguement d'elle : une jolie fille, avec la peau sombre et un grand sourire. »

Greg hocha la tête. Oui, un sourire éclatant, inoubliable. « Je voudrais juste son adresse. Pour lui écrire un mot.

— Je n'ai pas besoin de connaître vos motifs.

— Eh bien, c'est parfait », répliqua Greg. Les choses étaient-elles aussi simples que cela?

« Je prends dix dollars par jour. Sur la base d'un forfait minimum de deux jours. Plus les frais. »

C'était moins que ce à quoi Greg s'était attendu. Sortant son portefeuille, il tendit au détective un billet de vingt dollars.

« Merci.

— Eh bien, bonne chance! » répondit Greg.

2.

Ce samedi-là, comme il faisait chaud, Woody alla à la plage avec son frère Chuck.

La famille Dewar au grand complet se trouvait à Washington. Ils occupaient un appartement de neuf pièces, tout près de l'hôtel Ritz-Carlton. Chuck, qui était dans la marine, était en permission; Gus travaillait douze heures par jour à préparer cette réunion au sommet qu'il appelait déjà conférence de l'Atlantique, et Rosa était plongée dans la rédaction d'un nouveau livre sur les épouses de présidents.

Woody et Chuck, en shorts et polos, attrapèrent serviettes de bain, lunettes de soleil et journaux et prirent un train pour Rehoboth Beach, sur la côte du Delaware. Un trajet de presque deux heures, mais c'était le seul endroit où aller par un samedi d'été. Une vaste étendue de sable, une brise rafraîchissante soufflant de l'océan et un bon millier de filles en maillot de bain : que demander de plus?

Les deux frères ne se ressemblaient guère. Plus petit que Woody, Chuck était trapu et athlétique. Il avait la beauté de sa mère et son sourire conquérant. Il avait été un élève médiocre, malgré sa vive intelligence et ses idées originales, elles aussi héritées de sa mère. Il battait son frère dans tous les sports, sauf à la course où Woody l'emportait grâce à la longueur de ses jambes, et à la boxe où ses longs bras lui assuraient une meilleure défense.

Chuck n'avait pas raconté grand-chose de sa vie dans la marine à ses parents, sans doute parce qu'ils lui tenaient encore rigueur de ne pas être entré à Harvard. Mais seul avec Woody, il s'ouvrit un peu plus. « C'est chouette, Hawaï, mais franchement, je suis déçu d'avoir un poste à terre. Je suis entré dans la marine pour naviguer, moi.

— Qu'est-ce que tu fais, au juste ?

— Je suis dans la Signal Intelligence Unit. On écoute les messages radio, principalement ceux de la marine impériale japonaise.

— Ils ne sont pas codés ?

— Si, mais on peut apprendre pas mal de choses sans même les décoder. On appelle ça l'analyse du trafic. Une brusque augmentation des messages indique qu'il se prépare quelque chose. On finit aussi par repérer certains schémas. Les débarquements amphibies, par exemple, sont annoncés par une configuration de signaux tout à fait particulière.

— C'est fascinant. Tel que je te connais, je suis sûr que tu te débrouilles drôlement bien. »

Chuck haussa les épaules. « Je ne suis qu'un gratte-papier, chargé d'annoter et de classer les transcriptions. Mais, à force, je finis par avoir quelques bases.

— Et la vie à Hawaï, c'est comment ?

— Très amusant. Dans les bars à marins, c'est parfois assez chaud. Le Chat noir, c'est le plus sympa de tous. J'ai un bon copain, Eddie Parry, avec qui je fais du surf à Waikiki. On y va dès qu'on a un moment. Oui, je m'amuse bien, mais je regrette quand même de ne pas être en mer. »

Ils se baignèrent dans les eaux froides de l'Atlantique, s'offrirent des hot-dogs au déjeuner, se photographièrent réciproquement avec l'appareil de Woody et détaillèrent les maillots de bain, jusqu'à ce que le soleil commence à descendre. Au moment où ils repartaient, se frayant un chemin au milieu de la foule, Woody aperçut Joanne Rouzrokh.

Un coup d'œil lui suffit. Elle ne ressemblait à aucune autre fille de la plage, ni même du Delaware. Avec ses pommettes saillantes, son nez en cimeterre, son abondante chevelure noire et son teint d'une douce couleur café au lait, il était impossible de la confondre avec une autre.

D'un pas décidé, il marcha droit sur elle.

Elle était sensationnelle dans son maillot de bain noir une pièce. Les fines bretelles mettaient en valeur l'élégance de ses épaules, et sa coupe droite en haut des cuisses révélait la quasi-totalité de ses longues jambes brunes.

Dire qu'il avait tenu dans ses bras cette fabuleuse créature, qu'il l'avait embrassée comme si ce jour ne devait jamais finir !

Elle leva la tête vers lui en se protégeant les yeux du soleil. « Woody Dewar ! Je ne savais pas que tu étais à Washington ! »

Il n'avait pas besoin d'autre invitation pour s'agenouiller sur le sable. Le seul fait d'être aussi proche d'elle lui coupa le souffle. « Bonjour, Joanne. » Il jeta un bref regard à la fille bien en chair aux yeux marron installée à côté d'elle. « Ton mari n'est pas là ? »

Elle éclata de rire. « Qu'est-ce qui te fait croire que je suis mariée ? »

Woody se troubla. « Quand je suis venu à une soirée chez toi, il doit y avoir deux ans de ça, un été…

— Tu es venu ?

— Mais oui, je me rappelle ! intervint la compagne de Joanne. Je t'ai demandé ton nom et tu n'as pas voulu me le dire. »

Woody n'avait pas le moindre souvenir de cette scène. « Je suis désolé d'avoir été aussi impoli, dit-il. Je m'appelle Woody Dewar, et voici mon frère Chuck. »

La jeune fille aux yeux bruns serra la main des deux garçons. « Diana Taverner. » Chuck s'assit sur le sable à côté d'elle, ce qui parut lui plaire : Chuck était beau garçon, bien plus séduisant que Woody.

« Quoi qu'il en soit, poursuivit celui-ci, je t'ai cherchée partout, et dans la cuisine, je suis tombé sur un certain Bexforth

Ross qui s'est présenté comme ton fiancé. Je vous imaginais mariés maintenant. Est-ce que ce n'est pas extraordinairement long pour des fiançailles ?

— Que tu es bête ! dit-elle avec une pointe d'agacement, et il se souvint qu'elle n'aimait pas beaucoup les taquineries. Bexforth disait à tout le monde que nous étions fiancés parce qu'il vivait plus ou moins chez nous. »

Woody fut désarçonné. Voulait-elle dire que Bexforth couchait dans le même appartement ? Dans le lit de Joanne ? Ce n'était pas rare, bien sûr, mais peu de filles l'admettaient.

« C'était lui qui parlait mariage, poursuivit-elle. Moi, je n'ai jamais eu l'intention de l'épouser. »

Autrement dit, elle était seule. Woody n'aurait pas été plus heureux s'il avait gagné à la loterie.

Il fallait rester prudent, cela ne l'empêchait pas d'avoir peut-être un petit ami. Il faudrait qu'il se renseigne. Mais enfin… un petit ami, ce n'était pas comme un mari.

« Justement, j'étais l'autre jour à une réunion à laquelle il assistait, remarqua Woody. Bexforth, c'est quelqu'un, au Département d'État.

— Il ira loin et il se trouvera une femme qui remplira mieux que moi la fonction d'épouse d'un grand manitou du Département d'État. »

À en juger par son ton, Joanne ne gardait pas un excellent souvenir de son ancien amant. Woody en fut tout réjoui, sans savoir pourquoi.

Il prit appui sur son coude. Le sable était chaud. Si elle avait un petit ami auquel elle tenait, elle trouverait bien l'occasion de parler de lui avant longtemps, il en était certain.

« À propos du Département d'État, demanda-t-il, tu y travailles toujours ?

— Oui. Je suis l'assistante du sous-secrétaire chargé des affaires européennes.

— C'est intéressant ?

— À l'heure actuelle, très. »

Les yeux rivés sur la limite inférieure de son maillot de bain en haut de ses cuisses, Woody se disait que quelle que soit la surface de peau découverte, les pensées d'un homme se focalisent toujours sur les parties cachées d'une femme. Sentant venir une érection, il roula sur le ventre pour la dissimuler.

Joanne, à qui son regard n'avait pas échappé, lui demanda : « Tu aimes bien mon maillot de bain ? » Elle allait toujours droit au but. C'était un des traits de caractère qu'il appréciait le plus en elle.

Il décida d'être tout aussi franc. « C'est *toi* que j'aime bien, Joanne. Je t'ai toujours bien aimée. »

Elle éclata de rire. « Ne tourne pas autour du pot, Woody. Dis les choses carrément ! »

Autour d'eux, les gens rangeaient leurs affaires. « On ferait bien d'y aller, déclara Diana.

— Justement, on allait rentrer, dit Woody. On pourrait faire le voyage ensemble ? »

C'était le moment ou jamais pour elle de le rembarrer poliment en disant *Non, merci, les gars. Partez devant.* Mais elle répondit : « Bien sûr, pourquoi pas ? »

Les filles enfilèrent leur robe par-dessus leur maillot, entassèrent leurs affaires dans deux ou trois sacs et ils se dirigèrent tous vers la gare.

Le train était bondé de gens qui avaient passé l'après-midi à la plage comme eux, couverts de coups de soleil et mourant de faim et de soif. Woody acheta quatre Coca-Cola à la gare et les sortit dès que le train s'ébranla. « Une fois, à Buffalo, tu m'as acheté un Coca, dit Joanne, un jour de grande chaleur. Tu te rappelles ?

— Bien sûr ! C'était pendant la manifestation.

— On n'était que des gosses !

— Offrir un Coca, c'est une de mes techniques habituelles avec les jolies femmes. »

Elle rit. « Et ça te réussit ?

— Ça ne m'a pas encore valu un seul baiser. »

Elle leva sa bouteille comme pour porter un toast. « Persévère, ça marchera peut-être un jour ! »

Il décida d'y voir un encouragement et enchaîna : « Quand on sera rentrés en ville, tu veux aller prendre un hamburger quelque part ? Ou aller au ciné, peut-être ? »

C'était le moment ou jamais pour elle de dire *Non merci, je dois retrouver mon copain*. Diana ne lui en laissa pas le temps.

« Ce serait chouette, s'écria-t-elle. Qu'est-ce que tu en penses, Joanne ?

— Volontiers. »

Pas de petit ami donc, et en plus, un rendez-vous ! Woody fit de son mieux pour dissimuler sa joie. « On pourrait aller voir *Fiancée contre remboursement*, il paraît que c'est assez drôle.

— C'est avec qui ? demanda Joanne.

— James Cagney et Bette Davis.

— Ah oui, j'aimerais bien le voir.

— Moi aussi, renchérit Diana.

— Affaire conclue, lança Woody.

— Et toi, Chuck, ça te dit ?..., intervint Chuck. Oh oui, grand frère, c'est super ! Sympa à toi de me le proposer. »

Ce n'était pas franchement désopilant, mais Diana rit obligeamment.

Joanne s'endormit peu après, la tête sur l'épaule de Woody. Ses cheveux bruns lui chatouillaient le cou, il sentait son souffle chaud sur sa peau juste à l'endroit où s'arrêtait la manche courte de sa chemise. Il nageait dans le bonheur.

Ils se séparèrent à Union Station pour rentrer chez eux se changer avant de se retrouver dans un restaurant chinois du centre-ville.

Là, tout en dégustant des nouilles sautées arrosées de bière, ils discutèrent du Japon. Le sujet était sur toutes les lèvres. Chuck déclara : « On ne peut pas les laisser continuer, ce sont des fascistes !

— Peut-être, dit Woody.

— Ils sont militaristes et agressifs ; ils se comportent comme des racistes vis-à-vis des Chinois. Qu'est-ce qu'il te faut de plus pour les qualifier de fascistes ?

— Je peux te répondre, intervint Joanne. La différence, c'est la vision de l'avenir. Les vrais fascistes veulent éradiquer définitivement leurs ennemis pour édifier un type de société totalement nouveau. Les Japonais en font autant, mais pour défendre des groupes de pouvoir traditionnels : la caste militaire et l'empereur. En ce sens, l'Espagne n'est pas vraiment fasciste non plus : Franco assassine des gens pour maintenir en place l'Église catholique et la vieille aristocratie, pas pour créer un monde nouveau.

— N'empêche, intervint Diana. Il faut les arrêter.

— Je ne vois pas les choses comme ça, déclara Woody.

— Okay, Woody, fit Joanne. Et tu les vois comment ? »

Comme elle s'intéressait sérieusement à la politique, elle apprécierait une réponse réfléchie, se dit Woody : « Le Japon est une nation commerçante, mais dépourvue de ressources naturelles. Il n'a ni pétrole, ni fer. Uniquement des forêts. Il n'a qu'un moyen de gagner de l'argent : le commerce. Par exemple, en important du coton brut, en le tissant sur place et en le revendant à l'Inde et aux Philippines. Mais avec la Crise, les deux grands empires économiques – l'Angleterre et les États-Unis – ont dressé des barrières tarifaires pour protéger leur industrie. Ces mesures ont frappé le Japon très durement. Elles ont marqué la fin de son commerce avec l'Empire britannique, Inde comprise, et avec les pays qui relèvent de notre zone d'influence, Philippines comprises.

— Et ça leur donne le droit de conquérir le monde ? lança Diana.

— Non, mais ça les incite à croire que le seul moyen d'assurer leur sécurité économique, c'est de posséder un empire bien à eux, comme les Britanniques. Ou, du moins, de dominer leur hémisphère, comme le font les États-Unis. Parce que alors, personne ne pourra plus entraver leur activité économique. Voilà pourquoi ils veulent faire de l'Extrême-Orient leur terrain de jeux. »

Joanne acquiesça. « Et la faiblesse de notre politique, c'est que chaque fois que nous leurs imposons des sanctions économiques pour punir leurs menées agressives, nous ne faisons que renforcer leur conviction que l'autosuffisance est vitale pour eux.

— Peut-être, déclara Chuck, mais tout de même, il va bien falloir les stopper. »

Woody haussa les épaules. Il ne savait que répondre.

Après le dîner, ils allèrent au cinéma. Le film était excellent. Woody et Chuck raccompagnèrent ensuite les jeunes filles chez elles. En chemin, Woody saisit la main de Joanne. Elle lui sourit et serra ses doigts, ce qu'il prit pour un encouragement.

Arrivé devant leur immeuble, Woody prit Joanne dans ses bras. Du coin de l'œil, il vit Chuck en faire autant avec Diana.

Joanne effleura brièvement les lèvres de Woody, presque chastement, et dit : « Le traditionnel baiser du soir.

— Il n'y avait rien de traditionnel dans notre baiser, la dernière fois », murmura-t-il, et il inclina à nouveau la tête vers elle.

De l'index, elle repoussa son menton.

Woody en fut dépité. Il n'allait quand même pas se contenter de ce petit bécot !

« J'étais saoule, ce soir-là, lui rappela-t-elle.

— Je sais », répondit Woody. Il comprenait très bien : Joanne craignait qu'il ne la prenne pour une fille facile. C'est pourquoi il enchaîna : « Tu es encore plus séduisante quand tu n'as pas bu, tu sais. »

Elle resta pensive un moment. « Tu as gagné, dit-elle enfin, c'était la bonne réponse », et elle l'embrassa à nouveau, délicatement, sans hâte, non pas avec un élan passionné, mais avec une concentration pleine de tendresse.

Bien trop tôt pour son goût, Woody entendit Chuck chantonner : « Bonne nuit, Diana ! »

Joanne interrompit leur baiser.

« C'est un rapide, mon frère ! » maugréa Woody, contrarié.

Elle rit doucement. « Bonne nuit, Woody. » Puis elle se retourna et se dirigea vers l'immeuble.

Diana, déjà sur le seuil, avait l'air franchement déçue.

Woody lança : « On peut se revoir ? » Il se maudit aussitôt pour le ton suppliant de sa voix.

Joanne ne semblait pas l'avoir remarqué. « Appelle-moi », dit-elle avant d'entrer.

Woody suivit des yeux les deux jeunes filles jusqu'à ce qu'elles aient disparu. Il s'en prit ensuite à son frère plutôt vertement. « Tu ne pouvais pas l'embrasser plus longtemps, non ? dit-il avec humeur. Elle a pourtant l'air sympa, Diana.

— Pas mon genre !

— Vraiment ? répliqua Woody, plus surpris que contrarié. Des petits nénés bien ronds, une jolie frimousse, qu'est-ce qu'il te faut de plus ? Je n'aurais pas été avec Joanne, je l'aurais embrassée sans me faire prier.

— Chacun ses goûts ! »

Ils prirent le chemin du retour. « C'est quoi, ton type, alors ? voulut savoir Woody.

— Avant que tu te mettes à organiser une autre sortie à quatre, il faut probablement que je t'avoue quelque chose.

— Quoi donc ? »

Chuck s'arrêta, obligeant Woody à faire de même. « Mais d'abord, tu dois me jurer de ne jamais en parler aux parents, ni à Papa ni à Mama.

— Je le jure. » À la lumière jaune des réverbères, Woody scruta le visage de son frère. « C'est quoi, ce grand secret ?

— Je n'aime pas les filles.

— Ce sont des emmerdeuses, d'accord, mais qu'est-ce qu'on peut y faire ?

— Tu n'as pas compris. Je n'aime pas me coller contre une fille et l'embrasser.

— Quoi ? Ne dis pas de conneries.

— On n'est pas tous faits pareil, Woody.

— Ouais, mais alors, il faudrait que tu sois du genre pédé.

— Justement.

— Quoi, justement ?

— Justement, je suis du genre pédé.

— Tu te fous de ma gueule ?

— Ce n'est pas de la blague, Woody, je suis tout à fait sérieux.

— Toi, une pédale ?

— Eh oui, c'est comme ça. Je ne l'ai pas choisi. Quand on était petits et qu'on commençait à se branler, toi, tu pensais à des seins dodus et à des cons pleins de poils. Moi, je ne te l'ai jamais dit, mais c'étaient de grosses bites bien raides que je voyais dans ma tête.

— Chuck, c'est répugnant !

— Mais non ! Certains hommes sont comme ça. Et ils sont bien plus nombreux que tu le crois. Dans la marine, surtout.

— Il y a des pédés dans la marine ? »

Chuck hocha la tête avec force. « À la pelle !

— Ah oui… Et comment tu sais ça ?

— On se reconnaît entre nous. Un peu comme les Juifs entre eux si tu veux. Tiens, le serveur du restaurant chinois !

— Ah bon, tu crois ?

— Tu ne l'as pas entendu dire qu'il aimait bien mon veston ?

— Si, mais de là à en conclure que…

— C'est bien ce que je te disais.

— Tu crois que tu l'intéressais ?

— Je suppose.

— Pourquoi ?

— Pour la même raison que Diana, sans doute. Merde, je suis bien plus beau que toi !

— C'est bizarre.

— Allons, viens, on rentre ! »

Ils reprirent leur route. Woody était encore sous le choc. « Tu veux dire qu'il y a des pédés chinois ?

— Bien sûr ! dit Chuck, et il éclata de rire.

— Je ne sais pas, mais je n'aurais jamais pensé ça d'un Chinois !

— N'oublie pas, surtout : pas un mot à qui que ce soit et surtout aux parents. Dieu sait ce que Papa dirait. »

Au bout d'un moment, Woody passa le bras autour des épaules de son frère. « Et puis qu'est-ce que ça peut foutre, après tout ! Au moins, tu n'es pas républicain. »

3.

Greg Pechkov était à bord de l'*Augusta* dans la baie de Placentia, au large des côtes de Terre-Neuve, en compagnie de Sumner Welles et du président Roosevelt. En dehors de ce croiseur lourd, le convoi se composait du cuirassé l'*Arkansas*, du croiseur *Tuscaloosa* et de dix-sept destroyers.

Les navires avaient jeté l'ancre en formant deux longues rangées parallèles de façon à ménager un large passage au centre. En ce samedi 9 août, à neuf heures du matin, il faisait un soleil radieux. Les équipages de ces vingt navires, en uniforme blanc, étaient alignés le long du bastingage pour accueillir le cuirassé britannique *Prince of Wales* avec à son bord le Premier ministre Churchill, et les trois destroyers qui constituaient son escorte.

C'était la plus impressionnante démonstration de puissance que Greg ait jamais vue. Il était ravi d'être aux premières loges.

Il s'inquiétait pourtant à l'idée que les Allemands aient pu avoir vent de cette rencontre. Que se passerait-il alors ? Un U-Boot pouvait tuer les deux dirigeants de ce qui restait de la civilisation occidentale – et le tuer aussi, par la même occasion.

Avant de quitter Washington, Greg avait revu le détective Tom Cranmer, qui lui avait donné l'adresse d'un immeuble situé dans un quartier populaire, de l'autre côté de l'Union Station. « Elle est serveuse au University Women's Club, à côté du Ritz-Carlton. Ce qui explique que vous l'ayez rencontrée deux fois dans le coin, avait-il dit tout en empochant le solde de ses honoraires. Je suppose qu'elle n'a pas percé en tant qu'actrice. En tout cas, elle continue à se faire appeler Jacky Jakes. »

Greg lui avait écrit une lettre.

Chère Jacky,

Je voudrais simplement savoir pourquoi tu m'as plaqué comme ça, il y a six ans. Je croyais que nous étions heureux, je devais me tromper. La question me turlupine, voilà tout.

Tu as eu l'air effrayée de me revoir, mais tu n'as rien à craindre de moi. Je ne suis pas fâché, simplement curieux. Je ne ferai jamais rien qui puisse te blesser. Tu es la première fille que j'ai aimée.

Est-ce qu'on pourrait se voir, juste prendre un café ou autre chose, et bavarder un peu ?

Très sincèrement,

Greg Pechkov

Il avait ajouté son numéro de téléphone personnel et posté la lettre le jour même de son départ pour Terre-Neuve.

Le Président tenait absolument à ce que la conférence débouche sur une déclaration commune. Le patron de Greg, Sumner Welles, avait rédigé un projet, mais Roosevelt avait refusé de le présenter, disant qu'il valait mieux laisser Churchill soumettre la première version du texte.

Greg comprit immédiatement que Roosevelt était fin négociateur. Car la partie qui rédigerait la première ébauche du texte devrait nécessairement, par souci d'équité, évoquer certaines des exigences de la partie adverse en plus des siennes. De par leur seule mention dans le document, celles-ci deviendraient alors *de facto* une base de négociation minimale et incontournable, alors qu'aucune des revendications de la partie chargée de rédiger le texte n'aurait encore été négociée. Autrement dit, celui qui rédigeait la première mouture se trouvait toujours désavantagé. Greg se jura de ne jamais l'oublier.

Ce jour-là, le Président et le Premier ministre déjeunèrent ensemble à bord de l'*Augusta*. Le lendemain, un dimanche, ils assistèrent à un office religieux sur le pont du *Prince of Wales*. L'autel avait été drapé de rouge, de blanc et de bleu, couleurs des Stars and Stripes comme de l'Union Jack. Le lundi matin, les deux hommes, dès lors très bons amis, abordèrent les choses sérieuses.

Churchill avait élaboré un plan en cinq points qui enchanta Sumner Welles et Gus Dewar : il appelait en effet à la création d'une organisation internationale capable d'assurer efficacement la sécurité de tous les États. En d'autres termes, une Société des nations renforcée. À leur grande déception, ils découvrirent que cette proposition était trop radicale pour Roosevelt. Certes, il y était favorable, mais il craignait les isolationnistes, ceux qui croyaient encore que l'Amérique n'avait pas à s'occuper des problèmes du reste du monde. Très à l'écoute de l'opinion publique, Roosevelt mettait tout en œuvre pour ne pas susciter d'opposition.

Welles et Dewar n'abandonnèrent pas la partie, pas plus que les Britanniques. Ils se réunirent afin de trouver un compromis acceptable pour les deux dirigeants. Greg prenait des notes pour Welles. Le groupe rédigea une clause appelant au désarmement « en attendant la mise en place d'un système de sécurité globale, plus vaste et plus permanent ».

Elle fut soumise aux deux hommes d'État qui l'acceptèrent. Welles et Dewar étaient aux anges.

Greg ne comprenait pas pourquoi. « Ça paraît si peu de chose, remarqua-t-il. Tous ces efforts, les dirigeants de deux grands pays qui franchissent des milliers de kilomètres, des dizaines de collaborateurs, vingt-quatre navires et plusieurs jours de négociation, tout cela pour n'écrire que quelques mots sur une page, des mots, qui plus est, qui ne disent même pas tout à fait ce que nous voulons.

— Nous progressons pas à pas, répliqua Gus Dewar avec un sourire. C'est comme ça en politique, on ne fait jamais de grandes enjambées. »

4.

Cela faisait maintenant plus d'un mois que Woody et Joanne se voyaient régulièrement.

Le jeune homme serait volontiers sorti avec elle tous les soirs, mais il se retenait. Il avait quand même réussi à la voir quatre fois au cours de la semaine passée : dimanche à la plage, mercredi à dîner, vendredi au cinéma, et ce jour-là, samedi, ils passaient toute la journée ensemble.

Il ne se lassait pas de bavarder avec elle. Elle était drôle, intelligente, acerbe. Il aimait qu'elle ait des opinions aussi tranchées sur tout. Ils pouvaient discourir des heures entières sur ce qu'ils aimaient ou détestaient.

Les informations en provenance d'Europe étaient mauvaises. Les Allemands continuaient à tailler en pièces l'armée Rouge. À l'est de Smolensk, ils avaient anéanti les 16e et 20e armées soviétiques et fait trois cent mille prisonniers. Moscou n'était plus défendue que par des forces éparses. Pourtant ces mauvaises nouvelles ne parvenaient pas à tempérer l'allégresse de Woody.

Joanne n'éprouvait probablement pas pour lui des sentiments aussi exaltés que ceux qu'il avait pour elle, mais il était convaincu qu'elle l'aimait bien. Ils s'embrassaient toujours au moment de se quitter, et elle semblait le trouver sympathique même si elle ne manifestait pas la passion dont il la savait capable. Il faut dire qu'ils échangeaient toujours leurs baisers dans des lieux publics, au cinéma ou dans l'encoignure d'une porte d'immeuble près de chez elle. Quand ils se retrouvaient dans son appartement, il y avait toujours l'une ou l'autre de ses colocataires qui traînait au salon. Jusque-là, Joanne ne l'avait pas encore invité dans sa chambre.

Chuck avait regagné Hawaï depuis plusieurs semaines, sa permission achevée. Woody s'interrogeait toujours sur ce qu'il lui avait confié. Tantôt ces aveux le bouleversaient, tantôt il se disait qu'en réalité cela ne faisait aucune différence. Quoi qu'il en soit, il tenait parole et n'avait rien dit à personne, pas même à Joanne.

Sur ces entrefaites, le père de Woody quitta la ville avec le Président tandis que sa mère partait passer quelques jours à Buffalo auprès de ses parents. Woody avait donc leur grand appartement de Washington, neuf pièces, pour lui tout seul pendant quelques jours. Il décida de trouver une bonne occasion d'y faire venir Joanne Rouzrokh, dans l'espoir d'obtenir enfin d'elle un vrai baiser.

Ils déjeunèrent ensemble et visitèrent une exposition intitulée « L'Art nègre », descendue en flamme par les critiques conservateurs convaincus qu'une chose pareille n'existait pas – malgré l'incomparable génie d'artistes tels que le peintre Jacob Lawrence et la sculptrice Elizabeth Catlett.

Comme ils sortaient de l'exposition, Woody demanda à Joanne : « Tu veux prendre un verre, le temps qu'on décide où on va dîner ?

— Non, merci, dit-elle du ton résolu qui était le sien. Ce dont j'ai vraiment envie, c'est d'une tasse de thé.

— Du thé ? » Où pouvait-on boire un bon thé à Washington ? Et soudain, Woody eut une illumination. « On peut aller chez moi, si tu veux, ma mère boit du thé anglais.

— D'accord. »

C'était à quelques rues de là, dans la 22e Rue NW, près de L Street. Ils pénétrèrent dans le vestibule climatisé et respirèrent tout de suite mieux. Un portier les conduisit à l'étage par l'ascenseur.

Au moment d'entrer dans l'appartement, Joanne déclara : « Je tombe tout le temps sur ton père, ici ou là à Washington, mais je n'ai pas vu ta mère depuis des années. Je serai heureuse de la féliciter pour son dernier livre.

— Elle n'est pas là pour le moment, répondit Woody. Viens, on va aller à la cuisine. »

Il remplit la bouilloire au robinet et la posa sur la cuisinière. Puis, prenant Joanne dans ses bras, il soupira : « Enfin seuls !

— Où sont tes parents ?

— Ils sont partis, tous les deux.

— Et Chuck est à Hawaï.

— Oui. »

Elle s'écarta de lui. « Woody, comment oses-tu me faire une chose pareille ?

— Comment ça ? Je te fais du thé !

— Tu m'as entraînée ici sous de faux prétextes ! J'étais persuadée que tes parents étaient là.

— Je n'ai jamais prétendu ça.

— Pourquoi ne m'as-tu pas dit qu'ils étaient en voyage ?

— Tu ne me l'as pas demandé ! » répondit-il, indigné, tout en reconnaissant qu'elle n'avait pas tout à fait tort. Il ne lui aurait jamais menti, non. Il avait simplement profité de l'occasion et ne l'avait pas prévenue qu'il n'y avait personne chez lui.

« Tu m'as fait monter ici pour me faire des avances ! Tu me prends pour une fille légère.

— Pas du tout ! Mais on n'a jamais un moment d'intimité. Tout ce que j'espérais, c'est un vrai baiser. Rien de plus !

— Ne me raconte pas de salades ! »

Elle était vraiment injuste. Bien sûr qu'il espérait coucher avec elle un jour, mais sûrement pas ce jour-là. « Très bien, dit-il. Allons prendre le thé ailleurs. Le Ritz-Carlton est à deux pas. Ils en ont sûrement, avec tous les Anglais qui y descendent.

— Ça va, ne sois pas idiot. Maintenant qu'on est là, autant y rester. Tu ne me fais pas peur, je saurai me défendre. Je suis seulement furieuse contre toi. Je ne veux pas qu'un homme sorte avec moi parce qu'il me croit facile.

— Facile ? dit-il en haussant la voix. Mince alors ! J'ai attendu six ans pour que tu daignes sortir avec moi, et tout ce que je te demande, c'est un baiser ! Si c'est ça les filles faciles, qu'est-ce que ce serait si j'étais amoureux d'une fille difficile ! »

À son grand étonnement, Joanne éclata de rire.

« Qu'est-ce qu'il y a encore ? dit-il avec humeur.

— Excuse-moi, tu as raison. Si tu cherchais une fille facile, tu m'aurais laissée tomber depuis longtemps.

— Tu l'as dit !

— J'étais persuadée que tu te faisais une mauvaise image de moi parce que je t'avais embrassé ce fameux jour où j'avais trop bu. Je me disais que tu me courais après dans l'espoir de prendre un peu de plaisir à bon compte. Ces derniers temps, cette idée me tournicotait dans la tête. Pardonne-moi, je t'avais mal jugé. »

Woody était souvent déconcerté par ses rapides sautes d'humeur, mais il se dit que cette dernière phrase laissait entrevoir une amélioration. « Si tu crois que j'ai attendu de t'embrasser pour être fou de toi. Tu ne l'avais pas remarqué, je suppose.

— Je t'avais à peine remarqué *toi*.

— Ce n'est pas faute d'être grand.

— C'est bien ton seul trait physique intéressant. »

Il sourit. « Ah, ça ! Je ne risque pas d'attraper la grosse tête avec toi.

— Non, pas de danger de ce côté-là ! »

L'eau bouillait. Woody mit du thé dans une théière en porcelaine et versa l'eau dessus.

Joanne avait l'air pensif. « Tu as dit quelque chose tout à l'heure...

— Quoi donc ?

— Tu as dit : "Qu'est-ce que ça serait si j'étais amoureux d'une fille difficile !" Tu le pensais vraiment ?

— Quoi donc ?

— Le fait d'être amoureux.

— Ah ! Non, je ne voulais pas dire ça ! » Abandonnant toute prudence, il se reprit : « Et puis flûte ! Puisque tu veux savoir la vérité, oui, je suis amoureux de toi. Je crois que je t'aime depuis des années. Je t'adore. Je veux... »

Elle passa les bras autour de son cou et l'embrassa.

Un vrai baiser, cette fois. Sa bouche s'empara de la sienne avec avidité, la pointe de sa langue frôlant ses lèvres, son corps se serrant contre lui. C'était comme en 1935, sauf qu'elle ne sentait pas le whisky. C'était la jeune fille qu'il aimait, la vraie Joanne, pensa-t-il avec ravissement : une femme pleine de passion. Qui était dans ses bras et se livrait à lui sans retenue.

Elle glissa ses mains à l'intérieur de sa légère chemise de sport et caressa sa poitrine, lui enfonçant les doigts entre les côtes, lui pétrissant les mamelons de ses paumes, agrippant ses épaules comme si elle voulait pénétrer profondément dans sa chair. Il se rendit compte qu'elle était remplie du même désir refoulé que lui, un désir qui débordait maintenant comme l'eau d'un barrage qui rompt. Il lui rendit ses caresses, glissant les mains le long de ses flancs, saisissant ses seins avec un sentiment de libération et de bonheur proche de celui de l'écolier qui se voit offrir des vacances inattendues.

Quand il introduisit entre ses cuisses une main impatiente, elle se dégagea. Mais il ne s'attendait pas à ce qu'elle lui demande : « Tu as un préservatif ?

— Non ! Je suis désolé…

— Ça ne fait rien. En fait, je préfère. Ça prouve que tu n'avais pas monté ce guet-apens.

— Si seulement…

— Tant pis. Je connais une femme médecin qui me donnera ce qu'il faut lundi. En attendant, on va improviser. Embrasse-moi encore. »

Pendant qu'il s'exécutait, il la sentit déboutonner son pantalon.

« Oh, s'exclama-t-elle un instant plus tard. C'est formidable !

— Exactement ce que je pensais, murmura-t-il.

— Mais je vais avoir besoin de mes deux mains.

— Comment ?

— J'imagine que c'est proportionnel à la taille.

— Je ne comprends rien à ce que tu dis.

— Eh bien, je vais me taire et t'embrasser. »

Quelques minutes plus tard, elle demanda : « Un mouchoir ! »

Heureusement, il en avait un sur lui.

Il releva les paupières quelques instants avant la fin et vit qu'elle le regardait. Il lut dans ses yeux du désir, de l'excitation et aussi autre chose, qui pouvait bien être de l'amour.

Quand ce fut fini, il éprouva une béatitude pleine de sérénité. Je l'aime, pensa-t-il, je suis heureux. Que la vie est délicieuse ! Tout haut, il dit : « C'était merveilleux. Je voudrais te faire la même chose.

— Tu veux ? Vraiment ?

— Plutôt deux fois qu'une ! »

Ils étaient toujours debout dans la cuisine, appuyés contre la porte du réfrigérateur. Ni l'un ni l'autre n'avait envie de bouger. Elle lui prit la main, la guida sous sa robe d'été pour la glisser dans sa culotte en coton. Il sentit une peau chaude, des poils frisés et une fente humide. Il voulut enfoncer son doigt à l'intérieur. « Non », dit-elle. Elle saisit elle-même le bout de son doigt et le guida entre les lèvres souples, pour lui faire sentir, juste sous la peau, une petite protubérance toute dure, de la taille d'un pois. Elle fit remuer son doigt en cercle. « Oui, dit-elle en fermant les yeux. Comme ça. » Il étudia son visage avec adoration tandis qu'elle s'abandonnait au plaisir. Au bout d'une minute ou deux, elle poussa un petit cri, qu'elle répéta deux ou trois fois. Puis elle repoussa sa main et s'effondra contre lui.

« Ton thé va être froid », remarqua-t-il au bout d'un moment.

Elle se mit à rire. « Je t'aime, Woody.

— C'est vrai ?

— J'espère que ça ne te fait pas peur.

— Non, répondit-il avec un sourire. Ça me rend très heureux.

— Je sais que les filles ne sont pas censées être aussi directes. Mais je ne sais pas faire semblant. Une fois que j'ai décidé quelque chose, c'est pour de bon.

— Oui, j'avais cru remarquer. »

5.

Greg Pechkov habitait la suite que son père louait à l'année au Ritz-Carlton. Lev y venait de temps à autre, lorsqu'il s'arrêtait quelques jours à Washington entre Buffalo et Los Angeles. Pour l'heure, Greg avait la suite pour lui tout seul, à ce détail près que Rita Lawrence, la fille bien roulée du congressiste, y avait passé la nuit. Maintenant, elle était absolument adorable, ébouriffée dans une robe de chambre masculine en soie rouge.

Un garçon leur apporta le petit déjeuner, les journaux et une enveloppe contenant un message.

La déclaration commune de Roosevelt et Churchill avait provoqué plus de bruit que Greg ne s'y attendait. Une semaine plus tard, elle faisait encore la une de la presse, qui l'avait baptisée « charte de l'Atlantique ». Aux yeux de Greg, ce communiqué n'était qu'un conglomérat de phrases prudentes et de vagues promesses, mais le monde voyait les choses autrement. La nouvelle avait été saluée par certains comme une sonnerie de clairon en faveur de la liberté, de la démocratie et du commerce mondial. Hitler, en revanche, y avait vu l'équivalent d'une déclaration de guerre de la part des États-Unis. Il en avait été ulcéré, disait-on.

Des pays qui n'avaient pas participé à la conférence souhaitaient néanmoins signer cette charte et Bexforth Ross avait suggéré que les signataires prennent le nom de « Nations unies ».

Pendant ce temps, les Allemands envahissaient l'Union soviétique. Au nord, ils s'approchaient de Leningrad et tiendraient bientôt la ville en tenaille. Au sud, ils forçaient les Russes à battre en retraite. Au prix d'un sacrifice déchirant, ceux-ci avaient fait sauter le barrage sur le Dniepr, le plus grand complexe hydroélectrique du monde qui faisait leur joie et leur fierté, pour que les conquérants ne profitent pas de l'électricité qu'il produisait. « L'armée Rouge a réussi à freiner un peu l'invasion, annonça Greg à Rita, en lisant le *Washington Post*. Mais les Allemands continuent d'avancer de presque dix kilomètres par jour. Ils prétendent avoir déjà tué trois millions et demi de soldats soviétiques. Tu te rends compte ?

— Tu as de la famille en Russie ?

— En fait, oui. Un jour qu'il avait un peu trop bu, mon père m'a raconté qu'il avait laissé derrière lui une fille enceinte. »

Rita fit la grimace.

« C'est lui tout craché, malheureusement, continua Greg. C'est un grand homme, et les grands hommes n'obéissent pas aux règles. »

Rita resta muette, mais son expression était limpide : elle ne partageait pas ce point de vue, mais ne voulait pas se fâcher avec lui pour si peu.

« Toujours est-il que j'ai un demi-frère russe, tout aussi illégitime que moi, poursuivit Greg. Je ne sais qu'une chose à son sujet : il s'appelle Vladimir. Il est peut-être mort à l'heure qu'il est, il a l'âge de se battre. Il fait probablement partie de ces trois millions et demi de victimes. » Il tourna la page.

La lecture du journal achevée, il lut le message apporté par le garçon d'étage.

Il était de Jacky Jakes et contenait un numéro de téléphone accompagné de ces seuls mots : *Pas entre 1 et 3*.

Tout à coup, Greg n'eut plus qu'une envie : se débarrasser de Rita. « À quelle heure es-tu attendue chez toi ? » demanda-t-il sans grande délicatesse.

Elle regarda sa montre. « Oh, mon Dieu, il faut que je rentre avant que ma mère commence à me chercher partout. » La veille, elle avait dit à ses parents qu'elle restait dormir chez une amie.

Ils s'habillèrent en même temps et quittèrent l'hôtel dans deux taxis séparés.

Le numéro de téléphone indiqué par Jacky devait être celui de son lieu de travail, et les heures où elle ne voulait pas être dérangée celles où elle était particulièrement occupée. Greg décida de l'appeler en milieu de matinée.

Il se demanda pourquoi il s'emballait comme ça. Après tout, c'était pure curiosité de sa part. Rita Lawrence était une fille superbe, très attirante sexuellement. Et pourtant avec elle comme avec les autres, il n'avait jamais retrouvé la passion de sa première aventure avec Jacky. C'était sans doute parce qu'il n'avait plus quinze ans, et ne les aurait jamais plus.

Ayant rejoint son bureau dans le bâtiment de l'Old Executive Office, il s'attela à sa première tâche du jour : rédiger une note à destination des citoyens américains résidant en Afrique du Nord, où Britanniques, Italiens et Allemands se disputaient une bande côtière de trois mille kilomètres de long sur soixante de large, perdant et reprenant tour à tour une partie de ce territoire.

À dix heures et demie, il composa le numéro inscrit sur le message.

« University Women's Club », répondit une voix féminine. Greg n'y avait jamais mis les pieds. Les hommes ne se rendaient dans les cercles de femmes que sur invitation.

« Pourrais-je parler à Jacky Jakes ? demanda-t-il.

— Oui, elle attend justement un appel. Un instant, je vous prie. » Greg songea qu'elle devait probablement demander l'autorisation de recevoir un coup de téléphone pendant ses heures de travail.

Quelques instants plus tard, il entendit : « Ici Jacky. Qui est à l'appareil ?

— Greg Pechkov.

— Je m'en doutais. Comment as-tu eu mon adresse ?

— J'ai engagé un détective privé. On peut se voir ?

— J'ai l'impression que je n'ai pas le choix. Mais à une condition.

— Laquelle ?

— Tu dois me jurer sur ce que tu as de plus sacré de ne jamais le dire à ton père. Jamais.

— Pourquoi ?

— Je t'expliquerai plus tard. »

Il haussa les épaules. « D'accord.

— Tu le jures ?

— Bien sûr. »

Elle insista. « Dis-le.

— Je le jure. Ça te va ?

— Très bien. Tu peux m'inviter à déjeuner. »

Greg réfléchit. « Où est-ce qu'un Blanc et une Noire peuvent manger à la même table dans ce quartier ?

— L'Electric Diner, je ne vois que ça.

— Je suis déjà passé devant. » Il avait remarqué le nom de ce café-restaurant, mais n'y était jamais entré. C'était un snack bon marché, fréquenté par les concierges et les coursiers. « À quelle heure ?

— Onze heures et demie.

— Si tôt ?

— Parce que tu crois que ça mange à une heure, les serveuses ? »

Il sourit : « Toujours aussi insolente, à ce que je vois ! »

Elle raccrocha.

Greg acheva son communiqué et apporta les pages dactylographiées dans le bureau de son patron. Les ayant déposées dans la corbeille « arrivée », il demanda : « Est-ce que ça va si je pars déjeuner de bonne heure aujourd'hui, Mike ? Disons vers onze heures et demie ?

— Ouais, pas de problème », répondit l'autre sans lever les yeux de la page des chroniques et commentaires du *New York Times*.

Greg longea la Maison Blanche inondée de soleil et arriva au lieu du rendez-vous sur les coups de onze heures vingt. Le café était presque désert, il n'y avait que quelques personnes qui prenaient une pause en cette fin de matinée. Il s'installa dans un box et commanda un café.

Il se demanda ce que Jacky lui dirait. Il avait hâte de connaître le fin mot d'une histoire dont le mystère le taraudait depuis six ans.

Elle arriva à onze heures trente-cinq, vêtue d'une robe noire et chaussée de souliers plats. Son uniforme de serveuse, apparemment, mais sans le tablier blanc. Le noir lui allait bien et il se rappela le plaisir qu'il éprouvait, jadis, à la contempler, à admirer l'arc de sa bouche et ses grands yeux bruns. Elle s'assit en face de lui et commanda une salade et un Coca. Greg reprit du café : il était trop nerveux pour avaler quoi que ce soit.

Le visage de Jacky avait perdu les rondeurs enfantines qui étaient restées gravées dans sa mémoire. Elle avait seize ans à l'époque, elle en avait donc vingt-deux à présent. Ils étaient alors des enfants qui jouaient à être grands. Maintenant, ils étaient vraiment adultes. Il déchiffra sur son visage une histoire dont il n'y avait pas encore trace six ans auparavant : déception, souffrance, pauvreté.

« Je suis dans l'équipe de jour, expliqua-t-elle. J'arrive à neuf heures, je dresse le couvert, je prépare la salle à manger. Après, je sers le déjeuner, je débarrasse et je repars à cinq heures.

— La plupart des serveuses travaillent le soir.

— J'aime avoir mes soirées libres et mes week-ends aussi.

— Toujours partante pour faire la fête !

— Non, généralement, je reste chez moi et j'écoute la radio.

— Tu dois avoir quantité de soupirants.

— Autant que je veux. »

Il lui fallut un moment pour comprendre que cela pouvait être interprété de multiples façons.

Sa commande arriva. Elle but son Coca et picora la salade.

Greg se lança : « Dis-moi, pourquoi est-ce que tu es partie comme ça sans crier gare, en 1935 ? »

Elle soupira. « Je n'ai pas envie de te le dire, parce que ça ne va pas te faire plaisir.

— Il faut que je le sache.

— Ton père est venu me trouver. »

Greg hocha la tête. « Je me doutais qu'il n'était pas étranger à l'affaire.

— Il était accompagné d'un homme de main, un certain Joe quelque chose…

— Joe Brekhounov, un gangster, lâcha Greg, saisi d'une colère grandissante. Il t'a fait du mal ?

— Ce n'était pas la peine, Greg. Rien qu'à le regarder, j'étais morte de peur. J'aurais fait tout ce que ton père voulait. »

Greg réussit à contenir sa fureur. « Et il voulait quoi ?

— Que je parte sur-le-champ. Il m'a dit que je pouvais t'écrire un mot, mais qu'il le lirait. Il m'a obligée à revenir ici, à Washington. Ça m'a brisé le cœur de devoir te quitter.

— À moi aussi », murmura Greg en se rappelant son propre désespoir. Il faillit tendre la main par-dessus la table pour saisir la sienne, mais craignit qu'elle le repousse.

« Il m'a dit qu'il me verserait de l'argent toutes les semaines, reprit-elle, simplement pour que je ne te voie plus jamais. Il me paie toujours cette pension. Ce n'est pas grand-chose, quelques dollars à peine, mais ça paie mon loyer. J'ai donc promis. Je ne sais pas comment, mais j'ai tout de même eu le courage de poser une condition.

— Laquelle ?

— Qu'il ne chercherait jamais à me faire des avances. Que s'il le faisait, je te raconterais tout.

— Il a accepté ?

— Oui.

— Les gens qui s'en tirent après l'avoir menacé se comptent sur les doigts de la main. »

Elle repoussa son assiette. « Il m'a dit ensuite que si je ne tenais pas ma promesse, il enverrait Joe me lacérer le visage. Joe a sorti son rasoir. »

À présent, tout était clair. « Et tu continues à avoir peur. »

De fait, sa peau noire était exsangue. « Tu parles que j'ai la trouille, et pas qu'un peu !

— Oh, Jacky ! fit Greg d'une voix qui n'était plus qu'un chuchotement. Je suis tellement navré. »

Elle se força à sourire. « Est-ce qu'il a eu tort, finalement ? Tu n'avais que quinze ans. Ce n'est pas un âge pour se marier.

— Si au moins il m'en avait parlé, tout aurait été différent. Mais voilà, il est comme ça, il décide de l'avenir des autres et fait en sorte que tout se passe comme il l'entend. Et personne n'a son mot à dire.

— On a quand même passé de bons moments ensemble.

— Et comment !

— J'étais ton cadeau, tu te rappelles. »

Il se mit à rire. « Le plus beau que j'aie jamais reçu.

— Et toi, qu'est-ce que tu deviens ?

— Je travaille au bureau de presse du Département d'État. Un stage d'été. »

Elle fit la grimace. « Ça n'a pas l'air folichon.

— Tu te trompes ! C'est passionnant de regarder les puissants de ce monde prendre des décisions qui vont changer le cours de l'histoire. Comme ça, assis derrière leurs bureaux. Ils dirigent le monde !

— Oui, dit-elle sans grande conviction, c'est sûrement mieux que d'être serveuse. »

Il commença à entrevoir la distance qui les séparait désormais. « En septembre, je retourne à Harvard. Pour ma dernière année.

— Tu dois être la coqueluche des étudiantes !

— Il y a surtout des garçons, très peu de filles.

— Mais tu t'en sors sûrement très bien quand même, je me trompe ?

— Je ne vais pas te mentir. » L'image d'Emily Hardcastle lui traversa l'esprit et il se tut.

« Tu épouseras une de ces filles, tu auras de beaux enfants et tu habiteras une maison au bord d'un lac.

— J'aimerais faire une carrière politique, être secrétaire d'État ou bien sénateur comme le père de Woody Dewar. »

Elle détourna les yeux.

Greg se représenta cette maison au bord d'un lac. C'était probablement le rêve de Jacky. Il eut de la peine pour elle.

« Tu y arriveras, reprit-elle. Je le sais. Tu as ce qu'il faut pour ça. Tu l'avais déjà à quinze ans. Tu es comme ton père.

— Comment ? Arrête ! »

Elle haussa les épaules. « Réfléchis un peu, Greg. Tu savais que je ne voulais pas te voir, mais ça ne t'a pas empêché d'engager un détective privé pour me retrouver. *Il décide de l'avenir des autres et fait en sorte que tout se passe comme il l'entend. Et personne n'a son mot à dire.* C'est bien comme ça que tu viens de le décrire, non ?

— J'espère que je ne suis pas sa copie conforme », murmura Greg, consterné.

Elle le dévisagea comme si elle le jaugeait. « Le jury n'a pas encore rendu son verdict. »

La serveuse débarrassa son assiette. « Un dessert ? La tarte aux pêches est délicieuse. »

Ni l'un ni l'autre n'en voulait. La serveuse remit l'addition à Greg.

« J'espère que ta curiosité est satisfaite, déclara Jacky.

— Oui, je te remercie. C'était vraiment gentil de ta part.

— La prochaine fois que tu me croises dans la rue, passe ton chemin.

— Si c'est ce que tu veux… »

Elle se leva. « Sortons séparément. Je serai plus à l'aise.

— Comme tu voudras.

— Bonne chance, Greg.

— Bonne chance à toi !

— N'oublie pas de laisser un pourboire à la serveuse ! » Elle s'éloigna.

X

1941 (III)

1.

En ce mois d'octobre, la neige tombait et fondait aussitôt. À Moscou, les rues étaient froides et humides. En fouillant le placard de l'entrée à la recherche de ses *valenki*, ces traditionnelles bottes en feutre grâce auxquelles les Russes gardent les pieds au chaud en hiver, Volodia découvrit à sa grande surprise six caisses de vodka.

Ses parents n'étaient pas de gros buveurs. Ils se contentaient généralement d'un petit verre. Il arrivait à son père d'assister de temps à autre à l'un de ces dîners interminables donnés par Staline où l'alcool coulait à flots : il y retrouvait d'anciens camarades, et rentrait au petit matin saoul comme un Polonais. Mais à la maison, une bouteille de vodka durait au moins un mois.

Volodia entra à la cuisine, où ses parents prenaient le petit déjeuner : thé, sardines en boîte et pain noir. « Papa, demanda-t-il, pourquoi est-ce qu'on a des réserves de vodka pour six ans dans le placard ? »

Son père eut l'air étonné.

Les deux hommes se tournèrent vers Katerina qui rougit. Elle alla allumer la radio et en monta le son. Craignait-elle que l'appartement soit sur écoute ? se demanda Volodia.

D'une voix basse mais impatiente, elle déclara : « Qu'est-ce que vous aurez, comme monnaie d'échange, quand les Allemands seront là ? Nous n'appartiendrons plus à l'élite privilégiée, vous pouvez me croire ! Si nous n'avons pas de quoi acheter à manger au marché noir, nous mourrons de faim. Je suis trop vieille pour faire le tapin. La vodka vaudra plus cher que de l'or. »

Volodia fut médusé d'entendre sa mère parler ainsi. Quant à son père, il rétorqua : « Les Allemands n'arriveront jamais jusqu'ici, voyons. »

Volodia n'en était pas si sûr. Ils poursuivaient leur avance sur Moscou et les mâchoires de la tenaille se refermaient peu à peu autour de la ville. Ils étaient déjà à Kalinine au nord et à Kalouga au sud, deux villes situées à moins de deux cents kilomètres de la capitale. Les pertes dans le camp soviétique dépassaient l'imagination. Un mois plus tôt, huit cent mille hommes étaient sur le front. Aujourd'hui, il n'en restait plus que quatre-vingt-dix mille, selon les estimations parvenues à son bureau. « Qui diable va les arrêter ? demanda-t-il à son père.

— Leurs lignes de ravitaillement sont trop étirées et ils ne sont pas équipés pour nos hivers. Nous contre-attaquerons quand ils seront affaiblis.

— Dans ce cas, pourquoi le gouvernement quitte-t-il Moscou ? »

De fait, on avait commencé à transférer les administrations de la capitale à Kouïbychev, environ mille kilomètres à l'est, et le spectacle de tous ces fonctionnaires chargeant dans des camions des multitudes de cartons remplis de dossiers inquiétait grandement la population.

« Simple mesure de précaution, expliqua Grigori. Staline est toujours au Kremlin.

— Il existe pourtant une solution, fit remarquer Volodia : appeler en renfort nos centaines de milliers d'hommes qui se trouvent en Sibérie. »

Grigori secoua la tête. « On ne peut pas laisser la frontière orientale sans défense. Le Japon est toujours une menace.

— Le Japon ne nous attaquera pas, nous le savons parfaitement ! » Volodia jeta un coup d'œil à sa mère. Il n'était pas censé évoquer les questions d'espionnage devant elle, il le savait, mais passa outre. « Notre source à Tokyo, celle-là même qui nous a avertis à juste titre de l'invasion allemande imminente, nous annonce maintenant que les Japonais ne nous attaqueront pas. Nous n'allons quand même pas réitérer l'erreur de ne pas lui faire confiance !

— Évaluer la fiabilité d'un renseignement n'est jamais facile.

— Il faut faire revenir ces armées, c'est notre seule chance ! insista Volodia avec colère. Nous en avons douze là-bas, en réserve. Un million d'hommes ! Si on les déploie ici, Moscou tiendra peut-être. Dans le cas contraire, tout est fini pour nous. »

Grigori parut troublé. « Ne parle pas comme ça, même en privé.

— Pourquoi ? De toute façon, je serai sans doute bientôt mort. »

Sa mère fondit en larmes.

« Bravo ! s'écria son père. Tu peux être fier de toi ! »

Volodia sortit de la pièce. Tout en enfilant ses bottes, il se demanda ce qui l'avait poussé à s'en prendre à son père et à faire pleurer sa mère. En fait, c'était parce qu'il était désormais convaincu que l'Allemagne l'emporterait sur l'Union soviétique. Avec sa réserve de vodka, qui sous-entendait que les nazis risquaient d'occuper le pays, sa mère l'avait obligé à regarder la réalité en face. On va perdre la guerre, se dit-il. La fin de la révolution russe est proche.

Il mit son manteau et sa chapka et retourna dans la cuisine embrasser sa mère et son père.

« En quel honneur ? s'étonna Grigori. Tu ne fais jamais qu'aller au bureau.

— Au cas où on ne se reverrait plus ! » Et sur ces mots, il partit.

En traversant le pont pour gagner le centre-ville, Volodia constata que les transports en commun ne fonctionnaient plus. Le métro était fermé, les autobus et les trams ne circulaient pas.

Et les mauvaises nouvelles ne s'arrêtaient pas là.

Le bulletin du matin diffusé à la radio par le Sovinformburo et retransmis par les haut-parleurs noirs installés à tous les coins de rues était, pour une fois, conforme à la réalité. « Dans la nuit du 14 au 15 octobre, la situation sur le front ouest a empiré. De nombreux chars allemands ont franchi nos lignes de défense. » Tout le monde en avait conclu que la situation devait être bien plus grave, car le Sovinformburo mentait constamment.

Le centre-ville était encombré de réfugiés. Venus de l'est du pays, ils se déversaient par centaines dans la capitale, tirant des charrettes à bras contenant tous leurs biens, conduisant des troupeaux de vaches efflanquées, de cochons crasseux, de moutons trempés vers la campagne à l'est de la ville, fuyant le plus loin possible de ces Allemands que rien n'arrêtait.

Volodia essaya de faire du stop. Les civils ne circulaient plus guère en voiture, ces derniers temps. L'essence était réservée aux interminables convois militaires qui roulaient le long du Sadovoïé Koltso, le boulevard circulaire qui permettait de contourner le centre-ville.

Il réussit à monter dans un GAZ-64 flambant neuf, un véhicule tout terrain ouvert de type Jeep, depuis lequel il put découvrir une grande partie des dommages causés par les bombardements. À en croire des diplomates de retour d'Angleterre, ce n'était rien par rapport à ce que subissaient les Londoniens sous le Blitz, mais pour les Moscovites c'était largement suffisant. Il passa devant plusieurs immeubles en ruine et plusieurs dizaines de maisons en bois incendiées.

Grigori, chargé de la défense du ciel moscovite, avait fait installer des canons antiaériens sur les toits des bâtiments les plus hauts et lancé des ballons de barrage qui flottaient sous les nuages chargés de neige. Sa décision la plus insolite avait été de badigeonner les bulbes dorés des églises de peinture de camouflage verte et marron. Cela n'aurait sans doute guère d'effet sur la précision – ou l'imprécision – des tirs ennemis, avait-il admis devant Volodia, mais en tout cas, cela donnait aux citoyens l'impression d'être protégés.

Si les Allemands remportaient la guerre, si les nazis gouvernaient Moscou, alors, se disait Volodia, son neveu et sa nièce, les jumeaux de sa sœur, ne seraient pas élevés en patriotes communistes, mais en nazis serviles prompts à saluer Hitler. La Russie serait asservie comme la France, et peut-être même dirigée en partie par un gouvernement profasciste à la botte de l'envahisseur qui regrouperait les Juifs et les enverrait dans des camps de concentration. C'était une perspective qu'il se refusait à envisager. Dans l'avenir tel que Volodia se plaisait à l'imaginer, l'Union soviétique, libérée de la férule pernicieuse de Staline et de la brutalité de la police secrète, pourrait enfin édifier le vrai communisme.

Quand il arriva à son quartier général situé sur l'aérodrome de Khodynka, Volodia fut pris dans une bourrasque de flocons grisâtres qui n'étaient pas de la neige mais de la cendre : les services de renseignement de l'armée Rouge brûlaient leurs archives pour qu'elles ne tombent pas aux mains de l'ennemi.

Peu après, le colonel Lemitov entra dans son bureau. « Tu as envoyé un mémo à Londres à propos d'un physicien allemand appelé Wilhelm Frunze. C'était une excellente idée. Une piste de tout premier ordre. Bravo. »

Quelle importance, alors que les panzers n'étaient plus qu'à cent cinquante kilomètres de Moscou, pensa Volodia. Il était bien trop tard pour que les espions soient d'un quelconque secours. Il s'obligea pourtant à se concentrer. « Frunze, oui. J'étais au lycée avec lui à Berlin.

— Londres l'a contacté, il est disposé à collaborer. La rencontre s'est déroulée en lieu sûr. » Lemitov parlait tout en tripotant sa montre-bracelet. Cette agitation, tout à fait inhabituelle chez lui, montrait qu'il était tendu. Ces temps-ci, tout le monde était nerveux.

Volodia ne dit rien. À l'évidence, l'entrevue avait été fructueuse, sinon Lemitov ne lui en aurait pas parlé.

« D'après Londres, Frunze s'est d'abord montré méfiant. Il soupçonnait notre homme d'appartenir aux services secrets britanniques, ajouta Lemitov avec un sourire. En fait, après cette première rencontre, il est allé sonner à la porte de notre ambassade, à Kensington Palace Gardens, pour exiger confirmation que notre agent était bien celui qu'il prétendait être !

— Un vrai amateur, observa Volodia en souriant.

— Tu l'as dit. Un type qui aurait voulu jouer un double jeu et refiler de faux renseignements n'aurait jamais agi d'une manière aussi stupide. »

L'Union soviétique n'était pas encore vaincue, pas tout à fait. Volodia devait donc continuer à faire comme si ce Willi Frunze l'intéressait encore. « Qu'est-ce qu'il nous a donné, camarade colonel ?

— Il prétend qu'avec d'autres chercheurs il travaille avec les Américains sur un projet de superbombe. »

Volodia tressaillit, se rappelant ce que lui avait dit Zoïa Vorotsintseva. Apparemment, les pires craintes de la jeune femme se voyaient confirmées.

« Malheureusement, nous avons un problème avec les informations qu'il nous a transmises, reprit Lemitov.

— Oui ?

— Nous les avons traduites, mais nous n'y comprenons rien. » Lemitov tendit à Volodia une liasse de feuillets dactylographiés.

« Séparation isotopique par diffusion gazeuse, lut Volodia à haute voix.

— Tu vois ce que je veux dire ?

— J'ai étudié les langues à l'université, pas la physique.

— Tu n'as pas dit un jour que tu connaissais une physicienne ? demanda Lemitov toujours souriant. Une blonde superbe qui a refusé d'aller au cinéma avec toi, si je me souviens bien. »

Volodia rougit. Il avait raconté l'histoire à Kamen et celui-ci avait dû en faire des gorges chaudes. Le problème, quand on avait un espion pour patron, c'est qu'il était toujours au courant de tout. « C'est une amie de la famille. Elle m'a parlé d'un procédé d'explosion appelé fission. Vous voulez que je l'interroge ?

— Officieusement et discrètement. Je ne veux pas monter ce tuyau en épingle sans savoir auparavant de quoi il s'agit. On a peut-être affaire à un cinglé. Je ne tiens pas à être la risée de tous à cause de ce Frunze. Découvre de quoi traite ce rapport et si son contenu tient debout du point de vue scientifique. Si c'est le cas, il faudrait savoir si les Britanniques et les Américains sont vraiment en mesure de fabriquer une superbombe. Et les Allemands aussi.

— Je n'ai pas vu Zoïa depuis deux ou trois mois. »

Lemitov haussa les épaules. L'intimité des relations entre Volodia et Zoïa n'avait guère d'importance. En Union soviétique, répondre aux questions des autorités ne relevait pas d'un choix personnel.

« Je vais la retrouver. »

Lemitov hocha la tête. « Aujourd'hui même, c'est compris ? » Il sortit.

Volodia réfléchit. Zoïa était sûre et certaine que les Américains fabriquaient une superbombe. Elle avait su convaincre Grigori d'en parler à Staline, mais celui-ci avait balayé le sujet d'un revers de main. À présent, un espion en Angleterre confirmait les dires de Zoïa. Elle avait donc vu juste, apparemment. Et Staline avait eu tort, une fois de plus.

Les dirigeants de l'Union soviétique avaient une fâcheuse tendance à nier l'évidence quand les nouvelles n'étaient pas bonnes. La semaine précédente encore, une mission de reconnaissance aérienne avait repéré des blindés allemands à cent trente kilomètres de Moscou. Pour que l'état-major veuille bien y croire, il avait fallu que cette observation soit confirmée par deux fois. Il avait ensuite ordonné au NKVD d'arrêter et de torturer pour « provocation » l'officier d'aviation à l'origine du premier rapport.

Difficile de réfléchir à long terme avec les Allemands quasiment aux portes de la ville. Cependant, malgré l'extrême danger, on ne pouvait fermer les yeux sur l'existence possible d'une bombe capable de rayer Moscou de la carte. Si l'Allemagne écrasait l'URSS, celle-ci pouvait fort bien se faire attaquer ensuite par la Grande-Bretagne et les États-Unis : c'était plus ou moins ce qui s'était passé après la guerre de 1914-1918. Pouvait-on laisser l'URSS sans défense face à un impérialisme capitaliste doté d'une superbombe ?

Volodia confia à son adjoint, le lieutenant Biélov, la tâche de localiser Zoïa.

En attendant de connaître son adresse, il étudia le rapport de Frunze, dans l'original anglais et dans sa traduction russe. Ne pouvant sortir ces documents du bâtiment, il s'efforça d'en mémoriser les phrases qui lui paraissaient déterminantes. Au bout d'une heure, il en avait suffisamment compris pour être capable de poser les bonnes questions.

Biélov découvrit que Zoïa n'habitait ni dans les locaux de la faculté, ni dans l'immeuble voisin, réservé aux chercheurs. À l'université, on lui apprit que tous les assistants avaient été réquisitionnés pour la construction de nouvelles lignes de défense intra-muros, et on lui indiqua où il pourrait trouver Zoïa.

Volodia mit son manteau et sortit.

Il était tout excité. Parce qu'il allait revoir Zoïa ? Parce que cette superbombe existait bel et bien ? Peut-être pour les deux raisons à la fois.

Il réussit à obtenir une ZIS de l'armée et un chauffeur.

Devant la gare de Kazan d'où partaient les trains en direction de l'est, il aperçut ce qui ressemblait fort à une émeute. Impossible, semblait-il, de pénétrer dans la gare, et encore moins de monter dans un train. Des gens chargés de bagages se battaient pour atteindre la porte d'entrée, certains n'hésitant pas à faire usage de leurs poings et de leurs pieds, sous le regard de quelques policiers impuissants. Il aurait fallu une armée pour imposer l'ordre.

La scène scandalisa Volodia. Elle bouleversa son chauffeur au point de lui arracher un commentaire, ce qui n'était pas dans l'habitude des chauffeurs de l'armée, plutôt taciturnes d'ordinaire.

« Ces salauds de lâches ! Ils se tirent et nous laissent combattre les nazis tout seuls. Regardez-les, dans leurs manteaux de fourrure à la con ! »

Volodia en fut interloqué. Il était dangereux de proférer de telles critiques à l'encontre de l'élite dirigeante. Vous pouviez être dénoncé et aller passer une ou deux semaines dans les sous-sols du NKVD, place de la Loubianka, d'où vous risquiez de ressortir infirme à vie.

Volodia avait le sentiment troublant que le système rigide de hiérarchie et de déférence qui étayait le communisme soviétique commençait à donner des signes de faiblesse et à s'effriter.

Ils découvrirent l'équipe des barricades à l'endroit indiqué. Volodia descendit de voiture et demanda au chauffeur de l'attendre, avant d'observer le travail en cours.

Tout le long d'une artère principale s'étirait une ligne de « hérissons » antichars, composés de trois tronçons de rails de chemin de fer en acier d'un mètre de long entrecroisés et soudés en leur centre de manière à former une sorte d'astérisque reposant sur trois pieds. A priori, ces hérissons étaient censés empêcher les véhicules à chenilles de passer.

Au-delà de cette ligne, un fossé antichar avait été creusé à la pioche et à la pelle, et on était en train de le consolider par un mur en sacs de sable, pourvu de meurtrières. Un étroit passage en zigzag avait été ménagé entre ces différents obstacles pour que les Moscovites puissent continuer à circuler jusqu'à l'arrivée des Allemands.

Presque toute l'équipe occupée à creuser la tranchée ou à construire le mur était composée de femmes.

Volodia découvrit Zoïa à côté d'un immense tas de sable, en train de remplir des sacs à la pelle. Il la regarda de loin pendant une bonne minute, dans son manteau sale et ses bottes en feutre, les mains protégées par des mitaines en laine. Un fichu décoloré noué sous le menton recouvrait ses cheveux blonds tirés en arrière. Malgré la boue qui lui maculait le visage, elle était toujours aussi jolie. Elle travaillait avec efficacité, maniant son outil en cadence. Le chef de groupe donna un coup de sifflet, et tout le monde s'arrêta.

Zoïa se laissa tomber sur une pile de sacs de sable et sortit de sa poche de manteau un petit paquet enveloppé dans du papier journal. Volodia s'assit à côté d'elle. « Tu aurais pu obtenir une dispense !

— C'est ma ville. Il faut bien que je la défende !

— Autrement dit, tu ne pars pas pour l'est, toi.

— Il n'est pas question que je fuie devant ces salauds de nazis ! »

Sa véhémence le surprit. « Beaucoup le font.

— Je sais. Je te croyais d'ailleurs parti depuis longtemps.

— Je vois que tu as une bien piètre opinion de moi. Tu penses que j'appartiens à une élite égoïste, c'est ça ? »

Elle haussa les épaules. « En général, tous ceux qui le peuvent quittent la ville.

— Eh bien, tu as tort. Ma famille est toujours ici, à Moscou. Au grand complet.

— Je t'ai peut-être mal jugé. Tu veux un blini ? » Elle ouvrit son paquet. Il contenait quatre galettes ocre pâle, enveloppées dans des feuilles de chou. « Sers-toi. »

Il accepta et en prit une bouchée. Ce n'était pas franchement délicieux. « Tu fais ça avec quoi ?

— Des épluchures de pommes de terre. Tu peux en récupérer tout un seau gratuitement à la porte des cantines. Tu les râpes, tu les fais bouillir jusqu'à ce qu'elles soient tendres, tu ajoutes un peu de farine et de lait, du sel si tu en as, et tu les fais revenir dans du lard.

— Je ne savais pas que tu étais dans une situation aussi difficile, dit-il, gêné. Tu sais, tu peux toujours venir manger chez nous.

— Merci. Qu'est-ce qui t'amène ici ?

— Une question. La séparation isotopique par diffusion gazeuse, c'est quoi ? »

Elle le regarda fixement. « Oh, bon sang ! Qu'est-ce qui s'est passé ?

— Rien du tout. Je m'interroge simplement sur le sérieux de certaines informations.

— Aurions-nous enfin décidé de fabriquer une bombe atomique ? »

À sa réaction, Volodia se dit que les renseignements fournis par Frunze étaient certainement valables. Zoïa avait immédiatement compris leur importance. « Réponds-moi, s'il te plaît, dit-il sévèrement. C'est une affaire officielle, même si nous sommes amis.

— Très bien. Tu sais ce que c'est qu'un isotope ?

— Non.

— Certains éléments existent sous des formes légèrement différentes. Les atomes de carbone, par exemple, sont toujours constitués de six protons, mais certains peuvent avoir six neutrons et d'autres sept ou huit. C'est ça, les isotopes : les variantes d'un même élément qui diffèrent par leur masse atomique. On a ainsi le carbone 12, le carbone 13, le carbone 14.

— Jusque-là ça va. Même pour quelqu'un qui a fait des études de langues, déclara Volodia. Pourquoi est-ce que c'est important ?

— À cause de l'uranium. Il possède deux isotopes, l'uranium 235 et l'uranium 238. À l'état naturel, ces deux isotopes sont mélangés dans l'uranium, mais seul l'uranium 235 a des propriétés fissiles.

— Il faut donc les séparer.

— Oui. Théoriquement, on devrait pouvoir le faire par diffusion gazeuse. Quand un gaz diffuse à travers une membrane, les molécules les plus légères la traversent plus rapidement, de sorte que le gaz qui sort de la membrane est plus riche en isotope inférieur. Mais je n'ai jamais assisté à cette expérience, bien entendu. »

À en croire le rapport de Frunze, les Britanniques étaient en train de construire au pays de Galles, dans l'ouest du Royaume-Uni, une usine spécialisée dans ce processus de diffusion gazeuse, et les Américains en faisaient autant chez eux. « À quoi d'autre pourrait servir une usine de ce genre ?

— Aucune idée. » Elle secoua la tête. « À part séparer des isotopes, franchement, je ne vois pas. Pour mettre un processus pareil sur la liste des recherches prioritaires en temps de guerre, il faut être complètement cinglé, ou fabriquer une arme nouvelle. »

Volodia aperçut une KIM-10 qui s'approchait du passage ouvert à la circulation entre les barricades et commençait à en négocier les zigzags. C'était une petite voiture à deux portes, un modèle réservé aux privilégiés qui pouvait atteindre cent kilomètres à l'heure ; mais celle-ci ne dépassait sûrement pas les soixante-cinq tant elle était chargée.

Le conducteur, âgé d'une soixantaine d'années, portait un chapeau et un manteau en drap de laine de coupe occidentale. Une jeune femme en chapeau de fourrure était assise à côté de lui. La banquette arrière croulait sous les cartons et un piano avait été amarré sur le toit, en équilibre précaire.

À l'évidence, c'était un membre éminent de la classe dirigeante qui cherchait à quitter la ville avec sa femme, ou sa maîtresse, et tous les biens de valeur qu'il possédait. Un membre de la caste dans laquelle Zoïa le rangeait lui-même d'office, pensa Volodia, ce qui expliquait sans doute son refus de sortir avec lui. Si seulement elle pouvait revenir sur ses préjugés !

L'une des volontaires qui travaillaient à l'édification de la barricade saisit un hérisson qu'elle traîna devant la KIM-10. En la voyant faire, Volodia se dit qu'il y allait avoir du grabuge.

La voiture avança lentement jusqu'à ce que son pare-chocs heurte le hérisson. Le conducteur croyait peut-être pouvoir l'écarter. Plusieurs femmes s'approchèrent pour observer la scène. Mais ce dispositif avait été conçu pour résister à de plus fortes poussées. Les tronçons de rails s'enfoncèrent dans le sol, s'y ancrèrent solidement et n'en bougèrent pas. On entendit un vilain bruit de tôle froissée. Le conducteur enclencha la marche arrière et recula. Son pare-chocs était complètement embouti.

Passant la tête par la fenêtre, il cria sur le ton d'un homme habitué à être obéi : « Dégagez ça immédiatement !

— Tu n'as qu'à le faire toi-même, espèce de déserteur ! » rétorqua la volontaire en croisant les bras. C'était une solide femme d'âge moyen, coiffée d'une casquette d'homme en tissu à carreaux.

Le conducteur sortit, rouge de colère et Volodia eut la surprise de reconnaître le colonel Bobrov qu'il avait côtoyé en Espagne. Un Bobrov célèbre là-bas pour avoir abattu d'une balle dans la nuque ses propres soldats qui se repliaient. Il avait pour mot d'ordre : « Pas de pitié pour les lâches ! » À Belchite, Volodia l'avait vu de ses propres yeux tuer trois soldats des Brigades internationales qui avaient battu en retraite, faute de munitions. Bobrov était désormais en civil. Allait-il tirer sur la femme qui l'empêchait de passer ?

Bobrov s'avança vers le hérisson et empoigna une des tiges d'acier. L'objet était plus lourd qu'il ne s'y attendait, mais il réussit péniblement à dégager le passage.

Tandis qu'il revenait vers sa voiture, la femme à la casquette remit le hérisson en place.

À présent, toute l'équipe des volontaires s'était attroupée et, le sourire aux lèvres, suivait des yeux l'affrontement en échangeant des plaisanteries.

Bobrov marcha sur la femme tout en sortant son laissez-passer militaire de la poche de son manteau. « Libérez la voie ! Je suis le général Bobrov. » Il devait avoir obtenu cette promotion à son retour d'Espagne.

« Un soldat, vous ? ricana la femme. Ben qu'est-ce que vous fichez ici, au lieu de vous battre ? »

Bobrov rougit sous l'insulte, la sachant justifiée. Volodia se demanda si c'était sa jeune épouse qui avait réussi à convaincre ce vieux militaire brutal de s'enfuir.

« Un traître, oui ! poursuivait la volontaire à la casquette. Essayer de se tirer avec son piano et sa poule ! » Du plat de la main, elle fit voler le chapeau du général.

Volodia n'en revenait pas. Il n'avait jamais vu personne en Union soviétique défier ainsi l'autorité. À Berlin, oui. Avant l'arrivée des nazis au pouvoir, il avait été étonné de voir des Allemands ordinaires se quereller courageusement avec des policiers, mais ici, c'était nouveau.

La foule de femmes applaudit.

Bobrov avait toujours son abondante toison blanche coupée en brosse. Il suivit les virevoltes de son chapeau sur la route détrempée, esquissa un pas pour le rattraper et se ravisa.

Volodia n'avait pas envie d'intervenir. Que pouvait-il faire, tout seul, face à cette foule ? De toute façon, il n'éprouvait aucune sympathie pour Bobrov. Qu'on rende au général la monnaie de sa pièce : ce n'était que justice après tout.

Une autre femme, plus âgée et enveloppée dans une couverture crasseuse, ouvrit le coffre de la voiture. « Visez-moi ça ! » Il était rempli d'un monceau de bagages en cuir. Elle attrapa une valise, qu'elle exhiba avant d'en soulever le couvercle, faisant tomber son contenu : hauts en dentelle, jupons et chemises de nuit en satin, bas de soie et chemisettes, évidemment fabriqués à l'Ouest et plus délicats que tout ce que ces femmes russes avaient jamais vu et pouvaient s'offrir. Ils se répandirent dans la neige fondue et sale qui recouvrait la chaussée et y demeurèrent, éparpillés comme des pétales de fleurs sur un tas de fumier.

Certaines femmes entreprirent de les ramasser, d'autres se saisirent des bagages restants. Bobrov se précipita à l'arrière de sa voiture, bien décidé à repousser l'assaut. La situation allait dégénérer, se dit Volodia. À coup sûr, le général portait une arme et n'allait pas tarder à la sortir. Mais voilà que la vieille enveloppée d'une couverture s'empara d'une pelle qu'elle abattit dans un bruit retentissant sur le crâne de Bobrov avec toute la force d'une femme capable de creuser une tranchée. Le général s'effondra, et la femme lui balança un coup de pied.

La passagère descendit de voiture. Elle devait avoir une trentaine d'années. « Tu viens nous aider à creuser ? » l'interpella la femme à la casquette. Ses compagnes s'esclaffèrent.

Baissant la tête, la compagne du général s'engagea dans le passage par lequel la voiture était arrivée. La volontaire en casquette voulut la retenir, mais elle parvint à se faufiler entre les hérissons et se mit à courir. L'autre la prit en chasse. Dans ses souliers en daim beige à talons, la jeune femme dérapa sur la neige mouillée, perdant son chapeau en fourrure dans sa chute. S'étant relevée tant bien que mal, elle reprit ses jambes à son cou. La volontaire abandonna la poursuite pour faire main basse sur le chapeau.

Tout autour de l'automobile désertée, le sol était jonché de valises ouvertes. Les femmes avaient entrepris de tirer du véhicule les cartons entassés sur la banquette arrière, qu'elles renversaient et vidaient par terre. Des couverts s'en échappèrent, de la vaisselle en porcelaine, des verres, aussitôt brisés. Draps brodés et serviettes blanches furent traînés dans la boue. De jolies chaussures, une dizaine de paires, se retrouvèrent dispersées sur la chaussée.

Comme Bobrov s'était redressé sur les genoux et tentait de se relever, la femme à la couverture lui asséna un nouveau coup de pelle. Bobrov s'écroula. Elle se jeta aussitôt sur son beau manteau de laine pour le déboutonner et le lui arracher des épaules. Bobrov résistait, se débattait. Prise de fureur, la femme frappa à tour de bras jusqu'à ce que le général ne fasse plus un mouvement, ses cheveux blancs rouges de sang. Après quoi, elle se débarrassa de sa vieille couverture et revêtit le manteau convoité.

Volodia s'avança vers le corps immobile ; les yeux fixes du général ne voyaient plus. Il s'agenouilla près de lui pour prendre son pouls et voir s'il respirait toujours. Le cœur ne battait plus. Bobrov était mort.

« Pas de pitié pour les lâches », laissa tomber Volodia. Il lui ferma tout de même les yeux.

Le piano, que certaines libérèrent de ses cordes, glissa du toit de la voiture et s'écrasa au sol dans un bruit discordant. Des femmes se mirent alors à le fracasser à qui mieux mieux, à l'aide de pioches et de pelles, tandis que d'autres se disputaient âprement les biens éparpillés. Elles s'emparaient des couverts, se faisaient des ballots de draps, déchirant la lingerie fine dans leur lutte pour se l'approprier.

Une théière en porcelaine manqua de peu la tête de Zoïa. Volodia revint vers elle à la hâte. « Ça tourne à l'émeute. J'ai une voiture de l'armée et un chauffeur, je vais te sortir d'ici. »

Elle n'hésita qu'une seconde : « Merci. » Ils s'élancèrent vers la voiture, bondirent à l'intérieur et démarrèrent sur les chapeaux de roue.

2.

L'invasion de l'Union soviétique avait justifié la foi d'Erik von Ulrich dans le Führer. Tandis que l'armée allemande traversait l'immensité russe en balayant les soldats de l'armée Rouge comme des fétus de paille, Erik s'était félicité du génie stratégique du grand chef auquel il avait fait allégeance.

Pourtant les choses n'avaient pas été faciles, loin de là. En octobre, les pluies avaient transformé la campagne en bourbier. *Raspoutitsa*, tel était le nom donné à la période de l'année où les routes disparaissaient sous la boue. La terre détrempée s'accumulait peu à peu à l'avant des véhicules et ralentissait leur progression. Erik et Hermann avaient dû descendre de voiture et dégager la voie à la pelle. Partout, l'armée allemande devait faire face au même problème et la ruée sur Moscou s'était transformée en marche au pas. De plus, ces routes inondées entravaient la circulation des camions de ravitaillement. L'armée manquait de munitions, de carburant et de nourriture. L'unité d'Erik était menacée d'une pénurie de médicaments et de matériel médical.

Au début du mois de novembre, quand le froid s'installa, Erik commença par se réjouir. Ce gel était une bénédiction. Grâce à lui, les routes étaient redevenues praticables, et l'ambulance pouvait rouler à une allure normale. Il frissonnait cependant dans sa capote d'été et ses sous-vêtements en coton. Les tenues d'hiver n'étaient pas encore arrivées d'Allemagne, pas plus que l'antigel indispensable à son ambulance et aux autres engins militaires, camions, chars, pièces d'artillerie. La nuit, il devait se lever toutes les deux heures pour faire démarrer le moteur et le laisser tourner pendant cinq bonnes minutes. C'était le seul moyen d'empêcher l'huile de figer et le liquide de refroidissement de geler. Et encore, le matin, une heure avant de reprendre la route, il devait allumer un petit feu sous le moteur pour le réchauffer.

Des centaines de véhicules tombaient en panne et étaient abandonnés sur place. Les avions de la Luftwaffe, qui passaient la nuit dehors sur des aérodromes de fortune, étaient gelés au matin et ne décollaient plus. Aussi la couverture aérienne des troupes n'était-elle plus assurée.

Malgré les déboires de l'armée allemande, les Russes continuaient à céder du terrain. Ils avaient beau se battre avec acharnement, ils ne cessaient de reculer. L'unité d'Erik devait constamment s'arrêter pour dégager la route encombrée de cadavres ennemis, et tous les morts gelés empilés le long de la chaussée formaient un remblai macabre. Sans relâche, sans pitié, l'étau allemand se resserrait autour de Moscou.

On verrait bientôt les panzers défiler majestueusement sur la place Rouge, Erik en était sûr, et les bannières à croix gammée flotter gaiement sur les tours du Kremlin.

Mais pour le moment, la température avoisinait les moins dix et n'arrêtait pas de baisser.

L'hôpital de campagne auquel Erik était affecté s'était installé dans une petite localité dont il ignorait le nom, située à proximité d'un canal gelé, au cœur d'une forêt d'épicéas. Cette petite ville avait survécu à la retraite des Russes et était restée presque intacte, alors que bien souvent, ils ne laissaient rien derrière eux. L'hôpital local était moderne. Quand les Allemands s'en étaient emparés, le docteur Weiss avait donné l'ordre aux médecins qui y travaillaient de renvoyer les malades chez eux, quel que fût leur état.

À présent, Erik examinait un soldat allemand d'environ dix-huit ans atteint d'engelures. La peau gelée de son visage, d'un jaune cireux, était dure au toucher, et il avait les bras et les jambes couverts de cloques violettes, comme Erik et Hermann purent le constater après avoir découpé son mince uniforme d'été. Dans une pathétique tentative de se protéger du froid, il avait bourré de papier journal ses bottes éculées et déchirées. Quand Erik les lui retira, l'odeur putride caractéristique de la gangrène lui envahit les narines. Ils devraient pourtant réussir à éviter l'amputation. Il savait ce qu'il convenait de faire dans des cas pareils. Ces derniers temps, avec Hermann Braun, ils soignaient plus d'engelures que de blessures reçues au combat.

Erik remplit une baignoire d'eau tiède puis, aidé d'Hermann, il y plongea le soldat.

Erik l'examina pendant que son corps se réchauffait peu à peu. L'un des pieds était noir de gangrène, l'autre n'avait que les orteils atteints.

Quand l'eau commença à refroidir, ils sortirent le soldat du bain, le séchèrent en le tapotant délicatement, l'allongèrent sur un lit de camp sous des couvertures. Enfin, ils l'entourèrent de pierres chaudes enveloppées dans des serviettes.

Le patient, qui n'avait pas perdu conscience, voulut savoir si ses pieds pourraient être sauvés.

« Seul le médecin peut vous le dire, répondit Erik par automatisme. Nous ne sommes qu'infirmiers.

— Mais vous voyez beaucoup de malades, insista l'autre. Que pensez-vous de mon état ?

— À mon avis, vous devriez vous en sortir », répondit Erik. Si ce n'était pas le cas, il savait ce qui se passerait. Pour le pied le moins atteint, Weiss déciderait d'amputer les orteils, ce qu'il ferait à l'aide d'une grosse pince qui ressemblait à un coupe-boulons ; quant à l'autre jambe, il ordonnerait l'amputation au niveau du genou.

Weiss arriva quelques minutes plus tard. Il examina les pieds du soldat. « Préparez-le pour l'amputation », lança-t-il.

Erik en fut navré. Encore un jeune homme costaud qui passerait le restant de sa vie en fauteuil roulant. Quel gâchis !

Mais le patient voyait les choses différemment : « Dieu soit loué ! s'écria-t-il. Je ne retournerai pas au combat ! »

Tout en le préparant pour l'opération, Erik réfléchit à l'attitude défaitiste de ce soldat. Tant de gens la partageaient, jusque dans sa propre famille… Erik pensait souvent à son père décédé avec un mélange de colère et de tristesse. Walter n'aurait jamais rejoint la majorité pour célébrer le triomphe du IIIe Reich, pensa-t-il amèrement. Il aurait toujours trouvé à se plaindre d'une chose ou d'une autre, mettant en doute le jugement du Führer et sapant le moral des troupes. Pourquoi s'était-il toujours révolté ? Pourquoi avait-il été aussi attaché à la démocratie, une idéologie complètement dépassée ? La liberté n'avait rien apporté à l'Allemagne, alors que le fascisme avait sauvé le pays !

Non, la fureur d'Erik contre son père ne s'était pas apaisée et pourtant, les larmes lui vinrent aux yeux quand il repensa à la façon dont il était mort. Au début, Erik n'avait pas voulu admettre que c'était la Gestapo qui l'avait tué ; mais il avait fini par comprendre que c'était probablement la vérité. Ce n'étaient pas des enfants de chœur : ils tabassaient ceux qui racontaient des mensonges éhontés à propos du gouvernement. Son père s'était entêté à demander si le gouvernement tuait les enfants déficients. Il avait eu la sottise d'écouter son épouse anglaise et sa fille exagérément sentimentale. Erik souffrait d'autant plus de voir sa mère et sa sœur s'obstiner dans l'erreur qu'il les aimait profondément toutes les deux.

À Berlin, pendant sa permission, il était allé voir le père d'Hermann Braun. C'était cet homme qui, le premier, lui avait parlé de cette formidable philosophie nazie, alors qu'Hermann et lui étaient encore de jeunes garçons. Aujourd'hui, Herr Braun appartenait aux SS. Erik lui avait raconté que dans un bar, il avait entendu quelqu'un prétendre que le gouvernement éliminait les incurables dans des hôpitaux spéciaux. « Il est vrai qu'ils représentent un lourd fardeau financier sur la voie qui mène à la nouvelle Allemagne, lui avait expliqué Herr Braun. La race doit être purifiée, en prenant des mesures répressives contre les Juifs et autres dégénérés, et en interdisant les mariages mixtes, source d'abâtardissement. Mais l'euthanasie, non. Ça n'a jamais été la politique des nazis. Nous sommes déterminés, durs, parfois brutaux peut-être, mais nous n'assassinons personne. C'est un mensonge des communistes ! »

Les accusations de Vater avaient été erronées. Ce qui n'empêchait pas Erik de le pleurer quelquefois.

Heureusement, il n'avait pas une minute à lui. Le matin était toujours marqué par une arrivée massive de patients, pour la plupart des blessés de la veille. Il y avait ensuite une courte accalmie avant un second afflux de blessés : les premiers de la journée.

En milieu de matinée, après l'opération du soldat aux pieds gelés, il prit une pause en compagnie d'Hermann et du docteur Weiss dans la salle du personnel bondée.

Levant les yeux du journal qu'il était en train de lire, Hermann s'exclama soudain : « À Berlin, on dit que nous avons déjà gagné. Ils feraient bien de venir jeter un œil par ici ! »

Avec son cynisme habituel, le docteur Weiss répondit : « Le Führer a fait un discours des plus intéressants au Sportpalast. Il a évoqué l'infériorité bestiale des Russes. Ça m'a rassuré. J'avais l'impression que c'étaient les ennemis les plus coriaces que nous ayons rencontrés jusqu'ici. Ils se battent plus longtemps et plus durement que les Polonais, les Belges, les Hollandais, les Français ou les Britanniques. Ils sont sous-équipés, mal nourris et mal commandés, mais ça ne les empêche pas de se jeter contre nos mitrailleuses en agitant leurs vieux tromblons, comme s'ils se fichaient bien de vivre ou de mourir. Je suis heureux d'apprendre que ce trait de caractère n'est que la marque de leur bestialité. Je commençais à craindre qu'ils ne soient courageux et patriotes. »

À son habitude, Weiss feignait de partager les idées du Führer pour dire exactement le contraire. Sa réponse déconcerta Hermann, mais Erik, qui avait compris l'ironie, s'emporta. « Courageux ou pas, les Russes sont en train de perdre : nous sommes à soixante-cinq kilomètres de Moscou. Ça prouve que le Führer avait raison.

— Et qu'il est bien plus intelligent que Napoléon, renchérit le docteur Weiss.

— Du temps de Napoléon, le cheval était le moyen de locomotion le plus rapide, fit remarquer Erik. Aujourd'hui, nous avons des véhicules motorisés et la TSF. Les communications modernes nous ont permis de réussir là où Napoléon avait échoué.

— Enfin, elles nous *permettront* de réussir quand nous aurons pris Moscou.

— Ce qui sera chose faite dans quelques jours, pour ne pas dire quelques heures. Vous ne pouvez guère en douter !

— Ah bon ? Je crois savoir que certains de nos généraux ont proposé que nous nous arrêtions là où nous sommes pour construire une ligne de défense. Afin de sécuriser nos positions et de nous réapprovisionner pendant l'hiver, avant de reprendre l'offensive à l'arrivée du printemps.

— C'est de la lâcheté, pour ne pas dire de la trahison, si vous voulez mon avis ! répliqua Erik avec chaleur.

— C'est vous qui devez avoir raison, puisque c'est exactement ce que Berlin a répondu à ces généraux, si j'ai bien compris. Il va de soi que l'état-major a une meilleure vision des choses que les soldats du front.

— Nous avons presque éliminé l'armée Rouge !

— Sauf que Staline semble faire jaillir des armées de nulle part, comme un magicien. Au début de la campagne, nous pensions qu'il avait deux cents divisions. On estime maintenant qu'il en a plus de trois cents. Où a-t-il déniché cette centaine de divisions supplémentaire ?

— L'avenir prouvera une fois de plus que le Führer avait raison.

— Bien sûr, Erik.

— Il ne s'est encore jamais trompé !

— Un homme qui croyait pouvoir voler sauta du haut d'un immeuble de dix étages. Au cours de sa chute, alors qu'il passait devant le cinquième étage en battant vainement des bras, on l'entendit s'exclamer : "Jusqu'ici, tout va bien." »

À cet instant, un soldat fit irruption dans la salle. « Il y a eu un accident. Dans la carrière, au nord de la ville. Une collision entre trois véhicules. Plusieurs officiers SS sont blessés. »

À l'origine, les SS, les membres de la Schutzstaffel, avaient constitué la garde personnelle d'Hitler. Ils formaient à présent une puissante élite. Erik admirait leur discipline sans faille, leurs uniformes élégants et leur proximité avec Hitler.

« J'envoie une ambulance, répondit Weiss.

— Il s'agit de l'*Einsatzgruppe*, le groupe d'intervention spéciale », précisa le soldat.

Erik avait vaguement entendu parler de ces unités spéciales. Elles suivaient l'armée dans les territoires conquis et arrêtaient les fauteurs de troubles et autres saboteurs potentiels, comme les communistes. Ce groupe devait être en train d'installer un camp de prisonniers à l'extérieur de la ville.

« Combien de blessés ? demanda Weiss.

— Six ou sept. On essaie encore de les extraire des véhicules.

— Très bien. Braun, von Ulrich, allez-y ! »

Erik en fut ravi. Côtoyer les plus fervents partisans du Führer, quel bonheur ! S'il pouvait leur rendre service, sa joie serait sans bornes.

Le soldat lui tendit une fiche lui indiquant comment se rendre sur les lieux de l'accident.

Erik et Hermann se hâtèrent d'avaler leur thé, écrasèrent leur cigarette et sortirent. Erik enfila le manteau en mouton retourné qu'il avait récupéré sur le cadavre d'un officier russe, en prenant soin d'en laisser les pans ouverts pour qu'on voie son uniforme. Ils filèrent au garage. Hermann s'assit au volant tandis qu'Erik lui indiquait le chemin tout en scrutant les alentours à travers le fin rideau de neige qui tombait.

À la sortie de la ville, la route se mit à serpenter à travers la forêt. Ils croisèrent plusieurs autocars et camions venant en sens inverse. La neige était tassée et glissante, ce qui ralentissait leur allure. Qu'une collision ait pu se produire n'était pas difficile à imaginer.

On était déjà l'après-midi et les journées étaient courtes en cette période de l'année. Il n'y avait de lumière qu'entre dix heures du matin et cinq heures du soir. Ce jour-là, les nuages chargés de neige ne laissaient filtrer qu'une clarté grisâtre. Les grands conifères massés des deux côtés de la route obscurcissaient encore la chaussée. Erik avait l'étrange d'impression de suivre un chemin menant au plus profond des bois, là où règne le mal, comme s'ils avaient été plongés, Hermann et lui, au cœur d'un conte de Grimm.

Ils cherchaient une bifurcation sur la gauche : elle était gardée par un soldat qui leur indiqua le lieu de l'accident. Ils cahotèrent sur un chemin creux périlleux qui serpentait entre les arbres, jusqu'à ce qu'un autre garde leur fasse signe de s'arrêter. « Roulez au pas surtout ! C'est comme ça que s'est produit l'accident. »

Une minute plus tard, ils débouchèrent sur le lieu de la collision. Trois véhicules étaient comme soudés les uns aux autres : un autocar, une Jeep et une limousine Mercedes aux pneus pourtant équipés de chaînes. Erik et Hermann sautèrent de l'ambulance.

Il n'y avait personne dans l'autocar. Trois blessés étaient allongés par terre, peut-être les occupants de la Jeep. Près de la voiture prise en sandwich entre les deux autres véhicules, plusieurs soldats unissaient leurs efforts pour tenter d'extraire les passagers.

Une série de coups de fusil retentit. Erik se demanda l'espace d'un instant qui tirait, puis chassa cette pensée de son esprit pour se concentrer sur la tâche qui l'attendait.

Il examina tour à tour tous les blessés avec Hermann pour juger de leur état. Sur les trois victimes étendues au sol, l'un des hommes était mort, l'autre avait un bras cassé et le troisième ne souffrait apparemment que de contusions. À l'intérieur de la Mercedes, un homme avait succombé à une hémorragie et un autre avait perdu connaissance. Quant au troisième passager, il hurlait de douleur. Erik lui administra une piqûre de morphine. Il attendit que le médicament ait produit son effet pour le sortir du véhicule avec Hermann et le transporter dans l'ambulance.

L'espace qu'ils avaient libéré allait permettre aux soldats de dégager l'homme inconscient des tôles déformées qui le retenaient prisonnier. Il avait une blessure à la tête qui laissait peu d'espoir, mais Erik n'en dit rien. Il reporta son attention sur les passagers de la Jeep. Pendant qu'Hermann posait une attelle à celui qui avait le bras cassé, il aida l'autre blessé à rejoindre l'ambulance et à s'y asseoir. Puis il retourna à la Mercedes.

« Attendez quelques instants ! dit un capitaine. D'ici cinq ou dix minutes, on l'aura dégagé.

— D'accord », répondit Erik.

Comme une nouvelle salve de tirs retentissait, il céda à la curiosité et s'enfonça dans le sous-bois, à la recherche de l'*Einsatzgruppe*. Sous les arbres, la neige était piétinée et jonchée de mégots, de trognons de pommes, de vieux journaux et autres détritus, comme si une équipe d'ouvriers était passée par là.

Il déboucha dans une clairière, où étaient rangés des camions et des autocars. On avait apparemment transporté jusqu'ici un grand nombre de gens. Plusieurs autocars repartaient à vide en contournant le lieu de l'accident. Un autre arriva juste au moment où Erik traversait cette aire de stationnement. De l'autre côté, il découvrit un groupe d'une centaine de Russes de tous âges, des prisonniers apparemment. Chose curieuse, la plupart portaient des valises, des cartons ou des sacs auxquels ils se cramponnaient comme à leur bien le plus précieux. Un homme tenait un violon, une enfant sa poupée. À la vue de cette petite fille, Erik fut pris de malaise. Un pressentiment sinistre lui noua les tripes.

Ces prisonniers étaient sous la garde de policiers russes armés de matraques. De toute évidence, l'*Einsatzgruppe* faisait appel à des collaborateurs locaux. Les policiers le dévisagèrent mais gardèrent le silence en remarquant son uniforme allemand sous sa touloupe déboutonnée.

Il continua d'avancer dans la direction des tirs. Alors qu'il passait à la hauteur d'un prisonnier bien habillé, il s'entendit interpeller en allemand : « Monsieur, je suis le directeur de l'usine de pneumatiques de cette ville. Je n'ai jamais adhéré au communisme, ou seulement du bout des lèvres, comme tout responsable était tenu de le faire. S'il vous plaît, laissez-moi partir. Je peux vous aider, je connais tout et tout le monde ici. »

Erik passa son chemin.

Il arriva à la carrière. C'était une vaste dépression irrégulière, bordée sur son pourtour par de grands épicéas, semblables à des gardiens dont l'uniforme vert foncé disparaissait sous la neige. À une extrémité, une longue rampe menait au fond de la fosse. Une douzaine de prisonniers, marchant de front deux par deux, commençaient à descendre vers cette combe obscure, houspillés par des soldats.

Erik remarqua parmi eux trois femmes et un garçon d'une dizaine d'années. Le camp de prisonniers avait-il été installé quelque part dans cette carrière ? Mais ces détenus n'avaient plus leurs bagages. Les flocons de neige se déposaient sur leurs têtes nues comme une bénédiction.

« Sergent, qui sont ces gens ? demanda Erik à un SS debout à proximité.

— Des communistes de la ville, des commissaires politiques, de la racaille de ce genre.

— Ah bon ? Et ce petit garçon ?

— Il y a aussi des Juifs.

— Je ne comprends pas, ce sont des Juifs ou des communistes ?

— Qu'est-ce que ça change ?

— Ce n'est pas pareil.

— Quelles conneries ! La plupart des communistes sont juifs, et la plupart des Juifs sont communistes. Vous ne savez pas ça ? »

Pourtant, se dit Erik, le directeur de l'usine de pneus qui lui avait parlé n'avait l'air d'être ni l'un ni l'autre.

Les prisonniers avaient atteint le fond rocheux de la carrière. Jusque-là, ils avaient marché en silence comme un troupeau de moutons, sans regarder autour d'eux. Mais ils s'animèrent soudain, pointant le doigt vers le sol. À travers le rideau de neige, Erik distingua des formes qui ressemblaient à des corps, éparpillés au milieu des blocs de rochers, leurs vêtements saupoudrés de flocons.

C'est alors qu'il repéra douze soldats postés au bord du ravin, parmi les arbres. Douze prisonniers, douze tireurs. Il comprit. Une incrédulité mêlée d'horreur lui monta à la gorge, amère comme de la bile.

Les soldats épaulèrent leurs fusils, les braquèrent sur les prisonniers.

« Non ! cria Erik. Non, vous ne pouvez pas faire ça ! » Personne ne l'entendit.

Une femme se mit à hurler. Erik la vit saisir le petit garçon et le serrer contre elle comme si ses bras pouvaient arrêter les balles. C'était sans doute sa mère.

« Feu ! » cria un officier.

Les fusils claquèrent. Les prisonniers chancelèrent et s'écroulèrent. Le crépitement des balles fit tomber un peu de neige des épicéas, saupoudrant les soldats d'un blanc étincelant.

Erik vit le petit garçon et sa mère tomber ensemble, toujours enlacés. « Non, s'écria-t-il. Oh, non ! »

Le sergent se tourna vers lui. « Qu'est-ce qui vous prend ? demanda-t-il avec humeur. Et qui êtes-vous, d'abord ?

— Infirmier du bataillon, répondit Erik sans quitter des yeux l'effroyable scène qui s'était déroulée dans le fond du ravin.

— Qu'est-ce que vous venez faire ici ?

— Je suis venu avec l'ambulance chercher les officiers blessés dans l'accident. »

Douze autres prisonniers avaient été amenés et, déjà, ils descendaient la pente menant à la carrière. « Oh, mon Dieu, gémit-il. Mon père avait raison. Nous assassinons des gens.

— Vous avez fini de pleurnicher, bordel ? Retournez à votre ambulance !

— Oui, sergent. »

3.

À la fin du mois de novembre, Volodia demanda à être versé dans une unité de combat. Il ne voyait plus l'intérêt de son travail dans le Renseignement militaire. L'armée Rouge n'avait pas besoin d'espions à Berlin pour connaître les intentions d'une armée allemande qui se trouvait déjà aux portes de Moscou. Ce qu'il voulait, lui, c'était défendre sa ville, les armes à la main.

Ses doutes à l'égard du gouvernement n'avaient plus guère d'importance. La stupidité de Staline, la bestialité de la police secrète, le fait que rien ne marchait comme il l'aurait fallu en Union soviétique, tout s'effaçait devant la nécessité irrépressible de repousser un envahisseur qui menaçait de brutaliser, de violer, d'affamer et de tuer sa mère, sa sœur et les jumeaux de celle-ci, Dimka et Tania, ainsi que Zoïa.

En même temps, il n'ignorait pas que si tout le monde tenait le même raisonnement que lui, les espions comme ceux qu'il recrutait deviendraient une denrée rare. Ses informateurs allemands étaient des gens convaincus que la lutte contre l'abomination nazie devait passer avant leur patriotisme, avant leur loyauté. Volodia admirait leur courage et leur moralité sans faille. Il leur en était reconnaissant mais pour sa part, il ressentait les choses différemment.

Parmi les jeunes officiers du Renseignement militaire, il n'était pas le seul à penser ainsi. Ils furent tous rassemblés et incorporés à un bataillon d'infanterie légère au début du mois de décembre. Volodia embrassa ses parents, écrivit un mot à Zoïa pour lui dire qu'il espérait s'en sortir et la revoir, et partit pour la caserne.

Après de longs atermoiements, Staline avait fini par faire venir à Moscou des renforts de l'est. Treize divisions sibériennes étaient à présent déployées face aux Allemands qui se rapprochaient de plus en plus. Sur la route du front, plusieurs d'entre elles avaient fait halte dans la capitale et les Moscovites éberlués les avaient regardées passer dans leurs tenues blanches matelassées et leurs chaudes bottes en mouton retourné, équipées de leurs skis et de leurs lunettes protectrices, montés sur de rustiques poneys des steppes. Elles arrivèrent à temps pour prendre part à la contre-offensive.

Pour l'armée Rouge, c'était la bataille de la dernière chance. Au cours des cinq derniers mois, l'Union soviétique avait lancé à plusieurs reprises des centaines de milliers d'hommes contre les envahisseurs. Chaque fois, les Allemands avaient marqué une pause pour repousser l'attaque puis avaient repris leur progression implacable. Si la tentative se soldait par un échec, il n'y en aurait plus d'autre. Les Allemands prendraient Moscou et ensuite, l'URSS serait à eux. Sa mère n'aurait plus qu'à échanger sa vodka contre du lait au marché noir, pour nourrir Dimka et Tania.

Le 4 décembre, les bataillons soviétiques sortirent de la ville par le nord, l'ouest et le sud, pour prendre position en vue de cet ultime combat. Soucieux de ne pas alerter l'ennemi, ils marchaient dans le noir. Interdiction d'allumer torche ou cigarette.

Ce soir-là, des agents du NKVD vinrent inspecter la ligne de front. Volodia n'aperçut pas la tête de fouine de son beau-frère, Ilia Dvorkine, qui devait pourtant être parmi eux. Deux types qu'il ne connaissait pas entrèrent dans son bivouac alors qu'il était en train de nettoyer son fusil en compagnie d'une dizaine d'hommes. Ils voulaient savoir s'ils avaient entendu quelqu'un critiquer le gouvernement. Que disent les soldats à propos du camarade Staline ? demandèrent-ils. Qui, parmi vos camarades, s'interroge sur le bien-fondé de la stratégie et de la tactique de notre armée ?

Volodia n'en croyait pas ses oreilles. Quelle importance, au point où ils en étaient ? D'ici quelques jours, Moscou serait sauvée ou perdue. Les soldats râlaient contre les officiers ? Et alors ? Il coupa court à l'interrogatoire en affirmant que ses hommes et lui étaient tenus au silence en vertu d'un ordre impératif et que lui-même avait pour instruction de tirer sur quiconque l'enfreignait. Cependant, ajouta-t-il non sans témérité, il voulait bien fermer les yeux pour cette fois, à condition que la police secrète dégage immédiatement.

La ruse fut efficace, mais Volodia ne doutait pas que le NKVD sapait ainsi le moral des troupes sur toute la ligne de front.

Le vendredi 5 décembre au soir, l'artillerie russe donna de la voix. Le lendemain à l'aube, Volodia et son bataillon se mirent en marche dans le blizzard. Ils avaient reçu ordre de s'emparer d'une petite ville située de l'autre côté d'un canal.

Volodia ignora l'ordre d'attaquer les défenses allemandes de front, conformément à la vieille tactique russe. L'heure n'était plus à l'application obstinée de principes qui avaient fait la preuve de leur inefficacité. Avec sa compagnie de cent hommes, il remonta le canal pris par les glaces et le traversa au nord de la ville. Puis il opéra un mouvement tournant de façon à arriver sur le flanc des Allemands. Le fracas et le grondement de la bataille sur sa gauche lui indiquèrent qu'il avait effectivement dépassé les lignes ennemies.

Le blizzard était aveuglant. Au niveau du sol, la visibilité était de quelques mètres seulement même si, de temps à autre, les tirs illuminaient les nuages un court instant. Néanmoins, se dit Volodia avec optimisme, ce mauvais temps allait leur permettre de rester invisibles et de prendre les Allemands par surprise.

Il faisait un froid polaire, jusqu'à moins trente-cinq degrés par endroits. Les deux camps en souffraient, mais surtout les Allemands, insuffisamment équipés pour l'hiver.

À son grand étonnement, Volodia découvrit que leurs ennemis, d'habitude si efficaces, avaient négligé de consolider leur ligne de front. Pas de tranchées, pas de fossés antichars, pas de tranchées-abris. Leur front se limitait en fait à une série de points fortifiés espacés. Rien de plus facile que de le franchir en se glissant dans les brèches. Une fois entrés dans la petite ville, il ne restait qu'à localiser les cibles : casernes, cantines, dépôts de munitions.

Ses hommes abattirent trois sentinelles et s'emparèrent d'un terrain de football où étaient garés cinquante chars. Volodia n'en revenait pas. Cette puissance, qui avait conquis le quart de l'Union soviétique, était-elle aujourd'hui épuisée et à bout de souffle ?

Les cadavres de soldats soviétiques, tombés au cours d'escarmouches antérieures, ne portaient plus ni bottes ni capotes, dérobées sans doute par des Allemands transis.

Partout, des véhicules abandonnés, des camions aux portières ouvertes, des tanks couverts de neige aux moteurs gelés, des Jeep au capot relevé comme si des mécaniciens avaient vainement tenté de les réparer.

En traversant une grande rue, Volodia entendit un vrombissement. À travers la neige qui tombait, il distingua des phares qui se rapprochaient sur sa gauche. Il crut d'abord que c'était un véhicule soviétique qui avait franchi les lignes allemandes, mais bientôt son groupe fut pris pour cible et il cria à ses hommes de se mettre à couvert. C'était une *Kübelwagen*, une Jeep Volkswagen équipée d'une roue de secours sur le capot avant. Le système de refroidissement par air empêchait le moteur de geler. La *Kübelwagen* passa devant eux à toute vitesse, chargée d'Allemands qui tiraient depuis leurs sièges.

Volodia fut tellement ahuri qu'il en oublia de riposter. Pourquoi diable un véhicule rempli d'ennemis en armes fuyait-il la bataille ?

Il entraîna sa compagnie de l'autre côté de la rue. Il avait pensé qu'ils devraient progresser en se battant durement, immeuble après immeuble, mais ils ne rencontrèrent presque aucune résistance. Dans cette ville occupée, les bâtiments étaient fermés, volets tirés, sans lumière. Si des Russes y vivaient encore, ils se terraient sous leurs lits. C'était la seule chose sensée à faire.

D'autres véhicules passèrent. Volodia en conclut que les officiers fuyaient le champ de bataille. Il détacha une section munie d'une mitrailleuse Degtiarev DP-28 avec ordre de se mettre à couvert dans un café et de tirer sur ces voitures. Pas question que ces Allemands restent en vie et continuent à tuer des Russes !

Un peu en retrait de la rue principale, il repéra une bâtisse en brique d'un étage où des lumières brillaient derrière des rideaux étriqués. Profitant de la tempête de neige, il réussit à s'en approcher sans se faire voir de la sentinelle et à regarder à l'intérieur : des officiers allemands. C'était sûrement le quartier général du bataillon.

À voix basse, il donna ses instructions à ses sergents. Ils tirèrent dans les fenêtres avant de jeter des grenades à l'intérieur du bâtiment. Un groupe d'Allemands en sortit, les mains sur la tête. Une minute plus tard, Volodia avait pris l'immeuble.

Un autre bruit lui parvint, un bruit étrange qui le laissa perplexe. Il tendit l'oreille. On aurait dit une foule à un match de football. Il sortit sur le perron. Le bruit, de plus en plus fort, venait de la ligne de front.

Une mitrailleuse crépita. À une centaine de mètres de là, dans la grand-rue, un camion dérapa, partit en diagonale et percuta un mur de brique avant d'exploser, atteint sans doute par la DP-28 de la section que Volodia avait mise en place. Les deux véhicules qui le suivaient immédiatement parvinrent à s'esquiver.

Volodia se précipita vers le café. La mitrailleuse, en appui sur son bipied, était juchée sur une table. Cette arme était surnommée « tourne-disque » en raison du chargeur circulaire placé au-dessus du canon. Les hommes rigolaient. « Facile comme bonjour, camarade ! C'est comme de tirer les pigeons dans la cour. »

L'un d'eux avait fait une razzia dans la cuisine et découvert un grand bidon de crème glacée miraculeusement intact et, maintenant, ils l'engloutissaient à tour de rôle.

Volodia regarda par la fenêtre brisée du café. Un autre véhicule arrivait, une Jeep, lui sembla-t-il. Derrière, un groupe d'hommes qui couraient. Lorsqu'ils s'approchèrent, il reconnut des uniformes allemands. D'autres suivaient, des dizaines, des centaines peut-être. C'était donc eux qui faisaient ce bruit de match de football !

Le tireur pointa son canon sur la voiture. « Attends », ordonna Volodia en posant la main sur son épaule.

À travers la neige qui tombait sans relâche, il scruta la rue attentivement, au point d'en avoir les yeux qui piquaient. Tout ce qu'il voyait, c'était encore et encore des véhicules et un nombre toujours croissant d'hommes qui couraient. Quelques chevaux aussi.

Un soldat épaula. « Ne tire pas », fit Volodia. La foule se rapprochait. « On ne va pas pouvoir arrêter ce flot, on serait immédiatement submergé, expliqua-t-il. On va les laisser passer. Mettez-vous à couvert. » Les hommes se couchèrent au sol. Le tireur descendit le DP-28 de la table. Volodia s'assit par terre et jeta un coup d'œil au-dessus du rebord de fenêtre.

Le bruit s'était transformé en rugissement. Les hommes de tête arrivèrent à la hauteur du café et le dépassèrent. Ils couraient, trébuchant et boitant. Certains portaient des fusils, la plupart semblaient avoir perdu leurs armes ; quelques-uns avaient des capotes et des calots, d'autres ne portaient que leur tunique d'uniforme. Beaucoup étaient blessés. Volodia vit un homme à la tête bandée tomber, ramper sur quelques mètres et s'affaler. Personne n'y prêta attention. Un cavalier piétina un fantassin et poursuivit sa route au galop. Jeep et voitures de l'armée se frayaient dangereusement un chemin à travers la foule, dérapaient sur le verglas, klaxonnaient comme des forcenés, obligeant les soldats à pied à se rabattre sur les côtés.

C'était la déroute, comprit Volodia. Par milliers, des hommes passaient devant eux dans une fuite éperdue. Une vraie débandade.

Enfin, les Allemands battaient en retraite.

XI
1941 (IV)

1.

C'est à bord d'un hydravion Boeing B-314 de la Pan Am que Woody Dewar et Joanne Rouzrokh firent le voyage d'Oakland, en Californie, à Honolulu. Quatorze heures de vol.

Juste avant d'arriver, ils eurent une sérieuse dispute. Peut-être était-ce d'être restés aussi longtemps dans un espace confiné. Ce navire volant avait beau être l'un des plus grands avions au monde, les passagers y étaient répartis dans six petites cabines de deux rangées de quatre sièges se faisant face. « Je préfère le train », déclara Woody, en croisant maladroitement ses longues jambes, et Joanne eut la grâce de ne pas lui faire remarquer qu'on ne pouvait pas se rendre à Hawaï en chemin de fer.

Ce voyage était une idée des parents de Woody. Ils avaient décidé d'aller en vacances à Hawaï pour voir Chuck, leur fils cadet, stationné là-bas. Et ils avaient invité Woody et Joanne à les rejoindre pour la deuxième semaine.

Woody et Joanne étaient fiancés. Woody avait demandé Joanne en mariage à la fin de l'été, après quatre semaines d'amour passionné dans la chaleur de Washington. Joanne avait répondu que c'était trop tôt, mais Woody lui avait fait valoir qu'il était amoureux d'elle depuis six ans : n'avait-il pas suffisamment attendu ? Elle s'était inclinée. Ils devaient se marier au mois de juin suivant, dès que Woody serait diplômé d'Harvard. Entre-temps, ce statut de fiancés les autorisait à prendre des vacances ensemble en famille.

Elle l'appelait Woods, il l'appelait Jo.

En arrivant en vue d'Oahu, l'île principale, l'avion amorça sa descente. On voyait déjà des montagnes boisées, quelques villages éparpillés dans les plaines, une bande de sable et des vagues. « Je me suis acheté un nouveau maillot de bain », déclara Joanne. Ils étaient assis côte à côte, et les quatre moteurs à cylindres Wright Twin Cyclone 14 étaient si bruyants que leurs voisins ne risquaient pas de les entendre.

Woody, qui était plongé dans *Les Raisins de la colère*, reposa son livre. « Je meurs d'envie de te voir dedans », dit-il avec conviction. Avec sa silhouette parfaite, une fille comme elle était un rêve pour un fabricant de maillots de bain.

Elle lui jeta un coup d'œil sous ses paupières mi-closes. « Tu crois que tes parents nous auront réservé des chambres voisines à l'hôtel ? » demanda-t-elle, lui lançant un regard aguichant de ses yeux brun foncé.

Ce statut de fiancés ne les autorisait pas à passer la nuit ensemble, du moins pas officiellement, même si la mère de Woody, avec sa perspicacité habituelle, se doutait bien qu'ils étaient amants.

Woody déclara : « Je saurai te retrouver où que tu sois.

— Tu as intérêt !

— Ne me provoque pas, je t'en prie. Ce siège est déjà suffisamment inconfortable. »

Elle sourit de plaisir.

La base navale américaine était en vue. Une vaste lagune en forme de feuille de palmier offrait un abri naturel à une centaine de navires, la moitié de la flotte du Pacifique. Vus d'en haut, les réservoirs de carburant alignés sur le quai ressemblaient à des pions sur un échiquier.

Au milieu de cette rade, on apercevait une île équipée d'une piste d'atterrissage et, tout au bout à l'ouest, une bonne dizaine d'hydravions à l'amarre.

La base aérienne de Hickam se trouvait juste à droite de la lagune. Plusieurs centaines d'avions y étaient rangés sur le tarmac, aile contre aile, avec une précision toute militaire.

Prenant son virage sur l'aile, l'avion survola une plage plantée de palmiers et de gais parasols rayés – la plage de Waikiki, supposa Woody –, puis une petite ville qui devait être Honolulu, la capitale.

Joanne avait obtenu du Département d'État un congé qui lui était dû depuis longtemps, alors que Woody avait dû manquer une semaine de cours pour prendre ces vacances. « Je suis un peu surprise que ton père nous ait proposé de l'accompagner, déclara Joanne. En général, il n'est pas favorable à ce qui risque d'interrompre tes études.

— C'est vrai. Mais tu sais la vraie raison de ce voyage, Jo ? Il pense que c'est peut-être la dernière fois que nous verrons Chuck vivant.

— Oh, mon Dieu, vraiment ?

— Il est persuadé qu'il va y avoir la guerre. Et comme Chuck est dans la marine…

— Il a sûrement raison. La guerre est inévitable.

— D'où tires-tu cette certitude ?

— Le monde entier est hostile à la liberté. » Elle désigna le livre posé sur ses genoux, un best-seller intitulé *Journal de Berlin* dont l'auteur était un journaliste de radio du nom de William Shirer. « Les nazis tiennent l'Europe, les bolcheviks la Russie et maintenant, les Japonais sont en train de mettre la main sur l'Extrême-Orient. Je ne vois pas comment l'Amérique peut survivre dans un monde pareil. Il faut bien que nous ayons des partenaires commerciaux !

— Mon père pense à peu près la même chose. Il croit que nous entrerons en guerre contre le Japon l'année prochaine… Et en Russie, tu as des informations sur la manière dont ça se passe ? reprit Woody sur un ton pensif.

— Les Allemands semblent avoir du mal à prendre Moscou. Juste avant mon départ, on parlait d'une contre-offensive russe massive.

— Bonne nouvelle ! »

Regardant au-dehors, Woody aperçut l'aéroport d'Honolulu et supposa que l'hydravion allait se poser dans une crique protégée, parallèle à la piste.

« J'espère qu'il ne va rien se passer de grave pendant que je suis ici, murmura Joanne.

— Pourquoi ?

— Je brigue une promotion, Woods ! Alors il ne faudrait pas qu'une personne brillante et prometteuse profite de mon absence pour m'éclipser.

— De l'avancement ? Tu ne m'avais rien dit.

— Non, je ne t'en ai pas encore parlé, mais je vise le poste d'attachée de recherche. »

Il sourit. « Et tu veux monter jusqu'où comme ça ?

— Je me verrais bien ambassadrice, un jour. Dans un endroit compliqué et intéressant, comme Nankin ou Addis-Abeba.

— Vraiment ?

— Ne fais pas cette tête ! Frances Perkins est la première femme ministre du Travail, et elle fait un boulot du tonnerre. »

Woody hocha la tête. Perkins occupait ce poste depuis le début de la présidence Roosevelt, huit ans auparavant. Elle avait réussi à obtenir le soutien des syndicats pour le New Deal. À l'heure actuelle, une femme exceptionnelle pouvait aspirer à exercer presque n'importe quel métier, et Joanne était vraiment exceptionnelle. Pourtant, il était troublé de lui découvrir autant d'ambition. « Un ambassadeur, ça doit vivre à l'étranger, lui fit-il remarquer.

— Génial, non ? Une culture étrangère, un climat différent, des coutumes exotiques…

— Mais… comment tu concilies ça avec le mariage ?

— Pardon ? » demanda-t-elle durement.

Il haussa les épaules. « Ça me semble normal de poser la question, non ? »

L'expression de Joanne ne changea pas. Seules ses narines frémirent, un signe de colère imminente que Woody connaissait bien. « Est-ce que je me suis permis de te poser la question, à toi ?

— Non, mais…

— Eh bien ?

— Je m'interroge, Jo, c'est tout. Est-ce que tu t'attends à ce que je te suive où te conduira ta carrière ?

— J'essaierai de la faire concorder avec tes impératifs. Et il me paraîtrait normal que tu en fasses autant de ton côté.

— Ce n'est pas pareil.

— Ah bon ? Première nouvelle ! »

Cette fois, Joanne était vraiment fâchée. Woody se demanda comment la conversation avait pu dégénérer aussi vite. Se forçant à prendre un ton à la fois conciliant et raisonnable, il déclara : « On avait parlé d'avoir des enfants, non ?

— Tu en auras, tout comme moi.

— Pas exactement de la même manière.

— Si avoir des enfants doit faire de moi une citoyenne de seconde classe dans notre vie conjugale, dans ce cas, nous n'en aurons pas !

— Ce n'est pas ce que je veux dire !

— Qu'est-ce que tu veux dire, alors ?

— Si tu es nommée ambassadrice, est-ce que tu t'attends à ce que je laisse tout tomber pour te suivre je ne sais où ?

— Je m'attends à ce que tu me dises : "Ma chérie, c'est une merveilleuse chance pour toi. Tu peux compter sur moi pour ne pas te mettre des bâtons dans les roues." Est-ce déraisonnable ?

— Oui ! lâcha Woody, déconcerté et furieux. À quoi bon se marier, si ce n'est pas pour vivre ensemble ?

— Est-ce que tu t'engageras, si la guerre éclate ?

— Probablement, oui.

— Et l'armée t'enverra là où elle aura besoin de toi, en Europe, en Extrême-Orient ou ailleurs.

— Oui, évidemment.

— Et tu me laisseras à la maison pour suivre l'appel du devoir.

— S'il le faut, oui.

— Alors que moi, je ne peux pas en faire autant !

— Ça n'a rien à voir, voyons ! Tu t'en rends bien compte, non ?

— Figure-toi que ma carrière et ma volonté de servir mon pays comptent énormément pour moi. Tout autant que pour toi.

— Quelle mauvaise foi !

— Woods, je suis vraiment désolée que tu le prennes comme ça. Ce que je viens de te dire sur notre avenir commun était extrêmement sérieux, et je ne peux que m'interroger à présent sur la réalité de cet avenir.

— Comment peux-tu dire ça ! » répondit Woody. Il en aurait hurlé d'exaspération. « Mais comment en est-on arrivés là ? Comment ? »

Il y eut une secousse, des gerbes d'eau. L'hydravion s'était posé à Hawaï.

2.

Chuck Dewar était terrifié à l'idée que ses parents découvrent son secret.

À Buffalo, il n'avait jamais eu de véritable liaison, uniquement de brèves étreintes dans des ruelles obscures avec des garçons qu'il connaissait à peine. C'était une des raisons pour lesquelles il avait choisi la marine : pour vivre dans un milieu où il pourrait être lui-même sans que ses parents le sachent.

À Hawaï, tout avait été différent dès son arrivée. Ici, il faisait partie d'une communauté clandestine de gens comme lui. Dans les bars, les restaurants et les dancings où il allait, il n'avait pas besoin de faire semblant d'être hétérosexuel. Il avait eu plusieurs aventures et puis, il était tombé amoureux. Aujourd'hui, son secret n'en était plus un pour quantité de gens.

Et maintenant, ses parents étaient là.

Son père avait été convié à visiter la station HYPO, l'unité de Renseignement radio de la base navale. En sa qualité de membre de la commission des Affaires étrangères du Sénat, Gus Dewar était informé de nombreux secrets militaires. À Washington, il avait déjà visité l'Op-20-G, l'état-major du Renseignement radio.

Chuck alla le chercher à son hôtel d'Honolulu dans une limousine de la marine, une Packard LeBaron. Son père arborait un chapeau de paille blanche. Chuck lui fit faire le tour du port en voiture. « La flotte du Pacifique ! Quel spectacle magnifique, s'écria Gus avec un sifflement admiratif.

— Impressionnant, n'est-ce pas ? » acquiesça Chuck. De fait, les navires étaient magnifiques, surtout ceux de l'US Navy, qui, repeints et briqués, brillaient de tout leur éclat.

« Tous ces cuirassés parfaitement alignés ! s'émerveilla Gus.

— On appelle cette rangée de bâtiments le "Battleship Row", l'allée des cuirassés. Il y a le *Maryland*, le *Tennessee*, l'*Arizona*, le *Nevada*, l'*Oklahoma* et le *West Virginia*. » Les navires portaient le nom de différents États américains. « Le *California* et le *Pennsylvania* sont aussi au port, mais on ne les voit pas d'ici. »

En reconnaissant la voiture officielle, le fusilier marin en faction à l'entrée principale du bassin leur fit signe de passer. Ils rejoignirent la base des sous-marins et s'arrêtèrent sur le parking derrière le quartier général, l'Old Administration Building. Chuck conduisit son père dans la nouvelle aile qui venait d'être inaugurée.

Le capitaine Vandermeier les attendait.

C'était l'homme que le jeune marin redoutait le plus. Il avait pris Chuck en aversion et deviné son secret. Depuis, il le traitait à longueur de temps de gonzesse ou de tantouse. Pour peu qu'il en ait l'occasion, il se ferait un plaisir de lâcher le morceau devant son père.

Vandermeier était un petit homme trapu à la voix rocailleuse et à l'haleine fétide. Il fit le salut militaire avant de serrer la main de Gus. « Bienvenue, monsieur le sénateur. C'est un privilège pour moi de vous faire visiter l'unité de Renseignement des communications du 14e secteur de la marine. » Tel était le nom délibérément vague attribué à la cellule chargée de surveiller les signaux radio émis par la marine impériale japonaise.

« Merci, capitaine, répondit Gus.

— Tout d'abord un mot d'avertissement, monsieur. Il s'agit d'un groupe informel. Ce type de travail est souvent exécuté par des individus quelque peu excentriques réfractaires au port de l'uniforme. À en juger par sa veste en velours rouge, précisa Vandermeier avec un sourire de connivence à l'adresse du sénateur, vous pourriez penser que l'officier responsable, le commandant Rochefort, est un foutu homo. »

Chuck réprima de son mieux une grimace.

« Je n'en dirai pas plus jusqu'à ce que nous soyons à l'intérieur de la zone sécurisée, poursuivit Vandermeier.

— Très bien », approuva Gus.

Ils empruntèrent un escalier pour rejoindre le sous-sol où ils franchirent deux portes verrouillées.

La station HYPO comptait trente hommes. Elle occupait une cave sans fenêtres, éclairée au néon. En plus des bureaux et des chaises habituels, il y avait d'immenses comptoirs où déployer des cartes, des rangées d'incroyables machines IBM – imprimantes, trieuses et assembleuses –, ainsi que deux lits de camp sur lesquels les analystes pouvaient prendre un moment de repos pendant leurs marathons de séances de décryptage. Certains hommes portaient des uniformes soignés, mais d'autres, en civil comme l'avait annoncé Vandermeier, affichaient une tenue débraillée et n'étaient ni rasés ni lavés, à en juger par l'odeur.

« Comme toutes les marines du monde, la marine japonaise utilise différents codes. Elle réserve les plus simples aux messages les moins importants, tels que les rapports météorologiques, et les plus complexes aux messages à caractère sensible, expliqua Vandermeier. Ainsi, les signaux d'appel indiquant l'identité de l'expéditeur et du destinataire sont chiffrés à l'aide d'un code simple, alors que le texte lui-même peut être rédigé dans un code plus complexe. Celui des signaux d'appel a changé récemment, mais nous avons réussi à décrypter le nouveau en l'espace de quelques jours.

— Vous m'impressionnez, déclara Gus.

— Nous pouvons également déterminer la provenance d'un signal par triangulation. En nous fondant sur le lieu de l'émission et sur les signaux d'appels, nous pouvons nous faire une idée relativement précise de la position de la plupart des bâtiments, même si nous ne sommes pas capables de déchiffrer le contenu des messages.

— Autrement dit, nous savons où se trouvent les navires et dans quelle direction ils se déplacent, mais nous ignorons tout des ordres qu'ils ont reçus, résuma Gus.

— Oui, bien souvent.

— Mais s'ils voulaient passer inaperçus, il leur suffirait d'imposer le silence radio.

— C'est exact, admit Vandermeier. S'ils se taisent, notre travail perd toute utilité et on l'a vraiment dans l'os. »

Un homme en veste d'intérieur et en pantoufles s'approcha. Vandermeier le présenta comme le chef de l'unité. « Le commandant Rochefort n'est pas seulement un maître du décodage, il parle aussi japonais couramment.

— Jusqu'à ces derniers jours, nous avons fait des progrès appréciables dans le décryptage du chiffre le plus fréquemment utilisé, dit Rochefort. Mais ces salauds de Japonais viennent de le changer et tout est à refaire.

— Le capitaine Vandermeier me disait que vous pouviez déjà apprendre pas mal de choses sans vraiment lire les messages, fit remarquer Gus.

— C'est exact. » Rochefort désigna une carte murale. « À l'heure actuelle, la plus grande partie de la flotte japonaise a quitté les eaux territoriales et se dirige vers le sud.

— Mauvais signe.

— Ça, c'est sûr. Mais dites-moi, monsieur le sénateur, comment interprétez-vous les intentions du Japon ?

— Je crois qu'il va nous déclarer la guerre. À cause de l'embargo sur le pétrole qu'il subit de plein fouet. Comme les Britanniques et les Néerlandais refusent eux aussi de leur en fournir, ils essaient d'en faire venir d'Amérique du Sud. Mais ils ne pourront pas vivre comme ça indéfiniment.

— Que gagneraient-ils à nous attaquer ? intervint Vandermeier. Le Japon est un si petit pays ! Ils ne peuvent pas nous envahir !

— La Grande-Bretagne est un petit pays, elle aussi. Ça ne l'a pas empêchée de dominer le monde simplement en prenant le contrôle des mers, objecta Gus. Les Japonais n'ont pas besoin de conquérir l'Amérique, il leur suffira de nous infliger une défaite navale pour contrôler le Pacifique. Après, rien ne les empêchera de faire du commerce avec qui ils veulent.

— À votre avis, quel objectif poursuivent-ils, en se dirigeant vers le sud ?

— Les Philippines sont la cible la plus probable. »

Rochefort acquiesça de la tête. « Nous avons déjà renforcé notre base là-bas. Mais une chose me tracasse : voilà maintenant plusieurs jours que le commandant de leur flotte de porte-avions n'a pas reçu le moindre signal ! »

Gus se rembrunit. « Silence radio. Cela s'est-il déjà produit dans le passé ?

— Oui. En général, les porte-avions coupent toute communication quand ils regagnent leurs eaux territoriales. C'est sans doute ce qui se passe actuellement. »

Gus hocha la tête. « Cela me paraît fondé.

— Oui, dit Rochefort. Mais je préférerais en être certain. »

3.

À Honolulu, les guirlandes de Noël illuminaient Fort Street. En cette soirée du samedi 6 décembre, la rue grouillait de marins en tenue tropicale, uniforme blanc, casquette blanche et foulard croisé noir, bien décidés à s'offrir du bon temps.

La famille Dewar déambulait au milieu de cette joyeuse cohue, Rosa au bras de Chuck, Joanne entre Gus et Woody.

Les fiancés avaient oublié leur brouille. Woody s'était excusé de s'être fait de fausses idées sur la manière dont Joanne envisageait le mariage ; Joanne, de son côté, avait reconnu qu'elle s'était emportée. Rien n'avait vraiment été réglé, mais ils s'étaient suffisamment réconciliés pour s'arracher mutuellement leurs vêtements et se précipiter au lit.

À la suite de quoi, leur querelle avait perdu de sa gravité. Une seule chose comptait : ils s'aimaient. Ils s'étaient promis de discuter à l'avenir de ces problèmes avec tendresse et tolérance. Tandis qu'ils s'habillaient, Woody avait eu le sentiment d'avoir franchi une étape. Oui, ils s'étaient disputés, ils avaient eu une divergence d'opinion importante, mais ils avaient su passer l'éponge. C'était plutôt bon signe.

Maintenant, ils allaient tous dîner au restaurant. Woody, qui avait emporté son appareil photo, prenait des clichés tout en marchant. Ils n'avaient fait que quelques pas quand Chuck s'arrêta pour présenter un autre marin à sa famille. « Eddie Parry, un bon copain à moi. Eddie, je te présente le sénateur Dewar, Mrs. Dewar, mon frère Woody et miss Joanne Rouzrokh, sa fiancée.

— Ravie de faire votre connaissance, Eddie, dit Rosa. Chuck nous a parlé de vous dans ses lettres. Voulez-vous vous joindre à nous ? Un dîner tout simple, chez un Chinois. »

Woody s'étonna. Ce n'était pas le genre de sa mère d'inviter un inconnu à partager un repas de famille.

« Je vous remercie, madame. J'en serai très honoré », répondit Eddie. Il avait l'accent du Sud.

Ils entrèrent au restaurant Délices célestes et prirent place autour d'une table pour six. Eddie, un peu guindé, donnait à Gus du « monsieur » et du « madame » à Rosa, mais paraissait cependant plutôt à l'aise. Après qu'ils eurent passé la commande, il déclara : « J'ai tellement entendu parler de votre famille que j'ai presque l'impression de vous connaître tous. » Il avait des taches de rousseur et un grand sourire. Woody remarqua qu'il attirait tout de suite la sympathie.

Eddie interrogea Rosa sur sa première impression d'Hawaï. « Pour ne rien vous cacher, je suis un peu déçue, répondit-elle. J'ai l'impression de me trouver dans n'importe quelle petite ville américaine. Je m'attendais à découvrir quelque chose de plus asiatique.

— Je suis d'accord avec vous, déclara Eddie. Ça pullule de snacks, de motels et d'orchestres de jazz. »

Il demanda à Gus s'il allait y avoir la guerre. « Nous avons essayé par tous les moyens possibles et imaginables de trouver un modus vivendi acceptable avec le Japon, répondit le sénateur, et Woody se demanda si Eddie savait ce qu'était un modus vivendi. Au cours de l'été, poursuivit son père, le secrétaire d'État Cordell Hull a mené toute une série d'entretiens avec l'ambassadeur Nomura. Apparemment, sans résultat.

— Sur quoi butent les négociations ? s'enquit Eddie.

— Le commerce américain a besoin d'une zone de libre-échange en Extrême-Orient. Le Japon dit : d'accord, très bien, nous aussi nous aimons le libre-échange. Instaurons-le, pas seulement chez nous, mais dans le monde entier. Les États-Unis ne peuvent pas prendre cet engagement, quand bien même le voudraient-ils. Du coup, le Japon déclare : tant que d'autres pays auront leur propre zone économique, il nous en faudra une, à nous aussi.

— Je ne vois toujours pas pourquoi ils ont voulu envahir la Chine. »

Rosa, qui s'efforçait toujours d'être impartiale, déclara : « Les Japonais veulent avoir des troupes en Chine, en Indochine et dans les Indes néerlandaises pour protéger leurs intérêts, tout comme nous, les Américains, avons des troupes aux Philippines. Les Britanniques en ont en Inde, les Français en Algérie, et ainsi de suite.

— Vues sous cet angle, les exigences japonaises ne paraissent pas aussi déraisonnables !

— Elles ne sont pas déraisonnables, répliqua Joanne fermement, elles n'ont pas lieu d'être, voilà tout. Conquérir un empire est une solution du XIXe siècle. Le monde change. Le temps des empires et des zones économiques fermées est révolu. Céder aux Japonais, ce serait faire un pas en arrière. »

On apporta les plats. « Avant que j'oublie, intervint Gus : demain matin, petit déjeuner sur l'*Arizona*. À huit heures précises.

— Je ne suis pas invité, dit Chuck, mais j'ai été détaché pour vous conduire à bord. Je passerai vous chercher à l'hôtel à sept heures et demie et je vous conduirai à l'arsenal de la marine. Là, nous prendrons une vedette.

— Très bien. »

Woody plongea sa fourchette dans le riz frit. « C'est un délice. On devrait prévoir un repas chinois pour notre mariage. »

Gus se mit à rire. « Je ne suis pas sûr que ce soit une bonne idée.

— Pourquoi ? Ce n'est pas cher, et c'est excellent.

— Un mariage ne se résume pas à un repas, c'est une célébration. À ce sujet, Joanne, il faut que j'appelle votre mère.

— À propos du mariage ? demanda Joanne en fronçant les sourcils.

— À propos de la liste des invités. »

Joanne reposa ses baguettes. « Il y a un problème ? »

Au frémissement de ses narines, Woody comprit que l'orage menaçait.

« Oh, ce n'est pas vraiment un problème, répondit Gus. Il se trouve que j'ai un assez grand nombre d'amis et d'alliés politiques à Washington qui seraient vexés de ne pas être invités au mariage de mon fils. Je voudrais proposer à votre mère de partager les frais. »

Woody trouva que c'était délicat de la part de son père. La mère de Joanne n'avait peut-être pas les moyens d'offrir à sa fille un mariage princier. En effet, Dave, son mari, avait vendu son entreprise pour une bouchée de pain avant de mourir.

Mais l'idée que les parents puissent s'entendre sur l'organisation de son mariage sans lui demander son avis n'était pas du goût de Joanne. Et c'est sur un ton glacial qu'elle reprit : « Quand vous parlez d'amis et d'alliés, à qui pensez-vous au juste ?

— Principalement à des sénateurs et à des membres du Congrès. Il faudra inviter le Président, mais il ne viendra pas.

— Quels sénateurs et quels députés exactement ? »

Woody surprit le sourire discret de sa mère : visiblement, l'insistance de Joanne l'amusait. Peu de gens avaient le front de pousser Gus dans ses retranchements.

Le sénateur commença à dévider une liste de noms.

Joanne lui coupa la parole. « Cobb ? J'ai bien entendu ?

— Oui.

— Il a voté contre la loi antilynchage !

— C'est un type bien. Mais c'est un politicien du Mississippi. Nous vivons dans une démocratie, Joanne : nous devons représenter nos électeurs. Les gens du Sud ne soutiendront jamais une loi antilynchage… Sans vouloir vous offenser, Eddie, ajouta-t-il en se tournant vers l'ami de Chuck.

— Ne vous croyez pas obligé de prendre des gants, monsieur, répondit le marin. J'ai beau être texan, j'ai honte quand je pense à la politique en vigueur dans le Sud. Je déteste les préjugés. Un homme est un homme, quelle que soit sa couleur. »

Woody regarda son frère : Chuck se rengorgeait, très fier d'Eddie. En cet instant, il comprit qu'Eddie était plus qu'un ami pour lui.

Que c'était étrange ! Trois couples étaient réunis autour de cette table : son père et sa mère, Joanne et lui, mais aussi Chuck et Eddie.

Il regarda fixement le jeune homme. L'amant de Chuck, se dit-il.

Bizarre, quand même.

Eddie soutint son regard en souriant aimablement.

Woody détourna les yeux. Dieu merci, Papa et Mama ne se doutaient de rien !

Quoique… pour quelle raison Mama avait-elle invité Eddie à ce dîner en famille ? Savait-elle quelque chose ? Approuvait-elle cette liaison ? Non, c'était impensable !

« Quoi qu'il en soit, Cobb n'avait pas le choix, insistait Gus. Dans tous les autres domaines en revanche, c'est un libéral.

— Il n'y a rien de démocratique là-dedans ! lança Joanne avec véhémence. Cobb ne représente pas tous les gens du Sud, puisque seuls les Blancs ont le droit de vote là-bas.

— Rien n'est parfait sur cette terre, répliqua Gus. Cobb a soutenu le New Deal de Roosevelt.

— Ce n'est pas une raison pour que je l'invite à mon mariage. »

Woody s'interposa : « Papa, moi non plus je ne souhaite pas qu'il vienne, tu sais. Il a du sang sur les mains.

— C'est injuste.

— C'est comme ça que nous voyons les choses.

— Eh bien, la décision ne dépend pas entièrement de vous. C'est la mère de Joanne qui donne cette réception et j'en partagerai les frais si elle m'y autorise. Je suppose que cela nous permet d'avoir notre mot à dire sur la liste des invités ! »

Woody se cala dans son fauteuil. « Zut, c'est quand même nous qui nous marions ! »

Joanne se tourna vers Woody. « Nous devrions peut-être opter pour un mariage tout simple à la mairie, avec quelques amis seulement. »

Woody haussa les épaules. « Ça me convient parfaitement.

— Cela choquerait beaucoup de gens ! réagit Gus d'un ton sévère.

— Pas nous ! répliqua Woody. La personne la plus importante ce jour-là, c'est bien la mariée. Je veux que tout se déroule comme elle l'entend.

— C'est bon, vous tous, écoutez-moi ! intervint Rosa. Ne nous emballons pas. Gus, mon chéri, tu peux dire un mot à Peter Cobb, lui expliquer, délicatement, que tu as la chance d'avoir un fils idéaliste, qui va épouser une fille merveilleuse, tout aussi idéaliste, et que les jeunes mariés refusent obstinément malgré ton insistance que le député Cobb soit de la noce. Dis-lui que tu en es vraiment désolé mais que, sur ce sujet, tu ne peux pas davantage suivre tes propres inclinations que lui-même n'a pu suivre les siennes lors du vote sur le décret antilynchage. Il te répondra avec le sourire qu'il comprend très bien et qu'il a toujours apprécié ton franc-parler. »

Gus resta muet un long moment avant de s'avouer vaincu. Ce qu'il fit avec grâce. « Je suppose que tu as raison, ma chère. » Il ajouta avec un sourire à l'adresse de Joanne : « Je serais le dernier des imbéciles de me quereller avec ma délicieuse belle-fille à cause de Peter Cobb !

— Je vous remercie… Puis-je vous appeler Papa dès à présent ? »

Woody en fut ébahi. Quelle fille intelligente ! Elle avait trouvé exactement le mot à dire !

« Cela me ferait le plus grand plaisir », répondit Gus. Et Woody crut voir briller une larme dans l'œil de son père.

« Eh bien, merci, Papa. »

Woody n'en revenait pas. Joanne avait tenu tête à son père, et avait eu gain de cause. Quelle fille épatante !

4.

Le dimanche matin, Eddie voulut accompagner Chuck pour chercher le reste de la famille à l'hôtel.

« Je ne sais pas si c'est une bonne idée, mon chou, déclara Chuck. On est censés être amis, toi et moi. Pas inséparables. »

Ils étaient au lit dans un motel, et il faisait encore nuit noire. Ils devaient rentrer à la caserne en catimini, avant le lever du soleil.

« Tu as honte de moi, répliqua Eddie.

— Comment peux-tu dire ça ? Tu es venu dîner avec ma famille !

— Ce n'est pas toi qui en as eu l'idée, c'est ta mère. Mais j'ai l'impression d'avoir fait bon effet à ton père aussi, je me trompe ?

— Tout le monde t'a adoré. Comment pourrait-il en être autrement ? Mais ils ne savent pas que tu es un sale homo.

— Je ne suis pas un sale homo, je suis un homo très propre.

— C'est vrai.

— S'il te plaît, emmène-moi. J'ai envie de mieux les connaître. C'est vraiment important pour moi. »

Chuck soupira. « D'accord.

— Merci. » Eddie l'embrassa. « Est-ce qu'on a le temps…

— À condition de faire vite », répondit Chuck avec un grand sourire.

Deux heures plus tard, ils attendaient devant l'hôtel dans la Packard de la marine. À sept heures et demie tapantes, les quatre passagers firent leur apparition. Rosa et Joanne étaient gantées et chapeautées, Gus et Woody étaient en costume de lin blanc.

Woody, son appareil photo autour du cou, tenait Joanne par la main.

« Regarde mon frère, murmura Chuck à Eddie. Regarde comme il est heureux !

— C'est une bien jolie fille. »

Ils leur tinrent les portières ouvertes. Les Dewar montèrent à l'arrière de la limousine. Woody et Joanne rabattirent les strapontins. Chuck démarra et prit la direction de la base navale.

Il faisait un temps splendide. La KGMB, la station de radio d'Hawaï, diffusait des cantiques. Le soleil brillait sur la lagune et se reflétait sur les vitres des hublots et sur les cuivres étincelants d'une centaine de navires. Chuck s'exclama : « N'est-ce pas une vue magnifique ? »

Ils pénétrèrent dans l'enceinte de la base et roulèrent jusqu'à l'arsenal. Une dizaine de navires étaient à quai ou en cale sèche pour être réparés, entretenus ou réapprovisionnés en carburant. Chuck s'arrêta devant le débarcadère des officiers. Tout le monde descendit de voiture et admira les fiers et puissants cuirassés de l'autre côté de la lagune, dans la lumière du matin. Woody les prit en photo.

Il était presque huit heures. Chuck entendit tinter les cloches des églises voisines, dans la ville de Pearl. À bord des navires, un coup de sifflet donna le signal du petit déjeuner aux hommes du quart de la matinée. Les équipes chargées d'envoyer les couleurs se rassemblaient, attendant qu'il soit huit heures précises pour hisser les pavillons. Sur le pont du *Nevada*, un orchestre jouait le « Star-Spangled Banner ».

Ils se dirigèrent vers la jetée où les attendait une vedette amarrée. Elle pouvait transporter une douzaine de passagers. Eddie lança le moteur, qui se trouvait à la poupe, sous une trappe, tandis que Chuck donnait la main aux invités pour les aider à monter. Le petit moteur glouglouta gaiement. Chuck alla à l'avant. Eddie, resté à l'arrière, libéra les amarres et dirigea l'embarcation vers les cuirassés. La proue s'éleva sous l'effet de l'accélération et bientôt la vedette fila sur les flots, laissant derrière elle deux courbes d'écume jumelles, semblables aux ailes d'une mouette.

Un avion vrombit dans le ciel. Chuck leva les yeux. L'appareil arrivait de l'ouest et volait si bas qu'on aurait dit qu'il allait s'écraser. Il devait être sur le point de se poser sur la piste de la marine sur Ford Island.

Assis à l'avant à côté de Chuck, Woody demanda en plissant les yeux : « C'est quoi, cet avion ? »

Chuck, qui connaissait tous les appareils de l'armée de l'air et de la marine, avait du mal à identifier celui-ci. « On dirait presque un Type 97 », dit-il. C'étaient les bombardiers-torpilleurs embarqués sur les porte-avions japonais.

Woody leva son appareil photo.

L'avion se rapprochait, Chuck aperçut de grands soleils rouges peints sous les ailes. « Un avion japonais ! »

Eddie, qui barrait à la poupe, l'entendit. « Un simulacre d'attaque, probablement, une simple manœuvre. Une bonne blague pour enquiquiner tout le monde un dimanche matin.

— Tu dois avoir raison », renchérit Chuck.

Mais un second appareil surgit dans le sillage du premier.

Puis un troisième.

Il entendit son père s'exclamer d'une voix anxieuse : « Mais qu'est-ce qui se passe, nom d'un chien ? »

Les avions amorcèrent leur virage à hauteur de l'arsenal et, dans un rugissement qui évoquait les chutes du Niagara, survolèrent la vedette à basse altitude. Chuck en dénombra au moins dix, non, vingt, non, plus même.

Vingt appareils qui fonçaient droit sur Battleship Row.

Woody qui continuait à photographier s'interrompit pour crier : « Ils ne sont pas en train de nous attaquer pour de bon, quand même ?! » Sa voix exprimait autant d'inquiétude que d'étonnement.

« Des avions japonais, c'est impossible ! lâcha Chuck interloqué. Le Japon est à plus de six mille kilomètres d'ici ! Aucun avion ne peut franchir une telle distance. »

Il se souvint alors du silence radio des porte-avions japonais depuis plusieurs jours. L'unité de Renseignement avait supposé qu'ils avaient regagné leurs eaux territoriales, mais n'avait pas pu le confirmer.

Il croisa le regard de son père et comprit que celui-ci avait eu la même idée.

Subitement, tout lui parut d'une clarté limpide. Son incrédulité se changea en peur.

L'avion de tête était en train de passer en rase-mottes au-dessus du *Nevada*, le premier bâtiment de Battleship Row. Un bruit de canon retentit. Sur le pont, les marins se dispersèrent, l'orchestre s'éparpilla dans un diminuendo de notes éperdues.

À bord de la vedette, Rosa poussa un cri.

« Dieu du ciel, s'écria Eddie, c'est vraiment une attaque. »

Chuck crut que son cœur s'arrêtait de battre : les Japonais bombardaient Pearl Harbor, et il se trouvait dans une coque de noix au beau milieu de la rade. En voyant les visages terrifiés de ses parents, de son frère et d'Eddie, il prit conscience que tous les êtres qu'il aimait sur terre se trouvaient dans la vedette avec lui.

De longues torpilles commencèrent à pleuvoir des entrailles des avions, transformant les eaux tranquilles de la lagune en geysers.

Chuck hurla : « Eddie ! Fais demi-tour ! » Celui-ci s'y employait déjà et le bateau prit un virage serré.

Au cours de ce mouvement tournant, Chuck aperçut un autre groupe d'avions aux ailes décorées de disques rouges au-dessus de la base aérienne de Hickham. Ces bombardiers fondaient tels des oiseaux de proie sur les avions américains parfaitement alignés sur les pistes.

Combien y en avait-il dans le ciel, de ces salauds de Japonais ? À croire que la moitié de leur aviation s'était donné rendez-vous au-dessus de Pearl City !

Woody, pour sa part, continuait à prendre photo sur photo.

Chuck entendit un bruit sourd, comme une explosion souterraine, immédiatement suivi d'un autre. Il se retourna : une flamme jaillit du pont de l'*Arizona*, et tout de suite après, de la fumée s'éleva du navire.

Eddie mit les gaz, l'arrière de la vedette s'enfonça plus profondément dans l'eau. « Plus vite, plus vite ! » hurla Chuck bien inutilement.

À bord d'un des bâtiments, un hululement d'alerte insistant appelait les soldats du quartier général à rejoindre les postes de combat. Chuck prit conscience qu'il s'agissait bel et bien d'une bataille et que sa famille était prise au milieu des tirs. L'instant d'après, la sirène d'alarme du terrain d'aviation de Ford Island retentit, débutant par un gémissement sourd avant de monter progressivement dans les aigus jusqu'à atteindre la stridence de son point culminant.

Une longue série d'explosions retentit depuis Battleship Row : des torpilles avaient touché leurs cibles. « Le *Wee-Vee* ! hurla Eddie en désignant le *West Virginia*. Il gîte sur bâbord ! »

De fait, un énorme trou s'ouvrait dans le flanc du navire le plus exposé aux attaques des avions. Des millions de tonnes d'eau avaient dû s'y engouffrer en l'espace de quelques secondes, se dit Chuck, pour qu'un bateau aussi monumental se couche à ce point.

Juste à côté, l'*Oklahoma* avait subi le même sort et Chuck vit, horrifié, les marins impuissants déraper et glisser sur le pont de plus en plus incliné, avant de passer par-dessus bord.

Les vagues provoquées par les explosions malmenaient la vedette et tout le monde se cramponnait aux rebords.

Chuck vit une pluie de bombes s'abattre sur la base des hydravions, tout au bout de Ford Island. Ces fragiles appareils étaient amarrés tout près les uns des autres, et des fragments d'ailes et de fuselages déchiquetés se mirent à voler en l'air, semblables à des feuilles d'arbres prises dans un ouragan.

Formé à l'école du Renseignement, Chuck s'efforçait d'identifier les appareils japonais. Il venait de repérer un troisième modèle, le terrible « Zéro » Mitsubishi, le meilleur avion de chasse du monde qu'un porte-avions puisse accueillir. Il ne transportait que deux petites bombes, mais était équipé de mitrailleuses jumelles et de deux canons de 20 mm. Il devait être chargé d'escorter les bombardiers et de les défendre contre les chasseurs américains. Mais ceux-ci étaient encore tous au sol et un grand nombre d'entre eux avaient déjà été détruits. Ce qui laissait aux Zéro toute liberté pour mitrailler les navires, les installations et les troupes.

Pour ne rien dire de la famille perdue au milieu de la lagune et qui essayait désespérément de regagner le rivage, se dit Chuck terrifié.

Enfin, les Américains commencèrent à riposter. Sur Ford Island et sur les ponts des navires encore intacts, les canons antiaériens et les mitrailleuses entrèrent en action, ajoutant leur vacarme à la cacophonie meurtrière. Des obus antiaériens éclatèrent dans le ciel, telle une éclosion de fleurs noires. Presque immédiatement, un mitrailleur à terre atteignit un bombardier. Le cockpit s'enflamma et l'avion heurta l'eau dans un jaillissement d'écume. Involontairement, Chuck se mit à pousser des acclamations sauvages, brandissant les poings.

Le *West Virginia* qui gîtait revint peu à peu à l'horizontale, mais continua de s'enfoncer. Chuck se dit que le commandant avait dû ouvrir les vannes à tribord pour que le navire se redresse et que l'équipage puisse plus facilement s'en sortir. Hélas, l'*Oklahoma* ne connut pas cette chance. Stupéfaits et horrifiés, tous les passagers de la vedette regardèrent l'immense navire se retourner. « Oh, mon Dieu, murmura Joanne, l'équipage ! » Dans une tentative désespérée pour sauver leur peau, des marins s'efforcèrent d'escalader le pont désormais presque vertical, ou d'agripper le bastingage à tribord. C'étaient les plus chanceux, comprit Chuck en voyant le puissant navire se retourner, coque en l'air, dans un terrible fracas et se mettre à couler. Combien d'hommes étaient pris au piège sous les ponts ? Des centaines, assurément.

« Cramponnez-vous ! » cria-t-il en voyant déferler la vague gigantesque produite par le chavirement de l'*Oklahoma*. Gus attrapa Rosa, Woody saisit Joanne. La vague souleva la vedette à une hauteur vertigineuse. Chuck vacilla mais tint bon, agrippé au bastingage. Leur bateau se maintint à flot. Suivirent d'autres vagues moins violentes, qui ballottèrent les passagers en tous sens, mais personne ne fut blessé. Pour autant, ils étaient encore loin d'être en sécurité. Le rivage se trouvait à quatre cents mètres au moins, constata Chuck avec consternation.

Chose étonnante, le *Nevada*, mitraillé dans les premières minutes de l'attaque, commença à s'éloigner. Un officier devait avoir eu la présence d'esprit d'ordonner à tous les bâtiments de prendre le large. S'ils arrivaient à quitter la rade et à se disperser, ils ne seraient plus aussi faciles à atteindre.

Depuis Battleship Row retentit soudain une explosion dix fois plus assourdissante que toutes celles qu'ils avaient entendues jusque-là. Le souffle fut si violent que Chuck crut recevoir un coup de poing en pleine poitrine. Heureusement, le drame se déroulait à presque six cents mètres d'eux. Des flammes jaillirent de la tourelle numéro deux de l'*Arizona* et, une fraction de seconde plus tard, l'avant du navire éclata. Les débris volèrent dans les airs, des poutres d'acier tordues et déformées, des plaques de tôle à la dérive dans un brouillard de fumée se mirent à traverser le ciel avec une lenteur cauchemardesque, comme des morceaux de papier carbonisés échappés d'un feu de joie. L'avant du navire disparut au milieu des flammes et de la fumée, et le grand mât chavira, comme pris d'ivresse, pour s'abattre vers l'avant.

« C'était quoi, ça ? s'écria Woody.

— Le magasin de munitions du navire a dû sauter », expliqua Chuck, et il songea avec un profond chagrin que des centaines de ses compagnons marins avaient sans doute péri dans cette gigantesque explosion.

Une colonne de fumée rouge sombre s'éleva dans les airs au-dessus du bateau transformé en bûcher funéraire.

On entendit un violent craquement et la frêle embarcation fit une embardée comme si elle avait été heurtée par quelque chose. Tous les passagers rentrèrent la tête. Chuck, qui était tombé à genoux, crut qu'une bombe venait de les toucher. Mais non, ce n'était pas possible puisqu'il était toujours en vie. Ayant recouvré ses esprits, il vit qu'un énorme morceau de métal d'un mètre de long s'était fiché dans le bois du pont juste au-dessus du moteur. Un miracle que personne n'ait été atteint ! En revanche, le moteur était mort.

La vedette ralentit puis s'immobilisa, ballottée sur les vagues nées de la pluie d'enfer que les avions japonais continuaient de déverser sans relâche sur la rade.

« Chuck, on ne peut pas rester ici ! lâcha Woody d'une voix angoissée.

— Je sais. » Avec Eddie, il inspecta les dommages et ils tentèrent de déloger le morceau d'acier profondément enfoncé dans le pont en teck.

« Ne perdez pas votre temps avec ça ! protesta Gus.

— D'autant que le moteur est fichu, Chuck ! » renchérit Woody.

Ils étaient encore loin du rivage, mais la vedette était équipée pour les situations d'urgence. Chuck détacha une paire de pagaies, en garda une et tendit l'autre à Eddie.

Le bateau était trop large pour leur permettre de ramer avec efficacité et ils n'avançaient pas vite. Heureusement, une accalmie se produisit. Le ciel n'était plus noir d'avions, mais des colonnes de fumée s'élevaient toujours des navires endommagés. L'une d'elles, au-dessus de l'*Arizona* mortellement atteint, était monumentale. Elle s'élevait bien jusqu'à trois cents mètres de haut. Les explosions s'étaient interrompues et le *Nevada*, avec un courage remarquable, se dirigeait maintenant vers l'embouchure du port.

Autour des bâtiments, l'eau était couverte de radeaux de sauvetage, de chaloupes à moteur, de marins qui nageaient ou s'accrochaient à des morceaux d'épave. La noyade n'était pas le seul danger qui les menaçait. En effet, le gasoil qui s'échappait des flancs perforés des navires s'était enflammé et le feu se propageait à la surface de l'eau. Les hurlements effroyables des brûlés se mêlaient maintenant aux appels à l'aide de ceux qui ne savaient pas nager.

Chuck jeta un rapide coup d'œil à sa montre. Trente minutes à peine s'étaient écoulées, alors que l'attaque lui avait paru durer des heures.

À l'instant même où il se faisait cette remarque, débuta la deuxième offensive, venue de l'est cette fois.

Des avions prirent en chasse le *Nevada* ; d'autres choisirent pour cible l'arsenal de la marine, où les Dewar avaient embarqué. Presque aussitôt, le destroyer *Shaw* qui se trouvait dans un dock flottant explosa dans un jaillissement de flammes et de tourbillons de fumée. Le gasoil répandu sur l'eau s'enflamma. Et ce fut au tour du cuirassé *Pennsylvania* d'être touché dans la plus grande cale sèche. Deux destroyers amarrés dans la même cale explosèrent également, leurs magasins de munitions touchés.

Chuck et Eddie ramaient de toutes leurs forces, suant comme des chevaux de course.

Des fusiliers marins venus probablement de casernes voisines déployèrent des engins de lutte contre l'incendie.

Enfin, la vedette atteignit le quai d'embarquement des officiers. Chuck sauta à terre et noua rapidement les amarres pendant qu'Eddie aidait les passagers à sortir. Et tout le monde s'élança vers la voiture.

Chuck bondit au volant et fit démarrer le moteur. L'autoradio s'alluma automatiquement. « Tout le personnel de l'armée, de la marine et des fusiliers marins est tenu de se présenter immédiatement à son poste », annonça le speaker de la KGMB. Chuck n'avait pas encore pu se présenter où que ce soit, mais il était persuadé que sa première mission était d'assurer la sécurité des quatre civils à sa charge dont deux femmes et un sénateur.

Il démarra dès que tout le monde eut pris place dans la limousine.

La deuxième vague d'attaque semblait toucher à sa fin et la plupart des avions japonais s'éloignaient du port. Chuck accéléra tout de même : une troisième vague était toujours à craindre.

Par bonheur, la barrière de l'entrée principale était levée, ce qui évita à Chuck de la forcer.

Aucune voiture en face.

Fuyant le port à toute allure, il prit la grand-route de Kamehameha. Plus vite il s'éloignerait de Pearl Harbor, plus vite il mettrait les siens à l'abri, se disait-il.

C'est alors qu'il repéra dans le ciel un Zéro solitaire qui se dirigeait vers lui, volant à faible attitude au-dessus de la route. Chuck mit un moment à comprendre qu'il visait la voiture.

Les canons se trouvaient dans les ailes et il y avait de bonnes chances qu'ils manquent cette cible étroite. En revanche, les mitrailleuses de part et d'autre du moteur étaient très rapprochées. Si le pilote avait un peu de jugeote, il ne manquerait pas de s'en servir.

Chuck scruta fébrilement les bas-côtés. Rien que des champs de canne à sucre, à droite comme à gauche. Pas le moindre abri.

Il se mit à rouler en zigzag. Le pilote, qui ne cessait de se rapprocher, eut l'intelligence de ne pas chercher à le garder dans sa ligne de mire. La route n'était pas large, et il était impensable que Chuck pénètre dans le champ de canne, ce qui l'aurait obligé à rouler au pas. Comprenant que la vitesse était sa seule chance de survie, il appuya à fond sur l'accélérateur.

Il était trop tard. L'avion était si proche maintenant que Chuck distinguait sous ses ailes les bouches noires des canons. Comme il s'y attendait, le pilote ouvrit le feu à la mitrailleuse. Les balles, en touchant terre, firent voler la poussière.

Chuck se déporta sur la gauche, vers le milieu de la chaussée, puis au lieu de continuer dans la même direction, il donna un brutal coup de volant. Le pilote corrigea sa trajectoire. Des balles atteignirent le capot de la Packard. Le pare-brise éclata en mille morceaux. Eddie poussa un cri de douleur ; à l'arrière, l'une des femmes poussa un hurlement.

Le Zéro disparut.

La voiture se mit à zigzaguer toute seule : un des pneus avant avait dû éclater. Cramponné au volant, Chuck tenta désespérément de rester sur la route. La voiture fit un tête-à-queue, dérapa sur le bitume et alla s'enfoncer dans les tiges de canne à sucre qu'elle percuta et qui l'arrêtèrent. Des flammes s'élevèrent du capot.

« Tout le monde dehors ! hurla Chuck en sentant une odeur d'essence. Le réservoir va exploser ! » Il bondit à l'extérieur et courut ouvrir la portière arrière. Son père sauta du véhicule, tirant sa mère derrière lui. Chuck vit les autres s'échapper par l'autre côté. « Filez ! » cria-t-il encore, mais c'était superflu. Eddie se précipitait dans le champ en boitant, vraisemblablement blessé. Woody s'occupait de Joanne, la traînant et la portant tant bien que mal. Elle semblait avoir été touchée, elle aussi. Quant à ses parents, ils s'élancèrent au milieu de la canne à sucre, apparemment sains et saufs. Chuck les rejoignit. Ils coururent tous ensemble sur une centaine de mètres et se jetèrent à terre.

Il y eut un moment de silence. Le bruit des avions n'était plus qu'un bourdonnement lointain. Levant les yeux, Chuck vit des nuages de fumée épaisse s'élever dans le ciel au-dessus du port sur plusieurs centaines de mètres. Plus haut, à très haute altitude, les derniers bombardiers s'éloignaient vers le nord.

Une déflagration soudaine lui creva les tympans. Même à travers ses paupières closes, il vit l'éclair de lumière : le réservoir avait explosé. Une vague de chaleur passa au-dessus de lui.

Il releva la tête. La Packard était en flammes.

Il bondit sur ses pieds. « Mama, ça va ?

— Je n'ai rien, par miracle », répondit-elle calmement tandis que son père l'aidait à se relever.

Balayant du regard le champ de canne à sucre, il repéra les autres. Eddie, assis tout droit, serrait sa cuisse entre ses mains. Il s'élança vers lui. « Tu es blessé ?

— Putain, ça fait un mal de chien ! Heureusement, ça ne saigne pas beaucoup. » Il réussit à sourire. « Touché en haut de la cuisse, je pense, mais aucun organe vital ne devrait être atteint.

— On va te conduire à l'hôpital. »

À ce moment-là, Chuck entendit un bruit qui le fit tressaillir.

Son frère pleurait.

Woody ne pleurait pas comme un bébé, mais comme un enfant perdu : à gros sanglots, les sanglots d'un désespoir sans fond.

Les sanglots d'un cœur brisé, comprit Chuck immédiatement.

Il se précipita vers lui. Woody était à genoux, le corps agité de frissons, la bouche grande ouverte, les yeux ruisselants de larmes. Son costume de lin blanc était couvert de sang alors que lui-même n'était apparemment pas blessé. Ses sanglots étaient entrecoupés de gémissements : « Non, non. »

Joanne gisait sur le dos devant lui.

Elle était morte, cela ne faisait aucun doute. Son corps était immobile, ses yeux ouverts fixaient le vide. Le devant de sa robe en coton rayé était imprégné d'un sang rouge vif qui, par endroits, s'assombrissait déjà. La plaie n'était pas visible, mais Chuck comprit qu'une balle avait touché Joanne à l'épaule, sectionnant l'artère axillaire. Elle avait dû mourir en quelques instants, vidée de son sang.

Il resta muet.

Les autres le rejoignirent : Rosa, Gus et Eddie. Puis Rosa s'agenouilla à côté de Woody et le prit dans ses bras. « Mon pauvre garçon », murmura-t-elle comme si c'était un enfant.

Eddie passa le bras autour des épaules de Chuck et le serra discrètement contre lui.

Gus s'accroupit près du corps et prit la main de Woody.

Les sanglots de son fils s'apaisèrent un peu.

Il dit : « Ferme-lui les yeux, Woody. »

Non sans mal, Woody réussit à contrôler le tremblement de sa main. Il tendit les doigts vers les paupières de Joanne.

Et, avec une douceur infinie, il lui ferma les yeux.

XII

1942 (I)

1.

Le premier jour de l'année 1942, Daisy reçut une lettre de son ancien fiancé, Charlie Farquharson.

Elle l'ouvrit au petit déjeuner dans sa maison de Mayfair, ayant pour seule compagnie le vieux majordome en train de lui verser son café et la soubrette de quinze ans qui lui apportait des toasts chauds de la cuisine.

Charlie n'écrivait pas de Buffalo, mais de Duxford, une base de la RAF située dans l'est de l'Angleterre. Daisy en avait entendu parler. C'était près de Cambridge, la ville où elle avait rencontré son mari Boy Fitzherbert, et aussi Lloyd Williams, l'homme qu'elle aimait.

Elle était contente de recevoir des nouvelles de Charlie. Certes il l'avait plaquée autrefois et elle lui en avait voulu, mais le temps avait passé. Elle était une autre femme aujourd'hui. En 1935, elle était miss Pechkov, une riche héritière ; à présent, elle était une aristocrate anglaise, la vicomtesse d'Aberowen. Quoi qu'il en soit, elle était contente que Charlie se souvienne d'elle. Pour une femme rien n'est pire que l'oubli.

Charlie avait écrit sa lettre à l'encre noire, avec un stylo à grosse plume. L'écriture était brouillonne, les lettres épaisses et mal formées. Elle lut :

Avant tout, je dois te présenter mes excuses pour mon comportement à ton égard, à Buffalo. J'en frémis encore de honte chaque fois que j'y repense.

Eh bien, se dit Daisy, en voilà un qui a mûri.

Que nous étions snobs, tous autant que nous étions ! Et que j'étais faible, quant à moi, de laisser feu ma mère diriger ma vie.

« Feu ma mère. » Cette vieille carne était donc morte. Voilà qui expliquait peut-être le changement.

J'ai rejoint le 133e Eagle Squadron. C'est une escadrille d'avions de chasse. Pour l'heure, nous volons sur des Hurricane mais nous aurons bientôt des Spitfire.

Les Eagle Squadrons, les « escadrilles des aigles », au nombre de trois, étaient des unités de l'aviation britannique pilotées par des équipages de volontaires américains. Daisy fut surprise d'apprendre que Charlie s'était engagé. À l'époque où elle le connaissait, il ne s'intéressait qu'aux chiens et aux chevaux. Il avait vraiment mûri.

Si tu peux trouver au fond de ton cœur la force de me pardonner ou du moins d'oublier le passé, je serai ravi de te revoir et de faire la connaissance de ton mari.

La référence au mari devait être une manière délicate de lui faire comprendre que sa requête n'avait rien de sentimental.

Je serai à Londres en permission le week-end prochain. Puis-je vous inviter à dîner tous les deux? S'il te plaît, ne refuse pas.

Avec mes sentiments les meilleurs et les plus affectueux

Charlie H.B. Farquharson.

Boy serait absent ce week-end-là, mais Daisy accepta quant à elle l'invitation. Dans ce Londres en guerre, la compagnie des hommes lui manquait affreusement, comme à bien d'autres femmes. Lloyd était parti pour l'Espagne et avait disparu. Il avait prétendu avoir été nommé attaché militaire à l'ambassade britannique de Madrid. Quand elle s'était étonnée qu'on envoie un jeune officier valide remplir des fonctions de gratte-papier en pays neutre, il lui avait expliqué qu'il fallait empêcher coûte que coûte l'Espagne d'entrer en guerre aux côtés des fascistes. Mais son sourire contrit avait été suffisamment éloquent. Daisy redoutait qu'il n'ait franchi la frontière clandestinement pour gagner la France et rejoindre la Résistance. La nuit, elle faisait des cauchemars, l'imaginait prisonnier, torturé.

Cela faisait plus d'un an qu'elle ne l'avait pas vu et elle avait l'impression d'être amputée d'une partie d'elle-même. C'était une douleur de chaque instant. À l'idée de passer une soirée avec un homme, fût-ce ce Charlie Farquharson pataud, rondouillard et peu attirant, elle était ravie.

Charlie avait réservé une table au Grill Room du Savoy.

Dans le hall de l'hôtel, tandis qu'un serveur lui prenait son manteau de vison, un homme de haute taille, vêtu d'une élégante veste du soir, s'approcha d'elle. Son visage lui rappelait vaguement quelqu'un. Il lui tendit la main en disant timidement : « Bonsoir Daisy, quel plaisir de te revoir après tant d'années. »

Elle ne le reconnut qu'au son de sa voix. « Bon sang, Charlie ! Comme tu as changé !

— J'ai un peu fondu, avoua-t-il.

— Tu peux le dire ! » Il avait dû perdre vingt ou vingt-cinq kilos. Ça lui allait bien. À présent, ses traits paraissaient plus anguleux que laids.

« Toi, en revanche, tu n'as pas changé du tout ! » s'écria-t-il en l'examinant de haut en bas.

Elle avait fait des efforts de toilette. Elle ne s'était rien acheté depuis longtemps, à cause de l'austérité des temps. Pour l'occasion, elle avait ressorti une robe du soir en soie bleu saphir qui lui dénudait les épaules ; elle l'avait achetée chez Lanvin lors d'un de ses derniers voyages à Paris avant la guerre.

« J'aurai bientôt vingt-six ans, j'ai du mal à croire que je ressemble à la jeune fille que j'étais à dix-huit ans !

— Crois-moi, tu es toujours la même », insista-t-il en rougissant après avoir glissé un regard vers son décolleté.

Ils entrèrent dans le restaurant et prirent place à leur table. « J'ai eu peur que tu n'aies changé d'avis, dit-il.

— Ma montre s'était arrêtée, je suis désolée d'être en retard.

— Vingt minutes seulement, j'aurais attendu une heure. »

Un serveur leur proposa un apéritif. « C'est l'un des rares endroits d'Angleterre où les martinis sont buvables, remarqua Daisy.

— Alors deux, s'il vous plaît, demanda Charlie.

— Pour moi, sec et avec une olive.

— Pour moi aussi. »

Elle l'observait, intriguée de le voir aussi changé. Sa maladresse d'antan s'était adoucie en une timidité charmante. Tout de même, elle avait du mal à l'imaginer en pilote de combat, en train d'abattre des avions allemands. À Londres, le Blitz avait pris fin depuis six mois et il n'y avait plus de combats aériens dans le sud de l'Angleterre. « Quelle sorte de missions accomplis-tu ? lui demanda-t-elle.

— Des missions de jour, principalement. Ce que nous appelons des opérations circus sur le nord de la France.

— Ça consiste en quoi ?

— Une attaque de bombardier soutenue par une forte escorte de chasseurs qui cherchent à attirer les avions ennemis dans un combat aérien où ils se trouveront en infériorité numérique.

— J'ai horreur des bombardiers. J'en ai trop vu pendant le Blitz. »

Surpris, il répliqua : « J'aurais cru que tu serais contente de savoir qu'on rendait aux Allemands la monnaie de leur pièce.

— Pas du tout ! » Elle avait beaucoup réfléchi à la question. « Je pourrais pleurer des jours entiers sur tous ces civils innocents, ces femmes et ces enfants brûlés ou blessés pendant le Blitz. Savoir que des femmes et des enfants allemands subissent le même sort ne me fait pas plaisir du tout.

— Je n'avais pas envisagé la chose sous cet angle. »

Ils commandèrent à dîner. Selon les réglementations en vigueur, les repas devaient se limiter à trois plats et ne pas coûter plus de cinq shillings. Le menu comprenait du « similicanard », fait de saucisse de porc, ou de la « tourte Woolton », sans un gramme de viande, recettes spécialement élaborées au Savoy en ces temps de disette.

« Quel plaisir d'entendre une femme parler américain, tu ne peux pas savoir ! Je n'ai rien contre les Anglaises, j'en ai même fréquenté une, mais l'accent de chez nous me manque beaucoup.

— À moi aussi, dit-elle. Je suis ici chez moi, maintenant, et je ne crois pas que je retournerai un jour en Amérique. Mais je comprends très bien ce que tu veux dire.

— Je regrette de n'avoir pas pu faire la connaissance du vicomte d'Aberowen.

— Il est pilote, comme toi. Instructeur. Il rentre de temps en temps à la maison, mais pas ce week-end. »

Daisy accueillait à nouveau Boy dans son lit quand il était de passage, bien qu'elle se soit juré de ne plus jamais coucher avec lui après l'avoir surpris avec ces abominables femmes d'Aldgate. Mais il avait insisté lourdement, lui faisant valoir qu'un combattant avait besoin de réconfort quand il rentrait chez lui. Il lui avait aussi juré de ne plus fréquenter de prostituées. Elle n'y croyait pas vraiment, mais avait cédé de guerre lasse. Ne l'avait-elle pas épousé pour le meilleur et pour le pire ?

Elle n'éprouvait cependant plus aucun plaisir à coucher avec Boy, malheureusement. Elle pouvait avoir des rapports sexuels avec lui, mais son amour pour lui n'était plus qu'un souvenir. Elle devait utiliser une crème lubrifiante pour pallier son manque de désir. Elle avait essayé de se rappeler les tendres sentiments qu'il lui avait inspirés autrefois quand elle voyait en lui un jeune aristocrate plein de gaieté qui avait le monde à ses pieds et savait tirer le meilleur de la vie. Mais elle avait dû se rendre à l'évidence : Boy n'avait pas tant de charme que cela, finalement. C'était un homme égoïste et sans grand intérêt, dont le seul atout était son titre. Quand il s'allongeait sur elle, elle ne pensait qu'à la sale maladie qu'il risquait de lui transmettre.

Charlie lui demanda alors avec doigté : « Tu n'as peut-être pas très envie de parler des Rouzrokh…

— En effet.

— … mais as-tu appris que Joanne est morte ?

— Non, répondit-elle, bouleversée. Comment ?

— À Pearl Harbor. Elle était fiancée avec Woody Dewar, ils étaient allés voir son frère Chuck qui était en poste là-bas. Leur voiture a été attaquée par un Zéro, un chasseur japonais. Joanne a été touchée.

— Oh mon Dieu ! Pauvre Joanne, pauvre Woody ! »

On leur apporta leur repas, accompagné d'une bouteille de vin. Ils mangèrent en silence pendant un moment. Daisy constata que le similicanard n'avait pas vraiment goût de canard.

« Joanne est l'une des deux mille quatre cents personnes qui ont trouvé la mort à Pearl Harbor, reprit Charlie. Nous avons perdu huit cuirassés et dix autres navires. Ces sournois de Japonais, quels salauds quand même !

— Même si on ne le dit pas, tout le monde est bien content à Londres que les États-Unis soient entrés en guerre. Dieu seul sait pourquoi Hitler a eu la bêtise de déclarer la guerre aux

États-Unis. Avec les Russes et les Américains de leur côté, les Anglais pensent maintenant qu'ils ont de bonnes chances de l'emporter.

— Les Américains sont furieux à cause de Pearl Harbor.

— Ici, les gens ne comprennent pas pourquoi.

— Les Japonais ont continué à négocier jusqu'au bout, comme si de rien n'était, alors qu'ils avaient certainement déjà décidé de nous attaquer depuis longtemps ! C'est de la duplicité pure et simple ! »

Daisy cherchait à comprendre : « Je trouve ça plutôt raisonnable, moi. Si les négociations avaient abouti, ils auraient annulé l'ordre d'attaquer.

— Il n'y avait pas eu de déclaration de guerre !

— Et alors ? Nous nous attendions à ce qu'ils donnent l'assaut aux Philippines. Pearl Harbor nous aurait pris par surprise de toute façon. »

Charlie écarta les mains dans un geste d'incompréhension. « Mais pourquoi nous attaquer ?

— On leur avait piqué leur argent.

— Gelé leurs avoirs.

— Pour eux, cela revenait au même. Nous leur avions coupé tout approvisionnement en pétrole, nous les avions poussés à bout. Ils étaient au bord de la ruine, que pouvaient-ils faire d'autre ?

— Capituler et évacuer la Chine !

— Oui, sans doute ! Mais que dirais-tu si un autre pays prétendait obliger l'Amérique à faire ceci ou cela ? Tu voudrais qu'on cède ?

— Non, probablement pas. » Il sourit. « Tout à l'heure, j'ai prétendu que tu n'avais pas changé. Je retire ce que j'ai dit !

— Pourquoi ?

— Tu ne parlais pas comme ça, avant. Tu ne parlais jamais politique.

— C'est vrai. Mais si tu ne t'intéresses pas à ce qui se passe autour de toi, alors tu es responsable de ce qui se passe !

— Je crois qu'on a tous compris la leçon. »

Ils commandèrent des desserts. « Que va-t-il advenir du monde, Charlie ? demanda Daisy. Tout le continent européen est fasciste, les Allemands ont conquis une grande partie de la

Russie ! L'Amérique est un aigle à l'aile brisée. Parfois, je me réjouis de ne pas avoir d'enfants.

— Ne sous-estime pas les États-Unis. Nous sommes blessés, mais nous sommes encore debout. Pour le moment, le Japon parade mais viendra le jour où il versera des larmes amères pour le crime commis à Pearl Harbor.

— Puisses-tu avoir raison !

— Quant aux Allemands, les choses ne se passent plus exactement comme ils le souhaitaient. Ils n'ont pas réussi à prendre Moscou, ils battent en retraite. La bataille de Moscou a été la première vraie défaite d'Hitler, tu te rends compte ?

— Une défaite ou un simple revers ?

— Qu'importe ! Pour lui, c'est le pire échec qu'il ait essuyé à ce jour. Les bolcheviks lui ont donné une sacrée leçon. »

Charlie avait découvert le vieux porto, tant apprécié des Anglais. Traditionnellement, les hommes en buvaient une fois que les femmes avaient quitté la table. Une tradition que Daisy exécrait et qu'elle avait essayé d'abolir chez elle, sans succès. Ils en burent un verre chacun. Après le martini et le vin, Daisy se sentit un peu pompette… et très heureuse.

Ils évoquèrent leur adolescence à Buffalo et rirent des bêtises qu'ils avaient commises, les uns ou les autres. « Tu nous avais dit que tu allais à Londres danser avec le roi, et tu l'as fait ! s'exclama Charlie.

— J'espère bien que les autres en ont été vertes de jalousie !

— Et comment ! Dot Renshaw a failli en faire une attaque. »

Daisy rit joyeusement.

« Je suis content qu'on ait renoué, murmura Charlie, j'apprécie tant ta compagnie.

— Oui, ça me fait plaisir, à moi aussi. »

Ils quittèrent le restaurant et reprirent leurs manteaux. Le portier héla un taxi. « Je te raccompagne chez toi », dit Charlie.

Alors qu'ils longeaient le Strand, il passa le bras autour de ses épaules. Elle faillit protester puis se dit pourquoi pas, et se blottit contre lui.

« Quel imbécile j'ai été ! J'aurais vraiment dû t'épouser…

— Tu aurais fait un bien meilleur mari que Boy Fitzherbert », remarqua-t-elle.

Mais dans ce cas, elle n'aurait jamais rencontré Lloyd. Elle se rendit compte alors qu'elle n'avait pas parlé de lui de toute la soirée.

Au moment où le taxi bifurquait dans sa rue, Charlie l'embrassa.

Que c'était agréable d'être enlacée par un homme et de l'embrasser sur les lèvres. Bien sûr, l'alcool n'était pas étranger au plaisir qu'elle ressentait car en vérité, il n'y avait qu'un homme qu'elle eût vraiment envie d'embrasser, Lloyd. Néanmoins, elle ne repoussa Charlie que lorsque la voiture s'arrêta.

« Un dernier verre ? » suggéra-t-il.

Elle fut tentée un instant. Il y avait si longtemps qu'elle n'avait touché le corps musclé d'un homme ! En même temps, elle n'avait pas vraiment envie de lui. « Non, dit-elle. Je suis désolée Charlie, mais j'aime quelqu'un d'autre.

— On n'est pas obligés de coucher ensemble, murmura-t-il. On pourrait juste se caresser un peu… »

Elle descendit de voiture avec un sentiment de culpabilité. Tous les jours, Charlie risquait sa vie pour elle, et elle n'était même pas prête à lui donner un peu de plaisir. « Bonne nuit, Charlie, bonne chance. » Elle claqua la portière et entra chez elle avant d'avoir changé d'avis.

Elle monta directement dans sa chambre. Quelques minutes plus tard, seule dans son lit, elle était rongée de remords. Elle avait trompé deux hommes : Lloyd, en embrassant Charlie ; Charlie, en le quittant insatisfait.

Elle passa presque tout son dimanche au lit avec la gueule de bois.

Le lendemain soir, elle reçut un coup de téléphone. « Je m'appelle Hank Bartlett, annonça une voix jeune à l'accent américain. Je suis un ami de Charlie Farquharson, à Duxford. »

Elle crut que son cœur allait s'arrêter de battre : « Pourquoi m'appelez-vous ?

— Je crains d'avoir une mauvaise nouvelle à vous annoncer… Charlie est mort aujourd'hui, il a été abattu au-dessus d'Abbeville.

— Non !

— C'était sa première sortie dans son nouveau Spitfire.

— Il m'en avait parlé, balbutia-t-elle hébétée.

— J'ai voulu vous prévenir.

— Oui, merci, murmura-t-elle.

— Il vous trouvait du tonnerre.

— Vraiment ?

— Vous auriez dû l'entendre parler de vous !

— Que c'est triste ! Oh mon Dieu, que c'est triste… » Elle raccrocha, incapable d'ajouter un mot.

2.

Chuck Dewar lisait par-dessus l'épaule du lieutenant Bob Strong. Contrairement à tant d'autres officiers du chiffre, Strong était du genre soigneux. Il n'y avait rien d'autre sur son bureau que cette feuille où était écrit :

YO—LO—KU—TA—WA—NA

« Impossible à déchiffrer, lâcha Strong sur un ton excédé. Si cette transcription est juste, ça signifie qu'ils comptent attaquer yolokutawana. Ça ne rime à rien, ce mot n'existe pas ! »

Chuck examina les six syllabes japonaises. Malgré sa connaissance rudimentaire du japonais, il était convaincu qu'elles avaient un sens. Comme il lui échappait, il se remit à son travail.

Dans les bureaux de l'Old Administration Building, l'atmosphère était lourde.

Plusieurs semaines après le raid, la mer de Pearl Harbor était encore couverte de mazout et des cadavres boursouflés remontaient des profondeurs où gisaient les navires coulés. En même temps, les messages interceptés faisaient état d'attaques japonaises encore plus dévastatrices. Trois jours seulement après Pearl Harbor, l'aviation japonaise avait frappé la base américaine de Luzon aux Philippines et réduit à néant la réserve de torpilles de toute la flotte du Pacifique. Le même jour, dans la mer de Chine méridionale, elle avait coulé deux navires britanniques, le *Repulse* et le *Prince of*

Wales, privant les Britanniques de leurs forces de défense en Extrême-Orient.

Rien ne semblait pouvoir arrêter les Japonais. Les mauvaises nouvelles se succédaient sans interruption. Ces premiers mois de l'année 1942 avaient vu le Japon écraser les troupes américaines aux Philippines, et les troupes britanniques à Hong Kong, Singapour et Rangoon, capitale de la Birmanie.

Nombre de ces lieux n'évoquaient rien pour personne, même pour des marins avertis tels que Chuck et Eddie. Pour la majorité des Américains, ces noms faisaient penser aux planètes lointaines d'un roman de science-fiction : Guam, Wake, Bataan. En revanche, les mots de retraite, soumission et reddition étaient clairs pour tout le monde.

Chuck était perplexe. Le Japon réussirait-il vraiment à vaincre les États-Unis ? Il avait du mal à le croire.

Au mois de mai, les Japonais avaient atteint leur objectif : posséder un territoire qui leur fournissait du caoutchouc, de l'étain et – chose essentielle – du pétrole. À en croire les renseignements qui arrivaient au compte-gouttes, les Japonais gouvernaient leur empire avec une brutalité qui aurait fait rougir Staline.

Néanmoins, la marine américaine continuait à leur donner du fil à retordre. Cette seule pensée remplissait Chuck de fierté. Les Japonais avaient espéré anéantir Pearl Harbor et contrôler le Pacifique : ils avaient échoué. Les porte-avions et les lourds croiseurs de l'US Navy étaient toujours opérationnels. À en croire les services de renseignement, les commandants de l'armée japonaise étaient furieux que les Américains n'aient pas baissé les bras. Après les pertes subies à Pearl Harbor, ils se trouvaient en situation d'infériorité tant en matière d'effectifs humains que d'armement. Mais au lieu de prendre la fuite, ils avaient lancé des raids éclairs contre les bâtiments japonais. Certes, ils n'avaient fait subir que des dégâts minimes à la marine impériale, mais cela avait suffi à redonner le moral aux troupes et à prouver aux Japonais que la victoire était loin d'être acquise.

Le 25 avril, des appareils avaient décollé d'un porte-avions pour aller bombarder le centre de Tokyo, infligeant une blessure profonde à l'orgueil des militaires japonais. À Hawaï, la nouvelle avait été accueillie avec enthousiasme et force réjouissances. Cette nuit-là, Chuck et Eddie s'étaient même saoulés.

Mais une autre épreuve se préparait. À l'Old Administration Building, tout le monde s'accordait à penser que les Japonais allaient lancer une grande offensive au début de l'été afin d'obliger les navires américains à sortir en force livrer une ultime bataille navale. Les Japonais comptaient sur la supériorité numérique de leur marine pour anéantir définitivement la flotte américaine du Pacifique. Les Américains ne disposaient que d'un moyen pour remporter la victoire : être mieux préparés que l'ennemi, posséder un meilleur service de renseignement, être plus mobiles et plus intelligents.

Tout au long de ces mois, la station HYPO travailla jour et nuit pour décrypter le code JN-25b, le dernier en date de la marine impériale japonaise. Aux alentours du mois de mai, des progrès avaient été faits.

La marine américaine possédait des stations d'interception sans fil tout le long de la ceinture du Pacifique, de Seattle jusqu'en Australie. Dans ces stations, des hommes munis de casques et de récepteurs écoutaient les transmissions radio des Japonais. Les membres du « gang des Toits » – l'On The Roof Gang –, comme on l'appelait, avaient pour mission de balayer les ondes et de transcrire sur des blocs-notes tout ce qu'ils entendaient.

Les signaux étaient en morse, mais les points et les traits des messages maritimes étaient retranscrits en groupes de nombres à cinq chiffres, chacun correspondant à une lettre, un mot ou une expression extraits du manuel de chiffrage.

Ces chiffres apparemment aléatoires étaient retransmis par câble sécurisé aux téléscripteurs situés au sous-sol de l'Old Administration Building. Débutait alors la phase la plus difficile : le décryptage du code.

Ils commençaient toujours par de petites choses. Comme les signaux s'achevaient souvent par le mot OWARI, qui signifiait « fin », ils recherchaient des groupes similaires à l'intérieur du texte et les remplaçaient par le mot « fin ».

Une négligence, rarissime chez les Japonais, leur fournit une aide inattendue.

Comme la livraison du nouveau manuel de chiffrage pour le JN-25b à des unités stationnées très loin avait pris du retard, le haut commandement japonais transmit certains messages en utilisant simultanément les *deux* codes et ce, pendant plusieurs semaines. Les Américains, qui avaient réussi à déchiffrer une grande partie du code initial – le JN-25 –, furent ainsi en mesure de traduire ces messages grâce à l'ancienne clé. Ils purent ensuite les comparer aux messages transmis dans le nouveau code et en déduire le sens des groupes de cinq chiffres de celui-ci. Pendant un certain temps, ils progressèrent à pas de géant.

Aux huit officiers du chiffre engagés au départ vinrent s'adjoindre quelques rescapés de la fanfare du cuirassé *California* qui avait été coulé. Pour des raisons incompréhensibles, les musiciens se révélaient de bons décrypteurs.

Le moindre signal était conservé, et toutes les transcriptions classées. Comparer signaux et transcriptions était un travail primordial. Les analystes pouvaient demander à voir tous les signaux reçus au cours de telle ou telle journée, ou bien tous ceux qui avaient été envoyés à un bâtiment en particulier, ou encore tous ceux où figurait le nom d'Hawaï.

Pour leur venir en aide, Chuck et ses compagnons développèrent des systèmes de croisement de données de plus en plus complexes.

Son unité put ainsi prédire que les Japonais attaqueraient Port Moresby, la base alliée en Papouasie, au cours de la première semaine de mai. L'avenir lui donna raison, et la marine américaine fut à même d'intercepter la flotte d'invasion dans la mer de Corail. Les deux camps se déclarèrent vainqueurs, mais les Japonais ne prirent pas Port Moresby et l'amiral Nimitz, commandant en chef des forces du Pacifique, commença à faire confiance à ses décrypteurs.

Les Japonais ne désignaient pas les villes du Pacifique par leur nom habituel. Tout lieu de quelque importance était représenté par deux lettres – ou plus exactement par deux caractères, ou kanas, de l'alphabet japonais, que les Américains transcrivaient généralement par un équivalent latin, allant de A à Z. Bien souvent, dans leur sous-sol, les hommes s'arrachaient les cheveux pour comprendre ce que représentaient les deux lettres qu'ils avaient sous les yeux. Ils avançaient lentement. Ils avaient compris que MO correspondait à Port Moresby et AH à Oahu, mais combien de noms leur demeuraient inconnus !

En mai, il apparut avec évidence que les Japonais s'apprêtaient à lancer une offensive de grande envergure contre un lieu inconnu, qu'ils désignaient par les lettres AF.

De toutes les suppositions concernant cet endroit, la plus vraisemblable semblait être les îles Midway, un atoll situé à l'extrémité ouest de l'archipel qui partait d'Hawaï et s'étirait sur plus de deux mille quatre cents kilomètres à mi-chemin entre Los Angeles et Tokyo.

Mais l'amiral Nimitz n'avait que faire des suppositions. Les Japonais bénéficiant d'une évidente supériorité numérique, il avait besoin de certitudes.

Jour après jour, les compagnons de Chuck s'employaient à percer à jour le sinistre plan de bataille ennemi : de nouveaux appareils avaient été livrés aux porte-avions ; une « force d'occupation » avait été embarquée : à l'évidence, les Japonais avaient bien l'intention de conserver tous les territoires sur lesquels ils mettraient la main.

Ces préparatifs laissaient présager que la fameuse grande offensive était pour bientôt. Mais d'où l'attaque viendrait-elle ?

Un jour, les hommes parvinrent à décoder un message ennemi réclamant instamment à Tokyo d'*accélérer l'envoi des conduites de carburant*. Ils n'en furent pas peu fiers. D'abord, parce que ce message comportait un terme technique spécialisé ; ensuite, parce qu'il prouvait qu'une manœuvre majeure allait avoir lieu sous peu en plein océan.

Mais le haut commandement américain avait plutôt envisagé une attaque contre Hawaï tandis que l'armée craignait un débarquement sur la côte ouest des États-Unis. L'équipe de Pearl Harbor pour sa part croyait à une attaque contre l'île Johnston et son terrain d'aviation, à mille six cents kilomètres au sud de Midway.

Il fallait qu'ils en soient sûrs à cent pour cent.

Chuck avait bien une petite idée sur la question, mais il hésitait à dire quoi que ce soit. Les décrypteurs étaient des hommes tellement intelligents ! Lui-même n'avait jamais été bon élève. En neuvième, un de ses camarades de classe l'avait surnommé Chucky le Bêta. Et ses larmes n'avaient servi qu'à mieux ancrer ce surnom dans l'esprit de tous, lui-même compris, puisque aujourd'hui il se considérait toujours comme Chucky le Bêta.

À l'heure du déjeuner, il alla s'installer avec Eddie sur les docks, face à la rade, pour manger les sandwichs et boire le café qu'ils avaient pris à la cantine. La situation redevenait peu à peu normale. Le plus gros du mazout avait disparu et plusieurs épaves avaient été dégagées.

Pendant qu'ils mangeaient, un porte-avions endommagé apparut du côté de Hospital Point et entra lentement dans le port, laissant derrière lui une nappe de gasoil qui s'étirait jusqu'au large. Chuck reconnut le *Yorktown*. Sa coque était noire de suie et un grand trou s'ouvrait dans le pont d'envol, vraisemblablement causé par une bombe japonaise pendant la bataille de la mer de Corail. Sirènes et klaxons firent au navire un accueil triomphal tandis qu'il faisait route vers l'arsenal. Les remorqueurs se rassemblèrent pour l'escorter et l'aider à franchir les portes ouvertes de la cale sèche numéro un.

« Il paraît qu'il y en a au moins pour trois mois de boulot, mais qu'il reprendra la mer dans trois jours ! » annonça Eddie. Il travaillait dans le même bâtiment que Chuck, mais à un autre étage, au bureau du Renseignement maritime, où les ragots étaient plus nombreux.

« Et ils vont faire comment ?

— Ils ont déjà commencé. Le maître charpentier est à bord : il a rejoint le bateau en avion avec toute son équipe. Et regarde la cale sèche ! »

Chuck tourna la tête. La cale grouillait d'hommes et de matériel. On ne comptait plus le nombre de postes à souder alignés sur le quai.

« De toute façon, reprit Eddie, ils vont seulement le rafistoler : réparer le pont et le remettre en état de naviguer. Pour le reste, il faudra attendre. »

Quelque chose dans le nom de ce bateau troublait Chuck. Il lui revenait incessamment à l'esprit. *Yorktown*… Ce mot le titillait. Le siège de Yorktown avait été la dernière grande bataille de la guerre d'Indépendance.

Le capitaine Vandermeier passa à côté d'eux. « Allez, au boulot, les gonzesses ! »

Eddie chuchota : « Un de ces jours, je vais lui casser la gueule.

— Attends que la guerre soit finie, Eddie », lui conseilla Chuck.

De retour au sous-sol, en apercevant Bob Strong toujours penché sur sa feuille, Chuck se rendit compte qu'il avait résolu son énigme.

Regardant de nouveau par-dessus l'épaule du décrypteur, il relut les mêmes six syllabes japonaises.

YO—LO—KU—TA—WA—NA

Diplomatiquement, pour donner à Strong l'impression qu'il avait résolu le problème tout seul, il s'écria : « Mais vous avez trouvé, lieutenant !

— Ah bon ? répondit Strong décontenancé.

— Mais oui, c'est un nom anglais. Transcrit phonétiquement en japonais.

— Yolokutawana, c'est un nom anglais ?

— Mais oui, lieutenant. C'est comme cela qu'on prononce *Yorktown* en japonais.

— Comment ? » s'écria Strong. Il n'en revenait pas.

L'espace d'un moment, Chuck le Bêta se demanda avec épouvante s'il ne se trompait pas du tout au tout. Mais Strong déclara : « Oh mon Dieu, mais vous avez raison ! Yolokutawana, *Yorktown* avec l'accent japonais ! » Il éclata de rire, ravi. « Merci ! ajouta-t-il. Bien joué ! »

Chuck hésita un peu. Une autre idée le taraudait. Devait-il l'exprimer ? Le déchiffrage, ce n'était pas sa partie. D'un autre côté, l'Amérique était à deux doigts de la défaite. Oui, il fallait oser. « Puis-je vous suggérer autre chose ?

— Allez-y.

— C'est à propos de l'indicatif AF. Ce que nous voulons, c'est avoir la certitude que ça se rapporte à Midway, c'est bien ça ?

— Ouais.

— Et si nous expédions nous-mêmes un message suffisamment intéressant à propos de Midway pour que les Japonais le retransmettent en code ? Ensuite, au moment où nous intercepterons leur message à leur état-major, nous pourrions voir comment ils codent ce nom. »

Strong n'avait pas l'air convaincu. « Mieux vaudrait peut-être envoyer notre message en clair, pour être sûrs qu'ils l'ont bien compris.

— Éventuellement. Mais dans ce cas, il faudrait un message qui ne soit pas très confidentiel, du style : "Épidémie de chaude-pisse à Midway, envoyez des médicaments", ou quelque chose dans le genre.

— Pourquoi les Japonais retransmettraient-ils un message pareil ?

— D'accord, un renseignement d'ordre militaire, alors, mais pas hautement confidentiel. Le temps qu'il fait, par exemple.

— De nos jours, même les bulletins météo sont top secret.

— Vous pourriez évoquer une pénurie d'eau potable, intervint l'analyste assis au bureau voisin. S'ils ont l'intention d'occuper ce site, ça pourrait les intéresser.

— En effet.

— C'est sûr, ça pourrait marcher ! s'écria Strong qui commençait à s'exciter. Imaginons que Midway nous envoie ici un message en clair disant que l'usine de désalinisation est en panne.

— Et qu'Hawaï réponde qu'ils envoient une péniche d'eau ! compléta Chuck.

— Si les Japonais ont bien l'intention d'attaquer Midway, ils retransmettront forcément ce message, car il faudra bien qu'ils organisent une expédition d'eau potable.

— Et ils l'émettront en langage codé pour qu'on ne sache pas qu'ils s'intéressent à Midway. »

Strong se leva. « Venez avec moi, dit-il à Chuck. Allons proposer ça au patron et voir ce qu'il en pense. »

Les messages furent échangés le jour même.

Le lendemain, un message japonais signalait une pénurie d'eau douce sur AF.

Midway était bien la cible visée !

L'amiral Nimitz commença à tisser sa toile.

3.

Dans la soirée, alors que des milliers d'ouvriers s'activaient sur le porte-avions *Yorktown* paralysé, réparant les dégâts à la lumière de projecteurs, Chuck et Eddie se rendirent au Band Round The Hat, un bar situé dans une ruelle peu éclairée d'Hawaï. L'endroit était bondé, comme toujours. Marins et gens du cru mélangés. Presque tous les clients étaient des hommes, mais il y avait également quelques couples d'infirmières. Chuck et Eddie aimaient bien venir là, retrouver des hommes comme eux. Les lesbiennes aussi appréciaient ce lieu où les hommes ne les draguaient pas.

Rien ne se faisait au grand jour, bien sûr, car on pouvait se faire renvoyer de l'armée et jeter en prison pour homosexualité.

L'endroit n'en était pas moins sympathique. Le chef d'orchestre était maquillé et le chanteur hawaïen était si convaincant en femme que certains ne se rendaient pas compte que c'était un travesti. Quant au patron, c'était une vraie folle. Les clients pouvaient danser entre hommes et personne ne se serait fait traiter de chochotte en commandant un Vermouth.

Chuck aimait Eddie plus passionnément encore depuis la mort de Joanne. Qu'Eddie puisse mourir, il l'avait toujours su, en théorie, mais jusque-là le danger lui avait semblé abstrait. Depuis l'attaque de Pearl Harbor, il ne se passait pas un jour sans que l'image de cette jolie fille baignant dans une mare de sang et de son frère sanglotant à ses côtés ne lui revienne à l'esprit. Il aurait très bien pu être à la place de Woody, pleurant Eddie avec un chagrin aussi inconsolable. Certes, ils avaient défié la mort ensemble le 7 décembre, mais désormais c'était vraiment la guerre. La vie ne valait pas grand-chose dans des périodes pareilles. Chaque jour passé ensemble était précieux, car ce pouvait être le dernier.

Chuck était accoudé au bar, un verre de bière à la main, Eddie était perché sur un tabouret. Ils se moquaient d'un pilote de la marine, Trevor Paxman, surnommé Trixie, qui racontait sa première nuit avec une fille. « J'ai été horrifié ! disait Trixie. J'avais toujours cru que là en bas, c'était lisse et doux, comme sur les tableaux, mais figurez-vous qu'elle était plus poilue que moi ! » Tout le monde se tordit de rire. « On aurait dit un gorille ! » Juste à ce moment-là, Chuck aperçut du coin de l'œil la silhouette trapue du capitaine Vandermeier se dessiner sur le seuil.

Peu d'officiers fréquentaient ce genre d'endroit. Ce n'était pas interdit, mais cela ne se faisait pas, tout simplement, comme d'aller déjeuner au Ritz-Carlton dans des bottes crottées. Eddie s'empressa de se retourner en espérant que Vandermeier ne l'avait pas remarqué.

Pas de chance ! Le capitaine se dirigea droit sur eux. « Alors, toutes les gonzesses sont là, à ce que je vois ! »

Trixie leur tourna le dos et se fondit dans la foule.

« Où est-ce qu'il a filé, l'autre ? » s'écria Vandermeier. Il était déjà assez saoul pour ne plus articuler correctement.

Chuck vit Eddie se renfrogner. Il dit sèchement : « Bonsoir, capitaine. Puis-je vous offrir une bière ?

— Un scotch *onna rocks*. »

Chuck passa la commande. Vandermeier but une gorgée et dit : « Alors ici, tout se passe par-derrière, à ce que je vois. Je me trompe ? ajouta-t-il en regardant Eddie.

— Je n'en sais rien, répondit froidement celui-ci.

— Allons, à d'autres ! insista Vandermeier. Entre nous… » Il tapota le genou d'Eddie.

« Bas les pattes ! » Eddie se leva d'un bond en repoussant son tabouret.

« Eddie, calme-toi ! intervint Chuck.

— Aucune loi de la marine ne m'oblige à me laisser peloter par cette vieille tantouse !

— Tu m'as traité de quoi ? marmonna Vandermeier d'une voix pâteuse.

— S'il me touche encore une fois, je jure que je lui casse la gueule !

— Capitaine Vandermeier, je connais un endroit bien plus agréable que celui-ci. Voulez-vous y aller ? »

Vandermeier parut un peu désorienté. « Quoi ?

— Une autre boîte, plus petite, plus tranquille, improvisa Chuck. Un peu comme ici, mais en plus intime. Vous voyez ce que je veux dire ?

— Ça m'a l'air sympa ! » Le capitaine vida son verre.

Saisissant le capitaine par le bras droit, Chuck fit signe à Eddie de le prendre par le bras gauche. Ensemble, ils l'entraînèrent dehors.

Par chance, un taxi stationnait dans la ruelle obscure. Chuck en ouvrit la portière.

Vandermeier en profita pour enlacer Eddie et tenter de l'embrasser en murmurant : « Je t'aime. »

Chuck s'affola. Cette histoire allait mal finir, c'était sûr.

Eddie balança un coup de poing dans l'estomac de Vandermeier. Le capitaine grogna, le souffle coupé. Eddie le frappa à nouveau, au visage cette fois. Chuck s'interposa. Il réussit à rattraper le capitaine avant qu'il ne s'écroule et l'enfourna sur le siège arrière du taxi.

Il se pencha par la vitre du chauffeur et lui donna dix dollars. « Conduisez-le chez lui et gardez la monnaie ! »

Le taxi s'éloigna.

Chuck se tourna vers Eddie. « Eh bien, ce coup-là, on est dans la merde ! »

4.

Eddie Parry ne fut jamais poursuivi pour agression contre la personne d'un officier.

Le lendemain matin, le capitaine Vandermeier arriva à l'Old Administration Building avec un œil au beurre noir, mais il n'accusa personne. Probablement aurait-il pu dire adieu à sa carrière s'il avait admis avoir pris part à une rixe au Band Around The Hat, se dit Chuck. Ce qui n'empêchait pas tout le monde de parler de son cocard. « Vandermeier prétend avoir glissé sur une flaque d'huile dans son garage et s'être rétamé sur sa tondeuse à gazon, déclara Bob Strong. Je crois plutôt que sa femme l'a tabassé. Vous l'avez vue ? On dirait un catcheur. »

Ce même jour, les déchiffreurs du sous-sol avertirent l'amiral Nimitz que les Japonais attaqueraient Midway le 4 juin. Ou, plus précisément, qu'à sept heures du matin, les forces japonaises se trouveraient à cent soixante-quinze milles au nord de l'atoll.

Ils étaient presque prêts à en mettre leur main au feu.

Eddie était d'humeur morose. « Que peut-on faire ? » demanda-t-il à Chuck quand ils se retrouvèrent pour déjeuner. Travaillant lui aussi au Renseignement maritime, il n'ignorait rien de la puissance navale japonaise grâce aux rapports des décrypteurs. « Les Japonais ont deux cents navires en mer – la quasi-totalité de leur flotte – et nous, combien en avons-nous ? Trente-cinq ! »

Chuck n'était pas aussi pessimiste. « Mais leur force de frappe ne représente que le quart de leur puissance navale. Les trois autres quarts se partagent entre forces d'occupation, forces de diversion et réserves.

— Et alors ? Le quart de leur puissance maritime, c'est déjà plus que notre flotte du Pacifique tout entière !

— Cette force de frappe japonaise ne dispose que de quatre porte-avions.

— Et nous n'en avons que trois. » De son sandwich au jambon, Eddie désigna le porte-avions noir de suie en cale sèche et les ouvriers qui s'activaient dessus. « Et encore, en comptant le *Yorktown* tout déglingué.

— Nous avons l'avantage de savoir qu'ils arrivent. Eux ne savent pas que nous sommes à l'affût.

— J'espère que ça suffira à faire la différence, comme Nimitz le pense.

— Ouais ! J'espère aussi. »

Quand Chuck retourna au sous-sol, on lui apprit qu'il n'y travaillait plus. Il avait été réaffecté… sur le *Yorktown*.

« C'est un coup de Vandermeier, sa manière à lui de me punir, déclara Eddie en larmes, ce soir-là. Il pense que tu n'en reviendras pas.

— Ne sois pas pessimiste. Rien ne dit que nous ne gagnerons pas la guerre ! » répliqua Chuck.

Quelques jours avant l'attaque, les Japonais changèrent à nouveau de code. Au sous-sol, les hommes poussèrent de gros soupirs et retroussèrent leurs manches, mais ils ne furent pas en mesure de recueillir beaucoup d'informations nouvelles avant la bataille. Nimitz dut se contenter de ce qu'il savait déjà et espérer que les Japonais n'avaient pas révisé l'ensemble de leur plan à la dernière minute.

Les Japonais s'attendaient à attaquer Midway par surprise et à s'en emparer facilement. Ils espéraient que les Américains jetteraient toutes leurs forces dans la bataille pour reprendre l'atoll. À ce moment-là, la flotte de réserve japonaise se précipiterait et anéantirait la totalité de la flotte américaine. Le Japon serait alors maître du Pacifique.

Les Américains réclameraient des pourparlers de paix.

Un beau projet. Sauf que Nimitz avait prévu de l'étouffer dans l'œuf en tendant une embuscade à la force japonaise avant qu'elle n'ait réussi à prendre Midway.

Et Chuck participait à ce guet-apens.

Il fit son sac, embrassa Eddie et ils se dirigèrent tous deux vers les docks.

Là, ils tombèrent sur Vandermeier.

« On n'a pas eu le temps de réparer les compartiments étanches. Si le bâtiment est touché, il coulera comme un cercueil de plomb », leur annonça-t-il.

Retenant Eddie par l'épaule, Chuck répliqua : « Et votre œil, capitaine ? Ça va mieux ? »

La bouche de Vandermeier se tordit en une grimace mauvaise. « Bonne chance, pédé ! » Et il s'éloigna.

Chuck échangea une poignée de main avec Eddie et embarqua.

Dès qu'il fut à bord, il oublia Vandermeier. Enfin, son rêve se réalisait : il naviguait. Et en plus, sur l'un des plus beaux bâtiments de guerre au monde !

Le *Yorktown* était le vaisseau-amiral de la flotte de porte-avions. D'une longueur égale à deux terrains de football, il nécessitait un équipage de deux mille marins. Il transportait quatre-vingts avions : de vieux bombardiers-torpilleurs Douglas Devastator, aux ailes repliables ; des Douglas Dauntless plus récents, spécialisés dans les bombardements en piqué ; et des chasseurs Grumman Wildcat, servant d'escorte aux bombardiers.

Tous les équipements étaient situés sous le pont, à l'exception du château, sorte de tour de contrôle qui s'élevait à plus de dix mètres au-dessus de la piste d'envol. Il abritait le poste de commandement du navire, cœur des communications avec le pont, la salle de radio juste en dessous, la salle des cartes et la salle d'attente des pilotes. Derrière cet îlot, trois conduits placés l'un derrière l'autre formaient la cheminée extérieure.

Quand le *Yorktown* quitta la cale sèche et sortit dans la rade de Pearl Harbor, il y avait encore des ouvriers à bord qui terminaient leur travail. Chuck frissonna d'émotion en sentant ronfler les moteurs colossaux du navire tandis qu'il prenait la mer. Quand il fut entré en eaux profondes et commença à monter et descendre au gré de la houle du Pacifique, Chuck eut l'impression de danser.

Il avait été nommé à la salle de radio, poste sensible où il pouvait mettre à profit son expérience du traitement des signaux. Le bâtiment faisait route vers son lieu de rendez-vous, au nord-est de Midway, ses pièces rafistolées et soudées crissant comme des chaussures neuves. Le bateau avait une buvette, le Gedunk, où l'on servait aussi des glaces faites sur place. C'est là, le premier jour, qu'il tomba sur Trixie Paxman et il se réjouit d'avoir un ami à bord.

Le mercredi 3 juin, veille du jour de l'attaque présumée, un bateau de reconnaissance, croisant à l'ouest de Midway, repéra un convoi de transporteurs japonais, amenant probablement les troupes qui étaient censées se déployer sur l'atoll après la bataille. La nouvelle fut retransmise à tous les bâtiments américains. Chuck, qui était dans la salle de radio du *Yorktown*, fut l'un des premiers avisés. En voyant ainsi confirmées les prédictions de ses camarades de la station HYPO, il éprouva du soulagement. Soulagement un peu paradoxal, car si les gars s'étaient trompés dans leurs pronostics et si les Japonais s'étaient trouvés ailleurs, il n'aurait pas couru un tel danger.

Cela faisait un an et demi qu'il était dans la marine, et il n'avait encore jamais pris part à une bataille. Le *Yorktown*, retapé à la hâte, allait être la cible des torpilles et des bombes japonaises. Le navire faisait route vers des adversaires décidés à tout pour l'envoyer par le fond avec son équipage, et Chuck par la même occasion. C'était une sensation étrange. La plus grande partie du temps, il éprouvait un calme bizarre, mais parfois une envie folle le prenait de plonger par-dessus bord pour regagner Hawaï à la nage.

Ce soir-là, il écrivit à ses parents, tout en sachant que sa lettre et lui risquaient de couler en même temps que le navire, le lendemain. Il le fit néanmoins. Il passa sous silence les raisons de sa nouvelle affectation. Il eut envie un instant de leur avouer qu'il était homosexuel mais se contenta de leur dire qu'il les aimait et qu'il leur était reconnaissant de tout ce qu'ils avaient fait pour lui. « Si je meurs en me battant pour un pays démocratique contre une dictature militaire cruelle, je n'aurai pas vécu en vain », écrivit-il. À la relecture, la phrase lui parut un peu pompeuse, mais il ne la supprima pas.

La nuit fut courte. Les équipages des avions furent tirés du lit à une heure trente du matin pour le petit déjeuner. En contrepartie de ce réveil aux aurores, on servit aux aviateurs un steak et des œufs. Chuck alla souhaiter bonne chance à Trixie Paxman.

Les avions montèrent par d'énormes ascenseurs depuis les hangars situés sous le pont. Ils roulèrent ensuite jusqu'à leur emplacement pour être ravitaillés en carburant et en munitions. Quelques pilotes décollèrent pour effectuer des vols de reconnaissance. Les autres, vêtus de leur combinaison, restèrent dans la salle de briefing, attendant les nouvelles.

Chuck prit son service dans la salle de radio. Un peu avant six heures, il capta un message d'un des appareils envoyés en repérage :

NOMBREUX AVIONS ENNEMIS
VOLANT VERS MIDWAY.

Quelques minutes plus tard il reçut un signal partiel :

TRANSPORTEURS ENNEMIS

Les choses se précisaient.

La totalité du message arriva une minute plus tard. Il situait la force de frappe japonaise presque exactement à l'endroit prévu par les hommes du chiffre. Chuck en éprouva fierté et terreur à la fois.

Les trois porte-avions américains, le *Yorktown*, l'*Enterprise* et le *Hornet*, fixèrent un cap permettant aux avions embarqués de ne pas se trouver à portée des navires japonais.

L'amiral Frank Fletcher, un homme au long nez, âgé de cinquante-sept ans et décoré de la croix de la marine pendant la Première Guerre mondiale, arpentait la passerelle. Alors qu'il apportait un message sur le pont, Chuck l'entendit dire : « Nous n'avons pas encore aperçu d'avions japonais. Cela veut dire qu'ils ignorent encore notre présence ici. »

Chuck savait que l'avantage des Américains ne tenait qu'à cela : ils disposaient d'un meilleur service de renseignement.

À coup sûr, les Japonais s'attendaient à découvrir un atoll endormi et à rééditer leur exploit de Pearl Harbor. Mais cette fois-ci, à Midway, les choses ne se passeraient pas comme ça, grâce aux hommes du chiffre : ici, les avions américains n'étaient pas des cibles immobiles, gentiment rangées sur la piste. À l'arrivée des bombardiers japonais, ils étaient déjà tous dans les airs, prêts à en découdre.

Les officiers et les hommes présents dans la salle de radio du *Yorktown* écoutaient de toutes leurs oreilles les messages entrecoupés de grésillements provenant de Midway comme des bateaux japonais, et étaient convaincus qu'un combat gigantesque était en train de se dérouler sur ce bout d'atoll minuscule. Qui gagnait ? Ils n'en savaient rien.

Peu après, d'autres appareils américains décollèrent de Midway et attaquèrent les porte-avions japonais, portant la bataille au cœur des forces ennemies. Chuck se disait, sans être bien certain de tout comprendre, que les batteries antiaériennes jouaient un rôle capital dans les deux camps. Côté américain, elles avaient évité de gros dégâts à la base de Midway ; côté japonais, elles empêchaient presque toutes les bombes et torpilles lancées contre la flotte japonaise d'atteindre leurs cibles. Un grand nombre d'avions n'en fut pas moins abattu et les pertes furent sensiblement identiques de part et d'autre.

Chuck s'en inquiétait, sachant que les Japonais avaient d'autres forces en réserve.

Juste avant sept heures, le *Yorktown*, l'*Enterprise* et le *Hornet* durent changer de cap et se diriger vers le sud-est, ce qui les éloignait de l'ennemi car il fallait bien que les avions décollent face au vent, qui soufflait du sud-est.

Le puissant *Yorktown* tremblait de toutes parts dans le vacarme des moteurs poussés à plein régime : les avions filaient sur le pont à la suite les uns des autres et étaient catapultés dans les airs.

Chuck remarqua que les Wildcat avaient tendance à lever l'aile droite et à dévier un peu sur la gauche pendant qu'ils prenaient de la vitesse sur le pont, ce dont se plaignaient les pilotes.

À huit heures et demie, les trois porte-avions américains avaient lancé cent cinquante-cinq avions contre les forces ennemies.

Les premiers appareils atteignirent les environs de la cible juste au moment où les avions japonais, de retour de Midway, refaisaient le plein en carburant et en munitions. Synchronisation parfaite : les ponts d'envol étaient jonchés de caisses de munitions, de tuyaux d'essence enchevêtrés, le tout prêt à exploser en un instant. Il aurait dû se produire un carnage.

Celui-ci n'eut pas lieu.

Presque tous les avions américains de cette première vague d'assaut furent détruits.

Les Devastator étaient vétustes. Les Wildcat qui les escortaient étaient plus performants, mais ne soutenaient pas la comparaison avec les Zéro japonais, rapides et maniables. Les avions qui avaient réussi à lâcher leurs projectiles furent décimés par un barrage antiaérien tiré depuis les porte-avions.

Larguer une bombe d'un avion en mouvement sur une cible elle aussi en mouvement, tout comme lâcher une torpille droit sur un navire, était un exercice extrêmement difficile, d'autant plus que les pilotes essuyaient eux-mêmes des tirs venus d'en haut et d'en bas.

La plupart d'entre eux y laissèrent la vie.

Et aucun ne réussit à faire mouche.

Pas une bombe, pas une torpille américaine n'atteignit sa cible. Les trois premières vagues d'assaut lancées depuis les trois porte-avions américains ne causèrent aucun dégât à la force de frappe japonaise. Les munitions éparpillées sur les ponts d'envol n'explosèrent pas, les tuyaux de carburant ne s'enflammèrent pas. Les Japonais étaient indemnes.

Chuck, toujours à l'écoute des messages radio, commençait à désespérer.

Il saisissait avec d'autant plus d'acuité tout le génie de l'attaque contre Pearl Harbor, sept mois plus tôt. Les navires américains, à l'ancre et agglutinés les uns contre les autres, formaient des cibles statiques d'autant plus faciles à atteindre que les chasseurs, censés les protéger, avaient été détruits sur leurs terrains d'aviation. Le temps que les Américains arment et déploient leurs batteries antiaériennes, l'attaque était presque terminée.

La bataille en cours continuait à faire rage. Tous les avions américains n'avaient pas encore atteint les environs de la cible. Chuck entendit un officier de l'air de l'*Enterprise* hurler à la radio : « Attaquez ! Attaquez ! » La réponse laconique du pilote lui parvint : « Bien reçu, dès que je repère ces salauds ! »

Il y avait au moins une bonne nouvelle : le commandant japonais n'avait pas encore envoyé d'avions contre les navires américains. Il devait s'en tenir au plan établi et se concentrer sur Midway. Il avait dû comprendre désormais qu'il était attaqué par des avions décollant de porte-avions, mais n'avait sans doute pas encore déterminé où se trouvaient ces bâtiments.

C'était un avantage, et pourtant les Américains n'arrivaient pas à prendre le dessus.

Soudain, la situation changea. Une escadrille de trente-sept bombardiers Dauntless, partis de l'*Enterprise*, aperçut les Japonais. Or, dans leur combat contre la première vague d'assaut, les Zéro chargés de protéger les navires étaient presque descendus au ras de l'eau. De ce fait, les bombardiers se retrouvèrent au-dessus des chasseurs ce qui leur permit de piquer sur eux, le soleil dans le dos. Quelques minutes plus tard, dix-huit autres Dauntless, venant du *Yorktown*, atteignirent la zone de bataille ; Trixie était aux commandes de l'un d'eux.

Des voix surexcitées hurlaient dans la radio. Fermant les yeux pour mieux se concentrer, Chuck essaya de tirer un sens de tout ce brouhaha. Il ne reconnaissait pas la voix de Trixie.

Ensuite, il commença à entendre en bruit de fond le vrombissement caractéristique des bombardiers qui piquaient. L'attaque avait bel et bien commencé.

Tout à coup, pour la première fois, les pilotes se mirent à pousser des cris de triomphe.

« Je t'ai eu, mon salaud !

— Merde, j'ai cru qu'il allait exploser !

— Prends ça dans ta gueule, fils de pute !

— En plein dans le mille !

— Regarde-le, avec son cul en feu ! »

Dans la salle de radio, les hommes se mirent à applaudir à tout rompre sans vraiment savoir de quoi il retournait.

Cela ne dura que quelques minutes, mais il fallut un bon moment aux opérateurs radio pour avoir une vision claire des événements. Les hurlements de victoire des pilotes euphoriques étaient complètement incohérents. Ce ne fut qu'un peu plus tard, tandis que les pilotes regagnaient leur base, un peu calmés, qu'à bord du vaisseau-amiral, on put se faire une idée de la situation.

Trixie Paxman était sain et sauf.

Cette fois-ci encore, la plupart de leurs bombes avaient raté leur cible. Cependant, quelques-unes avaient fait mouche, une dizaine environ, et ce petit nombre avait causé d'énormes dégâts. Trois imposants porte-avions japonais étaient en feu : le *Kaga*, le *Soryu* et le vaisseau-amiral l'*Akagi*. L'ennemi n'avait plus qu'un porte-avions, le *Hiryu*.

« Trois sur quatre ! s'exclama Chuck, fou de joie. Et ils ne se sont même pas encore approchés de nos navires. »

Très vite, la situation bascula de nouveau.

L'amiral Fletcher avait expédié dix Dauntless en reconnaissance, pour repérer le dernier porte-avions japonais. En fait, ce fut le radar du *Yorktown* qui détecta à quatre-vingts kilomètres de distance une escadrille ennemie, venant probablement du *Hiryu*. À midi, Fletcher fit préparer douze Wildcat pour intercepter les agresseurs. Ordre fut donné aux autres avions encore sur le pont de décoller pour ne pas être vulnérables au moment de l'attaque. En même temps, les conduites de carburant du *Yorktown* furent remplies de dioxyde de carbone, pour parer à tout risque d'incendie.

L'escadrille d'attaque comprenait quatorze « Val », des bombardiers Aichi D3A, et des chasseurs Zéro.

Ça y est ! pensa Chuck, je vais vivre mon premier combat. Il faillit en vomir et dut se forcer à déglutir.

Avant même qu'on ait aperçu les attaquants, les canonniers du *Yorktown* ouvrirent le feu. Le bateau possédait quatre paires de grosses batteries antiaériennes, pourvues de canons de treize centimètres de diamètre, capables d'envoyer des obus à plusieurs kilomètres de distance. Ayant déterminé la position de l'ennemi grâce au radar, les canonniers tirèrent une salve d'énormes obus de vingt-cinq kilos en direction des avions à l'approche, leurs minuteurs réglés pour exploser au contact de leur cible.

Les Wildcat passèrent au-dessus des attaquants et descendirent six bombardiers et trois chasseurs, à en croire les messages radio des pilotes.

Chuck courut à la passerelle avec un message annonçant que l'escadrille piquait droit sur eux. « Eh bien, réagit calmement l'amiral Fletcher, j'ai mon casque sur la tête. Qu'est-ce que je peux faire de plus ? »

Regardant par le hublot, Chuck vit des bombardiers jaillir dans un hurlement strident et fondre sur lui en dessinant un angle tellement fermé qu'ils avaient l'air de tomber à la verticale. Il dut se retenir pour ne pas se jeter au sol.

Brusquement, le navire vira à bâbord. Tout ce qui pouvait détourner de leur trajectoire ces avions d'attaque méritait d'être tenté.

La batterie antiaérienne déployée sur le pont du *Yorktown* comptait quatre « pianos de Chicago » – des armes antiaériennes plus petites, de moindre portée et munies de quatre canons chacune. Elles ouvrirent le feu en même temps que les canons des croiseurs qui escortaient le *Yorktown*.

Chuck, resté sur la passerelle, regardait droit devant lui, terrifié, sans savoir comment se protéger. Un canonnier de pont visa un Val et le toucha. L'avion parut se casser en trois morceaux : deux tombèrent en mer, le troisième vint s'écraser sur le flanc du navire. Puis un autre Val explosa en vol. Chuck applaudit.

Il en restait six.

Et ces Val déterminés à poursuivre le vaisseau-amiral bravaient l'orage meurtrier qui jaillissait de ses canons.

Le *Yorktown* effectua un brusque virage à tribord.

Alors que les Val revenaient à la charge, les mitrailleuses placées sur les passerelles des deux côtés du poste de commandement crépitèrent à leur tour. À présent, toute l'artillerie du *Yorktown* exécutait une symphonie mortelle où les explosions graves des canons de treize centimètres répondaient aux sons plus aigus des pianos de Chicago et aux pétarades forcenées des mitrailleuses.

Chuck vit tomber la première bombe japonaise.

Nombre d'entre elles étaient munies d'un système à retardement et n'explosaient pas au moment de l'impact, mais une ou deux secondes plus tard. Elles devaient passer à travers le pont et s'écraser plus bas afin de se déclencher dans les entrailles du navire, causant un maximum de dégâts.

Or cette bombe-ci se mit à rouler sur le pont du *Yorktown*.

Chuck la suivit des yeux, pétrifié d'horreur. L'espace d'un instant, on put la croire inoffensive mais elle explosa soudain dans un fracas terrible et dans un jaillissement de flammes, pulvérisant les deux pianos de Chicago arrière. Plusieurs petits brasiers apparurent sur le pont et à l'intérieur des tourelles de la cheminée. À la stupéfaction de Chuck, les hommes debout à côté de lui n'avaient pas cillé. À croire qu'ils assistaient à une démonstration de stratégie militaire dans une salle de conférences. L'amiral Fletcher continua à donner ses ordres tout en traversant d'un pas mal assuré le pont branlant de la passerelle de commandement. Quelques instants plus tard, des équipes de secours envahirent la passerelle, les unes chargées de lances d'incendie, les autres de brancards pour évacuer les blessés par les escaliers raides descendant aux postes de secours.

Les feux n'étaient pas importants, le dioxyde de carbone contenu dans les canalisations de carburant ayant empêché leur propagation. Et la piste d'envol était déserte. Il ne restait plus un seul avion chargé de bombes susceptible d'exploser.

Un instant plus tard, un autre Val plongea sur le *Yorktown*. Une bombe toucha la cheminée. Le puissant navire trembla sous l'effet de l'explosion. Un énorme panache de fumée grasse s'échappa des conduits. Les moteurs avaient dû être endommagés, se dit Chuck, car le navire perdit de la vitesse immédiatement.

D'autres projectiles ratèrent leurs cibles et s'écrasèrent en mer en soulevant des geysers. Sur le pont, l'eau de mer se mêlait au sang des blessés.

Le *Yorktown* ralentit et s'arrêta. Le bateau désemparé n'était plus qu'un corps mort ballotté par les flots quand une troisième bombe japonaise l'atteignit, passant à travers une des cages d'ascenseur pour exploser quelque part sous le pont.

Et tout à coup, ce fut fini. Les Val encore indemnes prirent de la hauteur et disparurent dans le ciel bleu du Pacifique.

Je suis vivant ! pensa Chuck.

Le bateau n'était pas perdu. Les équipes anti-incendie furent à pied d'œuvre avant même que les avions japonais soient hors de vue. En bas, les ingénieurs firent savoir qu'ils en avaient pour une heure à réparer les chaudières. Des charpentiers de marine bouchèrent le trou dans le pont d'envol à l'aide de planches en pin Douglas.

Mais les appareils de radio avaient été détruits. Désormais, l'amiral Fletcher était aveugle et sourd. Transféré avec ses plus proches collaborateurs à bord du croiseur *Astoria*, il confia le commandement tactique à Spruance, à bord de l'*Enterprise*.

« Va te faire foutre, Vandermeier, je m'en suis tiré », marmonna Chuck dans sa barbe.

Mais il avait parlé trop vite.

Les moteurs reprirent vie dans un soubresaut. À présent, sous le commandement du capitaine Buckmaster, le *Yorktown* recommença à fendre les flots du Pacifique. Plusieurs de ses avions avaient trouvé refuge sur l'*Enterprise*, d'autres étaient encore en l'air. Aussi se plaça-t-il face au vent. Les appareils se posèrent et se ravitaillèrent. Les radios étant en panne, Chuck et ses camarades organisèrent une équipe de sémaphores pour communiquer avec les autres bâtiments à l'aide des pavillons, comme on le faisait autrefois.

À deux heures et demie, le radar du croiseur qui escortait le *Yorktown* repéra des avions arrivant par l'ouest à faible altitude. Probablement une escadrille qui avait décollé du *Hiryu*. Le croiseur en informa le porte-avions. Buckmaster fit aussitôt partir une patrouille d'interception composée de douze Wildcat.

Ils furent sans doute incapables de remplir leur mission car dix bombardiers-torpilleurs surgirent au ras des flots et se dirigèrent droit sur le *Yorktown*.

Chuck les voyait parfaitement. C'étaient des Nakajima B5N, surnommés « Kate » par les Américains. Ils portaient sous leur fuselage une torpille presque aussi longue que la moitié de l'avion.

Les quatre croiseurs lourds appartenant à l'escorte du *Yorktown* se mirent à bombarder la mer tout autour du vaisseau-amiral, créant un véritable écran d'eau bouillonnante. Les pilotes japonais ne se découragèrent pas pour si peu et traversèrent le rideau d'écume.

Chuck vit l'avion de tête lâcher sa torpille. Elle s'écrasa dans l'eau, pointée sur le *Yorktown*. L'appareil survola le navire à la vitesse de l'éclair, passant si près de Chuck qu'il put voir le visage du pilote : il portait un bandeau rouge et blanc autour de la tête, ainsi que son casque de pilote. Il brandit un poing triomphant en direction de l'équipage sur le pont et disparut.

D'autres avions passèrent en rugissant. Les torpilles étaient lentes et les bâtiments parvenaient parfois à les éviter, mais le *Yorktown* éclopé était trop lourd pour arriver à zigzaguer. Un bruit terrible retentit, ébranlant le vaisseau tout entier. Les torpilles étaient bien plus puissantes que les bombes ordinaires. Chuck eut l'impression que le bateau avait été touché à l'arrière. Une autre explosion suivit de près, qui souleva littéralement le navire, projetant la moitié de l'équipage sur le pont. Les puissants moteurs faiblirent dans l'instant.

Une fois de plus, les charpentiers se mirent au travail sans attendre que les attaquants soient hors de vue. Mais cette fois-ci, les dégâts étaient irréparables. Chuck, qui s'était joint à l'équipe de pompage, vit que la coque d'acier du grand bâtiment s'était ouverte comme une boîte de conserve. Un Niagara d'eau salée se déversait dans l'entaille. En l'espace de quelques instants, Chuck sentit le pont pencher sous ses pieds. Le *Yorktown* gîtait par bâbord.

Face à la quantité d'eau qui s'engouffrait, les pompes étaient impuissantes, d'autant que les compartiments étanches, endommagés sur la mer de Corail, n'avaient pas été réparés pendant les quelques jours de cale sèche.

Combien de temps le *Yorktown* mettrait-il à chavirer ?

À trois heures, Chuck entendit l'ordre tomber : « Abandonnez le navire ! »

Les marins firent descendre des cordages depuis la partie supérieure du pont de plus en plus incliné. D'autres membres d'équipage, sur le pont inférieur, attrapèrent des filins et, en quelques secousses, libérèrent une averse de plusieurs milliers de gilets de sauvetage arrimés au-dessus de leurs têtes. Les bâtiments d'escorte se rapprochèrent et mirent leurs chaloupes à la mer. Les marins du *Yorktown* enlevèrent leurs chaussures et se regroupèrent sur le flanc du navire, non sans avoir rangé au préalable ces centaines de paires de brodequins en lignes irréprochables, comme s'ils sacrifiaient à quelque rituel. Les blessés furent descendus dans les canots de sauvetage sur des brancards. Chuck se retrouva dans l'eau. Il se mit à nager de toute la puissance de ses membres pour s'éloigner du *Yorktown* avant qu'il ne se retourne. Une vague le prit par surprise et lui arracha son calot. Il se réjouit d'être au milieu du Pacifique, dans une eau plus chaude que celle de l'Atlantique où il serait mort de froid en attendant les secours.

Un canot de sauvetage le récupéra et continua à repêcher des hommes, comme le faisaient des dizaines d'autres bateaux. De nombreux membres d'équipage descendaient encore du pont principal, situé sous le pont d'envol. Le *Yorktown* se maintenait toujours à flot.

Dès que tout l'équipage fut en sécurité, il fut réparti à bord des navires d'escorte.

Chuck, debout sur le pont, regarda le soleil se coucher derrière le *Yorktown* toujours en train de sombrer. Il lui vint à l'esprit qu'il n'avait pas vu de la journée un seul bâtiment de la marine japonaise. La bataille s'était déroulée dans les airs, du début à la fin. Il se demanda si c'était une première dans les annales des batailles navales. Le cas échéant, les porte-avions étaient la clé de l'avenir. Rien ne pourrait les surpasser.

Trixie Paxman fut bientôt à ses côtés. Chuck fut si content de le voir vivant qu'il le serra dans ses bras.

Trixie lui apprit qu'au cours de leur dernière sortie, des bombardiers Dauntless de l'*Enterprise* et du *Yorktown* avaient mis le feu au dernier porte-avions japonais, le *Hiryu*, et l'avaient détruit.

« Si je comprends bien, les quatre porte-avions japonais sont hors de combat, résuma Chuck.

— Exactement. On les a tous eus et ils n'ont abattu qu'un seul des nôtres.

— Ça veut dire qu'on a gagné, alors ?

— On dirait bien », répondit Trixie.

5.

Après la bataille de Midway, il fallut se rendre à l'évidence : la bataille du Pacifique serait remportée par des avions décollant de haute mer. Le Japon et les États-Unis se lancèrent dans un programme intensif de construction de porte-avions.

Entre 1943 et 1944, le Japon arma sept de ces énormes vaisseaux coûteux.

Au cours de la même période, les États-Unis en mirent à l'eau quatre-vingt-dix.

XIII

1942 (II)

1.

L'infirmière Carla von Ulrich poussa un chariot dans la réserve et referma la porte derrière elle.

Elle devait se dépêcher. Si elle se faisait prendre, c'était le camp de concentration.

Dans un placard, elle fit main basse sur des pansements de différentes sortes, des bandages et un pot de pommade antiseptique, avant de passer à l'armoire où étaient conservés les médicaments. Elle tourna la clé et prit de la morphine, des sulfamides contre les infections et de l'aspirine contre la fièvre. À cela, elle ajouta une seringue hypodermique toute neuve, encore dans sa boîte.

Cela faisait plusieurs semaines déjà qu'elle reportait de fausses indications dans le registre pour donner à croire que ce qu'elle dérobait maintenant avait été utilisé de façon légitime. Mieux valait falsifier les livres avant de subtiliser quoi que ce soit plutôt que de le faire après, car en cas de contrôle, un surplus ferait croire à une négligence alors qu'un manque serait preuve de vol.

Elle avait déjà fait cela deux fois. Sa frayeur n'avait pas diminué pour autant.

Elle ressortit de la réserve en poussant son chariot de l'air innocent de l'infirmière qui apporte les fournitures nécessaires au chevet d'un malade. Du moins l'espérait-elle.

Elle entra dans la salle. À son grand désarroi, elle aperçut le docteur Ernst, assis auprès d'un patient et lui prenant le pouls.

À cette heure-là, tous les médecins étaient censés être partis déjeuner.

Trop tard pour faire demi-tour ! Avec une assurance qu'elle était loin d'éprouver, elle releva la tête et traversa la salle avec son chariot.

Le docteur Ernst la suivit du regard et sourit.

Berthold Ernst faisait battre le cœur de toutes les infirmières. Chirurgien de talent, chaleureux avec les malades, grand, beau et célibataire, il avait flirté avec la plupart des jolies infirmières et – à en croire les ragots – couché avec nombre d'entre elles.

Elle le salua de la tête et s'éloigna rapidement.

Elle sortit de la salle, poussant toujours son chariot, et entra prestement dans le vestiaire des infirmières.

Son imperméable était suspendu à une patère, couvrant un panier en osier qui contenait un vieux foulard de soie, un chou et un paquet de serviettes hygiéniques dans un sac en papier brun. L'ayant vidé à la hâte, elle y déposa les articles dérobés et les recouvrit d'un foulard aux motifs géométriques bleu et or que sa mère avait dû acheter dans les années 1920. Elle y replaça ensuite le chou et les serviettes hygiéniques, puis remit le panier en place, disposant les pans de son manteau par-dessus.

Ouf, je m'en suis tirée ! se dit-elle. Se rendant compte qu'elle tremblait un peu, elle prit une profonde inspiration, se ressaisit et ressortit dans le couloir, où elle tomba nez à nez avec le docteur Ernst.

L'avait-il suivie ? Allait-il l'accuser de vol ? Il n'avait pas l'air hostile, au contraire, même. Elle n'avait peut-être aucune raison de s'inquiéter.

« Bonjour, docteur. Puis-je faire quelque chose pour vous ? »

Il lui sourit. « Comment allez-vous, mademoiselle ? Tout se passe bien ?

— À la perfection, je crois. » Et d'ajouter d'une voix que la culpabilité rendait doucereuse : « Mais c'est à vous, docteur, de dire si tout se passe bien.

— Pas de plainte, en ce qui me concerne », répondit-il avec condescendance.

De quoi s'agissait-il alors ? se demanda Carla. Serait-il en train de jouer au chat et à la souris, en attendant le moment de m'accuser ?

Elle se tut et attendit, s'efforçant de maîtriser sa peur.

Il regarda le chariot : « Pourquoi l'avez-vous emporté dans le vestiaire ?

— J'avais quelque chose à chercher… dans mon manteau », improvisa-t-elle tant bien que mal. Faisant de son mieux pour réprimer le tremblement de sa voix, elle ajouta : « Un mouchoir que j'avais dans ma poche. » Cesse donc de bredouiller ! se sermonna-t-elle. C'est un médecin, pas un agent de la Gestapo. Pourtant, il la terrorisait.

Le médecin avait l'air de s'amuser de sa nervosité. « Et le chariot ?

— Je vais le remettre en place.

— L'ordre est une vertu essentielle. Vous êtes une excellente infirmière… Fräulein von Ulrich… Ou est-ce Frau ?

— Fräulein.

— Je regrette que nous n'ayons pas davantage d'occasions de nous parler. »

Elle comprit à son sourire qu'il ne soupçonnait pas le moins du monde un vol de matériel : il allait lui demander de sortir avec lui. De quoi faire mourir de jalousie des dizaines d'infirmières, si elle acceptait.

Mais il ne l'intéressait pas. Peut-être parce qu'elle avait déjà aimé un don Juan, Werner Franck, qui s'était révélé lâche et égoïste. Elle soupçonnait Berthold Ernst d'être du même acabit.

Toutefois, ne voulant pas courir le risque de lui déplaire, elle se contenta de sourire sans rien dire.

« Vous aimez Wagner ? »

Elle voyait clair dans son jeu. « Je n'ai pas de temps pour la musique, répondit-elle fermement. Je dois m'occuper de ma vieille mère. » Maud n'avait que cinquante et un ans et était en parfaite santé.

« J'ai deux billets pour un concert demain soir. On donne *Siegfried-Idyll*.

— Ah, de la musique de chambre ! s'exclama-t-elle. C'est inhabituel. » Wagner était plus connu pour ses opéras grandioses.

Il eut l'air ravi. « Je vois que vous vous y connaissez en musique. »

Elle regretta ses paroles, qui n'avaient fait que l'encourager. « Nous sommes une famille de mélomanes. Ma mère donne des leçons de piano.

— Alors, vous devez absolument m'accompagner. Je suis sûr que quelqu'un pourra s'occuper de votre mère pour un soir.

— Ce n'est vraiment pas possible. Je vous remercie tout de même pour l'invitation. » Elle vit de la colère dans ses yeux. Il n'avait pas l'habitude de s'entendre dire non. Elle s'éloigna en poussant son chariot.

« Une autre fois peut-être, alors ? lui lança-t-il.

— C'est très aimable à vous », répondit-elle sans ralentir.

Elle eut peur qu'il ne la suive. Mais sa réponse ambiguë semblait l'avoir adouci et quand elle regarda par-dessus son épaule, elle constata qu'il était parti.

Elle rangea le chariot à sa place et respira plus librement.

Elle retourna à son travail, fit la tournée de tous les malades de la salle et rédigea ses rapports. Puis il fut l'heure de passer le relais à l'équipe de nuit.

Elle enfila son imperméable, et glissa les anses de son panier sur son épaule. Il fallait maintenant quitter le bâtiment avec son butin et la peur l'envahit à nouveau.

Frieda Franck quittait l'hôpital au même moment. Elles sortirent ensemble. Frieda ne pouvait pas savoir que Carla transportait des objets dérobés et elles rejoignirent paisiblement l'arrêt de tram. Si Carla portait un manteau par ce beau soleil de juin, c'était surtout pour ne pas salir son uniforme.

Persuadée d'avoir l'air parfaitement normale, Carla fut très surprise d'entendre Frieda lui demander : « Il y a quelque chose qui te tracasse ?

— Non, pourquoi ?

— Tu as l'air inquiète.

— Non, non, tout va bien… » Pour changer de sujet, elle désigna une affiche : « Oh, tu as vu ça ? »

Le gouvernement venait d'inaugurer au Lustgarten, le parc situé en face de la cathédrale, une exposition consacrée à la vie sous le régime communiste intitulée ironiquement « Le paradis soviétique ». Le bolchevisme y était dépeint comme une ruse concoctée par les Juifs, et les Russes décrits comme des sous-hommes slaves. Mais à l'évidence, tout n'allait pas pour le mieux dans le meilleur des mondes nazi non plus, car quelqu'un avait collé dans tout Berlin des affiches satiriques sur lesquelles on pouvait lire :

Exposition permanente
LE PARADIS NAZI
Guerre Famine Mensonges Gestapo
Combien de temps encore ?

L'une d'elles avait été apposée dans l'abri du tram et sa vue réchauffa le cœur de Carla. « Qui peut bien coller ça ? » demanda-t-elle.

Frieda haussa les épaules en signe d'ignorance.

« Il faut être drôlement courageux, reprit Carla. Ils risquent leur vie. » Elle se rappela ce qu'elle transportait dans son panier. Elle aussi pouvait être tuée si elle se faisait prendre.

« C'est sûr », répondit Frieda laconiquement. Elle avait l'air à son tour sur des charbons ardents. Se pouvait-il qu'elle appartienne à ce groupe de colleurs d'affiches ? Probablement pas. Son petit ami, peut-être. Heinrich. Il était de ces gens passionnés et moralisateurs, capables de faire ce genre de choses. Carla demanda : « Comment va Heinrich ?

— Il veut qu'on se marie.

— Pas toi ? »

Frieda baissa la voix. « Je ne veux pas d'enfant. » C'était une déclaration parfaitement séditieuse, car les jeunes femmes étaient supposées se réjouir d'enfanter pour la plus grande gloire du Führer. Désignant l'affiche interdite, elle poursuivit : « Je n'ai pas envie de mettre un enfant au monde dans ce paradis-là.

— Moi non plus, certainement », renchérit Carla. C'était peut-être aussi, sans qu'elle en eût vraiment conscience, une des raisons qui l'avaient poussée à éconduire le docteur Ernst.

Le tram arriva. Une fois assise, Carla posa négligemment son panier sur ses genoux, comme si de rien n'était. Elle détailla les autres passagers et fut soulagée de ne pas apercevoir d'uniforme.

« Viens à la maison, proposa Frieda, on écoutera du jazz. On mettra les disques de Werner.

— J'aimerais bien, mais je ne peux pas, répondit Carla. Je dois voir quelqu'un. Tu te souviens des Rothmann ? »

Frieda jeta autour d'elle des regards méfiants, car ce patronyme pouvait très bien être juif. Par bonheur, personne ne se trouvait assez près pour les entendre. Elle souffla : « Bien sûr, c'était notre médecin, avant.

— Il n'est plus censé exercer. Eva, sa fille, s'est installée à Londres avant la guerre. Elle a épousé un militaire écossais. Les parents ne peuvent pas quitter l'Allemagne, évidemment. Leur fils, Rudi, qui était luthier, un excellent artisan, dit-on, a perdu son travail, lui aussi. Il ne peut plus fabriquer d'instruments, il se contente de les réparer et d'accorder des pianos. » Autrefois, il venait chez les Ulrich quatre fois par an pour accorder leur Steinway de concert. « J'ai promis de passer les voir ce soir.

— Ah », dit Frieda, laissant traîner cette voyelle d'un air entendu.

Carla réagit aussitôt : « Comment ça : ah ?

— Je comprends maintenant pourquoi tu te cramponnes à ton panier comme s'il contenait le Saint Sacrement ! »

Frieda avait percé son secret ! Carla en fut pétrifiée. « Comment as-tu deviné ?

— Tu as dit qu'il n'était pas censé exercer, ce qui donne à entendre qu'il le fait quand même. »

Carla comprit qu'elle avait vendu la mèche. Elle aurait dû dire : le docteur Rothmann n'est pas *autorisé* à exercer. Heureusement, ce n'était que Frieda ! Elle se justifia : « Qu'est-ce qu'il y peut, si les gens continuent de frapper à sa porte pour le supplier de les aider. Il ne peut tout de même pas renvoyer des malades. En plus, il ne gagne pas un sou. Ce sont tous des Juifs ou de pauvres gens qui le paient en pommes de terre ou en œufs.

— Tu n'as pas besoin de plaider sa cause, dit Frieda, chuchotant toujours. Je sais qu'il est courageux. Quant à toi, tu es héroïque. C'est la première fois que tu voles du matériel médical pour lui ? »

Carla secoua la tête. « La troisième. Mais je m'en veux tellement de t'avoir laissé deviner. Quelle idiote je suis !

— Mais non. C'est simplement que je te connais trop bien. »

Le tram s'immobilisa à l'arrêt de Carla. « Souhaite-moi bonne chance ! » dit-elle en descendant.

En entrant dans la maison, elle fut accueillie par des notes de piano maladroites provenant du premier étage. Sa mère avait un élève. Carla s'en réjouit : ça lui remonterait le moral et ça rapporterait un peu d'argent à la famille.

Elle retira son imperméable et se rendit à la cuisine pour dire bonjour à Ada. Après la mort de Walter, quand Maud lui avait annoncé qu'elle n'avait plus les moyens de lui payer ses gages, Ada avait insisté pour rester à son service. Le soir, elle faisait le ménage dans des bureaux et pendant la journée, elle travaillait chez les von Ulrich en échange du vivre et du couvert.

Carla retira ses chaussures sous la table et frotta ses pieds endoloris l'un contre l'autre. Ada lui versa une tasse d'ersatz de café.

Maud fit son entrée, l'œil brillant. « Un nouvel élève ! » Elle montra à Carla une poignée de billets. « Et il veut venir tous les jours ! » Elle lui avait demandé de faire des gammes, et c'étaient ses doigts novices qui produisaient les notes maladroites qu'on entendait et qui évoquaient un chat déambulant sur le clavier.

« Épatant, s'exclama Carla. Et qui est-ce ?

— Un nazi, forcément. Mais on ne peut pas cracher sur son argent.

— Comment s'appelle-t-il ?

— Joachim Koch. C'est un jeune homme timide. Si tu le croises, par pitié, retiens ta langue et sois polie !

— Promis. »

Maud ressortit.

Carla but son café avec plaisir. Comme presque tout le monde, elle s'était habituée à ce goût de glands grillés.

Elle bavarda tranquillement de choses et d'autres avec Ada pendant quelques minutes. Ada, qui avait été grassouillette, était toute maigre à présent. Dans l'Allemagne d'alors, les gros n'étaient pas légion, il est vrai, mais dans le cas d'Ada, les pénuries n'étaient pas seules en cause. La mort de Kurt, son fils handicapé, l'avait cruellement frappée, et plus rien ne l'intéressait. Elle effectuait son travail consciencieusement mais ensuite, elle s'asseyait près de la fenêtre et restait des heures à regarder au-dehors, le visage inexpressif. Carla, qui l'aimait beaucoup et comprenait sa douleur, ne savait comment la réconforter.

La musique s'interrompit. Peu après, deux voix se firent entendre dans l'entrée, celle de Maud et une autre, masculine. Sa mère devait raccompagner Herr Koch à la porte, se dit Carla.

Et c'est avec horreur qu'elle la vit, un instant plus tard, pénétrer dans la cuisine, suivie de près par un jeune homme en uniforme de lieutenant impeccable.

« Ma fille, lança Maud d'un ton jovial. Carla, je te présente le lieutenant Koch, un nouvel élève. »

C'était un jeune homme d'une vingtaine d'années, plutôt séduisant mais à l'air timide. Avec sa moustache blonde, il rappela à Carla des photos de son père jeune.

Son cœur se mit à battre la chamade. Le panier contenant les fournitures médicales dérobées était là, dans la cuisine, sur la chaise à côté d'elle. Allait-elle se trahir une nouvelle fois, comme tout à l'heure avec Frieda ?

« Ravie de faire votre connaissance », bredouilla-t-elle.

Maud la dévisagea, surprise par la nervosité de sa fille. Quel mal y avait-il à faire entrer un officier dans la cuisine ? Elle ne demandait qu'une chose à Carla : d'être aimable avec ce nouvel élève pour qu'il continue à prendre des leçons.

Koch s'inclina cérémonieusement. « Tout le plaisir est pour moi.

— Et voici Ada, notre bonne. »

Celle-ci jeta un regard hostile au nouveau venu, mais il ne le remarqua pas : les domestiques ne méritaient pas son attention. Voulant paraître à l'aise, il prit appui sur une jambe et n'en parut que plus emprunté.

Il se conduisait avec une étrange puérilité, la naïveté d'un petit garçon trop couvé. Il n'en représentait pas moins un danger.

Changeant de position, il s'appuya des deux mains au dossier de la chaise sur laquelle Carla avait posé son panier. « Vous êtes infirmière à ce que je vois.

— Oui », répondit Carla en essayant de réfléchir posément. Koch savait-il qui étaient les von Ulrich ? Peut-être était-il trop jeune pour avoir entendu parler des sociaux-démocrates. Cela faisait déjà neuf ans que le parti avait été interdit. Peut-être la mort de Walter avait-elle effacé en partie l'infamie attachée au nom de von Ulrich. Koch semblait les prendre pour une famille allemande respectable, tombée dans la misère après la disparition de leur seul soutien. C'était une situation fréquente parmi les femmes de bonne éducation.

Il n'y avait aucune raison pour qu'il s'inquiète du contenu de son panier.

Carla se força à être aimable. « Comment vous en sortez-vous au piano ?

— Je crois que je fais des progrès rapides... Du moins, c'est ce qu'affirme mon professeur, ajouta-t-il en se tournant vers Maud.

— Il n'en est qu'aux tout débuts, mais manifeste déjà un talent certain ! » déclara celle-ci.

C'était une phrase standard qu'elle répétait systématiquement à tous ses nouveaux élèves pour les encourager à revenir. Mais aujourd'hui, Carla eut l'impression qu'elle faisait assaut d'amabilité. Pourquoi pas ? se dit-elle. Sa mère avait bien le droit de faire du charme, cela faisait plus d'un an qu'elle était veuve. Mais elle ne pouvait évidemment pas éprouver de sentiment pour un jeune homme deux fois plus jeune qu'elle.

« De toute façon, j'ai décidé de ne rien dire à mes amis avant de jouer correctement, précisa Koch. Ce jour-là, ils seront drôlement épatés.

— Ce sera très amusant, en effet, approuva Maud. Lieutenant, prenez donc un siège, si vous avez un instant. » Elle désigna la chaise sur laquelle était posé le panier de Carla.

Celle-ci se précipita pour le prendre, mais Koch fut plus rapide. « Permettez ! dit-il en le soulevant. Votre dîner je suppose ? ajouta-t-il en voyant le chou.

— Oui », répondit Carla d'une voix proche du glapissement.

Il s'assit sur la chaise et posa le panier par terre à ses pieds, loin de Carla. « Je me suis toujours imaginé avoir un don pour la musique. J'ai décidé qu'il était temps de voir si c'était vrai. » Il croisa les jambes, pour les décroiser aussitôt.

Carla se demanda pourquoi il était aussi nerveux. Il n'avait rien à craindre, ici. L'idée que cette agitation puisse être d'ordre sexuel lui traversa l'esprit : il était seul au milieu de trois femmes. Comment savoir ce qui lui passait par la tête ?

Ada posa une tasse de café devant lui. Il sortit ses cigarettes. Il fumait maladroitement, comme un adolescent. Ada lui donna un cendrier.

« Le lieutenant Koch travaille au ministère de la Guerre, Bendlerstrasse, dit Maud.

— Exactement ! »

C'était le siège de l'Oberkommando, le haut commandement de la Wehrmacht, là où étaient conservés les secrets militaires de la plus haute importance. Mieux valait que le lieutenant ne parle pas là-bas de ses leçons de piano. Si Koch ignorait tout de la famille von Ulrich, certains de ses collègues se souvenaient certainement que Walter von Ulrich avait été un antinazi convaincu. Et Frau von Ulrich perdrait un élève.

« C'est un grand honneur de travailler dans ce service, ajouta Koch.

— Mon fils est en Russie, l'informa Maud. Nous sommes mortes d'inquiétude pour lui.

— Pour une mère, c'est bien naturel. Mais ne soyez pas pessimiste, s'il vous plaît ! La récente contre-offensive russe a été repoussée définitivement. »

Ce n'était que mensonge. La machine de propagande allemande n'avait pu dissimuler que les Soviétiques avaient remporté la bataille et repoussé les lignes allemandes de plus de cent cinquante kilomètres.

« À l'heure actuelle, continuait Koch, nous sommes en mesure de reprendre notre progression.

— En êtes-vous sûr ? » demanda Maud, visiblement inquiète. Carla partageait ses craintes. Elles redoutaient, l'une comme l'autre, qu'il n'arrive quelque chose à Erik.

Koch esquissa un sourire condescendant. « Croyez-moi, Frau von Ulrich, je suis sûr de ce que je dis. Je ne peux évidemment pas vous confier tout ce que je sais, mais je peux vous assurer qu'une autre offensive très agressive est planifiée.

— Je suis certaine que nos troupes ne manquent ni de nourriture ni de rien… » Elle posa la main sur le bras de Koch. « Pourtant, je ne peux m'empêcher d'être inquiète. Je ne devrais pas vous le dire, je sais, mais je sens que je peux vous faire confiance, lieutenant.

— Bien entendu.

— Cela fait des mois que je suis sans nouvelles de mon fils. Je ne sais même pas s'il est mort ou vivant. »

Koch sortit de sa poche un crayon et un petit carnet. « Je peux certainement me renseigner, proposa-t-il.

— C'est vrai ? » demanda Maud en écarquillant les yeux.

Voilà donc pourquoi sa mère se mettait en frais pour lui, pensa Carla pendant que Koch répliquait : « Mais certainement. Je suis à l'état-major, vous savez, même si je n'y tiens qu'un rôle insignifiant. » Il s'efforçait de jouer les modestes. « Je peux me renseigner au sujet de…

— Erik.

— Erik von Ulrich.

— Oh, ce serait merveilleux ! Il est infirmier. Il avait commencé ses études de médecine, voyez-vous, et s'est tout de suite porté volontaire pour se battre pour le Führer. »

C'était vrai. Erik était un nazi convaincu, bien qu'il ait donné l'impression, dans ses dernières lettres, de déchanter un peu.

Koch nota le nom dans son calepin.

« Vous êtes merveilleux, lieutenant Koch ! s'exclama Maud.

— Ce n'est rien.

— Je suis si contente que nous soyons sur le point de contre-attaquer. Vous ne pouvez évidemment pas me dire quand l'offensive aura lieu. Je meurs pourtant d'envie de le savoir, vous vous en doutez. »

Carla ne comprenait pas pourquoi sa mère cherchait à soutirer des informations à ce jeune homme. Que pouvait-elle en faire ?

Koch baissa la voix comme si un espion risquait d'être tapi sous la fenêtre ouverte. « C'est pour très bientôt ! » Il posa les yeux sur les trois femmes tour à tour, prenant manifestement plaisir à les voir tout ouïe. Peut-être ne lui arrivait-il pas souvent d'avoir des femmes suspendues à ses lèvres, se dit Carla.

« L'opération Fall Blau est imminente », lâcha-t-il enfin, après avoir visiblement savouré leur attente.

Maud darda les yeux sur lui. « L'opération Fall Blau, que c'est passionnant ! s'écria-t-elle sur le ton d'une femme qu'un homme invite à passer une semaine au Ritz, à Paris.

— Le 28 juin », ajouta-t-il dans un murmure.

Maud posa la main sur son cœur. « Déjà ! C'est merveilleux !

— Je n'aurais pas dû vous le dire. »

Maud couvrit sa main de la sienne. « Oh, je suis si contente que vous l'ayez fait, je me sens tellement mieux ! »

Il gardait le regard rivé sur la main de Maud et Carla se rendit compte qu'il n'était pas habitué aux caresses féminines. Il releva les yeux sur Maud. Elle lui souriait avec chaleur, tant de chaleur que Carla eut du mal à croire que ce sourire était entièrement feint.

Maud retira sa main. Koch écrasa sa cigarette et se leva. « Il faut que j'y aille. »

Pas trop tôt ! pensa Carla.

Il s'inclina devant elle : « J'ai été très heureux de faire votre connaissance, Fräulein.

— Au revoir, lieutenant », répondit-elle d'une voix neutre.

Maud le raccompagna jusqu'à la porte. « À demain, alors. À la même heure. »

De retour à la cuisine, elle s'écria : « Un garçon stupide qui travaille à l'état-major, c'est le gros lot !

— Je ne vois pas ce qui te met dans cet état, réagit Carla.

— Il est très beau, intervint Ada.

— Il vient de nous révéler une information secrète !

— Et ça nous sert à quoi ! Nous ne sommes pas des espionnes, répliqua Carla.

— Nous connaissons la date de la prochaine offensive. Il y a sûrement un moyen de transmettre ce renseignement aux Soviétiques, non ?

— Je ne vois pas comment.

— La ville est censée être truffée d'espions.

— Pure propagande. Dès que quelque chose se passe mal, on accuse les menées subversives des agents secrets judéo-bolcheviques au lieu de reconnaître l'incompétence des nazis.

— Il doit bien y avoir des espions, quand même. Des vrais.

— Comment comptes-tu entrer en contact avec eux ? »

Maud réfléchit un instant. « Je demanderai à Frieda.

— À Frieda ? Et pourquoi ?

— Une intuition. »

Carla se remémora le silence de son amie à l'arrêt du tram, au moment où elle s'était demandé qui pouvait bien placarder ces affiches antinazies. Sa mère n'avait sans doute pas tort.

Ce n'était cependant pas la seule difficulté. « Même si nous le pouvions, accepterions-nous de trahir notre pays ? »

La réponse de Maud fut catégorique : « Il faut renverser les nazis.

— Je les déteste moi aussi plus que tout au monde, mais je suis allemande.

— Je comprends ce que tu veux dire. L'idée de trahir ne me plaît pas plus qu'à toi, même si je suis anglaise de naissance. Mais le seul moyen de nous débarrasser des nazis, c'est de perdre la guerre.

— Admettons que le renseignement livré aux Russes leur permette de remporter une bataille, Erik pourrait trouver la mort dans ces combats ! Ton fils, mon frère ! Il pourrait mourir à cause de nous. »

Maud voulut répondre, mais les larmes l'en empêchèrent. Carla se leva et la prit dans ses bras.

Une minute plus tard, Maud murmura : « De toute façon, il risque de mourir. De mourir en se battant pour les nazis. Mieux vaut qu'il meure en perdant la bataille plutôt qu'en la gagnant. »

Carla n'en était pas convaincue.

Elle s'écarta de sa mère. « La prochaine fois, j'aimerais mieux que tu me préviennes avant de faire entrer quelqu'un de ce genre dans la cuisine. » Elle ramassa son panier. « Heureusement qu'il n'a pas cherché à savoir ce que j'avais là-dedans !

— Pourquoi ? Tu as quoi ?

— Des médicaments que j'ai volés pour le docteur Rothmann. »

Maud sourit fièrement à travers ses larmes. « Je te reconnais bien là !

— Quand il a pris le panier, j'ai failli mourir de peur.

— Je suis désolée.

— Tu ne pouvais pas deviner. Bon, je file me débarrasser de tout ça.

— Excellente idée. »

Carla remit son imperméable par-dessus son uniforme et sortit.

Elle partit à pied, d'un pas rapide. Les fenêtres de la maison des Rothmann étaient désormais barricadées et une pancarte grossière avait été clouée sur la porte : « Cabinet médical fermé. »

Le docteur comptait jadis parmi sa clientèle quelques patients fortunés, mais il soignait surtout des pauvres gens qui le payaient quand ils le pouvaient. Aujourd'hui, seuls les pauvres s'adressaient encore à lui.

Carla passa discrètement par-derrière.

Elle comprit immédiatement qu'il s'était passé quelque chose. La porte de service était restée ouverte et une guitare traînait par terre, sur le carrelage, le manche brisé. Il n'y avait personne, mais elle entendit du bruit ailleurs dans la maison.

Elle rejoignit le vestibule. Le rez-de-chaussée comportait deux pièces en plus de la cuisine : l'une servait jadis de salle d'attente, l'autre de cabinet de consultation. À présent, la salle d'attente faisait office de salon et le cabinet abritait l'atelier de Rudi. Il y avait là un établi, ses outils pour travailler le bois, et une demi-douzaine de mandolines, violons et contrebasses à divers stades de réparation. Tout l'équipement médical était rangé, invisible, dans des placards fermés à clé.

Ce n'était plus le cas aujourd'hui, comme Carla le constata depuis le seuil.

Les placards avaient été fracturés et leur contenu éparpillé sur le sol, lequel était jonché d'éclats de verre, de pilules, de poudres et de liquides. Carla reconnut un stéthoscope et un tensiomètre parmi les débris. Des morceaux d'instruments gisaient çà et là, fracassés ou écrasés.

Carla fut bouleversée. Quel gâchis ! se dit-elle, dégoûtée.

Elle passa dans la pièce voisine. Rudi Rothmann y était, allongé par terre. Ce jeune homme grand et athlétique de vingt-deux ans gémissait de douleur dans un coin, les yeux fermés.

Sa mère, Hannelore, était agenouillée à côté de lui. Mme Rothmann, qui avait été une belle blonde, était aujourd'hui une femme émaciée aux cheveux tout gris.

« Que s'est-il passé ? » s'écria Carla. Mais elle avait déjà compris.

« La police, expliqua Hannelore. Ils ont accusé mon mari de soigner des patients aryens. Ils l'ont emmené. Rudi a voulu s'interposer quand ils ont commencé à saccager la pièce. Ils l'ont… » Sa voix s'étrangla.

Carla posa son panier et vint s'accroupir près d'elle. « Qu'est-ce qu'ils lui ont fait ?

— Ils lui ont cassé les mains », lâcha Hannelore dans un murmure.

Carla se tourna vers Rudi. Ses mains étaient rouges et affreusement tordues. Les policiers avaient dû lui briser les doigts l'un après l'autre. Il souffrait certainement le martyre. C'était un spectacle qui lui soulevait le cœur, mais Carla voyait des horreurs chaque jour et avait appris à contrôler ses sentiments pour prodiguer les soins nécessaires. « Il lui faut de la morphine ! »

Hannelore désigna le fouillis qui les entourait. « Si nous en avions, il n'en reste rien. »

Carla fut folle de rage à l'idée que la police, dans son orgie de destruction, ait pu gâcher de précieux médicaments dont on manquait même dans les hôpitaux. « Je vous en ai apporté, justement. » Elle prit dans son panier une ampoule de liquide transparent et la seringue toute neuve, qu'elle dégagea de son étui et remplit adroitement. Puis elle fit une piqûre à Rudi.

L'effet fut quasi immédiat. Ses gémissements s'arrêtèrent, Rudi ouvrit les yeux et regarda Carla. « Tu es un ange ! » murmura-t-il, puis il ferma les yeux et parut s'endormir.

« Il faut lui remettre les doigts en place, dit Carla, pour que les os se ressoudent correctement. » Elle toucha délicatement la main gauche de Rudi, qui ne réagit pas. Elle saisit alors sa main et la souleva. Pas de réaction.

« Je n'ai jamais réduit de fracture, fit Hannelore. Mais j'ai vu plusieurs fois comment on faisait.

— Moi aussi, reprit Carla. De toute façon, il faut essayer. Je prends la main gauche, occupez-vous de la droite. Il faut qu'on ait fini avant que la morphine cesse de faire de l'effet. Il aura déjà assez mal comme ça.

— Entendu », acquiesça Hannelore.

Carla attendit un instant, plongée dans ses réflexions. Sa mère avait raison : elles devaient faire tout ce qui était en leur pouvoir, même trahir leur pays, pour en finir avec ce régime nazi. Tous ses doutes s'étaient dissipés.

« Allons-y ! » dit-elle.

Délicatement, avec un soin extrême, les deux femmes entreprirent de remettre en place les doigts brisés de Rudi.

2.

Tous les vendredis après-midi, Thomas Macke allait au bar de Tannenberg.

L'endroit n'avait rien de grandiose : au mur, la photographie du propriétaire, Fritz, en uniforme de la Première Guerre mondiale, sans sa bedaine et avec vingt ans de moins. À l'en croire, il avait tué neuf Russes à la bataille de Tannenberg. Dans la salle, quelques tables et des chaises, mais les habitués se regroupaient au bar. Le menu, sous sa reliure en cuir, était pure fiction, car on ne servait que deux plats : saucisses avec pommes de terre et saucisses sans pommes de terre.

Ce café, situé juste en face du commissariat de police de Kreuzberg, était essentiellement fréquenté par des policiers, ce qui permettait d'y enfreindre la loi en toute impunité. Le jeu s'y pratiquait ouvertement, les filles faisaient des pipes dans les toilettes, et aucun inspecteur des services sanitaires ne mettait jamais les pieds dans les cuisines. Le bar ouvrait à l'heure où Fritz se levait et il fermait une fois le dernier buveur parti.

Des années auparavant, alors qu'il n'était que policier subalterne, Macke avait travaillé au commissariat de Kreuzberg. C'était avant que les nazis ne prennent le pouvoir, offrant une seconde chance à des hommes comme lui. Au Tannenberg, il était sûr de rencontrer un ou deux visages connus, car plusieurs de ses anciens collègues venaient encore y boire un coup. Et Macke aimait bien discuter avec les vieux amis, même s'il s'était élevé bien au-dessus d'eux en devenant commissaire et membre des SS.

« Tu as fait une belle carrière, Thomas, je te l'accorde, déclara le sergent Bernhardt Engel qui, sous ce grade déjà, avait été le supérieur de Macke en 1932. Bonne chance à toi, mon vieux. » L'éternel sergent porta à ses lèvres la pinte de bière que Macke lui avait offerte.

« Je ne vais pas te contredire, répondit Macke. Mais je t'avouerai que c'est bien plus dur de travailler avec le commissaire principal Kringelein qu'avec toi.

— J'étais trop gentil avec vous, les gars », admit Bernhardt.

Un autre vieux camarade, Franz Edel, s'esclaffa non sans mépris : « Gentil ? Je ne dirais pas ça ! »

En regardant par la fenêtre, Macke vit une moto se garer. Elle était pilotée par un jeune homme en blouson bleu clair des officiers de l'armée de l'air qu'il eut l'impression d'avoir déjà rencontré. Oui, il l'avait déjà vu quelque part, ce type aux cheveux blond vénitien un peu trop longs qui retombaient sur un front patricien. Il le suivit des yeux pendant qu'il traversait le trottoir et entrait au Tannenberg.

Macke se souvint de son nom : c'était Werner Franck, le fils à papa du fabricant de radios Ludwig Franck.

Werner s'avança vers le comptoir et commanda un paquet de Kamel. Évidemment ! Un garçon pareil ne pouvait fumer que des américaines, pensa Macke, même si ce n'était qu'une imitation de fabrication allemande.

Werner paya, ouvrit le paquet, en sortit une cigarette et demanda du feu à Fritz. Se tournant pour s'éloigner, la cigarette aux lèvres d'un air canaille, il croisa le regard de Macke. « Commissaire Macke ? » lâcha-t-il après un bref instant d'hésitation.

Tous les clients regardèrent Macke, attendant sa réaction.

Celui-ci opina tranquillement. « Comment allez-vous, jeune Werner ?

— Très bien, commissaire, merci. »

Macke apprécia le ton déférent de Werner mais fut un peu surpris. Il gardait de lui le souvenir d'un freluquet arrogant, peu respectueux de l'autorité.

« Je reviens d'une visite sur le front de l'Est en compagnie du général Dorn », ajouta Werner.

Macke sentit les policiers dresser l'oreille : un homme qui avait été sur le front de l'Est méritait le respect. Et ses anciens collègues étaient épatés de découvrir qu'il évoluait dans des cercles aussi élevés. Involontairement, il en éprouva un certain plaisir.

Werner lui tendit son paquet de cigarettes, et Macke en prit une. « Une bière, commanda Werner à Fritz, puis il se tourna vers Macke : Puis-je vous offrir un verre, commissaire ?

— La même chose, merci. »

Fritz remplit deux bocs. Werner leva le sien en disant : « Je voudrais vous remercier. »

Nouvelle surprise. « Et de quoi ? » demanda Macke.

Ses amis ouvraient grand leurs oreilles.

« De la bonne leçon que vous m'avez donnée, il y a un an, répondit Werner.

— À l'époque, vous ne débordiez pas de gratitude.

— Et je m'en excuse. Mais j'ai beaucoup réfléchi à ce que vous m'avez dit, et j'ai compris que vous aviez raison. L'émotion m'avait obscurci l'esprit. Vous m'avez remis dans le droit chemin, je ne l'oublierai jamais. »

Macke fut touché. Werner lui avait inspiré une profonde antipathie, et il l'avait traité avec rudesse. Et voilà que le jeune homme avait tenu compte de ses paroles et changé d'attitude. Savoir qu'il était à l'origine d'une telle conversion lui fit chaud au cœur.

« Justement, poursuivit Werner, j'ai pensé à vous l'autre jour. Le général Dorn évoquait le problème des espions et se demandait s'il était possible de les repérer à l'aide de leurs signaux radio. Malheureusement, j'ai été bien en peine de lui répondre.

— Vous auriez dû me poser la question, c'est ma spécialité.

— Vraiment ?

— Venez, on va s'asseoir. »

Ils prirent leurs verres et s'installèrent à une table d'une propreté douteuse.

« Ces hommes sont tous des policiers, chuchota Macke, mais on ne sait jamais. Ces sujets ne sont pas censés être abordés en public.

— Bien sûr, dit Werner et il ajouta en baissant le ton : Je sais que je peux vous faire confiance. Voyez-vous, des commandants sur le terrain ont confié au général Dorn qu'ils ont souvent eu l'impression que l'ennemi était informé de nos intentions à l'avance.

— Ah ! soupira Macke, je le craignais.

— Que puis-je dire à Dorn sur la détection des signaux radio ?

— Le mot correct est goniométrie. » Macke rassembla ses idées. C'était une occasion inespérée d'impressionner un puissant général, fût-ce indirectement. Il fallait qu'il soit clair et insiste sur l'importance de ses missions, sans pour autant exagérer ses succès. Il imaginait déjà le général Dorn glissant incidemment au Führer un mot à son sujet : « Il y a un type très compétent à la Gestapo. Un certain Macke. Il n'est encore que commissaire, mais il est tout à fait remarquable… »

« Nous disposons d'un instrument qui nous indique la direction d'où vient le signal, commença-t-il. En prenant les coordonnées de trois lieux différents, assez éloignés les uns des autres, et en les reportant sur la carte, on peut tracer trois lignes. L'émetteur se trouve à l'intersection de ces lignes.

— C'est fantastique ! »

Macke leva la main dans un geste de mise en garde. « En théorie, oui. En pratique, c'est plus compliqué, car le pianiste, puisque que c'est ainsi qu'on appelle l'opérateur radio, ne reste généralement pas au même endroit assez longtemps pour que nous puissions lui mettre le grappin dessus. Un bon pianiste n'émet jamais deux fois de suite à partir du même site. Et comme notre instrument se trouve dans une camionnette munie d'une antenne sur le toit, ils nous voient arriver de loin.

— Mais avez-vous déjà réussi ?

— Bien sûr. Vous devriez nous accompagner un soir, dans la camionnette. Vous pourriez assister au déroulement des opérations et transmettre au général Dorn un rapport de visu.

— Quelle bonne idée ! » s'exclama Werner.

3.

À Moscou, le mois de juin était chaud et ensoleillé. À l'heure du déjeuner, Volodia attendit Zoïa près d'une fontaine dans les jardins Alexandre, au pied du Kremlin. Des centaines de badauds s'y promenaient, souvent en couple, profitant du beau temps. La vie était dure, on avait coupé l'eau des fontaines par souci d'économie, mais le ciel était bleu, les arbres couverts de feuilles et l'armée allemande à cent cinquante kilomètres.

Volodia débordait d'orgueil chaque fois qu'il pensait à la bataille de Moscou. La redoutable armée allemande, pourtant spécialisée dans les attaques éclairs, était arrivée aux portes de la ville et avait été repoussée. Les Soviétiques s'étaient battus comme des lions pour défendre leur capitale.

Malheureusement, la contre-offensive russe, qui avait permis de reconquérir un vaste territoire et de redonner confiance aux Moscovites, avait tourné court en mars. L'armée allemande avait désormais pansé ses plaies et se préparait à repartir à l'attaque.

Et Staline était encore au pouvoir.

Volodia vit Zoïa se diriger vers lui à travers la foule. Elle portait une robe à carreaux rouge et blanc. Sa démarche était pleine d'allant et ses cheveux blond pâle dansaient au rythme de ses pas. Elle attirait les regards de tous les hommes.

Volodia était sorti avec beaucoup de jolies filles avant Zoïa. Pourtant, il s'étonnait encore de sa chance. Pendant des années, elle l'avait traité avec une indifférence glaciale, ne parlant avec lui que de physique nucléaire. Et puis un jour, peu après l'émeute qui avait coûté la vie au général Bobrov, elle lui avait demandé, à sa grande surprise, de l'accompagner au cinéma.

Depuis lors, elle avait changé d'attitude à son égard, sans qu'il comprenne très bien pourquoi. Sans doute, cette expérience partagée les avait-elle rapprochés. Quoi qu'il en soit, ils étaient allés voir *Let George do it* avec l'Anglais George Formby, un joueur de banjolélé. Cette comédie britannique, à l'affiche depuis des mois, connaissait un immense succès. L'intrigue, d'un irréalisme achevé, retraçait les péripéties d'un musicien qui envoyait à son insu des messages aux U-Boots allemands par le truchement de son instrument. C'était si bête qu'ils en avaient ri aux larmes.

Depuis, ils se voyaient régulièrement.

Aujourd'hui, comme ils devaient déjeuner avec son père, Volodia avait donné rendez-vous à Zoïa près de la fontaine pour pouvoir passer un moment en tête à tête avec elle.

Elle lui adressa un sourire radieux et se haussa sur la pointe des pieds pour l'embrasser. Elle était grande, mais il était plus grand encore. Il lui rendit son baiser. Ses lèvres étaient douces et humides. Ce fut trop court à son goût.

Volodia n'était pas certain des sentiments qu'elle éprouvait pour lui. Pour le moment, ils ne faisaient que « se fréquenter », comme disait la vieille génération. Ils s'embrassaient souvent, mais ils n'avaient pas encore fait l'amour. Ils n'étaient pourtant pas trop jeunes puisqu'il avait vingt-sept ans et elle vingt-huit. Volodia savait que Zoïa ne coucherait avec lui que lorsqu'elle serait prête.

C'était un rêve auquel il ne croyait pas vraiment. Elle était trop blonde, trop intelligente, trop grande, trop sûre d'elle, trop séduisante pour se donner à lui. Il n'aurait sûrement jamais l'occasion de la regarder se déshabiller, de la contempler nue, de caresser son corps, de s'étendre sur elle…

Ils traversèrent cette place tout en longueur, bordée d'un côté par une rue très fréquentée, de l'autre par les hautes murailles derrière lesquelles se dressaient les tours du Kremlin. « Quand on voit ces murs, on serait tenté de penser que nos dirigeants y sont retenus prisonniers par la volonté du peuple, dit Volodia.

— C'est vrai, acquiesça Zoïa. Alors que c'est l'inverse. »

Il jeta un coup d'œil derrière lui pour s'assurer que personne ne l'avait entendue. Il était imprudent de s'exprimer ainsi. « Je comprends que mon père te trouve dangereuse.

— Avant, je pensais que tu étais comme lui.

— Si seulement ! C'est un héros : il a pris le palais d'Hiver. Pour ma part, ça m'étonnerait que je change un jour le cours de l'histoire.

— Oh, je sais, mais il est tellement étroit d'esprit, tellement conservateur. Tu n'es pas comme ça, toi. »

Volodia trouvait qu'il ressemblait assez à son père, mais il n'avait pas envie de discuter.

« Tu es libre ce soir ? demanda-t-elle, je voudrais te faire un bon petit plat.

— Et comment ! » Elle ne l'avait encore jamais invité chez elle.

« J'ai un bon morceau de steak.

— Formidable ! » Du bon bœuf était un régal même pour les privilégiés comme la famille de Volodia.

« Les Kovalev ne sont pas là », ajouta Zoïa.

Encore mieux ! Comme beaucoup de Moscovites, Zoïa vivait dans un appartement communautaire. Elle y occupait une chambre, partageant la cuisine et la salle de bains avec un autre scientifique, Kovalev, sa femme et leur enfant. En leur absence, Zoïa et Volodia auraient donc l'appartement pour eux. Son pouls se mit à battre plus vite. « Je dois apporter ma brosse à dents ? »

Elle lui adressa un sourire énigmatique et ne répondit pas.

Ils quittèrent le parc et traversèrent la rue pour rejoindre un restaurant. Un grand nombre d'établissements étaient fermés, mais comme il y avait encore beaucoup de petits bureaux dans le centre-ville et que leurs employés devaient bien déjeuner quelque part, plusieurs bars et cafés avaient survécu à l'évacuation.

Grigori était assis à une table sur le trottoir. Il aimait bien être vu dans des lieux fréquentés par le commun des mortels pour montrer qu'il ne se croyait pas au-dessus du peuple sous prétexte qu'il portait un uniforme de général. Il avait cependant choisi une table un peu à l'écart pour préserver une certaine intimité.

S'il n'approuvait pas les façons de Zoïa, il n'était pas insensible à son charme. Il se leva pour l'embrasser sur les deux joues.

Ils commandèrent des galettes de pommes de terre et de la bière. L'autre menu au choix proposait harengs marinés et vodka.

« Aujourd'hui, je ne vais pas vous parler de physique nucléaire, général, annonça Zoïa. Prenez pour acquis que je ne retire rien de ce que j'ai dit la dernière fois que nous avons évoqué le sujet. Mais je ne veux pas vous ennuyer.

— Ouf, quel soulagement ! »

Elle éclata de rire, découvrant des dents blanches. « En échange, vous pouvez peut-être me dire combien de temps encore va durer la guerre ? »

Volodia secoua la tête, feignant le désespoir. Elle ne pouvait pas s'empêcher de provoquer son père ! Tout autre que lui l'aurait déjà fait arrêter depuis longtemps ! Mais Grigori répondit : « Les nazis sont vaincus, mais ils ne l'admettront pas.

— À Moscou, tout le monde se demande ce qui va se passer cet été. Vous le savez sûrement, tous les deux !

— Si je le savais, je n'en dirais rien à ma petite amie, intervint Volodia. Même si j'en étais amoureux fou. » En dehors de toute autre considération, ça pourrait valoir à Zoïa une balle dans la tête, mais cela, il ne le dit pas.

On apporta les plats. Comme à son habitude, Zoïa se jeta sur la nourriture. Volodia adorait le plaisir évident qu'elle prenait à manger. Pour sa part, il n'aimait pas beaucoup les galettes. « Elles ont un drôle de goût, ces pommes de terre. On dirait du navet. »

Son père lui jeta un regard noir.

« Je ne m'en plains pas, loin de là ! » s'empressa d'ajouter Volodia.

Quand ils eurent fini, Zoïa se rendit aux toilettes. À peine se fut-elle éloignée que Volodia déclara : « Nous pensons que l'offensive d'été des Allemands est imminente.

— C'est aussi mon avis.

— On est prêts ?

— Bien sûr ! répondit Grigori, mais il avait l'air inquiet.

— Ils vont attaquer au sud. Pour s'emparer des champs de pétrole du Caucase. »

Grigori secoua la tête. « Non, ils vont revenir à Moscou. C'est la seule chose qui compte pour eux.

— Stalingrad est tout aussi symbolique. La ville porte le nom de notre dirigeant suprême.

— Je t'en foutrais des symboles ! S'ils prennent Moscou, la guerre est finie. Dans le cas contraire, ils n'auront pas gagné, même s'ils remportent la victoire sur d'autres fronts.

— Tu lances juste des hypothèses, lança Volodia agacé.

— Toi aussi.

— Non. J'ai des preuves. » Il regarda autour de lui : personne à proximité. « L'offensive a pour nom de code Fall Blau, et doit débuter le 28 juin. » Il tenait ce renseignement du réseau d'espions que dirigeait Werner Franck, à Berlin. « On a également trouvé des détails partiels sur cette opération dans la mallette d'un officier allemand qui s'est écrasé dans son avion de reconnaissance du côté de Kharkov.

— Les officiers en mission de reconnaissance ne se promènent pas avec des plans de bataille dans un porte-documents, objecta Grigori. Le camarade Staline pense que c'est un stratagème et je l'approuve. Les Allemands essaient tout simplement de nous inciter à dégarnir notre front central et de nous faire envoyer des troupes dans le sud pour repousser cette prétendue offensive, qui ne sera en fait qu'une diversion. »

C'était toujours le même problème ! pensa Volodia, excédé. Vous aviez beau détenir des renseignements parfaitement solides, les vieillards obtus qui dirigeaient le pays ne croyaient que ce qu'ils voulaient bien croire.

Zoïa traversait la terrasse dans leur direction, suivie de tous les regards. « Que faudrait-il pour te convaincre ? demanda-t-il à son père avant qu'elle ne les rejoigne.

— D'autres preuves.

— De quel genre ? »

Grigori réfléchit un moment, prenant la question très au sérieux. « Le plan de bataille. »

Volodia soupira. Werner Franck n'avait pas encore réussi à s'emparer de ce document. « Si j'arrive à l'avoir, est-ce que Staline reverra sa position ?

— Je le lui demanderai. Encore faut-il que tu l'obtiennes !

— Marché conclu », dit Volodia.

La réponse était pour le moins hâtive, puisqu'il n'avait aucune idée de la façon dont il pourrait y parvenir. Werner, Heinrich, Lili et les autres prenaient déjà des risques insensés, et il allait devoir leur imposer une pression supplémentaire.

Zoïa avait rejoint leur table et Grigori se leva. Ils partaient tous dans des directions différentes et se dirent au revoir.

« À ce soir donc », lança Zoïa à Volodia.

Il l'embrassa. « Je serai là à sept heures.

— N'oublie pas ta brosse à dents ! »

Il s'éloigna, au comble du bonheur.

4.

Une fille sait toujours quand sa meilleure amie lui cache quelque chose. Elle ne sait pas forcément de quoi il s'agit, mais pour elle, ça se voit comme le nez au milieu de la figure. À ses réponses prudentes et réticentes quand elle lui pose une question parfaitement anodine, elle devine que son amie fréquente quelqu'un qu'elle ne devrait pas. Elle ignore encore le nom de ce mystérieux amant, mais elle sait déjà que c'est forcément un homme marié, un étranger à la peau trop sombre, ou une autre femme. Elle admire un collier au cou de son amie et comprend à sa réaction en demi-teinte que ce collier est lié à une action honteuse. Peut-être n'apprendra-t-elle que des années plus tard que son amie l'avait dérobé dans la boîte à bijoux de sa grand-mère sénile.

Voilà à quoi pensait Carla en songeant à Frieda.

Son amie avait un secret, et ce secret était lié à la résistance contre les nazis. Frieda participait peut-être à un mouvement qui se livrait à des actions criminelles. Peut-être fouillait-elle tous les soirs dans la mallette de son frère Werner pour recopier des documents secrets qu'elle transmettait à un espion russe. Ou, sans aller jusque-là, elle pouvait prêter son aide à l'impression et à la distribution de ces affiches interdites et de ces pamphlets qui critiquaient le gouvernement.

Oui, elle allait parler de ce Joachim Koch à Frieda. L'occasion ne se présenta pas immédiatement, car les deux amies ne travaillaient pas dans le même service de l'immense hôpital où elles étaient infirmières. Leurs emplois du temps ne coïncidaient pas toujours, elles ne se voyaient pas quotidiennement.

Alors que Joachim, lui, venait tous les jours prendre sa leçon de piano. Il n'avait pas fait d'autre révélation à Maud, mais celle-ci continuait à flirter avec lui. Carla l'avait entendue dire : « Est-ce que vous vous rendez compte que j'ai presque quarante ans ? » En fait, elle en avait cinquante et un. Joachim se pâmait d'amour et Maud prenait visiblement plaisir à constater qu'elle n'avait rien perdu de son pouvoir de séduction auprès des beaux jeunes gens, fussent-ils un peu naïfs. L'idée que sa mère éprouvait peut-être des sentiments plus profonds pour ce jeune homme à la moustache blonde qui ressemblait vaguement à Walter jeune traversa bien l'esprit de Carla, mais elle la repoussa. C'était ridicule.

Joachim faisait tout ce qu'il pouvait pour lui plaire et lui apporta rapidement des nouvelles de son fils. Erik était vivant et en bonne santé. « Son unité est en Ukraine, c'est tout ce que je peux vous dire.

— J'aimerais tant qu'il ait une permission et puisse rentrer à la maison », soupira Maud avec mélancolie.

Le jeune officier hésita.

« Une mère se fait tant de soucis... Si je pouvais le voir, ne serait-ce qu'un jour, je serais tellement rassurée.

— Je pourrais peut-être arranger ça. »

Maud joua l'étonnement. « Vraiment ? Vous avez un tel pouvoir ?

— Je ne peux rien promettre, mais je peux essayer.

— Merci, ce serait déjà beaucoup ! » Elle saisit sa main et y posa les lèvres.

Une semaine s'écoula avant que Carla ne revoie Frieda. Elle lui parla alors de Joachim Koch en présentant les choses comme une anecdote amusante mais constata que Frieda ne prenait pas l'histoire à la légère. « Tu te rends compte ? enchaîna-t-elle, il nous a donné le nom de code de la prochaine opération et même sa date ! » Elle ménagea une pause pour observer la réaction de Frieda.

« Il pourrait se faire fusiller ! répondit Frieda.

— Et comment ! Si nous connaissions quelqu'un à Moscou, ça pourrait changer le cours de la guerre ! continua Carla comme si elle n'avait qu'une idée en tête : la gravité du délit commis par Joachim.

— Peut-être », acquiesça Frieda.

La preuve était faite. Normalement, une histoire pareille aurait dû inspirer à Frieda une réaction de vive surprise, un intérêt passionné. Elle aurait dû poser une foule de questions. Or elle n'avait répondu que par des phrases neutres et des acquiescements évasifs. De retour à la maison, Carla annonça à sa mère que son intuition ne l'avait pas trompée.

Le lendemain à l'hôpital, Frieda vint la trouver dans son service. « Il faut que je te parle de toute urgence ! » Elle avait l'air dans tous ses états.

Carla était en train de changer le pansement d'une jeune femme gravement brûlée dans l'explosion d'une usine de munitions. « Va m'attendre au vestiaire, je te rejoins dès que je peux. »

Cinq minutes plus tard, elle retrouvait son amie dans la petite pièce. Elle fumait devant la fenêtre ouverte.

« Que se passe-t-il ? »

Frieda éteignit sa cigarette. « C'est à propos de ton lieutenant Koch.

— Je m'en doutais.

— Tu dois lui soutirer plus d'informations !

— Je *dois* ? De quoi parles-tu ?

— Il a accès à tout le plan de bataille de l'opération Fall Blau. Nous savons déjà certaines choses, mais Moscou a besoin de renseignements plus précis. »

Les hypothèses échafaudées par Frieda étaient un peu improbables mais Carla ne chercha pas à la détromper. « Je peux éventuellement lui demander…

— Non. *Il faut* qu'il apporte le plan de bataille chez vous.

— Je ne vois pas comment. Il n'est pas complètement idiot. Tu ne crois pas que… »

Frieda n'écoutait même pas. « Et ensuite, tu photographieras ce plan. » Elle sortit de sa poche d'uniforme un petit boîtier métallique ressemblant à un paquet de cigarettes en un peu plus long et plus étroit. « C'est un appareil photo miniature, conçu spécialement pour photographier des documents. » Carla remarqua le nom gravé sur le côté : *Minox*. « Une pellicule te permet de prendre onze photos. Je t'en donne trois. » Elle sortit trois cassettes en forme d'haltères, assez petites pour s'insérer dans l'appareil. « Voilà comment on le charge. » Frieda fit une démonstration. « Pour prendre une photo, tu regardes par ce petit trou. Si tu n'es pas sûre, je te laisse le mode d'emploi. »

Carla n'aurait jamais imaginé que Frieda puisse se montrer aussi autoritaire. « Il faut que j'y réfléchisse, tu sais.

— On n'a pas le temps. C'est ton manteau ?

— Oui, mais… »

Frieda glissa dans la poche appareil photo, pellicules et mode d'emploi. « Bon, j'y vais ! » dit-elle, manifestement soulagée de s'en être débarrassée. Elle se dirigea vers la porte.

« Frieda, attends ! »

Frieda s'arrêta et regarda Carla droit dans les yeux. « Qu'y a-t-il ?

— Heu… je trouve que pour une amie, tu as une drôle de façon d'agir !

— C'est plus important que tout le reste.

— Tu ne me laisses même pas le choix !

— C'est toi qui l'as cherché en me parlant de ton Joachim Koch. Ne me dis pas que tu ne t'attendais pas à ce que je fasse quelque chose de cette information ! »

C'était vrai. Carla ne pouvait s'en prendre qu'à elle-même. Mais elle n'avait jamais imaginé en arriver là. « Et s'il refuse ?

— Eh bien, tu vivras probablement sous le régime nazi jusqu'à la fin de tes jours ! » Sur ces mots, Frieda sortit.

« Merde ! » s'exclama Carla.

Restée seule dans le vestiaire, elle réfléchit. Se débarrasser de l'appareil sans prendre de risques, oui, mais comment ? Elle ne pouvait quand même pas le jeter dans la poubelle de l'hôpital ! Elle allait devoir quitter les lieux avec l'appareil dans sa poche et trouver un endroit où s'en défaire discrètement.

Mais était-ce bien ce qu'elle voulait ?

Joachim avait beau être naïf, imaginer qu'il puisse accepter de sortir en douce du ministère de la Guerre une copie du plan de bataille simplement pour faire plaisir à Maud, c'était pousser le bouchon un peu loin ! D'un autre côté, si quelqu'un pouvait l'en persuader, c'était bien elle.

Carla était terrifiée. Si elle se faisait prendre, il ne lui serait accordé aucune pitié. Elle serait torturée. Elle revit Rudi Rothmann, gémissant de douleur, les mains brisées. Elle revit son père après sa libération : il avait été tellement maltraité qu'il n'avait pas survécu à ses blessures. Son crime à elle serait considéré comme bien pire que les leurs, et elle devait donc s'attendre à un châtiment encore plus inhumain. Elle serait exécutée, cela ne faisait aucun doute, mais après de très longues souffrances.

Elle se dit cependant qu'elle était prête à courir ce risque.

Mais il y avait une chose qu'elle ne pouvait accepter : provoquer la mort de son frère.

Or il était bien là-bas, sur le front est, Joachim l'avait confirmé. Il allait donc participer à l'opération Fall Blau. Si, grâce à son intervention, les Soviétiques gagnaient la bataille et qu'Erik trouvait la mort au cours de ces combats, elle ne se le pardonnerait jamais.

Elle retourna à son travail. Elle était distraite et commit des erreurs. Heureusement, les médecins ne s'en aperçurent pas. Quant aux patients, ils n'étaient pas en état de le signaler. Enfin, son service s'acheva, et elle partit précipitamment. Elle avait l'impression que l'appareil brûlait le fond de sa poche. Hélas, elle ne trouva pas d'endroit sûr où s'en débarrasser.

Où Frieda avait-elle bien pu se le procurer ? Certes, son amie avait beaucoup d'argent, mais pour acheter un appareil de ce genre, elle aurait dû inventer toute une histoire. Sans doute lui avait-il été remis par un Russe, avant la fermeture de l'ambassade, l'année précédente.

Carla rentra chez elle, l'appareil toujours dans sa poche.

Aucune note de piano ne venait de l'étage. La leçon de Joachim devait avoir lieu plus tard ce jour-là. Sa mère était à la cuisine, le visage radieux. « Regarde qui est là ! »

Erik !

Carla dévisagea son frère : il avait beau être d'une maigreur effroyable, il était apparemment entier. Il portait un uniforme crasseux et déchiré, mais s'était lavé le visage et les mains.

Il se leva et la prit dans ses bras. Elle le serra contre elle, se moquant bien de salir son propre uniforme, immaculé. « Tu es sain et sauf », s'exclama-t-elle. Erik était tellement efflanqué qu'elle sentait ses os à travers le tissu, ses côtes, ses hanches, ses épaules, ses vertèbres.

« Pour le moment », dit-il.

Elle relâcha son étreinte. « Comment vas-tu ?

— Mieux que la plupart de mes compagnons.

— Tu avais quand même autre chose sur le dos, pendant l'hiver russe, que cet uniforme de rien du tout ?

— Oui, un manteau en mouton retourné volé sur un cadavre russe. »

Elle s'assit à la table où Ada avait déjà pris place. « Tu avais raison ! murmura Erik. Je veux dire à propos des nazis, tu avais raison.

— En quel sens ? demanda-t-elle, contente, mais se demandant à quoi il faisait allusion.

— Ils assassinent des gens, comme tu me l'avais dit ! Et Vater aussi, et puis Mutter. Quand je pense que je ne vous ai pas crus ! Je m'en veux tellement ! Excuse-moi, Ada, de n'avoir pas cru qu'ils avaient tué ton pauvre petit Kurt. Maintenant, je sais que c'est vrai ! »

Pour un revirement, il était de taille !

« Qu'est-ce qui t'a fait changer d'avis ? voulut savoir Carla.

— Je les ai vus à l'œuvre en Russie. Quand ils arrivent dans une ville, ils rassemblent toutes les personnalités sous prétexte que ce sont des communistes ; ils embarquent aussi les Juifs... Et pas seulement les hommes : les femmes et les enfants avec, y compris les vieux bien trop faibles pour faire du tort à qui que ce soit ! » Les larmes ruisselaient sur ses joues. « Ce ne sont pas les soldats ordinaires qui font ça, mais des unités spéciales... Ils font sortir les prisonniers de la ville. Des fois, il y a une carrière ou une fosse quelconque. Quand il n'y en a pas, ils obligent les plus jeunes à creuser un grand trou, et ensuite, ils... »

Il s'étrangla.

« Ensuite quoi ? insista Carla.

— Ils les tuent. Douze à la fois, deux fois six. Parfois, les maris et les femmes descendent dans la fosse en se tenant par la main, les mères portent leurs enfants. Les tireurs attendent qu'ils soient au bon endroit, et puis, ils les abattent. » Erik essuya ses larmes avec la manche de son uniforme sale. « Pan ! »

Un long silence se fit dans la cuisine. Ada pleurait, Carla était atterrée. Seule Maud avait gardé un visage de pierre.

Finalement, Erik se moucha et sortit une cigarette. « J'ai été drôlement surpris d'obtenir une permission et un billet de train pour rentrer.

— Quand dois-tu repartir ?

— Demain. Je ne peux passer que vingt-quatre heures à Berlin. Mais j'ai fait l'envie de tous mes camarades. Ils donneraient n'importe quoi pour passer une journée chez eux. Le docteur Weiss m'a dit que j'avais sûrement des amis haut placés.

— C'est vrai, reconnut Maud. Joachim Koch. C'est un jeune lieutenant qui travaille au ministère de la Guerre et à qui je donne des leçons de piano. Je lui ai demandé s'il pouvait t'obtenir une permission. » Elle regarda sa montre. « Il ne devrait plus tarder, d'ailleurs. Il a un petit faible pour moi. Il doit avoir besoin d'une image maternelle, sans doute. »

Mutti, arrête ! pensa Carla. Il n'y avait rien de maternel dans les relations entre Maud et Joachim.

Mais celle-ci continuait : « Il est très naïf. Il nous a raconté qu'il allait y avoir une nouvelle offensive sur le front de l'Est, le 28 juin. Il nous en a même donné le nom de code : Fall Blau.

— De quoi se faire fusiller ! remarqua Erik.

— Joachim n'est pas le seul à risquer sa vie, murmura Carla. J'ai raconté à quelqu'un ce que j'avais appris et maintenant, on me demande de persuader Joachim de me donner les plans de bataille.

— Nom de Dieu ! s'exclama Erik abasourdi. C'est de l'espionnage pur et simple ! Tu risques bien plus gros que moi sur le front.

— Ne t'inquiète pas. Je n'imagine pas un instant que Joachim accepte de faire ça, le rassura Carla.

— N'en sois pas si sûre ! » réagit Maud.

Tous les yeux se tournèrent vers elle.

« Il le ferait peut-être pour moi. Si je le lui demandais habilement.

— Il est aussi naïf que ca ? » s'ébahit Erik.

Elle le défia du regard. « Il a le béguin pour moi.

— Oh. » L'idée même qu'un homme puisse être amoureux de sa mère l'embarrassait. Carla intervint : « De toute façon, on ne peut pas lui demander une chose pareille !

— Et pourquoi ? demanda Erik.

— Parce que si les Russes remportaient la bataille, tu pourrais y laisser ta peau.

— Je mourrai de toute façon, c'est presque sûr.

— Mais nous aiderions les Russes à te tuer ! répliqua Carla d'une voix que l'exaltation poussait dans les aigus.

— Je veux quand même que tu le fasses ! » insista Erik fermement. Il fixait la toile cirée à carreaux, l'esprit à des milliers de kilomètres de cette table de cuisine.

Carla était déchirée. Si son frère l'exigeait… « Mais pourquoi ?

— Je pense à tous ces gens qu'on fait descendre dans des fosses et qui se tiennent par la main. » Ses mains à lui, posées sur la table, étaient crispées dans une douloureuse étreinte. « Je suis prêt à donner ma vie pour que ça cesse. C'est même ce que je veux. Je me sentirai mieux vis-à-vis de moi-même et de mon pays. S'il te plaît Carla, si tu en as les moyens, transmets ce plan de bataille aux Russes ! »

Elle hésitait encore. « Tu es sûr ?

— Je t'en supplie.

— C'est bon, je le ferai. »

5.

Thomas Macke demanda à ses hommes, Wagner, Richter et Schneider, de se conduire irréprochablement. « Werner Franck n'est que lieutenant, mais il travaille pour le général Dorn. Je veux qu'il ait aussi bonne impression que possible de notre équipe et du travail que nous effectuons. Pas de jurons, pas de plaisanteries, pas de boustifaille et pas de tabassage sauf en cas d'absolue nécessité. Si on prend un espion communiste, vous pouvez lui filer un bon coup de pied, mais si on n'attrape personne, il n'est pas question de ramasser le premier venu, histoire de rigoler. » D'ordinaire, il fermait les yeux sur ce genre de dérapages qui entretenaient dans la population la peur de déplaire aux nazis. Mais allez savoir si Franck ne ferait pas le délicat ?

À l'heure dite, celui-ci arriva à moto au quartier général de la Gestapo, Prinz-Albrecht-Strasse. Ils montèrent tous à bord de la camionnette de surveillance, au toit équipé d'une antenne rotative. Avec tout l'équipement radio qui s'y trouvait déjà, ils étaient un peu à l'étroit. Richter se mit au volant et ils parcoururent les rues de la ville. On était en début de soirée. C'était apparemment l'heure préférée des espions pour envoyer leurs messages à l'ennemi. Werner s'en étonna.

« C'est parce que la plupart d'entre eux ont un emploi régulier qui leur sert de couverture, expliqua Macke. Dans la journée, ils sont au bureau ou à l'usine.

— Évidemment, je n'y avais pas pensé. »

Macke s'inquiétait à l'idée d'être bredouille ce soir-là. Sa terreur, c'était d'être blâmé pour les revers qu'essuyait l'armée allemande en Russie. Il faisait de son mieux, mais le IIIe Reich ne récompensait pas les efforts.

Certains soirs, il arrivait que son unité ne capte aucun signal. D'autres fois, elle en captait deux ou trois, et il était obligé de faire un choix. Il était convaincu qu'il existait en ville plusieurs réseaux d'espionnage distincts, qui ignoraient tout les uns des autres. Il s'efforçait d'accomplir une mission impossible et avec des moyens insuffisants.

Ils étaient tout près de la Potzdamer Platz quand ils captèrent un signal. Macke en reconnut immédiatement le son caractéristique. « Un pianiste ! » s'écria-t-il, soulagé. Au moins, il aurait démontré à Werner que le matériel fonctionnait. Quelqu'un était en train d'envoyer des signaux à cinq chiffres, l'un après l'autre. « Les services secrets soviétiques utilisent un code où les lettres sont représentées par deux chiffres, expliqua Macke en s'adressant à Werner. Par exemple, 11 peut signifier A. Transmettre les nombres par groupes de cinq n'est qu'une convention d'usage. »

L'opérateur radio, un ingénieur électricien du nom de Mann, lut à haute voix une série de coordonnées. À l'aide d'une règle et d'un crayon, Wagner traça une ligne sur une carte. Richter embraya et redémarra.

Le pianiste émettait toujours, et ses bips résonnaient bruyamment dans la camionnette. « Salopard de communiste ! s'exclama Macke, laissant libre cours à sa haine. Un de ces jours, il se retrouvera dans notre sous-sol et il me suppliera de le laisser mourir pour abréger ses souffrances. »

En voyant Werner pâlir, Macke se dit qu'il n'avait pas l'habitude du travail de la police.

Le jeune homme se ressaisit rapidement. « Tel que vous le décrivez, le code soviétique n'a pas l'air bien difficile à décrypter, dit-il pensivement.

— C'est vrai ! approuva Macke, heureux que Werner ait compris aussi vite. Mais j'ai un peu simplifié les choses. Ils ont des “astuces”. Après avoir codé son message sous forme d'une série de chiffres, le pianiste ajoute à plusieurs reprises un mot clé en dessous, par exemple Kurfürstendamm, et il code. Ensuite, il soustrait les seconds chiffres des premiers et il envoie le résultat.

— Presque impossible à déchiffrer si on ne connaît pas le mot clé !

— Exactement. »

Ils s'arrêtèrent à nouveau, près des ruines incendiées du Reichstag, pour tracer une autre ligne sur la carte. Les deux lignes se croisaient à Friedrichshain, à l'est du centre-ville.

Macke demanda au chauffeur de prendre en direction du nord-est, vers un point qui leur permettrait vraisemblablement de tracer une troisième ligne à partir d'un lieu différent. « L'expérience démontre qu'il vaut mieux prendre trois positions, expliqua Macke. Le matériel manque de précision. Plus on a de mesures, plus on réduit la marge d'erreur.

— Vous arrivez chaque fois à attraper le pianiste ? demanda Werner.

— Oh non, loin de là ! C'est même rare. Souvent, c'est parce qu'on n'est pas assez rapides. Mais il peut aussi changer de fréquence au beau milieu de son message, et alors nous le perdons. Quelquefois, il interrompt sa transmission pour la reprendre ailleurs. Il peut aussi avoir des guetteurs qui le préviennent quand ils nous voient arriver. Du coup, il prend la fuite.

— Ça fait beaucoup d'obstacles.

— Oui, mais on finit toujours par les attraper, tôt ou tard. »

Richter coupa le moteur. Macke enregistra une troisième position. Les trois lignes de crayon sur la carte de Wagner formaient un petit triangle près de l'Ostbahnhof, la gare de l'Est. Le pianiste devait se trouver quelque part entre la ligne de chemin de fer et le canal.

Macke indiqua le lieu à Richter en ajoutant : « Fonce !! »

Macke remarqua que Werner était en nage. Il faisait un peu chaud dans la camionnette, indéniablement, mais surtout, le jeune lieutenant n'était pas habitué à l'action. Il apprenait ce qu'était la vie à la Gestapo. Une bonne chose, pensa Macke.

Richter suivit la Warschauerstrasse en direction du sud, traversa la voie ferrée et s'engagea dans un quartier industriel fait de hangars, de dépôts et de petites usines. Devant une des entrées de la gare, des soldats mettaient des bardas en tas. Sans doute embarquaient-ils pour le front de l'Est. Et pendant ce temps, tapi dans une cachette toute proche, un de leurs compatriotes cherchait à les trahir, songea Macke furieux.

Wagner désigna une rue étroite qui partait de la gare. « Il se trouve dans les premières centaines de mètres, mais je ne sais pas de quel côté. Si on se rapproche, il va nous voir.

— C'est bon, les gars. Vous connaissez la musique, dit Macke. Wagner et Richter, vous prenez le côté gauche, Schneider et moi, le droit. » Chacun d'eux s'empara d'une masse à long manche. « Suivez-moi, Franck. »

Il n'y avait pas grand-monde dans la rue : un homme en casquette d'ouvrier qui se hâtait vers la gare et une vieille en vêtements miteux qui allait probablement faire le ménage dans des bureaux. En apercevant la Gestapo, l'un et l'autre accélérèrent le pas, ne voulant pas attirer son attention.

L'équipe de Macke visita un à un chacun des immeubles. Ils progressaient en se dépassant à tour de rôle. La plupart des bureaux étant fermés, ils devaient réveiller le gardien. Si celui-ci mettait plus d'une minute à arriver, ils enfonçaient la porte. Une fois à l'intérieur, ils visitaient rapidement les étages, vérifiant toutes les pièces.

Pas de pianiste dans le premier pâté de maisons.

Au fronton du premier immeuble à droite du pâté suivant, une vieille enseigne indiquait « Fourrures à la mode ». C'était un atelier de deux étages qui s'étirait sur toute la longueur de la rue transversale. Il avait l'air désaffecté, mais était muni d'une porte d'entrée blindée et de barreaux aux fenêtres, un dispositif de protection normal pour ce type d'établissement.

Cherchant comment y pénétrer, Macke entraîna Werner dans la rue perpendiculaire. L'immeuble voisin, endommagé par une bombe, tombait en ruine. Les gravats avaient été dégagés et il y avait un panneau écrit à la main : « Danger - Défense d'entrer ». Un fragment d'enseigne laissait supposer qu'à cet endroit s'élevait auparavant un entrepôt de meubles.

Ils franchirent l'amas de pierres et de poutres aussi vite qu'ils le pouvaient, tout en regardant où ils posaient les pieds. Un mur intact cachait l'arrière du bâtiment. Macke passa derrière et découvrit un trou qui permettait d'entrer dans l'atelier de fourrures.

Il était persuadé que le pianiste se trouvait à l'intérieur.

Il se faufila par la brèche, Werner sur ses talons.

Ils débouchèrent dans une salle déserte meublée d'un vieux bureau métallique sans sa chaise et d'un classeur sur le mur d'en face. Le calendrier accroché au mur datait de 1939, la dernière année sans doute où les Berlinoises avaient encore eu les moyens de s'offrir des vêtements aussi frivoles que des manteaux de fourrure.

Macke entendit des pas à l'étage.

Il sortit son pistolet.

Werner n'avait pas d'arme.

Ils ouvrirent la porte et s'engagèrent dans un couloir. Macke remarqua plusieurs portes ouvertes, un escalier qui montait à l'étage et, en dessous, une porte qui devait mener à la cave.

Se déplaçant sans bruit, Macke s'avança le long du couloir jusqu'au pied de l'escalier. Il vit que Werner gardait les yeux rivés sur la porte de la cave.

« J'ai cru entendre du bruit en bas. » Werner tourna la poignée. La porte était munie d'une serrure bon marché. Il recula et leva le pied droit.

« Non ! fit Macke.

— Si, si ! Je les entends ! » répliqua Werner et il enfonça la porte bruyamment.

L'écho résonna à travers tout l'atelier.

Werner se jeta en avant et disparut. Une lumière révéla un escalier en pierre. « Pas un geste ! hurla Werner. Vous êtes en état d'arrestation ! »

Macke descendit derrière lui.

Il arriva au sous-sol. Werner se trouvait au pied de l'escalier, apparemment déçu.

La pièce était vide.

Elle ne contenait que des tringles suspendues au plafond auxquelles on devait autrefois accrocher les manteaux, et un énorme rouleau de papier brun dans un coin, probablement destiné aux emballages. Pas l'ombre d'une radio ni d'un espion en train de transmettre un message à Moscou.

« Espèce d'imbécile ! » lâcha Macke à l'adresse de Werner.

Il pivota sur les talons et s'empressa de remonter au rez-de-chaussée, Werner courant à sa suite, puis ils grimpèrent à l'étage.

Des rangées de tables à ouvrage s'alignaient sous une verrière. Un grand nombre d'ouvrières avaient dû travailler sur des machines à coudre dans cet atelier maintenant désert.

Une porte vitrée menait à la sortie de secours, mais elle était fermée à clé. Macke regarda dehors. Personne.

Il rangea son arme et s'appuya à un établi, le souffle court.

Par terre, il remarqua des mégots récents, dont l'un portait des traces de rouge à lèvres. « Ils étaient ici ! s'exclama Macke en montrant le plancher. Avec votre raffut, vous leur avez donné l'alerte et ils se sont enfuis.

— Quelle maladresse ! Je suis désolé, je n'ai pas l'habitude de ce genre d'opération. »

Macke s'approcha de la fenêtre d'angle. En bas, dans la rue, un jeune homme et une femme s'éloignaient à grands pas. L'homme avait sous son bras un porte-documents en cuir fauve. Macke les vit s'engouffrer à l'intérieur de la gare.

« Merde ! s'exclama-t-il.

— Vous croyez vraiment que c'étaient des espions ? » demanda Werner en désignant le plancher. Macke aperçut un préservatif chiffonné. « Utilisé, mais vide, fit remarquer Werner. Je crois que nous les avons surpris en pleine action.

— Espérons que vous avez raison. »

6.

Joachim Koch avait promis d'apporter le plan de la bataille. Le jour dit, Carla n'alla pas à l'hôpital.

Elle aurait probablement pu assurer sa garde du matin et être rentrée à temps chez elle, mais ce « probablement » était trop incertain. Un grave incendie ou un accident de la route était toujours à craindre, ce qui l'aurait obligée à rester après son service pour faire face à l'afflux de blessés. Elle préféra donc ne pas sortir de chez elle de toute la journée.

En fin de compte, Maud n'avait pas eu à demander à Joachim d'apporter le plan. Il l'avait prévenue qu'il devait annuler sa leçon, puis, incapable de résister à la tentation de se faire valoir, il avait expliqué qu'il avait été chargé de porter une copie du plan à l'autre bout de la ville. « Arrêtez-vous donc en chemin pour prendre votre leçon », avait proposé Maud, et il avait accepté.

Le déjeuner se déroula dans une atmosphère tendue. Carla et Maud avalèrent un maigre potage de pois secs agrémenté d'un os de jambon. Carla ne chercha pas à savoir ce que sa mère avait fait, ou promis de faire, pour convaincre Koch. Peut-être lui avait-elle dit qu'il faisait de remarquables progrès au piano et ne pouvait pas se permettre de rater une leçon. Peut-être avait-elle laissé entendre qu'il devait occuper un poste bien subalterne pour être surveillé en permanence. Pareille remarque l'aurait certainement piqué à vif et incité à lui démontrer le contraire, car il aimait à se faire passer pour plus important qu'il n'était. Cependant, le stratagème le plus susceptible d'avoir réussi était évidemment celui auquel Carla s'interdisait de penser : le sexe. Sa mère faisait ouvertement du charme à Koch, et celui-ci réagissait à ses avances par une dévotion d'esclave. Carla soupçonnait que c'était cette tentation irrésistible qui avait poussé Joachim à rester sourd à la voix de la raison qui lui enjoignait de ne pas être aussi stupide.

Mais peut-être se trompait-elle. Peut-être n'était-il pas du tout tombé dans le panneau. Dans ce cas, ce n'était pas avec une copie carbone du plan de bataille dans son sac qu'il se présenterait cet après-midi-là, mais avec une escouade de la Gestapo et une paire de menottes.

Carla chargea l'appareil photo et le rangea en même temps que les deux pellicules restantes dans le tiroir d'un buffet bas de la cuisine, sous des torchons. Ce meuble se trouvant à côté de la fenêtre, elle pourrait poser le document dessus pour le photographier à la lumière.

Elle n'avait pas la moindre idée de la façon dont ces photos parviendraient à Moscou. Frieda lui avait simplement juré que ce serait fait. Carla imaginait un représentant – en produits pharmaceutiques ou en bibles allemandes, par exemple – autorisé à vendre ses marchandises en Suisse et remettant discrètement ces films à un membre de l'ambassade soviétique à Berne.

L'après-midi n'en finissait pas. Maud alla se reposer dans sa chambre, Ada fit la lessive, Carla s'installa pour lire dans la salle à manger qui ne servait quasiment plus ces derniers temps. Elle était incapable de se concentrer. Les journaux ne racontaient que des mensonges. Elle aurait dû réviser ses cours pour son prochain examen d'infirmière, mais les termes médicaux dansaient devant ses yeux. Elle lut un vieil exemplaire d'*À l'ouest, rien de nouveau*, un roman allemand sur la Première Guerre mondiale qui avait connu un grand succès mais était désormais interdit, parce que les épreuves subies par les soldats y étaient décrites de manière trop réaliste. Elle se retrouva bientôt, le livre à la main, en train de contempler par la fenêtre les toits poussiéreux de la ville dans la chaleur du soleil de juin.

Il arriva enfin. Elle entendit des pas dans l'allée et bondit sur ses pieds pour jeter un coup d'œil au-dehors. Joachim Koch était seul, sans agents de la Gestapo, dans son uniforme bien repassé et ses bottes cirées, son visage de jeune premier plein d'impatience, comme un petit garçon arrivant à une fête d'anniversaire. Il portait en bandoulière sa sacoche en toile, comme d'habitude. Avait-il tenu parole ? Contenait-elle une copie du plan de l'opération Fall Blau ?

Il sonna.

À partir de cet instant, tout avait été soigneusement préparé entre la mère et la fille. Elles avaient décidé que ce n'était pas Carla qui ouvrirait la porte. Quelques instants plus tard, elle vit sa mère traverser le couloir dans une robe d'intérieur en soie mauve, chaussée de mules à talons, et se dit, honteuse et gênée, qu'elle ressemblait presque à une prostituée. Carla entendit ensuite la porte d'entrée s'ouvrir et se refermer. Suivirent un froufrou de soie et un échange de mots doux suggérant une étreinte. La robe mauve et l'uniforme kaki passèrent devant la porte de la salle à manger et disparurent au premier étage.

Maud devait en premier lieu s'assurer que Joachim avait bien le document. Elle le regarderait, ferait part au jeune homme de son admiration, puis le reposerait. Elle conduirait ensuite Joachim au piano et plus tard, sous quelque prétexte auquel Carla préférait ne pas songer, sa mère ferait passer le lieutenant du salon au cabinet de travail contigu, une pièce beaucoup plus intime avec ses rideaux rouges et son grand canapé un peu avachi. Dès qu'ils y seraient entrés, Maud donnerait le signal à Carla.

Comme il était difficile de prévoir avec précision l'enchaînement de leurs faits et gestes, elles avaient imaginé plusieurs signaux ayant tous le même sens. Le plus simple consistait à faire claquer la double porte séparant le salon du cabinet de travail, mais Maud pouvait aussi appuyer sur le bouton de sonnette, vestige de l'ancien système permettant d'appeler les serviteurs. Situé près de la cheminée, il déclenchait une sonnerie dans la cuisine. De toute façon, n'importe quel bruit ferait l'affaire, avaient-elles décidé. En dernier ressort, Maud pourrait toujours renverser le buste en marbre de Goethe ou briser un vase « malencontreusement ».

Carla fit un pas dans le couloir et s'arrêta, les yeux levés vers le palier supérieur. Pas un bruit.

Elle alla jeter un œil dans la cuisine. Ada était en train de laver la marmite dans laquelle elle avait fait la soupe, la frottant avec une énergie qui révélait sa nervosité. Carla lui adressa un sourire qui se voulait encourageant. Elle aurait bien voulu, et Maud aussi, dissimuler toute cette affaire à Ada, non par manque de confiance – bien au contraire, car elle était une antinazie farouche – mais pour la protéger, car le seul fait d'en être informée faisait d'elle une complice et pouvait lui valoir le châtiment suprême. Mais elles partageaient une trop grande intimité avec Ada pour pouvoir lui cacher quoi que ce soit.

Carla perçut faiblement un rire léger qu'elle connaissait bien : le rire artificiel de sa mère poussant à l'extrême son pouvoir de séduction.

Joachim avait-il le document, oui ou non ?

Une minute plus tard, quelques notes de piano parvinrent à leurs oreilles. C'était Joachim, sans aucun doute. « ABC, die Katze lief im Schnee ». Une comptine à propos d'un chat dans la neige, que son père lui avait chantée des centaines de fois. À ce souvenir, sa gorge se noua. Comment les nazis osaient-ils jouer ces airs-là, alors que tant d'enfants étaient orphelins par leur faute ?

La mélodie s'interrompit brutalement. Que se passait-il ? Carla tendit l'oreille à l'affût du moindre bruit : voix, pas, n'importe quoi. Elle n'entendait rien.

Une minute s'écoula, une autre.

Il y avait sûrement un problème, mais lequel ?

Par la porte de la cuisine, elle regarda Ada. Celle-ci s'était arrêtée de récurer sa marmite et lui fit un geste de la main signifiant *Je n'y comprends rien*.

Carla décida d'en avoir le cœur net.

Elle gravit l'escalier, marchant sans bruit sur le tapis élimé, et s'avança jusqu'à la porte du salon.

Pas un bruit ne sortait de la pièce. Ni musique, ni voix, ni mouvement.

Elle tourna la poignée aussi doucement que possible.

Glissa un œil à l'intérieur. Personne. Elle entra et regarda autour d'elle. La pièce était déserte.

Pas trace de la sacoche en toile de Joachim.

Elle regarda la double porte qui menait au cabinet de travail. Un des battants était ouvert.

Carla traversa le salon sur la pointe des pieds. Il n'y avait pas de tapis, juste un parquet ciré, et on pouvait l'entendre. Mais il fallait qu'elle prenne ce risque.

Des murmures lui parvinrent à mesure qu'elle se rapprochait de la porte.

Arrivée sur le seuil, elle s'aplatit contre le mur et s'aventura à jeter un coup d'œil à l'intérieur.

Ils étaient debout, enlacés, et s'embrassaient. Dos à la porte, Joachim ne pouvait pas la voir. Maud avait dû veiller à ce qu'il se place ainsi. Sa mère interrompit alors ce baiser, regarda pardessus l'épaule de Joachim et aperçut Carla. Elle retira sa main posée sur la nuque du jeune homme et pointa le doigt avec insistance.

Carla repéra la sacoche sur une chaise.

En un instant, elle comprit tout. Maud avait réussi à attirer Joachim dans le cabinet de travail, mais il ne leur avait pas facilité la tâche en laissant son sac dans le salon, comme prévu. Dans son inquiétude, il l'avait emporté avec lui.

Et Carla allait devoir le récupérer !

Le cœur battant, elle fit un pas à l'intérieur de la pièce.

Maud murmura : « Oh, oui, continuez, continuez… »

Joachim gémit : « Je vous aime, ma chérie. »

Carla fit encore deux pas, attrapa le sac, fit demi-tour et sortit silencieusement de la pièce.

Le sac était léger.

Elle retraversa le salon à la hâte, dévala l'escalier, le souffle court.

Dans la cuisine, elle posa le sac sur la table et en défit les attaches. Il contenait le numéro du jour de *Der Angriff*, un journal berlinois, un paquet de cigarettes Kamel tout neuf et une chemise en carton beige. Elle la sortit d'une main tremblante et l'ouvrit. C'était le carbone d'un document.

La première page portait en titre :

DIRECTIVE N° 41

La dernière portait une ligne en pointillés destinée à la signature. Cette ligne était vierge, parce qu'il s'agissait d'une copie, bien sûr. Mais le nom dactylographié en marge de ces points était : Adolf Hitler.

Entre ces deux pages, le plan de l'opération Fall Blau.

Carla éprouva un sentiment d'allégresse, qui se mêla à l'immense tension et à la peur terrible de se faire prendre qui l'étreignaient déjà.

Elle plaça le document sur le buffet bas près de la fenêtre, en ouvrit vivement le tiroir et en sortit l'appareil Minox et les deux pellicules. Ayant positionné soigneusement les pages, elle entreprit de les photographier l'une après l'autre.

Cela ne lui prit pas longtemps, le dossier ne comptant que dix pages. Elle n'eut même pas besoin de recharger l'appareil. Fini ! Elle avait volé le plan de bataille.

Je l'ai fait pour toi, Vati.

Elle remit l'appareil à sa place, referma le tiroir, glissa le document dans sa chemise en carton et celle-ci dans la sacoche en toile qu'elle referma en prenant soin d'en attacher les sangles.

Le plus doucement possible, elle rapporta le sac à l'étage.

Au moment où elle se glissait dans le salon, elle entendit la voix de Maud, haute et claire comme si elle tenait à être entendue. Carla comprit tout de suite que sa mère cherchait à la mettre en garde. « Ne t'en fais pas, disait-elle. C'est parce que tu étais trop excité. Nous l'étions tous les deux ! »

La voix de Joachim répondit, sourde, gênée : « Je m'en veux tellement. Vous n'avez fait que me toucher et tout était fini. »

Carla devina ce qui s'était passé. Elle n'en avait pas personnellement fait l'expérience mais les filles bavardaient, et les conversations entre infirmières pouvaient être assez crues. Joachim avait dû avoir une éjaculation précoce. Frieda lui avait dit que c'était arrivé à Heinrich plusieurs fois au tout début de leur liaison et qu'il en avait été très mortifié. Il s'en était cependant vite remis. C'était un signe de nervosité, lui avait expliqué Frieda.

La fin prématurée des étreintes de Maud et Joachim n'allait pas lui faciliter la tâche, se dit Carla. Joachim serait à présent plus vigilant, moins sourd et aveugle à ce qui se passait autour de lui. Qu'importe, Maud allait faire son possible pour que Joachim continue à tourner le dos à la porte. Si elle parvenait à se glisser dans la pièce juste une seconde et à reposer le sac sur la chaise sans se faire voir de Joachim, elles pourraient encore s'en sortir.

Le cœur battant, Carla traversa le salon et s'arrêta près de la porte ouverte.

Maud disait d'un ton rassurant : « C'est très fréquent, tu sais. Le corps s'impatiente, ce n'est rien. »

Carla glissa la tête dans l'embrasure de la porte. Le couple était toujours à la même place, debout, enlacé.

Apercevant Carla, Maud posa la main sur la joue du jeune homme, l'obligeant à garder les yeux posés sur elle. « Embrasse-moi encore, dis-moi que tu ne me détestes pas pour ce petit incident. »

Carla franchit le seuil.

Joachim dit alors : « J'ai envie d'une cigarette. »

Comme il se retournait, elle se hâta de reculer.

Elle resta près de la porte. Avait-il des cigarettes sur lui ? Devrait-il prendre le paquet neuf dans sa sacoche ?

La réponse lui parvint une seconde plus tard. « Où est ma sacoche ? »

Le cœur de Carla s'arrêta de battre.

La voix de Maud lui parvint distinctement. « Tu l'as laissée au salon.

— Mais non ! »

Carla traversa la pièce en toute hâte, laissant tomber le sac sur une chaise au passage. Arrivée sur le palier, elle s'arrêta, aux aguets.

Elle les entendit passer du cabinet de travail dans le grand salon.

« La voilà ! Tu vois bien ! s'écria Maud.

— Je ne l'avais pas laissée là, s'obstina Joachim. Je m'étais juré de ne pas la lâcher des yeux ; le seul moment où je l'ai perdue de vue, c'est quand nous nous sommes embrassés.

— Mon chéri, tu es contrarié à cause de ce qui s'est passé. Essaie de te détendre.

— Quelqu'un est entré dans cette pièce pendant que j'avais la tête ailleurs.

— Voyons, c'est absurde !

— Pas du tout !

— Allons nous asseoir au piano, l'un à côté de l'autre, comme tu aimes le faire, suggéra-t-elle, avec une nuance d'affolement dans la voix.

— Qui d'autre y a-t-il dans la maison ? »

Devinant la suite, Carla dévala l'escalier et fila dans la cuisine. Ada la regarda, apeurée, mais elle n'avait plus le temps de lui donner d'explication.

Les marches résonnaient déjà sous les bottes de Joachim.

Il surgit dans la cuisine, brandissant sa sacoche en toile. Il avait le visage déformé par la colère. Les yeux rivés sur Carla et Ada, il déclara : « L'une de vous deux a fouillé dans mon sac !

— Je ne vois pas ce qui vous permet de dire ça, Joachim », déclara Carla de sa voix la plus calme.

Maud apparut dans le dos du jeune homme. Elle passa devant lui. « Ada, s'il vous plaît, faites-nous donc du café, dit-elle d'une voix enjouée. Joachim, vous devriez vous asseoir un instant. »

Il l'ignora et promena lentement les yeux tout autour de la cuisine. Son regard s'illumina en se posant sur le buffet près de la fenêtre. Carla s'aperçut avec consternation qu'elle avait oublié de ranger les pellicules dans le tiroir, avec l'appareil photo.

« Mais ce sont des pellicules huit millimètres ! s'écria Joachim. Vous avez un appareil photo miniature ? »

Tout à coup, il n'avait plus rien d'un petit garçon.

« Ah bon, c'est ça ! Je me demandais ce que ça pouvait bien être, s'écria Maud. C'est un autre élève qui les a oubliées, un officier de la Gestapo, en fait. »

L'improvisation était habile, mais Joachim ne s'y laissa pas prendre. « J'imagine qu'il a aussi laissé son appareil, n'est-ce pas ? » Il ouvrit le tiroir.

Le petit appareil en acier chromé était là, posé sur un torchon blanc, aussi compromettant qu'une tache de sang.

Joachim demeura pétrifié. Peut-être ne s'était-il pas réellement cru victime de trahison jusqu'à ce moment-là ; peut-être avait-il seulement cherché à compenser son échec sexuel en jouant les fanfarons ; maintenant la vérité s'étalait sous ses yeux. Consterné, la main sur la poignée du tiroir, il fixait l'appareil photo comme s'il était hypnotisé. En ce bref instant, Carla comprit que tous les rêves d'amour du jeune homme venaient d'être profanés et que sa colère serait implacable.

Il releva enfin les yeux. Regardant les trois femmes qui l'entouraient, il s'arrêta sur Maud. « C'est vous ! Vous m'avez trompé mais vous allez le payer cher. » Il saisit appareil et pellicules et les fourra dans sa poche. « Vous êtes en état d'arrestation, Frau von Ulrich. » Il marcha sur elle et la saisit par le bras. « Vous allez me suivre au siège de la Gestapo. »

Maud se dégagea et recula d'un pas.

Joachim leva le bras et la frappa de toutes ses forces. Il était grand, fort, jeune. Le coup atteignit Maud au visage et la fit tomber.

La dominant de toute sa taille, Joachim se mit à hurler : « Vous vous êtes moquée de moi ! Vous m'avez menti et moi, je vous ai crue ! » Il était au bord de l'hystérie. « La Gestapo nous torturera tous les deux, nous l'avons mérité ! » Il se mit à bourrer Maud de coups de pied. Elle tenta de se soustraire à ses coups mais en fut empêchée par le fourneau. La botte droite du jeune homme s'enfonçait avec un bruit sourd dans ses côtes, ses cuisses, son ventre.

Ada se rua sur Joachim et lui griffa le visage. Il l'écarta du bras et recommença à frapper Maud du pied, à la tête maintenant.

Carla intervint.

Elle avait appris que le corps pouvait se remettre de bien des contusions, mais qu'une blessure à la tête risquait de provoquer des dégâts irréversibles. Ce ne fut qu'une pensée fugace et elle agit sans réfléchir. Elle attrapa la marmite qu'Ada avait posée sur la table de la cuisine après l'avoir si bien récurée. La tenant par les poignées, elle la leva aussi haut qu'elle le pouvait et l'abattit de toutes ses forces sur la tête de Joachim.

Il chancela, assommé.

Elle le frappa à nouveau, plus fort encore.

Il s'écroula, inconscient. Maud s'écarta au moment où le corps s'affalait par terre et s'assit contre le fourneau, les mains sur la poitrine.

Carla leva à nouveau la marmite.

« Non ! Arrête ! » hurla Maud.

Carla la reposa sur la table.

Joachim remua, essaya de se lever.

Ada empoigna alors la marmite et le frappa avec fureur. Carla essaya vainement de s'interposer. En proie à une rage folle, Ada continuait de taper sur la tête du jeune homme désormais évanoui, s'acharnant sur lui encore et encore jusqu'à ne plus avoir la force de soulever la marmite qui tomba par terre avec fracas.

Se redressant tant bien que mal sur les genoux, Maud regarda Joachim. Ses yeux fixes étaient grands ouverts et il avait le nez tordu. Son crâne semblait déformé et du sang coulait d'une de ses oreilles. Apparemment, il ne respirait plus.

Carla s'agenouilla près de lui et posa le bout des doigts sur le côté de son cou. « Il est mort, dit-elle. Mon Dieu, nous l'avons tué. »

Maud fondit en larmes : « Pauvre gosse, quel petit idiot !

— Qu'est-ce qu'on fait maintenant ? » demanda Ada, encore essoufflée.

Carla comprit alors qu'elles allaient devoir se débarrasser du corps.

Maud se releva péniblement, la joue gauche tuméfiée. « Bon sang, ça fait mal ! » gémit-elle en appuyant la main sur son flanc. Elle devait avoir une côte cassée, songea Carla.

« On pourrait le cacher dans le grenier, suggéra Ada en baissant les yeux sur Joachim.

— Jusqu'à ce que les voisins se plaignent de l'odeur ! répliqua Carla.

— Alors, enterrons-le dans le jardin.

— Et que diront les gens en voyant trois femmes creuser un trou de deux mètres de long dans le jardin d'une maison berlinoise ? Que nous cherchons de l'or ?

— Il n'y a qu'à creuser de nuit.

— Tu crois que ça aurait l'air moins suspect ? »

Ada se gratta la tête.

« Il va falloir le sortir d'ici et le balancer dans un parc ou un canal, dit Carla.

— Mais comment allons-nous le porter ? demanda Ada.

— Il n'est pas bien lourd, remarqua Maud tristement. Si mince et si fort en même temps.

— Le problème, ce n'est pas le poids. À nous deux, Ada et moi, on arrivera bien à le porter. Le problème, c'est de le faire sans éveiller les soupçons.

— Si seulement on avait une voiture », soupira Maud.

Carla secoua la tête. « De toute façon, il n'y a plus d'essence. »

Le silence s'installa. Dehors la nuit tombait. Ada entoura la tête de Joachim d'une serviette pour que le sang ne tache pas le sol. Maud pleurait tout bas. Les larmes ruisselaient sur son visage crispé d'angoisse. Carla aurait bien voulu la réconforter mais il fallait d'abord régler ce problème.

« On pourrait le mettre dans une caisse », dit-elle.

À quoi Ada répondit : « La seule caisse de cette taille, c'est un cercueil.

— Et un meuble ? Un buffet ?

— Trop lourd… Il y aurait bien la penderie de ma chambre, reprit Ada après un moment de réflexion. Elle n'est pas très lourde. »

Carla approuva un peu gênée : il y avait effectivement dans la chambre d'Ada une petite penderie en bois blanc, légère car une domestique n'était pas censée posséder une vaste garde-robe et n'avait pas besoin d'une armoire en acajou. « Allons la chercher », approuva-t-elle.

Au début, Ada avait logé au sous-sol, mais cette partie de la maison ayant été transformée en abri antiaérien, elle avait désormais une chambre à l'étage. Ada et Carla y montèrent. Ada débarrassa l'armoire de tous ses vêtements. Elle n'en avait pas beaucoup : deux tenues de service, quelques robes, un manteau d'hiver, le tout très usé. Elle les déposa soigneusement sur son lit étroit.

Carla fit basculer l'armoire et la retint pendant qu'Ada la soulevait de l'autre côté. Elle n'était pas lourde mais encombrante, et il leur fallut un moment pour arriver à la faire passer par la porte et pour la descendre au rez-de-chaussée.

Enfin, elles la couchèrent sur le sol du vestibule et Carla ouvrit la porte. La penderie ressemblait maintenant à un cercueil muni d'un couvercle à charnières.

Carla retourna dans la cuisine et se pencha sur le corps. Elle retira de la poche d'uniforme l'appareil photo et les pellicules qu'elle rangea dans le tiroir du buffet.

Les deux femmes soulevèrent Joachim, Carla par les bras, Ada par les pieds, le sortirent de la cuisine et le portèrent jusque dans l'entrée où elles l'allongèrent à l'intérieur de l'armoire. Ada réarrangea la serviette autour de la tête de Joachim, mais il ne saignait plus.

Fallait-il lui retirer son uniforme ? se demanda Carla. Cela rendrait l'identification plus difficile ; en revanche, elles auraient alors à faire disparaître ses habits en plus du corps. Mieux valait s'en abstenir.

Elle déposa la sacoche dans l'armoire avec le cadavre.

Elle ferma à clé la porte de la penderie pour s'assurer qu'elle ne s'ouvrirait pas en chemin et glissa la clé dans la poche de sa robe.

Elle gagna la salle à manger et regarda par la fenêtre. « Il commence à faire nuit. Tant mieux. »

Maud s'inquiéta : « Que penseront les gens ?

— Que nous déménageons un meuble. Pour le vendre, par exemple. Pour avoir de quoi acheter à manger.

— Deux femmes déménageant une armoire ?

— Ce n'est pas si rare, tu sais, maintenant que tant d'hommes sont sur le front, ou morts. Difficile de trouver un camion de déménagement, il n'y a même plus d'essence !

— Mais pourquoi déménager à la tombée de la nuit ? »

Carla ne put dissimuler son exaspération. « Je n'en sais rien, Mutti. Si on nous le demande, j'inventerai quelque chose. De toute façon, ce cadavre ne peut pas rester ici.

— Quand on le retrouvera, on saura tout de suite qu'il a été assassiné. Il suffira d'examiner les blessures. »

Cette question inquiétait également Carla. « Que veux-tu qu'on y fasse ? répliqua-t-elle.

— La police voudra savoir où il est allé aujourd'hui.

— Il nous a bien dit qu'il n'avait parlé à personne de ses leçons de piano, il voulait faire la surprise à ses amis. Avec un peu de chance, personne ne sait qu'il est venu ici. » Autrement, c'est la mort pour nous trois ! songea Carla.

« Comment expliqueront-ils cet assassinat ?

— Est-ce qu'ils trouveront des traces de sperme dans son caleçon ? »

Maud se détourna, confuse. « Oui.

— Dans ce cas, ils imagineront une rencontre sexuelle qui a mal tourné. Peut-être avec un autre homme.

— Pourvu que tu aies raison. »

Carla était loin d'en être persuadée. Mais que faire ? « Le canal », dit-elle. Le corps remonterait à la surface et serait découvert tôt ou tard ; une enquête criminelle serait ouverte. Restait à espérer que cette enquête ne conduise pas jusqu'à elles.

Carla ouvrit la porte d'entrée.

Elle se plaça devant, légèrement sur la gauche de l'armoire, Ada derrière, sur la droite. Elles se baissèrent.

Ada, qui avait indéniablement plus l'habitude que ses employeuses de soulever de lourdes charges, lui conseilla : « Incline-la un peu et glisse tes mains dessous. »

Carla obtempéra.

« Maintenant, soulève de ton côté. »

Carla obéit.

Ada glissa elle aussi les mains sous l'armoire et dit : « Plie les genoux et prends la mesure de son poids avant de te relever. »

Elles soulevèrent l'armoire à hauteur de genoux. Ada se courba et la cala sur sa hanche, Carla l'imita.

Les deux femmes se redressèrent.

Elles descendirent ainsi les marches du perron, tout le poids de l'armoire reposant sur Carla, mais c'était supportable. Arrivées dans la rue, elles prirent la direction du canal, quelques rues plus loin.

Il faisait noir, la nuit était sans lune, mais de rares étoiles projetaient une faible lueur. Grâce au couvre-feu, il y avait peu de risque que quelqu'un les aperçoive en train de faire basculer l'armoire dans l'eau. L'inconvénient, c'était que Carla voyait à peine où elle posait les pieds. Elle était terrifiée à l'idée de trébucher, de tomber, et que la penderie se brise en mille morceaux, révélant son contenu : le cadavre d'un homme assassiné !

Une ambulance passa, ses phares masqués par des écrans ajourés. Elle se rendait probablement sur le lieu d'un accident. Les collisions étaient fréquentes pendant le couvre-feu. Ce qui voulait dire qu'il y aurait des voitures de police dans le voisinage.

Carla se remémora un célèbre assassinat commis au tout début du couvre-feu : un homme avait tué sa femme et enfermé son corps dans une valise qu'il avait transportée de nuit à travers toute la ville sur sa bicyclette, pour aller la jeter dans la Havel. La police se rappellerait-elle cette affaire en voyant quelqu'un transporter un objet encombrant ?

À l'instant précis où ce souvenir lui revenait, un véhicule de police les dépassa. Un policier regarda fixement les deux femmes chargées d'une armoire, mais la voiture ne s'arrêta pas.

La penderie commençait à peser. De plus, la nuit était chaude. Carla fut bientôt en nage. Le bois lui coupait les mains. Elle aurait dû mettre un mouchoir plié ou des gants.

Bifurquant dans une rue, elles se retrouvèrent sur les lieux de l'accident.

Un camion-remorque à huit roues chargé de bois était entré en collision de plein fouet avec une berline Mercedes qu'il avait gravement endommagée. La voiture de police et l'ambulance éclairaient le sinistre de leurs gyrophares. Dans un petit cercle de lumière, un groupe d'hommes entourait le véhicule. L'accident devait s'être produit quelques minutes auparavant, car trois personnes se trouvaient encore dans la Mercedes. Penché par la portière arrière, un ambulancier examinait probablement les blessés pour s'assurer qu'ils étaient transportables.

L'espace d'un instant, Carla se figea d'effroi et de culpabilité. Puis, comme personne ne prêtait attention à elle, à Ada ou à l'armoire qu'elles transportaient, elle se dit qu'elles n'avaient qu'à tourner les talons, revenir sur leurs pas et rejoindre le canal par un autre chemin.

Elle allait faire demi-tour quand un policier vigilant braqua sa torche sur elle.

Elle faillit lâcher l'armoire et s'enfuir à toutes jambes. Mais elle garda son sang-froid.

« Qu'est-ce que vous faites là ? demanda le policier.

— Nous déménageons une armoire, monsieur. » Retrouvant ses esprits, elle dissimula sa nervosité compromettante sous une curiosité macabre. « Qu'est-ce qui s'est passé ? demanda-t-elle. Il y a des morts ? »

Étant du métier, elle savait que secouristes et policiers détestaient ces manifestations de voyeurisme malsain. Comme prévu, le policier chercha à se débarrasser d'elle : « Dégagez, il n'y a rien à voir ! » et il lui tourna le dos, braquant sa torche sur la voiture accidentée.

Il rejoignit ses collègues au milieu de la rue. Il n'y avait plus qu'elles sur le trottoir et Carla décida finalement de continuer à avancer en direction des voitures accidentées.

Elle gardait les yeux rivés sur le groupe de secouristes dans le petit halo de lumière. Ils étaient concentrés sur leur tâche. Personne ne les remarqua quand elles dépassèrent la voiture.

Elle eut l'impression qu'elles mettaient une éternité à longer le camion et sa remorque. Quand elles arrivèrent enfin au bout, Carla eut une inspiration soudaine.

Elle s'arrêta.

« Qu'est-ce qui se passe ? murmura Ada.

— Par ici. » Carla descendit sur la chaussée, juste au niveau des roues arrière du camion. « Pose l'armoire. Sans faire de bruit ! » souffla-t-elle tout bas.

Elles déposèrent précautionneusement la penderie sur les pavés.

« Tu veux la laisser ici ? » chuchota Ada.

Carla sortit la clé de sa poche et la fit tourner dans la serrure de l'armoire. Puis elle releva les yeux : apparemment, les hommes étaient toujours autour de la voiture à six ou sept mètres, de l'autre côté du camion.

Elle ouvrit la porte de l'armoire.

Joachim Koch regardait droit devant lui sans voir, la tête enveloppée d'une serviette tachée de sang.

« Fais-le tomber, dit Carla. Près des roues ! » Elles inclinèrent la penderie. Le corps roula par terre et s'arrêta contre les pneus.

Carla ramassa la serviette pleine de sang et la jeta dans le meuble. Puis elle posa la sacoche en toile près du corps allongé, bien contente de s'en débarrasser. Ayant refermé l'armoire à clé, elles la soulevèrent à nouveau et repartirent.

Elle était nettement plus facile à transporter désormais.

Elles avaient parcouru une cinquantaine de mètres quand une voix s'éleva au loin dans le noir : « Oh, mon Dieu, une autre victime ! On dirait qu'un piéton s'est fait écraser. »

Carla et Ada tournèrent au coin de la rue et un sentiment de soulagement submergea Carla tel un raz de marée. Elle s'était débarrassée du cadavre ! Si elle parvenait à rentrer à la maison sans attirer l'attention, sans que personne ne vienne regarder à l'intérieur de l'armoire et ne remarque la serviette ensanglantée, alors elle serait sauvée. Il n'y aurait pas d'enquête pour meurtre, Joachim serait un piéton fauché par un camion pendant le couvre-feu. Car si les roues du camion l'avaient traîné sur cette chaussée pavée, il aurait très bien pu avoir des blessures identiques à celles qu'Ada lui avait infligées avec la marmite. Un médecin légiste expérimenté pourrait certainement voir la différence, mais qui songerait à réclamer une autopsie ?

Carla envisagea de se débarrasser de l'armoire, puis se ravisa. Même si on en retirait la serviette, le meuble était taché de sang, ce qui pourrait attirer l'attention de la police. Non, mieux valait la rapporter à la maison et la nettoyer à fond.

Elles regagnèrent la maison sans rencontrer personne.

Elles déposèrent la penderie dans l'entrée. Ada en sortit la serviette, la mit dans l'évier de la cuisine et fit couler de l'eau froide.

Carla éprouva un sentiment d'allégresse teinté de tristesse. Elle avait réussi à dérober le plan de bataille des nazis, mais avait tué un jeune homme, plus bête que méchant. Cette pensée continuerait à la tarauder pendant des jours, peut-être des années. Il lui faudrait longtemps pour voir clair dans ses sentiments. Pour le moment, elle était trop fatiguée.

Elle raconta à sa mère leur périple avec l'armoire. Maud avait la joue gauche tellement enflée que son œil était presque fermé. Elle tenait la main contre son flanc gauche comme pour tenter d'atténuer la douleur. Elle avait une mine épouvantable.

« Tu as été vraiment courageuse, Mutti. Je t'admire infiniment pour ce que tu as fait aujourd'hui.

— Eh bien moi, je ne me sens pas admirable du tout ! soupira Maud d'une voix lasse. J'ai tellement honte, je me méprise.

— Parce que tu ne l'aimais pas ?

— Non, parce que je l'aimais. »

XIV
1942 (III)

1.

Greg Pechkov obtint son diplôme de Harvard avec les félicitations du jury. Il aurait très bien pu poursuivre ses études et passer son doctorat de physique, ce qui lui aurait évité le service militaire. Mais il ne voulait pas être chercheur, il ambitionnait un autre type de pouvoir. La guerre finie, une carte d'ancien combattant serait un atout de taille pour une étoile montante de la politique. Il s'engagea donc dans l'armée.

Cela dit, il ne tenait pas vraiment à se battre.

Tout en suivant la guerre en Europe avec un intérêt accru, il faisait pression sur tous les gens qu'il connaissait à Washington – et ils étaient nombreux – pour obtenir un poste de bureau au ministère de la Guerre.

Le 28 juin, les Allemands avaient lancé leur offensive d'été. Ils avaient progressé vers l'est rapidement, sans rencontrer de véritable opposition jusqu'à la ville de Stalingrad, anciennement Tsaritsyne. Là, ils avaient été arrêtés par une farouche résistance. Maintenant, ils étaient dans l'impasse. Leurs lignes de ravitaillement étaient trop étirées, et il semblait de plus en plus évident que l'armée Rouge les avait attirés dans un guet-apens.

Greg venait de commencer sa formation militaire quand il fut convoqué chez le colonel. « L'Army Corps of Engineers, le Corps des ingénieurs de l'armée, réclame un jeune et brillant officier à Washington. Personnellement, je ne vous aurais jamais placé en tête de liste, malgré tous vos stages à Washington. Regardez-vous, vous n'êtes même pas foutu de garder votre uniforme propre ! Mais il se trouve que le travail en question requiert des connaissances en physique, et dans ce domaine, la concurrence est plutôt limitée.

— Je vous remercie, mon colonel.

— Vous feriez mieux d'éviter ce genre d'ironie avec votre nouveau supérieur, le colonel Groves. J'ai fait West Point avec lui. C'est le pire fils de pute que j'aie rencontré de ma vie, à l'armée comme dans le civil. Bonne chance. »

Greg appela Mike Penfold, au bureau de presse du Département d'État, et apprit que jusqu'à récemment, Leslie Groves était responsable des travaux publics pour toute l'armée américaine. Il avait dirigé notamment la construction du nouvel état-major de l'armée à Washington, un immense complexe à cinq côtés qu'on commençait à appeler le Pentagone. Depuis, il avait été transféré sur un nouveau projet dont on ne savait pas grand-chose. Les uns disaient qu'il avait si souvent offusqué ses supérieurs qu'on avait fini par le rétrograder, d'autres que sa nouvelle affectation était encore plus importante, mais ultra-secrète. Tout le monde s'accordait cependant à dire que c'était un homme égoïste, arrogant et impitoyable.

« Est-ce que vraiment *tout le monde* le déteste ? demanda Greg.

— Oh, non, répondit Mike. Seulement ceux qui l'ont rencontré. »

C'est donc empli d'appréhension que le lieutenant Greg Pechkov se présenta dans le bureau de Groves, au New War Department Building, spectaculaire palais de style Art déco de couleur ocre clair, situé à l'angle de la 21e Rue et de Virginia Avenue. Il y apprit dans l'instant qu'il ferait partie d'un groupe baptisé Manhattan Engineer District. Sous ce vocable délibérément abscons de « secteur d'ingénierie de Manhattan » se cachait une équipe de chercheurs qui essayait d'inventer un nouveau type de bombe se servant d'uranium comme explosif.

Greg était intrigué. Il savait, pour avoir lu plusieurs articles sur le sujet dans diverses revues scientifiques, que l'U-235, l'isotope léger de l'uranium, renfermait une masse d'énergie phénoménale. Mais ces deux dernières années, aucune information nouvelle n'avait été publiée sur le sujet. Il comprenait à présent pourquoi.

Il apprit que le projet n'avançait pas assez vite au gré du président Roosevelt et que celui-ci avait nommé Groves à sa tête, pour lui donner un coup de fouet.

Greg prit ses fonctions six jours après la nomination de Groves. Sa première tâche consista à aider son nouveau chef à piquer des étoiles dans le col de sa chemise kaki : Groves venait d'être promu général de brigade. « C'est surtout pour impressionner tous ces chercheurs civils avec qui nous sommes obligés de travailler, grommela Groves. Je suis attendu chez le ministre de la Guerre dans dix minutes. Vous feriez bien de venir avec moi, ça vous mettra au parfum. »

Groves était un homme massif. Il devait peser dans les cent vingt, cent trente kilos pour un mètre quatre-vingts. Il portait son pantalon d'uniforme sanglé haut, et sa bedaine dessinait une grosse bosse sous son ceinturon. Il avait des cheveux châtains qui auraient pu friser s'il les avait laissés pousser un peu, le front étroit, les mâchoires épaisses et le menton carré. Sa petite moustache était presque invisible. C'était un homme dénué de séduction à tous points de vue, et Greg n'était pas enchanté de travailler sous ses ordres.

Groves et son entourage, Greg compris, sortirent du bâtiment et descendirent Virginia Avenue en direction du National Mall. En chemin, Groves dit à Greg : « On m'a donné ce poste en me disant que ça pouvait nous faire gagner la guerre. Je ne sais pas si c'est vrai, mais je compte bien faire comme si. Je vous conseille d'en faire autant.

— Oui, mon général. »

Le Pentagone n'étant pas encore achevé, le ministère de la Guerre occupait toujours le Munitions Building, un vieux bâtiment « provisoire » tout en longueur situé sur Constitution Avenue.

Le ministre de la Guerre, Henry Stimson, appartenait au parti républicain. Il avait été nommé par Roosevelt pour empêcher ce parti de compromettre l'effort de guerre en lui mettant des bâtons dans les roues au Congrès. À soixante-quinze ans, Stimson était un politicien chevronné, un vieux monsieur soigné à la moustache blanche et aux yeux gris pétillants d'intelligence.

Cette réunion étant des plus officielles, la salle grouillait de personnalités en uniforme, parmi lesquelles George Marshall, le chef d'état-major de l'armée. Quelque peu intimidé, Greg admira le calme imperturbable de Groves qui, la veille encore, n'était que simple colonel.

Celui-ci commença son discours en expliquant comment il comptait imposer un minimum de discipline aux centaines de scientifiques civils et dans les dizaines de laboratoires de physique impliqués dans le projet Manhattan. Il ne feignit pas un instant de s'en remettre aux sommités qui auraient pu se croire responsables de ce projet. Il exposa son programme sans s'embarrasser de formules de politesse telles que « avec votre permission » ou « si vous n'y voyez pas d'inconvénient ». Au point que Greg se demanda s'il ne cherchait pas à se faire virer.

Greg apprenait tant d'informations nouvelles qu'il aurait volontiers pris des notes. Comme personne ne le faisait, il se dit que ce serait mal vu.

Quand Groves eut terminé son discours, quelqu'un demanda : « Si je ne me trompe, l'approvisionnement en uranium est un élément essentiel à la bonne marche du projet. En avons-nous suffisamment ? »

Groves répondit : « Un entrepôt de Staten Island abrite mille deux cent cinquante tonnes de pechblende. C'est le nom du minerai qui contient l'oxyde d'uranium.

— Dans ce cas, nous ferions bien d'en acheter un peu, reprit l'autre.

— J'ai acheté la totalité du stock vendredi dernier.

— Vendredi ? Au lendemain de votre nomination ?

— Oui. »

Le ministre de la Guerre réprima un sourire. Greg, tout d'abord effaré par l'arrogance de Groves, commença à admirer son sang-froid.

Un homme en tenue d'amiral prit la parole : « Quel est l'indice de priorité accordé à ce projet ? Vous devez régler cette question avec le War Production Board.

— J'ai vu Donald Nelson samedi dernier, amiral », dit Groves. C'était le responsable du Bureau de production de guerre, le WPB, dirigé par des civils. « Je lui ai demandé de le remonter d'un cran.

— Qu'a-t-il répondu ?

— Non.

— C'est un problème.

— Plus maintenant. Je lui ai dit que dans ce cas, je me verrais dans l'obligation de recommander au Président d'abandonner le projet Manhattan, parce que le War Production Board refusait de coopérer. Il nous a accordé un triple A.

— Bien », déclara le ministre de la Guerre.

Groves n'avait vraiment pas froid aux yeux, se dit Greg, de plus en plus impressionné.

Stimson intervint : « Dorénavant, vous serez placé sous le contrôle d'un comité qui en référera directement à moi. Neuf membres ont été proposés…

— Nom d'une pipe, pas question ! coupa Groves.

— Pardon ? » demanda le ministre de la Guerre.

Cette fois, Groves était vraiment allé trop loin, songea Greg. Son chef continuait pourtant : « Je ne peux pas être sous les ordres d'un comité de neuf membres, monsieur le ministre. Je les aurais continuellement sur le dos. »

Stimson sourit. Il avait trop d'expérience pour s'offusquer de ce genre de discours, semblait-il. Il dit avec douceur : « Quel nombre proposez-vous, général ? »

De toute évidence, Groves aurait volontiers répondu : « Zéro », mais la réponse qui sortit de ses lèvres fut : « Trois, ce serait parfait » et, à la stupéfaction de Greg, le ministre de la Guerre déclara : « Très bien. Autre chose ?

— Pour l'usine d'enrichissement d'uranium et les installations annexes, nous allons avoir besoin d'un site très étendu. Environ trente mille hectares, je dirais. Il y a un endroit qui conviendrait. Oak Ridge, dans le Tennessee. C'est une vallée encaissée. En cas d'accident, l'explosion serait limitée.

— En cas d'accident ? intervint l'amiral. Pareille éventualité est-elle probable ? »

Groves ne chercha pas à dissimuler qu'il trouvait la question stupide. « De quoi parle-t-on, nom de Dieu ? Il s'agit de fabriquer une bombe expérimentale ! Une bombe si puissante qu'elle pourrait rayer de la carte une ville de taille moyenne en un clin d'œil. Il faudrait être franchement crétin pour ne pas envisager cette éventualité ! »

L'amiral fit mine de protester, mais Stimson lui coupa l'herbe sous le pied : « Continuez, général.

— Les terrains sont bon marché dans le Tennessee, reprit Groves. L'électricité aussi. Notre usine va avoir besoin de quantités d'énergie considérables.

— Vous proposez donc d'acheter ce terrain.

— Je me propose d'aller le voir aujourd'hui même, répliqua Groves en regardant sa montre. En fait, il faut que je parte tout de suite si je ne veux pas rater mon train pour Knoxville. » Il se leva. « Si vous voulez bien m'excuser, messieurs, je n'ai pas de temps à perdre. »

L'assistance resta bouche bée et Stimson lui-même eut l'air ébahi. Personne à Washington n'aurait imaginé quitter le bureau d'un ministre avant que celui-ci ne lève la séance. C'était là un manquement au protocole de taille. Mais Groves ne semblait guère s'en soucier.

Son culot paya.

« Très bien, déclara Stimson. Nous ne vous retenons pas.

— Je vous remercie, monsieur », répondit Groves, et il sortit.

Greg se précipita à sa suite.

2.

Avec ses grands yeux noirs et sa bouche large et sensuelle, Margaret Cowdry était la plus jolie de toutes les secrétaires civiles qui travaillaient au ministère de la Guerre. Il suffisait de l'apercevoir derrière sa machine à écrire et de la voir lever les yeux en souriant pour avoir l'impression d'être déjà en train de faire l'amour avec elle.

Son père s'était lancé dans la pâtisserie industrielle. « Les biscuits Cowdry sont aussi croustillants que ceux de Maman ! » Margaret n'avait pas besoin de gagner sa vie, mais c'était sa manière de participer à l'effort de guerre. Avant de l'inviter à déjeuner, Greg lui fit discrètement savoir qu'il était lui aussi fils de millionnaire. En général, les riches héritières préféraient sortir avec des jeunes gens fortunés ; au moins, elles savaient qu'ils ne couraient pas après leur argent.

C'était le mois d'octobre et il faisait froid. Margaret portait un élégant manteau bleu marine aux épaules rembourrées et cintré à la taille. Son béret assorti lui donnait un petit air militaire.

Ils se rendirent au Ritz-Carlton. Là, en entrant dans la salle à manger, Greg aperçut son père qui déjeunait avec Gladys Angelus. Il n'avait aucune envie de se joindre à eux. Il l'expliqua à Margaret, qui répondit : « Pas de problème. Allons à l'University Women's Club, c'est au coin de la rue. Je suis membre. »

Greg n'y était jamais allé. Pourtant il avait l'impression d'avoir déjà entendu parler de cet endroit. Mais où, cela lui échappait et il chassa cette pensée de son esprit.

Au club, Margaret retira son manteau et apparut dans une robe en cachemire bleu roi qui lui allait à ravir. Elle conserva son chapeau et ses gants, comme toute femme chic qui se respectait lors d'un déjeuner au restaurant.

Greg était toujours heureux de parader en public au bras d'une jolie fille. Il n'y avait qu'une poignée d'hommes dans la salle à manger de ce cercle féminin, mais tous l'envièrent. Greg ne l'aurait probablement jamais avoué, mais ces moments lui procuraient autant de plaisir que de coucher avec une femme.

Il commanda du vin que Margaret coupa d'eau minérale, à la mode française, en expliquant : « Je ne tiens pas à passer l'après-midi à corriger mes fautes de frappe. »

Il lui parla du général Groves. « C'est un bulldozer. À certains égards, c'est la copie de mon père, en plus mal habillé.

— Tout le monde le déteste, déclara Margaret.

— Il faut dire qu'il prend les gens à rebrousse-poil.

— Votre père est comme ça ?

— Ça lui arrive. Mais en général, il préfère faire du charme.

— Le mien aussi ! C'est peut-être un trait commun à tous les hommes qui réussissent dans la vie. »

Le repas ne s'éternisa pas. Dans tous les restaurants de Washington, le service s'était accéléré. La nation était en guerre, les gens étaient pressés de se remettre au travail.

Une serveuse leur apporta la carte des desserts. En levant les yeux, Greg eut la surprise de reconnaître Jacky Jakes. « Bonjour, Jacky !

— Salut, Greg, tu vas bien ? répondit-elle avec une familiarité qui dissimulait sa nervosité. Qu'est-ce que tu deviens ? »

Greg se rappela que le détective lui avait appris qu'elle travaillait à l'University Women's Club. Voilà comment il en avait entendu parler. « Je vais très bien. Et toi ?

— En pleine forme.

— Rien de neuf? demanda-t-il, cherchant à savoir par ces mots si son père lui versait toujours une allocation.

— Pas grand-chose. »

Lev avait dû charger un avocat de s'occuper de cette affaire et l'avoir complètement oubliée, se dit Greg.

« Eh bien, tant mieux.

— Puis-je vous proposer un dessert ? reprit Jacky, se rappelant ses fonctions.

— Oui, volontiers. »

Margaret commanda une salade de fruits, Greg une glace.

Quand Jacky s'éloigna, Margaret se tourna vers lui d'un air interrogateur : « Elle est très jolie…

— Oui, en effet.

— Elle ne porte pas d'alliance. »

Greg soupira. Les femmes étaient vraiment trop perspicaces. « Vous vous demandez comment je peux être l'ami d'une jolie serveuse noire qui n'est pas mariée ? Autant vous dire la vérité. J'ai eu une aventure avec elle quand j'avais quinze ans. J'espère que ça ne vous choque pas.

— Bien sûr que si. Je suis outrée », fit-elle mi-figue, mi-raisin. Elle n'était pas véritablement scandalisée, il en était certain, mais peut-être ne voulait-elle pas lui donner l'impression de prendre ces choses-là à la légère, du moins lors d'un premier déjeuner.

Jacky apporta les desserts et leur demanda s'ils voulaient un café. Ils n'en avaient pas le temps, l'armée n'appréciait pas les longues pauses repas. Margaret demanda l'addition en expliquant : « Dans ce cercle, les invités ne sont pas autorisés à payer. »

Lorsque Jacky fut repartie, elle enchaîna : « Visiblement, vous éprouvez encore beaucoup de tendresse pour elle. Je trouve ça bien.

— Ah bon ? s'étonna Greg. Je garde de bons souvenirs, c'est vrai. Ça ne me dérangerait pas de retrouver mes quinze ans.

— Pourtant, elle a peur de vous.

— Mais non !

— Si. Elle est terrifiée.

— Ça m'étonnerait.

— Vous pouvez me croire sur parole. Les hommes sont aveugles, mais les femmes voient très bien ces choses-là. »

Quand Jacky posa la note sur la table, Greg la dévisagea attentivement. Margaret avait raison : Jacky n'était pas rassurée. Chaque fois qu'elle le revoyait, elle devait se rappeler Joe Brekhounov et son rasoir.

Cela le mit en colère. Jacky avait quand même le droit de vivre en paix.

Il allait devoir s'occuper de ça.

Fine mouche, Margaret insista : « Je crois même que vous savez parfaitement de quoi elle a peur.

— De mon père. Il lui a fichu la trouille pour qu'elle me plaque. Il craignait que je veuille l'épouser.

— Votre père est-il effrayant ?

— Il aime que les choses se passent comme il l'entend.

— Le mien est pareil. Doux comme un agneau jusqu'au moment où vous vous mettez en travers de sa route. Là, il devient carrément mauvais.

— Je suis tellement heureux que vous compreniez. »

Ils retournèrent à leur travail. La colère ne quitta pas Greg de tout l'après-midi. D'une façon ou d'une autre, la menace de son père continuait à peser sur Jacky comme une malédiction. Mais que pouvait-il faire ?

Que ferait son père à sa place ? se demanda-t-il. Oui, c'était la bonne façon d'aborder la question. Lev n'aurait qu'une idée en tête : parvenir à ses fins, coûte que coûte ! Le général Groves ferait exactement pareil. Eh bien moi, aussi, se dit Greg. Tel père, tel fils !

Un semblant de plan commença à prendre forme dans son esprit.

Il passa l'après-midi à lire et à résumer un rapport provisoire du laboratoire de métallurgie de l'université de Chicago. Cette équipe de chercheurs comptait dans ses rangs Leo Szilard, celui qui, le premier, avait conçu l'idée de réaction nucléaire en chaîne. C'était un Juif hongrois qui avait fait ses études à l'université de Berlin jusqu'à la funeste année 1933. Il y avait aussi

Enrico Fermi, un physicien italien dont la femme était juive et qui avait quitté son pays après la publication du « Manifeste de la race » de Mussolini. C'était lui qui dirigeait l'équipe de recherches de Chicago.

Greg se demanda si les fascistes avaient pris conscience qu'avec leur racisme, ils avaient eux-mêmes fourni à l'ennemi une manne de brillants savants.

Il comprit sans difficulté de quoi traitait ce rapport. Selon la théorie de Fermi et Szilard, quand un neutron heurtait un atome d'uranium, la collision pouvait engendrer deux neutrons, lesquels pouvaient à leur tour entrer en collision avec d'autres atomes d'uranium et produire alors quatre neutrons, puis huit, et ainsi de suite. Szilard avait donné à ce processus le nom de réaction en chaîne, une découverte absolument géniale.

Ainsi, une tonne d'uranium pouvait produire autant d'énergie que trois millions de tonnes de charbon. Théoriquement.

En pratique, cela n'avait encore jamais été démontré.

Fermi et son équipe travaillaient à la construction d'une pile à uranium à Stagg Field, un stade de football désaffecté appartenant à l'université de Chicago. Pour éviter tout risque d'explosion spontanée, ils avaient enfermé l'uranium dans du graphite, qui absorbait les neutrons et empêchait la réaction en chaîne. Leur objectif était d'augmenter la radioactivité très progressivement jusqu'au niveau où la quantité de neutrons créés serait supérieure à la quantité absorbée – prouvant ainsi la réalité de la réaction en chaîne –, et de l'interrompre immédiatement, avant que ladite réaction n'ait fait sauter la pile, le stade, le campus et très probablement la ville de Chicago tout entière.

Jusqu'à présent, ils n'avaient pas réussi.

Greg rédigea un résumé favorable de ce rapport et demanda à Margaret Cowdry de le taper immédiatement. Il le porta ensuite à Groves.

Le général lut le premier paragraphe et demanda : « Ça va marcher ?

— Eh bien, mon général…

— C'est vous le scientifique, nom de Dieu ! Alors, ça va marcher, oui ou non ?

— Oui, mon général, ça va marcher, répondit Greg.

— Très bien », dit Groves, et il balança le dossier dans sa corbeille à papier.

De retour dans son bureau, Greg resta un long moment le regard rivé sur le tableau périodique des éléments accroché en face de lui. Il était presque sûr que la pile nucléaire fonctionnerait. Il l'était moins de réussir à convaincre son père de laisser Jacky tranquille.

Dès le départ, il avait résolu de régler le problème à la manière de Lev. Il en était maintenant aux détails pratiques. Il allait devoir jouer une petite pièce de théâtre.

Son plan prenait forme.

Mais aurait-il le courage d'affronter son père ?

Il quitta son bureau à cinq heures.

Sur le chemin du retour, il s'arrêta chez un coiffeur pour acheter un rasoir, un instrument à lame escamotable comme les couteaux à cran d'arrêt. « Avec votre barbe, vous en serez bien plus satisfait que de votre rasoir de sécurité », lui dit le coiffeur.

Greg le réservait à un autre usage.

Il vivait dans la suite que son père louait à l'année au Ritz-Carlton. Lorsqu'il arriva à l'hôtel, il tomba sur Lev et Gladys qui prenaient des cocktails.

Sa première rencontre avec Gladys, sept ans auparavant, lui revint en mémoire. Elle était assise sur ce même canapé recouvert de soie jaune. Aujourd'hui, c'était une immense vedette. Lev l'avait imposée dans quantité de films honteusement va-t-en-guerre où elle défiait des nazis sardoniques, déjouait les plans pervers de Japonais sadiques et soignait des pilotes américains à la mâchoire carrée. Elle n'était plus aussi belle qu'à vingt ans, remarqua Greg. Son teint était moins lisse, ses cheveux moins flamboyants et elle portait un soutien-gorge, ce qu'elle n'aurait certainement pas fait autrefois. Cependant, elle avait toujours ces yeux d'un bleu profond, irrésistiblement aguicheurs.

Greg accepta un martini. Allait-il vraiment défier son père ? Il ne l'avait jamais fait au cours des sept années qui s'étaient écoulées depuis qu'il avait serré la main de Gladys pour la première fois. Peut-être le temps était-il venu ?

Je vais agir exactement comme il le ferait à ma place, pensa Greg.

Il sirota son cocktail et reposa le verre sur un guéridon aux pieds fuselés. Se tournant vers Gladys, il déclara comme si de rien n'était : « Quand j'avais quinze ans, mon père m'a présenté à une actrice qui s'appelait Jacky Jakes. »

Les yeux de Lev s'écarquillèrent.

« Je ne crois pas la connaître », répondit Gladys.

Greg sortit le rasoir de sa poche, sans l'ouvrir, le tenant dans sa main comme pour en sentir le poids. « Je suis tombé amoureux d'elle.

— Pourquoi est-ce que tu nous sors cette vieille histoire ? » intervint Lev.

Gladys perçut la tension entre le père et le fils et manifesta quelques signes d'inquiétude.

Greg poursuivit : « Mon père a eu peur que je l'épouse. »

Lev partit d'un rire moqueur. « Cette putain au rabais ?

— Parce que c'était une putain au rabais ? demanda Greg en fixant Gladys. Je croyais que c'était une actrice. »

Gladys rougit sous l'insulte à peine voilée.

Greg enchaîna : « Mon père lui a rendu visite, avec un de ses collaborateurs. Joe Brekhounov. Vous le connaissez, Gladys ?

— Je ne crois pas.

— Vous avez de la chance. Joe possède un rasoir identique à celui-ci. » Greg fit jaillir du manche une lame étincelante et acérée.

Gladys retint son souffle.

Lev déclara : « À quoi tu joues, on peut savoir ?

— Un instant, répliqua Greg. Gladys a envie d'entendre la fin de l'histoire. » Il lui sourit. Elle avait l'air terrifiée. Il poursuivit : « Mon père a dit à Jacky que si jamais elle me revoyait, Joe lui lacérerait le visage. »

Il agita juste un peu le rasoir. Gladys poussa un petit cri.

« Ça suffit maintenant ! » lança Lev et il fit un pas vers Greg. Celui-ci brandit le rasoir. Lev se figea.

Oserait-il porter la main sur son père ? Il n'en savait rien. Mais Lev non plus.

« Jacky vit toujours ici, poursuivit Greg. À Washington.

— Et tu as recommencé à la baiser ? demanda son père crûment.

— Non. Je ne baise personne. Mais j'ai quelqu'un en vue. Margaret Cowdry.

— L'héritière des biscuits ?

— Pourquoi ? Tu veux que Joe lui fiche la trouille, à elle aussi ?

— Ne sois pas ridicule !

— Jacky travaille comme serveuse maintenant. Elle n'a jamais obtenu le rôle qu'elle espérait. Il m'arrive de la croiser dans la rue. Aujourd'hui, au restaurant, c'est elle qui m'a servi. Chaque fois qu'elle me voit, elle a peur de Joe et de son rasoir.

— Elle est cinglée, déclara Lev. Jusqu'à ce que tu racontes tout ça, elle m'était complètement sortie de la tête.

— Je peux le lui dire ? lança Greg. Il me semble qu'elle a le droit d'avoir la paix, après tout ce temps.

— Tu peux lui dire ce que tu veux. Pour moi, elle n'existe pas.

— Eh bien, c'est parfait. Elle sera ravie de le savoir.

— Et maintenant, remballe ton putain de rasoir !

— Une chose encore. Une mise en garde.

— Une mise en garde ! s'écria Lev avec colère.

— Si jamais il devait arriver quelque chose à Jacky… N'importe quoi… » Greg fit rouler la lame d'un côté puis de l'autre.

Lev jeta avec mépris : « Tu t'en prendras à Joe Brekhounov, c'est ça ?

— Non.

— À moi ? » insista Lev, avec un soupçon d'inquiétude.

Greg secoua la tête.

« À qui, alors, bon sang ? »

Greg regarda Gladys.

L'actrice mit un instant à comprendre l'allusion. Elle se rejeta alors en arrière dans le canapé en soie jaune, les deux mains plaquées sur ses joues comme pour se protéger. Elle poussa un petit cri strident.

« Espèce de connard ! » lança Lev.

Greg referma son rasoir et se leva. « C'est comme ça que tu aurais réglé la question, mon cher père. »

Sur ces mots, il sortit.

Il claqua la porte et s'appuya contre le mur. Il était aussi essoufflé qu'après une course à pied. De toute sa vie, il n'avait jamais eu aussi peur ! En même temps, il éprouvait un sentiment de triomphe. Il avait tenu tête au vieux, il lui avait rendu la monnaie de sa pièce et l'avait même légèrement effrayé.

Tout en rangeant le rasoir dans sa poche, il se dirigea vers l'ascenseur. Il respirait plus librement. Il se retourna en direction du long couloir de l'hôtel, s'attendant vaguement à voir son père arriver en courant. Mais la porte de la suite ne s'ouvrit pas. Greg entra dans l'ascenseur et descendit au rez-de-chaussée.

Au bar de l'hôtel, il commanda un martini dry.

3.

Le dimanche suivant, Greg décida d'aller voir Jacky.

Il voulait lui annoncer la bonne nouvelle. Il se souvenait de l'adresse. C'était le seul renseignement qu'il ait jamais obtenu en recourant aux services d'un détective privé. À moins qu'elle n'ait déménagé, Jacky habitait de l'autre côté d'Union Station. Il lui avait promis de ne jamais aller chez elle, mais il pouvait lui expliquer à présent que cette prudence n'était plus nécessaire.

Il s'y rendit en taxi. Tout en traversant la ville, il se réjouit de mettre un point final à cette aventure. Certes, il éprouvait toujours une certaine tendresse pour sa première maîtresse, mais ne voulait en aucune manière être mêlé à sa vie. Il serait soulagé de ne plus avoir ce poids sur la conscience. Au moins, la prochaine fois qu'il croiserait Jacky, elle n'aurait plus cet air terrifié. Ils pourraient se dire bonjour, bavarder un moment et s'en aller chacun de son côté.

Son taxi le conduisit dans un quartier pauvre de petites maisons sans étage aux jardinets entourés de grillage. Il se demanda comment elle vivait aujourd'hui. Que faisait-elle de ses soirées, qu'elle tenait tant à garder pour elle ? Elle devait aller au cinéma avec des copines. Allait-elle applaudir les Redskins, l'équipe de football de Washington, ou voir des matchs de baseball avec l'équipe des Nats ? Quand il lui avait demandé si elle avait un petit ami, elle était restée dans le vague. Peut-être était-elle mariée et n'avait pas les moyens de s'offrir une alliance. Elle devait avoir désormais vingt-quatre ans. Si elle était à la recherche du prince charmant, elle l'avait certainement trouvé à l'heure qu'il était. Pourtant, elle n'avait pas parlé de mari. Le détective non plus.

Il régla la course. La voiture s'était arrêtée devant une petite maison proprette avec une cour en ciment égayée de pots de fleurs – plus soignée que ce à quoi il s'était attendu. Dès qu'il ouvrit le portillon, il entendit un chien aboyer. Rien d'étonnant à cela : une femme seule se sentait plus en sécurité avec un chien. Il avança jusqu'au perron et sonna. Les aboiements s'intensifièrent. Un gros chien, probablement…

Personne ne vint ouvrir.

Quand le chien se tut pour reprendre son souffle, Greg reconnut le silence caractéristique d'une maison déserte.

Il y avait un banc en bois sur le perron. Il s'y assit et attendit quelques minutes. Personne ne se montra, aucun voisin serviable ne vint lui annoncer que Jacky serait absente quelques minutes, toute la journée, ou deux semaines.

Il partit acheter l'édition du dimanche du *Washington Post* à quelques rues de là et revint le lire sur le banc. Le chien, qui sentait sa présence, continuait d'aboyer par intermittence. C'était le 1er novembre, et il faisait un temps hivernal. Sa capote vert olive et sa casquette d'uniforme étaient les bienvenues. Les élections de mi-mandat devaient avoir lieu le mardi suivant. À en croire le journal, les démocrates prendraient une déculottée à cause de Pearl Harbor. Cet événement avait transformé l'Amérique ; pourtant, se dit Greg avec étonnement, il ne remontait même pas à un an. Aujourd'hui, des Américains de son âge mouraient sur une île dont personne n'avait jamais entendu parler et qui avait pour nom Guadalcanal.

Entendant le portillon s'ouvrir, il releva les yeux.

Jacky ne le remarqua pas tout de suite, ce qui lui permit de l'observer un moment. Avec son manteau sombre un peu démodé et son chapeau en feutre uni, elle avait l'air tout à fait respectable. Elle tenait à la main un gros livre à couverture noire. S'il ne l'avait pas connue, Greg aurait pu croire qu'elle revenait de l'église.

Elle était accompagnée d'un petit garçon en manteau de tweed et casquette, qui la tenait par la main. Ce fut lui qui aperçut l'inconnu assis sur le banc.

« Regarde, Mommy, un soldat ! »

En reconnaissant Greg, Jacky porta la main à sa bouche.

Il se leva tandis qu'ils montaient les marches du perron. Un enfant ! Elle lui avait caché ça. Voilà pourquoi elle voulait être chez elle le soir. Cette pensée ne lui avait pas traversé l'esprit.

« Je t'avais dit de ne pas venir ici ! protesta-t-elle en mettant la clé dans la serrure.

— Je voulais t'annoncer que tu n'avais plus rien à redouter de mon père. Je ne savais pas que tu avais un fils. »

Elle entra dans la maison avec le petit garçon. Greg demeura sur le seuil, avec l'air d'attendre quelque chose. Le chien, un berger allemand, lui montra les crocs puis tourna la tête vers Jacky en quête d'instructions. La jeune femme jeta un regard noir à Greg, manifestement prête à lui claquer la porte au nez. Au bout d'un moment, elle laissa échapper un soupir exaspéré et tourna les talons, laissant la porte ouverte.

Greg entra. Il tendit sa main gauche au chien qui la renifla avec méfiance et voulut bien lui accorder une approbation provisoire. Greg suivit Jacky à la cuisine.

« C'est la Toussaint, remarqua Greg, c'est pour ça que tu es allée à l'église ? » Il n'était pas porté sur la religion mais avait dû apprendre les noms de toutes les fêtes chrétiennes quand il était au pensionnat.

« Nous y allons tous les dimanches, répondit-elle.

— Décidément, je vais de surprise en surprise ! » murmura Greg.

Elle aida le garçonnet à retirer son manteau, le fit asseoir à table et lui servit un verre de jus d'orange. Greg prit place en face de lui. « Comment tu t'appelles ?

— Georgy », répondit-il tout bas mais sans crainte. Ce n'était pas un timide, pensa Greg, et il l'observa plus attentivement. L'enfant était aussi beau que sa mère. Il avait la même bouche en forme d'arc de Cupidon, mais le teint plus clair, café au lait, et les yeux verts, chose inhabituelle pour un Noir. Greg lui trouva quelque chose de Daisy, sa demi-sœur. De son côté, le petit garçon le dévisageait avec un regard intense, presque intimidant.

« Et quel âge as-tu, Georgy ? » demanda Greg.

Il se tourna vers sa mère comme pour lui demander de l'aide. Elle jeta un regard bizarre à Greg et dit : « Il a six ans.

— Six ans ! répéta Greg. Tu es un grand garçon, dis-moi ! Mais… »

Une pensée venait de lui traverser l'esprit et il s'interrompit. Georgy était né six ans plus tôt… Son aventure avec Jacky remontait à sept ans. Il sentit son cœur flancher.

Il dévisagea Jacky. « Ne me dis pas… »

Elle hocha la tête.

« Il est né en 1936, déclara Greg.

— En mai, précisa-t-elle. Huit mois et demi après mon départ de Buffalo.

— Mon père est au courant ?

— Tu penses bien que non ! Ça lui aurait donné encore plus de pouvoir sur moi. »

Son hostilité avait disparu, laissant apparaître toute sa vulnérabilité. Dans ses yeux, il lut une supplique mais n'aurait su dire quelle en était le sens.

Il se remit à examiner Georgy. D'un œil neuf, cette fois : sa peau claire, ses yeux verts, cette étrange ressemblance avec Daisy. Serais-tu de moi ? pensa-t-il. Est-ce possible ?

Il connaissait déjà la réponse.

Une curieuse émotion l'envahit. Subitement, Georgy lui apparaissait dans toute sa fragilité, comme un enfant sans défense dans un monde cruel, un enfant dont il devait absolument s'occuper, sur lequel il devait veiller pour qu'il ne lui arrive jamais rien. Il mourait d'envie de le prendre dans ses bras, mais se retint, se rendant compte que cela risquait de l'effrayer.

Georgy reposa son jus d'orange. Il descendit de sa chaise et fit le tour de la table pour s'approcher de Greg. Avec un regard direct, il demanda : « Et toi, qui tu es ? »

Faites confiance aux enfants pour poser la question à laquelle il vous est le plus difficile de répondre, pensa Greg. Que dire ? La vérité était trop brutale pour un petit garçon de six ans. Je suis un ancien ami de ta mère, c'est tout, envisagea-t-il de répliquer. Je suis passé devant chez vous et je me suis dit que j'allais vous dire bonjour. C'est tout. On se reverra peut-être, mais ce n'est pas sûr.

Se tournant vers Jacky, il lut sur son visage une prière encore plus implorante. Il comprit ce qui la bouleversait : l'idée qu'il puisse rejeter Georgy.

« Je vais te dire quelque chose », répondit Greg. Il souleva le petit garçon et le posa sur ses genoux. « Et si tu m'appelais Oncle Greg ? »

4.

Greg grelottait sur le balcon réservé aux spectateurs d'une salle de squash non chauffée. C'était ici, sous la tribune ouest de ce stade désaffecté, en bordure du campus de l'université de Chicago, que Fermi et Szilard avaient construit leur pile atomique. Greg en était à la fois impressionné et effrayé.

La pile, un cube de briques grises qui montait jusqu'au plafond, se dressait juste devant le mur du fond qui portait encore les traces de centaines de balles de squash. Cette pile avait coûté un million de dollars et pouvait faire exploser toute la ville !

Le graphite était le matériau qu'on utilisait pour fabriquer les mines de crayons. L'horrible poussière qu'il dégageait recouvrait tout, sol et murs. Quiconque séjournait un moment dans cette salle en ressortait le visage aussi noir qu'un mineur. Toutes les blouses des chercheurs étaient maculées de crasse.

Le graphite n'était pas le matériau explosif, bien au contraire. Il servait à faire baisser le taux de radioactivité. Certaines briques de la pile avaient quand même été percées de petits trous et bourrées d'oxyde d'uranium, le matériau qui émettait les neutrons. L'intérieur de la pile était traversé de dix canaux destinés à accueillir les barres de contrôle, des tiges de huit mètres de long, fabriquées en cadmium, un métal dont les propriétés d'absorption des neutrons étaient encore supérieures à celles du graphite. Pour le moment, ces barres de contrôle maintenaient le calme dans la pile. C'était quand on les retirerait qu'on commencerait à rigoler.

L'uranium émettait déjà ses radiations mortelles, mais le graphite et le cadmium les absorbaient encore. Les radiations étaient jaugées par une batterie de compteurs qui cliquetaient de façon menaçante et par un cylindre enregistreur à stylet qui, lui, par bonheur, était silencieux. Tous ces instruments de contrôle et de mesure disposés près de Greg étaient bien les seules choses à dégager un peu de chaleur.

Un vent violent et un froid glacial régnaient à Chicago en ce mercredi 2 décembre, date de la visite de Greg. Pour la première fois ce jour-là, on devait atteindre la masse critique. Greg avait fait le déplacement pour assister à l'expérience en lieu et place de son supérieur, et il expliquait gaiement à qui voulait l'entendre que le général Groves, craignant une explosion, avait préféré l'envoyer en première ligne. En réalité, il était chargé d'une mission pernicieuse : établir une première évaluation des chercheurs, afin de déterminer si certains d'entre eux représentaient un danger pour la sûreté du pays.

Ce projet Manhattan posait en effet un problème de sécurité cauchemardesque. Les plus grands savants étaient étrangers. Quant aux autres, de gauche pour la plupart, c'étaient soit des communistes, soit des libéraux frayant avec des communistes. Si l'on devait renvoyer tous les suspects, il ne resterait plus un chercheur, ou presque, sur le projet. Greg cherchait donc à repérer les plus dangereux.

Enrico Fermi avait environ quarante ans. De petite taille, chauve et doté d'un long nez, il supervisait cette expérience terrifiante en souriant d'un air débonnaire dans son élégant costume trois-pièces.

Il donna l'ordre de démarrer le processus en milieu de matinée, demandant à un technicien de retirer toutes les barres de contrôle de la pile sauf une. « Comment ! Toutes les barres à la fois ? » s'exclama Greg. La décision lui semblait terriblement précipitée.

« On l'a déjà fait la nuit dernière, répliqua le chercheur qui se trouvait à côté de lui, un certain Barney McHugh. Tout a très bien marché.

— Je suis ravi de l'apprendre », dit Greg.

Ce McHugh barbu et corpulent se trouvait tout en bas sur la liste des suspects qu'avait établie Greg. Il était américain de naissance et ne s'intéressait pas du tout à la politique. Le seul grief retenu contre lui était qu'il avait épousé une étrangère, une Anglaise en l'occurrence. Si ce n'était jamais bon signe, ce n'était pas non plus, en soi, une preuve de trahison.

Greg avait supposé qu'un mécanisme sophistiqué permettait d'insérer ou de retirer les barres de contrôle. Apparemment, les scientifiques se contentaient d'un système beaucoup plus simple : un technicien juché sur une échelle appuyée contre le flanc de la pile était chargé de les retirer à la main.

McHugh poursuivit sur le ton de la conversation : « Au départ, on avait prévu de mener l'expérience dans l'Argonne Forest.

— Où est-ce ?

— À une trentaine de kilomètres au sud-ouest de Chicago. Un endroit assez isolé. Moins de victimes. »

Greg frissonna. « Qu'est-ce qui vous a fait changer d'avis et convaincus de transporter l'expérience ici, dans la 57e Rue ?

— Les ouvriers qu'on avait embauchés se sont mis en grève. Alors on a dû construire ce foutu machin tout seuls, et on ne pouvait pas être trop loin de nos labos.

— Et vous avez donc pris le risque de tuer toute la population de Chicago.

— Ça ne devrait pas arriver. »

Jusque-là, Greg avait été du même avis. Mais à quelques mètres de la pile, il se sentait moins rassuré.

Fermi comparait les relevés de ses moniteurs à la grille de radiation qu'il avait établie d'après ses propres calculs pour chaque étape de l'expérience. Apparemment, la première phase s'était déroulée conformément à ses prévisions car il ordonna de retirer la dernière barre jusqu'à la moitié.

Des mesures de sécurité avaient été prévues. Une barre lestée était suspendue en l'air, prête à être insérée automatiquement dans la pile si la radiation atteignait un niveau trop élevé. En cas de pépin, une autre barre, identique, était accrochée par une corde à la balustrade du balcon. Un jeune physicien, manifestement conscient du ridicule de sa position, se tenait à côté, une hache à la main, prêt à trancher la corde qui la retenait. Enfin, trois autres chercheurs – « le commando suicide », comme ils se surnommaient – avaient pris place au ras du plafond sur la plateforme du monte-charge utilisé pour la construction de la pile, munis de grands bidons remplis d'une solution de sulfate de cadmium à verser sur la pile, comme on éteint un feu de joie.

Greg savait que les générations de neutrons se multipliaient en un millième de seconde. Toutefois, Fermi avait fait valoir que pour certains neutrons, ça pouvait prendre plus longtemps, quelques secondes peut-être. S'il avait raison, tout irait bien. S'il se trompait, l'équipe aux bidons et le physicien à la hache seraient pulvérisés en un clin d'œil.

Greg nota que les cliquetis s'accéléraient. Il jeta un regard anxieux à Fermi, qui faisait des décomptes à l'aide d'une règle à calcul. Il avait l'air content. De toute façon, se dit Greg, s'il y a une catastrophe, elle se produira si rapidement qu'aucun d'entre nous ne s'en rendra compte.

La cadence des cliquetis se stabilisa. Fermi sourit et ordonna de dégager la barre de quinze centimètres de plus.

D'autres chercheurs arrivèrent, gravissant l'escalier menant au balcon, vêtus de gros manteaux d'hiver, de chapeaux, d'écharpes et de gants. Greg fut consterné par l'indigence des mesures de sécurité. Personne ne demanda leurs papiers à ces hommes, dont n'importe lequel aurait parfaitement pu être un espion à la solde des Japonais.

Parmi eux, il reconnut la haute et solide silhouette de Szilard, son visage rond et ses épais cheveux bouclés. Leo Szilard était un idéaliste qui avait rêvé de voir l'énergie nucléaire libérer la race humaine du fardeau du travail. C'était d'un cœur lourd qu'il avait rejoint une équipe chargée de concevoir une bombe atomique.

Encore quinze centimètres. La cadence des cliquetis augmenta.

Greg regarda sa montre. Onze heures et demie.

Soudain, un grand fracas retentit. Tout le monde sursauta. McHugh s'écria : « Eh merde !

— Que se passe-t-il ? s'inquiéta Greg.

— Oh, je vois ! soupira McHugh. C'est simplement la barre de contrôle de secours ! Le taux de radiation a dû déclencher le système de sécurité. »

Fermi annonça avec un fort accent italien : « J'ai faim. Allons déjeuner. »

Comment pouvait-on penser à manger en pareil instant ! s'étonna Greg en son for intérieur. Mais personne ne discuta. « On ne sait jamais combien de temps peut prendre une expérience, lui expliqua McHugh. Une heure ou la journée entière. Il vaut mieux manger quand on peut. » Greg en aurait hurlé d'exaspération.

Toutes les barres de contrôle furent réinsérées dans la pile et soigneusement assujetties, et tout le monde quitta les lieux.

Ils se retrouvèrent presque tous à la cantine du campus. Greg choisit un sandwich jambon-fromage grillé et s'assit pour le manger à côté d'un physicien à l'air grave, du nom de Wilhelm Frunze. La majorité des chercheurs s'habillaient n'importe comment, mais ce Frunze battait des records dans son costume vert orné d'innombrables parements en cuir fauve : aux boutonnières, au revers du col, aux coudes et jusque sur les rabats des poches. Cet Allemand, qui s'était installé à Londres au milieu des années 1930, occupait une des toutes premières places sur la liste des suspects de Greg. Antinazi, il n'était pas communiste mais social-démocrate. Il avait épousé une Américaine, une artiste. En bavardant avec lui pendant le déjeuner, Greg ne vit aucune raison de le soupçonner. Apparemment, il appréciait la vie en Amérique et s'intéressait à peu de choses en dehors de son travail. Mais avec ces étrangers, il était bien difficile de savoir envers qui finalement ils étaient loyaux.

De retour dans le stade abandonné après le déjeuner, face à ces gradins déserts, Greg se mit à penser à Georgy. Il n'avait confié à personne qu'il avait un fils, pas même à Margaret Cowdry avec qui il entretenait désormais de délicieuses relations charnelles. Il avait terriblement envie d'en parler à sa mère. Il éprouvait une étrange fierté, alors qu'il n'avait rien fait pour mettre Georgy au monde, sinon coucher avec Jacky. Ce qui était sans doute une des choses les plus faciles qu'il ait jamais faites. Il n'en était pas moins fou de joie. Il était au seuil d'une sorte d'aventure : Georgy grandirait, s'instruirait, changerait, et deviendrait un homme un jour et Greg suivrait toutes ces étapes, émerveillé.

Les chercheurs se rassemblèrent à deux heures. À présent, il y avait bien une quarantaine de personnes sur ce balcon où étaient installés les appareils de mesure. L'expérience reprit exactement là où ils l'avaient laissée, Fermi vérifiant constamment ses instruments. « Cette fois, on retire la barre de trente centimètres », annonça-t-il.

Les cliquetis s'accélérèrent. Greg s'attendait à ce que la cadence se stabilise, comme auparavant, mais ce fut le contraire qui se produisit : ils se firent de plus en plus rapides jusqu'à devenir un grondement continu.

Le niveau de radiations avait dépassé le maximum indiqué par les compteurs. Greg s'en rendit compte en voyant que tout le monde avait les yeux braqués sur le cylindre enregistreur. La graduation étant réglable, on la modifiait à mesure que le niveau montait, encore et encore.

Fermi leva la main. Tout le monde se tut. « La pile a dépassé le seuil critique », dit-il. Il sourit… et ne fit rien.

Greg faillit hurler : « Éteignez donc cette saloperie de machine ! » Mais Fermi, immobile et muet, observait le stylet avec une telle autorité que personne ne protesta. La réaction en chaîne se produisait, mais tout était sous contrôle. Il laissa l'expérience se poursuivre pendant une minute entière, puis une autre.

« Nom de Dieu ! » lâcha McHugh dans un murmure.

Greg n'avait aucune envie de mourir. Il voulait devenir sénateur. Il voulait coucher encore avec Margaret Cowdry. Il voulait voir Georgy entrer à l'université. Je n'en suis même pas à la moitié de ma vie, pensa-t-il.

Enfin Fermi ordonna de réinsérer les barres de contrôle.

Le bruit des compteurs redevint un cliquetis et ralentit peu à peu avant de s'arrêter enfin.

Greg recommença à respirer normalement.

McHugh ne se tenait plus de joie. « Ça y est ! On l'a prouvé ! La réaction en chaîne est une réalité !

— Et on peut la contrôler, ce qui est encore mieux ! renchérit Greg.

— Oui, c'est sans doute mieux. D'un point de vue pratique, s'entend. »

Greg sourit. C'était bien une remarque de scientifique ! À Harvard, il avait eu le temps de les connaître ! Pour eux, la réalité, c'était la théorie et le monde, un modèle plutôt imprécis.

Quelqu'un fit surgir une bouteille de vin italien entourée d'un panier d'osier ainsi que des gobelets en carton. Les chercheurs ne firent qu'y tremper les lèvres. Ces gens-là ne savaient pas faire la fête. Encore une raison qui avait dissuadé Greg d'embrasser une carrière scientifique.

Quelqu'un demanda à Fermi de signer le panier. Il s'exécuta et tous les autres l'imitèrent.

Les techniciens éteignirent les moniteurs, et tout le monde commença à s'éparpiller. Greg resta sur place, à observer ce qui se passait. Au bout d'un moment, il se retrouva seul sur le balcon en compagnie de Fermi et de Szilard. Il vit ces deux géants intellectuels – un grand costaud au visage rond et un petit homme à la stature d'elfe – se serrer la main, et l'image de Laurel et Hardy s'imposa à lui.

Puis il entendit Szilard s'exclamer : « Mon ami, je pense que ce jour est à marquer d'une pierre noire dans l'histoire de l'humanité. »

Greg se demanda ce qu'il voulait dire.

5.

Greg voulait que ses parents fassent bon accueil à Georgy.

Ce ne serait pas facile. Ils ne seraient certainement pas ravis qu'on leur ait caché pendant six ans qu'ils avaient un petit-fils. Ils seraient peut-être même furieux et risquaient de regarder Jacky de haut. Ils étaient pourtant mal placés pour jouer les moralisateurs, se dit Greg avec ironie, ayant eux-mêmes engendré un bâtard. Mais les gens n'étaient pas rationnels.

Greg était incapable de dire si le fait que Georgy fût noir pèserait pour beaucoup dans leur réaction. Ses parents étaient tolérants, il ne les avait jamais entendus proférer d'horreurs sur les nègres ou les youpins comme d'autres membres de leur génération. Mais de là à accueillir un Noir dans leur famille…

Devinant que la conversation serait plus difficile avec son père, Greg décida d'annoncer d'abord la nouvelle à sa mère.

Ayant obtenu une permission de quelques jours pour Noël, il alla la voir à Buffalo. Marga possédait un grand appartement dans le plus bel immeuble de la ville. Elle y vivait seule la plupart du temps, mais avait une cuisinière, deux bonnes et un chauffeur, un coffre-fort rempli de bijoux et une garde-robe grande comme un garage pour deux voitures. Toutefois, elle n'avait pas de mari.

Lev était à Buffalo lui aussi, mais traditionnellement, c'était avec Olga qu'il passait le réveillon de Noël. Ils étaient toujours légalement mari et femme, même si Lev n'avait pas passé une nuit sous le toit de son épouse légitime depuis des années. Ils n'avaient que haine l'un pour l'autre, mais pour une raison quelconque continuaient à se voir une fois par an.

Ce soir-là, Greg et Marga dînèrent ensemble dans l'appartement. Il revêtit un smoking pour lui faire plaisir car elle disait souvent : « J'aime voir mes hommes sur leur trente et un. » Ils mangèrent de la soupe de poisson, du poulet rôti et de la tarte aux pêches, le dessert préféré de Greg quand il était enfant.

« J'ai une nouvelle à t'annoncer, Maman », lança-t-il pendant que la bonne versait le café. Il était dans ses petits souliers. Ce n'était pas pour lui qu'il avait peur mais pour Georgy, et il se demanda si le fait d'être parent ne consistait pas à s'inquiéter davantage pour autrui que pour soi-même.

« Une bonne nouvelle ? »

Elle s'était un peu empâtée ces dernières années, mais était encore superbe à quarante-six ans. Si des fils d'argent parsemaient ses cheveux noirs, ils avaient été soigneusement camouflés par son coiffeur. Ce soir-là, elle portait une robe noire très sobre et un collier de diamants.

« Très bonne, mais un peu inattendue sans doute. Alors, s'il te plaît, ne monte pas sur tes grands chevaux. »

Elle haussa un sourcil, mais ne dit rien.

Il plongea la main dans la poche intérieure de son veston de smoking et en sortit une photo. Elle représentait Georgy juché sur un vélo rouge au guidon décoré d'un ruban. La roue arrière était munie de deux stabilisateurs. Le visage du petit garçon était tout illuminé de joie. Accroupi à côté de lui, Greg débordait de fierté.

Il tendit la photo à sa mère.

Elle l'examina d'un air pensif avant de déclarer : « Je suppose que c'est toi qui as offert cette bicyclette à ce petit garçon pour Noël.

— Oui. »

Elle releva les yeux. « Est-ce que tu es en train de me dire que tu as un fils ? »

Greg fit oui de la tête. « Il s'appelle Georgy.

— Tu es marié ?

— Non. »

Elle posa brutalement la photo sur la table. « Pour l'amour du ciel ! s'écria-t-elle avec colère. Mais qu'est-ce que vous avez, vous autres, les Pechkov ?

— Qu'est-ce qui te prend ? demanda Greg, consterné.

— Encore un enfant illégitime ! Encore une femme qui devra l'élever seule ! »

Il se rendit compte qu'elle se revoyait elle-même en Jacky. « Maman, j'avais quinze ans…

— Vous ne pouvez pas vous conduire normalement ? s'emporta-t-elle. Pour l'amour de Dieu, qu'y a-t-il de mal à avoir une famille normale ? »

Greg baissa les yeux. « Rien du tout. »

Il avait honte. Jusqu'à cet instant, il s'était considéré comme un acteur passif de ce drame, une victime même. Tout ce qui s'était passé avait été orchestré par son père et Jacky. À l'évidence, sa mère ne voyait pas les choses ainsi et il ne pouvait que lui donner raison. Il n'avait pas réfléchi à deux fois avant de coucher avec Jacky ; il ne l'avait pas interrogée quand elle lui avait dit d'un ton désinvolte qu'il n'avait pas à s'inquiéter de contraception ; et il n'avait pas affronté son père quand Jacky était partie. Il était très jeune, certes, mais assez âgé pour coucher avec elle. Assez âgé pour être responsable de ses actes et de leurs conséquences.

Sa mère fulminait toujours. « Ne me dis pas que tu as oublié que tu pleurnichais sans cesse quand tu étais petit : “Où il est mon papa ? Pourquoi est-ce qu'il ne dort pas ici ? Pourquoi est-ce qu'on ne peut pas aller avec lui chez Daisy ?” Et plus tard, toutes tes bagarres à l'école, quand les autres garçons te traitaient de bâtard. Et ta fureur quand tu n'as pas été accepté dans ce fichu Yacht-Club.

— Je m'en souviens, bien sûr ! »

Elle donna un coup de poing sur la table de sa main alourdie de bagues, faisant trembler les verres en cristal. « Alors, comment peux-tu imposer les mêmes tourments à un autre petit garçon ?

— Il y a deux mois encore, j'ignorais tout de son existence. Papa a terrorisé sa mère pour qu'elle ne me donne plus signe de vie.

— Qui est-ce ?

— Elle s'appelle Jacky Jakes. Elle est serveuse. » Il sortit une deuxième photo.

Sa mère soupira. « Une jolie Noire. » Elle commençait à se calmer.

« Elle espérait être actrice, mais je crois qu'elle y a renoncé à cause de Georgy. »

Marga hocha la tête. « Un bébé ruine une carrière plus vite qu'une chaude-pisse. »

Sa mère croyait probablement qu'une actrice devait coucher avec les gens qu'il fallait pour obtenir des rôles. Qu'en savait-elle, d'abord ? Il se rappela alors qu'elle était chanteuse de cabaret à l'époque où son père l'avait rencontrée…

Il garda le silence, préférant ne pas s'engager sur cette voie.

Elle reprit : « Qu'est-ce que tu lui as offert pour Noël ?

— Une assurance médicale.

— C'est très bien. C'est mieux qu'un colifichet. »

Greg entendit des pas dans le vestibule. Son père était rentré. « Maman, tu voudras bien voir Jacky ? demanda-t-il précipitamment. Accepteras-tu Georgy comme ton petit-fils ? »

Elle porta la main à sa bouche. « Oh, mon Dieu, je suis grand-mère ! » Elle ne savait pas si elle devait s'en réjouir ou s'en offusquer.

Greg se pencha en avant. « Je ne veux pas que Papa le rejette. Je t'en prie ! »

Lev entra dans la salle à manger avant qu'elle n'ait eu le temps de lui répondre.

« Bonjour, mon chéri, dit Marga, comment s'est passée ta soirée ? »

Il s'assit à la table d'un air grincheux. « J'ai eu droit à la liste détaillée de tous mes défauts. Je suppose donc que c'est ce qu'on appelle une soirée fructueuse.

— Mon pauvre. Est-ce que tu as assez mangé, au moins ? Je peux te faire une omelette si tu veux.

— Côté nourriture, il n'y a rien à redire. »

Les photographies étaient toujours sur la table, mais Lev ne les avait pas encore remarquées.

La bonne entra. « Voulez-vous un café, monsieur ?

— Non, merci.

— Apportez la vodka, dit Marga. Peut-être monsieur aura-t-il envie d'un verre plus tard.

— Bien, madame. »

Voyant sa mère aussi attentive au confort et au plaisir de son père, Greg songea que cette sollicitude expliquait qu'il ait toujours préféré passer la nuit là plutôt que chez Olga.

La domestique apporta une bouteille et trois petits verres sur un plateau d'argent. Lev buvait toujours la vodka à la russe, pure et à température ambiante.

« Papa, commença Greg, tu connais Jacky Jakes…

— Encore elle ? lança Lev irrité.

— Oui, il y a une chose que tu ne sais pas à son sujet. »

La phrase retint l'attention de Lev. Il détestait ignorer ce que d'autres savaient. « Quoi donc ?

— Elle a un enfant. » Greg poussa les photographies à travers la table vernie.

« Il est de toi ?

— Il a six ans. À ton avis ?

— Elle a drôlement bien tenu sa langue.

— Tu l'avais terrorisée.

— Qu'est-ce qu'elle pensait que j'allais lui faire, à son moutard ? Le passer à la broche et le bouffer ?

— Je ne sais pas, Papa. Pour ce qui est de terrifier les gens, c'est toi l'expert. »

Lev lui jeta un regard dur. « Tu marches sur mes traces, on dirait. »

Il voulait parler de la scène du rasoir. Et Greg ne put s'empêcher de se dire que peut-être, en effet, il en avait pris de la graine.

« Je peux savoir pourquoi tu me montres ces photos ? enchaîna Lev.

— Je me disais que tu serais peut-être content de savoir que tu as un petit-fils.

— D'une actrice de pacotille qui rêvait de piéger un type plein aux as !

— Chéri, je t'en prie ! intervint Marga. Rappelle-toi ! Moi aussi j'ai été une chanteuse de pacotille qui rêvait de piéger un type plein aux as. »

La fureur se peignit sur les traits de Lev et il jeta un regard noir à Marga. Puis son expression changea. « Tu sais quoi ? Tu as raison. Qui suis-je pour juger Jacky Jakes, d'abord ? »

Greg et Marga le regardèrent, abasourdis par cette soudaine humilité.

« Je suis comme elle après tout. Je n'étais qu'un voyou des bas quartiers de Saint-Pétersbourg avant d'épouser Olga Vialov, la fille de mon patron. »

Greg croisa le regard de sa mère qui esquissa un haussement d'épaules presque imperceptible, l'air de dire : « Avec lui, il faut s'attendre à tout. »

Lev baissa les yeux vers la photo. « À part la couleur, il me rappelle mon frère Grigori. Et moi qui croyais que tous ces négrillons avaient la même bouille ! Pour une surprise, c'est une surprise. »

Greg retint son souffle. « Tu veux le voir, Papa ? Tu m'accompagnerais pour faire la connaissance de ton petit-fils ?

— Plutôt deux fois qu'une ! » Lev déboucha la bouteille, remplit les trois verres de vodka et en tendit un à chacun. « Comment s'appelle-t-il, à propos ?

— Georgy. »

Lev leva son verre. « À Georgy ! »

Ils trinquèrent tous.

XV

1943 (I)

1.

Lloyd Williams fermait la marche derrière un groupe de fugitifs aux abois qui gravissaient un étroit sentier escarpé.

Son souffle était régulier. Il était rompu à cet exercice. Cela faisait plusieurs fois qu'il passait les Pyrénées. Il était chaussé d'espadrilles dont les semelles de corde lui assuraient une meilleure adhérence sur le terrain rocailleux et portait un lourd manteau au-dessus de son bleu de travail. Pour le moment, le soleil était chaud, mais plus tard, lorsque le soir tomberait et qu'ils prendraient de l'altitude, la température descendrait au-dessous de zéro.

Devant lui s'avançaient deux robustes chevaux, trois habitants de la région et huit rescapés débraillés et épuisés, tous chargés d'un lourd paquetage. Il y avait là trois aviateurs américains, survivants de l'équipage d'un bombardier B-24 Liberator abattu au-dessus de la Belgique, deux officiers britanniques évadés de l'Oflag 65, près de Strasbourg, auxquels s'ajoutaient un communiste tchèque, une violoniste juive et un énigmatique Anglais du nom de Watermill, probablement un espion.

Tous avaient fait un long chemin et subi quantité d'épreuves. Cette étape était la dernière de leur périple, la plus périlleuse aussi. S'ils se faisaient prendre, ils seraient soumis à des tortures qui les obligeraient à trahir les femmes et les hommes courageux qui les avaient aidés à arriver jusque-là.

Teresa ouvrait la marche. L'ascension était dure pour ceux qui n'avaient pas l'habitude, mais ils ne devaient pas traîner, de crainte de se faire repérer et, ainsi que Lloyd l'avait constaté, les réfugiés avaient moins tendance à lambiner lorsque c'était une femme, menue et jolie, qui les menait.

Le sentier cessa de monter au débouché d'une petite clairière. Soudain, une voix cria « Halte ! » dans un français fortement teinté d'allemand.

La colonne se figea.

Deux soldats allemands surgirent de derrière un rocher. Chacun d'eux était armé d'une carabine Mauser à verrou, dont le magasin contenait cinq cartouches.

Instinctivement, Lloyd porta la main sur le Luger 9 mm chargé dissimulé dans la poche de son manteau.

Il était de plus en plus difficile de fuir l'Europe et sa tâche devenait sans cesse plus dangereuse. L'année précédente, les Allemands avaient occupé la moitié sud de la France sans tenir compte du gouvernement de Vichy, révélant aux yeux de tous le piètre simulacre qu'il avait toujours été. Les quinze kilomètres bordant la frontière espagnole avaient été décrétés zone interdite. Or Lloyd et sa bande se trouvaient à présent justement dans ce secteur.

Teresa s'adressa aux soldats en français. « Bonjour, messieurs. Tout va comme vous voulez ? » Lloyd, qui la connaissait bien, perçut un frémissement d'inquiétude dans sa voix, mais il espérait que les deux sentinelles ne le remarqueraient pas.

Les rangs de la police française comptaient quantité de fascistes et de rares communistes, mais aucun n'était zélé au point de traquer les réfugiés dans les cols glacials des Pyrénées. On ne pouvait pas en dire autant des Allemands. Ceux-ci avaient envoyé dans les villes frontalières des troupes chargées de patrouiller sur les sentiers que Lloyd et Teresa utilisaient pour franchir la frontière. Rien à voir avec les soldats d'élite dépêchés sur le front russe, qui avaient récemment dû capituler devant Stalingrad à l'issue de combats meurtriers. Il s'agissait le plus souvent de vieillards, d'adolescents et d'invalides. Mais ils n'en semblaient que plus décidés à prouver leur valeur. Contrairement aux Français, ils n'étaient pas du genre à fermer les yeux.

L'aîné des deux, un type à la moustache grise d'une maigreur cadavérique, demanda à Teresa : « Où allez-vous ?

— À Lamont. Nous avons des provisions pour vos camarades et vous. »

Une unité allemande avait investi ce village de montagne reculé, en chassant tous les habitants. Les soldats avaient ensuite pris conscience des difficultés d'approvisionnement et Teresa avait eu une idée de génie en décidant de leur livrer de la nourriture moyennant finances, ce qui lui permettait de circuler dans une zone théoriquement interdite.

Le soldat examina d'un œil soupçonneux les hommes lourdement chargés. « Toutes ces provisions sont pour les Allemands ?

— J'espère bien, répliqua Teresa. Il n'y a personne d'autre à qui les vendre. » Elle sortit un feuillet de sa poche. « Voici notre laissez-passer, signé par le sergent Eisenstein. »

L'homme l'examina avec soin et le lui rendit. Puis il dévisagea le lieutenant-colonel Will Donelly, un pilote américain au visage rougeaud. « C'est un Français, lui ? »

Lloyd serra la main sur la crosse de son pistolet.

L'aspect des fugitifs était parfois problématique. Dans cette partie de l'Europe, on trouvait surtout des hommes petits et noirauds, Français comme Espagnols. Et tout le monde était maigre. Lloyd et Teresa correspondaient tous deux à ce physique, ainsi que le Tchèque et la violoniste. Mais les Britanniques étaient blonds et pâles, les Américains immenses.

« Guillaume est né en Normandie, expliqua Teresa. Il a été nourri au beurre et à la crème fraîche ! »

Le cadet des deux soldats, un jeune homme blafard portant des lunettes, lui adressa un sourire. Elle attirait toujours la sympathie. « Vous avez du vin ? demanda-t-il.

— Bien sûr. »

Le visage des deux sentinelles s'épanouit.

« Vous en voulez un peu ? demanda Teresa.

— C'est qu'il fait chaud en plein soleil », dit le plus âgé.

Lloyd ouvrit un panier porté par un des chevaux, en sortit quatre bouteilles de vin blanc du Roussillon et les tendit aux Allemands. Chacun en prit deux. Soudain, ce ne fut plus que sourires et poignées de main. « Passez, les amis », dit l'aîné.

Les fugitifs se remirent en marche. Lloyd n'avait pas eu vraiment peur qu'il y ait un problème, mais on n'était jamais sûr de rien et il était ravi d'avoir franchi cet obstacle.

Il leur fallut deux heures de plus pour rallier Lamont. Ce hameau misérable, composé d'une poignée de bâtisses et d'enclos à moutons déserts, était situé au bord d'un petit plateau où l'herbe printanière commençait tout juste à pousser. Lloyd avait pitié des gens qui avaient vécu là. Ils ne possédaient presque rien, et leurs maigres biens eux-mêmes leur avaient été confisqués.

Le petit groupe gagna la place du village et tous se défirent de leur paquetage en soupirant de soulagement. Ils étaient entourés de soldats allemands.

C'était le moment le plus risqué aux yeux de Lloyd.

Le sergent Eisenstein commandait une section d'une vingtaine d'hommes, qui les aidèrent tous à décharger les provisions : pain, saucisses, poisson frais, lait concentré, conserves. Les soldats étaient ravis d'avoir de quoi manger et de voir de nouvelles têtes. Ils tentèrent d'engager la conversation avec leurs bienfaiteurs.

Les fugitifs devaient en dire le moins possible. Le moindre faux pas aurait pu les trahir. Certains Allemands maîtrisaient suffisamment le français pour repérer un accent anglais ou américain. Et même ceux dont la prononciation était passable, comme Teresa et Lloyd, n'étaient pas à l'abri d'une faute de grammaire. Jamais un Français ne dirait *sur le table* au lieu de *sur la table*, par exemple.

Pour compenser, les Français du groupe se donnaient du mal pour faire preuve de volubilité. Chaque fois qu'un soldat s'adressait à un fugitif, l'un d'eux s'empressait de se mêler à la conversation.

Teresa présenta une facture au sergent, qui l'examina longuement avant de compter ses billets.

Enfin ils purent reprendre la route, leur cargaison livrée et le cœur léger.

Ils redescendirent vers la vallée sur sept ou huit cents mètres puis se séparèrent. Teresa poursuivit avec les Français et les chevaux. Lloyd et les fugitifs empruntèrent un sentier conduisant vers la montagne.

Les sentinelles qu'ils avaient croisées seraient sans doute trop éméchées pour se rendre compte qu'ils étaient moins nombreux au retour qu'à l'aller. Si elles posaient des questions, Teresa leur répondrait que quelques-uns de leurs compagnons jouaient aux cartes avec les soldats et redescendraient plus tard. Une fois que les sentinelles auraient été relevées, tout le monde aurait oublié ce détail.

Lloyd imposa à son petit groupe une marche forcée de deux heures, puis lui accorda une pause de dix minutes. Chacun des fugitifs s'était vu remettre une bouteille d'eau et des figues sèches. On les avait dissuadés d'emporter autre chose. Par expérience, Lloyd savait que les livres, l'argenterie, les bijoux et les disques auraient été jetés dans un ravin bien avant que les voyageurs épuisés n'aient franchi le col.

Le plus dur les attendait. À partir de là, il allait faire plus froid, plus sombre et le chemin serait plus rocailleux.

Juste avant les neiges éternelles, il leur conseilla de remplir leurs gourdes à un ruisseau limpide et glacial.

La nuit tomba, mais ils poursuivirent leur route. Il aurait été dangereux de les laisser dormir car ils risquaient de mourir de froid. Sous l'effet de la fatigue, ils ne cessaient de glisser et de trébucher sur les rochers gelés. Leur allure se ralentit, ce qui était inévitable. Lloyd devait les surveiller de près : celui qui s'écartait du groupe pouvait s'égarer, voire tomber dans une crevasse. Jusqu'à présent, il n'avait jamais eu de perte à déplorer.

La plupart des fugitifs étant des officiers, il n'était pas rare qu'ils contestent son autorité à cette étape du voyage, refusant tout net de poursuivre leur route. On l'avait promu au grade de commandant pour lui éviter ce genre de désagrément.

Au cœur de la nuit, alors que le moral de tous était au plus bas, Lloyd leur annonça : « Vous êtes en Espagne, en pays neutre ! » et il eut droit à quelques vivats. En fait, il ignorait l'emplacement exact de la frontière et faisait toujours cette déclaration quand il estimait que ses protégés avaient besoin d'un coup de fouet.

L'aube leur redonna courage. Il y avait encore du chemin à faire, mais le sentier descendait et le soleil réchauffait leurs membres engourdis.

Au lever du jour, ils contournèrent un village surmonté d'une église couleur de poussière juché au sommet d'une colline. Un peu plus loin, ils trouvèrent une grange en bord de route. Elle abritait un camion Ford vert équipé d'une bâche crasseuse. Il était suffisamment spacieux pour que tous y prennent place. Il était conduit par le capitaine Silva, un quadragénaire anglais d'origine espagnole qui travaillait avec Lloyd.

À sa grande surprise, était également présent sur les lieux le commandant Lowther, le responsable de la formation des officiers de Renseignement à Tŷ Gwyn, qui avait semblé désapprouver son amitié avec Daisy – à moins qu'il n'ait été tout simplement jaloux.

Lloyd, qui savait que Lowthie était affecté à l'ambassade d'Angleterre à Madrid, se doutait qu'il travaillait pour le MI6, les services secrets, mais jamais il n'aurait imaginé le croiser aussi loin de la capitale.

Lowther était vêtu d'un coûteux costume de flanelle blanc froissé et sali. Il se tenait campé près du camion avec des airs de propriétaire. « À partir d'ici, c'est moi qui prends les choses en main, Williams. » Il examina les fugitifs. « Lequel d'entre vous est Watermill ? »

Il pouvait s'agir d'un patronyme aussi bien que d'un nom de code.

Le mystérieux Anglais s'avança d'un pas et lui serra la main.

« Commandant Lowther. Je vous conduis directement à Madrid. » Se tournant vers Lloyd, il ajouta : « J'ai bien peur que votre groupe ne doive se diriger vers la gare la plus proche.

— Un instant, fit Lloyd. Ce camion appartient à mon réseau. » Il l'avait acheté avec les fonds fournis par le MI9, le service chargé d'aider les prisonniers évadés. « Et son chauffeur travaille pour moi.

— J'en suis navré, répliqua sèchement Lowther, mais Watermill est prioritaire. »

Les agents des services secrets se considéraient toujours comme prioritaires. « Je proteste, insista Lloyd. Je ne vois pas pourquoi nous n'irions pas tous à Barcelone dans ce camion, comme prévu. Une fois là-bas, vous pourrez emmener Watermill à Madrid en train.

— Je ne vous ai pas demandé votre avis, mon petit gars. Faites ce qu'on vous dit. »

Watermill intervint d'une voix conciliante : « Je serais ravi de partager ce camion avec mes compagnons.

— Laissez-moi régler cette affaire, je vous prie, objecta Lowther.

— Tous ces gens viennent de franchir les Pyrénées à pied, expliqua Lloyd. Ils sont épuisés.

— Raison de plus pour qu'ils se reposent avant de poursuivre leur route. »

Lloyd secoua la tête. « C'est trop dangereux. Le maire de la ville la plus proche est de notre côté – c'est pour ça que nous avons choisi ce lieu comme point de chute. Mais un peu plus bas dans la vallée, c'est une autre histoire. La Gestapo est partout, vous le savez bien – et la plupart des gardes civils sont dans leur camp, pas dans le nôtre. Comme nous sommes entrés clandestinement en Espagne, nous sommes en danger. Et vous savez à quel point il est difficile de sortir des geôles de Franco, même si on est innocent.

— Je n'ai pas l'intention de perdre mon temps à discuter avec vous. Je vous rappelle que je suis votre supérieur hiérarchique.

— Absolument pas.

— Pardon ?

— J'ai le grade de commandant. Et cessez de me donner du "mon petit gars" si vous ne voulez pas recevoir mon poing dans la gueule.

— Ma mission est urgente !

— Dans ce cas, pourquoi n'êtes-vous pas venu avec votre propre véhicule ?

— Parce que celui-ci était disponible !

— Sauf qu'il ne l'est pas. »

Will Donelly, le grand Américain, s'avança d'un pas. « Je soutiens le commandant Williams, déclara-t-il d'une voix traînante. Il vient de me sauver la vie. Alors que vous, commandant Lowther, vous n'avez encore rien fait.

— Je ne vois pas le rapport, protesta Lowther.

— En ce qui me concerne, la situation est on ne peut plus claire, reprit Donelly. Ce camion est placé sous l'autorité du commandant Williams. Le commandant Lowther voudrait bien le réquisitionner, mais il ne peut pas. Point final.

— Ne vous mêlez pas de ça, gronda Lowther.

— Il se trouve que j'ai le grade de lieutenant-colonel, ce qui fait de moi votre supérieur à tous deux.

— Ce n'est pas de votre ressort.

— Pas plus que du vôtre, manifestement. » Donelly se tourna vers Lloyd. « On y va ?

— J'insiste ! » bafouilla Lowther.

Donelly se tourna vers lui. « Commandant Lowther, dit-il. Fermez votre gueule. C'est un ordre.

— Allez, tout le monde, dit Lloyd. Montez. »

Lowther le gratifia d'un regard furibond. « Tu me revaudras ça, sale petit bâtard gallois », dit-il.

2.

Les jonquilles fleurissaient à Londres le jour où Daisy et Boy allèrent passer leur visite médicale.

C'était Daisy qui avait eu cette idée. Elle en avait assez que Boy lui reproche son infécondité. Il ne cessait de la comparer à May, l'épouse de son frère Andy, qui avait déjà trois enfants. « Il y a sûrement quelque chose qui cloche chez toi, déclarait-il d'un ton agressif.

— J'ai déjà été enceinte. » Elle cilla au douloureux souvenir de sa fausse couche puis, se rappelant comment Lloyd avait pris soin d'elle, elle éprouva une autre sorte de souffrance.

« Peut-être est-il arrivé quelque chose qui t'a rendue stérile depuis, insista Boy.

— Ou bien c'est toi.

— Comment ça ?

— C'est peut-être chez toi qu'il y a quelque chose qui ne va pas.

— Ne sois pas ridicule.

— Tiens, je te propose un marché. » L'espace d'un instant, elle songea qu'elle engageait une négociation un peu comme l'aurait fait son père Lev. « Je suis prête à subir un examen médical… à condition que tu en fasses autant. »

Il avait été surpris, et avait hésité avant de répondre : « Entendu. Vas-y d'abord. Si tu n'as rien, j'irai ensuite.

— Non, dit-elle. Toi d'abord.

— Pourquoi ?

— Parce que je n'ai pas confiance en toi.

— Très bien. Nous irons ensemble. »

Daisy se demandait pourquoi elle se donnait tout ce mal. Elle n'aimait pas Boy – elle ne l'aimait plus depuis longtemps. L'homme qu'elle aimait était Lloyd Williams, parti en Espagne pour une mission dont il n'avait pas dit grand-chose. Mais elle était la femme de Boy. Ce dernier lui avait été infidèle, bien sûr, à plusieurs reprises. Mais elle avait commis l'adultère elle aussi, même si c'était avec un seul homme. Elle pouvait difficilement lui faire la morale, et se sentait pieds et poings liés. Si elle accomplissait son devoir d'épouse, se disait-elle, peut-être parviendrait-elle à conserver un peu d'estime pour elle-même.

Le cabinet médical se trouvait dans Harley Street, non loin de leur domicile mais dans un quartier moins huppé. Daisy trouva l'examen extrêmement déplaisant. Le médecin était un homme et il lui reprocha vertement ses dix minutes de retard. Il lui posa quantité de questions sur son état de santé, sur ses règles et ce qu'il appelait ses « rapports » avec son époux, sans la regarder dans les yeux mais en se concentrant sur les notes qu'il rédigeait avec son stylo à plume. Puis il lui inséra une série d'instruments métalliques et froids dans le vagin. « Je fais ça tous les jours, inutile de vous inquiéter », affirma-t-il, avant d'esquisser une grimace qui semblait prouver le contraire.

Lorsqu'elle regagna la salle d'attente, elle n'aurait pas été surprise de voir Boy trahir sa parole et refuser de subir l'examen. Mais il se leva, visiblement contrarié, et entra à son tour dans le cabinet.

En l'attendant, Daisy relut une lettre de Greg, son demi-frère. Il venait de découvrir qu'il avait un fils, fruit de ses amours avec une jeune Noire quand il avait quinze ans. Au grand étonnement de Daisy, ce play-boy de Greg était fou de joie et impatient de participer à la vie de son enfant, même si c'était officiellement en tant qu'oncle et non en tant que père. Plus surprenant encore, Lev avait fait la connaissance du petit et affirmait qu'il était drôlement malin.

Quel paradoxe, songea-t-elle : Greg avait un fils, alors qu'il n'en avait pas voulu, tandis que Boy, qui en désirait ardemment un, en était privé.

Une heure s'était écoulée lorsque Boy sortit du cabinet médical. Le docteur leur promit les résultats une semaine plus tard. Ils le quittèrent à midi.

« J'ai besoin d'un verre, déclara Boy.

— Moi aussi », approuva Daisy.

Ils passèrent en revue les rangées de maisons identiques bordant la chaussée. « Ce quartier est un désert. Pas un seul pub à l'horizon.

— Pas question que je mette les pieds dans un pub, dit Daisy. J'ai envie d'un martini et dans les pubs, personne ne sait les préparer. » Elle parlait d'expérience. Lorsqu'elle avait commandé un martini dry au King's Head de Chelsea, on lui avait servi un verre de vermouth tiède. « Emmène-moi au Claridge, s'il te plaît. À pied, on en a seulement pour cinq minutes.

— Ça, c'est ce que j'appelle une bonne idée. »

Ils retrouvèrent une foule de connaissances au bar du Claridge. En raison des restrictions, les menus proposés par les restaurants étaient soumis à des règles, mais le Claridge avait trouvé une astuce : les lois concernaient la *vente* de nourriture, donc l'établissement offrait un buffet gratuit, facturant simplement encore plus cher les boissons déjà hors de prix.

Assis à une des tables du bar, splendeur Art déco, Daisy et Boy sirotèrent des cocktails parfaits et Daisy commença à se sentir mieux.

« Le docteur m'a demandé si j'avais eu les oreillons, lui raconta Boy.

— Tu les as eus, effectivement. » C'était une maladie infantile, mais Boy l'avait contractée deux ans plus tôt. Affecté dans l'est du pays, il logeait chez un pasteur dont les trois fils en bas âge avaient attrapé les oreillons et l'avaient contaminé. L'épreuve avait été plutôt douloureuse. « Il t'a expliqué pourquoi il te posait cette question ?

— Non. Tu sais comment sont les hommes de l'art. Motus et bouche cousue, c'est leur devise. »

Daisy songea qu'elle était bien moins insouciante que naguère. Quelques années plus tôt, jamais elle n'aurait broyé du noir à propos de son couple comme elle le faisait maintenant. À l'instar de Scarlett O'Hara dans *Autant en emporte le vent*, elle était du genre à se dire : « On verra demain. » Mais c'était fini désormais. Peut-être devenait-elle enfin adulte.

Boy commandait un second cocktail lorsque Daisy, en se tournant vers la porte, aperçut le marquis de Lowther qui faisait son entrée, vêtu d'un uniforme froissé et taché.

Daisy le détestait. Depuis qu'il avait deviné la nature de ses relations avec Lloyd, il la traitait avec une familiarité mielleuse, comme s'ils partageaient un secret qui faisait d'eux des intimes.

Il s'assit à leur table sans y avoir été invité, faisant choir les cendres de son cigare sur son pantalon kaki, et commanda un manhattan.

Daisy remarqua tout de suite qu'il ne mijotait rien de bon. Dans ses yeux luisait un éclat de joie malveillante que la perspective d'un bon cocktail ne suffisait pas à expliquer.

« Ça doit faire plus d'un an que je ne vous ai pas vu, Lowthie, dit Boy. Où étiez-vous passé ?

— À Madrid, répondit Lowthie. Mais je ne peux pas t'en dire plus. Secret militaire et tout ça. Et toi ?

— Je passe mes journées à former des pilotes, mais j'ai accompli moi-même quelques missions ces derniers temps, depuis qu'on a accéléré la fréquence des bombardements sur l'Allemagne.

— Excellente nouvelle. Au tour des Allemands de le sentir passer.

— Sans doute, mais beaucoup de pilotes sont mécontents.

— Ah bon ? Pourquoi ?

— Cette histoire d'objectifs militaires, c'est de la foutaise pure et simple. Bombarder les usines ne sert à rien, les Allemands les reconstruisent aussitôt. Alors nous visons les quartiers ouvriers. La main-d'œuvre est moins facile à remplacer. »

Lowther eut l'air choqué. « Autrement dit, nous avons pour politique de bombarder la population civile.

— Exactement.

— Pourtant, le gouvernement nous assure…

— Mensonges, coupa Boy. Les équipages l'ont compris. La plupart des hommes s'en fichent royalement, mais certains ont du mal à l'accepter. Selon eux, si nous estimons que cette tactique est justifiée, nous devrions le dire ; et dans le cas contraire, nous devrions en changer. »

Lowther paraissait gêné. « Je ne sais pas s'il est bien raisonnable d'aborder ce genre de sujets ici.

— Vous avez sans doute raison. »

On leur servit de nouveaux cocktails. Lowther se tourna vers Daisy. « Et comment va la petite madame ? demanda-t-il. Vous participez certainement à l'effort de guerre, vous aussi. Après tout, l'oisiveté est mère de tous les vices, comme dit le proverbe. »

Daisy répondit d'une voix neutre. « À présent que le Blitz est fini, on a moins besoin d'ambulancières, alors je travaille pour la Croix-Rouge américaine. Nous avons des bureaux dans Pall Mall. Nous essayons de venir en aide aux soldats américains affectés à Londres.

— Des hommes seuls, qui ont grand besoin de compagnie féminine, c'est ça ?

— Ils ont surtout le mal du pays. Ça leur fait du bien d'entendre parler anglais avec l'accent américain. »

Lowthie esquissa un sourire salace. « Je suis sûr que vous êtes très douée pour les consoler.

— Je fais ce que je peux.

— Je n'en doute pas un instant.

— Dites donc, Lowthie, intervint Boy, vous n'auriez pas un peu trop bu, par hasard ? Cette conversation me semble terriblement déplacée, vous savez. »

L'expression de Lowther se fit franchement venimeuse. « Allons, Boy, ne me dis pas que tu n'es pas au courant. Tu es aveugle, ou quoi ?

— S'il te plaît, Boy, ramène-moi à la maison », supplia Daisy.

Mais Boy l'ignora et demanda à Lowther : « Que voulez-vous dire au juste ?

— Tu devrais lui demander de te parler de Lloyd Williams.

— Qui diable est Lloyd Williams ?

— Si tu ne veux pas m'emmener, intervint Daisy, je rentre seule.

— Tu connais un Lloyd Williams, Daisy ? »

C'est ton frère, songea Daisy ; et elle éprouva une envie presque irrépressible de lui révéler ce secret pour le simple plaisir de le désarçonner, mais elle résista à la tentation. « Tu le connais, toi aussi. Il était à Cambridge en même temps que toi. Il y a quelques années, il nous a emmenés dans un music-hall de l'East End.

— Ah oui ! » se rappela Boy. Puis, toujours intrigué, il se tourna vers Lowther : « C'est de lui que vous parlez ? » Il avait peine à considérer un homme tel que Lloyd comme un rival. De plus en plus incrédule, il insista : « Un type qui n'a même pas les moyens de se payer un habit de soirée à lui ?

— Il y a trois ans, reprit Lowther, il faisait partie de mes élèves à Tŷ Gwyn, au moment où Daisy y résidait. À l'époque, si je me souviens bien, tu risquais ta vie aux commandes d'un Hawker Hurricane au-dessus de la France. Pendant ce temps, elle filait le parfait amour avec ce satané Gallois – dans ta demeure familiale, par-dessus le marché ! »

Le visage de Boy vira à l'écarlate. « Si vous inventez ça de toutes pièces, Lowthie, je vous étrillerai, je vous le jure.

— Demande à ta femme ! » répliqua Lowther avec un sourire suffisant.

Boy se tourna vers Daisy.

Elle n'avait pas couché avec Lloyd à Tŷ Gwyn. Ils s'étaient retrouvés dans son lit à lui, dans la maison de sa mère, pendant le Blitz. Mais elle ne pouvait pas le dire à Boy en présence de Lowther, et de toute façon, ce n'était qu'un détail. L'accusation d'adultère était fondée et elle ne comptait pas la nier. Le secret était révélé. L'important, désormais, c'était de conserver un semblant de dignité.

« Je te dirai tout ce que tu veux savoir, Boy – mais pas devant ce mufle concupiscent. »

Boy éleva la voix. « Ainsi, tu ne nies pas ? »

Les gens assis à la table voisine se tournèrent vers eux, l'air gêné, puis se concentrèrent sur leurs verres.

Daisy haussa le ton à son tour. « Je refuse de subir un interrogatoire au bar du Claridge.

— Donc, tu avoues ? » cria-t-il.

Le silence se fit dans la salle.

Daisy se leva. « Je n'avoue rien et je ne nie rien. Je te raconterai tout ce qu'il y a à raconter chez nous, en tête à tête, car c'est ainsi que les couples civilisés règlent ce genre de problème.

— Mon Dieu, tu as fait ça, tu as couché avec lui ! » rugit Boy.

Les garçons eux-mêmes avaient cessé de servir et s'étaient figés.

Daisy se dirigea vers la porte.

« Espèce de traînée ! » hurla Boy.

Daisy n'était pas disposée à lui laisser le dernier mot – surtout celui-là. Elle se retourna. « Il est vrai que tu t'y connais en matière de traînées. J'ai eu le malheur d'en rencontrer deux qui t'étaient particulièrement chères, tu te rappelles ? » Elle parcourut la salle du regard. « Joanie et Pearl, lança-t-elle avec dédain. Combien d'épouses auraient supporté cela ? » Et elle sortit sans lui laisser le temps de répliquer.

Elle monta dans un taxi qui attendait. Comme il démarrait, elle vit Boy sortir de l'hôtel et s'engouffrer dans le suivant.

Elle donna son adresse au chauffeur.

En un sens, cette révélation la soulageait. Mais elle était aussi infiniment triste. Quelque chose venait de prendre fin, elle le savait.

Leur maison était à cinq cents mètres à peine. Alors qu'elle arrivait, le taxi de Boy se gara derrière le sien.

Son mari la suivit dans le vestibule.

Elle ne pouvait plus habiter ici avec lui, comprit-elle. C'était fini. Plus jamais elle ne partagerait sa maison, ni son lit. « Apportez-moi une valise, s'il vous plaît, dit-elle au majordome.

— Bien, madame. »

Elle parcourut les lieux du regard. C'était une demeure du XVIIIe siècle, aux proportions parfaites, agrémentée d'un escalier aux courbes élégantes, mais elle ne regrettait pas vraiment de la quitter.

« Où vas-tu ? demanda Boy.

— À l'hôtel, je suppose. Sans doute pas au Claridge.

— Pour retrouver ton amant !

— Non, il est sur le continent. En effet, j'aime cet homme. Je suis navrée, Boy. Tu n'as pas le droit de me juger – tes fautes sont plus graves que les miennes –, mais je me juge moi-même.

— C'est décidé, dit-il. Je vais demander le divorce. »

C'étaient les mots qu'elle attendait. Ils avaient été prononcés, tout était fini. Sa nouvelle vie commençait à présent.

« Dieu merci », soupira-t-elle.

3.

Daisy loua un appartement à Piccadilly. Il était équipé d'une grande salle de bains à l'américaine, avec une douche. On y trouvait deux cabinets, dont un réservé aux invités – une extravagance ridicule aux yeux de la plupart des Anglais.

Heureusement, elle n'avait pas de problèmes d'argent. Son grand-père Vialov lui avait légué une fortune dont elle jouissait à part entière depuis l'âge de vingt et un ans. Une fortune en dollars américains.

Comme il était difficile d'acheter des meubles neufs, elle se rabattit sur des antiquités, que l'on trouvait à profusion pour des prix modiques. Elle accrocha aux murs des tableaux d'art moderne afin de composer un décor jeune et gai. Elle engagea une vieille lingère et une jeune femme de chambre et constata qu'il était facile de faire tourner une maison sans majordome ni cuisinière, en particulier quand on n'avait pas de mari à chouchouter.

Les domestiques de la maison de Mayfair rangèrent ses vêtements dans des cartons, qui lui furent livrés par un camion de déménagement. Daisy et sa lingère passèrent un après-midi à ouvrir les cartons et à tout ranger proprement.

Elle avait essuyé une humiliation mais gagné sa liberté. Le bilan était largement positif. Sa blessure finirait par guérir et au moins, elle était débarrassée de Boy à jamais.

Au bout de huit jours, elle se demanda quels étaient les résultats de l'examen médical. Le docteur avait dû les transmettre à Boy, naturellement, puisqu'il était le chef de famille. Comme elle n'avait pas envie de les lui demander et n'y attachait plus grande importance, elle cessa d'y penser.

Elle prit grand plaisir à aménager son nouvel appartement. Les quinze premiers jours, elle fut trop occupée pour fréquenter du monde. Mais une fois que tout fut prêt, elle décida de voir tous les amis qu'elle avait négligés.

Elle en avait beaucoup à Londres. Cela faisait sept ans qu'elle vivait là. Durant les quatre dernières années, Boy était plus souvent en mission qu'à la maison, aussi avait-elle pris l'habitude de sortir seule. L'absence d'un mari ne changerait donc pas grand-chose à sa vie, se dit-elle. Bien sûr, elle ne serait plus jamais invitée chez les Fitzherbert, mais la haute société londonienne ne se limitait pas à eux.

Elle fit des provisions de whisky, de gin et de champagne, achetant légalement tout ce qu'elle pouvait mais n'hésitant pas à recourir également au marché noir. Puis elle envoya des invitations pour sa pendaison de crémaillère.

Tout le monde lui répondit promptement – par la négative.

En larmes, elle appela Eva Murray. « Pourquoi personne ne veut-il venir à ma réception ? » sanglota-t-elle.

Dix minutes plus tard, Eva frappait à sa porte.

Elle était accompagnée de trois enfants et d'une nounou. Jamie avait six ans, Anna quatre et la petite Karen deux.

Daisy lui fit visiter l'appartement puis demanda du thé pendant que Jamie transformait le divan en char d'assaut, avec ses sœurs en guise d'équipage.

Eva parlait l'anglais avec un mélange d'accent allemand, américain et écossais. « Daisy, ma chère, Londres n'est pas Rome, tu sais.

— Je sais. Tu es sûre que tu es bien installée ? »

Eva était enceinte de son quatrième enfant et approchait du terme. « Ça ne te dérange pas si j'allonge mes jambes ?

— Bien sûr que non. » Daisy attrapa un coussin.

« La haute société londonienne est très à cheval sur les principes, reprit Eva. Ne crois pas que je l'approuve. J'ai souvent été mise en quarantaine, et ce pauvre Jimmy se fait parfois snober parce qu'il a épousé une Allemande, à moitié juive qui plus est.

— C'est affreux.

— Je ne le souhaiterais à personne, quelle qu'en soit la cause.

— Il y a des moments où je déteste les Anglais.

— Tu as oublié comment sont les Américains. Tu ne te rappelles pas m'avoir dit que toutes les filles de Buffalo étaient des bêcheuses ? »

Daisy éclata de rire. « Ça paraît si loin !

— Tu as quitté ton mari. En plus, tu l'as fait de façon spectaculaire, en l'insultant au bar du Claridge.

— Et je n'avais bu qu'un martini ! »

Un large sourire éclaira le visage d'Eva. « Je regrette d'avoir raté ça !

— Moi, je regrette un peu de l'avoir fait.

— Inutile de te dire que tout Londres ne parle que de ce scandale depuis trois semaines.

— J'aurais dû m'y attendre.

— Dans l'esprit des gens, assister à ta pendaison de crémaillère reviendrait à approuver l'adultère et le divorce. Moi-même, je ne serais pas tellement contente que ma belle-mère sache que je suis venue prendre le thé chez toi.

— Mais c'est vraiment injuste : Boy a été le premier à me tromper !

— Parce que tu pensais que les femmes étaient traitées à égalité avec les hommes ? »

Daisy se rappela qu'Eva avait d'autres motifs d'inquiétude, bien plus graves que les rebuffades de quelques snobs. Ses proches vivaient toujours dans l'Allemagne nazie. Fitz avait cherché à avoir de leurs nouvelles par l'entremise de l'ambassade de Suisse et avait appris que son père médecin était interné dans un camp de concentration et que son frère luthier avait été tabassé par la police qui lui avait brisé les doigts. « Quand je pense aux ennuis de ta famille, j'ai honte de me plaindre, dit-elle.

— Il ne faut pas. Mais annule ta réception. »

C'est ce que fit Daisy.

Elle en fut cependant très malheureuse. Grâce à son travail pour la Croix-Rouge, ses journées étaient bien remplies, mais, le soir venu, elle n'avait rien à faire et personne à voir. Elle allait au cinéma deux fois par semaine. Elle tenta de lire *Moby Dick* mais trouva ce roman indigeste. Un dimanche, elle décida d'aller à l'église. Celle de St James's, œuvre de Christopher Wren située juste en face de son immeuble de Piccadilly, ayant été bombardée, elle se rendit à St Martin's-in-the-Fields. Boy n'était pas dans l'assistance, mais Daisy reconnut Fitz et Bea, et passa toute la cérémonie les yeux fixés sur la nuque de son beau-père, songeant qu'elle était tombée amoureuse de deux de ses fils. Boy avait hérité de la beauté de sa mère et du caractère résolument égoïste de son père. Lloyd avait hérité de la beauté de Fitz et du grand cœur d'Ethel. Pourquoi ai-je mis tout ce temps à m'en rendre compte ? s'interrogea-t-elle.

Quantité de ses connaissances fréquentaient cette église, mais, après l'office, personne ne lui adressa la parole. Elle était seule et presque sans amis dans un pays étranger, alors que la guerre faisait rage.

Un soir, elle prit un taxi pour se rendre à Aldgate et frappa à la porte des Leckwith. Lorsque Ethel lui ouvrit, Daisy lui déclara :

« Je suis venue vous demander la main de votre fils. » Ethel partit d'un grand éclat de rire et la serra dans ses bras.

Elle avait apporté un cadeau, du jambon en conserve américain que lui avait donné un navigateur de l'US Air Force. Pour une famille anglaise soumise au rationnement, c'était un luxe. Elle s'assit dans la cuisine en compagnie d'Ethel et de Bernie, qui écoutaient des chansons à la radio. Tous reprirent en chœur *Underneath the Arches*, le succès du duo Flanagan et Allen. « Bud Flanagan est né ici même, dans l'East End, déclara Bernie avec fierté. Il s'appelle en réalité Chaim Reuben Weintrop. »

Les Leckwith se passionnaient pour le rapport Beveridge, un document gouvernemental qui était devenu un véritable best-seller. « Commandé par un Premier ministre conservateur et rédigé par un économiste libéral, lui expliqua Bernie. Et il propose de faire ce que le parti travailliste réclame depuis toujours ! En politique, vous savez que vous avez gagné la partie quand vos adversaires vous piquent vos idées.

— Son principe, enchaîna Ethel, c'est que toute personne en âge de travailler devrait verser une cotisation hebdomadaire, ce qui lui permettrait de toucher une pension quand elle sera malade, au chômage, à la retraite ou si elle perd son conjoint.

— Une proposition simple comme bonjour, mais qui transformera notre pays en profondeur, conclut Bernie avec enthousiasme. Du berceau à la tombe, personne ne pourra plus sombrer dans la misère.

— Et le gouvernement a accepté de l'appliquer ? demanda Daisy.

— Non, répondit Ethel. Clem Attlee a fortement insisté auprès de Churchill, mais Churchill a refusé d'y souscrire. Le Trésor estime que ça coûterait trop cher.

— Si nous voulons le faire appliquer, nous devrons d'abord gagner les élections », ajouta Bernie.

Millie, la fille d'Ethel et de Bernie, vint les rejoindre. « Je ne peux pas rester longtemps, annonça-t-elle. Abie peut garder les enfants une demi-heure, pas plus. » Elle avait perdu son emploi – les femmes n'achetaient pas de robes de luxe ces temps derniers, même celles qui pouvaient se le permettre – mais, heureusement, l'entreprise de vente de cuir de son époux était florissante et ils avaient déjà deux enfants, Lennie et Pammie.

Tout en buvant un chocolat chaud, ils parlèrent du jeune homme que tous adoraient. Ils avaient très peu de nouvelles de Lloyd. Tous les six ou huit mois, Ethel recevait une lettre sur papier à en-tête de l'ambassade britannique à Madrid, où il lui assurait qu'il se portait bien et participait activement à la lutte contre le fascisme. On l'avait promu commandant. Il n'écrivait jamais à Daisy, de peur que Boy n'intercepte ses lettres, mais cette précaution n'était plus nécessaire. Daisy donna sa nouvelle adresse à Ethel et nota celle de Lloyd, une boîte postale de l'armée britannique.

Ils ignoraient quand il pourrait venir en permission.

Daisy leur parla de Greg, son demi-frère, et de son fils Georgy. Elle savait que les Leckwith n'étaient pas du genre à s'offusquer et accueilleraient la nouvelle avec joie.

Elle leur raconta aussi les malheurs de la famille d'Eva à Berlin. Bernie était juif et il eut les larmes aux yeux en apprenant que Rudi avait eu les mains brisées. « Ils auraient dû affronter ces salopards de fascistes dans la rue, tant qu'il était temps, dit-il. C'est ce que nous avons fait.

— J'ai encore dans le dos les cicatrices de la vitrine de Gardiner's, renchérit Millie. J'en ai eu honte pendant longtemps : la première fois qu'Abie a vu mon dos, nous étions mariés depuis six mois. Mais il dit qu'il est fier de mes blessures.

— La bataille de Cable Street a été un sale moment, c'est sûr, reprit Bernie. Mais on a mis un terme à leurs conneries. » Il ôta ses lunettes et s'essuya les yeux avec son mouchoir.

Ethel posa le bras sur ses épaules. « Ce jour-là, j'avais conseillé aux gens de rester chez eux, murmura-t-elle. J'avais tort et c'est toi qui avais raison. »

Il sourit d'un air contrit. « Ce n'est pas fréquent.

— Mais c'est le Public Order Act, la loi adoptée par le Parlement après Cable Street, qui a définitivement balayé les fascistes britanniques, remarqua Ethel. Il a suffi d'interdire le port d'uniformes politiques sur la voie publique pour qu'ils disparaissent de la circulation. S'ils ne pouvaient plus parader en chemise noire, ils n'étaient plus rien. C'est aux conservateurs que nous le devons – reconnaissons-leur ce mérite. »

Toujours passionnés de politique, les Leckwith préparaient déjà la réforme de l'Angleterre d'après-guerre. Clement Attlee, le brillant dirigeant du parti travailliste, occupait le poste de vice-Premier ministre aux côtés de Churchill, et Ernie Bevin, un héros des luttes syndicales, celui de ministre du Travail. Grâce à eux, Daisy envisageait l'avenir avec optimisme.

Millie prit congé et Bernie alla se coucher. Lorsqu'elles furent seules, Ethel demanda à Daisy : « Vous avez vraiment l'intention d'épouser mon Lloyd ?

— C'est ce que je désire le plus au monde. Vous croyez que ça peut marcher ?

— Bien sûr. Pourquoi pas ?

— Nous venons de milieux si différents. Vous êtes des gens tellement épatants. Vous ne vivez que pour servir l'intérêt public.

— À l'exception de Millie. Elle tient du frère de Bernie – seul l'argent l'intéresse.

— Elle a tout de même gardé des cicatrices de Cable Street.

— C'est vrai.

— Lloyd est comme vous. Pour lui, la politique n'est ni un loisir ni un passe-temps – c'est l'essence de sa vie. Alors que moi, je ne suis qu'une millionnaire égoïste.

— Il existe, je crois, deux sortes de couples, murmura Ethel d'un air pensif. Le premier est une sorte d'association confortable, dont les deux partenaires partagent les mêmes espoirs, les mêmes craintes, élèvent leurs enfants ensemble et s'offrent mutuellement aide et réconfort. » Elle parlait de Bernie et d'elle-même, comprit Daisy. « Le second est une union passionnée, faite de folie, de joie et de sexe, souvent avec quelqu'un qui ne vous convient pas, voire qu'on n'admire pas et qu'on aime encore moins. » Ethel parlait de sa liaison avec Fitz, Daisy en était sûre. Elle retint son souffle : elle savait que ce qu'Ethel lui disait là était la vérité sans fard. « J'ai eu de la chance, j'ai connu les deux, conclut Ethel. Et si j'ai un conseil à vous donner, c'est le suivant : si vous avez l'occasion de vivre un amour fou, saisissez-la à bras-le-corps, et au diable les conséquences.

— Comptez sur moi ! » fit Daisy.

Elle prit congé quelques minutes plus tard. Elle était émue et honorée qu'Ethel se soit livrée avec une telle franchise. Mais en regagnant son appartement désert, elle se sentit déprimée. Elle se prépara un cocktail qu'elle vida dans l'évier. Elle mit la bouilloire à chauffer puis éteignit la cuisinière. Le programme de la radio s'acheva. Étendue entre des draps glacés, elle regretta l'absence de Lloyd.

Elle compara la famille de Lloyd à la sienne. Toutes deux avaient une histoire chaotique, mais Ethel avait su forger une famille forte et solide à partir de matériaux disparates, chose dont la mère de Daisy avait été incapable – même si la faute en incombait à Lev plus qu'à Olga. Ethel était une femme remarquable et Lloyd avait hérité nombre de ses qualités.

Où était-il à présent, que faisait-il ? Quelle que fût la réponse à ces questions, il était sûrement en danger. Et s'il se faisait tuer, alors qu'elle était enfin libre de l'aimer en toute liberté et peut-être même de l'épouser un jour ? Que ferait-elle s'il mourait ? Sa propre vie n'aurait plus aucun sens : pas de mari, pas d'amant, pas d'amis, pas de pays. Elle s'endormit enfin à l'aube, bercée par ses larmes.

Le lendemain, elle fit la grasse matinée. À midi, elle prenait son café dans sa petite salle à manger, vêtue d'un peignoir de soie noire, lorsque sa jeune femme de chambre entra et annonça : « Le commandant Williams, madame.

— Comment ? s'écria-t-elle. Ce n'est pas possible ! »

Il apparut sur le seuil, son sac sur l'épaule.

Il semblait épuisé, ses joues étaient mangées de barbe et, de toute évidence, il avait dormi dans son uniforme.

Elle le serra dans ses bras et embrassa ses joues râpeuses. Il l'embrassa en retour, un peu gêné par le sourire radieux qu'il n'arrivait pas à effacer de ses lèvres. « Je dois empester, dit-il entre deux baisers. Ça fait huit jours que je ne me suis pas changé.

— Tu pues comme une cave à fromages, acquiesça-t-elle. Et j'adore ça. » Elle l'entraîna dans sa chambre et commença à le déshabiller.

« Je vais prendre une douche en vitesse, dit-il.

— Non. » Elle le poussa sur le lit. « Je suis trop pressée. » Le désir qu'il lui inspirait tenait de la frénésie. Et, à vrai dire, cette forte odeur lui tournait la tête. Loin de la dégoûter, elle exerçait sur elle un effet diamétralement opposé. C'était lui, l'homme dont elle avait redouté la mort, et ces effluves lui emplissaient les narines et les poumons. Elle en aurait pleuré de joie.

Pour lui ôter son pantalon, elle devait d'abord lui retirer ses bottes, ce qui risquait d'être compliqué, aussi ne prit-elle pas cette peine. Elle se contenta de déboutonner sa braguette. Puis elle se débarrassa de son peignoir de soie noire et retroussa sa chemise de nuit sur ses hanches, sans cesser de fixer avec avidité le sexe pâle qui pointait hors de la toile kaki. Puis elle l'enfourcha, se laissa descendre sur lui, inclina la tête et l'embrassa. « Ô mon Dieu ! fit-elle. Si tu savais comme j'ai eu envie de toi. »

Elle resta ainsi, presque sans bouger, l'embrassant encore et encore. Il lui prit le visage au creux des mains et la regarda fixement. « C'est pour de vrai, hein ? chuchota-t-il. Ce n'est pas un rêve ?

— C'est pour de vrai.

— Tant mieux. Je n'aimerais pas me réveiller en ce moment.

— Je voudrais rester comme ça pour toujours.

— Excellente idée, mais je ne vais pas pouvoir me tenir tranquille plus longtemps. » Il se mit à bouger en elle.

« Si tu fais ça, je vais jouir », lui dit-elle tout bas.

Ce qu'elle fit.

Ils restèrent ensuite allongés un long moment à bavarder.

On lui avait accordé deux semaines de permission. « Installe-toi ici, proposa-t-elle. Tu peux aller voir tes parents tous les jours, mais je te veux ici toutes les nuits.

— Je ne voudrais pas te compromettre.

— Trop tard. Tout Londres m'évite déjà.

— Je sais. » Il avait téléphoné à Ethel depuis la gare de Waterloo, et elle lui avait appris que Daisy s'était séparée de Boy et lui avait donné sa nouvelle adresse.

« Nous devrions prendre quelques précautions, remarqua-t-il. Je me procurerai des capotes. Mais peut-être devrais-tu songer à te faire poser un de ces machins, tu sais. Qu'en penses-tu ?

— Tu veux être sûr que je ne tombe pas enceinte ? » demanda-t-elle.

Elle se rendit compte qu'il y avait une note de tristesse dans sa voix, et il la perçut lui aussi. « Ne te méprends pas », dit-il. Il se redressa sur le coude. « Je suis un enfant illégitime. On m'a raconté des mensonges sur mes origines et, quand j'ai découvert la vérité, je peux te dire que le choc a été terrible. » Sa voix tremblait un peu sous l'effet de l'émotion. « Jamais je ne ferai subir cela à mes enfants. Jamais !

— Rien ne nous obligerait à leur mentir.

— Nous leur dirions que nous ne sommes pas mariés ? Que tu es en fait l'épouse d'un autre ?

— Pourquoi pas ?

— Imagine les moqueries qu'on leur infligerait à l'école. »

Elle n'était pas convaincue, mais de toute évidence, c'était une question fondamentale pour lui. « Alors, que proposes-tu ? demanda-t-elle.

— Je veux que nous ayons des enfants. Mais seulement quand nous serons mariés. Ensemble.

— J'avais compris. Autrement dit…

— Nous devrons attendre. »

Décidément, les hommes n'étaient pas très vifs pour saisir les allusions. « Je ne suis pas très à cheval sur les traditions, dit-elle. Mais il y a quelques fondamentaux… »

Il comprit enfin où elle voulait en venir. « Oh ! D'accord. Une minute. » Il se redressa pour se mettre à genoux sur le lit. « Daisy, ma chérie… »

Elle éclata de rire. Comme il avait l'air drôle, encore en uniforme, son sexe fripé dépassant de sa braguette. « Je peux te prendre en photo ? » demanda-t-elle.

Il baissa la tête. « Oh, pardon…

— Non, surtout, ne change rien ! Reste comme tu es et dis-moi ce que tu as à me dire. »

Il sourit. « Daisy, ma chérie, veux-tu être mon épouse ?

— Tout de suite », répondit-elle.

Ils s'allongèrent à nouveau et s'étreignirent.

Elle se lassa bientôt de son étrange odeur et ils prirent une douche ensemble. Elle le savonna de la tête aux pieds, s'amusant de sa gêne lorsqu'elle lava ses parties intimes. Elle lui shampouina les cheveux et lui décrassa les pieds avec une brosse.

Une fois propre, il insista pour la laver à son tour, mais il n'en était qu'aux seins lorsque la passion les emporta à nouveau. Ils firent l'amour debout, dans la cabine de douche, sous une cascade d'eau délicieusement chaude. De toute évidence, il avait oublié son aversion pour les grossesses illégitimes et elle se garda de la lui rappeler.

Ensuite, il se planta devant le miroir pour se raser. Enveloppée dans un drap de bain, elle s'assit sur le couvercle des toilettes pour le regarder faire. « Combien de temps te faudra-t-il pour obtenir le divorce ? demanda-t-il.

— Je n'en sais rien. Il faudrait que j'aille discuter avec Boy.

— Pas aujourd'hui. Aujourd'hui, je te veux toute à moi.

— Quand comptes-tu aller voir tes parents ?

— Demain, peut-être.

— Alors, j'en profiterai pour aller voir Boy. Je veux en finir le plus vite possible.

— Bien. C'est décidé, alors. »

4.

Daisy éprouva une étrange impression en entrant dans la maison où elle avait vécu avec Boy. Un mois plus tôt, cette demeure était la sienne. Elle était libre d'y aller et venir comme bon lui semblait, d'entrer dans chacune de ses pièces sans demander la permission. Aujourd'hui, elle était une étrangère en visite. Elle garda ses gants et son chapeau et suivit le vieux majordome qui la conduisit au petit salon.

Boy ne daigna ni lui serrer la main ni l'embrasser sur la joue. Toute son attitude exprimait une vertueuse indignation.

« Je n'ai pas encore engagé d'avocat, dit Daisy en s'asseyant. Je voulais d'abord te parler en tête à tête. J'espère que nous réussirons à régler cette affaire sans nous détester. Après tout, nous n'avons pas d'enfants à nous disputer et nous sommes riches tous les deux.

— Tu m'as trahi ! » lança-t-il avec colère.

Daisy soupira. De toute évidence, elle avait nourri de faux espoirs. « Nous avons commis l'adultère l'un comme l'autre, reconnut-elle. Mais c'est toi qui as commencé.

— Tu m'as humilié publiquement. Tout Londres est au courant !

— J'ai fait mon possible pour t'empêcher de faire un esclandre au Claridge – mais tu étais trop occupé à me traîner dans la boue ! J'espère que tu as réglé son compte à ce détestable marquis.

— Et pourquoi l'aurais-je fait ? Il m'a rendu service.

— Il aurait été mieux inspiré de te faire des confidences au club.

— Je ne comprends pas comment tu as pu t'enticher d'un péquenaud comme ce Williams. J'en ai appris de belles à son sujet. Sa mère n'était qu'une gouvernante !

— C'est sans doute la femme la plus admirable que j'aie jamais rencontrée.

— J'espère que tu te rends compte que personne ne sait qui est son père ? »

L'ironie était de taille, se dit Daisy. « Je le sais, moi, répliqua-t-elle.

— Ah oui ? Qui est-ce ?

— Ne compte pas sur moi pour te le dire.

— Voilà qui a le mérite d'être clair.

— Nous n'irons pas bien loin comme ça, tu ne crois pas ?

— En effet.

— Peut-être vaut-il mieux qu'un avocat t'écrive de ma part. » Elle se leva. « Je t'ai aimé, Boy, ajouta-t-elle avec tristesse. Tu étais drôle. Je regrette de ne pas avoir suffi à te combler. Je te souhaite d'être heureux. J'espère que tu épouseras une femme qui te conviendra mieux et qu'elle te donnera de nombreux fils. J'en serais heureuse pour toi.

— Inutile d'y compter », murmura-t-il.

Elle se dirigeait déjà vers la porte mais elle se retourna. « Pourquoi dis-tu ça ?

— J'ai reçu les résultats de la visite médicale. »

Elle lui était complètement sortie de la tête depuis leur séparation. « Alors ?

— En ce qui te concerne, tout va bien – tu peux avoir une kyrielle de marmots. Quant à moi, je ne serai jamais père. Chez l'adulte, les oreillons peuvent être cause de stérilité, et j'ai tiré le mauvais numéro. » Il partit d'un rire amer. « Tous ces Allemands ont passé des années à me canarder, et je me suis fait avoir par trois morveux de fils de pasteur. »

Elle sentit son cœur se serrer. « Oh ! Boy, j'en suis vraiment peinée.

— Eh bien tu vas l'être plus encore, car j'ai décidé de ne pas divorcer. »

Un frisson la parcourut. « Que veux-tu dire ? Pourquoi fais-tu ça ?

— Pourquoi ne le ferais-je pas ? Je n'ai pas l'intention de me remarier. Après tout, je ne peux pas avoir d'enfants. C'est le fils d'Andy qui héritera du titre.

— Mais je veux épouser Lloyd !

— Qu'est-ce que ça peut me faire ? Pourquoi aurait-il des enfants et pas moi ? »

Daisy était atterrée. Le bonheur allait-il lui être refusé alors qu'il semblait enfin à sa portée ? « Boy, tu ne parles pas sérieusement !

— Je n'ai jamais été aussi sérieux de ma vie.

— Mais Lloyd veut des enfants, lui aussi ! insista-t-elle, effondrée.

— Il aurait dû y penser avant de baiser la femme d'un autre.

— Très bien, lança-t-elle d'un air de défi. C'est moi qui demanderai le divorce.

— Pour quel motif ?

— Adultère, bien sûr.

— Tu n'as aucune preuve. » Elle allait lui répondre qu'il ne devrait pas être difficile d'en obtenir lorsqu'il ajouta avec un sourire méchant : « Et je veillerai à ce que tu n'en aies pas. »

C'était réalisable s'il était discret, comprit-elle avec horreur. « Mais tu m'as chassée du domicile conjugal ! protesta-t-elle.

— Je dirai au juge que tu y es toujours la bienvenue. »

Elle tenta de refouler ses larmes. « Jamais je n'aurais cru que tu me haïrais à ce point, murmura-t-elle, bouleversée.

— Ah bon ? fit Boy. Eh bien, maintenant, tu es fixée. »

5.

Lloyd Williams frappa à la porte de Boy Fitzherbert en milieu de matinée, espérant qu'à cette heure-là il serait à jeun, et se présenta au majordome comme le commandant Williams, un parent éloigné. Il estimait qu'une conversation d'homme à homme s'imposait. Boy ne comptait sûrement pas savourer sa vengeance jusqu'à la fin de ses jours. Lloyd avait revêtu son uniforme, jugeant préférable de parler entre soldats. Le bon sens finirait sûrement par l'emporter.

On le fit entrer dans le petit salon, où Boy lisait le journal en fumant un cigare. Il lui fallut quelque temps pour le reconnaître. « Vous ! s'exclama Boy quand il comprit à qui il avait affaire. Foutez le camp.

— Je suis venu vous demander d'accorder le divorce à Daisy, déclara Lloyd.

— Sortez. » Boy se leva.

« Je vois que vous êtes tenté d'en venir aux mains, dit Lloyd, aussi permettez-moi de vous avertir que vous allez au-devant de certaines difficultés. Je suis certes un peu plus petit que vous, mais je boxe dans la catégorie des poids welter et j'ai remporté un certain nombre de combats.

— Je ne vais certainement pas me salir les mains en vous touchant.

— Je suis ravi de l'entendre. Mais êtes-vous prêt à revenir sur votre décision ?

— En aucun cas.

— Il y a un détail que vous ignorez, poursuivit Lloyd. Je me demande s'il ne serait pas susceptible de vous faire changer d'avis.

— J'en doute. Mais puisque vous êtes ici, essayez toujours. » Il se rassit, sans offrir de siège à son visiteur.

Tu l'auras voulu, songea Lloyd.

Il sortit de sa poche une vieille photographie sépia. « Si vous voulez bien examiner ce portrait de moi. » Il le posa sur la table basse, à côté du cendrier.

Boy ramassa le cliché. « Ce n'est pas vous. Cet homme vous ressemble, mais il porte un uniforme victorien. Votre père, sans doute.

— Mon grand-père, plutôt. Retournez donc ce portrait. »

Boy lut l'inscription figurant au dos. « Comte Fitzherbert ? dit-il avec dédain.

— Oui. Le précédent comte, votre grand-père – et le mien. Daisy a trouvé cette photo à Tŷ Gwyn. » Lloyd inspira profondément. « Personne ne connaît mon père, lui avez-vous dit. Eh bien, je peux vous détromper. Il s'agit du comte Fitzherbert. Nous sommes frères, vous et moi. » Il attendit sa réaction.

Boy s'esclaffa. « Ridicule !

— C'est exactement ce que j'ai dit la première fois.

— Eh bien, vous m'étonnez, je dois vous l'avouer. J'aurais cru que vous auriez trouvé des arguments plus sérieux que cette invention grotesque. »

Lloyd avait espéré que cette révélation mettrait Boy dans de meilleures dispositions, mais visiblement, ce n'était pas le cas. Néanmoins, il tenta encore de le raisonner. « Allons, Boy... Est-ce tellement improbable ? Ces choses-là arrivent dans les meilleures familles, comme on dit. Une jolie femme de chambre, un jeune noble vigoureux, et la nature fait le reste. Quand le bébé vient au monde, on étouffe l'affaire. Ne me dites pas que vous n'avez jamais entendu parler de ce genre d'incidents.

— Ils sont fréquents, je vous l'accorde. » Boy avait perdu de sa morgue, sans rendre les armes pour autant. « Mais il est tout aussi fréquent que des roturiers s'inventent des liens avec l'aristocratie.

— Oh ! Je vous en prie, fit Lloyd avec mépris. Je ne souhaite certainement pas de telles attaches. Je ne suis pas un vendeur d'une boutique de nouveautés, rêvant de grandeur. Je suis issu d'une éminente famille de politiciens socialistes. Mon grand-père maternel a été l'un des fondateurs de la Fédération des mineurs de Galles du Sud. La dernière chose dont j'aie besoin, c'est d'être apparenté à une famille de pairs conservateurs. C'est terriblement embarrassant pour moi. »

Boy éclata de rire, avec moins de conviction cependant. « Embarrassant *pour vous* ! Si ce n'est pas du snobisme à l'envers...

— À l'envers ? J'ai plus de chances que vous de devenir Premier ministre un jour. » Lloyd comprit que la discussion tournait à la dispute de cour de récréation, et ce n'était pas ce qu'il souhaitait. « Peu importe. Ce dont je veux vous convaincre, c'est que vous ne pouvez pas passer le reste de votre vie à vous venger de moi – ne serait-ce que parce que nous sommes frères.

— Je n'en crois rien », s'obstina Boy, qui reposa la photo sur la table pour reprendre son cigare.

« Moi aussi, j'ai eu du mal à me rendre à l'évidence. » Lloyd persista dans ses efforts : son avenir était en jeu. « Mais on m'a fait remarquer que ma mère travaillait à Tŷ Gwyn lorsqu'elle est tombée enceinte; qu'elle s'était toujours montrée très évasive à propos de l'identité de mon père; et que, peu de temps avant ma naissance, elle a obtenu les fonds nécessaires à l'achat d'une petite maison à Londres. Je lui ai fait part de mes soupçons et elle m'a avoué la vérité.

— C'est risible.

— Mais c'est vrai et vous le savez, n'est-ce pas ?

— En aucun cas.

— Oh si ! Au nom de nos liens familiaux, n'allez-vous pas agir en honnête homme ?

— Sûrement pas. »

Lloyd comprit, accablé, qu'il ne gagnerait pas. Boy avait le pouvoir de lui gâcher la vie et il était résolu à le faire.

Il ramassa la photographie et la remit dans sa poche. « Vous poserez la question à notre père. Vous ne pourrez pas vous en empêcher. Il faudra que vous en ayez le cœur net. »

Boy émit un reniflement de mépris.

Lloyd se dirigea vers la porte. « Je pense qu'il vous dira la vérité. Adieu, Boy. »

Il sortit, refermant la porte derrière lui.

XVI

1943 (II)

1.

Le colonel Albert Beck reçut une balle russe dans le poumon droit à Kharkov en mars 1943. Il eut de la chance : un chirurgien lui plaça un drain pour évacuer son pneumothorax, lui sauvant la vie de justesse. Affaibli par l'hémorragie et l'infection quasiment inévitable, Beck fut renvoyé en train à Berlin et se retrouva dans l'hôpital de Carla.

C'était un quadragénaire sec et noueux, au crâne prématurément dégarni, dont le menton en galoche évoquait la proue d'un drakkar. La première fois qu'il adressa la parole à Carla, les médicaments et la fièvre lui firent perdre toute prudence. « Nous sommes en train de perdre la guerre », dit-il.

Toute son attention fut aussitôt en éveil. Un officier mécontent était une source d'information potentielle. « Selon les journaux, répondit-elle d'un ton insouciant, nous réduisons nos lignes sur le front de l'Est.

— En d'autres termes, nous battons en retraite », fit-il avec un petit rire désabusé.

Elle l'encouragea à poursuivre. « Et en Italie, ça se présente mal. » Le dictateur Benito Mussolini – le plus puissant des alliés d'Hitler – venait de tomber.

« Vous vous rappelez 1939, et 1940 ? demanda Beck avec nostalgie. Les victoires foudroyantes se succédaient. C'était le bon temps. »

De toute évidence, il n'était pas animé par des buts idéologiques, ni même politiques. C'était un soldat ordinaire, un patriote, qui avait cessé de se bercer d'illusions.

Carla l'aiguillonna. « On raconte que l'armée est à court de tout, de cartouches comme de caleçons, mais je ne peux pas le croire. » Ce genre de rumeur, qu'il était risqué de propager, était désormais monnaie courante à Berlin.

« C'est la pure vérité. » Bien que totalement désinhibé, Beck s'exprimait clairement. « L'Allemagne n'a tout simplement pas les moyens de produire autant d'armes et de chars que l'Union soviétique, l'Angleterre et les États-Unis réunis – d'autant que nous subissons des bombardements incessants. Et nous avons beau tuer des Russes en masse, l'armée Rouge semble avoir des réserves inépuisables de nouvelles recrues.

— Que va-t-il se passer à votre avis ?

— Les nazis ne s'avoueront jamais vaincus, c'est évident. Et les gens continueront à mourir. Des millions de victimes, tout ça parce que leur fierté leur interdit de se rendre. C'est de la folie. De la folie. » Il sombra doucement dans le sommeil.

Il fallait être malade – ou dément – pour exprimer de telles opinions, mais elles étaient de plus en plus partagées, Carla en était persuadée. En dépit de la propagande intensive du gouvernement, personne ne pouvait plus se cacher qu'Hitler allait perdre la guerre.

La mort de Joachim Koch n'avait provoqué aucune enquête policière. Les journaux avaient parlé d'un accident de la circulation. Carla s'était remise de son choc initial, mais, de temps à autre, elle se rappelait qu'elle avait tué un homme et revivait sa mort en esprit. Elle était alors prise de tremblements si violents qu'elle était obligée de s'asseoir. Heureusement, cela ne s'était produit qu'une fois pendant ses heures de service, et elle avait évoqué un malaise passager dû à la faim – explication plausible dans Berlin en temps de guerre. Sa mère était plus durement atteinte. Étrange que Maud ait pu aimer un être faible et stupide comme Joachim ; mais l'amour ne s'explique pas. Carla elle-même s'était trompée au sujet de Werner Franck : celui qu'elle prenait pour un homme fort et courageux s'était révélé un lâche et un égoïste.

Elle parla longuement avec Beck durant son hospitalisation, le sondant pour se faire une idée de sa personnalité. Une fois remis, il s'abstint de tout nouveau commentaire sur la guerre. Elle apprit que c'était un militaire de carrière, veuf, dont la fille était mariée et vivait à Buenos Aires. Son père avait fait partie du conseil municipal de Berlin : comme il ne précisait pas à quelle formation politique il avait appartenu, elle en avait conclu qu'il ne s'agissait sûrement pas du parti nazi ni d'un de ses alliés. S'il ne disait jamais de mal d'Hitler, il n'en disait pas de bien non plus, et ne tenait jamais de propos désobligeants sur les Juifs et les communistes. Par les temps qui couraient, une telle attitude relevait presque de l'insubordination.

Son poumon finirait par guérir, mais il n'aurait plus la capacité physique de se battre, et il confia à Carla qu'il allait être affecté à l'état-major. Peut-être pourrait-il être une mine de précieux secrets. Sans doute risquait-elle sa vie en tentant de le recruter, mais le jeu en valait la chandelle.

Elle savait qu'il aurait oublié leur première conversation. « Vous avez été très franc », lui dit-elle à voix basse. Il n'y avait personne dans les parages. « Nous sommes en train de perdre la guerre, m'avez-vous dit. »

Une lueur de peur traversa son regard. Elle n'avait plus affaire à un malade en chemise de nuit, aux joues barbues. C'était un convalescent propre et rasé de frais, assis bien droit sur son lit dans son pyjama bleu marine boutonné jusqu'au cou. « Vous allez me dénoncer à la Gestapo, n'est-ce pas ? s'inquiéta-t-il. Il me semble tout de même qu'un homme en proie à la fièvre ne peut pas être tenu pour responsable de ses divagations.

— Il ne s'agissait pas de divagations. Vous étiez on ne peut plus clair. Mais je ne vous dénoncerai à personne, rassurez-vous.

— Pourquoi ?

— Parce que vous avez raison. »

Il était surpris. « Peut-être est-ce moi qui devrais vous dénoncer.

— Si vous faites ça, je dirai que vous avez insulté le Führer dans votre délire et inventé cette histoire pour vous protéger quand j'ai menacé de vous signaler à qui de droit.

— Si je vous dénonce, vous me dénoncez. Nous serons quittes.

— Mais vous n'allez rien faire. Je vous connais assez pour le savoir. Je vous ai soigné. Vous êtes quelqu'un de bien. Vous vous êtes engagé par amour pour la patrie, mais vous détestez la guerre et vous détestez les nazis. » Elle était sûre à quatre-vingt-dix-neuf pour cent de ce qu'elle avançait.

« C'est très dangereux de parler ainsi.

— Je sais.

— Il ne s'agit donc pas d'une conversation ordinaire.

— En effet. Vous m'avez dit que des millions de gens vont mourir uniquement parce que les nazis sont trop fiers pour se rendre.

— J'ai dit cela ?

— Vous pouvez nous aider à sauver ces malheureux.

— Comment ? »

Carla hésita. C'était maintenant qu'elle mettait sa vie en danger. « Toute information dont vous disposez pourra être transmise par mes soins à des personnes intéressées. » Elle retint son souffle. Si elle s'était trompée sur Beck, elle était condamnée à mort.

Elle perçut de l'étonnement dans son regard. Il avait peine à imaginer que cette jeune infirmière efficace puisse être une espionne. Mais elle vit qu'il la croyait. « Je pense vous avoir comprise », dit-il.

Elle lui tendit une chemise en carton verte de l'hôpital, vide.

Il la prit. « C'est pour quoi faire ? demanda-t-il.

— Vous êtes un soldat. Vous connaissez la notion de camouflage. »

Il acquiesça. « Vous risquez votre vie », observa-t-il et elle lut dans ses yeux ce qui ressemblait à de l'admiration.

« Vous aussi, à présent.

— Oui, approuva le colonel Beck. Mais moi, j'ai l'habitude. »

2.

De bon matin, Thomas Macke emmena le jeune Werner Franck à la prison de Plötzensee, à Charlottenburg, la banlieue ouest de Berlin. « Il faut que vous voyiez ça, dit-il. Ensuite, vous pourrez dire au général Dorn que nous sommes vraiment efficaces. »

Il se gara sur le Königsdamm et conduisit Werner derrière le bâtiment principal. Ils entrèrent dans une pièce de huit mètres de long sur quatre de large, où les attendait un homme portant une queue-de-pie, un haut-de-forme et des gants blancs. Werner fronça les sourcils devant cette étrange tenue. « Je vous présente Herr Reichhart, dit Macke. Le bourreau. »

Werner déglutit. « Nous allons assister à une exécution ?

— Oui. »

D'un air faussement décontracté, Werner demanda : « Pourquoi cette tenue de soirée ? »

Macke haussa les épaules. « C'est la tradition. »

Un rideau noir divisait la pièce en deux. Macke le tira, révélant huit crochets de boucher fixés à une poutre métallique courant sur le plafond.

« Pour les pendaisons ? » demanda Werner.

Macke acquiesça.

Il y avait aussi une table en bois avec des sangles permettant d'y maintenir un individu. L'une de ses extrémités était occupée par un appareil en hauteur, d'une forme caractéristique. Un lourd panier était posé sur le sol.

Le jeune lieutenant était pâle. « Une guillotine.

— Exactement », confirma Macke. Il consulta sa montre. « Ils ne devraient plus tarder. »

Plusieurs hommes entrèrent alors. Certains saluèrent Macke comme une vieille connaissance. Se penchant vers Werner, le commissaire lui dit à l'oreille : « Le règlement exige la présence des juges, des huissiers, du directeur de la prison et de l'aumônier. »

Werner déglutit encore. Il n'aimait pas ça, Macke le voyait bien.

C'était le but recherché. Si Macke l'avait fait venir là, ce n'était pas pour impressionner le général Dorn. Werner l'inquiétait. Il y avait chez lui quelque chose qui sonnait faux.

Werner travaillait pour Dorn, c'était incontestable. Il avait accompagné Dorn lors d'une visite du quartier général de la Gestapo et, peu après, Dorn avait rédigé une note louant le travail du contre-espionnage berlinois et citant nommément Macke. Durant les semaines suivantes, ce dernier s'était pavané de façon écœurante.

Macke n'arrivait pourtant pas à oublier le comportement de Werner le soir où, près d'un an auparavant, ils avaient failli prendre un espion dans un atelier de fourrures désaffecté près de l'Ostbahnhof. Werner avait paniqué – du moins en apparence. Par accident ou à dessein, il avait alerté le pianiste qui avait eu le temps de prendre la fuite. Macke le soupçonnait d'avoir joué la comédie, d'avoir délibérément cherché à sauver leur proie.

Macke n'était pas assez sûr de lui pour arrêter et torturer Werner. Il aurait pu le faire, certes, mais Dorn risquait de mal le prendre et Macke aurait des ennuis. Son supérieur, le commissaire principal Kringelein, qui ne l'appréciait guère, lui demanderait de produire des preuves matérielles contre Werner. Or il n'en avait aucune.

Les minutes suivantes devraient cependant lui révéler la vérité.

La porte s'ouvrit à nouveau sur deux gardiens escortant une jeune femme. Reconnaissant Lili Markgraf, Werner ne put réprimer un hoquet d'effroi.

« Qu'y a-t-il ? demanda Macke.

— Vous ne m'aviez pas dit que le condamné était une jeune fille.

— Vous la connaissez ?

— Non. »

Lili avait vingt-deux ans, mais paraissait plus jeune. Le matin même, ses cheveux blonds avaient été coupés ras. Elle boitait et avançait le dos voûté, comme si elle souffrait de douleurs abdominales. Elle était vêtue d'une robe sans col taillée dans une grossière cotonnade bleue. Ses yeux étaient rougis par les larmes. Les gardiens la tenaient fermement, ne voulant courir aucun risque.

« Cette femme a été dénoncée par l'une de ses proches qui a trouvé un manuel de chiffrage caché dans sa chambre, expliqua Macke. Le code russe à cinq chiffres.

— Pourquoi marche-t-elle comme ça ?

— Les conséquences de l'interrogatoire. Mais nous n'en avons rien tiré. »

Le visage de Werner demeurait impassible. « Dommage, dit-il. Elle aurait pu nous mener à d'autres espions. »

S'il jouait la comédie, il était très fort, songea Macke. « Elle ne connaissait son complice que sous le nom d'Heinrich – patronyme inconnu –, et peut-être est-ce un pseudonyme. En général, il ne sert à rien d'arrêter des femmes – elles ne savent presque rien.

— Au moins, vous avez récupéré le manuel de chiffrage.

— Pour ce que ça vaut… Le mot clé est changé régulièrement, si bien que le décryptage des transmissions n'est jamais simple.

— Dommage. »

L'un des hommes s'éclaircit la gorge et prit la parole d'une voix de stentor. Il se présenta comme le président de la cour et lut la sentence de mort.

Les gardes escortèrent Lili jusqu'à la table. Ils lui donnèrent la possibilité de s'y allonger volontairement, mais comme elle reculait d'un pas, ils l'empoignèrent pour la coucher de force. Elle ne leur résista pas. Ils lui plaquèrent le visage contre le bois et l'attachèrent.

L'aumônier récita une prière.

Lili se mit à supplier. « Non, non, dit-elle sans élever la voix. Non, je vous en prie. Laissez-moi. » Elle s'exprimait de façon cohérente, comme si elle demandait simplement un service.

L'homme au chapeau haut de forme se tourna vers le président, qui secoua la tête et dit : « Pas encore. Attendez que la prière soit finie. »

La voix de Lili monta dans les aigus, se faisant implorante. « Je ne veux pas mourir ! J'ai peur ! Ne me faites pas ça, je vous en supplie ! »

Le bourreau se tourna à nouveau vers le président. Cette fois-ci, celui-ci l'ignora.

Macke observait Werner avec attention. Il semblait avoir le cœur au bord des lèvres, mais on pouvait en dire autant de toute l'assistance. Cela ne prouvait pas que Werner était un traître. Tout ce qu'on pouvait déduire de sa réaction, c'est qu'il était sensible. Macke allait devoir trouver un autre moyen de le percer à jour.

Lili se mit à hurler.

Macke lui-même commençait à s'impatienter.

Le pasteur s'empressa d'achever sa prière.

Lorsqu'il dit *Amen*, la jeune fille cessa de crier, comme si elle avait compris que tout était fini.

Le président fit un signe de tête.

Le bourreau actionna un levier et la lame s'abattit.

Elle émit comme un murmure en tranchant le cou pâle de Lili. Sa tête aux cheveux ras tomba dans une gerbe de sang. Elle atterrit dans le panier avec un bruit sourd qui sembla résonner dans toute la pièce.

C'était absurde, mais Macke se demanda si la tête avait souffert.

3.

Carla tomba sur le colonel Beck dans un couloir de l'hôpital. Il était en uniforme. Elle lui jeta un regard épouvanté. Depuis qu'il était sorti, elle vivait dans la terreur, imaginant qu'il l'avait dénoncée et que la Gestapo allait l'arrêter d'un jour à l'autre.

Mais il sourit et lui dit : « Le docteur Ernst m'a convoqué pour un examen de contrôle. »

C'était tout ? Avait-il oublié leur conversation ? Ou faisait-il semblant ? Une Mercedes noire de la Gestapo était-elle garée dans la rue ?

Beck tenait dans ses mains une chemise en carton verte.

Un cancérologue en blouse blanche passa près d'eux. Comme il s'éloignait, Carla demanda à Beck d'une voix enjouée : « Comment vous sentez-vous ?

— Aussi bien que possible. Plus jamais je ne conduirai de troupes au combat, mais je mènerai une vie normale à condition de renoncer à l'athlétisme.

— Je suis ravie de l'entendre. »

On allait et venait tout autour d'eux. Carla craignait que Beck ne trouve pas l'occasion de lui parler en privé.

Il demeurait pourtant impassible. « Je voulais juste vous remercier pour votre gentillesse et votre professionnalisme.

— Je vous en prie.

— Au revoir, mademoiselle.

— Au revoir, colonel. »

Lorsque Beck s'éloigna, c'était Carla qui tenait la chemise en carton.

Elle gagna d'un pas vif le vestiaire des infirmières. Il était désert. Elle resta derrière la porte en la bloquant du talon pour que personne ne puisse entrer.

La chemise contenait une enveloppe de papier beige bon marché, comme on en utilisait dans les bureaux du monde entier. Carla l'ouvrit. Il s'y trouvait plusieurs feuillets tapés à la machine. Elle examina le premier sans le sortir de l'enveloppe. Il portait le titre suivant :

Ordre de mission n° 6
Code Zitadelle

Le plan de bataille pour l'offensive d'été sur le front de l'Est. Son cœur battit plus vite. C'était une mine d'or.

Elle devait transmettre cette enveloppe à Frieda. Malheureusement, celle-ci ne se trouvait pas à l'hôpital ce jour-là : c'était sa journée de congé. Carla envisagea de déserter son poste pendant ses heures de service pour se rendre chez son amie, mais y renonça aussitôt. Mieux valait se conduire normalement, ne pas attirer l'attention.

Elle glissa l'enveloppe dans le sac à bandoulière accroché à sa patère. Elle recouvrit celui-ci de l'écharpe de soie bleu et or qui lui servait toujours à dissimuler les objets compromettants. Puis elle resta sans bouger quelques instants, le temps que son souffle reprenne un rythme normal. Elle retourna à son poste.

Elle effectua le reste de son service comme si de rien n'était puis enfila son manteau, sortit de l'hôpital et se dirigea vers la station de métro. En passant devant un immeuble bombardé, elle vit des graffitis sur les ruines. Un vaillant patriote avait écrit : « Nos murs peuvent se briser, nos cœurs jamais. » En guise de commentaire, quelqu'un d'autre avait cité le slogan d'Hitler au moment de l'élection de 1933 : « Donnez-moi quatre ans et vous ne reconnaîtrez plus l'Allemagne. »

Elle acheta un ticket pour Zoologischer Garten.

Dans la rame, elle se fit l'effet d'une étrangère. Au sein de tous ces Allemands loyaux, elle était celle qui détenait dans son sac des secrets qu'elle allait livrer à Moscou. C'était un sentiment très déplaisant. Personne ne lui prêtait attention, mais cela ne faisait que la persuader que tous évitaient son regard. Il lui tardait de confier l'enveloppe à Frieda.

La station Zoologischer Garten se trouvait à la lisière du Tiergarten. Les arbres semblaient tout petits par rapport à l'immense tour de défense antiaérienne. On avait édifié à Berlin trois blocs de béton comme celui-ci, hauts de plus de trente mètres. Aux quatre angles du toit étaient placés des canons de cent vingt-huit millimètres pesant vingt-cinq tonnes chacun. On avait peint le béton en vert dans une vaine tentative pour rendre cette monstruosité plus agréable à l'œil au milieu de ce parc.

En dépit de la laideur de ces installations, les Berlinois étaient enchantés. Quand les bombes tombaient, le tonnerre qui montait de ces tours les rassurait car ils savaient que, dans leur camp, quelqu'un répliquait.

Les nerfs toujours tendus, Carla se dirigea vers la maison de Frieda. Comme on était en milieu d'après-midi, ses parents seraient certainement sortis, Ludwig à l'usine et Monika chez une amie, peut-être la mère de Carla. La moto de Werner était garée dans l'allée.

Ce fut le majordome qui lui ouvrit. « Fräulein Frieda n'est pas là, mais elle ne devrait pas tarder, lui dit-il. Elle est allée au KaDeWe s'acheter des gants. Herr Werner est au lit avec un gros rhume.

— J'attendrai Frieda dans sa chambre, comme d'habitude. »

Carla se défit de son manteau et monta, son sac à la main. Une fois dans la chambre de Frieda, elle ôta ses souliers et s'allongea sur le lit pour lire le plan de bataille de l'opération Zitadelle. Elle se sentait tendue comme un ressort, mais cela irait mieux dès qu'elle serait débarrassée de ce brûlot.

Un bruit de sanglots lui parvint de la chambre voisine.

Elle tendit l'oreille, interloquée. C'était la chambre de Werner. Carla avait peine à imaginer le tombeur de ces dames en larmes.

C'était pourtant bien un homme qui pleurait, et il semblait avoir peine à contenir son chagrin.

Carla fut prise de pitié, bien malgré elle. Sans doute s'était-il fait plaquer par une femme moins aveugle que les autres. Mais elle ne put s'empêcher de s'émouvoir de cette manifestation de détresse.

Elle se leva, rangea le plan de bataille dans son sac et sortit dans le couloir.

Elle colla l'oreille à la porte de Werner. Les sanglots étaient encore plus distincts. Elle était trop sensible pour rester indifférente à son chagrin. Elle ouvrit la porte et entra.

Werner était assis sur son lit, la tête entre les mains. En entendant la porte s'ouvrir, il sursauta et leva les yeux. Ses joues empourprées étaient inondées de larmes. Il avait ouvert son col et dénoué sa cravate. Les yeux qu'il tourna vers Carla étaient rougis par le désespoir. Il était sincèrement bouleversé, bien trop effondré pour penser à sauvegarder les apparences.

« Qu'y a-t-il ? demanda Carla, pleine de compassion.

— Je n'ai plus la force de continuer », dit-il.

Elle referma la porte. « Que s'est-il passé ?

— Ils ont guillotiné Lili Markgraf – et j'ai été obligé d'assister à ça. »

Carla en resta bouche bée. « Qu'est-ce que tu racontes ?

— Elle n'avait que vingt-deux ans. » Il prit un mouchoir dans sa poche et s'essuya les yeux. « Tu es déjà en danger, mais si je te raconte ça, ce sera encore pire. »

Les spéculations les plus folles se bousculaient dans son esprit. « Je crois que j'ai déjà deviné, mais parle », lança-t-elle.

Il acquiesça. « De toute façon, tu aurais fini par le découvrir. Lili aidait Heinrich à transmettre des informations à Moscou. Ça va plus vite si quelqu'un lit les séries de chiffres à l'opérateur radio. Et plus ça va vite, moins on court le risque de se faire pincer. Mais la cousine de Lili a passé quelques jours chez elle et elle a découvert ses manuels de chiffrage. Salope de nazie. »

Ces paroles confirmaient les soupçons les plus fous de Carla. « Tu es au courant pour le réseau ? » demanda-t-elle.

Il lui adressa un sourire ironique. « C'est moi qui le dirige.

— Bon Dieu !

— C'est pour ça que j'ai cessé de m'intéresser à toutes ces histoires de meurtres d'enfants déficients. Ordre de Moscou. Ils avaient raison. Si j'avais perdu mon poste au ministère de l'Air, j'aurais cessé d'avoir accès à des dossiers confidentiels, ainsi qu'aux gens susceptibles de me livrer des secrets. »

Les jambes coupées, elle se percha au bord du lit à côté de lui. « Pourquoi ne m'as-tu rien dit ?

— Nous partons du principe que tout le monde parle sous la torture. Celui qui ne sait rien ne peut pas trahir les autres. Cette pauvre Lili a été torturée, mais elle ne connaissait que Volodia, qui est reparti à Moscou, ainsi qu'Heinrich, mais elle ignorait son nom de famille et ne savait rien sur lui. »

Carla était transie jusqu'à la moelle. *Tout le monde parle sous la torture.*

« Je regrette de t'avoir dit tout ça, mais après m'avoir vu dans cet état, tu aurais fini par comprendre, acheva Werner.

— Autrement dit, je t'ai vraiment mal jugé.

— Ce n'est pas de ta faute. Je t'ai délibérément induite en erreur.

— Je me sens quand même bête. Dire que ça fait deux ans que je te méprise.

— Alors que je brûlais du désir de tout te dire. »

Elle lui passa un bras autour des épaules.

Il lui prit la main et l'embrassa. « Peux-tu me pardonner ? »

Elle n'était pas très sûre de ses sentiments, mais ne voulant pas le repousser dans cet état, elle lui répondit : « Oui, bien sûr.

— Pauvre Lili, soupira-t-il dans un murmure. Ils l'avaient tellement martyrisée qu'elle pouvait à peine marcher jusqu'à la guillotine. Et jusqu'au bout, elle les a suppliés de l'épargner.

— Mais pourquoi étais-tu là ?

— Je me suis lié d'amitié avec un type de la Gestapo, le commissaire Thomas Macke. C'est lui qui m'a emmené.

— Macke ? Je ne l'ai pas oublié, c'est lui qui a arrêté mon père. » Elle avait gardé le souvenir vivace d'un homme au visage rond, barré d'une petite moustache noire ; en repensant à son arrogance quand il avait embarqué son père et au chagrin qu'elle avait éprouvé quand son Vati chéri avait succombé aux blessures que cet ignoble individu et ses acolytes lui avait infligées, une rage intacte l'envahit.

« J'ai l'impression qu'il me soupçonne et qu'il a voulu me mettre à l'épreuve. Peut-être espérait-il que je perdrais mon sang-froid et chercherais à empêcher l'exécution. Quoi qu'il en soit, je pense avoir réussi l'examen.

— Mais si tu te fais arrêter… »

Werner acquiesça. « Tout le monde parle sous la torture.

— Et tu sais tout.

— Je connais tous les agents, tous les codes… La seule chose que j'ignore, c'est le lieu depuis lequel ils émettent. Je leur laisse le soin de le choisir et ils évitent de me le communiquer. »

Ils restèrent main dans la main en silence. Après un long moment, Carla dit : « Je suis venue apporter quelque chose à Frieda, mais autant que je te le donne.

— Quoi donc ?

— Le plan de bataille pour l'opération Zitadelle. »

Werner fit un bond. « Mais ça fait des semaines que j'essaie de mettre la main dessus ! Comment te l'es-tu procuré ?

— Grâce à un officier de l'état-major. Peut-être ne devrais-je pas te dire son nom.

— Surtout pas. Mais ce document est-il authentique ?

— Tu ferais mieux d'y jeter un œil. » Elle retourna dans la chambre de Frieda et en revint avec l'enveloppe en papier beige. Il ne lui était pas venu à l'idée de douter de son authenticité. « Il m'a l'air correct, mais je n'y connais rien. »

Werner s'empara des feuillets dactylographiés. « C'est du solide. Fantastique ! s'écria-t-il au bout d'une minute.

— Je suis bien contente ! »

Il se leva. « Il faut que j'apporte ça à Heinrich sur-le-champ. Nous devons le coder et le transmettre cette nuit même. »

Carla fut déçue – sans savoir vraiment ce qu'elle en attendait – que ces instants d'intimité s'achèvent aussi vite. Elle suivit Werner dans le couloir. Après avoir récupéré son sac dans la chambre de Frieda, elle descendit au rez-de-chaussée.

La main sur la poignée de la porte d'entrée, Werner lui dit : « Ça me fait plaisir que nous soyons à nouveau amis.

— Moi aussi.

— Crois-tu que nous pourrons oublier cette période de brouille ? »

Elle ignorait où il voulait en venir. Souhaitait-il redevenir son amant – ou voulait-il lui faire comprendre qu'il n'en était pas question ? « Il me semble que nous pouvons tourner la page, dit-elle sans trop s'engager.

— Bien. » Il lui déposa un bref baiser sur les lèvres. Puis il ouvrit la porte.

Ils sortirent ensemble et il enfourcha sa moto.

Carla gagna la rue et se dirigea vers la station de métro. L'instant d'après, Werner la dépassa et la salua d'un coup de klaxon et d'un geste de la main.

Restée seule, elle avait enfin le loisir de réfléchir à cette révélation. Que ressentait-elle exactement ? Elle l'avait détesté deux ans durant. Mais, au cours de cette période, jamais elle n'avait eu de petit ami. Était-elle toujours amoureuse de lui ? À tout le moins, elle ne l'avait pas entièrement banni de ses pensées. Tout à l'heure, devant la détresse qui l'habitait, elle avait senti se dissiper l'hostilité qu'il lui inspirait. Désormais, l'affection qu'elle lui portait lui réchauffait le cœur.

L'aimait-elle encore ?

Elle n'en savait rien.

4.

Macke avait pris place aux côtés de Werner sur la banquette arrière de la Mercedes noire. Il avait au cou un sac ressemblant à un cartable d'écolier, à cette différence près qu'il le portait devant et non sur son dos. Il était suffisamment petit pour être dissimulé sous son manteau boutonné. Un mince fil le reliait à un minuscule écouteur. « C'est le dernier cri, expliqua Macke. Plus on s'approche de l'émetteur, plus le volume augmente.

— C'est plus discret qu'une fourgonnette avec une antenne sur le toit, approuva Werner.

— Les deux nous sont utiles – la fourgonnette pour repérer la zone d'émission, cet appareil pour localiser l'émetteur avec précision. »

Macke était dans ses petits souliers. L'opération Zitadelle avait tourné à la catastrophe. Bien avant le début de l'offensive, l'armée Rouge avait attaqué les aérodromes où étaient rassemblées les forces de la Luftwaffe. Zitadelle avait été annulée au bout d'une semaine, mais l'armée allemande avait subi des pertes irréparables.

Quand les choses tournaient mal, les dirigeants allemands étaient prompts à blâmer les conspirateurs judéo-bolcheviques, mais cette fois, ils avaient raison. Selon toute apparence, l'armée Rouge avait eu connaissance à l'avance de l'ensemble du plan. Et, à en croire le commissaire principal Kringelein, Thomas Macke était le seul responsable. C'était lui le chef du contre-espionnage berlinois. Sa carrière était en jeu. Il risquait son poste – sinon pire.

Son seul espoir était de réussir un coup d'éclat, de capturer l'ensemble des espions qui sabotaient l'effort de guerre allemand. Ce soir-là, il avait tendu un piège à Werner Franck.

Si celui-ci était innocent, il n'avait pas de plan de rechange.

Un talkie-walkie posé à l'avant de la voiture se mit à grésiller. Le cœur de Macke battit plus fort. Le chauffeur saisit l'appareil. « Ici Wagner. » Il démarra. « Nous arrivons, dit-il. Terminé. »

Les dés étaient jetés.

« Où allons-nous ? demanda Macke.

— À Kreuzberg. » C'était un quartier populaire situé au sud du centre-ville.

Comme ils se mettaient en route, une sirène annonça un raid aérien.

Ce n'est vraiment pas le moment, se dit Macke en jetant un coup d'œil au-dehors. Les projecteurs s'allumèrent, oscillant comme des lances de lumière. Sans doute leur arrivait-il parfois de repérer des avions, mais Macke n'avait jamais vu ce cas de figure. Lorsque les sirènes cessèrent de hurler, il entendit le rugissement des bombardiers qui approchaient. Au début de la guerre, les escadrilles britanniques se composaient de quelques dizaines d'avions – ce qui était déjà beaucoup –, mais ils en envoyaient désormais plusieurs centaines à la fois. Le bruit était terrifiant avant même qu'ils ne larguent leurs bombes.

« Il vaudrait mieux annuler la mission, suggéra Werner.

— Vous voulez rire ? » fit Macke.

Le rugissement des avions s'accentua.

Fusées éclairantes et bombes incendiaires se mirent à pleuvoir alors que la voiture approchait de Kreuzberg. Ce quartier était une cible idéale pour la RAF, dont la stratégie actuelle consistait à tuer le plus de civils possible, des ouvriers surtout. Avec une hypocrisie éhontée, Churchill et Attlee prétendaient n'attaquer que des objectifs militaires et regretter les pertes civiles occasionnées par leurs raids. Les Berlinois n'étaient pas dupes.

Wagner négociait à vive allure les rues éclairées par intermittence par les flammes. On n'y voyait personne excepté le personnel de la défense antiaérienne : tous les citoyens étaient tenus de gagner les abris. Les seuls véhicules qu'ils croisaient étaient des ambulances, des camions de pompiers et des voitures de police.

Macke observait discrètement Werner. Le jeune homme était tendu, ne tenait pas en place, jetait au-dehors des regards anxieux, tapait du pied sans s'en rendre compte.

Macke n'avait confié ses soupçons à personne hormis à ses subordonnés. Il aurait eu du mal à justifier le fait d'avoir révélé le fonctionnement des opérations de la Gestapo à un homme qu'il soupçonnait d'espionnage. Cela aurait pu le conduire dans sa propre salle de torture. Il ne pouvait agir qu'à coup sûr. Sa seule planche de salut était de livrer un espion à ses supérieurs.

Si ses soupçons étaient fondés, en revanche, il ne se contenterait pas d'arrêter Werner mais embarquerait aussi sa famille et ses amis, ce qui lui permettrait d'annoncer l'élimination de tout un réseau. Voilà qui changerait la donne. Peut-être même aurait-il droit à une promotion.

Les bombes changèrent de nature à mesure que le raid progressait, et Macke entendit le bruit sourd et profond des explosifs. Une fois la cible éclairée, la RAF aimait lâcher un mélange de bombes incendiaires de gros calibre qui faisaient partir des brasiers que des explosifs attisaient et qui gênaient les services de secours. C'était une tactique cruelle, mais Macke savait que la Luftwaffe en faisait autant.

Le son s'amplifia dans l'écouteur de Macke lorsqu'ils s'engagèrent précautionneusement dans une rue bordée d'immeubles de cinq étages. Le quartier était soumis à un terrible pilonnage et plusieurs bâtiments s'étaient déjà effondrés. « Nous sommes en plein dans leur ligne de mire, bon sang ! » s'exclama Werner.

Macke s'en fichait. Bombardement ou pas, sa vie était en jeu. « Tant mieux, se réjouit-il. Le pianiste s'imaginera qu'il n'a rien à craindre de la Gestapo pendant un raid aérien. »

Wagner s'arrêta près d'une église en flammes et désigna une rue latérale. « C'est ici », dit-il.

Macke et Werner descendirent de voiture.

Macke s'avança d'un pas vif, Werner à ses côtés et Wagner sur leurs talons. « Vous êtes sûr que c'est un espion ? lui demanda Werner. Et si c'était autre chose ?

— Qui d'autre lancerait des signaux radio ? » répliqua Macke.

Il entendait toujours du bruit dans son écouteur, mais faiblement, à cause de la cacophonie due au raid : les avions, les bombes, les canons antiaériens, les bâtiments qui s'effondraient et le rugissement des flammes.

Ils passèrent devant des écuries où des chevaux hennissaient de terreur, et le signal s'accentua. Werner jetait autour de lui des regards de plus en plus anxieux. Si c'était bien un espion, il devait redouter qu'un de ses collaborateurs ne se fasse prendre par la Gestapo – et se demander comment l'éviter. Tenterait-il la même ruse que la dernière fois, ou trouverait-il un nouveau moyen de donner l'alerte ? Si ce n'était pas un espion, pensa Macke, toute cette mascarade n'était qu'une perte de temps.

Macke saisit son écouteur et le tendit à Werner. « Tenez, écoutez », dit-il sans cesser d'avancer.

Werner opina. « Le signal est de plus en plus net. » Une lueur affolée traversa son regard. Il rendit l'écouteur à Macke.

Cette fois, je te tiens, se dit le commissaire, triomphant.

Un fracas retentissant ébranla l'air au moment où une bombe s'abattait sur un immeuble tout proche. Ils se retournèrent et virent les flammes dévorer une boulangerie. « Bon Dieu, ce n'est pas passé loin », s'exclama Wagner.

Ils s'arrêtèrent devant une école, un bâtiment de brique qui s'élevait dans une cour asphaltée. « C'est ici, je crois », dit Macke.

Les trois hommes montèrent une volée de marches donnant sur une entrée latérale. La porte était ouverte. Ils la poussèrent.

Ils se trouvaient à l'extrémité d'un couloir. À l'autre bout, une porte donnant sans doute sur une salle de classe. « Allons-y », s'écria Macke.

Il dégaina son pistolet, un Luger 9 mm.

Werner n'était pas armé.

Ils entendirent un coup de tonnerre, puis un grondement sourd suivi du rugissement d'une explosion très proche. Toutes les vitres du couloir se brisèrent et des éclats de verre s'éparpillèrent sur le sol carrelé. Une bombe avait dû toucher la cour de récréation.

« Dégagez, tout le monde ! hurla Werner. Le bâtiment va s'effondrer. »

Il n'y avait aucun danger qu'il s'écroule, remarqua Macke. Werner ne cherchait qu'à alerter le pianiste.

Werner se mit à courir, mais, au lieu de fuir par où ils étaient venus, il se précipita vers l'autre bout du couloir.

Pour alerter ses amis, se dit Macke.

Wagner dégaina son arme, mais Macke cria : « Non ! Ne tire pas ! »

Arrivé au bout du corridor, Werner ouvrit la porte donnant sur une salle. « Barrez-vous tous ! » hurla-t-il. Puis il se figea et se tut.

Devant lui, Mann, l'opérateur radio de Macke, tapait des signaux sans queue ni tête sur son émetteur.

Il était flanqué de Schneider et de Richter, l'arme au poing.

Macke arbora un sourire triomphant. Werner était tombé dans le panneau.

Wagner avança jusqu'au jeune homme et lui colla le canon de son arme sur la tempe.

« Tu es en état d'arrestation, ordure bolchevique », lui lança Macke.

Werner réagit au quart de tour. Il releva la tête pour se dégager, empoigna le bras de Wagner et l'entraîna dans la salle. L'espace d'un instant, il s'en fit un bouclier pour se protéger des pistolets de Schneider et Richter. Puis il se débarrassa de lui sans ménagement, le repoussant brutalement au fond de la salle. Wagner trébucha et tomba, entraînant les autres dans sa chute. Et, avant que personne n'ait eu le temps de réagir, Werner attrapa une chaise et ressortit de la salle dont il claqua la porte avant d'en bloquer la poignée avec le dossier de la chaise. Macke et Werner se retrouvèrent seuls, face à face, dans le couloir. Werner s'avança.

Macke pointa son Luger sur lui. « Arrête ou je tire.

— Ça m'étonnerait. » Werner continua de marcher sur lui. « Il faudra bien que vous m'interrogiez si vous voulez savoir qui sont mes complices. »

Macke visa les jambes de Werner. « Je peux t'interroger avec une balle dans le genou », rétorqua-t-il, et il tira.

Il manqua son coup.

Werner bondit sur lui, lutta et lui fit lâcher son arme qui alla glisser plus loin. Comme Macke se baissait pour la ramasser, Werner prit ses jambes à son cou.

Macke récupéra son pistolet.

Werner arriva devant la porte. Macke visa à nouveau ses jambes et tira.

Ses trois premières balles manquèrent leur cible et Werner franchit la porte.

Macke tira une quatrième fois. Werner poussa un cri et s'effondra.

Macke se mit à courir dans le couloir. Derrière lui, il entendit les autres qui enfonçaient la porte de la salle dans laquelle ils étaient enfermés.

Puis le toit s'effondra dans un bruit de tonnerre, et une averse de feu s'abattit sur lui. Macke poussa un cri de terreur, qui se transforma en cri de souffrance lorsque ses vêtements s'embrasèrent. Il tomba à terre, et ce fut le silence, puis les ténèbres.

5.

Les médecins triaient les patients dans le hall de l'hôpital. Les blessés légers étaient dirigés vers la salle d'attente, où les infirmières débutantes les pansaient et leur administraient de l'aspirine. Les cas les plus graves avaient droit à un traitement d'urgence sur place, avant d'être orientés vers les spécialistes aux étages. Quant aux morts, on les évacuait dans la cour où on les allongeait sur le pavé en attendant que leurs proches viennent les chercher.

Le docteur Ernst examina un brûlé qui hurlait, auquel il prescrivit de la morphine. « Déshabillez-le et mettez de la pommade sur ses brûlures », ordonna-t-il, et il passa au suivant.

Carla prépara une seringue tandis que Frieda découpait les vêtements calcinés du blessé. Il était grièvement brûlé sur le flanc droit, mais le gauche était quasiment indemne. Carla localisa une zone de peau intacte sur sa cuisse gauche. Elle allait lui administrer une piqûre lorsqu'elle examina son visage et se figea.

Elle connaissait cette tête ronde, cette moustache pareille à une tache de saleté sous le nez. Deux ans plus tôt, cet homme avait fait irruption chez eux pour arrêter son père qu'elle n'avait revu qu'à l'agonie. Ce grand brûlé était le commissaire Thomas Macke, de la Gestapo.

Tu as tué mon père, songea-t-elle.

Et à présent, ta vie est entre mes mains.

Rien de plus simple. Quatre fois la dose maximale de morphine, et on n'en parlerait plus. Personne ne remarquerait rien, surtout par une nuit comme celle-ci. Il sombrerait aussitôt dans l'inconscience et mourrait au bout de quelques minutes. Un médecin épuisé attribuerait son décès à une défaillance cardiaque. Personne ne contesterait son diagnostic, personne ne poserait de question embarrassante. Il ferait partie des milliers de victimes de ce raid aérien. *Requiescat in pace.*

Werner craignait que Macke ne l'ait démasqué, elle le savait. Il risquait d'être arrêté d'un jour à l'autre. *Tout le monde parle sous la torture.* Werner leur livrerait Frieda, Heinrich et les autres – elle ferait partie du lot. Elle avait la possibilité de les sauver tous.

Elle hésitait.

Elle se demanda pourquoi. Macke était un assassin et un tortionnaire. Il méritait mille fois la mort.

Carla avait déjà tué Joachim, ou du moins contribué à le tuer. Mais cela s'était passé au cours d'une lutte. Joachim était en train de frapper brutalement sa mère lorsqu'elle lui avait fracassé le crâne avec une marmite. Cette fois, ce n'était pas la même chose : Macke était un patient.

Si Carla n'était pas croyante, certaines choses n'en étaient pas moins sacrées à ses yeux. Elle était infirmière et les patients lui faisaient confiance. Macke n'hésiterait pas à la torturer et à la tuer, bien sûr… Mais elle n'était pas comme lui. Ce n'était pas lui qui était en cause, c'était elle.

Si elle tuait un patient, elle devrait renoncer au métier d'infirmière, renoncer à soigner les autres. Elle serait comme un banquier qui vole les épargnants, un homme politique qui accepte des pots-de-vin, un prêtre pervertissant des premières communiantes. Elle se trahirait elle-même.

« Qu'est-ce que tu attends ? lui demanda Frieda. Je ne peux rien faire tant qu'il n'est pas anesthésié. »

Carla plongea la seringue dans le bras de Thomas Macke, qui cessa aussitôt de hurler.

Frieda entreprit d'enduire son épiderme brûlé de pommade.

« Celui-ci souffre d'une simple commotion, disait le docteur Ernst à propos d'un autre patient. Mais il a reçu une balle dans le derrière. » Il leva la voix pour s'adresser au patient. « Qui vous a tiré dessus ? La RAF nous balance de tout, sauf des balles. »

Carla se retourna pour voir. Le patient était allongé sur le ventre. On avait coupé son pantalon, mettant ses fesses à nu. Il avait la peau blanche et le creux des reins couvert d'un duvet blond. À moitié dans le brouillard, le blessé réussit néanmoins à bredouiller quelques mots.

« Un policier vous a tiré dessus par erreur, c'est ça ? » demanda Ernst.

Le patient répondit distinctement : « Oui.

— Je vais vous extraire la balle. Ça va faire mal, mais nous sommes à court de morphine et il y a des gens plus gravement atteints que vous.

— Allez-y. »

Carla nettoya la plaie. Ernst sélectionna une longue paire de pinces. « Mordez votre oreiller », conseilla-t-il.

Il inséra les pinces dans la plaie. Le patient poussa un cri de douleur étouffé.

« Essayez de ne pas contracter vos muscles, dit le docteur Ernst. Vous ne faites qu'aggraver les choses. »

Quelle remarque stupide, se dit Carla. Comment ne pas être crispé quand on triturait votre plaie ?

« Ah, merde ! rugit le patient.

— Je la tiens, annonça le docteur Ernst. Restez tranquille ! »

Le patient obtempéra, et Ernst réussit à extraire la balle, qu'il jeta dans une cuvette.

Carla nettoya la plaie et y appliqua un pansement.

Le patient se retourna.

« Non, non, lui dit-elle. Vous devez rester allongé sur le… »

Elle se tut. C'était Werner.

« Carla ? fit-il.

— C'est moi, dit-elle, aux anges. En train de te panser les fesses.

— Je t'aime. »

Elle l'étreignit d'une façon qui n'avait vraiment rien de professionnel et s'écria : « Oh ! mon chéri. Je t'aime, moi aussi. »

6.

Thomas Macke revint lentement à lui. Il resta d'abord prisonnier de ses rêves. Puis reprenant progressivement conscience, il saisit qu'il était à l'hôpital et sous calmants. Il comprit pourquoi : sa peau était douloureuse, notamment son flanc droit. Les médicaments atténuaient la sensation de brûlure sans l'éliminer entièrement.

Peu à peu, il se rappela comment il était arrivé là. Le bombardement. Il aurait sûrement péri s'il n'avait pas tenté de fuir le bâtiment pour rattraper sa proie. Tous ses collaborateurs devaient être morts. Mann, Schneider, Richter et le jeune Wagner. Son équipe au complet.

Mais il avait pincé Werner.

L'avait-il vraiment pincé ? Il lui avait tiré dessus et Werner était tombé, et puis la bombe avait explosé. Macke avait survécu, alors peut-être Werner s'en était-il tiré, lui aussi.

Macke était désormais le seul à savoir que Werner était un espion. Il devait parler immédiatement à son supérieur, le commissaire principal Kringelein. Il tenta de se redresser, mais s'aperçut qu'il n'en avait pas la force. Il décida d'appeler l'infirmière avant de constater qu'il en était incapable. Ce simple effort l'épuisa et il sombra dans le sommeil.

Lorsqu'il se réveilla, il devina que la nuit était tombée. Tout était calme autour de lui, personne ne bougeait. Il ouvrit les yeux et découvrit un visage penché sur lui.

Werner.

« Vous allez quitter cet endroit maintenant », déclara Werner.

Macke voulut appeler à l'aide mais aucun son ne sortit de sa gorge.

« Vous allez partir ailleurs, reprit Werner. Vous ne serez plus tortionnaire – là-bas, c'est vous qui serez torturé. »

Macke ouvrit la bouche pour hurler.

Il vit un oreiller descendre sur lui, se plaquer sur son visage et sur son nez. Il ne pouvait plus respirer. Il tenta de se débattre, mais ses membres étaient sans force. Il essaya vainement de reprendre son souffle. La panique le gagna. Il réussit à agiter la tête de droite à gauche, mais l'oreiller se plaqua plus fermement encore sur son visage. Lorsque enfin il réussit à émettre un son, ce ne fut qu'un pitoyable bruit de gorge.

L'univers devint un disque de lumière qui se réduisit peu à peu à un point.

Puis ce fut le noir.

XVII

1943 (III)

1.

« Veux-tu m'épouser ? demanda Volodia Pechkov en retenant son souffle.

— Non, répondit Zoïa Vorotsintseva. Mais merci quand même. »

Elle était souvent incroyablement directe dans ses rapports avec les gens mais là, même de sa part, la réponse était un peu abrupte.

Ils étaient allongés après l'amour dans une chambre du luxueux hôtel Moskva. Zoïa avait joui deux fois. Sa pratique préférée était le cunnilingus. Elle n'aimait rien tant que se laisser aller sur une pile de coussins tandis qu'il s'agenouillait entre ses cuisses tel un adorateur. Il ne se faisait jamais prier et elle lui rendait la pareille avec enthousiasme.

Ils étaient ensemble depuis plus d'un an et tout semblait aller à merveille. Son refus le déconcerta.

« Est-ce que tu m'aimes ? demanda-t-il.

— Oui. Je t'adore. Et je te remercie de m'aimer au point de me proposer le mariage. »

Voilà qui était mieux. « Alors pourquoi ne veux-tu pas m'épouser ?

— Je ne veux pas avoir d'enfants dans un monde en guerre.

— D'accord, je comprends ça.

— Repose-moi la question quand nous aurons gagné.

— Mais peut-être que je ne voudrai plus t'épouser à ce moment-là.

— Si tu es aussi inconstant, j'ai bien fait de refuser aujourd'hui.

— Pardon. J'avais oublié que tu ne comprenais pas la taquinerie.

— Il faut que j'aille faire pipi. » Elle se leva et traversa la chambre toute nue. Volodia avait du mal à croire à sa chance. Elle avait le corps d'un mannequin ou d'une vedette de cinéma. Sa peau était blanche comme le lait et ses cheveux – ainsi que sa toison – blonds comme les blés. Elle s'assit sur le siège sans fermer la porte de la salle de bains et il l'écouta. Son absence de pudeur était un délice sans cesse renouvelé.

Il était censé être en train de travailler.

Les milieux du Renseignement moscovites étaient complètement désorganisés à chaque visite des dirigeants alliés et la conférence des ministres des Affaires étrangères, inaugurée le 18 octobre, avait bouleversé la routine de Volodia.

L'Union soviétique recevait Cordell Hull, le secrétaire d'État américain, et Anthony Eden, son homologue britannique. Ils avaient en vue un projet d'alliance quadripartite insensé incluant la Chine. Staline, qui jugeait cette idée ridicule, ne comprenait pas pourquoi on lui faisait perdre son temps. Hull, l'Américain, était âgé de soixante-douze ans et crachait du sang – son médecin personnel l'avait suivi à Moscou –, mais il n'en était pas moins déterminé à faire aboutir ce projet.

Il y avait tant à faire pendant la conférence que le NKVD était obligé de coopérer avec ses rivaux des services de renseignement de l'armée Rouge, pour lesquels travaillait Volodia. Il fallait dissimuler des micros dans les hôtels – il y en avait un dans cette chambre, mais Volodia l'avait débranché. Il fallait placer les ministres et leurs assistants sous une surveillance constante. Il fallait ouvrir et fouiller clandestinement leurs bagages. Il fallait enregistrer, transcrire, traduire en russe, lire et résumer leurs conversations téléphoniques. La plupart des gens qu'ils croisaient, serveurs et femmes de chambre compris, étaient des agents du NKVD, mais tous ceux à qui ils adressaient la parole, quels qu'ils soient, dans l'hôtel ou au-dehors, devaient faire l'objet d'un contrôle, qui risquait de déboucher sur une arrestation et un interrogatoire pouvant aller jusqu'à la torture. Tout cela représentait beaucoup de travail.

Volodia était sur un petit nuage. Ses espions berlinois lui fournissaient de précieux renseignements. Ils lui avaient communiqué le plan de bataille de l'opération Zitadelle, la plus grande offensive allemande de l'été, et l'armée Rouge avait infligé à l'ennemi une cuisante défaite.

Zoïa était heureuse, elle aussi. L'Union soviétique avait relancé ses recherches nucléaires et Zoïa appartenait à l'équipe chargée de concevoir une bombe atomique. Ils avaient pris du retard sur l'Ouest à cause du scepticisme de Staline, mais, en compensation, ils recevaient une aide inestimable des espions communistes opérant en Angleterre et en Amérique, parmi lesquels Willi Frunze, le vieux camarade de classe de Volodia.

Elle revint au lit. « Lorsqu'on s'est rencontrés, dit Volodia, j'ai eu l'impression que tu ne m'appréciais pas beaucoup.

— Je n'aimais pas les hommes, répondit-elle. Je ne les aime toujours pas. La plupart d'entre eux sont des ivrognes, des brutes et des imbéciles. Il m'a fallu quelque temps pour comprendre que tu étais différent.

— Merci. Mais tu trouves vraiment les hommes aussi affreux que ça ?

— Regarde autour de toi. Regarde notre pays. »

Il tendit le bras pour allumer le poste de radio posé sur la table de chevet. Bien qu'il ait débranché le micro dissimulé derrière la tête de lit, on n'était jamais trop prudent. Lorsque la radio eut chauffé, elle se mit à diffuser une marche militaire. Assuré de ne pas être entendu, Volodia reprit : « Tu penses à Staline et à Beria. Mais ils ne sont pas éternels.

— Sais-tu comment mon père est mort ? demanda-t-elle.

— Non. Mes parents ne me l'ont jamais dit.

— Il y a une bonne raison à cela.

— Je t'écoute.

— À en croire ma mère, des élections ont été organisées à l'usine de mon père afin d'envoyer un député au soviet de Moscou. Un candidat menchevique s'est présenté contre le bolchevik, il a organisé une réunion et mon père est allé l'écouter. Il ne soutenait pas ce menchevik, il n'a pas voté pour lui ; mais tous ceux qui ont assisté à cette réunion ont été licenciés et, quelques semaines plus tard, mon père était arrêté et conduit à la Loubianka. »

Elle parlait du quartier général de la police secrète, place Loubianka.

« Ma mère est allée voir ton père et l'a supplié d'intervenir, poursuivit-elle. Il l'a aussitôt accompagnée à la Loubianka. Ils sont arrivés juste à temps ; tous les hommes arrêtés avec mon père ont été fusillés, mais ton père a pu sauver le mien in extremis. Grigori l'a cru tiré d'affaire, mais ils ont exécuté mon père en douce quelques jours plus tard.

— C'est horrible, murmura Volodia. Mais c'est Staline qui…

— Non. Ça s'est passé en 1920. À l'époque, Staline était commandant de l'armée Rouge et participait à la guerre soviéto-polonaise. Le dirigeant, c'était Lénine.

— C'est arrivé du temps de Lénine ?

— Oui. Tu vois, il n'y a pas que Staline et Beria. »

Cela ébranlait sérieusement la vision qu'avait Volodia de l'histoire du communisme. « Mais pourquoi ? »

La porte s'ouvrit.

Volodia tendit le bras pour prendre le pistolet caché dans le tiroir de la table de chevet.

Ce n'était qu'une jeune fille vêtue d'un manteau de fourrure, sans rien dessous apparemment.

« Pardon, Volodia, dit-elle. Je ne savais pas que tu avais de la compagnie.

— Qui diable est cette fille ? lança Zoïa.

— Natacha, comment as-tu fait pour ouvrir cette porte ?

— Tu m'as donné un passe. Il ouvre toutes les portes de l'hôtel.

— Tu aurais pu frapper, quand même !

— Désolée. Je venais t'annoncer une mauvaise nouvelle.

— Hein ?

— Je suis allée dans la chambre de Woody Dewar, comme tu me l'avais ordonné. Mais ça n'a pas marché.

— Qu'est-ce que tu as fait ?

— Ça. » Natacha ouvrit son manteau révélant son corps nu. Elle avait des formes voluptueuses et une toison pubienne noire et luxuriante.

« C'est bon, je vois le tableau, lança Volodia. Qu'est-ce qu'il a dit ? »

Elle lui répondit en anglais. « Il a simplement dit :"Non." J'ai demandé : "Comment ça, non ?" Et il m'a répondu : "Le contraire de oui." Puis il a ouvert la porte et attendu que je sois sortie.

— Merde, fit Volodia. Il va falloir que je trouve autre chose. »

2.

Chuck Dewar comprit qu'il allait y avoir du grabuge lorsque le capitaine Vandermeier débarqua dans la section Territoire ennemi en milieu d'après-midi, le visage empourpré par un déjeuner bien arrosé à la bière.

Le service du Renseignement militaire de Pearl Harbor avait pris de l'ampleur. Jadis baptisé station HYPO, il avait reçu l'appellation ronflante de Joint Intelligence Center, Pacific Ocean Area (JICPOA), Centre mixte du renseignement pour la zone de l'océan Pacifique.

Vandermeier était accompagné d'un sergent des marines. « Hé ! les deux tantes, lança le capitaine. Je vous amène un client mécontent. »

Depuis le développement des opérations, tout le monde s'était spécialisé dans un domaine précis, et Chuck et Eddie étaient chargés de cartographier les zones de débarquement des forces américaines envoyées reprendre une par une les îles du Pacifique.

« Voici le sergent Donegan », reprit Vandermeier. Le marine était un homme de haute taille, raide comme un fusil. Connaissant les penchants sexuels un peu troubles de Vandermeier, Chuck se demanda s'il avait le béguin pour lui.

Il se leva. « Enchanté de faire votre connaissance, sergent. Je suis le premier maître Dewar. »

Chuck et Eddie étaient tous deux montés en grade. À mesure que des milliers de conscrits rejoignaient l'armée américaine, les officiers commençaient à manquer, et les soldats qui s'étaient engagés avant le conflit et savaient se débrouiller gravissaient rapidement les échelons. Chuck et Eddie n'étaient plus tenus de loger sur la base. Ils avaient loué un petit appartement.

Chuck tendit la main mais Donegan ne la serra pas.

Il se rassit. Son grade faisait de lui le supérieur hiérarchique de ce sergent et il pouvait lui aussi se montrer grossier. « Que puis-je pour vous, capitaine Vandermeier ? »

Un capitaine de marine ne manquait pas de moyens pour mener la vie dure à ses subalternes, et Vandermeier les connaissait tous. Il modifiait le tableau de service afin que Chuck et Eddie ne soient jamais en congé en même temps. Il notait leurs rapports « passable », sachant parfaitement que toute appréciation inférieure à « excellent » équivalait à un blâme. Il envoyait des messages contradictoires au chef comptable afin que la solde de Chuck et d'Eddie soit versée en retard, voire diminuée, ce qui les obligeait à perdre des heures à régulariser leur situation. Bref, c'était un enquiquineur patenté. Quel mauvais tour avait-il encore imaginé ?

Donegan sortit de sa poche une feuille de papier fort abîmée et la déplia. « C'est votre boulot, ça ? » demanda-t-il d'un air agressif.

Chuck prit le papier. C'était une carte de la Nouvelle-Géorgie, une des îles Salomon. « Faites-moi voir », dit-il. Il s'agissait bien de son œuvre, il le savait, mais il cherchait à gagner du temps.

Il s'approcha d'un classeur et ouvrit un tiroir. Il en sortit le dossier de la Nouvelle-Géorgie puis referma le tiroir d'un coup de genou. De retour à son bureau, il s'assit et ouvrit la chemise en carton. Elle contenait un double de la carte de Donegan. « Oui, confirma Chuck. C'est mon boulot.

— Eh bien, je suis venu vous dire que c'est de la merde, déclara Donegan.

— Ah bon ?

— Regardez ici. D'après vous, la jungle s'étend jusqu'à la mer. En réalité, il y a une plage large de quatre cents mètres.

— Je suis navré de l'apprendre.

— Navré ! » Donegan avait bu à peu près autant de bière que Vandermeier et il cherchait la bagarre. « Cinquante de mes gars sont morts sur cette plage. »

Vandermeier rota et demanda : « Comment avez-vous pu commettre une aussi grossière erreur, Dewar ? »

Chuck était atterré. S'il était effectivement responsable d'une erreur qui avait coûté la vie à cinquante hommes, il méritait qu'on lui passe un sacré savon. « On n'avait que ça pour travailler », se justifia-t-il. Le dossier contenait une carte approximative de l'archipel, datant sans doute de l'époque victorienne, et une carte nautique plus récente qui détaillait les profondeurs au large de l'île sans révéler grand-chose de sa topographie. Ils ne disposaient ni de description de première main, ni de messages japonais décryptés. Pour compléter le dossier, une photographie noir et blanc un peu floue obtenue par les services de reconnaissance aérienne. Posant l'index sur celle-ci, Chuck déclara : « On dirait pourtant que les arbres vont jusqu'à la ligne de haute mer. Y a-t-il une marée ? Dans le cas contraire, il n'est pas impossible que le sable ait été recouvert d'algues le jour où on a pris cette photo. Les algues, ça apparaît et disparaît en un rien de temps.

— Vous feriez preuve de moins de désinvolture si vous deviez vous battre sur des îles comme celle-ci. »

C'était sans doute vrai, se dit Chuck. Donegan était grossier et agressif, et cette ordure de Vandermeier lui avait bien monté la tête, mais ça ne voulait pas dire qu'il avait tort.

« Ouais, Dewar, renchérit Vandermeier. Vous deux, les tapettes, vous feriez peut-être bien d'accompagner les marines, lors du prochain assaut. Pour voir ce que valent vos cartes sur le terrain. »

Chuck cherchait une repartie cinglante lorsqu'il lui vint l'idée de prendre cette suggestion au sérieux. Peut-être devrait-il effectivement aller au feu. Il était facile d'être blasé derrière un bureau. Les griefs de Donegan méritaient d'être pris en considération.

D'un autre côté, il allait risquer sa vie.

Chuck regarda Vandermeier droit dans les yeux. « Ça me semble une bonne idée, capitaine, acquiesça-t-il. J'aimerais me porter volontaire pour cette mission. »

Donegan parut surpris, comme s'il commençait à comprendre qu'il avait mal jugé la situation.

Eddie prit la parole à son tour. « Moi aussi. Je suis prêt.

— Parfait, dit Vandermeier. Vous apprendrez sûrement des choses qui vous seront utiles à votre retour… Si vous revenez. »

3.

Volodia n'arrivait pas à saouler Woody Dewar.

Au bar de l'hôtel Moskva, il posa un verre de vodka devant le jeune Américain et lui dit dans un anglais scolaire : « Vous allez adorer celle-ci : c'est la meilleure de toutes.

— Je vous remercie infiniment, répondit Woody. C'est très aimable de votre part. » Et il ne toucha pas au verre.

Woody était grand, dégingandé et semblait d'une franchise confinant à la naïveté, raison pour laquelle Volodia l'avait pris pour cible.

S'exprimant par le truchement de son interprète, Woody lui demanda : « Pechkov est-il un patronyme courant en Russie ?

— Pas tellement, répondit Volodia en russe.

— Je suis de Buffalo et il y a là-bas un homme d'affaires bien connu qui s'appelle Lev Pechkov. Je me demandais si vous étiez apparentés. »

Volodia sursauta. Le frère de son père s'appelait Lev Pechkov et avait émigré à Buffalo avant la Première Guerre mondiale. Mais il préféra se montrer prudent. « Il faudra que je pose la question à mon père, dit-il.

— J'étais à Harvard avec Greg, le fils de Lev Pechkov. C'est peut-être votre cousin.

— Possible. » Volodia jeta un coup d'œil inquiet aux agents de la police secrète assis autour de la table. Woody l'ignorait, mais à leurs yeux, tout citoyen soviétique entretenant des liens avec un Américain devenait automatiquement suspect. « Vous savez, Woody, dans ce pays, refuser un coup à boire passe pour une insulte. »

Woody esquissa un sourire affable. « Pas en Amérique », dit-il.

Volodia attrapa son verre et jeta un regard circulaire aux espions déguisés en fonctionnaires et en diplomates. « Portons un toast ! À l'amitié entre les États-Unis et l'Union soviétique ! »

Les autres levèrent leurs verres. Woody les imita. « À l'amitié ! » répétèrent-ils.

Tous burent hormis Woody, qui reposa son verre sans y avoir touché.

Volodia se demanda s'il était aussi naïf qu'il en avait l'air.

Woody se pencha au-dessus de la table. « Il faut que vous sachiez une chose, Volodia : je ne connais aucun secret. Je n'occupe pas un poste assez important pour cela.

— Moi non plus », répondit Volodia. Ce n'était pas vrai, loin de là.

« Ce que je veux dire, poursuivit Woody, c'est que vous n'avez qu'à me poser des questions. Si je peux répondre, je le ferai. Rien ne m'en empêche puisque je ne détiens aucun secret. Inutile de m'enivrer ou d'envoyer des prostituées dans ma chambre. Interrogez-moi, c'est tout. »

C'était sûrement une ruse, se dit Volodia. Personne n'était ingénu à ce point. Mais il décida de se prêter au petit jeu de Woody. Pourquoi pas ? « Très bien, approuva-t-il. J'aimerais bien savoir ce que vous voulez. Pas vous personnellement, bien entendu. Votre délégation, le secrétaire d'État Hull et le président Roosevelt. Qu'attendez-vous de cette conférence ?

— Nous voulons que vous souteniez l'alliance quadripartite. »

C'était la réponse courante, mais Volodia décida d'aller plus loin. « Voilà précisément ce que nous ne comprenons pas. » C'était à lui d'être sincère cette fois, trop peut-être, mais son instinct lui soufflait de baisser un peu sa garde. « Qui peut bien s'intéresser à une alliance avec la Chine ? Nous devons vaincre les nazis en Europe. Et nous avons besoin de votre aide.

— Vous l'aurez.

— C'est ce que vous dites. Mais vous aviez promis d'envahir l'Europe cet été.

— Eh bien, nous avons débarqué en Italie.

— Ça ne suffit pas.

— L'année prochaine, c'est au tour de la France. Nous l'avons promis.

— Alors pourquoi avez-vous besoin de cette alliance ?

— Comment dire… » Woody marqua une pause, cherchant à rassembler ses idées. « Nous devons montrer au peuple américain qu'il est dans son intérêt d'envahir l'Europe.

— Pourquoi ?

— Pourquoi quoi ?

— Pourquoi devez-vous expliquer ça au peuple ? Roosevelt est président, non ? Il n'a qu'à le faire !

— Il y a des élections l'année prochaine. Il veut être réélu.

— Et alors ?

— Les Américains ne voteront pas pour lui s'ils estiment qu'il les a entraînés sans raison valable dans la guerre en Europe. Il veut donc leur présenter celle-ci comme un élément de son plan général pour une paix mondiale. Si nous mettons sur pied l'alliance quadripartite, cela montrera que nous sommes décidés à créer une Organisation des Nations unies. Alors les Américains seront plus nombreux à admettre que le débarquement en France est une étape sur la voie d'un monde plus pacifique.

— Je n'en reviens pas, lança Volodia. C'est lui le président, et il est obligé de justifier tous ses actes !

— Il y a de ça, acquiesça Woody. C'est ce qu'on appelle la démocratie. »

Volodia avait la vague impression que cette histoire incroyable était vraie. « Autrement dit, cette alliance est nécessaire pour convaincre les électeurs américains de soutenir votre intervention en Europe ?

— Exactement.

— Mais pourquoi avons-nous besoin de la Chine ? » Staline n'avait que mépris pour l'insistance avec laquelle les Alliés tenaient à inclure la Chine dans cette alliance.

« La Chine est un allié faible.

— Alors, oubliez-la.

— Si les Chinois sont tenus à l'écart, cela va les décourager, et ils se battront peut-être avec moins d'énergie contre les Japonais.

— Et alors ?

— Et alors, nous devrons renforcer nos troupes dans le Pacifique, ce qui nous empêchera de le faire en Europe. »

Volodia s'alarma. L'Union soviétique n'avait pas la moindre envie que des forces alliées soient retirées d'Europe pour être déployées dans le Pacifique. « Donc, vous faites une fleur à la Chine dans le seul but de pouvoir concentrer vos forces sur le débarquement en Europe ?

— Oui.

— À vous entendre, ça paraît si simple.

— Pour une bonne raison : ça l'est », conclut Woody.

4.

Le 1er novembre au matin, Chuck et Eddie mangèrent un steak pour le petit déjeuner en compagnie de la 3e division de marines au large de Bougainville, dans les mers du Sud.

Cette île mesurait deux cents kilomètres de long. Les Japonais y avaient construit deux bases aériennes, une au nord, l'autre au sud. Les marines se préparaient à débarquer à mi-chemin de la côte ouest, faiblement défendue. Leur objectif était d'établir une tête de pont et de s'emparer de suffisamment de terrain pour construire une piste d'atterrissage d'où ils pourraient attaquer les bases japonaises.

Chuck était sur le pont à sept heures vingt-six, lorsque les fusiliers marins, casqués et chargés de leur barda, descendirent le long des filets accrochés sur la coque du navire pour sauter dans les barges de débarquement. Ils étaient accompagnés d'un petit nombre de chiens, des dobermans aux qualités de sentinelles infatigables.

Comme les barges approchaient de la côte, Chuck repéra aussitôt une erreur sur la carte qu'il avait dressée. De fortes vagues s'écrasaient sur une plage escarpée. Sous ses yeux, une barge secouée par le ressac chavira. Les marines durent nager jusqu'au rivage.

« Il faut indiquer le ressac sur la carte », dit Chuck à Eddie, qui se tenait près de lui.

« Mais comment le repérer ?

— La reconnaissance aérienne doit envoyer un appareil à assez basse altitude pour que les gerbes d'écume soient visibles sur les photos qu'il prend.

— Ils ne peuvent pas prendre un tel risque à proximité d'une base aérienne ennemie. »

Eddie avait raison. Mais il fallait trouver une solution. Chuck enregistra mentalement le problème, premier enseignement de cette mission.

Pour celle-ci, ils avaient bénéficié d'informations plus précises qu'à l'ordinaire. Outre les cartes peu fiables et les photographies aériennes difficiles à déchiffrer, ils disposaient du rapport d'un groupe d'éclaireurs débarqués par sous-marin six semaines plus tôt. Ils avaient identifié douze plages propices à un débarquement sur une longueur de côte de six kilomètres. Mais ils n'avaient pas parlé du ressac. Peut-être était-il moins violent ce jour-là.

Pour l'instant, la carte de Chuck était relativement exacte. Une plage de sable de cent mètres de large, puis un fouillis de palmiers et d'autres végétaux. Derrière les fourrés, à en croire la carte, s'étendait un marais.

La côte n'était pas totalement sans défense. Chuck entendit le rugissement de l'artillerie et vit un obus se perdre dans les hauts-fonds. Un coup dans l'eau, mais le canonnier ajusterait son tir. Les marines semblaient galvanisés lorsqu'ils sautèrent sur le sable et foncèrent en direction du couvert végétal.

Chuck se félicitait d'être là. Jamais il ne s'était montré négligent dans le tracé de ses cartes, mais il était salutaire de constater par soi-même qu'une carte précise était capable de sauver des vies et que la moindre erreur pouvait être fatale. Avant même leur embarquement, Eddie et lui s'étaient montrés plus exigeants. Ils avaient demandé à ce qu'on reprenne des clichés quand les photos aériennes n'étaient pas nettes, interrogé les éclaireurs par téléphone et demandé un peu partout des cartes nautiques plus fiables.

Il se félicitait pour une autre raison. Il avait pris la mer, ce qui le ravissait. Il partageait un navire avec sept cents jeunes gens et il appréciait leur camaraderie, leurs blagues, leurs chansons et l'intimité des cabines bondées et des douches communes. « C'est un peu ce que doit ressentir un hétérosexuel dans une école de filles, dit-il un soir à Eddie.

— Sauf que ça ne leur arrive jamais, mais à nous si », répondit Eddie. Il était aussi enchanté que Chuck. Tous deux s'aimaient sincèrement, ce qui ne les empêchait pas de reluquer les matelots en tenue d'Adam.

À présent, les sept cents fusiliers marins cherchaient à débarquer le plus rapidement possible. La même scène se reproduisait en huit autres points de la côte. Dès qu'une barge était vide, elle s'empressait de faire demi-tour pour embarquer un nouveau contingent, mais la procédure n'en semblait pas moins d'une lenteur désespérante.

L'artilleur japonais, tapi quelque part dans la jungle, parvint enfin à ajuster son tir et, sous les yeux horrifiés de Chuck, un obus explosa au milieu d'un groupe de marines, projetant une gerbe de corps, de fusils et de membres arrachés qui retombèrent sur la plage et vinrent rougir le sable.

Chuck contemplait cette scène de carnage lorsqu'il entendit le rugissement d'un avion et vit un chasseur Zéro longer la côte à basse altitude. Les soleils rouges peints sur ses ailes le terrifièrent. La dernière fois qu'il les avait vus, c'était à la bataille de Midway.

Le Zéro mitrailla la plage. Les marines qui sortaient de la barge se retrouvèrent sans défense. Certains se jetèrent à plat ventre dans l'eau, d'autres tentèrent de s'abriter derrière la coque, d'autres encore de courir vers la jungle. Pendant quelques secondes, on vit jaillir le sang et tomber les hommes.

Puis l'avion repartit, laissant la plage jonchée de cadavres américains.

Quelques instants plus tard, Chuck l'entendit faire feu sur la plage suivante.

Il reviendrait.

Des avions américains auraient dû les protéger, mais il n'en voyait aucun. Le soutien aérien n'était jamais là où il aurait dû être, c'est-à-dire au-dessus des troupes.

Lorsque tous les marines eurent rejoint la rive, les barges y transportèrent médecins et brancardiers. Puis ce fut au tour du matériel : munitions, eau potable, nourriture, médicaments et pansements. Au retour, la barge évacuait les blessés jusqu'au navire.

Chuck et Eddie, dont le rôle n'était pas essentiel pour l'opération, débarquèrent avec le matériel.

Le pilote de la barge s'était adapté au ressac et parvint à la stabiliser, la rampe abaissée sur le sable et la poupe encaissant les vagues, tandis qu'on déchargeait les caisses et que Chuck et Eddie gagnaient le rivage en pataugeant.

Ils arrivèrent ensemble sur le sable.

À ce moment-là, une mitrailleuse ouvrit le feu.

Elle devait être dissimulée dans la jungle, à environ quatre cents mètres. S'y trouvait-elle depuis le début, le tireur ayant patiemment attendu son heure, ou bien venait-on de la mettre en position ? Baissant la tête, Chuck et Eddie coururent vers les palmiers.

Un marin qui transportait une caisse de munitions poussa un cri de douleur et s'effondra, laissant choir son fardeau.

Puis Eddie hurla.

Chuck fit encore deux enjambées avant de pouvoir s'arrêter. Quand il se retourna, ce fut pour voir Eddie se rouler par terre, le genou entre les mains, en criant : « Ah, merde ! »

Chuck revint près de lui. « Tout va bien, je suis là ! » hurla-t-il. Eddie avait les yeux clos, mais il était vivant et Chuck ne lui vit de blessure qu'au genou.

Il leva les yeux. La barge qui les avait transportés était toujours en cours de déchargement près du rivage. Il lui suffirait de quelques minutes pour y ramener Eddie. Mais la mitrailleuse tirait toujours.

Il s'accroupit. « Ça va faire mal, prévint-il. Hurle autant que tu veux. »

Il passa le bras droit sous les épaules d'Eddie, puis glissa le gauche sous ses cuisses. Il le souleva et se redressa. Eddie poussa un cri de douleur lorsque sa jambe blessée se balança dans le vide. « Tiens bon, mon vieux », dit Chuck. Il se tourna vers l'océan.

Soudain, il sentit une douleur insoutenable lui poignarder les jambes, le dos et enfin la tête. Pendant la fraction de seconde qui suivit, il se dit qu'il ne devait pas lâcher Eddie. Puis il comprit qu'il ne pourrait pas faire autrement. Une explosion de lumière derrière ses yeux l'aveugla.

Et le monde prit fin.

5.

Carla travaillait à l'hôpital juif pendant son jour de congé.

C'était le docteur Rothmann qui l'en avait persuadée. Les nazis l'avaient libéré du camp de concentration – ils étaient les seuls à savoir pourquoi, et ils n'en dirent rien à personne. Le médecin juif avait perdu un œil et boitait, mais il était vivant et capable d'exercer son métier.

L'hôpital se trouvait à Wedding, un quartier ouvrier du nord de la ville, mais son architecture n'avait rien de prolétaire. On l'avait édifié avant la Première Guerre mondiale, à une époque où les Juifs berlinois étaient fiers et prospères. Il consistait en sept élégants bâtiments disposés dans un vaste jardin. Les différents services étaient reliés par des tunnels, si bien que patients et personnel pouvaient se déplacer de l'un à l'autre sans craindre les intempéries.

C'était un miracle qu'il existe encore un hôpital juif. Il ne restait presque plus de Juifs à Berlin. On les avait raflés par milliers et déportés dans des trains spéciaux. Nul ne savait où ils avaient été conduits ni ce qu'ils étaient devenus. De folles rumeurs circulaient sur des camps d'extermination.

S'ils tombaient malades, les rares Juifs berlinois ne pouvaient pas se faire soigner par des médecins et des infirmières aryens. Aussi la logique tordue du racisme nazi imposait-elle le maintien de cet hôpital. Son personnel était composé en majorité de Juifs et d'autres malheureux auxquels on refusait le statut d'Aryens. Slaves d'Europe de l'Est, individus d'origines mixtes, époux et épouses de Juifs. Comme il n'y avait pas assez d'infirmières, Carla venait donner un coup de main.

L'hôpital était constamment en butte aux tracasseries de la Gestapo, à court de fournitures médicales et plus particulièrement de médicaments ; son fonctionnement était encore entravé par des effectifs insuffisants et des fonds presque inexistants.

Carla violait la loi lorsqu'elle prenait la température d'un garçon de onze ans au pied broyé lors d'un raid aérien. Et elle commettait aussi un crime en chapardant des fournitures médicales dans l'hôpital où elle travaillait quotidiennement pour les apporter là. Mais elle tenait à prouver, ne fût-ce qu'à elle-même, que certains n'avaient pas plié l'échine devant les nazis.

Alors qu'elle finissait sa tournée, elle aperçut Werner devant la porte, dans son uniforme de la Luftwaffe.

Plusieurs jours durant, ils avaient vécu l'un comme l'autre dans la terreur, se demandant si quelqu'un avait survécu au bombardement de l'école et risquait de dénoncer Werner, mais il était clair à présent que tout le monde était mort et que Macke n'avait confié ses soupçons à personne. Une fois encore, ils s'en étaient tirés.

Werner s'était rapidement remis de sa blessure.

Et ils étaient amants. Werner avait emménagé dans la grande maison à moitié vide des von Ulrich et couchait toutes les nuits dans le lit de Carla. Leurs parents n'y avaient vu aucune objection : tout le monde savait que la mort pouvait frapper d'un jour à l'autre et qu'il fallait profiter de tous les instants de joie dérobés à cette vie de souffrance et de privations.

Mais Werner avait l'air plus solennel que d'habitude lorsqu'il fit signe à Carla à travers la vitre de la porte du service. Elle l'invita à entrer et l'embrassa. « Je t'aime », dit-elle. Elle ne se lassait pas de prononcer ces mots.

Et il était toujours heureux de lui répondre : « Moi aussi, je t'aime.

— Que viens-tu faire ici ? demanda-t-elle. Tu avais envie d'un petit baiser ?

— J'ai reçu de mauvaises nouvelles. On m'envoie sur le front de l'Est.

— Oh, non ! » Elle sentit les larmes lui monter aux yeux.

« C'est un miracle que ce ne soit pas arrivé plus tôt. Mais le général Dorn ne peut plus me garder. La moitié de nos troupes se compose de vieillards et d'adolescents, et je suis un officier de vingt-quatre ans en excellente forme.

— Ne meurs pas, je t'en prie, chuchota-t-elle.

— Je ferai mon possible. »

Toujours dans un murmure, elle poursuivit : « Mais que va devenir le réseau ? Tu es le seul à être au courant de tout. Qui d'autre pourrait le diriger ? »

Il la regarda sans rien dire.

Elle comprit à quoi il pensait. « Oh, non, pas moi !

— Tu es la plus qualifiée. Frieda est un bon agent, mais ce n'est pas un chef. Tu as montré que tu étais capable de recruter de nouveaux éléments et de les motiver. La police ne s'est jamais intéressée à toi et on ne te connaît aucune activité politique. Tout le monde ignore le rôle que tu as joué dans l'affaire de l'Aktion T4. Aux yeux des autorités, tu es une infirmière irréprochable.

— Mais Werner, j'ai peur !

— Tu n'es pas obligée de le faire. Mais si tu ne le fais pas, personne ne le fera. »

À ce moment-là, un bruit parvint à leurs oreilles.

Le service voisin était réservé aux malades mentaux, et il n'était pas rare d'entendre des cris et des hurlements, mais ce bruit-là sortait de l'ordinaire. Une voix cultivée protestait d'un ton irrité. Puis une seconde voix s'éleva, marquée par un fort accent berlinois et par un ton brutal, typique des gens de la capitale à en croire les provinciaux.

Carla s'avança dans le couloir, Werner sur ses talons.

Le docteur Rothmann, dont la veste portait l'étoile jaune, discutait ferme avec un SS en uniforme. Derrière eux, la double porte du service psychiatrique, normalement fermée à double tour, était grande ouverte. Les patients sortaient. Deux policiers et quelques infirmières escortaient dans l'escalier une file désordonnée d'hommes et de femmes, pour la plupart en pyjama ; certains se tenaient droits et paraissaient normaux, alors que d'autres avaient le dos voûté et traînaient les pieds en marmonnant.

Carla se rappela aussitôt Kurt, le fils d'Ada, Axel, le frère de Werner, et le prétendu hôpital d'Akelberg. Elle ignorait où se rendaient ces patients mais elle était sûre qu'on les y tuerait.

« Ces gens sont malades ! vitupérait le docteur Rothmann. Ils ont besoin de soins.

— Ce ne sont pas des malades, répondit l'officier SS, ce sont des aliénés, et nous les conduisons là où ils devraient déjà se trouver.

— Dans un hôpital ?

— Vous en serez informé en temps utile.

— Cela ne me suffit pas. »

Carla savait qu'elle ne devait pas intervenir. Si on s'apercevait qu'elle n'était pas juive, elle aurait de graves ennuis. Elle n'avait pas vraiment l'air d'une Aryenne, avec ses cheveux noirs et ses yeux verts. Si elle se tenait tranquille, personne sans doute ne l'inquiéterait. Mais si elle protestait contre les agissements du SS, elle serait arrêtée et interrogée, et on découvrirait qu'elle travaillait dans l'illégalité. Elle décida donc de serrer les dents.

L'officier éleva la voix : « Dépêchez-vous – faites monter ces crétins dans le bus. »

Rothmann insista. « Je dois savoir où ils se rendent. Ce sont mes patients. »

Ce n'était pas tout à fait vrai car il n'était pas psychiatre.

« Si vous vous faites du souci pour eux, vous n'avez qu'à les accompagner », répliqua le SS.

Le médecin pâlit. Il irait vers une mort presque certaine, il le savait.

Carla pensa à son épouse Hannelore, à son fils Rudi et à sa fille Eva, réfugiée en Angleterre, et la peur lui serra la gorge.

L'officier sourit. « On se sent soudain moins concerné, hein ? » railla-t-il.

Rothmann se redressa. « Vous vous trompez. J'accepte votre proposition. Il y a bien des années, j'ai fait serment de m'efforcer d'aider les malades. Je ne vais pas trahir ma parole aujourd'hui. J'espère mourir en paix avec ma conscience. » Il descendit l'escalier en boitant.

Une vieille femme passa, vêtue d'une simple robe de chambre grande ouverte qui ne laissait rien ignorer de sa nudité.

Carla ne put rester silencieuse plus longtemps. « Nous sommes en novembre ! s'écria-t-elle. Ils ne sont pas habillés pour sortir ! »

L'officier la gratifia d'un regard mauvais. « Ils seront très bien dans le bus.

— Je vais chercher des vêtements chauds. » Carla se tourna vers Werner. « Aide-moi. Récupère des couvertures où tu pourras. »

Ils firent le tour du service psychiatrique à présent désert, ramassant des couvertures sur les lits et dans les armoires. Puis ils s'empressèrent de descendre, chacun avec sa pile.

Dans le jardin, la terre avait gelé. Un autocar gris était garé devant la porte principale ; son chauffeur fumait au volant en faisant tourner le moteur. Carla vit qu'il portait un épais manteau, un chapeau et des gants, ce qui signifiait que le car n'était pas chauffé.

Un petit groupe de SS et de membres de la Gestapo surveillait les opérations.

Les derniers patients montaient dans le véhicule. Carla et Werner les suivirent et entreprirent de distribuer les couvertures.

Le docteur Rothmann se tenait debout au fond du bus. « Carla, dit-il. Tu… tu diras à Hannelore ce qui s'est passé. Je dois accompagner les patients. Je n'ai pas le choix.

— Bien sûr. » Elle avait la gorge nouée par l'émotion.

« Peut-être arriverai-je à protéger ces malheureux. »

Carla acquiesça, mais elle ne le croyait pas vraiment.

« Quoi qu'il en soit, je ne peux pas les abandonner.

— Je le lui dirai.

— Dis-lui aussi que je l'aime. »

Carla ne put retenir ses larmes.

« Dis-lui que c'est la dernière chose que je t'aie dite. Je l'aime. »

Carla acquiesça.

Werner lui prit le bras. « Allons-nous-en. »

Ils descendirent du car.

Un SS lança à Werner : « Hé, vous, vous portez l'uniforme de la Luftwaffe, qu'est-ce que vous faites ici ? »

Werner était si furieux que Carla craignit de le voir déclencher un pugilat. Mais ce fut d'une voix posée qu'il répondit : « Je distribue des couvertures à des vieillards qui ont froid. C'est contraire à la loi à présent ?

— Vous devriez être sur le front de l'Est.

— J'y pars dès demain. Et vous ?

— Attention à ce que vous dites.

— Si vous aviez l'obligeance de m'arrêter, vous me sauveriez peut-être la vie. »

L'homme se détourna.

On entendit grincer les vitesses du car et le bruit du moteur s'amplifia. Carla et Werner se retournèrent. À chaque carreau se plaquait un visage, et ils étaient tous différents : celui-ci bredouillait, celui-là bavait, un autre riait de façon hystérique, celui-là était distrait, le dernier en proie à une détresse spirituelle – tous étaient déments. Les occupants d'une unité psychiatrique évacués par des SS. Les fous guidant les fous.

Le car s'éloigna.

6.

« Peut-être aurais-je aimé la Russie si on m'avait permis de la voir, dit Woody à son père.

— Je partage ton sentiment.

— Je n'ai même pas pu prendre une photo correcte. »

Ils étaient assis dans le grand hall de l'hôtel Moskva, près de l'entrée de la station de métro. Leurs valises étaient bouclées et ils rentraient chez eux.

« Il faut que je dise à Greg Pechkov que j'ai rencontré un certain Volodia Pechkov, reprit Woody. Encore que celui-ci n'ait guère été ravi de l'apprendre. J'imagine qu'avoir un parent à l'Ouest suffirait à le rendre suspect.

— Tu peux en être sûr.

— Enfin, nous ne repartons pas les mains vides – c'est l'essentiel. Les Alliés se sont engagés à créer l'Organisation des Nations unies.

— Oui, approuva Gus d'un air satisfait. Staline s'est fait tirer l'oreille, mais il a fini par admettre que c'était une idée sensée. Tu nous as aidés, je crois, en parlant franchement à Pechkov comme tu l'as fait.

— Toute ta vie tu t'es battu pour ça, Papa.

— C'est un grand moment, c'est sûr. »

Une idée inquiétante traversa l'esprit de Woody. « Tu ne vas pas prendre ta retraite, au moins ? »

Gus s'esclaffa. « Non. Nous n'avons obtenu qu'un accord de principe : le travail ne fait que commencer. »

Cordell Hull avait déjà quitté Moscou, mais certains de ses assistants s'y trouvaient encore et l'un d'eux s'approcha des Dewar. Woody le connaissait – un jeune homme du nom de Ray Baker. « J'ai un message pour vous, monsieur le sénateur », annonça-t-il. Il paraissait nerveux.

« Eh bien, vous avez failli me manquer – je suis sur le départ, dit Gus. Qu'y a-t-il ?

— C'est à propos de votre fils Charles – Chuck. »

Gus pâlit : « Quel est le message, Ray ? »

Le jeune homme avait peine à s'exprimer. « Ce sont de mauvaises nouvelles, monsieur. Il participait à une bataille dans les îles Salomon.

— Il est blessé ?

— Non, monsieur, pire.

— Oh Seigneur », murmura Gus, et il fondit en larmes.

Woody n'avait jamais vu son père pleurer.

« Je suis désolé, monsieur, reprit Ray. Le message dit qu'il est mort. »

XVIII
1944

1.

Woody se tenait devant le miroir de sa chambre, dans l'appartement de ses parents à Washington. Il était vêtu de son uniforme de sous-lieutenant du 510e régiment de parachutistes de l'armée des États-Unis.

Il l'avait fait confectionner par un excellent tailleur de Washington, mais n'avait pas fière allure pour autant. La couleur kaki lui donnait un teint cireux, les écussons et les insignes de la veste faisaient curieusement négligé.

Sans doute aurait-il pu éviter de rejoindre le contingent, mais il s'en serait voulu. En un sens, il aurait bien aimé continuer à travailler avec son père, qui aidait le président Roosevelt à établir un nouvel ordre mondial destiné à éviter d'autres guerres planétaires. Ils avaient remporté une victoire à Moscou, mais Staline se montrait capricieux et semblait prendre un malin plaisir à leur mettre des bâtons dans les roues. Lors de la conférence de Téhéran, au mois de décembre précédent, le dirigeant soviétique avait cherché à imposer une nouvelle fois la solution de compromis des comités régionaux, et Roosevelt avait dû batailler ferme pour le faire renoncer à cette idée. De toute évidence, l'Organisation des Nations unies exigerait une vigilance de tous les instants.

Gus pouvait cependant se passer de Woody. Et Woody s'en voulait de plus en plus de laisser d'autres faire la guerre à sa place.

Décidant qu'il ne pourrait jamais paraître plus fringant, il gagna le salon pour montrer son uniforme à sa mère.

Rosa avait un invité, un jeune homme en uniforme de l'US Navy, et Woody reconnut les taches de rousseur et le visage avenant d'Eddie Parry. Assis sur le canapé à côté de Rosa, il tenait une canne. Il se leva non sans difficulté pour saluer Woody.

Mama avait l'air triste. « Eddie était en train de me raconter comment Chuck est mort », dit-elle.

Eddie se rassit et Woody prit place en face de lui. « J'aimerais le savoir, moi aussi, murmura-t-il.

— Il n'y a pas grand-chose à raconter, commença Eddie. On était sur la plage de Bougainville depuis à peine cinq secondes quand une mitrailleuse a ouvert le feu depuis les marais. On s'est mis à courir, mais j'ai reçu deux balles dans le genou. Chuck aurait dû foncer se mettre à l'abri sous les arbres. C'est ce qu'on nous apprend à l'entraînement – il faut laisser aux toubibs le soin de récupérer les blessés. Évidemment, il a désobéi à la consigne. Il a fait demi-tour pour venir me chercher. »

Eddie marqua une pause. Une tasse de café était posée sur la table basse devant lui et il en but une gorgée.

« Il m'a pris dans ses bras, reprit-il. Satané idiot. Il faisait une cible idéale. Je suppose qu'il voulait me ramener jusqu'à la barge. Ces embarcations ont une coque assez haute, en acier qui plus est. On aurait été parfaitement à l'abri et j'aurais pu me faire soigner tout de suite. Mais il n'aurait pas dû faire ça. Dès qu'il s'est redressé, il a reçu une rafale de mitrailleuse – qui l'a atteint aux jambes, au dos et à la tête. Je pense qu'il est mort avant de toucher le sol. En tout cas, quand j'ai réussi à relever les yeux vers lui, il n'était déjà plus de ce monde. »

Woody vit que sa mère avait peine à rester maîtresse d'elle-même. Si elle se mettait à pleurer, il craignait d'en faire autant.

« Je suis resté allongé près de lui pendant une heure, poursuivit Eddie. Je ne lui ai pas lâché la main de tout ce temps-là. Puis des brancardiers sont venus me chercher. Je ne voulais pas le quitter. Je savais que je ne le reverrais plus jamais. » Il enfouit son visage entre ses mains. « Je l'aimais tellement », sanglota-t-il.

Rosa passa un bras autour de ses larges épaules et l'étreignit. Il se laissa aller contre elle, pleurant comme un enfant. Elle lui caressa les cheveux. « Allons, allons, fit-elle. Allons, allons. »

Woody comprit que sa mère n'ignorait rien de la nature des relations de Chuck et d'Eddie.

Au bout d'une minute, ce dernier reprit contenance. Il se tourna vers Woody. « Vous savez ce que c'est », dit-il.

Il faisait allusion à la mort de Joanne. « Oui, acquiesça Woody. C'est la pire chose qui puisse vous arriver – mais chaque jour qui passe apaise un peu la douleur.

— J'espère que c'est vrai.

— Vous êtes toujours à Hawaï ?

— Oui. Chuck et moi, on travaille à la section Territoire ennemi. On travaillait, je veux dire. » Il déglutit. « Chuck avait décidé de participer à une opération pour vérifier si nos cartes étaient vraiment utiles sur le terrain. C'est pour ça qu'on a débarqué à Bougainville avec les marines.

— Vous devez faire du bon boulot, approuva Woody. Apparemment, nous faisons reculer les Japs dans le Pacifique.

— Peu à peu, oui », confirma Eddie. Il examina l'uniforme de Woody. « Où êtes-vous affecté ?

— Jusqu'ici, j'ai suivi un entraînement de parachutiste à Fort Benning, en Géorgie. Mais je pars demain pour Londres. »

Il surprit le regard de sa mère et la trouva soudain vieillie. Il s'aperçut qu'elle avait des rides. C'était en toute discrétion qu'on avait célébré son cinquantième anniversaire. L'évocation de la mort de Chuck à un moment où son aîné venait de revêtir l'uniforme de l'armée avait dû la secouer, devina-t-il.

Cette douloureuse coïncidence échappa à Eddie. « Il paraît qu'on va envahir la France cette année, dit-il.

— C'est sans doute pour ça que j'ai eu droit à une formation accélérée, répondit Woody.

— Vous irez sûrement au feu. »

Rosa étouffa un sanglot.

« J'espère me montrer aussi courageux que mon frère, fit Woody.

— J'espère surtout que vous n'aurez pas l'occasion de le faire », répliqua Eddie.

2.

Greg Pechkov invita Margaret Cowdry à un concert symphonique en matinée. Margaret avait des yeux noirs et une bouche aux lèvres pulpeuses, faites pour les baisers. Mais Greg avait autre chose en tête.

Il avait pris Barney McHugh en filature.

Un agent du FBI nommé Bill Bicks en faisait autant.

Barney McHugh était un jeune et brillant physicien. Le laboratoire secret de l'armée américaine installé à Los Alamos au Nouveau-Mexique lui avait accordé un congé et il faisait visiter Washington à son épouse britannique.

Le FBI avait été informé que McHugh assisterait à ce concert, et l'agent spécial Bicks avait obtenu deux places pour Greg, quelques rangées derrière McHugh. Une salle de concert où se pressaient plusieurs centaines d'inconnus était un lieu idéal pour un rendez-vous clandestin, et Greg voulait savoir ce que mijotait McHugh.

Quel dommage qu'ils se soient déjà rencontrés ! Greg avait discuté avec McHugh à Chicago le jour où on avait testé la pile atomique. Même si cela s'était produit dix-huit mois plus tôt, peut-être McHugh s'en souvenait-il malgré tout. Greg devait donc veiller à ne pas se faire voir.

Lorsqu'il arriva en compagnie de Margaret, les places réservées par McHugh étaient inoccupées. Deux couples d'aspect fort banal étaient assis de part et d'autre : à gauche un quinquagénaire en complet gris anthracite bon marché et sa femme grassouillette, à droite deux dames d'un certain âge. Greg espérait que McHugh ne tarderait pas. Si c'était un espion, il était résolu à le coincer.

Ils allaient entendre la *Symphonie n° 1* de Tchaïkovski. « Alors, comme ça, tu aimes la musique classique », remarqua Margaret pendant que les musiciens s'accordaient. Elle n'avait aucune idée de la vraie raison de leur présence ici. Elle savait que Greg travaillait dans la recherche militaire, un champ d'activité ultrasecret, mais, à l'instar de la majorité des Américains, elle ignorait tout de la bombe atomique. « Je croyais que tu n'écoutais que du jazz, ajouta-t-elle.

— Les compositeurs russes me fascinent – je trouve leur musique bouleversante. Je dois avoir ça dans le sang.

— J'ai été élevée dans la musique classique. Mon père aime bien faire venir un orchestre de chambre pour ses réceptions. » La famille de Margaret était incomparablement plus riche que celle de Greg. Mais il n'avait toujours pas rencontré ses parents, et se doutait que ceux-ci n'accueilleraient pas à bras ouverts le fils illégitime d'un célèbre don Juan hollywoodien. « Qu'est-ce que tu regardes comme ça ? demanda-t-elle.

— Rien. » Les McHugh venaient d'arriver. « Comment s'appelle ton parfum ?

— *Chichi*, de Renoir.

— J'adore. »

Les McHugh semblaient ravis, un jeune couple prospère en vacances. Greg se demanda s'ils étaient en retard parce qu'ils avaient fait l'amour dans leur chambre d'hôtel.

Barney McHugh s'assit à côté de l'homme au complet gris anthracite. La raideur des épaulettes prouvait que ce n'était pas un costume de qualité. Son propriétaire ne daigna même pas se tourner vers les nouveaux venus. Les McHugh ouvrirent un journal et commencèrent à faire les mots croisés, tête contre tête. Quelques minutes plus tard, le chef d'orchestre fit son entrée.

Le premier morceau était de Saint-Saëns. La popularité des compositeurs allemands et autrichiens avait décliné depuis la déclaration de guerre, et les mélomanes élargissaient leurs horizons. Sibelius connaissait une véritable renaissance.

McHugh était probablement communiste. Greg le savait, parce que J. Robert Oppenheimer le lui avait dit. Oppenheimer, un prestigieux physicien de l'université de Californie, dirigeait le laboratoire de Los Alamos et était à la tête de l'ensemble du projet Manhattan. Malgré ses liens étroits avec les communistes, il affirmait énergiquement n'avoir jamais adhéré au Parti.

L'agent spécial Bicks avait demandé à Greg : « Pourquoi l'armée fait-elle appel à tous ces rouges ? J'ignore sur quoi vous travaillez au fin fond de ce désert, mais vous ne pourriez pas recruter de brillants jeunes savants américains conservateurs ?

— Il n'y en a pas, lui avait répondu Greg. Sinon, on les aurait déjà embauchés. »

Les communistes étaient parfois plus dévoués à leur cause qu'à leur patrie et ne voyaient aucun inconvénient à partager les secrets de la recherche nucléaire avec l'Union soviétique. On ne pouvait pas les accuser de livrer des informations à l'ennemi, après tout. Les Soviétiques étaient les alliés des Américains dans la lutte contre les nazis – en fait, ils s'étaient battus plus vaillamment que tous les autres alliés réunis. De telles initiatives n'en étaient pas moins dangereuses. Une information destinée à Moscou risquait de se retrouver à Berlin. Et quiconque réfléchissait ne fût-ce qu'une minute au monde d'après-guerre devait bien conclure que les États-Unis et l'URSS ne resteraient pas éternellement amis.

Le FBI, qui considérait Oppenheimer comme un élément à risque, ne cessait de réclamer son renvoi au général Groves, le patron de Greg. Mais Oppenheimer était le scientifique le plus brillant de sa génération, et le général ne voulait rien savoir.

Désireux de prouver sa loyauté, Oppenheimer avait dénoncé McHugh comme un communiste probable, raison pour laquelle Greg le surveillait.

Le FBI restait sceptique. « Oppenheimer se fiche de vous, avait dit Bicks.

— Ça m'étonnerait, avait répliqué Greg. Ça fait un an que je le connais maintenant.

— C'est un enfoiré de coco, comme sa femme, son frère et sa belle-sœur.

— C'est un type qui travaille dix-neuf heures par jour pour améliorer l'armement des soldats américains – vous appelez ça un traître ? »

Greg espérait confondre McHugh, car cela innocenterait Oppenheimer et assoirait la crédibilité du général Groves tout en le mettant lui-même en avant.

Il ne quitta pas McHugh des yeux durant la première partie du concert. Le physicien n'accorda pas un regard à ceux qui l'entouraient. Il semblait concentré sur la musique et l'attention qu'il portait à l'orchestre ne se relâchait que lorsqu'il jetait des regards enamourés à Mrs. McHugh, une pâle fleur anglaise. Et si Oppenheimer s'était trompé sur son compte ? À moins qu'il n'ait ainsi cherché, plus subtilement, à détourner les soupçons qui pesaient sur lui ?

Bicks était sur le qui-vive, lui aussi. Il avait pris une place au premier balcon. Peut-être avait-il repéré quelque chose.

Lorsque arriva l'entracte, Greg suivit les McHugh et fit la queue derrière eux pour commander un café. Ni le couple de quinquagénaires ni les deux vieilles dames ne les avaient suivis.

Greg était déçu. Il ne savait que conclure. Ses soupçons étaient-ils infondés ? Ou la présence des McHugh dans cette salle de concert ce jour-là était-elle parfaitement anodine ?

Alors que Margaret et lui regagnaient leurs places, Bill Bicks s'approcha d'eux. L'agent fédéral était un homme d'une quarantaine d'années bedonnant, au crâne déjà dégarni. Il portait un complet gris clair taché de sueur sous les bras. « Vous aviez raison, dit-il à voix basse.

— Comment le savez-vous ?

— Le type assis à côté de McHugh.

— Le costume gris anthracite ?

— Ouais. C'est Nikolaï Ienkov, attaché culturel à l'ambassade soviétique.

— Merde alors ! » s'exclama Greg.

Margaret se retourna. « Qu'y a-t-il ?

— Rien », fit Greg.

Bicks s'éloigna.

« Il y a quelque chose qui te préoccupe, observa-t-elle comme ils s'asseyaient. Je parie que tu n'as pas écouté une seule mesure de Saint-Saëns.

— Je pensais à mon boulot, c'est tout.

— Dis-moi que ce n'est pas une autre femme et je n'en parle plus.

— Ce n'est pas une autre femme. »

Pendant toute la seconde partie du concert, il fut sur le qui-vive. Il ne remarqua aucun échange entre McHugh et Ienkov. Ils ne prononcèrent pas un mot, et Greg ne les vit échanger ni bout de papier, ni rouleau de pellicule – rien.

Le concert s'acheva et le chef d'orchestre salua le public. Celui-ci quitta lentement la salle. La traque de Greg tournait au fiasco.

Une fois dans le foyer, Margaret gagna les toilettes pour dames. Pendant que Greg l'attendait, Bicks s'approcha de lui.

« Rien à signaler, lui dit Greg.

— Moi non plus.

— C'est peut-être une coïncidence. Le fait que McHugh ait été assis à côté d'Ienkov, je veux dire.

— Les coïncidences, ça n'existe pas.

— Ou alors il y a eu un pépin. Il n'avait pas le bon mot de passe, par exemple. »

Bicks fit non de la tête. « Ils ont échangé quelque chose. On n'a rien vu, c'est tout. »

Mrs. McHugh se rendit elle aussi aux toilettes, et, tout comme Greg, McHugh l'attendit à proximité. Dissimulé derrière un pilier, Greg l'étudia avec attention. Il ne portait ni porte-documents, ni imperméable sous lequel il aurait pu dissimuler une chemise ou un dossier. Quelque chose l'intriguait pourtant, sans qu'il puisse mettre le doigt dessus. Qu'était-ce ?

Soudain, il comprit. « Le journal ! s'exclama-t-il.

— Quoi ? sursauta Bicks.

— En arrivant, Barney avait un journal. Ils ont fait les mots croisés, sa femme et lui, en attendant le début du concert. Où est passé ce journal ?

— Soit il l'a jeté... soit il l'a passé à Ienkov, en planquant quelque chose dedans.

— Ienkov et sa femme sont déjà partis.

— Ils ne doivent pas être bien loin. »

Les deux hommes foncèrent vers la sortie.

Bicks bouscula les mélomanes qui se pressaient vers les portes, Greg sur ses talons. Une fois dehors, ils regardèrent tout autour d'eux. Greg ne localisa pas Ienkov, mais Bicks avait les yeux d'un chasseur. « En face ! » s'écria-t-il.

Debout sur le trottoir, l'attaché culturel et son épouse grassouillette attendaient une limousine noire qui se dirigeait lentement vers eux.

Ienkov tenait un journal à la main.

Greg et Bicks traversèrent la rue, ventre à terre.

La limousine s'arrêta.

Plus rapide que l'agent fédéral, Greg arriva le premier sur le trottoir.

Ienkov ne les avait pas remarqués. Sans se presser, il ouvrit la portière de la voiture puis s'écarta pour laisser monter son épouse.

Greg se jeta sur lui et ils tombèrent tous les deux. Mrs. Ienkov hurla.

Greg se releva en hâte. Le chauffeur, qui était descendu de voiture, se précipitait vers lui, mais Bicks s'écria : « FBI ! » et brandit son insigne.

Ienkov avait lâché le journal. Il voulut le ramasser. Mais Greg fut plus prompt. Il s'en empara, recula d'un pas et l'ouvrit.

Une feuille de papier était glissée entre ses pages. Greg reconnut tout de suite le diagramme qui y figurait. Il représentait le mécanisme d'implosion d'une bombe au plutonium. « Bon Dieu de merde ! s'écria-t-il. Ils viennent juste de le mettre au point ! »

Ienkov s'engouffra dans la limousine, referma la portière et la verrouilla de l'intérieur.

Le chauffeur remonta en voiture et fonça.

3.

C'était un samedi soir et l'appartement de Daisy était bondé. Il devait y avoir une bonne centaine de personnes, songea-t-elle avec satisfaction.

Elle avait pris la tête d'un groupe d'amis qui s'était constitué autour de la Croix-Rouge américaine de Londres. Tous les samedis, elle donnait une réception pour les militaires américains, et invitait les infirmières de St Bartholomew. Les pilotes de la RAF étaient également les bienvenus. Ils éclusaient ses réserves de scotch et de gin, et dansaient au son de ses disques de Glenn Miller. Sachant que c'était peut-être la dernière fête de ces hommes, elle faisait tout ce qu'elle pouvait pour leur faire plaisir – excepté les embrasser, mais les infirmières y pourvoyaient largement.

Daisy ne buvait jamais d'alcool lors de ces soirées. Elle avait bien trop de choses à régler. Tous ces couples qui s'enfermaient dans les toilettes et qu'elle devait chasser pour que ceux qui voulaient les utiliser normalement puissent en disposer. Tous ces généraux qu'il fallait raccompagner chez eux en état d'ivresse. Il lui arrivait souvent de se retrouver à court de glaçons – en dépit de tous ses efforts, elle ne parvenait pas à convaincre ses domestiques anglais qu'une quantité importante de glaçons était indispensable à une soirée digne de ce nom.

Pendant un certain temps après sa rupture avec Boy Fitzherbert, elle n'avait eu d'autres amis que la famille Leckwith. Ethel, la mère de Lloyd, ne l'avait jamais jugée. Bien qu'elle soit devenue une femme éminemment respectable, elle avait commis son content d'erreurs par le passé, ce qui l'avait rendue plus tolérante. Daisy lui rendait toujours visite tous les mercredis dans sa maison d'Aldgate pour déguster un chocolat chaud en écoutant la radio. C'était sa sortie préférée de la semaine.

Cela faisait deux fois maintenant qu'elle avait été mise au ban de la société, une fois à Buffalo et une fois à Londres, et elle se demandait non sans accablement si elle n'en était pas responsable. Peut-être n'avait-elle pas vraiment sa place dans ce milieu snob à l'extrême, où les règles de conduite étaient d'une rigidité impitoyable. Elle était stupide de chercher à fréquenter ces gens-là.

Malheureusement, elle adorait les réceptions, les pique-niques et toutes ces occasions de s'amuser et de faire assaut d'élégance.

Elle avait tout de même fini par comprendre qu'elle n'avait pas besoin des aristocrates britanniques ni des nouveaux riches américains pour prendre du bon temps. Elle avait créé autour d'elle son propre cercle, bien plus excitant à ses yeux. Parmi ceux qui lui avaient battu froid après qu'elle avait quitté Boy, certains lui faisaient désormais comprendre qu'ils aimeraient bien être invités à l'une de ses fameuses soirées du samedi. Et quantité de gens n'hésitaient pas à se présenter chez elle pour se détendre à l'issue d'un dîner aussi grandiose qu'assommant dans un hôtel particulier de Mayfair.

Ce soir-là, elle était aux anges, car Lloyd était en permission.

Il s'était installé chez elle et ne s'en cachait pas. Au diable le qu'en-dira-t-on, songeait Daisy : de toute façon, elle était perdue de réputation. Par ailleurs, la guerre encourageait bien des gens à faire fi des convenances. Les domestiques pouvaient se montrer aussi collet monté que les duchesses, mais ceux de Daisy l'adoraient, si bien que Lloyd et elle ne feignaient même plus de faire chambre à part.

Elle ne se lassait pas de coucher avec lui. Quoique moins expérimenté que Boy, il compensait cela par son enthousiasme… et était fort réceptif à la nouveauté. Chaque nuit était pour eux l'occasion d'explorer de nouveaux plaisirs.

Alors qu'ils contemplaient leurs invités affairés à rire et à bavarder, à boire et à fumer, à danser et à flirter, Lloyd lui sourit et lui dit : « Tu es heureuse ?

— Presque, rétorqua-t-elle.

— Presque ? »

Elle soupira. « Je veux des enfants, Lloyd. Que nous soyons mariés ou non, je m'en fiche. Enfin, pas tout à fait, mais je veux quand même un bébé. »

Le visage de Lloyd s'assombrit. « Tu sais ce que je pense des naissances illégitimes.

— Oui, tu me l'as expliqué. Mais je voudrais avoir un peu de toi à chérir s'il devait t'arriver quelque chose.

— Je ferai tout ce que je peux pour rester en vie.

— Bien sûr. » Mais si elle ne se trompait pas, il opérait clandestinement en territoire ennemi, et risquait donc d'être exécuté, comme l'étaient les espions allemands appréhendés sur le sol britannique. Il quitterait ce monde sans rien lui laisser. « Des millions de femmes sont dans le même cas que moi, mais je n'arrive pas à imaginer la vie sans toi. Je crois que j'en mourrais.

— Si je le pouvais, je convaincrais Boy de t'accorder le divorce.

— Arrête, ce n'est pas le moment de parler de ça. » Elle parcourut les lieux du regard. « Hé, mais tu sais quoi ? Je crois bien que c'est Woody Dewar ! » s'écria-t-elle en apercevant un jeune homme en uniforme de lieutenant. Elle se dirigea vers lui pour le saluer. C'était bizarre de le revoir au bout de neuf ans – mais il n'avait pas beaucoup changé, il paraissait juste un peu plus vieux.

« Il y a maintenant des milliers de soldats américains en Angleterre, remarqua Daisy tandis qu'ils dansaient le fox-trot sur « Pennsylvania Six-Five Thousand ». Nous sommes sûrement sur le point d'envahir la France. Qu'en dis-tu ?

— Ce que j'en dis, c'est que le haut commandement ne me fait pas de confidences, répliqua Woody. Mais je suis comme toi, je ne vois pas d'autre raison à ma présence ici. On ne peut quand même pas laisser les Russes se taper tout le boulot.

— C'est pour quand, selon toi ?

— En général, on attend l'été pour lancer une offensive. Fin mai ou début juin, c'est ce que disent les gens bien informés.

— Déjà !

— Mais personne ne sait où ça se fera.

— La traversée de Douvres à Calais est la plus courte.

— C'est pour ça que les Allemands ont renforcé leurs défenses autour de Calais. Alors on va peut-être tenter de les prendre par surprise – en débarquant à Marseille, par exemple.

— Et la guerre sera enfin finie.

— Ça m'étonnerait. Une fois que nous aurons établi une tête de pont, il nous restera à libérer la France et ensuite à conquérir l'Allemagne. Ça fait un sacré bout de chemin.

— Je veux bien te croire. » Woody avait visiblement besoin qu'on lui remonte le moral. Et Daisy connaissait la fille idéale pour ça. Isabel Hernandez, une étudiante de Rhodes College à Memphis dans le Tennessee, était venue au St Hilda's College d'Oxford pour faire une thèse d'histoire. C'était une beauté, mais son intelligence impressionnante lui avait valu le surnom de « casse-couilles » parmi les garçons. Woody n'était pas du genre à se laisser impressionner. « Viens par ici, dit-elle à Isabel. Woody, je te présente mon amie Bella. Elle vient de San Francisco. Bella, je te présente Woody Dewar, de Buffalo. »

Ils se serrèrent la main. Grande, les cheveux noirs, le teint mat, Bella ressemblait beaucoup à Joanne Rouzrokh. Woody lui sourit et lui dit : « Que faites-vous à Londres ? » Daisy les laissa.

Elle servit le souper à minuit. Quand elle arrivait à se procurer des provisions américaines, ses invités avaient droit à des œufs et à du jambon ; sinon, ils se contentaient de sandwichs au fromage. Cela leur permettait de faire une pause pour bavarder un peu, comme durant l'entracte au théâtre. Elle remarqua que Woody Dewar était toujours en grande conversation avec Bella Hernandez. Après avoir vérifié que tout le monde était servi, elle alla rejoindre Lloyd dans un coin.

« J'ai décidé ce que j'allais faire après la guerre, si je suis encore en vie, déclara-t-il. À part t'épouser, bien entendu.

— Et quoi donc ?

— Je vais me présenter aux élections. »

Daisy frémit de joie. « Lloyd, c'est merveilleux ! » Elle lui sauta au cou et l'embrassa.

« Il est un peu tôt pour me féliciter. Je me suis porté candidat pour représenter le parti travailliste dans la circonscription de Hoxton, juste à côté de celle de Mam. Mais la section locale n'est pas obligée de me choisir. Et si elle le fait, je ne serai pas forcément élu. Le député actuel, un libéral, est solidement implanté.

— J'aimerais tant t'aider, dit-elle. Je pourrais être ton bras droit. J'écrirais tes discours – je suis sûre que je suis douée pour ça.

— J'en serais ravi.

— Alors c'est décidé ! »

Les plus âgés des invités partirent après le souper, mais les disques continuèrent de tourner et l'alcool de couler à flots, et la fête devint de plus en plus effrénée. Woody dansait le slow avec Bella. Daisy se demanda si c'était son premier flirt depuis Joanne.

L'ambiance se relâcha encore et on vit des couples s'éclipser dans les deux chambres. Il était impossible d'en fermer les portes – Daisy avait caché les clés –, aussi y trouvait-on parfois plusieurs couples à la fois, mais personne ne s'en offusquait. Un soir, Daisy avait surpris deux personnes dormant enlacées dans un placard à balais.

Son mari débarqua à une heure du matin.

Elle ne l'avait pas invité, mais il était accompagné de deux pilotes américains. Elle le laissa donc entrer avec un haussement d'épaules. Déjà passablement éméché, il dansa avec plusieurs infirmières puis l'invita poliment.

Était-il saoul ou était-il revenu à de meilleurs sentiments à son égard ? Le cas échéant, se laisserait-il convaincre de lui accorder le divorce ?

Elle accepta son invitation et ils entamèrent un swing. La plupart des personnes présentes ne savaient rien de leur situation, mais ceux qui en étaient informés étaient stupéfaits.

« J'ai lu dans la presse que tu as acheté un nouveau cheval de course, dit-elle pour faire la conversation.

— Lucky Laddie, confirma-t-il. Il m'a coûté huit mille quatre cents livres – un record.

— J'espère qu'il les vaut. » Elle adorait les chevaux et avait espéré qu'ils en élèveraient ensemble, mais il n'avait pas daigné partager cette passion avec elle. Un des nombreux motifs de frustration de leur vie conjugale.

Il dut lire dans son esprit. « Je t'ai déçue, n'est-ce pas ?

— Oui.

— Toi aussi, tu m'as déçu. »

Voilà qui était nouveau pour elle. Après quelques instants de réflexion, elle demanda : « En refusant de fermer les yeux sur tes infidélités ?

— Exactement. » L'ivresse le rendait honnête.

Elle décida de saisir l'occasion. « À ton avis, combien de temps allons-nous encore nous châtier mutuellement ?

— Nous châtier ? s'étonna-t-il. Comment ça ?

— En restant mariés, chacun de nous punit l'autre. Nous devrions divorcer, comme des gens raisonnables.

— Tu as peut-être raison, dit-il. Mais le lieu et le moment sont mal choisis pour en discuter. »

Elle reprit espoir. « Et si je venais te voir ? suggéra-t-elle. Quand nous serons tous les deux frais et dispos… et dégrisés ? »

Il hésita. « Entendu. »

Elle profita de son avantage. « Demain matin, par exemple ?

— Entendu.

— Je passe chez toi après l'office religieux. Disons à midi ?

— Entendu », répéta Boy.

4.

Woody raccompagnait Bella à l'appartement d'une de ses amies, à South Kensington, et elle l'embrassa alors qu'ils traversaient Hyde Park.

Il n'avait pas embrassé une seule femme depuis la mort de Joanne. Tout d'abord, il se figea. Il aimait beaucoup Bella : c'était la fille la plus intelligente qu'il ait connue depuis Joanne. Et à la façon dont elle se serrait contre lui pendant leur slow, il avait compris qu'il pourrait l'embrasser s'il le voulait. Il était resté sur la réserve. Il ne cessait de penser à Joanne.

Bella prit alors l'initiative.

Elle ouvrit les lèvres et il sentit le goût de sa langue, mais cela lui rappela encore Joanne et ses baisers. Il ne s'était écoulé que deux ans et demi depuis sa mort.

Son esprit cherchait des paroles de rejet courtois, mais son corps ne lui en laissa pas le temps. Il fut soudain brûlant de désir et se mit à l'embrasser goulûment.

Elle réagit avec fièvre à cet accès de passion. Lui saisissant les deux mains, elle les plaqua sur ses seins, aussi opulents que moelleux. Il ne put s'empêcher de gémir.

Il faisait noir et on y voyait à peine, mais, à en juger par les bruits étouffés montant des buissons alentour, ils n'étaient pas le seul couple à profiter de la nuit.

Elle colla son corps contre le sien et il sut qu'elle devait sentir son érection. Il était tellement excité qu'il craignit d'éjaculer d'une seconde à l'autre. Apparemment, le désir de Bella était égal au sien. Il sentit des doigts impatients déboutonner sa braguette puis des mains fraîches enserrer son sexe ardent. Elle le dégagea du pantalon puis, à sa grande surprise et à son immense plaisir, s'agenouilla devant lui. Dès que ses lèvres se refermèrent sur son gland, il se répandit dans sa bouche. Elle le suça et le lécha avec frénésie.

Une fois qu'il eut joui, elle continua d'embrasser son sexe jusqu'à ce qu'il s'amollisse. Puis elle le remit doucement en place et se redressa.

« C'était très excitant, murmura-t-elle. Merci. »

Il ne savait trop comment la remercier, lui aussi : alors il la prit dans ses bras et la serra très fort. Il était à deux doigts de pleurer de reconnaissance. Il n'avait pas compris à quel point il avait besoin, ce soir-là, de l'affection d'une femme. C'était comme si on l'avait libéré des ténèbres qui pesaient sur lui. « Je ne peux pas te dire… », commença-t-il, mais il était incapable de trouver les mots.

« Alors ne dis rien, répliqua-t-elle. De toute façon, je sais. Je l'ai senti. »

Ils reprirent la direction de son immeuble. Devant la porte, il demanda : « Est-ce qu'on pourrait… »

Elle lui posa l'index sur les lèvres pour le faire taire. « Va vite gagner la guerre », murmura-t-elle.

Puis elle entra.

5.

Lorsque Daisy se rendait à l'office, ce qui ne lui arrivait pas souvent, elle évitait désormais les églises huppées du West End, dont les fidèles l'avaient snobée, et prenait le métro pour se rendre à la chapelle évangélique du Calvaire, à Aldgate. Les différences doctrinales étaient considérables, mais elle s'en fichait. On chantait beaucoup mieux dans l'East End.

Lloyd et elle arrivèrent séparément. Les habitants d'Aldgate savaient qui était Daisy et appréciaient de voir sur leurs bancs une aristocrate en rupture avec son milieu, mais ils n'auraient sûrement pas toléré qu'une femme mariée entre dans l'église au bras de son amant. Pour citer Billy, le frère d'Ethel : « Jésus n'a pas condamné la femme adultère, mais il lui a dit de ne plus pécher. »

Elle pensa à Boy pendant le service. Était-il sincère la veille au soir, lorsqu'il avait prononcé ces paroles conciliantes, ou bien simplement attendri par l'alcool ? Il était allé jusqu'à serrer la main de Lloyd en partant. Cela voulait dire qu'il lui pardonnait, non ? Elle ne voulait pas s'emballer trop vite. Boy était l'individu le plus narcissique qu'elle ait jamais connu, plus encore que son propre père et son demi-frère Greg.

Après l'office, Daisy allait souvent déjeuner chez Eth Leckwith mais, ce jour-là, elle laissa Lloyd à sa famille et se hâta de regagner le West End. Elle frappa à la porte de la maison de Mayfair. Le majordome l'introduisit dans le salon.

Boy fit son entrée, furibond. « Qu'est-ce que ça veut dire, bon sang ? » rugit-il en jetant un journal à ses pieds.

Elle l'avait souvent vu d'humeur massacrante et il avait cessé de lui faire peur. Il n'avait osé lever la main sur elle qu'une seule fois. S'emparant d'un lourd chandelier, elle l'avait menacé. Il n'avait plus jamais recommencé.

Si elle n'était pas effrayée, elle n'en était pas moins déçue. Il avait été de si bonne humeur la veille au soir ! Mais peut-être lui ferait-elle tout de même entendre raison.

« Qu'est-ce qui te met dans cet état ? demanda-t-elle posément.

— Regarde cette feuille de chou. »

Elle se pencha pour ramasser le journal. C'était le *Sunday Mirror*, un journal socialiste très populaire. En première page figurait une photo de Lucky Laddie, le nouveau cheval de Boy, avec la légende suivante :

LUCKY LADDIE
SON PRIX :
28 MINEURS DE FOND

Les quotidiens de la veille s'étaient fait l'écho de la dépense somptuaire de Boy, mais le *Mirror* lui consacrait un violent éditorial, qui soulignait que le prix du cheval, soit huit mille quatre cents livres, représentait vingt-huit fois le montant de la prime versée à la veuve d'un mineur ayant péri dans un accident.

Or la fortune des Fitzherbert venait des mines de charbon.

« Mon père est furieux, dit Boy. Il comptait bien obtenir le poste de ministre des Affaires étrangères dans le gouvernement d'après-guerre. Ce torchon a probablement ruiné tous ses espoirs.

— Boy, peux-tu m'expliquer en quoi je suis responsable de cet article ? demanda Daisy d'une voix exaspérée.

— Regarde qui l'a signé ! »

Daisy s'exécuta.

Billy Williams
Député d'Aberowen

« L'oncle de ton amant !

— Penses-tu qu'il me consulte avant de prendre la plume ? »

Il agita un index menaçant. « Pour une raison qui m'échappe, cette famille nous déteste !

— Ils jugent injuste que tu gagnes des millions grâce au charbon alors que les mineurs sont aussi mal traités. C'est la guerre, tu sais.

— Toi aussi, tu vis de tes rentes, rétorqua-t-il. Et hier soir, dans ton appartement de Piccadilly, on ne peut pas dire que l'ambiance était à l'austérité.

— Tu as raison, admit-elle. Mais moi, je donne des réceptions pour remonter le moral de nos soldats. Toi, tu dépenses une fortune pour un cheval.

— C'est mon argent !

— Mais il provient du charbon.

— À force de coucher avec ce bâtard de Williams, tu es devenue une bolchevik comme lui.

— C'est une des choses qui nous séparent, une parmi tant d'autres. Boy, veux-tu vraiment rester mon époux ? Tu n'aurais aucun mal à trouver une compagne qui te convienne mieux que moi. La moitié des jeunes filles de Londres seraient ravies de devenir vicomtesse d'Aberowen.

— Il n'est pas question que je fasse ce plaisir à la famille Williams. À propos, j'ai appris hier soir que ton joli cœur voulait se présenter aux élections.

— Il fera un excellent député.

— Pas en s'encombrant de toi. Il sera forcément battu. C'est un foutu socialiste. Et toi, tu es une ex-fasciste.

— J'y ai pensé et je suis consciente que ça peut poser un léger problème…

— Un léger problème ? Une barrière insurmontable, oui ! Attends que les journaux soient au courant. Ils te cloueront au pilori comme je l'ai été aujourd'hui.

— J'imagine que tu te feras un malin plaisir d'informer le *Daily Mail*.

— Inutile – ses adversaires s'en chargeront. Retiens bien ce que je te dis. Si tu restes avec lui, Lloyd Williams n'a aucune chance d'être élu. »

6.

Durant les cinq premiers jours de juin, le lieutenant Woody Dewar et sa section de parachutistes furent cantonnés, en même temps qu'un millier d'autres soldats, sur un aérodrome du nord-ouest de Londres. On avait converti un hangar en dortoir géant où s'alignaient des centaines de lits de camp. Pour tromper leur ennui, les soldats pouvaient voir des films et écouter des disques de jazz.

Leur objectif était la Normandie. Grâce à de faux plans soigneusement élaborés, les Alliés s'étaient efforcés de convaincre l'état-major allemand que leur cible n'était autre que Calais, trois cents kilomètres plus loin au nord-est. Si la ruse avait pris, les forces d'invasion ne rencontreraient que peu de résistance, du moins dans un premier temps.

Les parachutistes constitueraient la première vague d'assaut, déclenchée en pleine nuit. La deuxième vague était composée de cent trente mille hommes, qu'une flotte de cinq mille navires allait débarquer à l'aube sur les plages normandes. À ce moment-là, les paras auraient déjà détruit les bastions situés à l'intérieur des terres et pris le contrôle de nœuds de communication stratégiques.

La section de Woody avait pour mission de s'emparer d'un pont qui traversait un cours d'eau dans une bourgade du nom d'Église-des-Sœurs, à quinze kilomètres du littoral. Ils devraient ensuite en assurer le contrôle et empêcher le passage de toutes les unités allemandes envoyées en renfort sur la côte, jusqu'à ce que la force d'invasion ait opéré la jonction avec eux. Et ils devaient à tout prix empêcher les Allemands de faire sauter le pont.

Pendant qu'ils attendaient le feu vert, Ace Webber avait entamé un tournoi de poker qui tournait au marathon ; il avait gagné un millier de dollars pour les reperdre aussitôt. Lefty Cameron ne cessait de nettoyer et de graisser de façon obsessionnelle son fusil semi-automatique M1 ultraléger à crosse pliante. Lonnie Callaghan et Tony Bonanio, qui ne s'appréciaient guère, allaient à la messe ensemble tous les jours. Sneaky Pete Schneider aiguisait le poignard de commando qu'il avait acheté à Londres, obtenant un fil aussi affûté que celui d'un rasoir. Patrick Timothy, qui ressemblait à Clark Gable, moustache comprise, jouait sans se lasser le même air sur son ukulélé, ce qui portait sur les nerfs de son entourage. Le sergent Defoe écrivait inlassablement à sa femme des lettres qu'il déchirait les unes après les autres. Mack Trulove et Smoking Joe Morgan se rasaient mutuellement le crâne, persuadés que cela faciliterait le traitement d'éventuelles blessures à la tête.

La plupart d'entre eux avaient un surnom. Woody avait découvert que le sien était Scotch.

Le jour J était prévu pour le dimanche 4 juin, mais fut repoussé pour cause de mauvais temps.

Le soir du lundi 5 juin, le colonel s'adressa à ses troupes. « Les gars ! s'écria-t-il. Cette nuit, nous envahissons la France ! »

On entendit un rugissement approbateur. Woody savoura l'ironie de la situation. Tous ces hommes bien au chaud et à l'abri étaient impatients de sauter d'un avion pour atterrir dans les bras de soldats ennemis décidés à les tuer.

Ils eurent droit à un repas de gala : steaks, côtes de porc, poulet, frites et crème glacée à volonté. Woody préféra s'abstenir. Contrairement à la plupart des hommes, il savait ce qui l'attendait et préférait ne pas avoir l'estomac plein. Il se contenta d'un café et d'un beignet. C'était du café américain, délicieux et odorant, bien différent de l'épouvantable mixture que servaient les Anglais, quand ils réussissaient à s'en procurer.

Il ôta ses bottes et s'allongea sur son lit de camp. Il songea à Bella Hernandez, à son sourire en coin et à la douceur de ses seins.

Soudain, une sirène mugit.

L'espace d'un instant, Woody crut se réveiller d'un cauchemar où il se voyait partir au combat pour tuer d'autres soldats. Puis il se rendit compte que c'était la réalité.

Tous enfilèrent leur tenue de saut et rassemblèrent leur paquetage, qui était bien trop lourd. Certaines choses étaient essentielles : une carabine, cent cinquante chargeurs de calibre 30, des grenades antichar, une petite bombe baptisée grenade gammon, des rations alimentaires, des pastilles pour purifier l'eau, une trousse de premiers secours contenant de la morphine. D'autres semblaient moins indispensables : une pelle pliante, un nécessaire de rasage, un manuel de conversation en français. Ils étaient tellement chargés que les plus chétifs eurent du mal à gagner les avions alignés sur le tarmac.

Ils allaient être transportés par des C-47 Skytrain. À sa grande surprise, Woody remarqua dans la pénombre qu'on avait peint sur leur carlingue des rayures noires et blanches. Son pilote, le capitaine Bonner, un type du Midwest mauvais coucheur, lui expliqua : « C'est pour éviter que les nôtres nous canardent. »

On pesa les hommes avant l'embarquement. Callaghan et Bonanio transportaient chacun un bazooka en pièces détachées dans des sacs attachés à leurs jambes, ce qui leur ajoutait quarante kilos. Au fur et à mesure que le total augmentait, le capitaine Bonner devenait plus râleur. « Vous m'alourdissez ! gronda-t-il en se tournant vers Woody. Jamais je ne pourrai faire décoller mon zinc !

— Je n'y suis pour rien, mon capitaine, dit Woody. Adressez-vous au colonel. »

Le sergent Defoe monta le premier et prit place à l'avant de l'appareil, sur un siège proche de la porte du cockpit. Il serait le dernier à sauter. Si l'un des hommes hésitait avant de plonger dans la nuit, Defoe l'y aiderait d'une solide bourrade.

Il fallut aider Callaghan et Bonanio à gravir la passerelle tant ils étaient lourds. En qualité que chef de section, Woody fut le dernier à embarquer. Il serait le premier à sauter et le premier à atterrir.

La cabine toute en longueur n'était équipée que de sièges métalliques alignés contre les cloisons. Les hommes eurent des difficultés à boucler leur ceinture vu les dimensions de leur barda, et certains ne prirent même pas cette peine. La porte se referma et les moteurs se mirent à vrombir.

Woody était partagé entre la peur et l'excitation. Paradoxalement, il était impatient de se battre. Il lui tardait de poser le pied par terre, de localiser l'ennemi et de tirer dessus. Il n'en pouvait plus d'attendre.

Il se demanda s'il reverrait jamais Bella Hernandez.

Il crut sentir l'avion peiner sur la piste, prenant laborieusement de la vitesse. On aurait dit qu'il se traînait sur le sol. Woody s'interrogea sur la longueur de cette foutue piste d'envol. Puis l'appareil s'éleva enfin dans les airs. Comme il n'avait pas l'impression de voler, Woody crut qu'il restait à quelques mètres du sol. Puis il regarda au-dehors. Assis près du septième et dernier hublot, juste à côté de la porte, il distingua les lumières embrumées de l'aérodrome qui s'éloignaient. Ils avaient décollé.

Le ciel était couvert, mais les nuages étaient légèrement lumineux, sans doute parce que la lune s'était levée derrière eux. Un feu de position bleu équipait la pointe de chaque aile et Woody vit son avion se mettre en formation en un V géant avec un grand nombre d'autres appareils.

La cabine était si bruyante qu'il fallait crier dans l'oreille de son voisin pour se faire entendre, et toute conversation cessa bientôt. Les hommes s'agitaient sur leurs sièges, cherchant en vain une position confortable. Certains fermèrent les yeux, mais Woody ne pensait pas qu'ils arrivaient à dormir.

Ils volaient à basse altitude, mille pieds à peine, et Woody apercevait parfois l'éclat terne d'une rivière ou d'un lac. À un moment, il entrevit une foule de gens, des centaines de visages tournés vers les avions rugissant dans le ciel. Ainsi qu'il le savait, plus de mille avions survolaient le sud de l'Angleterre en cet instant, offrant probablement un spectacle remarquable. Ces badauds regardaient l'histoire en train de s'écrire, songea-t-il, une histoire dont il ferait partie.

Au bout d'une demi-heure, ils passèrent au-dessus des plages de l'Angleterre et entreprirent la traversée de la Manche. L'espace d'un instant, la lune apparut entre les nuages et Woody découvrit les navires. Il n'en crut pas ses yeux. On aurait dit une ville flottante, composée de milliers de bateaux de toutes tailles progressant à la file, telles des maisons toutes différentes bien rangées le long des rues, des milliers de bâtiments à perte de vue. Avant qu'il ait pu attirer l'attention de ses camarades, la lune se cacha derrière les nuages et cette vision s'évanouit comme un rêve.

Les avions obliquèrent lentement vers la droite, comptant atteindre la France à l'ouest de la zone de largage puis remonter le littoral vers l'est, afin de repérer le terrain et de s'assurer que les parachutistes atterriraient à l'endroit prévu.

Les îles Anglo-Normandes, possessions britanniques bien que proches de la France, étaient occupées par l'Allemagne depuis la fin de la bataille de France, en 1940, et la défense anti-aérienne allemande ouvrit le feu lorsque l'armada d'avions les survola. En raison de leur faible altitude, les Skytrain étaient terriblement vulnérables. Woody songea qu'il risquait de se faire tuer avant même d'être arrivé sur le champ de bataille. L'idée qu'il puisse mourir pour rien le fit frémir.

Le capitaine Bonner se mit à zigzaguer pour éviter les tirs. Woody s'en réjouit, mais les hommes en subirent les conséquences. Tous furent pris du mal de l'air, Woody compris. Patrick Timothy fut le premier à vomir. La puanteur n'arrangea pas les choses. Ce fut ensuite au tour de Sneaky Pete, puis de plusieurs de ses camarades. Ils s'étaient empiffrés de steak et de crème glacée, qu'ils régurgitaient à présent dans leur totalité. L'odeur était atroce et le sol devint glissant.

Ils reprirent une trajectoire rectiligne en s'éloignant des îles. Quelques minutes plus tard, la côte française était en vue. L'avion vira sur l'aile gauche. Le copilote se leva pour aller parler au sergent Defoe, qui se tourna vers le peloton et leva les deux mains en montrant ses dix doigts. Largage dans dix minutes.

L'avion ralentit, passant de sa vitesse de croisière, deux cent soixante kilomètres à l'heure, à la vitesse optimale pour le saut en parachute, soit environ cent soixante kilomètres à l'heure.

Soudain, ils entrèrent dans un banc de brouillard, assez épais pour occulter le feu de position bleu à la pointe de l'aile. Le pouls de Woody s'accéléra. Les avions volant en formation serrée, ils étaient en danger. Et s'il mourait dans un accident d'avion sans jamais être allé au feu ! Bonner n'avait pas le choix et ne pouvait que continuer sur sa trajectoire en espérant que tous en feraient autant. Au moindre changement de cap, ce serait la collision.

L'appareil émergea du brouillard aussi soudainement qu'il y avait plongé. À droite comme à gauche, les autres étaient toujours en formation, ce qui tenait du miracle.

Presque aussitôt, les batteries antiaériennes ouvrirent le feu, et les obus se mirent à éclore autour des avions ainsi que des fleurs meurtrières. Dans de telles circonstances, Woody le savait, les pilotes avaient ordre de filer droit devant en maintenant leur vitesse. Mais Bonner n'en fit qu'à sa tête et quitta la formation. Le rugissement des moteurs devint assourdissant. L'avion se remit à zigzaguer. Il piqua du nez en cherchant à prendre de la vitesse. En jetant un coup d'œil par le hublot, Woody constata que plusieurs autres pilotes se montraient tout aussi indisciplinés. Ils n'avaient plus qu'une idée en tête : sauver leur peau.

Un voyant rouge s'alluma au-dessus de la porte : plus que quatre minutes.

Woody était certain que ce signal était prématuré, que Bonner voulait se débarrasser d'eux pour gagner un abri sûr. Toutefois c'était le capitaine qui avait les cartes en main, et il ne pouvait pas contester sa décision.

Il se leva. « Debout et accrochez-vous ! » hurla-t-il. La plupart des hommes ne pouvaient pas l'entendre, mais ils savaient ce qu'il disait. Ils se levèrent et chacun d'eux attacha sa sangle d'ouverture automatique au câble courant le long du plafond, afin de ne pas être projeté accidentellement par la porte. Celle-ci s'ouvrit et le vent s'y engouffra. L'avion volait encore trop vite. Ils seraient désagréablement secoués en sautant à cette vitesse, mais ce n'était pas le plus grave. Les hommes se poseraient trop loin les uns des autres et Woody perdrait du temps à rassembler sa section. Du coup, il atteindrait son objectif avec un certain retard, ce qui affecterait tout le déroulement de sa mission. Il pesta contre Bonner.

Celui-ci ne cessait de virer d'un côté et de l'autre pour éviter les obus. Les hommes avaient le plus grand mal à ne pas glisser sur le sol couvert de vomissures.

Woody jeta un coup d'œil au-dehors. Bonner avait perdu de l'altitude en essayant de gagner de la vitesse, et l'avion volait à présent à cinq cents pieds – beaucoup trop bas. Les parachutes n'auraient peut-être pas le temps de s'ouvrir complètement avant que les hommes touchent le sol. Il hésita, puis fit signe à son sergent de s'approcher.

L'ayant rejoint, Defoe regarda à l'extérieur puis secoua la tête. Il colla ses lèvres à l'oreille de Woody et cria : « La moitié des gars se casseront une patte si on saute à cette altitude. Et ceux qui portent les bazookas sont sûrs d'y passer. »

Woody prit sa décision.

« Veillez à ce que personne ne saute ! » cria-t-il à Defoe.

Puis il décrocha sa sangle et se dirigea vers le cockpit, se frayant un chemin entre les deux rangées de parachutistes. Le pilote était assisté de deux hommes d'équipage. « Grimpez ! Prenez de la hauteur ! hurla Woody à s'en époumoner.

— Retournez dans la cabine et sautez !

— Personne ne sautera à cette altitude ! » Woody se pencha et pointa l'index sur l'altimètre, qui affichait quatre cent quatre-vingts pieds. « C'est du suicide !

— Sortez de mon cockpit, lieutenant. C'est un ordre. »

Woody aurait dû obéir à un supérieur, mais il tint bon. « Pas avant que vous n'ayez repris de l'altitude.

— Si vous ne sautez pas tout de suite, la zone cible sera bientôt derrière nous ! »

Woody perdit son calme. « Grimpez, espèce de con ! Grimpez, je vous dis ! »

Bonner avait l'air furieux, mais Woody ne bougea pas. Il savait que le pilote ne voudrait pas regagner sa base avec un avion encore plein. On ordonnerait une enquête pour savoir ce qui s'était passé. Et Bonner avait enfreint plusieurs ordres cette nuit-là. Poussant un juron, il tira sur le manche à balai. Le nez de l'avion se redressa et il regagna de l'altitude en perdant de la vitesse.

« Ça y est ? Vous êtes content ? gronda Bonner.

— Vous rigolez ? » Si Woody retournait dans la cabine, Bonner s'empresserait d'inverser la manœuvre. « On sautera à mille pieds, pas avant. »

Bonner mit les gaz. Woody ne quittait pas l'altimètre des yeux.

Quand le cadran afficha mille pieds, il retourna auprès de ses hommes. Arrivé devant la porte, il se raccrocha, regarda au-dehors, lança le signal attendu – les deux pouces levés – puis sauta.

Son parachute s'ouvrit aussitôt. Il tomba à toute vitesse pendant que sa corolle se déployait, puis sa chute se ralentit. Quelques secondes plus tard, il plongeait dans l'eau. Il éprouva un instant de panique, se demandant si ce poltron de Bonner les avait largués en pleine mer. Puis ses pieds touchèrent la terre, ou plutôt la boue, et il comprit qu'il se trouvait dans un pré inondé.

La soie du parachute le recouvrit. Il s'en extirpa et se défit de son harnachement.

Debout dans cinquante centimètres d'eau, il jeta autour de lui un regard circulaire. Il avait atterri dans une prairie inondable ou dans un champ noyé par les Allemands pour ralentir une éventuelle armée d'invasion. Il ne distingua personne dans la pénombre, ni ami ni ennemi, ni homme ni animal.

Il consulta sa montre – trois heures quarante – puis sortit sa boussole et s'orienta.

Il sortit ensuite son M1 de son paquetage et en déplia la crosse. Il mit un chargeur en place puis arma le fusil. Et pour finir, il débloqua le levier de sûreté.

Plongeant une main dans sa poche, il en sortit un cricket, un petit objet en tôle ressemblant à un jouet d'enfant. Quand on appuyait dessus, il émettait un cliquetis caractéristique. Tous les soldats en avaient reçu un afin de pouvoir se reconnaître dans le noir sans risquer de se trahir en prononçant un mot de passe en anglais.

Une fois prêt, il jeta un nouveau regard alentour.

Puis il se décida et pressa deux fois sur le cricket. Au bout de quelques instants, il perçut un cliquetis en réponse droit devant lui.

Il s'avança dans l'eau. Une odeur de vomi lui chatouilla les narines. « Qui va là ? demanda-t-il à voix basse.

— Patrick Timothy.

— Lieutenant Dewar. Suis-moi. »

Timothy avait été le deuxième à sauter, et Woody en déduisit que s'il continuait dans la même direction, il avait de bonnes chances de trouver les autres.

Cinquante mètres plus loin, il butait sur Mack et Smoking Joe, qui s'étaient posés très près l'un de l'autre.

Ils émergèrent du champ inondé pour s'engager sur une route étroite, où ils localisèrent leurs premières pertes. Lonnie et Tony, alourdis par les bazookas, avaient atterri trop brutalement. « Je crois bien que Lonnie est mort », dit Tony. Woody vérifia : il ne se trompait pas. Lonnie ne respirait plus. Apparemment, il s'était brisé la nuque. Quant à Tony, il était incapable de bouger et Woody en déduisit qu'il s'était cassé la jambe. Il lui administra de la morphine puis le traîna dans un pré voisin. Il ne lui restait qu'à attendre les secours.

Woody ordonna à Mack et à Smoking Joe de dissimuler le cadavre de Lonnie de crainte qu'il ne conduise les Allemands à Tony.

Il tenta de distinguer le paysage qui l'entourait, dans l'espoir de reconnaître un repère figurant sur sa carte. La tâche semblait impossible, surtout en pleine nuit. Comment conduire ses hommes à leur objectif s'il ne savait pas où il était ? Il n'était sûr que d'une chose : ils ne se trouvaient pas dans la zone de largage prévue.

Il entendit un bruit suspect et aperçut presque immédiatement une lumière.

Il fit signe aux autres de se jeter à terre.

Les parachutistes n'étant pas censés utiliser de lampe torche et les Français étant soumis au couvre-feu, ils avaient probablement affaire à un soldat allemand.

Woody distingua une bicyclette dans l'obscurité.

Il se leva, empoigna sa carabine. Il était prêt à abattre le cycliste, mais ne put se résoudre à tirer sans sommation. Il s'écria : « Halte ! *Arrêtez !* »

Le cycliste obtempéra et dit : « Salut, mon lieutenant. » Woody reconnut la voix d'Ace Webber.

Il baissa son arme. « Où as-tu trouvé ce vélo ? demanda-t-il d'une voix incrédule.

— Devant une ferme », répondit Ace, laconique.

Woody fit prendre à sa petite troupe la direction dont venait Ace, jugeant qu'il avait plus de chances de récupérer les autres du même côté. Il ne cessait d'examiner le terrain dans l'espoir de trouver des repères figurant sur sa carte, mais il faisait bien trop noir. Il se sentait stupide et inutile. Pourtant, c'était lui l'officier et c'était à lui de résoudre de tels problèmes.

Il retrouva d'autres membres de sa section et ils arrivèrent devant un moulin à vent. Décidant qu'il ne pouvait pas continuer à errer à l'aveuglette, il tambourina sur la porte de la maison du meunier.

Une fenêtre s'ouvrit à l'étage et une voix demanda en français : « *Qui est là ?*

— Les Américains, répondit Woody. *Vive la France !*

— Que voulez-vous ?

— *Vous libérer*, dit Woody dans son français de collège. *Mais d'abord, je voudrais bien savoir où je suis.* »

Le meunier s'esclaffa et lança : « *Je descends.* »

Une minute plus tard, dans la cuisine, Woody étalait sa carte en soie sur une table bien éclairée. Le meunier lui montra où il se trouvait. La situation n'était pas aussi grave que l'avait craint Woody. Malgré l'affolement du capitaine Bonner, ils n'étaient qu'à six ou sept kilomètres d'Église-des-Sœurs. Le meunier lui indiqua sur la carte l'itinéraire le plus pratique.

Une jeune fille d'environ treize ans en chemise de nuit s'avança d'un air timide. « Maman dit que vous êtes américain, dit-elle à Woody.

— C'est exact, *mademoiselle.*

— Vous connaissez Gladys Angelus ? »

Woody éclata de rire. « Il se trouve que je l'ai rencontrée un jour, chez le père d'un de mes amis.

— Elle est vraiment belle ?

— Encore plus belle que dans les films.

— J'en étais sûre ! »

Le meunier lui offrit du vin. « Non, merci, dit Woody. Quand nous aurons gagné, peut-être. » L'homme l'embrassa sur les deux joues.

Woody ressortit et conduisit sa section sur la route d'Église-des-Sœurs. Sur ses dix-huit parachutistes, il en avait déjà rassemblé neuf, lui-même compris. Ses pertes s'élevaient à deux hommes, un mort et un blessé, plus sept qu'il n'avait pas encore retrouvés. Il avait ordre de ne pas consacrer trop de temps à regrouper ses troupes. Il devait gagner son objectif dès qu'il aurait suffisamment d'hommes pour accomplir sa mission.

L'un des sept manquants se manifesta aussitôt. Sneaky Pete sortit d'un fossé et rejoignit ses camarades en leur lançant un « Salut, les gars », comme si c'était la chose la plus naturelle au monde.

« Qu'est-ce que tu fichais là-dedans ? lui demanda Woody.

— Je vous avais pris pour des Allemands, répondit Pete. Je me planquais. »

Woody avait aperçu un bout de parachute dans le fossé. Pete devait se cacher là depuis son atterrissage. De toute évidence, il avait paniqué et était resté blotti au fond de la rigole. Mais Woody feignit de croire sa version des faits.

Celui qu'il tenait vraiment à retrouver était le sergent Defoe. Il avait compté s'appuyer sur cet homme d'expérience. Or il demeurait invisible.

Comme ils approchaient d'un carrefour, ils entendirent des voix. Woody identifia le bruit d'un moteur tournant au ralenti et dénombra deux ou trois interlocuteurs distincts. Il ordonna à ses hommes de se mettre à plat ventre et d'avancer en rampant.

Un motocycliste s'était arrêté pour discuter avec deux hommes à pied. Tous trois étaient en uniforme. Ils parlaient allemand. Un petit bâtiment se dressait au carrefour, une taverne ou une boulangerie.

Il décida d'attendre. Peut-être finiraient-ils par partir. Il voulait que son groupe se fasse repérer le plus tard possible.

Au bout de cinq minutes, sa patience était à bout. Il se retourna. « Patrick Timothy ! » souffla-t-il.

Une voix lança tout bas : « Pat ! Scotch te demande. »

L'intéressé rampa jusqu'à lui.

Woody l'avait vu jouer au base-ball et savait que c'était un excellent lanceur. « Balance une grenade sur cette moto », lui ordonna-t-il.

Timothy attrapa une grenade dans son barda, la dégoupilla et la lança.

On entendit un bruit métallique. « Qu'est-ce que c'est ? » demanda l'un des soldats en allemand. Puis la grenade explosa.

Il y eut en fait deux déflagrations. La première projeta les trois Allemands à terre. Puis le réservoir de la moto explosa dans un jet de flammes qui brûla les soldats, répandant une odeur de chair carbonisée.

« Restez où vous êtes ! » ordonna Woody à sa section. Il observa le bâtiment. Y avait-il quelqu'un à l'intérieur ? Cinq minutes passèrent sans que personne n'ouvre ni porte ni fenêtre. Soit la maison était déserte, soit ses occupants s'étaient cachés sous leurs lits.

Woody se leva et fit signe à ses hommes de poursuivre. Il ressentit une étrange sensation en enjambant les cadavres brûlés des trois Allemands. Il avait décidé la mort de ces hommes – qui avaient un père et une mère, une épouse ou une fiancée, peut-être même des fils ou des filles. Il ne restait plus d'eux que des poupées de sang et de chair calcinée. Woody aurait dû éprouver un sentiment de triomphe. C'était son premier contact avec l'ennemi et il l'avait vaincu. Or il se sentait un peu nauséeux.

Passé le carrefour, il fit avancer ses hommes à marche forcée, leur interdisant de bavarder comme de fumer. Pour garder ses forces, il mangea une barre chocolatée qui avait un goût de mastic sucré.

Au bout d'une demi-heure, il entendit un véhicule et donna l'ordre à ses hommes de se cacher dans les champs. La voiture roulait à vive allure, tous feux allumés. Sans doute était-elle allemande, mais les Alliés comptaient transporter des Jeep par planeur, ainsi que des canons antichar et autres pièces d'artillerie, de sorte qu'il avait peut-être affaire à des camarades. Il se tapit derrière un buisson et ouvrit les yeux.

Le véhicule roulait trop vite pour être identifié. Il se demanda s'il aurait dû ordonner de faire feu sur lui. Non, décida-t-il ; tout bien considéré, mieux valait se concentrer sur leur mission.

Ils traversèrent trois hameaux que Woody réussit à localiser sur sa carte. Quelques chiens aboyèrent, mais personne ne vint voir ce qui se passait. De toute évidence, les Français avaient appris sous l'Occupation à se mêler de leurs affaires. Il trouvait un peu sinistre de progresser ainsi en terre inconnue, dans les ténèbres, armé jusqu'aux dents, en marchant devant des maisons paisibles dont les habitants dormaient sans se douter de la présence de soldats dans les parages.

Ils arrivèrent enfin dans les faubourgs d'Église-des-Sœurs. Woody décréta une brève pause. Ils se dirigèrent vers un bosquet et s'assirent dans l'herbe. Ils burent à leurs gourdes et mangèrent quelques rations. Woody leur interdisait toujours de fumer : la braise d'une cigarette pouvait être visible de très, très loin.

La route qu'ils suivaient devait mener droit au pont, estima-t-il. Ils ne disposaient d'aucune information sur les effectifs qui le gardaient. Puisque les Alliés le jugeaient important, il supposait qu'il en allait de même des Allemands et qu'il devait s'attendre à y trouver une ou plusieurs sentinelles ; mais il ignorait tout de l'armement qu'elles pourraient leur opposer. Pour planifier son assaut, il devait d'abord examiner sa cible.

Dix minutes plus tard, il donna le signal. Inutile d'imposer le silence aux gars : tous sentaient le danger à présent. Ils avancèrent à pas de loup, longeant des maisons, une église et des boutiques, rasant les murs et ouvrant l'œil, sursautant au moindre bruit suspect. En entendant une quinte de toux derrière une fenêtre ouverte, Woody faillit lâcher une rafale.

Église-des-Sœurs était un gros village plutôt qu'une petite ville, et Woody aperçut l'éclat de la rivière plus tôt qu'il ne l'aurait cru. Il leva la main pour arrêter ses hommes. La grand-rue descendait en pente douce jusqu'au pont, ce qui lui offrait une vue parfaite sur celui-ci. La rivière faisait une trentaine de mètres de large et le pont n'avait qu'une arche. Il devait être très vieux, car il était si étroit que deux voitures n'auraient pas pu s'y croiser.

Malheureusement, les Allemands avaient édifié un blockhaus sur chaque rive, un dôme de béton où s'ouvraient des meurtrières horizontales. Deux sentinelles allaient et venaient de l'un à l'autre. Pour le moment, elles se trouvaient chacune à une extrémité du pont. La plus proche, le visage collé à une meurtrière, discutait probablement avec un soldat posté dans le blockhaus. Puis les deux hommes s'avancèrent l'un vers l'autre et, arrivés au milieu du pont, contemplèrent les eaux noires de la rivière. Comme ils ne semblaient pas très inquiets, Woody en déduisit qu'ils ne savaient pas encore que l'invasion avait commencé. Leur vigilance ne s'était pas relâchée pour autant : ils semblaient attentifs au moindre bruit suspect et ne cessaient de regarder à droite et à gauche.

Impossible de deviner le nombre des soldats en place dans les blockhaus et la nature de leur armement. Auraient-ils à affronter des fusils ou des mitrailleuses ? Cela ferait une sacrée différence.

Woody regrettait amèrement son manque d'expérience. Qu'était-il censé faire en pareil cas ? Sans doute étaient-ils des milliers de jeunes officiers comme lui contraints d'improviser sur le champ de bataille. Si seulement le sergent Defoe était à ses côtés.

La meilleure méthode pour neutraliser un blockhaus, c'était de s'en approcher discrètement et de jeter une grenade par une meurtrière. Un commando particulièrement habile n'aurait aucun mal à gagner le plus proche sans se faire repérer. Mais il fallait que Woody neutralise les deux en même temps, faute de quoi l'attaque du premier alerterait les occupants du second.

Comment atteindre le blockhaus le plus éloigné à l'insu des deux sentinelles ?

Il sentit ses hommes s'impatienter. Ils n'aimaient pas que leur chef fasse preuve d'indécision.

« Sneaky Pete, appela-t-il. Rampe jusqu'au blockhaus le plus proche et balance une grenade par la meurtrière. »

Pete avait l'air mort de peur, mais il obtempéra : « À vos ordres, mon lieutenant. »

Woody appela ensuite les deux meilleurs tireurs du peloton. « Smoking Joe, Mack. Choisissez chacun une sentinelle. Dès que Pete aura lancé sa grenade, abattez-les. »

Les deux hommes acquiescèrent et levèrent leurs armes.

En l'absence de Defoe, il décida de faire d'Ace Webber son second. Il appela trois autres soldats et leur dit : « Suivez Ace. Dès que ça commencera à tirer, traversez le pont en courant et arrosez le blockhaus sur l'autre rive. Si vous êtes assez rapides, vous les prendrez par surprise.

— À vos ordres, mon lieutenant, répondit Ace. Ces salopards ne vont rien piger à ce qui leur arrive. » Sa gouaille masquait sa peur, devina Woody.

« Les autres, suivez-moi jusqu'au blockhaus le plus proche. »

Woody avait des scrupules à confier à Ace et à son trio la mission la plus dangereuse, se réservant celle qui semblait relativement inoffensive, mais on lui avait dit et répété qu'un officier ne devait jamais risquer sa vie sans nécessité, de peur de laisser ses hommes sans commandement.

Ils se dirigèrent vers le pont, Pete ouvrant la marche. L'instant était périlleux. Même de nuit, dix hommes avançant de concert dans une rue ne pouvaient pas passer inaperçus. Il suffirait qu'une sentinelle regarde dans leur direction pour percevoir un mouvement.

Si l'alarme était donnée trop tôt, Sneaky Pete risquait de ne pas atteindre le blockhaus et la section perdrait l'avantage de la surprise.

Ce fut une longue marche.

Pete fit halte à un coin de rue. Woody comprit qu'il attendait que la sentinelle la plus proche s'éloigne du blockhaus et lui tourne le dos.

Les deux tireurs d'élite s'abritèrent et prirent position.

Woody mit un genou à terre et fit signe aux autres de l'imiter. Tous avaient les yeux rivés sur la sentinelle.

L'homme tira une longue bouffée de sa cigarette, la lâcha, l'écrasa sous sa botte et exhala un nuage de fumée. Puis il se redressa, ajusta la lanière de son fusil sur son épaule et se mit en marche.

L'autre sentinelle en fit autant.

Pete courut jusqu'au pâté de maisons suivant et arriva au bout de la rue. Il se jeta à quatre pattes et traversa la chaussée en rampant. Arrivé près du blockhaus, il se releva.

Personne ne l'avait repéré. Les deux sentinelles s'approchaient l'une de l'autre.

Pete prit une grenade et la dégoupilla. Puis il attendit quelques secondes. Il ne voulait pas laisser aux soldats ennemis le temps de la lui renvoyer, comprit Woody.

Tendant la main vers la meurtrière, il laissa doucement tomber la grenade à l'intérieur du blockhaus.

Les carabines de Joe et de Mack crépitèrent. La sentinelle la plus proche s'effondra, mais l'autre était indemne. Au lieu de s'enfuir, le soldat mit courageusement un genou à terre et saisit son fusil. Mais il fut trop lent : les carabines tirèrent à nouveau, presque simultanément, et il tomba sans avoir eu le temps de riposter.

La grenade de Pete explosa dans le premier blockhaus avec un bruit étouffé.

Woody courait déjà, ses gars sur les talons. En quelques secondes, il atteignit le pont.

Le blockhaus était muni d'une porte en bois assez basse. Woody l'ouvrit et entra. Trois corps vêtus de l'uniforme allemand gisaient sur le sol.

Il se posta devant l'une des meurtrières. Ace et son trio traversaient le pont en courant sans cesser de tirer comme des forcenés. Le pont ne faisait que trente mètres, mais c'était quinze de trop. Comme ils arrivaient au milieu, une mitrailleuse ouvrit le feu. Les Américains étaient pris au piège dans un étroit couloir, sans couverture. Une rafale sèche, et ils tombèrent tous les quatre. L'ennemi continua de les arroser par acquit de conscience – achevant du même coup les deux sentinelles, si elles n'étaient pas déjà mortes.

Les balles cessèrent de voler et le calme revint.

Silence total.

« Dieu tout-puissant », souffla Lefty Cameron tout près de Woody.

Il en aurait pleuré. Il avait causé la mort de neuf hommes, quatre Américains et cinq Allemands, sans pour autant parvenir à ses fins. L'ennemi tenait toujours l'autre rive et était en mesure d'arrêter l'avance des Alliés.

Il lui restait cinq hommes. S'ils tentaient eux aussi de franchir le pont, ils n'y survivraient pas. Il fallait qu'il trouve une autre stratégie.

Il observa le village. Que faire ? Si seulement il avait eu un char.

Il fallait agir vite. Peut-être y avait-il dans ce petit bourg d'autres soldats ennemis, que les détonations n'auraient pas manqué d'alerter. Ils ne tarderaient pas à réagir. Il pourrait leur résister en tenant les deux blockhaus. Autrement, il était dans le pétrin.

Si ses hommes ne pouvaient pas traverser le pont, peut-être parviendraient-ils à franchir la rivière à la nage. Il décida de jeter un coup d'œil au rivage. « Mack et Smoking Joe, dit-il. Tirez sur l'autre blockhaus. Essayez de viser les meurtrières. Occupez-les pendant que je pars en reconnaissance. »

Les carabines parlèrent et il sortit.

Il réussit à s'abriter derrière le blockhaus pendant qu'il jetait un coup d'œil au rivage en amont. Puis il lui fallut traverser la chaussée pour examiner l'autre rive. Par bonheur, l'ennemi s'abstint de faire feu.

Il n'y avait pas de quai. La berge se réduisait à une pente qui descendait vers l'eau. L'autre rive lui parut identique, mais l'obscurité ne lui permettait pas d'en être sûr. Un bon nageur traverserait sans problème. Une fois sous le pont, il ne serait pas facile à repérer depuis la position ennemie. Et arrivé sur l'autre rive, il lui suffirait de répéter la manœuvre de Sneaky Pete et de jeter une grenade dans le blockhaus.

En examinant le pont de plus près, il eut une meilleure idée. Une corniche de trente centimètres de large courait sous le parapet. Un homme qui aurait le cran de ramper dessus resterait hors de vue.

Il regagna le blockhaus pris à l'ennemi. Lefty Cameron était le plus petit des survivants. Et puis c'était un bagarreur, pas du genre à se dégonfler. « Lefty, dit Woody. Il y a une corniche qui longe le pont par l'extérieur, sous le parapet. Elle sert sans doute aux ouvriers qui font des réparations. Tu vas passer par là pour rejoindre l'autre rive et balancer une grenade dans le blockhaus.

— Ça gaze », fit Lefty.

Décidément, il avait du cran. Il venait pourtant de voir périr quatre de ses camarades.

Woody se tourna vers Mack et Smoking Joe. « Couvrez-le. » Ils ouvrirent le feu.

« Et si je me retrouve dans la flotte ? demanda Lefty.

— Tu tomberas de cinq ou six mètres à peine, répondit Woody. Il n'y a aucun danger.

— Okay », dit Lefty. Il se dirigea vers la porte. « Seulement, je ne sais pas nager », ajouta-t-il. Et il sortit.

Woody le vit traverser la rue en courant. Il jeta un coup d'œil par-dessus le parapet, l'enjamba et se laissa glisser, jusqu'à être hors de vue.

« C'est bon, lança-t-il aux deux autres. Cessez le feu. Il est en route. »

Ils restèrent tous aux aguets. Rien ne bougeait. Woody remarqua que le jour se levait : le village lui apparaissait de plus en plus nettement. Mais aucun villageois ne montra le bout de son nez : ils étaient trop avisés pour ça. Peut-être des soldats allemands se rassemblaient-ils dans une rue voisine, mais il n'entendait rien. Il se rendit compte qu'il tendait l'oreille en quête d'un bruit d'éclaboussures, redoutant que Lefty ne tombe dans la rivière.

Un chien traversa le pont en trottinant, un bâtard de taille moyenne à la queue fièrement dressée vers le ciel. Il renifla les cadavres avec curiosité puis poursuivit sa route d'un air décidé, comme s'il avait un rendez-vous de la plus haute importance.

Woody le regarda passer près de l'autre blockhaus et disparaître sur l'autre rive.

Puisque le jour se levait, les forces alliées avaient dû commencer à débarquer sur les plages. À en croire un expert, c'était l'opération amphibie la plus importante de l'histoire. Il se demanda si les Alliés rencontraient une forte résistance. Il n'y a rien de plus vulnérable qu'un fantassin chargé comme une mule s'avançant dans les vagues, en direction d'une plage constituant une zone de tir idéale pour les soldats ennemis planqués dans les dunes. Woody se sentait bien plus en sécurité dans son blockhaus.

Lefty prenait son temps. Était-il tombé dans l'eau sans faire de bruit ? Avait-il rencontré un obstacle imprévu ?

Enfin Woody le vit, mince silhouette en kaki se hissant sur le parapet près de l'autre rive. Il retint son souffle. Lefty se laissa tomber à genoux, rampa jusqu'au blockhaus et se redressa, le dos collé au mur de béton. Il attrapa une grenade de la main gauche. Il la dégoupilla, attendit deux ou trois secondes puis bondit et la jeta à travers la meurtrière.

Woody entendit le fracas d'une explosion et aperçut un violent éclair derrière les meurtrières. Lefty leva les bras au-dessus de sa tête, prenant une pose de champion.

« Planque-toi, connard », dit Woody, sachant pourtant que l'autre ne pouvait pas l'entendre. Un soldat allemand se cachait peut-être dans un bâtiment tout proche, brûlant du désir de venger ses amis.

Mais aucune balle ne fendit l'air et, après avoir esquissé un pas de danse, Lefty entra dans le blockhaus et Woody poussa un soupir de soulagement.

Cependant, il n'était pas encore tout à fait rassuré. Il suffirait d'une attaque surprise d'une vingtaine d'Allemands pour reprendre le pont. Et tout ce qu'ils avaient fait n'aurait servi à rien.

Il s'obligea à patienter une minute de plus pour s'assurer qu'aucun soldat ennemi n'apparaissait. Toujours rien. Apparemment, les seuls Allemands présents à Église-des-Sœurs étaient ceux qui gardaient le pont : sans doute étaient-ils relevés toutes les douze heures par les soldats d'une caserne distante de quelques kilomètres.

« Smoking Joe, dit-il. Débarrasse-nous de ces trois cadavres. Balance-les à la flotte. »

Joe traîna les corps à l'extérieur du blockhaus et les fit basculer dans l'eau. Les cadavres des deux sentinelles suivirent le même chemin.

« Pete et Mack, ordonna Woody. Allez rejoindre Lefty dans l'autre blockhaus. Restez sur le qui-vive, tous les trois. On n'a pas encore tué tous les Allemands de France. Si vous voyez des troupes ennemies approcher de votre position, pas d'hésitation, pas de négociation : tirez. »

Les deux hommes sortirent et traversèrent le pont au petit trot.

Il y avait désormais trois Américains dans chaque blockhaus. Si les Allemands tentaient de reprendre le pont, ils auraient de sacrées difficultés, notamment en plein jour.

Woody se rendit compte que les cadavres américains qui gisaient sur le pont risquaient d'apprendre aux forces ennemies que les blockhaus avaient été pris. Mieux valait conserver l'effet de surprise.

Autrement dit, il fallait les jeter à l'eau, eux aussi.

Il donna ses instructions aux autres et sortit.

L'air matinal était frais et clair.

Arrivé au milieu du pont, il prit le pouls de chacun des hommes tombés, mais aucun doute ne subsistait : ils étaient tous morts.

Un par un, il souleva ses camarades et les jeta par-dessus le parapet.

Ace Webber était le dernier. Comme il disparaissait dans les eaux, Woody déclara : « Reposez en paix, mes potes. » Il resta un moment immobile, la tête basse et les yeux clos.

Lorsqu'il se retourna, le soleil se levait.

7.

Les Alliés craignaient par-dessus tout que les Allemands envoient des renforts massifs en Normandie et lancent une contre-attaque suffisamment puissante pour les rejeter à la mer, rééditant ainsi le fiasco de Dunkerque.

Lloyd Williams faisait partie de ceux qui devaient s'assurer que cela n'arriverait pas.

Depuis le débarquement, l'évacuation des prisonniers évadés était moins prioritaire, et il travaillait désormais en liaison avec la Résistance française.

Dès la fin mai, la BBC avait transmis des messages codés pour déclencher une campagne de sabotage dans la France occupée. Les premiers jours de juin, des centaines de fils téléphoniques furent sectionnés, le plus souvent dans des lieux difficiles d'accès. Des dépôts de carburant furent incendiés, des arbres abattus pour bloquer les routes, les pneus de véhicules allemands crevés.

Lloyd assistait des cheminots majoritairement communistes, qui avaient baptisé leur réseau « Résistance-Fer ». Des années durant, ils avaient harcelé les nazis par leurs activités subversives. Il arrivait que les transports de troupes allemands soient détournés sur des lignes secondaires et échouent à des kilomètres de leur destination. Les locomotives accumulaient les pannes inexplicables, quand ce n'était pas les convois qui déraillaient. La situation était si grave que l'occupant avait fait venir des employés des chemins de fer allemands. Mais les nazis n'étaient pas au bout de leurs peines. Au printemps 1944, les résistants s'attaquèrent à l'infrastructure du réseau. Ils firent sauter les voies ferrées et sabotèrent les grues nécessaires pour déplacer les wagons accidentés.

Les nazis ne se laissèrent pas faire. Ils fusillèrent des centaines de cheminots et en déportèrent des milliers dans les camps. Mais la campagne de sabotage ne fit que s'intensifier et, quand vint le jour J, le trafic était totalement paralysé dans certaines régions de France.

Le 7 juin, Lloyd était allongé sur le ventre au sommet d'un talus près de la ligne menant à Rouen, la capitale de la Normandie, juste avant l'entrée d'un tunnel. De son poste d'observation, il voyait les trains approcher à plus d'un kilomètre de distance.

Il était accompagné de deux hommes ayant pour noms de code Légionnaire et Cigare. Le premier était le chef local de la Résistance, le second un cheminot. Lloyd avait apporté la dynamite. La fourniture d'armes était la principale contribution des Britanniques à la Résistance.

Les trois hommes se dissimulaient derrière de hautes herbes parsemées de fleurs sauvages. Le genre d'endroit où on aurait plaisir à amener une jolie fille par une aussi belle journée, se dit Lloyd. Daisy aurait adoré.

Un train apparut dans le lointain. Cigare l'observa attentivement tandis qu'il approchait. C'était un petit homme d'une soixantaine d'années, sec comme une trique, au visage ridé de gros fumeur. Le train était encore à quatre ou cinq cents mètres d'eux lorsqu'il secoua la tête en signe de dénégation. Ce n'était pas celui qu'ils attendaient. La locomotive passa en crachant sa fumée et disparut dans le tunnel. Elle tractait quatre wagons bondés de passagers, un mélange de civils et de militaires. Lloyd traquait une proie plus importante.

Légionnaire consulta sa montre. Avec son teint basané et son épaisse moustache noire, il devait avoir un ancêtre nord-africain dans son arbre généalogique. Il était nerveux : ils étaient particulièrement exposés sur ce talus. Plus ils s'attardaient, plus ils couraient le risque de se faire repérer. « Il y en a encore pour longtemps ? » interrogea-t-il d'une voix anxieuse.

Cigare haussa les épaules. « On verra bien.

— Vous pouvez partir maintenant si vous voulez, intervint Lloyd en français. Tout est en place. »

Légionnaire ne répondit pas. Il n'avait pas l'intention de manquer ça. Son prestige et son autorité étaient en jeu. Il fallait qu'il puisse dire par la suite : « J'y étais. »

Cigare se crispa et plissa les yeux, creusant encore les rides de ses joues. « Ah », fit-il sans autre précision. Il se redressa sur ses genoux.

Lloyd avait peine à distinguer le train, sans parler de l'identifier, mais Cigare était aux aguets. La motrice allait bien plus vite que la précédente, constata Lloyd. Comme elle approchait, il remarqua aussi que la rame était plus longue : vingt-quatre wagons au bas mot.

« C'est lui », annonça Cigare.

Le pouls de Lloyd s'accéléra. Si le cheminot ne se trompait pas, ce transport de troupes convoyait vers le front de Normandie plus de mille soldats et officiers – et ce n'était peut-être que le premier d'une longue série. Lloyd devait s'assurer qu'aucun d'eux ne franchirait le tunnel.

Son regard fut soudain attiré par autre chose. Un avion avait pris le train en chasse. Sous ses yeux, l'appareil adapta sa vitesse à celle du convoi et perdit de l'altitude.

C'était un avion britannique.

Lloyd reconnut un Hawker Typhoon, un chasseur-bombardier monoplace surnommé le Tiffy. Leurs pilotes effectuaient souvent des missions dangereuses, pénétrant profondément en territoire ennemi pour attaquer les voies de communication. L'homme qui était aux manettes était un brave, se dit-il.

Mais son intervention n'était pas prévue. Il ne fallait pas que le train soit détruit avant de s'engager dans le tunnel.

« Merde », souffla-t-il.

Le Tiffy lâcha une rafale sur les wagons.

« Qu'est-ce que ça veut dire ? demanda Légionnaire.

— Je n'en sais foutre rien », répliqua Lloyd.

Il remarqua alors que la rame était composée de wagons de passagers et de wagons à bestiaux. Mais ceux-ci devaient également contenir des soldats.

Le chasseur accéléra et mitrailla les wagons l'un après l'autre. Il était équipé de quatre canons à chargement par bande qui tiraient des projectiles de 20 mm, et faisaient un bruit d'enfer, couvrant celui du moteur et du train. Lloyd ne put s'empêcher de plaindre les soldats pris au piège dans les wagons, incapables d'échapper au tir meurtrier du Tiffy. Il se demanda pourquoi le pilote ne lâchait pas ses obus. Même s'ils rataient souvent leur cible, ils étaient capables de détruire des véhicules en marche. Peut-être les avait-il tous utilisés lors d'un précédent engagement.

Des Allemands particulièrement hardis passèrent la tête par les fenêtres pour riposter à coups de fusil ou de pistolet – sans résultat.

Lloyd aperçut alors une batterie antiaérienne légère sur un wagon plat situé juste derrière la motrice. Deux artilleurs s'activaient à la déployer. Le canon pivota sur son socle et se pointa sur le chasseur anglais.

Sans doute le pilote ne l'avait-il pas vu, car il maintint son cap, continuant à mitrailler les toits des wagons.

Le canon tira et manqua son coup.

Lloyd se demanda s'il connaissait ce pilote de chasse. Ils n'étaient que cinq mille en service actif dans tout le Royaume-Uni et un grand nombre d'entre eux fréquentaient les soirées de Daisy. Lloyd pensa à Hubert St John, un brillant diplômé de Cambridge avec qui il avait échangé des souvenirs de fac quelques semaines plus tôt ; à Dennis Chaucer, qui venait de la Trinité et se plaignait de la fadeur de la cuisine anglaise, en particulier de la sempiternelle purée de pommes de terre ; et aussi à Brian Mantel, un Australien affable auquel il avait fait franchir les Pyrénées lors de sa dernière mission d'évacuation. Oui, il connaissait sans doute le courageux pilote de ce Tiffy.

Le canon antiaérien tira de nouveau, sans toucher sa cible.

Le pilote ne l'avait toujours pas vu, ou bien il se croyait invulnérable ; au lieu de tenter de lui échapper, il continua à voler suffisamment bas pour arroser le convoi.

La motrice n'était qu'à quelques secondes du tunnel lorsque l'avion fut touché.

Les flammes jaillirent de son moteur, un panache de fumée noire monta dans le ciel. Le pilote voulut virer, mais il était trop tard.

Le train entra dans le tunnel et Lloyd vit défiler les wagons devant sa position. Chacun d'eux était rempli de dizaines, de centaines de soldats allemands.

Le Tiffy passa juste au-dessus de lui. L'espace d'un instant, il crut qu'il allait lui tomber dessus. Il était déjà plaqué au sol, mais ne put s'empêcher de se protéger la tête des deux mains, ce qui n'aurait servi à rien.

Le Tiffy volait en rugissant à trente mètres d'altitude.

Légionnaire actionna le détonateur.

Un coup de tonnerre ébranla le tunnel, suivi d'un terrible grincement d'acier torturé lorsque le train s'écrasa contre la roche.

Les wagons continuèrent d'avancer quelques secondes encore puis s'arrêtèrent net. Deux d'entre eux s'élevèrent dans les airs, dessinant un V inversé. Lloyd entendit hurler leurs passagers. Tous les wagons déraillèrent et s'éparpillèrent telles des allumettes autour de la gueule noire du tunnel. Le fer se froissa comme du papier, des éclats de verre s'abattirent sur les trois saboteurs qui observaient la scène depuis leur talus. Comprenant qu'ils étaient eux aussi en danger de mort, ils se levèrent d'un bond et s'enfuirent en courant.

Lorsqu'ils furent en lieu sûr, tout était fini. Un nuage de fumée sortait du tunnel : au cas bien improbable où des hommes auraient survécu à la collision, ils seraient brûlés vifs.

Le plan de Lloyd était une réussite. Non seulement il avait tué plusieurs centaines de soldats ennemis et détruit un transport de troupes, mais il avait bloqué en outre une voie ferrée stratégique. Il faudrait plusieurs semaines pour dégager ce tunnel. Les Allemands auraient un peu plus de difficulté à renforcer leurs défenses en Normandie.

Un sentiment d'horreur l'envahit.

Il avait déjà assisté à des scènes de carnage en Espagne, mais rien de comparable à cette boucherie. Et c'était lui le responsable.

Un nouveau bruit retentit et, en se tournant vers sa source, il vit que le Tiffy venait de s'écraser. Il brûlait mais son fuselage était intact. Le pilote était peut-être vivant.

Il courut vers l'avion, Cigare et Légionnaire sur les talons.

Le chasseur gisait sur le ventre. L'une de ses ailes s'était brisée en deux. Son moteur fumait toujours. Le capot en plexiglas du cockpit était noirci par la suie et Lloyd ne distinguait pas le pilote.

Il monta sur une aile et débloqua le capot. Cigare en fit autant de l'autre côté. Ensemble, ils le firent coulisser sur ses rails.

Le pilote était inconscient. Il portait un casque et des lunettes de vol ; un masque à oxygène lui couvrait le nez et la bouche. Lloyd était incapable de dire s'il le connaissait ou pas.

Il se demanda où était la réserve d'oxygène et si elle avait déjà explosé.

Légionnaire avait eu la même idée. « Il faut le sortir de là avant que l'avion prenne feu », dit-il.

Lloyd déboucla la ceinture du pilote. Puis il l'attrapa par les aisselles et tira. L'homme resta inerte. Impossible d'évaluer la gravité de ses blessures. Peut-être même était-il déjà mort.

Il extirpa le pilote du cockpit, puis le cala sur ses épaules comme le font les pompiers, ne faisant halte qu'à bonne distance de l'épave en feu. Puis, avec d'infinies précautions, il l'allongea dans l'herbe, le visage tourné vers le ciel.

Il entendit un bruit – à la fois un souffle et un bruit sourd –, et vit que l'avion était en feu.

Se penchant sur le pilote, il lui retira doucement ses lunettes et son masque à oxygène, découvrant, bouleversé, un visage familier.

C'était Boy Fitzherbert.

Il respirait encore.

Lloyd essuya son nez et sa bouche maculés de sang.

Boy ouvrit les yeux. Il sembla tout d'abord ne pas comprendre ce qui se passait. Puis, au bout d'une minute, son expression s'altéra : « Vous ?

— On a fait sauter le train », expliqua Lloyd.

Boy semblait ne pouvoir remuer que les yeux et les lèvres. « Le monde est petit, murmura-t-il.

— En effet.

— Qui est-ce ? » demanda Cigare.

Lloyd hésita un instant avant de répondre : « Mon frère.

— Mon Dieu. »

Boy ferma les yeux.

« Il faut aller chercher un médecin », dit Lloyd à Légionnaire.

Celui-ci secoua la tête. « Nous ne pouvons pas traîner ici. Dans quelques minutes, les Allemands viendront voir ce qui est arrivé à leur train. »

Il avait raison, Lloyd le savait. « Il va falloir l'emmener. »

Boy ouvrit les paupières : « Williams.

— Oui, Boy ? »

On aurait dit qu'il souriait. « Tu peux épouser cette salope, maintenant. »

Et il mourut.

8.

Daisy pleura en apprenant la nouvelle. Boy était un sale type et il l'avait fait souffrir, mais elle l'avait aimé et lui devait son éducation amoureuse ; sa mort l'attristait.

Son frère Andy était désormais vicomte et héritier du titre ; May, son épouse, devenait du coup vicomtesse ; quant à Daisy, compte tenu des usages byzantins de l'aristocratie anglaise, elle était la vicomtesse douairière d'Aberowen – jusqu'à ce qu'elle épouse Lloyd, ce qui ferait d'elle, pour son plus grand soulagement, Mrs. Williams tout court.

Et elle devrait pourtant attendre encore un moment. Au fil de l'été, on perdit tout espoir de voir le conflit s'achever rapidement. Le 20 juillet, un complot de généraux allemands décidés à assassiner Hitler échoua. La Wehrmacht battait en retraite sur le front de l'Est et les Alliés reprirent Paris en août, mais Hitler était résolu à lutter jusqu'au bout, coûte que coûte. Daisy ignorait quand elle reverrait Lloyd, sans parler de l'épouser.

Un mercredi de septembre, comme elle allait passer la soirée à Aldgate, elle fut accueillie par une Eth Leckwith folle de joie. « Grande nouvelle ! annonça-t-elle lorsque Daisy entra dans la cuisine. Lloyd a été choisi comme candidat travailliste de la circonscription de Hoxton ! »

Millie, la sœur de Lloyd, était là ainsi que ses deux enfants, Lennie, quatre ans, et Pammie, deux ans. « C'est formidable, non ? lança-t-elle. Je parie qu'il finira Premier ministre.

— Oui, fit Daisy en s'asseyant lourdement sur une chaise.

— Eh bien, ça n'a pas l'air de vous enchanter, remarqua Ethel. Comme dirait mon amie Mildred, ça a tout l'air de vous rester en travers du gosier. Que se passe-t-il ?

— Eh bien, si nous nous marions, ça compromettra ses chances d'être élu. » Elle l'aimait, voilà pourquoi elle en était toute désemparée. Comment pourrait-elle se résoudre à gâcher ses perspectives d'avenir ? Mais comment se résoudre aussi à renoncer à lui ? Lorsque de telles pensées l'agitaient, elle avait le cœur gros et la situation lui paraissait sans issue.

« Parce que vous êtes riche ? demanda Ethel.

— Pas seulement. Avant de partir pour sa dernière mission, Boy m'a dit que Lloyd ne serait jamais élu à cause du passé fasciste de son épouse. » Elle se tourna vers Ethel, qui n'était pas femme à mâcher ses mots. « Il avait raison, n'est-ce pas ?

— Pas forcément », répondit Ethel. Elle mit la bouilloire à chauffer puis s'assit en face de Daisy. « Je ne vais pas prétendre que ça n'a aucune importance. En même temps, il ne faut pas désespérer. »

Tu es comme moi, songea Daisy. Tu dis ce que tu penses. Pas étonnant qu'il soit tombé amoureux de moi : je ressemble à sa mère, en plus jeune !

« L'amour triomphe de tout, non ? » lança Millie avant de remarquer que Lennie tapait sur Pammie avec un soldat de bois. « Arrête de frapper ta sœur ! » s'écria-t-elle. Se retournant vers Daisy, elle reprit : « Et mon frère est fou de toi. À vrai dire, je crois qu'il n'a jamais aimé personne d'autre.

— Je sais, murmura Daisy au bord des larmes, mais il est résolu à changer le monde et je ne supporte pas l'idée d'être un obstacle sur sa route. »

Ethel prit sur ses genoux la fillette en pleurs, qui se calma aussitôt. « Je vais vous dire ce qu'il faut faire, déclara-t-elle à Daisy. Préparez-vous à ce qu'on vous pose des questions déplaisantes, et à ce que certains se montrent hostiles à votre égard, mais ne cherchez pas à éluder le débat, ni à dissimuler votre passé.

— Que dois-je dire ?

— Que vous avez cédé aux illusions du fascisme, comme des millions d'autres ; mais que vous avez conduit une ambulance pendant le Blitz et que vous espérez vous être ainsi rachetée. Préparez vos arguments avec Lloyd. Ayez confiance en vous, déployez votre charme irrésistible et ne vous laissez pas démonter.

— Est-ce que ça marchera ? »

Ethel hésita. « Je ne sais pas, dit-elle enfin. Vraiment pas. Mais il faut tenter le coup.

— S'il devait, à cause de moi, renoncer à ce qu'il aime le plus au monde, je ne me le pardonnerais pas. Il y a de quoi détruire n'importe quel couple. »

Daisy espérait vaguement qu'Ethel la détromperait, mais celle-ci n'en fit rien. « Je ne sais pas », répéta-t-elle.

XIX
1945 (I)

1.

Woody eut vite fait de s'habituer à ses béquilles.

Il avait été blessé en Belgique, à la fin de l'année 1944, au cours de la bataille des Ardennes. Alors qu'ils fonçaient vers la frontière allemande, les Alliés avaient été surpris par une contre-attaque. En compagnie d'autres soldats de la 101e division aéroportée, Woody avait tenu Bastogne, une ville qui occupait une position stratégique. Lorsque les Allemands les avaient sommés de se rendre, le général McAuliffe leur avait répondu d'un seul mot, destiné à passer à la postérité – *Nuts !* –, les traitant en fait de cinglés.

Le jour de Noël, une rafale de mitrailleuse avait démoli la jambe droite de Woody. Ça faisait un mal de chien. Pis encore, il avait dû attendre un mois avant d'être évacué de la ville assiégée vers un hôpital digne de ce nom.

Ses os finiraient par se ressouder, et peut-être cesserait-il un jour de boiter, mais plus jamais il ne sauterait en parachute.

La bataille des Ardennes avait été la dernière offensive allemande sur le front occidental. Désormais, l'armée hitlérienne ne lancerait plus aucune contre-attaque.

Woody retourna à la vie civile, ce qui lui permit de s'installer chez ses parents, à Washington, et de se faire dorloter par sa mère. Une fois son plâtre enlevé, il retourna travailler au bureau de son père.

Le jeudi 12 avril 1945, il se trouvait au sous-sol du Capitole, le bâtiment abritant le Sénat et la Chambre des représentants, et discutait avec son père du sort des personnes déplacées. « Selon nos estimations, environ vingt et un millions de personnes ont été chassées de leur foyer en Europe, expliqua Gus. L'Administration des Nations unies pour le secours et la reconstruction est prête à les aider.

— Elle ne tardera pas à se mettre au travail, je pense, observa Woody. L'armée Rouge est aux portes de Berlin.

— Et l'armée américaine en est à moins de cent kilomètres.

— Combien de temps Hitler peut-il encore tenir ?

— Un homme sain d'esprit se serait déjà rendu. »

Woody baissa la voix. « On m'a dit que les Russes avaient découvert une sorte de camp d'extermination. Les nazis y tuaient plusieurs centaines de personnes par jour. C'est un endroit qui s'appelle Auschwitz, en Pologne. »

Gus acquiesça d'un air sombre. « C'est exact. Le public n'en a pas encore été informé, mais il le sera tôt ou tard.

— Les coupables devront être jugés et condamnés.

— Depuis deux ans déjà, la Commission des crimes de guerre des Nations unies dresse une liste des criminels de guerre et rassemble des preuves. Il y aura un procès, à condition bien sûr que nous arrivions à maintenir les Nations unies sur pied après le conflit.

— Mais nous y arriverons, c'est évident ! protesta Woody. Roosevelt a fait campagne sur ce thème l'année dernière et il a remporté l'élection. La conférence des Nations unies s'ouvre à San Francisco dans quinze jours. » San Francisco avait une signification toute particulière pour Woody, car c'était là qu'habitait Bella Hernandez, mais il n'avait pas encore parlé d'elle à son père. « Le peuple américain exige une coopération internationale afin d'éviter de nouveaux conflits comme celui-ci. Qui pourrait s'opposer à cela ?

— Tu serais surpris ! Tu sais, la plupart des membres du parti républicain sont des hommes de bonne volonté qui ont une vision du monde différente de la nôtre, c'est tout. Mais il y a parmi eux un noyau dur de connards et de cinglés. »

Woody sursauta. Il n'était pas dans les habitudes de son père de parler ainsi.

« Les types qui ont fomenté une insurrection contre Roosevelt durant les années 1930, poursuivit Gus, les hommes d'affaires comme Henry Ford, qui considéraient Hitler comme un excellent chef d'État, un anticommuniste résolu. Ils ont adhéré à des groupes de droite comme America First. »

Woody ne se rappelait pas l'avoir jamais vu aussi furieux.

« Si ces imbéciles ont leur mot à dire, nous aurons une troisième guerre mondiale et elle sera encore pire que les deux premières, ajouta Gus. La guerre m'a pris un fils et, si je dois avoir un petit-fils un jour, je ne veux pas le perdre, lui aussi. »

Woody sentit son cœur se serrer. Si elle avait vécu, Joanne aurait donné des petits-enfants à Gus.

Woody ne fréquentait personne pour le moment, donc la paternité n'était qu'une perspective lointaine – à moins qu'il ne retrouve la trace de Bella à San Francisco…

« Pour les crétins finis, il n'y a rien à faire, reprit Gus. Mais peut-être arriverons-nous à quelque chose avec le sénateur Vandenberg. »

Arthur Vandenberg était un républicain du Michigan, un conservateur qui s'était opposé au New Deal. Il siégeait à la commission des Affaires étrangères du Sénat aux côtés de Gus.

« C'est l'homme le plus dangereux pour notre camp, remarqua celui-ci. Un homme vaniteux et imbu de sa personne, sans doute, mais respecté de tous ou presque. Le Président lui a fait du charme et il a fini par se ranger à notre point de vue, mais il peut encore changer d'avis.

— Pourquoi le ferait-il ?

— Il est farouchement anticommuniste.

— Il n'y a pas de mal à ça. Nous le sommes aussi.

— Oui, mais Arthur est assez rigide sur ce point. S'il estime que nous plions l'échine devant Moscou, il ne décolérera pas.

— Que veux-tu dire au juste ?

— Dieu sait à quels compromis nous devrons nous résoudre à San Francisco. Nous avons déjà accepté de considérer l'Ukraine et la Biélorussie comme des États distincts, ce qui revient à donner trois voix à Moscou à l'Assemblée générale. Nous ne pouvons pas nous passer des Soviétiques – mais, s'il juge que nous allons trop loin, Arthur risque de se braquer contre le projet d'Organisation des Nations unies. Dans ce cas, le Sénat refusera de le ratifier, comme il a rejeté la Société des nations en 1919.

— Autrement dit, notre mission à San Francisco consistera à contenter les Soviétiques sans offusquer le sénateur Vandenberg.

— Exactement. »

Ils entendirent un bruit de pas précipités, inhabituel en ce lieu empreint de dignité. Ils se retournèrent d'un même mouvement. À sa grande surprise, Woody vit Harry Truman, le vice-président, qui courait dans le couloir. Il était vêtu de façon ordinaire, costume trois-pièces gris et cravate à pois, mais ne portait pas de chapeau. Apparemment, il avait perdu son escorte d'assistants et de gardes du corps. Il courait à un rythme soutenu, le souffle court, en regardant droit devant lui, comme un homme terriblement pressé.

Woody et Gus, ainsi que toutes les personnes présentes, ouvrirent des yeux étonnés.

Lorsque Truman eut disparu, Woody lança : « Que diable… ?

— Je pense que le Président est mort », murmura Gus.

2.

Volodia Pechkov entra en Allemagne à bord d'un camion militaire Studebaker US6 à dix roues. Fabriqué à South Bend dans l'Indiana, ce camion avait été transporté par rail à Baltimore, expédié par navire dans le golfe Persique via l'océan Atlantique et le cap de Bonne-Espérance, puis acheminé par voie ferrée jusqu'au centre de la Russie. Comme le savait Volodia, c'était l'un des deux cent mille camions Studebaker offerts à l'armée Rouge par le gouvernement américain. Les Russes les adoraient : ils étaient aussi robustes que fiables. À en croire les soldats, le sigle *USA* peint au pochoir sur leurs flancs signifiait *Ubil Sukina sina Adolfa*, c'est-à-dire « Ce camion a tué un salopard d'Adolf ».

Ils appréciaient tout autant la nourriture envoyée par les Américains, notamment le SPAM, cette viande en conserve d'un étrange rose vif, mais très riche en matières grasses.

Volodia avait été envoyé en Allemagne parce que les renseignements transmis par ses espions berlinois étaient désormais moins pertinents que les informations que l'on pouvait tirer des prisonniers de guerre allemands. Comme il parlait allemand couramment, il avait été chargé de les interroger.

En franchissant la frontière, il avait vu une affiche du gouvernement soviétique proclamant : « Soldat de l'armée Rouge ! Te voici en territoire allemand. L'heure de la vengeance a sonné ! » C'était l'un des exemples les plus modérés de propagande. Depuis quelque temps, le Kremlin attisait la haine des soldats soviétiques contre les Allemands, dans l'espoir de les rendre plus combatifs. Les commissaires politiques avaient calculé, ou prétendaient l'avoir fait, le nombre d'hommes morts sur le champ de bataille, de maisons incendiées, de civils massacrés dans les villes conquises par l'ennemi, et dont le seul crime était d'être juifs, slaves ou communistes. Les soldats du front étaient nombreux à connaître par cœur les chiffres concernant leurs propres villages et ils brûlaient du désir de rendre à l'ennemi la monnaie de sa pièce.

L'armée Rouge était arrivée sur les rives de l'Oder, un fleuve dont les méandres traversaient la Prusse du sud au nord, le dernier obstacle avant Berlin. Un million de soldats soviétiques se massaient à moins de cent kilomètres de la capitale, prêts à lui porter un coup fatal. Volodia se trouvait avec la 5e armée de choc. En attendant le début des combats, il lisait attentivement *L'Étoile rouge*, le journal de l'armée.

Son contenu l'horrifia.

La propagande était plus virulente que tout ce qu'il avait pu voir jusque-là. « Si tu n'as pas tué au moins un Allemand par jour, tu as perdu ta journée, lut-il. Si tu attends que la bataille commence, tue un Allemand pour passer le temps. Si tu tues un Allemand, tues-en un autre – rien ne nous amuse autant qu'un monceau de cadavres allemands. Tue les Allemands – c'est la prière de ta vieille mère. Tue les Allemands – c'est la supplique de tes enfants. Tue les Allemands – c'est le cri de ta terre russe. N'hésite pas. Ne recule pas. Tue. »

Cela donnait un peu la nausée, se dit Volodia. Mais certains sous-entendus étaient encore pires. Le rédacteur se montrait indulgent avec les pillards : « Les manteaux de fourrure et les cuillers en argent que vous volez aux Allemandes, elles les ont volés à d'autres. » Et le viol faisait l'objet d'une plaisanterie de mauvais goût : « Un soldat soviétique ne refuse jamais les avances d'une femme allemande. »

Les soldats n'étaient pas au départ les hommes les plus civilisés du monde et le comportement des envahisseurs allemands en 1941 avait indigné tous les Russes. Le gouvernement mettait de l'huile sur le feu en parlant ainsi de vengeance. Et voilà que l'organe officiel de l'armée incitait les soldats à se déchaîner sur les Allemands vaincus.

Autant appeler l'Apocalypse de ses vœux.

3.

Erik von Ulrich espérait de tout cœur que la guerre serait bientôt finie.

En compagnie de son ami Hermann Braun et de leur supérieur, le docteur Weiss, Erik avait installé un hôpital de campagne dans un petit temple protestant ; puis ils s'étaient assis dans la nef, désœuvrés, pour attendre l'arrivée des ambulances à cheval transportant leur cargaison d'hommes affreusement mutilés et brûlés.

L'armée allemande avait renforcé ses positions sur les hauteurs de Seelow, qui dominaient l'Oder à son point le plus proche de Berlin. L'hôpital de campagne se trouvait dans un village à quinze cents mètres du front.

Selon le docteur Weiss, dont un ami travaillait pour le Renseignement militaire, cent dix mille soldats allemands étaient censés défendre Berlin contre un million de soldats soviétiques. Toujours sarcastique, il avait ajouté : « Mais comme le moral des troupes est au beau fixe et qu'Adolf Hitler est le plus grand génie de l'histoire militaire, nous sommes sûrs de gagner. »

Les soldats allemands continuaient à se battre férocement, malgré leur situation désespérée. Sans doute était-ce à cause des rumeurs sur les exactions de l'armée Rouge qui étaient parvenues jusqu'à eux, se disait Erik. Prisonniers exécutés, maisons pillées et incendiées, femmes violées et clouées aux portes des granges. Les Allemands étaient persuadés de protéger leurs familles des brutalités communistes. La propagande haineuse du Kremlin se retournait contre elle.

Erik priait pour la défaite. Il fallait mettre fin à ce massacre. Il voulait rentrer chez lui.

Son vœu serait bientôt exaucé – ou alors il serait mort.

Il s'était endormi sur un banc et fut réveillé à trois heures du matin par l'artillerie russe. En ce lundi 16 avril, ce n'était pas la première fois qu'il entendait le son du canon, mais celui-ci était dix fois plus fort ce jour-là que tout ce dont il avait été témoin auparavant. Pour les hommes du front, il devait être littéralement assourdissant.

Les premiers blessés arrivèrent à l'aube et ils se mirent au travail en dépit de leur fatigue, amputant les membres, réduisant les fractures, extrayant les balles, nettoyant et pansant les plaies. Ils manquaient de tout, aussi bien de médicaments que d'eau pure, et n'administraient de la morphine qu'à ceux qui hurlaient de souffrance.

Les hommes encore capables de marcher et de tenir une arme étaient renvoyés au combat.

Les troupes allemandes résistèrent plus longtemps que ne l'avait pensé le docteur Weiss. Au soir du premier jour, elles n'avaient pas quitté leur position, et l'afflux de blessés se tarit à la nuit tombée. L'unité médicale réussit à voler quelques heures de sommeil.

Le lendemain de bonne heure, Werner Franck fit son apparition, le poignet horriblement broyé.

Il avait à présent le grade de capitaine. On lui avait confié la responsabilité d'une section du front avec trente canons de DCA de 88 mm. « Nous n'avions que huit obus par canon, expliqua-t-il tandis que les doigts habiles du docteur Weiss soignaient méticuleusement ses os fracassés. Nous avions ordre d'en tirer sept sur les chars russes puis de détruire chaque canon avec le huitième pour éviter qu'il tombe entre les mains des rouges. » Il se tenait à côté d'un de ces canons lorsqu'un projectile russe l'avait frappé de plein fouet, le renversant sur lui. « J'ai eu de la chance que ce soit la main qui prenne. Ça aurait pu être ma foutue tête. »

Une fois son poignet bandé, il demanda à Erik : « Tu as des nouvelles de Carla ? »

Erik savait que sa sœur et Werner étaient ensemble. « Ça fait des semaines que je n'ai pas reçu de lettre.

— Moi non plus. Il paraît que la situation est très dure à Berlin. J'espère qu'elle va bien.

— Je suis inquiet, moi aussi », dit Erik.

À la surprise générale, les Allemands tinrent les hauteurs de Seelow pendant un jour et une nuit de plus.

Personne n'avertit l'hôpital de campagne que la ligne de front avait été enfoncée. Ils triaient un nouvel arrivage de blessés lorsque sept ou huit soldats soviétiques firent irruption dans le temple. L'un d'eux tira une rafale de mitraillette en direction du plafond voûté et Erik se jeta à terre, imité par tous ceux qui pouvaient encore bouger.

Constatant que personne n'était armé, les Russes se détendirent. Ils firent le tour de la salle pour s'emparer des montres et des alliances de ceux qui en possédaient. Puis ils repartirent.

Erik se demanda ce qui allait se passer à présent. C'était la première fois qu'il était piégé derrière les lignes ennemies. Devaient-ils abandonner leur hôpital de campagne pour tenter de rejoindre leur armée en déroute ? Ou leurs patients étaient-ils plus en sécurité ici ?

Le docteur Weiss avait déjà pris sa décision. « Que tout le monde se remette au travail », dit-il.

Quelques minutes plus tard, ils virent arriver un soldat soviétique portant un camarade sur son épaule. Il braqua son arme sur Weiss et prononça un flot de paroles en russe. Il était visiblement affolé et son compagnon était couvert de sang.

Weiss ne perdit pas son sang-froid. Il lui répondit dans un russe hésitant : « Inutile de me menacer. Posez votre ami sur cette table. »

Le soldat s'exécuta et l'équipe se mit à l'œuvre. Pendant tout ce temps, le soldat garda son fusil braqué sur le médecin.

Plus tard dans la journée, les patients allemands furent évacués, à pied ou à bord d'un camion qui prit la direction de l'est. Erik vit Werner Franck disparaître : il était prisonnier de guerre. Quand Erik était petit, on lui avait souvent raconté l'histoire de son oncle Robert, qui avait été pris par les Russes lors de la Première Guerre mondiale, et était rentré à pied au pays depuis la Sibérie, un périple de six mille kilomètres. Il se demanda où l'on conduisait Werner.

On leur amena d'autres blessés russes, et les Allemands les soignèrent comme ils l'auraient fait pour leurs propres compatriotes.

Plus tard, sombrant dans un sommeil épuisé, Erik se rendit compte qu'il était, lui aussi, prisonnier de guerre.

4.

Alors que les armées alliées s'approchaient de Berlin, les pays vainqueurs commencèrent à se chamailler à la conférence des Nations unies de San Francisco. Woody en aurait été découragé s'il n'avait pas consacré tous ses efforts à tenter de retrouver la trace de Bella Hernandez.

Elle n'avait pas quitté ses pensées durant tous les combats, du jour J à la conquête de la France, de son entrée à l'hôpital à la fin de sa convalescence. Un an auparavant, elle s'apprêtait à terminer ses études à Oxford et comptait aller soutenir sa thèse à Berkeley, c'est-à-dire à San Francisco. Sans doute habitait-elle chez ses parents, à Pacific Heights, à moins qu'elle n'ait loué un appartement près du campus.

Malheureusement, il n'arrivait pas à la joindre.

Toutes ses lettres restaient sans réponse. Quand il composa le numéro qu'il avait trouvé dans l'annuaire, une femme d'un certain âge qui devait être sa mère lui répondit avec une politesse glaciale : « Elle n'est pas là pour le moment. Voulez-vous lui laisser un message ? » Bella ne rappela jamais.

Elle devait sortir avec quelqu'un. Dans ce cas, il voulait qu'elle le lui dise en face. Mais peut-être sa mère interceptait-elle son courrier et ne lui transmettait-elle pas ses messages.

Il aurait été raisonnable de renoncer. Il allait finir par se ridiculiser. Mais ce n'était pas dans sa nature. Il se rappela Joanne et la longue cour qu'il s'était entêté à lui faire. On dirait que l'histoire se répète, songea-t-il ; est-ce moi qui suis bizarre ?

Pendant ce temps, chaque matin, il accompagnait son père dans la vaste suite au dernier étage de l'hôtel Fairmont, où Edward Stettinius, le secrétaire d'État, réunissait l'équipe américaine. Stettinius remplaçait Cordell Hull, qui venait d'être hospitalisé. Les États-Unis avaient également un nouveau Président, Harry Truman, qui avait prêté serment après le décès du grand Franklin D. Roosevelt. Dommage, avait commenté Gus Dewar, qu'en ces circonstances décisives pour l'avenir du monde, les États-Unis soient dirigés par deux hommes sans expérience.

Les choses avaient mal commencé. Le président Truman avait maladroitement heurté Molotov, le ministre soviétique des Affaires étrangères, lors d'une rencontre préliminaire à la Maison Blanche. Du coup, Molotov arriva à San Francisco de fort méchante humeur. Il repartirait sur-le-champ à Moscou, affirma-t-il, si la conférence n'admettait pas immédiatement la Biélorussie, l'Ukraine et la Pologne.

Personne ne souhaitait le départ de l'URSS. Sans les Soviétiques, les Nations unies ne seraient pas les Nations unies. La majorité de la délégation américaine était favorable à un compromis avec les communistes, mais le sénateur Vandenberg, célèbre pour son nœud papillon, insista d'un air pincé pour qu'aucune décision ne soit prise sous la pression de Moscou.

Un matin, alors que Woody disposait de deux heures de liberté, il se rendit chez les parents de Bella.

Ils habitaient Nob Hill, un quartier huppé relativement proche de l'hôtel Fairmont, mais comme Woody marchait toujours avec une canne, il prit un taxi. Il en descendit dans Gough Street, devant une grande demeure victorienne à la façade jaune. La femme qui lui ouvrit la porte était trop bien habillée pour être une domestique. Devant son sourire en coin, identique à celui de Bella, il comprit que c'était sa mère. « Bonjour, madame, lui dit-il poliment. Je m'appelle Woody Dewar. J'ai fait la connaissance de Bella Hernandez l'année dernière à Londres et j'aimerais bien la revoir, si cela est possible. »

Le sourire de la femme s'effaça. Elle le fixa longuement du regard et dit : « Ainsi, c'est vous. »

Woody interloqué ne sut que répondre.

« Je suis Caroline Hernandez, la mère d'Isabel. Vous feriez mieux d'entrer.

— Je vous remercie. »

Elle ne prit pas la main qu'il lui tendait, affichant clairement une hostilité dont il ignorait le motif. Mais il était dans la place.

Mrs. Hernandez conduisit Woody dans un vaste salon confortable jouissant d'une vue splendide sur l'océan. Elle lui désigna un fauteuil, l'invitant à s'y asseoir d'un geste à peine poli. Elle prit place en face de lui et lui décocha un nouveau regard noir. « Combien de temps avez-vous passé avec Bella en Angleterre ? demanda-t-elle.

— Quelques heures à peine. Mais je n'ai cessé de penser à elle depuis. »

Suivit un silence pesant, puis elle reprit : « Quand elle est partie pour Oxford, Bella était fiancée à Victor Rolandson, un jeune homme charmant qu'elle connaissait depuis toujours, ou presque. Les Rolandson sont d'excellents amis de mon époux et de moi-même – ou plutôt ils l'étaient, jusqu'à ce que Bella rentre à la maison et rompe brutalement ses fiançailles. »

Le cœur de Woody fit un bond dans sa poitrine.

« Tout ce qu'elle a consenti à nous dire, c'est qu'elle avait compris qu'elle n'aimait pas Victor. Je me suis doutée qu'elle avait rencontré quelqu'un d'autre. Je sais maintenant qui c'est.

— J'ignorais qu'elle était fiancée, murmura Woody.

— Elle portait une bague en diamant qu'il était difficile de ne pas voir. Vos piètres facultés d'observation ont provoqué une tragédie.

— Je suis navré », dit Woody. Puis il se reprocha d'être trop timoré. « En fait, non, reprit-il. Je suis ravi qu'elle ait rompu ses fiançailles, parce que je trouve que c'est une fille sensationnelle et que je la veux pour moi. »

Mrs. Hernandez n'apprécia guère. « Vous êtes bien impertinent, jeune homme. »

Mais Woody ne supportait plus sa condescendance. « Madame Hernandez, vous venez de prononcer le mot de "tragédie". Ma fiancée Joanne est morte dans mes bras à Pearl Harbor. Mon frère Chuck a été fauché par une mitrailleuse sur la plage de Bougainville. Le jour J, j'ai envoyé quatre jeunes Américains à la mort pour prendre un pont dans un trou perdu de Normandie. Je sais ce que c'est qu'une tragédie, madame, croyez-moi, et ça n'a rien à voir avec une rupture de fiançailles. »

Elle fut manifestement désarçonnée. Sans doute n'avait-elle pas l'habitude que des jeunes gens lui tiennent tête. Elle ne répondit pas, mais pâlit légèrement. Au bout d'un moment, elle se leva et sortit sans un mot. Woody ne savait pas ce qu'elle attendait de lui, mais, comme il n'avait pas encore vu Bella, il décida de ne pas bouger.

Cinq minutes plus tard, Bella était là.

Woody se leva, le cœur battant. Il lui suffit de l'apercevoir pour avoir le sourire aux lèvres. Elle portait une robe jaune pâle très sobre qui faisait ressortir ses cheveux d'un noir lustré et sa peau couleur café. Les tenues les plus simples étaient toujours les plus seyantes sur elle, devina-t-il ; comme sur Joanne. Il mourait d'envie de la serrer dans ses bras, de presser son corps souple contre le sien, mais il attendit un signe de sa part.

Elle paraissait inquiète et mal à l'aise. « Que fais-tu ici ? demanda-t-elle.

— Je suis venu te voir.

— Pourquoi ?

— Parce que je n'arrête pas de penser à toi.

— On ne se connaît même pas.

— Il faut y remédier dès aujourd'hui. Puis-je t'inviter à dîner ce soir ?

— Je ne sais pas. »

Il traversa la pièce pour la rejoindre.

Elle fut surprise de le voir s'appuyer sur une canne. « Qu'est-ce qui t'est arrivé ?

— Quelques balles dans le genou, en Europe. Ça va mieux, mais c'est long.

— Je suis désolée.

— Bella, je te trouve merveilleuse. J'ai l'impression que tu m'aimes bien. Nous sommes libres tous les deux. Qu'est-ce qui t'inquiète ? »

Elle lui adressa ce sourire en coin qu'il aimait tant. « Disons que je suis gênée. À cause de ce que j'ai fait cette nuit-là à Londres.

— C'est tout ?

— C'était un peu beaucoup, pour une première soirée ensemble.

— Ces choses-là arrivaient tout le temps. Pas à moi personnellement, mais j'en ai entendu parler. Tu pensais que j'allais mourir. »

Elle acquiesça. « Je n'avais jamais fait ça de ma vie, même avec Victor. Je ne sais pas ce qui m'a pris. Et dans un parc public, en plus ! J'ai l'impression d'être une putain.

— Je sais exactement ce que tu es, dit Woody. Tu es une jeune femme intelligente, belle et généreuse. Alors oublions ce moment de folie à Londres et apprenons à nous connaître comme les jeunes gens respectables et bien élevés que nous sommes, veux-tu ? »

Elle s'adoucit. « Tu crois que c'est possible ?

— Bien sûr.

— Alors c'est d'accord.

— Je viens te chercher à sept heures.

— Entendu. »

C'était le moment de prendre congé, mais il hésita. « Je ne peux pas te dire à quel point je suis heureux de t'avoir retrouvée ».

Pour la première fois, elle le regarda droit dans les yeux. « Oh ! Woody, moi aussi. Je suis si heureuse ! » Puis elle lui passa les bras autour de la taille et le serra contre elle.

C'était ce qu'il avait espéré. Il lui rendit son étreinte et plongea le visage dans ses cheveux splendides. Ils restèrent ainsi durant une longue minute.

Puis elle s'écarta de lui. « Rendez-vous à sept heures, dit-elle.

— Compte sur moi. »

Il sortit de la maison aux anges.

De là, il se rendit directement à une réunion du comité de pilotage qui se tenait dans une salle du Veterans Building, juste à côté de l'Opéra. Quarante-six personnes avaient pris place autour de la longue table, flanquées de leurs assistants. Gus Dewar faisait partie de ces derniers et, en tant qu'assistant d'un assistant, Woody s'assit sur une chaise contre le mur.

Molotov, le ministre soviétique des Affaires étrangères, fut le premier à prendre la parole. Il n'était pas très impressionnant, songea Woody. Avec son crâne dégarni, ses lunettes et sa petite moustache bien taillée, on aurait dit un vendeur de grand magasin, profession que le père de Molotov avait d'ailleurs exercée. Mais il avait survécu aux intrigues bolcheviques durant de longues années. Ami de Staline bien avant la révolution, il était l'architecte du pacte germano-soviétique de 1939. C'était un travailleur acharné, que l'on surnommait Cul-de-Pierre à cause des heures qu'il passait assis à son bureau.

Il proposa que l'Ukraine et la Biélorussie soient admises au rang des membres fondateurs des Nations unies. Ces deux républiques soviétiques avaient particulièrement souffert de l'invasion nazie, souligna-t-il, et chacune d'elles avait fourni plus d'un million de soldats à l'armée Rouge. Certains laissaient entendre qu'elles n'étaient pas totalement indépendantes de Moscou, mais on pouvait en dire autant du Canada et de l'Australie, deux dominions de l'Empire britannique auxquels on avait accordé un statut de membre à part entière.

La proposition fut adoptée à l'unanimité. Comme le savait Woody, tout avait été décidé d'avance. Les pays d'Amérique latine avaient menacé de se retirer si l'Argentine, l'alliée d'Hitler, voyait sa candidature refusée, et on leur avait accordé cette concession en échange de leurs voix.

L'intervention suivante fit l'effet d'une bombe. Jan Masaryk, le ministre tchèque des Affaires étrangères, se leva. Ce célèbre libéral, adversaire farouche des nazis, avait fait la couverture du *Time* en 1944. Il proposa que la Pologne soit, elle aussi, admise à l'ONU.

Les Américains refusaient que cet État soit membre tant que Staline n'y aurait pas autorisé la tenue d'élections libres et Masaryk, un démocrate, aurait dû soutenir leur démarche, d'autant plus qu'il s'efforçait lui aussi d'édifier une démocratie dans son pays sous l'œil vigilant de Staline. Molotov avait dû exercer de fortes pressions pour qu'il trahisse ainsi ses idéaux. De fait, en se rasseyant, Masaryk affichait l'expression de quelqu'un qui a avalé de travers.

Gus Dewar était tout aussi sombre. Grâce au compromis auquel on était parvenu sur la Biélorussie, l'Ukraine et l'Argentine, cette réunion aurait dû se dérouler sans anicroche. Et voilà que Molotov leur infligeait ce coup bas.

Le sénateur Vandenberg, assis avec les représentants américains, était outré. Attrapant un stylo et un bloc-notes, il se mit à griffonner furieusement. Au bout d'une minute, il arracha le feuillet, fit signe à Woody et le lui donna en disant : « Apportez ceci au secrétaire d'État. »

Woody s'approcha de la table, se pencha au-dessus de l'épaule de Stettinius et posa la feuille de papier devant lui : « De la part du sénateur Vandenberg, monsieur.

— Merci. »

Woody regagna sa place contre le mur. Ma contribution à l'histoire, songea-t-il. Il avait jeté un coup d'œil à la note avant de la transmettre. Vandenberg avait rédigé un discours bref mais passionné pour rejeter la proposition tchèque. Stettinius suivrait-il l'initiative du sénateur ?

Si Molotov réussissait son coup, Vandenberg risquait de saboter les Nations unies au Sénat. Mais si Stettinius allait dans le sens de Vandenberg, Molotov risquait de quitter les négociations et de rentrer à Moscou, ce qui marquerait également la fin de l'ONU.

Woody retint son souffle.

Stettinius se leva, la note de Vandenberg à la main. « Nous venons d'honorer les engagements pris envers la Russie à la conférence de Yalta », déclara-t-il. Les États-Unis avaient alors accepté de soutenir les candidatures de l'Ukraine et de la Biélorussie. « D'autres engagements de Yalta attendent encore d'être tenus. » Il répétait le discours de Vandenberg. « Parmi ceux-ci figure la mise en place en Pologne d'un gouvernement provisoire représentatif. »

Un murmure choqué parcourut l'assistance. Stettinius s'opposait de front à Molotov. Woody jeta un coup d'œil à Vandenberg. Il buvait du petit-lait.

« Tant que cela ne sera pas fait, poursuivit Stettinius, cette conférence ne pourra, en toute conscience, reconnaître le gouvernement de Lublin. » Il regarda Molotov en face et conclut en citant mot pour mot le texte de Vandenberg. « Ce serait là une déplorable démonstration de mauvaise foi. »

Molotov semblait sur le point d'exploser.

Anthony Eden, le ministre britannique des Affaires étrangères, déplia sa longue carcasse et se leva pour apporter son soutien à Stettinius. Sa voix était d'une courtoisie irréprochable, mais son discours était virulent. « Mon gouvernement n'a aucun moyen de savoir si le peuple polonais soutient son gouvernement provisoire, dit-il, puisque nos alliés soviétiques refusent l'entrée de la Pologne aux observateurs britanniques. »

Woody sentit que le vent tournait contre Molotov. De toute évidence, le Russe partageait cette impression. Il consulta ses assistants d'une voix si forte que Woody perçut nettement la colère qui l'animait. Mais irait-il jusqu'à claquer la porte ?

Le ministre belge des Affaires étrangères, un homme chauve et bedonnant au visage alourdi par un double menton, proposa un compromis, une motion exprimant l'espoir que le nouveau gouvernement polonais puisse être constitué à temps pour être représenté à San Francisco avant la fin de la conférence.

Tous les regards se tournèrent vers Molotov. On lui offrait une possibilité de sauver la face. La saisirait-il ?

Il était toujours furieux. Mais il hocha légèrement la tête en signe d'assentiment.

La crise était terminée.

Eh bien, se dit Woody, deux victoires en un jour. Ça s'arrange.

5.

Carla sortit faire la queue à la pompe.

Cela faisait deux jours qu'il n'y avait plus d'eau aux robinets. Par bonheur, les Berlinoises avaient découvert l'existence d'antiques pompes publiques installées toutes les deux ou trois rues, désaffectées depuis des lustres mais toujours reliées à des réservoirs souterrains. Elles étaient mangées par la rouille et grinçaient terriblement, mais elles fonctionnaient encore. Tous les matins, les femmes du quartier s'y retrouvaient avec leurs cruches et leurs seaux.

Les attaques aériennes s'étaient interrompues, sans doute parce que l'ennemi s'apprêtait à entrer dans la ville. Pourtant, il était toujours dangereux de s'aventurer dans les rues, car l'artillerie de l'armée Rouge poursuivait son pilonnage. Carla ne comprenait pas pourquoi. La quasi-totalité de la ville était détruite. Des quartiers entiers n'étaient plus que des champs de ruines. Tous les services publics étaient interrompus. Il ne circulait plus ni bus ni trains. Les sans-abri se comptaient par milliers, voire par millions. La capitale n'était plus qu'un gigantesque camp de réfugiés. Cependant, les obus continuaient de tomber. La plupart des gens passaient la journée dans leur cave, ou dans des abris antiaériens publics, mais il fallait bien sortir chercher de l'eau.

Peu avant que l'électricité ne soit définitivement coupée, la BBC avait annoncé que l'armée Rouge avait libéré le camp de concentration de Sachsenhausen. Comme celui-ci se trouvait au nord de Berlin, cela voulait dire que les Soviétiques, venant de l'est, avaient choisi d'encercler la capitale au lieu de foncer droit sur elle. Maud, la mère de Carla, en avait déduit que les Russes souhaitaient arrêter au plus vite l'avancée des troupes américaines, britanniques, françaises et canadiennes qui arrivaient de l'ouest. Elle avait cité Lénine : « Qui tient Berlin tient l'Allemagne, qui tient l'Allemagne tient l'Europe. »

Mais l'armée allemande n'avait pas renoncé. Inférieurs en nombre et en armes, manquant cruellement de munitions et de carburant, ses soldats à moitié morts de faim tenaient bon. Leurs chefs lançaient sans désemparer de nouveaux assauts contre les forces ennemies, et les hommes leur obéissaient sans broncher, se battant vaillamment et mourant par centaines de milliers. Parmi eux figuraient les deux hommes que Carla aimait le plus au monde : son frère Erik et Werner, son amant. Elle ignorait où ils se trouvaient, ignorait même s'ils étaient encore en vie.

Carla avait mis fin aux activités de son réseau d'espionnage. Les combats tournaient au chaos. Les plans de bataille ne voulaient plus rien dire, ou si peu. Pour les Soviétiques, les renseignements en provenance de Berlin avaient de moins en moins de valeur. Inutile, dans ces conditions, de courir des risques. Les espions avaient brûlé leurs manuels de chiffrage et enfoui leurs émetteurs radio dans les gravats des immeubles bombardés. Ils étaient convenus de ne jamais parler à quiconque de leurs activités. Ils s'étaient montrés courageux, ils avaient abrégé la guerre et sauvé des vies, mais ils ne devaient pas s'attendre aux remerciements du peuple allemand vaincu. Leur bravoure demeurerait à jamais un secret.

Pendant que Carla attendait son tour devant la pompe, un groupe de membres de la Jeunesse hitlérienne chargés de la lutte antichars longea la file d'attente, se dirigeant vers la zone des combats à l'est. Une douzaine d'adolescents à bicyclette étaient encadrés par deux quinquagénaires. Ils avaient attaché à leur guidon deux lance-grenades Panzerfaust du dernier modèle. Perdus dans leurs uniformes et leurs casques bien trop grands pour eux, ils auraient paru du plus haut comique si l'épreuve qui les attendait avait été moins terrible. Ils partaient affronter l'armée Rouge.

Ils marchaient vers une mort certaine.

Carla détourna les yeux à leur passage : elle ne voulait pas se rappeler leurs visages.

Comme elle remplissait son seau, Frau Reichs, qui patientait derrière elle, lui demanda tout bas afin de ne pas être entendue : « Vous êtes bien une amie de la femme du docteur ? »

Carla se crispa. Frau Reichs parlait certainement d'Hannelore Rothmann. Le médecin avait disparu en même temps que tous les malades mentaux de l'hôpital juif. Rudi, le fils des Rothmann, avait arraché son étoile jaune pour rejoindre les Juifs qui vivaient dans la clandestinité, comme des non-Juifs : les *U-Boot* ainsi que les surnommaient les Berlinois. Mais Hannelore, qui n'était pas juive, vivait toujours dans la vieille maison familiale.

Pendant douze ans, une question comme celle-ci – vous êtes une amie de la femme d'un Juif? – avait été une accusation. Comment l'interpréter aujourd'hui ? se demanda Carla. Frau Reichs n'était pour elle qu'une vague connaissance : elle ne pouvait pas lui faire confiance.

Carla ferma le robinet. « Le docteur Rothmann était notre médecin de famille quand j'étais petite, répondit-elle, sur ses gardes. Pourquoi ? »

Frau Reichs prit à son tour la pompe et commença à remplir un vieux bidon d'huile. « Des gens ont emmené Frau Rothmann, dit-elle. J'ai pensé que ça vous intéresserait peut-être. »

Cela n'avait rien d'extraordinaire. On « emmenait » des gens tous les jours. Mais quand cela arrivait à un proche, c'était comme un coup au cœur.

Inutile de tenter de découvrir ce qui leur était arrivé – c'était même dangereux : celui qui s'inquiétait d'une disparition avait vite fait de disparaître à son tour. Carla ne put pourtant s'empêcher de demander : « Vous savez où elle est allée ? »

Cette fois-ci, elle obtint une réponse. « Au camp de transit de la Schulstrasse. » Carla sentit l'espoir renaître. « C'est au vieil hôpital juif de Wedding. Vous connaissez ?

— Oui. » Comme il lui arrivait d'y travailler, officieusement et en toute illégalité, elle savait que le gouvernement avait réquisitionné un des bâtiments, le laboratoire de pathologie, qui était désormais entouré de barbelés.

« J'espère qu'elle va bien, dit Frau Reichs. Elle a été très bonne avec moi quand ma Steffi est tombée malade. » Elle ferma le robinet et s'éloigna avec son bidon rempli d'eau.

Carla s'empressa de rentrer chez elle.

Elle devait faire quelque chose pour Hannelore. Jusqu'ici, il avait été quasiment impossible de faire sortir quelqu'un d'un camp, mais comme tout allait à vau-l'eau, peut-être trouverait-elle un moyen d'y parvenir.

Arrivée à la maison, elle confia le seau à Ada.

Maud était partie faire la queue avec ses tickets de rationnement. Carla se changea et enfila son uniforme d'infirmière, jugeant que cela lui faciliterait peut-être les choses. Elle dit à Ada où elle se rendait et ressortit.

Elle dut gagner Wedding à pied, ce qui faisait près de cinq kilomètres de marche. Elle se demanda si cela en valait la peine. Même si elle retrouvait Hannelore, elle ne pourrait sans doute pas l'aider. Puis elle pensa à Eva, réfugiée à Londres, et à Rudi, caché quelque part dans Berlin : quelle tragédie s'ils perdaient leur mère pendant les dernières heures de la guerre. Il fallait tenter le coup.

La police militaire avait envahi les rues et tout le monde devait montrer patte blanche. Les policiers travaillaient par groupes de trois et s'intéressaient surtout aux hommes en âge de se battre qui risquaient une exécution sommaire s'ils cherchaient à éviter le combat. Grâce à son uniforme, Carla ne fut pas inquiétée.

Dans ce paysage de désolation, il était étrange de voir fleurir les pommiers et les cerisiers et d'entendre les oiseaux gazouiller entre deux explosions, aussi gaiement que par un printemps ordinaire.

Horrifiée, elle vit plusieurs hommes pendus à des réverbères, dont certains en uniforme. La plupart d'entre eux portaient au cou un écriteau annonçant « Lâche » ou « Déserteur ». Des victimes de la justice sommaire appliquée par la police militaire.

Les nazis n'avaient donc pas étanché leur soif de massacres ? Elle en aurait pleuré.

Trois bombardements successifs l'obligèrent à chercher un abri. Lors du dernier, alors qu'elle ne se trouvait plus qu'à quelques centaines de mètres de l'hôpital, il lui sembla qu'Allemands et Soviétiques s'affrontaient deux ou trois rues plus loin. Le bruit des détonations était si violent qu'elle fut tentée de faire demi-tour. Hannelore était sans doute condamnée, voire déjà exécutée : pourquoi sacrifier sa propre vie ? Elle poursuivit tout de même sa route dans le soir tombant.

L'hôpital se trouvait dans l'Iranische Strasse, au coin de la Schulstrasse. Les arbres de la rue arboraient un feuillage tout neuf. Le laboratoire transformé en camp de transit était étroitement surveillé. Carla envisagea de s'adresser à un gardien pour lui exposer sa mission, mais cette stratégie ne semblait pas très prometteuse. Elle se demanda si elle pourrait s'introduire à l'intérieur du bâtiment par le tunnel qui le desservait.

Elle entra dans le bâtiment principal. L'hôpital était toujours en activité. Tous les patients avaient été transférés au sous-sol et dans les tunnels. Médecins et infirmiers travaillaient à la lueur des lampes à pétrole. À en juger par l'odeur, les toilettes ne fonctionnaient plus. On allait chercher l'eau au vieux puits du parc.

Surprise, elle vit des soldats amener des camarades blessés. Que tout le personnel soit juif leur était désormais indifférent.

Elle emprunta le tunnel qui passait sous le parc et rejoignait le sous-sol du laboratoire. Comme elle s'y attendait, la porte d'accès était gardée. Mais, en voyant son uniforme, le jeune membre de la Gestapo en faction la laissa passer sans rien dire. Peut-être ne voyait-il plus l'utilité de sa mission.

Elle était maintenant à l'intérieur du camp. Elle se demanda s'il serait aussi facile d'en ressortir.

L'odeur était atroce et elle comprit vite pourquoi. Le sous-sol était bondé. Plusieurs centaines de prisonniers étaient enfermés dans quatre réserves du laboratoire. Ils étaient assis ou allongés à même le sol, et seuls les plus chanceux disposaient d'un bout de mur pour s'y adosser. Ils empestaient ; sales et épuisés, ils la regardèrent d'un œil morne et indifférent.

Elle trouva Hannelore au bout de quelques minutes.

La femme du médecin avait jadis eu un corps sculptural et un visage séduisant couronné de cheveux blonds. Comme la plupart des gens, elle était aujourd'hui maigre à faire peur, et ses cheveux étaient gris et ternes. L'angoisse lui avait ridé les traits et creusé les joues.

Elle parlait à une adolescente, à la physionomie encore enfantine mais aux formes exagérément voluptueuses, les hanches larges et la poitrine plantureuse, comme en présentent certaines jeunes filles de cet âge-là. Elle pleurait, assise par terre, et Hannelore s'était agenouillée près d'elle pour lui tenir la main tout en la consolant à voix basse.

En apercevant Carla, elle se leva et s'écria : « Mon Dieu ! Que viens-tu faire ici ?

— J'ai pensé que si je leur disais que vous n'êtes pas juive, ils vous relâcheraient peut-être.

— Tu es bien courageuse !

— Votre mari a sauvé tant de vies. Il est normal que quelqu'un sauve la vôtre. »

L'espace d'un instant, Carla crut qu'Hannelore allait fondre en larmes. Son visage se décomposa. Puis elle battit des cils et secoua la tête. « Je te présente Rebecca Rosen, dit-elle d'une voix ferme. Ses parents sont morts aujourd'hui dans un bombardement.

— Oh ! Ma pauvre Rebecca », s'exclama Carla.

La jeune fille resta sans réaction.

« Quel âge as-tu, Rebecca ? insista Carla.

— Presque quatorze ans.

— Il va falloir que tu te conduises en adulte à présent.

— Pourquoi est-ce que je ne suis pas morte, moi aussi ? J'étais juste à côté d'eux. J'aurais dû mourir. Maintenant, je suis toute seule.

— Non, tu n'es pas toute seule, répliqua Carla vivement. Nous sommes avec toi. » Elle se tourna vers Hannelore. « Qui est le responsable de ce camp ?

— Walter Dobberke.

— Je vais lui dire qu'il doit vous libérer.

— Il est parti pour la journée. Et son adjoint est un sergent complètement borné. Mais attends, voilà Gisela. Elle a une liaison avec Dobberke. »

Une jolie jeune femme, aux longs cheveux blonds et à la peau laiteuse, venait d'entrer. Tous détournèrent les yeux en la voyant. Elle affichait un air hautain.

« Elle couche avec lui sur le lit d'examen de la salle d'électrocardiographie à l'étage, expliqua Hannelore. En échange, elle a droit à des rations supplémentaires. Je suis la seule à lui adresser la parole. J'estime que nous n'avons pas à juger les autres pour les compromis qu'ils acceptent. Nous sommes en enfer, après tout. »

Carla était sceptique. Jamais elle n'accepterait d'être l'amie d'une Juive qui couchait avec un nazi.

Croisant le regard d'Hannelore, Gisela se dirigea vers elle. « Il a reçu de nouveaux ordres », dit-elle, si bas que Carla dut tendre l'oreille pour l'entendre. Puis elle hésita.

« Oui ? fit Hannelore. Et quels sont ces ordres ? »

La voix de Gisela se réduisit à un murmure. « D'abattre tous ceux qui sont ici. »

Carla sentit une étreinte glacée lui serrer le cœur. Tous – y compris Hannelore et la petite Rebecca.

« Walter ne veut pas le faire, reprit Gisela. Au fond, ce n'est pas un mauvais bougre.

— Quand est-il censé nous exécuter ? demanda Hannelore sur un ton fataliste.

— Immédiatement. Mais il veut d'abord détruire les archives. Hans-Peter et Martin sont en train de les brûler dans la chaudière. Comme ça risque d'être long, nous avons quelques heures de répit. Peut-être l'armée Rouge arrivera-t-elle à temps pour nous sauver.

— Et peut-être pas, rétorqua Hannelore. Y a-t-il un moyen de le convaincre de désobéir aux ordres ? Pour l'amour de Dieu, la guerre est presque finie !

— Avant, j'arrivais à le convaincre de faire tout ce que je voulais, dit tristement Gisela. Mais il commence à se lasser de moi. Vous savez comment sont les hommes.

— Il ferait bien de penser à l'avenir. D'un jour à l'autre, les Alliés vont prendre le contrôle du pays. Les crimes des nazis ne resteront pas impunis.

— Si nous mourons tous, qui l'accusera ? demanda Gisela.

— Moi », déclara Carla.

Les deux autres la regardèrent, et se turent.

Carla comprit alors qu'on la fusillerait avec les autres, même si elle n'était pas juive, pour faire disparaître tous les témoins.

Cherchant désespérément une solution, elle lança : « Si Dobberke nous épargnait, ça l'aiderait peut-être auprès des Alliés.

— Bonne idée, approuva Hannelore. Nous pourrions attester par écrit qu'il nous a sauvé la vie à tous. »

Carla se tourna vers Gisela. « Peut-être accepterait-il », dit-elle d'un ton dubitatif.

Hannelore jeta un regard autour d'elle. « Voici Hilde. C'est elle qui lui sert de secrétaire. » Elle appela la jeune femme et lui exposa leur plan.

« Je vais taper des ordres de libération pour tous les prisonniers, proposa Hilde. Nous lui demanderons de les signer avant de lui donner notre attestation de bonne conduite. »

Comme il n'y avait pas de gardiens au sous-sol, excepté la sentinelle de faction devant la porte d'accès du tunnel, les prisonniers avaient toute liberté de mouvement. Hilde se rendit dans la pièce qui servait de bureau à Dobberke. Elle commença par dactylographier l'attestation. Hannelore et Carla firent ensuite le tour des prisonniers pour leur expliquer leur projet et recueillir leurs signatures. Pendant ce temps, Hilde tapa les ordres de libération.

Il était près de minuit lorsqu'elles eurent fini. Elles ne pouvaient rien faire de plus avant le retour de Dobberke, prévu pour la matinée.

Carla s'allongea sur le sol près de Rebecca Rosen. Il n'y avait pas d'autre endroit où dormir.

Au bout d'un moment, Rebecca se mit à pleurer doucement.

Carla ne savait pas quoi faire. Elle aurait voulu la réconforter, mais les mots lui manquaient. Que dire à une enfant qui vient de voir ses parents se faire tuer ? Ses sanglots étouffés ne s'apaisaient pas. Finalement, Carla s'approcha d'elle et la prit dans ses bras.

Elle comprit aussitôt que c'était la chose à faire. Rebecca se blottit contre elle, la tête sur sa poitrine. Carla lui tapota le dos comme à un bébé. Peu à peu, ses sanglots s'espacèrent et elle s'endormit.

Carla fut incapable de fermer l'œil. Elle passa la nuit à composer des discours imaginaires à l'intention du commandant du camp. Tantôt elle faisait appel à sa bonté d'âme, tantôt elle le menaçait des foudres de la justice alliée, tantôt encore elle lui suggérait de veiller à ses propres intérêts.

Elle s'efforça de ne pas penser à l'exécution qui l'attendait probablement. Erik lui avait raconté comment les nazis massacraient les Russes par douzaines. Sans doute disposeraient-ils ici d'un système tout aussi efficace. Elle avait peine à l'imaginer. Cela valait peut-être mieux.

Elle pourrait sans doute échapper à la mort en s'enfuyant tout de suite, ou dès le lever du jour. Elle n'était ni prisonnière ni juive, et ses papiers étaient en règle. Il lui suffirait de sortir par où elle était entrée, toujours vêtue de son uniforme d'infirmière. Mais cela l'obligerait à abandonner Hannelore et Rebecca. Elle ne pouvait s'y résoudre, malgré la tentation.

Les combats firent rage dans les rues jusqu'au petit matin, puis il y eut une brève accalmie. Ils reprirent dès le lever du soleil. Ils étaient désormais suffisamment proches pour qu'on entende les fusils-mitrailleurs en plus des canons.

En début de matinée, les gardiens leur apportèrent une marmite de soupe fade et un sac rempli de quignons de pain rassis. Après avoir bu et mangé, Carla se rendit à contrecœur dans des toilettes d'une indicible saleté.

En compagnie d'Hannelore, de Gisela et d'Hilde, elle monta au rez-de-chaussée pour attendre Dobberke. Les bombardements avaient repris et elles risquaient leur vie à chaque seconde, mais elles tenaient à le voir dès son arrivée.

Il n'était pas à l'heure. Pourtant, il était généralement ponctuel, affirma Hilde. Peut-être avait-il été retardé par les combats. Peut-être même était-il mort. Carla espérait que non. Son adjoint, le sergent Ehrenstein, était tellement stupide qu'il était impossible de discuter avec lui.

Au bout d'une heure, Carla commença à perdre espoir.

Le commandant arriva après un deuxième tour d'horloge.

« Que se passe-t-il ? demanda-t-il en découvrant les quatre femmes qui l'attendaient dans le hall. Une réunion de mères de famille ?

— Tous les prisonniers ont signé une attestation affirmant que vous leur avez sauvé la vie, répondit Hannelore. Et cela peut sauver la *vôtre*, si vous acceptez nos conditions.

— Ne soyez pas ridicule. »

Carla intervint. « D'après ce que dit la BBC, les Nations unies détiennent une liste nominative des officiers nazis qui ont pris part à des crimes de guerre. Dans huit jours, vous serez peut-être sur le banc des accusés. N'aimeriez-vous pas avoir entre les mains une attestation signée prouvant que vous avez épargné des condamnés ?

— Écouter la BBC est un crime.

— Moins grave que le meurtre. »

Hilde tenait une chemise en carton. « J'ai dactylographié un ordre de libération pour chaque prisonnier, expliqua-t-elle. Si vous les signez, nous vous remettrons l'attestation.

— Qu'est-ce qui m'empêche de vous la prendre de force ?

— Si nous sommes tous morts, personne ne croira à votre innocence. »

Dobberke avait beau être furieux, il n'était pas assez sûr de lui pour refuser de les écouter. « Je pourrais vous faire fusiller toutes les quatre pour insolence, remarqua-t-il.

— Voilà à quoi ressemble la défaite, lança Carla avec impatience. Autant vous y habituer tout de suite. »

Le visage de Dobberke s'empourpra et elle comprit qu'elle était allée trop loin. Si seulement elle pouvait retirer ses paroles ! Elle ne le quitta pas du regard, s'efforçant de ne pas laisser transparaître sa peur.

À cet instant, un obus explosa tout près du bâtiment. Les portes vibrèrent et l'une des fenêtres se brisa. Tous se baissèrent instinctivement, mais personne ne fut blessé.

Lorsqu'ils se redressèrent, l'expression de Dobberke avait changé. La rage avait cédé la place à une sorte de résignation écœurée. Le cœur de Carla battit plus vite. Allait-il céder ?

Le sergent Ehrenstein fit irruption. « Aucune perte à déplorer, mon commandant.

— Très bien, sergent. »

Ehrenstein allait ressortir quand Dobberke le rappela. « Ce camp est désormais fermé », annonça-t-il.

Carla retint son souffle.

« Fermé, mon commandant ? » Le sergent hésitait entre surprise et agressivité.

« J'ai reçu de nouveaux ordres. Dites aux hommes… » Dobberke hésita. « Dites-leur de se présenter au bunker de la gare de la Friedrichstrasse. »

Dobberke improvisait, Carla le savait, et Ehrenstein sembla s'en douter. « À quelle heure, mon commandant ?

— Immédiatement.

— Immédiatement. » Ehrenstein resta cloué sur place, comme s'il avait besoin qu'on lui explique le sens de ce mot.

Dobberke le fit plier d'un regard.

« À vos ordres, mon commandant, acquiesça le sergent. Je vais rassembler les hommes. » Il sortit.

Carla triomphait, tout en se disant qu'elles n'étaient pas encore tirées d'affaire.

« Montrez-moi cette attestation », demanda Dobberke à Hilde.

Hilde ouvrit la chemise. Elle contenait une douzaine de feuillets, tous portant le même texte dactylographié et une série de signatures. Elle les lui tendit.

Dobberke les plia et les fourra dans sa poche.

Hilde lui tendit ensuite les ordres de libération. « Veuillez signer ces documents, s'il vous plaît.

— Vous n'avez pas besoin de cette paperasse, répliqua-t-il. Et je n'ai pas le temps de signer cent fois le même formulaire.

— Il y a des policiers partout, protesta Carla. Ils pendent des gens aux réverbères. Nous avons besoin de ces papiers. »

Il tapota sa poche. « C'est moi qui serai pendu si on trouve cette attestation. » Il se dirigea vers la porte.

« Emmène-moi avec toi, Walter ! » s'écria Gisela.

« Hein ? fit-il en se tournant vers elle. Et que dirait ma femme ? » Il sortit en claquant la porte.

Gisela éclata en sanglots.

Carla se dirigea vers la porte, l'ouvrit et vit Dobberke s'éloigner d'un pas vif. Pas d'autres membres de la Gestapo en vue : ils avaient obéi à ses ordres et abandonné le camp.

Arrivé dans la rue, Dobberke se mit à courir.

Il avait laissé le portail ouvert.

À côté de Carla, Hannelore ouvrait des yeux incrédules.

« Nous sommes libres, je crois, murmura Carla.

— Il faut prévenir les autres.

— Je m'en charge », fit Hilde. Elle redescendit au sous-sol.

D'un pas craintif, Carla et Hannelore s'engagèrent dans l'allée conduisant du laboratoire au portail ouvert. Elles échangèrent un regard hésitant.

« La liberté est effrayante », remarqua Hannelore.

Derrière elles, une petite voix s'écria : « Carla, ne pars pas sans moi ! » C'était Rebecca qui se dépêchait dans l'allée, ses seins ballottant sous son chemisier crasseux.

Carla soupira. Me voilà avec un enfant sur les bras, songea-t-elle. Je ne me sens pas vraiment prête à assumer un rôle de mère. Mais que faire ?

« Viens, dit-elle. Mais il va falloir courir, tu sais. » Ses inquiétudes étaient infondées : Rebecca était bien plus rapide qu'Hannelore et elle-même.

Elles traversèrent le parc en direction de la porte principale où elles s'arrêtèrent pour inspecter l'Iranische Strasse. Tout semblait tranquille. Elles traversèrent la chaussée et accélérèrent le pas jusqu'au coin de la rue. Comme Carla jetait un coup d'œil dans la Schulstrasse, elle entendit une rafale de mitraillette et remarqua un échange de coups de feu un peu plus loin. Des soldats allemands battaient en retraite dans leur direction, poursuivis par ceux de l'armée Rouge.

Jetant autour d'elle un regard circulaire, elle n'aperçut aucun endroit où se cacher, sinon derrière les arbres, qui ne leur offraient aucune protection.

Un obus tomba sur la chaussée à une cinquantaine de mètres et explosa. Carla en sentit le souffle mais en fut quitte pour la peur.

Sans avoir besoin de se concerter, les trois fugitives regagnèrent l'hôpital.

Elles retournèrent dans le laboratoire. Quelques prisonniers se tenaient derrière les barbelés, comme s'ils redoutaient de sortir.

« Je sais que le sous-sol empeste, mais c'est le seul refuge qui nous reste », leur dit Carla. Elle entra dans le bâtiment, descendit l'escalier, et la plupart des autres la suivirent.

Elle se demanda combien de temps elle allait devoir rester là. L'armée allemande finirait par se rendre, mais quand ? Hitler n'accepterait certainement pas la défaite, quelles que soient les circonstances. Cet homme avait fondé toute son existence sur l'affirmation arrogante de sa supériorité de chef. Comment pourrait-il reconnaître ses erreurs, sa stupidité, sa monstruosité ? Admettre qu'il avait massacré des millions de gens et entraîné son pays à la ruine ? Affronter le jugement de l'Histoire, qui le présenterait comme l'être le plus malfaisant de tous les temps ? C'était impossible. Il allait devenir fou, mourir de honte ou s'enfoncer un pistolet dans la bouche et presser la détente.

Mais quand ? Dans combien de temps ? Un jour ? Une semaine ? Davantage encore ?

On entendit un cri à l'étage. « Ils sont là ! Les Russes arrivent ! »

Puis un bruit de lourds brodequins fit vibrer la cage d'escalier. Où les Russes avaient-ils trouvé des godillots aussi robustes ? Était-ce un cadeau des Américains ?

Ils entrèrent dans la salle, quatre, six, huit, neuf hommes au visage sale, armés de mitraillettes à chargeur tambour, prêts à tirer sur tout ce qui bougeait. Ils semblaient occuper tant d'espace ! Les gens se recroquevillaient d'effroi devant eux, devant ces hommes censés venir les libérer.

Les soldats évaluèrent la situation et décidèrent manifestement que les prisonniers – en majorité des femmes – étaient parfaitement inoffensifs. Ils baissèrent leurs armes. Certains allèrent inspecter les pièces voisines.

Un soldat de haute taille releva sa manche gauche. Il portait six ou sept montres-bracelets. Il aboya un ordre en russe en les désignant de la crosse de son arme. Carla crut comprendre ce qu'il voulait, mais elle n'en revenait pas. L'homme saisit alors une vieille femme, s'empara de sa main et désigna son alliance.

« Vont-ils nous prendre le peu que les nazis nous ont laissé ? » demanda Hannelore.

C'était bien cela. L'air agacé, le soldat tenta d'arracher l'alliance au doigt de la vieille dame. Comprenant ce qu'il voulait, celle-ci l'ôta d'elle-même pour la lui donner.

Il la prit, hocha la tête puis balaya la pièce du canon de son arme.

Hannelore s'avança vers lui. « Tous ces gens sont des prisonniers ! dit-elle en allemand. Des Juifs et des proches de Juifs, persécutés par les nazis ! »

Qu'il l'ait comprise ou non, il se contenta de désigner à nouveau les montres passées à son bras.

Les quelques détenus qui avaient réussi à conserver des objets plus ou moins précieux les remirent aux soldats.

L'arrivée de l'armée Rouge libératrice ne ressemblait guère à la fête que certains avaient espérée.

Mais le pire était encore à venir.

Le soldat pointa le doigt vers Rebecca.

Celle-ci se fit toute petite et tenta de se cacher derrière Carla.

Un second soldat, un blond, moins grand que le premier, agrippa Rebecca par l'épaule et l'attira contre lui. La jeune fille hurla et le visage de l'homme se fendit d'un large sourire, comme si ce bruit lui plaisait.

Un affreux pressentiment étreignit Carla.

Le petit blond immobilisa Rebecca pendant que le grand soldat lui tripotait les seins brutalement, lançant un commentaire qui les fit s'esclaffer tous les deux.

Des cris de protestation s'élevèrent du groupe de prisonniers.

Le grand soldat pointa son arme sur eux. Terrorisée, Carla crut qu'il allait tirer. Dans cette pièce surpeuplée, une rafale de mitraillette pouvait faire des dizaines de victimes.

Tous comprirent le danger et se turent.

Les deux soldats reculèrent vers la porte, emmenant Rebecca. Elle hurlait et se débattait, sans parvenir à se dégager.

Comme ils arrivaient sur le seuil, Carla s'avança d'un pas et cria : « Attendez ! »

Quelque chose dans sa voix les fit obéir.

« Elle est trop jeune, dit Carla. Elle n'a que treize ans ! » Avaient-ils compris ? Elle leva les mains, leur montrant ses dix doigts puis en ajoutant trois. « Treize ans ! »

Le plus grand des deux soldats parut comprendre. Un large sourire aux lèvres, il lança : « *Frau ist Frau.* » Une femme est une femme.

Carla s'entendit répondre : « C'est une vraie femme qu'il vous faut. » Elle s'avança d'un pas lent. « Prenez-moi à sa place. » Elle s'efforça d'esquisser un sourire enjôleur. « Je ne suis pas une enfant. Je sais comment on fait. » Elle s'approcha encore, assez près pour humer l'odeur nauséabonde d'un homme qui ne s'était pas lavé depuis des mois. S'efforçant de dissimuler son dégoût, elle baissa la voix et ajouta : « Je sais ce que désire un homme. » Elle frôla ses seins d'un geste aguicheur. « Oubliez cette enfant. »

Le grand soldat regarda Rebecca. Elle avait les yeux rougis par les larmes et la morve au nez, ce qui, par bonheur pour elle, la faisait ressembler à une gamine bien plus qu'à une femme.

Puis il se retourna vers Carla.

« Il y a un lit à l'étage, ajouta-t-elle. Vous voulez que je vous montre ? »

Ignorant toujours s'il comprenait ce qu'elle disait, elle le prit par la main et il la suivit au rez-de-chaussée.

Le blond lâcha Rebecca et gravit les marches derrière eux.

Elle avait réussi, pensa Carla tout en regrettant amèrement son audace. Elle n'avait qu'une envie : prendre ses jambes à son cou. Mais alors les Russes l'abattraient et s'en prendraient à Rebecca. Carla pensa à l'enfant bouleversée qui avait perdu ses parents la veille. Subir un viol aussitôt après la briserait à jamais. Carla devait la sauver.

Ça ne me détruira pas, se jura-t-elle. Je m'en remettrai. Je redeviendrai moi-même.

Elle les conduisit dans la salle d'électrocardiographie. Glacée, elle avait l'impression d'avoir le cœur gelé, le cerveau engourdi. Près du lit, elle aperçut un bidon de vaseline, dont les médecins se servaient pour améliorer la conductivité électrique de leurs appareils. Elle ôta sa culotte et préleva une bonne quantité de vaseline qu'elle s'enfonça dans le vagin. Cela lui éviterait peut-être de saigner.

Elle devait continuer à jouer la comédie. Se retournant vers les deux soldats, elle constata avec horreur que trois autres les avaient suivis. Elle s'efforça de sourire, en vain.

Elle s'allongea sur le dos et écarta les jambes.

Le grand soldat s'agenouilla entre ses cuisses. Il déchira son uniforme d'infirmière pour dénuder ses seins. Elle vit qu'il se tripotait pour hâter son érection. Il s'allongea sur elle et la pénétra. Elle se dit et se répéta que cela n'avait rien à voir avec les étreintes qu'elle partageait avec Werner.

Elle voulut détourner le visage, mais le soldat lui prit le menton et l'obligea à le regarder pendant qu'il la besognait. Elle ferma les yeux. Elle le sentit qui l'embrassait, cherchait à insinuer la langue entre ses lèvres. Il avait une haleine de viande pourrie. Lorsqu'elle serra les dents, il la frappa au visage. Elle poussa un cri et ouvrit sa bouche tuméfiée. Elle s'efforça de se rappeler qu'une vierge de treize ans aurait bien plus souffert.

Le soldat éjacula dans un grognement. Elle fit tout son possible pour dissimuler son dégoût.

Il descendit et le blond prit sa place.

Carla tenta de fermer son esprit, de le détacher de son corps, de transformer celui-ci en machine, sans aucun rapport avec elle. Le blond n'avait pas envie de l'embrasser, mais il la téta et lui mordit les mamelons, et, lorsqu'elle poussa un cri de douleur, il sembla ravi et la mordit de plus belle.

Le temps passa, et il éjacula.

Un autre prit sa place.

Elle songea soudain que lorsque cette épreuve serait enfin finie, elle ne pourrait ni se baigner ni se doucher, car il n'y avait plus d'eau courante dans la capitale. Ce fut cette perspective qui la fit basculer. Elle garderait leur semence en elle, leur odeur sur sa peau, leur salive dans sa bouche, et n'aurait aucun moyen de se laver. Pour une raison qui lui échappait, cela l'accabla plus encore que le reste. Perdant tout courage, elle se mit à pleurer.

Le troisième soldat prit son plaisir en elle, un quatrième le remplaça.

XX

1945 (II)

1.

Le lundi 30 avril 1945, Adolf Hitler mettait fin à ses jours dans son bunker de Berlin. Une semaine plus tard exactement, à dix-neuf heures quarante, à Londres, le ministère de l'Information annonçait la capitulation de l'Allemagne. Le lendemain, mardi 8 mai, fut déclaré férié.

Assise à la fenêtre de son appartement de Piccadilly, Daisy assistait aux réjouissances. Il y avait tellement de monde dans la rue que les voitures et les autobus ne pouvaient plus passer. Les filles embrassaient tous les hommes en uniforme, et des milliers d'heureux militaires en profitaient de bon cœur. Au début de l'après-midi, bien des gens étaient ivres. Par la vitre ouverte, Daisy entendait chanter au loin, et elle devina que la foule, massée devant Buckingham Palace, chantait « Land of Hope and Glory ». Elle prenait part à la liesse générale, mais elle n'avait envie d'embrasser qu'un soldat, Lloyd, mais elle ne savait pas exactement où il était, en France ou en Allemagne. Elle priait pour qu'il n'ait pas été tué pendant les dernières heures de la guerre.

Millie arriva avec ses deux enfants. Son mari, Abe Avery, était aussi sous les drapeaux, quelque part. Elle avait amené les petits dans le West End pour participer aux festivités, et ils venaient souffler un peu à l'abri de la foule, chez Daisy. La maison des Leckwith à Aldgate avait longtemps été un refuge pour Daisy, et elle était heureuse d'avoir l'occasion de rendre la pareille à Millie. Elle alla préparer du thé – les domestiques étaient sortis faire la fête – et servit du jus d'orange aux enfants. Lennie avait cinq ans, à présent, et Pammie trois.

Depuis qu'Abe avait été enrôlé, c'était Millie qui dirigeait son affaire de négoce de cuir en gros. Elle s'occupait de la partie commerciale tandis que sa belle-sœur, Naomi Avery, tenait la comptabilité. « Les choses vont changer, forcément, remarqua Millie. Ces cinq dernières années, on avait besoin de grosses peaux épaisses pour faire des chaussures et des bottes. Maintenant, la demande va se tourner vers des cuirs plus souples, de l'agneau et du porc, pour des sacs à main et des mallettes. Quand le marché du luxe reprendra, il y aura enfin moyen de gagner correctement sa vie. »

Daisy se rappela que son père, Lev, avait la même façon de voir les choses que Millie. Comme elle, il regardait toujours vers l'avenir, à l'affût de toutes les possibilités.

Eva Murray vint ensuite, accompagnée de ses quatre enfants. Jamie, qui avait huit ans, organisa une partie de cache-cache, et l'appartement se transforma en cour d'école maternelle. Le mari d'Eva, Jimmy, qui avait été promu colonel, se trouvait également en France ou en Allemagne, et Eva connaissait les mêmes affres que Daisy et Millie.

« Nous allons avoir des nouvelles d'un jour à l'autre, vous allez voir, dit Millie. Et alors, ce sera vraiment fini. »

Eva attendait aussi avec angoisse des nouvelles de sa famille à Berlin. Mais dans le chaos de l'après-guerre, elle pensait qu'il s'écoulerait peut-être des semaines voire des mois avant qu'on puisse savoir ce qu'était devenu tel ou tel Allemand.

« Je me demande si mes enfants connaîtront jamais mes parents », soupira-t-elle tristement.

À cinq heures, Daisy prépara une cruche de martini. Millie alla faire un tour à la cuisine et, toujours aussi efficace, revint bientôt avec une assiette de canapés aux sardines à grignoter en buvant un verre. Eth et Bernie arrivèrent juste au moment où Daisy en confectionnait une seconde tournée.

Bernie annonça à Daisy que Lennie savait déjà lire, et que Pammie connaissait l'hymne national. « Le grand-père dans toute sa splendeur, commenta Ethel. Il pense qu'il n'y a jamais eu d'enfants intelligents avant. » Mais Daisy voyait bien qu'au fond de son cœur, elle était tout aussi fière d'eux.

Alors qu'elle buvait son deuxième martini, heureuse et détendue, elle parcourut du regard le groupe disparate réuni chez elle.

Ils lui avaient fait honneur en passant la voir sans être invités, sachant qu'ils seraient les bienvenus. Ils étaient les siens comme elle était des leurs. Ils étaient sa famille, songea-t-elle.

Elle se sentait très privilégiée.

2.

Assis devant la porte du bureau de Leo Shapiro, Woody Dewar contemplait pensivement un paquet de photos. Celles qu'il avait prises à Pearl Harbor, pendant l'heure qui avait précédé la mort de Joanne. Après avoir laissé la pellicule dans son appareil pendant des mois, il avait fini par la développer et avait tiré les clichés. Les regarder l'avait tellement attristé qu'il les avait rangés dans un tiroir de sa chambre, dans son appartement de Washington, et ils y étaient restés.

Mais le temps du changement était venu.

Il n'oublierait jamais Joanne, et pourtant, il était amoureux à nouveau. Il adorait Bella, et elle le lui rendait bien. Quand ils s'étaient séparés, à la gare d'Oakland à la périphérie de San Francisco, il lui avait dit qu'il l'aimait, et elle avait répondu : « Moi aussi, je t'aime. » Il allait lui demander de l'épouser. Il l'aurait déjà fait si ça ne lui avait semblé un peu précipité – moins de trois mois ; il ne voulait pas donner aux parents de la jeune fille, hostiles à cette union, une bonne raison d'élever des objections.

Et puis, il avait une décision à prendre ; une décision concernant son avenir.

Il ne voulait pas se lancer dans la politique.

Ses parents en seraient bouleversés, il le savait. Ils avaient toujours pensé qu'il suivrait les traces de son père, devenant ainsi le troisième sénateur Dewar. Il s'était engagé sur cette voie sans réfléchir. Mais au cours de la guerre, et surtout pendant son séjour à l'hôpital, il s'était demandé ce qu'il voulait *vraiment* faire, s'il s'en sortait, et la réponse avait été : « Pas de politique ».

C'était le bon moment pour se retirer. Son père avait réalisé l'ambition de sa vie. Le Sénat avait débattu du projet des Nations unies. C'était à un point similaire de l'histoire que la

SDN, l'ancienne Société des nations, avait fait naufrage, un souvenir pénible pour Gus Dewar. Mais le sénateur Vandenberg avait défendu avec passion ce qu'il avait appelé le « rêve le plus précieux de l'humanité », et la charte des Nations unies avait été ratifiée par quatre-vingt-neuf voix contre deux. C'était gagné. Même si Woody renonçait maintenant à cette carrière, il n'aurait pas l'impression de laisser tomber son père.

Il espérait que Gus verrait les choses comme lui.

Shapiro ouvrit la porte de son bureau et lui fit signe. Woody se leva et entra.

Shapiro était plus jeune que Woody ne s'y attendait. Il devait avoir une trentaine d'années. C'était le chef du bureau de Washington de la National Press Agency. Il retourna s'asseoir et lui demanda :

« Que puis-je faire pour le fils du sénateur Dewar ?

— Je voudrais vous montrer quelques photos, si vous permettez.

— Je vous en prie. »

Woody étala ses photos sur le bureau de Shapiro.

« C'est Pearl Harbor ? demanda Shapiro.

— Oui. Le 7 décembre 1941.

— Mon Dieu. »

Woody voyait les clichés à l'envers, mais il en avait les larmes aux yeux. Il y avait Joanne, si belle, et Chuck, arborant un grand sourire heureux, heureux d'être en compagnie de sa famille avec Eddie. Et puis les avions qui surgissaient dans le ciel, les bombes, les torpilles qui tombaient de leur ventre, les explosions de fumée noire sur les navires, les marins qui grimpaient sur les bastingages, se jetaient à la mer et nageaient avec l'énergie du désespoir.

« C'est votre père, là, remarqua Shapiro. Et votre mère. Je les reconnais.

— Et ma fiancée, qui est morte quelques minutes plus tard. Mon frère, qui a été tué à Bougainville. Et le meilleur ami de mon frère.

— Ce sont des clichés sensationnels ! Combien en voulez-vous ?

— Je ne veux pas d'argent », répondit Woody.

Shapiro leva les yeux, surpris.

« Ce que je veux, dit Woody, c'est du travail. »

3.

Deux semaines après la fête de la Victoire en Europe, Winston Churchill annonça l'organisation d'élections législatives.

La famille Leckwith fut prise au dépourvu. Comme la plupart des gens, Ethel et Bernie pensaient que Churchill attendrait la capitulation japonaise. Clement Attlee, le chef du parti travailliste, avait envisagé des élections en octobre. Churchill les avait tous pris à contre-pied.

Le commandant Lloyd Williams fut démobilisé pour pouvoir être candidat du parti travailliste à Hoxton, dans l'East End de Londres. Le parti proposait une vision d'avenir qui lui inspirait un enthousiasme ardent. Le fascisme avait été vaincu, et le peuple britannique pouvait désormais fonder une société alliant liberté et mieux-être social. Les travaillistes avaient un programme bien conçu pour éviter les catastrophes des vingt dernières années : l'assurance chômage intégrale et universelle afin d'aider les familles à traverser les périodes difficiles, la planification économique censée empêcher une nouvelle crise, et l'Organisation des Nations unies pour maintenir la paix.

« Tu n'as aucune chance de gagner », lui dit son beau-père, Bernie, dans la cuisine de la maison d'Aldgate, le lundi 4 juin. Son pessimisme était d'autant plus convaincant qu'il lui ressemblait bien peu. « Ils vont voter conservateur parce que Churchill a gagné la guerre, poursuivit-il d'un ton funèbre. C'est ce qui s'est passé avec Lloyd George en 1918. »

Lloyd s'apprêtait à répliquer, mais Daisy le prit de vitesse. « La guerre n'a pas été gagnée par le libéralisme et le capitalisme, protesta-t-elle. Elle a été gagnée par des gens qui ont uni leurs forces pour supporter les épreuves, chacun faisant sa part. C'est ça, le socialisme ! »

Cette passion, voilà ce que Lloyd aimait plus que tout chez elle. Quant à lui, il avait tendance à être plus mesuré : « Nous avons déjà appliqué des mesures que les anciens conservateurs auraient jugées dignes des bolcheviks, comme le contrôle gouvernemental des chemins de fer, des mines et du transport maritime, autant de mesures instaurées par Churchill. Et Ernie Bevin a été chargé de la planification économique pendant toute la durée de la guerre. »

Bernie secoua la tête d'un air entendu, une attitude de vieil homme qui avait le don d'agacer Lloyd. « Les gens votent avec leur cœur, pas avec leur cerveau, objecta-t-il. Ils vont vouloir exprimer leur reconnaissance.

— Eh bien, ça ne sert à rien de rester ici à discuter avec toi, dit Lloyd. Je préfère discuter avec les électeurs. »

Ils prirent le bus, Daisy et lui, pour se rendre quelques stations plus au nord, au Black Lion, un pub de Shoreditch où ils retrouvèrent une équipe de militants du parti travailliste de la circonscription de Hoxton qui faisait du porte-à-porte. En réalité, Lloyd savait bien que le porte-à-porte n'avait pas pour but d'essayer de convaincre les électeurs. Son principal objectif était d'identifier les sympathisants, afin que, le jour de l'élection, la machine du parti s'assure qu'ils se rendaient bien tous aux urnes. Ils cochaient les sympathisants déclarés du parti travailliste ; ceux des autres partis étaient purement et simplement rayés. Seuls les indécis méritaient qu'ils leur consacrent plus de quelques secondes : on leur donnait l'occasion de parler au candidat.

Lloyd s'attira quelques réactions négatives. « Vous êtes commandant, hein ? lui lança une femme. Mon Alf est caporal. Il dit que les officiers ont bien failli perdre la guerre. »

S'ajoutaient quelques accusations de népotisme : « Vous êtes le fils de la députée d'Aldgate, non ? Alors, c'est quoi, une monarchie héréditaire ? »

Il se remémora le conseil de sa mère : « On ne gagne jamais une élection en prouvant à l'électeur qu'il est idiot. Sois charmant, sois modeste et ne perds jamais ton calme. Si un électeur se montre hostile et grossier, remercie-le de t'avoir donné un peu de son temps et prends congé. Il se demandera après ton départ s'il ne t'a pas mal jugé. »

Les électeurs de la classe ouvrière étaient majoritairement favorables aux travaillistes. Beaucoup firent remarquer à Lloyd qu'Attlee et Bevin avaient fait du bon boulot pendant le conflit. Les indécis appartenaient surtout à la petite bourgeoisie. Quand ils lui disaient que c'était Churchill qui avait gagné la guerre, Lloyd leur rappelait l'aimable critique d'Attlee : « Cela n'a pas été un gouvernement d'un seul homme, et cela n'a pas été la guerre d'un seul homme. »

Churchill avait présenté Attlee comme un homme modeste qui avait d'excellentes raisons de l'être. L'humour d'Attlee était moins agressif, et de ce fait même plus percutant ; tel était du moins l'avis de Lloyd.

Deux électeurs lui annoncèrent qu'ils voteraient pour l'actuel député de Hoxton, un libéral, parce qu'il les avait aidés à résoudre un problème. Les gens faisaient souvent appel aux membres du Parlement lorsqu'ils avaient l'impression d'être injustement traités par le gouvernement, un employeur ou un voisin. C'était un travail qui prenait du temps, mais qui rapportait des voix.

L'un dans l'autre, Lloyd aurait été incapable de dire de quel côté penchait l'opinion.

Un seul électeur lui parla de Daisy. Un homme qui vint lui ouvrir la porte, la bouche pleine. « Bonsoir, monsieur Perkinson, commença Lloyd. Je crois savoir que vous aviez quelque chose à me demander.

— Votre fiancée a été fasciste », répondit l'homme en mastiquant.

Lloyd devina qu'il avait lu le *Daily Mail*, qui avait publié un article fielleux sur Lloyd et Daisy intitulé LE SOCIALISTE ET LA VICOMTESSE.

« Elle a été brièvement aveuglée par le fascisme, convint Lloyd. Comme beaucoup d'autres.

— Comment un socialiste peut-il épouser une fasciste ? »

Lloyd chercha Daisy du regard, et lui fit signe d'approcher.

« Monsieur Perkinson ici présent m'interroge sur ma fiancée, à laquelle il reproche d'avoir été fasciste.

— Enchantée, monsieur Perkinson, dit Daisy en serrant la main de l'homme. Je comprends tout à fait que cette question vous préoccupe. Mon premier mari a été fasciste dans les années 1930, et je l'ai effectivement soutenu. »

Perkinson hocha la tête. Il jugeait probablement qu'une femme devait avoir les mêmes opinions politiques que son mari.

« Nous étions de jeunes idiots, poursuivit Daisy. Mais quand la guerre a éclaté, mon premier mari s'est engagé dans l'armée de l'air et a combattu les nazis aussi courageusement que n'importe qui.

— C'est vrai ?

— L'année dernière, il pilotait un Typhoon au-dessus de la France, et mitraillait en rase-mottes un convoi de troupes allemand quand son appareil a été abattu. Il est mort. Je suis donc veuve de guerre. »

Perkinson déglutit. « Ma foi, je vous présente toutes mes condoléances. »

Mais Daisy n'avait pas fini. « Quant à moi, je suis restée à Londres durant le conflit. J'ai conduit une ambulance pendant toute la durée du Blitz.

— C'était très courageux de votre part, en tout cas.

— Je voudrais seulement vous convaincre que nous avons payé notre dette, mon défunt mari et moi.

— Ça, c'est une autre affaire, marmonna Perkinson.

— Nous ne voudrions pas vous retenir davantage, intervint Lloyd. Merci de m'avoir fait part de votre point de vue. Bonne soirée. »

Comme ils s'éloignaient, Daisy remarqua : « Je n'ai pas l'impression que nous l'ayons gagné à notre cause.

— On n'y arrive jamais, répondit Lloyd. Mais au moins, maintenant, il a entendu les deux sons de cloche, et quand il parlera de nous au pub, plus tard, dans la soirée, il sera peut-être un peu moins virulent.

— Hum. »

Lloyd sentit qu'il n'avait pas réussi à rassurer Daisy.

Le porte-à-porte s'acheva de bonne heure, parce que ce soir-là, la BBC diffusait sa première émission sur les élections et que tous les militants voulaient l'écouter. Churchill avait le privilège d'ouvrir les débats.

Dans le bus qui les ramenait chez eux, Daisy murmura :

« Je suis inquiète. Je suis un obstacle pour ton élection.

— Aucun candidat n'est parfait, répondit Lloyd. C'est la façon de gérer ses faiblesses qui compte.

— Je ne veux pas être une faiblesse pour toi. Il vaudrait peut-être mieux que je ne m'en mêle pas.

— Au contraire, je veux que tout le monde sache tout à ton sujet, de A à Z. Si tu es un obstacle, c'est moi qui laisserai tomber la politique.

— Ah non ! Je ne me pardonnerais jamais de t'avoir fait renoncer à tes ambitions.

— On n'en arrivera pas là. » Mais il se rendait bien compte, encore une fois, qu'il n'avait pas su apaiser ses craintes.

Quand ils regagnèrent Nutley Street, la famille Leckwith était assise autour du poste de radio Marconi, dans la cuisine. Daisy prit Lloyd par la main. « Je suis souvent venue ici pendant ton absence, lui dit-elle. Nous écoutions de la musique, du swing, et nous parlions de toi. »

À cette pensée, Lloyd songea qu'il avait bien de la chance.

La voix de Churchill se fit entendre sur les ondes. Son timbre rauque si familier avait quelque chose de galvanisant. Pendant cinq années noires, cette voix avait donné aux gens espoir, force et bravoure. Lloyd perdit courage : il était tenté, lui-même, de voter pour lui.

« Mes amis, commença le Premier ministre, je me dois de vous dire qu'une politique socialiste est incompatible avec la notion britannique de liberté. »

Bon, c'était l'argumentation habituelle. Toutes les idées nouvelles étaient condamnées comme des importations étrangères. Mais qu'est-ce que Churchill avait à offrir au peuple ? Les travaillistes avaient un programme précis. Et les conservateurs, que proposaient-ils ?

« Le socialisme est intrinsèquement lié au totalitarisme, affirma Churchill.

— Il ne va quand même pas prétendre que nous ne valons pas mieux que les nazis ? s'indigna Ethel.

— J'ai bien peur que si, répondit Bernie. Il va dire que nous avons vaincu l'ennemi à l'étranger, et que maintenant nous devons le vaincre chez nous. La bonne vieille tactique conservatrice.

— Les gens ne vont jamais avaler ça, objecta Ethel.

— Chut ! fit Lloyd.

— Un État socialiste, une fois solidement et précisément établi sous tous ses aspects, poursuivait Churchill, ne tolérera aucune opposition.

— C'est révoltant ! s'exclama Ethel.

— Mais j'irai plus loin, continuait Churchill. Je vous le déclare du fond du cœur, aucun système socialiste ne peut se passer de police politique.

— Une police politique ? répéta Ethel, indignée. Où est-il allé chercher ça ?

— En un sens, c'est bien, intervint Bernie. Comme il ne trouve rien à critiquer dans notre programme, il nous attaque sur des propositions qui ne sont pas les nôtres. Satané menteur !

— Écoutez ! s'écria Lloyd.

— Ils seraient obligés d'établir à leur tour une sorte de Gestapo », disait Churchill.

Ils bondirent sur leurs pieds comme un seul homme, poussant des hurlements indignés qui couvrirent la voix du Premier ministre.

« Ah, le salaud ! hurla Bernie en agitant le poing vers le poste de radio. Le salaud, le salaud ! »

Lorsqu'ils se furent calmés, Ethel reprit la parole : « Ça va être ça, leur campagne ? Une kyrielle de mensonges ?

— Ça m'en a tout l'air, répondit Bernie.

— Mais est-ce que les gens vont le croire ? » demanda Lloyd.

4.

Au sud du Nouveau-Mexique, non loin d'El Paso, s'étend un désert appelé Jornada del Muerto, le voyage du mort. Toute la journée, un soleil meurtrier accable les buissons épineux de prosopis et les yuccas aux feuilles lancéolées. Les seuls habitants sont des scorpions, des serpents à sonnettes, des fourmis rouges et des tarentules. C'est là que les hommes du projet Manhattan testaient l'arme la plus effroyable jamais conçue par l'humanité.

Greg Pechkov accompagnait les chercheurs qui suivaient l'essai à dix kilomètres de distance. Il espérait deux choses : d'abord, que la bombe fonctionnerait, ensuite que dix kilomètres seraient suffisants pour qu'ils soient à l'abri.

Le compte à rebours commença à cinq heures neuf du matin, le lundi 16 juillet. Le jour se levait, barrant le ciel de traînées d'or à l'est.

L'essai portait le nom de code Trinity. Quand Greg avait demandé pourquoi, le responsable de l'équipe de chercheurs,

J. Robert Oppenheimer, un Juif de New York aux oreilles pointues, avait cité un poème de John Donne : « Bats, mon cœur, Dieu de Trinité. »

« Oppie » était l'homme le plus intelligent que Greg ait jamais rencontré, le physicien le plus brillant de sa génération. Il parlait six langues, il avait lu *Le Capital* de Karl Marx dans sa version originale allemande, et pour se divertir, il apprenait entre autres le sanscrit. Greg l'aimait et l'admirait. La plupart des physiciens étaient des asociaux binoclards, mais Oppie, comme Greg, d'ailleurs, faisait exception à la règle : grand, beau, charmant, un véritable bourreau des cœurs.

Il avait ordonné au Corps des ingénieurs de l'armée de construire au milieu du désert, sur des fondations de béton, une tour en poutrelles d'acier de trente mètres de haut qui soutenait une plate-forme de chêne. La bombe avait été treuillée au sommet le samedi.

Les chercheurs n'utilisaient jamais le terme « bombe », ils parlaient du « gadget ». Son cœur était formé par une boule de plutonium, un métal qui n'existait pas à l'état naturel mais était un sous-produit créé dans les piles nucléaires. La boule pesait dix livres et contenait tout le plutonium du monde. Quelqu'un avait estimé son prix à un milliard de dollars.

Trente-deux détonateurs placés à la surface de celle-ci devaient se déclencher simultanément, créant une pression interne si puissante qu'elle accroissait la densité du plutonium, jusqu'à ce qu'il atteigne la masse critique.

Ensuite, personne ne savait ce qui se passerait.

Les chercheurs avaient ouvert les paris, à un dollar la mise, sur la force de l'explosion mesurée en tonnes d'équivalent de TNT. Edward Teller avait misé sur quarante-cinq mille tonnes, Oppie sur trois cents. La prévision officielle était de vingt mille tonnes. La veille au soir, Enrico Fermi avait proposé un autre pari : l'explosion allait-elle, oui ou non, rayer de la carte tout l'État du Nouveau-Mexique ? Le général Groves n'avait pas trouvé ça drôle.

Les physiciens avaient eu un débat extrêmement sérieux sur les conséquences de l'explosion : et si elle enflammait toute l'atmosphère de la Terre et détruisait la planète ? Ils étaient arrivés à la conclusion que cela n'arriverait pas. S'ils se trompaient, Greg ne pouvait qu'espérer que ce serait rapide.

L'essai avait été prévu au départ pour le 4 juillet. Mais chaque fois qu'ils avaient testé un composant, l'expérience avait été un échec, et le grand jour avait été retardé plusieurs fois. À Los Alamos, le samedi, une maquette absolument identique au véritable spécimen avait refusé de se déclencher. Ce qui avait relancé les paris : Norman Ramsey avait pronostiqué zéro, convaincu que la bombe ferait un flop.

Ce jour-là, l'explosion avait été programmée pour deux heures du matin, mais à l'heure prévue, un orage avait éclaté – en plein désert ! La pluie aurait précipité les retombées radioactives sur la tête des observateurs. Aussi la mise à feu avait-elle été retardée.

L'orage s'était calmé à l'aube.

Greg était au niveau d'un bunker appelé le S-10000, qui abritait le centre de commandement. Comme la plupart des membres de l'équipe scientifique, il était sorti pour mieux voir. Il était partagé entre l'espoir et la peur. Si la bombe faisait long feu, les efforts de centaines de personnes – sans parler de près de deux milliards de dollars – auraient été investis en pure perte. Si elle explosait, ils seraient peut-être tous morts quelques minutes plus tard.

À côté de lui se tenait Wilhelm Frunze, un jeune physicien allemand dont il avait fait la connaissance à Chicago.

« Que se serait-il passé, Will, si la foudre était tombée sur la bombe ? »

Frunze haussa les épaules. « Personne n'en sait rien. »

Une fusée éclairante verte fila dans le ciel, faisant sursauter Greg.

« Mise à feu dans cinq minutes », annonça Frunze.

Les services de sécurité n'avaient pas été très méthodiques. Santa Fe, la ville la plus proche de Los Alamos, grouillait d'agents du FBI trop bien habillés. Nonchalamment adossés aux murs avec leurs vestes de tweed et leurs cravates, ils tranchaient sur la population locale en blue jeans et en bottes de cow-boy.

Le FBI avait aussi mis sur écoute, en toute illégalité, les lignes téléphoniques de centaines de personnes mêlées au projet Manhattan. Greg n'en revenait pas. Comment l'institution, responsable au premier chef de l'application de la loi, pouvait-elle commettre systématiquement des actes délictueux ?

Les services de sécurité de l'armée et du FBI avaient tout de même identifié quelques espions, comme Barney McHugh, et les avaient exclus en douceur du projet. Mais les avaient-ils tous débusqués ? Greg n'en savait rien. Groves avait été obligé de prendre des risques. S'il avait viré tous ceux dont le FBI lui demandait de se débarrasser, il ne serait plus resté assez de chercheurs pour fabriquer la bombe.

Malheureusement, la plupart des scientifiques étaient des radicaux, des socialistes et des libéraux. On aurait eu du mal à trouver un conservateur parmi eux. Ils étaient convaincus que les découvertes scientifiques devaient être partagées avec toute l'humanité, et que le savoir ne devait jamais être tenu secret, au profit d'un régime ou d'un pays unique. C'est ainsi que, pendant que le gouvernement américain conservait un silence absolu sur ce projet colossal, les chercheurs organisaient des groupes de discussion sur le partage de la technologie nucléaire avec toutes les nations du monde. Oppie lui-même était suspect. La seule raison pour laquelle il n'était pas affilié au parti communiste était qu'il n'avait jamais été membre d'aucun cercle, d'aucune association.

Pour le moment, Oppie était allongé par terre à côté de son jeune frère, Frank, qui était lui aussi un physicien exceptionnel, également communiste. À l'image de Greg et de Frunze, ils tenaient tous les deux des boucliers de soudeur à travers lesquels ils observeraient l'explosion. Certains chercheurs portaient des lunettes de soleil.

Une autre fusée fut lancée.

« Une minute », annonça Frunze.

Greg entendit Oppie dire : « Bon sang, ça met le cœur à rude épreuve, des machins pareils. »

Il se demanda si ce seraient ses dernières paroles.

À plat ventre sur le sol sablonneux, leur plaque de verre opaque devant les yeux, ils regardaient tous en direction de la zone d'essai.

Face à la mort, Greg pensa à sa mère, à son père et à sa sœur Daisy, à Londres. Il se demanda à quel point il leur manquerait. Il songea, avec une pointe de regret, à Margaret Cowdry, qui l'avait plaqué pour un type disposé à l'épouser. Mais surtout, il pensa à Jacky Jakes et Georgy, qui avait maintenant neuf ans. Il aurait tant voulu le voir grandir ! Il se rendit compte que Georgy était la principale raison pour laquelle il espérait rester en vie. En catimini, l'enfant s'était glissé dans son âme, s'emparant de son amour. La puissance de ce sentiment le surprenait.

Un carillon tinta, un bruit pour le moins insolite dans le désert.

« Dix secondes. »

Greg fut pris de l'envie subite de se lever et de prendre ses jambes à son cou. C'était évidemment stupide – jusqu'où pourrait-il aller en dix secondes ? – mais il dut faire un effort sur lui-même pour ne pas bouger.

La bombe explosa à cinq heures vingt-neuf minutes et quarante-cinq secondes.

Il y eut d'abord un éclair effroyable, d'une luminosité incroyable, la lumière la plus intense que Greg aie jamais vue, plus vive que le soleil.

Et puis un dôme de feu maléfique sembla surgir du sol. Il s'éleva monstrueusement, à une vitesse terrifiante, atteignit le niveau des montagnes et continua à monter ; en comparaison, les sommets devinrent minuscules.

Greg murmura :

« Bon Dieu… »

Le dôme changea de forme et devint cubique. La lumière était encore plus vive qu'en plein midi, et les montagnes lointaines étaient si vivement illuminées que Greg en distinguait la moindre faille, le moindre plissement, la moindre roche.

Et puis la forme se modifia à nouveau. Une colonne apparut en dessous et sembla s'élever à plusieurs milliers de mètres dans le ciel, tel le poing de Dieu. Le nuage de feu bouillonnant qui surmontait la colonne se déploya en parapluie, jusqu'à ce que le tout ressemble à un champignon de plus de dix kilomètres de hauteur. Un champignon de nuages teinté de vert, d'orange et de violet démoniaques.

Greg fut heurté par une vague de chaleur comme si le Tout-Puissant avait ouvert un four géant. Au même moment, le bang de l'explosion atteignit ses oreilles, pareil à un coup de tonnerre infernal. Mais ce n'était qu'un début. Un bruit pareil au grondement d'un orage d'une puissance surnaturelle roula sur le désert, noyant tous les autres sons.

Le nuage incandescent commença à diminuer alors que le tonnerre rugissait encore et encore, se prolongeant insupportablement, au point que Greg se demanda si ce n'était pas le bruit de la fin du monde.

Et puis il finit par s'estomper, et le nuage en forme de champignon se dissipa peu à peu.

Greg entendit Frank Oppenheimer murmurer : « Ça a marché.

— Oui, ça a marché », renchérit Oppie.

Les deux frères se serrèrent la main.

Le monde est toujours là, pensa Greg.

Mais il a changé à jamais.

5.

Le matin du 26 juillet, Lloyd Williams et Daisy se rendirent à l'hôtel de ville de Hoxton pour assister au décompte des voix.

Si Lloyd perdait, Daisy était décidée à rompre leurs fiançailles.

Il refusait d'admettre qu'elle puisse représenter un obstacle à sa carrière politique, mais elle savait à quoi s'en tenir. Les adversaires de Lloyd mettaient un point d'honneur à l'appeler « Lady Aberowen ». Certains électeurs prenaient un air indigné face à son accent américain, comme si elle n'avait pas le droit de prendre part à la politique britannique. Les membres du parti travailliste eux-mêmes la traitaient différemment, lui demandant si elle voulait du café alors qu'ils buvaient tous du thé.

Comme l'avait prévu Lloyd, elle réussissait souvent à vaincre l'hostilité initiale dont elle était l'objet par son charme, sa spontanéité et sa serviabilité. Mais cela suffirait-il ? Seul le résultat de l'élection apporterait une réponse définitive.

Elle refuserait de l'épouser si leur mariage devait obliger Lloyd à renoncer à ce qui était toute sa vie. Il se déclarait prêt à le faire, mais ce serait un bien mauvais départ pour un jeune couple. Daisy frémissait d'horreur lorsqu'elle l'imaginait faire autre chose – travailler dans une banque ou dans l'administration –, affreusement malheureux et essayant de faire comme si elle n'y était pour rien. Mieux valait ne pas y penser.

Malheureusement, tout le monde était convaincu que les conservateurs remporteraient l'élection.

Au cours de la campagne, certains éléments avaient joué en faveur du parti travailliste. Le discours sur la « Gestapo » de Churchill s'était retourné contre lui. Les conservateurs eux-mêmes avaient été consternés. Clement Attlee, qui s'était exprimé à la radio le lendemain, au nom des travaillistes, avait été d'une ironie glacée : « Hier soir, en entendant le Premier ministre caricaturer ainsi le programme du parti travailliste, j'ai immédiatement saisi son objectif. Il voulait que les électeurs comprennent l'immense différence qu'il y a entre Winston Churchill, grand chef de guerre d'une nation unie, et Mr. Churchill, chef du parti conservateur. Il craignait que ceux qui avaient accepté qu'il les dirige pendant la guerre soient tentés, par gratitude, de continuer à le suivre. Je le remercie de les avoir si puissamment désillusionnés. » Le dédain magistral d'Attlee avait fait passer Churchill pour un fomentateur de troubles. Les gens avaient eu leur dose de passion rouge sang, se disait Daisy. En temps de paix, ils préféreraient sûrement le bon sens et la modération.

Un sondage effectué la veille de l'élection annonçait la victoire des travaillistes, mais personne n'y croyait. George Gallup, un Américain, avait fait un pronostic erroné lors de la dernière élection présidentielle. L'idée selon laquelle on pouvait prévoir le résultat en interrogeant un petit nombre d'électeurs paraissait un peu irréaliste. Le *News Chronicle*, qui avait publié le sondage, voyait les deux partis dans un mouchoir de poche.

Tous les autres journaux prédisaient la victoire des conservateurs.

Auparavant, jamais Daisy ne s'était intéressée aux rouages de la démocratie, mais son destin était dans la balance à présent, et elle regarda, fascinée, les bulletins de vote sortir des urnes. Elle observa les scrutateurs qui les triaient, les comptaient, les rassemblaient en petits paquets et les recomptaient. Le président du bureau de vote – appelé *returning officer*, comme s'il venait de revenir après une absence – était en réalité le secrétaire de mairie. Les observateurs de chacun des partis surveillaient la procédure de dépouillement afin de s'assurer qu'il n'y avait ni fraude ni erreur. Toutes ces opérations prenaient du temps, et l'attente était une torture pour Daisy.

À dix heures et demie, on annonça les premiers résultats d'un autre bureau. Harold Macmillan, un protégé de Churchill qui avait fait partie du cabinet du Premier ministre pendant la guerre, avait été battu par les travaillistes à Stickton-on-Tees. Un quart d'heure plus tard, on apprenait que Birmingham avait nettement basculé du côté du parti travailliste. Les radios n'étaient pas autorisées dans la salle de dépouillement, si bien que Daisy et Lloyd en étaient réduits à se fier aux rumeurs qui filtraient du dehors, et Daisy ne savait pas trop ce qu'elle devait croire.

Vers le milieu de la journée, le président du bureau de vote appela les candidats et leurs agents électoraux dans un coin de la salle pour leur communiquer les résultats avant de procéder à l'annonce publique. Daisy aurait bien voulu suivre Lloyd, mais on ne l'y autorisa pas.

L'homme leur parla à voix basse. En dehors de Lloyd et du député sortant, il y avait un conservateur et un communiste. Daisy tenta de déchiffrer leur expression, sans réussir à deviner qui avait gagné. Ils montèrent tous sur l'estrade, et le silence se fit dans la pièce. Daisy fut prise d'une vague nausée.

« Moi, Michael Charles Davis, agissant en tant que président du bureau de vote de la circonscription de Hoxton…

Daisy se leva avec les observateurs du parti travailliste, les yeux rivés sur Lloyd. Allait-elle le perdre ? À cette pensée, l'angoisse lui serra le cœur et lui coupa la respiration. Par deux fois, dans sa vie, elle avait choisi un homme, et ce choix s'était révélé désastreux. Charlie Farquharson était l'opposé de son père, gentil mais faible. Boy Fitzherbert, qui ressemblait beaucoup plus à son père, était un homme à poigne, mais égoïste. Enfin, elle avait trouvé Lloyd, qui était à la fois fort et gentil. Elle ne l'avait pas choisi pour son prestige social, ni pour ce qu'il pouvait lui apporter, mais tout simplement parce qu'il était d'une rare bonté. Il était doux, intelligent, solide, et il l'adorait. Elle avait mis longtemps à comprendre qu'il était exactement ce qu'elle cherchait. Ce qu'elle avait pu être bête !

Le président du bureau de vote annonça le nombre de suffrages exprimés pour chacun des candidats. Ils étaient classés par ordre alphabétique, et Williams était donc le dernier. Daisy était tellement tendue qu'elle n'arrivait pas à retenir les chiffres. « Reginald Sidney Blenkinsop, cinq mille quatre cent vingt-sept… »

Il en arriva enfin au résultat de Lloyd. Une ovation délirante s'éleva parmi les membres du parti travailliste qui l'entouraient. Daisy mit un moment à comprendre qu'il avait gagné. Puis elle vit l'expression solennelle de Lloyd se muer en un large sourire ; elle se mit alors à l'applaudir et à l'acclamer plus fort que tous les autres. Il avait été élu ! Elle ne serait pas obligée de le quitter ! C'était comme si on venait de lui rendre la vie.

« Je déclare donc Lloyd Williams officiellement élu député de Hoxton. »

Lloyd était membre du Parlement. Daisy le regarda fièrement s'avancer et prononcer son discours de remerciement. Elle constata que l'exercice comportait une part de rituel : il enchaîna les formules fastidieuses, remercia le président du bureau de vote et son équipe, puis les adversaires qui s'étaient livrés à une bataille loyale. Elle était impatiente de le serrer sur son cœur. Il finit par quelques phrases sur la tâche qui les attendait, reconstruire l'Angleterre dévastée par la guerre et créer une société plus juste, et quitta la scène sous un nouveau tonnerre d'applaudissements.

Redescendu de l'estrade, il s'avança droit vers Daisy, la prit dans ses bras et l'embrassa.

« Bravo, mon chéri », dit-elle, avant de prendre conscience qu'elle n'arrivait plus à parler.

Au bout d'un moment, ils sortirent et prirent l'autobus pour Transport House, le quartier général du parti travailliste. Ils y apprirent que les travaillistes avaient déjà remporté cent six sièges.

C'était un véritable raz-de-marée.

Les experts s'étaient bel et bien trompés, et les prévisions de tout un chacun étaient bouleversées. Quand l'ensemble des résultats fut connu, le parti travailliste disposait de trois cent quatre-vingt-treize sièges, les conservateurs de deux cent dix, les libéraux de douze et les communistes d'un seul – Stepney. Les travaillistes l'avaient emporté à une écrasante majorité.

À sept heures du soir, Winston Churchill, le grand chef de guerre de l'Angleterre, se rendit à Buckingham Palace donner sa démission de Premier ministre.

Daisy pensa à la façon dont Churchill s'était moqué d'Attlee : « Une voiture vide s'est approchée et Clem en est descendu. » L'homme qu'il considérait comme inexistant l'avait écrasé.

À sept heures et demie, Clement Attlee gagna le palais dans sa propre voiture conduite par sa femme, Violet, et le roi George VI le pria de devenir Premier ministre.

Dans la maison de Nutley Street, lorsqu'ils eurent tous écouté les nouvelles à la radio, Lloyd se tourna vers Daisy et lui dit : « Bon, eh bien, voilà. On peut se marier, maintenant ?

— Oui, répondit Daisy. Dès que tu voudras. »

6.

Volodia et Zoïa fêtèrent leur mariage dans l'une des plus petites salles de banquet du Kremlin.

La guerre avec l'Allemagne était terminée, mais l'Union soviétique étant encore ravagée, exsangue, l'heure ne se prêtait pas à une fête fastueuse. Zoïa avait une robe neuve, mais Volodia portait son uniforme. Cela dit, il y avait de la nourriture à profusion, et la vodka coulait à flots.

Le neveu et la nièce de Volodia étaient là, les jumeaux de sa sœur, Ania, et de son affreux mari, Ilia Dvorkine. Ils n'avaient pas encore six ans. Contredisant le comportement que l'on était en droit d'attendre d'un garçon et d'une fille, le petit Dimka aux cheveux noirs était tranquillement assis en train de lire un livre, tandis que l'espiègle Tania aux yeux bleus courait partout dans la pièce, heurtait les tables et asticotait les invités.

Zoïa était tellement séduisante en rose que Volodia aurait aimé l'entraîner sans attendre vers leur chambre à coucher. C'était hors de question, bien sûr. Le cercle d'amis de son père comprenait certains des généraux et des politiciens les plus en vue du pays, et ils étaient nombreux à être venus porter un toast au jeune et heureux couple. Grigori laissait entendre qu'un invité extrêmement prestigieux pourrait arriver plus tard. Volodia espérait que ce n'était pas Beria, le chef dépravé du NKVD.

Le bonheur de Volodia ne lui faisait pas entièrement oublier les horreurs qu'il avait vues, et les profondes réticences qu'avait fini par lui inspirer le communisme soviétique. L'indicible brutalité de la police secrète, les erreurs de Staline qui avaient coûté des millions de vies et la propagande qui avait encouragé l'armée Rouge à se comporter comme une meute de bêtes sauvages en Allemagne, tout cela l'avait conduit à douter des valeurs les plus fondamentales que son éducation lui avait inculquées. Il se demandait avec un certain malaise dans quel pays Dimka et Tania allaient grandir. Enfin, ce n'était pas le moment de penser à tout cela.

Les élites soviétiques étaient de bonne humeur. Elles avaient gagné la guerre et vaincu l'Allemagne. Quant à leur vieil ennemi, le Japon, les États-Unis étaient en train de l'écraser. Le code d'honneur insensé des dirigeants japonais leur rendait la capitulation extrêmement difficile, mais ce n'était plus qu'une question de temps. Chose tragique, pendant qu'ils se cramponnaient à leur orgueil, des soldats japonais et américains continueraient à mourir, des femmes et des enfants japonais continueraient à être chassés de leurs maisons par les bombardements, et tout cela pour quoi ? Le résultat final serait le même. Le plus triste était que les Américains paraissaient incapables d'accélérer les choses et d'empêcher toutes ces morts inutiles.

Le père de Volodia, saoul et ravi, fit un discours. « La Pologne est occupée par l'armée Rouge, déclara-t-il. Plus jamais ce pays ne servira de tremplin à l'invasion de la Russie par l'Allemagne. »

Tous ses vieux camarades l'acclamèrent en tapant sur les tables.

« En Europe occidentale, les partis communistes sont soutenus par les masses comme ils ne l'ont encore jamais été. À Paris, aux élections municipales de mars dernier, le parti communiste a remporté le plus fort pourcentage de voix. Je félicite nos camarades français. »

Ils l'acclamèrent de plus belle.

« Quand j'observe le monde d'aujourd'hui, je vois que la révolution russe, pour laquelle tant de braves ont combattu et ont donné leur vie… »

Il s'interrompit alors que des larmes d'ivresse lui montaient aux yeux. Un silence plana dans la salle. Il réussit à se dominer. « Je vois que jamais la révolution n'a été plus solide qu'aujourd'hui ! »

Ils levèrent leurs verres. « Révolution ! Révolution ! »

Ils les vidèrent cul sec.

Les portes s'ouvrirent à la volée, et le camarade Staline fit son entrée.

Tout le monde bondit sur ses pieds.

Il avait les cheveux gris et l'air fatigué. Il avait près de soixante-cinq ans, et avait été malade : selon certaines rumeurs, il avait fait plusieurs crises cardiaques, ou des attaques cérébrales sans gravité. Mais ce jour-là, il était d'humeur joviale. « Je suis venu embrasser la mariée ! » s'exclama-t-il.

Il s'approcha de Zoïa et posa les mains sur ses épaules. Elle mesurait dix bons centimètres de plus que lui, mais réussit à se baisser discrètement. Il l'embrassa sur les deux joues, laissant ses lèvres surmontées d'une moustache grise s'attarder sur sa peau juste le temps d'attiser légèrement la jalousie de Volodia, puis il recula et dit : « Alors, on me donne à boire ? »

Plusieurs personnes s'empressèrent de lui tendre un verre de vodka. Grigori insista pour lui céder sa place au milieu de la table d'honneur. Le brouhaha des conversations reprit, un peu atténué : les convives étaient transportés de joie par sa présence, mais devaient maintenant surveiller leurs paroles et leurs gestes. Cet homme pouvait faire tuer n'importe qui d'un claquement de doigts ; il l'avait souvent fait.

On apporta encore de la vodka, l'orchestre se mit à jouer des danses folkloriques russes, et peu à peu l'assistance se détendit. Volodia, Zoïa, Grigori et Katerina s'engagèrent dans un quadrille, une danse à quatre destinée à faire rire les spectateurs. Après cela, d'autres couples commencèrent à se trémousser, et les hommes entamèrent une barinia endiablée, une danse qui s'exécutait accroupi, en lançant les jambes devant soi. Beaucoup d'entre eux tombèrent. Comme tous les autres convives, Volodia surveillait toujours Staline du coin de l'œil : il semblait bien s'amuser et tapait avec son verre sur la table au rythme des balalaïkas.

Zoïa et Katerina dansaient une troïka avec le patron de Zoïa, Vassili, un physicien d'un certain âge qui travaillait sur le projet de bombe atomique, et Volodia les contemplait lorsque soudain, l'atmosphère changea.

Un aide de camp en civil entra, fit précipitamment le tour de la pièce et se dirigea vers Staline. Sans cérémonie, il se pencha sur l'épaule du dirigeant et lui parla à l'oreille, d'un ton pressant.

Staline sembla d'abord intrigué. Il posa sèchement une question, puis une autre et son visage s'altéra. Il pâlit et donna l'impression de regarder les danseurs sans les voir.

Volodia chuchota : « Bon sang, qu'est-ce qui se passe ? »

Les danseurs n'avaient encore rien remarqué, mais les convives assis à la table d'honneur avaient l'air épouvanté.

Staline se leva. Ceux qui l'entouraient l'imitèrent par déférence. Volodia remarqua que son père dansait toujours. Des gens s'étaient fait fusiller pour moins que ça.

Mais Staline n'avait plus d'yeux pour les invités de la noce. Flanqué de son aide de camp, il quitta la table et se dirigea vers la porte en traversant la piste de danse. Apeurés, les fêtards s'écartèrent précipitamment de son chemin. Un couple tomba. Staline ne parut pas s'en apercevoir. L'orchestre s'interrompit. Sans un mot, sans un regard, Staline quitta la pièce.

Certains des généraux lui emboîtèrent le pas, visiblement inquiets.

Un second aide de camp apparut, suivi de deux autres. Chacun chercha son supérieur pour lui glisser quelques mots. Un jeune homme en veste de tweed s'approcha de Vassili. Zoïa, qui semblait le connaître, l'écouta attentivement. Elle se figea, bouleversée.

Vassili et son aide de camp quittèrent la pièce. Volodia s'approcha de Zoïa et lui demanda : « Bon sang, que se passe-t-il ? »

Elle répondit d'une voix tremblante : « Les Américains ont largué une bombe atomique sur le Japon. » Son beau visage pâle semblait avoir encore blêmi. « Le gouvernement japonais n'a pas tout de suite compris. Ils ont mis des heures à réaliser de quoi il s'agissait.

— On en est sûr ?

— Mille trois cents hectares de bâtiments ont été rasés. On estime que soixante-quinze mille personnes ont été tuées instantanément.

— Combien de bombes ?

— Une.

— Une seule bombe ?

— Oui.

— Nom d'un chien ! Pas étonnant que Staline ait pâli. »

Ils demeurèrent un instant silencieux. La nouvelle se répandait dans la salle comme une traînée de poudre. Certains restèrent assis, pétrifiés. D'autres se levèrent et sortirent, pressés de retrouver leurs bureaux, leurs téléphones, leurs collaborateurs.

« Ça change tout, dit Volodia.

— Y compris nos projets de voyage de noces, soupira Zoïa. Je peux dire adieu à mon congé.

— Nous qui pensions être en sécurité en Union soviétique…

— Ton père vient de dire, dans son discours, que la révolution n'avait jamais été aussi solide.

— Rien ne l'est plus aujourd'hui.

— Non. Rien ne le sera plus tant que nous n'aurons pas notre propre bombe. »

7.

Jacky Jakes et Georgy étaient à Buffalo. C'était leur premier séjour chez Marga. Greg et Lev étaient là, eux aussi, et le jour de la victoire sur le Japon – le mercredi 15 août –, ils se rendirent tous au Humboldt Park. Les allées regorgeaient de couples radieux, et des centaines d'enfants pataugeaient dans le lac.

Greg était heureux et fier. La bombe avait explosé. Les deux engins largués sur Hiroshima et Nagasaki avaient provoqué des dégâts épouvantables, mais avaient hâté la fin de la guerre, sauvant ainsi des milliers de vies américaines. Greg avait joué un rôle dans ce triomphe. Grâce à tout ce qu'ils avaient fait, Georgy grandirait dans un monde libre.

« Il a neuf ans », remarqua Greg en se tournant vers Jacky.

Ils étaient assis sur un banc et bavardaient, pendant que Lev et Marga avaient emmené Georgy acheter une glace.

« Je n'arrive pas à le croire !

— Je me demande ce qu'il fera plus tard.

— En tout cas, il ne sera sûrement pas acteur ou trompettiste, répondit Jacky avec véhémence. Il est incroyablement intelligent.

— Tu voudrais qu'il soit professeur d'université, comme ton père ?

— Oui.

— Dans ce cas… » C'était là que Greg voulait en venir, et il craignait un peu la réaction de Jacky. « … Il faut qu'il aille dans une bonne école.

— À quoi penses-tu ?

— Que dirais-tu d'une pension ? Celle où je suis allé, par exemple.

— Il serait le seul élève noir.

— Pas forcément. Quand j'y étais, nous avions un garçon de couleur, un Indien de Delhi qui s'appelait Kamal.

— Un seul.

— Oui.

— Les autres élèves ne l'embêtaient pas ?

— Si. On l'appelait Chamelle. Mais les garçons ont fini par s'habituer à lui, et il s'est fait des amis.

— Qu'est-ce qu'il est devenu ? Tu le sais ?

— Il est pharmacien. J'ai entendu dire qu'il était déjà propriétaire de deux drugstores à New York. »

Jacky hocha la tête. Greg voyait qu'elle n'était pas hostile à son projet. Elle était issue d'une famille cultivée. Bien qu'elle se soit rebellée et ait abandonné ses études, elle croyait à la valeur de l'éducation. « Et les frais de scolarité ?

— Je pourrais demander à mon père.

— Tu crois qu'il accepterait de payer ?

— Regarde-les. »

Greg tendit le doigt vers le chemin. Lev, Marga et Georgy revenaient de la voiture du marchand de glaces. Lev et Georgy marchaient main dans la main, dégustant leurs cornets.

« Mon père conservateur, tenant un enfant de couleur par la main dans un parc public. Fais-moi confiance, il paiera les frais de scolarité.

— Georgy n'est vraiment à sa place nulle part, observa Jacky, l'air troublé. C'est un petit garçon noir qui a un papa blanc.

— En effet.

— Dans l'immeuble de ta mère, les gens me prennent pour la bonne – tu le savais ?

— Oui.

— Je me suis bien gardée de les détromper. S'ils apprennent qu'il y a des Noirs dans l'immeuble, pas comme domestiques mais comme invités, ça pourrait faire des histoires. »

Greg soupira.

« Ça me navre, mais tu as raison.

— La vie va être rude pour Georgy.

— Je sais, acquiesça Greg. Mais il nous a, nous.

— Ouais, répondit-elle avec un de ses rares et précieux sourires. Et ça, c'est quelque chose. »

TROISIÈME PARTIE

La paix froide

XXI
1945 (III)

1.

Après le mariage, Volodia et Zoïa s'installèrent dans un appartement à eux. Rares étaient les jeunes couples russes à avoir cette chance. Pendant quatre ans, toute la puissance industrielle de l'Union soviétique avait été mobilisée dans l'armement. On avait construit peu d'immeubles d'habitation, et beaucoup avaient été détruits. Mais Volodia était commandant dans les services de renseignement de l'armée Rouge, il était fils de général, et savait faire jouer ses relations.

L'espace était néanmoins exigu : un salon avec une table de salle à manger, une chambre si petite que le lit l'occupait presque entièrement, une cuisine dans laquelle on tenait tout juste à deux, un minuscule cabinet de toilette avec un lavabo et une douche, et une entrée étroite avec une penderie. Lorsque la radio était allumée dans le salon, on l'entendait dans tout l'appartement.

Ils en firent vite leur nid. Zoïa acheta un dessus-de-lit jaune vif. La mère de Volodia leur offrit un service qu'elle avait acheté en 1940, en prévision du mariage de son fils, et mis de côté pendant toute la durée de la guerre. Volodia accrocha un tableau au mur, une photo de fin d'année de sa promotion à l'Académie militaire, section Renseignement.

Ils faisaient plus souvent l'amour, maintenant. Le fait d'être seuls était plus appréciable que Volodia ne l'aurait cru. Il ne s'était jamais senti particulièrement gêné quand il couchait avec Zoïa chez ses parents, ou dans l'appartement qu'elle habitait en colocation, mais il se rendait compte à présent que ce n'était pas sans importance. Il fallait parler à voix basse, éviter de faire grincer le lit, et il y avait toujours un risque, faible mais réel, que quelqu'un entre dans la chambre. Chez les autres, on n'avait jamais vraiment d'intimité.

Ils se réveillaient souvent tôt, faisaient l'amour, puis restaient allongés à s'embrasser et à bavarder pendant une heure avant de s'habiller et d'aller travailler. Par un matin comme ceux-là, allongé la tête sur les cuisses de Zoïa, respirant encore l'odeur de leurs étreintes, Volodia proposa : « Tu veux du thé ?

— Oui, je veux bien s'il te plaît. »

Elle s'étira voluptueusement, adossée aux oreillers.

Volodia enfila une robe de chambre et traversa l'étroit couloir menant à la petite cuisine, où il alluma le gaz sous le samovar. Il fut agacé de voir les casseroles et les assiettes du dîner de la veille encore entassées dans l'évier.

« Zoïa ! cria-t-il. C'est la pagaille à la cuisine ! »

L'appartement était assez petit pour qu'elle l'entende sans difficulté. « Je sais », répondit-elle.

Il retourna dans la chambre. « Pourquoi n'as-tu pas fait la vaisselle, hier soir ?

— Et toi ? Pourquoi ne l'as-tu pas faite ? »

Cela ne lui était même pas venu à l'esprit. Mais il répondit : « J'avais un rapport à rédiger.

— Et moi, j'étais fatiguée. »

L'idée qu'il puisse être tenu pour responsable en quelque façon que ce soit l'irrita.

« Je déteste trouver de la vaisselle sale quand j'entre dans la cuisine.

— Moi aussi. »

Pourquoi se montrait-elle aussi bornée ? « Dans ce cas, tu n'as qu'à la laver !

— Faisons-la ensemble, tout de suite. » Elle sauta du lit, passa devant lui avec un sourire aguichant et gagna la cuisine.

Volodia la suivit.

« Tu laves, j'essuie », proposa-t-elle.

Elle prit un torchon propre dans un tiroir.

Elle était encore nue et il ne put s'empêcher de sourire. Elle était svelte et longiligne, et avait la peau très blanche. Elle avait des petits seins plats aux mamelons pointus, et des poils pubiens fins et blonds. L'une des joies du mariage, pour lui, était cette habitude qu'elle avait de se promener toute nue dans l'appartement. Il pouvait admirer son corps autant qu'il le voulait. Elle paraissait aimer ça. Lorsqu'elle surprenait son regard, elle n'était pas gênée, elle se contentait de sourire.

Il retroussa les manches de son peignoir et commença à laver les assiettes qu'il passait à Zoïa pour qu'elle les essuie. Faire la vaisselle n'était pas une activité très virile – Volodia n'avait jamais vu son père la faire –, mais Zoïa semblait penser qu'ils devaient partager ce genre de tâches. C'était une idée excentrique. Zoïa avait-elle une vision exagérée de l'égalité dans le mariage ? Ou se laissait-il émasculer ?

Il crut entendre du bruit au-dehors ; il jeta un coup d'œil dans le couloir : la porte de l'appartement n'était qu'à trois ou quatre pas de l'évier de la cuisine. Soudain, on enfonça le battant.

Zoïa poussa un cri.

Volodia attrapa le couteau à découper qu'il venait de laver. Passant devant Zoïa, il se campa sur le seuil de la cuisine. Un policier en uniforme brandissant une masse se trouvait juste devant la porte fracturée.

Volodia bouillonnait de peur et de colère. « Et merde ! Qu'est-ce que ça veut dire ? » s'écria-t-il.

Le policier s'écarta, et un petit homme mince au visage sournois entra dans l'appartement. C'était le beau-frère de Volodia, Ilia Dvorkine, un agent de la police secrète. Il portait des gants en cuir.

« Ilia ! s'exclama Volodia. Espèce de sale fouine !

— Un peu de respect », répondit Ilia.

Volodia était à la fois déconcerté et furieux. La police secrète n'était pas censée arrêter les agents du Renseignement de l'armée Rouge. Autrement, cela aurait tourné à la guerre des gangs. « Bon sang ! Pourquoi as-tu enfoncé la porte ? Je t'aurais ouvert ! »

Deux autres agents s'engagèrent dans le couloir et prirent position derrière Ilia. Ils portaient leurs manteaux de cuir caractéristiques, malgré la douceur de cette fin d'été.

Ilia dit d'une voix tremblante : « Repose ce couteau, Volodia.

— Il n'y a pas de quoi avoir peur, répondit Volodia. J'étais en train de le laver, c'est tout. »

Il tendit le couteau à Zoïa, debout derrière lui. « Je t'en prie, passons au salon. On pourra discuter pendant que Zoïa s'habille.

— Parce que tu crois que c'est une visite de courtoisie ? rétorqua Ilia, indigné.

— Quel que soit le motif de ta visite, tu préféreras sûrement t'éviter la gêne de voir ma femme dans cette tenue.

— Je suis ici en mission officielle !

— Dans ce cas, pourquoi envoient-ils mon beau-frère ? »

Ilia baissa la voix. « Tu ne comprends pas que ç'aurait été bien pire pour toi avec un autre ? »

Voilà qui semblait présager de gros ennuis. Volodia s'efforça de conserver une attitude bravache.

« Que voulez-vous au juste, cette bande de connards et toi ?

— Le camarade Beria a repris la direction du programme de physique nucléaire. »

Volodia le savait. Staline avait mis en place un nouveau comité pour diriger les travaux et en avait confié la présidence à Beria. Celui-ci n'avait aucune notion de physique et n'était absolument pas qualifié pour piloter un projet de recherche scientifique. Mais Staline lui faisait confiance. C'était le problème récurrent du gouvernement soviétique : des gens loyaux mais incompétents étaient promus à des fonctions qu'ils étaient incapables d'assumer.

« Et le camarade Beria a besoin que ma femme rejoigne son laboratoire pour mettre la bombe au point, poursuivit Volodia. Tu es venu la chercher pour la conduire au travail en voiture ?

— Les Américains ont fabriqué leur bombe atomique avant les Soviétiques.

— En effet. Auraient-ils accordé aux recherches en physique nucléaire une plus grande priorité que nous ?

— La science capitaliste ne peut pas être supérieure à la science communiste !

— Cela va de soi. »

Volodia était perplexe. Où voulait-il en venir ? « Alors, qu'est-ce que tu en conclus ?

— Il y a forcément eu sabotage. »

C'était exactement le genre de fantasme grotesque que la police secrète était capable d'imaginer.

« Quel genre de sabotage ?

— Certains chercheurs ont délibérément retardé la mise au point de la bombe soviétique. »

Volodia commençait à comprendre… et à avoir peur. Il n'en continua pas moins à répondre sur un ton belliqueux ; trahir sa faiblesse devant ces gens-là était toujours une erreur.

« Bon sang, pourquoi feraient-ils une chose pareille ?

— Parce que ce sont des traîtres – et ta femme est du nombre !

— J'espère que tu n'es pas sérieux, espèce de petit merdeux…

— Je suis venu l'arrêter.

— Comment ? fit Volodia, estomaqué. Mais c'est de la folie !

— Telle n'est pas l'opinion de mon organisation.

— Vous n'avez aucune preuve.

— Si tu veux des preuves, va à Hiroshima ! »

Zoïa prit la parole pour la première fois depuis le cri de surprise qui lui avait échappé.

« Il va bien falloir que je les suive, Volodia. Il vaudrait mieux que tu ne te fasses pas arrêter, toi aussi. »

Volodia pointa le doigt sur Ilia.

« Tu n'as pas idée de la merde dans laquelle tu es en train de te fourrer.

— Je ne fais qu'exécuter les ordres.

— Pousse-toi. Il faut que ma femme aille s'habiller.

— Pas le temps, objecta Ilia. Qu'elle vienne comme elle est.

— Ne sois pas ridicule ! »

Ilia releva le menton. « Une citoyenne soviétique respectable ne se promènerait pas chez elle dans le plus simple appareil. »

Volodia eut une pensée pour sa sœur et se demanda fugitivement comment elle supportait d'être mariée à cette vermine.

« Parce que la police secrète désapprouve moralement la nudité ?

— C'est la preuve de sa dépravation. Nous l'emmènerons dans la tenue où elle se trouve.

— Sûrement pas, bordel !

— Écarte-toi.

— Écarte-toi toi-même. Elle va s'habiller. »

Volodia s'avança dans le couloir et se dressa devant les trois agents, bras écartés, pour laisser Zoïa passer derrière lui.

Comme elle s'avançait, Ilia se précipita derrière Volodia et attrapa la jeune femme par le bras.

Le poing de Volodia partit tout seul. Un coup, un autre. Ilia poussa un cri et recula en titubant. Les deux hommes en manteau de cuir se précipitèrent. Volodia essaya d'en frapper un, malheureusement l'homme esquiva. Puis ils empoignèrent Volodia par les bras. Il se débattit, mais ils étaient costauds, et la manœuvre leur était visiblement familière. Ils le plaquèrent contre le mur.

Pendant qu'ils le maintenaient, Ilia lui donna un, puis deux coups de poing en plein visage avec ses poings gantés de cuir, puis trois, puis quatre, après quoi il lui martela l'estomac, encore et encore, jusqu'à ce que Volodia se mette à cracher du sang. Zoïa essaya de s'interposer, mais Ilia la repoussa brutalement, et elle tomba à la renverse en hurlant.

Les pans du peignoir de Volodia s'écartèrent. Ilia lui donna des coups de pied dans les parties, et dans les genoux. Volodia s'affaissa sur lui-même, incapable de tenir debout, mais les deux hommes en manteau de cuir le hissèrent sur ses pieds et Ilia lui asséna encore plusieurs coups de poing.

Ilia se détourna enfin en se frottant les jointures. Ses deux acolytes lâchèrent Volodia qui s'écroula. Il arrivait à peine à respirer et était incapable de bouger, mais il était conscient. Du coin de l'œil, il vit les deux gros bras empoigner Zoïa et la faire sortir, toujours nue, de l'appartement. Ilia leur emboîta le pas.

Plusieurs minutes s'écoulèrent. La souffrance aiguë se changea en une douleur sourde, profonde, et Volodia retrouva une respiration normale.

Il récupéra peu à peu l'usage de ses membres et réussit à se mettre debout. Il parvint à atteindre le téléphone, et composa le numéro de son père, en espérant qu'il n'était pas encore parti au travail. Il fut soulagé d'entendre sa voix.

« Ils ont arrêté Zoïa, annonça-t-il.

— Les ordures ! s'exclama Grigori. Qui ça ?

— Ilia.

— Quoi ?

— Appelle tous les gens que tu peux, supplia Volodia. Essaie de savoir ce qui se passe. Il faut que je me lave, il y a du sang partout.

— Du sang ? Comment ça du sang ? »

Volodia raccrocha. Il n'avait que quelques pas à faire pour rejoindre la salle de bains. Il laissa tomber son peignoir ensanglanté et entra dans la douche. L'eau chaude soulagea son corps endolori. Ilia était vicieux, mais pas très costaud ; il ne lui avait rien cassé.

Refermant le robinet, il se regarda dans le miroir de la salle de bains. Il avait le visage couvert de plaies et d'ecchymoses.

Il ne prit pas la peine de se sécher. Il enfila, au prix d'efforts considérables, son uniforme de l'armée Rouge. Ce symbole d'autorité pourrait lui être utile.

Son père arriva alors qu'il essayait de nouer les lacets de ses chaussures.

« Qu'est-ce que c'est que ce foutoir ? » rugit Grigori.

Volodia répondit : « Ils cherchaient la bagarre, et j'ai été assez idiot pour leur donner satisfaction. »

Son père ne lui témoigna d'abord aucune compassion. « Je t'aurais cru plus malin.

— Elle n'était pas encore habillée et ils ont absolument tenu à l'emmener comme ça.

— Les fumiers !

— Tu as trouvé quelque chose ?

— Pas encore. J'ai passé quelques coups de fil. Personne ne sait rien. »

Grigori avait l'air préoccupé.

« Soit quelqu'un a fait une vraie connerie… Soit, pour une raison ou une autre, ils sont drôlement sûrs d'eux.

— Conduis-moi à mon bureau. Lemitov va être fou de rage. Il ne les laissera pas s'en tirer comme ça. S'ils se permettent de me faire ça à moi, aucun membre des services de renseignement de l'armée Rouge n'est à l'abri. »

La voiture avec chauffeur de Grigori attendait devant l'immeuble. Ils se rendirent au terrain d'aviation de Khodynka. Grigori resta dans la voiture pendant que Volodia entrait en boitant dans les bureaux des services de renseignement de l'armée

Rouge. Il se dirigea droit vers le bureau de son supérieur, le colonel Lemitov.

Il frappa à la porte, entra et dit :

« Ces salauds de la police secrète ont arrêté ma femme.

— Je sais, répondit Lemitov.

— Vous êtes au courant ?

— J'ai donné mon accord. »

Volodia en resta bouche bée.

« Bordel de merde…

— Assieds-toi.

— Mais que se passe-t-il ?

— Assieds-toi et boucle-la. Je vais t'expliquer. »

Volodia se laissa tomber dans un fauteuil.

« Il faut qu'on fabrique une bombe atomique, et vite, reprit Lemitov. Pour le moment, Staline tient la dragée haute aux Américains parce qu'on est à peu près sûrs qu'ils n'ont pas un arsenal nucléaire suffisant pour nous rayer de la surface de la planète. Mais ils sont en train de constituer des stocks, et ils vont bien finir par s'en servir – à moins que nous n'ayons les moyens de riposter. »

Ça n'avait aucun sens. « Parce que vous croyez que ma femme pourra concevoir une bombe pendant que la police secrète la tabasse. C'est complètement crétin !

— Tais-toi, bon sang ! Notre problème, c'est qu'il y a plusieurs conceptions possibles. Les Américains ont mis cinq ans à comprendre laquelle était la bonne. Or le temps presse. Nous devons leur voler leurs résultats.

— Mais nous aurons encore besoin de chercheurs russes pour reproduire le modèle – et pour ça, il faut qu'ils soient dans leurs laboratoires, et pas au fond des sous-sols de la Loubianka.

— Tu connais un certain Wilhelm Frunze ?

— Oui, j'étais en classe avec lui. À Berlin. À l'Académie de garçons.

— Il nous a communiqué de précieuses informations sur les recherches nucléaires britanniques. Ensuite, il est parti pour les États-Unis, où il a travaillé sur le projet de bombe atomique. L'équipe du NKVD en poste à Washington l'a approché. Ils ont été tellement maladroits qu'ils l'ont effrayé et il a rompu tout contact. Il faut qu'on le récupère.

— Et qu'est-ce que ça a à voir avec moi ?

— Il te fait confiance.

— Ça, c'est vous qui le dites. Il y a douze ans que je ne l'ai pas vu.

— Il faut que tu ailles en Amérique lui parler.

— Mais pourquoi avez-vous arrêté Zoïa ?

— Pour être sûrs que tu reviendras. »

2.

Volodia se dit que c'était dans ses cordes. À Berlin, avant la guerre, il avait déjoué la filature d'agents de la Gestapo, il avait rencontré des espions potentiels, les avait recrutés et en avait fait des sources de renseignement fiables. Ce n'était jamais simple – le moment le plus difficile était celui où il fallait convaincre quelqu'un de trahir – mais il était expert en la matière.

Cette fois, tout de même, c'était l'Amérique.

Les pays de l'Ouest où il s'était rendu – l'Allemagne et l'Espagne dans les années 1930 et 1940 – n'avaient rien à voir avec ça.

Il était dépassé. Toute sa vie, on lui avait raconté que les films hollywoodiens donnaient une impression exagérée de prospérité et qu'en réalité, la plupart des Américains vivaient dans la misère. Or depuis que Volodia avait mis les pieds aux États-Unis, il avait pu se convaincre que les films n'exagéraient pas, au contraire. Et les pauvres ne couraient pas les rues.

New York grouillait de voitures, dont les conducteurs n'avaient pas franchement l'allure de hauts fonctionnaires : des jeunes, des hommes en tenue de travail, même des femmes qui faisaient leurs courses. Et tout le monde était tellement bien habillé ! Tous les hommes paraissaient porter leur plus beau costume. Les femmes avaient les jambes gainées de bas d'une finesse arachnéenne. Et apparemment, toutes les chaussures étaient neuves.

Il devait faire un effort pour se rappeler les mauvais côtés de l'Amérique. Il y avait forcément de la pauvreté quelque part. Les Noirs étaient persécutés, et dans le Sud, ils n'avaient pas le droit de vote. Le taux de criminalité était considérable –

les Américains disaient eux-mêmes qu'elle était endémique –, et pourtant étrangement, Volodia n'en voyait aucune trace et se sentait tout à fait en sécurité quand il marchait dans les rues.

Il passa quelques jours à explorer New York. Il s'efforça de perfectionner son anglais, qui n'était pas fameux, mais ça n'avait pas d'importance : la ville était pleine de gens qui parlaient un anglais approximatif, avec un accent épouvantable. Il apprit à reconnaître les visages de certains des agents du FBI affectés à sa filature, et identifia plusieurs endroits commodes pour les semer.

Par un matin ensoleillé, il quitta le consulat soviétique de New York, tête nue, vêtu d'un pantalon gris et d'une chemise bleue, comme s'il allait faire quelques courses. Un jeune homme en costume noir et cravate lui emboîta le pas.

Il alla acheter des sous-vêtements et une chemise avec un petit motif à carreaux marron dans le grand magasin Saks de la Cinquième Avenue. Celui qui le filait devait penser qu'il se contentait de faire des achats.

Le chef du NKVD au consulat l'avait prévenu qu'une équipe soviétique le suivrait à chaque instant de son séjour en Amérique pour s'assurer de sa bonne conduite. Volodia avait eu du mal à contenir sa colère contre l'organisation qui avait emprisonné Zoïa, et il avait dû résister à la tentation de sauter à la gorge du type et de l'étrangler. Il avait tout de même réussi à garder son calme et avait répondu d'un ton sarcastique que, pour remplir sa mission, il devrait échapper à la surveillance du FBI, et qu'il risquait ainsi de déjouer, par inadvertance, la filature du NKVD ; mais il leur souhaitait bonne chance. La plupart du temps, il s'en débarrassait en cinq minutes.

Le jeune homme qui l'avait pris en filature était certainement un agent du FBI, ce que confirmait son costume d'un classicisme irréprochable.

Portant ses emplettes dans un sachet en papier, Volodia quitta le magasin par une porte latérale et héla un taxi. Il laissa le gars du FBI en plan au bord du trottoir, en train d'agiter le bras. Après avoir fait tourner le taxi à deux coins de rue successifs, Volodia jeta un billet au chauffeur et descendit de voiture. Il se précipita dans une bouche de métro, ressortit par une autre issue et attendit cinq minutes dans l'embrasure de la porte d'un immeuble de bureaux.

Le jeune homme en costume sombre avait disparu.

Volodia gagna Penn Station à pied.

Là, il vérifia deux fois qu'il n'était pas suivi, et prit un billet. Avec pour tout bagage son sachet en papier, il monta dans un train.

Le trajet jusqu'à Albuquerque dura trois jours.

Le train traversait à toute allure des kilomètres et des kilomètres : fermes opulentes, usines puissantes vomissant des panaches de fumée et grandes villes aux gratte-ciel insolemment dressés vers les nuages. L'Union soviétique était plus vaste, mais l'Ukraine mise à part, elle était surtout constituée de forêts de pins et de steppes glacées. Il n'aurait jamais imaginé une richesse d'une telle ampleur.

Et la richesse n'était pas tout. Il y avait plusieurs jours qu'une arrière-pensée le titillait, une singularité propre à la vie en Amérique. Il finit par comprendre ce que c'était : personne ne lui demandait ses papiers. Depuis qu'il avait franchi la barrière des services d'immigration à New York, il n'avait plus jamais montré son passeport. Apparemment, dans ce pays, n'importe qui pouvait entrer dans une gare de chemin de fer ou d'autobus et prendre un billet pour n'importe quelle destination sans avoir besoin d'obtenir une autorisation ou d'expliquer le but de son voyage à un fonctionnaire. Cela lui donnait une impression de liberté dangereusement exaltante. Il pouvait aller où il voulait !

L'opulence de l'Amérique lui fit aussi prendre une conscience plus aiguë du danger auquel son pays était exposé. Les Allemands avaient bien failli détruire l'Union soviétique alors qu'elle était trois fois plus peuplée et dix fois plus riche que l'Allemagne. La pensée que la Russie pourrait devenir une nation subalterne, soumise par la peur de la bombe atomique, tempéra ses doutes à propos du communisme, malgré ce que le NKVD leur avait fait, à sa femme et à lui. S'il avait des enfants, il ne voulait pas qu'ils grandissent dans un monde dominé par une Amérique tyrannique.

Il traversa Pittsburgh et Chicago sans attirer l'attention de qui que ce soit. Il portait des vêtements américains, et on ne remarquait pas son accent pour la simple raison qu'il ne parlait à personne. Il achetait des sandwichs et du café en les désignant du doigt et en tendant l'argent. Il feuilletait les revues et les journaux que les autres voyageurs abandonnaient derrière eux, et essayait de comprendre la signification des gros titres en regardant les photos.

La dernière partie du trajet lui fit traverser un paysage désertique d'une beauté désolée. Le soleil couchant maculait de rouge les pics enneigés dressés dans le lointain, ce qui expliquait probablement leur nom : les montagnes du Sang du Christ.

Il s'enferma dans les toilettes pour changer de sous-vêtements et enfiler la chemise neuve achetée chez Saks.

Il se doutait que le FBI ou la sécurité militaire surveillerait la gare d'Albuquerque et repéra effectivement un jeune homme dont la veste à carreaux – trop chaude pour le climat du Nouveau-Mexique en septembre – ne réussissait pas tout à fait à dissimuler la bosse que faisait son pistolet dans son holster. Mais l'agent s'intéressait probablement aux passagers venus de loin, de New York ou de Washington. Volodia, qui n'avait ni chapeau, ni veston, ni bagages, avait l'air d'un gars du coin revenant d'un court trajet. Il se rendit sans qu'on le suive jusqu'à la gare routière et monta à bord d'un bus Greyhound pour Santa Fe.

Il arriva à destination en fin d'après-midi. Il remarqua deux hommes du FBI à la gare autoroutière de Santa Fe qui le détaillèrent. Mais ils ne pouvaient pas suivre tous ceux qui descendaient de l'autobus, et encore une fois, il se fondait si bien dans la foule qu'ils le laissèrent partir sans l'inquiéter.

Feignant dans la mesure du possible de savoir où il allait, il se promena dans les rues. Les maisons à toit plat, de style pueblo, et les églises trapues qui rôtissaient au soleil lui rappelaient l'Espagne. Les devantures des boutiques formaient des avancées sur les trottoirs, créant des arcades agréablement ombragées.

Il évita le grand hôtel La Fonda, sur la place centrale, à côté de la cathédrale, et prit une chambre au St Francis. Il paya en espèces et s'inscrivit sous le nom de Robert Pender, un nom qui pouvait aussi bien être américain que de plusieurs nationalités d'Europe. « Ma valise arrive demain, expliqua-t-il à la jolie réceptionniste. Si je suis sorti quand on la livrera, pourriez-vous la faire monter dans ma chambre, s'il vous plaît ?

— Mais bien sûr, sans problème, répondit-elle.

— Merci », dit-il. Et il ajouta une phrase qu'il avait entendue à plusieurs reprises dans le train : « C'est très aimable à vous.

— Si je ne suis pas là, quelqu'un d'autre s'occupera de votre bagage, pourvu qu'il soit étiqueté.

— Il l'est. »

Il n'avait pas de bagage, mais elle ne le saurait jamais.

Elle regarda ce qu'il avait inscrit dans le registre.

« Alors comme ça, monsieur Pender, vous venez de New York ? »

Il y avait une nuance de scepticisme dans sa voix, sans doute parce qu'il ne s'exprimait pas comme un New-Yorkais.

« Je suis d'origine suisse, expliqua-t-il, choisissant un pays neutre.

— Ah, voilà qui explique votre accent. C'est la première fois que je rencontre quelqu'un qui vient de Suisse. C'est comment là-bas ? »

Volodia n'avait jamais mis les pieds en Suisse, mais il avait vu des photos.

« Il y a beaucoup de neige.

— Eh bien, j'espère que vous apprécierez le climat du Nouveau-Mexique !

— Sûrement. »

Il ressortit cinq minutes plus tard.

Certains chercheurs vivaient au laboratoire de Los Alamos, lui avaient appris ses collègues du consulat soviétique, mais ce n'était guère qu'un amas de baraquements qui ne disposaient pas de tout le confort moderne. Aussi ceux qui en avaient les moyens préféraient-ils louer des maisons ou des appartements dans le coin. Will Frunze pouvait se le permettre : il avait épousé une artiste à succès, auteur d'une bande dessinée publiée dans tous les journaux sous le titre de *Slack Alice*. Sa femme – Alice, de son prénom –, pouvait travailler n'importe où ; ils habitaient donc dans le centre-ville historique.

C'était le bureau new-yorkais du NKVD qui lui avait fourni cette information. Ils avaient fait des recherches poussées sur Frunze, et Volodia disposait de son adresse, de son numéro de téléphone et de la description de sa voiture, une Plymouth décapotable d'avant-guerre aux pneus à flancs blancs.

Une galerie d'art occupait le rez-de-chaussée de l'immeuble des Frunze. L'appartement, à l'étage, avait une grande baie vitrée orientée au nord, qui devait faire un excellent atelier d'artiste. Une Plymouth décapotable était garée devant.

Volodia préféra ne pas entrer : l'endroit était peut-être truffé de micros.

Les Frunze étaient un couple aisé, sans enfants, et il se dit qu'un vendredi soir, il était peu probable qu'ils restent chez eux à écouter la radio. Il décida d'attendre à proximité pour voir s'ils sortaient.

Il passa un moment dans la galerie, à regarder les tableaux à vendre ; il aimait la peinture lumineuse, vivante, et n'était pas très séduit par ces croûtes bâclées. Il trouva un café un peu plus loin et s'installa devant la fenêtre, à un endroit d'où il pouvait voir la porte des Frunze. Il partit au bout d'une heure, acheta un journal et se planta à un arrêt de bus en faisant semblant de lire.

Cette longue attente lui permit de constater que personne ne surveillait l'appartement des Frunze. Ce qui voulait dire que le FBI et les services de sécurité de l'armée ne considéraient pas Frunze comme un sujet hautement suspect. Il était étranger, comme beaucoup d'autres scientifiques, et l'on n'avait probablement rien d'autre à retenir contre lui.

C'était un quartier commercial du centre-ville, pas un quartier résidentiel, et il y avait beaucoup de monde dans les rues. Au bout de quelques heures, Volodia commença tout de même à craindre de se faire remarquer s'il continuait à traîner dans les parages.

C'est alors que les Frunze sortirent.

Frunze s'était empâté depuis la dernière fois qu'il l'avait vu, douze ans plus tôt – les restrictions alimentaires n'existaient pas en Amérique. Ses tempes commençaient à se dégarnir, alors qu'il n'avait que trente ans, et il avait toujours le même air compassé. Il portait une chemise de sport et un pantalon kaki, une tenue courante aux États-Unis.

Sa femme était habillée de façon moins conventionnelle. Ses cheveux blonds étaient retenus par des épingles sous un béret et elle portait une robe de coton informe d'un brun incertain, mais elle avait une collection de bracelets aux deux poignets, et d'innombrables bagues. C'était le genre de tenue que portaient les artistes en Allemagne, avant Hitler, se rappela Volodia.

Les Frunze s'engagèrent dans la rue, et Volodia les suivit.

Il se demandait quelles pouvaient être les idées politiques de la femme, et si sa présence influerait sur la conversation délicate qu'il était sur le point d'engager. À l'époque où il vivait en Allemagne, Frunze était un social-démocrate pur et dur, et il était peu probable qu'il ait épousé une conservatrice, mais cette spéculation ne reposait que sur son allure générale. D'un autre côté, elle ignorait probablement qu'il avait transmis des secrets aux Russes, quand il était à Londres. Elle représentait donc une inconnue.

Il aurait préféré traiter avec Frunze en tête à tête, et envisagea de les laisser là et de retenter sa chance le lendemain. Mais la réceptionniste de l'hôtel avait remarqué son accent étranger, et il n'était pas impossible que le FBI lui mette un homme aux trousses dès le jour suivant. Il devrait pouvoir s'en débarrasser, mais ce serait moins facile dans cette petite ville qu'à New York ou Berlin. Le lendemain étant en outre un samedi, le couple passerait sans doute la journée ensemble. Combien de temps Volodia devrait-il attendre avant de pouvoir rencontrer Frunze seul ?

Ce n'était jamais une décision facile à prendre. Tout bien pesé, il décida de tenter sa chance le soir même.

Les Frunze entrèrent dans un snack.

Passant devant, Volodia jeta un coup d'œil par les vitres. C'était un restaurant sans prétention, dont la salle était divisée en box. Il envisagea un instant d'entrer et de s'asseoir avec eux, mais préféra les laisser manger tranquillement. Ils seraient de meilleure humeur le ventre plein.

Il attendit une demi-heure en surveillant la porte de loin. Puis, n'y tenant plus, il entra.

Ils finissaient de dîner. Frunze leva les yeux sur Volodia qui traversait le restaurant, puis détourna le regard. Il ne l'avait pas reconnu.

Volodia se glissa dans le box à côté d'Alice et dit tout bas, en allemand : « Alors, Willi, tu ne reconnais plus un vieux copain d'école ? »

Le regard sévère de Frunze se posa sur lui pendant plusieurs secondes, puis son visage s'éclaira. « Pechkov ? Volodia Pechkov ? C'est vraiment toi ? » demanda-t-il avec un grand sourire.

Volodia fut envahi par une vague de soulagement. Frunze était toujours amical. Il n'y aurait pas de barrière d'hostilité à surmonter. « Eh oui, c'est bien moi », fit Volodia. Il lui tendit la main et l'autre la lui serra. Il se tourna vers Alice et dit en anglais : « Désolé, je parle très mal votre langue.

— Ça n'a aucune importance, répondit-elle dans un allemand parfait. Mes parents sont des immigrés bavarois. »

Frunze reprit avec étonnement : « Figure-toi que j'ai pensé à toi il n'y a pas longtemps, parce que je connais un type qui porte le même nom que toi : un certain Greg Pechkov.

— Ah oui ? Mon père avait un frère qui est venu en Amérique vers 1915. Mais il s'appelait Lev.

— Non, le lieutenant Pechkov est beaucoup plus jeune que ça. Enfin, peu importe. Quel bon vent t'amène ?

— Je suis venu te voir », annonça Volodia en souriant.

Sans laisser à Frunze le temps de demander pourquoi, il poursuivit : « La dernière fois qu'on s'est vus, tu étais secrétaire du parti social-démocrate de l'arrondissement de Neukölln. »

C'était la deuxième étape. Ayant établi des bases amicales, il rappelait à Frunze ses idéaux de jeunesse.

« Cette expérience m'a convaincu que le socialisme démocratique ne marchait pas, répondit Frunze. Contre les nazis, nous étions complètement désarmés. Il a fallu que l'Union soviétique intervienne pour les arrêter. »

C'était vrai, et Volodia se réjouit que Frunze en ait conscience ; mais surtout, ce commentaire montrait que les idées politiques de Frunze n'avaient pas changé au contact de la vie américaine facile.

« Nous pensions aller prendre un verre dans un bar, au coin de la rue, intervint Alice. Beaucoup de chercheurs y vont le vendredi soir. Vous voulez vous joindre à nous ? »

Volodia ne voulait surtout pas être vu en public avec les Frunze. « Je ne sais pas », éluda-t-il. En réalité, il était déjà resté trop longtemps avec eux dans ce restaurant. Le moment de l'étape numéro trois était arrivé : rappeler à Frunze sa terrible responsabilité. Se penchant en avant, il baissa la voix : « Willi, tu savais que les Américains allaient larguer des bombes nucléaires sur le Japon ? »

Il y eut un long silence. Volodia retint son souffle. Il avait fait le pari que Frunze serait rongé de remords.

L'espace d'un instant, il eut peur d'être allé trop loin. Frunze semblait sur le point de fondre en larmes.

Puis il inspira profondément et reprit son empire sur lui-même. « Non, répondit-il. Je ne le savais pas. Aucun de nous ne le savait. »

Alice intervint, rageusement : « Nous pensions que l'armée américaine allait se livrer à une démonstration de force quelconque, ça oui, qu'elle allait brandir la menace de la bombe pour pousser les Japonais à capituler plus vite. » Elle avait donc été informée de l'existence de la bombe avant l'explosion, nota Volodia. Il n'en était pas surpris. Les hommes avaient du mal à cacher ce genre de choses à leur femme. « Nous nous attendions donc à une explosion quelque part, à un moment ou un autre, poursuivit-elle. Mais nous supposions qu'ils détruiraient une île inhabitée, ou peut-être une installation militaire contenant beaucoup d'armes et très peu de soldats.

— Ça, ç'aurait été justifiable, reprit Frunze. Mais… Personne n'imaginait qu'ils la largueraient sur une ville et qu'ils tueraient quatre-vingt mille hommes, femmes et enfants », poursuivit-il d'une voix réduite à un souffle.

Volodia hocha la tête. « Je me disais bien que tu verrais sans doute les choses comme ça. » Il l'avait même espéré de tout son cœur.

« Comment faire autrement ? répliqua Frunze.

— Laisse-moi te poser une question encore plus importante. »

Il abordait l'étape numéro quatre. « Est-ce qu'ils vont recommencer ?

— Je n'en sais rien, soupira Frunze. Ils en sont bien capables. Que Dieu nous pardonne, ils en sont bien capables. »

Volodia dissimula sa satisfaction. Il avait amené Frunze à se sentir responsable de l'usage futur de l'arme nucléaire autant que de son utilisation passée.

Volodia hocha la tête. « C'est ce que nous pensons, nous aussi.

— Qui ça, *nous* ? » demanda sèchement Alice.

Elle était futée, et avait probablement plus de bon sens que son mari. Volodia aurait du mal à la duper ; il décida de ne même pas essayer. Mieux valait jouer franc jeu avec elle.

« Bonne question, répondit-il, et je ne suis pas venu jusqu'ici pour mentir à un vieil ami. Je suis commandant des services de renseignement de l'armée Rouge. »

Ils le regardèrent en ouvrant de grands yeux. Cette possibilité les avait sûrement déjà effleurés, mais ils étaient désarçonnés par la franchise de l'aveu.

« J'ai quelque chose à vous dire, poursuivit Volodia. Une chose d'une extrême importance. Y a-t-il un endroit où nous pourrions parler tranquillement ? »

Ils se regardèrent, indécis. « Chez nous ? proposa Frunze.

— Votre appartement a probablement été mis sur écoute par le FBI. »

Frunze avait une certaine expérience de l'action clandestine, mais Alice fut scandalisée. « Vous croyez ? demanda-t-elle, incrédule.

— Oui. Le mieux serait de prendre votre voiture et de sortir de la ville.

— Il y a un endroit où nous allons parfois, à peu près à cette heure-ci, pour admirer le coucher de soleil, suggéra Frunze.

— Parfait. Allez chercher votre voiture et attendez-moi à l'intérieur. Je vous rejoins dans une minute. »

Frunze régla l'addition et sortit avec Alice, Volodia ne tarda pas à leur emboîter le pas. Le court trajet pour les rejoindre lui permit de s'assurer qu'il n'était pas filé. Arrivé à la Plymouth, il monta dedans. Ils s'assirent tous les trois sur la banquette avant, à l'américaine, Frunze au volant.

Ils quittèrent la ville et s'engagèrent sur une route de terre battue qui gravissait une petite colline. Frunze arrêta la voiture. Volodia fit signe aux deux autres de descendre et de le suivre. Ils s'éloignèrent d'une centaine de mètres : la voiture aurait pu être équipée de micros, elle aussi.

Ils regardèrent le soleil se coucher sur le désert caillouteux, ponctué de buissons d'épineux, et Volodia aborda l'étape numéro cinq : « Nous pensons que la prochaine bombe atomique pourrait être lâchée au-dessus de l'Union soviétique. »

Frunze hocha la tête. « Dieu nous en préserve, mais tu as probablement raison.

— Et nous sommes complètement impuissants, insista Volodia, enfonçant le clou. Nous ne pouvons prendre aucune précaution, dresser aucune barrière, nous ne pouvons rien faire pour préserver notre peuple. Il n'existe aucun moyen de défense contre la bombe atomique – la bombe que tu as fabriquée, Willi.

— Je sais », répondit Frunze, consterné. Il était clair que si l'URSS était victime d'une attaque nucléaire, il se sentirait responsable.

Sixième étape : « Notre seul moyen de protection serait d'avoir notre propre arme nucléaire. »

Frunze tiqua. « Ce n'est pas un moyen de défense, objecta-t-il.

— Non, mais c'est un moyen de dissuasion.

— Peut-être, convint-il.

— Nous ne voudrions pas que ces bombes fleurissent un peu partout, protesta Alice.

— Moi non plus, approuva Volodia, mais le seul moyen sûr d'empêcher les Américains de raser Moscou comme ils ont rasé Hiroshima serait que l'Union soviétique ait sa propre bombe atomique, pour pouvoir les menacer de représailles.

— Il a raison, Willi, murmura Alice. Bon sang, on le sait tous. »

C'était elle la plus solide des deux, comprit Volodia.

Il adopta un ton léger pour la septième étape : « Combien de bombes les Américains ont-ils pour le moment ? »

C'était le moment décisif. Si Frunze répondait à cette question, il aurait franchi une ligne. Jusque-là, leur conversation s'était limitée à des généralités. Maintenant, Volodia demandait des informations secrètes.

Frunze hésita longuement. Il se tourna enfin vers Alice.

Volodia vit qu'elle lui adressait un imperceptible hochement de tête.

« Une seule », répondit Frunze.

Volodia dissimula son sentiment de triomphe. Frunze avait trahi un secret. C'était la première démarche, la plus difficile. Le deuxième secret viendrait plus aisément.

« Mais ils en auront bientôt d'autres, ajouta Frunze.

— C'est une course, et si nous la perdons, nous sommes morts, reprit Volodia d'un ton pressant. Nous devons fabriquer au moins une bombe avant que les États-Unis en aient suffisamment pour nous rayer de la carte.

— Vous y arriverez ? »

Pour Volodia, c'était le signal de la huitième étape :

« Nous avons besoin d'aide. »

Il vit le visage de Frunze se fermer, et devina qu'il pensait à l'incident, quel qu'il fût, qui l'avait conduit à refuser de coopérer avec le NKVD.

« Et si nous répondons que nous ne pouvons pas vous aider ? avança Alice. Que c'est trop dangereux ? »

Volodia suivit son instinct. Il leva les mains dans un geste fataliste. « Je rentrerai chez moi et je dirai que j'ai échoué, répondit-il. Je ne peux pas vous obliger à faire quelque chose que vous ne voulez pas faire. Je n'ai pas l'intention d'exercer de pression ni de contrainte sur vous.

— Pas de menaces ? » insista Alice.

Ce qui confirma à Volodia son intuition : les agents du NKVD avaient cherché à intimider Frunze. Ils ne pouvaient pas s'en empêcher : ils ne savaient rien faire d'autre. « Je n'essaie même pas de te convaincre, expliqua Volodia à Frunze. Je me contente de t'exposer des faits. Le reste dépend de toi. Si tu veux nous aider, je suis ici, comme contact. Si tu vois les choses autrement, point final. Vous êtes intelligents, tous les deux. Je ne pourrais pas vous tromper, même si je voulais. »

Une fois encore, ils échangèrent un regard. Volodia espérait qu'ils le trouvaient très différent des derniers agents soviétiques qui les avaient approchés.

Leur silence s'éternisa.

Ce fut Alice qui reprit enfin la parole. « De quel genre d'aide auriez-vous besoin ? »

Ce n'était pas un « oui », mais c'était toujours mieux qu'un refus, et cette question menait logiquement à l'étape numéro neuf.

« Ma femme est physicienne ; elle fait partie de l'équipe de recherche », leur confia Volodia, espérant se donner ainsi un visage plus humain à un moment où il risquait d'être perçu comme un manipulateur.

« Elle me dit qu'il y a plusieurs pistes qui devraient permettre de fabriquer une bombe atomique, et nous n'avons pas le temps de les explorer toutes. Nous pourrions gagner des années si nous savions ce qui a marché pour vous.

— Ça se tient », répondit Willi.

Dixième étape, la plus critique. « Il faudrait que nous sachions quel type de bombe a été larguée sur le Japon. »

Frunze avait l'air torturé. Il regarda sa femme. Cette fois, elle ne lui donna pas le feu vert en opinant du chef, mais elle ne secoua pas la tête non plus. Elle avait l'air aussi partagée que lui.

Frunze soupira. « Deux types de bombes ont été utilisés, murmura-t-il enfin.

— De conceptions différentes ? » demanda Volodia, surpris et captivé.

Frunze hocha la tête. « Pour Hiroshima, ils ont utilisé une bombe à l'uranium avec un dispositif de mise à feu par charge explosive. Nous l'avions appelée *Little Boy*. La bombe lâchée sur Nagasaki, *Fat Man*, était une bombe au plutonium qui recourt à la technique de l'implosion. »

Volodia retint son souffle. C'étaient des informations de toute première importance.

« Et quelle est la plus efficace ?

— À l'évidence, elles ont toutes les deux fonctionné. Mais Fat Man est plus facile à fabriquer.

— Pourquoi ?

— Il faut des années pour produire suffisamment d'uranium enrichi pour fabriquer une bombe. Le plutonium est plus rapide à obtenir, à partir du moment où on a une pile atomique.

— L'URSS aurait donc intérêt à s'inspirer de Fat Man.

— Sans aucun doute.

— Il y a encore une chose que tu pourrais faire pour éviter le pire à l'Union soviétique, poursuivit Volodia.

— Quoi donc ? »

Volodia le regarda bien en face. « Me procurer les plans de fabrication. »

Willi blêmit. « Je suis citoyen américain, murmura-t-il. Ce que tu me demandes, c'est une trahison. Un crime passible de la peine de mort. Je pourrais me retrouver sur la chaise électrique. »

Et ta femme aussi, pensa Volodia. Elle est complice. Heureusement, cette idée ne t'a pas effleuré.

« J'ai demandé à bien des gens de mettre leur vie en jeu au cours de ces dernières années, dit-il. À des gens comme toi. Des Allemands qui détestaient les nazis, des hommes et des femmes qui ont pris des risques terribles pour nous communiquer les informations qui nous ont aidés à gagner la guerre. Et je vais te dire la même chose qu'à eux : beaucoup de gens mourront si tu ne le fais pas. » Il se tut. Il avait abattu son dernier atout. Il n'avait plus rien à offrir.

Frunze consulta sa femme du regard.

« Tu as fait la bombe, Willi, souffla Alice.

— Je vais y réfléchir, Volodia », répondit Frunze.

3.

Deux jours plus tard, il remettait les plans à Volodia, lequel les rapporta à Moscou.

Zoïa fut libérée. Elle était moins furieuse que lui d'avoir été emprisonnée. « Ils l'ont fait pour protéger la révolution, expliqua-t-elle. Et ils ne m'ont fait aucun mal. C'était comme un séjour dans un très mauvais hôtel. »

Le jour de son retour à la maison, après qu'ils eurent fait l'amour, il lui annonça : « J'ai quelque chose à te montrer, quelque chose que j'ai rapporté d'Amérique. »

Il roula à bas du lit, ouvrit un tiroir et en sortit un livre. « Ça s'appelle le catalogue Sears Roebuck », dit-il.

Il s'assit à côté d'elle sur le lit et ouvrit le volume.

« Regarde ça. »

Le catalogue s'ouvrit à la page des robes. Les mannequins étaient d'une minceur invraisemblable, mais les modèles étaient coupés dans des tissus lumineux et gais, avec des rayures, des carreaux, des couleurs unies, des ruchés, des plis et des ceintures. « Elle est drôlement jolie celle-là, fit Zoïa en posant le doigt sur une robe. Mais deux dollars quatre-vingt-dix-huit, ça fait beaucoup d'argent, non ?

— Pas tant que ça, répondit Volodia. Le salaire moyen est d'une cinquantaine de dollars par semaine, et le loyer du tiers, à peu près.

— Vraiment ? s'exclama Zoïa, stupéfaite. Alors la plupart des femmes pourraient sans difficulté s'offrir ce genre de robe ?

— Eh oui. Peut-être pas les paysannes. Encore que ces catalogues aient été destinés au départ aux fermiers qui vivaient à cent cinquante kilomètres du magasin le plus proche.

— Comment ça marche ?

— Tu choisis ce que tu veux dans le livre, tu leur envoies l'argent, et deux semaines plus tard, le facteur t'apporte ce que tu as commandé.

— On doit avoir l'impression d'être une tsarine. »

Zoïa lui prit le livre des mains et tourna la page. « Oh ! Regarde, il y en a encore ! » La page suivante présentait des tailleurs à quatre dollars quatre-vingt-dix-huit. « Ça aussi, c'est élégant, remarqua-t-elle.

— Regarde plus loin », suggéra Volodia.

Des pages et des pages de vêtements pour femmes, manteaux, chapeaux, chaussures, sous-vêtements, bas, pyjamas, s'étalaient sous les yeux stupéfaits de Zoïa.

« Les gens peuvent avoir tout ça ? demanda-t-elle.

— Mais oui.

— On trouve plus de choix sur une seule de ces pages que dans la plupart de nos magasins !

— C'est vrai. »

Elle continua à feuilleter lentement le catalogue. Il y avait une variété similaire de vêtements pour hommes, et même pour enfants. Zoïa montra à Volodia un épais manteau d'hiver en laine qui coûtait quinze dollars.

« À ce prix-là, j'imagine que tous les garçons d'Amérique en ont un.

— Probablement. »

Après les vêtements venaient les meubles. On pouvait acheter un lit pour vingt-cinq dollars. Tout était abordable, quand on gagnait cinquante dollars par semaine. Et c'était la même chose, page après page. Le catalogue contenait des centaines d'articles tout simplement impossibles à acheter en Union soviétique, même quand on avait de l'argent : des jeux et des jouets, des produits de beauté, des guitares, des fauteuils élégants, des appareils électriques, des romans sous des jaquettes en couleur, des décorations de Noël et des grille-pain.

Même un tracteur. « Tu crois, fit Zoïa, qu'en Amérique, n'importe quel fermier qui veut un tracteur peut l'avoir *tout de suite* ?

— À condition d'avoir l'argent pour l'acheter, répondit Volodia.

— Il n'a pas besoin de s'inscrire sur une liste et d'attendre plusieurs années ?

— Non. »

Zoïa referma le volume et regarda Volodia d'un air grave.

« Pourquoi des gens qui peuvent avoir tout ça voudraient-ils être communistes ? demanda-t-elle.

— Bonne question », répondit Volodia.

XXII
1946

1.

Les petits Berlinois avaient inventé un nouveau jeu appelé *Komm, Frau* – « Viens, femme ». Les jeux de course-poursuite entre garçons et filles étaient nombreux – une dizaine au moins – mais Carla remarqua que celui-ci avait une particularité. Les garçons faisaient équipe et prenaient une des filles pour cible. Quand ils l'avaient attrapée, ils criaient : « *Komm, Frau !* » et la jetaient par terre. Ensuite, ils la maintenaient pendant que l'un d'eux s'allongeait sur elle et mimait l'acte sexuel. Des enfants de sept et huit ans qui n'auraient pas dû savoir ce qu'était un viol jouaient ainsi parce qu'ils avaient vu ce que les soldats de l'armée Rouge faisaient aux Allemandes. Tous les Russes connaissaient au moins cette phrase en allemand : « *Komm, Frau.* »

Mais qu'est-ce qu'ils avaient, ces hommes russes ? Carla n'avait jamais rencontré de femme violée par un soldat français, anglais, américain ou canadien, et pourtant elle supposait que ça devait bien arriver. A contrario, elle ne connaissait pas une seule femme âgée de quinze à cinquante-cinq ans qui n'ait été violée par au moins un soldat russe : sa mère, Maud ; son amie Frieda ; la mère de Frieda, Monika ; Ada, leur domestique ; elles avaient toutes été violées.

En même temps, elles avaient eu de la chance, parce qu'elles étaient encore en vie. Certaines, violées par des dizaines d'hommes, des heures durant, n'avaient pas survécu. Carla avait entendu parler d'une fille qui avait été mordue à mort.

Rebecca Rosen y avait échappé, cependant. Depuis que Carla l'avait prise sous sa protection, le jour de la libération de l'hôpital juif, Rebecca s'était installée chez les von Ulrich. La maison se trouvait dans le secteur soviétique, mais elle n'avait pas d'autre point de chute. Elle s'était cachée pendant des mois dans le grenier, comme une criminelle, ne sortant qu'en pleine nuit, quand ces barbares de Russes avaient sombré dans un sommeil aviné. Carla passait quelques heures là-haut avec elle, quand elle pouvait, et elles jouaient aux cartes ou se racontaient leur vie. Carla aurait voulu être une espèce de grande sœur pour elle, mais Rebecca la considérait comme une mère.

Et puis Carla avait découvert qu'elle allait vraiment être mère.

Maud et Monika, qui avaient une cinquantaine d'années, étaient trop âgées pour tomber enceintes, heureusement ; et Ada avait eu de la chance. Mais Carla et Frieda avaient été engrossées par leurs violeurs.

Frieda s'était fait avorter.

C'était illégal, et une loi nazie qui punissait l'avortement de mort était encore en vigueur. Mais Frieda était allée voir une « sage-femme » d'un certain âge, qui le lui avait fait pour cinq cigarettes. Frieda avait contracté une grave infection, et serait morte si Carla n'avait pas réussi à voler de la pénicilline, un médicament encore extrêmement rare, à l'hôpital.

Carla avait décidé de garder son bébé.

Ses sentiments à ce sujet passaient avec violence d'un extrême à l'autre. Quand elle souffrait de nausées matinales, elle fulminait contre les sauvages qui avaient profané son corps et lui avaient infligé ce fardeau. À d'autres moments, elle se surprenait à rester assise, les mains sur le ventre, le regard perdu dans le vide, rêvant à de petits vêtements de bébé. Elle se demandait alors si les traits de l'enfant lui rappelleraient l'un des hommes et lui feraient haïr sa propre progéniture. Il aurait certainement aussi quelque chose des von Ulrich, non ? Elle était terrifiée et angoissée.

En janvier 1946, elle était enceinte de huit mois. Comme la plupart des Allemands, elle était également gelée, affamée et complètement démunie. Lorsque sa grossesse avait commencé à se voir, elle avait dû renoncer à son poste d'infirmière et rejoindre les millions de chômeurs. Les rations alimentaires étaient distribuées tous les dix jours. Pour ceux qui ne bénéficiaient pas de privilèges particuliers, la ration quotidienne était de mille cinq cents calories. Encore fallait-il avoir de quoi payer, évidemment. Et même quand on avait de l'argent et des tickets de rationnement, il arrivait qu'on ne trouve tout simplement aucune denrée alimentaire à acheter.

Carla avait envisagé de demander aux Soviétiques un traitement spécial parce qu'elle leur avait transmis de précieuses informations pendant la guerre. Mais Heinrich, qui s'y était risqué, avait vécu une expérience épouvantable. Les services de renseignement de l'armée Rouge s'étaient attendus à ce qu'il continue à espionner pour leur compte et lui avaient demandé d'infiltrer l'armée américaine. Quand il s'était dérobé, les choses avaient mal tourné, et ils avaient menacé de l'envoyer en camp de travail. Il s'en était sorti en disant qu'il ne parlait pas anglais et ne pouvait donc pas leur être utile. Ainsi avertie, Carla avait préféré se faire oublier.

Ce jour-là, Carla et Maud étaient contentes parce qu'elles avaient vendu un meuble. C'était une commode Jugendstil en loupe de chêne clair que les parents de Walter avaient achetée pour leur mariage, en 1889. Carla, Maud et Ada l'avaient chargée sur une charrette à bras qu'elles avaient empruntée.

Il n'y avait toujours pas d'homme chez elles. Erik et Werner faisaient partie des millions de soldats allemands portés disparus. Peut-être étaient-ils morts. Le colonel Beck avait confié à Carla que près de trois millions d'Allemands avaient été tués au cours des combats sur le front est, et que davantage encore avaient péri – de faim, de froid ou de maladie – dans des camps de prisonniers. Deux autres millions étaient toujours vivants et trimaient dans des camps de travail en Union soviétique. Quelques-uns étaient rentrés : certains s'étaient évadés, d'autres avaient été libérés parce qu'ils étaient trop malades pour accomplir une quelconque besogne, et ils avaient rejoint les milliers de personnes déplacées qui arpentaient les routes de toute l'Europe, s'efforçant de rentrer chez elles. Carla et Maud leur avaient adressé des lettres aux bons soins de l'armée Rouge, mais n'avaient jamais eu de réponse.

Carla se torturait à la perspective du retour de Werner. Elle l'aimait toujours, et espérait de tout son cœur qu'il était sain et sauf, ce qui ne l'empêchait pas de redouter leurs retrouvailles alors qu'elle portait le bébé d'un violeur. Elle n'y était pour rien, mais elle éprouvait une honte irrationnelle.

Les trois femmes poussaient donc leur charrette dans les rues. Elles avaient laissé Rebecca à la maison. Après avoir atteint un point culminant cauchemardesque, l'orgie de viols et de pillages de l'armée Rouge semblait être retombée, et Rebecca avait quitté son grenier. Néanmoins, il n'était pas encore très sûr pour une jolie fille de se promener dans les rues.

D'immenses portraits de Lénine et de Staline dominaient maintenant Unter den Linden, l'ancien lieu de promenade préféré de la société élégante. La plupart des rues de Berlin avaient été déblayées, et les décombres des immeubles détruits formaient des tas de gravats espacés d'une centaine de mètres, prêts à être réutilisés, peut-être, si les Allemands arrivaient un jour à reconstruire leur pays. Des hectares d'habitations avaient été rasés, souvent par pâtés de maisons entiers. Il faudrait des années pour nettoyer la ville. Des milliers de cadavres pourrissaient dans les ruines, et tout l'été, l'odeur de chair humaine corrompue avait plané dans l'air. Maintenant, cela n'empestait plus qu'après la pluie.

La ville avait été provisoirement divisée en quatre secteurs : russe, américain, anglais et français. Beaucoup de bâtiments encore debout avaient été réquisitionnés par les troupes d'occupation et les Berlinois vivaient où ils pouvaient, cherchant parfois un abri précaire dans les pièces restantes de maisons à moitié démolies. L'eau courante avait été rétablie et l'électricité fonctionnait par intermittence, mais on avait du mal à trouver du combustible pour se chauffer et faire la cuisine. La commode serait peut-être presque aussi précieuse une fois débitée en bois de chauffage.

Elles l'avaient transportée à Wedding, dans le secteur français, où elles l'avaient vendue à un charmant colonel contre une cartouche de Gitanes. La monnaie d'occupation n'avait plus aucune valeur, parce que les Russes en imprimaient trop, et tout se troquait contre des cigarettes.

À présent, elles revenaient triomphantes. Maud et Ada tiraient la charrette vide pendant que Carla, qui l'avait poussée à l'aller et avait mal partout, marchait à côté. Elles étaient riches : avec toute une cartouche de cigarettes, on pouvait acheter bien des choses.

La nuit tomba ; la température frôlait zéro degré. Pour rentrer chez elles, elles devaient traverser une petite portion de secteur britannique. Carla se demandait parfois si les Anglais auraient aidé sa mère s'ils avaient su ce qu'elle endurait. Mais après tout, Maud était citoyenne allemande depuis vingt-six ans. Son frère, le comte Fitzherbert, était un homme fortuné et influent, mais il avait refusé de l'aider après son mariage avec Walter von Ulrich, et c'était un homme buté : il y avait peu de chances qu'il change d'attitude.

Elles tombèrent sur un attroupement : trente ou quarante personnes dépenaillées massées devant une demeure réquisitionnée par la puissance occupante. Les trois femmes s'arrêtèrent, intriguées, et virent que les badauds observaient une fête donnée à l'intérieur. Par les fenêtres, on distinguait des pièces vivement éclairées dans lesquelles des domestiques allaient et venaient avec des plateaux d'amuse-gueules qu'ils présentaient aux invités, des hommes et des femmes qui riaient, un verre à la main. Carla regarda autour d'elle. Il y avait surtout des femmes et des enfants – il ne restait plus beaucoup d'hommes à Berlin, ni même en Allemagne, à vrai dire –, aux yeux brillants d'envie, comme des pécheurs refoulés aux portes du paradis. C'était un spectacle pathétique.

« C'est indécent », déclara Maud.

Elle remonta l'allée qui menait à la porte de la maison. Une sentinelle britannique s'interposa en disant : « *Nein*, *nein* », probablement les seuls mots d'allemand que ce jeune homme connaissait.

Maud lui répondit dans l'anglais châtié qu'elle avait appris dans son enfance : « Je désire voir immédiatement votre commandant. Veuillez le prévenir. »

Carla admira, comme toujours, l'aplomb et la dignité de sa mère.

Le planton contempla d'un air dubitatif le manteau élimé de Maud, mais finit par frapper à la porte, qui s'ouvrit tandis qu'une tête se penchait à l'extérieur.

« Il y a une dame anglaise qui veut voir le commandant », annonça la sentinelle.

Un instant plus tard, le battant s'écarta devant deux personnes. On aurait dit une caricature d'officier britannique flanqué de son épouse : lui en tenue de soirée avec un nœud papillon noir, elle en robe longue et collier de perles.

« Bonsoir, dit Maud. Veuillez m'excuser de perturber ainsi votre fête. »

Ils la regardèrent, stupéfaits de voir une femme en guenilles s'exprimer ainsi.

Maud poursuivit : « Vous ne vous rendez certainement pas compte du spectacle que vous infligez à ces pauvres gens qui sont là, dehors. »

Le couple posa les yeux sur la foule.

Maud continua : « Pour l'amour du ciel, tirez au moins les rideaux. »

Après un instant de désarroi, la femme demanda : « Oh mon Dieu, George, aurions-nous fait preuve d'un terrible manque de délicatesse ?

— Peut-être, bougonna l'homme. Mais ce n'était pas intentionnel.

— Pensez-vous que nous pourrions nous racheter en leur apportant un peu de nourriture ?

— Oui, certainement, répondit Maud promptement. Ce serait gentil de votre part, et ce serait une excellente façon de présenter vos excuses. »

L'officier parut hésiter. Offrir des canapés à des Allemands qui mouraient de faim constituait probablement une violation d'un règlement quelconque.

La femme implora : « George chéri, tu veux bien ?

— Eh bien, c'est entendu », acquiesça son mari.

La femme se tourna vers Maud. « Nous ne l'avons vraiment pas fait exprès. Merci de nous avoir prévenus.

— Je vous en prie. » Et Maud redescendit l'allée.

Quelques minutes plus tard, les invités commencèrent à sortir de la maison chargés de plateaux de sandwichs et de gâteaux, qu'ils offrirent à la foule affamée. Carla sourit. L'impudence de sa mère avait été payante. Elle prit une grosse tranche de cake qu'elle engloutit avidement. Elle contenait plus de sucre qu'elle n'en avait mangé au cours des six derniers mois.

Les rideaux furent tirés, les invités regagnèrent la demeure et l'attroupement se dispersa. Maud et Ada reprirent les bras de leur charrette et recommencèrent à la pousser vers la maison. « Bravo, Mutti, lança Carla. Une cartouche de Gitanes *et* un repas gratuit, en un seul après-midi ! »

Les Russes mis à part, rares étaient les soldats d'occupation qui se montraient cruels envers les Allemands, se dit Carla. Cela l'étonnait. Les GI américains distribuaient des tablettes de chocolat. Les Français eux-mêmes, dont les propres enfants avaient connu la faim sous l'occupation allemande, faisaient souvent preuve de gentillesse. Après toutes les souffrances que nous, les Allemands, avons infligées à nos voisins, il est surprenant qu'ils ne nous détestent pas davantage, songeait-elle. D'un autre côté, entre les nazis, l'armée Rouge et les raids aériens, ils estiment peut-être que nous avons suffisamment souffert.

Il était tard lorsqu'elles arrivèrent chez elles. Elles rendirent la charrette aux voisines qui la leur avaient prêtée, et leur donnèrent un demi-paquet de Gitanes en dédommagement. Elles rentrèrent dans la maison qui était, par bonheur, encore intacte. Il n'y avait plus de vitres à la plupart des fenêtres, et la façade de pierre était criblée d'impacts de balles et d'éclats d'obus, mais l'immeuble n'avait pas subi de dégâts structurels et les abritait encore du froid.

Cela dit, les quatre femmes vivaient désormais dans la cuisine et dormaient sur des matelas qu'elles traînaient le soir depuis le couloir. Il était déjà assez difficile de garder la chaleur dans cette unique pièce, et elles n'avaient pas de quoi chauffer le reste de la maison. Dans le temps, le poêle de la cuisine était alimenté au charbon, mais il était maintenant presque impossible de s'en procurer. Elles avaient cependant découvert qu'on pouvait y faire brûler bien d'autres choses : des livres, des journaux, des meubles réduits en morceaux, et même des voilages.

Elles dormaient à deux par matelas, Carla avec Rebecca et Maud avec Ada. Rebecca s'endormait parfois épuisée d'avoir pleuré dans les bras de Carla, comme la nuit où ses parents avaient été tués.

Exténuée par leur longue marche, Carla s'allongea aussitôt. Ada alluma le poêle avec de vieilles revues que Rebecca avait descendues du grenier. Maud ajouta de l'eau au reste de la soupe aux haricots du déjeuner, et la réchauffa en guise de dîner.

En s'asseyant pour manger, Carla eut l'impression de recevoir un coup de poignard dans le ventre. Ce n'était pas parce qu'elle avait poussé la charrette, c'était autre chose. Elle vérifia la date, compta à rebours depuis la date de libération de l'hôpital juif.

« Mutti, murmura-t-elle inquiète, je crois que le bébé arrive.

— C'est trop tôt ! s'écria Maud.

— Je suis enceinte de trente-six semaines, et je commence à avoir des contractions.

— Dans ce cas, nous ferions mieux de nous préparer. »

Maud monta à l'étage chercher des serviettes.

Ada apporta un fauteuil en bois de la salle à manger. Elle avait récupéré sur un site bombardé un morceau de gros câble d'acier torsadé qui faisait un excellent marteau. Elle brisa le fauteuil et mit les fragments dans le poêle.

Carla posa les mains sur son ventre distendu. « Tu aurais pu attendre qu'il fasse un peu plus chaud, bébé », dit-elle.

Mais elle eut bientôt trop mal pour sentir le froid. Elle n'avait pas imaginé que l'on puisse souffrir à ce point.

Ni que cela pouvait durer aussi longtemps. Elle fut en travail toute la nuit. Maud et Ada lui tinrent la main à tour de rôle pendant qu'elle gémissait et criait. Rebecca assistait à la scène, le visage blême, épouvantée.

La lumière grise du matin filtrait à travers les journaux collés sur les fenêtres sans vitres de la cuisine quand enfin la tête du bébé apparut. Carla fut submergée par un sentiment de soulagement comme elle n'en avait jamais connu, et pourtant la douleur ne cessa pas immédiatement.

Après une dernière poussée atroce, Maud attrapa le bébé entre ses jambes.

« C'est un garçon », annonça-t-elle.

Elle lui souffla sur la figure. Il ouvrit la bouche et se mit à crier.

Elle donna le bébé à Carla et la redressa sur le matelas à l'aide de coussins qu'elle était allée chercher dans le salon.

Le bébé avait la tête couverte d'une tignasse noire.

Maud noua le cordon avec un morceau de coton et le coupa. Carla déboutonna son corsage et posa le bébé sur sa poitrine.

Elle craignait de ne pas avoir de lait. Ses seins auraient dû gonfler et suinter vers la fin de sa grossesse, or ils ne l'avaient pas fait, peut-être parce que le bébé était arrivé trop tôt, peut-être parce qu'elle était sous-alimentée. Mais après qu'il eut tété pendant quelques instants, elle éprouva une douleur curieuse, et le lait commença à couler.

Le petit ne tarda pas à s'endormir.

Ada apporta un bol d'eau chaude et un linge, et nettoya précautionneusement le visage, la tête, puis le corps du bébé.

« Comme il est beau, murmura Rebecca.

— Mutti, dit Carla, si on l'appelait Walter? Tu veux bien? »

Elle n'avait pas eu l'intention de donner à sa question un tour mélodramatique, mais Maud s'effondra. Son visage se crispa et elle se plia en deux, secouée de terribles sanglots. Elle reprit suffisamment le dessus pour balbutier « Pardon », avant d'être à nouveau submergée par le chagrin.

« Oh, Walter, mon Walter », bredouillait-elle, en larmes.

Ses pleurs finirent par se calmer.

« Pardon, répéta-t-elle encore. J'ai horreur de me donner en spectacle. » Elle s'essuya le visage avec sa manche. « J'aurais tellement voulu que ton père soit là pour voir le bébé, c'est tout. C'est tellement injuste. »

Ada les surprit en citant le Livre de Job : « Le Seigneur donne et le Seigneur reprend. Béni soit le nom du Seigneur. »

Carla ne croyait pas en Dieu – jamais un être saint digne de ce nom n'aurait permis l'existence des camps de la mort nazis –, mais cette citation la réconforta quand même. Elle voulait dire qu'il fallait tout accepter de la vie humaine, y compris la souffrance de la naissance et le chagrin de la mort. Maud sembla l'apprécier aussi, et s'apaisa peu à peu.

Carla regardait son petit Walter avec adoration. Elle se jura de s'occuper de lui, de lui procurer chaleur et nourriture quelles que puissent être les difficultés qui se dresseraient sur leur chemin. C'était le plus merveilleux bébé qui ait jamais vu le jour, elle l'aimerait et le chérirait pour toujours.

Il se réveilla, et Carla lui redonna le sein. Il téta avec satisfaction, faisant de petits bruits de succion sous le regard des quatre femmes. Et pendant un petit moment, dans la lumière crépusculaire et la chaleur de la cuisine, il n'y eut pas d'autre son.

2.

La première intervention d'un nouveau membre du Parlement, son discours inaugural, est généralement assommante. Certaines choses doivent être dites, des phrases toutes faites doivent être prononcées, et la règle veut que le sujet choisi ne prête surtout pas à controverse. Tout le monde, sympathisants comme opposants, félicite le nouvel élu, les traditions sont respectées et la glace est rompue.

Lloyd Williams prononça son premier *vrai* discours quelques mois plus tard, lors du débat sur le projet de loi concernant la protection sociale. C'était plus impressionnant pour lui.

Il le prépara en songeant à deux orateurs : son grand-père, Dai Williams, qui recourait au langage et au phrasé bibliques, non seulement au temple, mais aussi – et peut-être surtout – quand il évoquait les épreuves et les injustices de la vie des mineurs. Il adorait les mots courts, lourds de sens : labeur, péché, profit. Il parlait du foyer, de la mine et de la tombe.

Churchill en faisait autant, mais avec un humour qui manquait à Dai Williams. Ses longues périodes majestueuses s'achevaient souvent sur une image saugrenue, ou un détournement de sens. Il avait été rédacteur en chef de la *British Gazette*, le journal du gouvernement, pendant la grève générale de 1926, et avait prévenu les syndicalistes : « Que les choses soient bien claires entre nous : si vous nous infligez une nouvelle grève générale, nous vous infligerons une nouvelle *British Gazette.* » Lloyd estimait qu'un discours devait contenir des surprises de ce genre. C'était comme les raisins dans un pudding.

Mais quand il se leva pour prendre la parole, il trouva soudain que les phrases qu'il avait si soigneusement ciselées sonnaient creux. De toute évidence, son public partageait cette impression : il sentait bien que les cinquante ou soixante députés présents ne lui prêtaient qu'une oreille distraite. Il fut pris d'un moment de panique : comment pouvait-il ennuyer tout le monde avec un sujet tellement important pour ceux qu'il représentait ?

Sur le premier banc du gouvernement, il aperçut sa mère, devenue ministre des Écoles, et son oncle Billy, ministre du Charbon. Lloyd savait que Billy Williams était descendu pour la première fois dans la mine à l'âge de treize ans. Ethel avait le même âge quand elle avait commencé à briquer les parquets de Tŷ Gwyn. Ce débat n'était pas une histoire de belles formules ; c'était leurs vies dont il était question.

Au bout d'une minute, il abandonna son texte et commença à improviser. Il rappela les souffrances des familles de la classe ouvrière ruinées par le chômage ou l'invalidité, les drames auxquels il avait assisté de ses propres yeux dans l'East End à Londres et dans les mines de charbon de Galles du Sud. Sa voix trahissait son émotion, ce qui l'embarrassait, mais il poursuivit tout de même. Il sentit que son public commençait à l'écouter. Il parla de son grand-père et de bien d'autres, de tous ceux qui avaient lancé le mouvement travailliste en rêvant qu'une assurance chômage universelle bannirait à jamais le spectre de la misère. Quand il se rassit, son discours fut salué par un tonnerre d'acclamations.

Dans la galerie des visiteurs, sa femme, Daisy, souriait fièrement, le pouce levé en signe de victoire.

Il écouta le reste des débats, rouge de satisfaction. Il avait l'impression d'avoir passé avec succès sa première véritable épreuve de député.

Plus tard, dans le hall, un whip travailliste, un des parlementaires chargés de la discipline du parti, s'approcha de lui. Après l'avoir félicité pour son discours, il lui demanda s'il aimerait devenir attaché parlementaire.

Lloyd était aux anges. Chaque ministre et secrétaire d'État avait au moins un attaché. En réalité, ceux-ci n'étaient que des porteurs de serviette, mais ce poste était souvent un tremplin pour une nomination ministérielle.

« J'en serais très honoré, répondit Lloyd. Pour qui devrais-je travailler ?

— Ernie Bevin. »

Quelle chance ! Lloyd osait à peine à y croire. Bevin était ministre des Affaires étrangères, et le plus proche collaborateur du Premier ministre, Attlee. L'intimité entre ces deux hommes était un bon exemple d'attirance des contraires. Attlee était issu de la classe moyenne : fils d'avocat, diplômé d'Oxford, officier pendant la Première Guerre mondiale. Bevin, quant à lui, était le fils illégitime d'une femme de chambre, il n'avait jamais connu son père, avait commencé à travailler à onze ans et fondé la Transport and General Workers Union, le très puissant syndicat des transports et des travailleurs. Ils étaient aussi physiquement différents qu'il est possible de l'être : Attlee était mince et soigné de sa personne, calme et solennel, tandis que Bevin était un colosse aux éclats de rire tonitruants. Le ministre des Affaires étrangères appelait le Premier ministre « Petit Clem ». Ce qui ne les empêchait pas d'être des alliés indéfectibles.

Bevin était un héros aux yeux de Lloyd et de millions d'Anglais ordinaires.

« Rien ne pourrait me faire plus plaisir, répondit Lloyd. Mais Bevin n'a-t-il pas déjà un attaché parlementaire ?

— Il lui en faut deux, répondit le whip. Présentez-vous demain matin à neuf heures au Foreign Office. Vous commencerez tout de suite.

— Merci ! »

Lloyd se rua dans le couloir aux lambris de chêne, en direction du bureau de sa mère. Il devait y retrouver Daisy après la séance.

« Mam ! cria-t-il en entrant. Je viens d'être nommé attaché parlementaire d'Ernie Bevin ! »

C'est alors qu'il remarqua qu'Ethel n'était pas seule. Le comte Fitzherbert se trouvait avec elle.

Fitz posa sur Lloyd un regard chargé de surprise autant que d'aversion.

Malgré son émoi, Lloyd remarqua que son père portait un costume gris clair à veston croisé extrêmement bien coupé.

Il se tourna vers sa mère. Elle était parfaitement calme. Cette rencontre n'avait pas l'air de la surprendre. C'était elle qui avait dû la manigancer.

Le comte en vint à la même conclusion. « Qu'est-ce que cela veut dire, Ethel ? »

Lloyd avait les yeux rivés sur l'homme dont le sang coulait dans ses veines. Cette situation gênante n'empêchait pas Fitz de conserver son assurance et sa dignité. Il était beau, malgré la paupière tombante que lui avait laissée la bataille de la Somme. Il s'appuyait sur une canne, autre héritage de la guerre. À quelques mois de ses soixante ans, il était tiré à quatre épingles, ses cheveux gris soigneusement coupés, sa cravate argent impeccablement nouée, ses chaussures noires parfaitement lustrées. Lloyd aimait, lui aussi, être sur son trente et un. C'est de lui que je tiens ça, se dit-il.

Ethel s'approcha tout près du comte, supprimant toute distance physique entre eux. Lloyd connaissait suffisamment sa mère pour comprendre ce geste : elle n'hésitait pas à user de son charme quand elle voulait gagner un homme à sa cause. Tout de même, Lloyd n'aimait pas beaucoup la voir aussi chaleureuse envers celui qui avait profité d'elle avant de la laisser tomber.

« J'ai été vraiment navrée d'apprendre la mort de Boy, dit-elle à Fitz. Nos enfants sont notre bien le plus précieux, n'est-ce pas ?

— Je dois y aller », marmonna Fitz.

Avant cette rencontre, Lloyd n'avait fait que croiser son père. C'était la première fois qu'il passait autant de temps en sa présence, la première fois qu'il l'entendait prononcer autant de mots d'affilée. Il avait beau être mal à l'aise, il n'en était pas moins fasciné. Bien que bougon, il avait vraiment de l'allure.

« Je vous en prie, Fitz, reprit Ethel. Vous avez un fils que vous n'avez jamais reconnu – un fils dont vous devriez être fier.

— Vous n'auriez pas dû faire ça, Ethel, répliqua Fitz. Un homme a le droit d'oublier ses erreurs de jeunesse. »

Lloyd était mortifié, mais sa mère insista.

« Pourquoi voudriez-vous l'oublier ? Je sais bien que c'était une erreur, mais regardez-le, maintenant : il est député, il vient de faire un discours enthousiasmant, et d'être nommé attaché parlementaire du ministre des Affaires étrangères. »

Fitz refusait obstinément de poser les yeux sur Lloyd.

« Vous voudriez faire comme si notre liaison avait été une aventure sans importance, poursuivit Ethel, mais vous savez la vérité. Oui, nous étions jeunes et stupides, et fougueux bien sûr – vous comme moi –, mais nous nous aimions ; nous nous aimions vraiment, Fitz. Il va bien falloir que vous l'admettiez. Vous ne savez donc pas qu'en refusant de reconnaître la vérité sur vous-même, vous aliénez votre âme ? »

Le visage de Fitz avait perdu un peu de son impassibilité. Il avait visiblement du mal à se dominer. Lloyd comprit que sa mère avait mis le doigt sur le véritable problème. Ce n'était pas tant que Fitz avait honte de ce fils illégitime, c'était avant tout qu'il était trop fier pour reconnaître qu'il avait été amoureux d'une femme de chambre. Il aimait probablement Ethel plus que sa propre femme, subodora Lloyd. Et cela bouleversait ses convictions fondamentales sur la hiérarchie sociale.

Lloyd ouvrit la bouche pour la première fois. « J'étais aux côtés de Boy lors de ses derniers moments, monsieur. Il est mort en brave. »

Pour la première fois, Fitz le regarda. « Mon fils n'a que faire de votre approbation », lança-t-il.

Lloyd eut l'impression d'avoir été giflé.

Ethel elle-même fut outrée. « Fitz ! s'exclama-t-elle. Comment pouvez-vous être aussi mesquin ? »

C'est alors que Daisy fit son entrée.

« Bonjour, Fitz, dit-elle gaiement. Vous pensiez probablement avoir réussi à vous débarrasser de moi, et vous revoilà mon beau-père. C'est amusant, non ?

— J'essayais précisément de convaincre Fitz de serrer la main de Lloyd, lui expliqua Ethel.

— Dans toute la mesure du possible, j'évite de serrer la main à des socialistes », rétorqua Fitz.

Ethel livrait un combat perdu d'avance, mais elle n'était pas femme à s'avouer vaincue.

« Ne voyez-vous pas à quel point il tient de vous ! Il vous ressemble, il s'habille comme vous, il partage votre goût pour la politique – il finira probablement ministre des Affaires étrangères, ce que vous avez toujours voulu être ! »

L'expression de Fitz s'assombrit encore. « Il est fort improbable désormais que je devienne jamais ministre des Affaires étrangères. » Il se dirigea vers la porte. « Et rien ne saurait me déplaire davantage que de voir ce grand ministère occupé par mon bâtard bolchevique. » Sur ces mots, il sortit.

Ethel fondit en larmes.

Daisy passa son bras autour des épaules de Lloyd. « Quel mufle, s'écria-t-elle.

— Ne te frappe pas, répondit Lloyd. Je ne suis ni scandalisé, ni déçu. » Ce n'était pas vrai, mais il ne voulait surtout pas apitoyer les autres sur son sort. « Voici bien longtemps qu'il m'a rejeté. » Il regarda Daisy avec adoration. « Et par bonheur, il y a beaucoup d'autres gens qui m'aiment.

— Tout est de ma faute, murmura Ethel d'une voix entrecoupée de sanglots. Je n'aurais jamais dû lui demander de passer. Je pouvais me douter que ça tournerait mal.

— Ne vous en faites pas, voyons, la réconforta Daisy. J'ai une bonne nouvelle pour vous. »

Lloyd la regarda en souriant. « De quoi s'agit-il ? »

Elle se tourna vers Ethel. « Êtes-vous prête ?

— Je crois.

— Allez, insista Lloyd. Qu'est-ce que c'est ?

— Nous allons avoir un bébé », annonça Daisy.

3.

Erik, le frère de Carla, rentra à la maison cet été-là. Il était aux portes de la mort. Il avait attrapé la tuberculose en Russie, dans un camp de prisonniers, et avait été libéré quand il avait été trop malade pour travailler. Il avait dormi à la dure pendant des semaines, voyagé dans des trains de marchandises et demandé à des camionneurs de lui faire faire un bout de chemin. Il arriva chez les von Ulrich pieds nus et vêtu de haillons crasseux. Il avait un visage cadavérique.

Mais il ne mourut pas. Était-ce de se retrouver parmi des gens qui l'aimaient, le temps plus clément à l'approche du printemps, ou seulement le repos ? Toujours est-il qu'il commença à moins tousser et reprit assez de forces pour pouvoir effectuer de petits travaux dans la maison, clouer des planches sur les fenêtres brisées, réparer les tuiles du toit, déboucher la tuyauterie.

Par bonheur, au début de l'année, Frieda Franck avait déniché un filon en or.

Ludwig Franck avait été tué dans le raid aérien qui avait détruit son usine, et pendant un moment, Frieda et sa mère s'étaient retrouvées dans la même misère que les autres. Mais elle avait obtenu un poste d'infirmière dans le secteur américain, et peu après – ainsi qu'elle l'expliqua à Carla –, un petit groupe de médecins lui avait demandé d'écouler leurs surplus de vivres et de cigarettes au marché noir, en échange d'une partie du produit de la vente. Et depuis, elle arrivait chez Carla une fois par semaine avec un panier contenant des vêtements chauds, des bougies, des piles pour les torches électriques, des allumettes, du savon et des denrées alimentaires – bacon, chocolat, pommes, riz, pêches en conserve. Maud partageait les provisions en portions et en donnait une double ration à Carla. Laquelle acceptait sans scrupules, pas pour elle-même, mais pour pouvoir nourrir son bébé, le petit Walli.

Sans la manne de Frieda, Walli n'aurait peut-être pas survécu.

Il grandissait vite. La tignasse noire de sa naissance avait disparu et laissé place à des cheveux blonds, fins. À six mois, il avait les merveilleux yeux verts de Maud. Alors que son petit visage prenait forme, Carla nota un pli cutané au coin des yeux qui lui faisait des yeux bridés, et elle se demanda si son père n'était pas sibérien. Elle ne se souvenait pas de tous les hommes qui l'avaient violée. La plupart du temps, elle avait fermé les yeux.

Elle ne les détestait plus. C'était bizarre, mais elle était tellement heureuse d'avoir Walli qu'elle n'arrivait même plus à regretter ce qui s'était passé.

Rebecca était fascinée par Walli. Elle avait maintenant tout juste quinze ans, et était en âge d'éprouver les premiers frémissements d'instinct maternel. C'est avec empressement qu'elle aidait Carla à baigner et à habiller le bébé, elle jouait tout le temps avec lui, et il se mettait à glousser de ravissement dès qu'il la voyait.

Lorsque Erik fut suffisamment rétabli, il adhéra au parti communiste.

Carla n'en revenait pas. Après tout ce que les Soviétiques lui avaient fait endurer, comment pouvait-il prendre une décision pareille ? Mais elle se rendit compte qu'il parlait du communisme exactement comme du nazisme dix ans plus tôt. Elle espérait seulement que, cette fois, la désillusion serait plus rapide.

Les Alliés avaient hâte que la démocratie soit rétablie en Allemagne, et il devait y avoir des élections municipales à Berlin un peu plus tard dans l'année.

Carla était sûre que la ville ne retrouverait pas une vie normale tant que son propre peuple n'aurait pas repris son administration en main. Elle décida donc de défendre la cause du parti social-démocrate. Mais les Berlinois ne tarderaient pas à découvrir que les occupants soviétiques avaient une curieuse notion de la démocratie.

Les Russes avaient été déconcertés par le résultat des élections en Autriche, au mois de novembre précédent. Les communistes autrichiens s'attendaient à faire jeu égal avec les socialistes, or ils n'avaient remporté que quatre sièges sur cent soixante-cinq. Apparemment, les électeurs mettaient sur le dos du communisme la barbarie de l'armée Rouge. Le Kremlin, qui n'était pas habitué aux élections libres, ne l'avait pas prévu.

Pour éviter un résultat similaire en Allemagne, les Soviétiques suggérèrent que les communistes et les sociaux-démocrates fusionnent au sein d'un front commun. Les sociaux-démocrates refusèrent, malgré les pressions. En Allemagne de l'Est, les Russes commencèrent à arrêter les sociaux-démocrates, exactement comme les nazis l'avaient fait en 1933, et la fusion fut imposée de force. Mais les élections berlinoises étant supervisées par les quatre Alliés, les sociaux-démocrates réussirent à survivre.

Lorsque le temps se réchauffa, Carla put aller faire la queue pour le ravitaillement comme les autres membres de la maisonnée. Elle portait Walli enroulé dans une taie d'oreiller – elle n'avait pas de vêtements de bébé. Un matin où elle faisait la queue pour des pommes de terre, à quelques rues de chez elle, elle vit s'arrêter une Jeep américaine et eut la surprise de reconnaître Frieda, assise à côté du conducteur. Lequel conducteur, un homme d'un certain âge, aux tempes dégarnies, l'embrassa sur la bouche avant qu'elle descende d'un bond. Elle portait une robe bleue sans manches et des chaussures neuves. Elle s'éloigna rapidement et se dirigea vers la maison des von Ulrich, avec son petit panier.

En un éclair, Carla comprit tout : Frieda ne faisait pas de marché noir, il n'y avait pas de groupe de médecins ; elle était la maîtresse d'un officier américain.

Cela n'avait rien d'exceptionnel. Des milliers de jolies Allemandes s'étaient trouvées devant la même alternative : laisser mourir leur famille ou coucher avec un officier généreux. Les Françaises en avaient fait autant sous l'occupation allemande ; les épouses d'officiers restées en Allemagne s'en étaient suffisamment plaintes.

Carla n'en fut pas moins horrifiée. Elle qui croyait que Frieda aimait Heinrich ! Ils avaient l'intention de se marier dès que la vie aurait retrouvé un semblant de normalité. Elle en avait la mort dans l'âme.

Lorsqu'elle arriva au bout de la queue, elle acheta sa ration de pommes de terre et rentra précipitamment chez elle.

Elle trouva Frieda à l'étage, au salon. Erik avait fait le ménage dans la pièce et mis des journaux aux fenêtres, ce qui était la meilleure solution faute de verre. Les rideaux avaient été depuis longtemps recyclés en draps de lit, mais la plupart des fauteuils étaient toujours là, avec leur tissu fané et élimé, tout comme le piano à queue, miraculeusement. Un officier russe l'avait remarqué et avait annoncé qu'il reviendrait le lendemain avec une grue pour le faire sortir par la fenêtre, mais on ne l'avait jamais revu.

Frieda prit aussitôt Walli des bras de Carla et commença à lui chanter « A, B, C, die Katze lief im Schnee ». Carla avait remarqué que Rebecca et Frieda, qui n'avaient pas encore d'enfants, ne se lassaient jamais de jouer avec le bébé. Alors que Maud et Ada, qui en avaient eu, l'adoraient et s'en occupaient également, mais de façon plus pragmatique.

Frieda souleva le couvercle du piano et encouragea Walli à taper sur les touches pendant qu'elle chantait. Il y avait des années que personne ne l'avait ouvert : Maud n'avait pas joué une note depuis la mort de son dernier élève, Joachim Koch.

Au bout de quelques minutes, Frieda s'adressa à Carla :

« Tu as l'air bien sérieuse aujourd'hui. Qu'y a-t-il ?

— Je sais comment tu obtiens la nourriture que tu nous apportes, répondit Carla. Ce n'est pas au marché noir.

— Bien sûr que si, protesta Frieda. Qu'est-ce que tu racontes ?

— Je t'ai vue descendre d'une Jeep tout à l'heure.

— Le colonel Hicks m'a déposée en passant.

— Il t'a embrassée sur la bouche. »

Frieda détourna le regard. « J'aurais dû descendre plus tôt, je le savais bien. J'aurais pu venir à pied du secteur américain.

— Frieda, et Heinrich ?

— Il ne le saura jamais ! Je serai plus prudente dorénavant, je te le jure.

— Tu l'aimes toujours ?

— Mais bien sûr ! Nous allons nous marier.

— Alors, pourquoi… ?

— J'en ai assez des privations ! Je veux pouvoir mettre de jolis vêtements, aller en boîte de nuit et danser.

— Non, ce n'est pas ça, dit fermement Carla. Tu ne peux pas me raconter d'histoires à moi, Frieda – nous sommes amies depuis trop longtemps. Dis-moi la vérité.

— La vérité ?

— Oui, s'il te plaît.

— Tu es sûre ?

— Oui.

— Je l'ai fait pour Walli. »

Carla étouffa un hoquet de surprise. Cela ne lui serait jamais venu à l'esprit, mais cela tenait debout. Frieda était certainement capable de faire un tel sacrifice pour son bébé et pour elle.

Elle se sentit affreusement coupable. C'était en quelque sorte sa faute si Frieda se prostituait.

« C'est affreux ! s'exclama Carla. Tu n'aurais jamais dû faire ça. Nous aurions réussi à nous en sortir, d'une façon ou d'une autre. »

Frieda bondit du tabouret de piano, tenant toujours le bébé.

« Non, ce n'est pas vrai ! » s'emporta-t-elle.

Effrayé, Walli se mit à pleurer. Carla le prit dans ses bras, le berça, lui tapota le dos.

« Non, tu n'y serais pas arrivée, poursuivit Frieda, plus calmement.

— Comment peux-tu dire ça ?

— Pendant tout l'hiver dernier, des bébés sont arrivés à l'hôpital, nus, enveloppés dans des journaux, morts de faim et de froid. Ça me brisait le cœur de les voir.

— Oh mon Dieu, fit Carla en serrant Walli plus fort.

— Ils prennent une couleur bleue bien particulière quand ils meurent gelés.

— Arrête.

— Il faut que je te le dise, sans quoi tu ne comprendras pas ce que j'ai fait. Walli aurait été un de ces bébés bleus, morts de froid.

— Je sais, souffla Carla. Je sais.

— Percy Hicks est gentil. C'est un brave homme qui a une femme un peu popote à Boston, et je suis la créature la plus affriolante qu'il ait vue de sa vie. Il est tendre et rapide pendant l'amour, et il utilise toujours un préservatif.

— Tu devrais arrêter.

— Tu ne penses pas ce que tu dis.

— Non, avoua Carla. Je ne le pense pas, et c'est ce qu'il y a de pire. Je me sens tellement coupable. Je suis coupable.

— Mais non. C'est moi qui l'ai voulu. Les Allemandes ont des choix douloureux à faire. Nous payons les décisions faciles que les hommes de notre pays ont prises il y a quinze ans. Des hommes comme mon père, qui pensaient que l'arrivée d'Hitler au pouvoir serait bonne pour les affaires, ou comme le père d'Heinrich, qui ont voté la loi sur les pleins pouvoirs. Les péchés des pères retombent sur les filles. »

Un coup violent retentit à la porte d'entrée. Un instant plus tard, ils entendirent Rebecca grimper l'escalier quatre à quatre pour se cacher, craignant l'armée Rouge.

Et puis la voix d'Ada : « Oh, monsieur ! Bonjour ! » Elle avait l'air surprise et vaguement inquiète, mais pas effrayée. Carla se demanda qui pouvait bien inspirer cette réaction à leur domestique.

Un pas masculin monta pesamment les marches, et Werner apparut sur le seuil.

Il était sale, déguenillé et maigre comme un clou, mais son beau visage arborait un grand sourire : « C'est moi ! s'écria-t-il avec exubérance. Je suis rentré ! »

Et puis il vit le bébé. Sa mâchoire tomba, et son sourire radieux s'effaça.

« Oh, bafouilla-t-il. Que… qui… À qui est ce bébé ?

— À moi, mon chéri, répondit Carla. Laisse-moi t'expliquer.

— M'expliquer quoi ? répliqua-t-il, furieux. Il n'y a rien à expliquer ? Tu as eu un enfant d'un autre ! »

Il tourna les talons.

Frieda se dressa d'un bond : « Werner ! Dans cette pièce, il y a deux femmes qui t'aiment. Ne t'en va pas avant de nous avoir écoutées. Tu ne comprends pas.

— Je crois que je comprends très bien.

— Carla a été violée. »

Il blêmit. « Violée ? Par qui ?

— Je n'ai jamais su leurs noms, répondit Carla.

— Leurs noms ? répéta Werner en déglutissant péniblement. Il… Il y en a eu plusieurs ?

— Cinq. Des soldats de l'armée Rouge.

— Cinq ? » fit-il d'une voix réduite à un souffle.

Carla hocha la tête.

« Mais… tu n'aurais pas pu… Je veux dire…

— Werner, moi aussi, j'ai été violée, intervint Frieda. Maman également.

— Bon sang ! Mais qu'est-ce qui s'est passé, ici ?

— C'était l'enfer », répondit Frieda.

Werner se laissa tomber lourdement sur un fauteuil de cuir usé. « Moi qui croyais que l'enfer, c'était là où j'étais », murmura-t-il.

Il enfouit son visage dans ses mains.

Carla traversa la pièce, Walli dans ses bras, et se campa devant le fauteuil de Werner : « Regarde-moi, Werner. Je t'en prie. »

Il leva vers elle un visage ravagé.

« L'enfer, c'était hier, dit-elle.

— Vraiment ?

— Oui, répondit-elle fermement. La vie est dure, mais les nazis sont partis, la guerre est finie. Hitler est mort, et les violeurs de l'armée Rouge ont été ramenés à l'ordre, plus ou moins. Le cauchemar est terminé. Nous sommes tous les deux en vie, et ensemble. »

Il tendit le bras et lui prit la main.

« C'est vrai.

— Nous avons Walli, et dans une minute, tu vas faire la connaissance de Rebecca. Elle a quinze ans et c'est un peu ma fille à présent. Il va falloir fonder une famille avec ce que la guerre nous a laissé, exactement comme il va falloir reconstruire des maisons avec les décombres entassés dans les rues. »

Il hocha la tête en signe d'acquiescement.

« J'ai besoin de ton amour, poursuivit-elle. Rebecca et Walli aussi. »

Il se releva lentement. Elle le regardait d'un air plein d'espoir. Il ne répondit pas, mais au bout d'un long moment, il la serra dans ses bras, le bébé avec elle, les enlaçant tous les deux avec une infinie tendresse.

4.

Les réglementations de guerre étant toujours en vigueur, le gouvernement britannique avait le droit d'ouvrir des mines de charbon partout, avec ou sans l'assentiment du propriétaire du terrain. Des indemnités n'étaient versées qu'en cas de perte de revenus lorsqu'il s'agissait de terres agricoles ou de propriétés commerciales.

En tant que ministre du Charbon, Billy Williams autorisa l'exploitation d'une houillère à ciel ouvert sur le domaine de Tŷ Gwyn, propriété ancestrale du comte Fitzherbert, à la périphérie d'Aberowen.

Aucune compensation n'était due, puisqu'il s'agissait d'un parc d'agrément.

Un tollé s'éleva des bancs des conservateurs, à la Chambre des communes. « Votre crassier se trouvera juste sous les fenêtres de la chambre à coucher de la comtesse ! » protesta un tory indigné.

Billy Williams sourit. « Il y a cinquante ans que les fenêtres de ma mère donnent sur le crassier du comte », rétorqua-t-il.

Lloyd Williams et Ethel accompagnèrent Billy à Aberowen, la veille du jour où les ingénieurs devaient commencer les travaux d'excavation. Lloyd hésitait à laisser Daisy, qui devait accoucher deux semaines plus tard, mais c'était un moment historique et il tenait à y assister.

Ses deux grands-parents avaient maintenant près de quatre-vingts ans. Granda était presque aveugle, malgré ses lunettes en cul de bouteille, et Grandmam était toute voûtée. « Mes deux enfants sont là. C'est bien agréable », se réjouit-elle quand ils furent tous assis autour de la vieille table de la cuisine. Elle leur servit un ragoût de bœuf avec une purée de navets et de grosses tranches de pain qu'elle faisait elle-même, sur lesquelles était tartinée la graisse figée du jus de cuisson. Le tout arrosé de grandes tasses de thé au lait sucré.

Lloyd avait souvent fait ce genre de repas quand il était petit, mais trouvait maintenant que ce menu manquait de raffinement. Il savait que même dans les périodes les plus difficiles, en France et en Espagne, les femmes réussissaient à servir des plats savoureux, délicatement assaisonnés d'ail et agrémentés d'épices. Il avait honte de se montrer aussi difficile, et faisait semblant de se régaler.

« C'est quand même dommage pour les jardins de Tŷ Gwyn », lâcha abruptement Grandmam.

Billy fut piqué au vif. « Qu'est-ce que tu racontes ? L'Angleterre a besoin de charbon.

— Mais les gens aiment ce parc. Il est si beau. J'y vais au moins une fois par an depuis que je suis petite fille. Quel dommage qu'il disparaisse !

— Il y a un jardin public tout à fait agréable au centre d'Aberowen !

— Ce n'est pas pareil, s'obstina Grandmam.

— Les femmes ne comprendront jamais rien à la politique, observa Granda.

— C'est sans doute vrai », répondit Grandmam.

Lloyd croisa le regard de sa mère. Elle lui sourit sans rien dire.

Billy et Lloyd partagèrent la deuxième chambre, tandis qu'Ethel se fit un lit par terre dans la cuisine.

« J'ai dormi dans cette chambre toutes les nuits de ma vie, jusqu'à mon départ pour l'armée, confia Billy à Lloyd alors qu'ils s'allongeaient. Et tous les matins, en regardant par la fenêtre, j'ai vu ce putain de crassier.

— Pas si fort, oncle Billy, chuchota Lloyd. Tu ne voudrais pas que ta mère t'entende jurer.

— Zut, tu as raison ».

Le lendemain matin, après le petit déjeuner, ils montèrent tous à pied vers le château. Il faisait doux, et pour une fois, il ne pleuvait pas. Le profil de la crête montagneuse, à l'horizon, était adouci par l'herbe d'été. Alors que Tŷ Gwyn apparaissait à leurs regards, Lloyd ne put s'empêcher d'y voir une belle demeure plus qu'un symbole d'oppression. C'étaient les deux, bien sûr : rien n'était simple, en politique.

Le grand portail de fer forgé était ouvert. La famille Williams entra dans le parc où une foule s'était déjà rassemblée : les ouvriers de l'entreprise avec leurs machines, une centaine de mineurs accompagnés de leurs femmes et de leurs enfants, le comte Fitzherbert et son fils Andrew, une poignée de journalistes avec leurs calepins et une équipe de tournage.

Les jardins étaient d'une beauté à couper le souffle. L'allée de vieux châtaigniers formait un berceau de feuillage, des cygnes évoluaient sur le lac et les parterres de fleurs éclataient de mille couleurs. Lloyd se dit que le comte avait dû veiller à ce que le parc se présente sous son plus beau jour pour mieux dénoncer aux yeux du monde le vandalisme du gouvernement travailliste.

Lloyd ne put se défendre d'un sentiment de compassion envers Fitz.

Le maire d'Aberowen s'adressait à la presse : « Les gens de cette ville sont contre cette exploitation à ciel ouvert », disait-il. Lloyd fut étonné : le conseil municipal était travailliste, et ce n'était certainement pas de gaieté de cœur qu'il s'opposait ainsi à une décision gouvernementale. « Depuis plus de cent ans, la beauté de ces jardins rafraîchit l'âme de ceux qui vivent dans un sinistre paysage industriel », poursuivit le maire. Abandonnant son discours préparé pour se laisser aller à des souvenirs personnels, il ajouta : « J'ai demandé la main de ma femme sous ce cèdre. »

Il fut interrompu par un fracas assourdissant, semblable aux pas d'un géant de fer. Se retournant, Lloyd vit une énorme machine approcher dans l'allée. On aurait dit la plus grande grue du monde. Elle était munie d'une immense flèche d'une trentaine de mètres de long et d'un godet dans lequel on aurait pu aisément faire tenir un camion. Mais le plus surprenant était qu'elle avançait sur des soles d'acier rotatives qui ébranlaient le sol chaque fois qu'elles se posaient.

Billy dit fièrement à Lloyd : « C'est une excavatrice Monaghan à pattes et à bennes traînantes. Elle est capable de retirer six tonnes de terre à la fois. »

La caméra commença à filmer tandis que le monstrueux engin remontait l'allée.

Lloyd n'avait qu'un reproche à faire au parti travailliste : il y avait chez beaucoup de socialistes un fond d'autoritarisme puritain. C'était vrai chez son grand-père aussi bien que chez Billy. Les plaisirs des sens leur étaient étrangers. Le sacrifice et l'abnégation leur convenaient davantage. Peu importait à leurs yeux la splendeur de ces jardins. Ils avaient tort.

Ethel n'était pas comme eux, Lloyd non plus. Peut-être cette fibre rabat-joie avait-elle été éliminée de leur lignée. Il l'espérait.

Pendant que le conducteur de l'excavatrice manœuvrait sa machine pour la mettre en position, Fitz accorda une interview sur l'allée de gravier rose : « Le ministre du Charbon vous a dit que quand la mine serait épuisée, le parc ferait l'objet de ce qu'il a appelé un “programme global de restauration du site”. Croyez-moi, c'est une promesse en l'air. Il nous a fallu plus d'un siècle à mon grand-père, à mon père et à moi-même pour parvenir à ce sommet de beauté et d'harmonie. Il faudra encore cent ans pour restaurer ces jardins. »

La flèche de l'excavatrice s'abaissa pour former un angle de quarante-cinq degrés avec les parterres de fleurs et les arbustes du jardin ouest. Le godet était suspendu juste au-dessus du terrain de croquet. La foule attendait, silencieuse, quand Billy s'écria d'une voix forte : « Allez-y, pour l'amour du ciel ! »

Un ingénieur en chapeau melon donna un coup de sifflet.

Le godet s'abattit sur le sol dans un choc épouvantable. Les dents d'acier mordirent dans le gazon vert. Le câble de traction se raidit, on entendit un grincement assourdissant de rouages, et le godet commença à reculer. En raclant le sol, il arracha un parterre d'énormes tournesols jaunes, un massif de clèthre et de pavier, la roseraie et un petit magnolia. À la fin de son parcours, l'auge était pleine de terre, de fleurs et de plantes.

Le godet s'éleva ensuite à six mètres de hauteur, dans une pluie de terre et de végétaux.

La flèche pivota latéralement. Lloyd remarqua qu'elle était plus haute que le château. Il craignit un instant qu'elle fracasse les fenêtres de l'étage supérieur, mais le conducteur était habile, et s'arrêta juste à temps. Le câble se tendit, l'auge s'inclina, et six tonnes de jardin retombèrent à quelques mètres de l'entrée.

Le godet reprit sa position initiale, et le processus se renouvela.

Lloyd se tourna vers Fitz : il pleurait.

XXIII
1947

1.

Au début de l'année 1947, on pouvait se demander si toute l'Europe n'allait pas se convertir au communisme.

Volodia Pechkov ne savait s'il fallait l'espérer ou souhaiter le contraire.

L'Europe de l'Est était sous la botte de l'armée Rouge, et à l'Ouest, les communistes remportaient une élection après l'autre. Leur résistance au nazisme leur avait valu le respect. Aux élections d'après-guerre, cinq millions de Français avaient voté pour le parti communiste qui était ainsi devenu la formation politique la plus populaire. En Italie, une alliance entre socialistes et communistes avait recueilli quarante pour cent des suffrages. En Tchécoslovaquie, les communistes avaient remporté à eux seuls trente-huit pour cent des voix et dirigeaient un gouvernement démocratiquement élu.

Il en allait différemment en Autriche et en Allemagne, où les électeurs avaient été pillés et violés par l'armée Rouge. Aux élections municipales de Berlin, les sociaux-démocrates avaient obtenu soixante-trois sièges sur cent trente, et les communistes seulement vingt-six. Mais l'Allemagne était ruinée, affamée, et le Kremlin espérait encore que la population aux abois se tournerait vers le communisme, exactement comme elle s'était jetée dans les bras des nazis au moment de la grande Crise.

La vraie déception venait de Grande-Bretagne. Un seul communiste était entré au Parlement à la suite des élections d'après-guerre. Il faut dire que le gouvernement travailliste offrait tout ce que proposait le communisme : la sécurité sociale, la gratuité des soins médicaux, l'éducation pour tous, et même la semaine de cinq jours pour les mineurs des houillères.

Mais dans le reste de l'Europe, le capitalisme était impuissant à sortir les populations du bourbier de l'après-guerre.

Le vent soufflait en faveur de Staline, pensait Volodia alors que les couches de neige s'accumulaient sur les coupoles à bulbe. L'hiver 1946-1947 était le plus froid que l'Europe ait connu depuis plus d'un siècle. Il neigeait à Saint-Tropez. En Angleterre, les routes et les voies ferrées étaient impraticables et l'industrie était paralysée, ce qui n'était jamais arrivé, même durant le conflit. En France, les rations alimentaires étaient inférieures à ce qu'elles avaient été pendant la guerre. L'Organisation des Nations unies avait calculé que cent millions d'Européens devaient se contenter de mille cinq cents calories par jour – le seuil où les effets de la malnutrition commencent à se faire sentir. Alors que les outils de production fonctionnaient au ralenti, les gens commençaient à se dire qu'ils n'avaient rien à perdre, et la révolution semblait la seule issue.

Dès que l'URSS posséderait l'arme nucléaire, aucun pays ne pourrait plus se dresser sur son chemin. Zoïa et ses collègues avaient construit une pile atomique au laboratoire numéro deux de l'Académie des sciences – cette dénomination délibérément vague désignait le cœur de la recherche nucléaire soviétique. La masse critique avait été atteinte le jour de Noël, six mois après la naissance de Konstantin, qui dormait pour le moment dans la crèche du laboratoire. Zoïa avait chuchoté à l'oreille de Volodia que si l'expérience tournait mal, le fait de se trouver à deux ou trois kilomètres de là ne changerait pas grand-chose pour le petit Kotia : tout le centre de Moscou serait rasé.

Les sentiments conflictuels que l'avenir inspirait à Volodia avaient pris une nouvelle intensité avec la naissance de son fils. Il voulait que Kotia grandisse dans un pays fier et puissant. Il estimait que l'Union soviétique méritait de dominer l'Europe. C'était l'armée Rouge qui avait vaincu les nazis, après quatre cruelles années de guerre totale. Les autres Alliés étaient prudemment restés en coulisse, livrant de petits combats et ne rejoignant les Soviétiques que pendant les onze derniers mois. Toutes leurs pertes humaines additionnées ne constituaient qu'une fraction de celles que déplorait le peuple soviétique.

Mais il songeait aussi aux réalités du communisme : aux purges arbitraires, à la torture dans les sous-sols de la police secrète, aux soldats conquérants poussés à des excès de bestialité, à cet immense pays obligé d'obéir aux décisions erratiques d'un tyran plus puissant qu'un tsar. Volodia voulait-il vraiment que ce système impitoyable s'étende au reste du continent ?

Il se rappelait être entré dans Penn Station, à New York, et avoir acheté un billet pour Albuquerque sans demander l'autorisation à personne, sans avoir à présenter ses papiers d'identité, et n'avait pas oublié l'impression grisante de liberté absolue qu'il avait éprouvée. Il avait depuis longtemps brûlé le catalogue Sears Roebuck, mais celui-ci hantait toujours sa mémoire, avec ses centaines de pages d'articles à la disposition de tous. Le peuple russe croyait que les histoires de liberté et de prospérité occidentales n'étaient que propagande, mais Volodia savait à quoi s'en tenir. Au fond de lui-même, il avait du mal à ne pas espérer la défaite du communisme.

L'avenir de l'Allemagne, et donc de l'Europe, serait réglé à la conférence des ministres des Affaires étrangères organisée à Moscou en mars 1947.

Volodia, désormais colonel, était responsable de l'équipe du Renseignement affectée à la conférence. Les réunions avaient lieu dans une salle richement décorée de la Maison de l'industrie aéronautique, commodément située près de l'hôtel Moskva. Comme toujours, les délégués et leurs interprètes siégeaient autour d'une table, leurs assistants assis derrière eux, sur plusieurs rangées de chaises. Le ministre des Affaires étrangères soviétique, Viatcheslav Molotov, le vieux Cul-de-Pierre, exigeait de l'Allemagne dix milliards de dollars à titre de réparations. Les Américains et les Britanniques protestaient : cela porterait un coup mortel à l'économie allemande moribonde. C'était probablement ce que voulait Staline.

Volodia retrouva Woody Dewar, devenu photographe de presse et chargé de couvrir la conférence. Il était marié, lui aussi, et montra à Volodia une photo d'une femme brune, à la beauté saisissante, tenant un bébé dans ses bras. Assis à l'arrière d'une limousine ZIS-11OB, alors qu'il revenait d'une séance de photos officielles au Kremlin, Woody fit remarquer à Volodia :

« Vous vous rendez bien compte que l'Allemagne n'a pas de quoi vous payer ces réparations, n'est-ce pas ?

— Dans ce cas, comment nourrissent-ils leur population et avec quoi rebâtissent-ils leurs villes ? » rétorqua Volodia.

Il avait fait des progrès en anglais, et ils pouvaient discuter sans interprète.

« Grâce aux aides que nous leur versons, évidemment, répondit Woody. Ça nous coûte une fortune. Toutes les réparations que les Allemands vous verseraient seraient en réalité payées avec notre argent.

— Et alors ? Les États-Unis ont bien profité de la guerre. Mon pays a été dévasté. Il n'est peut-être pas anormal que vous mettiez la main à la poche.

— Ce n'est pas ce que pensent les électeurs américains.

— Les électeurs américains ont peut-être tort. »

Woody haussa les épaules. « C'est vrai… mais c'est leur argent. »

Et voilà, c'était reparti, songea Volodia : il ne comprendrait jamais cette complaisance à l'égard de l'opinion publique qu'il avait déjà eu l'occasion de relever en discutant avec Woody. Les Américains parlaient des électeurs comme les Russes de Staline : il fallait leur céder, qu'ils aient tort ou raison.

Woody baissa la vitre. « Ça ne vous ennuie pas que je prenne une photo de la ville ? La lumière est merveilleuse. » Il appuya sur le déclencheur.

Il savait qu'il était censé ne prendre que des photos autorisées. Mais il n'y avait rien de compromettant dans la rue, seulement des femmes qui pelletaient la neige. Volodia dit malgré tout : « Ne faites pas ça, s'il vous plaît. » Il se pencha par-dessus Woody et remonta sa vitre. « Des photos officielles, c'est tout. »

Il était sur le point de demander à Woody de lui remettre sa pellicule quand celui-ci changea de sujet : « Vous vous rappelez que je vous ai parlé de mon ami Greg Pechkov, qui porte le même nom que vous ? »

Volodia ne risquait pas de l'avoir oublié. Willi Frunze lui avait fait une réflexion similaire. Il s'agissait probablement du même homme. « Non, je ne m'en souviens pas. » Volodia mentait. Il ne voulait pas avoir affaire à un éventuel parent à l'Ouest.

Ce genre de lien ne valait généralement aux Russes que suspicion et ennuis.

« Il fait partie de la délégation américaine. Vous devriez lui parler. Voir si vous êtes de la même famille.

— Je n'y manquerai pas », répondit Volodia, bien résolu à éviter ce type à tout prix.

Il décida de ne pas réclamer la pellicule de Woody. Inutile de faire des histoires pour une innocente scène de rue.

À la conférence du lendemain, le secrétaire d'État américain, George Marshall, proposa que les quatre Alliés suppriment la division de l'Allemagne en zones d'occupation et la réunifient, afin qu'elle redevienne le cœur économique de l'Europe et qu'elle produise, exploite ses mines, achète et vende.

C'était la dernière chose que voulaient les Russes.

Molotov refusa d'aborder ce point tant que la question des réparations ne serait pas réglée.

La conférence était au point mort.

Ce qui, pensa Volodia, était exactement ce que désirait Staline.

2.

Décidément, la diplomatie internationale était un tout petit monde, se disait Greg Pechkov. L'un des jeunes assistants de la délégation britannique à la conférence de Moscou n'était autre que Lloyd Williams, le mari de Daisy, sa demi-sœur. Au début, l'allure de Lloyd et son élégance de gentleman anglais déplurent à Greg, mais il découvrit que c'était un brave type. « Molotov est un con », lui dit Lloyd, au bar de l'hôtel Moskva, après deux vodkas martinis.

« Sans doute, mais que faire ?

— Je n'en sais rien. Une chose est sûre : l'Angleterre ne supportera pas plus longtemps ces manœuvres dilatoires. L'occupation de l'Allemagne coûte des sommes astronomiques, et cet hiver rigoureux a changé le problème en crise.

— Vous voulez que je vous dise ? fit Greg, réfléchissant tout haut. Si les Russes ne veulent pas jouer le jeu, nous n'avons qu'à aller de l'avant sans eux.

— Mais comment faire ?

— Que voulons-nous ? Unifier l'Allemagne et organiser des élections, commença Greg en comptant sur ses doigts.

— Nous aussi.

— Retirer de la circulation le reichsmark qui n'a plus aucune valeur et introduire une nouvelle monnaie, pour que les Allemands puissent recommencer à faire du commerce.

— Oui.

— Et sauver le pays du communisme.

— C'est aussi la politique britannique.

— Nous ne pouvons pas faire ça à l'Est, parce que les Soviétiques refuseront de participer. Qu'ils aillent au diable ! Nous contrôlons les trois quarts de l'Allemagne – accomplissons tout ça dans nos zones, et que l'Est du pays aille se faire voir. »

Lloyd eut l'air songeur. « Vous en avez discuté en haut lieu ?

— Certainement pas ! Je me contente de cogiter tout haut. Cela dit, pourquoi pas ?

— Je pourrais en parler à Ernie Bevin.

— Et moi, j'en toucherai un mot à George Marshall. » Greg prit une petite gorgée de son verre. « La vodka, c'est la seule chose que les Russes sachent faire convenablement, remarqua-t-il. Alors, comment va ma sœur – Daisy ?

— Elle est enceinte de notre deuxième bébé.

— Et comment est-elle, en mère de famille ? »

Lloyd se mit à rire. « Vous devez penser qu'elle n'est vraiment pas faite pour ça.

— J'ai un peu de mal à la voir dans ce rôle, c'est vrai, répondit Greg avec un haussement d'épaule.

— Elle se montre pourtant patiente, calme, organisée.

— Elle n'a pas embauché six nounous pour faire tout le boulot à sa place ?

— Une seule, pour pouvoir m'accompagner à des soirées, des réunions politiques, le plus souvent.

— Ça alors ! Elle a bien changé.

— Pas complètement. Elle aime encore faire la fête. Et vous ? Toujours célibataire ?

— Il y a une fille, Nelly Fordham, avec qui ça devient vraiment sérieux. Et j'imagine que vous savez que j'ai un filleul.

— Oui, répondit Lloyd. Daisy m'a parlé de lui. Georgy. »

Greg eut la conviction, à en juger par l'air vaguement gêné de Lloyd, qu'il savait que Georgy était son fils.

« Je l'aime beaucoup.

— C'est formidable. »

Un membre de la délégation russe s'approcha du bar, et son regard se posa sur Greg. Il y avait quelque chose de familier dans son apparence. La trentaine, bel homme, malgré une coupe de cheveux militaire trop courte, un regard bleu presque intimidant. Il esquissa un hochement de tête amical, et Greg l'aborda : « On ne se serait pas rencontrés quelque part ?

— Peut-être, répondit le Russe. J'étais au lycée en Allemagne – à l'Académie de garçons, à Berlin. »

Greg secoua la tête. « Vous n'êtes jamais venu aux États-Unis ?

— Non.

— C'est l'homme qui porte le même nom que vous, Volodia Pechkov », intervint Lloyd.

Greg se présenta : « Il n'est pas impossible que nous soyons parents. Mon père, Lev Pechkov, a émigré en 1914, laissant derrière lui une petite amie qui attendait un enfant de lui et qui a ensuite épousé son frère aîné, Grigori Pechkov. Se pourrait-il que nous soyons demi-frères ? »

L'attitude de Volodia changea aussitôt : « En aucun cas, lança-t-il. Excusez-moi. » Il quitta le bar sans commander à boire.

« Pas très causant, commenta Greg.

— En effet, convint Lloyd.

— Il avait même l'air un peu retourné.

— Ça doit être quelque chose que vous avez dit. »

3.

C'était impossible, songeait Volodia.

À en croire Greg, Grigori aurait épousé une fille qui était déjà enceinte de Lev. Si tel était le cas, l'homme que Volodia avait toujours appelé Papa n'était pas son père mais son oncle.

Cette similitude de noms n'était peut-être qu'une coïncidence. À moins que l'Américain n'ait simplement voulu semer le trouble dans son esprit.

Quoi qu'il en soit, Volodia en était encore ébranlé.

Il rentra chez lui à l'heure habituelle. Ils gravissaient rapidement l'échelle sociale, Zoïa et lui, et on leur avait attribué un appartement dans la Maison du gouvernement, l'immeuble luxueux où habitaient ses parents. Comme presque tous les soirs, Grigori et Katerina arrivèrent chez eux au moment où Kotia prenait son dîner. Katerina donna le bain à son petit-fils, puis Grigori lui chanta des chansons et lui raconta des légendes russes. Kotia avait neuf mois, il ne parlait pas encore, mais prenait visiblement plaisir à écouter des histoires avant de s'endormir.

Volodia se conforma aux rituels habituels de la soirée comme un somnambule. Il avait beau essayer de se comporter normalement, il avait le plus grand mal à adresser la parole à ses parents. Il ne croyait pas un mot du récit de Greg, mais ne pouvait s'empêcher d'y penser.

Lorsque Kotia fut endormi et que les grands-parents firent mine de se lever pour rentrer chez eux, Grigori demanda à Volodia : « J'ai un bouton sur le nez ?

— Non.

— Alors, pourquoi est-ce que tu m'as regardé comme ça pendant toute la soirée ? »

Volodia décida de dire la vérité. « J'ai rencontré un type qui s'appelle Greg Pechkov. Il fait partie de la délégation américaine. Il pense que nous sommes apparentés.

— C'est possible, répondit Grigori d'un ton insouciant, comme si cela n'avait pas grande importance, mais Volodia remarqua que son cou avait rougi, signe révélateur d'une émotion réprimée. La dernière fois que j'ai vu mon frère, c'était en 1919. Depuis, je n'ai plus jamais eu de ses nouvelles.

— Le père de Greg s'appelle Lev, et Lev avait un frère appelé Grigori.

— Il se pourrait donc que ce Greg soit ton cousin.

— Mon frère, d'après lui. »

Grigori ne répondit pas, mais sa rougeur s'accentua.

« Que veux-tu dire ? intervint Zoïa.

— D'après ce Pechkov américain, répondit Volodia, Lev avait une petite amie à Saint-Pétersbourg, elle attendait un enfant de lui et aurait épousé son frère après son départ.

— Ridicule ! » s'exclama Grigori.

Volodia se tourna vers Katerina. « Tu n'as rien dit, Mamotchka ? »

Il y eut un long silence. Révélateur, se dit Volodia. Pourquoi hésitaient-ils si l'histoire de Greg ne comportait aucune part de vérité ? Un froid étrange s'empara de lui comme un brouillard glacé.

Sa mère se décida enfin : « J'étais une fille volage. Je n'avais pas la tête sur les épaules, comme ta femme », ajouta-t-elle avec un coup d'œil en direction de Zoïa.

Elle poussa un profond soupir et poursuivit : « Grigori Pechkov est tombé amoureux de moi. Il a eu le coup de foudre, le pauvre idiot, dit-elle avec un sourire attendri en regardant son mari. Mais voilà. Son frère, Lev, avait de beaux habits, des cigarettes, de l'argent pour acheter de la vodka, des amis un peu louches. Il me plaisait beaucoup plus. J'étais encore plus idiote que lui, comme tu vois.

— Alors, c'est vrai ? » demanda Volodia, stupéfait. Il ne pouvait s'empêcher d'espérer encore un démenti.

« Lev a fait ce que font tous les hommes de son genre, reprit Katerina. Il m'a mise enceinte et il m'a quittée.

— Autrement dit, mon père, c'est Lev. Et tu n'es que mon oncle ! » s'exclama Volodia en regardant Grigori. Il avait l'impression qu'il allait tomber à la renverse. Le sol s'effondrait sous ses pieds. C'était un véritable tremblement de terre.

Zoïa s'approcha de la chaise de Volodia et posa la main sur son épaule, debout à ses côtés, comme pour l'apaiser, ou peut-être le retenir.

Katerina poursuivit : « Et Grigori a fait ce que font toujours les hommes comme lui : il s'est occupé de moi. Il m'aimait, il m'a épousée, et il a subvenu à nos besoins, les miens et ceux de mes enfants. » Assise sur le canapé, à côté de Grigori, elle lui prit la main. « Ce n'était pas lui que je voulais, et je ne le mérite sûrement pas, mais c'est lui que le destin m'a donné.

— J'ai toujours redouté ce jour, murmura Grigori. Depuis ta naissance, je l'ai redouté.

— Alors, pourquoi as-tu gardé le secret ? demanda Volodia. Pourquoi ne pas m'avoir dit la vérité, tout simplement ? »

Grigori répondit d'une voix étranglée : « Je ne pouvais pas supporter de t'avouer que je n'étais pas ton père, articula-t-il avec difficulté. Je t'aimais trop.

— Permets-moi de te dire une chose, mon fils chéri, reprit Katerina. Tu vas m'écouter, et tant pis si c'est la dernière fois, mais je tiens à ce que tu entendes ceci : oublie l'étranger qui a séduit une fille stupide et s'est enfui en Amérique. Regarde l'homme qui est assis devant toi, les yeux pleins de larmes. »

Volodia se tourna vers Grigori, et son expression implorante lui perça le cœur.

Katerina poursuivit : « Cet homme t'a nourri, il t'a habillé et t'a aimé sans faillir pendant trois décennies. Si le mot *père* veut dire quelque chose, ton père, c'est lui.

— Oui, répondit Volodia. Je le sais. »

4.

Lloyd Williams s'entendait bien avec Ernie Bevin. Ils avaient beaucoup de points communs, malgré la différence d'âge. Au cours de leurs quatre jours de voyage en train à travers l'Europe enneigée, Lloyd lui avait confié que, comme lui, il était le fils illégitime d'une femme de chambre. Ils étaient tous les deux farouchement anticommunistes : Lloyd à cause de ce qu'il avait vécu en Espagne, Bevin parce qu'il avait vu les communistes à l'œuvre dans le mouvement syndicaliste. « Ils sont les esclaves du Kremlin et des tyrans pour tous les autres », déclara Bevin, et Lloyd ne pouvait que lui donner raison.

Lloyd n'avait pas beaucoup de sympathie pour Greg Pechkov : il donnait perpétuellement l'impression de s'être habillé à la hâte avec ses poignets de chemise déboutonnés, son col de veston tortillé et ses lacets de chaussures défaits. Greg était astucieux, et Lloyd essayait de l'apprécier, mais sentait que son charme facile dissimulait un noyau de dureté. Daisy lui avait dit que Lev Pechkov était un gangster, et Lloyd se demandait si Greg ne tenait pas un peu de lui.

Pourtant, Bevin adopta avec enthousiasme l'idée de Greg pour l'Allemagne : « Vous pensez qu'il s'exprimait au nom de

Marshall ? » lui demanda le ventripotent ministre des Affaires étrangères avec son accent traînant de l'ouest du pays.

« Il m'a dit que non, répondit Lloyd. Vous croyez que ça pourrait marcher ?

— Je crois que c'est la meilleure idée que j'aie entendue en trois putains de semaines dans cette putain de Moscou. S'il est sérieux, organisez un déjeuner informel, juste Marshall, ce jeune homme, vous et moi.

— Je m'en occupe tout de suite.

— Mais n'en parlez à personne. Pas question que les Soviétiques soient au courant. Ils nous accuseraient, à juste titre, de conspirer contre eux. »

Ils se retrouvèrent le lendemain à la résidence de l'ambassadeur des États-Unis, au 10, place Spassopeskovskaïa, une demeure néoclassique extravagante construite avant la Révolution. Marshall était grand et mince, le militaire dans toute sa splendeur, Bevin, ventripotent, myope, la cigarette souvent pendue aux lèvres, mais le courant passa immédiatement. C'étaient deux hommes qui ne mâchaient pas leurs mots. Staline lui-même avait accusé un jour Bevin de ne pas s'exprimer comme un gentleman, distinction dont le ministre des Affaires étrangères était très fier. Sous les plafonds peints et les lustres à pendeloques, ils s'attaquèrent à la tâche de remettre l'Allemagne debout sans l'aide de l'URSS.

Ils tombèrent rapidement d'accord sur les principes : la nouvelle monnaie, l'unification des zones britannique, américaine et – si possible – française, la démilitarisation de l'Allemagne de l'Ouest, l'organisation d'élections et la création d'une nouvelle alliance militaire transatlantique. C'est alors que Bevin lança sans ménagement : « Rien de tout ça ne marchera, vous savez. »

Marshall eut l'air ahuri. « Alors, je ne vois pas pourquoi nous discutons, remarqua-t-il sèchement.

— L'Europe est dans le pétrin. Ce plan échouera si les gens meurent de faim. La meilleure protection contre le communisme, c'est la prospérité. Staline le sait bien : c'est pour ça qu'il veut que l'Allemagne reste dans la misère.

— Je suis d'accord.

— Ce qui veut dire que nous devons reconstruire. Mais nous ne pouvons pas le faire à mains nues. Nous avons besoin de tracteurs, de bulldozers, de matériel roulant : autant de choses que nous n'avons pas les moyens de payer. »

Marshall comprit où il voulait en venir. « Les Américains n'ont pas l'intention de continuer à faire la charité aux Européens.

— C'est compréhensible. Mais il faudrait trouver un moyen pour que les États-Unis nous prêtent l'argent dont nous avons besoin pour leur acheter du matériel. »

Le silence se fit. Marshall n'était pas du genre à parler pour ne rien dire, mais l'interruption dura longtemps, même pour un homme comme lui.

Enfin, il reprit la parole : « Ça se tient. Je vais voir ce que je peux faire. »

La conférence dura six semaines, et quand chacun regagna son pays, rien n'avait été décidé.

5.

Eva Williams avait un an quand ses molaires poussèrent. Les autres dents avaient percé assez facilement, mais celles-là la firent souffrir. Lloyd et Daisy ne pouvaient pas faire grand-chose pour elle. Elle était malheureuse, elle n'arrivait pas à dormir, elle les empêchait de dormir, et ils étaient malheureux aussi.

Daisy avait beaucoup d'argent, mais ils vivaient sans ostentation. Ils avaient acheté à Hoxton une agréable maison et avaient pour voisins un commerçant et un entrepreneur en bâtiment. Ils avaient une petite voiture familiale, une Morris Eight neuve, qui pouvait atteindre cent kilomètres à l'heure. Daisy s'achetait encore des jolis vêtements, mais Lloyd n'avait que trois costumes : un habit de soirée, un costume rayé pour la Chambre des communes et un autre en tweed pour arpenter sa circonscription, le week-end.

Il était tard, et Lloyd était en pyjama en train de bercer la petite Evie grincheuse dans l'espoir de l'endormir tout en feuilletant le magazine *Life*, lorsqu'il tomba sur une photo remarquable prise à Moscou. On y voyait une Russe, un fichu sur la tête, son manteau ceinturé par une ficelle comme un paquet, son vieux visage profondément ridé, en train de pelleter de la neige dans la rue. Quelque chose dans la façon dont la lumière tombait sur elle lui donnait l'air d'avoir échappé au temps, d'être là depuis un millier d'années. Il chercha le nom du photographe : Woody Dewar, le type qu'il avait rencontré à la conférence.

Le téléphone sonna. Il décrocha. C'était Ernie Bevin. « Allumez votre radio, dit-il. Marshall est en train de faire un discours. » Il raccrocha sans lui laisser le temps de répondre.

Lloyd descendit au salon, Evie toujours dans les bras, et alluma la radio. L'émission s'appelait « American Commentary ». Le correspondant de la BBC à Washington, Leonard Miall, parlait depuis l'université de Harvard, à Cambridge, dans le Massachusetts. « Le secrétaire d'État a annoncé aux anciens étudiants que la reconstruction de l'Europe prendrait plus longtemps et exigerait plus d'efforts que prévu. »

Voilà qui était prometteur, s'enthousiasma Lloyd. « Chut, Evie, je t'en prie », dit-il. Et, miracle, elle se tut.

Lloyd entendit ensuite la voix grave, raisonnable, de George C. Marshall. « Les besoins de l'Europe pour les trois ou quatre années à venir, en vivres et autres produits de première nécessité importés de l'étranger – surtout d'Amérique – sont plus grands que sa capacité de paiement. Il est nécessaire d'envisager une aide supplémentaire, sous peine de s'exposer à une dislocation économique, sociale et politique très grave. »

Lloyd était galvanisé. *Une aide supplémentaire*, c'était ce que Bevin avait demandé.

« Il faut briser le cercle vicieux et rétablir la confiance du peuple européen dans l'avenir économique, poursuivit Marshall. Les États-Unis doivent faire tout ce qui est en leur pouvoir pour permettre le redressement de l'économie mondiale. »

« Il l'a fait ! s'écria triomphalement Lloyd, s'adressant à sa petite fille ébahie. Il a dit à l'Amérique qu'elle devait nous accorder une nouvelle aide ! Mais combien ? Comment ? Quand ? »

La voix changea et le journaliste reprit la parole : « Le secrétaire d'État n'a pas défini de plan détaillé d'aide à l'Europe ; il a déclaré que c'était aux Européens de rédiger ce programme.

— Veut-il dire que nous avons carte blanche ? » demanda Lloyd à Evie.

La voix de Marshall se fit à nouveau entendre : « Je pense que l'initiative doit venir de l'Europe. »

Le reportage s'acheva, et le téléphone sonna à nouveau. « Vous avez entendu ? demanda Bevin.

— Oui, mais qu'est-ce que ça veut dire ?

— Pas de questions ! s'exclama Bevin. Posez des questions, et vous obtiendrez des réponses qui ne vous plairont pas.

— Très bien, fit Lloyd, déconcerté.

— Peu importe ce qu'il a voulu dire. La vraie question est : "Qu'allons-nous faire ?" L'initiative doit venir de l'Europe, vous l'avez entendu. Autrement dit, de vous et moi.

— Que puis-je faire ?

— Votre valise. Nous allons à Paris. »

XXIV
1948

1.

Volodia était à Prague avec une délégation de l'armée Rouge venue discuter avec l'armée tchèque. Ils séjournaient à l'hôtel Impérial, splendeur Art déco de la ville.

Il neigeait.

Zoïa lui manquait, et le petit Kotia aussi. Son fils avait deux ans, et apprenait à parler à une vitesse stupéfiante. L'enfant changeait si rapidement qu'il semblait chaque jour différent de la veille. Et Zoïa était à nouveau enceinte. Volodia n'appréciait pas du tout de devoir passer deux semaines loin de sa famille. La plupart des hommes du groupe considéraient ce voyage comme une occasion d'échapper un moment à leurs épouses, de boire trop de vodka et peut-être de se laisser aller à quelques fredaines avec des femmes de petite vertu. Volodia, quant à lui, ne demandait qu'une chose : rentrer chez lui.

Les pourparlers militaires étaient bien réels, mais le rôle qu'il y jouait dissimulait sa véritable mission. Il avait été envoyé à Prague pour enquêter sur les activités de la police secrète soviétique – toujours aussi maladroite –, éternelle rivale des services de renseignement de l'armée Rouge.

Son travail ne lui inspirait plus grand enthousiasme. Tout ce en quoi il croyait jadis vacillait. Staline, le communisme, la bonté fondamentale du peuple russe ne lui inspiraient plus confiance. Son père lui-même n'était pas son père. Il serait passé à l'Ouest s'il avait pu trouver le moyen d'emmener Zoïa et Kotia.

Il n'en mettait pas moins tout son cœur dans sa mission à Prague. C'était une rare occasion de faire une chose en laquelle il avait encore foi.

Deux semaines auparavant, le parti communiste tchèque avait pris le contrôle intégral du gouvernement, éliminant les autres partenaires de la coalition. Le ministre des Affaires étrangères, Jan Masaryk, héros de guerre et démocrate anticommuniste, était désormais littéralement prisonnier à l'étage supérieur du palais Czernin, sa résidence officielle. La police secrète soviétique était de toute évidence derrière ce coup d'État. En réalité, le beau-frère de Volodia, le colonel Ilia Dvorkine, qui se trouvait lui aussi à Prague et séjournait dans le même hôtel, y avait certainement pris part.

Le supérieur de Volodia, le général Lemitov, considérait ce coup d'État comme une catastrophe pour les relations diplomatiques de l'Union soviétique. Masaryk avait représenté aux yeux du monde la preuve que les pays d'Europe de l'Est pouvaient vivre, libres et indépendants, à l'ombre de l'URSS. Il avait permis à la Tchécoslovaquie de se doter d'un gouvernement communiste ami de l'Union soviétique, avec toutes les apparences d'une démocratie bourgeoise. Un arrangement idéal : l'URSS avait tout ce qu'elle voulait et les Américains étaient rassurés. Et voilà que ce bel équilibre avait été bouleversé.

Pourtant, Ilia exultait. « Les partis bourgeois ont été écrasés ! lança-t-il un soir à Volodia au bar de l'hôtel.

— Tu es au courant de ce qui s'est passé au Sénat américain ? demanda doucement Volodia. Vandenberg, le vieil isolationniste, a prononcé un discours de quatre-vingts minutes en faveur du plan Marshall, et il a été accueilli par une ovation délirante. »

Les idées vagues de George Marshall étaient devenues un plan. Surtout grâce à l'habileté de ce renard d'Ernie Bevin, le ministre des Affaires étrangères britannique. D'après Volodia, Bevin était un anticommuniste de la pire espèce : un social-démocrate de la classe ouvrière. Malgré sa masse imposante, il était rapide. Il avait organisé à la vitesse de l'éclair une conférence à Paris qui avait réservé, au nom de l'Europe, un accueil collectif enthousiaste au discours que George Marshall avait fait à Harvard.

Volodia savait, grâce à ses espions en place au ministère des Affaires étrangères britannique, que Bevin était déterminé à faire entrer l'Allemagne dans le plan Marshall, et à en tenir l'URSS à l'écart. Et Staline était tombé tête baissée dans le piège tendu par Bevin en ordonnant aux pays de l'Est de refuser l'aide de Marshall.

La police secrète soviétique semblait maintenant faire tout son possible pour favoriser le passage de la loi au Congrès. « Le Sénat s'apprêtait à rejeter Marshall, poursuivit Volodia. Les contribuables américains ne voulaient pas payer l'addition. Mais le coup d'État, ici, à Prague, les a convaincus de le faire, parce que le capitalisme européen risque de s'effondrer.

— Les partis bourgeois tchèques étaient prêts à se laisser graisser la patte par les Américains, rétorqua Ilia, indigné.

— Nous aurions dû les laisser faire, répondit Volodia. C'était peut-être la façon la plus rapide de saboter le processus. Le Congrès aurait rejeté le plan Marshall : les Américains ne veulent pas donner d'argent aux communistes.

— Le plan Marshall est une manœuvre impérialiste !

— En effet, acquiesça Volodia. Et j'ai bien peur que cette manœuvre soit une réussite. Nos alliés de guerre sont en train de constituer un bloc antisoviétique.

— Ceux qui font obstacle à la marche en avant du communisme doivent être traités en conséquence.

— C'est certain. » Décidément, des types comme Ilia avaient un talent indéniable pour commettre les pires erreurs de jugement politique.

« Et moi, je dois aller me coucher. »

Il n'était que dix heures, mais Volodia remonta lui aussi dans sa chambre. Il resta allongé sur son lit sans pouvoir trouver le sommeil, pensant à Zoïa et à Kotia et regrettant de ne pas être auprès d'eux pour leur souhaiter bonne nuit.

Ses pensées divaguèrent vers sa mission. Il avait rencontré deux jours plus tôt Jan Masaryk, le symbole de l'indépendance tchèque à une cérémonie organisée sur la tombe de son père, Thomas Masaryk, le fondateur de la Tchécoslovaquie et son premier président. Vêtu d'un manteau à col de fourrure, tête nue sous la neige, le second Masaryk semblait accablé.

Si on réussissait à le convaincre de rester ministre des Affaires étrangères, certains compromis seraient possibles, se dit rêveusement Volodia. La Tchécoslovaquie pourrait avoir un authentique gouvernement communiste, mais rester neutre dans ses relations internationales, ou du moins aussi peu antiaméricaine que possible. Masaryk possédait à la fois les compétences diplomatiques et la crédibilité internationale nécessaires pour effectuer cet exercice de funambulisme.

Volodia décida d'en parler à Lemitov dès le lendemain.

Il dormit mal, d'un sommeil agité, et fut réveillé avant six heures. Une alarme sonnait dans sa tête : quelque chose qu'Ilia avait dit la veille au soir le tracassait. Volodia passa mentalement en revue leur conversation. Quand Ilia avait évoqué « ceux qui font obstacle à la marche en avant du communisme », il pensait à Masaryk ; et quand un agent de la police secrète parlait de traiter quelqu'un « en conséquence », cela voulait toujours dire « éliminer ».

Ilia était allé se coucher de bonne heure, ce qui suggérait un départ matinal le lendemain.

Quel idiot ! se reprocha Volodia. J'avais tous les indices en main, et il m'a fallu toute la nuit pour les déchiffrer.

Il se leva d'un bond. Peut-être n'était-il pas trop tard.

Il s'habilla rapidement et mit un gros pardessus, une écharpe et un chapeau. Il n'y avait pas de taxis devant l'hôtel – il était trop tôt. Il aurait pu appeler une voiture de l'armée Rouge, mais le temps qu'on réveille un chauffeur et que la voiture arrive, il fallait bien compter une heure.

Il décida d'y aller à pied. Le palais Czernin n'était qu'à deux ou trois kilomètres à l'ouest. Il quitta l'élégant centre-ville de Prague par le pont Saint-Charles et se dirigea d'un bon pas vers le château, en haut de la colline.

Masaryk ne l'attendait pas, et le ministre des Affaires étrangères n'était pas obligé d'accorder une audience à un colonel de l'armée Rouge. Mais Volodia était sûr qu'il serait assez intrigué pour le recevoir.

Il marcha rapidement dans la neige et arriva au palais Czernin à sept heures moins le quart. C'était une gigantesque bâtisse baroque, dont les trois étages supérieurs étaient ornés d'une grandiose rangée de demi-colonnes corinthiennes. Il constata avec étonnement que l'endroit n'était pas très sévèrement gardé. Une sentinelle était en faction devant la porte d'entrée. Volodia traversa un vestibule au décor chargé sans qu'on lui demande quoi que ce soit.

Il s'attendait à trouver l'éternel crétin de la police secrète assis à un bureau, dans l'entrée, mais il n'y avait personne. C'était mauvais signe. Un sombre pressentiment l'envahit.

Le vestibule donnait sur une cour intérieure. Il jeta un coup d'œil par une fenêtre et aperçut une silhouette allongée dans la neige. On aurait dit quelqu'un qui dormait. Peut-être un type qui s'était écroulé là, ivre mort. Dans ce cas, il risquait fort de mourir de froid.

Faisant tourner la poignée de la porte donnant sur la cour, Volodia s'aperçut qu'elle était ouverte.

Il traversa le quadrilatère en courant. Un homme en pyjama de soie bleue gisait par terre à plat ventre. Il n'y avait pas de neige sur ses vêtements, il ne devait pas être là depuis plus de quelques minutes. Volodia s'agenouilla à côté de lui. L'homme était parfaitement immobile. Apparemment, il ne respirait plus.

Volodia leva la tête. Des rangées de fenêtres identiques s'alignaient au-dessus de la cour tels des soldats à la parade. Elles étaient toutes fermées hermétiquement pour ne pas laisser entrer le froid glacial – toutes sauf une, très haut, au-dessus de l'homme en pyjama. Grande ouverte, celle-là.

Comme si quelqu'un avait été poussé dans le vide.

Volodia retourna la tête sans vie et regarda le visage de l'homme.

C'était Jan Masaryk.

2.

Trois jours plus tard, à Washington, le Comité des chefs d'état-major interarmées présentait au président Truman un plan de bataille d'urgence destiné à riposter à une invasion de l'Europe de l'Ouest par les Soviétiques.

Le risque d'une troisième guerre mondiale faisait la une des journaux. « Nous venons de gagner la guerre, fit remarquer Jacky Jakes à Greg Pechkov. Comment avons-nous pu en arriver là une nouvelle fois ?

— C'est ce que je ne cesse de me demander », répondit Greg.

Ils étaient assis sur un banc, dans un parc. Greg soufflait un peu après avoir joué au foot avec Georgy.

« Je suis contente qu'il soit trop jeune pour se battre, dit Jacky.

— Moi aussi. »

Ils avaient les yeux rivés sur leur fils, qui bavardait un peu plus loin avec une fille blonde à peu près du même âge que lui. Les lacets de ses Keds étaient dénoués, et sa chemise était sortie de son pantalon. Il avait douze ans, et avait beaucoup changé ces derniers temps. Quelques poils de duvet noir ombraient sa lèvre supérieure, et il donnait l'impression d'avoir pris au moins cinq bons centimètres depuis la semaine précédente.

« Ça fait un moment que nous rapatrions nos troupes aussi rapidement que possible, poursuivit Greg. Les Anglais et les Français aussi. Mais l'armée Rouge est restée sur place. Résultat : ils ont maintenant trois fois plus de soldats en Allemagne que nous.

— Les Américains ne veulent pas d'un nouveau conflit.

— Ça, tu peux le dire. Et Truman, qui espère remporter l'élection présidentielle en novembre, fera tout son possible pour l'éviter. Il se pourrait qu'elle éclate quand même.

— Tu seras bientôt démobilisé. Qu'est-ce que tu vas faire ? »

Un frémissement dans sa voix lui fit soupçonner que, contrairement à ce qu'elle aurait voulu lui faire croire, ce n'était pas une question de pure forme. Il se tourna vers elle, mais son expression était indéchiffrable. Il répondit : « Si l'Amérique n'est pas en guerre, je tenterai de me faire élire au Congrès en 1950. Mon père a accepté de financer ma campagne. Je commencerai juste après l'élection présidentielle. »

Elle détourna les yeux. « Pour quel parti ? » demanda-t-elle machinalement.

Il se demanda s'il avait dit quelque chose qui lui avait déplu. « Républicain, bien sûr.

— Et le mariage ? »

Greg fut pris de court. « Pourquoi cette question ? »

Elle le regardait durement, à présent.

« Tu vas te marier ? insista-t-elle.

— Eh bien oui, en effet. Elle s'appelle Nelly Fordham.

— C'est bien ce que je pensais. Quel âge a-t-elle ?

— Vingt-deux ans. Comment ça, c'est bien ce que tu pensais ?

— Un homme politique doit avoir une épouse.

— Mais je l'aime !

— Bien sûr. Sa famille est dans la politique ?

— Son père est avocat à Washington.

— Excellent choix. »

Greg était agacé. « Je te trouve bien cynique.

— Je te connais, Greg. Mon Dieu, quand j'ai couché avec toi, tu n'étais pas beaucoup plus vieux que Georgy aujourd'hui. Tu peux abuser tout le monde, sauf ta mère, et moi. »

Elle était fine mouche, comme toujours. Sa mère avait elle aussi critiqué ses fiançailles. Une union stratégique, destinée à servir sa carrière, elles avaient raison. Mais enfin, Nelly était jolie, charmante, et elle adorait Greg, alors où était le mal ? « Je la retrouve pour déjeuner près d'ici dans quelques minutes.

— Nelly est au courant, pour Georgy ? demanda Jacky.

— Non. Et il n'est pas question qu'elle l'apprenne.

— Tu as raison. Avoir un enfant illégitime, c'est déjà assez fâcheux ; un enfant noir pourrait ruiner ta carrière.

— Je sais.

— Presque aussi sûrement qu'une épouse noire. »

Greg fut tellement surpris que la question lui échappa : « Tu pensais que j'allais t'épouser, toi ?

— Oh, bon sang ! Non, Greg, répliqua-t-elle avec amertume. Si j'avais le choix entre Jack l'Éventreur et toi, je demanderais un peu de temps pour réfléchir. »

Il savait qu'elle mentait. L'espace d'un instant, il essaya d'imaginer sa vie s'il épousait Jacky. Les mariages interraciaux n'étaient pas fréquents, et suscitaient pas mal d'hostilité de la part des Noirs comme des Blancs, mais certains couples mixtes se mariaient quand même et en assumaient les conséquences. Il n'avait jamais rencontré de fille qui lui plaise autant que Jacky ; même pas Margaret Cowdry, avec qui il était sorti pendant deux ans, jusqu'à ce qu'elle en ait assez d'attendre une hypothétique demande en mariage. Jacky avait la langue acérée, mais il aimait ça, peut-être parce qu'elle lui rappelait sa mère. Et l'idée de passer tout leur temps ensemble tous les trois était terriblement tentante. Georgy apprendrait à l'appeler Papa. Ils pourraient acheter une maison dans un quartier habité par des gens aux idées larges, un endroit où il y aurait beaucoup d'étudiants et de jeunes professeurs, peut-être Georgetown.

Et puis il vit que la petite amie blonde de Georgy se faisait rappeler à l'ordre par sa mère, une mère blanche, fâchée, qui agitait un doigt en signe de reproche, et il se rendit à l'évidence : épouser Jacky était la plus mauvaise idée du monde.

Georgy regagna le banc où Greg et Jacky étaient assis.

« Alors, comment ça marche en classe ? demanda Greg.

— Ça me plaît mieux qu'avant, répondit le garçon. Je trouve les maths plus intéressantes cette année.

— J'étais bon en maths, remarqua Greg.

— Quelle coïncidence ! » lança Jacky.

Greg se leva. « Il faut que j'y aille. » Il serra la main de Georgy. « Continue à bosser les maths, mon pote.

— D'accord », répondit Georgy.

Greg s'éloigna en faisant un signe de la main à Jacky.

Elle avait songé au mariage elle aussi, cela ne faisait pas de doute. Elle savait que son départ de l'armée représentait un tournant pour lui et l'obligeait à réfléchir à son avenir. Elle ne pensait sûrement pas vraiment qu'il l'épouserait, mais quand même, elle devait avoir caressé des rêves secrets. Et voilà qu'il les avait brisés. Eh bien, tant pis. Même si elle avait été blanche, il ne l'aurait pas épousée. Il avait beaucoup d'affection pour elle, il adorait le gamin, mais il avait toute sa vie devant lui et avait besoin d'une femme qui lui apporterait des relations et des appuis. Le père de Nelly était un homme puissant dans le clan républicain.

Il alla à pied jusqu'au Napoli, un restaurant italien, à quelques rues du parc. Nelly était déjà là. Ses boucles cuivrées s'échappaient d'un petit chapeau vert.

« Tu es superbe ! s'exclama-t-il. J'espère que je ne suis pas en retard. » Il s'assit.

Nelly avait le visage fermé. « Je t'ai vu au parc », commença-t-elle.

Et merde, se dit Greg.

« J'étais un peu en avance, et je me suis assise un moment, poursuivit-elle. Tu ne m'as pas remarquée. Et puis j'ai eu l'impression d'être indiscrète, alors je suis partie.

— Tu as donc vu mon filleul ? demanda-t-il avec une jovialité forcée.

— Ah, parce que tu es son parrain ? C'est bizarre de t'avoir choisi pour ce rôle, toi qui ne vas jamais à l'église.

— Je suis gentil avec lui.

— Comment s'appelle-t-il ?

— Georgy Jakes.

— C'est la première fois que tu m'en parles.

— Vraiment ?

— Quel âge a-t-il ?

— Douze ans.

— Tu avais donc seize ans quand il est né. C'est bien jeune pour être parrain.

— Peut-être, oui.

— Et que fait sa mère ?

— Elle est serveuse. Il y a des années, elle était actrice. Son nom de scène était Jacky Jakes. Je l'ai connue quand elle était sous contrat dans le studio de mon père. » Ce qui était plus ou moins vrai, se dit Greg, mal à l'aise.

« Et son père ? »

Greg secoua la tête. « Jacky est célibataire. » Un serveur s'approcha. « Si on prenait un cocktail ? suggéra Greg, dans l'espoir de détendre l'atmosphère. Deux martinis, commanda-t-il au garçon.

— Tout de suite, monsieur. »

Dès que le serveur fut parti, Nelly reprit : « Tu es son père, n'est-ce pas ?

— Son parrain.

— Oh, arrête, fit-elle d'une voix chargée de mépris.

— Qu'est-ce qui te permet d'en être tellement sûre ?

— Il est peut-être noir, mais il te ressemble. Il est incapable de nouer les lacets de ses chaussures ou de rentrer sa chemise dans son pantalon, comme toi. Et il faisait du charme à cette petite blonde à qui il parlait. Il est de toi, ça ne fait aucun doute. »

Greg rendit les armes. « J'avais l'intention de t'en parler, soupira-t-il.

— Quand ?

— J'attendais le bon moment.

— Le bon moment aurait été avant de me demander en mariage.

— Je te demande pardon. » Il était ennuyé, mais pas vraiment penaud. Il trouvait qu'elle faisait beaucoup d'histoires pour pas grand-chose.

Le garçon leur apporta les menus et ils les consultèrent tous les deux. « Les spaghettis bolognaise sont excellents, suggéra Greg.

— Je vais prendre une salade. »

Leurs martinis arrivèrent. Greg leva son verre : « Au pardon dans le mariage. »

Nelly ne toucha pas à son cocktail. « Je ne peux pas t'épouser.

— Chérie, voyons, ne prends pas les choses comme ça. Je me suis excusé. »

Elle secoua la tête. « Tu ne comprends pas ?

— Qu'est-ce que je ne comprends pas ?

— Cette femme, qui était assise sur le banc, dans le parc, à côté de toi : elle t'aime.

— Tu crois ? »

Greg l'aurait encore nié la veille, mais après la conversation qu'ils venaient d'avoir, il n'en était plus aussi sûr.

« Bien sûr qu'elle t'aime. Pourquoi ne s'est-elle pas mariée ? Elle est très jolie. Depuis le temps, si elle l'avait vraiment voulu, elle aurait pu trouver un homme prêt à accepter un beau-fils. Mais elle est amoureuse de toi, espèce de brute.

— Je n'en suis pas si sûr.

— Et le petit t'adore, lui aussi.

— Je suis son oncle préféré.

— Sauf que tu n'es pas son oncle. » Elle repoussa son cocktail. « Tu peux prendre mon verre.

— Chérie, je t'en prie, détends-toi.

— Je m'en vais. »

Elle se leva.

Greg n'avait pas l'habitude que les filles le laissent en plan comme ça. Il n'en revenait pas. Son charme aurait-il cessé d'agir ?

« Je veux qu'on se marie ! lança-t-il d'une voix désespérée.

— Greg, tu ne peux pas m'épouser. » Elle retira le solitaire de son doigt et le posa sur la nappe à carreaux rouges. « Tu as déjà une famille. »

Et elle quitta le restaurant.

3.

La crise mondiale arriva à un paroxysme au moins de juin. Carla et sa famille étaient aux premières loges.

Le plan Marshall était devenu une loi signée par le président Truman, et les premiers envois d'aide humanitaire arrivaient en Europe, à la grande fureur du Kremlin.

Le vendredi 18 juin, les Alliés occidentaux avisèrent les Allemands qu'ils allaient faire une annonce importante à huit heures du soir. Carla et sa famille étaient réunies dans la cuisine, autour du poste réglé sur Radio Francfort, attendant avec impatience. La guerre était finie depuis trois ans, mais aucun d'eux ne savait ce que l'avenir leur réservait : le capitalisme ou le communisme, l'unité ou la division, la liberté ou l'oppression, la prospérité ou la misère.

Werner était assis à côté de Carla. Il tenait sur ses genoux Walli, qui avait maintenant deux ans et demi. Ils s'étaient mariés dans l'intimité un an plus tôt. Carla avait repris son travail d'infirmière. Elle était aussi conseillère municipale de Berlin pour les sociaux-démocrates. Tout comme le mari de Frieda, Heinrich.

En Allemagne de l'Est, les Russes avaient interdit le parti social-démocrate, mais Berlin était une oasis en pleine zone soviétique, dirigée par un conseil des quatre principaux Alliés appelé la Kommandantura, qui avait opposé son veto à cette interdiction. Avec pour conséquence la victoire des sociaux-démocrates aux élections, les communistes n'étant arrivés que troisièmes, derrière les chrétiens-démocrates conservateurs. Les Russes ne décoléraient pas et faisaient tout ce qui était en leur pouvoir pour mettre des bâtons dans les roues du conseil élu. Ce que Carla trouvait exaspérant, mais elle ne pouvait renoncer à un espoir d'indépendance vis-à-vis des Soviétiques.

Werner avait réussi à monter une petite entreprise. Il avait fouillé dans les ruines de l'usine de son père et récupéré un stock de matériel électrique et de pièces de radio. Les Allemands n'avaient pas les moyens d'acheter des postes neufs, mais tout le monde voulait faire réparer les vieux. Werner avait retrouvé certains des anciens techniciens de l'usine et les avait embauchés pour remettre en état les appareils en panne. Il était tout à la fois le directeur et le vendeur de sa société, allant frapper aux portes des maisons et des immeubles pour se constituer une clientèle.

Maud, qui était également assise à la table de la cuisine ce soir-là, travaillait comme interprète pour les Américains. C'était l'une des meilleures, et elle traduisait souvent les procès-verbaux des réunions de la Kommandantura.

Erik, le frère de Carla, portait un uniforme de policier. Il était entré au parti communiste – au grand désespoir de sa famille – et avait trouvé du travail dans la nouvelle police d'Allemagne de l'Est constituée par les occupants russes. Erik prétendait que les Alliés occidentaux cherchaient à couper l'Allemagne en deux. « Vous autres, les sociaux-démocrates, vous n'êtes que des séparatistes », disait-il, récitant le bréviaire communiste sans plus de réflexion qu'il n'avait répété autrefois la propagande nazie.

« Les Alliés occidentaux n'ont rien divisé du tout, répliquait Carla. Ils ont ouvert les frontières entre leurs zones. Pourquoi les Russes n'en font-ils pas autant ? Comme ça, nous formerions de nouveau un seul pays. » Il n'avait pas l'air de l'entendre.

Rebecca avait presque dix-sept ans. Carla et Werner l'avaient adoptée légalement. Elle allait au lycée, et était particulièrement bonne en langues.

Carla était à nouveau enceinte, mais ne l'avait pas encore dit à Werner. Elle était aux anges. Il avait déjà une fille adoptive et un beau-fils, et maintenant il allait avoir un enfant bien à lui. Elle savait qu'il serait enchanté quand elle le lui annoncerait et attendait encore un peu pour en être tout à fait sûre.

Mais elle avait hâte de savoir dans quel genre de pays ses trois enfants allaient vivre.

La voix de Robert Lochner, un officier américain, se fit entendre sur les ondes. Ayant grandi en Allemagne, il parlait allemand couramment. Il annonça qu'à partir du lundi suivant, à sept heures du matin, l'Allemagne de l'Ouest aurait une nouvelle monnaie, le deutsche mark.

Carla n'était pas surprise. Le reichsmark perdait de sa valeur tous les jours. Les gens qui avaient du travail étaient généralement payés en reichsmark, que l'on pouvait utiliser pour les dépenses de base comme les rations alimentaires et les tickets de bus, mais tout le monde préférait se faire rémunérer en cigarettes ou en produits d'épicerie. Werner facturait les clients de son entreprise en reichsmarks, mais il demandait cinq cigarettes pour les dépannages de nuit, et trois œufs pour une livraison en ville, quelle que soit la distance.

Carla savait, grâce à Maud, que la nouvelle monnaie avait fait l'objet de discussions à la Kommandantura. Les Russes avaient demandé des planches pour pouvoir l'imprimer. Mais ils avaient mis en circulation une telle quantité de l'ancienne monnaie qu'ils l'avaient complètement dévaluée, et il ne servirait à rien de tirer de nouveaux billets si le même phénomène devait se reproduire. Les Alliés occidentaux avaient donc refusé, et les Soviétiques l'avaient mal pris.

Autrement dit, l'Ouest avait décidé d'aller de l'avant en se passant de la coopération des Soviétiques. Carla était contente parce que la nouvelle monnaie serait bonne pour l'Allemagne, mais elle s'inquiétait de la réaction des Russes.

Les Allemands de l'Ouest pourraient échanger soixante anciens reichsmarks dévalués contre trois nouveaux deutsche marks et quatre-vingt-dix pfennigs, expliqua Lochner.

Puis il annonça que ces mesures ne s'appliqueraient pas à Berlin, du moins au début, déclaration qui fut accueillie par un gémissement collectif dans la cuisine.

Carla alla se coucher en se demandant comment les Soviétiques allaient réagir. Elle était allongée à côté de Werner, l'oreille tendue pour s'assurer que Walli ne pleurait pas dans la pièce voisine. Depuis quelques mois, les occupants soviétiques étaient de plus en plus hargneux. La police secrète soviétique avait enlevé un journaliste, un certain Dieter Friede, dans la zone américaine, et l'avait jeté en prison. Les Russes avaient prétendu ne rien savoir à propos de cette affaire, avant d'accuser Friede d'espionnage. Trois étudiants avaient été exclus de l'université pour avoir critiqué les Russes dans une revue. Pire encore, un avion de chasse soviétique avait frôlé un avion de ligne de la British European Airways qui se posait sur l'aéroport de Gatow : il lui avait arraché une aile, et les deux appareils s'étaient écrasés au sol, provoquant la mort de quatre membres de l'équipage de la BEA, de dix passagers et du pilote soviétique. Quand les Russes étaient en colère, on pouvait s'attendre à ce qu'ils le fassent payer aux autres.

Le lendemain matin, les Soviétiques décrétèrent que l'importation de deutsche marks en Allemagne de l'Est serait considérée comme un délit. Ils précisèrent que cela incluait Berlin « qui faisait partie de la zone soviétique ». Les Américains protestèrent aussitôt contre cette formule, rappelant que Berlin était une ville internationale. Mais les esprits s'échauffaient, et Carla n'était pas rassurée.

Le lundi, la nouvelle monnaie fut mise en circulation en Allemagne de l'Ouest.

Le mardi, un coursier de l'armée Rouge se présenta chez Carla et la convoqua à l'hôtel de ville.

Ce n'était pas la première fois, mais elle sortit quand même de chez elle la peur au ventre. Rien ne pouvait empêcher les Soviétiques de l'emprisonner. Les communistes jouissaient de tous les pouvoirs arbitraires qu'avaient détenus les nazis. Ils allaient jusqu'à utiliser les mêmes camps de concentration.

Le célèbre Rotes Rathaus – l'hôtel de ville rouge – avait été gravement endommagé par les bombardements, et le conseil municipal s'était installé dans le nouvel hôtel de ville de la Parochialstrasse. Les deux bâtiments se trouvaient dans le quartier du Mitte, où vivait Carla, en plein secteur soviétique.

Lorsqu'elle y arriva, elle découvrit que le maire par intérim, Louise Schroeder, et plusieurs autres, avaient également été convoqués à une réunion avec l'agent de liaison soviétique, le commandant Otchkine. Il les informa que la monnaie de l'Allemagne de l'Est serait réformée, et qu'à l'avenir, seul le nouvel ostmark serait légal dans la zone soviétique.

Louise Schroeder mit immédiatement le doigt sur le point critique : « Vous voulez dire que cette mesure s'appliquera à tous les secteurs de Berlin ?

— Oui. »

Frau Schroeder n'était pas femme à se laisser intimider.

« Conformément à la Constitution municipale, la force d'occupation soviétique ne peut imposer une telle règle aux autres secteurs, dit-elle fermement. Tous les Alliés doivent être consultés.

— Ils n'élèveront pas d'objection. C'est un décret du maréchal Sokolovski, ajouta-t-il en brandissant une feuille de papier. Vous le présenterez demain au conseil municipal. »

Plus tard, dans la soirée, alors que Carla et Werner allaient se coucher, elle lui raconta ce qui s'était passé : « Tu comprends la tactique des Russes ? Si le conseil municipal vote le décret, les Alliés occidentaux, qui se réclament de la démocratie, auront bien du mal à revenir dessus.

— Mais le conseil ne le votera pas. Les communistes sont en minorité, et personne d'autre ne veut de l'ostmark.

— Non. Voilà pourquoi je voudrais bien savoir quel atout le maréchal Sokolovski a dans la manche. »

Les journaux du lendemain annoncèrent qu'à partir du vendredi, deux monnaies concurrentes seraient en circulation à Berlin, l'ostmark et le deutsche mark. En fait, les Américains avaient secrètement transporté en avion deux cent cinquante millions de ces nouveaux deutsche marks dans des caisses en bois marquées « Clay » et « Bird Dog », qui étaient maintenant entreposées en lieu sûr un peu partout dans Berlin.

Dans la journée, Carla commença à entendre des rumeurs en provenance d'Allemagne de l'Ouest. La récente monnaie y avait fait des miracles. Du jour au lendemain, de nouvelles marchandises étaient apparues dans les devantures des magasins : des paniers de cerises, des bottes de carottes soigneusement attachées venant de la campagne environnante, du beurre, des œufs et des pâtisseries, des articles de luxe longtemps thésaurisés comme des chaussures et des sacs neufs, et même des bas à quatre deutsche marks la paire. Les gens avaient attendu de pouvoir les vendre contre de l'argent qui valait quelque chose.

Cet après-midi-là, Carla partit pour l'hôtel de ville. Elle devait participer à la réunion du conseil municipal prévue à quatre heures. En s'approchant, elle vit, rangés dans les rues voisines, des dizaines de camions de l'armée Rouge, dont les chauffeurs traînaient aux alentours, la cigarette au bec. C'étaient majoritairement des véhicules américains qui avaient dû être donnés à l'URSS pendant la guerre en vertu de l'accord de prêt-bail. Elle comprit la raison de leur présence en entendant les premiers échos révélateurs d'une foule en colère. L'atout que le gouvernement soviétique avait dans sa manche, se dit-elle, était probablement une matraque.

Devant l'hôtel de ville, des drapeaux rouges flottaient au-dessus d'un rassemblement de plusieurs milliers de personnes, dont la plupart arboraient des insignes du parti communiste. Des camions équipés de haut-parleurs diffusaient des discours rageurs, et la foule scandait : « À bas les séparatistes ! »

Carla se demanda comment atteindre l'entrée du bâtiment. Une poignée de policiers observait la scène d'un œil distrait sans chercher à aider les conseillers à passer. Ce spectacle rappela à Carla un pénible souvenir : l'attitude de la police le jour où les Chemises brunes avaient dévasté le bureau de sa mère, quinze ans plus tôt. Elle était absolument convaincue que les conseillers communistes étaient déjà à l'intérieur, et que si les sociaux-démocrates n'arrivaient pas à entrer, la minorité voterait le décret et proclamerait sa validité.

Elle prit une profonde inspiration et commença à jouer des coudes.

Elle réussit à avancer de quelques pas sans se faire remarquer, mais un homme la reconnut alors et cria « Putain américaine ! » en tendant le doigt vers elle. Elle poursuivit son chemin avec détermination. Quelqu'un d'autre lui cracha dessus, maculant sa robe de salive. Elle continua à se frayer un passage, tout en sentant la panique la gagner. Elle était entourée de gens qui la détestaient, une épreuve qu'elle n'avait jamais subie, et dut résister à la tentation de s'enfuir. On la bouscula, mais elle réussit à rester debout. Une main agrippa sa robe. Elle se dégagea, entendit un bruit d'étoffe déchirée et retint un hurlement. Qu'allaient-ils faire ? Lui arracher tous ses vêtements ?

Elle s'aperçut que quelqu'un d'autre essayait de traverser la foule derrière elle, et jeta un coup d'œil par-dessus son épaule. C'était Heinrich von Kessel, le mari de Frieda. Il arriva à son niveau et ils foncèrent ensemble. Heinrich était plus agressif, n'hésitant pas à marcher sur les pieds et à donner de vigoureux coups de coude à tous ceux qui se trouvaient à sa portée. À deux, ils avancèrent plus vite ; ils atteignirent enfin la porte et entrèrent.

Leur calvaire n'était pas terminé. Des manifestants communistes s'étaient rassemblés par centaines à l'intérieur du bâtiment. Carla et Heinrich durent se bagarrer pour traverser les couloirs. Dans la salle de réunion, les trublions étaient partout – non seulement dans la galerie des visiteurs, mais dans la chambre même. Leur comportement était tout aussi belliqueux qu'au-dehors.

Certains sociaux-démocrates étaient déjà là, d'autres arrivèrent après Carla. D'une façon ou d'une autre, les soixante-trois conseillers réussirent presque tous à se frayer un chemin à travers la foule. Elle fut soulagée. L'ennemi n'avait pas réussi à les terroriser au point de les mettre en fuite.

Quand le président de l'assemblée réclama le silence, un conseiller communiste debout sur un banc exhorta les manifestants à ne pas sortir. Voyant Carla, il cria : « Que les traîtres restent dehors ! »

Tout cela rappelait sinistrement 1933 : les intimidations, les brimades et le chahut qui entravaient l'exercice de la démocratie. Carla était désespérée.

Levant les yeux vers la galerie, elle reconnut avec consternation son frère, Erik, au milieu de la foule hurlante. « Tu es allemand ! lui cria-t-elle. Tu as vécu sous les nazis. La leçon ne t'a donc pas suffi ? »

Il n'eut pas l'air de l'entendre.

Frau Schroeder, debout sur l'estrade, lança un appel au calme. Les manifestants la huèrent et l'accablèrent de sarcasmes. Elle haussa le ton et hurla : « Si le conseil municipal ne peut pas siéger en bon ordre dans ce bâtiment, je transférerai le débat en secteur américain. »

Elle fut saluée par une nouvelle bordée d'injures, mais les vingt-six conseillers communistes durent se rendre à l'évidence : le désordre ne servait pas leur cause. Si le conseil tenait une réunion hors du secteur soviétique, rien ne l'empêcherait de le refaire, et même d'aller s'installer définitivement hors de portée de l'intimidation communiste. Après une brève discussion, l'un d'eux se leva et demanda aux manifestants de quitter la salle. Ils sortirent en rang, en chantant l'Internationale.

« Pas besoin de se demander à qui ils obéissent », lui chuchota Heinrich.

Enfin, le calme fut rétabli. Frau Schroeder exposa la revendication des Soviétiques, ajoutant qu'elle ne pouvait s'appliquer à l'extérieur du secteur soviétique de Berlin, à moins d'être ratifiée par les autres Alliés.

Un député communiste prononça un discours l'accusant de recevoir ses ordres directement de New York.

Les accusations et les injures fusèrent. Finalement, ils passèrent au vote. Les communistes soutinrent à l'unanimité le décret soviétique – après avoir accusé le camp adverse d'être contrôlé par l'étranger. Tous les autres votèrent contre et la motion fut rejetée. Berlin avait repoussé les manœuvres d'intimidation. Carla était à la fois triomphante et épuisée.

Mais ce n'était pas fini.

Le temps qu'ils quittent le bâtiment, il était sept heures du soir. L'essentiel de la foule s'était dispersé, mais un noyau dur rôdait encore autour de l'entrée, prêt à en découdre. En sortant, une conseillère municipale d'un certain âge reçut des coups de pied et des coups de poing sous les yeux indifférents de la police.

Carla et Heinrich s'éclipsèrent par une porte latérale, avec quelques amis, espérant passer inaperçus, mais un communiste à bicyclette surveillait la sortie. Il s'éloigna à grands coups de pédales.

Alors que les conseillers hâtaient le pas, il revint à la tête d'une petite bande. L'un des types fit un croche-pied à Carla, qui tomba par terre. Elle reçut un, deux, trois coups de pied, très douloureux. Terrifiée, elle se protégea le ventre des deux mains. Elle était enceinte de près de trois mois : le moment où se produisent la plupart des fausses couches, elle le savait. Le bébé de Werner va-t-il mourir, se demanda-t-elle, désespérée, tué à coups de pied dans une rue de Berlin par des brutes communistes ?

Et puis ils se dispersèrent.

Les conseillers se relevèrent. Aucun n'était gravement blessé. Ils s'éloignèrent ensemble, craignant de nouvelles agressions, mais apparemment les communistes avaient eu leur content de violences pour la journée.

Carla arriva chez elle à huit heures. Il n'y avait pas trace d'Erik.

En voyant ses ecchymoses et sa robe déchirée, Werner fut atterré. « Que s'est-il passé ? demanda-t-il. Tu vas bien ? »

Elle fondit en larmes.

« Tu es blessée, reprit Werner. Tu veux que je t'emmène à l'hôpital ? »

Elle secoua vigoureusement la tête. « Ce n'est pas ça, dit-elle. Ce ne sont que des bleus. J'ai connu pire. » Elle s'effondra dans un fauteuil. « Bon sang, ce que je suis fatiguée…

— Qui t'a fait ça ? insista-t-il, furieux.

— Toujours les mêmes, répondit-elle. Ils ne se disent plus nazis mais communistes, et pourtant ce sont les mêmes. C'est 1933 qui recommence. »

Werner la prit dans ses bras.

Elle était inconsolable. « Les brutes et les tyrans sont au pouvoir depuis si longtemps ! soupira-t-elle entre deux sanglots. Ça ne finira donc jamais ? »

4.

Cette nuit-là, l'agence de presse soviétique publia un communiqué. À partir de six heures du matin le lendemain, aucun transport de passagers ou de marchandises ne pourrait plus entrer à Berlin ou en ressortir : trains, voitures, péniches, plus rien ne passerait. Aucune livraison ne serait autorisée : plus de nourriture, de lait, de médicaments, de charbon. Comme les centrales électriques ne pourraient plus fonctionner, l'électricité serait en conséquence coupée – dans les secteurs occidentaux seulement.

La ville était en état de siège.

Lloyd Williams se trouvait au quartier général de l'armée britannique. Il y avait de brèves vacances parlementaires, et Ernie Bevin était parti se reposer à Sandbanks, sur la côte Sud de l'Angleterre. Mais il était suffisamment inquiet pour envoyer Lloyd à Berlin observer l'introduction de la nouvelle monnaie et le tenir au courant.

Daisy n'avait pas accompagné Lloyd. Leur nouveau-né, Davey, n'avait que six mois ; de plus, le centre de planning familial pour les habitantes de Hoxton qu'elle mettait sur pied avec Eva Murray était sur le point d'ouvrir ses portes.

Lloyd redoutait que cette crise ne mène à la guerre. Des guerres, il en avait fait deux, et était bien décidé à ne pas en voir une troisième. Il avait deux enfants en bas âge et souhaitait qu'ils grandissent dans un monde en paix. Il avait épousé la plus jolie, la plus séduisante, la plus adorable des femmes de la planète, et n'aspirait qu'à une chose : passer de longues et nombreuses décennies avec elle.

Le général Clay, le gouverneur militaire américain – réputé pour être un bourreau de travail – ordonna à ses collaborateurs de constituer un convoi blindé qui foncerait sur l'autoroute depuis Helmstedt, à l'ouest, et traverserait le territoire soviétique jusqu'à Berlin, balayant tout sur son passage.

Lloyd eut vent de ce projet en même temps que le gouverneur britannique, Brian Robertson, et il entendit Sir Brian dire, dans son phrasé militaire incisif : « Si Clay fait ça, ce sera la guerre. »

Mais il n'y avait guère d'autre solution. D'après ce que Lloyd apprit en discutant avec de jeunes assistants de Clay, les Américains avaient avancé d'autres idées : le ministre de l'Armée, Kenneth Royall, avait proposé de revenir sur la réforme monétaire. Clay avait objecté que le processus était trop engagé pour qu'on pût faire marche arrière. Royall avait suggéré ensuite d'évacuer tous les Américains. Clay avait rétorqué que c'était exactement ce que les Soviétiques voulaient.

Sir Brian envisageait, quant à lui, de ravitailler la ville grâce à un pont aérien. La plupart estimaient que c'était impossible. Quelqu'un calcula que Berlin avait besoin de quatre mille tonnes de carburant et de nourriture par jour. Y avait-il suffisamment d'avions au *monde* pour transporter autant de marchandises ? Personne ne le savait. Et pourtant, Sir Brian ordonna à la Royal Air Force de se mettre au travail.

Le vendredi après-midi, Sir Brian rendit visite à Clay, et Lloyd fut invité à l'accompagner. Sir Brian présenta le projet à Clay en ces termes : « Les Russes peuvent bloquer l'autoroute devant votre convoi, et attendre de voir si vous avez le cran de les attaquer ; mais je ne pense pas qu'ils aillent jusqu'à abattre des appareils en plein vol.

— Je ne vois pas comment nous pourrions livrer suffisamment de marchandises par avion, objecta Clay.

— Moi non plus, convint Sir Brian. C'est pourtant ce que nous allons faire en attendant de trouver une meilleure solution. »

Clay décrocha son téléphone. « Passez-moi le général LeMay à Wiesbaden », ordonna-t-il. Et au bout d'une minute : « Curtis, auriez-vous des avions susceptibles de transporter du charbon ? »

La ligne resta silencieuse.

« Du charbon », répéta Clay, plus fort.

Nouveau silence.

« Oui, c'est bien ce que j'ai dit : du charbon. »

Quelques instants plus tard, Clay se tourna vers Sir Brian. « Il dit que l'armée de l'air américaine peut livrer tout ce qu'on veut. »

Les Anglais regagnèrent leur quartier général.

Le samedi, Lloyd prit un chauffeur de l'armée et mit le cap sur le secteur soviétique pour une mission personnelle. Il se fit conduire à l'adresse à laquelle il avait rendu visite à la famille von Ulrich, quinze ans plus tôt.

Il savait que Maud y habitait toujours. Sa mère et elle avaient repris leur correspondance à la fin de la guerre. Dans ses lettres, Maud s'efforçait de faire bonne figure alors que la situation devait être très éprouvante. Elle ne demandait pas d'aide, et de toute façon, Ethel n'aurait rien pu faire pour elle : le rationnement était toujours en vigueur en Grande-Bretagne.

L'endroit avait beaucoup changé. En 1933, c'était un bel hôtel particulier, qui avait connu des jours meilleurs mais conservait son élégance. Aujourd'hui, ce n'était plus qu'un taudis. Les vitres de la plupart des fenêtres avaient été remplacées par des planches ou du papier. La maçonnerie était criblée d'impacts de balles, le mur du jardin était écroulé et les menuiseries n'avaient pas été repeintes depuis des lustres.

Lloyd resta quelques instants assis dans la voiture, les yeux rivés sur la maison. Lors de son dernier séjour, il avait dix-huit ans ; Hitler venait de devenir chancelier. Le jeune Lloyd n'aurait jamais imaginé les horreurs que le monde allait connaître. Personne, d'ailleurs, n'avait soupçonné que le fascisme serait aussi près de s'abattre sur toute l'Europe, n'avait deviné les sacrifices que tous devraient consentir pour le vaincre. Il se sentait un peu comme la demeure des von Ulrich, délabré, bombardé, criblé de balles, mais toujours debout.

Il s'avança dans l'allée et frappa.

La domestique qui lui ouvrit était la même qu'autrefois. « Bonjour, Ada. Me reconnaissez-vous ? demanda-t-il en allemand. Je suis Lloyd Williams. »

L'intérieur de la maison était en meilleur état que l'extérieur. Ada le fit monter au salon. Il y avait un vase de fleurs sur le piano et un plaid aux couleurs vives jeté sur le canapé, sans doute pour en dissimuler les trous. Les journaux qui obturaient les fenêtres laissaient filtrer une lumière étonnamment vive.

Un petit garçon de deux ans entra dans la pièce et examina Lloyd sans dissimuler sa curiosité. Il portait des vêtements visiblement confectionnés à la maison et avait l'air vaguement asiatique. « Qui tu es ? demanda-t-il.

— Je m'appelle Lloyd. Et toi, comment tu t'appelles ?

— Walli », répondit le gamin. Il ressortit en courant, et Lloyd l'entendit dire à quelqu'un, dehors : « Il parle drôle, le monsieur ! »

Au temps pour mon accent allemand, pensa Lloyd.

Et puis il entendit la voix d'une femme d'âge mûr. « Il ne faut pas faire de remarques de ce genre. C'est très impoli.

— Pardon, Oma. »

Aussitôt après, Maud entra dans la pièce.

Lloyd eut un choc en la voyant. Elle avait un peu plus de cinquante-cinq ans, mais on lui en aurait donné soixante-dix. Elle avait les cheveux gris, le visage émacié, et était vêtue d'une robe de soie bleue élimée. De ses lèvres fanées, elle posa un baiser sur sa joue. « Lloyd Williams ! Quelle joie de te revoir ! »

C'est ma tante, pensa Lloyd, avec un sentiment étrange. Mais elle ne le savait pas : Ethel avait gardé le secret.

Maud était suivie de Carla, méconnaissable, et de son mari. La dernière fois que Lloyd avait vu Carla, c'était une enfant de onze ans, d'une intelligence précoce ; il calcula qu'elle avait vingt-six ou vingt-sept ans. Elle avait l'air à moitié morte de faim – comme la plupart des Allemands –, ce qui ne l'empêchait pas d'être jolie et d'afficher une assurance qui surprit Lloyd. Quelque chose dans son attitude lui fit penser qu'elle devait être enceinte. Il savait, parce que Maud l'avait écrit à sa mère, que Carla avait épousé Werner, qui était un véritable séducteur en 1933 ; il n'avait pas changé.

Ils passèrent une heure à se raconter les événements des dernières années. La famille von Ulrich avait traversé des horreurs effroyables, et n'hésitait pas à en parler. Lloyd avait pourtant l'impression qu'ils éludaient les détails les plus sordides. Il leur parla de Daisy, d'Evie et du petit Davey. Pendant la conversation, une adolescente entra et demanda à Carla l'autorisation d'aller chez une de ses amies.

« C'est notre fille, Rebecca », expliqua Carla à Lloyd.

Comme elle devait avoir seize ou dix-sept ans, Lloyd supposa qu'elle était adoptée.

« Tu as fait tes devoirs ? demanda Carla à la jeune fille.

— Je les ferai demain matin.

— Non, fais-les tout de suite, s'il te plaît, répondit fermement Carla.

— Oh, Mutti !

— Et ne discute pas », ajouta Carla. Elle se retourna vers Lloyd, et Rebecca sortit bruyamment.

Ils parlèrent de la crise. En tant que conseillère municipale, Carla y était plongée jusqu'au cou. Elle était pessimiste quant à l'avenir de Berlin et pensait que les Russes allaient simplement affamer la population jusqu'à ce que l'Ouest baisse les bras et abandonne la ville au contrôle absolu des Soviétiques.

« Je vais vous montrer quelque chose qui pourrait vous faire changer d'avis, dit Lloyd. Vous voulez bien faire un tour en voiture avec moi ? »

Maud resta à la maison avec Walli, mais Carla et Werner accompagnèrent Lloyd. Il demanda au chauffeur de les conduire à Tempelhof, l'aéroport situé dans le secteur américain. Sur place, il les fit entrer dans le bâtiment et monter à l'étage. Ils s'approchèrent d'une fenêtre d'où ils avaient une vue plongeante sur les pistes.

Sur le tarmac, une dizaine d'avions, des C-47 Skytrain, étaient alignés l'un derrière l'autre, certains ornés de l'étoile américaine, d'autres de la cocarde de la RAF. Un camion était rangé devant chacun des appareils. Les portes de la soute étaient ouvertes, et des manutentionnaires allemands aidés d'aviateurs américains déchargeaient des sacs de farine, d'énormes barils de kérosène, des cartons de fournitures médicales et des caisses en bois contenant des milliers de bouteilles de lait.

Sous leurs yeux, les avions vides décollaient tandis que d'autres atterrissaient.

« C'est stupéfiant, murmura Carla, les yeux brillants. Je n'ai jamais rien vu de pareil.

— Il n'y a jamais rien eu de pareil, confirma Lloyd.

— Les Anglais et les Américains pourront-ils continuer comme ça ?

— Il va bien falloir, je crois.

— Mais pendant combien de temps ?

— Le temps qu'il faudra », répondit fermement Lloyd.

Et c'est ce qu'ils firent.

XXV

1949

1.

Le 29 août 1949 – approximativement le milieu du XXe siècle – Volodia Pechkov se trouvait dans le sud profond de l'Union soviétique, au Kazakhstan, à l'est de la mer Caspienne, très précisément sur le plateau d'Oustiourt. C'était un désert rocailleux, où les nomades gardaient des chèvres un peu comme aux temps bibliques. Volodia était ballotté dans un camion militaire qui bringuebalait sur une piste cahoteuse. Le petit jour commençait à poindre sur un paysage minéral, accidenté, fait de sable et de buissons d'épineux. Un chameau osseux, tout seul le long de la route, posa un regard malveillant sur le camion qui passait.

Dans le lointain, Volodia distingua, d'abord vaguement, la tour de la bombe, éclairée par une batterie de projecteurs.

Zoïa et ses collègues avaient construit leur première bombe atomique conformément aux plans que Volodia avait obtenus de Willi Frunze à Santa Fe. C'était une bombe au plutonium avec un dispositif de mise à feu par implosion. Il y avait d'autres solutions possibles, mais celle-là avait déjà fait la preuve de son efficacité à deux reprises, une fois au Nouveau-Mexique, une autre fois à Nagasaki.

Il n'y avait aucune raison pour que l'expérience de ce jour-là ne réussisse pas.

L'essai portait le nom de code RDS-1, mais entre eux ils l'appelaient « Premier Éclair ».

Le camion s'arrêta au pied de la tour. Volodia leva les yeux et aperçut, sur la plate-forme, un groupe de chercheurs qui s'affairaient autour d'un véritable enchevêtrement de câbles reliés aux détonateurs fixés sur le revêtement de la bombe. Une silhouette en combinaison bleue recula dans un jaillissement de cheveux blonds : Zoïa. Volodia éprouva une bouffée de fierté. Ma femme, se dit-il ; éminente physicienne *et* mère de deux enfants.

Elle discutait avec deux hommes, leurs trois têtes réunies dans une conversation animée. Volodia espérait que rien ne clochait.

C'était la bombe qui allait sauver Staline.

Tout le reste n'avait été qu'une série de déboires pour l'Union soviétique. L'Europe de l'Ouest avait définitivement opté pour la démocratie, détournée du communisme par les manœuvres d'intimidation du Kremlin et soudoyée par les pots-de-vin du plan Marshall. L'URSS n'avait même pas réussi à prendre le contrôle de Berlin : le pont aérien s'était poursuivi sans trêve ni relâche, jour après jour, pendant près d'un an, et l'Union soviétique avait fini par baisser les bras, rouvrant les routes et les voies ferrées. Staline n'avait réussi à garder le contrôle de l'Europe de l'Est qu'en usant de la force brutale. Truman avait été réélu président, et se prenait pour le maître du monde. Les Américains avaient constitué des stocks d'armes nucléaires et des bombardiers B-29 étaient stationnés en Grande-Bretagne, prêts à transformer l'Union soviétique en désert radioactif.

Mais tout allait changer ce jour-là.

Si la bombe explosait comme prévu, l'URSS et les États-Unis seraient à nouveau à égalité. L'Union soviétique pourrait brandir la menace du feu nucléaire, et c'en serait fini de la domination américaine sur le monde.

Volodia se demandait si ce serait une bonne ou une mauvaise chose.

Si la bombe n'explosait pas, Zoïa et Volodia seraient probablement victimes d'une nouvelle purge, envoyés dans des camps de travail en Sibérie, ou tout simplement fusillés. Volodia avait déjà parlé à ses parents, et ils avaient promis de s'occuper de Kotia et de Galina.

Ce qu'ils feraient également si Volodia et Zoïa trouvaient la mort lors de l'essai de la bombe.

Dans la lumière de plus en plus vive, Volodia vit près de la tour une juxtaposition hétéroclite de bâtiments : des maisons de brique et de bois, un pont qui n'enjambait rien, et l'entrée d'une espèce d'architecture souterraine. L'armée voulait probablement mesurer l'effet de l'explosion. En regardant plus attentivement, il remarqua des camions, des chars d'assaut et un vieil avion d'un modèle obsolète, placés là dans le même but, supposa-t-il. Les chercheurs voulaient également évaluer l'impact de la bombe sur des êtres vivants : il y avait des chevaux, du bétail, des moutons et des chiens dans des chenils.

Sur la plate-forme, le conciliabule s'acheva sur une décision. Les trois chercheurs hochèrent la tête et se remirent au travail.

Quelques minutes plus tard, Zoïa descendit et s'approcha de son mari.

« Tout va bien ? demanda-t-il.

— Nous pensons que oui, répondit Zoïa.

— Vous *pensez* ? »

Elle haussa les épaules. « Évidemment, c'est la première fois que nous faisons ça. »

Ils montèrent dans le camion et s'éloignèrent à travers ce qui était déjà un paysage de désolation, vers le bunker de commandement.

Les autres chercheurs étaient juste derrière eux.

Au bunker, ils mirent tous des lunettes de soudeurs alors que le compte à rebours commençait.

À soixante secondes, Zoïa prit Volodia par la main.

À dix secondes, il lui sourit et lui dit : « Je t'aime. »

À une seconde, il bloqua sa respiration.

Et puis ce fut comme si le soleil s'était levé d'un coup. Une lumière plus vive que celle du plein midi inonda le désert. À l'endroit où se trouvait la tour de la bombe, une boule de feu s'éleva à une hauteur invraisemblable, montant toujours comme si elle cherchait à atteindre la lune. Volodia fut stupéfait par ses couleurs éclatantes : vert, violet, orange et pourpre.

La boule se changea en champignon dont le chapeau continua à monter, monter, monter. Et puis, enfin, le bruit parvint à leurs oreilles, une déflagration aussi violente qu'un tir rapproché de la plus grosse pièce d'artillerie de l'armée Rouge, suivie d'un roulement de tonnerre qui rappela à Volodia le terrible bombardement des Hauteurs de Seelow.

Au bout d'un temps infini, le nuage commença à se dissiper, et le vacarme s'estompa, suivi d'un interminable silence abasourdi.

Une voix s'écria : « Bon sang, je ne m'attendais pas à *ça.* »

Volodia serra sa femme dans ses bras. « Tu l'as fait », dit-il.

Elle le regarda solennellement. « Je sais, murmura-t-elle. Mais *qu'est-ce que* nous avons fait ?

— Tu as sauvé le communisme », répondit Volodia.

2.

« La bombe russe était une réplique de Fat Man, celle que nous avons larguée sur Nagasaki, dit l'agent spécial Bill Bicks. Quelqu'un leur a donné les plans.

— Comment le savez-vous ? demanda Greg.

— Par un transfuge. »

Il était neuf heures du matin, et ils étaient assis dans le bureau de Bicks, au quartier général du FBI à Washington. Bicks avait tombé la veste. Il avait des auréoles de sueur sous les bras, alors que le bâtiment était agréablement climatisé.

« D'après ce gars, poursuivit Bicks, un colonel des services de renseignement de l'armée Rouge a obtenu les plans grâce à un des membres de l'équipe du projet Manhattan.

— Il a dit son nom ?

— Il ne sait pas de quel chercheur il s'agit. C'est pour ça que je vous ai appelé. Il faut démasquer le traître.

— Le FBI avait pourtant mené des enquêtes serrées sur toute l'équipe à l'époque.

— Et la plupart présentaient un risque en matière de sécurité ! Que pouvions-nous faire ? Mais vous, vous les avez tous côtoyés personnellement.

— Qui est le colonel de l'armée Rouge ?

— J'allais y venir. Ce n'est pas un inconnu pour vous. Il s'appelle Vladimir Pechkov.

— Mon demi-frère !

— Oui.

— À votre place, c'est moi que je soupçonnerais, lança Greg en riant pour dissimuler son malaise.

— Oh, mais nous l'avons fait, croyez-moi, répondit Bicks. Vous avez fait l'objet de l'enquête la plus approfondie que j'aie vue depuis vingt ans que je suis au FBI. »

Greg lui jeta un regard incrédule. « Sans blague ?

— Votre gamin se débrouille bien à l'école, n'est-ce pas ? »

Qui avait bien pu parler de Georgy au FBI ? se demanda Greg, atterré. « Vous voulez parler de mon filleul ? corrigea-t-il.

— Greg, j'ai dit *approfondie*. Nous savons que c'est votre fils. »

Greg fut ennuyé, mais réprima ce sentiment. Il avait lui-même enquêté sur la vie privée de nombreux suspects lorsqu'il était dans la sécurité militaire. Il était mal placé pour s'offusquer.

« J'ai le plaisir de vous annoncer que vous êtes au-dessus de tout soupçon, poursuivit Bicks.

— Vous m'en voyez fort aise.

— De toute façon, notre transfuge a été très clair : les plans venaient d'un chercheur et non d'un des membres du personnel militaire employés sur le projet. »

Greg dit pensivement : « Lorsque j'ai rencontré Volodia à Moscou, il m'a assuré qu'il n'était jamais venu aux États-Unis.

— Il vous a menti, répondit Bicks. Il est venu ici en septembre 1945. Il a passé une semaine à New York. Et puis nous avons perdu sa trace pendant huit jours. Il a refait surface brièvement avant de rentrer chez lui.

— Huit jours ?

— Ouais. Vous imaginez notre embarras.

— Ça suffit pour aller à Santa Fe, y rester deux jours et revenir. »

— Exactement. » Bicks se pencha au-dessus de son bureau. « Mais réfléchissez. Si le chercheur avait déjà été recruté comme espion, pourquoi n'a-t-il pas été contacté par son officier traitant habituel ? Pourquoi faire venir quelqu'un de Moscou pour lui parler ?

— Vous pensez que le traître aurait pu être recruté au cours de cette visite de deux jours ? Ça paraît un peu bref.

— Peut-être avait-il déjà travaillé pour les Soviétiques avant de les laisser tomber. Quoi qu'il en soit, nous pensons qu'ils ont dû envoyer quelqu'un que ce chercheur connaissait déjà. Ce qui veut dire qu'il existait un lien quelconque entre Volodia et l'un des membres de notre équipe scientifique. » Bicks désigna d'un geste une petite table qui disparaissait sous les chemises de papier kraft. « La réponse est là-dedans, quelque part. Ce sont nos dossiers sur chacun des chercheurs qui ont eu accès à ces plans.

— Que voulez-vous que je fasse ?

— Que vous y jetiez un coup d'œil.

— Ce n'est pas votre boulot ?

— Nous l'avons déjà fait. Nous n'avons rien trouvé. Nous comptons sur vous pour repérer quelque chose qui nous a échappé. Je vais rester ici et vous tenir compagnie en faisant de la paperasse.

— Ça risque de prendre du temps.

— Vous avez la journée devant vous. »

Greg fronça les sourcils.

Bicks déclara d'un ton sans réplique : « Vous n'avez pas d'autres projets pour aujourd'hui. »

Greg haussa les épaules. « Vous avez du café ? »

Il eut droit à du café et des beignets, et encore du café, puis un sandwich à l'heure du déjeuner, et une banane dans l'après-midi. Il lut tous les détails connus sur la vie des chercheurs, de leurs femmes et de leurs familles : leur enfance, leurs études, leur carrière, leurs relations amoureuses, leur mariage, leurs réussites, leurs excentricités et leurs vices.

Il avalait la dernière bouchée de sa banane lorsqu'il s'exclama : « Bordel de merde !

— Quoi, qu'est-ce qu'il y a ? demanda Bicks.

— Willi Frunze a fréquenté l'Académie de garçons, à Berlin, répondit Greg triomphant, en flanquant une claque sur le dossier posé sur la table.

— Et alors… ?

— Alors ? Volodia aussi. Il me l'a dit. »

Bicks frappa du poing sur son bureau, fou d'excitation. « Camarades de classe ! C'est ça ! On te tient, mon salaud !

— Ce n'est pas une preuve, objecta Greg.

— Oh, ne vous en faites pas, il avouera.

— Comment pouvez-vous en être sûr ?

— Ces scientifiques sont convaincus que le savoir ne doit pas rester secret, qu'il faut le partager avec tout le monde. Il va essayer de se justifier en prétendant avoir agi pour le bien de l'humanité.

— C'est peut-être vrai.

— N'empêche qu'il va se retrouver sur la chaise électrique », affirma Bicks.

Greg fut glacé. Il avait trouvé Willi Frunze sympathique. « Vraiment ?

— Vous pouvez en être sûr. Il va griller. »

Bicks avait dit vrai. Willi Frunze fut convaincu de trahison, condamné à mort et envoyé à la chaise électrique.

Sa femme aussi.

3.

Daisy regarda son mari nouer son nœud papillon blanc et enfiler la veste à queue-de-pie de son habit impeccablement coupé. « Tu as vraiment fière allure », dit-elle, et c'était sincère. Il aurait dû faire du cinéma.

Elle repensa à lui au bal de Trinity College, treize ans plus tôt, avec sa tenue de soirée d'emprunt, et éprouva un agréable frisson de nostalgie. Il était vraiment séduisant à l'époque, se rappelait-elle, et pourtant son costume était trop grand de deux tailles au moins.

Ils étaient descendus au Ritz-Carlton de Washington, dans la suite que louait à l'année le père de Daisy. Lloyd était maintenant secrétaire d'État au ministère des Affaires étrangères, et il était venu en visite diplomatique. Les parents de Lloyd, Ethel et Bernie, étaient enchantés de garder leurs deux petits-enfants pendant une semaine.

Ce soir-là, Daisy et Lloyd étaient invités à un bal à la Maison Blanche.

Elle portait une robe renversante de Christian Dior, en satin rose avec une longue jupe d'une ampleur spectaculaire, aux innombrables plis de tulle. Après toutes ces années d'austérité, elle était ravie de pouvoir recommencer à acheter ses robes à Paris.

Elle se revoyait au bal du Yacht-Club, en 1935, à Buffalo, à cette soirée dont elle avait pensé, sur le moment, qu'elle avait fichu sa vie en l'air. La Maison Blanche était évidemment bien plus prestigieuse, mais elle savait que rien de ce qui se passerait ce soir-là ne pourrait lui gâcher la vie. C'était ce qu'elle se disait pendant que Lloyd l'aidait à attacher le collier de diamants roses de sa mère, avec les boucles d'oreilles assorties. À dix-neuf ans, elle recherchait désespérément l'estime de la haute société. Maintenant, c'était le cadet de ses soucis. Tant que Lloyd l'admirait, les autres pouvaient bien penser ce qu'ils voulaient. La seule autre personne dont l'approbation comptait pour elle était sa belle-mère, Eth Leckwith, qui n'appartenait pas au grand monde, et n'avait assurément jamais porté une robe de grand couturier.

Est-ce que toutes les femmes se disaient, en se penchant sur leur passé, qu'elles avaient été des bécasses quand elles étaient jeunes ? Daisy repensa à Ethel, qui avait évidemment fait une bêtise – tout de même, se faire engrosser par son patron, un homme marié ! – mais qui n'en parlait jamais avec regret. C'était sans doute l'attitude la plus sensée. Daisy réfléchit à ses propres erreurs : se fiancer à Charlie Farquharson, repousser Lloyd, épouser Boy Fitzherbert. Elle n'arrivait pas encore tout à fait, malgré le recul, à évaluer tout le bénéfice de ces choix passés. En réalité, il avait fallu qu'elle soit définitivement mise au ban de la haute société et trouve du réconfort dans la cuisine d'Ethel à Aldgate pour que sa vie s'engage réellement sur la bonne voie. Elle avait cessé de courir après le prestige social, elle avait appris ce qu'était une vraie amitié, et depuis elle était heureuse.

Maintenant qu'elle s'en fichait, elle appréciait encore plus les soirées.

« Tu es prête ? » lui demanda Lloyd.

Elle était prête. Elle enfila le manteau de soirée assorti à sa robe Christian Dior. Ils descendirent par l'ascenseur, quittèrent l'hôtel et montèrent dans la limousine qui les attendait.

4.

Carla avait persuadé sa mère de se mettre au piano le soir de Noël.

Il y avait des années que Maud n'avait pas joué. Peut-être cela l'attristait-il, en ravivant les souvenirs de Walter : ils avaient toujours joué et chanté ensemble, et elle avait souvent raconté aux enfants comment elle avait essayé, sans succès, de lui apprendre le ragtime. Mais elle ne racontait plus cette histoire, et Carla se demandait si désormais, le piano ne lui rappelait pas surtout Joachim Koch, le jeune officier nazi qui était venu chez elle prendre des leçons, qu'elle avait séduit et dupé, et que Carla et Ada avaient tué dans la cuisine. Carla elle-même n'arrivait pas à effacer les souvenirs de cette soirée de cauchemar, et notamment de leur expédition nocturne pour se débarrasser du corps. Elle ne regrettait rien – elles avaient fait ce qu'il fallait –, mais quand même, elle aurait préféré oublier tout ça.

Maud accepta enfin de jouer « Douce Nuit » pendant qu'ils chantaient. Werner, Ada, Erik et les trois enfants – Rebecca, Walli et Lili, le nouveau bébé –, se réunirent dans le salon autour du vieux Steinway. Carla posa une bougie sur le piano et observa, à la lueur mouvante de la flamme, les visages des membres de sa famille alors qu'ils reprenaient en cœur le refrain familier.

Walli, dans les bras de Werner, aurait quatre ans quelques semaines plus tard, et il essayait de chanter avec eux, devinant promptement les paroles et la mélodie. Il avait les yeux bridés du violeur qui l'avait engendré : Carla avait décidé que sa vengeance serait d'élever un fils qui traiterait les femmes avec douceur et respect.

Erik chanta les paroles du cantique avec sincérité. Il soutenait le régime soviétique aussi aveuglément qu'il avait pris fait et cause pour les nazis. Au début, Carla en était folle de rage ; cela la dépassait. Mais elle y décelait à présent une triste logique. Erik était un de ces individus éternellement inadaptés, qui avaient peur de la vie au point de préférer vivre sous un régime autoritaire et implacable, ne tolérant aucune opposition et disant aux gens quoi faire et quoi penser. C'étaient des êtres stupides et dangereux, mais tragiquement nombreux.

Carla regarda son mari avec tendresse. Il était encore incroyablement séduisant à trente ans. Elle se rappelait l'avoir embrassé, et même un peu plus, sur la banquette avant de son irrésistible voiture garée dans la forêt de Grünewald, quand elle avait dix-neuf ans. Ses baisers lui faisaient toujours autant d'effet.

Quand elle songeait au temps qui s'était écoulé depuis, elle éprouvait mille regrets, mais le plus douloureux était la mort de son père. Il lui manquait constamment, et elle pleurait encore lorsqu'elle repensait à lui, gisant dans l'entrée, si cruellement torturé qu'il était mort avant l'arrivée du médecin.

Mais tout le monde mourait un jour, et son père avait donné sa vie pour un monde meilleur. S'il y avait eu plus d'Allemands aussi courageux que lui, jamais les nazis n'auraient triomphé. Elle voulait faire tout ce qu'il avait fait : donner une bonne éducation à ses enfants, jouer un rôle dans la politique de son pays, aimer et être aimée. Mais surtout, elle voulait qu'à sa mort, ses enfants puissent dire, comme elle disait de son père, que sa vie avait eu un sens et qu'elle laissait en partant un monde un peu meilleur.

Le chant s'acheva. Maud tint l'accord final ; et puis le petit Walli se pencha et souffla la bougie.

FIN

REMERCIEMENTS

Mon principal conseiller historique pour la trilogie du *Siècle* est Richard Overy. Je suis également redevable aux historiens Evan Mawdsley, Tim Rees, Matthias Reiss et Richard Toye d'avoir relu le tapuscrit de *L'Hiver du monde* et je les remercie pour les corrections qu'ils y ont apportées.

Comme toujours, j'ai bénéficié de l'aide inestimable de mes éditeurs et agents, et plus particulièrement d'Amy Berkower, de Leslie Gelbman, de Phyllis Gramm, de Neil Nyren, de Susan Opie et de Jeremy Treviathan.

J'ai rencontré mon agent, Al Zuckerman, vers 1975, et depuis tout ce temps, il est mon lecteur le plus exigeant, en même temps qu'une source d'inspiration.

Plusieurs amis m'ont fait des commentaires utiles : Nigel Dean, avec un souci du détail à nul autre pareil ; Chris Manners et Tony McWalter, avec leur regard perçant et observateur. Angela Spizig et Annemarie Behnke, qui m'ont épargné de nombreuses erreurs dans les passages situés en Allemagne.

On remercie toujours sa famille, à juste titre. Merci à Barbara Follett, Emanuele Follett, Jann Turner et Kim Turner, qui ont lu le premier jet et m'ont fait des remarques précieuses, tout en m'offrant le cadeau inestimable de leur amour.